T0134276

# Lasers and Electro-optics

Covering a broad range of topics in modern optical physics and engineering, this textbook is invaluable for undergraduate students studying laser physics, optoelectronics, photonics, applied optics, and optical engineering.

This new edition has been re-organized, so that it now covers many new topics such as the optics of stratified media, quantum-well lasers and modulators, free-electron lasers, diode-pumped solid-state and gas lasers, imaging and non-imaging optical systems, squeezed light, periodic poling in nonlinear media, very-short-pulse lasers, and new applications of lasers.

The textbook gives a detailed introduction to the basic physics and engineering of lasers, as well as covering the design and operational principles of a wide range of optical systems and electro-optic devices. It features full details of important derivations and results, and provides many practical examples of the design, construction, and performance characteristics of different types of lasers and electro-optic devices.

**Christopher C. Davis** is Minta Martin Professor of Engineering and Professor of Electrical and Computer Engineering at the University of Maryland. He is co-author of the best-selling text *Building Scientific Apparatus* (fourth edition, Cambridge University Press, 2009).

# Lasers and Electro-optics

## Second Edition

CHRISTOPHER C. DAVIS
University of Maryland

CAMBRIDGE
UNIVERSITY PRESS

CAMBRIDGE
UNIVERSITY PRESS

University Printing House, Cambridge CB2 8BS, United Kingdom

Published in the United States of America by Cambridge University Press, New York

Cambridge University Press is part of the University of Cambridge.

It furthers the University's mission by disseminating knowledge in the pursuit of education, learning and research at the highest international levels of excellence.

www.cambridge.org
Information on this title: www.cambridge.org/9780521860291

© C. Davis 2014

This publication is in copyright. Subject to statutory exception and to the provisions of relevant collective licensing agreements, no reproduction of any part may take place without the written permission of Cambridge University Press.

First published 2014

Printed in the United Kingdom by TJ International Ltd. Padstow Cornwall

*A catalog record for this publication is available from the British Library*

*Library of Congress Cataloging-in-Publication Data*

Davis, Christopher C., 1944-
Lasers and electro-optics / Christopher C. Davis, University of Maryland. – Second edition.
pages cm
Includes bibliographical references and index.
ISBN 978-0-521-86029-1 (hardback)
1. Lasers. 2. Electrooptics. I. Title.
TA1675.D38 2014
621.36–dc23 2013019087

ISBN 978-0-521-86029-1 Hardback

Cambridge University Press has no responsibility for the persistence or accuracy of URLs for external or third-party internet websites referred to in this publication, and does not guarantee that any content on such websites is, or will remain, accurate or appropriate.

# Contents

*Preface to the Second Edition* *page* xv

**1 Electromagnetic waves, light, and lasers** 1

1.1 The laser 1
1.2 Light and electromagnetic waves 2
1.3 Some basic electromagnetic theory 3
1.4 The polarization state of an electromagnetic wave 5
1.5 Electromagnetic waves and their propagation through matter 6
1.6 Spontaneous and stimulated transitions 7
1.7 Lasers as oscillators 8
1.8 The energy levels of atoms, molecules, and condensed matter 10
1.9 Amplification at optical frequencies 11
1.10 The relation between energy density and intensity 16
1.11 The intensity of a beam of electromagnetic radiation in terms of photon flux 20
1.12 Black-body radiation 20
1.13 The relation between the Einstein $A$ and $B$ coefficients 25
1.14 The effect of level degeneracy 27
1.15 The ratio of spontaneous and stimulated transitions 29
1.16 Problems 30
Further reading 31
References 31

**2 Optical frequency amplifiers** 33

2.1 Introduction 33
2.2 Homogeneous line broadening 33
2.3 Inhomogeneous broadening 37
2.4 Optical frequency amplification 43
2.5 Optical frequency oscillation – saturation 48
2.6 The power output from a laser amplifier 58
2.7 The electron oscillator model of a radiative transition 59
2.8 The classical oscillator explanation for stimulated emission 66
2.9 Problems 69
References 70

**3 An introduction to two practical laser systems** 72
3.1 Introduction 72
3.2 Three- and four-level lasers 72
3.3 The ruby laser 74
3.4 The helium–neon laser 79
References 84

**4 Optical resonators containing amplifying media** 85
4.1 Introduction 85
4.2 Optical resonators containing amplifying media 87
4.3 The oscillation frequency 91
4.4 Multimode laser oscillation 93
4.5 Mode-beating 100
4.6 The power output of a laser 103
4.7 Optimum coupling 106
4.8 Problems 107
References 108

**5 Laser radiation** 109
5.1 Introduction 109
5.2 Diffraction 109
5.3 Babinet's principle 111
5.4 Two parallel narrow slits 112
5.5 A single slit 112
5.6 Two-dimensional apertures 113
5.7 Laser modes 116
5.8 Beam divergence 122
5.9 The linewidth of laser radiation 122
5.10 Coherence properties 125
5.11 Interference 127
5.12 Problems 130
References 130

**6 Control of laser oscillators** 132
6.1 Introduction 132
6.2 Multimode operation 132
6.3 Single-longitudinal-mode operation 133
6.4 Mode-locking 137
6.5 Methods of mode-locking 140
6.6 Pulse compression 147
6.7 Attosecond pulses 149
References 150

**7 Optically pumped solid-state lasers** 153
7.1 Introduction 153

7.2 Optical pumping in three- and four-level lasers 153
7.3 Pulsed versus continuous-wave operation 157
7.4 Threshold population inversion and the stimulated-emission cross-section 159
7.5 Paramagnetic-ion solid-state lasers 160
7.6 The Nd:YAG laser 160
7.7 Continuous-wave operation of the Nd:YAG laser 168
7.8 The $Nd^{3+}$ glass laser 169
7.9 Geometrical arrangements for optical pumping 174
7.10 High-power pulsed solid-state lasers 182
7.11 Diode-pumped solid-state lasers 183
7.12 Fiber lasers 184
7.13 Relaxation oscillations (spiking) 188
7.14 Rate equations for relaxation oscillation 190
7.15 Undamped relaxation oscillations 194
7.16 Giant-pulse (*Q*-switched) lasers 195
7.17 A theoretical description of the *Q*-switching process 199
7.18 Problem 203
References 204

**8 Gas lasers** 206
8.1 Introduction 206
8.2 Excitation schemes 206
8.3 The argon-ion laser 211
8.4 Pumping saturation in gas-laser systems 213
8.5 Pulsed gas lasers 214
8.6 Continuous-wave ion lasers 217
8.7 "Metal"-vapor ion lasers 220
8.8 Gas discharges for exciting gas lasers 222
8.9 Rate equations for gas-discharge lasers 225
8.10 Problem 228
References 229

**9 Molecular gas lasers I** 232
9.1 Introduction 232
9.2 The energy levels of molecules 232
9.3 Vibrations of a polyatomic molecule 236
9.4 Rotational energy states 239
9.5 Rotational populations 240
9.6 The overall energy state of a molecule 242
9.7 The carbon dioxide laser 244
9.8 The carbon monoxide laser 249
9.9 Other gas-discharge molecular lasers 251
References 251

**10 Molecular gas lasers II** 252
10.1 Introduction 252
10.2 Gas transport lasers 252
10.3 Gas dynamic lasers 256
10.4 High-pressure pulsed gas lasers 260
10.5 Ultraviolet molecular gas lasers 264
10.6 Photodissociation lasers 269
10.7 Chemical lasers 270
10.8 Far-infrared lasers 272
10.9 Problem 274
References 275

**11 Tunable lasers** 276
11.1 Introduction 276
11.2 The titanium–sapphire laser 276
11.3 Organic dye lasers 277
11.4 Calculation of threshold pump power in dye lasers 285
11.5 Inorganic liquid lasers 289
11.6 Free-electron lasers 291
11.7 Problems 298
References 298

**12 Semiconductor lasers** 301
12.1 Introduction 301
12.2 Semiconductor physics background 301
12.3 Carrier concentrations 307
12.4 Intrinsic and extrinsic semiconductors 309
12.5 The p–n junction 311
12.6 Recombination and luminescence 316
12.7 Heterojunctions 321
12.8 Semiconductor lasers 328
12.9 The gain coefficient of a semiconductor laser 331
12.10 Threshold current and power–voltage characteristics 334
12.11 Longitudinal and transverse modes 335
12.12 Quantum-well lasers 336
12.13 Semiconductor laser structures 341
12.14 Surface-emitting lasers 348
12.15 Laser diode arrays and broad-area lasers 350
12.16 Blue–green lasers 352
12.17 Quantum cascade lasers 353
12.18 Silicon lasers 355
12.19 Modulation of semiconductor lasers 356
12.20 Problems 356

Further reading 357
References 357

**13 Passive optical systems** 361
13.1 Introduction 361
13.2 The propagation of rays and waves through isotropic media 362
13.3 Simple reflection and refraction analysis 363
13.4 Paraxial-ray analysis 366
13.5 Non-imaging light collectors 384
13.6 Generalized imaging systems 385
13.7 The numerical aperture 385
13.8 Apertures, stops, and pupils 386
13.9 Vignetting 389
13.10 Exact ray tracing and aberrations 389
13.11 Chromatic aberrations 391
13.12 Geometrical aberrations 392
13.13 Spot diagram 395
13.14 The modulation transfer function 395
13.15 The use of impedances in optics 398
13.16 Problems 406
References 407

**14 Periodic optical systems, resonators, and inhomogeneous media** 409
14.1 Introduction 409
14.2 Plane waves in media with complex refractive indices 409
14.3 Negative-refractive-index materials 411
14.4 Structures with periodically varying dielectric properties 411
14.5 The transfer matrix for a single slab 412
14.6 The etalon 415
14.7 A thin metal film 415
14.8 A stratified medium of multiple slabs 417
14.9 Fiber Bragg gratings 420
14.10 Analysis of multi-layer structures by impedance techniques 420
14.11 Photonic crystals 421
14.12 Periodic lens sequences 425
14.13 The identical-thin-lens waveguide 426
14.14 The propagation of rays in mirror resonators 428
14.15 The propagation of rays in isotropic media with refractive-index gradients 431
14.16 Media with a Gaussian radial index variation 433
14.17 The propagation of spherical waves 434
14.18 Problems 436
References 437

**15 The optics of Gaussian beams** 438
15.1 Introduction 438

15.2 Beam-wave solutions of the wave equation 438
15.3 Higher-order modes 445
15.4 Transformation of Gaussian beams by general optical systems 454
15.5 The transformation of a Gaussian beam by a lens 459
15.6 Gaussian beams in lens waveguides 462
15.7 The propagation of a Gaussian beam in a medium with a quadratic refractive-index profile 463
15.8 The propagation of Gaussian beams in media with spatial gain or absorption variations 464
15.9 Propagation in a medium with a parabolic gain profile 465
15.10 Gaussian beams in plane and spherical mirror resonators 467
15.11 Symmetrical resonators 469
15.12 Examples of resonator design 471
15.13 Diffraction losses 474
15.14 Unstable resonators 475
15.15 Other beam waves 477
15.16 Problems 478
References 479

**16 Optical fibers and waveguides** 481
16.1 Introduction 481
16.2 Ray theory of cylindrical optical fibers 481
16.3 Ray theory of a dielectric-slab guide 490
16.4 The Goos–Hänchen shift 491
16.5 Wave theory of the dielectric-slab guide 494
16.6 P waves in the slab guide 495
16.7 Dispersion curves and field distributions in a slab waveguide 498
16.8 S waves in the slab guide 502
16.9 Practical slab guide geometries 503
16.10 Cylindrical dielectric waveguides 504
16.11 Modes and field patterns 511
16.12 The weakly guiding approximation 512
16.13 Mode patterns 514
16.14 Cutoff frequencies 518
16.15 Multimode fibers 520
16.16 Fabrication of optical fibers 521
16.17 Dispersion in optical fibers 524
16.18 Holey fibers 527
16.19 Solitons 528
16.20 Erbium-doped fiber amplifiers 529
16.21 Coupling optical sources and detectors to fibers 532
16.22 Problems 535
References 536

**17 The optics of anisotropic media** 539
17.1 Introduction 539
17.2 The dielectric tensor 539
17.3 Stored electromagnetic energy in anisotropic media 542
17.4 Propagation of monochromatic plane waves in anisotropic media 543
17.5 The two possible directions of D for a given wave vector are orthogonal 545
17.6 Angular relationships involving D, E, H, k, and the Poynting vector S 547
17.7 The indicatrix 548
17.8 Uniaxial crystals 550
17.9 Index surfaces 553
17.10 Other surfaces related to the uniaxial indicatrix 555
17.11 Huygenian constructions 557
17.12 Retardation 560
17.13 Biaxial crystals 565
17.14 Intensity transmission through polarizer/waveplate/polarizer combinations 568
17.15 The Jones calculus 570
17.16 Mueller calculus 576
17.17 Problems 578
References 578

**18 The electro-optic and acousto-optic effects and modulation of light beams** 580
18.1 Introduction to the electro-optic effect 580
18.2 The linear electro-optic effect 581
18.3 The quadratic electro-optic effect 587
18.4 Longitudinal electro-optic modulation 588
18.5 Transverse electro-optic modulation 590
18.6 Electro-optic amplitude modulation 595
18.7 Electro-optic phase modulation 597
18.8 High-frequency waveguide electro-optic modulators 599
18.9 Other high-frequency electro-optic devices 603
18.10 Electro-absorption modulators 606
18.11 Electro-optic beam deflectors 607
18.12 Acousto-optic modulators 608
18.13 Applications of acousto-optic modulators 614
18.14 Construction and materials for acousto-optic modulators 617
18.15 Problem 620
References 620

**19 Introduction to nonlinear processes** 622
19.1 Introduction 622
19.2 Anharmonic potentials and nonlinear polarization 622
19.3 Nonlinear susceptibilities and mixing coefficients 626
19.4 Second-harmonic generation 629

19.5 The linear electro-optic effect 631
19.6 Parametric and other nonlinear processes 633
19.7 Macroscopic and microscopic susceptibilities 634
19.8 Problem 638
References 638

**20 Wave propagation in nonlinear media** 640
20.1 Introduction 640
20.2 Electromagnetic waves and nonlinear polarization 640
20.3 Second-harmonic generation 645
20.4 The effective nonlinear coefficient $d_{\text{eff}}$ 646
20.5 Phase matching 649
20.6 Beam walk-off and 90° phase matching 653
20.7 Second-harmonic generation with Gaussian beams 654
20.8 Up-conversion and difference-frequency generation 660
20.9 Optical parametric amplification 661
20.10 Parametric oscillators 665
20.11 Parametric-oscillator tuning 668
20.12 Phase conjugation 670
20.13 Optical bistability 674
20.14 Practical details of the use of crystals for nonlinear applications 679
20.15 Problems 680
References 681

**21 Detection of optical radiation** 684
21.1 Introduction 684
21.2 Noise 684
21.3 Detector performance parameters 692
21.4 Practical characteristics of optical detectors 693
21.5 Thermal detectors 715
21.6 Detection limits for optical detector systems 717
21.7 Coherent detection 725
21.8 The bit-error rate 730
21.9 Problems 733
References 734

**22 Coherence theory** 736
22.1 Introduction 736
22.2 Square-law detectors 736
22.3 The analytic signal 737
22.4 Correlation functions 741
22.5 Temporal coherence 744
22.6 Spatial coherence 748
22.7 Spatial coherence with an extended source 751

22.8 Propagation laws of partial coherence 753
22.9 Propagation from a finite plane surface 756
22.10 The van Cittert–Zernike theorem 761
22.11 The spatial coherence of a quasi-monochromatic, uniform, spatially incoherent circular source 763
22.12 Intensity-correlation interferometry 766
22.13 Intensity fluctuations 767
22.14 Photon statistics 771
22.15 The Hanbury Brown–Twiss interferometer 776
22.16 The Hanbury Brown–Twiss experiment with photon-count correlations 778
22.17 Squeezed light 780
References 781

**23 Laser applications** 783
23.1 Optical communication systems 783
23.2 Optical amplification and wavelength-division multiplexing 788
23.3 Holography 797
23.4 Medical applications of lasers 806
23.5 Laser weapons 809
23.6 Laser isotope separation 809
23.7 Laser plasma generation and fusion 813
References 816

**Appendix 1 Optical terminology** 821

**Appendix 2 The $\delta$-function** 824

**Appendix 3 Black-body radiation formulas** 826

**Appendix 4 *RLC* circuits** 828

**Appendix 5 Storage and transport of energy by electromagnetic fields** 831

**Appendix 6 The reflection and refraction of a plane electromagnetic wave at a boundary between two isotropic media of different refractive indices** 834

**Appendix 7 The vector differential equation for light rays** 837

**Appendix 8 Symmetry properties of crystals and the 32 crystal classes** 841

**Appendix 9 Tensors** 845

**Appendix 10 Bessel-function relations** 848

**Appendix 11 Green's functions** 849

**Appendix 12 Recommended values of some physical constants** 852
*Index* 853

# Preface to the Second Edition

The author of a text generally feels obligated to explain the reasons for his or her writing. This is a matter of tradition as it provides an opportunity for explaining the development and philosophy of the text, its subject matter and intended audience, and acknowledges the help that the author has received. In the case of a second edition of a text, as is the case here, a new preface provides an opportunity for the author to explain the revisions of the second edition and to further acknowledge help from colleagues. I hope to accomplish these tasks briefly here.

The first edition of this text grew over many years out of notes that I had developed for courses at the senior undergraduate and beginning graduate student level at the University of Manchester, Cornell University, and the University of Maryland, College Park. These courses covered many aspects of laser physics and engineering, the practical aspects of optics that pertain to an understanding of these subjects, and a discussion of related phenomena and devices whose importance has grown from the invention of the laser in 1960. These include nonlinear optics, electro-optics, acousto-optics, and the devices that take practical advantage of these phenomena. The names given to the fields that encompass such subject matter have included laser physics, optical electronics, optoelectronics, photonics, and quantum electronics. The fundamentals of these subjects have not changed significantly in the years that have intervened since the publication of the first edition. However, there have been important technological advances that need discussion, as well as new laser applications. A few important new topic areas now included are fiber lasers, aberrations in optical systems, stratified media, photonic crystals, periodic poling for phase matching, and new methods for obtaining laser pulses of very short duration. Expanded discussion is provided about diode-pumped solid-state lasers, free-electron lasers, semiconductor physics, semiconductor lasers, photon statistics, squeezed light, and analysis of polarized light.

This is a textbook, not a research monograph, although I have attempted to provide enough detail to help researchers obtain the background necessary for exploration of more specialized literature. After teaching this material for over 40 years I hope that I have found helpful ways to make the subject understandable to students. I have certainly benefited greatly from their feedback, which has contributed to the changes in this second edition.

Although a comprehensive understanding of the laser and associated phenomena requires a quantum-mechanical treatment, almost all aspects of laser operation and important related phenomena can be explained well classically. Therefore, this text requires no knowledge of quantum mechanics, although a background in electromagnetic theory at the undergraduate level is desirable for a greater understanding. Most electrical engineering majors do not take a course in quantum mechanics until they reach graduate level, and many physics majors will not have acquired sufficient quantum mechanics knowledge at the undergraduate level for this to make a meaningful contribution to better understanding in a study of lasers and electro-optics.

For all the above reasons this text should be suitable for senior undergraduate and beginning graduate students in electrical engineering or physics. It should also prove useful to other engineers or chemists who use lasers and electro-optic devices.

The text is broken up into two principal parts. Chapters 1–12 discuss the basic physics and engineering of lasers of all kinds, beginning with a discussion of the fundamental physics of the stimulated emission process and laser amplifiers. This is followed by chapters on laser resonators and the characteristics of laser radiation and methods for controlling it. There are succeeding chapters that cover optically pumped insulating crystal lasers, atomic gas lasers, and molecular gas lasers of various kinds including gas transport, gas dynamics and chemically pumped varieties, and tunable lasers. The first section of the book concludes with a chapter on semiconductor lasers that begins with a review of the basic physics necessary for their understanding.

The second part of the text covers various issues of relevance to lasers and electro-optics, including optical analysis and design techniques, the optics of Gaussian beams, laser resonators, and anisotropic crystalline materials. There are chapters on optical fibers, electro-optic and acousto-optic devices, the fundamentals of nonlinear optics, and application of nonlinear optics in harmonic generation, parametric processes, phase conjugation, and optical bistability. The text concludes with chapters on optical detectors and the detection process, coherence theory, and applications of lasers.

I have found that Chapters 1–12 provide sufficient material for a one semester course on lasers, with some applications from Chapter 23 included. Chapters 5 and 13–18, together with Chapter 22, form the basis of a one-semester course on optical design, electro-optic devices, and optical detectors. I draw on the somewhat more difficult material in Chapters 19–21 as reference material in both the one-semester courses just mentioned, and also as adjunct material in more advanced graduate courses.

I have been an experimentalist in the optics and laser business for over four decades. I have always found the laser itself a fascinating device that provides a teaching vehicle for discussing many fundamental physics concepts as well as practical aspects of optical design. I have always tried to introduce practical details of real lasers into my classes as early as a treatment of some of the associated fundamentals permits. This should be apparent in the current text where I digress in Chapter 3 into a fairly detailed practical discussion of two historically important lasers, even though contextually a fuller discussion of these devices could be left until later. I believe that this makes pedagogical sense as students get a glimpse of where they are headed. Throughout the text I have attempted to provide full details of important derivations, and provide practical examples from the literature on the design, construction, and performance characteristics of lasers and electro-optic devices.

In developing a sound approach to teaching the material in this text to many students over more than four decades I have drawn inevitably on the work of many others. There have been many other books that cover material that is shared in common with the current one. What is different between this and related texts is not so much the analytic treatment of common subject matter, but the specific choice of material presentation sequence, and assorted explanations. As it is said, "there is nothing new under the sun," so it is not the intent of this author to claim that the treatment of particular topics in the current text is necessarily unique. Different authors impart their own slant to the same subject matter: sometimes their treatments converge, particularly when there is one especially good way of explanation that is valid. I have attempted in every case to provide references to the original literature from which I have benefited in my writing, and I apologize for any inadvertent omissions. There is a vast literature on the subjects covered and would be it impossible to list all the important published work that has added to our knowledge. I have attemped to list key early work in each area covered, since use of the *Science Citations Index*[1] will in this way provide links to almost all the work in an area.

Over the course of many years I have learned much from my contacts in the classroom, office, and research laboratory, and at conferences. I am indebted to numerous past and present colleagues and students for their intellectual stimulation, collaboration in research, advice, provision of material, and feedback on early versions of the current text that have contributed greatly to the finished product. I would like to thank especially my past and present faculty colleagues Quirino Balzano, Kyuman Cho, Mario Dagenais, Julius Goldhar, Ping-Tong Ho, the late Urs Hochuli, Terry King, Chi Lee, Ross McFarlane, Stuart Milner, Ian Smith, Igor Smolyaninov; and finally George Wolga who gave me valuable help at the very beginning of this work. I am especially grateful to Professor Thomas Murphy for his helpful comments on how to treat the topic of semiconductor lasers, which just on its own could fill many volumes. My graduate students and post-doctoral research associates have over several years provided help and advice that have helped me greatly. I am particularly grateful to Navik Agrawal, David Coleman, Mohammed Eslami, Ehren Hwang, Jonathan Ko, Jaime Llorca, Billy Nelson, John Rzasa, Tommy Shen, and Chensheng Wu, for their current help with my research that has provided the time to finish this new edition. I am most grateful to Joan Hamilton and Nono Kusuma for drawing many of the diagrams and to Dave Mazzoni and Sarah Fish for help with additional diagrams.

I appreciate the patience of several editors at Cambridge University Press in waiting for completion of this second edition, and for accepting ongoing excuses as to why it was not completed earlier. I am most grateful to Patricia Keehn for her expert and careful computer typesetting work using TEX over enormous numbers of revisions of the first edition.[2]

Most of all, I am indebted to my family, especially my wife Mary, for their unfailing love and support over the course of my career.

Christopher C. Davis

College Park, Maryland
November 2013

[1] Part of the *Web of Science*, a subscription data base that is widely available: http://thomsonreuters.com/web-of-science/

[2] In most cases where curves showing the parametric variation of phenomena discussed in the text are given these have been calculated from scratch using Mathcad (©Mathsoft, Inc.) The entire text has been typeset using Donald Knuth's TEX: *The TEXbook*, Addison-Wesley, Reading, MA, 1984. TEX is a trademark of the American Mathematical Society.

# 1 Electromagnetic waves, light, and lasers

## 1.1 The laser

A laser is an oscillator that operates at *optical* frequencies. These frequencies of operation lie within a spectral region that extends from the very far infrared to the *vacuum-ultraviolet* (VUV) or soft-X-ray region. At the lowest frequencies at which they operate, lasers overlap with the frequency coverage of masers, to which they are closely related, and millimeter-wave sources using solid-state or vacuum-tube electronics, such as TRAPATT, IMPATT, and Gunn diodes, klystrons, gyroklystrons, and traveling-wave tube oscillators, whose principles of operation are quite different [1]. In common with electronic-circuit oscillators, a laser is constructed using an amplifier with an appropriate amount of positive feedback. The positive feedback is generally provided by mirrors that re-direct light back and forth through the laser amplifier. The acronym LASER, which stands for *light* **amplification** *by stimulated emission of radiation*, is in reality therefore a slight misnomer.[1]

### 1.1.1 A little bit of history

The basic physics underlying light emission and absorption by atoms and molecules was first expounded by Albert Einstein (1879–1955) in 1917 [2]. Richard Chace Tolman (1881–1948) observed that stimulated emission could lead to "negative absorption." In 1928 Rudolph Walther Landenburg (1889–1953) confirmed the existence of stimulated emission and negative absorption. It is interesting to note that a famous spectroscopist, Curtis J. Humphreys (1898–1986), who for most of his career worked at the U.S. Naval Ordnance Laboratory in Corona, California, might have operated the first gas laser without knowing it. Humphreys and his colleagues, notably William F. Meggers (1888–1966), studied the spectra emitted by low-pressure xenon lamps beginning in the later 1920s [3], and if they had had access to photodetectors operating in the near and middle infrared would almost certainly have observed the anomalously strong lines from the pure-xenon gas laser, notably at 2.062 μm and 3.507 μm. These laser lines are so characteristic of low-pressure xenon gas discharges that it is almost impossible to stop them from exhibiting laser oscillation, even in short laser tubes with no mirrors. Later, namely in 1939, Valentin Aleksandrovich Fabrikant (1907–1991) predicted the actual use of stimulated emission in gas discharges to amplify light. He subsequently claimed that he had observed

[1] The more truthful acronym LOSER was long ago deemed inappropriate.

"negative absorption" experimentally, and in 1951 he and his students filed a patent, but their work had little impact. Attempts to replicate Fabrikant's results experimentally showed that they had arisen from an experimental artefact (C. E. Webb, private communication). In the early 1950s, Charles Hard Townes (born in 1915) and his colleagues at Columbia University worked on the development of the maser, a microwave amplifier using stimulated emission, and published their first report of a working device in 1954 [4, 5]. Independent researchers Joseph Weber (1919–2000), of the University of Maryland [6], and Nikolai Basov (1922–2001) and Aleksandr Prokhorov (1916–2002), of the P. N. Lebedev Physics Institute in Moscow, proposed similar ideas [7]. On May 16, 1960 Theodore H. Maiman at the Hughes Research Laboratories reported the successful operation of the first laser – the pulsed ruby laser [8]. On December 13, 1961, Ali Javan (born in 1926), William R. Bennett Jr. (1930–2008), and Donald Heriott (1928–2007) demonstrated the first continuous-wave laser – the helium–neon laser [9]. In 1964, Nikolai Basov, Charles Townes, and Aleksandr Prokhorov received the Nobel Prize for "fundamental work in the field of quantum electronics, which has led to the construction of oscillators and amplifiers based on the maser–laser principle." Subsequently, in 1981, Arthur Schawlow (1921–1999), who was Charles Townes's brother-in-law, and Nicolaas Bloembergen (born in 1920) received the Nobel Prize for "their contribution to the development of laser spectroscopy."

## 1.2 Light and electromagnetic waves

Light is one form of electromagnetic radiation, the many categories of which make up the electromagnetic spectrum. Electromagnetic radiation, which transports energy from point to point at the velocity of light, can be described in terms of both *wave* and *particle* "pictures" or "models." This is the famous "wave–particle" duality of all fields or particles in our model of the Universe. In the electromagnetic-wave picture, waves are characterized by their frequency $\nu$, wavelength $\lambda$, and the velocity of light $c$, which are inter-related by $c = \nu\lambda$. A propagating electromagnetic wave is characterized by a number of field vectors, which vary in time and space. These include the electric field $\mathbf{E}$ (measured in volts/m), the magnetic field $\mathbf{H}$ (measured in amps/m), the displacement vector $\mathbf{D}$ (measured in coulombs/m$^2$), and the magnetic flux density $\mathbf{B}$ (measured in webers/m$^2$ or tesla). For a complete description the *polarization* state of the wave must also be specified. *Linearly polarized* waves have fixed directions for their field vectors, which do not re-orient themselves as the wave propagates. *Circularly* or *elliptically* polarized waves have field vectors that trace out circular, or elliptical, helical paths as the wave travels along. In the particle picture, electromagnetic energy is carried from point to point as quantized packets of energy called *photons*. The energy of a photon of frequency $\nu$ is $h\nu$, where $h$ is Planck's constant, namely $6.626 \times 10^{-34}$ J s. Photons have zero mass, and travel at the velocity of light, but carry both linear and angular momentum. The linear momentum of a photon of wavelength $\lambda$ is $p = h/\lambda$, and the angular momentum depends on the equivalent polarization state of the corresponding wave. Circularly polarized photons have angular momentum $h/(2\pi) = \hbar$.

Our everyday experience of "light" generally encompasses only the small part of the electromagnetic spectrum to which the human eye is sensitive, a wavelength range running roughly from 400 nm to 700 nm. The full electromagnetic spectrum, going from low to high frequencies, is divided into radiowaves (0–1 GHz), microwaves (1–300 GHz), infrared waves (of wavelength $\lambda = 0.7-1000$ μm; 300 GHz to 430 THz),[2] visible light ($\lambda = 400-700$ nm), ultraviolet light ($\lambda = 10-400$ nm), X-rays ($\lambda = 0.1$–$10$ nm), and $\gamma$ rays ($\lambda < 0.1$ nm).

## 1.3 Some basic electromagnetic theory

In this book we will frequently discuss the electromagnetic wave properties of laser radiation, so we begin with a brief sumary of some key concepts that are important in later chapters. A linearly polarized electromagnetic wave of angular frequency $\omega$ that is polarized in the $x$ direction and is traveling in the $z$ direction has electric and magnetic fields[3]

$$E_x = E_0 e^{j(\omega t - kz)} \tag{1.1}$$

$$H_y = H_0 e^{j(\omega t - kz)}. \tag{1.2}$$

A linearly polarized wave polarized orthogonally to this one would have field pairs $(E_y, H_x)$. The parameter $k$ is called the propagation constant of the wave. The corresponding vector $\mathbf{k}$, the *wave vector*, represents the direction of propagation of the wave and has magnitude $|\mathbf{k}| = k = 2\pi/\lambda$. The *phase* velocity of the wave is $c = \omega/k$. The important relations between the $\mathbf{E}$ and $\mathbf{H}$ fields and related vectors of the wave are

$$\mathbf{D} = \epsilon_0 \mathbf{E} + \mathbf{P} \tag{1.3}$$

and

$$\mathbf{B} = \mu_0(\mathbf{H} + \mathbf{M}), \tag{1.4}$$

where $\epsilon_0 = 8.854 \times 10^{-12}$ F m$^{-1}$ (farads/m) is the *permittivity* of free space, $\mathbf{P}$ is the polarization produced in the medium through which the wave is traveling, $\mu_0 = 4\pi \times 10^{-12}$ H m$^{-1}$ (henrys/m) is the *permeability* of free space, and $\mathbf{M}$ is the *magnetization* produced in the medium through which the wave is traveling. The polarization is produced by the electric field of the wave according to

$$\mathbf{P} = \epsilon_0 \chi \mathbf{E}, \tag{1.5}$$

[2] 1 GHz (gigahertz) = $10^9$ Hz; 1 THz (terahertz) = $10^{12}$ Hz. The infrared range can also be somewhat arbitarily divided up into the sub-millimeter region ($\lambda = 0.1$–$1$ mm), the far infrared ($\lambda = 20-100$ μm), the middle infrared ($\lambda = 3-20$ μm), and the near infrared ($\lambda = 0.7-3$ μm).

[3] The polarization direction of a wave is strictly speaking the direction of its $\mathbf{D}$ vector, which is parallel to $\mathbf{E}$ except in *anisotropic media*, which will be dealt with in Chapter 17.

where $\chi$ is the (electric) *susceptibility* of the medium. The magnetization produced by the magnetic field of the wave is

$$\mathbf{M} = \chi_{\mathrm{m}}\mathbf{H}, \tag{1.6}$$

where $\chi_{\mathrm{m}}$ is the magnetic susceptibility of the medium. Using Eqs. (1.5) and (1.6), Eqs. (1.3) and (1.4) can also be written, respectively, as

$$\mathbf{D} = \epsilon_0\epsilon_{\mathrm{r}}\mathbf{E} \tag{1.7}$$

and

$$\mathbf{B} = \mu_0\mu_{\mathrm{r}}\mathbf{H}, \tag{1.8}$$

where $\epsilon_{\mathrm{r}}$ is the relative permittivity[4] and $\mu_{\mathrm{r}}$ is the relative permeability of the medium. Clearly $\epsilon_{\mathrm{r}} = 1 + \chi$, and $\mu_{\mathrm{r}} = 1 + \chi_{\mathrm{m}}$. One of the triumphs of James Clerk Maxwell in his 1873 formulation of electromagnetic theory [10] in terms of the famous *Maxwell* equations was to show that the velocity of light was related to the fundamental permittivity and permeability quantities of a medium by

$$c = \frac{1}{\sqrt{\mu_0\mu_{\mathrm{r}}\epsilon_0\epsilon_{\mathrm{r}}}}, \tag{1.9}$$

which was first verified experimentally by Weber and Kohlrausch in 1857 [11]. In 1907 Rosa and Dorsey [12] obtained a value for the velocity of light of 299,788 $\mathrm{km\,s^{-1}}$ in this way. This was the most accurately known value at that time.[5] The propagation constant is $k = \omega\sqrt{\mu_0\mu_{\mathrm{r}}\epsilon_0\epsilon_{\mathrm{r}}}$.

An electromagnetic wave transports energy at the velocity of light.[6] The local energy flux is described in terms of the *Poynting vector* $\mathbf{S} = \mathbf{E} \times \mathbf{H}$. The time-averaged value of the Poynting vector at a point in space is frequently called the *intensity* and is measured in $\mathrm{W\,m^{-2}}$ (watts/m$^2$). The direction of the Poynting vector at a point in space is the direction of the *light ray*. If all the field components of an electromagnetic wave are the same in planes perpendicular to the $\mathbf{k}$ vector then this is a *plane wave*. An ideal *spherical wave* has equal field components on spherical surfaces, whose centers correpond to the point source of the waves. The Poynting vector of a spherical wave is directed radially outwards at any point on the surface of these spheres. A *beam wave*, of which a laser beam is an example, has its energy flux localized in a cone around some axial direction, and has field components that vary with the radial distance from this axis.

[4] This is frequently called the *dielectric constant* of the medium.

[5] The velocity of light has now been measured so accurately that it has been *defined* to have a value of $2.99792458 \times 10^8\,\mathrm{m\,s^{-1}}$. Consequently in the S.I. system of units the meter is now a derived unit: it is the distance traveled by light in vacuum in 3.335640952 ns.

[6] A more detailed discussion is provided in Appendix 5.

## 1.4 The polarization state of an electromagnetic wave

Description of the polarization state of an electromagnetic wave, whether this be a plane wave, spherical wave, or beam wave, involves specification of the magnitude, and phase, of orthogonal field components tranverse to the propagation direction **k**. For propagation in the $z$ direction linearly polarized light propagates with a fixed **D** direction, and circularly polarized light has orthogonal $D_x$ and $D_y$ components of equal amplitude, but 90° out of phase. The resultant **D** traces out a helical path of circular cross-section as the wave propagates. If the **D** vector rotates in a clockwise direction when viewed in the propagation direction, the light is said to be *left-hand* circularly polarized.[7] Elliptically polarized light has orthogonal $D_x$ and $D_y$ components of unequal amplitude and arbitrary phase difference. The polarization state can be specified by the use of the *Stokes*[8] parameters, $I, Q, U$, and $V$, which four parameters are often also written as $S_0, S_1, S_2$, and $S_3$. The parameter $I$ is related to the total intensity of the wave and can be written, in the case of a right-handed coordinate system with the wave propagating in the $+z$ direction, as

$$I = \langle E_x^2 \rangle + \langle E_y^2 \rangle,$$

where the $\langle\rangle$ indicate time-averaging. The $Q$ parameter is

$$Q = \langle E_x^2 \rangle - \langle E_y^2 \rangle,$$

so $Q$ describes the extent of linear polarization, since for pure $x$ or $y$ polarization $Q = I$. If we write the time-harmonic behavior of the electric-field components as

$$E_x = E_{0x}(t)\cos(\omega t + \phi_x)$$

and

$$E_y = E_{0y}(t)\cos(\omega t + \phi_y),$$

then

$$U = 2\langle E_{0x}E_{0y}\cos(\phi_x - \phi_y)\rangle$$

and

$$V = 2\langle E_{0x}E_{0y}\sin(\phi_x - \phi_y)\rangle.$$

$U$ also describes the linearly polarized character of the wave, while $V$ describes its circular polarization character, since $V$ is maximum for a phase difference of 90° between $E_x$ and $E_y$. The polarization state of an electromagnetic wave is often represented on a diagram called the *Poincaré sphere*.[9] Figure 1.1 shows a representation of the Poincaré sphere. The Stokes parameters $U, V$, and $Q$ are the Cartesian coordinates of a point on the sphere. At any point P on the surface of the sphere the intensity is $I = \sqrt{U^2 + V^2 + Q^2}$.

[7] For a picture see Fig. 17.16.

[8] Sir George Gabriel Stokes (1819–1903), British mathematician and physicist.

[9] This is named for the French mathematician and physicist Jules Henri Poincaré (1854–1912).

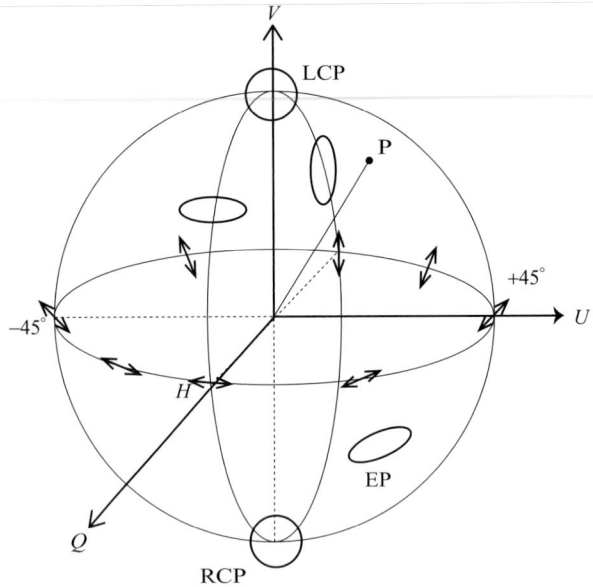


Fig. 1.1 The Poincaré sphere.

## 1.5 Electromagnetic waves and their propagation through matter

The values of the relative permittivity and relative permeability characterize a medium in terms of its difference from a vacuum. Most materials that are important in a discussion of lasers and optical devices are not strongly magnetic, and it is generally legitimate to assume that for such materials $\mu_r = 1$. It is important to note that the dielectric properties of a medium depend on frequency, even though this has not been explicitly stated in our preceding brief discussion. In addition, to include the properties of real, rather than ideal, matter in our discussion we allow the possibility that the dielectric can be represented by a complex number, so the frequency-dependent dielectric constant can be written as

$$\epsilon_r(\omega) = \epsilon'(\omega) - j\epsilon''(\omega). \tag{1.10}$$

Along with this goes a corresponding definition for the complex susceptibility, which is

$$\chi(\omega) = \chi'(\omega) - j\chi''(\omega). \tag{1.11}$$

Clearly, $\chi'(\omega) = \epsilon'(\omega) - 1$ and $\chi''(\omega) = \epsilon''(\omega)$. The minus sign in Eqs. (1.10) and (1.11) is a sign convention. We shall see that, with this sign convention, when $\chi''(\omega)$ is positive the medium absorbs energy from the wave.

Although a laser is a device that requires quantum mechanics for a complete description of its behavior, much can be learned from a classical analysis. In this context we describe a medium that has gain or absorption as one whose dielectric properties have been modified in a special way. When an electromagnetic wave propagates in the $z$ direction through a medium with complex dielectric constant $\epsilon_r = \epsilon' - j\epsilon''$ the variation of the fields of the wave is, taking the electric field as an example,

$$\mathbf{E} = \mathbf{E}_0 e^{j(\omega t - k'z)}, \tag{1.12}$$

where the new propagation constant is

$$k' = \omega\sqrt{\mu_0 \mu_r \epsilon_r \epsilon_0}. \tag{1.13}$$

Both $\epsilon'$ and $\epsilon''$ vary with frequency. If we make the usual assumption that optical media are generally not strongly magnetic then we can take $\mu_r = 1$ and write

$$k' = \frac{\omega}{c_0}\sqrt{\epsilon' - j\epsilon''}. \tag{1.14}$$

For a medium that does not have large absorption or gain, we can make the assumption that $\epsilon'' \ll \epsilon'$, and Eq. (1.14) becomes

$$k' = \frac{\omega}{c}\left(1 - j\frac{\epsilon''}{2\epsilon'}\right), \tag{1.15}$$

where $c = c_0/\sqrt{\epsilon'}$ is the modified velocity of light in the medium. The variation of the amplitude of the electric field of the wave as it propagates is now

$$\mathbf{E} = \mathbf{E}_0 e^{-\frac{\alpha}{2}z} e^{j(\omega t - kz)}, \tag{1.16}$$

where $k = \omega/c$ and $\alpha = k\epsilon''/\epsilon'$. If we write $n = \sqrt{\epsilon'}$ and note from our previous discussion that $\epsilon'' = \chi''$, then

$$\alpha = \frac{k\chi''}{n^2}. \tag{1.17}$$

If the complex susceptibility were to become *negative* then we could write

$$\gamma = -\frac{k\chi''}{n^2}, \tag{1.18}$$

and the wave would increase in field amplitude as it propagated according to

$$\mathbf{E} = \mathbf{E}_0 e^{\frac{\gamma}{2}z} e^{j(\omega t - kz)}, \tag{1.19}$$

meaning that the medium has become an amplifier.

## 1.6 Spontaneous and stimulated transitions

In this chapter we shall consider the fundamental processes whereby amplification at optical frequencies can be obtained. These processes involve the fundamental atomic nature of matter. At the atomic level matter is not a continuum, but is composed of discrete particles – atoms, molecules, or ions. These particles have energies that can have only certain discrete values. This discreteness, or *quantization*, of energy is intimately connected with the duality that exists in nature. Light sometimes behaves as if it were a wave and in other circumstances as if it were composed of particles. These particles, called *photons*, carry the discrete packets of energy associated with the wave. For light of frequency $\nu$ the energy of each photon is $h\nu$, where $h$ is Planck's constant, $6.6 \times 10^{-34}$ J s. The energy $h\nu$ is the *quantum* of energy associated with the frequency $\nu$. At the microscopic level the amplification of light within a laser involves the emission of these quanta. Thus, the term *quantum electronics* is often used to describe the branch of science that has grown from the development of the maser in 1954 and the laser in 1960.

The widespread practical use of lasers and optical devices in applications such as communications, and increasingly in areas such as signal processing and image analysis, has led to the use of the term *photonics*. Whereas electronics uses electrons in various devices to perform analog and digital functions, photonics aims to replace the electrons with photons. Because photons have zero mass, do not interact with each other to any significant extent, and travel at the speed of light, photonic devices promise small size and high speed.

## 1.7 Lasers as oscillators

In "conventional" electronics, where by the word "conventional" for the present purposes we mean frequencies at which solid-state devices such as transistors or diodes will operate, say below $10^{11}$ Hz, an oscillator is conveniently constructed by applying an appropriate amount of positive feedback to an amplifier. Such an arrangement is shown schematically in Fig. 1.2. The input and output voltages of the amplifier are $V_i$ and $V_o$, respectively. The voltage gain of the amplifier is $A_0$, and, in the absence of feedback, $A_0 = V_o/V_i$. The feedback circuit returns part of the amplifier output to the input. The feedback factor $\beta = |\beta|e^{j\phi}$ is in general a complex number with amplitude $|\beta| \leq 1$ and phase $\phi$:

$$V_o = A_0(V_i + \beta V_o), \tag{1.20}$$

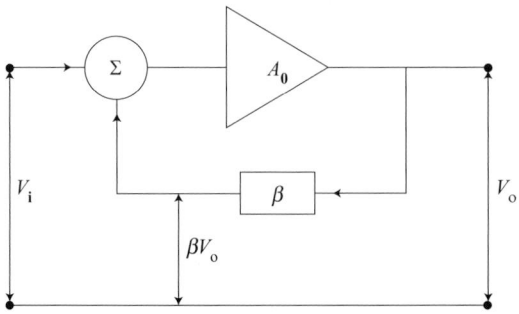


Fig. 1.2 The circuit diagram of a simple amplifier with feedback.

so

$$V_o = \frac{A_0 V_i}{1 - \beta A_0}, \tag{1.21}$$

and the overall voltage gain is

$$A = \frac{A_0}{1 - \beta A_0}. \tag{1.22}$$

As $\beta A_0$ approaches $+1$ the overall gain of the circuit goes to infinity, and the circuit would appear capable of generating a finite output without any input. This corresponds to the output of the amplifer being fed back being in phase with the input, which is referred to as *positive* feedback. In practice, electrical "noise," which is a random oscillatory voltage generated to a greater or lesser extent in all electrical components in any amplifier system, provides a finite input. Because $\beta A_0$ is generally a function of frequency the condition $\beta A_0 = +1$ is generally satisfied only at one frequency. The circuit oscillates at this frequency by amplifying the noise at this frequency which appears at its input. However, the output does not grow infinitely large, because, as the signal grows, $A_0$ falls – this process is called saturation. This phenomenon is fundamental to all oscillator systems. A laser (or maser) is an optical (microwave)-frequency oscillator constructed from an optical (microwave)-frequency amplifier with positive feedback, as shown schematically in Fig. 1.3. Light waves that become amplified on traversing the amplifier are returned through the amplifier by the reflectors and grow in intensity, but this intensity growth does not continue indefinitely because the amplifier saturates. The arrangement of mirrors (and sometimes other components) which provides the feedback is generally referred to as the laser *cavity* or *resonator*.

The characteristics of the device consisting of amplifying medium and resonator will be covered later; for the moment we must concern ourselves with the problem of how to construct an amplifier at optical frequencies, which range from $10^{11}$ Hz to beyond $10^{16}$ Hz. The operating frequencies of masers overlap this frequency range at the low-frequency end, the fundamental difference between the two devices being primarily one of scale. If the length of the resonant cavity that provides feedback is $L$, then for $L \gg \lambda$, where $\lambda$ is the wavelength at which oscillation occurs, we generally have a laser. For $L \sim \lambda$ we usually have a maser, although the development of *microlasers*, which have small cavity lengths, has removed this easy way of distinguishing lasers from masers.

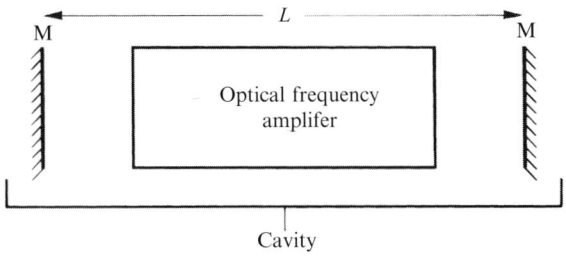


**Fig. 1.3** A schematic diagram of a basic laser structure incorporating an amplifying medium and two feedback mirrors, M.

## 1.8 The energy levels of atoms, molecules, and condensed matter

All particles in nature have distinct states[10] that they can occupy. These states in general have different energies, although it is possible for particles in different states to have the same energy. The term "energy level" is used to describe a particle with a specific, distinct energy, without implying any particular information about its (quantum) state. The lowest energy state, in which a particle is stable, is called the *ground state*. All higher energy states are called *excited* states. Excited states are intrinsically unstable, and a particle occupying one will eventually lose energy and fall to lower energy states. When a particle falls from a higher energy state to a lower, energy is conserved. The energy $\Delta E$ lost by the particle can be emitted as a photon with energy $h\nu = \Delta E$: this is radiative energy loss. The particle can also lose energy *non-radiatively*, in which case the energy is dissipated into heating. Atomic systems have only electronic states, which in the simple Bohr model of the atom correspond to different configurations of electron orbits. The types of energy state that exist in a molecular system are more varied, and include electronic, vibrational, and rotational states. In a molecule, changes in the internuclear separation of the constituent atoms give rise to *vibrational* energy states, which have quantized energies. The various characteristic vibrational motions of a molecule are called its *normal modes*, which for a molecule with $N$ atoms number $3N - 6$, unless the molecule is linear, in which case they number $3N - 5$. The quantized energies of a normal mode can be written as (see Chapter 10 for more details)

$$E_{\text{vib}} \simeq \left(n + \tfrac{1}{2}\right) h\nu_{\text{vib}}, \tag{1.23}$$

and form a ladder of (almost) equally spaced energy levels. Molecules also have quantized rotational energy levels, whose energies can be written as

$$E_{\text{rot}} \simeq BJ(J + 1), \tag{1.24}$$

where $B$ is a rotational energy constant, and $J$ is called the rotational quantum number. The overall energy state of a molecule thus has electronic, vibrational, and rotational components. A molecule in a particular combination of electronic and vibrational states is described as being in a *vibronic* state. A state with a specific combination of vibrational and rotational energies would be described as being in a *vibrot* state. As a rough rule of thumb, transitions between different vibronic states where the electronic state changes lie in the visible spectrum with energy spacings[11] $\sim$20,000 cm$^{-1}$, which corresponds to an energy spacing of $3 \times 10^{10} h/\lambda$ J. Transitions between vibrot states where the electronic energy does not change, but the vibrational state changes, are typically of energy $\sim$1,000 cm$^{-1}$. Transitions between different rotational states where the electronic and vibrational states do

[10] The term "state" in quantum mechanics corresponds to a configuration with a particular "state function," which often corresponds to a specific set of quantum numbers that identify the state.

[11] The cm$^{-1}$ unit is often used to describe energy spacings. A transition at wavelength $\lambda$ (cm) between two levels has an energy spacing characterized by $\lambda$ cm$^{-1}$.

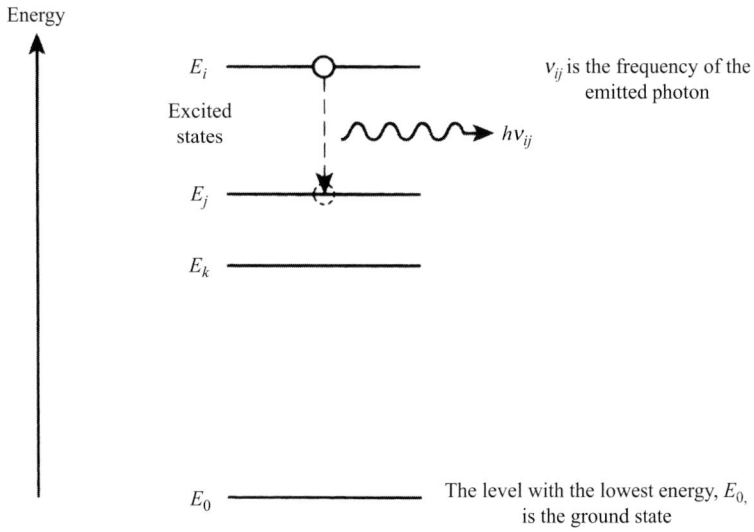


**Fig. 1.4** A simple schematic energy-level diagram for a particle.

not change are typically of energy $\sim 100\ \text{cm}^{-1}$. In practical terms vibrational transitions are typically in the 3–20-μm range, and rotational transitions are typically in the 50–1,000-μm range. In the gas phase the energy levels of atoms or molecules are quite sharp and distinct, as shown schematically in Fig. 1.4, although we shall see later that even these precise energies are "broadened." This broadening occurs for several reasons, but perhaps most importantly because of the interactions between neighboring particles. In condensed matter, whether this be in the solid or in the liquid state, there are very many particles close to any individual particle of interest, and inter-particle interactions are strong. Consequently, the allowed energies of particles in the medium occupy broad, continuous ranges of energy called energy "bands." The lowest-lying energy band, which is analogous to the ground state of an isolated particle, is called the *valence* band. The next highest band of allowed energies is called the *conduction* band. An energy band can be thought of as the result of very many sharp isolated energy states having their energies "smeared" out so that they overlap. We will reserve further discussion of the energy bands in condensed matter until a little later, and for the moment will consider the energy levels of particles as relatively sharp, and not strongly influenced by inter-particle interactions.

## 1.9 Amplification at optical frequencies

How do we build an amplifier at such high frequencies? We use the energy delivered as the particles which constitute the amplifying medium make jumps between their different energy levels. The medium may be gaseous, liquid, a crystalline or glassy solid, an insulating material, or a semiconductor. The electrons that are bound within the particles of the

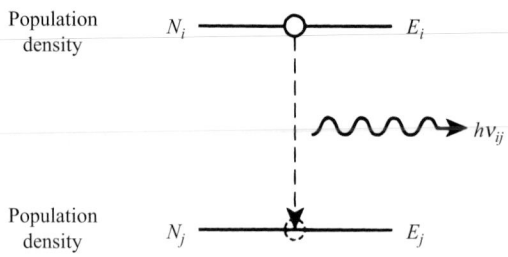


Fig. 1.5 A schematic representation of spontaneous emission between two levels of energy $E_i$ and $E_j$.

amplifying medium, whether these are atoms, molecules, or ions, can occupy only certain discrete energy levels. Consider such a system of energy levels, as shown schematically in Fig. 1.4. Electrons can make jumps between these levels in three ways: spontaneous emission, stimulated emission, and stimulated absorption.

### 1.9.1 Spontaneous emission

When an electron spontaneously falls from a higher energy level to a lower one as shown in Fig. 1.5, the emitted photon has frequency

$$\nu_{ij} = \frac{E_i - E_j}{h}. \tag{1.25}$$

This photon is emitted in a random direction with arbitrary polarization (except in the presence of magnetic fields, but this need not concern us here). The photon carries away momentum $h/\lambda = h\nu/c$, and the emitting particle (atom, molecule, or ion) recoils in the opposite direction. The probability of such a spontaneous jump is given quantitatively by the Einstein $A$ coefficient defined as $A_{ij}$ being the "probability" per second of a spontaneous jump from level $i$ to level $j$.

For example, if there are $N_i$ particles per unit volume in level $i$ then $N_iA_{ij}$ per second make jumps to level $j$. The total rate at which jumps between the two levels are made is

$$\frac{dN_i}{dt} = -N_iA_{ij}. \tag{1.26}$$

There is a negative sign because the population of level $i$ is decreasing.

Generally an electron can make jumps to more than one lower level, unless it is in the first (lowest) excited level. The total probability that the electron will make a spontaneous jump to any lower level is $A_i\ \mathrm{s}^{-1}$, where

$$A_i = \sum_j A_{ij}. \tag{1.27}$$

The summation runs over all levels $j$ that are lower in energy than level $i$, and the total rate at which the population of level $i$ changes by spontaneous emission is

$$\frac{dN_i}{dt} = -N_iA_i,$$

which has the solution

$$N_i = \text{ constant} \times e^{-A_i t}. \tag{1.28}$$

If, at time $t = 0, N_i = N_i^0$, then

$$N_i = N_i^0 e^{-A_i t}, \tag{1.29}$$

so the population of level $i$ falls exponentially with time as electrons leave by spontaneous emission. The time in which the population falls to $1/e$ of its initial value is called the natural lifetime of level $i, \tau_i$, which is given by $\tau_i = 1/A_i$. The magnitude of this lifetime is determined by the actual probabilities of jumps from level $i$ by spontaneous emission. Jumps that are likely to occur are called *allowed* transitions, whereas those which are unlikely are said to be *forbidden*. *Allowed* transitions in the visible region typically have $A_{ij}$ coefficients in the range $10^6$–$10^8$ s$^{-1}$. *Forbidden* transitions in this region have $A_{ij}$ coefficients below $10^4$ s$^{-1}$. These probabilities decrease as the wavelength of the jump increases. Consequently, levels that can decay by allowed transitions in the visible have lifetimes generally shorter than 1 μs, whereas similar forbidden transitions have lifetimes in excess of 10–100 μs. Although no jump turns out to be absolutely forbidden, some jumps are so unlikely that levels whose electrons can fall to lower levels only by such jumps are very long-lived. Levels with lifetimes in excess of an hour have been observed under laboratory conditions. Levels that can decay only slowly, and usually only by forbidden transitions, are said to be *metastable*.

When a particle changes its energy spontaneously the emitted radiation is not, as might perhaps be expected, all at the same frequency. Real energy levels are not infinitely sharp; they are smeared out or *broadened*. A particle in a given energy level can actually have any energy within a finite range. The frequency spectrum of the spontaneously emitted radiation is described by the *lineshape function*, $g(\nu)$. This function is usually normalized so that

$$\int_{-\infty}^{\infty} g(\nu) d\nu = 1. \tag{1.30}$$

Here $g(\nu)d\nu$ is the probability that a photon will be emitted spontaneously in the frequency range $\nu + d\nu$. The lineshape function $g(\nu)$ is a true probability function for the spectrum of emitted radiation, and is usually sharply peaked near some frequency $\nu_0$, as shown in Fig. 1.6. For this reason the function is frequently written as $g(\nu_0, \nu)$ to highlight this. Since negative frequencies do not exist in reality, the following question might properly be asked: "Why does the integral have a lower limit of minus infinity?" This is because $g(\nu)$ can be viewed as the Fourier transform of a real function of time, so negative frequencies have to be permitted mathematically. In practice $g(\nu)$ is only of significant value for some generally small range of positive frequencies, so

$$\int_{0}^{\infty} g(\nu) d\nu \simeq 1. \tag{1.31}$$

The amount of radiation emitted spontaneously by a collection of particles can be described quantitatively by their *radiant intensity* $I_e(\nu)$. The units of radiant intensity are watts per

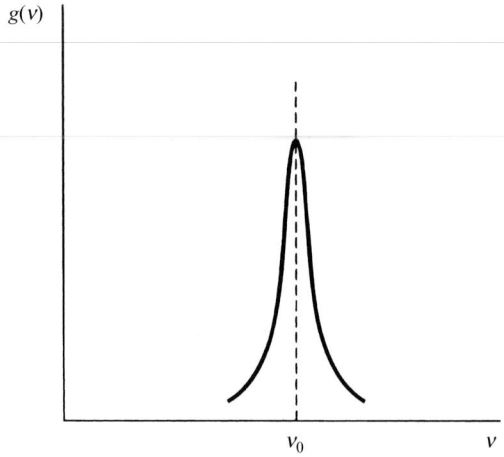


Fig. 1.6 A lineshape function $g(\nu_0, \nu)$.

steradian.[12] The total power (watts) emitted in a given frequency interval $d\nu$ is

$$W(\nu) = \int_S I_e(\nu) d\nu \, d\Omega, \tag{1.32}$$

where the integral is taken over a closed surface $S$ surrounding the emitting particles.

The total power emitted is

$$W_0 = \int_{-\infty}^{\infty} W(\nu) d\nu. \tag{1.33}$$

Here $W(\nu)$ is closely related to the lineshape function

$$W(\nu) = W_0 g(\nu). \tag{1.34}$$

For a collection of $N_i$ identical particles the total spontaneously emitted power per frequency interval is

$$W(\nu) = N_i A_i h\nu g(\nu). \tag{1.35}$$

Clearly this power decreases with time if the number of excited particles decreases.

For a plane electromagnetic wave we can introduce the concept of *intensity*, which has units of W m$^{-2}$. The intensity is the average amount of energy per second transported across unit area in the direction of travel of the wave. The spectral distribution of intensity, $I(\nu)$, is related to the total intensity, $I_0$, by

$$I(\nu) = I_0 g(\nu). \tag{1.36}$$

Although perfect plane waves do not exist, because such waves would have a unique propagation direction and infinite radiant intensity, they can be employed as a useful, simple idealization. To a very good degree of approximation we can treat the light from a small

[12] The steradian is the unit of solid angle, $\Omega$. The surface of a sphere encompasses a solid angle of $4\pi$ steradians.

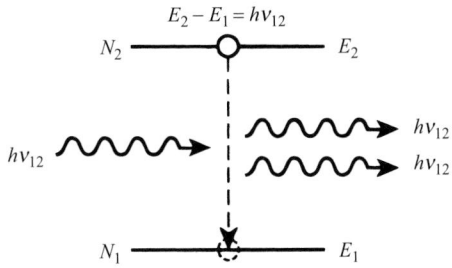


Fig. 1.7 A schematic representation of stimulated emission between two levels of energy $E_2$ and $E_1$.

source as a plane wave if we are far enough away from the source. The light coming from a star viewed outside the Earth's atmosphere is a good example.[13]

### 1.9.2 Stimulated emission

Besides being able to make transitions from a higher level to a lower one by spontaneous emission, electrons can also be stimulated to make these jumps by the action of an external radiation field, as shown in Fig. 1.7.

The probability of the external radiation field causing stimulated emission depends on its energy density, which is written as $\rho(\nu)$ and measured in J m$^{-3}$ Hz$^{-1}$. The rate at which stimulated emissions occur within a small band of frequencies $d\nu$ is

$$\frac{dN_2}{dt}(\nu)d\nu = N_2 B'_{21}(\nu)\rho(\nu)d\nu \text{ s}^{-1} \text{ m}^{-3}, \tag{1.37}$$

where $B'_{21}(\nu)$ is a function specific to the transition between levels 2 and 1, and $N_2$ is the number of particles per unit volume in the upper level of the transition. Stimulated emission will occur if the external radiation field contains energy in a frequency range that overlaps the lineshape function. The frequency dependence of $B'_{21}(\nu)$ is the same as the lineshape function

$$B'_{21}(\nu) = B_{21} g(\nu_0, \nu). \tag{1.38}$$

$B_{21}$ is called the Einstein coefficient for stimulated emission. The total rate of change of particle concentration in level 2 by stimulated emission is

$$\begin{aligned} \frac{dN_2}{dt} &= -N_2 \int_{-\infty}^{\infty} B'_{21}(\nu)\rho(\nu)d\nu \\ &= -N_2 B_{21} \int_{-\infty}^{\infty} g(\nu_0, \nu)\rho(\nu)d\nu. \end{aligned} \tag{1.39}$$

Note that, for the dimensions of the two sides of Eq. (1.39) to balance, $B_{21}$ must have units of m$^3$ J$^{-1}$ s$^{-2}$. To evaluate the integral in Eq. (1.39) we must consider how energy density is related to intensity and varies with frequency.

[13] The plane waves coming from a star are distorted by the atmosphere because of density and refractive-index fluctuations referred to as *atmospheric turbulence*.

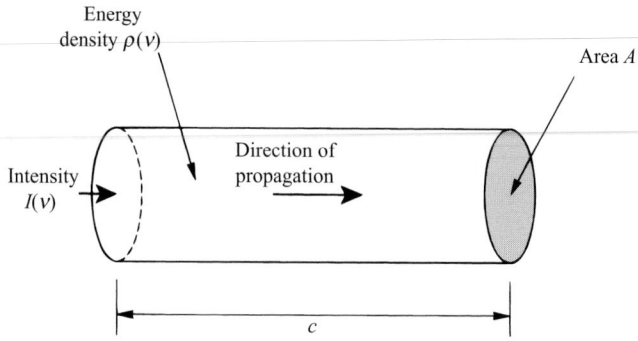


**Fig. 1.8** A volume of space swept through per second by part of a plane wave.

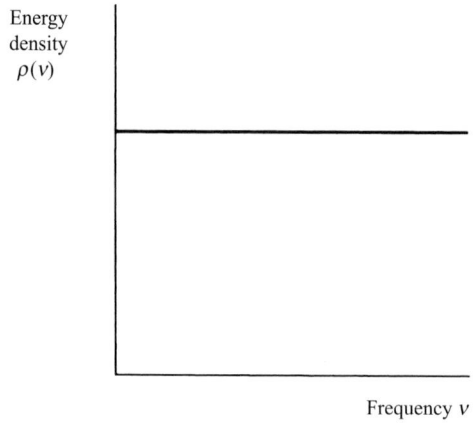


**Fig. 1.9** A "white" energy-density spectrum.

## 1.10 The relation between energy density and intensity

The energy density of a radiation field $\rho(\nu)$ (energy per unit volume per unit frequency interval) can be simply related to the intensity of a plane electromagnetic wave. If the intensity of the wave is $I(\nu)$ (for example, watts per unit area per frequency interval), then

$$\rho(\nu) = \frac{I(\nu)}{c}, \tag{1.40}$$

where $c$ is the velocity of light in the medium.[14] This is illustrated in Fig. 1.8. All the energy stored in the volume of length $c$ passes across the plane $A$ in one second, so

$$\rho(\nu) \times A \times c = I(\nu) \times A. \tag{1.41}$$

The energy density in a general radiation field $\rho(\nu)$ is a function of frequency $\nu$. If $\rho(\nu)$ is independent of frequency the radiation field is said to be *white*, as shown in Fig. 1.9.

[14] Here $c = c_0/n$, where $c_0$ is the velocity of light in a vacuum and $n$ is the *refractive index*.

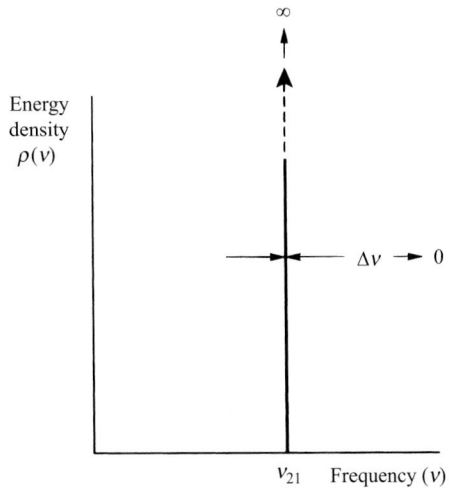


**Fig. 1.10** A monochromatic energy-density spectrum.

If the radiation field is *monochromatic* at frequency $\nu_{21}$, its spectrum is as shown in Fig. 1.10. The ideal *monochromatic* radiation field has an infinitely high energy density profile of zero frequency width at frequency $\nu_{21}$. This type of profile is called a $\delta$-function. The properties of this function are described in more detail in Appendix 2.

The $\delta$-function has the property

$$\delta(\nu - \nu_{21}) = 0 \text{ for } \nu \neq \nu_{21}, \tag{1.42}$$

and

$$\int_{-\infty}^{\infty} \delta(\nu - \nu_{21}) d\nu = 1. \tag{1.43}$$

For a general radiation field, such as that shown in Fig. 1.11, the total energy stored per unit volume between frequencies $\nu_1$ and $\nu_2$ is $\int_{\nu_1}^{\nu_2} \rho(\nu) d\nu$. For a monochromatic radiation field at frequency $\nu_{21}$

$$\rho(\nu) = \rho_{21}\delta(\nu - \nu_0). \tag{1.44}$$

So for a monochromatic radiation field the total stored energy per unit volume is

$$\int_{-\infty}^{\infty} \rho(\nu) d\nu = \int_{-\infty}^{\infty} \rho_{21}\delta(\nu - \nu_{21}) d\nu = \rho_{21}. \tag{1.45}$$

For such a monochromatic radiation field Eq. (1.39) can be written as

$$\begin{aligned} \frac{dN_2}{dt} &= -N_2 B_{21} \int_{-\infty}^{\infty} g(\nu_0, \nu)\rho_{21}\delta(\nu - \nu_{21}) d\nu \\ &= -N_2 B_{21} g(\nu_0, \nu_{21})\rho_{21}. \end{aligned} \tag{1.46}$$

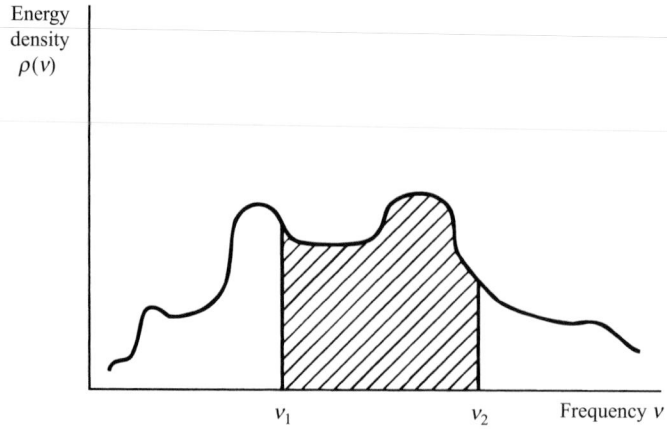


**Fig. 1.11** A generalized energy-density spectrum.

It is very important to note that the rate of stimulated emissions produced by this input monochromatic radiation is directly proportional to the value of the lineshape function at the input frequency. The maximum rate of stimulated emission is produced, all else being equal, if the input radiation is at the line center frequency $\nu_0$.

If the stimulating radiation field has a spectrum that is broad, we can assume that the energy density $\rho(\nu)$ is constant over the narrow range of frequencies for which $g(\nu_0, \nu)$ is significant. In this case Eq. (1.39) gives

$$\frac{dN_2}{dt} = -N_2 B_{21} \rho(\nu), \tag{1.47}$$

where $\rho(\nu) \simeq \rho(\nu_0)$ is the energy density in the frequency range where transitions take place.

### 1.10.1 Stimulated absorption

As well as making stimulated jumps in a downward direction, electrons may make transitions in an upward direction between energy levels of a particle by absorbing energy from an electromagnetic field, as shown in Fig. 1.12. The rate of such absorptions, and the rate at which electrons leave the lower level, is

$$\frac{dN_1}{dt} = N_1 \rho(\nu) B_{12} g(\nu_0, \nu) \text{ s}^{-1} \text{ Hz}^{-1} \text{ m}^{-3}, \tag{1.48}$$

which yields a result similar to Eq. (1.47):

$$\frac{dN_1}{dt} = -N_1 B_{12} \int_{-\infty}^{\infty} g(\nu_0, \nu) \rho(\nu) d\nu. \tag{1.49}$$

Once again $B_{12}$ is a constant specific to the jump, and is called the Einstein coefficient for stimulated absorption. Here again $\rho(\nu)$ is the energy density of the stimulating field. There is no analog in the absorption process to spontaneous emission. A particle cannot

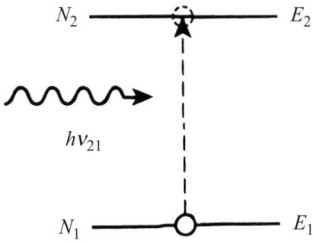


**Fig. 1.12** A schematic representation of stimulated absorption between two levels of energy $E_1$ and $E_2$.

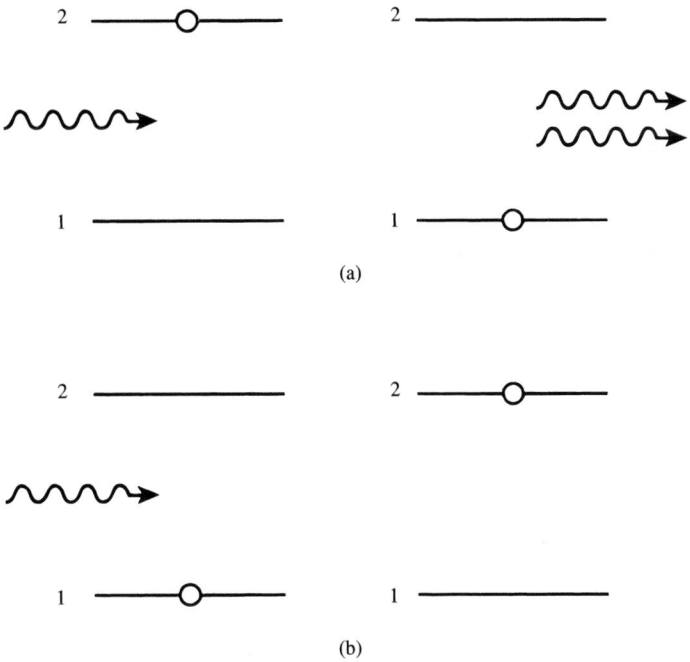


**Fig. 1.13** A photon–particle "collision" picture of the stimulated emission and absorption processes: (a) stimulated emission, and (b) absorption.

spontaneously *gain* energy without an external energy supply. Thus, it is unnecessary for us to continue to describe the absorption process as stimulated absorption.

It is interesting to view both stimulated emission and absorption as photon–particle collision processes. In *stimulated* emission the incident photon produces an identical photon by "colliding" with the electron in an excited level, as shown in Fig. 1.13(a). After the stimulated emission process, both photons are traveling in the same direction and with the same polarization as the incident photon originally had. When light is described in particle terms, polarization can be viewed as describing the angular motion or spin of individual photons. Left- and right-hand circularly polarized light corresponds in this particle picture to beams of photons that spin clockwise and counterclockwise, respectively, about their direction of propagation. Linearly polarized light corresponds to a beam of photons that have no net angular momentum about an axis parallel to their direction of propagation. In stimulated emission the stimulated photon has exactly the same frequency as the stimulating photon. In absorption the incident photon disappears, as shown in Fig. 1.13(b). In both stimulated emission and absorption the atom recoils to conserve linear momentum.

## 1.11 The intensity of a beam of electromagnetic radiation in terms of photon flux

If the intensity of a beam of light is $I(\nu)$ (W m$^{-2}$ per frequency interval) then the number of photons in the beam crossing unit area per unit time is

$$N_{\text{photons}} = \frac{I(\nu)}{h\nu} \text{ (photons s}^{-1}\text{ m}^{-1}\text{ per frequency interval).} \tag{1.50}$$

If the beam is monochromatic and has total intensity $I(\nu_{21})$ then

$$N_{\text{photons}} = \frac{I(\nu_{21})}{h\nu} \text{ photons s}^{-1}\text{ m}^{-2}. \tag{1.51}$$

## 1.12 Black-body radiation

A collection of particles in thermal equilibrium is described by a common temperature $T$ K. Although the collection of particles is described as being in "equilibrium," this is a dynamic equilibrium. The processes of spontaneous emission, stimulated emission, and absorption continuously occur, because there is always a radiation field present. Even though no external radiation field is supplied, the thermal background radiation is always present. This radiation is called *black-body* radiation, and constantly interacts with each particle. Black-body radiation is so called because of the special characteristics of the radiation emitted and absorbed by a *black body*. Such a body absorbs with 100% efficiency all the radiation falling on it, irrespective of the radiation frequency. A close approximation to a black body (absorber and emitter) is an enclosed cavity containing a small hole, as shown in Fig. 1.14. Radiation that enters the hole has very little chance of escaping. If the inside of this cavity is in thermal equilibrium, it must lose as much energy as it absorbs, and the emission from the hole is therefore characteristic of the equilibrium temperature $T$ inside the cavity. Thus this type of radiation is often called "thermal" or "cavity" radiation. Black-body radiation has a spectral distribution of the kind shown in Fig. 1.15.

In the early days of the quantum theory, the problem of describing theoretically the spectral profile of the emission from a black body was a crucial one. In the latter part of the nineteenth century experimental measurements of this spectral profile had already been obtained, and the data had even been fitted to an empirical formula. Attempts to explain the form of the data were based on treating the electromagnetic radiation as a collection of

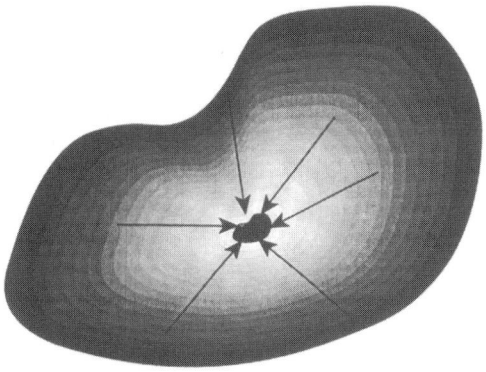

Fig. 1.14 A simple model of a black-body absorber/emitter – an enclosed cavity containing a small hole.

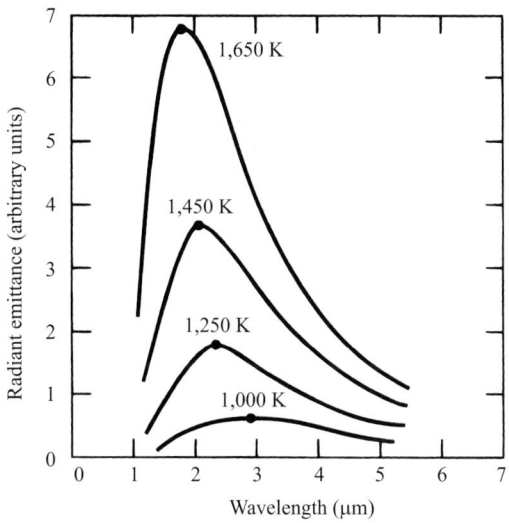


Fig. 1.15 The spectral distribution of black-body radiation at different temperatures.

oscillators, with each oscillator having its own characteristic frequency. The problem was to determine how many oscillations at a given frequency could be fitted inside a cavity.

Thermodynamically, the shape of the cavity for which the calculation is performed is arbitrary (provided that it is much larger than the wavelength of any of the oscillations), otherwise we could make a heat engine by connecting together cavities of different shapes. If, for example, two cavities of different shapes, but at the same temperature, were connected together with a reflective hollow pipe, we could imagine placing filters having different narrow-frequency bandpass characteristics in the pipe. Unless the radiation emitted in each elemental frequency band from both cavities was identical, one cavity could be made to heat up and the other cool down, thereby violating the second law of thermodynamics. For convenience, we choose a cubical cavity with sides of length $L$. A plane electromagnetic wave will "fit" inside this cavity if it satisfies appropriate *periodic* boundary conditions. If the wave has a spatial variation written in complex notation as $e^{j\mathbf{k}\cdot\mathbf{r}}$ these boundary conditions can be written as

$$e^{jk_x x} = e^{2k_x(x+L)}, \tag{1.52}$$

with similar equations for the $y$- and $z$-directed components of the wave. Equations like (1.52) are satisfied if

$$k_x = \frac{2\pi\ell}{L}, \qquad k_y = \frac{2\pi m}{L}, \qquad k_z = \frac{2\pi n}{L}, \tag{1.53}$$

where $k_x, k_y$, and $k_z$ are the components of the wave vector $\mathbf{k}$ of the oscillation, and $\ell, m$, and $n$ are integers:

$$\mathbf{k} = \frac{\omega}{c}\hat{\mathbf{k}}, \quad |\mathbf{k}| = 2\pi/\lambda, \tag{1.54}$$

where $\omega$ is the angular frequency of the wave, $\lambda$ is its wavelength, $c$ is the velocity of light, and $\hat{\mathbf{k}}$ is a unit vector in the direction of the wave vector.

In a three-dimensional space whose axes are $k_x, k_y$, and $k_z$ (called $\mathbf{k}$-space) the possible $k$ values that are periodic inside the cube form a lattice as shown in Fig. 1.16. The size of a unit cell of this lattice is $(2\pi/L)^3$. Each cell corresponds to one possible *mode*, characterized by its own values of $k_x, k_y$, and $k_z$, which are periodic inside the cube. Note that the spacing of adjacent modes, say in the $k_x$ direction, is $2\pi/L$. Thus the permitted values of $k_x$, namely $0, 2\pi/L, 4\pi/L$ etc., correspond to oscillation wavelengths $\infty, L, L/2, L/3$ etc. Thus, in counting the permitted values of $k_x$, all the intuitive values of $\lambda$ which "fit"

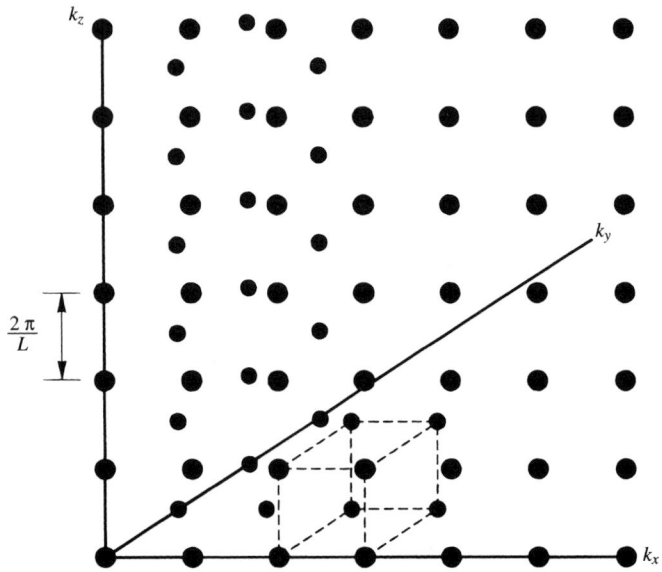


**Fig. 1.16** Allowed values of $k_x$, $k_y$, and $k_z$ in $\mathbf{k}$-space for a cubical cavity of side $L$.

between the walls of the cavity are included. The total number of modes of oscillation with $|\mathbf{k}| \leq k$ is

$$N_k = \frac{\text{total volume of } \mathbf{k}\text{-space with } |\mathbf{k}| \leq k}{\text{volume of unit cell}} \times 2$$
$$= 2 \times \frac{4}{3}\pi k^3/(2\pi/L)^3. \tag{1.55}$$

The factor of 2 enters because we must take account of the two distinct polarizations of the radiation field. Therefore,

$$N_k = \frac{k^3 L^3}{3\pi^2}. \tag{1.56}$$

Now, since $k = 2\pi\nu/c$, the number of modes with frequency $\leq \nu$ is

$$N_\nu = \frac{8\pi\nu^3}{3c^3}L^3. \tag{1.57}$$

Since $L^3$ is the volume of the enclosure, $V$,

$$\frac{N_\nu}{V} = \frac{8\pi\nu^3}{3c^3}. \tag{1.58}$$

The mode density (per unit volume per unit frequency interval) is

$$p(\nu) = \frac{1}{V}\frac{dN_\nu(\nu)}{d\nu} = \frac{8\pi\nu^2}{c^3}. \tag{1.59}$$

Rayleigh and Jeans attempted to use this type of frequency dependence to describe the spectral composition of black-body radiation. They assumed that the law of equipartition of energy held for the distribution of energy among the various modes of the radiation field. Consequently, they assigned an equal energy $k_B T$ to each mode of oscillation, and predicted that the intensity distribution of the black-body radiation would be

$$I(\nu) \propto \frac{8\pi\nu^2}{c^3}. \tag{1.60}$$

This did not fit experimental observations of black-body radiation except at relatively low frequencies (in the red and infrared regions of the spectrum). The predicted large increase in $I(\nu)$ at high frequencies was in dramatic conflict with observations of black-body emission intensities in the ultraviolet. This conflict between theory and experiment was called the *ultraviolet catastrophe*, as shown in Fig. 1.17. Planck resolved these difficulties with his quantum hypothesis. He proposed that each oscillation mode could only take certain quantized energies

$$E_{n_\nu} = \left(n + \frac{1}{2}\right)h\nu, \qquad n = 0, 1, 2, 3, ..., \tag{1.61}$$

where the contribution $\frac{1}{2}h\nu$ is called the zero-point energy. The probability of finding energy $E_n$ in a particular mode of oscillation is given by classical Maxwell–Boltzmann statistics, i.e.,

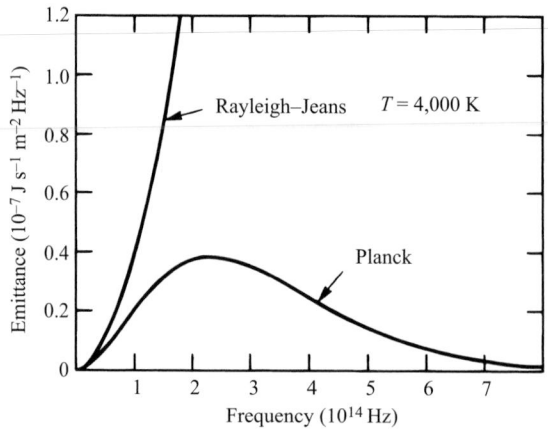


**Fig. 1.17** The Rayleigh–Jeans prediction of the spectral intensity of a black-body compared with the Planck formula.

$$\frac{P(n)}{P(0)} = \frac{e^{-E_n/(k_B T)}}{e^{-E_0/(k_B T)}} = e^{-nh\nu/(k_B T)}, \tag{1.62}$$

where $h$ is Planck's constant, $k_B$ is Boltzmann's constant, $T$ is the absolute temperature, and $P(0)$ is the probability of finding the lowest energy in the mode. Consequently the average energy of a mode is

$$\overline{E}_\nu = \sum_{n=0}^{\infty} \text{energy of excitation} \times \text{probability} = \sum_{n=0}^{\infty} P(n)E_n, \tag{1.63}$$

giving

$$\overline{E}_\nu = \sum_{n=0}^{\infty} P(0)e^{-nh\nu/(k_B T)} \left(n + \frac{1}{2}\right) h\nu. \tag{1.64}$$

Now, if a particular oscillation is excited, it must be in one of the quantized states, therefore

$$\sum_{n=0}^{\infty} P(n) = 1,$$

so

$$\sum_{n=0}^{\infty} P(0)e^{-nh\nu/(k_B T)} = 1. \tag{1.65}$$

Thus

$$P(0) = \frac{1}{\sum_{n=0}^{\infty} e^{-nh\nu/(k_B T)}}, \tag{1.66}$$

and

$$\overline{E}_\nu = \frac{\sum_{n=0}^{\infty}(n+\frac{1}{2})h\nu e^{-nh\nu/(k_\mathrm{B}T)}}{\sum_{n=0}^{\infty} e^{-nh\nu/(k_\mathrm{B}T)}} \tag{1.67}$$

$$= \frac{1}{2}h\nu + \frac{h\nu e^{-h\nu/(k_\mathrm{B}T)} + 2h\nu e^{-2h\nu/(k_\mathrm{B}T)} + \cdots + nh\nu e^{-nh\nu/(k_\mathrm{B}T)}}{1 + e^{-h\nu/(k_\mathrm{B}T)} + e^{-2h\nu/(k_\mathrm{B}T)} + \cdots + e^{-nh\nu/(k_\mathrm{B}T)}} \tag{1.68}$$

$$= \frac{1}{2}h\nu - \frac{[d/d(1/(k_\mathrm{B}T))]\left(e^{-h\nu/(k_\mathrm{B}T)} + e^{-2h\nu/(k_\mathrm{B}T)} + \cdots\right)}{1/\left(1 - e^{-h\nu/(k_\mathrm{B}T)}\right)} \tag{1.69}$$

$$= \frac{1}{2}h\nu - \frac{[d/d(1/(k_\mathrm{B}T))]\left[(1/(1 - e^{-h\nu/(k_\mathrm{B}T)})\right]}{1/\left(1 - e^{-h\nu/(k_\mathrm{B}T)}\right)} \tag{1.70}$$

$$= \frac{1}{2}h\nu + \frac{h\nu e^{-h\nu/(k_\mathrm{B}T)}}{\left(1 - e^{-h\nu/(k_\mathrm{B}T)}\right)^2}(1 - e^{-h\nu/(k_\mathrm{B}T)}). \tag{1.71}$$

We have used a mathematical "trick" here to sum the series in the numerator by recognizing that it is the derivative of a geometric series that we can sum easily. So

$$\overline{E}_\nu = \frac{1}{2}h\nu + \frac{h\nu}{e^{h\nu/(k_\mathrm{B}T)} - 1}. \tag{1.72}$$

This is the average energy per mode. Consequently, the stored energy in the black-body radiation field treated as a collection of quantized oscillators is

$$\rho(\nu) = p(\nu)\overline{E}_\nu = \frac{8\pi h\nu^3}{c^3}\left(\frac{1}{2} + \frac{1}{e^{h\nu/(k_\mathrm{B}T)} - 1}\right). \tag{1.73}$$

The factor of $\frac{1}{2}$ comes from zero-point energy that cannot be released, so the available stored energy in the field is

$$\rho(\nu) = \frac{8\pi h\nu^3}{c^3}\left(\frac{1}{e^{h\nu/(k_\mathrm{B}T)} - 1}\right). \tag{1.74}$$

This formula predicts exactly the observed spectral character of black-body radiation, and was the first spectacular success of quantum theory. Note that

$$\rho(\nu) = \frac{8\pi h\nu^3}{c^3}\left(\frac{1}{e^{h\nu/(k_\mathrm{B}T)} - 1}\right) = \frac{8\pi\nu^2}{c^2} \times h\nu \times \left(\frac{1}{e^{h\nu/(k_\mathrm{B}T)} - 1}\right), \tag{1.75}$$

which is the number of modes per volume per frequency interval $\times$ the photon energy $\times 1/(e^{h\nu/(k_\mathrm{B}T)} - 1)$. The quantity $1/(e^{h\nu/(k_\mathrm{B}T)} - 1)$ represents the average number of photons in each mode, this is called the *occupation number* of the modes of the field.

## 1.13 The relation between the Einstein *A* and *B* coefficients

We can derive a useful relationship between Einstein's $A$ and $B$ coefficients by considering a collection of atoms in thermal equilibrium inside a cavity at temperature $T$. The energy density of the radiation within the cavity is given by

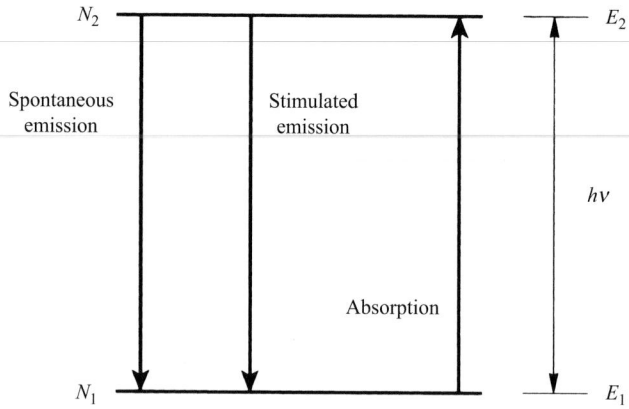


**Fig. 1.18** Radiative processes connecting two energy levels in thermal equilibrium at temperature $T$.

$$\rho(\nu) = \frac{8\pi h\nu^3}{c^3}\left(\frac{1}{e^{h\nu/(k_B T)} - 1}\right), \tag{1.76}$$

since the radiation in thermal equilibrium in the cavity will be black-body radiation. Although real atoms in such a cavity possess many energy levels, we can restrict ourselves to considering the dynamic equilibrium between any two of them, as shown in Fig. 1.18. The jumps which occur between two such levels as a result of interaction with radiation essentially occur independently of the energy levels of the system which are not themselves involved in the jump.

In thermal equilibrium the populations $N_2$ and $N_1$ of these two levels are constant, so

$$\frac{dN_2}{dt} = \frac{dN_1}{dt} = 0, \tag{1.77}$$

and the rates of transfer between the levels are equal. Since the energy density of a black-body radiation field varies very little over the range of frequencies where transitions between levels 2 and 1 take place we can use Eqs. (1.26) and (1.47), and write

$$\frac{dN_2}{dt} = -N_2 B_{21}\rho(\nu) - A_{21}N_2 + N_1 B_{12}\rho(\nu). \tag{1.78}$$

Therefore, substituting from Eq. (1.74) gives

$$N_2\left[B_{21}\frac{8\pi h\nu^3}{c^3(e^{h\nu/(k_B T)} - 1)} + A_{21}\right] = N_1\left[B_{12}\frac{8\pi h\nu^3}{c^3(e^{h\nu/(k_B T)} - 1)}\right]. \tag{1.79}$$

For a collection of particles that obeys Maxwell–Boltzmann statistics, in thermal equilibrium energy levels of high energy are less likely to be occupied than levels of low energy. In exact terms the ratio of the population densities of two levels whose energy difference is $h\nu$ is

$$\frac{N_2}{N_1} = e^{-h\nu/(k_B T)}. \tag{1.80}$$

So

$$\frac{8\pi h\nu^3}{c^3(e^{h\nu/(k_B T)} - 1)} = \frac{A_{21}}{B_{12}e^{h\nu/(k_B T)} - B_{21}}. \tag{1.81}$$

This equality can be satisfied only if

$$B_{12} = B_{21} \tag{1.82}$$

and

$$\frac{A_{21}}{B_{21}} = \frac{8\pi h\nu^3}{c^3}, \tag{1.83}$$

so a single coefficient $A_{21}$ (say) will describe both stimulated emission and absorption. Equations (1.81) and (1.83) are called the *Einstein relations*. The stimulated emission rate is $W_{21}$, where

$$W_{21} = B_{21}\rho(\nu) = \frac{c^3 A_{21}}{8\pi h\nu^3}\rho(\nu), \tag{1.84}$$

which is proportional to the energy density. The spontaneous emission rate is $A_{21}$, which is independent of external radiation.

If we ignored $A_{21}$ in describing the equilibration of the two energy levels previously considered, then in equilibrium we would have

$$N_2 B_{21} = N_1 B_{12}, \tag{1.85}$$

and if $B_{12} = B_{21}$ this would imply $N_1 = N_2$, which we know cannot be so.

Although spontaneous emission would appear to be a different kind of radiative process from stimulated emission, in fact it is not. The modes of the radiation field contain photons. Even a mode of the radiation field containing no photons has a zero-point energy of $\frac{1}{2}h\nu$. This zero-point energy cannot be extracted from the mode, but nonetheless it plays a role in producing spontaneous emission. Spontaneous emission is stimulated emission resulting from this zero-point energy.

## 1.14 The effect of level degeneracy

In real systems containing atoms, molecules, or ions, it frequently happens that different configurations of the system can have exactly the same energy. If a given energy level corresponds to a number of different arrangements specified by an integer $g$, we call $g$ the *degeneracy* of the level. We call the separate states of the system with the same energy *sub-levels*. The levels 2 and 1 that we have been considering may consist of a number of degenerate sub-levels, such that each sub-level has the same energy, as shown in Fig. 1.19, with $g_2$ sub-levels making up level 2 and $g_1$ sub-levels making up level 1. For each of the sub-levels of levels 1 and 2, with populations $n_1$ and $n_2$, respectively, the ratio of

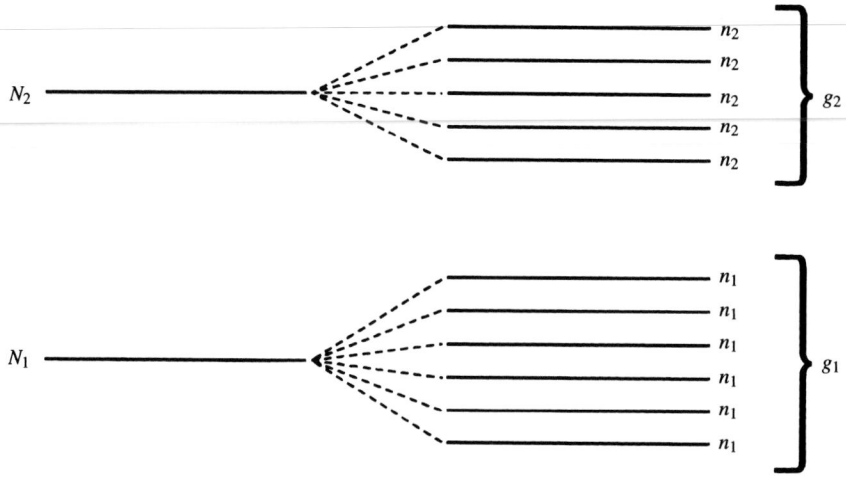


**Fig. 1.19** Two energy levels, each of which has a number of sub-levels of the same energy.

populations is

$$\frac{n_2}{n_1} = e^{-h\nu/(k_B T)}, \tag{1.86}$$

and

$$N_1 = g_1 n_1, \quad N_2 = g_2 n_2. \tag{1.87}$$

Therefore

$$\frac{n_2}{n_1} = \frac{g_1 N_2}{g_2 N_1}, \tag{1.88}$$

and

$$\frac{N_2}{N_1} = \frac{g_2}{g_1} e^{-h\nu/(k_B T)}. \tag{1.89}$$

From Eqs. (1.79) and (1.89) it follows that in this case, where degenerate levels are involved, the Einstein relations become

$$g_1 B_{12} = g_2 B_{21}, \tag{1.90}$$

and, as before,

$$\frac{A_{21}}{B_{21}} = \frac{8\pi h\nu^3}{c^3}. \tag{1.91}$$

Note that

$$A_{21} = B_{21}\frac{8\pi h\nu^3}{c^3} = B_{21}\frac{8\pi \nu^2}{c^3} h\nu, \tag{1.92}$$

which can be described as

$$B_{21} \times \text{no. of modes per unit volume per frequency interval} \times \text{photon energy}.$$

If there were only one photon in each mode of the radiation field, then the resulting energy density would be

$$\rho(\nu) = p(\nu)\overline{E}_\nu = \frac{8\pi\nu^2}{c^3}h\nu. \tag{1.93}$$

The resulting number of stimulated transitions would be

$$W_{21} = B_{21}\frac{8\pi\nu^2}{c^3}h\nu = A_{21}. \tag{1.94}$$

Thus, the number of spontaneous transitions per second is equal to the number of stimulated transitions per second that would take place if just one photon were excited in each mode.

## 1.15 The ratio of spontaneous and stimulated transitions

It is instructive to examine the relative rates at which spontaneous and stimulated processes occur in a system in equilibrium at temperature $T$. This ratio is

$$R = \frac{A_{21}}{B_{21}\rho(\nu)}. \tag{1.95}$$

We choose the $\rho(\nu)$ appropriate to a black-body radiation field, since such radiation is always present to interact with an excited atom that is contained within an enclosure at temperature $T$:

$$R = \frac{A}{B\rho(\nu)} = e^{h\nu/(k_B T)} - 1. \tag{1.96}$$

If we use $T = 300$ K and examine the *microwave region*, $\nu = 10^{10}$ (say), then

$$\frac{h\nu}{k_B T} = \frac{6.626 \times 10^{-34} \times 10^{10}}{1.38 \times 10^{-23} \times 300} = 1.6 \times 10^{-3},$$

so

$$R = e^{0.0016} - 1 \approx 0.0016,$$

and stimulated emission dominates over spontaneous. In particular, in any microwave laboratory experiment,

$$\rho(\nu)_{\text{laboratory created}} > \rho(\nu)_{\text{black-body}},$$

and spontaneous emission is negligible. However, spontaneous emission is still observable as a source of noise – the randomly varying component of the optical signal.

In the *visible region*

$$\nu \approx 10^{15} \frac{h\nu}{(k_B T)} \approx 160 \text{ and } A \gg B\rho(\nu).$$

So, in the visible and near-infrared region spontaneous emission generally dominates unless we can arrange for there to be several photons in a mode. The average number of photons in a mode in the case of black-body radiation is

$$\overline{n}(\nu) = \frac{1}{e^{h\nu/(k_B T)} - 1}, \tag{1.97}$$

which is very small in the visible and infrared.

## 1.16 Problems

1.1 In a *dispersive* medium the refractive index varies with wavelength. We can define a *group* refractive index by the relation

$$n_g = n - \lambda \frac{dn}{d\lambda}.$$

(i) Prove that $n_g = n + \nu\, dn/d\nu$.

(ii) Prove that if a black-body cavity is filled with such a dispersive material then the radiation mode density, $p_\nu$, satisfies

$$p_\nu = \frac{8\pi \nu^2 n^2 n_g}{c_0^3}.$$

(iii) Prove also that in such a situation

$$\frac{A_{21}}{B_{21}} = \frac{8\pi h \nu^3 n^2 n_g}{c_0^3}.$$

1.2 Calculate the photon flux (photons $m^{-2}$ $s^{-1}$) in a plane monochromatic wave of intensity 100 W $m^{-2}$ at wavelengths of (a) 100 nm and (b) 100 μm.

1.3 What is the total number of modes per unit volume for visible light?

1.4 Prove Wien's displacement law for black-body radiation, namely

$$\lambda_m T = 2898 \text{ μm K},$$

where $\lambda_m$ is the wavelength of peak emission of the black body at absolute temperature $T$ K.

1.5 What fraction of the total emission of a normalized Lorentzian lineshape of full width at half maximum (FWHM) $\Delta\nu$ occurs between $\nu_0 - \Delta\nu$ and $\nu_0 + \Delta\nu$? (See Eq. (2.15).)

1.6 Estimate the total force produced by the photon pressure of the Sun on an aluminum sheet of area $10^6$ $m^2$ situated on the surface of the Earth. Use sensible values for the parameters of the problem in order to produce a numerical result.

1.7 Calculate the total stored energy in a box of volume 1 $m^3$ that lies between the wavelengths 10.5 μm and 10.7 μm at a temperature of 3,000 K.

1.8 What would the spectral energy density distribution be if "black-body" radiation could occupy only *two-photon* states? Namely, the only allowed energies for the various modes of a cavity would be

$$E_n = (2n + 1)h\nu.$$

Hint: first calculate the mode density, and then average energy per mode.

1.9 At what temperature would the stimulated and spontaneous emission rates be equal for particles in a cavity and a transition at a wavelength of 1 μm?

1.10 A point source emits a Lorentzian line of FWHM 1 GHz at $\lambda_0 = 500$ nm and total radiated power of 1 W. A square bandpass optical filter with transmittance 0.8 between $\nu_0 - \Delta\nu$ and $\nu_0$ covers an optical detector of active surface area 10 $mm^2$ placed 1 m from the source. What light intensity reaches the detector?
Hint: $\int [1/(1 + x^2)]dx = \arctan x$.

## Further reading

W. T. Silfvast, *Laser Fundamentals*, 2nd edn., Cambridge: Cambridge University Press, 2004.

O. Svelto, *Principles of Lasers*, 5th edn., New York: Springer, 2010.

A. E. Siegman, *Lasers*, Mill Valley, CA: University Science Books, 1985.

P. W. Milonni and J. H. Eberly, *Lasers*, 2nd edn., New York: Wiley, 2010.

A. Yariv, *Optical Electronics in Modern Communications*, 5th edn., Oxford: Oxford University Press, 1997.

A. Yariv, *Quantum Electronics*, 3rd edn., New York: Wiley, 1989.

## References

[1] S. Y. Liao, *Microwave Solid-State Devices*, Englewood Cliffs, NJ: Prentice-Hall, 1985.

[2] A. Einstein, "Zur Quantentheorie der Strahlung," *Phys. Z.* **18**, 121–128, 1917.

[3] C. J. Humpreys and W. F. Meggers, "Further description and analysis of the first spectrum of xenon," *Bureau Standards J. Res.* **10**, 139–149, 1932.

[4] J. P. Gordon, H. J. Zeiger, and C. H. Townes, "Molecular microwave oscillator and new hyperfine structure in the microwave spectrum of $NH_3$," *Phys. Rev.* **95**, 282–284, 1954.

[5] J. P. Gordon, H. J. Zeiger, and C. H. Townes, "The maser – new type of microwave amplifier, frequency standard, and spectrometer," *Phys. Rev.* **99**, 1264–1274, 1955.

[6] J. Weber, "Amplification of microwave radiation by substances not in thermal equilibrium," *Trans. IRE Professional Group Electron Devices* **3**, June, 1–4, 1953.

[7] N. G. Basov and A. M. Prokhorov, "Theory of a molecular oscillator and a molecular power amplifier," *Dokl. Akad. Nauk SSSR* **101**, 47–49, 1955.

[8] T. Maiman, "Stimulated optical radiation in ruby," *Nature* **187**, 493–494, 1960.

[9] A. Javan, W. R. Bennett Jr., and D. R. Herriott, "Population inversion and continuous optical maser oscillation in a gas discharge containing a He–Ne mixture," *Phys. Rev. Lett.* **6**, 106–110, 1961.

[10] J. Clerk Maxwell, *A Treatise on Electricity and Magnetism*, Oxford: Clarendon Press, 1873 (now available in an edition edited by T. F. Torrance, Edinburgh: Scottish Academic Press, 1982).

[11] R. Kohlrausch and W. E. Weber, "Über die Elektrizitätsmenge, welche bei galvanischen Strömen durch den Querschnitt der Kette fließt," *Ann. Phys.* **99**, 10–25, 1856.

[12] E. B. Rosa and N. E. Dorsey, "A new determination of the ratio of the electromagnetic to the electrostatic unit of electricity," *Bull. Bureau of Standards* **3**, 433–540, 1907.

# 2 Optical frequency amplifiers

## 2.1 Introduction

When an electromagnetic wave propagates through a medium stimulated emissions increase the intensity of the wave, while absorptions diminish it. The overall intensity will increase if the number of stimulated emissions can be made larger than the number of absorptions. If we can create such a situation then we have built an amplifier that operates through the mechanism of stimulated emission. This *laser amplifier*, in common with electronic amplifiers, has useful gain only over a particular frequency bandwidth. Its operating frequency range will be determined by the lineshape of the transition, and we expect the frequency width of its useful operating range to be of the same order as the width of the lineshape. It is very important to consider how this frequency width is related to the various mechanisms by which transitions between different energy states of a particle are smeared out over a range of frequencies. This *line broadening* affects in a fundamental way not only the frequency bandwidth of the amplifier, but also its gain. A laser amplifier can be turned into an oscillator by supplying an appropriate amount of positive feedback. The level of oscillation will stabilize because the amplifier saturates. Laser amplifiers fall into two categories, which saturate in different ways. A *homogeneously* broadened amplifier consists of a number of amplifying particles that are essentially equivalent, whereas an *inhomogeneously* broadened amplifier contains amplifying particles with a distribution of amplification characteristics.

## 2.2 Homogeneous line broadening

All energy states, except the lowest energy state of a particle (the ground state) cover a range of possible energies. This is reflected in the lineshape function, which shows the range of energies for a transition between energy states of which one or two are broadened. At the fundamental level this smearing out of the energy is caused by the uncertainty involved in energy measurement. This gives rise to an intrinsic and unavoidable amount of line broadening called *natural broadening*.

### 2.2.1 Natural broadening

This most fundamental source of line broadening arises, as just mentioned, because of uncertainty in the exact energy of the states involved. This uncertainty in measured energy, $\Delta E$, arises from the time uncertainty, $\Delta t$, involved in making such a measurement. The

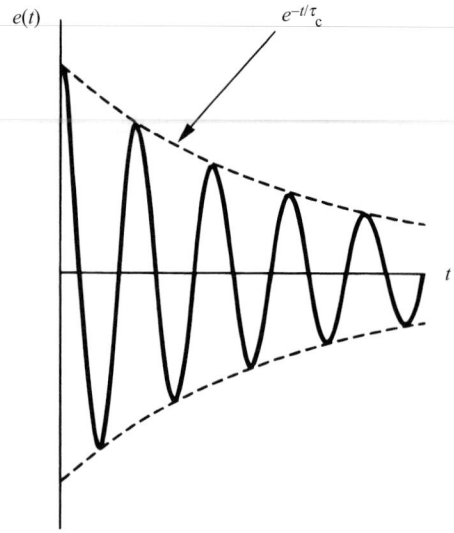


Fig. 2.1 The damped oscillation of the electric field produced by an excited particle as it decays.

product of these uncertainties is [1–3] $\Delta E\,\Delta t \sim \hbar$.[1] Because an excited particle can be observed only for a time that is of the order of its lifetime, the measurement time uncertainty $\Delta t$ is roughly the same as the lifetime, so

$$\Delta E \sim \hbar/\tau = A\hbar. \tag{2.1}$$

The uncertainty in emitted frequency $\Delta\nu$ is $\Delta E/h$, so

$$\Delta\nu \sim A/(2\pi). \tag{2.2}$$

When the decay of an excited particle is viewed as a photon-emission process, we can think of the particle, which was initially placed in the excited state at time $t' = 0$, emitting a photon at time $t$. The distribution of these times $t$ among many such particles varies as $e^{-t/\tau}$. Our knowledge of when the photon is likely to be emitted with respect to the time origin restricts our ability to be sure of its frequency. For example, if a photon is observed at time $t$, and is known to have come from a state with lifetime $\tau$, we know that the *probable* time $t'$ at which the atom became excited was $t - \tau < t' \leq t$. The longer the lifetime of the state the greater the uncertainty about when the particle acquired its original excitation. In the limit as $\tau \to \infty$, our knowledge of the time of excitation becomes infinitely uncertain and we can ascribe a very well-defined frequency to the emitted photon. In this limit the electromagnetic waveform emitted by the atom approaches infinite length and is undamped. We can put the above approximate determination of $\Delta\nu$ on a more exact basis by considering the exponential intensity decay of a group of excited particles: that is, the decay of each individual excited particle is modeled as an exponentially decaying (damped) cosinusoidal oscillation, as shown schematically in Fig. 2.1. It must be stressed that this is only a

[1] $\hbar = h/(2\pi)$.

convenient way of *picturing* how an excited particle decays and emits electromagnetic radiation. It would not be possible in practice to observe such an electromagnetic field by watching a single excited particle decay. We can observe a classical field only by watching many excited particles simultaneously. Within the framework of our model we can represent the electric field of a decaying excited particle as

$$e(t) = E_0 e^{-t/\tau_c} \cos(\omega_0 t). \tag{2.3}$$

We need to determine the time constant $\tau_c$ that applies to this damped oscillation. The instantaneous intensity $i(t)$ emitted by an individual excited atom is

$$i(t) \propto |e(t)|^2 = E_0^2 e^{-2t/\tau_c} \cos^2(\omega_0 t). \tag{2.4}$$

If we observe many such atoms the total observed intensity is

$$\begin{aligned} I(t) &= \sum_{\text{particles}} i(t) = \sum_i E_0^2 e^{-2t/\tau_c} \cos^2(\omega_0 t + \epsilon_i) \\ &= \sum_i \frac{E_0^2}{2} e^{-2t/\tau_c} [1 + \cos^2(\omega_0 t + \epsilon_i)], \end{aligned} \tag{2.5}$$

where $\epsilon_i$ is the phase of the wave emitted by atom $i$. In the summation the cosine term gets smeared out because individual atoms are emitting with random phases. So, $I(t) \propto e^{-2t/\tau_c}$. However, we know that $I(t) \propto e^{-t/\tau}$, where $\tau$ is the lifetime of the emitting state, so the time constant $\tau_c$ is in fact $2\tau$. Thus

$$e(t) = E_0 e^{-t/(2\tau)} \cos(\omega_0 t). \tag{2.6}$$

To find the frequency distribution of this signal we take its Fourier transform,

$$E(\omega) = \frac{1}{2\pi} \int_{-\infty}^{\infty} e(t) e^{-j\omega t}\, dt, \tag{2.7}$$

where

$$e(t) = \begin{cases} (E_0/2)(e^{j(\omega_0 + j/(2\tau))t} + e^{-j(\omega_0 - j/(2\tau))t}) \text{ for } t > 0, \\ 0 \text{ for } t < 0. \end{cases} \tag{2.8}$$

The start of the period of observation at $t = 0$, taken for example at an instant when all the particles are pushed into the excited state, allows the lower limit of integration to be changed to 0, so

$$\begin{aligned} E(\omega) &= \frac{1}{2\pi} \int_0^{\infty} e(t) e^{-j\omega t}\, dt \\ &= \frac{E_0}{4\pi} \left[ \frac{i}{\omega_0 - \omega + j/(2\tau)} - \frac{i}{\omega_0 + \omega - j/(2\tau)} \right]. \end{aligned} \tag{2.9}$$

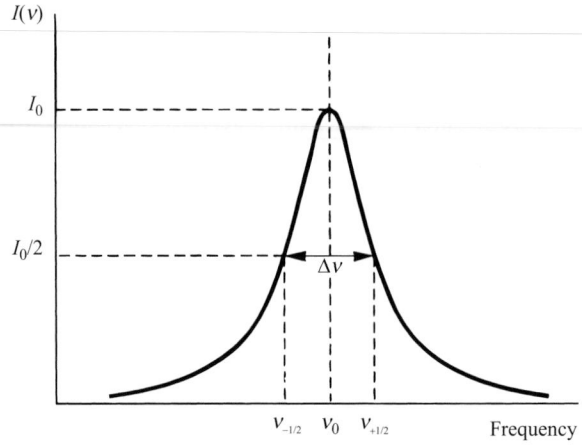


**Fig. 2.2** The Lorentzian lineshape function for natural broadening.

The intensity of emitted radiation is

$$I(\omega) \propto |E(\omega)|^2 = E(\omega)E^*(\omega) \propto \frac{1}{(\omega - \omega_0)^2 + (1/(2\tau))^2}. \tag{2.10}$$

Or, in terms of ordinary frequency,

$$I(\nu) \propto \frac{1}{(\nu - \nu_0)^2 + (1/(4\pi\tau))^2}. \tag{2.11}$$

The FWHM of this function is found from the half-intensity points of $I(\nu)$, which occur at frequencies $\nu_{\pm\frac{1}{2}}$ as shown in Fig. 2.2. This occurs where

$$\left(\frac{1}{4\pi\tau}\right)^2 = (\nu_{\pm\frac{1}{2}} - \nu_0)^2. \tag{2.12}$$

The FWHM is $\Delta\nu = \nu_{+\frac{1}{2}} - \nu_{-\frac{1}{2}}$, which gives[2]

$$\Delta\nu = \frac{1}{2\pi\tau} = \frac{A}{2\pi}. \tag{2.13}$$

So, from Eq. (2.11),

$$I(\nu) \propto \frac{1}{(\nu - \nu_0)^2 + (\Delta\nu/2)^2}. \tag{2.14}$$

The normalized form of this function is the lineshape function for natural broadening,

$$g(\nu_0, \nu)_N = \frac{2/(\pi\,\Delta\nu)}{1 + [2(\nu - \nu_0)/\Delta\nu]^2}. \tag{2.15}$$

This type of function is called a Lorentzian. Since natural broadening is the same for each particle, it is said to be a *homogeneous* broadening mechanism.

[2] A more exact treatment gives $\Delta\nu = (A_1 + A_2)/(2\pi)$, where $A_2$ and $A_1$ are the Einstein coefficients of the upper and lower levels of the transition.

### 2.2.2 Other homogeneous broadening mechanisms

Besides natural broadening other mechanisms of homogeneous broadening exist, for example the following.

(i) In a crystal the constituent particles of the lattice are in constant vibrational motion. This collective vibration can be treated as being equivalent to sound waves bouncing around inside the crystal. These sound waves, just like electromagnetic waves, can carry energy only in quantized amounts. These packets of acoustic energy are called *phonons*, and are analogous in many ways to photons. The principal differences between them are that phonons travel at the speed of sound and can exist only in a material medium. Collisions of phonons with the particles of the lattice perturb the phase of any excited, emitting particles present. This type of collision, which does not abruptly terminate the lifetime of the particle in its emitting state, is called a *soft* collision.

(ii) Particularly in the gasous and liquid phases, interaction of an emitting particle with its neighbors causes perturbation of its emitting frequency and subsequent *pressure broadening* of the transition. This interaction may arise in a number of ways.

  (a) Collisions with neutral particles, which may be *soft* or *hard*. A hard collision causes abrupt decay of the emitting species.

  (b) Collisions with charged particles. These collisions need not be very direct, but may involve a very small interaction that occurs when the charged particle passes relatively near, but perhaps as far as several tens of atomic diameters away from, the excited particle. In any case the relative motion of the charged and excited particles leads to a time-varying electric field at the excited particle that perturbs its energy states. This general effect in which an external electric field perturbs the energy levels of an atom (or molecule or ion) is called the *Stark* effect; hence line broadening caused by charged particles (ions or electrons) is called Stark broadening.

  (c) Van der Waals and resonance interactions (usually small effects). Resonance interactions occur when an excited particle can easily exchange energy with like neighbors. The effect is most important for transitions involving the ground state, since in this case there are generally many particles near an excited particle for which the possibility of energy exchange exists. Such a process is shown schematically in Fig. 2.3. Broadening occurs because the *possibility* of energy exchange exists, not because an actual emission/re-absorption process occurs.

## 2.3 Inhomogeneous broadening

When the environments or properties of particles in an emitting sample are non-identical, *inhomogeneous* broadening can occur. In this type of broadening the shifts and perturbations of emission frequencies differ from particle to particle. For example, in a real crystal

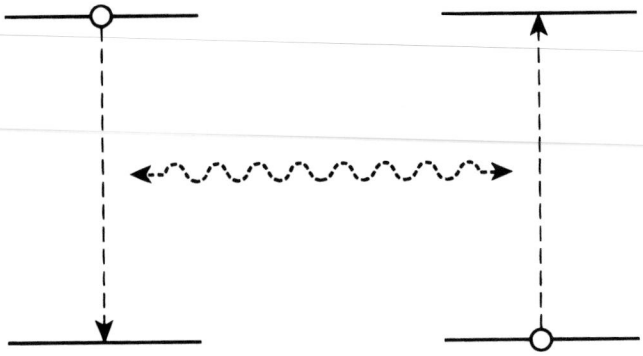

Fig. 2.3 A schematic illustration of how resonance and van der Waals broadening arise.

the presence of imperfections and impurities in the crystal structure alters the physical environment of atoms from one lattice site to another. The random distribution of lattice point environments leads to a distribution of particles whose center frequencies are shifted in a random way throughout the crystal. In a gas the random distribution of particle velocities leads to a distribution in the emission center frequencies of different emitting particles seen by a stationary observer. For an atom whose component of velocity towards the observer is $v_x$ the observed frequency of the transition whose stationary center frequency is $\nu_0$ is

$$\nu = \nu_0 + \frac{v_x}{c}\nu_0, \tag{2.16}$$

where $c$ is the velocity of light in the gas. The Maxwell–Boltzmann distribution of atomic velocities for particles of mass $M$ at absolute temperature $T$ is [4, 5]

$$f(v_x, v_y, v_z) = \left(\frac{M}{2\pi k_B T}\right)^{3/2} \exp\left[-\frac{M}{2k_B T}\left(v_x^2 + v_y^2 + v_z^2\right)\right]. \tag{2.17}$$

The number of particles per unit volume that have velocities simultaneously in the ranges $v_x \to v_x + dv_x$, $v_y \to v_y + dv_y$, and $v_z \to v_z + dv_z$ is $Nf(v_x, v_y, v_z)dv_x\, dv_y\, dv_z$, where $N$ is the total number of particles per unit volume. The $(M/(2\pi k_B T))^{3/2}$ factor is a normalization constant that ensures that the integral of $f(v_x, v_y, v_z)$ over all velocities is equal to unity, i.e.,

$$\iiint_{-\infty}^{\infty} f(v_x, v_y, v_z)dv_x\, dv_y\, dv_z = 1. \tag{2.18}$$

The normalized one-dimensional velocity distribution is

$$f(v_x) = \sqrt{\frac{M}{2\pi k_B T}} \exp\left(-\frac{Mv_x^2}{2k_B T}\right). \tag{2.19}$$

This Gaussian-shaped function is shown in Fig. 2.4. It represents the probability that the velocity of a particle towards an observer is in the range $v_x \to v_x + dv_x$. This is the same

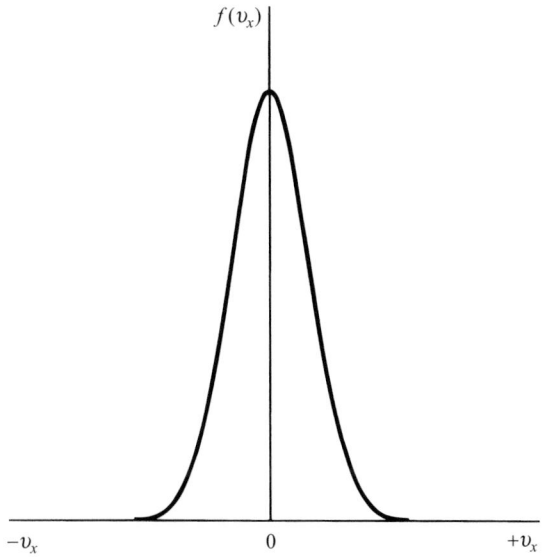


Fig. 2.4 The one-dimensional Gaussian distribution of velocities for the particles in a gas.

as the probability that the frequency be in the range

$$\nu_0 + \frac{v_x}{c}\nu_0 \to \nu_0 + \frac{v_x + dv_x}{c} = \nu_0 + \frac{v_x}{c}\nu_0 + \frac{dv_x}{c}\nu_0. \tag{2.20}$$

The probability that the frequency lies in the range $\nu \to \nu + d\nu$ is the same as the probability of finding the velocity in the range $(\nu - \nu_0)c/\nu_0 \to (\nu - \nu_0)c/\nu_0 + c/\nu_0\, d\nu$, so the distribution of the emitted frequencies is

$$g(\nu) = \frac{c}{\nu_0}\sqrt{\frac{M}{2\pi k_B T}}\exp\left[-\frac{M}{2k_B T}\frac{c^2}{\nu_0^2}(\nu - \nu_0)^2\right]. \tag{2.21}$$

This is already normalized (since $f(v_x)$ was normalized). It is called the normalized Doppler-broadened lineshape function. Its FWHM is

$$\Delta\nu_D = 2\nu_0\sqrt{\frac{2k_B T\,\ell n\,2}{Mc^2}}. \tag{2.22}$$

It can be seen that this increases with $\sqrt{T}$ and falls with atomic mass as $1/\sqrt{M}$. In terms of the Doppler-broadened linewidth $\Delta\nu_D$ the normalized Doppler lineshape function is

$$g(\nu_0, \nu) = \frac{2}{\Delta\nu_D}\sqrt{\frac{\ell n\,2}{\pi}}\, e^{-[2(\nu-\nu_0)/\Delta\nu_D]^2 \ell n 2}. \tag{2.23}$$

This is a Gaussian function. It is shown in Fig. 2.5 compared with a Lorentzian function of the same FWHM. The Gaussian function is much more sharply peaked, while the Lorentzian has considerable intensity far away from its center frequency, in its *wings*.

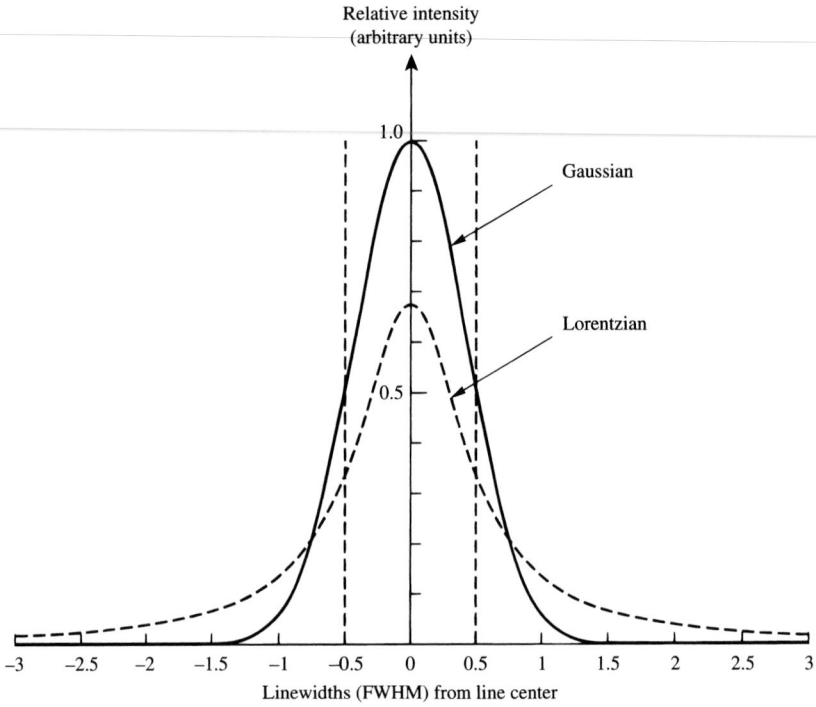


**Fig. 2.5** A comparison of normalized Lorentzian and Gaussian lineshapes.

**Example** The 632.8-nm transition of neon is the most important transition to show laser oscillation in the helium–neon laser. The relative atomic mass of neon is 20. Therefore, using $M = 20 \times 1.67 \times 10^{-27}$ kg, $\nu_0 = 3 \times 10^8/632.8 \times 10^{-9}$ Hz, and $T = 400$ K, which is an appropriate temperature to use for the gas in a He–Ne laser, the Doppler width is $\Delta\nu_D \sim$ 1.5 GHz.

Doppler broadening usually dominates over all other sources of broadening in gaseous systems, except occasionally in very heavy gases at high pressures and in highly ionized plasmas of light gases (in the latter case Stark broadening frequently dominates). Doppler broadening gives rise to a Gaussian lineshape, and is a common source of inhomogeneous broadening. The inhomogeneous lineshape covers a range of frequencies because many *different* particles are being observed. The particles are *different* in the sense that they have different velocities, and consequently different center frequencies. Homogeneous broadening also always occurs at the same time as inhomogeneous broadening, to a greater or lesser degree. To illustrate this, imagine a hypothetical experiment in which only those particles in a gas within a certain narrow velocity range are observed. The center frequencies of these particles are confined to a narrow frequency band, and in this sense there is no inhomogeneous broadening – all the *observed* particles are the same. However, a broadened

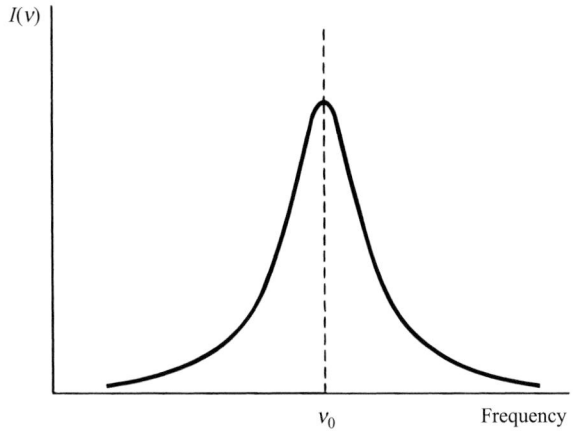


Fig. 2.6 Homogeneous broadening of a group of particles in a gas that have the same velocity.

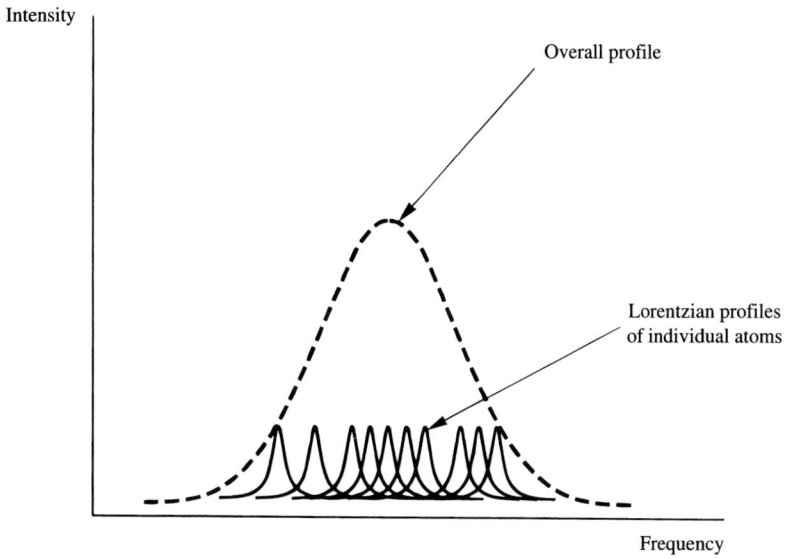


Fig. 2.7 A Doppler-broadened distribution of Lorentzian lineshapes.

lineshape would still be observed – the homogeneous lineshape resulting from natural and pressure broadening. This is illustrated in Fig. 2.6. When all particles are observed, irrespective of their velocity, an *overall* lineshape called a *Voigt* profile will be observed. This overall lineshape results from the superposition of Lorentzian lineshapes spread across the Gaussian distribution of Doppler-shifted center frequencies, as shown in Fig. 2.7. If the constituent Lorentzians have FWHM $\Delta\nu_L \ll \Delta\nu_D$ then the overall lineshape remains Gaussian and the system is properly said to be *inhomogeneously* broadened. On the other hand, if all observed particles are identical, or almost identical, so that $\Delta\nu_D \ll \Delta\nu_L$, then the system as a whole is said to be *homogeneously* broadened.

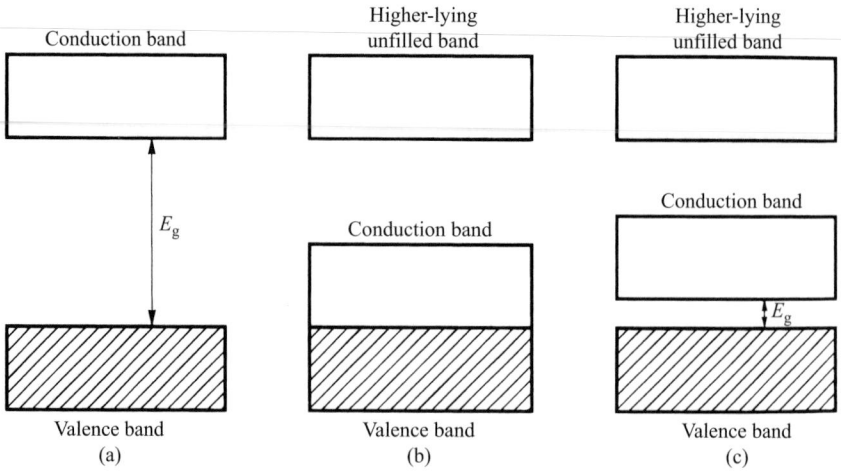


**Fig. 2.8** A schematic illustration of the different band structures of insulators, metals, and semiconductors.

In solid materials, inhomogeneous broadening, when it is important, results from lattice imperfections and impurities that cause the local environments of individual excited particles to differ in a random way. We shall assume that the broadening which thereby results also gives rise to a Gaussian lineshape of appropriate FWHM, which we shall also designate as $\Delta\nu_D$.[3]

Energy bands are the result of extensive line broadening in condensed matter, and also occur because the energy states of electrons in such material reflect the delocalization of electrons throughout the material. The outermost electrons of atoms in condensed matter, the valence electrons, can no longer be thought of as being bound to a particular atom, rather they "wander" throughout the material. Because of the Pauli exclusion principle, which says that no two electrons can occupy the same energy state, these valence electrons separate themselves into very many, closely spaced energy states. These closely spaced states overlap and form the valence band. At absolute zero all energy states in this band are filled with electrons. Some energy levels in condensed matter can remain relatively sharp if inter-particle interactions do not lead to extensive broadening. For example, for some levels of an impurity in a solid material, the spacing between adjacent impurity atoms can on average be as large as it might be in the gas phase, and broadening of the energy levels of these impurities need not lead to broad energy bands. The next highest band of energies results from the electrons in the lowest excited states of the individual atoms becoming delocalized, interacting strongly, and forming a band, called the conduction band. At absolute zero all the energy states in the conduction band are empty. At any temperature above absolute zero some electrons are thermally excited into the conduction band. It is the accessibility of the energy states in the conduction band which characterizes *insulators*, *metals*, and *semiconductors*. This is illustrated in Fig. 2.8. The energy spacing between the top of the valence band and the bottom of the conduction band is called the *band gap*.

[3] The designation $\Delta\nu_D$ originates from Doppler broadening, but is used generally to designate the inhomogeneous linewidth.

The gap in energy between the top of the valence band and the bottom of the conduction band is called the *energy gap*, $E_g$. If $E_g \gg k_B T$, where $k_B T$ is the characteristic thermal energy at temperature $T$, then electrons do not readily jump across the band gap into the conduction band. The probability of such a jump depends on the Boltzmann factor $e^{-E_g/(k_B T)}$. In a metal, however, $E_g$ is either a lot less than $k_B T$ or zero. In a semiconductor, $E_g \sim k_B T$. It is the presence of electrons in the conduction band, and also the absence of some electrons from the valence band (the "holes"), which gives rise to electrical conductivity. In a simple sense, a material conducts electricity if there are available energy states for electrons to move into, since conduction corresponds to directed electron motion, which implies that an electron has moved into a different energy state. In a laser using condensed matter, the laser transition involves a jump between energy bands or sometimes between relatively sharper levels that have not been extensively broadened. In a semiconductor laser the laser transition generally involves electrons jumping across the energy gap.

## 2.4 Optical frequency amplification

### 2.4.1 Optical frequency amplification with a homogeneously broadened transition

In an optical frequency amplifier we are generally concerned with the interaction of a monochromatic radiation field with a transition between two energy states whose center frequency is at, or near, the frequency of the monochromatic field. The magnitude of this interaction with each particle is controlled by the homogeneous lineshape function of the transition. In the general case, the monochromatic radiation field and the center frequency of the transition are not the same. This situation is shown schematically in Fig. 2.9. The stimulating radiation field is taken to be at frequency $\nu$, whilst the center frequency of the transition is at $\nu'$. The closer $\nu$ is to $\nu'$, the greater the number of transitions that can be stimulated. The stimulated transitions occur at frequency $\nu$, since this is the frequency of the stimulating radiation. The number of stimulated transitions is proportional to the homogeneous lineshape function $g(\nu', \nu)$, and can be written as

$$N_s = N_2 B_{21} \rho(\nu) g(\nu', \nu). \tag{2.24}$$

We have written $g(\nu', \nu)$ to indicate that this lineshape function has its center frequency at $\nu'$ but is being evaluated at frequency $\nu$. It is important to stress that the lineshape function $g(\nu', \nu)$ which is used here is the *homogeneous* lineshape function of the individual particles in the system, even though the contribution of homogeneous broadening to the overall broadening in the system may be small, for example when the overall broadening is predominantly inhomogeneous. The important point about the interaction of a particle with radiation is that an excited atom, molecule, or ion can interact only with a monochromatic radiation field that overlaps its homogeneous (usually Lorentzian-shaped) lineshape profile. For example, consider the case of two excited atoms with different center frequencies, which may be the different center frequencies of atoms with different velocities

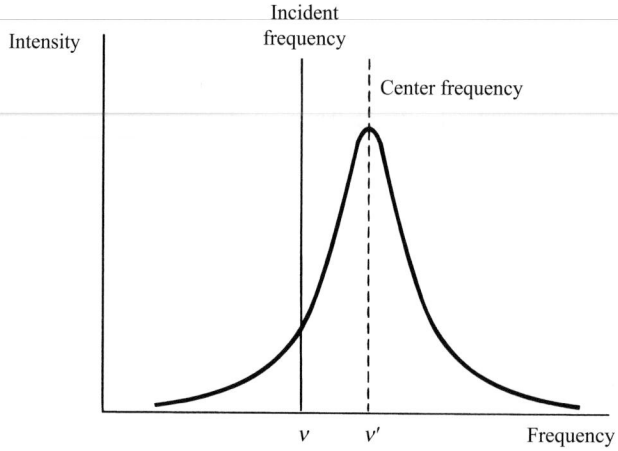


**Fig. 2.9** A monochromatic field interacting with a homogeneously broadened lineshape.

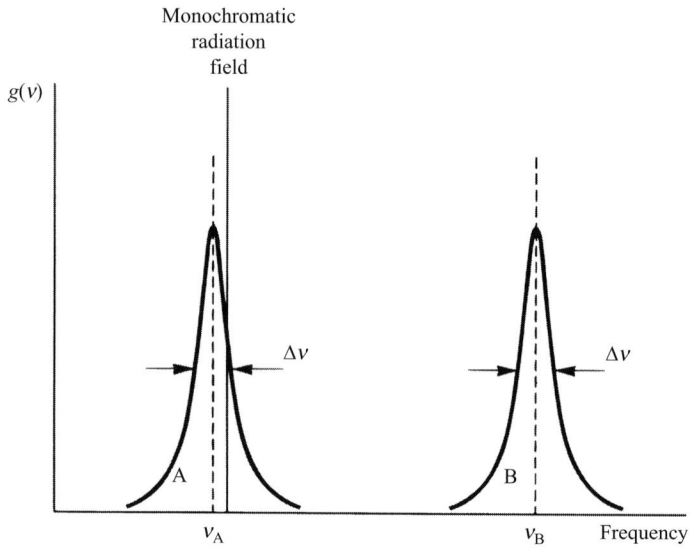


**Fig. 2.10** A monochromatic radiation field interacting with two homogeneously broadened lineshapes whose center frequencies are different.

relative to a fixed observer. The homogeneous lineshapes of these two atoms are shown in Fig. 2.10 together with a monochromatic radiation field at frequency $\nu$. Particle A with center frequency $\nu_A$ and *homogeneous width* $\Delta\nu$ can interact strongly with the field, while the interaction of particle B is negligible.

We can analyze the interaction between a plane monochromatic wave and a collection of homogeneously broadened particles with reference to Fig. 2.11. As the wave passes through the medium it grows in intensity if the number of stimulated emissions exceeds

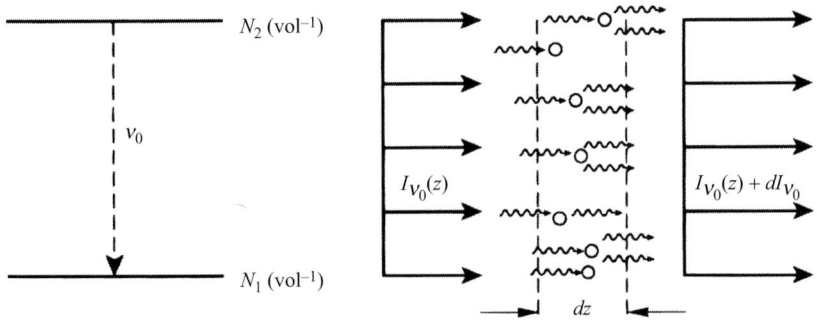


**Fig. 2.11** A plane wave interacting with a collection of homogeneously broadened particles.

the number of absorptions. The change in intensity of the wave in traveling a small distance $dz$ through the medium is

$$\begin{aligned} dI_\nu &= (\text{number of stimulated emissions} - \text{number of absorptions})/\text{volume} \\ &\quad \times h\nu \times dz \\ &= \left( N_2 B_{21} g(\nu', \nu) \frac{I_\nu}{c} - N_1 B_{12} g(\nu', \nu) \frac{I_\nu}{c} \right) h\nu \, dz. \end{aligned} \tag{2.25}$$

Use of the Einstein relations, Eqs. (1.80) and (1.81), gives

$$dI_\nu = \frac{I_\nu}{c} \left( N_2 - \frac{g_2}{g_1} N_1 \right) \frac{c^3 A_{21}}{8\pi h \nu^3} h\nu g(\nu', \nu) dz. \tag{2.26}$$

Therefore,

$$\frac{dI_\nu}{dz} = \left( N_2 - \frac{g_2}{g_1} N_1 \right) \frac{c^2 A_{21}}{8\pi \nu^2} g(\nu', \nu) I_\nu, \tag{2.27}$$

which has the solution

$$I_\nu = I_\nu(0) e^{\gamma(\nu) z}, \tag{2.28}$$

where $I_\nu(0)$ is the initial intensity at $z = 0$ and

$$\gamma(\nu) = \left( N_2 - \frac{g_2}{g_1} N_1 \right) \frac{c^2 A_{21}}{8\pi \nu^2} g(\nu', \nu); \tag{2.29}$$

$\gamma(\nu)$ is called the gain coefficient of the medium and has the same frequency dependence as $g(\nu', \nu)$. If $N_2 > (g_2/g_1)N_1$ then $\gamma(\nu) > 0$ and we have an optical frequency amplifier. If $N_2 < (g_2/g_1)N_1$ then $\gamma(\nu) < 0$ and net absorption of the incident radiation occurs. For a system in thermal equilibrium

$$\frac{N_2}{N_1} = \frac{g_2}{g_1} e^{-h\nu/(k_B T)}, \tag{2.30}$$

and, for $T > 0$, $e^{-h\nu/(k_B T)} < 1$, which implies that, in thermal equilibrium at positive temperatures, we cannot have positive gain. If we allow the formal existence of a negative temperature, at least for our system of two levels, then we can have $N_2 > (g_2/g_1)N_1$. Such a situation, which is essential for the construction of an optical frequency amplifier, is called a state of *population inversion* or *negative temperature*. This is not a true state of thermal equilibrium and can be maintained only by feeding energy into the system. In the above discussion we have neglected the occurrence of spontaneous emission: this is reasonable for a truly plane wave since the total number of spontaneous emissions into the zero solid angle subtended by the wave is zero. If the wave being amplified were diverging into a small solid angle $\delta\omega$, then $N_2 A_{21}\,\delta\omega/(4\pi)$ spontaneous emissions per second per unit volume would contribute to the increase in intensity of the wave. However, such emissions, being independent of the incident wave, are not in a constant phase relationship with this wave as are the stimulated emissions. These spontaneous emissions constitute a kind of "noise" superimposed on the beam of identical photons created by stimulated emission.

### 2.4.2 The stimulated emission rate in a homogeneously broadened system

The stimulated emission rate $W_{21}(\nu)$ is the number of stimulated emissions per particle per second caused by a monochromatic imput wave at frequency $\nu$:

$$W_{21}(\nu) = B_{21} g(\nu', \nu)\rho_\nu, \tag{2.31}$$

where $\rho_\nu$ (J m$^{-3}$) is the energy density of the stimulating radiation. Equation (2.31) can be rewritten in terms of more practical parameters as

$$W_{21}(\nu) = \frac{A_{21} c^2 I_\nu}{8\pi h \nu^3} g(\nu', \nu). \tag{2.32}$$

Here $W_{21}(\nu)$ has units of s$^{-1}$ per particle. Note that the frequency variation of $W_{21}(\nu)$ follows the lineshape function. The total number of stimulated emissions is

$$N_s = N_2 W_{21}(\nu). \tag{2.33}$$

### 2.4.3 Optical frequency amplification with inhomogeneous broadening

Although in our discussion so far we have been restricting our attention to homogeneous systems, we can show that Eqs. (2.25)–(2.29) hold generally, even in a system with inhomogeneous broadening, if we take $g(\nu', \nu)$ as the *total* lineshape function. In an inhomogeneously broadened system we can divide the atoms up into classes, each class consisting of particles with a certain range of center emission frequencies and the same homogeneous lineshape. For example, the class with center frequency $\nu''$ in the frequency range $d\nu''$ has $N g_D(\nu', \nu'')d\nu''$ particles in it, where $g_D(\nu', \nu'')$ is the normalized inhomogeneous distribution of center frequencies – the inhomogeneous lineshape function centered at $\nu'$.

This class of particles contributes to the change in intensity of a monochromatic wave at frequency $\nu$ as

$$\begin{aligned}\Delta(dI_\nu)&(\text{from the group of particles in the band } d\nu'')\\ &= \left(N_2 B_{21} g_{\mathrm{D}}(\nu', \nu'') d\nu''\, g_{\mathrm{L}}(\nu'', \nu)\frac{I_\nu}{c}\right.\\ &\quad \left. - N_1 B_{12} g_{\mathrm{D}}(\nu', \nu'') d\nu''\, g_{\mathrm{L}}(\nu'', \nu)\frac{I_\nu}{c}\right) h\nu\, dz, \end{aligned} \tag{2.34}$$

where $g_{\mathrm{L}}(\nu'', \nu)$ is the homogeneous lineshape function of a particle at center frequency $\nu''$. Equation (2.34) is equivalent to Eq. (2.25). The increase in intensity from all the classes of particles is found by integrating over these classes, that is, over the range of center frequencies $\nu''$, so Eq. (2.34) becomes

$$dI_\nu = \frac{I_\nu}{c}(N_2 B_{21} - N_1 B_{12})\left[\int_{-\infty}^{\infty} g_{\mathrm{D}}(\nu', \nu'') g_{\mathrm{L}}(\nu'', \nu) d\nu''\right] \nu\, dz, \tag{2.35}$$

which, in a similar fashion to Eqs. (2.25)–(2.29), gives

$$\gamma(\nu) = \left(N_2 - \frac{g_2}{g_1} N_1\right) \frac{c^2 A_{21}}{8\pi \nu^2} g(\nu', \nu), \tag{2.36}$$

where $g(\nu', \nu)$ is now the overall lineshape function defined by the equation

$$g(\nu', \nu) = \int_{-\infty}^{\infty} g_{\mathrm{D}}(\nu', \nu'') g_{\mathrm{L}}(\nu'', \nu) d\nu''. \tag{2.37}$$

In other words, the overall lineshape function is the convolution [6] of the homogeneous and inhomogeneous lineshape functions. The convolution integral in Eq. (2.37) can be put in a more familiar form if we measure frequency relative to the center frequency of the overall lineshape; that is, we put $\nu' = 0$, and Eq. (2.37) becomes

$$\begin{aligned} g(0, \nu) &= \int_{-\infty}^{\infty} g_{\mathrm{D}}(0, \nu'') g_{\mathrm{L}}(\nu'', \nu) d\nu'' \\ &= \int_{-\infty}^{\infty} g_{\mathrm{D}}(0, \nu'') g_{\mathrm{L}}(0, \nu - \nu'') d\nu'', \end{aligned} \tag{2.38}$$

which can be written in simple form as

$$g(\nu) = \int_{-\infty}^{\infty} g_{\mathrm{D}}(\nu'') g_{\mathrm{L}}(\nu - \nu'') d\nu''. \tag{2.39}$$

This is recognizable as the standard convolution integral of two functions $g_{\mathrm{D}}(\nu)$ and $g_{\mathrm{L}}(\nu)$. If $g_{\mathrm{D}}(\nu', \nu'')$ is indeed a normalized Gaussian lineshape as in Eq. (2.23) and $g_{\mathrm{L}}(\nu'', \nu)$ is a Lorentzian, then Eq. (2.37) can be written in the form

$$g(\nu', \nu) = \frac{2}{\Delta\nu_{\mathrm{D}}}\sqrt{\frac{\ell n\, 2}{\pi}}\, \frac{y}{\pi} \int_{-\infty}^{\infty} \frac{e^{-t^2}}{y^2 + (x - t)^2}\, dt, \tag{2.40}$$

where $y = \Delta\nu_{\mathrm{L}}\sqrt{\ell n\, 2}/\Delta\nu_{\mathrm{D}}$ and $x = 2(\nu - \nu')\sqrt{\ell n\, 2}/\Delta\nu_{\mathrm{D}}$. This is one way of writing a normalized Voigt profile [7, 8]. The integral in Eq. (2.40) cannot be evaluated analytically but must be evaluated numerically. For this purpose the Voigt profile is often written in terms of the error function for complex argument $W(z)$, which is available in tabulated form [9]:

$$g(\nu', \nu) = \frac{2}{\Delta\nu_D}\sqrt{\frac{\ell n\, 2}{\pi}}\mathcal{R}[W(z)], \tag{2.41}$$

where $z = x + jy$ and $\mathcal{R}[W(z)]$ denotes the real part of the function.

## 2.5 Optical frequency oscillation – saturation

If we can prepare a medium in a state of population inversion for a pair of its energy levels, then the transition between these levels can be used to make an optical frequency amplifier. To produce an oscillator we need to apply appropriate feedback by inserting the amplifying medium between a pair of suitable mirrors. If the overall gain of the medium exceeds the losses of the mirror cavity and ancillary optics then oscillation will result. The level at which this oscillation stabilizes is set by the way in which the amplifier saturates.

### 2.5.1 Homogeneous systems

Consider an amplifying transition at center frequency $\nu_0$ between two energy levels of a particle. We maintain this pair of levels in population inversion by feeding in energy. In equilibrium in the absence of an external radiation field, the rates $R_2$ and $R_1$ at which particles are fed into these levels must be balanced by spontaneous emission and non-radiative loss processes (such as collisions). The population densities and effective lifetimes of the two levels are $N_2$ and $\tau_2$ and $N_1$ and $\tau_1$, respectively, as shown in Fig. 2.12. These effective lifetimes include the effect of non-radiative deactivation. If $X_{2j}$ is the rate per particle per unit volume by which collisions depopulate level 2 and cause the particle to end up in a

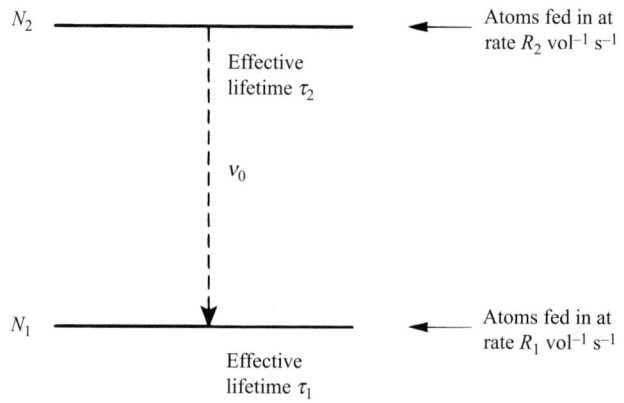


Fig. 2.12 The simple energy-level system used in a discussion of amplifier saturation.

lower state $j$ we can write

$$\frac{1}{\tau_2} = \sum_j (A_{2j} + X_{2j}). \tag{2.42}$$

In equilibrium

$$\frac{dN_2^{\mathrm{o}}}{dt} = R_2 - \frac{N_2^{\mathrm{o}}}{\tau_2} = 0, \tag{2.43}$$

where the term $N_2^{\mathrm{o}}/\tau_2$ is the total loss rate per unit volume from spontaneous emission and other deactivation processes, so

$$N_2^{\mathrm{o}} = R_2\tau_2, \tag{2.44}$$

where the superscript o indicates that the population is being calculated in the absence of a radiation field. Similarly, for the lower level of the transition, in equilibrium

$$\frac{dN_1^{\mathrm{o}}}{dt} = R_1 + N_2^{\mathrm{o}}A_{21} - \frac{N_1^{\mathrm{o}}}{\tau_1} = 0. \tag{2.45}$$

The term $N_1^{\mathrm{o}}/\tau_1$ is the total loss rate per unit volume from the level by spontaneous emission and other deactivation processes, while the term $N_2^{\mathrm{o}}A_{21}$ is the rate at which atoms are feeding into level 1 by spontaneous emission from level 2. So, from Eqs. (2.44) and (2.45),

$$N_1^{\mathrm{o}} = (R_1 + N_2^{\mathrm{o}}A_{21})\tau_1 = (R_1 + R_2\tau_2A_{21})\tau_1. \tag{2.46}$$

The population inversion is

$$N_2^{\mathrm{o}} - \frac{g_2}{g_1}N_1^{\mathrm{o}} = \Delta N^{\mathrm{o}} = R_2\tau_2 - \frac{g_2}{g_1}\tau_1(R_1 + R_2\tau_2A_{21}), \tag{2.47}$$

or, when the level degeneracies are equal,

$$\Delta N^{\mathrm{o}} = R_2\tau_2 - \tau_1(R_1 + R_2\tau_2A_{21}). \tag{2.48}$$

If we now feed in a monochromatic (or other) signal then stimulated emission and absorption processes will occur. We take the energy density of this signal at some point within the medium to be $\rho(\nu) = I(\nu)/c$. The rate at which this signal causes stimulated emissions is

$$W_{21}(\nu) = \int_{-\infty}^{\infty} B_{21}g(\nu_0, \nu)\rho(\nu)d\nu \quad \text{per particle per second,} \tag{2.49}$$

where $g(\nu_0, \nu)$ is the *homogeneous* lineshape function. If the input radiation were *white*, which in this context means that $\rho(\nu)$ is a constant over the range of frequencies spanned by the lineshape function, then

$$W_{21}(\nu) = B_{21}\rho(\nu)\int_{-\infty}^{\infty} g(\nu_0, \nu)d\nu = B_{21}\rho(\nu). \tag{2.50}$$

The total rate at which a monochromatic plane wave causes stimulated transitions is

$$\begin{aligned} W_{21}(\nu) &= \int_{-\infty}^{\infty} B_{21}g(\nu_0, \nu)\rho(\nu)d\nu = \frac{I_\nu}{c}B_{21}g(\nu_0, \nu)\int_{-\infty}^{\infty} \delta(\nu - \nu'')d\nu'' \\ &= B_{21}g(\nu_0, \nu)\frac{I_\nu}{c}. \end{aligned} \tag{2.51}$$

In the presence of a radiation field, in equilibrium the population densities of the pair of energy levels shown in Fig. 2.12 obey

$$\frac{dN_2}{dt} = R_2 - \frac{N_2}{\tau_2} - N_2 B_{21} g(\nu_0, \nu)\rho(\nu) + N_1 B_{12} g(\nu_0, \nu)\rho(\nu) = 0 \tag{2.52}$$

and

$$\frac{dN_1}{dt} = R_1 + N_2 A_{21} - \frac{N_1}{\tau_1} + N_2 B_{21} g(\nu_0, \nu)\rho(\nu) - N_1 B_{12} g(\nu_0, \nu)\rho(\nu) = 0. \tag{2.53}$$

If we write $B_{21} g(\nu_0, \nu)\rho(\nu) = W_{12}(\nu)$, the stimulated emission rate at frequency $\nu$ per particle, and neglect degeneracy factors so that we can assume that $W_{12}(\nu) = W_{21}(\nu) = W$ (which is equivalent to having $B_{12} = B_{21}$), then we have

$$R_2 - \frac{N_2}{\tau_2} - N_2 W + N_1 W = 0 \tag{2.54}$$

and

$$R_1 + N_2 A_{21} - \frac{N_1}{\tau_1} + N_2 W - N_1 W = 0. \tag{2.55}$$

From Eq. (2.54)

$$N_2 = \frac{N_1 W + R_2}{1/\tau_2 + W} \tag{2.56}$$

and from Eq. (2.55)

$$N_2 = \frac{N_1/\tau_1 + N_1 W - R_1}{A_{21} + W}, \tag{2.57}$$

so

$$N_1 = \frac{R_1/\tau_2 + R_2 A_{21} + W(R_1 + R_2)}{1/(\tau_1\tau_2) + W(1/\tau_1 + 1/\tau_2 - A_{21})}. \tag{2.58}$$

The population inversion in the system is now

$$N_2 - N_1 = \frac{N_1 W + R_2}{1/\tau_2 + W} - N_1 = \frac{R_2 - N_1/\tau_2}{1/\tau_2 + W}. \tag{2.59}$$

From Eq. (2.58) this gives

$$\begin{aligned} N_2 - N_1 &= \frac{(1/\tau_2)(R_2/\tau_1 - R_2 A_{21} - R_1/\tau_2) + W(R_2/\tau_1 - R_2 A_{21} - R_1/\tau_2)}{(1/\tau_2 + W)(1/(\tau_1\tau_2) + W/\tau_2 + W/\tau_1 - WA_{21})} \\ &= \frac{R_2/\tau_1 - R_1/\tau_2 - R_2 A_{21}}{1/(\tau_1\tau_2) + W(1/\tau_2 + 1/\tau_1 - A_{21})}. \end{aligned} \tag{2.60}$$

On multiplying the numerator and denominator of Eq. (2.60) by $\tau_1\tau_2$ we get

$$N_2 - N_1 = \frac{R_2\tau_2 - R_1\tau_1 - R_2\tau_1\tau_2 A_{21}}{1 + W\tau_2(1 + \tau_1/\tau_2 - A_{21}\tau_1)}. \tag{2.61}$$

The numerator of this expression is just the population inversion in the absence of any light signal, $\Delta N^{\mathrm{o}}$. Thus,

$$N_2 - N_1 = \frac{\Delta N^{\mathrm{o}}}{1 + W\tau_2(1 + \tau_1/\tau_2 - A_{21}\tau_1)} \tag{2.62}$$

or, with the substitution

$$\phi = A_{21}\tau_2\left[1 + (1 - A_{21}\tau_2)\frac{\tau_1}{\tau_2}\right], \tag{2.63}$$

we have

$$N_2 - N_1 = \frac{\Delta N^{\mathrm{o}}}{1 + \phi W/A_{21}}. \tag{2.64}$$

Now, from Eq. (2.32),

$$W = \frac{c^2 A_{21}}{8\pi h\nu^3} I_\nu g(\nu_0, \nu). \tag{2.65}$$

If we define

$$I_{\mathrm{s}}(\nu) = \frac{8\pi h\nu^3}{c^2\phi g(\nu_0, \nu)} \tag{2.66}$$

then

$$N_2 - N_1 = \frac{\Delta N^{\mathrm{o}}}{1 + I_\nu/I_{\mathrm{s}}(\nu)} \tag{2.67}$$

and $I_{\mathrm{s}}(\nu)$, called the saturation intensity, is the intensity of an incident light signal (power area$^{-1}$) that reduces the population inversion to half its value when no signal is present. Note that the value of the saturation intensity depends on the frequency of the input signal relative to the line center. Returning to our expression for the gain constant of a laser amplifier,

$$\gamma(\nu) = (N_2 - N_1)\frac{c^2 A_{21}}{8\pi\nu^2} g(\nu_0, \nu), \tag{2.68}$$

the gain as a function of intensity is, in a *homogeneously* broadened system,

$$\gamma(\nu) = \frac{\Delta N^{\mathrm{o}}}{1 + I_\nu/I_{\mathrm{s}}(\nu)} \frac{c^2 A_{21}}{8\pi\nu^2} g(\nu_0, \nu), \tag{2.69}$$

which is reduced, that is, *saturates*, as the strength of the amplified signal increases. A good optical amplifier should have a large saturation intensity, and, from Eq. (2.65), this implies that $\phi$ should be a minimum. In such systems often $A_{21} \simeq 1/\tau_2$, so $\phi \simeq 1$.

### 2.5.2 Inhomogeneous systems

The problem of gain saturation in inhomogeneous media is more complex. For example, in a gas a plane monochromatic wave at frequency $\nu$ interacts with a medium whose individual particles have Lorentzian homogeneous lineshapes with FWHM $\Delta\nu_{\mathrm{N}}$, but whose center frequencies are distributed over an inhomogeneous (Doppler) broadened profile of width (FWHM) $\Delta\nu_D$. The Lorentzian contribution to the overall lineshape is

$$g_L(\nu', \nu) = \frac{2/(\pi\,\Delta\nu_N)}{1 + [2(\nu - \nu')/\Delta\nu_N]^2}, \tag{2.70}$$

where $\nu'$ is the center frequency of a particle set by its velocity relative to the observer. The Doppler-broadened profile of all the particles is

$$g_D(\nu_0, \nu') = \frac{2}{\Delta\nu_D}\sqrt{\frac{\ell n\,2}{\pi}}\,e^{-[(2(\nu'-\nu_0)/\Delta\nu_D)^2 \ell n 2]}, \tag{2.71}$$

where $\nu_0$ is the center frequency of a particle at rest. The overall lineshape (from all the particles) is a sum of Lorentzian profiles spread across the particle velocity distribution, as shown in Appendix 1. If $\Delta\nu_N \gg \Delta\nu_D$, the overall profile remains approximately Lorentzian, and the observed behavior of the system will correspond to *homogeneous* broadening. Such a situation is likely to arise for long-wavelength transitions in a gas, particularly if this has a high relative atomic or molecular mass, at pressures for which pressure broadening (which is a homogeneous process) is important, and frequently in solid materials. If $\Delta\nu_D \gg \Delta\nu_N$ (as is often the case in gases), the overall lineshape remains Gaussian, and the system is *inhomogeneously* broadened. Once again we reduce the problem to a consideration of the interaction of a plane electromagnetic wave with a two-level system as shown in Fig. 2.12. As the wave passes through the system its intensity changes according to whether the medium is amplifying or absorbing. We take the intensity of the monochromatic wave to be $I(\nu, z)$ at plane $z$ within the medium. The individual particles of the medium have a distribution of emission center frequencies (or absorption center frequencies) because of their random velocities (or, for example, their different crystal environments). We take the population density functions (particles vol$^{-1}$ Hz$^{-1}$) in the upper and lower levels whose center frequency is at $\nu'$ to be $N_2(\nu', z)$ and $N_1(\nu', z)$, respectively, at plane $z$. Particles are fed into levels 2 and 1 at rates $R_2(\nu')$ and $R_1(\nu')$. These rates are assumed uniform throughout the medium. Then $N_2(\nu', z), N_1(\nu', z), R_2(\nu')$, and $R_1(\nu')$ are assumed to follow the Gaussian frequency dependence set by the particle velocity distribution, and are normalized so that, for example, the total pumping rate of level 2 is

$$R_2 = \int_{-\infty}^{\infty} R_2(\nu')d\nu' = R_{20}\int_{-\infty}^{\infty} e^{[2(\nu'-\nu_0)/\Delta\nu_D]^2 \ell n 2}\,d\nu'. \tag{2.72}$$

In practice, the primary pumping process need not have this Gaussian dependence, but, even when this is the case, the effect of collisions among particles that have been excited will be to *smear out* any non-Gaussian pumping process into a near-Gaussian form. This conclusion is justified by observations of Doppler-broadened lines under various excitation conditions, where deviations from a true Gaussian lineshape are found to be minimal. From Eq. (2.72),

$$R_2 = \sqrt{\frac{\pi}{\ell n\,2}}\,\frac{\Delta\nu_D}{2}R_{20}, \tag{2.73}$$

where $R_{20}$ is a pumping rate constant and the total population density of the level at plane $z$ is

$$N_2 = \int_{-\infty}^{\infty} N_2(\nu', z)d\nu'. \tag{2.74}$$

The rate equations for the atoms whose center frequencies are at $\nu'$ are

$$\begin{aligned}\frac{dN_2}{dt}(\nu', z) = R_2(\nu') - N_2(\nu', z)\left[\frac{1}{\tau_2} + B'_{21}(\nu', \nu)\frac{I(\nu, z)}{c}\right] \\ + N_1(\nu', z)B'_{12}(\nu', \nu)\frac{I(\nu, z)}{c}\end{aligned} \tag{2.75}$$

and

$$\begin{aligned}\frac{dN_1}{dt}(\nu', z) = R_1(\nu') + N_2(\nu', z)\left[A_{21} + B'_{21}(\nu', \nu)\frac{I(\nu, z)}{c}\right] \\ - N_1(\nu', z)\left[\frac{1}{\tau_1} + B'_{12}(\nu', \nu)\frac{I(\nu, z)}{c}\right].\end{aligned} \tag{2.76}$$

Here we have used the modified Einstein coefficients $B'_{21}(\nu', \nu)$ and $B'_{12}(\nu', \nu)$ that describe stimulated emission processes when the stimulating radiation is at frequency $\nu$ and the particle's center emission frequency is at $\nu'$. Written out in full,

$$B'_{21}(\nu', \nu) = B_{21}g(\nu', \nu), \tag{2.77}$$

where $g(\nu', \nu)$ is the *homogeneous* lineshape function. The rate of change of intensity of the incident wave due to atoms with center frequencies in a small range $d\nu'$ at $\nu'$ is

$$\left[\frac{dI}{dz}(\nu, z)\right]_{d\nu'} = h\nu\frac{I(\nu, z)}{c}[B'_{21}(\nu', \nu)N_2(\nu', z) - B'_{12}(\nu', \nu)N_1(\nu', z)]d\nu'. \tag{2.78}$$

The total rate of change of intensity due to all the particles, that is, from all possible center frequencies $\nu'$, is

$$\frac{dI(\nu, z)}{dz} = \frac{h\nu I(\nu, z)}{c}\int_{-\infty}^{\infty}[B'_{21}(\nu', \nu)N_2(\nu', z) - B'_{12}(\nu', \nu)N_1(\nu', z)]d\nu'. \tag{2.79}$$

In the steady state

$$\frac{dN_2}{dt}(\nu', z) = \frac{dN_1}{dt}(\nu', z) = 0,$$

so, from Eqs. (2.75) and (2.76),

$$\begin{aligned}&B'_{21}(\nu', \nu)N_2(\nu') - B'_{12}(\nu', \nu)N_1(\nu') \\ &= \frac{B'_{21}(\nu', \nu)\left[\dfrac{R_2(\nu')}{1/\tau_2} - \dfrac{g_2}{g_1}\dfrac{(R_2(\nu')A_{21} + R_1(\nu')/\tau_2)}{1/(\tau_1\tau_2)}\right]}{1 + \left[\dfrac{(g_1/g_2)(1/\tau_2 - A_{21})}{1/(\tau_1\tau_2)} + \tau_2\right]B'_{21}(\nu', \nu)\dfrac{I(\nu, z)}{c}}.\end{aligned} \tag{2.80}$$

We note that

$$R_2(\nu') = R_{20} e^{-[2(\nu'-\nu_0)/\Delta\nu_D]^2 \ell n2}. \tag{2.81}$$

Substituting into Eq. (2.79) from (2.80) and (2.81) and bearing in mind that $B'_{21}(\nu', \nu)$ has a Lorentzian form,

$$B'_{21}(\nu', \nu) = \frac{B_{21}[2/(\pi\,\Delta\nu_N)]}{1 + \left[\dfrac{2(\nu - \nu')}{\Delta\nu_N}\right]^2}, \tag{2.82}$$

where $\Delta\nu_N$ is the *homogeneous* FWHM of the transition, gives

$$\begin{aligned}\frac{1}{I(\nu,z)}\frac{dI(\nu,z)}{dz} &= \gamma(\nu)\\ &= \gamma_0 \int_{-\infty}^{\infty} d\nu' \frac{\dfrac{2}{\pi\,\Delta\nu_N}\{1 + [2(\nu-\nu')/\Delta\nu_N]^2\}^{-1} e^{-[\frac{2(\nu'-\nu)}{\Delta\nu_D}]^2 \ell n2}}{1 + \eta I(\nu,z)\dfrac{2}{\pi\,\Delta\nu_N}[1 + [2(\nu-\nu')/\Delta\nu_N]^2]^{-1}}.\end{aligned} \tag{2.83}$$

We have made the substitutions

$$\gamma_0 = \frac{h\nu}{c} B_{21}\left[R_{20}\tau_2 - \frac{g_2}{g_1}\left(\frac{R_{20}A_{21} + R_{10}A_2}{1/\tau_1\tau_2}\right)\right] \tag{2.84}$$

and

$$\eta = \left[\frac{g_2}{g_1}\frac{(1/\tau_2 - A_{21})}{1/(\tau_1\tau_2)} + \tau_2\right]\frac{B_{21}}{c}. \tag{2.85}$$

Equation (2.83) can be written

$$\gamma(\nu) = \frac{2\gamma_0\,\Delta\nu_N}{\pi}\int_{-\infty}^{\infty} \frac{e^{-[2(\nu'-\nu_0)/\Delta\nu_D]^2 \ell n2}\,d\nu'}{4(\nu-\nu')^2 + \Delta\nu_N^2[1 + 2\eta I(\nu,z)/(\pi\,\Delta\nu_N)]}. \tag{2.86}$$

Although it is fairly clear from Eq. (2.86) that the gain of the amplifier falls as $I(\nu, z)$ increases, it is not easy to see from the integral exactly how this occurs. If the intensity is small, the gain approaches its *unsaturated* value

$$\gamma_0(\nu) = \frac{2\gamma_0\,\Delta\nu_N}{\pi}\int_{-\infty}^{\infty} \frac{e^{-[2(\nu'-\nu_0)\Delta\nu_D]^2 \ell n2}\,d\nu'}{4(\nu-\nu')^2 + \Delta\nu_N^2}. \tag{2.87}$$

If Eq. (2.86) is examined closely, for frequencies $\nu'$ close to the input frequency $\nu$ the integrand can be written approximately as

$$\frac{e^{-[2(\nu'-\nu_0)/\Delta\nu_D]^2 \ell n2}}{\Delta\nu_N^2[1 + 2\eta I(\nu,z)/(\pi\,\Delta\nu_N)]}. \tag{2.88}$$

On the other hand, for frequencies $\nu'$ far from the input frequency, the integrand can be written approximately as

$$\frac{e^{-[2(\nu'-\nu_0)/\Delta\nu_D]^2 \ell n2}}{4(\nu-\nu')^2+\Delta\nu_N^2}, \tag{2.89}$$

which can be seen to be identical to the integrand in the unsaturated gain expression, Eq. (2.87). Thus, we conclude that, in making their contribution to the overall gain, particles whose frequencies are far from the input frequency are relatively unaffected by the input radiation, whereas particles whose frequencies are close to that of the input show strong saturation effects. The gain in the system comes largely from those particles whose frequencies are within (roughly) a homogeneous linewidth of the input radiation frequency. The consequences of this are best illustrated by considering a hypothetical experiment, shown schematically in Fig. 2.13, in which the small-signal gain of a predominantly inhomogeneously broadened amplifier is measured with and without a strong saturating signal simultaneously present. Without the presence of a strong signal at a fixed frequency $\nu_S$ the observed (small-signal) gain follows the Gaussian curve of the overall line profile of the amplifier, as shown in Fig. 2.14(a). However, if we perform this experiment again when a strong fixed-frequency field is also present, which causes saturation of the gain, we find that the gain is reduced locally by the saturating effect of the strong field as shown in

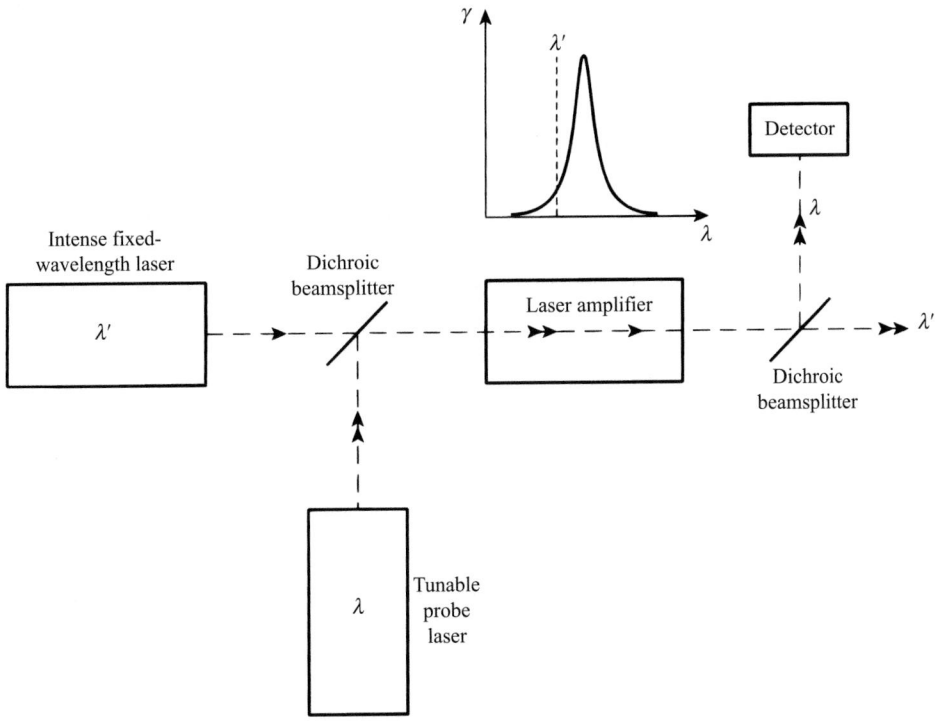


**Fig. 2.13** A schematic representation of the experimental arrangement for measuring the frequency dependence of the gain of an amplifier that is experiencing saturation from a strong fixed monochromatic signal.

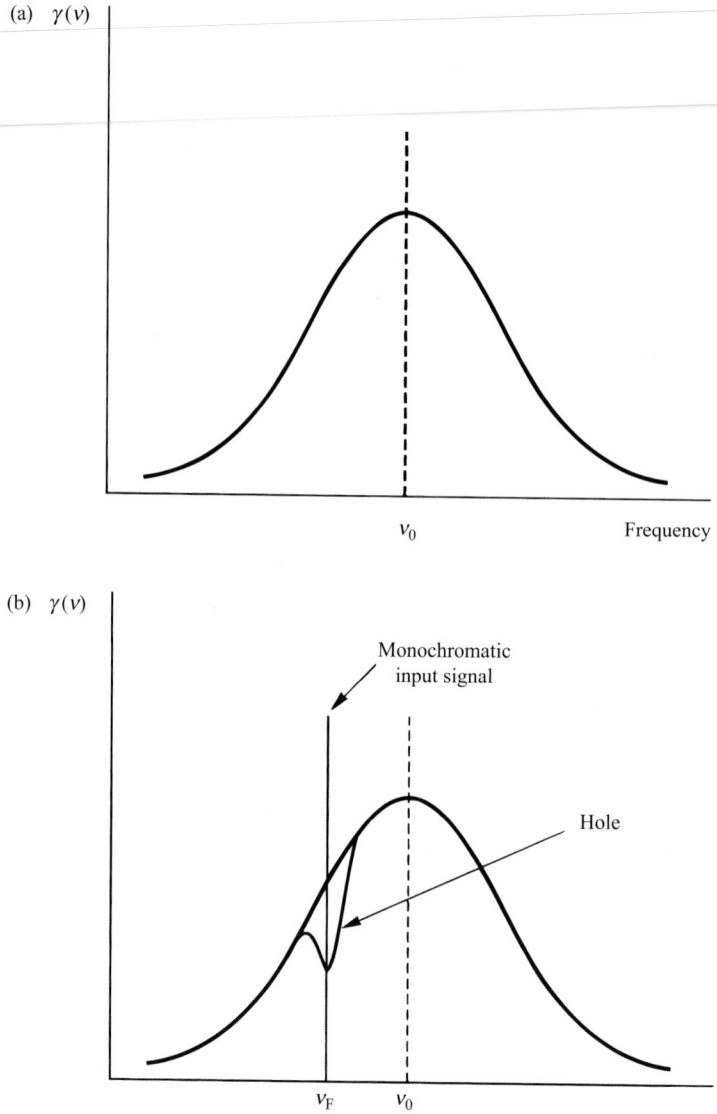


**Fig. 2.14** Gain as a function of frequency in an inhomogeneously broadened amplifier. (a) The small-signal situation when no saturation has occurred. (b) The production of a "hole" in the gain curve by a strong monochromatic input at frequency $\nu_F$.

Fig. 2.14(b). This phenomenon is called *hole burning* [10]. The width of the *hole* that is thus produced is determined by the quantity

$$\Delta\nu_N^2 \left[ 1 + \frac{2\eta I(\nu, z)}{\pi\, \Delta\nu_N} \right]. \tag{2.90}$$

If

$$\Delta\nu_{\mathrm{N}}\sqrt{1+\frac{2\eta I(\nu,z)}{\pi\,\Delta\nu_{\mathrm{N}}}} \ll \Delta\nu_{\mathrm{D}}, \tag{2.91}$$

for example, in a gaseous system where Doppler broadening is the largest contribution to the total observed line broadening, Eq. (2.86) can be integrated by bringing the much less sharply peaked exponential factor outside the integral. In this case, the sharply peaked Lorentzian lineshape makes the integrand largest for frequencies $\nu'$ near to $\nu$; over a small range of frequencies $\nu'$ near $\nu$ the exponential factor remains approximately constant so Eq. (2.86) can be written

$$\begin{aligned}\gamma(\nu) &= \frac{2\gamma_0\,\Delta\nu_{\mathrm{N}}}{\pi}e^{-[2(\nu-\nu_0)\Delta\nu_{\mathrm{D}}]^2\ell\mathrm{n}2}\\ &\quad\times\int_{-\infty}^{\infty}\frac{d\nu'}{4(\nu-\nu')^2+\Delta\nu_{\mathrm{N}}^2[1+2\eta I(\nu,z)/(\pi\,\Delta\nu_{\mathrm{N}})]}.\end{aligned} \tag{2.92}$$

Now the integral can be evaluated to give

$$\gamma(\nu)=\gamma_0\left[1+\frac{2\eta I(\nu,z)}{\pi\,\Delta\nu_{\mathrm{N}}}\right]^{-\frac{1}{2}}e^{-[2(\nu-\nu_0)/\Delta\nu_{\mathrm{D}}]^2\ell\mathrm{n}2}, \tag{2.93}$$

which gives

$$\gamma(\nu)=\gamma_0\left[1+\frac{I(\nu,z)}{I_{\mathrm{s}}'(\nu)}\right]^{-\frac{1}{2}}e^{-[2(\nu-\nu_0)/\Delta\nu_{\mathrm{D}}]^2\ell\mathrm{n}2}, \tag{2.94}$$

where $I_{\mathrm{s}}'(\nu)=\pi\,\Delta\nu_{\mathrm{N}}/(2\eta)$ is called the saturation intensity for inhomogeneous broadening. Note that $\gamma_0$ is the small-signal gain at line center of the inhomogeneously broadened line, which can be written in the form

$$\gamma_0=\frac{1}{4\pi}\sqrt{\frac{\ell\mathrm{n}\,2}{\pi}}\frac{\lambda^2A_{21}}{\Delta\nu_{\mathrm{D}}}\left(N_2-\frac{g_2}{g_1}N_1\right). \tag{2.95}$$

When Doppler broadening dominates in a system, incident radiation at frequency $\nu$ cannot interact with any particle whose Doppler-shifted frequency is different from $\nu$ by much more than $\Delta\nu_{\mathrm{N}}$. If the amplifier is homogeneously broadened, that is, if

$$\Delta\nu_{\mathrm{N}}\sqrt{1+\frac{\eta I(\nu,z)}{\pi\,\Delta\nu_{\mathrm{N}}}} \gg \Delta\nu_{\mathrm{D}}, \tag{2.96}$$

then Eq. (2.86) can be integrated by bringing the less-sharply peaked Lorentzian factor outside the integral to give

$$\begin{aligned}\gamma(\nu) &= \frac{\gamma_0\,\Delta\nu_{\mathrm{D}}}{\Delta\nu_{\mathrm{N}}\sqrt{\pi\,\ell\mathrm{n}\,2}}\left\{\left[\frac{2(\nu-\nu_0)}{\Delta\nu_{\mathrm{N}}}\right]^2+1+\frac{I(\nu,z)}{I_{\mathrm{s}}'(\nu)}\right\}^{-1}\\ &= \frac{\Delta\nu_{\mathrm{D}}}{2}\sqrt{\pi/\ell\mathrm{n}\,2}\,\frac{\gamma_0 g(\nu_0,\nu)}{1+I(\nu,z)/I_{\mathrm{s}}(\nu)},\end{aligned} \tag{2.97}$$

where $g(\nu_0, \nu)$ is the homogeneous lineshape function

$$g(\nu_0, \nu) = \frac{2/(\pi\, \Delta\nu_N)}{1 + [2(\nu - \nu_0)\Delta\nu_N]^2} \tag{2.98}$$

and $I_s(\nu)$ is the saturation intensity for homogeneous broadening given by

$$I_s(\nu) = \frac{2I'_s(\nu)}{\pi\, \Delta\nu_N\, g(\nu_0, \nu)} = \frac{1}{\eta g(\nu_0, \nu)}. \tag{2.99}$$

It can be seen from Eq. (2.85) that, for $g_2 = g_1$, $\eta$ reduces to the expression

$$\eta = \left[\frac{1/\tau_2 - A_{21}}{1/(\tau_1\tau_2)} + \tau_2\right]\frac{B_{21}}{c} \tag{2.100}$$

and $I_s(\nu)$ reduces to the expression obtained previously as the saturation intensity for homogeneous broadening, namely

$$I_s(\nu) = \frac{8\pi h\nu^3}{c^2\phi g(\nu_0, \nu)}, \tag{2.101}$$

where

$$\phi = A_{21}\tau_2\left[1 + (1 - A_{21}\tau_2)\frac{\tau_1}{\tau_2}\right]. \tag{2.102}$$

## 2.6 The power output from a laser amplifier

For a laser amplifier of length $\ell$ and gain coefficient $\gamma(\nu)$ the output intensity for a monochromatic input intensity of $I_0$ (W m$^{-2}$) at frequency $\nu$ is

$$I = I_0 e^{\gamma(\nu)\ell} \tag{2.103}$$

*if* saturation effects are neglected. If saturation effects cannot be neglected then the differential equation that describes how the intensity increases must be re-examined. This equation is

$$\gamma(\nu) = \frac{1}{I}\frac{dI}{dz}. \tag{2.104}$$

For a homogeneously broadened amplifier with saturation an explicit solution to this equation can be found. In this case, if $\gamma_0(\nu)$ is the small-signal gain,

$$\gamma(\nu) = \frac{\gamma_0(\nu)}{1 + I/I_s(\nu)} = \frac{1}{I}\frac{dI}{dz}, \tag{2.105}$$

which can be rewritten in the form

$$\frac{dI}{I} + \frac{dI}{I_s(\nu)} = \gamma_0(\nu)dz. \tag{2.106}$$

The solution to this equation is

$$I = I_0 e^{\gamma_0(\nu)\ell - (I - I_0)/I_s(\nu)}. \tag{2.107}$$

**Table 2.1** Iterative solution of Eq. (2.107)

| LHS | RHS |
|---|---|
| 10 | 2.57 |
| 8 | 5 |
| 7 | 6.98 |
| 6.9 | 7.21 |
| 6.99 | 7.00 |
| 6.993 | 6.994 |
| 6.9934 | 6.9934 |

Equation (2.107) must be solved iteratively. The solution will be somewhere between the input intensity and the output intensity when saturation is neglected.

We can best illustrate how this equation can be used to find the output of a saturated amplifier with a numerical example.

**Example** A homogeneously broadened laser has a saturation intensity of 3 W m$^{-2}$ and a small-signal gain at line center of 0.5 m$^{-1}$. Calculate the output intensity if a monochromatic signal of intensity 5 W m$^{-2}$ at line center enters an amplifier 2 m long.

If saturation were negligible the output intensity would be

$$I = I_0 e^{\gamma_0 \ell} = 5e^{0.5\times 2} = 13.59 \text{ W m}^{-2}.$$

We know the solution will be somewhere between the input intensity and the output when saturation is neglected, i.e., $5\,\text{W m}^{-2} < I < 13.59\,\text{W m}^{-2}$. We make an initial guess for $I$ (LHS) and evaluate the RHS of Eq. (2.107). The successive calculation of the LHS and RHS for different guesses for $I$ is shown in Table 2.1. The answer, which is obtained when the LHS and RHS agree, is 6.9934 W m$^{-2}$.

## 2.7 The electron oscillator model of a radiative transition

When a particle decays from a higher energy level to a lower, we can model the resultant electric field as a damped oscillation. There is an analogy between this decay of an excited particle and the damped oscillation of an electric circuit. For example, for the *RLC* circuit shown in Fig. 2.15 the resonance frequency is $\nu_0 = 1/(2\pi\sqrt{LC})$. If the sinusoidal driving voltage is disconnected from the circuit, then the oscillation of the circuit decays exponentially – provided that the circuit is underdamped. The power spectrum of the decaying electric current is Lorentzian, just as it is for a spontaneous transition. The FWHM of the circuit resonance is $\nu_0/Q$, where $Q$ is called the *quality factor* of the circuit, which is

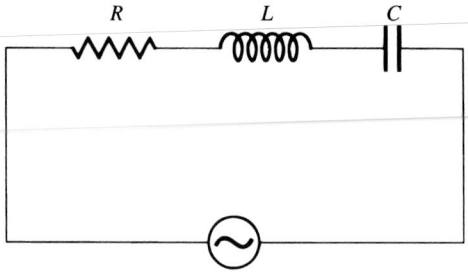


Fig. 2.15 An $RLC$ circuit.

analogous to the homogeneously broadened linewidth $\Delta\nu_N$. A transition between levels almost always has $\Delta\nu_N \ll \nu_0$, and hence clearly has a very high $Q$.

In the classical theory of how a particle responds to electromagnetic radiation each of the $n$ electrons attached to the particle is treated as a damped harmonic oscillator. For example, when an electric field acts on an atom, the nucleus, which is positively charged, moves in the direction of the field, while the electron cloud, which is negatively charged, moves in the opposite direction to the field. The resultant separation of the centers of positive and negative charge causes the atom to become an elemental dipole. If the separation of the nucleus and electron cloud is $d$, then the resultant dipole has magnitude $ed$ and points from the negative towards the positive charge.[4] As the frequency of the electric field that acts on the atom increases, the amount of nuclear motion decreases much more rapidly than does that of the electrons. At optical frequencies we generally neglect the motion of the nucleus; its great inertia compared with the electron cloud prevents it following the rapidly oscillating applied electric field. If the vector displacement of the $i$th electron on the atom from its equilibrium position is $\mathbf{x}_i$ then at any instant the atom has acquired a dipole moment

$$\mu = -\sum_{i=1}^{n} e\mathbf{x}_i, \tag{2.108}$$

where the summation runs over all the $n$ electrons on the atom. The magnitude of the displacement of each electron depends on the value of the electric field $\mathbf{E}_i$ at the electron,

$$k_i\mathbf{x}_i = -e\mathbf{E}_i, \tag{2.109}$$

where $k_i$ is a force constant. A time-varying field $\mathbf{E}$ leads to a time-varying dipole moment. This dipole moment can become large if there is a resonance between the applied field and a particular electron on the atom. This happens if the frequency of the field is near the natural oscillation frequency of a particular electron. Classically, the resonance frequency of electron $i$ is, by analogy with a mass attached to a spring, $\omega_i = \sqrt{k_i/m}$. If the applied electric field is near this frequency, one electron in the summation in Eq. (2.108) makes a dominant contribution to the dipole moment and we can treat the atom as a single-electron oscillator. The physical significance of the resonance frequencies of the electrons is that

[4] The magnitude of the electronic charge is $e \simeq 1.9 \times 10^{-19}$ C; the charge on an electron is $-e$.

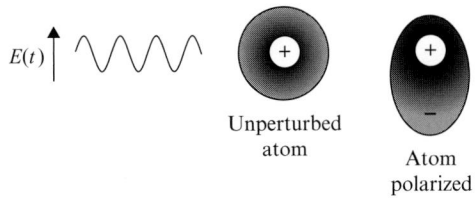


Fig. 2.16 The electron cloud and the nucleus are displaced in opposite directions by an applied field.

they correspond to the frequencies of transitions from one energy level to another that the electrons of the atom can make. If we confine our attention to one of these resonances then we can treat an $n$-electron atom as a one-electron classical oscillator. A time-varying electric field $E(t)$, which we assume *a priori* to be at a frequency near to an atomic resonance, perturbs only a single electron to a significant degree, thereby inducing a dipole moment that varies at the same frequency as the applied field. The electron and nucleus are perturbed in opposite directions by the field as shown in Fig. 2.16.

The motion of the electrons on each of the particles in the medium, which for simplicity can be assumed to be identical, obeys the differential equation

$$\frac{d^2x}{dt^2} + 2\Gamma\frac{dx}{dt} + \frac{k}{m}x = -\frac{e}{m}E(t). \tag{2.110}$$

The terms on the left-hand side of the equation represent, reading from left to right, the acceleration of the electron, a damping term proportional to the electron velocity, and a restoring force. These terms are balanced by the effect of the applied electric field $E(t)$. The restoring force is analogous to the restoring force acting on a mass suspended from a spring and given a small displacement from its equilibrium position. The damping can be regarded as a viscous drag that the moving electron experiences because of its interaction with the other electrons on the particle. We take $E(t) = \mathcal{R}(Ee^{j\omega t})$ and $x(t) = \mathcal{R}[X(\omega)e^{j\omega t}]$. The possibility that $E(t)$ and $x(t)$ are not in phase is taken into account by allowing the function $X(\omega)$ to include a phase factor. If we define a resonance frequency by $\omega_0 = \sqrt{k/m}$ then the differential equation (2.110) above becomes

$$(\omega_0^2 - \omega^2)X + 2j\omega\Gamma X = -\frac{e}{m}E, \tag{2.111}$$

giving

$$X(\omega) = \frac{-(e/m)E}{\omega_0^2 - \omega^2 + 2j\omega\Gamma}. \tag{2.112}$$

This is the amplitude of the displacement of the electron from its equilibrium position as a function of the frequency of the applied field. Near resonance $\omega \simeq \omega_0$, so

$$X(\omega \simeq \omega_0) = \frac{-(e/m)E}{2\omega_0(\omega_0 - \omega) + 2j\omega_0\Gamma}. \tag{2.113}$$

The dipole moment of a single particle is $\mu(t) = -e[x(t)]$, which arises from the separation of the electron-charge cloud and the nucleus. If there are $N$ electron oscillators per unit volume there results a net polarization (dipole moment per unit volume) of

$$P(t) = -Nex(t) = \mathcal{R}[P(\omega)e^{j\omega t}], \tag{2.114}$$

where $P(\omega)$ is the complex amplitude of the polarization:

$$\begin{aligned} P(\omega) = -NeX(\omega) &= \frac{(Ne^2/m)E_0}{2\omega_0(\omega_0 - \omega) + 2j\omega_0\Gamma} \\ &= \frac{-jNe^2/(2m\omega_0\Gamma)}{1 + j(\omega - \omega_0)/\Gamma}E_0. \end{aligned} \tag{2.115}$$

The electronic susceptibility $\chi(\omega)$ is defined by the equation

$$P(\omega) = \epsilon_0\chi(\omega)E_0, \tag{2.116}$$

where $\epsilon_0$ is the permittivity of free space.[5] The susceptibility $\chi(\omega)$ is complex, and can be written in terms of its real and imaginary parts as

$$\chi(\omega) = \chi'(\omega) - j\chi''(\omega), \tag{2.117}$$

where the use of the negative sign is a common convention. The polarization is

$$P(t) = \mathcal{R}[\epsilon_0\chi(\omega)E_0e^{j\omega t}] = \epsilon_0E_0\chi'(\omega)\cos(\omega t) + \epsilon_0E_0\chi''\sin(\omega t). \tag{2.118}$$

Therefore $\chi'(\omega)$, the real part of the susceptibility, is related to the in-phase polarization, while $\chi''(\omega)$, the complex part, is related to the out-of-phase (quadrature) component. From Eqs. (2.115) and (2.116),

$$\chi(\omega) = -j\frac{Ne^2}{2m\omega_0\Gamma\epsilon_0}\frac{1}{1 + j(\omega - \omega_0)/\Gamma}, \tag{2.119}$$

so

$$\chi'(\omega) = \frac{Ne^2}{2m\omega_0\Gamma\epsilon_0}\frac{(\omega_0 - \omega)/\Gamma}{1 + (\omega - \omega_0)^2/\Gamma^2}, \tag{2.120}$$

$$\chi''(\omega) = \frac{Ne^2}{2m\omega_0\Gamma\epsilon_0}\frac{1}{1 + (\omega - \omega_0)^2/\Gamma^2}. \tag{2.121}$$

On changing to conventional frequency, $\nu = \omega/(2\pi)$, and putting $\Delta\nu = \Gamma/\pi$, which is the FWHM of the Lorentzian shape that describes $\chi''(\omega)$, we obtain

$$\chi''(\nu) = \frac{Ne^2}{16\pi^2m\nu_0\epsilon_0}\frac{\Delta\nu}{(\Delta\nu/2)^2 + (\nu - \nu_0)^2}, \tag{2.122}$$

$$\chi'(\nu) = \frac{2(\nu_0 - \nu)}{\Delta\nu}\chi''(\nu) = \frac{Ne^2}{8\pi^2m\nu_0\epsilon_0}\frac{\nu_0 - \nu}{(\Delta\nu/2)^2 + (\nu - \nu_0)^2}. \tag{2.123}$$

[5] $\epsilon_0 = 8.854 \times 10^{-12}\,\mathrm{F\,m^{-1}}$.

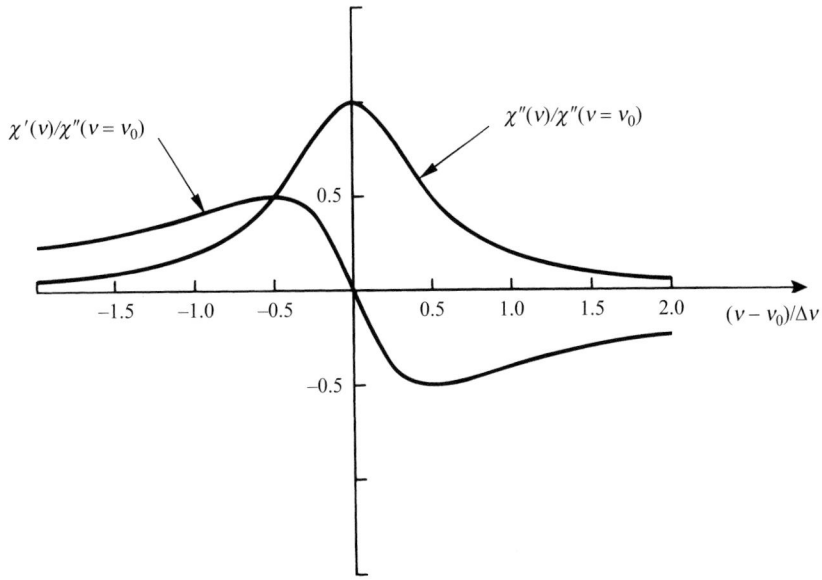


**Fig. 2.17** The frequency variation of the real, $\chi'(\nu)$, and imaginary, $\chi''(\nu)$, parts of the susceptibility calculated using the electron oscillator model.

Figure 2.17 is a graph of $\chi''$ and $\chi'$ normalized with respect to the peak value $\chi''_0$ of $\chi''$. Note that $\chi''$ has the characteristic Lorentzian shape common to the frequency response of *RLC* circuits and homogeneously broadened spectral lines.[6]

### 2.7.1 Relating the complex susceptibility, gain, and absorption

The complex susceptibility of a particle, or a medium containing a collection of particles, is closely related to the change in amplitude of an electromagnetic wave passing through such a medium. We can make this connection by considering some fundamental concepts in classical electromagnetic theory. The relationship between the applied electric field **E** and the electron displacement vector **D** is

$$\begin{aligned} \mathbf{D} &= \epsilon_0 \mathbf{E} + \mathbf{P} \\ &= \epsilon_0 (1 + \chi) \mathbf{E}, \end{aligned} \tag{2.124}$$

which, on introducing the dielectric constant $\epsilon_r = 1 + \chi$, can be written as

$$\mathbf{D} = \epsilon_0 \epsilon_r \mathbf{E}. \tag{2.125}$$

The refractive index $n$ of the medium is related to $\epsilon_r$ by $n = \sqrt{\epsilon_r}$.[7] When an external electric field interacts with a group of particles there are two contributions to the induced

[6] It can be shown that, although Eqs. (2.132) and (2.123) have the correct frequency dependence for the susceptibility of a real particle, they have incorrect magnitude. This arises because of inadequacies in the classical electron oscillator model, and these inadequacies should be borne in mind before the classical model is used to make detailed predictions about the behavior of a particle.

[7] In a magnetic material with relative magnetic permeability $\mu_r \neq 1$, $n = \sqrt{\mu_r \epsilon_r}$.

polarization, namely a macroscopic contribution $\mathbf{P}_m$ from the collective properties of the particles, for example their arrangement in a crystal lattice, and a contribution from the polarization $\mathbf{P}_t$ associated with transitions in the medium; so, in general,

$$\mathbf{P} = \mathbf{P}_m + \mathbf{P}_t. \tag{2.126}$$

Usually many transitions are possible for the particles of the medium but only one will be in near resonance with the frequency of an applied field. $\mathbf{P}_t$ is dominated by the contribution of this single transition near resonance. Far from any such resonance $\mathbf{P}_t$ is negligible and $\mathbf{P} = \mathbf{P}_m$, which allows us to define the macroscopic dielectric constant $\epsilon_r$ (far from resonance) from

$$\mathbf{P} = \epsilon_0\mathbf{E} + \mathbf{P}_m = \epsilon_0\mathbf{E} + \chi_m\epsilon_0\mathbf{E} = \epsilon_r\epsilon_0\mathbf{E}, \tag{2.127}$$

where $\chi_m$ is the macroscopic, non-resonant, susceptibility. However, if the frequency of the electric field is near the frequency of a possible transition within the medium then there is a significant contribution to $\mathbf{P}$ from this transition. Other possible transitions far from resonance do not contribute, and we can write

$$\mathbf{D} = \epsilon_0\mathbf{E} + \mathbf{P}_m + \mathbf{P}_t = \epsilon_r\epsilon_0\mathbf{E} + \mathbf{P}_t. \tag{2.128}$$

$\mathbf{P}_t$ is related to the complex susceptibility that results from the transition according to $\mathbf{P}_t = \epsilon_0\chi(\omega)\mathbf{E}$. Therefore we can rewrite Eq. (2.124) as

$$\mathbf{D} = \epsilon_0[\epsilon_r + \chi(\omega)]\mathbf{E} = \epsilon_0\epsilon_r^*\mathbf{E}, \tag{2.129}$$

so the complex susceptibility modifies the effective dielectric constant from $\epsilon_r$ to $\epsilon_r^*$. When an electromagnetic wave propagates through a medium with a complex susceptibility, both the amplitude and the phase velocity of the wave are affected. This can be illustrated easily for a plane wave propagating in the $z$ direction with a field variation $\sim e^{j(\omega t - kz)}$, where the propagation constant, $k$, is given by the expression

$$k = \omega\sqrt{\mu\epsilon} = \omega\sqrt{\mu_r\mu_0\epsilon_r\epsilon_0}. \tag{2.130}$$

The relative permeability $\mu_r$ is generally unity for optical materials. For a complex dielectric constant Eq. (2.130) can be rewritten as

$$k' = \omega\sqrt{\mu_0\epsilon_r\epsilon_0}\sqrt{1 + \frac{\chi(\omega)}{\epsilon_r}} = k\sqrt{1 + \frac{\chi(\omega)}{\epsilon_r}}, \tag{2.131}$$

where $k'$ is now the new propagation constant, which differs from the non-resonant propagation constant $k$ because of the complex susceptibility resulting from a transition. If $|\chi(\omega)| \ll \epsilon_r$ Eq. (2.131) can be simplified by use of the binomial theorem to give

$$k' = k\left[1 + \frac{\chi(\omega)}{2\epsilon_r}\right] = k\left[1 + \frac{\chi'(\omega)}{2\epsilon_r} - \frac{j\chi''(\omega)}{2\epsilon_r}\right]. \tag{2.132}$$

The wave now propagates through the medium as $e^{-jk'z}$. Written out in full, the electric field varies as

$$\begin{aligned} E &= E_0 \exp\left(j\left\{\omega t - k\left[1 + \frac{\chi'(\omega)}{2\epsilon_r} - \frac{j\chi''(\omega)}{2\epsilon_r}\right]z\right\}\right) \\ &= E_0 \exp\left(j\left\{\omega t - k\left[1 + \frac{\chi'(\omega)}{2\epsilon_r}\right]\right\}\right) \exp\left[-\frac{k\chi''(\omega)}{2\epsilon_r}z\right]. \end{aligned} \tag{2.133}$$

Clearly, this is a wave whose phase velocity is

$$c' = \frac{\omega}{k[1 + \chi'(\omega)/(2\epsilon_r)]} = \frac{\omega}{k + \Delta k} \tag{2.134}$$

and whose field amplitude changes exponentially with distance. If we write

$$\gamma = -\frac{k\chi''(\omega)}{\epsilon_r} \tag{2.135}$$

then the wave changes its electric-field amplitude with distance according to $e^{(\gamma/2)z}$. The intensity of the wave is $I \propto E(z,t)E^*(z,t)$, which changes as the wave passes through the medium as $I \propto e^{\gamma z}$. We can identify $\gamma$ as the familiar gain coefficient of the medium, which was calculated previously by considering the spontaneous and stimulated radiative jumps between two energy levels. Now, from Eq. (2.135),

$$\gamma(\nu) = -\frac{k\chi''(\nu)}{n^2} = -\frac{2\pi\nu_0 n}{c_0}\frac{\chi''(\nu)}{n^2}, \tag{2.136}$$

which, from Eq. (2.122), obtained by consideration of the system as a collection of classical oscillators, is

$$\gamma(\nu) = -\frac{2\pi\nu}{nc_0}\frac{Ne^2}{16\pi^2 m\nu_0\epsilon_0}\frac{\Delta\nu}{(\Delta\nu/2)^2 + (\nu - \nu_0)^2}, \tag{2.137}$$

which is *always negative*. Clearly, this is an incorrect result since we have previously shown that

$$\gamma(\nu) = \left(N_2 - \frac{g_2}{g_1}N_1\right)\frac{c^2 A_{21}}{8\pi\nu^2}g(\nu_0, \nu), \tag{2.138}$$

which can be either positive or negative, depending on the sign of $N_2 - (g_2/g_1)N_1$. Thus, the classical electron oscillator model appears to predict only absorption of incident radiation. It is possible, however, within the framework of the classical electron oscillator model to show that under certain conditions stimulated emission can occur. Although the classical electron oscillator model is instructive, it is not entirely adequate in describing the interaction between particles and radiation. It is better to accept that $\gamma(\nu) = -k\chi''(\nu)/n^2$ and use Eq. (2.138) as the expression for $\gamma(\nu)$. In this case we find that the imaginary part of the complex susceptibility of the medium is

$$\chi''(\nu) = -\frac{n^2\gamma(\nu)}{k} = -\frac{n^2 c}{2\pi\nu}\gamma(\nu), \tag{2.139}$$

which, from Eq. (2.136), gives

$$\chi''(\nu) = -\frac{[N_2 - (g_2/g_1)N_1]\, n^2 c^3 A_{21}}{8\pi^3 \nu^3 \Delta\nu} \frac{1}{1 + [2(\nu - \nu_0)/\Delta\nu]^2}. \tag{2.140}$$

This quantum-mechanical susceptibility is negative or positive depending on whether $N_2 - (g_2/g_1)N_1$ is positive or not. A *negative* value of $\chi''(\nu)$ corresponds to a system in population inversion.

The power expended in inducing polarization in a medium is given by $\mathbf{E} \dot{\sim} d\mathbf{P}/dt$ (see Appendix 5). If we write

$$|\mathbf{P}| = P(t) = \mathcal{R}(Pe^{j\omega t}), \tag{2.141}$$

which in an isotropic medium is parallel to $\mathbf{E}$, where

$$|\mathbf{E}| = \mathcal{R}(Ee^{j\omega t}), \tag{2.142}$$

then the rate of expenditure of energy is

$$\frac{\text{power}}{\text{volume}} = \mathcal{R}(Ee^{j\omega t}) \frac{\partial}{\partial t} \mathcal{R}(Pe^{j\omega t}). \tag{2.143}$$

On remembering that

$$\mathbf{P} = \epsilon_0 \chi \mathbf{E} \tag{2.144}$$

and averaging we have

$$\overline{\frac{\text{power}}{\text{volume}}} = \overline{\mathcal{R}(Ee^{j\omega t})\mathcal{R}(j\omega\epsilon_0\chi Ee^{j\omega t})} \tag{2.145}$$

$$= \frac{1}{2}\mathcal{R}(j\omega\epsilon_0\chi EE^*) \tag{2.146}$$

$$= \frac{\omega}{2}\epsilon_0 |E|^2 \mathcal{R}(j\chi). \tag{2.147}$$

Since $\chi = \chi' - j\chi''$,

$$\overline{\frac{\text{power}}{\text{volume}}} = \frac{\omega\epsilon_0\chi''}{2}|E|^2. \tag{2.148}$$

If our system is in a state of population inversion, that is, with a negative $\chi''$, then from Eq. (2.145) the average power per volume used in inducing polarization in the medium is *negative* – and the system provides us with more energy than is put in by the electromagnetic wave.

## 2.8 The classical oscillator explanation for stimulated emission

If there were no applied electric field acting on the electron, then, from Eq. (2.110), the position of the electron would satisfy the equation

$$x(t) = x_0 e^{-\Gamma t} \cos(\omega_0 t + \phi_0), \tag{2.149}$$

where $\omega_0$ is the resonance frequency and $\phi_0$ is a phase factor set by the initial conditions. If at time $t = 0$ the position and velocity of the electron are $a_0$ and $v_0$, respectively, then

$$\begin{aligned} a_0 &= x_0 \cos\phi_0, \\ v_0 &= -\Gamma x_0 \cos\phi_0 - \omega_0 x_0 \sin\phi_0. \end{aligned} \tag{2.150}$$

When the electric field is applied, we have already seen that in the steady state energy is apparently only absorbed. However, this impression is erroneous. It neglects the fact that in reality no electromagnetic field interacts indefinitely with an electron. Therefore, we must consider what happens when an electron, which can already be regarded as oscillating if it is in an excited state, is suddenly subjected to the additional perturbation of an applied field. We are going to be interested in the behavior of the electron over the first few cycles of the applied field, so we neglect damping and write

$$\frac{d^2x}{dt^2} + \omega_0^2 = -\frac{eE_0}{2m}(e^{j\omega t} + e^{-j\omega t}), \tag{2.151}$$

where the applied field is $E_0 \cos(\omega t)$ and has been written in its complex exponential form. On introducing a new variable $z = \dot{x} + j\omega_0 x$, where the overdot indicates differentiation with respect to time, Eq. (2.151) can be rewritten in the form

$$\frac{dz}{dt} - j\omega_0 z = -\frac{eE_0}{2m}(e^{j\omega t} + e^{-j\omega t}). \tag{2.152}$$

This equation can be solved by multiplying each term by $e^{-j\omega_0 t}$ and then integrating to give

$$ze^{-j\omega_0 t} = -\frac{eE_0}{2m}\int (e^{j(\omega-\omega_0)t} + e^{-j(\omega+\omega_0)t})dt. \tag{2.153}$$

This gives

$$\frac{dx}{dt} + j\omega_0 x = -\frac{eE_0}{2m}\left[-\frac{je^{j\omega t}}{\omega - \omega_0} + \frac{je^{-j\omega t}}{\omega + \omega_0} + Ae_0^{j\omega t}\right], \tag{2.154}$$

where $A$ is a constant of integration. By integrating a second time in a similar manner and introducing the initial values of position and velocity it is possible to show that the final solution is

$$x(t) = -\frac{eE_0}{m}\left[\frac{\cos(\omega_0 t) - \cos(\omega t)}{\omega^2 - \omega_0^2}\right] + \sqrt{\left(\frac{v_0}{\omega_0}\right)^2 + x_0^2}\cos(\omega_0 t + \phi), \tag{2.155}$$

where $\tan\phi = -v_0/(\omega_0 x_0)$. By use of the trigonometrical identity

$$\cos X - \cos Y = -2\sin\left(\frac{X+Y}{2}\right)\sin\left(\frac{X-Y}{2}\right) \tag{2.156}$$

and assuming that the applied frequency is close to resonance, Eq. (2.155) can be written

$$x(t) = -\frac{eE_0}{2m\omega_0}t\sin(\omega_0 t) + \sqrt{\left(\frac{v_0}{\omega_0}\right)^2 + x_0^2}\cos(\omega_0 t + \phi). \tag{2.157}$$

Thus, near resonance the amplitude of oscillation will increase linearly with time, which is, of course, a consequence of our neglect of damping. It is more interesting, however, to use Eq. (2.157) to calculate the work done during the first $n$ cycles of application of the field. This work is calculated as the work done by the electric field in polarizing the medium: the polarization $\mathbf{P}$ is proportional to the electron displacement. The work done is $\mathbf{E} \dot{\sim} \partial \mathbf{P}/\partial t$. During the first $n$ cycles the total work done by the field is

$$W = \int_0^{2n\pi/\omega_0} \mathbf{E} \dot{\sim} \frac{\partial \mathbf{P}}{\partial t} dt = -NeE_0 \int_0^{2n\pi/\omega_0} \cos(\omega_0 t)\dot{x}(t)dt, \tag{2.158}$$

where $N$ is the total number of particles per unit volume. Writing $(v_0/\omega_0)^2 + x_0^2 = a^2$ and substituting from Eq. (2.157) gives

$$W = -NeE_0 \int_0^{2n\pi/\omega_0} \left[ -\frac{eE_0}{m\omega_0} \sin(2\omega_0 t) - \frac{eE_0 t}{4m}(1 + \cos(2\omega_0 t)) \right.$$
$$\left. - \frac{a\omega_0}{2} \sin\phi + \frac{a\omega_0}{2} \sin(2\omega_0 t + \phi) \right] dt. \tag{2.159}$$

Clearly, the first and last terms of the integrand average to zero over a whole number of cycles. The remaining terms can be integrated to give

$$W = \frac{Ne^2 E_0^2}{m} \left( \frac{n^2\pi^2}{2\omega_0^2} + \frac{n\pi ma}{eE_0} \sin\phi \right). \tag{2.160}$$

This work done by the applied field is negative, implying that the oscillating electrons supply energy to the field if $\sin\phi < 0$ and $|\sin\phi| > eE_0 n\pi/(2ma\omega_0^2)$. This is the condition set by classical theory for stimulated emission to occur. Because the charge on the electron is negative, stimulated emission can result only if the electron velocity when the applied field is turned on is in the direction of the field. If the electron velocity is in the same direction as the field, the electron is decelerated by the field and consequently radiates energy. If the electron were accelerated by the field then absorption of energy from the field would occur. There is a maximum number of cycles of application of the field after which the oscillating electrons start, and continue indefinitely, to absorb energy. This is set by the condition

$$n < 2ma\omega_0^2/(eE_0\pi). \tag{2.161}$$

After a long enough time the motion of the electron is dominated by the first term in Eq. (2.157) and can be written

$$x(t) = -\frac{eE_0}{2m\omega_0} t \sin(\omega_0 t), \tag{2.162}$$

and the electron velocity is, for large enough $t$,

$$\dot{x}(t) \simeq -\frac{eE_0}{2m} t \cos(\omega_0 t). \tag{2.163}$$

The electron now has a velocity that is oppositely directed from the applied field; the electron is being accelerated and absorbs energy from the field. We can conclude by saying that, when an electric field near resonance is applied to an already oscillating electron, stimulated emission can occur at early times, provided that the initial velocity of the electron is in the same direction as the field.

## 2.9 Problems

2.1 A homogeneously broadened laser amplifier has a small-signal gain at line center of $0.6\ \mathrm{m}^{-1}$ and is at line center 3 m long. The saturation intensity of the medium at line center is $3\ \mathrm{W\ m}^{-2}$. An input signal of intensity $2.5\ \mathrm{W\ m}^{-2}$ enters the amplifier. What is the output intensity if (a) the input is at line center and gain saturation is neglected, (b) the input is one linewidth (FWHM) from line center and gain saturation is neglected, (c) as in (a) but gain saturation is included, and (d) as in (b) but gain saturation is included?

Hint: (c) and (d) need to be solved numerically.

2.2 A laser amplifier contains two groups of particles centered at frequencies $\nu_0 - a/2$ and $\nu_0 + a/2$. Each group is individually homogeneously broadened with homogeneous FWHM $\Delta\nu_\mathrm{L}$. (a) At what two frequencies $\nu_0 \pm b$ is the gain of the amplifier a maximum? Consider all three cases: (i) $a \gg \Delta\nu_\mathrm{L}$, (ii) $a \sim \Delta\nu_\mathrm{L}$, and (iii) $a \ll \Delta\nu_\mathrm{L}$. (b) What is the relative gain at frequency $\nu_0$ as a function of $a/\Delta\nu_\mathrm{L}$?

2.3 What is the overall lineshape function if

$$g_\mathrm{D}(\nu_0, \nu) = \begin{cases} 1/\Delta\nu_\mathrm{D} & \text{for } \nu_0 - \Delta\nu_\mathrm{D}/2 \le \nu \le \nu_0 + \Delta\nu_\mathrm{D}/2; \\ 0, & \text{elsewhere} \end{cases}$$

and

$$g_\mathrm{L}(\nu', \nu'') = \frac{2/(\pi\ \Delta\nu_\mathrm{L})}{1 + [2(\nu' - \nu'')/\Delta\nu_\mathrm{L}]^2}?$$

2.4 What is the stimulated emission rate if $A_{21} = 10^8\ \mathrm{s}^{-1}$, the line is homogeneously broadened with $\Delta\nu = 1$ GHz, and the input radiation is "white" with $\rho(\nu) = 1\ \mathrm{J\ m}^{-3}\ \mathrm{Hz}^{-1}$? What would the stimulated emission rate be if the input radiation were monochromatic, one FWHM from line center, and had intensity $1\ \mathrm{W\ m}^{-2}$? Take the line center frequency to be $10^{14}$ Hz.

2.5 A homogeneously broadened amplifier with FWHM 1 GHz and a small-signal gain at line center of $1\ \mathrm{m}^{-1}$ has a saturation intensity at line center of $1\ \mathrm{W\ m}^{-2}$. An input signal of intensity $2\ \mathrm{W\ m}^{-2}$ enters the amplifier, which is 0.5 m long. Calculate (a) the output intensity neglecting saturation if the input is at 500 MHz above line center, (b) the output intensity including gain saturation if the input signal is 500 MHz above line center, (c) the output intensity including gain saturation if the input signal is at line center, and (d) the output intensity 500 MHz above line center if an additional $1\ \mathrm{W\ m}^{-2}$ signal is also injected 500 MHz below line center.

2.6 What is the saturation intensity one FWHM from line center in a naturally broadened amplifier if the amplifier has the following parameters: $A_{21} = 10^8\ \mathrm{s}^{-1}$, $\tau_1 = 1$ ns, $\tau_2 = 5$ ns, and $\lambda_0 = 1\ \mu\mathrm{m}$?

2.7 The lifetime of a particular excited argon level varies with pressure as $\tau = 10^{-8}/(1+P)$ s at 2,000 K, where $P$ is the pressure measured in atmospheres. At what pressure will the collisional broadening be as great as the Doppler broadening? Neglect the effect of the lower-level lifetime. Take $\lambda_0 = 488$ nm.

2.8 A homogeneously broadened gas laser amplifier with $\Delta\nu = 1$ GHz working at 300 nm has a small-signal gain coefficient at line center of $1\ \mathrm{m}^{-1}$. Its susceptibility as a function of frequency varies as

$$\chi''(\nu) \propto \frac{1}{1 + [2(\nu - \nu_0)/\Delta\nu]^2}$$

and

$$\chi'(\nu) = \frac{2(\nu - \nu_0)}{\Delta\nu}\chi''(\nu).$$

Both $\chi'$ and $\chi'' \ll 1$. The refractive index of the gas is $n = 1$. Calculate the slowest phase velocity that will be observed for waves traveling through the amplifier.

2.9 A gaseous medium has a Lorentzian broadened lineshape of FWHM 1 GHz. Other parameters of the medium are $N_2 = 5 \times 10^{16}\ \mathrm{m}^{-3}$, $N_1 = 5 \times 10^{15}\ \mathrm{m}^{-3}$, $g_2 = g_1$, $A_{21} = 10^8\ \mathrm{s}^{-1}$, $\tau_2 = 5$ ns, $\tau_1 = 1$ ns, and $\lambda_0 = 1\ \mu\mathrm{m}$.
Calculate (i) $I_S(\nu_0)$, (ii) $\chi = \chi' - j\chi''$, and (iii) the change in phase velocity of a weak signal propagating one FWHM from line center relative to $c_0$. What would happen if $\tau_1 = 1\ \mu\mathrm{s}$ and the medium were pumped from $N_2 = 0$, $N_1 = 0$ at time 0 at a rate $R_2 = 10^{20}\ \mathrm{m}^{-3}\ \mathrm{s}^{-1}$, $R_1 = 0$?

2.10 How is the result given in Eq. (2.131) modified if $|\chi(\omega)|$ is not much smaller than $\epsilon_r$? Derive new expressions for the phase velocity and gain coefficient.

2.11 Write a computer program to plot the variation in phase velocity as a function of gain one FWHM on the low-frequency side of line center as the gain increases from 0 to $1/\lambda\ \mathrm{m}^{-1}$ for a medium with $\epsilon_r = 1.5$ and $\lambda = 1\ \mu\mathrm{m}$.

## References

[1] R. H. Dicke and J. P. Wittke, *Introduction to Quantum Mechanics*, Reading, MA: Addison-Wesley, 1960.

[2] R. Liboff, *Introductory Quantum Mechanics*, 4th edn., San Francisco, CA: Addison Wesley, Inc., 2003.

[3] R. L. White, *Basic Quantum Mechanics*, New York: McGraw-Hill, 1966.

[4] Sir J. H. Jeans, *An Introduction to the Kinetic Theory of Gases*, Cambridge: Cambridge University Press, 1940.

[5] R. D. Present, *Kinetic Theory of Gases*, New York: McGraw-Hill, 1958.

[6] D. C. Champeney, *Fourier Transforms and Their Physical Applications*, London: Academic Press, 1973.

[7] A. C. G. Mitchell and M. W. Zemansky, *Resonance Radiation and Excited Atoms*, Cambridge: Cambridge University Press, 1971.

[8] P. W. Milonni and J. H. Eberly, *Lasers*, 2nd edn., New York: Wiley, 2010.

[9] M. Abramowitz and I. A. Stegun, *Handbook of Mathematical Functions*, New York: Dover, 1972.

[10] W. R. Bennett, Jr., "Hole-burning effects in a He–Ne optical maser," *Phys. Rev.* **126**, 580–593, 1962.

# 3 An introduction to two practical laser systems

## 3.1 Introduction

To give a little more practical emphasis to some of the ideas we have dealt with so far, let us consider some of the details of the two laser systems in which population inversion and laser oscillation were first demonstrated. One of these lasers uses an amplifying medium that is a crystalline solid – the ruby laser; in the other the amplifying medium is a gas – a mixture of helium and neon. In each case, the amplifying medium is pumped into a state of population inversion by feeding energy into it in an appropriate way. Laser oscillation occurs when the amplifying medium is placed between a pair of suitable aligned mirrors that provide the necessary optical feedback to cause oscillation to occur. The ruby laser was the first operational laser, being demonstrated on May 16, 1960 by Theodore Maiman of the Hughes Aircraft Company in Malibu, California [1].

That the ruby laser was the first laser to be demonstrated surprised many in the scientific community.[1] This is because the ruby laser is a *three-level* laser, which was expected to be much more difficult to operate than a *four-level* laser. This is an important distinction, which we will examine before describing the first two lasers in detail.

## 3.2 Three- and four-level lasers

The distinction between three- and four-level lasers can be illustrtated with the aid of Fig. 3.2. Energy is fed into the system to move particles from the ground state, level 0, to a *pumping* energy level, level 3. This is called the pumping process. Particles that reach this level should, ideally, transfer with high efficiency to the upper laser level, level 2. The laser transition is from level 2 to level 1. Particles arriving in the lower laser level should, ideally, leave this level quickly and fall back to the ground state via the *drain* transition from level 1 to the ground state.

If the energy $E_1$ of the lower laser level is sufficiently small, such that $E_1 \ll k_B T$, or if level 1 is the ground state, the system is a three-level laser. In such a laser the threshold inversion is substantially higher than in a four-level system where $E_1 \gg k_B T$. It is this

[1] When Maiman submitted his manuscript describing the first laser to the journal *Physical Review Letters*, the editor at that time, Samuel Goudsmit, rejected it, saying it was just another "maser" paper. Consequently, the first article describing successful operation of a laser appeared in the periodical *Nature*.

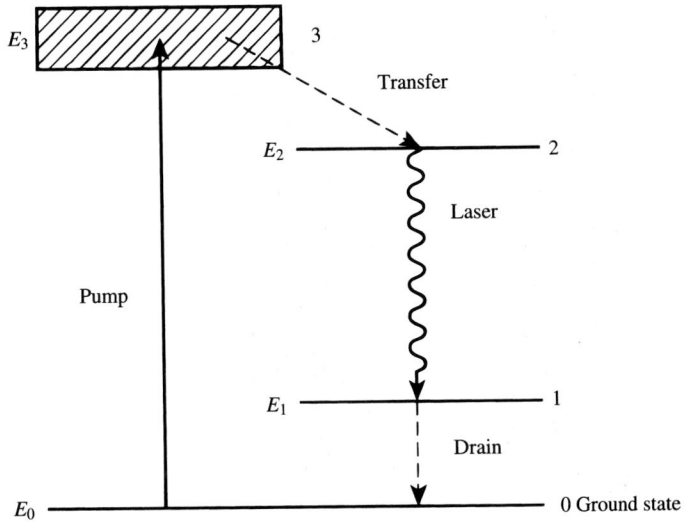


**Fig. 3.1** A schematic energy-level diagram of three- and four-level lasers.

increased inversion needed for oscillation that makes a three-level laser more difficult to operate than a four-level laser. Most efficient lasers fall into the class of four-level lasers (two-level lasers are also possible).

### 3.2.1 Threshold inversion in three- and four-level lasers

If the total number of particles per unit volume participating in the pumping and lasing process is $N$ we can write

$$N = N_3 + N_2 + N_1 + N_0, \tag{3.1}$$

where the subscripts indicate the level involved, and $N_0$ is the population density in the ground state. Equation (3.1) assumes that other levels of the system are not actively involved and that their populations are negligible.

In a three-level laser, level 1 is essentially the ground state, so we can write

$$N = N_3 + N_2 + N_1. \tag{3.2}$$

In a good laser, level 3 should transfer its excitation very rapidly to level 2, so its population can be neglected. In this case

$$N = N_2 + N_1. \tag{3.3}$$

The threshold inversion is

$$N_t = N_2 - \frac{g_2}{g_1} N_1. \tag{3.4}$$

From Eqs. (3.3) and (3.4),

$$(N_2)_{3\text{-level}} = \frac{(g_2/g_1)N + N_t}{g_2/g_1 + 1}, \tag{3.5}$$

which, if $g_2 = g_1$, gives $N_2 \simeq N/2$.

For a good four-level laser in which the lower laser level remains relatively depopulated we can write

$$(N_2)_{4\text{-level}} \simeq N_t.$$

Consequently, the relative rate at which level 2 must be excited to produce an inversion (all other factors such as decay rates being comparable) is

$$\frac{(N_2)_{3\text{-level}}}{(N_2)_{4\text{-level}}} \simeq \frac{(g_2/g_1)N + N_t}{(g_2/g_1 + 1)N_t}. \tag{3.6}$$

Usually, $N \gg N_t$, and, if we take the simple case $g_2 = g_1$, the ratio of pumping rates becomes $N/(2N_t)$. This number can easily be $10^4$ or more, demonstrating once again how surprising it was that a three-level laser was the first ever to be operated.

## 3.3 The ruby laser

The amplifying medium of this laser is sapphire (crystalline aluminum oxide) doped with $Cr^{3+}$ ions, typically in concentrations of about 0.05% by weight. The crystal structure of the undoped sapphire is shown in Fig. 3.3. When the crystal is doped with $Cr^{3+}$ ions some of the aluminum sites become occupied by chromium. It is these $Cr^{3+}$ ions that are the active particles in the amplification process. A schematic energy-level diagram for the $Cr^{3+}$ ions in such material is shown in Fig. 3.4. Population inversion is produced by irradiating such a ruby crystal with light that causes absorption transitions from the ground state into the broad absorption bands labeled $^4F_1$ and $^4F_2$. The absorption spectrum of a typical ruby laser crystal is illustrated in Fig. 3.5, which clearly shows these two broad absorption bands, one in the violet–blue and the other in the green–yellow region, which lead to the $^4F$ levels. These absorption bands are very broad because the very many sub-levels of the $^4F$ levels of slightly different energy are "smeared" together by line broadening. The sharper absorption peak due to the laser transition itself is also visible as a small peak around 700 nm in Fig. 3.5. Figure 3.6 shows in more detail the absorption spectrum of the laser transition that illustrates that the laser transition consists of two closely spaced lines, labeled $R_2$ and $R_1$ [3]. In Figs. 3.5 and 3.6 the amount of absorption depends on the relative linear polarization of the incident light and the orientation of the crystal.[2]

[2] A fuller discussion of the effect of crystal symmetry on the propagation of light is given in Chapter 17.

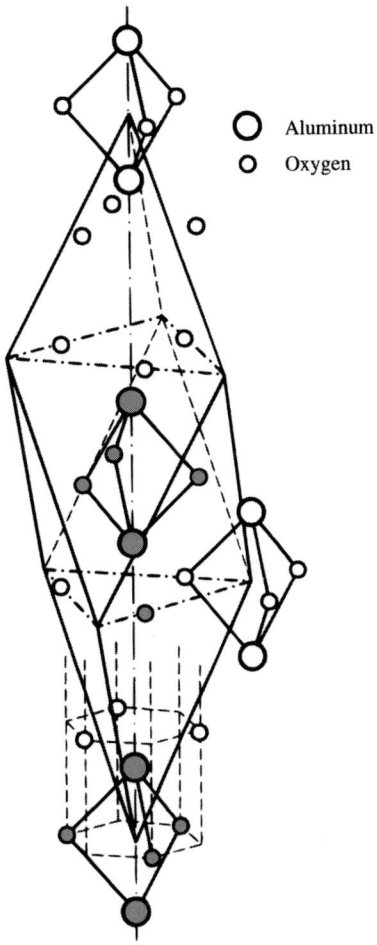


**Fig. 3.2** The crystal structure of sapphire: $\alpha$-$Al_2O_3$ (aluminum oxide). The shaded atoms make up a unit cell of the structure. The aluminum atom inside the dashed hexagonal prism experiences an almost cubic field symmetry from the oxygen atoms on the prism [1].

The $^4F_1$ and $^4F_2$ levels decay (with lifetimes ~50 ns) preferentially into the upper laser level $^2E$. This level actually consists of a pair of closely spaced levels separated by 29 $cm^{-1}$.[3]

[3] The unit $cm^{-1}$ is frequently used as a unit of energy. If two energy levels are 1 $cm^{-1}$ apart, this implies that the wavelength of the transition between them is 1 cm, and its frequency is $3 \times 10^{10}$ Hz, i.e.,

$$\Delta E = h\nu = \frac{hc}{\lambda} = 3 \times 10^{10} h.$$

If two levels are $k$ $cm^{-1}$ apart, then

$$\Delta E = \frac{hc}{\lambda} = \frac{hc}{1/k} = \left(k \times 3 \times 10^{10}\right) h.$$

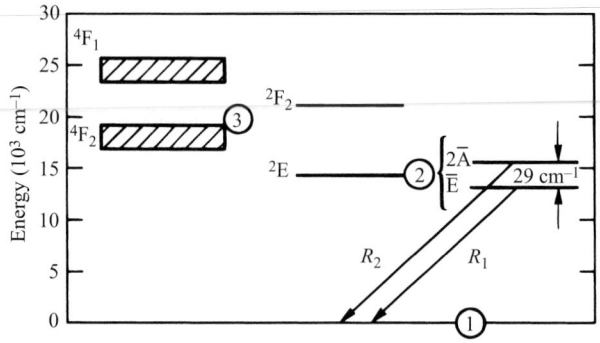


**Fig. 3.3** A schematic energy-level diagram for ruby – $Cr^{3+}$ ions in sapphire [2].

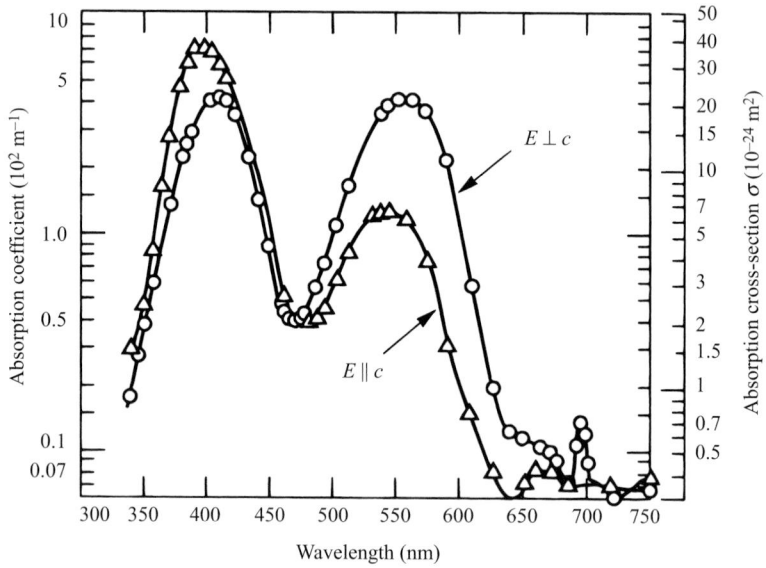


**Fig. 3.4** The absorption coefficient and absorption cross-section as functions of wavelength for pink ruby. These absorption spectra are slightly different depending on whether the incident polarized light being absorbed is linearly polarized with its electric vector parallel, or perpendicular, to the *c* symmetry axis of the crystal [3].

The splitting of this energy level arises because of the symmetry of the crystal lattice. The $Cr^{3+}$ ions in the aluminum oxide lattice have energy levels that are affected by the spatial arrangement of aluminum and oxygen ions around them. If this arrangement had exact octahedral symmetry around each $Cr^{3+}$ ion, no splitting would occur. In practice the symmetry of the arrangement is rhombohedral, and the small energy splitting results.

In practice, to produce a state of population inversion in the crystal, it is necessary to irradiate the crystal with a very intense light source. This is most easily accomplished with a flashlamp, which uses a tube, generally made of quartz and filled with a noble gas, through which energy stored in a capacitor charged to high voltage is discharged. Such a lamp can, for the short period of time during which current flows through it, produce

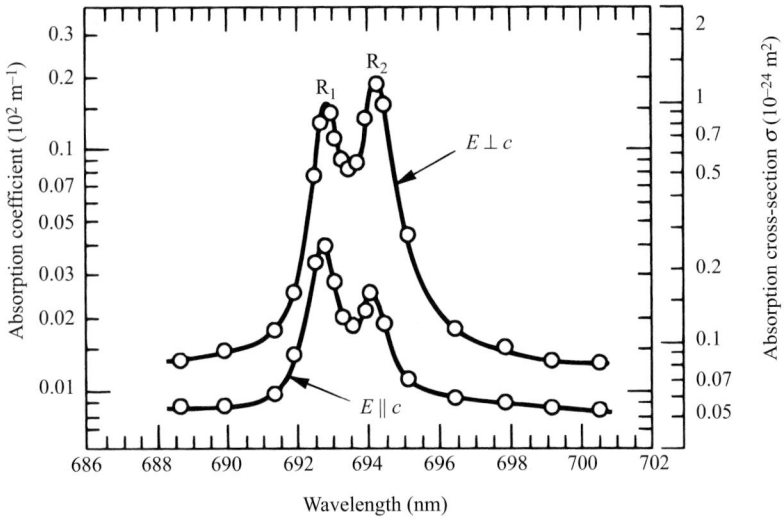


**Fig. 3.5** The detailed absorption spectrum of pink ruby in the 686–702-nm region showing the absorption peaks corresponding to the $R_1$ and $R_2$ components of the ruby laser transition [3].

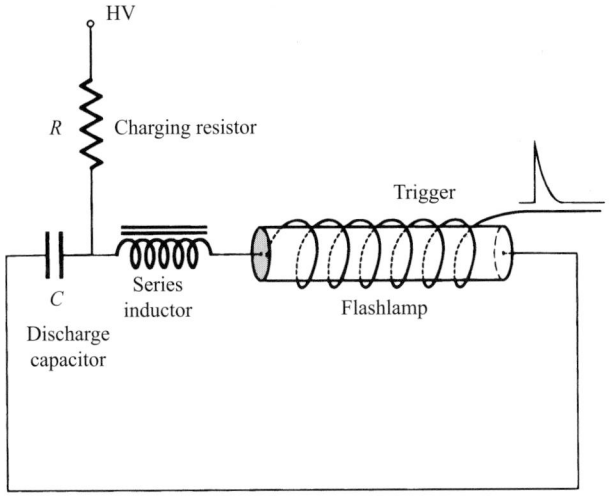


**Fig. 3.6** A simple electric circuit for driving a flashlamp.

a very high light intensity. A simple electric circuit for the operation of such a lamp is shown in Fig. 3.1. The spectral characteristics of the lamp will depend on a number of parameters; for example, the capacitor size, voltage, and tube diameter, and the filling gas and its pressure. Generally, if the discharge is of short duration, $\lesssim 10\ \mu s$, and high energy, the hot gas in the lamp will emit approximately as a black body, perhaps at an effective temperature of 20,000 K, at which much of its emission would be in the blue and ultraviolet. For ruby laser excitation the emission of the lamp is optimized in the region of the crystal's absorption bands by using slower flashlamp pulses (of duration 100 μs to 10 ms) and appropriate gas mixtures. If the flashlamp intensity is great enough, then sufficient ground state $^4A_2$ $Cr^{3+}$ ions can be excited to the $2\overline{A}$ and $\overline{E}$ levels via the broad absorption bands $^4F_2$ and $^4F_1$ that a population inversion exists between the $2\overline{A}$ and $\overline{E}$ levels and the split ground state. An efficient transfer occurs from the broad absorption bands to these upper laser levels because the radiative decay from these bands to the ground state has a lifetime of about 3 μs, against a value of about 50 ns for transfer into the upper laser levels. Consequently, about 99% of the ions that reach these bands transfer into the upper laser levels. These levels have an overall lifetime $\tau_2 \simeq 3$ ms, and lose their energy almost exclusively by spontaneous emission to the ground state, so $A_{21} \simeq 1/\tau_2$. Because these levels are relatively long-lived, if the exciting flash is shorter than their overall lifetime then most of the ions transferred into them will still be there when the flash is finished. The main laser transition that occurs is $R_1$, namely that between level $\overline{E}$ and the ground state; its wavelength is 694.3 nm (corresponding to a visible dark red color). Laser oscillation can also be obtained on the transition $R_2$, but only by taking special measures to reduce the gain of $R_1$, which otherwise operates preferentially.

Typical ruby lasers employ cylindrical crystals 1–20 mm in diameter and 20–200 mm in length. Usually the resonant cavity consists of the parallel polished faces of the crystal, one of which is coated to make it 100% reflecting. The Fresnel reflection from the other end is often sufficient to provide the optical feedback necessary for oscillation, although it may be made to have a higher reflection coefficient by giving it a suitable coating. Theodore Maiman's original ruby laser employed a very small crystal surrounded by a helical flashlamp, the whole being contained inside a coaxial cylindrical metal reflector, as shown in Fig. 3.7 [1]. In this arrangement the optical pumping of the laser is axisymmetric, which offers certain advantages in many applications, since the population inversion produced in the crystal is then itself axisymmetric. Prior to obtaining laser oscillation in the ruby system, Maiman and his colleagues had measured the absorption spectrum and the relative decay rate of the important energy levels [2]. It was these measurements that made them believe it would be possible to obtain population inversion in the ruby system.

Although helical flashlamp excitation is still sometimes used, other methods of optical pumping are more common. The most usual arrangement employs a cylindrical flashlamp and crystal placed along the two focal lines of an elliptical reflector, as shown in Fig. 3.8. It

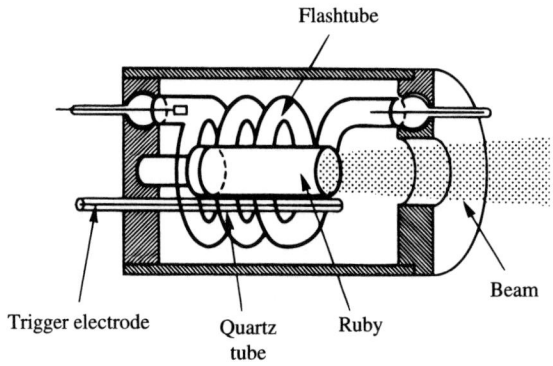


Fig. 3.7 A schematic diagram showing the arrangement of Maiman's original ruby laser.

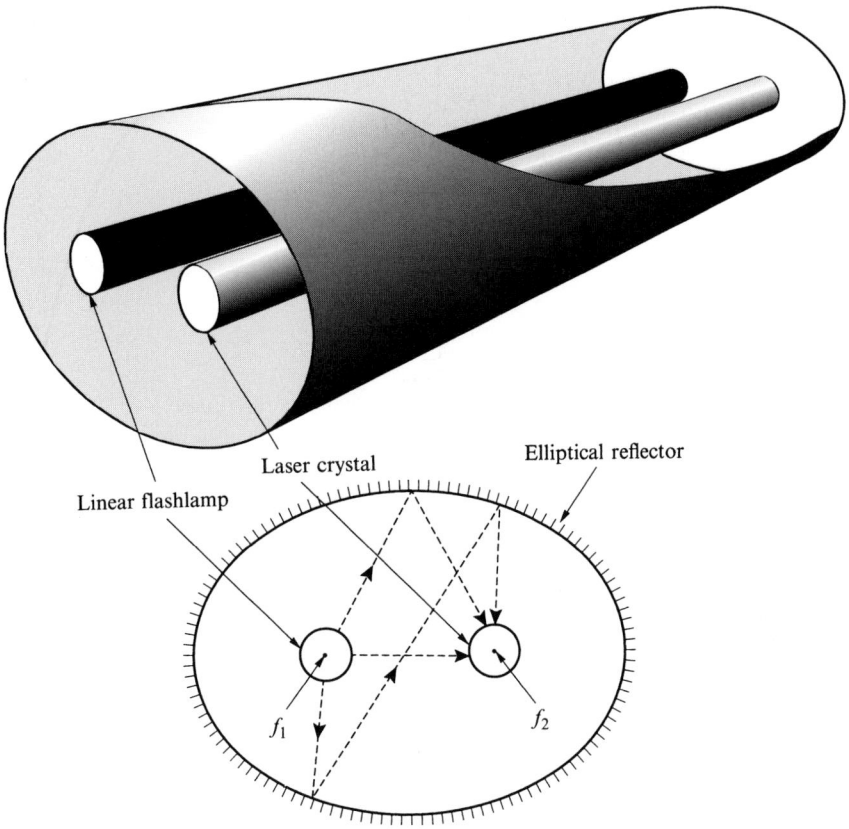


Fig. 3.8 An elliptical reflector arrangement for optical pumping of a laser crystal by a linear flashlamp.

is a property of such an elliptical reflector that rays of light from one focus, after reflection, pass through the other focus. Thus, light from the flashlamp is efficiently reflected towards the laser crystal.

Ruby lasers are rarely used these days, since they have been supplanted by other lasers with improved performance. Further discussion of their detailed properties can be found elsewhere [4].

## 3.4 The helium–neon laser

The helium–neon (He–Ne) laser was the first gas laser to be operated, and also the first continuous-wave (CW) laser. It is a four-level laser. The production of a population inversion in this system occurs as a result of energy-transfer processes between metastable helium atoms and ground-state neon atoms, which are as a result excited to various upper laser levels.

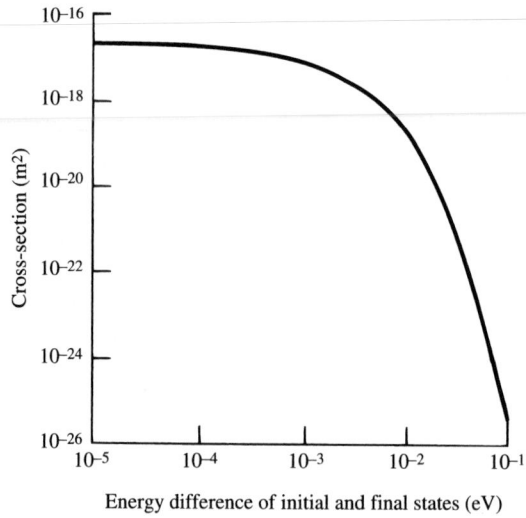


**Fig. 3.9** The calculated variation of the energy-transfer cross-section for a collision between two atomic species as a function of the energy discrepancy $\Delta E_\infty$ [3]. The probability of excitation transfer is linearly dependent on the cross-section.

In a collisional excitation transfer process of the kind

$$\mathrm{A} + \mathrm{B}^* \rightleftarrows \mathrm{A}^* + \mathrm{B} \pm \Delta E_\infty,$$

where the colliding particles are not molecular and an asterisk indicates that the particle is in an excited state, the probability of the reaction occurring becomes extremely small if the energy discrepancy $\Delta E_\infty$ of the reaction exceeds $k_\mathrm{B}T$ [5], as shown in Fig. 3.9. Here $\Delta E_\infty$ is the energy difference between the excited states $\mathrm{A}^*$ and $\mathrm{B}^*$ at large (theoretically infinite) separation of the reacting particles ($\Delta E$ is a function of the separation of the two atoms in an $\mathrm{A}^*$ and in a B state because of forces between them; the effect of this force goes to zero as the two atoms become infinitely far apart).

If either of the colliding particles is molecular, then the probability of reaction can remain large even if $\Delta E_\infty$ exceeds $k_\mathrm{B}T$, since one or other of the particles may be able to release or store energy in internal degrees of freedom, such as vibration.

In the He–Ne laser the excitation transfer processes occur in a glow discharge in a He–Ne mixture at a total pressure of about 1 torr.[4] The mixture is usually in the range 5:1 to 10:1 helium to neon. The discharge is produced either by DC or RF excitation in tubes usually of inside diameter 1–20 mm, and carries a current density from 0.1 to 1 $\mathrm{A\,cm^{-2}}$. By various collisional and radiative processes electrons excite ground-state helium atoms to the metastable $2\,^1\mathrm{S}_0$ and $2\,^3\mathrm{S}_1$ levels, as shown in Fig. 3.10:

$$\mathrm{He}\left(^1\mathrm{S}_0 = \text{ground state}\right) + \mathrm{e} \rightarrow \mathrm{He}\left(2\,^1\mathrm{S}_0 \text{ or } 2\,^3\mathrm{S}_1\right)$$

[4] The torr, named for the seventeenth-century Italian mathematician Torricelli, who invented the mercury manometer, is a unit of pressure equal to 1 mm of mercury. Standard atmospheric pressure is 760 torr; 1 torr $\equiv$ 133 pascals (1 Pa $\equiv$ 1 N $\mathrm{m^{-2}}$).

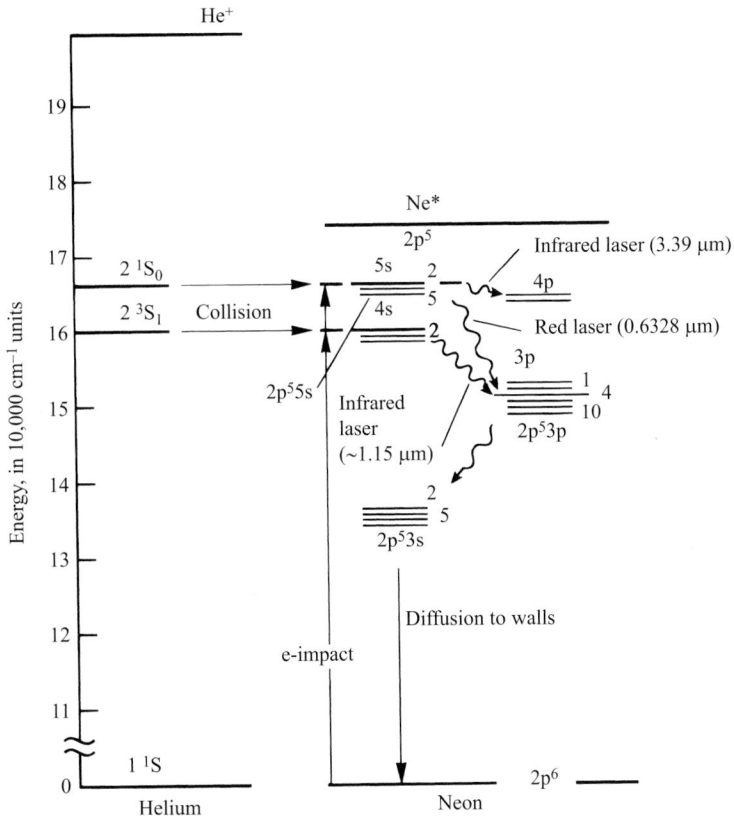


**Fig. 3.10** The energy-level diagram for the helium–neon laser.

or

$$\mathrm{He}(^1\mathrm{S}_0) + \mathrm{e} \rightarrow \mathrm{He}^*(\text{higher non-metastable state})$$
$$\mathrm{He}^* \rightarrow \mathrm{He}(2\,^1\mathrm{S}_0)(\text{or } 2\,^3\mathrm{S}_1) + h\nu.$$

Since the $2\,^1\mathrm{S}_0$ and $2\,^3\mathrm{S}_1$ states are metastable they do not lose energy readily by undergoing spontaneous emission (the radiative lifetime of the $2\,^3\mathrm{S}_1$ state is well in excess of an hour). However, these metastable states can lose energy in collisions, and in collisions with ground-state neon atoms energy transfer in reactions such as

$$\mathrm{He}(2\,^1\mathrm{S}_0) + \mathrm{Ne}(^1\mathrm{S}_0) \rightarrow \mathrm{He}(^1\mathrm{S}_0) + \mathrm{Ne}(2\mathrm{p}^5 4\mathrm{s}) + \Delta E_1$$

and

$$\mathrm{He}(2\,^3\mathrm{S}_1) + \mathrm{Ne}(^1\mathrm{S}_0) \rightarrow \mathrm{He}(^1\mathrm{S}_0) + \mathrm{Ne}(2\mathrm{p}^5 5\mathrm{s}) + \Delta E_2$$

is highly probable. The energy discrepancies $\Delta E_1$ and $\Delta E_2$ are small, $\Delta E_1 \sim 400\ \mathrm{cm}^{-1}$, which is comparable to, or less than, $k_\mathrm{B}T$ for the hot atoms in the discharge. The neon 4s and 5s levels which are excited in this way, under appropriate conditions,

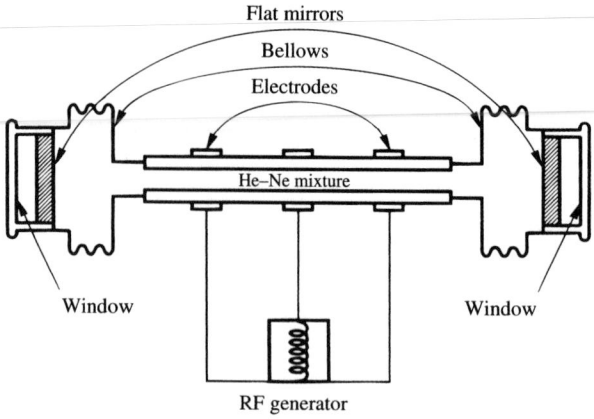


Fig. 3.11 A schematic diagram showing the arrangement of the first gas laser.

exhibit population inversion with respect to the 3p and 4p levels. The existence of such a population inversion does, however, require that the lower laser levels not be efficiently populated. This could occur either because the electrons in them are excited by electron collisions or because population builds up in them if they become long-lived.

One interesting laser transition that results because of population inversion is between one of the 4s and one of the 3p levels, which occurs at 1.15 μm. This transition (which is in the near-infrared and consequently invisible to the human eye) was the first transition to be observed in the He–Ne laser. When Javan, Bennett, and Herriott first obtained oscillation on the 1.15-μm transition in 1961 [6], their success built on solid theoretical predictions that population inversion could be achieved in this atomic four-level gaseous system. A schematic diagram of their laser, the first CW and gas laser, is shown in Fig. 3.11. The features of this system are of largely historic interest, since most of the technological features of the device have been superseded in modern He–Ne gas lasers. The glow discharge was produced in this first gas laser by coupling RF power to the tube with external electrodes; in most modern CW gas lasers direct DC input to the gas is preferred. The laser mirrors situated at each end of the amplifier tube were plane, and were fixed inside metal assemblies to allow them to be aligned parallel while retaining the vacuum integrity of the structure. Nowadays, Brewster-angle window assemblies allow the laser mirrors to be situated outside the tube. Such windows allow a plane-polarized laser oscillation to occur without the Fresnel reflection loss normally associated with in-cavity windows. It also turns out that plane-parallel mirror resonators are extremely sensitive to misalignment (misalignment of a few seconds of arc will prevent oscillation being obtained). In order to be sure that their laser mirrors would be aligned, Javan and his coworkers fixed one mirror to an electromechanical device that wobbled the mirror through a range of angular positions near the expected position where parallelism with the other mirror would occur. By using an image-converter tube, which made the infrared laser beam visible, they were able to detect the burst of oscillation as the laser mirror passed through the alignment position. It was subsequently discovered that the use of spherical mirrors of appropriate radius of curvature made the alignment tolerances of the cavity much less severe (oscillation

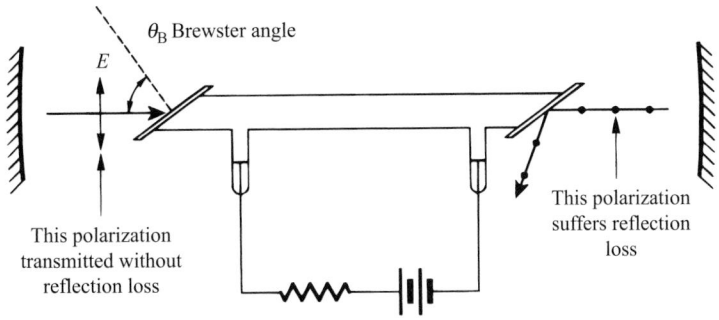


Fig. 3.12 A typical design of a modern gas laser.

could be sustained over perhaps a $\frac{1}{2}$-degree range near the coaxial position). A schematic diagram of a gas laser system with DC excitation, Brewster windows, and external spherical mirrors is shown in Fig. 3.12.

The 1.15-μm transition first observed in the He–Ne laser is now mainly of historical importance. Many other laser transitions between the (5s, 4s) and (4p, 3p) levels have been observed since, the most important of which are those between the 5s and 3p and the 5s and 4p levels. Two of these transitions are particularly noteworthy, namely those at 632.8 nm and 3.39 μm. The red laser transition at 632.8 nm is familiar because many thousands of commercial He–Ne lasers operating at this red wavelength are in widespread use.[5] Population inversion on this transition results from the energy transfer pumping from helium $2\,^1S$ metastables. The lower level of the transition has a radiative lifetime of about 10 ns, which is about a factor of ten less than that of the upper level. The desirability of such a favorable lifetime ratio in a laser system will become more apparent later. It can be seen from Fig. 3.10 that the lower level involved in the 632.8-nm transition decays to one of the 3s levels. This level is metastable and loses its energy mostly in collisions with the discharge tube walls. If the population of this level rises too high, repopulation of the lower laser level from it may destroy the population inversion. This repopulation may result from electron-collisional excitation of the 3p levels from the 3s levels, or by the re-absorption of photons emitted by atoms in the lower laser level (this phenomenon is called *radiation trapping* [7, 8]). If the rate of collisional deactivation of the 3s levels is increased (for example, by reducing the diameter of the laser tube), then the gain of the 632.8-nm transition (and also that of the 1.15-μm transition) is increased. The gain increases inversely with tube diameter, but cannot be increased too far by this means or the available volume of inverted gas falls, diffraction losses increase, and the laser's power output decreases.

The 3.39-μm laser transition shares a common upper level with the 632.8-nm transition but has a very much higher small-signal gain (as high as 50 dB $m^{-1}$). This high gain is partly due to a very favorable lifetime ratio for the 3.39-μm transition and also to its longer wavelength. In gas lasers of this kind the gain is proportional to at least $\lambda^2$ ($\lambda^3$ if the transition is predominantly Doppler-broadened). Because of its higher gain, the 3.39-μm transition has a lower oscillation threshold under conditions of comparable

[5] These lasers have largely been replaced in many applications by red emitting semiconductor lasers, which will be discussed later.

feedback than does the 632.8-nm transition. Under such conditions it will oscillate preferentially and prevent oscillation at 632.8 nm. This occurs because it builds up in oscillation first and "clamps" the population inversion (because of gain saturation) at a threshold level for its oscillation. This will be discussed in more detail in Chapter 5. The population inversion may be clamped at a level too low for laser oscillation at 632.8 nm. To prevent this occurring, the oscillation threshold at 3.39 μm must be raised. In short lasers with quartz or glass windows these windows introduce sufficient loss for this to occur; also the resonator mirrors will generally have lower reflectivity at 3.39 μm than at 632.8 nm. In long He–Ne lasers ($\gtrsim$1 m) the loss introduced by the windows and mirrors may fail to prevent parasitic oscillation at 3.39 μm. There are two common ways to lower the gain at 3.39 μm in such lasers. An in-cavity methane cell (which absorbs strongly at 3.39 μm but not at 632.8 nm) can be used, or a spatially inhomogeneous magnetic field can be applied to the discharge tube. Such a field introduces a Zeeman splitting[6] of the 3.39-μm laser transition that is different at different points along the discharge: this effectively increases the linewidth of the laser transition and reduces its gain. The effect of the magnetic field is much smaller at 632.8 nm because its Doppler-broadened linewidth is already large. The spatially inhomogeneous magnetic field can be supplied by a series of bar magnets placed outside the tube and spaced along its length.

## References

[1] T. H. Maiman, "Stimulated optical radiation in ruby," *Nature* **187**, 493–494, 1960.

[2] T. H. Maiman, R. H. Hoskins, I. J. D'Haenens, C. K. Asawa, and V. Evtuhov, "Stimulated emission in fluorescent solids II. Spectroscopy and stimulated emission in ruby," *Phys. Rev.* **123**, 1151–1157, 1961.

[3] D. C. Cronemeyer, "Optical absorption characteristics of pink ruby," *J. Opt. Soc. Am.* **56**, 1703–1706, 1966.

[4] V. Evtuhov and J. K. Neeland, "Pulsed ruby lasers," in *Laser: A Series of Advances*, Vol. I, ed. A. K. Levine, New York: Marcel Dekker, 1966, Chapter 1.

[5] N. F. Mott and H. S. W. Massey, *The Theory of Atomic Collisions*, Oxford: Oxford University Press, 1965.

[6] A. Javan, W. R. Bennett, Jr., and D. R. Herriott, "Population inversion and continuous optical maser oscillation in a gas discharge containing a He–Ne mixture," *Phys. Rev. Lett.* **6**, 106–110, 1961.

[7] T. Holstein, "Imprisonment of resonance radiation in gases," *Phys. Rev.* **72**, 1212–1233, 1947.

[8] T. Holstein, "Imprisonment of resonance radiation in gases II," *Phys. Rev.* **83**, 1159–1168, 1951.

[6] An effect whereby a magnetic field separates an energy level into a series of closely spaced sub-levels of slightly different energy. The transition changes from a single transition at a unique center frequency to a group of closely spaced transitions at center frequencies that depend on the strength of the applied magnetic field.

# 4 Optical resonators containing amplifying media

## 4.1 Introduction

A feedback structure is used to channel output emissions from a laser amplifier back through the amplifier. The simplest example consists of a pair of plane mirrors placed parallel to each other, but at opposite ends of the amplifying medium. The structure is called the laser resonator or optical cavity. It is often referred to as a Fabry–Perot resonator after Fabry and Perot.[1] The start of laser oscillation begins with spontaneous emission that occurs close to the axis of the resonator system, namely the direction perpendicular to both mirrors. This spontaneous emission is re-directed by the mirrors back through the amplifying medium and its intensity grows. The fact that only photons that are traveling close to the axis direction can make many passes through the amplifying medium, especially if this is of small width, explains the directional character of laser beams.

The frequency spectrum of laser radiation is controlled by the interaction between the lineshape function of the amplifying transition and the properties of the optical resonator. A plane wave bouncing back and forth between the two flat mirrors of an empty cavity will "resonate" if the wavelength of the radiation satisfies the simple condition

$$\frac{n\lambda}{2} = \ell, \tag{4.1}$$

where $\ell$ is the axial spacing between the two parallel mirrors. This is the simple condition for the bouncing waves to be in phase and develop a standing wave of maximum amplitude between the two mirrors, much like the acoustic resonances of an open or closed pipe. When an amplifying medium is present, this resonance condition must still be satisfied, and determines the precise frequency (or frequencies) at which laser oscillation can occur. In this case the effective "length" $\ell$ of the cavity is modified by the presence of the amplifier.

Before exploring this further, let us consider at what frequency a laser would oscillate if the feedback provided by the end "mirrors" did not interact with the gain profile of the amplifying medium in this way. This is called "non-resonant" feedback [1–4]. To make a non-resonant laser the end "mirrors" are made from rough material that scatters strongly. There is no specific length to include in the resonance condition given by Eq. (4.1). Suppose the amplifying medium is homogeneously broadened and has a gain profile (gain/frequency response) of Lorentzian form, as shown in Fig. 4.1.

[1] Marie P. A. C. Fabry and Jean B. G. G. A. Perot, the inventors of the plane-parallel mirror structure developed originally as an optical instrument for interferometry and first described in 1899.

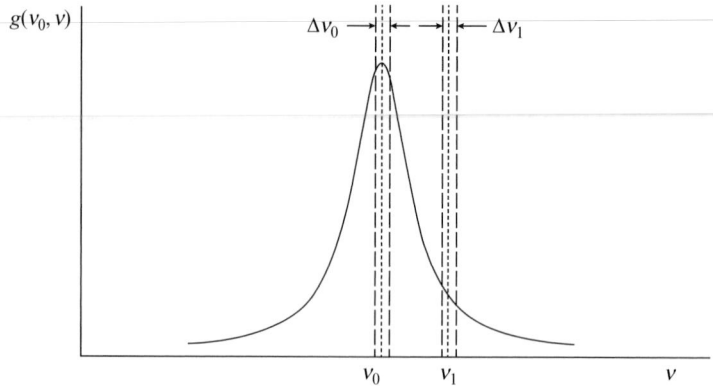


**Fig. 4.1** A gain profile with a Lorentzian shape interacting with two narrow bands of input light, one at the line center and one at a frequency that is off center.

The maximum gain of the medium is at frequency $\nu_0$, the line center, so it is perhaps logical to expect that oscillation will build up at this frequency rather than at any other. If we view the build-up of oscillation as a process triggered by spontaneous emission, we can see why this is so. A photon traveling in a direction that keeps it bouncing back and forth within the cavity is more likely to be emitted in a narrow band of frequencies $\Delta\nu_0$ near $\nu_0$ than in some other band $\Delta\nu_1$ at frequency $\nu_1$. As oscillation builds up, one can imagine photons spontaneously emitted at all points of the lineshape being amplified to some extent, but oscillation at $\nu_0$ builds up fastest. As the intracavity intensity grows it depletes the inversion by causing stimulated emission. If the medium is homogeneously broadened, then, since photons oscillating at $\nu_0$ can stimulate emission from all the particles in the medium, it is easy to see that oscillation at frequency $\nu_0$ can suppress oscillation at any other frequency under the gain profile. The possibility of additional oscillation at frequencies far away from $\nu_0$ in an inhomogeneously broadened gaseous amplifier is not precluded by this discussion; this in fact oftens happens, as we shall see later.

The monochromatic character of the oscillation can be predicted by simple consideration of the shape of the gain profile of the amplifier. In the early stages of oscillation, photons with a frequency distribution $g(\nu_0, \nu)$ (the total lineshape) are being amplified in a material whose gain/frequency response is $\gamma(\nu)$ (which is proportional to $g(\nu_0, \nu)$). The amplification process changes the lineshape of the emitted photons circulating in the cavity by a process that is dependent on the product of $g(\nu_0, \nu)$ and $\gamma(\nu)$, that is on $[g(\nu_0, \nu)]^2$. The resulting profile of the laser radiation is dependent on higher powers of $[g(\nu_0, \nu)]^2$ since the oscillation is dependent on many passes of photons back and forth through the amplifying medium. For Gaussian lineshapes like $e^{-[2(\nu-\nu_0)/\Delta\nu_D]^2 \ell n2}$, which is like $e^{-x^2/\sigma^2}$, the product of two lineshapes produces a narrower profile, for example

$$[e^{-x^2/\sigma^2}]^2 = e^{-2x^2/\sigma^2} = e^{-x^2/(\sigma/\sqrt{2})^2}, \tag{4.2}$$

a function that has a width that is smaller by a factor of $1/\sqrt{2}$ than that of the original. The same can also be shown to be true for Lorentzian profiles. Figure 4.2 shows the narrowing

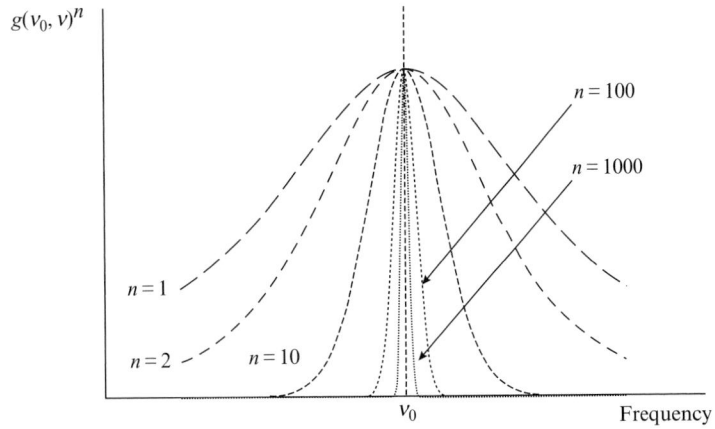


**Fig. 4.2** The narrowing of a Gaussian line as it is multiplied by itself $n$ times.

of a Gaussian line as it is multiplied by itself $n$ times. In all cases the gain of the medium causes a narrowing of the original spontaneously emitted lineshape. Thus, we can see that, in a non-interactive laser cavity, the laser oscillation will be highly monochromatic and at the line center.

## 4.2 Optical resonators containing amplifying media

When an optical resonator is filled with an amplifying medium, laser oscillation will occur at specific frequencies if the gain of the medium is large enough to overcome the loss of energy through the mirrors and by other loss mechanisms within the laser medium. The onset of laser oscillation and the frequency, or frequencies, at which it occurs are governed by threshold amplitude and phase conditions. Once laser oscillation has become established, it stabilizes at a level that depends on the saturation intensity of the amplifying medium and the reflectance of the laser mirrors.

Figure 4.3 represents an optical (Fabry–Perot) resonator, whose interior is filled with an amplifying medium and has plane mirrors. We consider the complex amplitudes of the waves bouncing backwards and forwards normally between the resonator mirrors. These waves result from an incident beam with electric vector $E_0$ at the first mirror, as shown in Fig. 4.3, where $E_0$ is the complex amplitude at some reference point. The reflection and transmission coefficients in the various directions at the mirrors are as shown in Fig. 4.3. For example, $t$ is the transmission coefficient for electromagnetic fields passing through the left-hand reflector, and $r_1$ is the reflection coefficient for fields striking the left-hand reflector from inside the resonator. $A$, $A_1$, and $A_2$ are absorption coefficients, which lead to energy dissipation in the resonator reflectors. These coefficients are not used explicitly in the analysis which follows, but they modify the values of the reflection and transmission coefficients. If the mirrors are lossless, then their reflectances, $R_1$ and $R_2$, and transmittances,

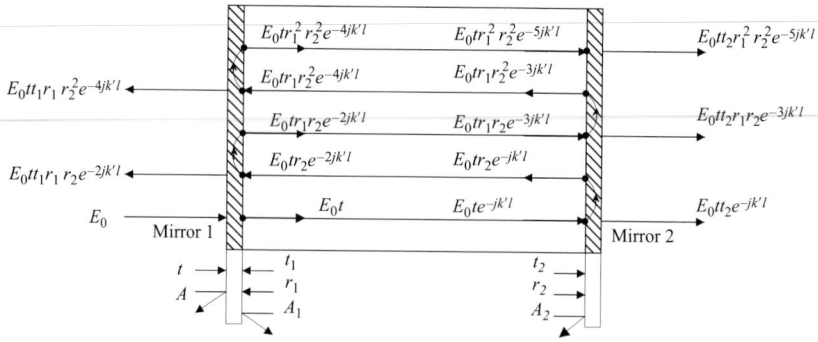


**Fig. 4.3** The amplitude of the electric field vectors of the successively transmitted, amplified, and reflected waves in a Fabry–Perot system containing an amplifying (or absorbing) medium. The absorption factors $A$, $A_1$, and $A_2$ are not used explicitly in the analysis given in the text, but they modify the values of $t$, $t_1$, and $t_2$.

$T_1$ and $T_2$, obey $R_1 + T_2 = 1$, etc. and $|r_1|^2 = R_1$, $|r_2|^2 = R_2$, $|t|^2 = |t_1|^2 = T_1$, and $|t_2|^2 = T_2$. The absorption coefficients $A, A_1$, and $A_2$ at the two mirrors include any reflection losses or scattering that will send energy out of the resonator. We do not at this stage include diffraction losses, which result from the finite lateral dimensions of the mirrors or medium.

If there were nothing inside the resonator then a wave propagating between the mirrors would propagate as $E_0 e^{j(\omega t - kz)}$ to the right and $E_0 e^{j(\omega t + kz)}$ to the left. The presence of a gain medium changes the otherwise passive propagation factor $k$ to

$$k'(\omega) = k + \Delta k = k + \frac{k\chi'(\omega)}{2n^2}, \tag{4.3}$$

and the gain coefficient $\gamma(\omega) = -k\chi''(\omega)/n^2$ causes the complex amplitude of the wave to change with distance as $e^{(\gamma/2)z}$. We allow for the possible existence of a distributed loss per pass given by an absorption coefficient $\alpha$. Such absorption causes a fractional change in intensity for a single pass through the medium of $e^{-\alpha\ell}$. Such a distributed loss could, for example, arise from scattering by crystal imperfections in a rod-shaped crystal laser medium. This distributed loss modifies the complex amplitude by a factor of $e^{-j\alpha\ell/2}$ per pass. Therefore, the full propagation constant of the wave in the presence of both gain and loss is

$$k'(\omega) = k + k\frac{\chi'(\omega)}{2n^2} - \frac{jk\chi''(\omega)}{2n^2} - \frac{j\alpha}{2}, \tag{4.4}$$

and the wave propagates as $e^{j(\omega t \pm k'z)}$.

A wave traveling to the right with complex amplitude $E_0$ at plane $z = 0$ in the resonator, the left-hand mirror, has at plane $\ell$, the right-hand mirror, become

$$E = E_0 e^{j(\omega t - k\ell)} = E_0 e^{-jk\ell} e^{j\omega t} = E_0' e^{j\omega t}.$$

This wave then begins to propagate to the left as

$$E = E_0' e^{j(\omega t + kz)}.$$

At plane $-\ell$, the left-hand mirror, with the right-hand mirror now taken as the origin, it has become once more a wave traveling to the right:

$$E = E_0 e^{jk\ell} e^{j(\omega t - k\ell)} = E_0 e^{-2jk\ell} e^{j\omega t}.$$

In this way we can write down the complex amplitudes of successive rays traveling at normal incidence between the two reflectors, as shown in Fig. 4.3.

The output beam through the right-hand mirror arises from the transmission of waves traveling to the right: its total electric field amplitude is

$$\begin{aligned} E_t &= E_0 t t_2 e^{-jk'\ell} + E_0 t t_2 r_1 r_2 e^{-3jk'\ell} + \cdots \\ &= E_0 t t_2 e^{-jk'\ell}(1 + r_1 r_2 e^{-2jk'\ell} + r_1^2 r_2^2 e^{-4jk'\ell} + \cdots) \\ &= \frac{E_0 t t_2 e^{-jk'\ell}}{1 - r_1 r_2 e^{-2jk'\ell}} \\ &= \frac{E_0 t t_2 e^{-j(k+\Delta k)\ell} e^{(\gamma-\alpha)\ell/2}}{1 - r_1 r_2 e^{-2j(k+\Delta k)\ell} e^{(\gamma-\alpha)\ell}}, \end{aligned} \tag{4.5}$$

where

$$\gamma(\nu) = \left(N_2 - \frac{g_2}{g_1} N_1\right) \frac{c^2 A_{21}}{8\pi\nu^2} g(\nu_0, \nu).$$

The ratio of input to output intensities is

$$\frac{E_t}{E_0} = \frac{I_t}{I_0} = \frac{t^2 t_2^2 e^{(\gamma-\alpha)\ell}}{(1 - r_1 r_2 e^{-2j(k+\Delta k)\ell} e^{(\gamma-\alpha)\ell})(1 - r_1 r_2 e^{2j(k+\Delta k)\ell} e^{(\gamma-\alpha)\ell})}, \tag{4.6}$$

which becomes

$$\frac{I_t}{I_0} = \frac{t^2 t_2^2 e^{(\gamma-\alpha)\ell}}{1 + r_1^2 r_2^2 e^{2(\gamma-\alpha)\ell} - 2 r_1 r_2 e^{(\gamma-\alpha)\ell} \cos[2(k + \Delta k)\ell]}. \tag{4.7}$$

In a passive resonator, which has no gain $\gamma$ or loss $\alpha$, $\Delta k = 0$, and, if $r_1 = r_2 = R$,

$$\frac{I_t}{I_0} = \frac{T^2}{1 + R^2 - 2R\cos(2k\ell)}.$$

The resonances of this structure satisfy $2k\ell = m\pi$, and correspond to wavelengths that satisfy the condition $m\lambda/2 = \ell$. If $R$ is close to 1, then the resonances are very sharp. It is easy to show that they approximate a Lorentzian shape with a FWHM defined as

$$\Delta\nu_{1/2} = \Delta\nu/F,$$

where $\Delta\nu = c/(2\ell)$ and $F = \sqrt{R}/(1 - R)$; $\Delta\nu$ is called the *free spectral range* of the resonator, and $F$ is called its *finesse*.

In a resonator containing an active medium, as $\gamma - \alpha$ increases from zero, the denominator of Eq. (4.5) approaches zero, and the whole expression blows up when

$$r_1 r_2 e^{-2j(k+\Delta k)\ell} e^{(\gamma-\alpha)\ell} = 1. \tag{4.8}$$

When this happens we have an infinite-amplitude transmitted wave for a finite-amplitude incident wave. In other words, a finite-amplitude transmitted wave for zero incident wave – *oscillation*. Physically, Eq. (4.8) is the condition that must be satisfied in order for a wave to make a complete round trip inside the resonator and return to its starting point with the same amplitude and, apart from a multiple of $2\pi$, the same phase.

Equation (4.8) provides an amplitude condition for oscillation that gives an expression for the threshold gain constant, $\gamma_t(\nu)$,

$$r_1 r_2 e^{[\gamma_t(\nu)-\alpha]\ell} = 1. \tag{4.9}$$

To satisfy Eq. (4.8), $e^{-2j(k+\Delta k)\ell}$ must be real, which provides us with the phase condition

$$2[k + \Delta k(\nu)]\ell = 2\pi m, \quad m = 1, 2, 3, ... \tag{4.10}$$

The threshold gain coefficient can be written

$$\gamma_t(\nu) = \alpha - \frac{1}{\ell}\ell n(r_1 r_2), \tag{4.11}$$

which, from the gain equation (2.68), gives the population inversion needed for oscillation:

$$\left(N_2 - \frac{g_2}{g_1}N_1\right)_t = \frac{8\pi}{g(\nu_0, \nu)A_{21}\lambda^2}\left(\alpha - \frac{1}{\ell}\ell n(r_1 r_2)\right). \tag{4.12}$$

For a homogeneously broadened transition the parametric variation of Eq. (4.12) that depends on the gain medium can be written as

$$\left(N_2 - \frac{g_2}{g_1}N_1\right)_t \propto \frac{\Delta\nu}{A_{21}\lambda^2},$$

whereas, for an inhomogeneously broadened transition, since $\Delta\nu_D \propto 1/\lambda$,

$$\left(N_2 - \frac{g_2}{g_1}N_1\right)_t \propto \frac{1}{A_{21}\lambda^3}.$$

Clearly, lower inversions are needed in order to achieve laser oscillation at longer wavelengths. It is much easier to build lasers that oscillate in the infrared than at visible, ultraviolet, or X-ray wavelengths. For example, in an inhomogeneously broadened laser, a population inversion $10^6$ times greater would be required for oscillation at 200 nm rather than at 20 μm (all other factors such as $A_{21}$ being equal). In practice, since $A_{21}$ factors generally increase at shorter wavelengths, the difference in population inversion might not need to be as great as this.

In a resonator such as is shown in Fig. 4.3, if $R_1 = r_1^2 \simeq 1, R_2 = r_2^2 \simeq 1$, and distributed losses are small, then a wave starting with intensity $I$ inside the resonator will, after one complete round trip, have intensity $IR_1R_2e^{-2\alpha\ell}$, and the change in intracavity intensity after one round trip is

$$dI = (R_1R_2e^{-2\alpha\ell} - 1)I. \tag{4.13}$$

This loss occurs in a time $dt = 2\ell/c$. So

$$\frac{dI}{dt} = cI[R_1R_2e^{-2\alpha\ell} - 1]/(2\ell). \tag{4.14}$$

This equation has the solution

$$I = I_0 \exp\{-[1 - R_1 R_2 e^{-\alpha\ell}]ct/(2\ell)\}, \tag{4.15}$$

where $I_0$ is the intensity at time $t = 0$. The time constant for intensity (energy) loss is

$$\tau_0 = \frac{2\ell}{c(1 - R_1 R_2 e^{-2\alpha\ell})}. \tag{4.16}$$

Now, if $R_1 R_2 e^{-2\alpha\ell} \simeq 1$, with $\alpha$ small as we have assumed here, then

$$1 - R_1 R_2 e^{-2\alpha\ell} \simeq -\ell \mathrm{n}(R_1 R_2 e^{-2\alpha\ell}) = -\ell \mathrm{n}(R_1 R_2) + 2\alpha\ell, \tag{4.17}$$

and we get

$$\tau_0 = \frac{2\ell}{c(2\alpha\ell - \ell \mathrm{n}(R_1 R_2))} = \frac{1}{c[\alpha - (1/\ell)\ell \mathrm{n}(r_1 r_2)]}. \tag{4.18}$$

Thus, the threshold population inversion can be written

$$N_\mathrm{t} = \frac{8\pi}{A_{21}\lambda^2 g(\nu) c\tau_0}. \tag{4.19}$$

**Example** *Threshold population inversion – numerical example.* For the 488-nm transition in the argon-ion laser $\lambda = 488$ nm, $c = 3 \times 10^8$ m s$^{-1}$, $A_{21} \simeq 10^9$ s$^{-1}$, and $\Delta\nu_\mathrm{D} \sim 3$ GHz. Take $\ell = 1$ m, and $R_1 = 100\%$, and $R_2 = 90\%$ (typical values for a practical device). Since this is a gas laser, internal losses are easily kept small, so $\alpha \simeq 0$. In this case

$$\begin{aligned}\tau_0 &= 2\ell/[c(1 - R_1 R_2)] \\ &= 66.67 \text{ ns}.\end{aligned}$$

For oscillation at, or near, line center

$$g(\nu_0, \nu_0) = \frac{2}{\Delta\nu_\mathrm{D}}\sqrt{\frac{\ell \mathrm{n}\, 2}{\pi}} = \frac{0.94}{\Delta\nu_\mathrm{D}} \sim \frac{1}{\Delta\nu_\mathrm{D}}.$$

The threshold inversion is, from Eq. (4.19),

$$N_\mathrm{t} = \frac{8\pi \times 3 \times 10^9}{10^9 \times (488 \times 10^{-9})^2 \times 3 \times 10^8 \times 66.67 \times 10^{-9}} = 1.58 \times 10^{13} \text{ m}^{-3}.$$

## 4.3 The oscillation frequency

To determine the frequency at which laser oscillation can occur, we return to the phase condition, Eq. (4.10). This phase condition was

$$(k + \Delta k)\ell = m\pi, \tag{4.20}$$

which from Eq. (4.3) gives

$$k\ell\left[1 + \frac{\chi'(\nu)}{2n^2}\right] = m\pi. \tag{4.21}$$

Now, from Eq. (2.118)

$$\chi'(\nu) = \frac{2(\nu_0 - \nu)}{\Delta\nu}\chi''(\nu), \tag{4.22}$$

where $\nu_0$ is the line center frequency and $\Delta\nu$ is its *homogeneous* FWHM, and

$$\gamma(\nu) = -\frac{k\chi''(\nu)}{n^2}. \tag{4.23}$$

So we must have

$$\frac{2\pi\nu\ell}{c}\left[1 - \frac{\nu_0 - \nu}{\Delta\nu}\frac{\gamma(\nu)}{k}\right] = m\pi, \tag{4.24}$$

and, on rearranging,

$$\nu\left[1 - \frac{\nu_0 - \nu}{\Delta\nu}\frac{\gamma(\nu)}{k}\right] = \frac{mc}{2\ell} = \nu_m, \tag{4.25}$$

where $\nu_m$ is the $m$th resonance of the passive laser resonator in normal incidence as calculated previously. Equation (4.25) can be rewritten as

$$\nu = \nu_m - (\nu - \nu_0)\frac{\gamma(\nu)c}{2\pi\Delta\nu}. \tag{4.26}$$

We expect the actual oscillation frequency $\nu$ to be close to $\nu_m$, so we can write $\nu - \nu_0 \simeq \nu_m - \nu_0$ and $\gamma(\nu) \simeq \gamma(\nu_m)$, to give

$$\nu = \nu_m - (\nu_m - \nu_0)\frac{\gamma(\nu_m)c}{2\pi\Delta\nu}. \tag{4.27}$$

At threshold

$$\gamma_t(\nu_m) = \alpha - \frac{1}{\ell}\ell n(r_1 r_2),$$

and, if $\alpha \simeq 0$ and $r_1 = r_2 = \sqrt{R}$,

$$\gamma_t(\nu_m) = \frac{1 - R}{\ell}. \tag{4.28}$$

Now the FWHM of the passive resonances (the transmission intensity maxima of the Fabry–Perot resonator) is

$$\Delta\nu_{1/2} = \frac{\Delta\nu_{\mathrm{FSR}}}{F} = \frac{c(1 - R)}{2\pi\ell\sqrt{R}}, \tag{4.29}$$

which, with $R \simeq 1$, gives

$$\Delta\nu_{1/2} = \frac{c(1 - R)}{2\pi\ell}, \tag{4.30}$$

and, finally,

$$\nu = \nu_m - (\nu_m - \nu_0)\frac{\Delta\nu_{1/2}}{\Delta\nu}. \tag{4.31}$$

Thus, if $\nu_m$ coincides with the line center, oscillation occurs at the line center. If $\nu_m \neq \nu_0$, oscillation takes place near $\nu_m$ but is shifted slightly towards $\nu_0$. This phenomenon is called "mode-pulling" and is illustrated in Fig. 4.4.

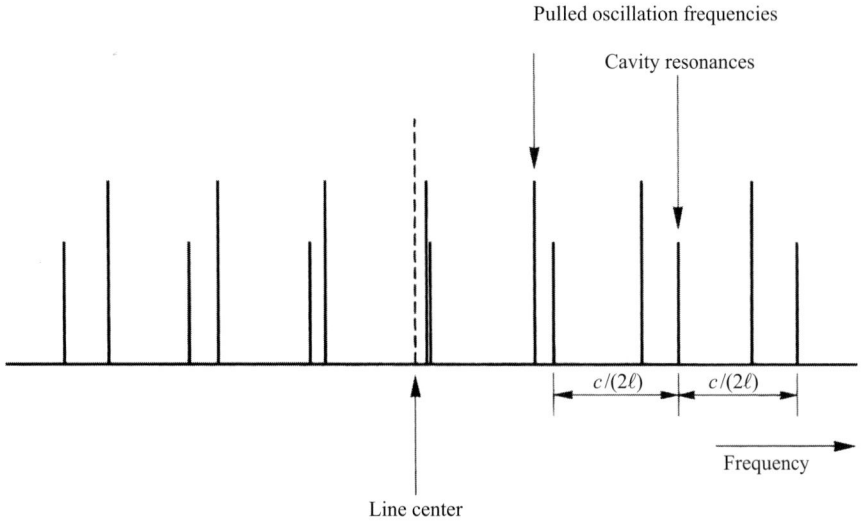


**Fig. 4.4** The relative positions of the line center, Fabry–Perot resonances, and pulled oscillation frequencies that satisfy the phase condition (4.20).

## 4.4 Multimode laser oscillation

We have seen that for oscillation to occur in a laser system the gain must reach a threshold value $\gamma_t(\nu) = \alpha - (1/\ell)\ell n(r_1 r_2)$. For gain coefficients greater than this, oscillation can occur at, or near (because of mode-pulling effects), one or more of the passive resonance frequencies of the Fabry–Perot laser cavity. The resulting oscillations of the system are called longitudinal modes. As oscillation at a particular one of these mode frequencies builds up, the growing intracavity energy density depletes the inverted population and gain saturation sets in. The reduction in gain continues until

$$\gamma(\nu) = \gamma_t(\nu) = \alpha - \frac{1}{\ell}\ell n(r_1 r_2). \tag{4.32}$$

Further reduction of $\gamma(\nu)$ below $\gamma_t(\nu)$ does not occur, otherwise the oscillation would cease. Therefore, the gain is stabilized at the loss

$$\alpha - \frac{1}{\ell}\ell n(r_1 r_2).$$

Usually $\alpha$, $r_1$, and $r_2$ are nearly constant over the frequency range covered by typical amplifying transitions, so, over such moderate frequency ranges, $10^{11}$ Hz say, $\alpha - (1/\ell)\ell n(r_1 r_2)$ as a function of frequency is a straight line parallel to the frequency axis. This line is called the *loss line*. At markedly different frequencies $\alpha$, $r_1$, and $r_2$ can be expected to change: for example, a laser mirror with high reflectivity in the red region of the spectrum could have quite low reflectivity in the blue.

In a homogeneously broadened laser, because the reduction in gain caused by a monochromatic field is uniform across the whole gain profile, the clamping of the gain at

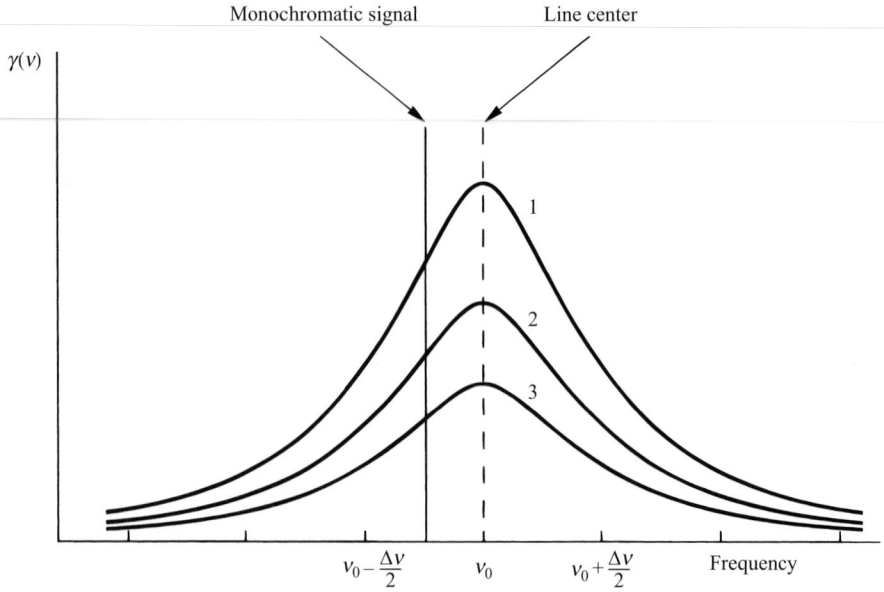


**Fig. 4.5** Saturation of the gain of a homogeneously broadened transition by a monochromatic signal whose intensity increases in the sequence 1 → 2 → 3.

$\gamma_t(\nu)$ leads to final oscillation at only one of the cavity resonance frequencies, namely the one for which the original unsaturated gain was highest. We can show this schematically by plotting $\gamma(\nu)$ at various stages as oscillation builds up. Remember first the effect on $\gamma(\nu)$ produced by a monochromatic light signal of increasing intensity as shown in Fig. 4.5. Note that the gain profile is depressed uniformly even though the saturating signal is not at the line center, as predicted by Eq. (2.69).

In a laser, as oscillation begins, several such monochromatic fields start to build up at those cavity resonances where gain exceeds loss, as shown in Fig. 4.6. The oscillation stabilizes when the highest (small-signal) gain has been reduced to the loss line by saturation as shown in Figs. 4.7 and 4.8. Thus, in a *homogeneously* broadened laser, oscillation occurs only at *one* longitudinal mode frequency.

In an inhomogeneously broadened laser, the onset of gain saturation due to a monochromatic signal merely reduces the gain locally over a region that is of the order of a homogeneous width. Only particles whose velocities (or environments in a crystal) make their center emission frequencies lie within a homogeneous width of the monochromatic field can interact strongly with it. Schematically, the effect of an increasing intensity monochromatic field on the gain profile is as shown in Fig. 4.9. A localized dip, or *hole*, in the gain profile occurs. If only one cavity resonance has a small-signal gain above the loss line then only this longitudinal mode oscillates. The stabilization of the oscillation might be expected to occur as shown schematically in Fig. 4.10. However, the situation is not quite as simple as this! Oscillation at this single longitudinal-mode frequency implies waves traveling in both directions inside the laser cavity. These waves can be represented by

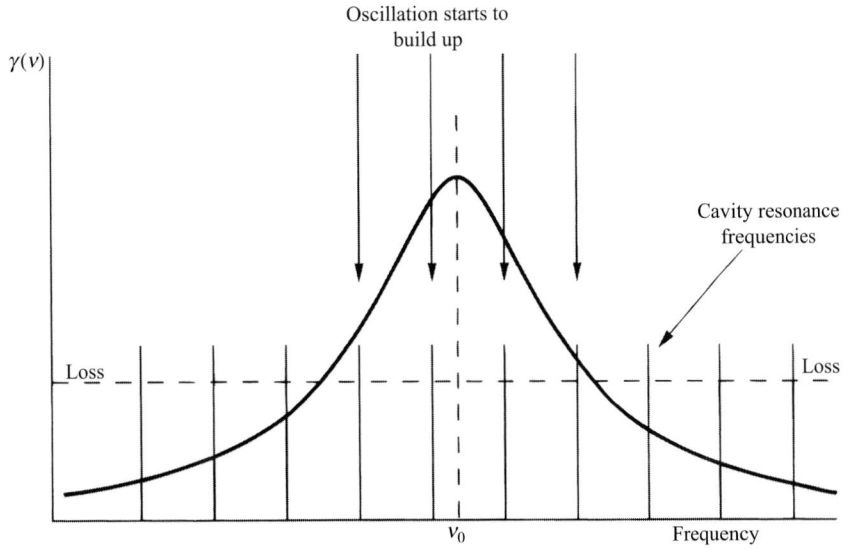


**Fig. 4.6** A schematic illustration of the onset of oscillation at cavity resonances that lie above the loss line in a homogeneously broadened laser.

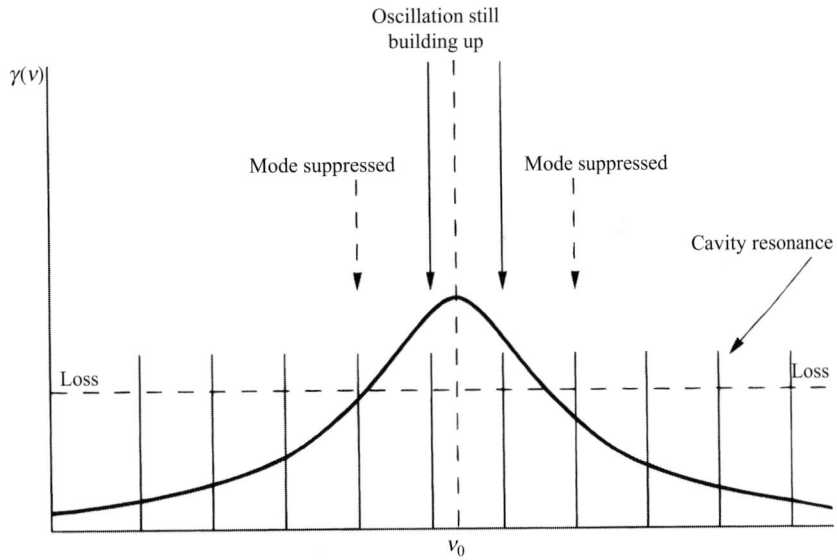


**Fig. 4.7** Oscillation building up in a homogeneously broadened laser. Gain saturation has already suppressed oscillation at two of the cavity modes that were above the loss line in Fig. 4.6.

(a) the wave traveling to the right, $\sim E_0 e^{j(\omega t - kz)}$, and
(b) the wave traveling to the left, $\sim E_0 e^{j(\omega t + kz)}$,

where we choose for convenience that $\omega = 2\pi\nu < 2\pi\nu_0$. Wave (a) can interact with particles whose center frequency is near $\nu$. These particles are, insofar as their Doppler shifts are concerned, moving away from an observer looking into the laser from right to left. Their

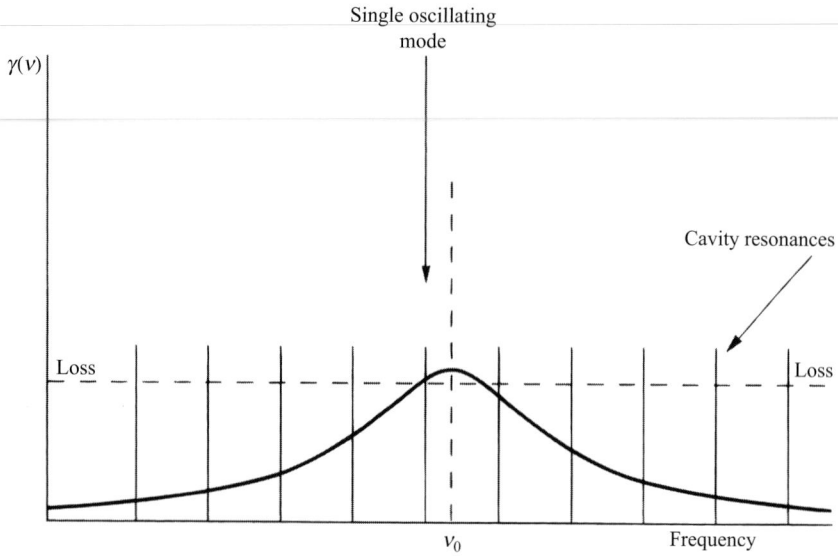


**Fig. 4.8** Oscillation stabilized in a homogeneously broadened laser. The gain has been uniformly saturated until only one mode remains at the loss line.

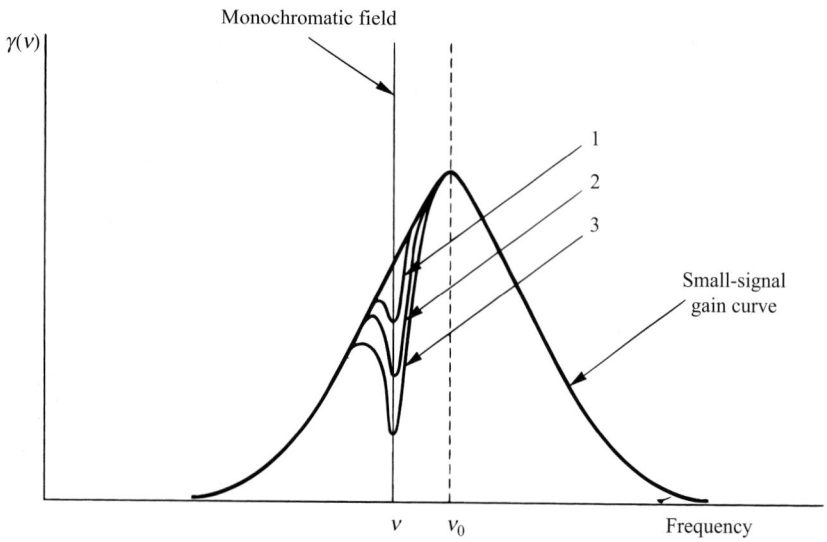


**Fig. 4.9** Localized gain saturation in an inhomogeneously broadened amplifier produced by a monochromatic signal whose intensity increases in the sequence 1→2→3.

center frequencies satisfy $\nu = \nu_0 - |\nu|\nu_0/c$, where positive atom velocities correspond to particles moving from left to right. Wave (b), which is traveling in the opposite direction (to the left) and is monitored, still at frequency $\nu < \nu_0$, by a second observer looking into the laser from left to right cannot interact with the same velocity group of particles as wave (a). The particles which interacted with wave (a) were moving away from the first

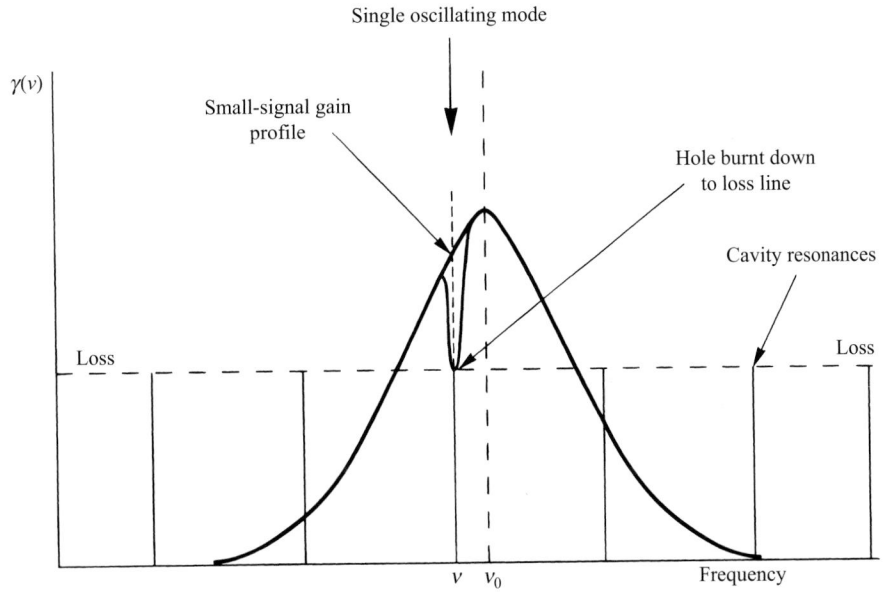


**Fig. 4.10** A simplified illustration of stabilization of oscillation at a single longitudinal mode in an inhomogeneously broadened laser.

observer and were Doppler-shifted to lower frequencies so as to satisfy

$$\nu = \nu_0 - \frac{|v|}{c}\nu_0. \tag{4.33}$$

The second observer sees these particles approaching and their center frequency as

$$\nu = \nu_0 + \frac{|v|}{c}\nu_0, \tag{4.34}$$

so they cannot interact with wave (b). Wave (b) interacts with particles moving away from the second observer so that their velocity would be the solution of

$$\nu = \nu_0 - \frac{|v|}{c}\nu_0. \tag{4.35}$$

These particles would be monitored by the first observer at center frequency

$$\nu = \nu_0 + \frac{|v|}{c}\nu_0. \tag{4.36}$$

So the oscillating waves interact with two velocity groups of particles as shown in Fig. 4.11. This leads to saturation of the gain by a single laser mode in an inhomogeneously broadened laser both at the frequency of the mode $\nu$ and at a frequency $\nu_0 + (\nu_0 - \nu)$, which is equally spaced on the opposite side of the line center, as shown in Fig. 4.12. The power output of the laser (strictly the intracavity power) comes from those groups of particles that

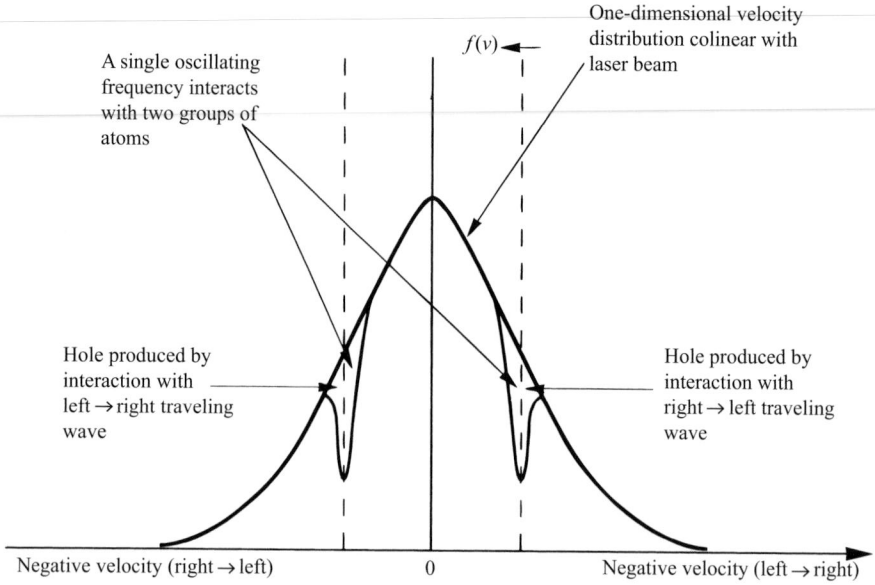


**Fig. 4.11** The production of two holes in the velocity distribution of a collection of amplifying particles by a single cavity mode.

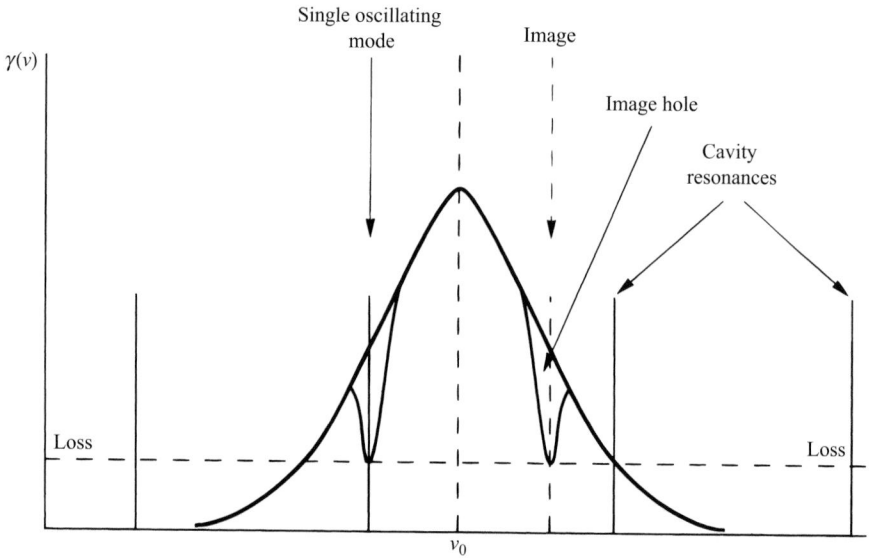


**Fig. 4.12** Stabilization of a single longitudinal mode in an inhomogeneously broadened laser.

have gone into stimulated emission and left the two holes. The combined area of these two holes gives a measure of the laser power.

If the frequency of the oscillating mode is moved in towards the line center, the main hole and image hole begin to overlap. This corresponds physically to the left and right traveling waves within the laser cavity beginning to interact with the same velocity group

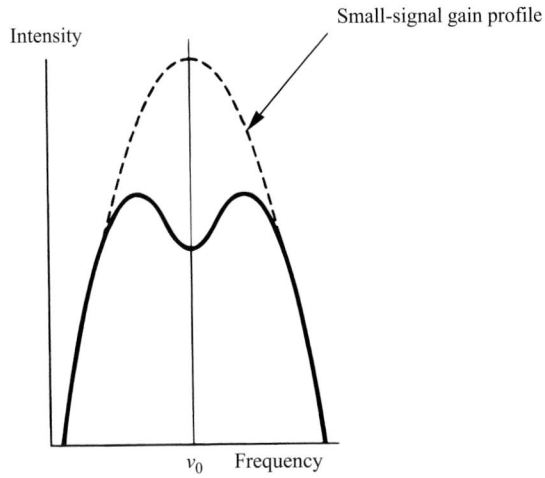


**Fig. 4.13** The Lamb dip – a reduction in the intensity of a single oscillating longitudinal mode in an inhomogeneously broadened laser as its frequency is scanned through line center.

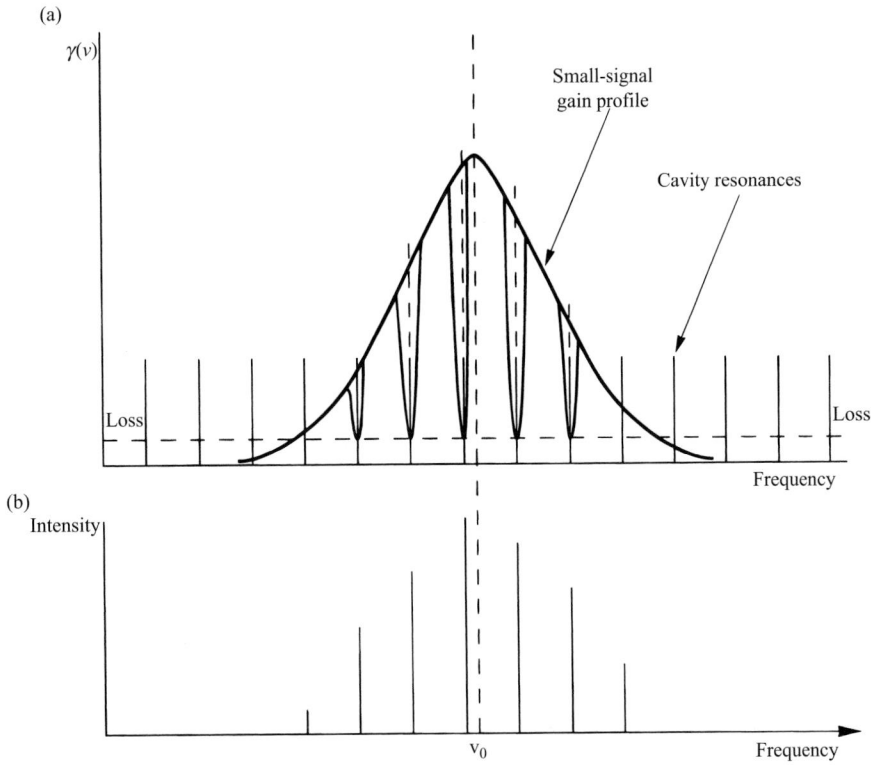


**Fig. 4.14** Multi-longitudinal-mode oscillation in an inhomogeneously broadened laser. (a) Only the primary holes are shown burnt down to the loss line. The image holes are not shown. (b) A schematic laser output spectrum.

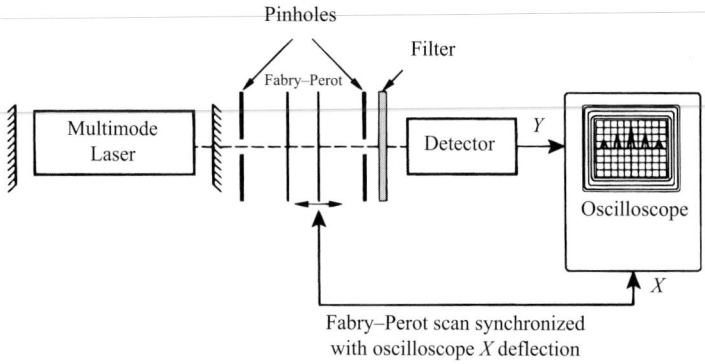


**Fig. 4.15** An experimental arrangement with a scanning Fabry–Perot interferometer for observing multimode laser oscillation.

of particles. As the oscillating mode moves in towards the line center, the holes overlap further, the combined area decreases, and the laser output power falls, reaching a minimum at the line center. This phenomenon is called the Lamb dip [5]. It was named after Willis E. Lamb, Jr., who first predicted the effect, and was first observed experimentally by Szoke and Javan [6]. The phenomenon is illustrated in Fig. 4.13. When the cavity resonance is at the line center frequency $\nu_0$, both traveling waves are interacting with the same group of atoms – those with near-zero directed velocity along the laser resonator axis.

Because hole-burning in gain saturation in inhomogeneously broadened lasers is localized near the frequency of a cavity mode, one oscillating mode does not reduce the gain at other cavity modes, so simultaneous oscillation at several longitudinal modes is possible. If several such modes have small-signal gains above the loss line the oscillation stabilizes in the manner shown in Fig. 4.14(a). The output frequency spectrum from the laser would appear as shown in Fig. 4.14(b). This simultaneous oscillation at several closely spaced frequencies ($c/(2\ell)$ apart) can be observed with a high-resolution spectrometer – for example a scanning Fabry–Perot interferometer as shown in Fig. 4.15. The multiple modes are almost exactly $c/(2\ell)$ apart in frequency, but are not exactly equally spaced because of mode-pulling. This effect can be observed in the beat spectrum observed with a square-law optical detector (which means most optical detectors). Such a detector responds to the intensity, not the electric field, of an incident light signal.

## 4.5 Mode-beating

Suppose we shine the light from a two-mode laser on a square-law detector. The incident electric field is

$$E_i = \mathcal{R}(E_1 e^{j\omega t} + E_2 e^{j(\omega+\Delta\omega)t}), \tag{4.37}$$

where $E_1$ and $E_2$ are the complex amplitudes of the two modes and $\Delta\omega$ is the frequency spacing between them. Using real notation for these fields, the output current $i$ from the detector is

$$\begin{aligned} i &\propto \{|E_1|\cos(\omega t + \phi_1) + |E_2|\cos[(\omega + \Delta\omega)t + \phi_2]\}^2 \\ &\propto |E_1|^2 \cos^2(\omega t + \phi_1) + |E_2|^2 \cos^2[(\omega + \Delta\omega)t + \phi_2] \\ &\quad + 2|E_1||E_2|\cos(\omega t + \phi_1)\cos[(\omega + \Delta\omega)t + \phi_2] \\ &\propto |E_1|^2 \cos^2(\omega t + \phi_1) + |E_2|^2 \cos^2[(\omega + \Delta\omega)t + \phi_2] \\ &\quad + |E_1||E_2|\cos[(2\omega + \Delta\omega)t + \phi_1 + \phi_2] \\ &\quad + |E_1||E_2|\cos(\Delta\omega t + \phi_2 - \phi_1). \end{aligned} \tag{4.38}$$

Furthermore, since, for example,

$$|E_1|^2 \cos^2(\omega t + \phi_1) = \frac{1}{2}|E_1|^2[1 + \cos[2(\omega t + \phi_1)]], \tag{4.39}$$

the output frequency spectrum of the detector appears to contain the frequencies $2\omega, 2(\omega + \Delta\omega), 2\omega + \Delta\omega$, and $\Delta\omega$. However, the first three of these frequencies are very high, particularly for light in the visible and infrared regions of the spectrum, and do not appear in the output of the detector. It is as if the high-frequency terms are averaged to zero by the detector time response to give

$$i \propto \frac{|E_1|^2}{2} + \frac{|E_2|^2}{2} + |E_1||E_2|\cos(\Delta\omega t + \phi_2 - \phi_1), \tag{4.40}$$

so only the difference-frequency beat $\Delta\omega$ is observed. This result can be derived directly using the *analytic* signal of the incident electric field. The *analytic* signal of a field that is $E(t) = \mathcal{R}(Ee^{j\omega t})$ is

$$V(t) = Ee^{j\omega t}, \tag{4.41}$$

so that $E(t) = \mathcal{R}[V(t)]$. The response of a square-law detector can be found directly from $i \propto V(t)V^*(t)$. The significance of this is discussed further in Chapter 23.

If the output from the square-law detector is analyzed with a radio-frequency spectrum analyzer (because this is the frequency range in which the difference frequencies between longitudinal laser modes are usually observed) different displays are obtained according to how many longitudinal modes of a multimode laser are simultaneously oscillating. Figure 4.16 gives some examples. Because Eq. (4.31) is not quite exact, the beat frequencies can split as shown because of nonlinear mode-pulling. This splitting will be observed only if the nonlinear mode-pulling is large and the spectrum analyzer that analyzes the output of the photodetector has high resolution.

If a predominantly inhomogeneously broadened laser also has a significant amount of homogeneous broadening, the holes burnt in the gain curve can start to overlap, for example, when $\Delta\nu \gtrsim c/(2\ell)$. If $\Delta\nu$ is large enough this causes neighboring oscillating modes to

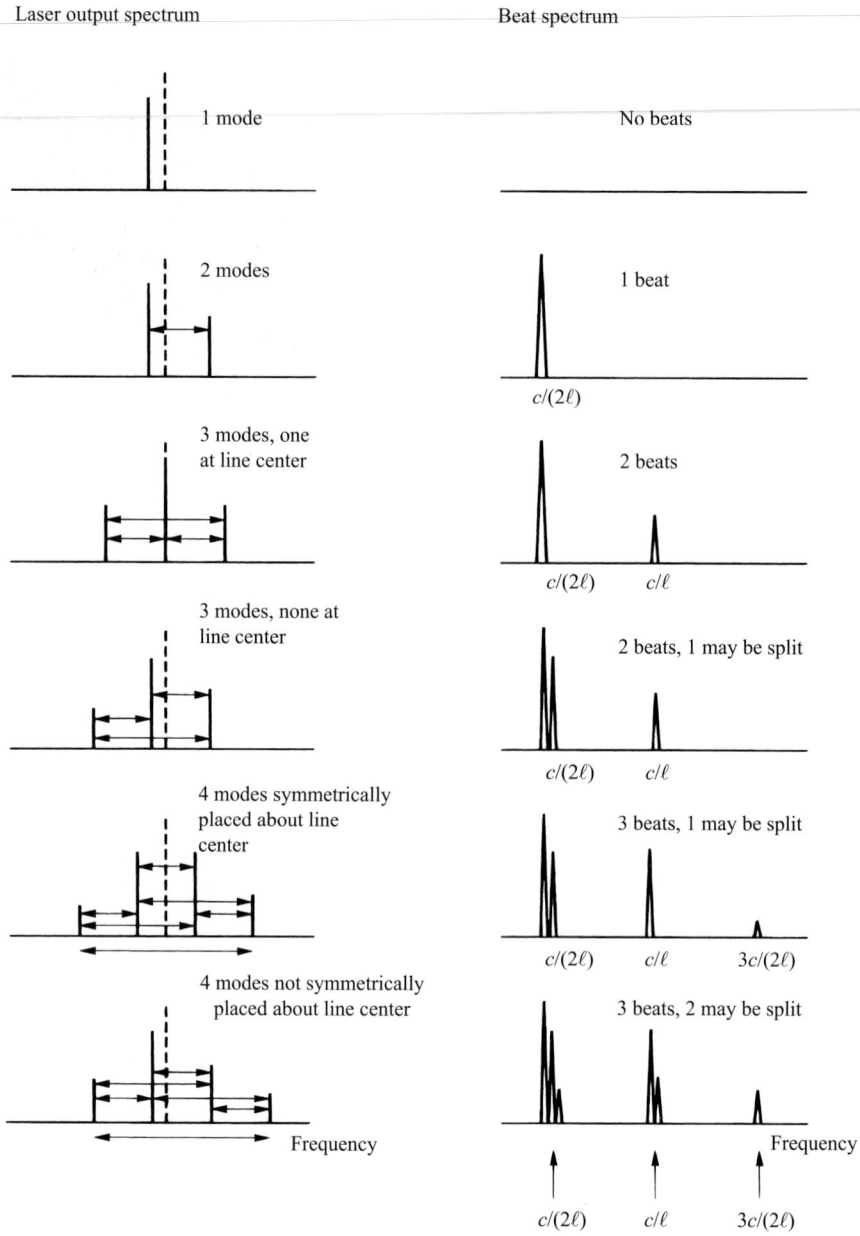


**Fig. 4.16** Schematic mode-beating spectra observed with a square-law detector and a multimode laser.

compete, and may lead to oscillation on a strong mode suppressing its weaker neighbors, as shown in Fig. 4.17. This effect has been observed in several laser systems, for example in the argon-ion laser, where an increase in the strength of the oscillation can lead to the successive disappearance first of every other mode, then of two modes out of every three, and so on.

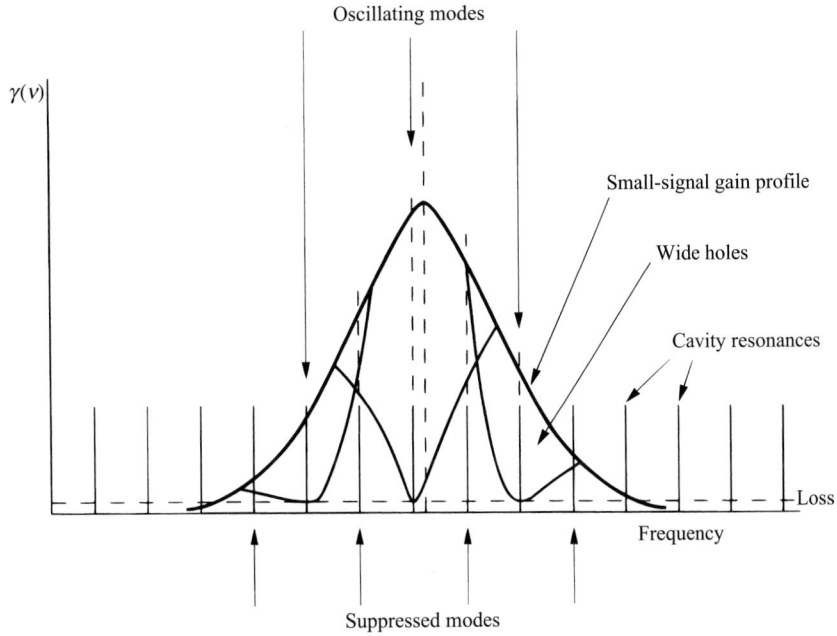


**Fig. 4.17** A schematic illustration of mode competition in an inhomogeneously broadened laser in which there is significant homogeneous broadening.

## 4.6 The power output of a laser

When a laser oscillates, the intracavity field grows in amplitude until saturation reduces the gain to the loss line for each oscillating mode. What this means in practice can be best illustrated with reference to Fig. 4.18.

For an asymmetrical resonator, whose mirror reflectances are not equal, the distribution of standing-wave energy within the resonator is not symmetrical. For example, in Fig. 4.18, if $R_2 > R_1$ the distribution of intracavity traveling wave intensity will be as shown, and

$$\frac{I_3}{I_2} = R_2, \qquad \frac{I_1}{I_4} = R_1. \tag{4.42}$$

The left traveling wave, of intensity $I_-$, grows in intensity from $I_3$ to $I_4$ on a single pass. The right traveling wave, of intensity $I_+$, grows in intensity from $I_1$ to $I_2$ on a single pass. The total output intensity is

$$I_{\text{out}} = T_2 I_2 + T_1 I_4. \tag{4.43}$$

However, calculation of $I_2$ and $I_4$ is not straightforward in the general case. Actually $I_2$ grows from $I_1$ through a gain process that depends in a complex way on $I_+ + I_-$, as does the growth of $I_3$ to $I_4$. We can identify at least three scenarios in which the calculation proceeds differently:

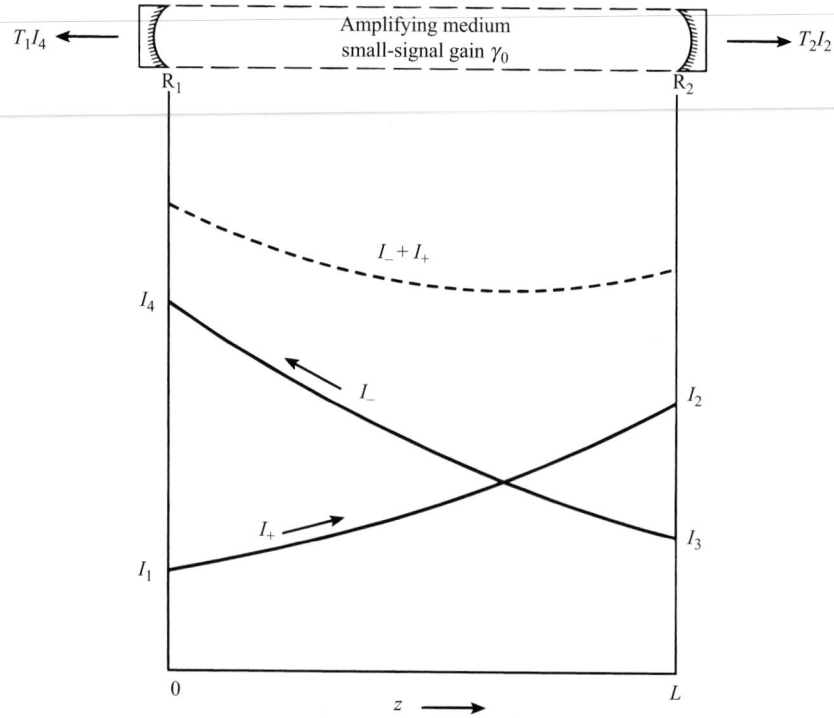


**Fig. 4.18** Distribution of wave intensities in an oscillating laser cavity with unequal laser intensities.

(a) a homogeneously broadened amplifier and single-mode operation;
(b) an inhomogeneously broadened amplifier and single-mode operation; and
(c) an inhomogeneously broadened amplifier and multimode operation.

In cases (b) and (c) the calculation of the output power becomes more complicated as the homogeneous contribution to the broadening grows more significant compared with $\Delta\nu_D$. This additional complexity arises because each oscillating mode burns both a primary and an image hole in the gain curve. The resultant distribution of overlapping holes makes the gain for each mode dependent not only on its own intensity but also on the intensity of the other simultaneously oscillating modes. The presence of distributed intracavity loss presents additional complications. We shall not attempt to deal with these complex situations here but will follow Rigrod [7–9] in dealing with a homogeneously broadened amplifier in which the primary intensity loss occurs at the mirrors. Inhomogeneously broadened systems and multimode operation have been discussed elsewhere by Smith [10].

In a purely homogeneously broadened system the saturated gain in Fig. 4.18 is

$$\gamma(z) = \frac{\gamma_0}{1 + (I_+ + I_-)/I_s}. \tag{4.44}$$

Both $I_-$ and $I_+$ grow according to $\gamma(z)$:

$$\frac{1}{I_+}\frac{dI_+}{dz} = -\frac{1}{I_-}\frac{dI_-}{dz} = \gamma(z). \tag{4.45}$$

Consequently,

$$I_+ I_- = \text{constant} = C. \tag{4.46}$$

From Eq. (4.46)

$$I_4 I_1 = I_2 I_3 = C, \tag{4.47}$$

and therefore, from Eq. (4.42),

$$I_2/I_4 = \sqrt{R_1/R_2}. \tag{4.48}$$

For the right traveling wave, using Eqs. (4.45) and (4.46) gives

$$\frac{1}{I_+}\frac{dI_+}{dz} = \frac{\gamma_0}{1 + (I_+ + C/I_+)/I_s}, \tag{4.49}$$

which can be integrated to give

$$\gamma_0 L = \ell n\left(\frac{I_2}{I_1}\right) + \frac{I_2 - I_1}{I_s} - \frac{C}{I_s}\left(\frac{1}{I_2} - \frac{1}{I_1}\right). \tag{4.50}$$

In a similar way, for the left traveling wave

$$\gamma_0 L = \ell n\left(\frac{I_4}{I_3}\right) + \frac{I_4 - I_3}{I_s} - \frac{C}{I_s}\left(\frac{1}{I_4} - \frac{1}{I_3}\right). \tag{4.51}$$

Adding Eqs. (4.50) and (4.51), and using Eqs. (4.42), (4.47), and (4.48), gives

$$I_2 = \frac{I_s\sqrt{R_1}(\gamma_0 L + \ell n\sqrt{R_1 R_2})}{(\sqrt{R_1} + \sqrt{R_2})(1 - \sqrt{R_1 R_2})}. \tag{4.52}$$

From Eq. (4.48)

$$I_4 = I_2\sqrt{\frac{R_2}{R_1}}. \tag{4.53}$$

Now

$$T_1 = 1 - R_1 - A_1, \tag{4.54}$$

$$T_2 = 1 - R_2 - A_2, \tag{4.55}$$

so, from Eqs. (4.43) and (4.52), if $A_1 = A_2 = A$, then

$$I_{\text{out}} = I_s\frac{1 - A - \sqrt{R_1 R_2}}{1 - \sqrt{R_1 R_2}}(\gamma_0 L + \ell n\sqrt{R_1 R_2}). \tag{4.56}$$

If one mirror is made perfectly reflecting, say $T_1 = 0$ and $R_1 = 1$, then

$$I_{\text{out}} = T_2 I_2 = \frac{T_2 I_s[\gamma_0 L + \frac{1}{2}\ell n(1 - A_2 - T_2)]}{A_2 + T_2}. \tag{4.57}$$

For a symmetrical resonator, defined by

$$R_1 R_2 = R^2,$$

$$R = 1 - A - T, \tag{4.58}$$

the output intensity at each mirror is

$$\frac{I_{\text{out}}}{2} = \frac{I_s}{2}\frac{1-A-R}{1-R}(\gamma_0 L + \ell\text{n}\, R). \tag{4.59}$$

## 4.7 Optimum coupling

To maximize the output intensity from the symmetrical resonator we must find the value of $R$ such that $\partial I_{\text{out}}/\partial R = 0$, which gives

$$\frac{T_{\text{opt}}}{A} = \frac{1-A-T_{\text{opt}}}{A+T_{\text{opt}}}[\gamma_0 L + \ell\text{n}(1-A-T_{\text{opt}})]. \tag{4.60}$$

For small losses, such that $A + T_{\text{opt}} \ll 1$, Eq. (4.58) gives

$$\frac{T_{\text{opt}}}{A} = \sqrt{\frac{\gamma_0 L}{A}} - 1. \tag{4.61}$$

Figure 4.19 shows the calculated optimum coupling for various values of the loss parameter $A$ and the unsaturated gain in dB ($4.343\gamma_0 L$).

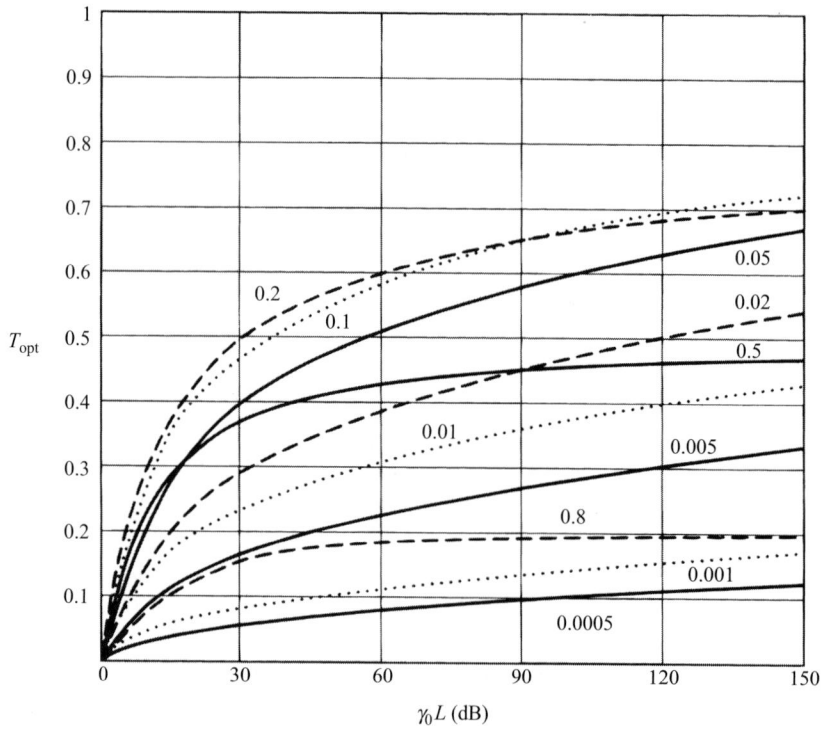


**Fig. 4.19** The calculated optimum coupling for a symmetrical resonator for various values of the loss parameter $A$ and the unsaturated gain.

In practice, it should be pointed out, the optimum mirror transmittance in a laser system is generally determined empirically. For example, for the CW $CO_2$ laser, whose unsaturated gain varies roughly inversely with the tube diameter $d$, the optimum mirror transmittance has been determined to be $T \simeq L/(500d)$.

## 4.8 Problems

4.1 A four-level laser is pumped into its pump band at a rate of $10^{24}$ m$^{-3}$ s$^{-1}$, and the transfer efficiency to the upper level is 0.5. The lifetime of the upper laser level is $7 \times 10^{-4}$ s. For the laser transition $A_{21} = 10^3$ s$^{-1}$, and $\lambda_0 = 1$ μm. The laser is homogeneously broadened, with $\Delta\nu = 1$ GHz. Assume $n = 1.6$. The amplifying medium is 20 mm long. Neglect population of the lower laser level. (a) What is the gain at line center? (b) What minimum value of $R_2$ is needed in order to get oscillation if mirror 1 has $R_1 = 1$? Assume $\alpha_{\text{distributed loss}} = 0$.

4.2 How many longitudinal modes will oscillate in an inhomogeneously broadened gas laser with $\ell = 1$ m, $\gamma(\nu_0) = 1$ m$^{-1}$, $R_1 = R_2 = 99\%$, $\alpha = 0.001$ m$^{-1}$, $\lambda_0 = 500$ nm, and $\Delta\nu_D = 3$ GHz?

4.3 A gas laser is operating simultaneously on five modes, none of which is at line center. The laser beam illuminates a square-law detector. Draw the RF beat spectrum that will be observed. Is the beat signal near $c/(2\ell)$ split?

4.4 Derive an expression for the peak transmittance of a Fabry–Perot filter that has a round-trip in-cavity absorption that causes the intensity to change from $I_0$ to $AI_0$ on a round trip between the two mirrors.

4.5 Derive the amplitude condition and the laser oscillation frequencies for a laser whose cavity is of length $L$ and whose amplifying medium is of length $\ell$, where $\ell < L$.

4.6 A laser is exactly 1 m long and operates at the wavelength $\lambda_0 = 632.8$ nm. The mirrors of the laser have $R = 99\%$. The index of refraction of the amplifying medium is exactly 1.0001, for $\Delta\nu = 100.000$ MHz. The laser operates just on the two modes nearest to the line center. The laser output illuminates a photodiode whose output is mixed with a 150-MHz local oscillator. What is the frequency of the lowest beat signal observed? Take the velocity of light in free space to be $2.997 \times 10^8$ m s$^{-1}$.

4.7 A gas laser with $\lambda_0 = 325$ nm has $\gamma_0(\nu_0) = 0.1$ m$^{-1}$, and $\Delta\nu_D = 3$ GHz. The laser medium fills the space between two mirrors 500 mm apart. The refractive index of the laser medium can be assumed to be 1, $\Delta\nu_{\text{homogeneous}} = 10$ MHz. The distributed loss parameter is $\alpha = 0.01$ m$^{-1}$. What minimum equal mirror reflectance is needed in order to allow ten longitudinal modes to oscillate? What would happen if $\Delta\nu_{\text{homogeneous}} = 500$ MHz?

4.8 A Fabry–Perot interferometer has two mirrors of reflectance $R$ spaced by a distance $\ell$, and the space between its plates is filled with a material of absorption coefficient $\alpha$ and refractive index $n$. Calculate its contrast ratio $C = (I_t/I_0)_{\max}/(I_t/I_0)_{\min}$.

4.9 A single-mode homogeneously broadened laser system with a small-signal gain at the oscillation frequency of 0.001 m$^{-1}$ has a total power output of 1 mW when operated in a symmetrical resonator with $R_1 = R_2 = 0.99$ and $L = 0.5$ m. The loss in each mirror is $10^{-4}$. Calculate the saturation intensity.

4.10 Write a computer program to solve Eq. (4.60) for the optimum mirror transmittance for arbitrary values of $L$, $\gamma_0$, and $A$. Find the value of $T_{\text{opt}}$ for $L = 1$ m, $\gamma_0 = 1\ \text{m}^{-1}$, and $A = 10^{-4}$.

## References


[1] R. V. Ambartsumian, N. G. Basov, P. G. Kryukov, and V. S. Letokhov, "A laser with a nonresonant feedback," *Pis'ma Zh. Éksp. Teor. Fiz.* **3**, No. 6, 262–264, 1966.

[2] R. V. Ambartsumian, N. G. Basov, P. G. Kryukov, and V. S. Letokhov, "Lasers with non-resonant feedback," *IEEE J. Quant. Electron.* **2**, No. 9, 442–446, 1966.

[3] R. V. Ambartsumian, P. G. Kryukov, and V. S. Letokhov, "Emission spectrum narrowing dynamics of a laser oscillator with a nonresonant feedback," *Zh. Éksp. Teor. Fiz.* **51**, No. 6(12), 1669–1675, 1966 (in Russian); *English translation Sov. Phys. JETP* **24**, 1129–1136, 1966.

[4] R. V. Ambartsumian, N. G. Basov, P. G. Kryukov, and V. S. Letokhov, "Non-resonant feedback in lasers," in *Progress in Quantum Electronics*. Vol. 1, Part 3, eds. J. H. Sanders and K. W. H. Stevens, London: Pergamon Press, 1970, pp. 105–193.

[5] W. E. Lamb, "Theory of an optical maser," *Phys. Rev.* **134A**, 1429–1450, 1964.

[6] A. Szoke and A. Javan, "Isotope shift and saturation behavior of the 1.15 μm transition of neon," *Phys. Rev. Lett.* **10**, 521–524, 1963.

[7] W. W. Rigrod, "Saturation effects in high-gain lasers," *J. Appl. Phys.* **36**, 2487–2490, 1965.

[8] W. W. Rigrod, "Gain saturation and output power of optical masers," *J. Appl. Phys.* **34**, 2602–2609, 1963.

[9] W. W. Rigrod, "Homogeneously broadened CW lasers with uniform distributed loss," *IEEE J. Quant. Electron.* **14**, 377–381, 1978.

[10] P. W. Smith, "The output power of a 6328 Å He–Ne gas laser," *IEEE J. Quant. Electron.* **2**, 62–68, 1966.

# 5 Laser radiation

## 5.1 Introduction

In this chapter we shall examine some of the characteristics of laser radiation that distinguish it from ordinary light. Our discussion will include the monochromaticity and directionality of laser beams, and a preliminary discussion of their *coherence* properties. Coherence is a measure of the temporal and spatial phase relationships which exist for the fields associated with laser radiation.

The special nature of laser radiation is graphically illustrated by the ease with which the important optical phenomena of interference and diffraction are demonstrated using it. This chapter includes a brief discussion of these two phenomena with some examples of how they can be observed with lasers. Interference effects demonstrate the coherence properties of laser radiation, while diffraction effects are intimately connected with the beam-like properties that make this radiation special.

## 5.2 Diffraction

Diffraction of light results whenever a plane wave has its lateral extent restricted by an obstacle. By definition, a plane wave traveling in the $z$ direction has no field variations in planes orthogonal to the $z$ axis, so the derivatives $\partial/\partial x$ or $\partial/\partial y$ operating on any field component give zero. Clearly this condition cannot be satisfied if the wave strikes an obstacle: at the edge of the obstacle the wave is obstructed, and there must be variations in field amplitude in the lateral direction. In other words, the derivative operations $\partial/\partial x$ and $\partial/\partial y$ do not give zero, and the wave after passing the obstacle is no longer a plane wave. When the wave ceases to be a plane wave, its phase fronts are no longer planes, and there is no unique propagation direction associated with the wave. The range of $\mathbf{k}$ vectors associated with the new wave gives rise to lateral variations in intensity observed behind the obstacle – the *diffraction pattern*.

We can show that at sufficiently great distances from the obstacle, or aperture, the diffraction pattern can be obtained as a Fourier transform of the amplitude distribution at the obstacle or aperture. There is a close parallel between this phenomenon, where light travels within a range of angular directions determined by the size of an obstacle, and the frequency spread associated with a waveform that is restricted in time. We saw in Chapter 2 that the frequency spectrum of a waveform is given by the Fourier transform of

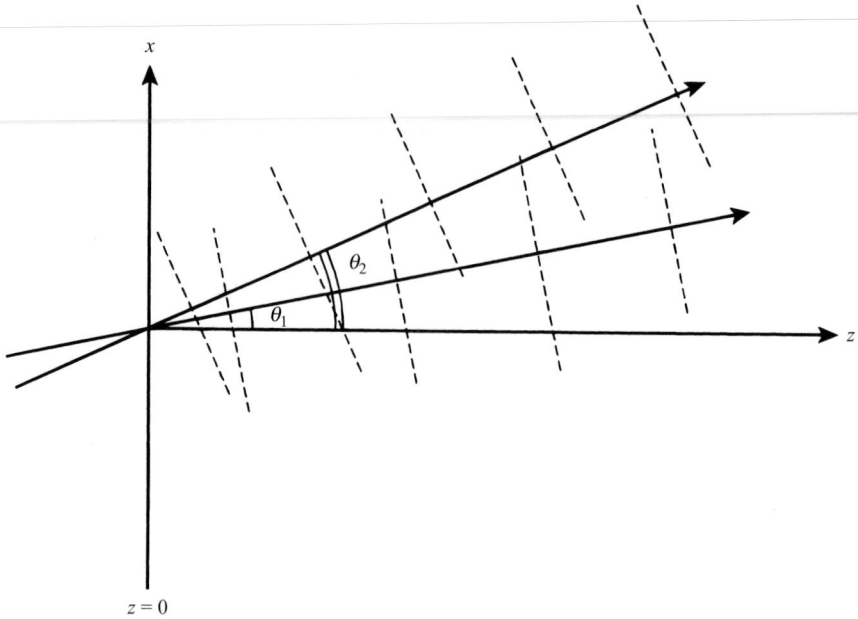


**Fig. 5.1** Superposition of two plane waves traveling at different angles in the *xz* plane.

its waveform. It is perhaps not surprising to learn that the angular spectrum of a plane wave transmitted through an aperture is a Fourier transform over the shape and transmission of the aperture.

Consider two plane waves whose field variations are of the general form $e^{j(\omega t - \mathbf{k}\cdot\mathbf{r})}$ traveling at slightly different small angles and giving rise to a resultant field in the $z = 0$ plane, as shown in Fig. 5.1.

If the first wave has field amplitudes of the form

$$V_1 = V_0 e^{j(\omega t - k\sin\theta_1 x - k\cos\theta_1 z)} \tag{5.1}$$

and the second has

$$V_2 = V_0 e^{j(\omega t - k\sin\theta_2 x - k\cos\theta_2 z)}, \tag{5.2}$$

the resultant disturbance at $z = 0$ is

$$V_1 + V_2 = V_0 e^{j\omega t}\left(e^{-jk_1 x} + e^{-jk_2 x}\right). \tag{5.3}$$

Because $\theta$ is a small angle we can assume that the $V_1$ and $V_2$ vectors are parallel, and we have written

$$k_1 = k\sin\theta_1, \qquad k_2 = k\sin\theta_2, \quad k = 2\pi/\lambda. \tag{5.4}$$

The assumption that the field vectors are parallel becomes more valid the further from the obstacle the resultant field amplitude is being calculated.

The resultant disturbance due to many such waves can be written as

$$V = V_0 e^{j\omega t} \sum_n e^{-jk_n x}. \tag{5.5}$$

In the limit of infinitely many waves, this summation can be written as an integral over a continuous distribution of waves for which the amplitude distribution is $a(k_x)$. For example, the total amplitude of the group of waves traveling in the small range of angles corresponding to a range $dk_x$ at $k_x$ is $a(k_x)dk_x$. The total disturbance in the plane $z = 0$ is

$$V_0(x) = \int_{-\infty}^{\infty} a(k_x) e^{-jk_x x}\, dk_x. \tag{5.6}$$

The limits on the integral in Eq. (5.6) are set to $\pm\infty$. Implicit in this is the physical reality that $a(k_x) = 0$ for any $|k_x| > |\mathbf{k}|$.

On recognizing Eq. (5.6) as a Fourier transform, we can write

$$a(k_x) = \frac{1}{2\pi} \int_{-\infty}^{\infty} V_0(x) e^{jk_x x}\, dx. \tag{5.7}$$

Thus, if we know the field distribution at the aperture or obstacle, given by $V(x)$, we can calculate $a(k_x)$, which gives a measure of the contribution of various plane waves in the diffraction pattern. This is because an observer cannot distinguish between light coming from an aperture and a collection of plane waves that combine in space to give non-zero field amplitudes only in a region corresponding to the aperture.

## 5.3 Babinet's principle

If an aperture of whatever shape and size is filled with an obstacle of exactly the same shape, size, and characteristics as the screen containing the aperture, then it is obvous that there will be zero transmitted wave. This observation can be interpreted to mean *diffraction pattern of an aperture + diffraction pattern of corresponding obstacle = 0*. In other words, the diffraction pattern of an obstacle is exactly the same as the diffraction pattern of the corresponding indentical aperture, except for a phase difference of $\pi$. This is called *Babinet's principle.*[1] An interesting historical story is connected with this. In 1818 Augustine Fresnel (1788–1827) submitted a paper on diffraction in a competition sponsored by the French Academy. His paper analyzed light as a wave rather than as a stream of small particles, as originally postulated by Sir Isaac Newton. Simon Poisson (1781–1840), who was one of the judges of the competition, was not a fan of wave theory. He pointed out that, according to Fresnel's analysis, close to a circular obstacle there should be a bright spot in the center of the shadow, so the wave theory had to be nonsense. Dominique Arago (1786–1853), another of the judges, promptly showed experimentally that the spot existed, much to Poisson's chagrin. Fresnel won the competition. Consequently, the bright spot is

[1] After the French scientist Jacques Babinet, 1794–1872.

often called "Poisson's bright spot," as a reminder of Poisson's error. If a circular aperture is illuminated then close to the aperture there is a dark spot in the center of the diffraction pattern, but this is harder to observe experimentally.

## 5.4 Two parallel narrow slits

As an example, let us look at the case of a pair of slits, which we will represent as a pair of $\delta$-function sources. If the slits are at $x = \pm a$ in the plane $z = 0$ and are parallel to the $y$ axis,

$$V_0(x) = \delta(x - a) + \delta(x + a). \tag{5.8}$$

So

$$\begin{aligned} a(k_x) &= \frac{1}{2\pi} \int_{-\infty}^{\infty} [\delta(x - a) + \delta(x + a)] e^{jk_x x}\, dx \\ &= \frac{1}{2\pi} (e^{jk_x a} + e^{jk_x a}) = \frac{\cos(k_x a)}{\pi}. \end{aligned} \tag{5.9}$$

Note that

$$|a(k_x)|^2 \propto \cos^2(k_x a) = \cos^2\left(\frac{2\pi a \sin\theta}{\lambda}\right). \tag{5.10}$$

In the center of the diffraction pattern $|a(k_0)|^2 \propto 1$. Thus

$$\frac{|a(k_x)|^2}{|a(k_0)|^2} = \cos^2\left(\frac{2\pi a \sin\theta}{\lambda}\right). \tag{5.11}$$

We can interpret $|a(k_x)|^2$ as the intensity in the diffraction pattern at angle $\theta$ relative to the intensity in the center of the pattern. The conditions under which Eq. (5.7) can be used to calculate the relative intensity in the diffraction pattern can be roughly stated as $z \gg D^2/\lambda$ and $\theta^2 \ll z\lambda/D^2$, where $D$ is the maximum lateral dimension of the obstacle or aperture. When these conditions are satisfied the diffraction pattern is referred to as a *Fraunhofer* diffraction pattern. Diffraction patterns observed too close to an obstacle or aperture for the above conditions to be satisfied are called *Fresnel* diffraction patterns [1–7].

## 5.5 A single slit

The diffraction pattern for a single slit of width $2d$ can be calculated in the same way as above. The Fourier transform of the slit is

$$a(k_x) = \frac{1}{\Delta\pi} \int_{-d}^{d} e^{jk_x x}\, dx = \frac{\sin(k_x d)}{\pi k_x}, \tag{5.12}$$

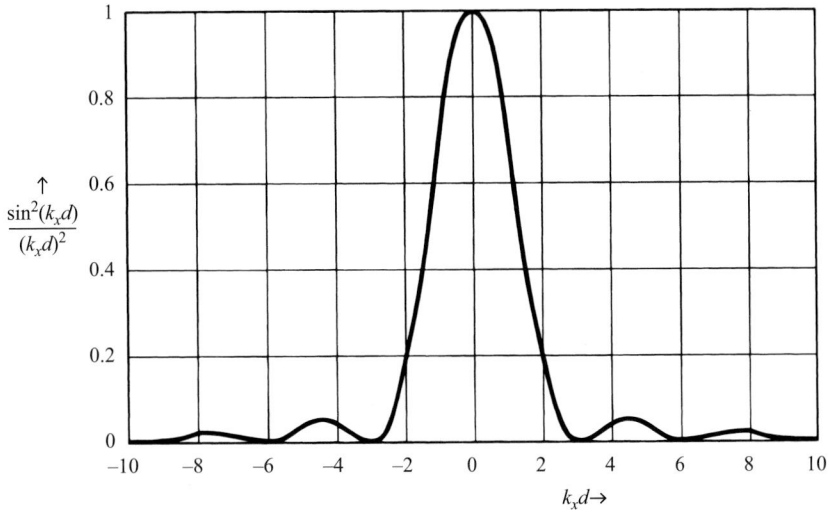


**Fig. 5.2** The diffraction pattern of a single slit of width $2d$.

and the relative intensity at angle $\theta$ is

$$I = \frac{|a(k_x)|^2}{|a(k_0)|^2} = \frac{\sin^2(k_x d)}{(k_x d)^2} = \frac{\sin^2((2\pi/\lambda)d\sin\theta)}{((2\pi d\sin\theta)/\lambda)}, \tag{5.13}$$

which is shown graphically in Fig. 5.2. The center of the diffraction pattern is an intensity maximum. The first minimum occurs when

$$\frac{2\pi d\sin\theta}{\lambda} = \pi, \tag{5.14}$$

which gives

$$\theta_{\min} = \frac{\lambda}{2d} = \frac{\lambda}{w}, \tag{5.15}$$

where $w$ is the width of the slit.

## 5.6 Two-dimensional apertures

We can generalize Eq. (5.7) for the case of a two-dimensional aperture by writing

$$a(k_x, k_y) = \frac{1}{4\pi^2}\int_{-\infty}^{\infty} \xi(x,y)e^{j(k_x x + k_y y)}\,dx\,dy, \tag{5.16}$$

where $\xi(x,y)$ is the aperture function, $k_x = |\mathbf{k}|\sin\theta$, and $k_y = |\mathbf{k}|\sin\phi$, where $\theta$ and $\phi$ are the angles the wave vector of the contributing plane waves make with the $z$ axis when projected onto the $xz$ and $zy$ planes, respectively.

### 5.6.1 A circular aperture

For a circular aperture of radius $R$ illuminated normally with a coherent plane wave (so that all points in the aperture have the same phase) we can write

$$a(k_x, k_y) = \frac{1}{4\pi^2} \int_{-R}^{R} \int_{-\sqrt{R^2-x^2}}^{\sqrt{R^2-x^2}} [\cos(k_x x + k_y y) + j\sin(k_x x + k_y y)] dx\, dy, \tag{5.17}$$

which can be written

$$a(k_x, k_y) = \frac{1}{4\pi^2} \int_{-R}^{R} \int_{-\sqrt{R^2-x^2}}^{\sqrt{R^2-x^2}} \big[\cos(k_x x)\cos(k_y y) - \sin(k_x x)\sin(k_y y) + j\sin(k_x x)\cos(k_y y) + j\cos(k_x x)\sin(k_y y)\big]\, dx\, dy. \tag{5.18}$$

All the terms in the integrand which contain sine terms give zero because sine is an odd function: $\sin(k_x x) = -\sin(-k_x x)$. Thus

$$a(k_x, k_y) = \frac{1}{4\pi^2} \int_{-R}^{R} \int_{-\sqrt{R^2-x^2}}^{\sqrt{R^2-x^2}} \cos(k_x x)\cos(k_y y) dx\, dy. \tag{5.19}$$

Because the aperture is rotationally symmetric we can choose $k_y = 0$, which gives

$$a(k_x, k_y) = \frac{1}{2\pi^2} \int_{-R}^{R} \sqrt{R^2 - x^2}\cos(k_x x) dx. \tag{5.20}$$

Putting $x = R\cos\chi$ and $\rho = k_x R$ gives

$$a(k_x, k_y) = \frac{R^2}{2\pi^2} \int_0^{\pi} \sin^2\chi \cos(\rho\cos\chi) d\chi. \tag{5.21}$$

The Bessel function of order 1 is defined by [8]

$$J_1(\rho) = \frac{\rho}{\pi} \int_0^{\pi} \sin^2\chi \cos(\rho\cos\chi) d\chi, \tag{5.22}$$

so

$$a(k_x, k_y) = \frac{R^2}{2\pi} \frac{J_1(\rho)}{\rho}. \tag{5.23}$$

The intensity corresponding to the direction $k_x, k_y$ is

$$I = I_0 \frac{|a(k_x, k_y)|^2}{|a(k_0, k_0)|^2} = 4\left[\frac{J_1(\rho)}{\rho}\right]^2, \tag{5.24}$$

where $I_0$ is a constant, and we have used the fact that $\lim_{\rho\to 0}[J_1(\rho)/\rho] = 1/2$ [8]. Remembering that $\rho = k_x R = |\mathbf{k}| R\sin\theta$:

$$I = 4I_0\left[\frac{J_1(2\pi R\sin\theta/\lambda)}{(2\pi R\sin\theta/\lambda)}\right]^2. \tag{5.25}$$

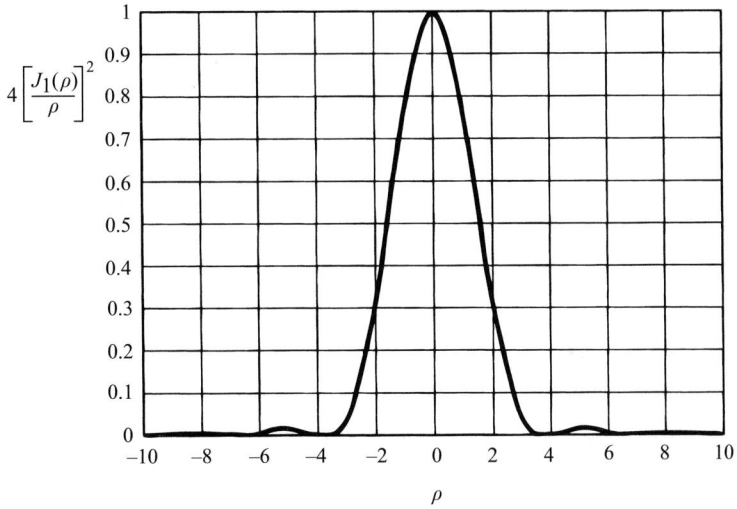


**Fig. 5.3** The Airy diffraction pattern of a circular aperture.

This diffraction pattern is shown in Fig. 5.3. The central disk is called Airy's disk, after Sir G. B. Airy (1801–1892), who first solved this problem. The minima occur when $J_1(\rho) = 0$, which is satisfied for $\rho = 1.220\pi, 2.233\pi, 2.635\pi, 2.679\pi, 3.238\pi, 3.699\pi$, etc. The first minimum satisfies $\rho = 1.220\pi$, which gives

$$\frac{2\pi R \sin\theta}{\lambda} = 1.22\pi \tag{5.26}$$

and

$$\sin\theta = \frac{1.22\lambda}{2R} = \frac{1.22\lambda}{D}, \tag{5.27}$$

where $D$ is the diameter of the circular aperture.

The peak intensity of the diffraction pattern at $\theta = 0$ can be determined easily for small angles that obey $\theta = \sin\theta$. In this case the intensity on a screen a distance $L$ in front of the circular aperture is

$$I(r) = 8\pi I_0 \left[\frac{J_1(krR/L)}{(krR/L)}\right]^2, \tag{5.28}$$

where $k = 2\pi/\lambda$. If the total power passing through the circular aperture is $P$ then

$$P = 8\pi I_0 \int_0^\infty \left[\frac{J_1(krR/L)}{(krR/L)}\right]^2 r\,dr. \tag{5.29}$$

The integral in Eq. (5.29) can be simplified by writing $krR/L = \rho$, which gives

$$P = 2I_0 \frac{\lambda^2 L^2}{\pi R^2} \int_0^\infty \frac{[J_1(\rho)]^2}{\rho}\,d\rho. \tag{5.30}$$

The integral $\int_0^\infty ([J_1(\rho)]^2/\rho)d\rho = 1/2$, so, finally,

$$I_0 = \frac{PA}{\lambda^2 L^2}, \tag{5.31}$$

where $A = \pi R^2$ is the area of the aperture.

For small angles the diffraction angle in Eq. (5.27) can be written as $\theta \simeq 1.22\lambda/D$. This angle provides a measure of when diffraction effects are important. Diffraction effects become negligible when $D \gg \lambda$. The value of $\theta$ sets limits to the ability of optical systems to produce images, limits the ability of a focusing lens to focus plane waves to small spots, and represents a fundamental deviation of the performance of an optical system from that represented by geometrical optics. We will examine these points in further detail in Chapters 14, 15, and 16. A fundamental observation (see Section 14.6) is that the smallest focal spot that can be produced with a plane wave and lens, and also the smallest object that can be imaged without excessive diffraction, are both about the size of the wavelength of the light used.

For the transmission of light through very small apertures, which are smaller than the wavelength, the Fraunhofer diffraction analysis that has been presented here is no longer sufficient to explain what happens [9]. In this case, the nature of the aperture must be carefully taken into account. For example, small apertures in thick metal sheets should prevent the transmission of any radiation, although anomalously large transmisison through sub-wavelength holes has been observed [10].

## 5.7 Laser modes

In the last chapter we saw that when a laser oscillates it emits radiation at one or more frequencies that lie close to passive resonance frequencies of the cavity. These frequencies are called *longitudinal* modes. In our initial discussion of these modes we treated them as plane waves reflecting back and forth between two plane laser mirrors. In practice, laser mirrors are not always plane. Usually at least one of the laser mirrors will have concave spherical curvature. The use of spherical mirrors relaxes the alignment tolerance that must be maintained in order for adequate feedback to be achieved. Even if the laser mirrors are plane, the waves reflecting between them cannot be plane, since true plane waves can exist only if there is no lateral restriction of the wave fronts. Practical laser mirrors are of finite size, so any wave reflecting from them will spread out because of diffraction. We have just shown that this diffractive spreading of a plane wave results when the plane wave is restricted laterally, for example by passing the wave through an aperture. Reflection from a finite-sized mirror produces equivalent effects. We can explain this phenomenon qualitatively by introducing the concept of Huygens secondary wavelets.[2] If a plane mirror is illuminated by a plane wave then each point on the mirror can be treated as a source of a spherical wave called a secondary wavelet. The overall reflected wave is the envelope

[2] Christiaan Huygens (1629–1695) was a Dutch astronomer who was the first to suggest the concept of secondary wavelets.

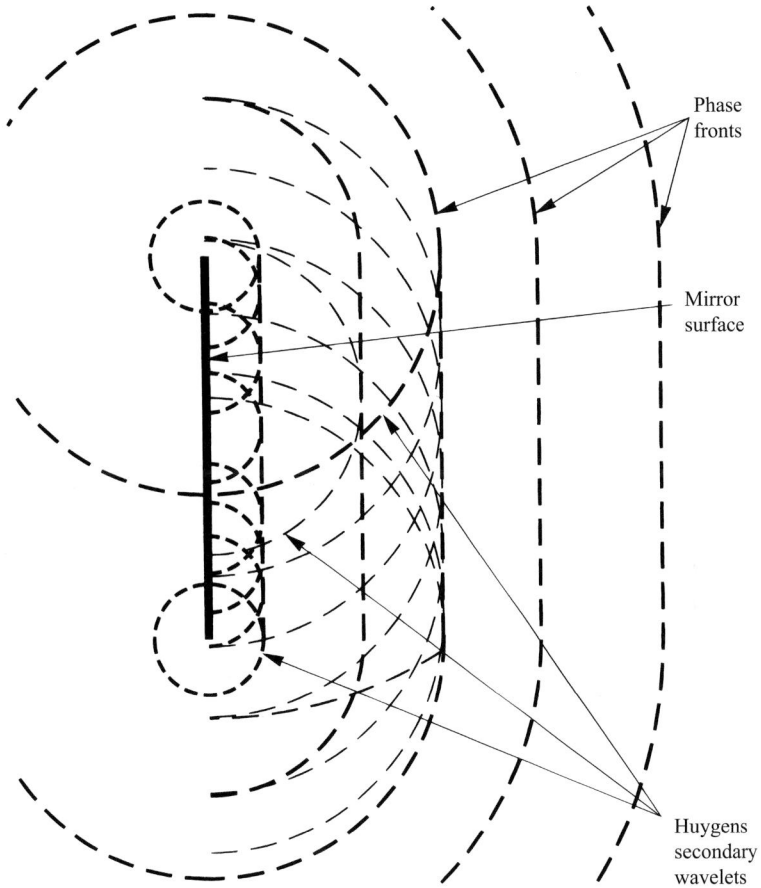


Fig. 5.4 Secondary wavelets originating from a finite plane mirror.

of the sum total of secondary wavelets originating from every point on the mirror surface, as shown in Fig. 5.4. This construction shows that the reflected wave from a finite-sized mirror is not a plane wave.

The existence of diffraction in a laser resonator places restrictions on the minimum size of mirrors that can be used at a given wavelength $\lambda$ and spacing $\ell$. According to Eq. (5.27) a plane wave that reflects from a mirror of diameter $d$ will spread out into a range of directions characterized by a half angle $\theta$, where $\theta = \lambda/d$.

Consider the resonator shown in Fig. 5.5, whose reflectors have diameters $d_1$ and $d_2$, respectively, and spacing $\ell$. A wave diffracting from mirror $M_1$ towards mirror $M_2$ will lose substantial energy past the edges of $M_2$ if the diffraction angle $\theta \leq d_1/\ell$. This gives us the Fresnel condition

$$\frac{d_1 d_2}{\lambda \ell} \geq 1. \tag{5.32}$$

The actual waves that reflect back and forth between the mirrors of a laser resonator are not plane waves, but they do have characteristic spatial patterns of electric (and magnetic)

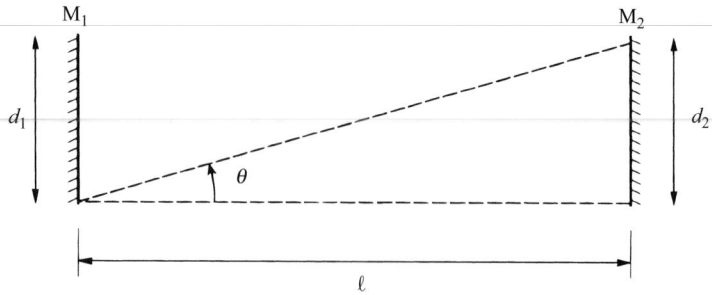


Fig. 5.5 The geometry for calculating the magnitude of diffraction effects in a laser resonator.

field amplitude; they are called *transverse modes*. To be amplified effectively, such modes must contain rays that undergo substantial numbers of specular reflections before being lost from the cavity. A transverse mode is a field configuration on the surface of one reflector that propagates to the other reflector and back, returning in the same pattern, apart from a complex amplitude factor that gives the total phase shift and loss of the round trip. To each of these transverse modes there corresponds a set of longitudinal modes spaced by approximately $c/(2\ell)$.

The electromagnetic field distributions that propagate back and forth between the mirrors of a laser resonator are called *Gaussian beam waves*. These waves have a radial field and intensity distribution that is Gaussian, for example the electric field could be of the form

$$E_x = E_0 e^{-r^2/w_0^2}. \tag{5.33}$$

Why this should be so can be illustrated by the use of Eq. (5.16). If we had a field distribution like Eq. (5.29) on one mirror of a resonator, then this field would propagate away from the mirror in the same way as if an incoming plane wave were being transmitted though an aperture with a Gaussian aperture function. The diffracted field in this case is

$$a(k_x, k_y) = \frac{1}{4\pi^2} \int_{-\infty}^{\infty} \int_{-\infty}^{\infty} e^{-(x^2+y^2)/w_0^2} e^{j(k_x x + k_y y)}\, dx\, dy. \tag{5.34}$$

The integral in Eq. (5.30) can be written as the product of two integrals of the form

$$a'(k_x x) = \frac{1}{2\pi} \int_{-\infty}^{\infty} e^{-x^2/w_0^2} (\cos(k_x x) + j \sin(k_x x)) dx. \tag{5.35}$$

Because $\sin(k_x x)$ is an odd function, the imaginary part of each integral vanishes. Each remaining integral is of the form

$$a'(k_x x) = \frac{1}{2\pi} \int_{-\infty}^{\infty} e^{-x^2/w_0^2} \cos(k_x x) dx. \tag{5.36}$$

This is a well-known integral, whose value is [8]

$$a'(k_x x) = \frac{w}{2\sqrt{\pi}} e^{-k_x w_0^2/4}. \tag{5.37}$$

The Gaussian field distribution propagates from the mirror as a Gaussian function of the wave vector. The relative intensity at angle $\theta$ to the axis is, by analogy with Eq. (5.24),

$$I(\theta) = e^{-|\mathbf{k}|^2 w_0^2/4}. \tag{5.38}$$

If we introduce the *beam divergence angle*

$$\theta_{\text{beam}} = \lambda/(\pi w_0) \tag{5.39}$$

then Eq. (5.34) can be written for small angles as

$$I(\theta) = e^{-\theta^2/\theta_{\text{beam}}^2}. \tag{5.40}$$

A distance $L$ away the radial illumination pattern on a second resonator mirror would be

$$I(r) = e^{-\frac{r^2}{L^2\theta_{\text{beam}}^2}}. \tag{5.41}$$

So a Gaussian intensity distribution on one resonator mirror has become a Gaussian intensity distribution on the second mirror. It is clear, however, that the Gaussian pattern on the second mirror will not diffract back to the first mirror and replicate the initial Gaussian distribution without adding some focusing character to one or more of the resonator mirrors. This explains why laser resonators do not use two flat mirrors if a stable, self-replicating Gaussian intensity distribution is required.

A more detailed treatment of these Gaussian beams and their use to describe the transverse modes of resonators is given in Chapter 15. A few of their important additional properties can be summarized as follows.

(a) The nature of these transverse modes is a function of the mirror sizes, their radii of curvature, the presence of additional limiting apertures between the mirrors, and the resonator length.
(b) These modes are affected by the existence of spatial variations of gain in the laser medium.
(c) Only certain configurations of laser mirrors allow propagating transverse modes that do not suffer substantial diffraction loss from the laser cavity to exist, as illustrated in Fig. 5.6. Our brief discussion above has already demonstrated why this is so for plane–plane mirror resonators.
(d) The transverse modes, because they are propagating-beam solutions of Maxwell's equations, are analogous to the confined-beam TEM modes in waveguides, and are labeled accordingly. A laser mode of order $m, n$ would be labeled $\text{TEM}_{mn}$.
(e) The transverse modes can have Cartesian (rectangular) or polar (circular) symmetry. Cartesian symmetry usually arises when some element in the laser cavity imposes a preferred direction on the direction of the electric and magnetic field vectors (for example, Brewster windows in the laser cavity). If distance along the resonator axis is measured by the coordinate $z$ then a transverse mode with its electric field in the $x$ direction would be a function of the form $E_{mn}^x(x,y,z)e^{j(\omega t \pm kz)}$ with a magnetic field of the form $H_{mn}^y(x,y,z)e^{j(\omega t \pm kz)}$.
(f) Depending on the radii of the resonator mirrors and their separation, the longitudinal modes associated with each transverse mode may have the same or different

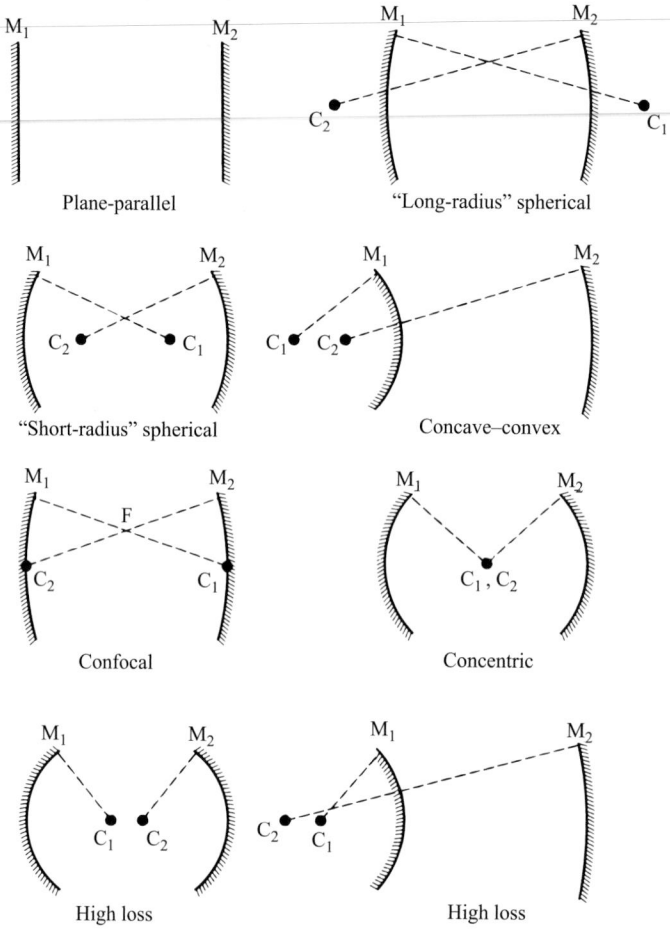


**Fig. 5.6** Laser resonator configurations that use two spherical mirrors. Here $C_1$ and $C_2$ are the centers of curvature of mirrors $M_1$ and $M_2$, respectively.

frequencies. Each individual longitudinal mode associated with a transverse mode $TEM_{mn}$ is labeled $TEM_{mnq}$. In the general case the longitudinal mode frequencies of two different TEM modes will be different, as shown in Fig. 5.7.

(g) Since the field distribution of the transverse modes is propagating in both directions inside the laser resonator, this field distribution is maintained in the output beam from the laser, and the resultant intensity pattern shows an $xy$ spatial dependence in a plane perpendicular to the direction of propagation of the laser beam. These patterns have an intensity distribution

$$I(x, y, z) \propto [E^x_{mn}(x, y, z)]^2 \tag{5.42}$$

and are called mode patterns. Some examples of simple transverse mode patterns having Cartesian symmetry are shown in Fig. 5.8. It can be seen that the number of $xy$ nodal lines in the intensity pattern determines the designation $TEM_{mn}$.

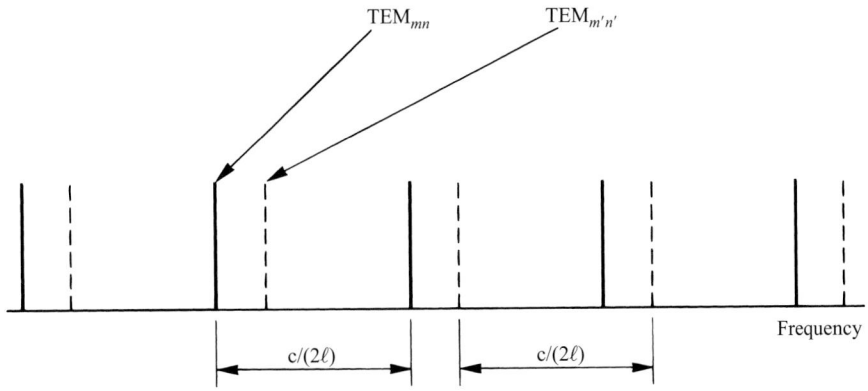


**Fig. 5.7** Schematic frequencies of longitudinal modes belonging to two different transverse modes, TEM$_{mn}$ and TEM$_{m'n'}$.

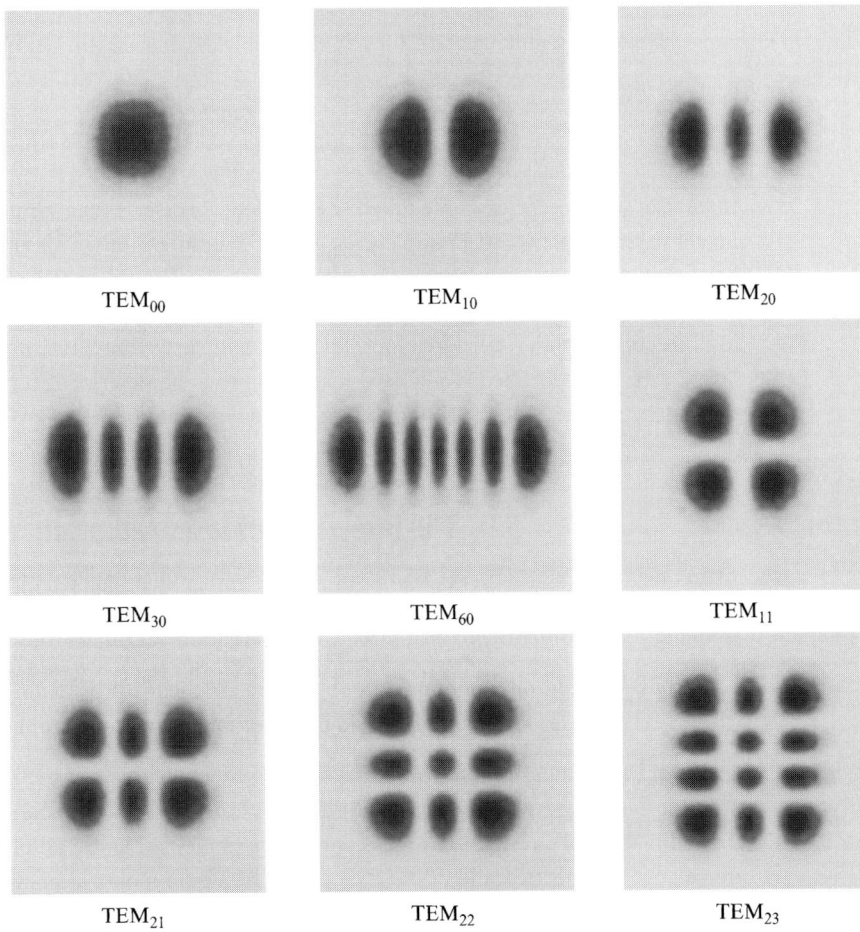


**Fig. 5.8** Intensity patterns for different TEM modes possessing Cartesian symmetry.

## 5.8 Beam divergence

Since the oscillating field distributions inside a laser are not plane waves, when they propagate through the mirrors as output beams they spread by diffraction.

The semivertical angle of the cone into which the output beam diverges is, from Eq. (5.34),[3]

$$\theta_{\text{beam}} \approx \frac{\lambda}{\pi w_0}, \tag{5.43}$$

where $\lambda$ is the wavelength of the output beam and $w_0$ is a parameter called the "minimum spot size" that characterizes the transverse mode. For the special case of two mirrors of equal radii in a confocal arrangement, as shown in Fig. 5.6, the value of $w_0$ is

$$(w_0)_{\text{conf}} = \sqrt{\frac{\lambda \ell}{2\pi}}, \tag{5.44}$$

so

$$\theta_{\text{beam(conf)}} = \frac{\lambda}{\pi}\sqrt{\frac{2\pi}{\lambda \ell}} = \sqrt{\frac{2\lambda}{\pi \ell}}. \tag{5.45}$$

If we take the specific example of a 632.8-nm He–Ne laser with a symmetric confocal resonator 0.3 m long, $w_0 = 0.17$ mm and $\theta_{\text{beam}} = 1.2$ mrad $= 0.66°$. This is a highly directional beam, but the beam does become wider as it travels away from the laser. Such a beam is very useful in providing a perfect straight-line reference. For this reason lasers are widely used in construction – for tunneling, leveling, and surveying. Over a 100 m distance the laser beam just described would have expanded to a diameter of 230 mm. After traveling the distance to the Moon ($\sim$390,000 km) the beam would be $\simeq$900 km in diameter.

## 5.9 The linewidth of laser radiation

A single longitudinal mode of a laser is an oscillation resulting from the interaction of a broadened gain curve with a passive resonance of the Fabry–Perot laser cavity. The frequency width of the gain curve is $\Delta\nu$, and the frequency width of the passive cavity resonance is $\Delta\nu_{1/2} = \Delta\nu_{\text{FSR}}/F$. We expect the linewidth of the resulting oscillation to be narrower than either of these widths, as shown schematically in Fig. 5.9.

In the ideal case, we might expect the linewidth of laser oscillation to approach zero, since the stimulated emission process produces photons that are all at the same identical frequency. The electric field of the laser would be a pure monochromatic wave of the form

$$E(t) = E_0 \cos(\omega t - \phi). \tag{5.46}$$

[3] For a more detailed derivation of this result see Chapter 15.

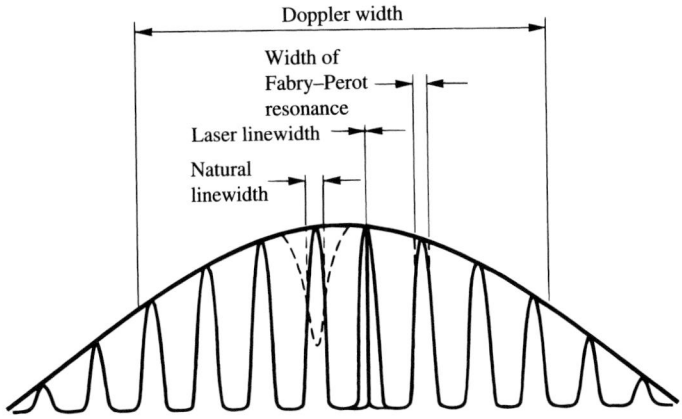


Fig. 5.9 Linewidth factors in a laser.

But we must not forget about spontaneous emission, which also produces photons, and some small fraction of these spontaneously emitted photons will emerge from the laser traveling in the same direction as the laser beam. The effect of these photons is to add to the pure monochromatic wave represented by Eq. (5.46) and have the effect of producing phase fluctuations $\Delta\phi(t)$, since the spontaneously emitted photons do not have the identical phase or frequency of the stimulated emission. We could represent the overall wave now as

$$E(t) = E_0 \cos[\omega t - \phi + \Delta\phi(t)]. \tag{5.47}$$

This wave has a broader frequency spectrum than the idealized $\delta$-function of a pure laser. We can describe the frequency width of this new (laser plus spontaneous emission) spectrum by considering the relative amounts of stimulated emission in a laser.

Consider a laser of length $\ell$ and cross-sectional area $A$, with an output power $P$. To keep our analysis simple we assume that the laser has a symmetric resonator with two identical mirrors of reflectance $R$. If the laser is oscillating just above threshold, then its gain is close to the threshold value, and the intracavity laser intensity can be assumed to be uniform between the two resonator mirrors. Consequently, there is an intracavity laser intensity $P/[2A(1-R)]$ propagating in both directions inside the cavity. The rate of stimulated emissions produced by this monochromatic field is

$$R_{\text{stim}} = B_{21}\left(N_2 - \frac{g_2}{g_1}N_1\right)_{\text{t}} g(\nu_0, \nu) V \frac{P}{A(1-R)c}, \tag{5.48}$$

where $V$ is the volume of the laser cavity, and $\nu$ is the laser oscillation frequency. Spontaneous emissions that add to the laser ouput lie within a frequency band $\Delta\nu_{1/2}$. Only these spontaneous emissions emerge from the laser cavity, which in this context acts like a narrow-band filter for the spontaneous emission. Moreover, since spontaneous emission occurs into all modes of the electromagnetic field, the number of spontaneous emissions into modes corresponding to the same propagation direction as the laser field is

$$R_{\text{spont}} = N_2 A_{21} g(\nu_0, \nu) \frac{c^3}{8\pi \nu^2} \Delta\nu_{1/2}. \tag{5.49}$$

The ratio of stimulated emissions to spontaneous emissions in the laser beam direction is

$$R_{\text{ss}} = \frac{8\pi \nu^2}{c^3} \frac{B_{21}}{A_{21}} \frac{\left(N_2 - (g_2/g_1)N_1\right)_{\text{t}}}{N_2} \frac{g(\nu_0, \nu)PV}{A(1-R)c\,\Delta\nu_{1/2}}, \tag{5.50}$$

which, on substituting for the ratio $B_{21}/A_{21}$ from Eq. (1.83) gives

$$R_{\text{ss}} = \frac{P\ell}{c(1-R)h\nu\,\Delta\nu_{1/2}} \frac{\left(N_2 - (g_2/g_1)N_1\right)_{\text{t}}}{N_2}. \tag{5.51}$$

We note that in this expression, for values of $R$ near 1,

$$\Delta\nu_{1/2} = \frac{c(1-R)}{2\pi\ell}. \tag{5.52}$$

It is plausible to take the spectral purity of the laser emission, including the co-directional spontaneous emission as

$$\Delta\nu_{\text{laser}} = \Delta\nu_{1/2}/R_{\text{ss}}, \tag{5.53}$$

which gives finally

$$\Delta\nu_{\text{laser}} = \frac{2\pi(\Delta\nu_{1/2})^2 h\nu}{P} \frac{N_2}{\left(N_2 - (g_2/g_1)N_1\right)_{\text{t}}}, \tag{5.54}$$

which is called the Schawlow–Townes formula [11–14].

Equation (5.54) predicts very narrow linewidths for many lasers. For a typical He–Ne laser with 99%-reflectance mirrors and a cavity 30 cm long

$$\Delta\nu_{1/2} = \frac{c(1-R)}{2\pi\ell} = 1.59 \text{ MHz}.$$

The factor $N_2/[N_2 - (g_2/g_1)N_1]_{\text{t}}$ is close to unity for a typical low-power, say 1-mW, laser. Consequently,

$$\Delta\nu_{\text{laser}} = \frac{\pi \times 6.626 \times 10^{-34} \times 3 \times 10^8 \times 1.59^2 \times 10^{12}}{10^{-3} \times 632.8 \times 10^{-9}} = 2.5 \times 10^{-3} \text{ Hz}.$$

Such a small linewidth is never observed in practice because thermal instabilities and acoustic vibrations lead to variations in resonator length that further broaden the output radiation lineshape. The best observed minimum linewidths for highly stabilized gas lasers operating in the visible region of the spectrum are around $10^3$ Hz. Even if macroscopic thermal and acoustic vibrations could be eliminated from the system, a fundamental limit to the resonator length stability would be set by Brownian motion of the mirror assemblies. For example, consider two laser mirrors mounted on a rigid bar. The mean amount of energy stored in the Brownian motion of the whole bar is $\overline{E} = k_{\text{B}}T$.

The frequency spread of the laser output that thereby results is

$$\Delta\nu_{\text{Brownian}} = \nu\sqrt{\frac{2k_{\text{B}}T}{YV}}, \tag{5.55}$$

where $Y$ is Young's modulus of the bar material and $V$ is the volume of the mounting bar. Typical values of $\Delta\nu_{\text{Brownian}}$ are $\sim$2 Hz. The linewidths of most lasers are *much* larger than this, as we will see in our discussion of specific lasers.

## 5.10 Coherence properties

Because of its extremely narrow output linewidth the output beam from a laser exhibits considerable *temporal coherence* (longitudinal coherence). To illustrate this concept, consider two points A and B a distance $L$ apart in the direction of propagation of a laser beam, as shown in Fig. 5.10. If there exists a definite and fixed phase relationship between the wave amplitudes at A and B, then the wave shows temporal coherence for a time $c/L$. The further apart A and B can be, while still maintaining a fixed phase relation with each other, the greater is the temporal coherence of the output beam. The maximum separation at which the fixed phase relationship is retained is called the *coherence length*, $L_c$, which is a measure of the length of the continuous uninterrupted wave trains emitted by the laser. The coherence length is related to the *coherence time* $\tau_c$ by $L_c = c\tau_c$. The coherence time itself is a direct measure of the monochromaticity of the laser, since by Fourier transformation, done in an analogous manner to the treatment of natural broadening,

$$\tau_c \simeq \frac{1}{\Delta\nu_L} \text{ and } L_c \simeq \frac{c}{\Delta\nu_L}. \tag{5.56}$$

The coherence length and time of a laser source are considerably better than a conventional monochromatic source (a spontaneously emitted line source). The greatly increased coherence can be demonstrated in a Michelson interferometer experiment, which allows interference between waves at longitudinally different positions in a wave front to be studied, as shown in Fig. 5.11.

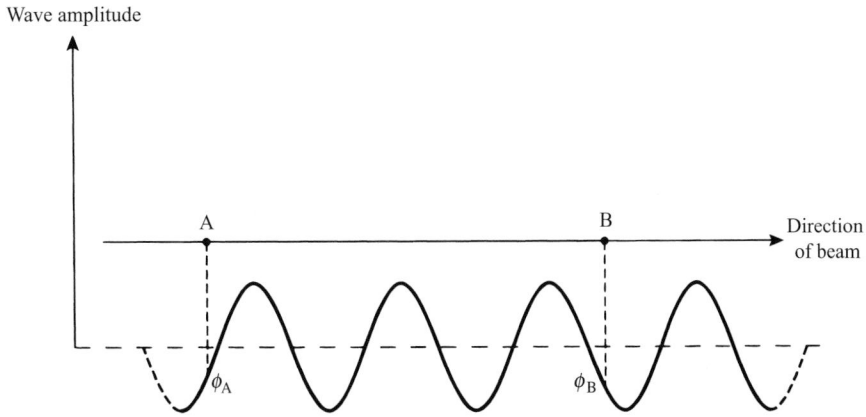


**Fig. 5.10** To illustrate the concept of temporal coherence. If an unbroken wave train connects the points A and B then the phase difference ($\phi_B - \phi_A$) will have a constant value.

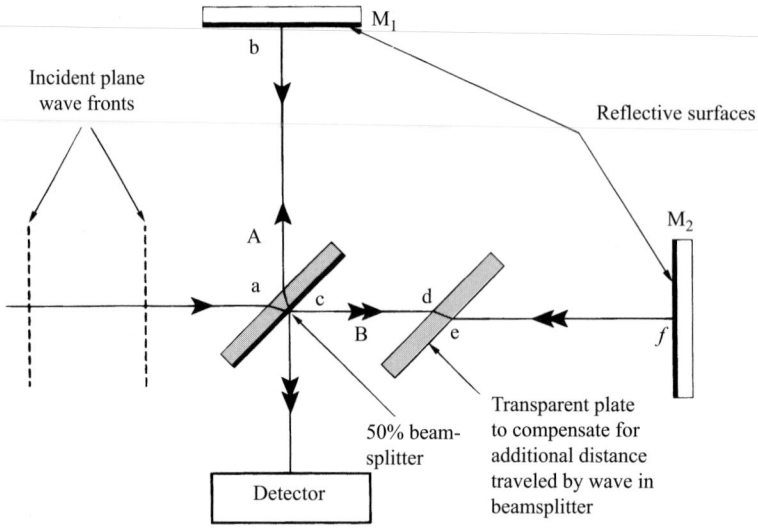


**Fig. 5.11** The Michelson interferometer. Only the primary reflections are shown. Weak reflections from one surface of the beamsplitter and both faces of the compensating plate also occur unless these faces are antireflection-coated. (See Chapter 13.)

The operation of this instrument can be described as follows. Incident waves divide at the input beamsplitter. One part of the wave, A, takes the path c → a → b → a → c → detector. The other part of the incident wave, B, takes the path c → d → e → f → e → d → c → detector. We can write the electric field of wave A, at the detector, as

$$E_A = E_0 e^{j(\omega t + \phi_A)}, \tag{5.57}$$

where $\phi_A$ is the phase shift experienced along the path c → a → b → a → c → detector. Similarly, for wave B

$$E_B = E_0 e^{j(\omega t + \phi_B)}. \tag{5.58}$$

The signal from the detector can be written as

$$i(t) \propto |E_A + E_B|^2, \tag{5.59}$$

since the detector responds to the intensity of the light, which gives

$$i(t) \propto 2E_0^2[1 + \cos(\phi_B - \phi_A)]. \tag{5.60}$$

We expect maximum signal (corresponding to maximum observed illumination) if waves A and B are in phase, that is if $\phi_B - \phi_A = 2n\pi$. Minimum signal results if $\phi_B - \phi_A = (2n+1)\pi$, when the two waves are out of phase. We can write

$$\phi_B - \phi_A = \frac{2\pi L}{\lambda}, \tag{5.61}$$

where $L$ is the difference in path length for the two waves A and B. If $L$ is altered, for example by moving mirror $M_2$ to the right in Fig. 5.11, we expect the detector signal to go up and down between its maximum and minimum values. However, this interference phenomenon will be observed only provided that $L \lesssim L_c$. For a laser with $\Delta\nu \sim 1$ kHz this would require $L \lesssim 300$ km. In practice, it is not feasible to build laboratory interferometers with path differences as large as this, although large path differences can be obtained by incorporating a long optical fiber into one of the paths in Fig. 5.11. If the linewidth of the source is large then it is quite easy to demonstrate the disappearance of any interference effects of $L > L_c$. For example, with a "white" light source that covers the spectrum from 400–700 nm

$$\Delta\nu = 3 \times 10^8 \left( \frac{1}{400 \times 10^{-9}} - \frac{1}{700 \times 10^{-9}} \right) = 3.2 \times 10^{14} \text{ Hz},$$

and therefore $L_c \sim 10^{-3}$ mm. In this case interference effects will be observed only with careful adjustment of the interferometer so that the two paths are almost equal. Interference fringes will be detected due to interference between the two parts of the split wave, A and B, if the coherence length of the incident wave is greater than the distance $2(\text{cd} + \text{ef}) - 2(\text{ab})$.

It should be noted that failure to observe fringes in a Michelson interferometer does not imply that interference between waves is not occurring, but merely that the continuing change in phase relationship between these waves shifts the positions of intensity maxima and minima around in a time $\sim\tau_c$, so apparent uniform illumination results.

A laser also possesses *spatial* (lateral) coherence, which implies a definite fixed phase relationship between points separated by a distance $L$ transverse to the direction of beam propagation. The transverse coherence length, which has a similar physical meaning to the longitudinal coherence length, is

$$L_{tc} \sim \frac{\lambda}{\theta_{\text{beam}}} \sim \pi\omega_0 \tag{5.62}$$

for a laser source.

## 5.11 Interference

The existence of spatial coherence in a wave front and the limit of its extent can be demonstrated in a classic Young slits interference experiment. In this experiment a pair of thin, parallel slits or a pair of pinholes is illuminated normally with a spatially coherent monochromatic plane wave, as illustrated in Fig. 5.12. A laser beam operating in a circularly symmetric Gaussian mode, and expanded so that its central portion illuminates both slits or pinholes, serves as an ideal source to illustrate this classic interference phenomenon and at the same time test the spatial coherence properties of the laser beam.

The field amplitudes produced by the plane wave at the slits are equal and in phase, and can be represented as

$$V = V_0 \cos(\omega t + \phi), \tag{5.63}$$

where $\phi$ is a phase factor.

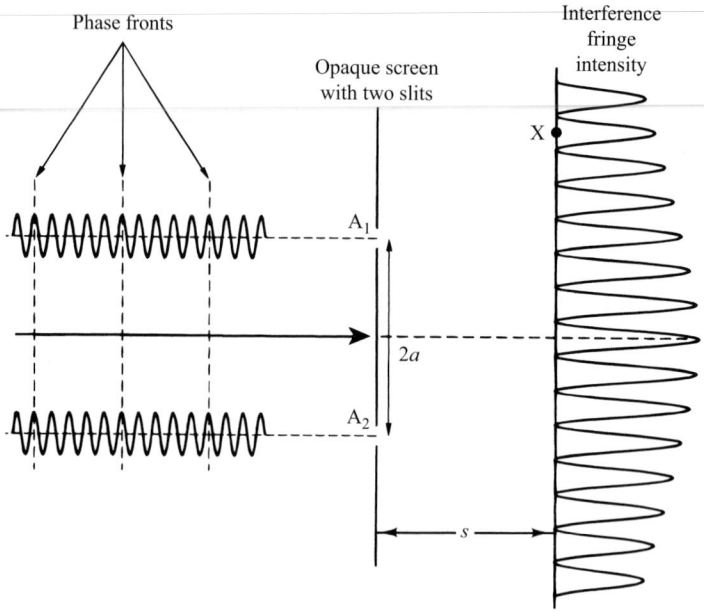


**Fig. 5.12** Young's slits interference experiment to demonstrate the existence of spatial coherence in a phase front.

Each slit acts as a source of secondary wavelets that propagate towards the observation screen. The resultant field amplitude at a point such as X will result from the superposition of the waves that have propagated from $A_1$ and $A_2$. The field amplitude at X from $A_1$ is

$$V_1 = b_1 V_0 \cos(\omega t + \phi - kR_1), \tag{5.64}$$

where

$$R_1 = A_1X = \sqrt{(x-a)^2 + s^2}, \tag{5.65}$$

$k = 2\pi/\lambda$ is the wavelength of the plane wave, and $b_1$ is a constant that depends on the geometry of the experiment. The field amplitude at X due to $A_2$ is

$$V_2 = b_2 V_0 \cos(\omega t + \phi - kR_2), \tag{5.66}$$

where

$$R_2 = A_2X = \sqrt{(x+a)^2 + s^2}. \tag{5.67}$$

If $s \gg a, x$ then the magnitudes of $V_1$ and $V_2$ can be taken to be equal, but they differ in phase. The total field at X is

$$V_X = V_1 + V_2 = bV_0[\cos(\omega t + \phi - kR_1) + \cos(\omega t + \phi - kR_2)], \tag{5.68}$$

where $b$ is a constant. If $s \gg a, x$ then $\theta$ is a small angle and $V_1$ and $V_2$ can be taken to be parallel, irrespective of the polarization of the input wave. From Eq. (5.64) we get

$$V_X = 2bV_0 \cos\left[\omega t + \phi - \frac{k(R_1 + R_2)}{2}\right] \cos\left[\frac{k(R_2 - R_1)}{2}\right] \tag{5.69}$$

and the intensity at $X$ is

$$I \propto V_X^2 \propto \cos^2\left[\omega t + \phi - \frac{k(R_1 + R_2)}{2}\right]\cos^2\left[\frac{k(R_2 - R_1)}{2}\right], \tag{5.70}$$

which gives

$$I \propto \frac{1}{2}\left\{1 + \cos^2\left[\omega t + \phi - \frac{k(R_1 + R_2)}{2}\right]\right\}\cos^2\left[\frac{k(R_2 - R_1)}{2}\right]. \tag{5.71}$$

Because $\omega$ is a very high frequency ($\sim 10^{15}$ Hz in the visible region), the detector does not see the term at frequency $2\omega$. It is as if the detector averages this term, which is as often positive as it is negative, to zero.[4] The observed intensity is therefore

$$I \propto \cos^2\left[\frac{k(R_2 - R_1)}{2}\right]. \tag{5.72}$$

Now, if $s \gg a, x$,

$$R_2 = s\left[1 + \left(\frac{x+a}{s}\right)^2\right]\sqrt{1/2} = s\left[1 + \frac{1}{2}\left(\frac{x+a}{s}\right)^2\right], \tag{5.73}$$

$$R_1 = s\left[+\frac{1}{2}\left(\frac{x-a}{s}\right)^2\right], \tag{5.74}$$

so

$$R_2 - R_1 = 2ax/s, \tag{5.75}$$

which gives

$$I \propto \cos^2\left(\frac{kax}{s}\right). \tag{5.76}$$

We can write $\theta = x/s$ for $x \ll s$, giving

$$I \propto \cos^2(ka\theta) = \cos^2\left(\frac{2\pi a\theta}{\lambda}\right). \tag{5.77}$$

Note that this is the same result as Eq. (5.10). The intensity is a maximum whenever

$$\frac{2\pi a\theta}{\lambda} = m\pi, \tag{5.78}$$

giving

$$\theta_{\max} = \frac{m\lambda}{2a}, \tag{5.79}$$

whereas, for the minima

$$\frac{2\pi a\theta}{\lambda} = \frac{2m+1}{2}, \tag{5.80}$$

[4] For a further discussion of this point see Chapter 22.

giving

$$\theta_{\min} = \left(\frac{2m+1}{4\pi a}\right)\lambda. \tag{5.81}$$

The interference pattern on the screen appears as a series of equally spaced alternate bright and dark bands.

The appearance of these bright and dark bands does, however, depend crucially on the spatial coherence of the waves illuminating the two slits. If the slit variation were increased to beyond the lateral coherence length, so that $2a > L_{\text{tc}}$, then the fringe pattern would disappear.

The classic diffraction pattern of a circular aperture can also be easily observed by illuminating a small circular hole with the central portion of a $\text{TEM}_{00}$-mode laser beam. The cleanest patterns can be obtained by focusing the laser beam with a lens and placing a small circular aperture (of size smaller than the focused laser beam) in the focal plane of the lens.

## 5.12 Problems

5.1 A pair of narrow slits is illuminated with a monochromatic plane wave with $\lambda_0 = 488$ nm. At a distance of 500 mm behind the slits the dark-band spacing is 5 mm. What is the spacing of the slits?

5.2 Calculate the interference pattern produced by four narrow slits, equally spaced by $2a$. Extend your analysis to $N$ equally spaced slits. What happens as $N \to \infty$?

5.3 Calculate the Fraunhofer diffraction pattern of a rectangular aperture of dimension $a \times b$.

5.4 A Michelson interferometer with identical arms shows sharp interference effects when illuminated by a point source. A glass slab of thickness 10 mm and refractive index 1.6 is placed in one arm. No interference effects are then observed. What can you say about the coherence properties of the source?

5.5 A gas laser with $\lambda_0 = 446$ nm has a resonator 0.5 m long. One of the laser mirrors is randomly vibrating with an amplitude of 10 nm. Estimate how much this will effectively broaden the linewidth of the emitted radiation.

## References

[1] M. Born and E. Wolf, *Principles of Optics*, 6th edn., Oxford: Pergamon Press, 1980.
[2] R. W. Ditchburn, *Light*, 3rd edn., New York: Academic Press, 1975.
[3] E. Hecht and A. Zajac, *Optics*, 2nd edn., Reading, MA: Addison-Wesley, 1987.
[4] F. A. Jenkins and H. E. White, *Fundamentals of Optics*, 3rd edn., New York: McGraw-Hill, 1957.
[5] M. V. Klein and T. E. Furtak, *Optics*, 2nd edn., New York, Wiley, 1970.

[6] R. S. Longhurst, *Geometrical and Physical Optics*, 3rd edn., London: Longman, 1973.

[7] F. G. Smith and J. H. Thompson, *Optics*, London: Wiley, 1971.

[8] A. Abramowitz and I. A. Stegun, eds., *Handbook of Mathematical Functions*, New York: Dover Publications, 1968.

[9] H. A. Bethe, "Theory of diffraction by small holes," *Phys. Rev.* **66**, 63–182, 1944.

[10] T. W. Ebbesen, H. J. Lezec, H. F. Ghaemi, T. Thio, and P. A. Wolff, "Extraordinary optical transmission through sub-wavelength hole arrays," *Nature* **391**, 667–669, 1998.

[11] P. W. Milonni and J. H. Eberly, *Lasers*, New York: Wiley, 1988.

[12] A. E. Siegman, *Lasers*, Mill Valley, CA: University Science Books, 1985.

[13] A. Yariv, *Introduction to Optical Electronics*, 4th edn., New York: Holt, Rinehart and Winston, 1991.

[14] A. Yariv, *Quantum Electronics*, 3rd edn., New York: Wiley, 1989.

# 6 Control of laser oscillators

## 6.1 Introduction

During laser oscillation one or more distinct mode frequencies can be emitted, which in the general case behave as independent monochromatic oscillations. When more than one mode oscillates this multimode operation usually corresponds to simultaneous oscillation on more than one longitudinal mode of a single transverse mode of the cavity. It is also possible for more than one transverse mode to oscillate simultaneously, but this is not of great importance or interest to us here. In this chapter we will examine how multimode operation can be suppressed in favor of single-mode operation. We shall also see how many simultaneously oscillating modes can become *locked* together in phase to produce a form of laser oscillation in which the laser output becomes a train of very short pulses. This *mode-locked* operation is of considerable fundamental and practical interest, and we shall examine the various ways in which the phenomenon can be induced.

## 6.2 Multimode operation

We have seen that in an inhomogeneously broadened laser simultaneous oscillation is possible on more than one longitudinal mode. These oscillations can be coupled to each other by applying an external perturbation, as we shall see shortly. However, in the absence of such perturbation, each mode oscillates independently of the others. We can represent the total electric field of the laser beam as

$$E(t) = \sum_{n=1}^{N} E_n(t)\cos[\omega_n t + \phi_n(t)], \tag{6.1}$$

where $\omega_n$ is the frequency of the $n$th of $N$ modes and $\phi_n(t)$ is its phase. The frequencies $\omega_n$ and $\omega_{n+1}$ are separated by approximately $\pi c/\ell$, but successive adjacent modes are not exactly evenly spaced because of mode-pulling. The phase term fluctuates with time in a random way because the coherence time of the beam is finite. These fluctuations can be attributed to three main causes:

(i) spontaneous emission, of which there is always some, even into the narrow solid angle occupied by the laser beam;
(ii) fluctuation in the index of refraction of the amplifying medium; and
(iii) vibrations of the mechanical structure of the laser resonator.

The amplitude $E_n(t)$ changes with time if the frequency $\omega_n$ drifts relative to the center of the gain profile. Such drift can be made very small by stabilizing the frequency of the laser. This is most simply done by stabilizing the optical length $L$ of the resonator. The optical length can be defined in the most general way as

$$L = \int_0^{\ell} n(x)dx, \tag{6.2}$$

where $n(x)$ is the refractive index of the medium (or media) between the resonator mirrors placed at the geometrical spatial locations 0 and $\ell$.

If the laser intensity is monitored with a fast optical detector, it will be observed to fluctuate with time because of the beating together of the various longitudinal modes. We saw in Chapter 4 that these beats allow the number of oscillating modes to be determined. Although these beats might generally be expected to give only high-frequency fluctuation, this is not so. The beat signals from different pairs of modes also beat with each other to give fluctuations down to quite low frequencies. The laser intensity also fluctuates because of variations in the mode amplitudes $E_n(t)$ caused by fluctuations in the gain of the amplifying medium.

Fluctuations in amplitude and laser frequency can be controlled in several ways. The simplest approach to ideal behavior is to cause the laser to oscillate in only a single longitudinal and transverse mode. The transverse mode most desirable in this context is the so-called *fundamental* mode in which the distribution of intensity across the laser beam is Gaussian.[1] Oscillation in the fundamental transverse mode can be controlled by appropriate selection of the radii of curvature of the laser mirrors and the size of apertures in the laser cavity – for example, the apertures provided by the lateral size of the amplifying medium itself.

## 6.3 Single-longitudinal-mode operation

Single-longitudinal-mode operation in a homogeneously broadened laser should be automatic. Surprisingly, sometimes this is not so. The strongest mode should suppress all its neighbors, but because of the standing-wave character of the in-cavity field this might not occur if the maxima of field amplitudes of two different modes do not overlap in space. This phenomenon is called *spatial hole-burning*. Figure 6.1 shows schematically how it happens. Mode 1 draws its gain from, for example, particles located near a field maximum such as A. Mode 2 draws its gain from particles located near field maximum B. The two modes will not compete strongly unless particles can move rapidly from A to B. This phenomenon is called *spatial cross-relaxation*. Clearly, such effects are more important in a gas than a solid-state laser.

[1] This will be discussed in detail in Chapter 15.

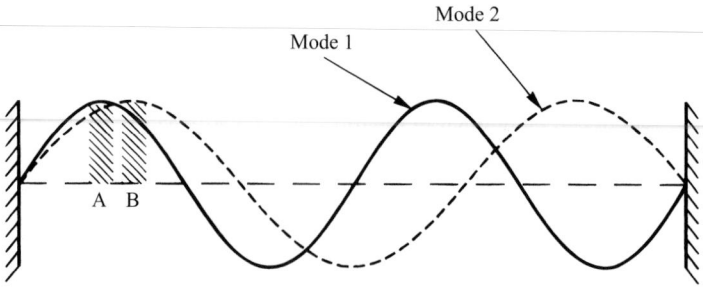


**Fig. 6.1** Standing-wave patterns of two adjacent oscillating longitudinal modes in a laser resonator.

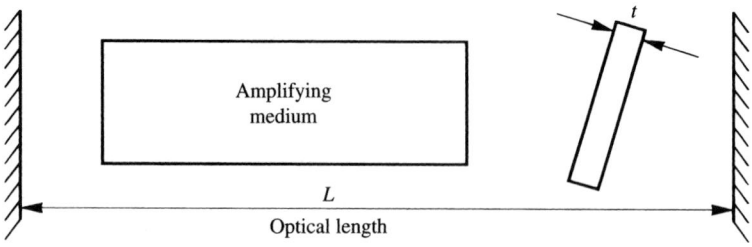


**Fig. 6.2** Use of a tilted intracavity etalon to obtain single-longitudinal-mode laser oscillation. The optical length includes both the amplifying medium and the etalon.

Several methods have been used to force a laser into single-longitudinal-mode operation.

(i) Making the laser sufficiently short that only one cavity resonance lies under the gain profile and above the loss line. This generally requires $c/(2L) > \Delta\nu_D$.

(ii) Use of an intracavity etalon of thickness $t \ll L$, as shown in Fig. 6.2. The etalon is usually tilted slightly to prevent complications being caused by reflections between the etalon faces and laser mirrors. The transmission maxima of such an etalon are relatively wide in frequency compared with the transmission maxima of the main cavity, but they are separated by a much greater frequency, $\Delta\nu_{\text{etalon}} \gg \Delta\nu_{\text{cavity}}$,

$$\Delta\nu_{\text{etalon}} = \frac{c_0}{2nd}, \tag{6.3}$$

where $n$ is the refractive index of the etalon. For example, for a 10-mm-thick fused silica etalon $\Delta\nu_{\text{etalon}} \simeq 10$ GHz. If this etalon is used inside the cavity of a laser with a gain linewidth of 1 GHz (say), then it is likely that only one of the main cavity resonances will be within a transmission maximum of the in-cavity etalon. This situation is shown schematically in Fig. 6.3.

(iii) Placing a very thin absorbing film inside the cavity. The position of the film, which could be a thin layer of metal evaporated onto a glass substrate, is adjusted to be at a node of the intracavity standing wave of a strong longitudinal mode. Other modes suffer loss at the film and can be suppressed.

(iv) Making the laser homogeneously broadened. This can sometimes be done in the case of gas lasers by making them operate at high gas pressure.

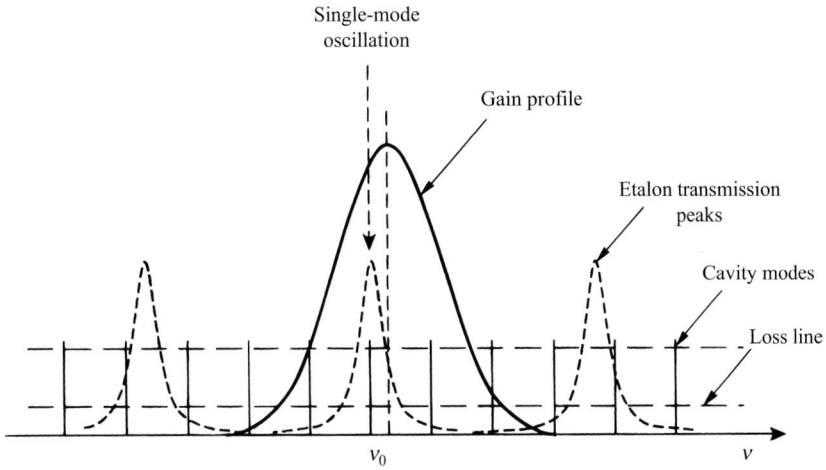


**Fig. 6.3** Intracavity etalon transmission peaks superimposed on the laser gain profile. Main cavity resonances span the gain curve. In this figure four such resonances lie above the loss line when the etalon is absent.

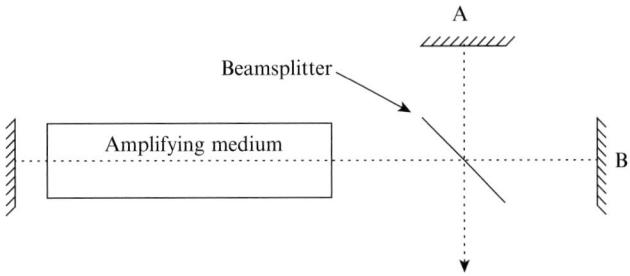


**Fig. 6.4** A laser resonator incorporating a Fox–Smith interferometer for encouraging single-longitudinal-mode laser oscillation.

(v) Use of a Fox–Smith interferometer as shown in Fig. 6.4, which acts as a frequency-dependent [1, 2] cavity mirror. Laser oscillation can use either mirror A or mirror B, the combination of which behaves like a Michelson interferometer. A longitudinal mode that divides into two parts at the beamsplitter will suffer high loss from the cavity unless both parts reflect in phase towards the amplifying medium.

(vi) If one of the main cavity mirrors is an etalon, both faces of which are reflective, then a frequency-dependent cavity mirror results and can discriminate between longitudinal modes.

(vii) In homogeneously broadened lasers, multimode oscillation frequently results because the standing waves corresponding to different modes deplete the population inversion at different spatial locations. This is the spatial hole-burning already mentioned. Single-longitudinal-mode operation can be obtained if the periodicity of the intracavity field can be smoothed out. This requires the wave inside the cavity to look like a traveling rather than a standing wave, which can be accomplished by building

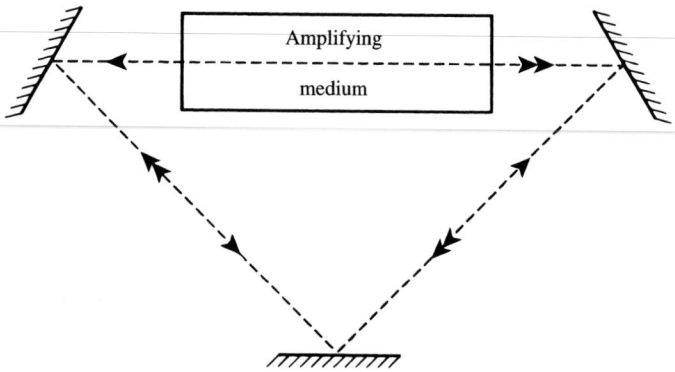


**Fig. 6.5** A ring laser configuration with counterpropagating traveling wave fields. Unidirectional laser oscillation in such a cavity suppresses spatial hole-burning and will encourage single-mode oscillation.

the laser in the form of a ring. In Fig. 6.5 waves can travel around the three-mirror cavity in both directions; if one of these waves can be suppressed, then a traveling wave exists inside the laser medium and no spatial hole-burning results. To explain how this might be done, it is actually simplest to describe a clever scheme that uses a conventional cavity with two mirrors but with elements inside the cavity that change the polarization state of the wave. The laser configuration shown in Fig. 6.6 has a Brewster window in the laser cavity and two quarter-wave plates. These optical components, which will be described in detail in Chapter 17, change linearly polarized light passing through them into circularly polarized light. At point A inside the cavity the wave that oscillates with minimum loss will be linearly polarized in the direction shown. With respect to $x$ and $y$ axes oriented at 45° to the plane of the diagram we can represent a wave traveling to the right as

$$(E_x)_{\mathrm{A}} = E_0 \cos(\omega t - kz),$$
$$(E_y)_{\mathrm{A}} = E_0 \cos(\omega t - kz). \tag{6.4}$$

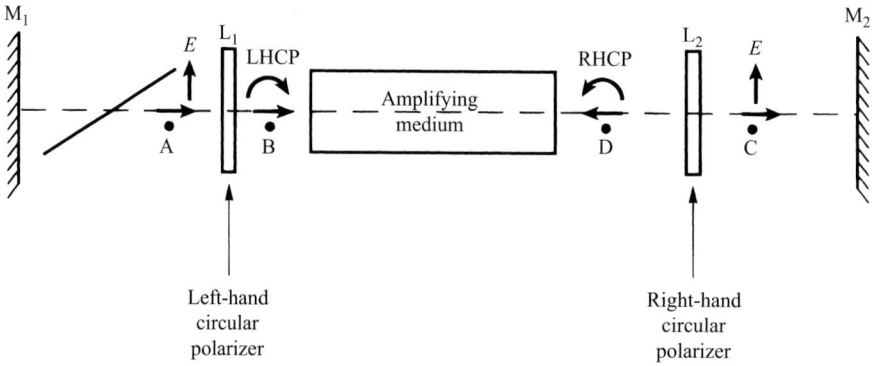


**Fig. 6.6** The "twisted-mode" arrangement for obtaining single-mode operation of a homogeneously broadened laser by eliminating standing waves in the amplifying medium.

After passage through quarter-wave plate $L_1$, which is oriented appropriately, the electric fields at point B are

$$(E_x)_B = E_0 \cos(\omega t - kz + \pi/2),$$
$$(E_y)_B = E_0 \cos(\omega t - kz). \tag{6.5}$$

This is left-hand circularly polarized light (LHCP). The second quarter-wave plate $L_2$ is oriented to produce a $\pi/2$ phase shift of the $y$-directed electric vector, so at point C the right traveling wave can be written as

$$(E_x)_C = E_0 \cos(\omega t - kz + \pi/2),$$
$$(E_y)_C = E_0 \cos(\omega t - kz + \pi/2), \tag{6.6}$$

which is once again linearly polarized. This wave reflects from mirror $M_2$ and returns through $L_2$. At point D this left traveling wave can be written as

$$(E_x)_D = E_0 \cos(\omega t + kz + \pi/2),$$
$$(E_y)_D = E_0 \cos(\omega t + kz + \pi), \tag{6.7}$$

which is right-hand circularly polarized light (RHCP). After a second pass through $L_1$ this wave becomes linearly polarized in its original direction once more.

Inside the amplifying medium the total electric field resulting from the right and left traveling waves is

$$E_x = E_0 \cos(\omega t - kz + \pi/2) + E_0 \cos(\omega t + kz + \pi/2)$$
$$= -2E_0 \sin(\omega t)\cos(kz), \tag{6.8}$$
$$E_y = E_0 \cos(\omega t - kz) + E_0 \cos(\omega t + kz + \pi)$$
$$= 2E_0 \sin(\omega t)\sin(kz). \tag{6.9}$$

The intracavity intensity is proportional to

$$E_x^2 + E_y^2 = 4E_0^2 \sin^2(\omega t), \tag{6.10}$$

which is independent of $z$. No standing-wave distribution of energy density exists in the cavity and single-longitudinal-mode operation can be obtained.

## 6.4 Mode-locking

So far in this chapter we have discussed ways in which to cause a laser to generate one or more CW oscillating frequencies corresponding to cavity modes. However, it is also possible to cause the laser to generate a train of regularly spaced, generally very short, pulses. A laser that operates in this way is said to be *mode-locked*. This kind of behavior often occurs spontaneously, and is then referred to as *self-mode-locking*. A fundamental mode-locked pulse train consists of a series of pulses separated by the cavity round-trip time $2\ell/c$. That this kind of behavior could occur should not seem surprising. Early in the development of the laser, Fleck [3] showed that as a laser oscillation built up from spontaneous emission there was a natural tendency for the laser output to show strong fluctuations on a time scale that corresponded to the round-trip time in the cavity. A single photon spontaneously emitted into a high-$Q$ mode of the cavity will bounce back and forth in the resonator, and be amplified, leading to a pulse train of spacing $2\ell/c$. Whether such a group of photons will succeed in competing with other growing oscillations will determine whether the laser ends up operating in CW mode or mode-locked.[2]

There are two ways of looking at mode-locking. The more common, but less physically satisfying, is to treat the independent oscillating cavity modes whose electric fields are given in Eq. (6.1) as modes that have had their phases locked together, so that each mode has $\phi_n(t) = \phi_0$. By a shift of our choice of the time origin we can arbitrarily set $\phi_0 = 0$. The total electric field of the laser can then be represented as

$$E(t) = \mathcal{R}\left[\sum_{n=1}^{N} E_n(t)e^{j\omega_n t}\right]. \tag{6.11}$$

To simplify the analysis we take $E_n(t) = 1$ and

$$\omega_n = \omega_0 + \left(n - \frac{N+1}{2}\right)\Delta\omega_c, \tag{6.12}$$

where $\Delta\omega_c = \pi c/\ell$.

Substituting Eq. (6.12) into Eq. (6.11) and summing the resultant geometric series gives

$$E(t) = \cos(\omega_0 t)\frac{\sin(N\,\Delta\omega_c\,t/2)}{\sin(\Delta\omega_c\,t/2)}. \tag{6.13}$$

This is a pure oscillation at frequency $\omega_0$ modulated with the *envelope* function

$$f(t) = \frac{\sin(N\,\Delta\omega_c\,t/2)}{\sin(\Delta\omega_c\,t/2)}. \tag{6.14}$$

The average power corresponding to this envelope is

$$P(t) \propto \frac{\sin^2(N\,\Delta\omega_c\,t/2)}{\sin^2(\Delta\omega_c\,t/2)}. \tag{6.15}$$

Equation (6.15) represents a periodic train of pulses that have the following properties:

(a) the pulse spacing is $\Delta T = 2\pi/\Delta\omega_c = 2\ell/c$;
(b) the peak power in the train is $N$ times the average power, and the peak field is $N$ times the average for a single mode; and
(c) the pulses within the train become narrower as $N$ increases, and for large $N$ approach a value $\tau = \Delta T/N$.

[2] In this context the term CW is used to include lasers in which the laser excitation is pulsed, but in which the output radiation consists of one or more waves at longitudinal mode frequencies that last for the period of excitation.

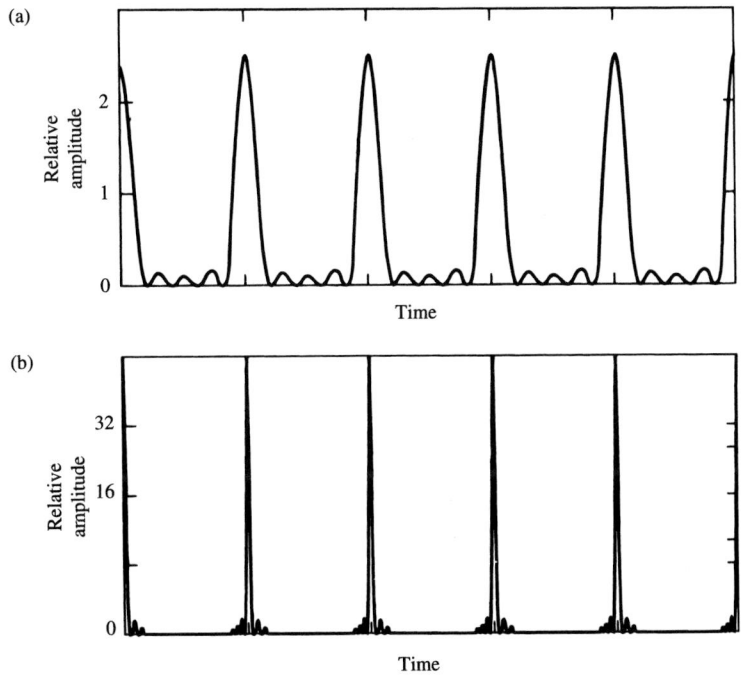


Fig. 6.7 Mode-locked pulse trains calculated from Eq. (6.15): (a) $N = 5$ and (b) $N = 20$.

These characteristics are evident from Fig. 6.7, which is a plot of Eq. (6.15). For an inhomogeneously broadened laser with linewidth $\Delta\nu_D$, the number of independent oscillating modes will be

$$N \sim \frac{2\pi\,\Delta\nu_D}{\Delta\omega_c}, \tag{6.16}$$

so the mode-locked pulse width will approach a value $\tau \simeq 1/\Delta\nu_D$.

For a homogeneously broadened laser, which ideally oscillates in a single longitudinal mode, the frequency explanation of the generation of short "mode-locked" pulses seems less than satisfactory. This brings us to a more internally self-consistent model of mode-locking as a mode of operation that the laser finds *energy advantageous*, and which does not require the *a priori* assumption of independent oscillating cavity modes that become locked. This argument is illuminated by considering one of the most common ways in which mode-locking is accomplished – passive locking with a saturable dye cell in the laser cavity as shown in Fig. 6.8. The dye cell is frequently placed in close proximity to one of the cavity mirrors.

The oscillation threshold for an individual cavity mode in this arrangement will be high because of the intracavity loss presented by the absorbing dye. However, if the laser oscillates in a "bouncing-pulse" mode in which the intracavity intensity is concentrated, then this pulse can "bleach" the dye.[3]

[3] A dilute dye becomes less absorbing at high light intensities, because the stimulated absorption and emission of dye molecules become balanced as the ground state of the dye is depleted by excitation.

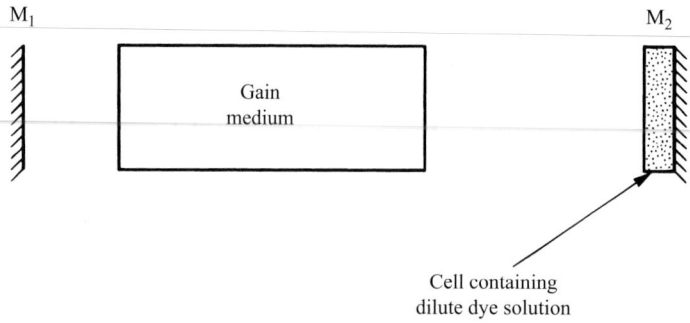


**Fig. 6.8** A schematic diagram for passive mode-locking with a saturable dye.

In this situation a bouncing pulse is more energy efficient in overcoming the intracavity loss presented by the dye. The actual shape of the pulse will be determined by a self-consistency condition. The pulse train $f(t)$ will have a Fourier transform (FT) $F(\omega)$ that describes the frequency content of the bouncing pulse. Each of the frequency components of this effective pulse train will be amplified by the gain medium according to a saturated gain profile $g_s(\omega)$. It is clear that the *narrowest* width that the bouncing pulse can have is $\sim 1/\Delta\nu$, where $\Delta\nu$ is the width of the gain profile. In common with any amplifier, frequencies outside the gain bandwidth are not amplified. Schematically, we could write a self-consistency equation

$$\mathrm{FT}[f(t)] = F(\omega) g_s(\omega) \alpha_d(\omega), \tag{6.17}$$

where $\alpha_d(\omega)$ is the effective absorption spectrum of the bleached dye. This explanation does not impose an inhomogeneously broadened description of the gain medium. Any broadband gain profile can, in principle, amplify a short pulse. Of course, once we accept the bouncing-pulse model of mode-locking we can examine $F(\omega)$ and will find that the principal components of the Fourier transform will correspond to longitudinal mode frequencies.[4] Mode-locked pulses frequently have a $\mathrm{sech}^2$ shape.

## 6.5 Methods of mode-locking

In general, mode-locked behavior is caused to occur by modulating the gain or loss of the laser cavity in a periodic way, usually at a frequency $f_m = c/(2L)$. In amplitude-modulation (AM) mode-locking the magnitude of the cavity loss (or gain) is modulated; in phase (or frequency) modulation (FM) mode-locking only the complex part of the gain is modulated. In *active* mode-locking the modulation is introduced by external means, for example, with an intracavity modulator. In *passive* or *self*-mode-locking the modulation is created by the bouncing mode-locked pulse itself.

[4] In a sense this is an example of what comes first – *the chicken or the egg?*

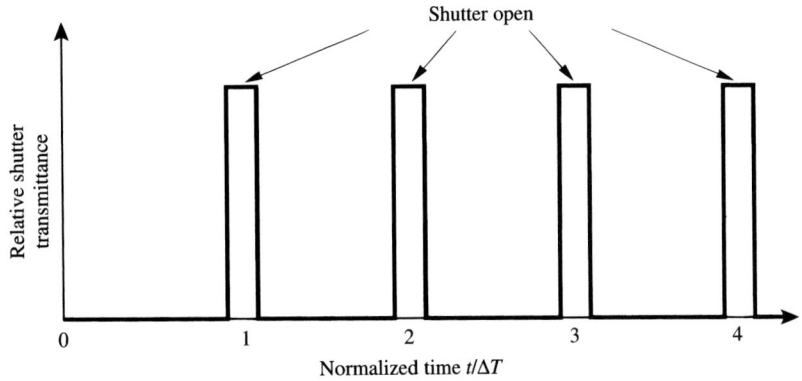


Fig. 6.9 The time sequence of opening of an intracavity shutter used to produce AM mode-locking.

### 6.5.1 Active mode-locking

In the simplest form of AM mode-locking the intracavity loss is switched periodically with an intracavity shutter as shown schematically in Fig. 6.9. The precise time variation of the periodic intracavity loss is not very important, except that its period should equal the round-trip time, $\Delta T$, for an optical pulse bouncing back and forth in the cavity. It is energy advantageous for a bouncing pulse to develop, rather than one or more CW longitudinal modes, since the pulse will adjust its arrival time at the intracavity modulator to correspond to the time of maximum transmission. An alternative way of viewing what happens is to realize that the modulator generates new frequency components in a longitudinal mode of frequency $\nu_m$ passing through it. These new frequencies are called *side-bands*[5] and have values

$$\nu_k = \nu_m + kf_{\mathrm{m}} \quad k = 0, \pm 1, \pm 2 \ldots. \tag{6.18}$$

Each of these side-bands also corresponds to another longitudinal cavity mode with frequency $\nu_{(m+k)}$.

The interaction between the fundamental longitudinal modes and side-bands causes the phases of the longitudinal modes to lock together and generate a mode-locked pulse train. In an inhomogeneously broadened laser the existence of several independent longitudinal modes whose phases can become locked is clear. In a homogeneously broadened laser in which only one such mode *should* oscillate, the side-bands of the dominant mode are not suppressed by gain saturation: they compete for gain and a self-consistent bouncing pulse results. Almost all of the lasers that are used to generate very short mode-locked pulses are (ideally) homogeneously broadened: for example Nd:YAG, Nd:glass, dye, and semiconductor lasers.

A common method of intracavity AM mode-locking is to use an acousto-optic modulator.[6] This was the method used in the first experimental observation of mode-locking using

[5] For a more detailed discussion see Chapter 18.

[6] A full discussion of these devices is given in Chapter 18.

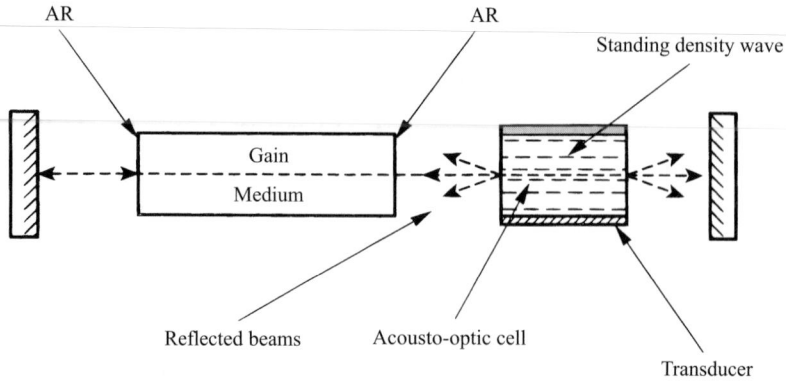


**Fig. 6.10** The arrangement for mode-locking with an intracavity acousto-optic modulator (AOM). AR, antireflection-coated face.

a helium–neon laser [4]. The acousto-optic modulator can be used to deflect light from the cavity. The undeflected beam is thereby amplitude-modulated.

In mode-locking applications the acousto-optic modulator is used in a standing-sound-wave configuration; the acousto-optic modulator material is designed to resonate for sound waves at frequency $f_s$.[7] For an acousto-optic medium used in this way the variation of refractive index is

$$n(z,t) = n_0 + \Delta n \cos(2\pi f_s t)\cos(k_s z). \tag{6.19}$$

This standing wave acts like a phase-diffraction grating that deflects light. The light traveling orthogonally to the sound wave is deflected at frequency $2f_s$, because the maximum index variation in Eq. (6.19) occurs twice per cycle of the standing sound wave. A typical experimental arrangement for mode-locking with an intracavity acoustic cell is shown in Fig. 6.10.

Mode-locking can also be induced through periodic variation of the gain of the amplifying medium. In a semiconductor laser (see Chapter 12) this can be simply done by modulating the drive current to the laser at frequency $f_m$. Alternatively, one periodically modulated laser can be used to mode-lock another. This approach is called *synchronous* pumping. For example, an argon laser that is itself mode-locked, but does not generate very short mode-locked pulses, can be used to periodically pump a dye laser (an optically excited laser that uses a dilute organic dye solution as its gain medium – see Chapter 11). The dye laser generates much shorter mode-locked pulses than were injected from the pump laser. The success of this scheme requires that the cavities of pump and sychronously mode-locked lasers be matched so that

$$f_m(\text{pump}) = \frac{1}{T}, \tag{6.20}$$

where $T$ is the cavity round-trip time in the pumped laser. Synchronous pumping allows the use of convenient mode-locked pumps, such as Nd:YAG or argon-ion lasers, which do not

[7] This is in contrast to acousto-optic frequency shifter applications, where one end of the crystal is terminated in a matched acoustic load or absorber so that only a unidirectional traveling sound wave results.

themselves intrinsically generate the shortest mode-locked pulses, to pump dyes or doped glasses and crystals that support a large gain bandwidth.

In the ideal case the length of the mode-locked pulses generated by any laser, whether it be homogeneously or inhomogeneously broadened, can approach a value $\tau \sim 1/\Delta\nu$, where $\Delta\nu$ is the width of the gain profile. Consequently gain media with large values of $\Delta\nu$ are desirable for the shortest mode-locked pulse generation; these include dye lasers [5, 6], Nd:glass lasers [7, 8], and titanium–sapphire lasers [9–12]. The gain bandwidth in the titanium–sapphire laser is about 300 nm (128 THz) wide around 800 nm. Typical pulse lengths generated by such lasers are in the picosecond (ps) range (1 ps = $10^{-12}$ s), although special configurations achieve sub-picosecond values. Figure 6.11 shows a good example of a mode-locked laser system that can generate sub-picosecond pulses. In this scheme a CW argon-ion laser pumps an acousto-optically mode-locked Nd:phosphate glass

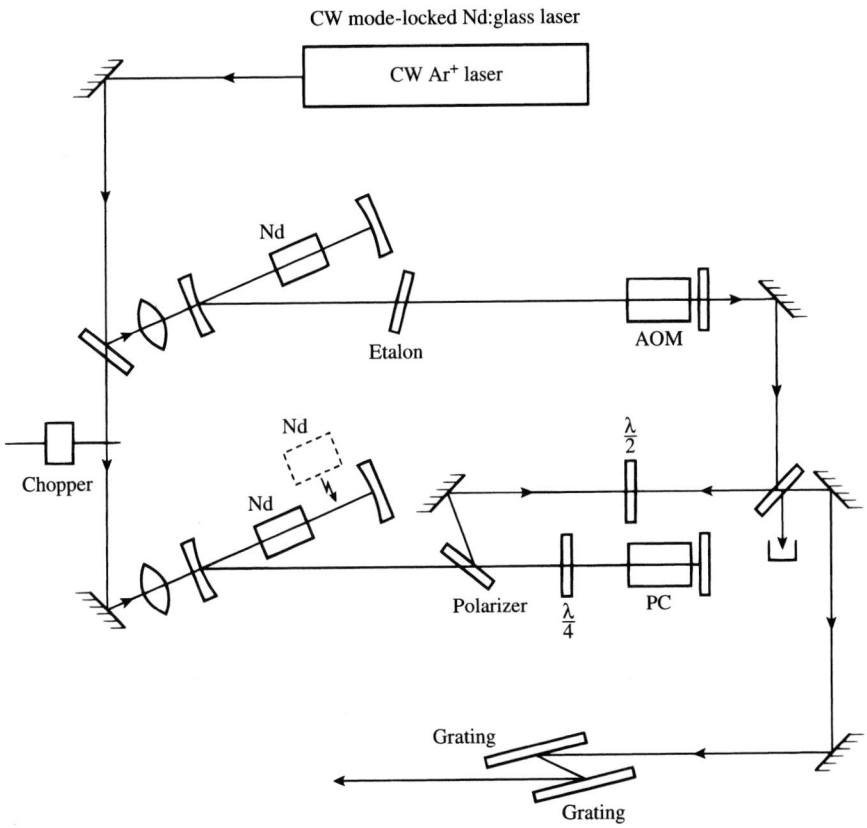


**Fig. 6.11** Generating ultrashort pulses from a neodymium–glass laser system [7]. The CW $Ar^+$ laser pumps an Nd:phosphate glass laser that is mode-locked with an acousto-optic modulator (AOM). The mode-locked pulses are further amplified in a *regenerative* amplifier, where they may make at least 60 round trips before being switched out of the cavity with the Pockels cell (PC) (an electro-optic crystal that can switch the polarization state of the beam). (Courtesy of Professor Chi H. Lee.)

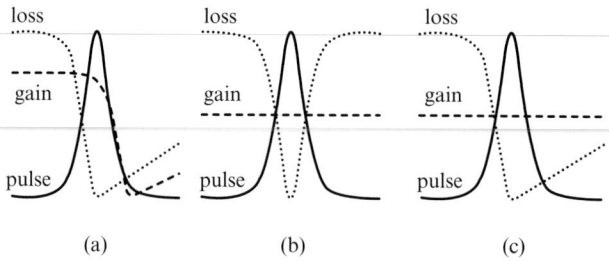


**Fig. 6.12** The three interaction mechanisms that lead to pulse shortening in passive mode-locking. (a) Passive mode-locking with dynamic gain saturation and a slow saturable absorber. (b) With a fast saturable absorber. The shorter the mode-locked pulse becomes the faster the absorber introduces loss. The gain stays approximately constant because gain saturation is weak. (c) With a slow saturable absorber, where the loss increases rapidly at the leading edge of the pulse, but then recovers more slowly.

oscillator. The mode-locked pulses are then further amplified by causing them to make multiple passes inside a Nd:phosphate glass *regenerative* amplifier [8].[8]

### 6.5.2 Passive mode-locking with saturable absorbers

Passive mode-locking has become the dominant technique for the generation of the shortest laser pulses. It produces much shorter pulses than active mode-locking, whose ability to produce short pulses is limited by the speed of external electronic modulators. In passive mode-locking a saturable absorber or nonlinear medium allows the generation of pulses whose duration is controlled by the pulse itself. The first observations of passive mode-locking used saturable absorbers such as intracavity absorbing dye cells whose absorbance was intensity-dependent [13, 14], and these techniques evolved over time to ever shorter pulse generations and cavity configurations. When the dye is bleached it recovers its absorption as the excited population relaxes. This process may be slower than the length of the mode-locked pulse. Femtosecond pulses cannot be generated using a saturable absorber alone. This also requires dynamic gain saturation. In essence, the saturable dye suddenly bleaches to create the leading edge of the mode-locked pulse and the pulse is shortened as it passes through the gain medium, whose gain is saturated by the intense mode-locked pulse and thereby shortens the falling edge of the pulse [15, 16]. The different mode-locking concepts are illustrated in Fig. 6.12.

It is usual in a mode-locked system using a linear resonator to place the intracavity modulator or absorbing dye close to one end mirror to ensure that the laser pulse passes the modulator only once in every round trip. If this is not done, then more than one bouncing pulse can develop within the cavity, and the output will not be a regular pulse train of spacing $2\ell/c$. In laser cavities configured in the form of a ring, the placement of the intracavity modulator or absorber is not so important. In this case it is possible for two counterpropagating pulses to develop in the ring. In ring and linear resonators where two bouncing

[8] Synchronous mode-locking can also be achieved if the optical lengths of the *pump* laser cavity and the *pumped* laser cavity have a ratio that is of the form $L_{\text{pump}}/L_{\text{pumped}} = p/p'$, where $p$ and $p'$ are small integers.

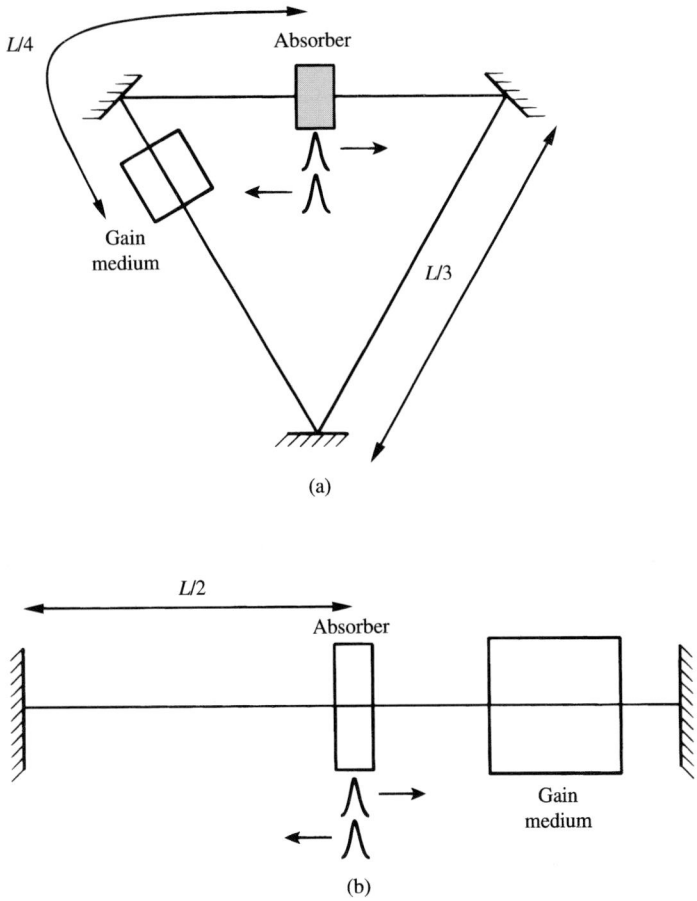


Fig. 6.13 Schematic placement of a saturable absorber for CPM mode-locking: (a) ring cavity and (b) linear cavity.

pulses develop, it is possible to cause the pulses to overlap or *collide* as they pass through an intracavity saturable absorber (dye). Because of the nonlinear variation of the absorption of the dye with intensity the superimposed counterpropagating short pulses actually become shorter in order to maximize bleaching of the dye. In this scheme, called colliding pulse mode-locking (CPM), the typical placement of the absorber is at the center of a linear cavity or one quarter perimeter away from the gain medium in a ring cavity as shown in Fig. 6.13. CPM lasers can directly generate mode-locked pulses of sub-picosecond generation [6, 17].

The shortest pulses generated to date in a ring-cavity CPM are 27 femtoseconds (fs) long [18] (1 fs = $10^{-15}$ s).

### 6.5.3 Passive mode-locking with nonlinear media

Attempts to reduce the length of mode-locked pulses from dye-based saturable absorbers into the few-femtosecond and attosecond range (1 attosecond = $10^{-18}$ s) eventually proved

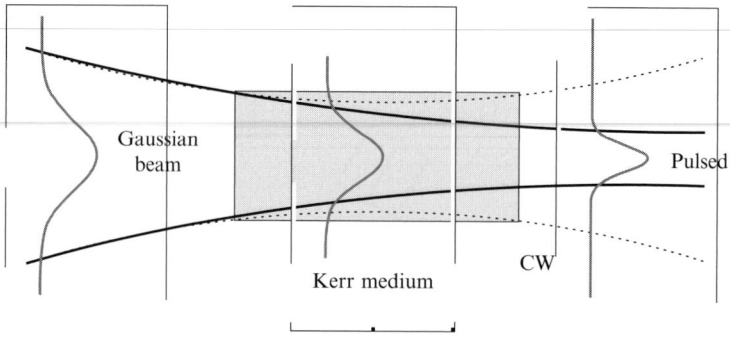


Fig. 6.14 The arrangement for hard-aperture Kerr mode-locking.

difficult and led to the development of Kerr-lens mode-locking (KLM) [11], additive-pulse mode-locking (APM) [19–21], and the semiconductor saturable absorbed mirror (SESAM) [22, 23]. The systems used in KLM and APM act like saturable absorbers using intracavity nonlinear media.[9] They do not rely on inter-level excitation, and thus can act as very fast saturable absorbers. In KLM an intracavity nonlinear element, whose refractive index is intensity-dependent, can lead to mode-locking in a suitable designed cavity. The nonlinear element may be the gain medium itself in, for example, a titanium–sapphire [11] or Nd:YAG laser [24], or a slab of glass that exhibits a sufficient Kerr effect. The Kerr effect, or quadratic electro-optic effect, occurs in all materials to some extent. One consequence of the effect is that an intense light beam changes the refractive index of a medium though which it passes according to $n = n_0 + \gamma I$, where $I$ is the intensity. The coefficient $\gamma$ is very small, typically on the order of $10^{-20}$ m$^2$ W$^{-1}$ for glasses. If a Gaussian beam passes through such a medium, the index is increased most on the beam axis, and the Gaussian beam focuses itself. This *self-focusing* effect can allow an intense beam to pass through an aperture placed inside the laser cavity, which favors mode-locked over CW operation. This is called *hard-aperture* KLM, which is illustrated in Fig. 6.14. In *soft-aperture* (SA) KLM the self-focusing effect can provide better overlap between the cavity mode and the pump beam in a laser-pumped laser.[10] Figure 6.15 shows the SAKLM arrangement used by Juang *et al.* [25]. This figure introduces a number of resonator concepts that will be dealt with again later. The gain medium is a titanium–sapphire laser pumped by the beam from an argon-ion laser. The laser rod has Brewster-angled ends and lies inside a folded resonator consisting of mirrors $M_1$, $M_2$, $M_3$, and $M_4$. The angles and distances in the resonator are chosen carefully so as to avoid astigmatism.[11] The two prisms $P_1$ and $P_2$ provide *group-velocity dispersion* [26]. Without this compensation for the variation of refractive index with frequency of the various transparent components and mirror coatings in the laser cavity different frequency components of a short-pulse laser pulse would have different round-trip times in the cavity.

[9] See Chapter 20.

[10] See Sections 7.5 and 7.11.

[11] Astigmatism in this case is the optical aberration that results when light reflects at an angle from a spherical mirror.

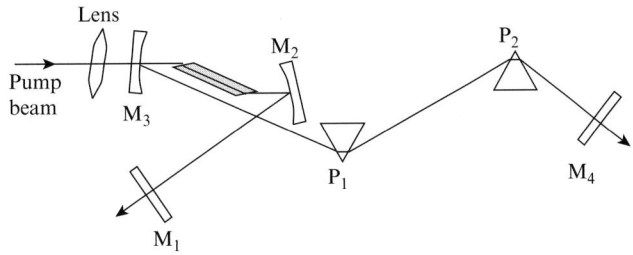


Fig. 6.15 The arrangement for soft-aperture Kerr mode-locking [25].

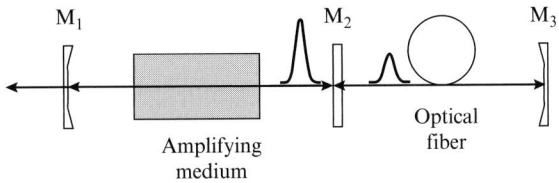


Fig. 6.16 Coupled cavities for additive-pulse mode-locking.

Additive-pulse mode-locking (sometimes called coupled-cavity mode-locking) uses an external cavity coupled to the main laser cavity, which contains a length of single-mode optical fiber (see Chapter 16), as shown in Fig. 6.16. The fiber produces nonlinear phase shifts so that pulses returning from the external cavity interfere with pulses in the main cavity. The lengths both of the main cavity and of the external cavity are matched so that the round-trip times in them are equal. With correct cavity adjustment, two coherent pulses with a relative phase shift interfere so that their wings cancel out but their fields add near the peak of the pulse. This reduces the width of the pulses emerging from the system.

SESAM mode-locking uses a mirror with multiple alternating thin layers of two different semiconductor materials such as AlAs and AlGaAs, which acts as a solid-state saturable absorber [27]. At the same time a properly designed SESAM device suppresses unwanted $Q$-switching in mode-locked solid-state lasers. SESAM mode-locking has allowed GHz pulse-repetition frequencies from mode-locked diode-pumped solid-state (DPSS) lasers.[12]

## 6.6 Pulse compression

Any short pulse can be represented by its Fourier transform

$$F(\omega) = \mathrm{FT}\{f(t)\}. \tag{6.21}$$

[12] For further discussion of DPSS lasers see Section 7.11.

The Fourier transform describes the spectral composition of the pulse $f(t)$ in both amplitude and phase. For example,

$$F(\omega) = A(\omega)e^{j\phi(\omega)}. \tag{6.22}$$

If we pass our original short pulse through a system in which the optical delay varies with frequency as $\Delta\phi(\omega)$, then the emerging pulse will have a Fourier transform

$$F'(\omega) = F(\omega)e^{j\Delta\phi(\omega)}. \tag{6.23}$$

If the lower-frequency components suffer a greater phase shift than the higher-frequency components, this is called a negative group-velocity delay. In this case the output pulse, $g(t)$, will be

$$g(t) = \mathrm{FT}\{F'(\omega)\}. \tag{6.24}$$

If the original pulse is written as

$$f(t) = a(t)e^{-j[\omega_0 t+\phi(t)]}, \tag{6.25}$$

where $a(t)$ is a smooth, slowly varying function of time (the pulse envelope), then the *instantaneous* frequency of the pulse is

$$\omega(t) = \omega_0 + \frac{d\phi}{dt}. \tag{6.26}$$

If $\phi(t)$ is nonlinear, the pulse is said to be *chirped*. If

$$\phi(t) = bE^2, \tag{6.27}$$

the frequency chirping is linear, and, with $b > 0$, is positive. The leading edge of the pulse has a lower instantaneous frequency than the trailing edge – it is red-shifted relative to the blue-shifted trailing edge. In a medium with a negative group-velocity delay characteristic, the lower frequencies in the leading edge of the pulse travel more slowly than the higher frequencies in the trailing edge. The back of the pulse catches up with the front and the pulse is compressed, as shown schematically in Fig. 6.17.

A simple pulse compressor can be constructed with two diffraction gratings as shown in Fig. 6.18 [28]. The first-order-diffraction angle $\theta$ satisfies

$$\sin\theta = \frac{\lambda}{d} - \sin\theta_{\mathrm{in}}. \tag{6.28}$$

The longer of two wavelengths diffracts at a larger angle and takes a longer path between the two gratings – thereby acquiring a larger phase shift and satisfying the condition for negative group delay.

An optical fiber can also be used to produce pulse compression. Because of the Kerr effect the refractive index of the fiber varies with intensity.

Accordingly, the peak of a pulse is slowed relative to its tail. There are additional effects on the pulse because of the dispersion in the fiber $n(\lambda)$. If $n(\lambda)$ lies in a region of negative group-velocity dispersion (which for silica fibers lies beyond 1.3 μm) then the index

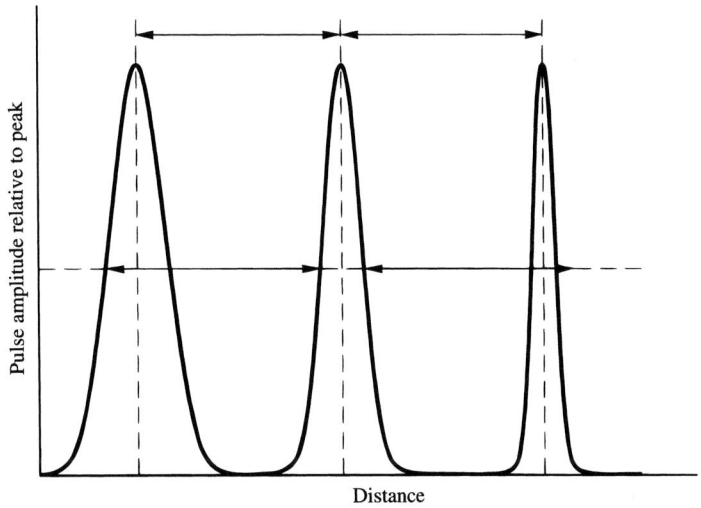


Fig. 6.17 Pulse compression resulting from the traveling edge of a pulse catching up with its leading edge. The arrows in the figure correspond to equal time intervals.

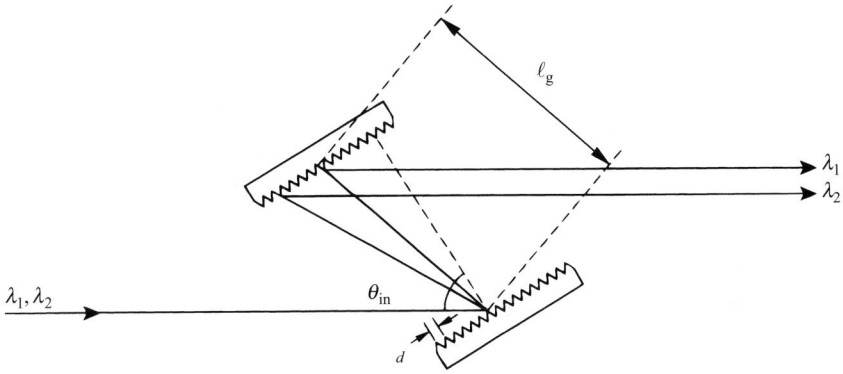


Fig. 6.18 A diffraction-grating pair used to produce pulse compression of a "chirped" pulse.

nonlinearity and dispersion work together to make the pulse shorter.[13] To date pulses have been compressed to as little as 6 fs [29].

## 6.7 Attosecond pulses

The shortest pulses that have been directly generated by a mode-locked Ti–sapphire laser are on the order of 5 fs [30], which corresponds to only about two oscillations of the optical electric field. New techniques involving shorter wavelengths have steadily reduced the

[13] An optical soliton (see Chapter 17) develops for similar reasons.

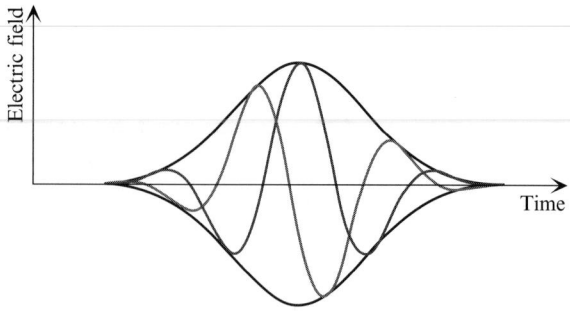


Fig. 6.19 Under the envelope of an attosecond optical pulse there are very few oscillations of the electric field.

length of optical pulses, so that now pulses with durations in the attosecond (as) ($10^{-18}$ s) range are being produced. At 400 nm one optical oscillation takes only 1,334 as. Under the pulse envelope in the few-femtosecond range there are only a few oscillations of the electric field of the light, as shown in Fig. 6.19. To generate shorter pulses the laser wavelength must be reduced, for example by harmonic generation (see Chapter 20). If a short pulse is converted to its second harmonic, then the pulse length is shortened and the wavelength is halved. For example, if a Gaussian pulse of envelope $f(t) = e^{-t^2/\sigma^2}$ is multiplied by itself the result is $f_2(t) = e^{-2t^2/\sigma^2}$, which has an FWHM that is $\sqrt{2}$ times smaller. Pulses as short as 130 as at a photon energy of 36 eV have been reported [31]. These pulses were generated by high-order harmonic generation from a 5-fs 750-nm pulse.

## References

[1] P. W. Smith, "Mode selection in lasers," *Proc. IEEE* **60**, 422–440, 1972.

[2] A. G. Fox and P. W. Smith, "Mode-locked laser and the 180° pulse," *Phys. Rev. Lett.* **18**, 826–828, 1967.

[3] J. A. Fleck, "Mode-locked pulse generation in passively switched lasers," *Appl. Phys. Lett.* **12**, 178–181, 1968.

[4] L. E. Hargrove, R. L. Fork, and M. A. Pollack, "Locking of He–Ne laser modes induced by synchronous intracavity modulation," *Appl. Phys. Lett.* **5**, 4–5, 1964.

[5] C. V. Shank and E. P. Ippen, "Subpicosecond kilowatt pulses from a mode-locked CW dye laser," *Appl. Phys. Lett.* **24**, 373–375, 1974.

[6] R. L. Fork, B. I. Greene, and C. V. Shank, "Generation of optical pulses shorter than 0.1 psec by colliding pulse mode-locking," *Appl. Phys. Lett.* **38**, 671–672, 1981.

[7] D. J. Bradley and W. Sibbett, "Streak-camera studies of picosecond pulses from a mode-locked Nd:glass laser," *Opt. Commun.* **9**, 17–20, 1973.

[8] L. Yan, P.-T. Ho, C. H. Lee, and G. L. Burdge, "Generation of ultrashort pulses from a neodymium glass laser system," *IEEE J. Quant. Electron.* **26**, 2431–2440, 1989.

[9] J. D. Kafka, M. L. Watts, D. J. Roach, M. S. Keirstead, H. W. Schaaf, and T. Baer, "Pulse compression of a mode-locked Ti:sapphire laser," in *Ultrafast Phenomena VII*,

eds. C. B. Harris, E. R. Ippen, G. A. Mourou, and A. H. Zewail, New York: Springer-Verlag, 1990, pp. 66–68.

[10] N. Sarnkara, Y. Ishida, and H. Nakano, "Generation of 50-fsec pulses from a pulse-compressed, CW, passively mode-locked Ti:sapphire laser," *Opt. Lett.* **16**, 153–155, 1991.

[11] D. E. Spence, P. N. Kean, and W. Sibbett, "60-fsec pulse generation from a self-mode-locked Ti:sapphire laser," *Opt. Lett.* **16**, 42–44, 1991.

[12] J. Squier, F. Salin, and G. Mourou, "100-fs pulse generation and amplification in Ti:$Al_2O_3$," *Opt. Lett.* **16**, 324–326, 1991.

[13] A. J. DeMaria, D. A. Stetsen, and H. Heyman, "Self mode locking of lasers with saturable absorbers," *Appl. Phys. Lett.* **8**, 174–176, 1966.

[14] H. W. Mocker and R. J. Collins, "Mode competition and self locking effects in a *Q*-switched ruby laser," *Appl. Phys. Lett.* **7**, 270–273, 1965.

[15] G. H. C. New, "Pulse evolution in mode-locked quasi-continuous lasers," *IEEE J. Quant. Electron.* **10**, 115–124, 1974.

[16] H. A. Haus, "Theory of mode-locking with a slow saturable absorber," *IEEE J. Quant. Electron.* **11**, 736–746, 1975.

[17] P. M. W. French and J. R. Taylor, "Generation of sub-100 fsec pulses tunable near 497 nm from a colliding-pulse mode-locked ring dye laser," *Opt. Lett.* **13**, 470–472, 1988.

[18] J. A. Valdmanis, R. L. Fork, and J. P. Gordon, "Generation of optical pulses as short as 28 femtoseconds directly from a laser balancing self-phase modulation, group-velocity dispersion, saturable absorption, and saturable gain," *Opt. Lett.* **10**, 131–133, 1985.

[19] E. P. Ippen, H. A. Haus, and C. Y. Liu, "Additive pulse mode-locking," *J. Opt. Soc. Am.* **B6**, 1736–1745, 1989.

[20] K. J. Blow and D. Wood, "Mode-locked lasers with nonlinear external cavities," *J. Opt. Soc. Am.* **B5**, 629–632, 1988.

[21] U. Keller, D. A. B. Miller, G. D. Boyd, T. H. Chiu, J. F. Ferguson, and M. T. Asom, "Solid-state low-loss intracavity saturable absorber for Nd–YLF laser – an antiresonant semiconductor Fabry–Perot saturable absorber," *Opt. Lett.* **17**, 505–507, 1992.

[22] U. Keller, K. J. Weingarten, F. X. Kärtner, D. Kopf, B. Braun, I. D. Jung, R. Fluck, C. Hönninger, N. Matuschek, and J. Aus der Au, "Semiconductor saturable absorber mirrors (SESAM's) for femtosecond to nanosecond pulse generation in solid-state lasers," *IEEE J. Selected Topics Quantum Electron.* **2**, 435–453, 1997.

[23] J. Mark, L. Y. Liu, K. L. Hall, H. A. Haus, and E. P. Ippen, "Femtosecond pulse generation in a laser with a nonlinear external resonator," *Opt. Lett.* **14**, 48–50, 1989.

[24] K. X. Liu, C. J. Flood, D. R. Walker, and H. M. van Driel, "Kerr lens mode locking of a diode-pumped Nd:YAG laser," *Opt. Lett.* **17**, 1361–1363, 1992.

[25] D.-G. Juang, Y.-C. Chen, S.-H. Hsu, K.-H. Lin, and W.-F. Hsieh, "Differential gain and buildup dynamics of self starting Kerr lens mode-locked Ti:sapphire laser without an internal aperture." *J. Opt. Soc. Am.* **B14**, 2216–2121, 1996.

[26] N. Langford, R. S. Grant, C. I. Johnston, K. Smith, and W. Sibbett, "Group-velocity-dispersion compensation of a passively mode-locked ring LiF:$F_2^+$ color-center laser," *Opt. Lett.* **14**, 45–47, 1989.

[27] U. Keller, "Ultrafast solid-state laser oscillators: a success story for the last 20 years with no end in sight," *Appl. Phys.* **B100**, 15–28, 2010.

[28] E. B. Treacy, "Optical pulse compression with diffraction gratings," *IEEE J. Quantum Electron.* **5**, 454–458, 1969.

[29] R. L. Fork, C. H. Brito Cruz, P. C. Becker, and C. V. Shank, "Compression of optical pulses to six femtoseconds by using cubic phase compensation," *Opt. Lett.* **12**, 483–485, 1987.

[30] D. H. Sutter, G. Steinmeyer, L. Gallmann, N. Matuschek, F. Morier-Genoud, U. Keller, V. Scheuer, G. Angelow, and T. Tschudi, "Semiconductor saturable-absorber mirror-assisted Kerr-lens mode-locked Ti:sapphire laser producing pulses in the two-cycle regime," *Opt. Lett.* **24**, 631–633, 1999.

[31] G. Sansone, E. Benedetti, F. Calegari, V. Vozzi, L. Avaldi, R. Flammini, L. Poletto, P. Villoresi, C. Altucci, R. Velotta, S. Stagira, S. De Silvestri, and M. Nisoli, "Isolated single-cycle attosecond pulses," *Science* **314**, 443–446, 2006.

# 7 Optically pumped solid-state lasers

## 7.1 Introduction

In this chapter we shall discuss in some detail the operating principles, characteristics, and design features of solid-state lasers in which the laser medium is an insulating or glassy solid. In many of these lasers the active particles are impurity ions doped into a host matrix. These lasers are pumped optically, with a pulsed or continuous lamp, and most commonly by another laser. Our discussion will build on the brief introduction to one laser of this class, namely the ruby laser, given in Chapter 3. The chapter will conclude with a discussion of the characteristics of the radiation emitted by such lasers and how this radiation can be modified and controlled in time.

## 7.2 Optical pumping in three- and four-level lasers

Optical pumping in an insulating solid-state laser is illustrated schematically in Fig. 7.1. Light from the pumping lamp(s) excites ground-state particles into an absorption band, labeled 3 in the figure. Ideally, particles that reach this state should transfer rapidly into the upper laser level, level 2. If transfer occurs preferentially to level 2 rather than to level 1, a population inversion will result between levels 2 and 1, and laser action can be obtained. The drain transition from level 1 back to the ground state should be fast, in order to keep level 1 from becoming a "bottleneck." The performance of the laser will be influenced by several factors.

### 7.2.1 The effective lifetimes of the levels involved

The length of time a particle can remain in an excited level is governed by its *effective* lifetime. This lifetime is influenced both by radiative and by non-radiative processes. The transition from level 3 to level 2 in Fig. 7.1 is generally non-radiative. Particles make the transition by dumping their excess energy into the lattice – they literally heat up the medium. Such a non-radiative process involving transfer from level $i$ to level $j$ can be described by a rate coefficient $X_{ij}$, which is similar to a spontaneous emission coefficient $A_{ij}$. The overall rate at which particles leave level $i$ is

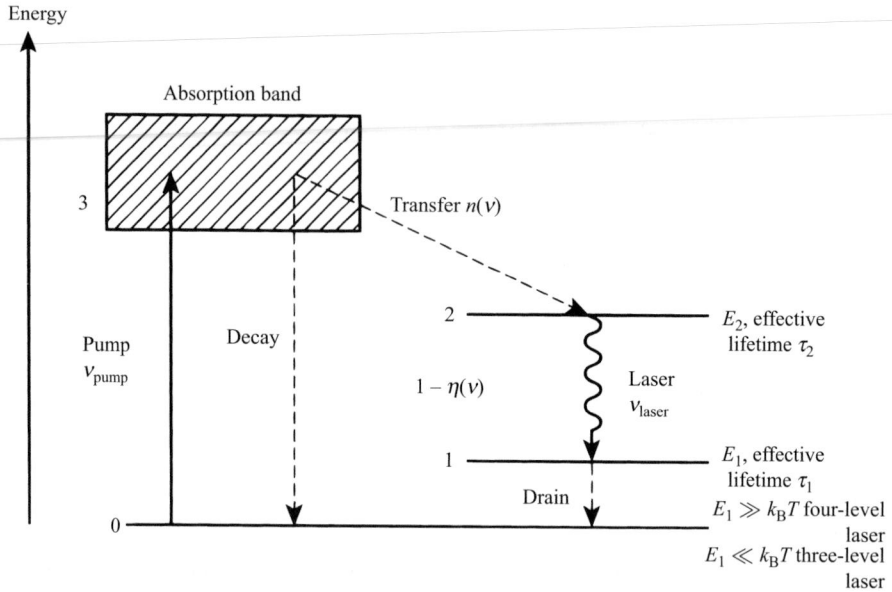


**Fig. 7.1** A schematic diagram of a general optically pumped laser system. In a good laser system the transfer coefficient should be close to unity and transfer should occur rapidly.

$$\frac{dN_i}{dt} = -\sum_j N_i A_{ij} - \sum_j N_i X_{ij}, \tag{7.1}$$

where the summation in $j$ runs over all levels below level $i$ in energy. The effective lifetime of level $i$ is

$$\tau_i = 1 \Big/ \sum_j (A_{ij} + X_{ij}). \tag{7.2}$$

The non-radiative decay of an excited state can be thought of as a collisional process, in a similar way to stimulated emission, in which a quantized packet of acoustic energy within the solid, called a *phonon*, collides with the excited particle. These quantized acoustic energy packets always exist in a solid, even at absolute zero: they correspond closely to the waves we considered in our discussion of black-body radiation, except that the modes being counted are now vibrational motions of the particles constituting the solid. The perturbation of excited states by these phonons leads to line broadening, which can be substantial. Broad absorption bands in a solid result from the line-broadened "smearing" together of levels that would be sharp and distinct in the vapor phase. The vastly greater number of neighboring particles in a solid (or liquid) than in a gas causes large amounts of line broadening, although the extent of the broadening can vary greatly from level to level.

### 7.2.2 Threshold inversion in three- and four-level lasers

If the lower laser level is very close to the ground state ($E_1 \ll k_B T$), or is the ground state, the system is a three-level laser. In such a laser the threshold inversion is substantially higher than in a four-level system ($E_1 \gg k_B T$). If the total number of particles per unit volume participating in the pumping and lasing process is $N$ we can write

$$N = N_3 + N_2 + N_1 + N_{gs}, \tag{7.3}$$

where the subscripts indicate the level involved, and $N_{gs}$ is the population density in the ground state. Equation (7.3) assumes that other levels of the system are not actively involved and that their populations are negligible.

In a three-level laser level 1 is essentially the ground state, so we can write

$$N = N_3 + N_2 + N_1. \tag{7.4}$$

In a good solid-state laser, level 3 transfers its excitation very rapidly to level 2, so its population can be neglected. In this case

$$N = N_2 + N_1. \tag{7.5}$$

The threshold inversion is

$$N_t = N_2 - \frac{g_2}{g_1} N_1. \tag{7.6}$$

From Eqs. (7.5) and (7.6)

$$(N_2)_{3\text{-level}} = \frac{(g_2/g_1)N + N_t}{g_2/g_1 + 1}, \tag{7.7}$$

which, if $g_2 = g_1$, gives $N_2 \simeq N/2$.

For a good four-level laser in which the lower laser level remains relatively depopulated we can write

$$(N_2)_{4\text{-level}} \simeq N_t.$$

Consequently, the relative rate at which level 2 must be excited to produce an inversion (all other factors such as decay rates being comparable) is

$$\frac{(N_2)_{3\text{-level}}}{(N_2)_{4\text{-level}}} \simeq \frac{(g_2/g_1)N + N_t}{(g_2/g_1 + 1)N_t}. \tag{7.8}$$

Usually, $N \gg N_t$, and, if we take the simple case $g_2 = g_1$, the ratio of pumping rates becomes $N/(2N_t)$. This number can easily be $10^4$ or more, demonstrating once again how surprising it was that a three-level laser was the first ever to be operated.

### 7.2.3 Quantum efficiency

If the average energy of photons from the pumping lamp is $h\overline{\nu}_{\text{pump}}$ and the laser photon is $h\nu$, the intrinsic quantum efficiency of the pumping process is

$$\eta_i = \frac{h\nu}{h\overline{\nu}_{\text{pump}}}. \tag{7.9}$$

### 7.2.4 Pumping power

The rate at which particles must be excited to the upper laser level to sustain a population inversion is

$$R_2 = N_2/\tau_2 \text{ (particles m}^{-3}\text{ s}^{-1}). \tag{7.10}$$

If the average probability factor for a particle in level 3 transferring to level 2 rather than elsewhere is $\overline{\eta}$ (called the branching factor) then the rate at which level 3 must be pumped is

$$R_3 = \frac{N_2}{\overline{\eta}\tau_2}. \tag{7.11}$$

The corresponding absorbed pump power is

$$P_{\text{A}} = \frac{N_2 h\overline{\nu}_{\text{pump}}}{\overline{\eta}\tau_2}. \tag{7.12}$$

### 7.2.5 Threshold lamp power

To determine the power and spectral characteristics of the lamp(s) needed to create an inversion, we must relate the threshold pumping rate $R_2$ to the absorption coefficient in the pump band, and electrical and geometrical factors that determine how efficiently the lamp generates pump light and couples this into the laser medium.

The absorption band has a lineshape function $g_3(\nu)$. If the energy density of the pump radiation in the laser medium is $\rho_{\text{P}}(\nu)$, then the rate at which level 3 is excited is

$$R_3 = \int N_0 B_{03} g_3(\nu)\rho_{\text{P}}(\nu)d\nu. \tag{7.13}$$

If we assume plane-wave illumination of the laser medium then we can write $\rho(\nu) = I(\nu)/c$, and Eq. (7.13) becomes

$$R_3 = \int N_0 \frac{c^2 A_{30}}{8\pi\nu^2}\frac{I(\nu)}{h\nu} g_3(\nu)d\nu. \tag{7.14}$$

From Eq. (7.14) we can recognize $N_0 c^2 A_{30} g_3(\nu)/(8\pi\nu^2)$ as the absorption (negative gain) coefficient of the laser medium $\alpha(\nu)$. So

$$R_3 = \int \frac{I(\nu)\alpha(\nu)d\nu}{h\nu}. \tag{7.15}$$

The integration covers the range of frequencies encompassed by the absorption band.

If we allow for the possibility of a frequency-dependent branching factor, the rate of excitation of level 2 is

$$R_2 = \int \frac{I(\nu)\alpha(\nu)\eta(\nu)}{h\nu}\, d\nu. \tag{7.16}$$

If the absorption band over which the integration is carried out is narrow, of width $\Delta\nu_3$, we can replace the quantities in (7.16) by averages and write

$$R_2 = \bar{I}(\nu)\overline{\alpha}(\nu)\overline{\eta(\nu)}\Delta\nu_3/(h\overline{\nu}_{\text{pump}}). \tag{7.17}$$

## 7.3 Pulsed versus continuous-wave operation

If the pumping lamp is a flashlamp producing pulses of duration $t_{\text{p}} \ll \tau_2$, then, at the end of the flash, all the excited particles will still be in level 2. In this case

$$N_2 = t_{\text{p}}\overline{I(\nu)\alpha(\nu)\eta(\nu)}\Delta\nu_3/(h\overline{\nu}_{\text{pump}}). \tag{7.18}$$

The flashlamp energy needed to achieve $N_2 \geq N_{\text{t}}$ depends additionally on the electrical efficiency $e$ of the lamp, which is defined as

$$e = \frac{\text{joules of light energy out}}{\text{capacitor joules in}}, \tag{7.19}$$

(for xenon flashlamps this efficiency factor can be as high as 80%) [1], the fraction $f$ of this light in the spectral region that will pump the absorption band,

$$f = \frac{\text{light energy within absorption band}}{\text{total light energy}}, \tag{7.20}$$

and the geometrical efficiency $g$ with which the light is coupled to the laser medium, which usually takes the form of a cylindrical rod or flat slab:

$$g = \frac{\text{light energy within absorption band reaching laser medium}}{\text{total light energy within absorption band}}. \tag{7.21}$$

The total energy that must reach the laser crystal, in the right spectral region, to reach threshold can be written as

$$U_{\text{t}} = t_{\text{p}}\overline{I(\nu)} = \frac{N_{\text{t}}h\overline{\nu}_{\text{pump}}}{\overline{\alpha}(\nu)\overline{\eta(\nu)}}. \tag{7.22}$$

### 7.3.1 Threshold for pulsed operation of a ruby laser

In the ruby laser the typical $Cr^{3+}$-ion concentration is ~0.05% by weight, equivalent to about $10^{25}$ $Cr^{3+}$ ions $m^{-3}$. Since this is a three-level laser, roughly half of these ions must be excited to the upper laser level to reach threshold. Therefore, we can take $N_{\text{t}} = 5 \times 10^{24}\ \text{m}^{-3}$. Figure 3.5 shows that the approximate average absorption coefficient

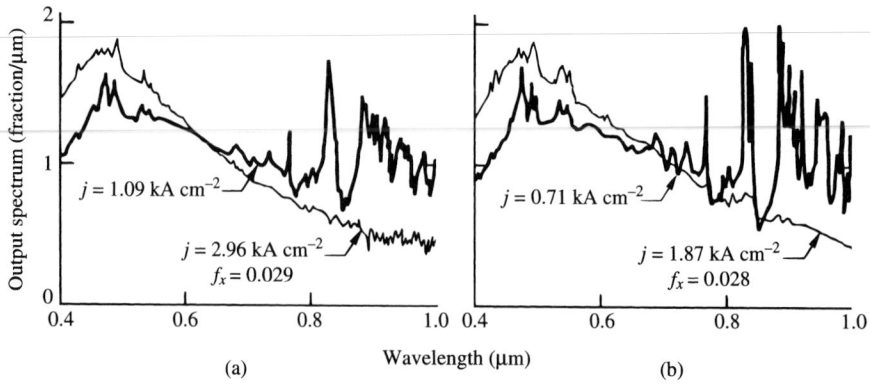


**Fig. 7.2** Examples of output spectra from various flashlamps at different flashlamp current densities; $f_x$ is the fraction of explosion energy at which each lamp was operated: (a) 1.5-cm-bore lamp and (b) 4.2-cm-bore lamp [2].

of ruby in the 350–600-nm region is about 100 $m^{-1}$. Assuming efficient transfer from the pump bands, we take $\overline{\eta(\nu)} = 1$. The average pump wavelength is $\simeq$475 nm. Therefore,

$$U_t \simeq \frac{5 \times 10^{24} \times 6.66 \times 10^{-34} \times 3 \times 10^8}{100 \times 475 \times 10^{-9}} = 21 \text{ kJ m}^{-2}.$$

For a cylindrical ruby crystal 20 mm long and 10 mm in diameter, if the lamp energy reaches the crystal uniformly over its curved surface, the threshold input energy reaching the crystal in the appropriate spectral band is about 13 J. For a modern flashlamp the electrical efficiency $e$ can be 50% or more. The spectral output of the lamp will approximate a black body with superimposed spectral features, as shown in Fig. 7.2. It is reasonable to assume that 25% of the emitted light is in the ruby pump bands. If the optical arrangement of lamp(s) and crystal is efficient, the geometrical factor $g$ should be 50% or better. The electrical input to reach threshold will therefore be about 200 J. Small ruby crystals in an efficient pumping arrangement such as the axisymmetric ellipsoidal cavity shown in Fig. 7.3 can have thresholds as low as 34 J.

To achieve CW operation in a solid-state insulating laser the pumping rate of the upper laser level must be sufficient to maintain $N_2 = N_t$ in the face of all the spontaneous relaxation processes. The absorbed power per unit volume must be

$$U_t = \frac{N_t h\overline{\nu}_{\text{pump}}}{\tau_2 \overline{\eta(\nu)}}. \tag{7.23}$$

### 7.3.2 The threshold for CW operation of a ruby laser

Although the ruby laser is not commonly operated in a CW mode, it is interesting to calculate its threshold pump rate. We use the same values for the parameters in Eq. (7.23) as before, with the addition of the value for $\tau_2$, 3 ms, to get

$$U_t = \frac{5 \times 10^{24} \times 6.626 \times 10^{-34} \times 3 \times 10^8}{3 \times 10^{-3} \times 475 \times 10^{-9}} \simeq 7 \times 10^8 \text{ W m}^{-3}.$$

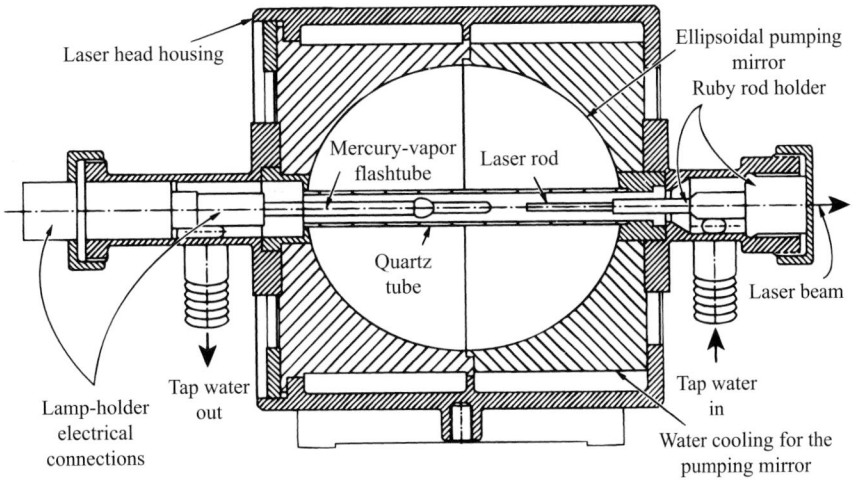


**Fig. 7.3** An ellipsoidal cavity for solid-state laser pumping that provides axisymmetric illumination of a cylindrical laser rod by a cylindrical lamp [6, 7].

In the first CW ruby-laser experiments, Nelson and Boyle [3, 4] obtained threshold pump rates of this order. They had to use an 850-W mercury xenon lamp focused onto a ruby crystal 11.5 mm long by 0.61 mm in diameter kept at liquid-nitrogen temperature inside a Dewar flask to achieve 300 mW of power actually absorbed in the crystal.

## 7.4 Threshold population inversion and the stimulated-emission cross-section

The small-signal gain at the center of the line can be written in the form

$$\gamma_0 = \left(N_2 - \frac{g_2}{g_1} N_1\right) \sigma_0, \tag{7.24}$$

where $\sigma_0$ is called the stimulated-emission cross-section. It serves as a useful parameter for comparing different laser media. From Eq. (2.30) it is clear that, for a homogeneously broadened line, for which

$$g(\nu_0, \nu_0) = \frac{2}{\pi \, \Delta\nu}, \tag{7.25}$$

one has

$$\sigma_0 = \frac{c^2 A_{21}}{4\pi^2 \nu^2 \, \Delta\nu}. \tag{7.26}$$

For the ruby laser $\sigma_0 \simeq 1.2 \times 10^{-24}$ m$^2$.

To achieve threshold the minimum inversion, $N_t$, is

$$N_t = \left(N_2 - \frac{g_2}{g_1}N_1\right)_t = \frac{1}{\sigma_0}\left(\alpha - \frac{1}{\ell}\,\ell n(r_1 r_2)\right). \tag{7.27}$$

Therefore, knowledge of the stimulated-emission cross-section and the parameters of the laser cavity allows quick estimation of the threshold inversion. We shall meet some examples of this later.

## 7.5 Paramagnetic-ion solid-state lasers

Many paramagnetic ions from the iron, rare-earth (lanthanide), and actinide groups of the periodic table exhibit laser action when doped into a large number of host crystals or glasses. In a separate class of solid-state lasers – the so-called *stoichiometric* lasers – the active ion is not an impurity dopant but is an intrinsic part of the lattice. These lasers have not to date achieved significant importance, and we will not discuss them further. The large diversity of doped crystalline lasers is demonstrated by Table 7.1, which lists some of the most important such lasers, together with some of their relevant operating parameters. These are all optically pumped lasers; most are four-level systems. By far the most important is the $Nd^{3+}$ laser, which is operated successfully in a variety of host materials, for example in yttrium aluminum garnet ($Y_3Al_5O_{12}$, YAG), calcium tungstate ($CaWO_4$), lithium yttrium fluoride ($LiYF_4$, YLF), $Ca_5(PO_4)_3F$ (FAP), $YAlO_3$ (YALO), gadolinium gallium garnet (GGG), scandium-substituted GGG (GSGG), and various glasses. Another important laser of this type is the holmium ($Ho^{3+}$) laser in $LiYF_4$ (YLF), frequently with erbium ions ($Er^{3+}$) or thulium ions ($Tm^{3+}$) added to the host to assist in the absorption of flashlamp energy and transfer into the $Ho^{3+}$ upper laser level.

The chromium ($Cr^{3+}$)-in-$BeAl_2O_4$ (alexandrite) laser is interesting since its laser wavelength can be tuned continuously from 700 to 800 nm.

The $Ti^{3+}$ in $Al_2O_3$, the titanium–sapphire laser, is a very attractive source for the generation of tunable near-infrared radiation and has largely replaced dye lasers in this application. Figure 7.4 shows the absorption and emission spectroscopic properties of the $Ti^{3+}$:$Al_2O_3$ material. When pumped by a high-power argon-ion laser, a power output in excess of 1 W can be obtained [5].

Although specific details of their operation differ from one doped insulating crystal or glass laser to other, many of the general features of their construction and operation are similar. The Nd:YAG laser serves as a benchmark for this discussion.

## 7.6 The Nd:YAG laser

YAG has a combination of desirable properties as a host medium for $Nd^{3+}$ ions: it has relatively high thermal conductivity, which allows it to disperse the waste heat from the

**Table 7.1** Characteristics of some optically pumped solid-state laser systems

| Ion | Host | Laser wavelength (μm) | Lower laser level energy ($cm^{-1}$) | Laser transition | Laser ion concentration (%) | $\tau_2$ (ms) | $\Delta\nu$ (GHz) | Pulsed (P) or CW | Operating temperature (K) | Method of pumping |
|---|---|---|---|---|---|---|---|---|---|---|
| Cerium, $Ce^{3+}$ | $LiYF_4$ | 0.286 | Variable (this is a tunable laser) | $5d \rightarrow {}^2F_{7/2}$ | 0.05 | 18 ns | | P | 300 | KrF |
| Neodymium, $Nd^{3+}$ | $Bi_4Ge_3O_{12}$ | 1.0638 | ~2,000 | ${}^4F_{3/2} \rightarrow {}^4I_{11/2}$ | 0.3 | | | P (low threshold) | 77 | Xe |
| | $CaF_2$ | 1.0448 | 2,034 | | 0.4–0.6 | 1.1 | 270 | P | 120 | Xe |
| | $CaF_2$–$YF_3$ | 1.0461 | 2,032 | | 0.2–12 | 0.48 | 900 | P | 300 | Xe |
| | $CaF_2$–$YF_3$ | 1.0632 | 2,170 | | 0.2–12 | 0.48 | 2,700 | P | 300 | Xe |
| | $CaMoO_4$ | 1.061 | ~2,000 | | 1.8 | 0.12 | – | P or CW | 300 | Xe |
| | $Ca_5(PO_4)_3F$ | 1.063 | 1,900 | | 1 | 0.25 | 180 | P or CW | 300 | Xe |
| | $CaWO_4$ | 0.9145 | 471 | ${}^4F_{3/2} \rightarrow {}^4I_{9/2}$ | 0.14–3 | 0.18 | 450 | P | 77 | Xe |
| | | 1.0649 | 2,016 | ${}^4F_{3/2} \rightarrow {}^4I_{11/2}$ | 0.14–3 | 0.18 | 210 | P or CW | 85 | Hg |
| | | 1.3340 | ~3,970 | ${}^4F_{3/2} \rightarrow {}^4I_{13/2}$ | 0.14–3 | 0.18 | 600 | P | 300 | Xe |
| | | 1.3370 | ~3,925 | | 0.14–3 | 0.18 | 690 | CW | 300 | Xe |
| | $CaY_2Mg_2Ge_3O_{12}$ | 1.05986 | 2,008 | ${}^4F_{3/2} \rightarrow {}^4F_{11/2}$ | 2–6 | 0.305 | 1,109 | CW (high efficiency) | 300 | $Ar^+$ laser |
| | $CeF_3$ | 1.0638 | 2,189 | ${}^4F_{3/2} \rightarrow {}^4I_{11/2}$ | ~4 | ~0.27 | ~1,050 | P | 300 | Xe |
| | $GdGa_5O_{12}$ | 1.0621 | 2,064 | | 1 | 0.27 | 216 | P | 300 | Xe |
| | $Gd_3Sc_2Al_3O_{12}$ | 1.06 | 1,978 | | 1 | 0.275 | 345 | CW | 300 | W |
| | $Gd_3Sc_2Ga_3O_{12}$ | 1.0612 | 2,070 | | 1 | 0.26 | 420 | P | 300 | Xe |
| | $LaF_3$ | 1.04065 | 1,983 | | 1–2 | 0.6–0.7 | 750 | P or CW | 300 | Xe |
| | | 1.3310 | 4,070 | ${}^4F_{3/2} \rightarrow {}^4I_{13/2}$ | 2 | 0.6–0.7 | 630 | P | 300 | Xe |

**Table 7.1** (Cont.)

| Ion | Host | Laser wavelength (μm) | Lower laser level energy ($cm^{-1}$) | Laser transition | Laser ion concentration (%) | $\tau_2$ (ms) | $\Delta\nu$ (GHz) | Pulsed (P) or CW | Operating temperature (K) | Method of pumping |
|---|---|---|---|---|---|---|---|---|---|---|
| | $LiYF_4$ | 1.0471 | 2,042 | $^4F_{3/2} \rightarrow {}^4I_{11/2}$ | 2 | 0.5 | 360 | P | 300 | Xe |
| | | 1.0530 | | | 2 | 0.5 | 375 | P | 300 | Xe |
| | $Lu_3Al_5O_{12}$ | 1.06425 | 2,099 | | ~0.6 | 0.245 | 160 | P or CW | 300 | Xe |
| | $SrF_2$ | 1.0370 | 2,008 | | 0.8 | 1.1 | 600 | P | 300 | Xe |
| | $YAlO_3$ | 0.930 | 670 | $^4F_{3/2} \rightarrow {}^4I_{9/2}$ | 1 | 0.18 | 900 | P | 300 | XeF laser |
| | | 1.0795 | 2,157 | $^4F_{3/2} \rightarrow {}^4I_{11/2}$ | 1–3 | 0.18 | 330 | P or CW (low threshold) | 300 | Xe |
| | | 1.0645 | 2,026 | | 1–3 | 0.18 | 285 | P or CW | 300 | Xe |
| | $Y_3Al_5O_{12}$ | 0.8910 | 200 | $^4F_{3/2} \rightarrow {}^4I_{9/2}$ | 1 | 0.255 | | P | 300 | $Ar^+$ laser |
| | | 0.8999 | 311 | | 1 | 0.255 | | P | 300 | $Ar^+$ laser |
| | | 0.9385 | 852 | | 1 | 0.255 | | P | 300 | $Ar^+$ laser |
| | | 0.09460 | 852 | | 1 | 0.255 | 270 | P | 300 | Xe |
| | | 1.06415 | 2,110 | $^4F_{3/2} \rightarrow {}^4I_{11/2}$ | ~1 | 0.255 | 195 | P or CW (low threshold) | 300 | Xe |
| | | 1.0682 | 2,146 | | ~1 | 0.255 | ~300 | P | 300 | Xe |
| | | 1.3188 | 3,222 | $^4F_{3/2} \rightarrow {}^4I_{13/2}$ | ~1 | 0.255 | 195 | CW | 300 | Kr |
| | | 1.3382 | 4,034 | | ~1 | 0.255 | 231 | CW | 300 | Kr |
| | | 1.3564 | ~4,000 | | ~1 | 0.255 | – | CW | 300 | Kr |
| | | 1.4140 | ~4,000 | | ~1 | 0.255 | – | CW | 300 | Kr |

| | | | | | | | | | | |
|---|---|---|---|---|---|---|---|---|---|---|
| | $Y_2O_3$ | ~1.0746 | 1,895 | $^4F_{3/2} \rightarrow {}^4I_{11/2}$ | 1.5 | 0.34 | 240 | CW (very low threshold) | 300 | $Kr^+$ laser |
| | | ~1.358 | 3,837 | $^4F_{3/2} \rightarrow {}^4I_{13/2}$ | 1.5 | 0.34 | 360 | CW (very low threshold) | 300 | $Kr^+$ laser |
| | $Y_3Sc_2Ga_3O_{12}$ | 1.0583 | ~2,000 | $^4F_{3/2} \rightarrow {}^4I_{11/2}$ | 1 | | 240 | P | 300 | Xe |
| | | 1.3310 | ~4,000 | $^4F_{3/2} \rightarrow {}^4I_{13/2}$ | 1 | | 420 | P | 300 | Xe |
| | $YVO_4$ | 1.0625 | 1,964 | | ~1 | 0.092 | | P | 300 | Xe |
| | | 1.0634 | ~2,000 | | ~1 | 0.092 | 210 | CW | 300 | $Ar^+$ laser |
| | | 1.3425 | 3,913 | | 0.1 | 0.033 | 240 | P | 300 | Xe |
| Holmium, $Ho^{3+}$ | $CaF_2$–$ErF_3$ (3%) | 2.06 | ~250 | $^4I_7 \rightarrow {}^5I_8$ | 0.5–1 | | | P | 298 | Xe |
| | $LiYF_4$ | 0.7505 | ~5,300 | $^5S_2 \rightarrow {}^5I_7$ | 2 | 0.09 | 360 | P | 300 | Xe |
| | | 0.9794 | ~5,300 | $^5F_5 \rightarrow {}^5I_7$ | 2 | 0.1 | 360 | P | 90 | Xe |
| | | 1.3960 | ~11,300 | $^5S_2 \rightarrow {}^5I_5$ | 2 | 0.05 | | P | 300 | Xe |
| | $LaYF_4$ (+5% Er) | 2.066 | 300 | $^5I_7 \rightarrow {}^5I_8$ | 2 | | | P | 300 | Xe |
| | $Li(Y, Er)F_4$ (+6.7% Tm) | 2.0654 | ~300 | $^5I_7 \rightarrow {}^5I_8$ | 1.7 | 12 | | P | 300 | Xe |
| | $Lu_3Al_5O_{12}$ (+10% Yb, 0.3% Cr) | 2.9460 | ~5,400 | $^5I_6 \rightarrow {}^5I_7$ | 10 | | 270 | P | 300 | Xe |
| | $Y_3Al_5O_{12}$ | 2.0914 | 532 | $^5I_7 \rightarrow {}^5I_8$ | ~4 | 4–5 | | P | 77 | Xe |
| | | 2.9403 | ~5,400 | $^5I_6 \rightarrow {}^5I_7$ | 10 | | 270 | P | 300 | Xe |
| | $Y_3Al_5O_{12}$ (+50% Er) (+6–7% Tm) | ~2.13 | 532 | $^5I_7 \rightarrow {}^5I_8$ | 1.65 | 5 | | P | 300 | Xe |
| | | | | | | | | CW | 77 | W |

**Table 7.1** (Cont.)

| Ion | Host | Laser wavelength (μm) | Lower laser level energy ($cm^{-1}$) | Laser transition | Laser ion concentration (%) | $\tau_2$ (ms) | $\Delta\nu$ (GHz) | Pulsed (P) or CW | Operating temperature (K) | Method of pumping |
|---|---|---|---|---|---|---|---|---|---|---|
| Erbium, $Er^{3+}$ | $CaF_2$ | 2.7307 | ~6,500 | $^4I_{11/2} \rightarrow {}^4I_{13/2}$ | 4 | | | P | 300 | Xe |
| | $LiYF_4$ | 0.85 | 6,714 | $^4S_{3/2} \rightarrow {}^4I_{13/2}$ | 2 | 0.2 | 300 | P | 300 | Xe |
| | | ~1.2308 | 10,313 | $^4S_{3/2} \rightarrow {}^4I_{11/2}$ | 2 | 0.2 | | P | 300 | Xe |
| | | 1.7320 | 12,572 | $^4S_{3/2} \rightarrow {}^4I_{9/2}$ | 2 | 0.2 | | P | 300 | Xe |
| | | 2.870 | 6,738 | $^4I_{11/2} \rightarrow {}^4I_{13/2}$ | 2 | 10 | | P | 300 | Xe |
| | $Lu_3Al_5O_{12}$ | 0.86325 | 6,818 | $^4S_{3/2} \rightarrow {}^4I_{13/2}$ | 1.5 | 0.13 | 240 | P | 300 | Xe |
| | | 1.6525 | 534 | $^4I_{13/2} \rightarrow {}^4I_{15/2}$ | 2–5 | 6–7 | 195 | P | 77 | Xe |
| | | 1.7762 | 12,772 | $^4S_{3/2} \rightarrow {}^4I_{9/2}$ | 1.5 | 0.11 | | P | 300 | Xe |
| | | 2.8298 | 6,885 | $^4I_{11/2} \rightarrow {}^4I_{13/2}$ | ~33 | 0.12 | | P | 300 | Xe |
| | (+5%Yb, 0.3%Cr) | 2.8298 | ~6,885 | $^4I_{11/2} \rightarrow {}^4I_{13/2}$ | 33 | 0.12 | | P (low threshold) | 300 | Xe |
| | $Y_3Al_5O_{12}$ | 0.8627 | ~6,800 | $^4S_{3/2} \rightarrow {}^4I_{13/2}$ | 2.5 | 0.12 | | P | 300 | Xe |
| | | 2.8302 | ~6,800 | $^4I_{11/2} \rightarrow {}^4I_{13/2}$ | 33 | 0.09 | | P (low threshold) | 300 | Xe |
| Thulium, $Tm^{3+}$ | $LiYF_4$ | 0.4526 | 5,969 | $^1D_2 \rightarrow {}^3F_4$ | 10 | 0.001 | 8,990 | P | 300 | XeF laser |
| | $YAlO_3$ (+1% Cr) | 2.34 | 8,250 | $^3F_4 \rightarrow {}^3H_5$ | 1 | ~0.1 | | P | 300 | Xe |
| | $Y_3Al_5O_{12}$ (+0.1% Cr) | 2.324 | ~8,300 | $^3F_4 \rightarrow {}^3H_5$ | 1 | | | CW | 300 | W |

| | | | | | | | | | | |
|---|---|---|---|---|---|---|---|---|---|---|
| Chromium, $Cr^{3+}$ | $Al_2O_3$ (sapphire) | 0.6943 ($R_1$) | 0 | $^2E(\bar{E}) \rightarrow {}^4A_2$ | ~0.05 | 3 | ~300 | P | 300 | Xe |
| | | 0.6929 ($R_2$) | 0 | $^2E(2\bar{A}) \rightarrow {}^4A_2$ | ~0.05 | 3 | 225 | P | 300 | Xe |
| | | 0.6943 ($R_1$) | 0 | $^2E(2\bar{A}) \rightarrow {}^4A_2$ | ~0.05 | 3 | ~300 | CW | 300 | Hg |
| | $BeAl_2O_4$ (alexandrite) | 0.701–0.818 | Varies | $^4T_2 \rightarrow {}^4A_2$ | | 0.26 | | P | 300 | Xe |
| | | 0.744–0.788 | Varies | $^4T_2 \rightarrow {}^4A_2$ | 0.27 | 0.26 | | CW | 300 | Hg |
| Cobalt, $Co^{2+}$ | $MgF_2$ | 1.63–2.11, tunable | Varies | $^4T_2 \rightarrow {}^4T_1$ | 0.9 | | | CW | 80 | Nd:YAG laser |
| Nickel, $N_i^{2+}$ | $MgF_2$ | 1.674–1.676 | Varies | $^3T_2 \rightarrow {}^3A_2$ | 0.5 | 11.5 | | P | 82–100 | |
| | | 1.731–1.756 | Varies | | | | | | 100–192 | |
| | | 1.785–1.797 | Varies | | | | | | 198–240 | |
| Titanium, $Ti^{3+}$ | $Al_2O_3$ | 0.66–1.1 | Varies | | | | | CW | 300 | $Ar^+$ laser |
| | | 0.66–1.38 | | | | | | P | | Frequency-doubled Nd:YAG laser |

Abstracted from a more comprehensive listing in M. J. Weber, ed., *Handbook of Laser Science and Technology, Vol. I, Lasers and Masers*, Boca Raton, FL: CRC Press, 1982.

$LiYF_4$ is known as YLF, $GdGa_5O_{12}$ as GGG, and $Y_3Al_5O_{12}$ as YAG.

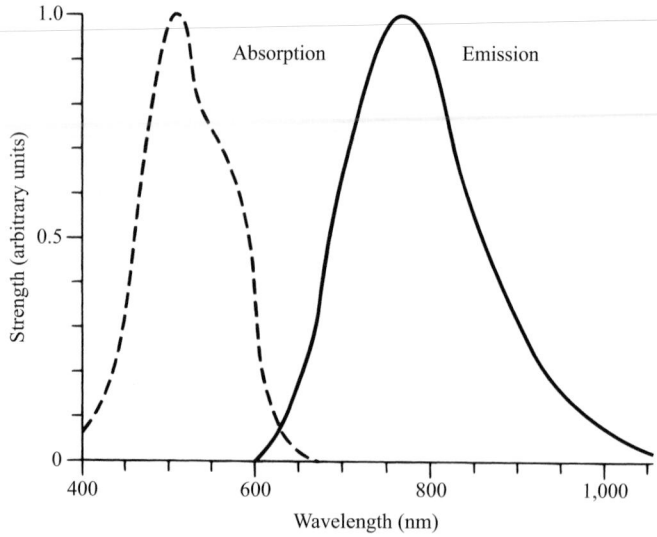


**Fig. 7.4** Absorption and emission properties of titanium-doped sapphire.

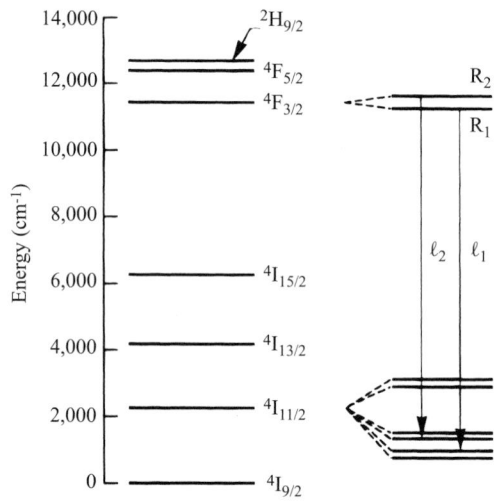


**Fig. 7.5** The energy-level diagram for $Nd^{3+}$:YAG. Emissions $\ell_1$ and $\ell_2$ correspond to the two transitions that make up the laser line [7].

optical pumping process; it has high mechanical strength; and it can be grown as crystals of large size with good optical quality. The $Nd^{3+}$ ions substitute within the YAG lattice in a single site so the emission and absorption lines are homogeneously broadened. Typical $Nd^{3+}$ doping densities range up to 1%.

An energy-level diagram for the $Nd^{3+}$ ions in YAG is shown in Fig. 7.5. The primary pumping process is absorption of lamp energy from the ground state $^4I_{9/2}$ into the $^4F_{5/2}$ level, followed by transfer to the upper laser level $^4F_{3/2}$. This consists of a number of

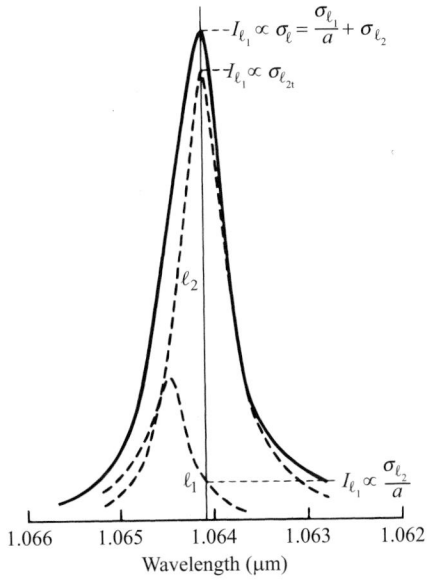


**Fig. 7.6** The spontaneous-emission spectrum of the 1.06-μm transition in $Nd^{3+}$:YAG at room temperature. The two Lorentzian lineshapes that contribute to the laser transition are shown by dashed lines [7]. The factor $a$ is a lineshape factor that gives the contribution to the intensity at the center of $\ell_2$ from the line $\ell_1$.

closely spaced levels. Two of these, labeled $R_2$ and $R_1$ in Fig. 7.5, serve as the upper levels of the closely spaced transitions $\ell_2$ and $\ell_1$. The stronger of these, $\ell_2$, has $\lambda_2 = 1.06415$ μm; the other, $\ell_1$, has $\lambda_1 = 1.0646$ μm. Each of these transitions is substantially homogeneously broadened, with $\Delta\nu \sim 2 \times 10^{11}$ Hz, and because they are so close in frequency, $\sim 10^{11}$ Hz apart, they form a single asymmetric lineshape. This lineshape, shown in Fig. 7.6, contributes to the effective overall gain of the laser, which sees its peak gain near 1.06415 μm. It is interesting in a case like this to compute the effective spontaneous-emission coefficient describing the transition.

### 7.6.1 The effective spontaneous-emission coefficient

At ambient temperature the populations of the two levels $R_2$ and $R_1$ are in thermal equilibrium. The ratio of their populations is the ratio of their degeneracy factors.

$$\frac{N_{R_2}}{N_{R_1}} = \frac{g_{R_2}}{g_{R_1}} = \frac{2}{3}.$$

The effective spontaneous emission coefficient $A_{21}$ is related to the coefficients for $\ell_2$ and $\ell_1$ separately according to

$$A_{21} = A_{\ell_1}\frac{N_{R_1}}{N_{R_2}} + A_{\ell_2}. \tag{7.28}$$

For values $A_{\ell_2} = 1{,}440\ \mathrm{s}^{-1}$ and $A_{\ell_1} \simeq 250\ \mathrm{s}^{-1}$; $A_{21} = 1{,}815\ \mathrm{s}^{-1}$. With these parameters the stimulated-emission cross-section for the Nd:YAG laser is $\simeq 9 \times 10^{-23}\ \mathrm{m}^2$.

### 7.6.2 Example – the threshold pump energy of a pulsed Nd:YAG laser

We take for an example a crystal 50 mm long by 5 mm in diameter. One end of the crystal has a reflectance of 100%, while the other has a reflectance of 92%. For a good-quality crystal $\alpha \simeq 0$. The threshold gain is

$$\gamma_t = \alpha - \frac{1}{\ell}\,\ell n(r_1 r_2) = 0.833\ \mathrm{m}^{-1}.$$

Therefore, from Eq. (7.27) the threshold inversion is $N_t \simeq 9.3 \times 10^{21}\ \mathrm{m}^{-3}$ – more than 500 times smaller than that for the ruby laser. The energy that must be absorbed in order to reach threshold is

$$U_t = \frac{N_t h\overline{\nu}_{\mathrm{pump}}}{\overline{\eta(\nu)}}, \tag{7.29}$$

which, with $\overline{\eta(\nu)} \simeq 1$ and a pump wavelength of 810 nm, gives $U_t = 2{,}280\ \mathrm{J\ m}^{-3}$. For the crystal specified this gives a threshold absorbed energy of 2.24 mJ.

Overall conversion efficiencies from electrical energy to radiant emission in the $Nd^{3+}$ absorption bands may range as high as 15%, so the necessary electrical input to reach threshold may be as small as 15 mJ. With these small thresholds it is not surprising that high-energy pulsed operation of Nd:YAG lasers is routine. Single transverse-mode outputs up to 50 J are readily available. Although 1.06 μm is the commonest wavelength produced by these lasers, a laser transition at 1.3188 μm that has the $^4I_{13/2}$ level shown in Fig. 7.5 as its lower level is also easy to obtain.

## 7.7 Continuous-wave operation of the Nd:YAG laser

Continuous-wave laser operation of Nd:YAG lasers is possible because of their low threshold pump requirements. Outputs of tens of watts are readily obtained in the $TEM_{00}$ mode. To operate the Nd:YAG crystal described above in a CW mode requires a steady pumping power ($\mathrm{W\ m}^{-3}$) of

$$P_t = \frac{N_t h\overline{\nu}_{\mathrm{pump}}}{\tau_2\overline{\eta(\nu)}}, \tag{7.30}$$

where, if spontaneous emission is the dominant loss process from the upper laser level, $\tau_2 = 1/A_{21}$. With the parameters used above

$$P_t = 4.15 \times 10^6\ \mathrm{W\ m}^{-3},$$

and, for the crystal of dimensions 50 mm long × 5 mm diameter, $P_t = 4.07$ W.

Because CW arc lamps are not as efficient at converting electrical energy to light as are flashlamps, the factors $e$, $f$, and $g$ can be taken to be on the order of 10%, 10%, and 50%,

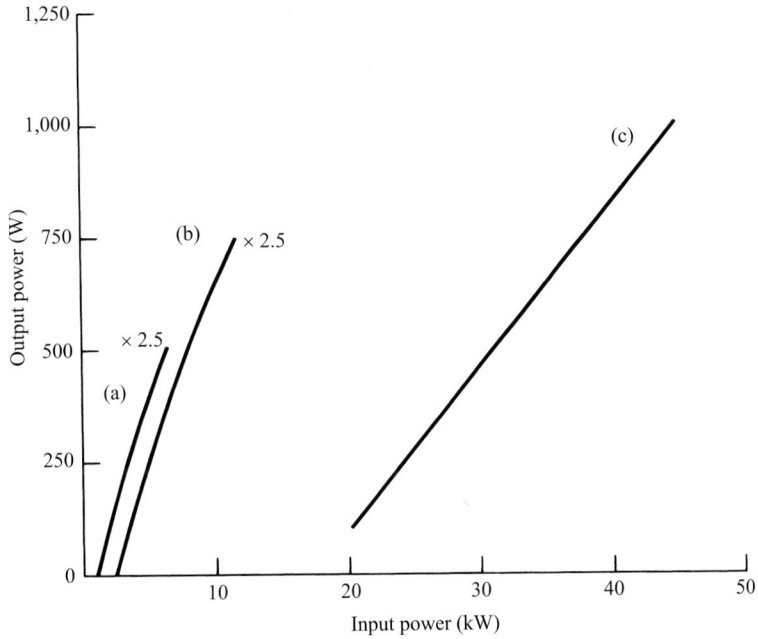


**Fig. 7.7** The performance of CW Nd:YAG lasers at 1.06 μm pumped by krypton arc lamps: (a) 6.3-mm × 100-mm rod in a single-elliptical pump cavity, (b) 6.3-mm × 100-mm rod in a double-elliptical cavity, and (c) eight 6-mm × 75-mm rods in series, each pumped with an individual lamp [8].

respectively. The lamp power required in order to reach threshold would be $\simeq 800$ W. The overall efficiency for producing laser power is on the order of 1%–3%, as can be seen from Fig. 7.7, which shows the actual performance characteristics of some CW Nd:YAG lasers pumped by krypton arc lamps.

## 7.8 The $Nd^{3+}$ glass laser

The construction of doped crystalline lasers of high power or energy output, particularly in configurations where the output of a laser oscillator is amplified by successive crystals, requires the growth of crystals of substantial size. This must be accomplished without the introduction of optical inhomogeneities into the crystal. In practice, most such crystals are grown by using a small seed crystal and pulling a larger crystal from the melt – the so-called Czochralski method, which is illustrated in Fig. 7.8. Such methods become difficult to use if very large crystals are needed. However, the use of glasses to serve as host media, particularly for $Nd^{3+}$ ions, circumvents this difficulty since extremely large pieces of doped glass are easily fabricated.

In a glass the environments of the $Nd^{3+}$ ions vary much more than in a crystalline material because of the random structural character of the glass matrix. Therefore, not

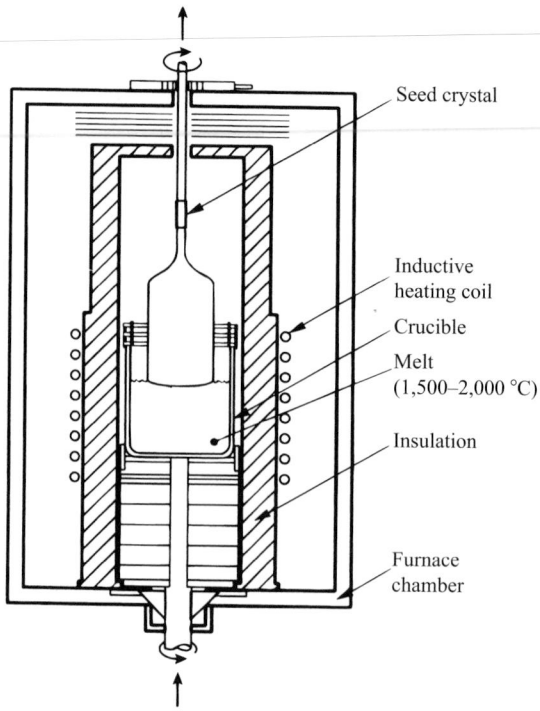


**Fig. 7.8** The arrangement of an apparatus for pulling a crystal boule from the melt by the Czochralski method.

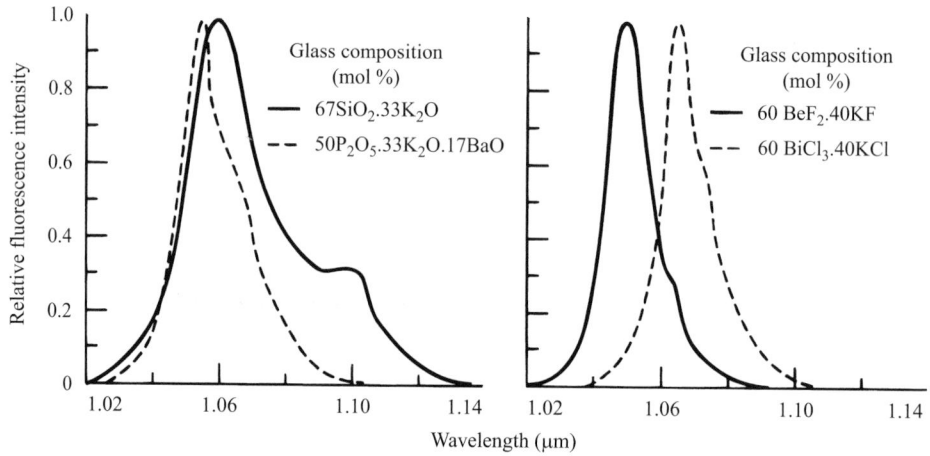


**Fig. 7.9** The spontaneous-emission lineshape of the 1.06-μm neodymium laser transition in various glasses at 295 K [9].

only is the laser substantially inhomogeneously broadened, but also it has a much broader linewidth than, for example, Nd:YAG. The laser transition linewidth is typically on the order of 300 $cm^{-1}$, $\sim 10^{13}$ Hz, and its actual shape varies from one glass to another, as can be seen from Fig. 7.9 and Table 7.2. The lifetime of the upper laser level also depends on the type of glass and the $Nd^{3+}$ concentration, as can be seen from Fig. 7.10. Figure 7.11

**Table 7.2** Spectroscopic properties of the $^4F_{3/2} \rightarrow {}^4I_{11/2}$ transition of $Nd^{3+}$ ions in different glasses at 295 K [9]

| Glass | Refractive index $n$ | Cross-section ($10^{24}$ $m^2$) | Wavelength $\lambda_p$ (nm) | Line-width $\Delta_{eff}$ (nm) | Lifetime $\tau_R$ (μs) |
|---|---|---|---|---|---|
| Oxides | | | | | |
| Silicates | 1.46–1.75 | 0.9–3.6 | 1,057–1,088 | 34–55 | 170–1,090 |
| Phosphates | 1.49–1.63 | 2.0–4.8 | 1,052–1,057 | 22–35 | 280–530 |
| Borates | 1.51–1.69 | 2.1–3.2 | 1,054–1,063 | 34–38 | 270–450 |
| Germanates | 1.61–1.71 | 1.7–2.5 | 1,060–1,063 | 36–43 | 300–460 |
| Tellurites | 2.0–2.1 | 3.0–5.1 | 1,056–1,063 | 26–31 | 140–240 |
| Halides | | | | | |
| Fluoroberyllates | 1.28–1.38 | 1.6–4.0 | 1,046–1,050 | 19–29 | 460–1,030 |
| Fluoroaluminates | 1.41–1.48 | 2.2–2.9 | 1,049–1,051 | 30–33 | 420–570 |
| Fluorozirconates | 1.52–1.56 | 2.9–3.0 | 1,049 | 26–27 | 430–450 |
| Chlorides | 1.67–1.91 | 6.0–6.3 | 1,062–1,064 | 19–20 | 180–220 |

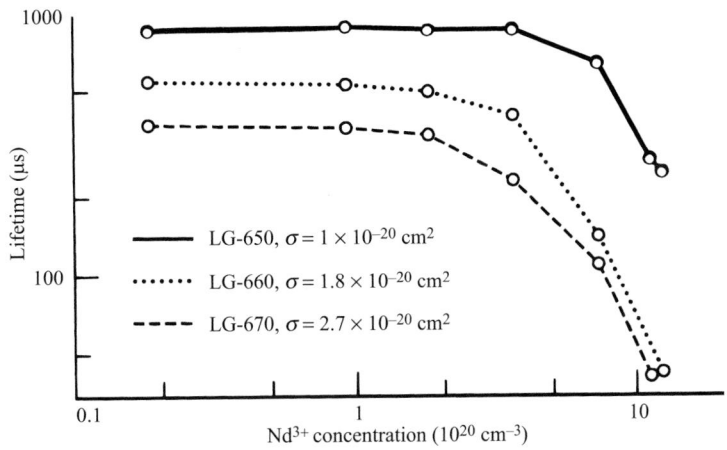


**Fig. 7.10** The lifetime of the upper laser level as a function of $Nd^{3+}$ concentration for different silicate-glass laser materials. The $\sigma$ values quoted are the peak values of the stimulated-emission cross-sections of the $^4F_{3/2} \rightarrow {}^4I_{11/2}$ transition [10].

shows a typical energy-level diagram for $Nd^{3+}$ ions in glass together with the absorption spectrum of the material.

The world's most powerful lasers are based on this material. Because their linewidths are much larger than in YAG, their inversion threshold is higher; however, once they are pumped sufficiently above threshold they offer comparable (energy-out)/(energy-in) performance to Nd:crystalline lasers. The relatively long upper laser level lifetime, ~100 μs,

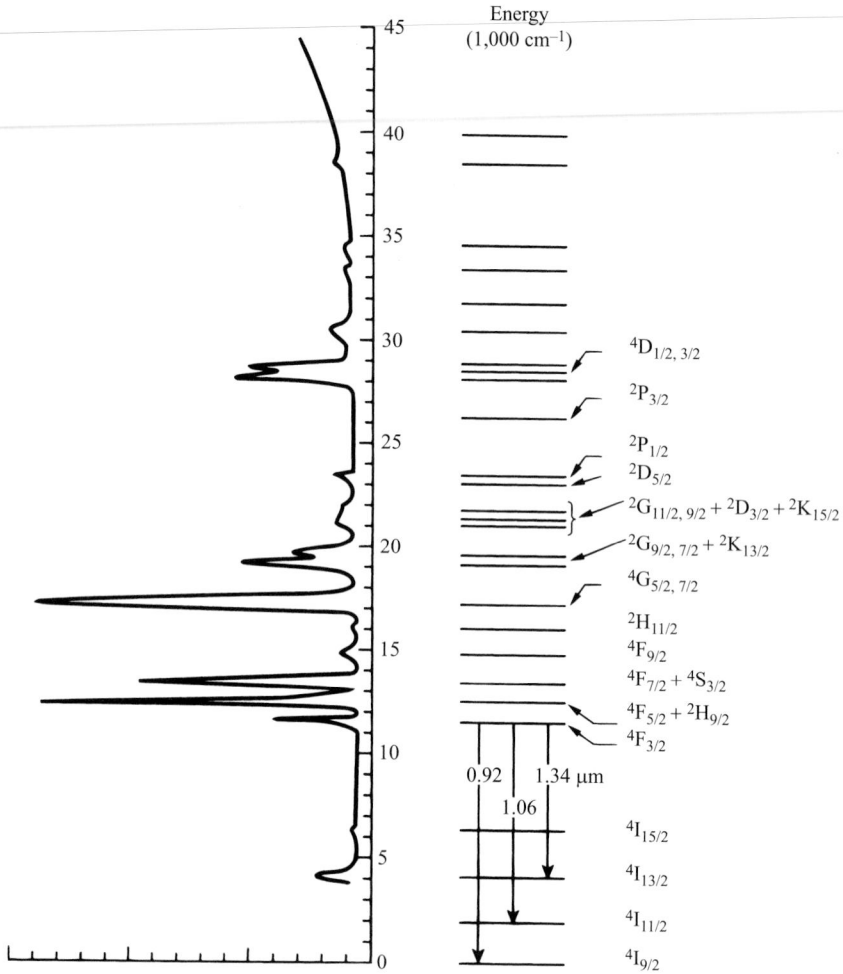


**Fig. 7.11** The absorption spectrum and energy levels of $Nd^{3+}$ in glass [11].

allows them to store larger quantities of energy – up to 0.5 MJ $m^3$. The best example of their capability is their use in research into inertial-confinement fusion (ICF), which will be discussed in more detail in Chapter 24. In such high-energy applications the output of a well-controlled Nd:YAG crystal laser oscillator will be amplified up to a million times by a series of laser amplifiers of progressively larger aperture, as shown schematically in Fig. 7.12. These amplifiers utilize large slabs of Nd:glass material placed at Brewster's angle and pumped through their flat faces with arrays of linear flashlamps. A large oscillator/amplifier(s) chain of this kind will contain many additional components over and above the oscillator, its mirror feedback system, and the successively larger amplifiers. There will be Faraday isolators (see Chapter 13), which allow light to pass in one direction but not the other, beam expanders, spatial filters for smoothing the transverse spatial profile

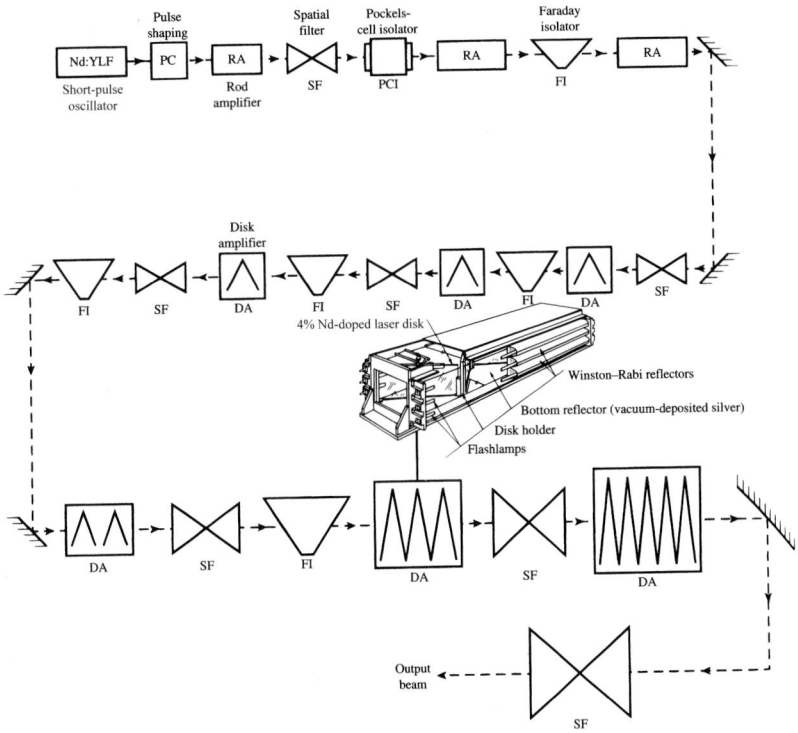


**Fig. 7.12** A schematic diagram showing the arrangement of an oscillator–amplifier chain for the production of very-high-energy laser pulses from an $Nd^{3+}$ laser. The isolators allow radiation to pass only in one direction. Spatial filters clean up the radial intensity profile of the beam. As the beam increases in intensity down this chain the size of the beam increases, as does the size of the successive rod amplifiers (RA) and disk amplifiers (DA).

of the laser beam (see Chapter 15), electro-optic switches (see Chapter 18), and nonlinear crystals for harmonic generation (see Chapter 20).

The large Nd:glass laser that is the source for the National Ignition Facility (NIF) at the Lawrence Livermore National Laboratory delivers a final amplified pulse energy of several MJ at 1.06 μm in 1 ns – corresponding to a peak power of more than $10^{15}$ W (1 petawatt). By the use of nonlinear techniques a substantial part (>1 MJ) of this laser energy can be converted to its third harmonic (0.35 μm) for irradiation of nuclear-fusion targets. In such multi-megajoule laser systems the advantages of Nd:glass become apparent: it has a high damage threshold, so it can amplify high intensity pulses; and because of its large linewidth its gain coefficient is smaller than that for Nd:YAG, which helps to eliminate the occurrence of spontaneous parasitic oscillations in amplifier sections.

Amplification of very short optical pulses can be achieved with a glass laser amplifier, because of the broad linewidth $\sim 10^{13}$ Hz, the amplification of pulses ~1 ps in length is feasible. In mode-locked operation the broad linewidth of the Nd:glass laser allows it to generate pulses in the 2–20-ps range.

## 7.9 Geometrical arrangements for optical pumping

In Chapter 3 we have already briefly discussed the optical pumping arrangement in which the laser crystal is pumped by a helical lamp wrapped around the crystal, and the elliptical-reflector scheme in which flashlamp and crystal lie along the two focal lines of a cylindrical reflector of elliptical cross-section. Other optical pumping arrangements are also commonly used.

In the elliptical-reflector arrangement, all the light from the lamp (along one focal line) passes through the crystal (along the other focal line). In this arrangement the linear lamp and the laser rod are quite separate, which makes access and replacement of either a simple matter. High-purity, deionized water coolant can be circulated directly inside the reflector to keep both lamp and laser rod cool. This arrangement is very efficient and convenient, but the illumination it provides is not axisymmetric. The side of the laser crystal nearest to the flashlamp receives its illumination directly, and the greater intensity of this direct illumination produces a population inversion in the system that is greater on the side of the crystal nearest to the lamp. The resultant laser beam has an asymmetric intensity distribution that is skewed towards the side of the laser rod facing the lamp.

The close-coupling optical pumping arrangements shown in Fig. 7.13 are simple and convenient to manufacture, and provide comparable efficiency to elliptical cavities, although again without axial symmetry (unless very many closely coupled lamps are used). However, the close proximity of flashlamp and laser can lead to thermal dissipation problems. If the laser crystal is heated by the pumping lamp, its oscillation threshold will rise. This can occur for several reasons: because of an increase in the linewidth of the laser transition, thermal population of the lower laser level, or, particularly, the production of refractive-index inhomogeneities in the laser crystal. In very-high-power solid-state laser systems, especially $Nd^{3+}$:glass, the laser medium is frequently fabricated as a series of flat slabs placed at Brewster's angle and optically pumped through their flat faces by arrays of linear flashlamps. Special reflector arrangements have been developed for efficiently coupling light from these flashlamp arrays into the laser slabs, as shown in Fig. 7.14.

### 7.9.1 Axisymmetric optical pumping of a cylindrical rod

An ingenious method for obtaining axisymmetric irradiation of the laser crystal is to use an ellipsoidal mirror (an ellipsoid of revolution) as shown in Fig. 7.3. Practically all the light from the flashlamp passes through the laser crystal. However, even in this case where the illumination of the laser crystal is axisymmetric, the volume illumination of the crystal is not uniform. This can be illustrated by considering a simple two-dimensional model where a cylindrical transparent rod is uniformly illuminated around its circumference as shown in Fig. 7.15. Note that groups of rays that do not strike the surface of the rod normally will be focused through the curved surface by refraction. We assume that the illumination consists of many, incoherent plane waves of equal intensity that have their electric vectors parallel to the axis of the cylinder [13, 14] (it can be shown that the final result is the same if the electric vectors of the incident waves are perpendicular to the cylinder axis). If we assume

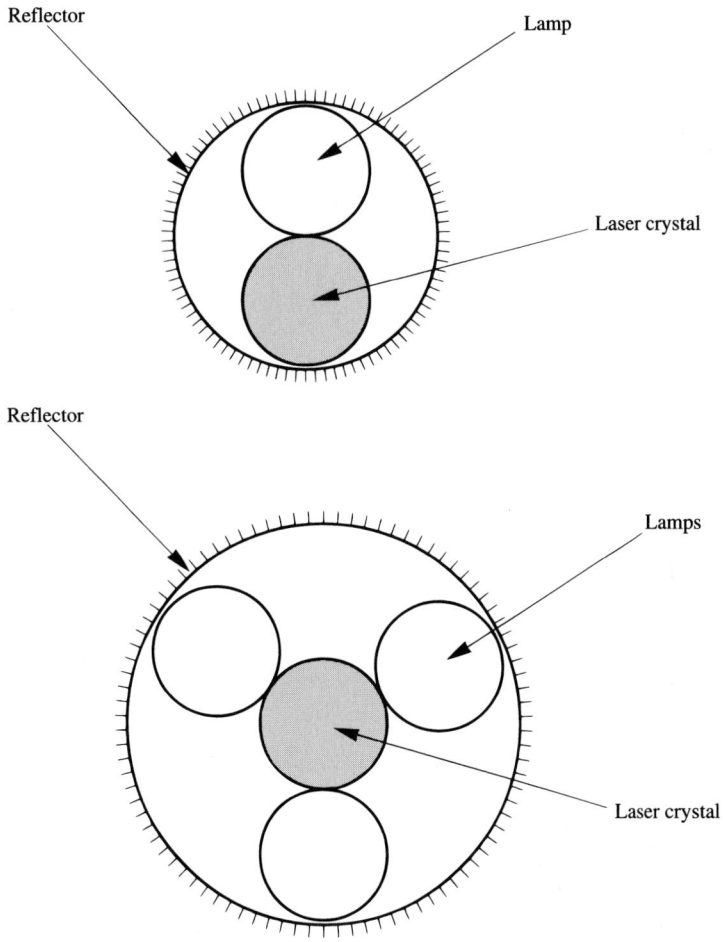


Fig. 7.13 Lamp–reflector configurations used in close optical coupling arrangements for solid-state laser pumping.

that interference effects between these waves are "smeared out," then the energy density in the rod due to all the illuminating waves can be taken as the sum of their individual energy densities.

We consider the group of waves whose propagation vectors lie within a small range of angles $d\alpha$. Their contribution to the total average energy density $\rho_0$, at a point outside the rod, is

$$\Delta\rho_0 = \frac{\rho_0}{2\pi}\, d\alpha, \tag{7.31}$$

since the total energy density comes from waves covering the total circular angle of $2\pi$. Now, if there are $m$ waves altogether (where in the limit we will let $m$ become infinitely large), the energy density associated with each wave is $\rho_0/m$. If the electric field of each

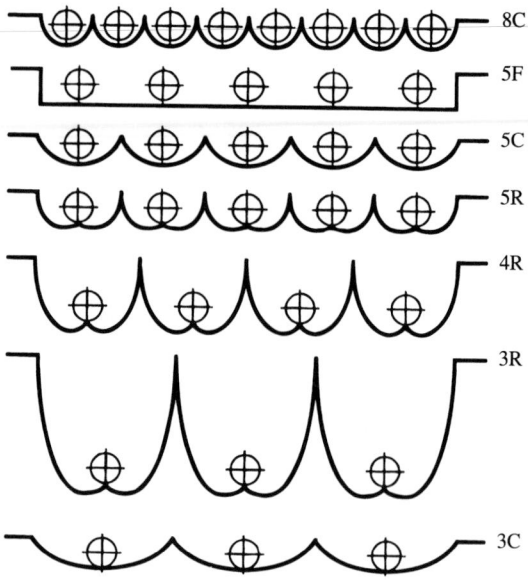


**Fig. 7.14** Reflector configurations used in face pumping of slab amplifiers. The classification numbers indicate the number of lamps and reflector type used by the Lawrence Livermore National Laboratory: C, cylindrical reflector; F, flat; and R, Rabl [12].

wave is of the form $E\cos(\omega t + \phi)$ (where $E$ becomes infinitely small as $m \to 0$) then the time-averaged energy density outside the rod due to an individual wave is

$$d\rho_0 = \overline{\epsilon_0 E^2 \cos^2(\omega t + \phi)}, \tag{7.32}$$

where the bar indicates time averaging, which gives

$$d\rho_0 = \frac{1}{2}\epsilon_0 E^2. \tag{7.33}$$

The total energy density due to the $m$ waves is

$$\rho_0 = \sum_m d\rho_0 = \frac{1}{2}\epsilon_0 \sum_m E^2 = \frac{1}{2}\epsilon_0 E_0^2, \tag{7.34}$$

where $E_0$ is the root-mean-square value of the total electric field due to all the waves. Thus, from (7.31) and (7.34)

$$\Delta\rho_0 = \frac{\epsilon_0 E_0^2}{4\pi}\, d\alpha. \tag{7.35}$$

The group of waves which make this contribution to the energy density strike the outer wall of the cylindrical rod at angle $\alpha$, as shown in Fig. 7.16. Their incident intensity is

$$I = c_0\, \Delta\rho, \tag{7.36}$$

and the intensity transmitted through the surface, from Eq. (4.30) with $n' = 1$, is

$$I_t = nt^2 I, \tag{7.37}$$

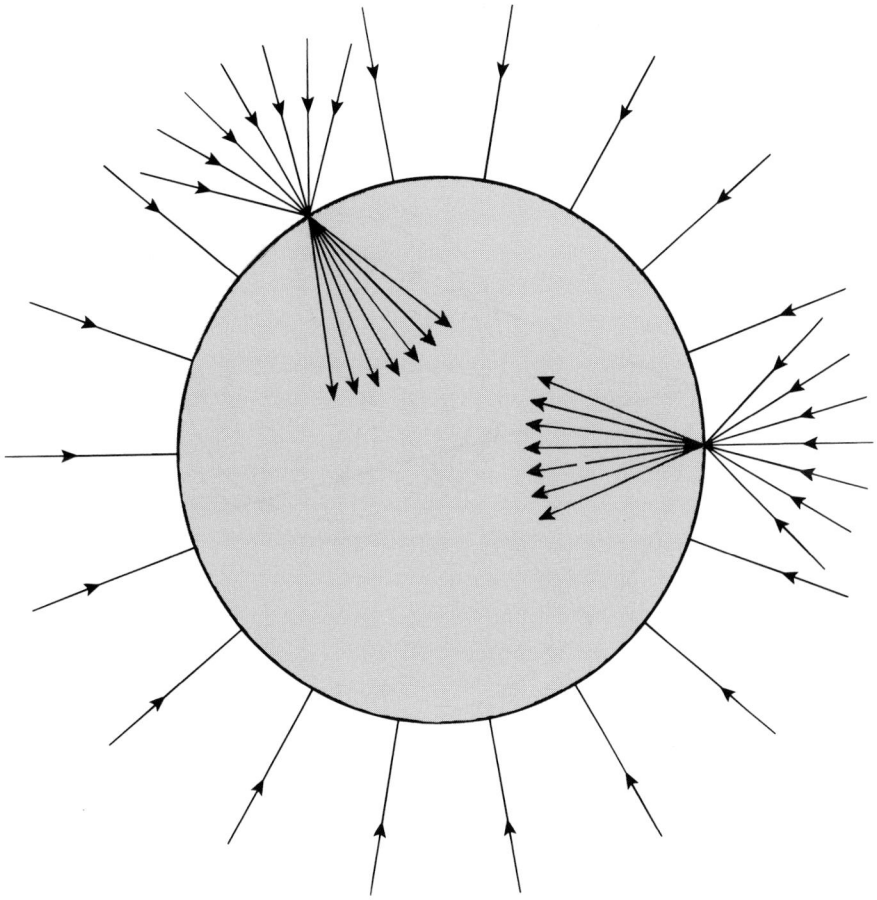

Fig. 7.15 Circularly symmetric illumination of a laser rod showing the refraction of rays through the cylindrical surface.

where $t$ is the transmission coefficient, which in this case is given by[1]

$$t = \frac{2\cos\alpha}{\cos\alpha + n\cos\beta}. \tag{7.38}$$

The angle of refraction $\beta$ is related to the angle of incidence $\alpha$ by Snell's law

$$\frac{\sin\alpha}{\sin\beta} = n, \tag{7.39}$$

where $n$ is the refractive index of the cylinder. The total internal flux transmitted through a small area of surface $dA$ of unit length along the cylinder axis is, from (7.37),

$$F^0_{\text{int}} = nt^2\, dA\, I\cos\beta, \tag{7.40}$$

[1] See Appendix 4.

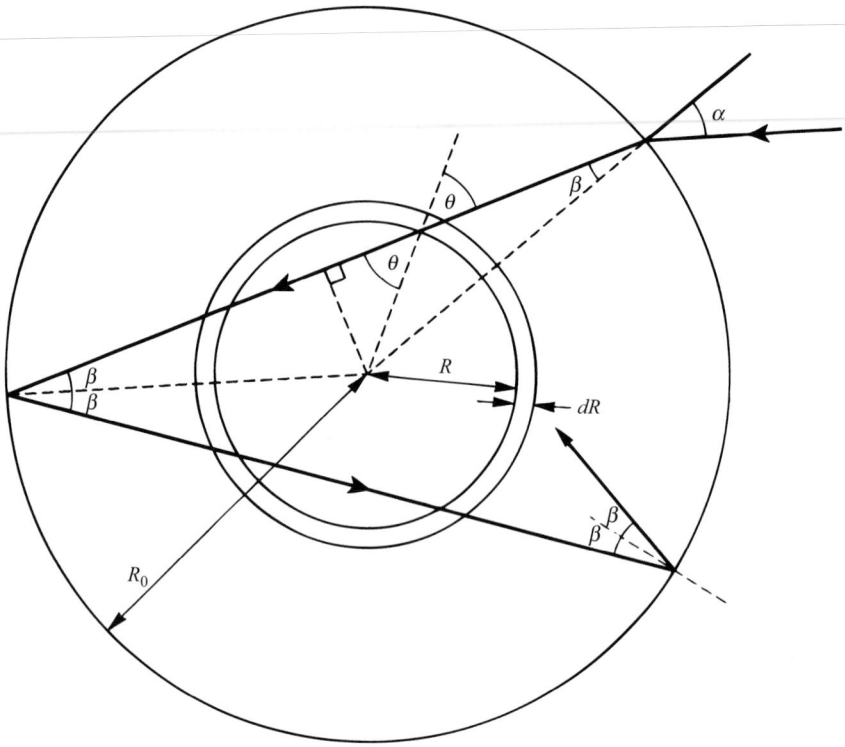


Fig. 7.16 The geometry of irradiation of a cylindrical laser rod in a cylinder cross-section.

since $dA \cos\beta$ is the effective area of the element of surface normal to the direction of propagation of the wave.

Once this group of waves has entered the cylinder, they reflect around inside, maintaining internal angles of incidence and reflection of $\beta$, as shown in Fig. 7.16. Each time they strike the wall their intensity is diminished by a factor $r^2$, where $r$ is the reflection coefficient given by[2]

$$r = \frac{n\cos\beta - \cos\alpha}{n\cos\beta + \cos\alpha}. \tag{7.41}$$

We consider the contribution that these waves make to the energy density in a small annulus of the cylinder of radius $R$ and thickness $dR$. On any given pass across the cylinder, between internal reflections, the waves strike this region at an angle $\theta$, where

$$\theta = \arcsin(R_0 \sin\beta / R), \tag{7.42}$$

where, of course, if $R < R_0 \sin\beta$, the particular group of waves does not pass through the part of the cylinder of radius $R$. If the flux that strikes this annular region is $F^j_{\text{int}}$, where $F^j_{\text{int}}$

[2] See Appendix 4.

is the internal flux after $j$ internal reflections, then the stored energy within the annulus due to this passage of the internally reflected wave is

$$du^j(r) = \frac{2F^j_{\text{int}} n\, dR}{c_0 \cos\theta}, \tag{7.43}$$

which gives

$$du^j(r) = \frac{2r^{2j} F^0_{\text{int}} nR\, dR}{c_0\sqrt{R^2 - R_0^2 \sin^2\beta}}. \tag{7.44}$$

The total stored energy in the annulus is

$$du(r) = \sum_{j=0}^{\infty} \frac{2r^{2j} F^0_{\text{int}} nR\, dR}{c_0\sqrt{R^2 - R_0^2 \sin^2\beta}}, \tag{7.45}$$

which, on substituting from (7.40) and summing the geometric series, gives

$$du(r) = \frac{2n^2 t^2 I \cos\beta\, dA\, R\, dR}{c_0\sqrt{R^2 - R_0^2 \sin^2\beta(1 - r^2)}}. \tag{7.46}$$

The energy density in the annulus, since its volume is $2\pi R\, dR$, is

$$d\rho(r) = \frac{n^2 t^2 I \cos\beta\, dA}{\pi c_0\sqrt{R^2 - R_0^2 \sin^2\beta(1 - r^2)}}. \tag{7.47}$$

The total energy density at radius $r$ due to all the waves that strike an area $2\pi R_0$ of the outer cylinder in a small angular range $d\alpha$ at $\alpha$ is

$$\Delta\rho(r) = \frac{2n^2 t^2 I R_0 \cos\beta}{c_0\sqrt{R^2 - R_0^2 \sin^2\beta(1 - r^2)}}. \tag{7.48}$$

Now, from (7.38) and (7.41),

$$\frac{t^2}{1 - r^2} = \frac{\cos\alpha}{n\cos\beta}, \tag{7.49}$$

and, since $I = c_0\, \Delta\rho_0$ and $\Delta\rho_0 = [\rho_0/(2\pi)]d\alpha$,

$$\Delta\rho(r) = \frac{n\rho_0 R_0}{\pi} \frac{\cos\alpha\, d\alpha}{\sqrt{R^2 - R_0^2 \sin^2\alpha/n}}. \tag{7.50}$$

The total energy density at radius $R$, $\rho(R)$, is obtained by integrating (7.50) over the range of angles of incidence $\alpha$ that allow waves to reach radius $R$,

$$\rho(R) = \frac{n\rho_0 R_0}{\pi} \int_{-\alpha_0}^{\alpha_0} \frac{\cos\alpha\, d\alpha}{\sqrt{R^2 - R_0 \sin^2\alpha/n}}, \tag{7.51}$$

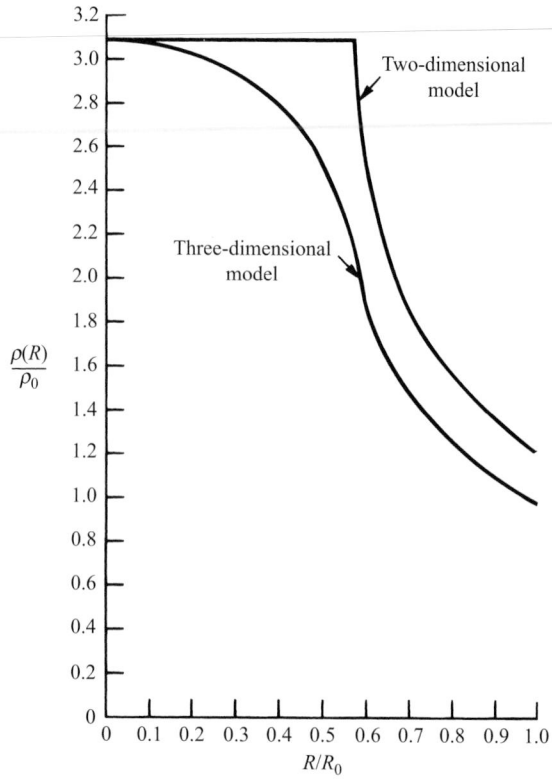


**Fig. 7.17** The normalized energy density as a function of radius in a transparent cylindrical rod illuminated uniformly in a cylinder cross-section. In the two-dimensional model all illuminating rays travel in planes orthogonal to the cylinder axis. In the three-dimensional model the illuminating rays take all possible directions [13, 14].

where $\sin\alpha$ is the smaller of 1 and $nR/R_0$. The solution of (7.51) is

$$\frac{\rho(R)}{\rho_0} = \begin{cases} n^2 & \text{for } 0 \le R \le R_0/n, \\ (2n^2/\pi)\arcsin(R_0/nR) & \text{for } R_0/n \le R \le R_0, \end{cases} \tag{7.52}$$

which is shown in Fig. 7.17 for the case $n = 1.76$.

The energy density within the cylinder is a maximum, and constant, for all radii $\le R_0/n$. Figure 7.17 shows also the energy density as a function of radius inside a cylindrical rod when the incident radiation consists of all possible polarizations in three dimensions. If the cylinder absorbs the incident radiation then these energy-density functions are modified as illustrated in Fig. 7.18, which shows both the two- and the three-dimensional calculated energy-density functions for different values of $d$, where $d = 2R_0 \times$ absorption coefficient.

It is clear from Figs. 7.17 and 7.18 that, if a cylindrical laser crystal is made up as a composite rod with an inner absorbing core, for example $Cr^{3+}$-doped $\alpha$-$Al_2O_3$ (pink

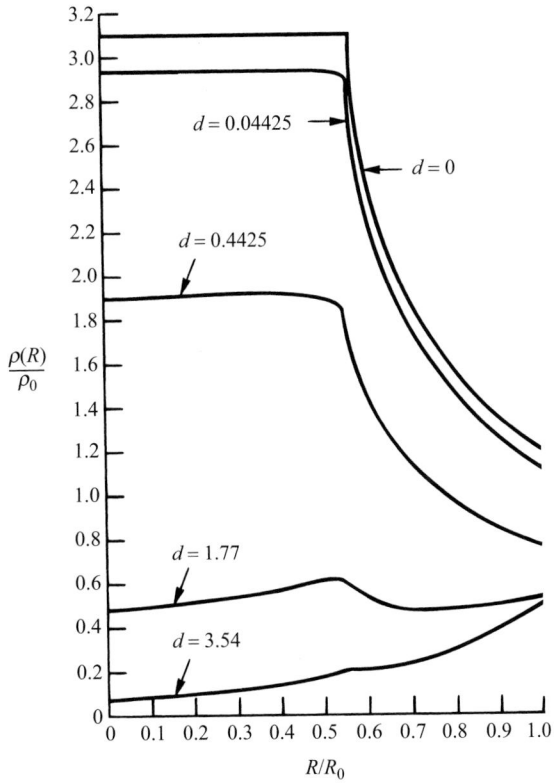


**Fig. 7.18** The normalized energy density inside a cylindrical laser rod uniformly illuminated with linearly polarized light polarized along the cylinder axis. The energy-density distributions are shown for different amounts of rod absorption, where $d$ is the product of the cylinder diameter and the absorption coefficient [13, 14].

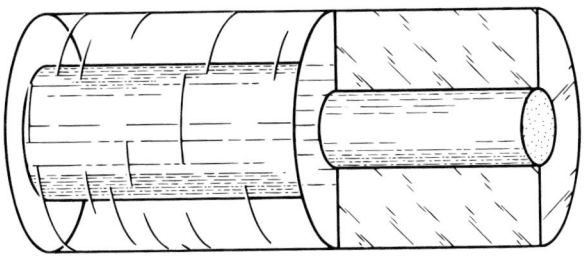

**Fig. 7.19** A composite cylindrical laser rod with a doped center [15].

ruby) of radius $R_0/n$, and an outer sheath of transparent material of the same refractive index, for example $\alpha$-$Al_2O_3$ (sapphire) as shown in Fig. 7.19, then very close to uniform illumination of the amplifying medium will be achieved. Such composite laser rods do, in fact, have lower oscillation thresholds than those of comparable uniform rods, although the manufacture of these composite rods is more difficult.

## 7.10 High-power pulsed solid-state lasers

The input energies to the flashlamps used in pumping high-energy solid-state lasers can be very large, ranging up to several thousand joules for each lamp. Typical flash durations in these systems are in the range 100 μs to 1 ms. These large energy inputs can cause severe heating of the laser rod, and consequently restrict operation to pulsed operation, unless special cooling arrangements are incorporated.

For this reason, high-power solid-state lasers cannot be built just by increasing the diameter and length of a cylindrical laser rod. Large-diameter rods cannot eliminate waste heat effectively, and are also difficult to fabricate.

The highest available energy outputs are obtained from $Nd^{3+}$:glass lasers, which at high power levels are efficiently pumped in a face-pumped slab configuration. One example of such a system, which contains some interesting features, is shown in Fig. 7.20. In this arrangement the thermal and optical pumping inhomogeneity problems associated with large-diameter cylindrical laser crystals are considerably reduced. Irradiation of the crystal is efficient over a large surface/volume ratio and large-sized glass slabs, placed at Brewster's angle, can be used.

Development of high-power $Nd^{3+}$:glass systems of this kind is continuing in many laboratories as part of a program to obtain laser-induced fusion. Current attention is focused on the development of lasers with output energies of up to 1 MJ to be delivered to pellets of nuclear fuel on time scales from 100 ps to 1 ns. This corresponds to peak powers of $10^{14}$–$10^{15}$ W.

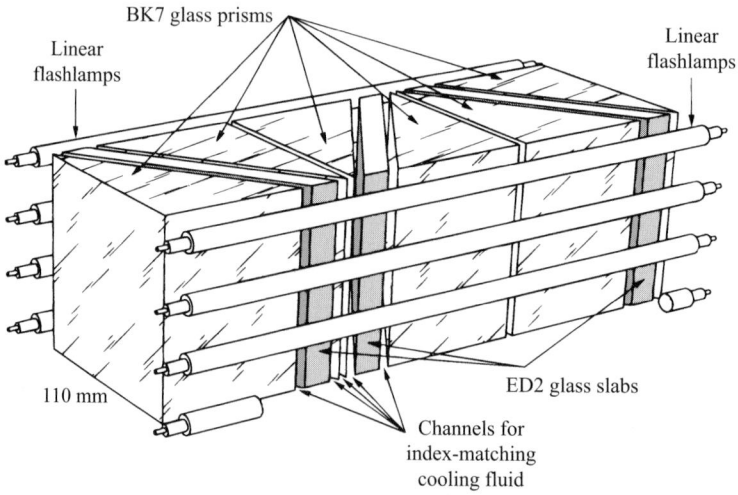


**Fig. 7.20** The arrangement of a "zig-zag" face-pumped laser amplifier module that incorporates prisms and index-matching coolant for reducing Fresnel-reflection losses at the neodymium–glass (ED2) slabs [16].

## 7.11 Diode-pumped solid-state lasers

The development of efficient, high-power semiconductor lasers using GaAlAs has created the most important and efficient optically pumped lasers in which a crystalline laser is optically pumped by a semiconductor laser.

Because of a fortuitous coincidence between the output of GaAlAs semiconductor lasers working near 809 nm and a strong absorption of the neodymium ions ($Nd^{3+}$) in YAG, YLF, or $YVO_4$, it is very efficient to optically pump $Nd^{3+}$ lasers with diode lasers. These lasers have become widely available commercially. They can also be frequency doubled efficiently to 530 nm, and have largely replaced ion lasers in applications where intense green light is required. Green output powers of hundreds of watts are available. The principal advantages of diode-pumped solid-state (DPSS) lasers are reliablity, narrow linewidth, and single-frequency operation.

We shall discuss the characteristics of the semiconductor pump lasers in detail in Chapter 12. Suffice it to say that, because of their high electrical efficiency, and directional outputs, they can be efficiently coupled to solid-state laser slabs, rods, or fibers. The most common DPSS lasers are the Nd:YAG and Nd:$YVO_4$ lasers operating at 1.064 μm, 1.32 μm, or 946 nm, although other hosts for the $Nd^{3+}$ ions, such as YLF and $La_2Be_2O_5$ (BEL) are also used. The coincidence between GaAlAs laser emission at ~809 nm and the absorption of $Nd^{3+}$ ions into the $^4F_{5/2}$ pump band (see Fig. 7.5) allows very efficient conversion from 809 nm to laser oscillation at 1.06 μm; over 10% conversion is easily obtained with pump powers of ~200 mW. The 1.06-μm and 946-nm laser transitions just mentioned can be frequency doubled with nonlinear crystals (see Chapter 21) to give green light at 540 nm, or blue light at 473 nm.

Table 7.3 summarizes some important DPSS laser transitions that have been obtained [17]. Two more lasers in this list are worthy of note. The 2.1-μm Ho:YAG laser has optical radar (LIDAR) applications because its wavelength makes it far less of an eye hazard than lasers operating at 1.06 μm and shorter wavelengths. The erbium laser emits very close to a strong absorption of water, so it has become important in medical applications involving absorption of laser energy for removing tissue.

Diode-pumped lasers in which the semiconductor lasers are arranged to inject their pump radiation through the cylindrical faces of a laser rod are in many ways akin to lamp-pumped lasers of the same kind. However, because of the high electrical efficiency of GaAlAs laser diodes (>10%), the overall electrical-to-optical conversion efficiency of these lasers is very high, approaching 10%.

If the diode-laser pump radiation is injected along the axis and matched to the transverse-mode geometry of the solid-state laser then very stable, narrow-linewidth laser oscillation can be obtained: Fig. 7.21 shows the clever monolithic Nd:YAG laser design of Kane and Byer [18]. In this laser the magneto-optical properties of YAG are used to obtain unidirectional amplification of a traveling wave in the laser cavity. This prevents the spatial hole-burning that can occur in a standing-wave homogeneously broadened laser, which can allow multiple longitudinal modes to oscillate.

**Table 7.3** Diode-pumped solid-state lasers

| Ion | Transition | Wavelength (μm) | Operating temperature (K) |
|---|---|---|---|
| $Nd^{3+}$ | $^4F_{3/2} \rightarrow {}^4I_{11/2}$ | 1.06 | 300 |
| | $^4F_{3/2} \rightarrow {}^4I_{13/2}$ | 1.32 | 300 |
| | $^4F_{3/2} \rightarrow {}^4I_{9/2}$ | 0.95 | 300 |
| $Ho^{3+}$ | $^5I_7 \rightarrow {}^5I_8$ | 2.1 | 300 |
| $Er^{3+}$ | $^4I_{13/2} \rightarrow {}^4I_{13/2}$ | 2.8 | 300 |
| | $^4I_{13/2} \rightarrow {}^4I_{9/2}$ | 1.6 | 300 |
| $Tm^{3+}$ | $^3F_4 \rightarrow {}^3H_5$ | 2.3 | 300 |
| $U^{3+}$ | $^4I_{11/2} \rightarrow {}^4I_{9/2}$ | 2.61 | 4.2 |
| $Dy^{2+}$ | $^5I_7 \rightarrow {}^5I_8$ | 2.36 | 1.9 |
| $Yb^{3+}$ | $^2F_{7/2} \rightarrow {}^2F_{5/2}$ | 1.03 | 77 |

Frequency doubling of a DPSS laser is generally carried out with an intracavity nonlinear crystal such as β-barium borate (BBO) (β-$BaB_2O_4$), BiBO (bismuth triborate, $BiB_3O_6$), KTP (potassium titanium oxide phosphate ($KTiOPO_4$), PPLN (periodically poled lithium niobate), or PPKTP (periodically poled KTP), or with an external resonant cavity containing a nonlinear crystal. Figure 7.22 shows how intracavity frequency doubling is done with an Nd:$YVO_4$ Coherent Verdi laser. This use of a ring laser resonator with unidirectional laser oscillation provides single-longitudinal-mode operation and a narrow linewidth. Conversion efficiency to green light can reach 35%. Frequency-tripled operation of diode-pumped $Nd^{3+}$ lasers at 355 nm is available at powers in excess of 3 W; frequency-quadrupled operation at 266 nm provides powers up to 500 mW. Frequency doubling of Nd:$YVO_4$ has become so routine that this is used to make green laser pointers, in which the 1.06-μm laser is frequency doubled with a KTP crystal. Other laser wavelengths can be generated from DPSS lasers using harmonic-mixing processes (see Chapter 20), for example mixing 1.064-μm and 1.342-μm light gives 593.5 nm.

## 7.12 Fiber lasers

Cylindrical optical fibers will be discussed in detail in Chapter 16, and are most important in optical communications. In their simplest form they have a central core region of a higher index than that of the surrounding medium – the *cladding*. Light in the core region

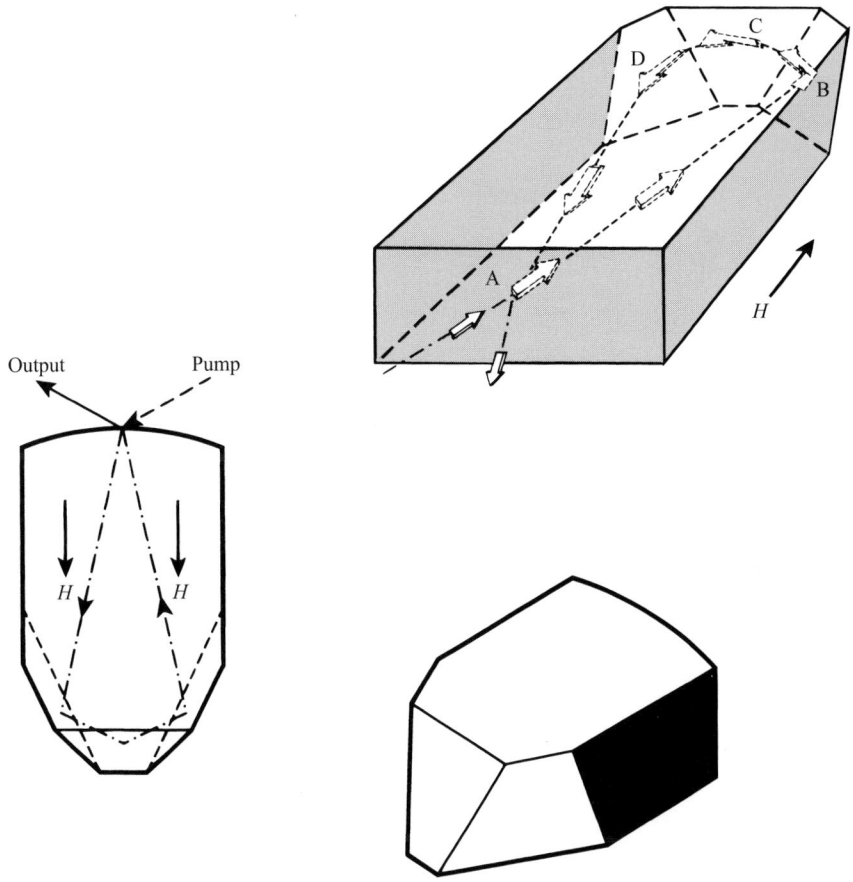


Fig. 7.21 A monolithic integrated diode-pumped Nd:YAG laser – the MISER laser [18]. Faraday rotation within the laser crystal, produced by an integral permanent magnet, leads to unidirectional oscillation and single-longitudinal-mode oscillation as described in Chapter 6. The curved face is partially transmitting and selects a specific polarization state for oscillation. Total internal reflection occurs at points B, C, and D. Faraday radiation takes place along the paths AB and DA.

traveling at a small enough angle to the axis is trapped because of total internal reflection at the core–cladding boundary (see Fig. 16.2). Fiber lasers use a gain medium that is an optical fiber doped with rare-earth ions such as erbium ($Er^{3+}$), neodymium ($Nd^{3+}$), ytterbium ($Yb^{3+}$), thulium ($Tm^{3+}$), or praseodymium ($Pr^{3+}$). Research on fiber lasers based on ($Yb^{3+}$)has been particularly intense because these lasers have a simple energy-level structure, as shown in Fig. 7.23, which facilitates diode-laser pumping. They have electrical efficiency and scalability to high powers [19]. They are pumped by one or more semiconductor laser diodes [20, 21].

A simple schematic diagram of one way in which they are constructed is shown in Fig. 7.24. The guiding properties of the higher-index core guarantee that light stays in the fiber and resonates between the end mirrors. The end mirrors can be external to the

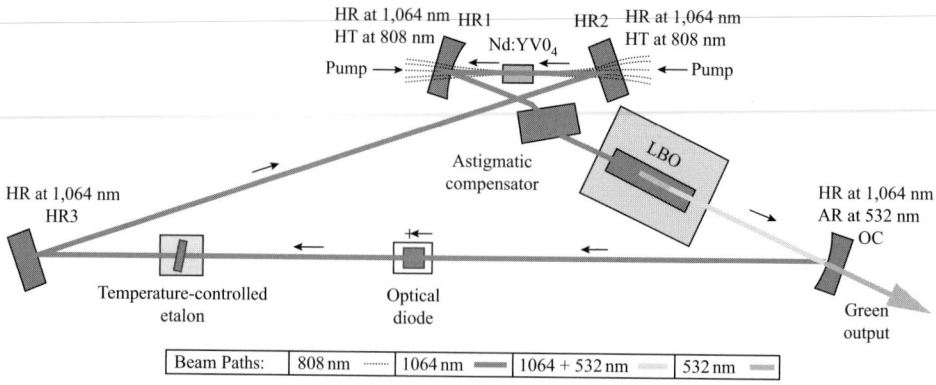


**Fig. 7.22** The layout of the Coherent, Inc. "Verdi" laser, wherein intracavity frequency doubling with a lithium triborate ($LiB_3O_5$, LBO) crystal is used to generate 530 nm from 1.06 μm. The initial laser oscillation is at 1.064 μm using a neodymium-doped yttrium vanadate ($Nd:YVO_4$) crystal pumped with an 808-nm diode laser. HR, high-reflectance mirror; OC, output-coupling mirror.

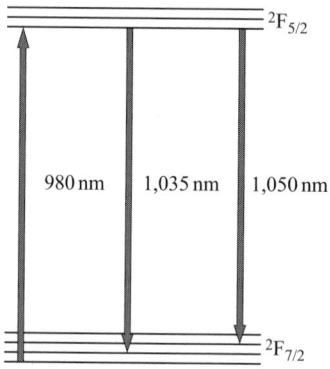


**Fig. 7.23** A simplified energy-level diagram for $Yb^{3+}$ in glass.

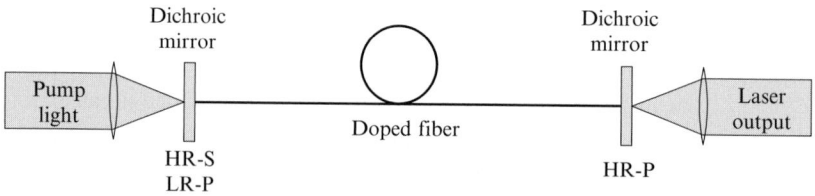


**Fig. 7.24** A simple schematic diagram of how a fiber laser is constructed (HR-P, high-reflectance mirror for P polarization; LR-P, low-reflectance mirror for P polarization; HR-S, high-reflectance mirror for S polarization).

polished end of the fiber, may consist of a fiber loop, or may use sections of fiber in which the refractive index varies periodically in the axial direction. These *Bragg reflectors* reflect strongly at specific wavelengths, as will be discussed in more detail in Section 14.4. In

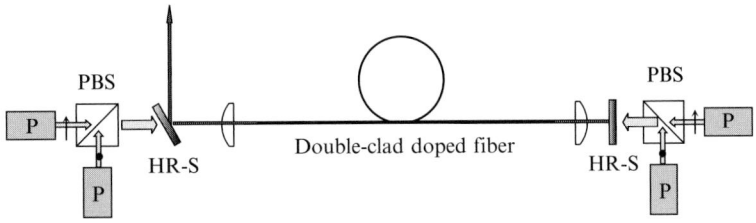


Fig. 7.25 A fiber laser using a double-clad fiber. HR-S, high reflectance at fiber laser signal wavelength.

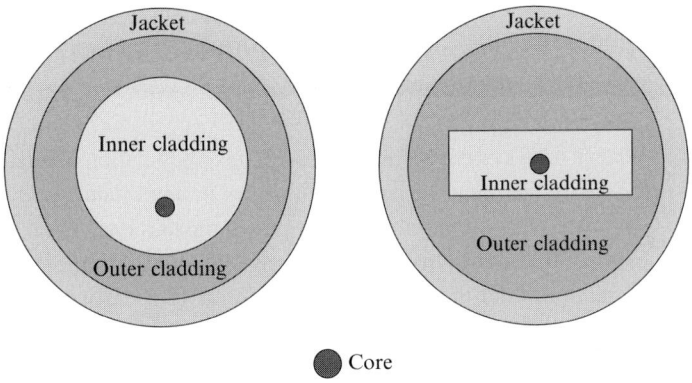


Fig. 7.26 The double-clad fiber design used in fiber lasers.

the most recently developed versions of these lasers the fiber has a double cladding, and the pump laser light is directed into the cladding. Figure 7.25 shows how a high-power fiber laser is constructed. Pump light from diode lasers operating near 980 nm (in the case of a $Yb^{3+}$-doped fiber laser) is directed into the cladding of a double-clad fiber. Various double-cladding designs are used, as shown in Fig. 7.26. The pump light from laser diodes enters over a large acceptance angle and is internally reflected within the other cladding as shown in Fig. 7.27. The laser light is confined by the inner cladding. The various refractive indices obey $n_{\text{core}} > n_{\text{innercladding}} > n_{\text{outercladding}}$. The inner cladding may be circular or rectangular, and the core may be offset from the axis of symmetry of the fiber. Part of the fiber may take the form of a photonic crystal (see Section 14.11). In this type of fiber the central core is surrounded by a periodic structure of air holes that run along parallel to the axis. These structures confine the laser light to a single transverse mode, but still provide the injection of pump light in a multimode structure. Very-high-power fiber lasers with powers above 1 kW are increasingly being used in industrial applications because they have high electrical efficiency and good output beam quality. Fiber lasers can be mode-locked to generate very short pulses (<1 ps) in the infrared. An important application of erbium-doped fiber is in erbium-doped optical fiber amplifiers (EDFAs), which are widely used in fiber-optic communication networks to amplify 1.55-μm semiconductor laser radiation.

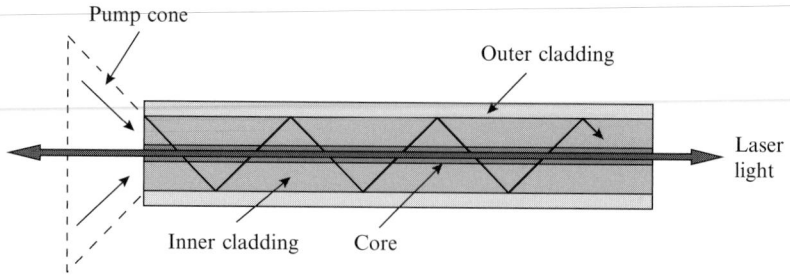


Fig. 7.27 The pumping of a double-clad fiber laser.

## 7.13 Relaxation oscillations (spiking)

When a laser system is pushed into a state of population inversion, as soon as its gain passes the threshold value the oscillation grows in amplitude and, by reducing the gain through stimulated emission, stabilizes the gain at the loss line. The approach to a stable oscillation may occur smoothly, as if the approach to the equilibrium situation were damped. However, oscillation about the equilibrium position is frequently observed. This oscillation appears as a time-varying output laser intensity following the attainment of a population inversion sufficient to start the oscillation. It is called "spiking." It results from the attempt of the intracavity field and the population inversion to become balanced, and can be visualized as an oscillatory process of energy exchange between the intracavity intensity and the population inversion. As the intracavity intensity increases, the upper-level population is reduced, so the gain falls – this reduces the intracavity intensity. This reduction of the intracavity intensity causes the gain to increase again. The intracavity intensity then increases again, and so on.

If we observe the output intensity from, for example, a pulsed ruby laser, we generally see the sort of behavior illustrated in Fig. 7.28. The laser emission in successive spikes tends to shift between different longitudinal and transverse modes. Suppose the emission in one spike is in a longitudinal mode of order $m$ so that the oscillation frequency is

$$\nu = \frac{mc}{2\ell} \qquad \left(\lambda = \frac{2\ell}{m}\right),$$

then the standing-wave field in the laser cavity has nodes and antinodes of intensity in a certain pattern, as shown in Fig. 7.29. Following laser emission in the spike, the inverted population is depleted near the antinodes of the field relative to the nodes; this causes a longitudinal mode of different order to oscillate in the next spike. This mode will have higher gain because its field antinodes will be near the field nodes of the previously oscillating mode as shown schematically in Fig. 7.29.

One phenomenon that can suppress this longitudinal mode jumping is called *spatial cross-relaxation*, whereby an excited particle at position A in Fig. 7.29 can rapidly move to a position B. This smears out any spatial sinusoidal variation of gain within the laser

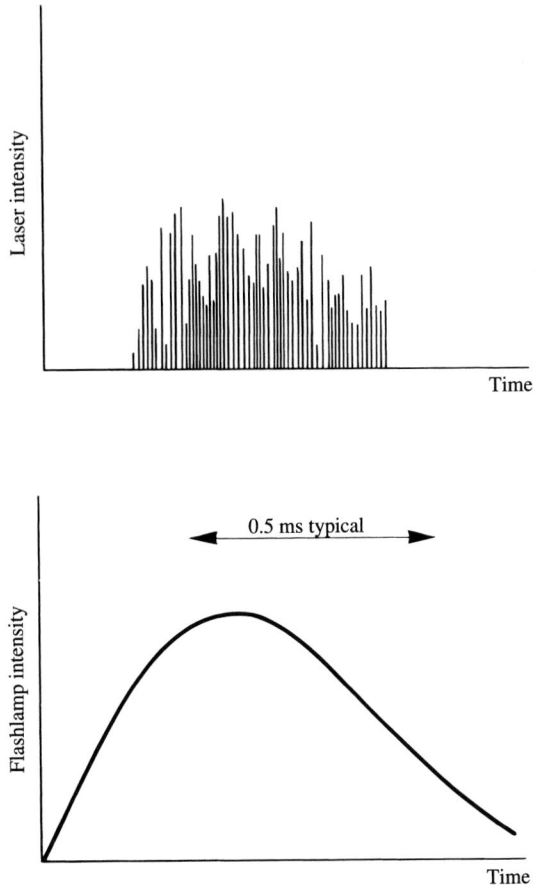


Fig. 7.28 A schematic illustration of the "spiking" phenomenon in the output of a pulsed solid-state laser.

medium, and sustains oscillation on a particular high-gain longitudinal mode. In a crystal lattice, because the excited particles are fixed, it is relatively difficult for them to transfer excitation from one position to another. However, the transfer can still occur radiatively, the relative efficiency of this process depending on the Einstein coefficient $A_{21}$ for the laser transition and on the absorption coefficient of the laser crystal for this transition (which, of course, depends on the atom concentration and the linewidth of the transition).

Jumping from one transverse mode to another in successive spikes is understandable in much the same way. Oscillation in one TEM mode depletes a certain region of the crystal preferentially. This favors subsequent oscillation on a different TEM mode, which has its maxima of intracavity intensity in the least-depleted spatial region of the laser medium.

Frequency and mode jumping of the above kinds also tend to be enhanced by non-uniformity of optical pumping, and heating of the crystal during the flash, which produces spatial refractive-index variations. Simultaneous oscillation on different longitudinal and transverse modes may occur even though the laser transition is homogeneously broadened. The simultaneously oscillating modes arrange to use different spatial regions of the crystal,

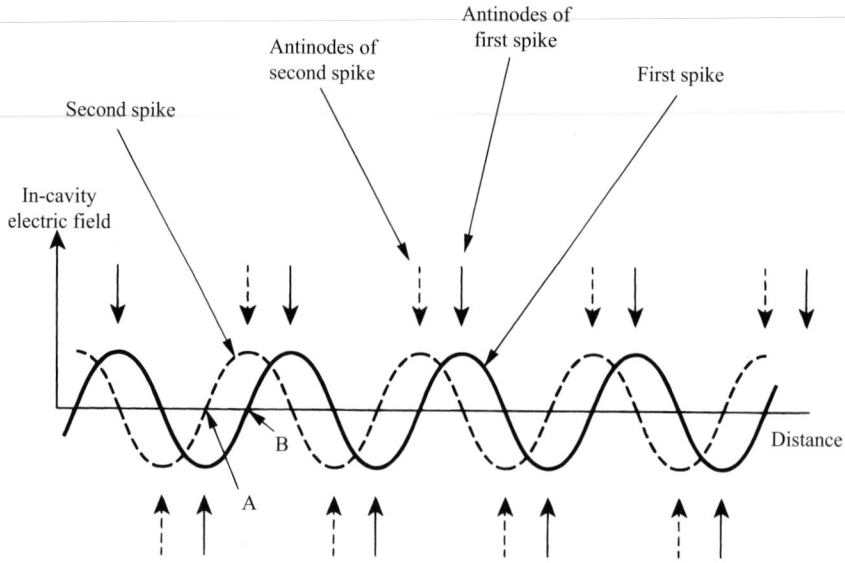


Fig. 7.29 An illustration of how oscillation can jump from one longitudinal mode to another in successive "spikes" because of population depletion.

so, although in theory they can compete for the same group of atoms, in fact they do not. A simple illustration of how this can happen is given in Fig. 7.30. In gaseous lasers spatial cross-relaxation is fast, and in *homogeneously* broadened gas lasers the above type of process, which would lead to multimode oscillation, is generally suppressed.

## 7.14 Rate equations for relaxation oscillation

In describing relaxation oscillation in a laser we follow the approach first given by Statz and deMars [22]. We assume that laser oscillation builds up in a single mode of the optical resonator.

Let $q$ be the number of photons in the mode per unit volume. The mode corresponds to a monochromatic radiation field at frequency $\nu'$ whose energy density is a $\delta$-function,

$$\rho(\nu) = qh\nu'\delta(\nu - \nu'). \tag{7.53}$$

The upper and lower laser levels have population densities $N$ and $N_1$, and effective lifetimes $\tau$ and $\tau_1$, respectively. We assume that the laser is a *good* one such that $\tau \gg \tau_1$, and $N \gg N_1$. We neglect $N_1$. The population inversion in this case is the same as $N$ and obeys the differential equation

$$\frac{dN}{dt} = R - NW - \frac{N}{\tau}, \tag{7.54}$$

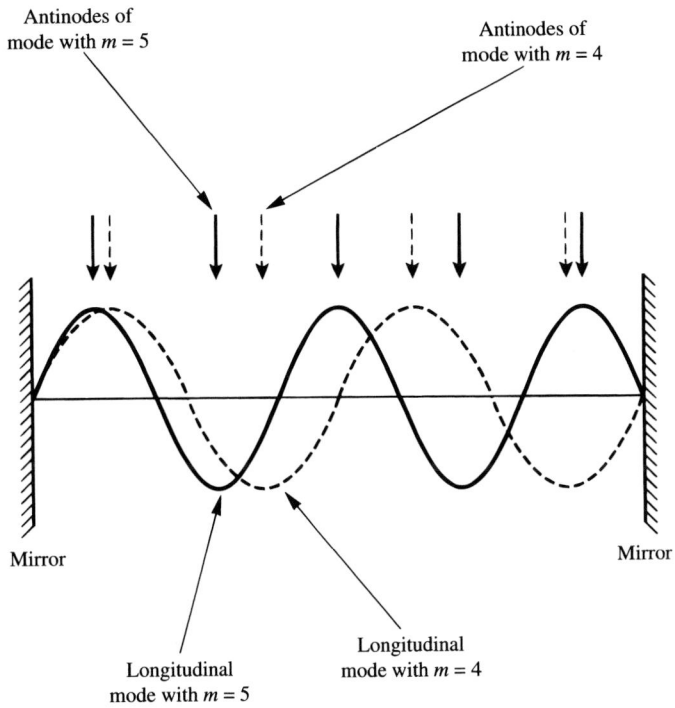


**Fig. 7.30** An illustration of how two different longitudinal modes partially avoid competition with each other by using different spatial regions of inverted population in a gain medium.

where $R$ is the rate per unit volume at which particles are fed into the upper laser level, and $W$ is the stimulated emission rate. In this case

$$W = \int qh\nu'\delta(\nu - \nu')B_{21}g(\nu_0, \nu)d\nu, \tag{7.55}$$

which gives

$$W = qh\nu' B_{21}g(\nu_0, \nu') = qB. \tag{7.56}$$

The constant $B$ is

$$B = h\nu' B_{21}g(\nu_0, \nu') = \frac{c^3}{8\pi\nu'^2}A_{21}g(\nu_0, \nu'), \tag{7.57}$$

which is the ratio of the spontaneous emission rate per frequency interval to the mode density. Equation (7.54) therefore becomes

$$\frac{dN}{dt} = R - qBN - \frac{N}{\tau}. \tag{7.58}$$

The equation describing the rate of change of the number of photons in the mode is

$$\frac{dq}{dt} = qBN - \frac{q}{\tau_0} + N\epsilon A_{21}, \tag{7.59}$$

where $\tau_0$ is the time constant of the laser cavity, for example given by Eq. (4.11), and $\epsilon$ is the probability that a photon will be emitted spontaneously into the particular mode we are considering. Given the very large number of available modes for spontaneous emission, we neglect the term containing $\epsilon$ in Eq. (7.59). We cannot solve Eqs. (7.58) and (7.59), which are called the *Statz–deMars* equations, exactly, although it is easy to examine their solution numerically using a computer. To find an analytic solution we consider small departures from the equilibrium situation in which $dN/dt = 0$ and $dq/dt = 0$.

In equilibrium, $N = N_0$, where, from Eq. (7.59),

$$N_0 = \frac{1}{B\tau_0} \tag{7.60}$$

and, from Eqs. (7.58) and (7.60),

$$q_0 = R\tau_0 - \frac{1}{B\tau}. \tag{7.61}$$

There are no photons in the mode when $q_0 = 0$. This occurs when the pumping rate is at its threshold value, $R_t$, where

$$R_t = \frac{1}{B\tau\tau_0}. \tag{7.62}$$

We define a pumping factor $r$, where

$$r = \frac{R}{T_t}, \tag{7.63}$$

so Eq. (7.61) can be written

$$q_0 = \frac{r-1}{B\tau}. \tag{7.64}$$

We solve Eqs. (7.58) and (7.59) approximately by considering small oscillations about the equilibrium position. We write

$$\begin{aligned} N(t) &= N_0 + N'(t), \qquad N' \ll N_0, \\ q(t) &= q_0 + q'(t), \qquad q' \ll q_0. \end{aligned} \tag{7.65}$$

Substitution into Eqs. (7.58) and (7.59) gives

$$\frac{dN'}{dt} = R - B(q_0 + q')(N_0 + N') - \left(\frac{N_0 + N'}{\tau}\right), \tag{7.66}$$

$$\frac{dq'}{dt} = B(q_0 + q')(N_0 + N') - \left(\frac{q_0 + q'}{\tau_0}\right). \tag{7.67}$$

Neglecting the product of small quantities gives

$$\frac{dN'}{dt} = \left(R - Bq_0N_0 - \frac{N_0}{\tau}\right) - N'\left(Bq_0 + \frac{1}{\tau}\right) - Bq'N_0, \tag{7.68}$$

$$\frac{dq'}{dt} = \left(Bq_0N_0 - \frac{q_0}{\tau_0}\right) + Bq_0N' + q'\left(BN_0 - \frac{1}{\tau_0}\right). \tag{7.69}$$

By virtue of Eqs. (7.58) and (7.59), the first term in Eq. (7.68) and the first and third terms in Eq. (7.69) are zero. That is,

$$\frac{dN'}{dt} = -N\left(Bq_0 + \frac{1}{\tau}\right) - q'BN_0 \tag{7.70}$$

and

$$\frac{dq'}{dt} = NBq_0. \tag{7.71}$$

Differentiation of Eq. (7.70) followed by substitution from Eq. (7.71) gives

$$\frac{d^2N'}{dt^2} + \frac{dN'}{dt}\left(Bq_0 + \frac{1}{\tau}\right) + \frac{N'q_0B}{\tau_0} = 0. \tag{7.72}$$

Use of Eq. (7.61) gives us

$$\frac{d^2N'}{dt^2} + \frac{dN'}{dt}(RB\tau_0) + N'\left(RB - \frac{1}{\tau\tau_0}\right) = 0. \tag{7.73}$$

Similarly,

$$\frac{d^2q'}{dt^2} - Bq_0\frac{dN'}{dt} = 0. \tag{7.74}$$

Substitution from (7.70) gives

$$\frac{d^2q'}{dt^2} + Bq_0\left(Bq_0 + \frac{1}{\tau}\right)N' + q'Bq_0BN_0 = 0, \tag{7.75}$$

which becomes

$$\frac{d^2q'}{dt^2} + \frac{dq'}{dt}\left(Bq_0 + \frac{1}{\tau}\right) + q'\left(\frac{q_0B}{\tau_0}\right) = 0. \tag{7.76}$$

Finally, from Eq. (7.61)

$$\frac{d^2q'}{dt^2} + \frac{dq'}{dt}(RB\tau_0) + q'\left(RB - \frac{1}{\tau\tau_0}\right) = 0. \tag{7.77}$$

So, $q'$ and $N'$ both obey the same differential equation, which is logical since the fluctuations in photon density are closely coupled to fluctuations in population inversion.

For a trial solution of Eq. (7.77) we try $q' = q_0e^{-(\eta - j\xi)t}$. Remembering that

$$q_0 = \frac{RB\tau_0 - 1/\tau}{B},$$

we find

$$\eta = \frac{1}{2}\left(q_0B + \frac{1}{\tau}\right) = \frac{RB\tau_0}{2} = \frac{r}{2\tau} \tag{7.78}$$

and

$$\xi = \sqrt{\frac{q_0B}{\tau_0} - \frac{q_0^2B^2}{4} - \frac{q_0B}{2\tau} - \frac{1}{4\tau^2}} = \sqrt{\frac{r-1}{\tau\tau_0} - \left(\frac{q_0B}{2} + \frac{1}{2\tau}\right)^2}. \tag{7.79}$$

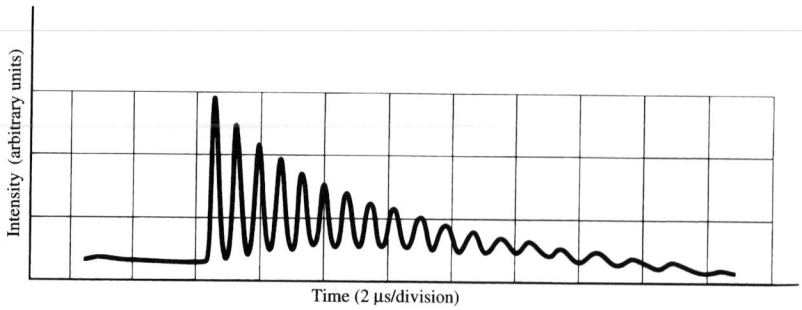


Fig. 7.31 Almost idealized relaxation oscillations obtained from an $Nd^{3+}$:$CaWO_4$ laser.

Equation (7.79) can be simplied by using Eqs. (7.61)–(7.63) to give

$$\xi = \sqrt{\frac{r-1}{\tau\tau_0} - \left(\frac{r}{2\tau}\right)^2}. \tag{7.80}$$

The photon density is a damped oscillatory function of angular frequency $\xi$, provided that

$$\frac{r-1}{\tau\tau_0} > \left(\frac{r}{2\tau}\right)^2,$$

of the form $q' = q_0 e^{-\eta t} e^{j\xi t}$.

If we take the specific example of a ruby laser with $\tau = 3 \times 10^{-3}$ s, $\tau_0 = 10^{-8}$ s, and $r = 2$, then Eq. (7.79) gives the frequency of the relaxation oscillation as $1.82 \times 10^5$ s$^{-1}$ and its period (i.e., $2\pi/\xi$) as $3.45 \times 10^{-5}$ s. Such a damped oscillation in the output of an $Nd^{3+}$:calcium tungstate laser is shown in Fig. 7.31 [23]. It would appear from Eq. (7.80) that for pumping far enough above threshold, when $r \gg 1$ and $r > 4\tau/\tau_0$, the output laser intensity would cease to be oscillatory. Such an assumption cannot be made with any certainty, however, because under these circumstances, $q'$ and $N'$ no longer satisfy the conditions $q' \ll q_0$ and $N' \ll N_0$, and our solution of the Statz–deMars equations breaks down.

## 7.15 Undamped relaxation oscillations

Under normal circumstances relaxation oscillations are damped, although the relaxation oscillation may become undamped if the pumping rate is varying with time. Suppose that $R = R_0 + r'(t)$, in which case Eq. (7.68) becomes

$$\frac{dN'}{dt} = r' - R_0 B \tau_0 N' - \frac{q'}{\tau_0} \tag{7.81}$$

and Eq. (7.61) takes the form

$$\frac{d^2q'}{dt^2} + \frac{r}{\tau}\frac{dq'}{dt} + \frac{1}{\tau\tau_0}(r-1)q' = -\frac{1}{\tau}(r-1)r'. \tag{7.82}$$

This is now the differential equation describing a damped oscillator that is being driven with a "force" $-(1/\tau)(r-1)r'$, and as such can have an oscillatory solution of large amplitude if the driving force has a frequency component near the natural frequency $\xi$.

Suppose that $r'(t)$ is oscillatory and can be written as

$$r'(t) = R'e^{j\xi t},$$

then Eq. (7.82) becomes

$$\frac{d^2q'}{dt} + \frac{r}{\tau}\frac{dq'}{dt} + \frac{1}{\tau\tau_0}(r-1)q' = -\frac{1}{\tau}(r-1)r_0e^{j\xi' t}. \tag{7.83}$$

We try as a solution $q' = Qe^{j\xi' t}$.

Substitution into Eq. (7.83) gives

$$-\xi'^2Q + j\xi'\frac{r}{\tau}Q + \frac{1}{\tau\tau_0}(r-1)Q = -\frac{1}{\tau}(r-1)r_0 \tag{7.84}$$

and

$$Q = \frac{(r-1)r_0}{\tau\xi'^2 - jr\xi' - (r-1)/\tau_0}. \tag{7.85}$$

From Eqs. (7.78) and (7.80), Eq. (7.85) can be written as

$$Q = \frac{(r-1)r_0/\tau}{(\xi' - \xi - j\eta)(\xi' + \xi - j\eta)}. \tag{7.86}$$

This amplitude becomes large as $\xi' \to \xi$, particularly if the damping factor $\eta$ is small.

Undamped relaxation oscillations are quite common in real lasers. A component in the pumping process that fluctuates at a natural resonance frequency of the laser system will tend to drive a relaxation oscillation at, or near, the resonance frequency. Figure 7.32 shows just such an oscillation, and its frequency spectrum observed in the output of a Nd:YAG laser pumped by a semiconductor laser diode. Such relaxation oscillations can be suppressed by minimizing fluctuations in the pump through feedback control, by increasing the damping factor $\eta$, or by channeling the fluctuations in the pump into a frequency region far from a natural resonance frequency of the laser.

## 7.16 Giant-pulse (Q-switched) lasers

In a pulsed solid-state laser the time dependence of the optical pumping leads naturally to "spiking" in the laser output. The pseudo-random nature of this light output is generally undesirable, but can be controlled by a technique called *Q-switching* [8–10]. If the feedback process within the oscillator is blocked, for example by preventing reflection from one of

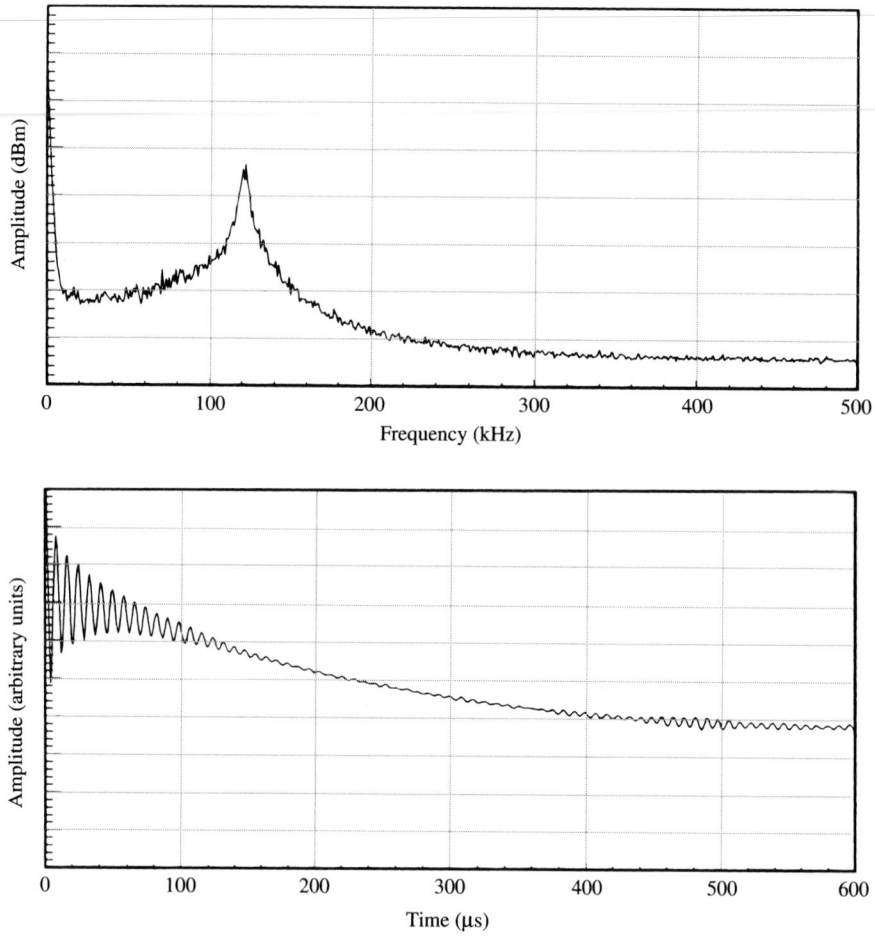


**Fig. 7.32** Relaxation oscillations in the output of a semiconductor-diode-laser-pumped Nd:YAG laser. (Courtesy of Dr. Simon P. Bush.)

the laser mirrors, then the cavity has low $Q$. If the lamp is fired, inversion builds up within the laser, but oscillation does not start. If the obstacle to feedback is suddenly removed, the $Q$ of the cavity switches to a higher value and oscillation begins. Because the laser now finds itself with an inversion much above threshold, the intracavity intensity grows very rapidly and the laser delivers a giant or *Q-switched* pulse of high intensity and short duration. This duration is typically on the order of the natural lifetime $\tau_0$ of the laser cavity described in Section 4.4:

$$\tau_0 = \frac{\ell}{c(1-R)}. \tag{7.87}$$

This giant pulse may so deplete the inversion in the laser that no subsequent laser emission occurs for that particular firing of the flashlamp. There are several ways of achieving

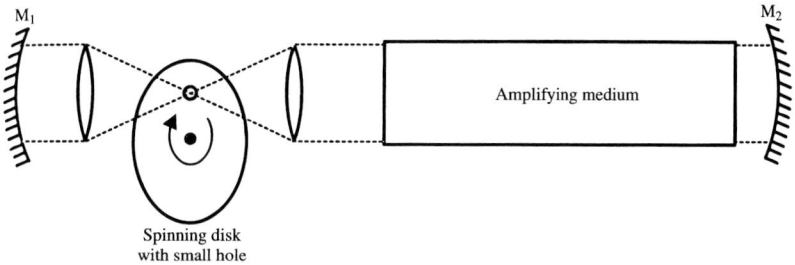


Fig. 7.33 A rotating intracavity wheel with a hole for $Q$-switching.

$Q$-switching [24–26]. They all rely on removal of an obstacle to reflection from one laser mirror after a substantial population has been stored in the upper laser level.

(a) Use of a spinning wheel with a hole or slot placed inside the laser cavity [27]. This method is largely of historic interest, but it is the easiest to understand. The lamp is synchronized to fire before the spinning wheel brings the hole into alignment. To provide a rapid transition from low to high $Q$, the hole in the wheel should be of small size, with a pair of lenses to focus the intracavity laser beam, as shown in Fig. 7.33.

(b) One of the end reflectors is spun about an axis orthogonal to the alignment axis of the laser cavity. This is most easily accomplished if one of the laser mirrors is replaced with a spinning roof prism, as shown in Fig. 7.34(a). A roof prism has the property of reflecting any ray incident orthogonal to the ridge line of the roof back parallel to itself. Therefore, if such a prism is spun about an axis perpendicular to the ridge, alignment of the laser cavity is guaranteed twice per rotation. Alternatively, the roof prism can remain fixed, or be fabricated directly onto the end of the laser rod. A conventional laser mirror spun about its diameter perpendicular to the ridge of the roof prism as shown in Fig. 7.34(b) is guaranteed to align twice per rotation.

(c) If a glass cell containing a dilute solution of absorbing dye such as cryptocyanine is placed inside the cavity then the oscillation threshold predicted by Eq. (5.9) is raised to

$$\gamma_t = \alpha - \frac{1}{\ell}[\alpha_d d + \ell n(r_1 r_2)], \tag{7.88}$$

where $\alpha_d$ is the absorption coefficient of the dye and $d$ is its thickness. The laser will not start to oscillate until this threshold has been reached. When oscillation does start, the increasing intracavity intensity "bleaches" the dye by exciting ground-state dye molecules into excited states. Solid materials with absorbing impurities can be used in a similar way [28]. This is the reverse of the gain-saturation process described in Chapter 2. In describing this process theoretically, provided that the $Q$-switched pulse is of short duration, spontaneous emission of the dye molecules can be neglected [29].

(d) With an optically active element in the laser cavity that changes the polarization state of the light traveling through it by virtue of the Kerr effect,[3] the Faraday effect,[4] or

[3] See Chapter 19.
[4] See Chapter 13.

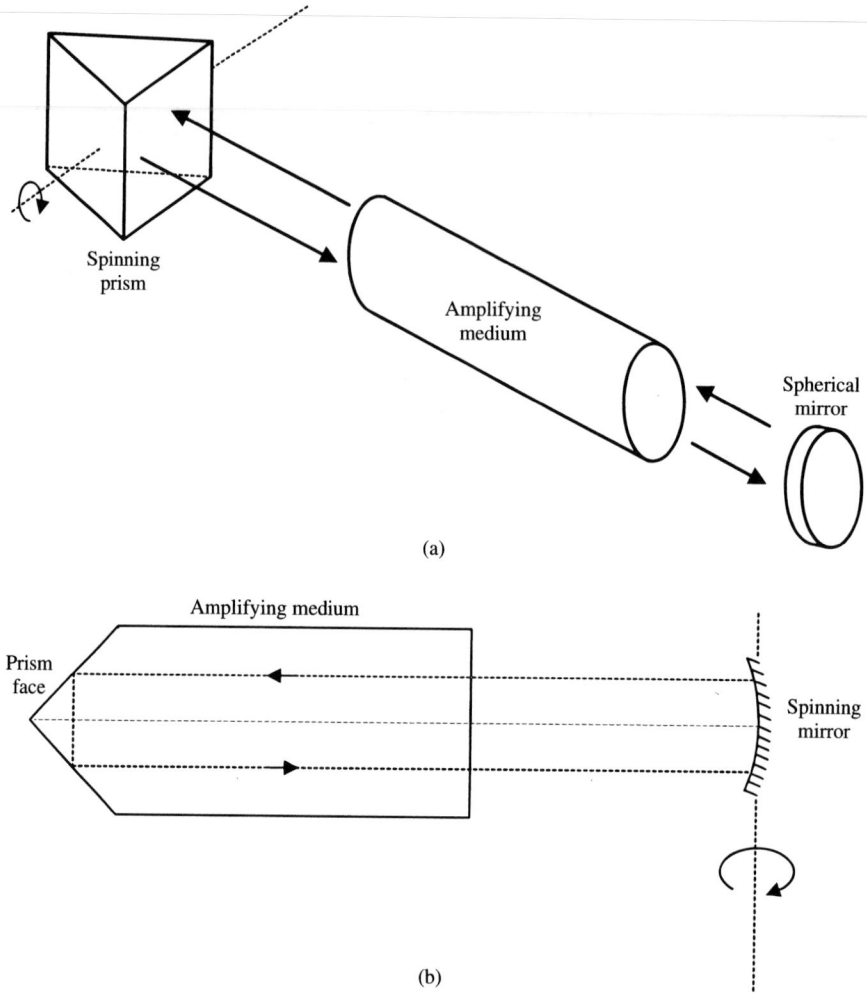


**Fig. 7.34** (a) *Q*-switching with one fixed cavity mirror and a rotating roof prism. (b) A laser rod with a roof prism fashioned on one end, where *Q*-switching is accomplished by spinning the other cavity reflector.

electro-optic activity in a crystal.[5] In a Kerr-active medium or electro-optic crystal a transverse electric field of sufficient magnitude will change linearly polarized light to circularly polarized or rotate its plane of linear polarization by 90°. In a Faraday-active medium an axial magnitude field causes the plane of linear polarization of a wave traveling along the field to rotate. The general way in which these phenomena can be used to build a *Q*-switch is best illustrated with Fig. 7.35, which shows an electro-optic crystal used with a transverse electric field as the switching element. The crystal has sufficient voltage applied to it that the electric field causes input linearly polarized light to be converted to circularly polarized light on output. Exactly why this

[5] See Chapter 18.

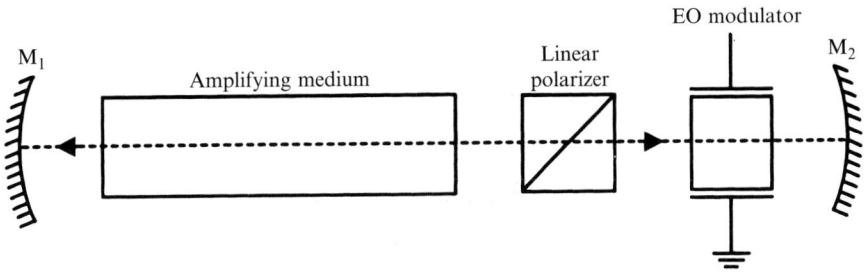


Fig. 7.35 An arrangement for $Q$-switching with an electro-optic (EO) crystal.

occurs in certain materials is explained in detail in Chapters 18 and 19. The circularly polarized light strikes mirror $M_2$, and returns through the crystal. In this second pass through the crystal the circularly polarized light is converted back to linearly polarized, but with its polarization direction orthogonal to its original direction. This light cannot pass the linear polarizer, and reflection from mirror $M_2$ is effectively blocked. Fast removal of the voltage applied to the crystal opens up mirror $M_2$ for reflection, and laser oscillation occurs.

## 7.17 A theoretical description of the $Q$-switching process

In a simple model of the $Q$-switching process, the $Q$-switched pulse is assumed to be short enough that spontaneous emission processes during the pulse can be neglected and the changeover from low $Q$ to high $Q$ is instantaneous [30]. For a resonator of volume $V$ the total number of photons is

$$\phi = Vq \tag{7.89}$$

and the total inversion is

$$n = \left(N_2 - \frac{g_2}{g_1}N_1\right)V. \tag{7.90}$$

The gain of the amplifying medium is $\gamma$. We neglect the additional complexity of gain saturation. Radiation of intensity $I$ passing through the amplifier grows according to

$$\frac{dI}{dz} = \gamma I \tag{7.91}$$

and, as a function of time,

$$\frac{dI}{dt} = \frac{dI}{dz}\frac{dz}{dt} = c\gamma I, \tag{7.92}$$

where $c$ is the velocity of light within the laser medium. For a laser cavity of length $\ell$ containing an amplifying medium of length $L$, only a fraction $L/\ell$ of all the light is being amplified. We can describe the average intensity increase within the cavity by Eq. (7.92) with a scale factor of $L/\ell$. The total number of photons changes according to

$$\frac{d\phi}{dt} = \phi\left(\frac{c\gamma L}{\ell} - \frac{1}{\tau_0}\right), \tag{7.93}$$

where the cavity time constant $\tau_0$ includes all passive loss effects such as mirror transmission and scattering within the amplifying medium. If we introduce a normalized time unit $\tau = t/\tau_0$, Eq. (7.93) becomes

$$\frac{d\phi}{d\tau} = \phi\frac{\gamma}{\ell/(cLT_0)} - 1. \tag{7.94}$$

Oscillation will not occur if $\gamma < \gamma_t = \ell/(cL\tau_0)$; in this case $d\phi/d\tau = 0$. Therefore we can write

$$\frac{d\phi}{d\tau} = \phi\left(\frac{\gamma}{\gamma_t} - 1\right), \tag{7.95}$$

and, since the gain is proportional to the population inversion,

$$\frac{d\phi}{d\tau} = \phi\left(\frac{n}{n_t} - 1\right). \tag{7.96}$$

Each stimulated emission reduces the population of the upper laser level by one and increases the lower level by the same number. Consequently,

$$\frac{dn}{d\tau} = -2\phi\frac{n}{n_t}. \tag{7.97}$$

Equations (7.96) and (7.97) can be solved numerically very easily, for example by a Runge–Kutta [31] method. Before giving examples of such solutions we can learn quite a lot about the way the population inversion and photon density behave by examining the equations. If we divide Eq. (7.96) by Eq. (7.97) we get

$$\frac{d\phi}{dn} = \frac{1}{2}\left(\frac{n_t}{n} - 1\right), \tag{7.98}$$

which has the solution

$$\phi = \frac{1}{2}(n_t\,\ell\mathrm{n}\,n - n) + \text{constant}. \tag{7.99}$$

If the initial photon density and inversion are $\phi_0$ and $n_0$, respectively, then

$$\phi_0 = \frac{1}{2}(n_t\,\ell\mathrm{n}\,n_0 - n_0) + \text{constant}. \tag{7.100}$$

Combining Eqs. (7.99) and (7.100) gives

$$\phi - \phi = \frac{1}{2}\left[n_t\,\ell\mathrm{n}\left(\frac{n}{n_0}\right) - (n - n_0)\right]. \tag{7.101}$$

Since there is negligible photon density before laser action starts, we set $\phi_0 = 0$. Therefore,

$$\phi = \frac{1}{2}\left[n_t\,\ell\mathrm{n}\left(\frac{n}{n_0}\right) - (n - n_0)\right]. \tag{7.102}$$

After the $Q$-switched pulse is over $\phi \to 0$ and the final inversion $n_f$ satisfies

$$n_t \, \ell n\left(\frac{n_f}{n_0}\right) - (n_f - n_0) = 0, \tag{7.103}$$

or

$$\frac{n_f}{n_0} = e^{-(n_0 - n_f)/n_t}. \tag{7.104}$$

The quantity $(n_0 - n_f)/n_0$ is the fraction of the original inversion that is converted to laser energy, which is given by

$$\frac{n_0 - n_f}{n_0} = 1 - e^{-(n_0 - n_f)/n_t}. \tag{7.105}$$

This approaches unity as $n_0/n_t \to \infty$.

If $n_f$ is very small, this means that the $Q$-switched pulse has very efficiently depleted the inversion. The value of $n_f$ can swing below the threshold value $n_t$, which is analogous to the underdamped discharge of a capacitor.

The output power of the laser is

$$P = \frac{\phi h\nu}{\tau_0}, \tag{7.106}$$

since the rate at which instantaneous energy stored is being lost from the cavity depends only on the cavity lifetime. Therefore, from Eq. (7.102),

$$P = \frac{h\nu}{2\tau_0}\left[n_t \, \ell n\left(\frac{n}{n_0}\right) - (n - n_0)\right]. \tag{7.107}$$

Maximum power output results when $d\phi/d\tau = 0$, which, from Eq. (7.96), occurs when $n = n_t$. The maximum power output is

$$P_{max} = \frac{h\nu}{2\tau_0}\left[n_t \, \ell n\left(\frac{n_t}{n_0}\right) - \left(n_t - n_0\right)\right]. \tag{7.108}$$

If the laser is $Q$-switched from far above threshold then $n_0 \gg n_t$ and

$$P_{max} \simeq \frac{n_0 h\nu}{2\tau_0}, \tag{7.109}$$

and the maximum number of photons stored in the cavity is

$$\phi_{max} = \frac{P_{max} h\nu}{\tau_0} = \frac{n_0}{2}. \tag{7.110}$$

All these predictions are borne out by actual numerical solution of Eqs. (7.96) and (7.97), as can be seen from Figs. 7.36 and 7.37. To summarize: these simulations demonstrate the following.

(i) The $Q$-switched pulse has higher amplitude and turns on faster as the initial inversion increases from the threshold value – see Fig. 7.36.
(ii) The inversion falls more rapidly at the onset of $Q$-switching as the initial inversion increases from the threshold value – see Fig. 7.36.

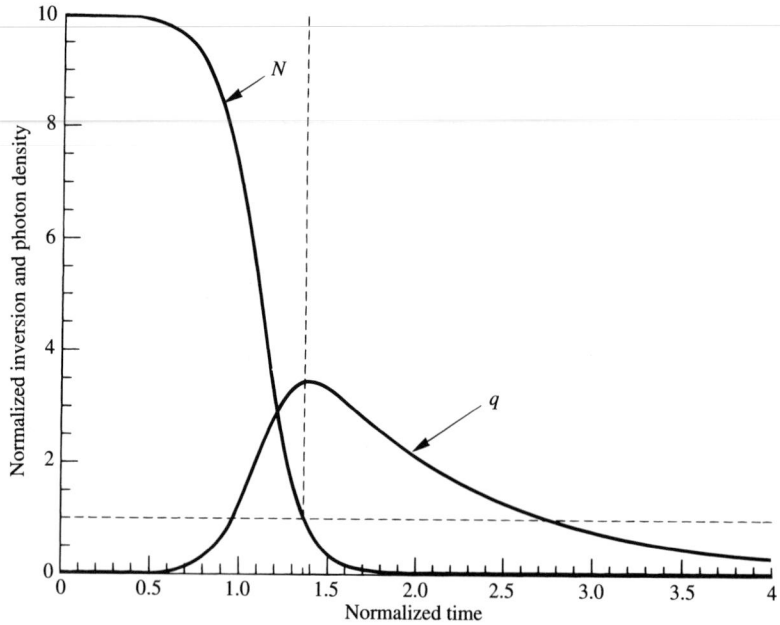


**Fig. 7.36** A numerical simulation of $Q$-switched behavior from Eqs. (7.105) and (7.106) for a system that is pumped initially to ten times above threshold.

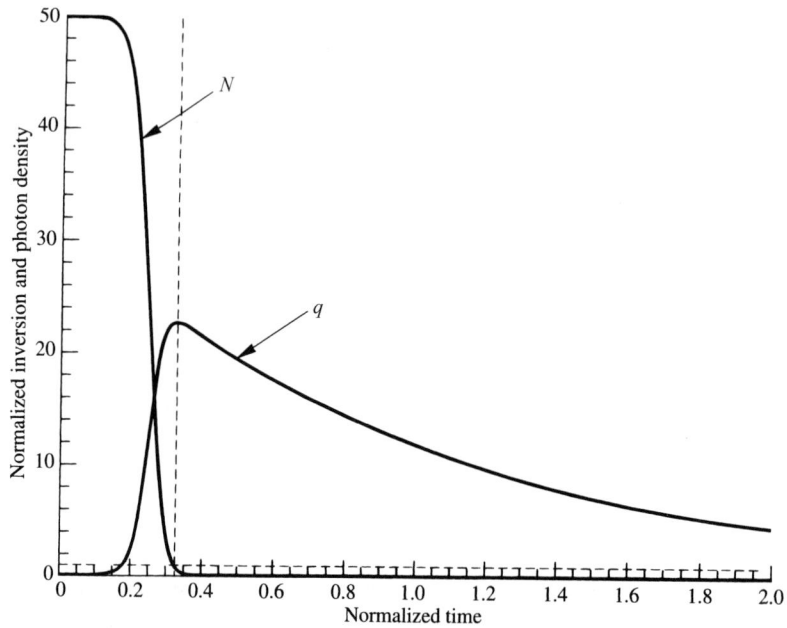


**Fig. 7.37** A numerical simulation of $Q$-switched behavior from Eqs. (7.105) and (7.106) for a system that is pumped initially to 50 times above threshold.

(iii) The peak photon density occurs when the inversion passes through the threshold value – see Figs. 7.36 and 7.37.
(iv) After the inversion has fallen to a low value the remaining photon density decays with the characteristic lifetime of the cavity.
(v) The final inversion falls increasingly below threshold as the initial inversion rises from the threshold value – see Fig. 7.36.

### 7.17.1 An example calculation of $Q$-switched pulse characteristics

We take as an example a Nd:YAG laser with a 1% $Nd^{3+}$ doping density. The density of YAG is 4,550 kg $m^{-3}$ so a 1% by weight doping level of $Nd^{3+}$ corresponds to $1.9 \times 10^{26}$ ions $m^{-3}$. For the Nd:YAG laser crystal considered earlier in this chapter the threshold inversion was $9.3 \times 10^{21}$ ions $m^{-3}$, which corresponds to an excited $Nd^{3+}$ ion population within the crystal volume ($\simeq 10^{-6}$ $m^3$) of $9.3 \times 10^{15}$. For this laser, with $R_{avg} = 0.96$ the cavity lifetime is

$$t_0 = \frac{\ell}{c(1 - R_{avg})} = \frac{50 \times 10^{-3}}{2.997 \times 10^8 \times 0.04} = 4.17 \text{ ns}.$$

If we can pump this crystal to 100 times above threshold then the small-signal gain will be 88.3 $m^{-1}$. The peak $Q$-switched power will be, from Eq. (7.109),

$$P_{max} \simeq \frac{9.3 \times 10^{17} \times 6.626 \times 10^{-34} \times 2.947 \times 10^8}{2 \times 4.17 \times 10^{-9} \times 1.06 \times 10^{-6}} = 2.09 \times 10^7 \text{ W}.$$

The total output pulse energy is

$$U \simeq \frac{n_0 h\nu}{2} \simeq 87 \text{ mJ},$$

where the factor of 2 in the denominator results from the "bottleneck" imposed by the lower laser level. Since the $Q$-switched pulse is very fast, ions that reach the lower laser level can be assumed to remain there during the pulse, so the inversion will be lost when half the initially excited ions have emitted.

## 7.18 Problem

7.1 Consider a ruby laser rod whose pump band absorption characteristics correspond to the data in Fig. 3.5, i.e., the concentration of $Cr^{3+}$ is $1.88 \times 10^{25}$ $m^{-3}$. Use the data given for $E \perp c$. Assume that the laser rod is rectangular (10 mm × 10 mm × 100 mm) and is closed-coupled to the flashlamp so that 25% of the flashlamp radiation enters the rod. Calculate the optical flux requirement from the lamp (W $m^{-2}$ $Hz^{-1}$) to exceed laser threshold by 100% on the $R_1$ transition.

(a) The two cavity mirror reflectances are 90% and 100%.
(b) The flashlamp pulse is a square pulse of $10^{-4}$ s duration with a flat spectral output from 300 nm to 700 nm.
(c) The efficiency $\eta$ for transfer of energy to the upper laser level is 50% for both pump bands.
(d) For the laser transition $\tau_2 \simeq 1/A_{21} = 3 \times 10^{-3}$ s, $\Delta\nu = 200$ GHz (homogeneously broadened).
(e) $T = 300$ K.

Take both the $2\overline{A}$ and the $\overline{E}$ upper ruby laser states into consideration. The ratio of degeneracy factors for the upper/lower levels is 1/2. Make a suitable approximation in evaluating the integral for the population of the upper laser levels.

## References

[1] Laser Program Annual Report, C. D. Hendricks, ed., Lawrence Livermore National Laboratory, Livermore, CA, UCRL-50021-83, 1984, 6–19.

[2] Laser Program Annual Report, C. D. Hendricks, ed., Lawrence Livermore National Laboratory, UCRL-50021-83, 1984, 6–22.

[3] D. F. Nelson and W. S. Boyle, "A continuously operating ruby optical maser," *Appl. Opt.* **1**, 181–183, 1962.

[4] D. F. Nelson and W. S. Boyle, "A continuously operating ruby optical maser," *Appl. Opt. Suppl.* **1**, 99–101, 1962.

[5] A. Sanchez, R. E. Fahey, A. J. Strauss, and R. L. Aggrawal, "Room temperature continuous-wave operation of a Ti:$Al_2O_3$ laser," *Opt. Lett.* **11**, 363–364, 1986.

[6] D. Röss, *Lasers, Light Amplifiers and Oscillators*, New York: Academic Press, 1968.

[7] D. Röss and G. Zeidlov, "Pumping new life into ruby lasers," *Electronics*, **39**, 115–118, 1966.

[8] T. Kushida, H. M. Marcos, and J. E. Geusic, "Laser transition cross section and fluorescence branching ratio for $Nd^{+3}$ in yttrium aluminium garnet," *Phys. Rev.* **167**, 289–291, 1968.

[9] Laser Program Annual Report, E. V. George, ed., Lawrence Livermore National Laboratory, Livermore, CA, UCRL-50021-81, 1982.

[10] Laser Program Annual Report, C. D. Hendricks, ed., Lawrence Livermore National Laboratory, Livermore, CA, UCRL-50021-83, 1984.

[11] S. E. Stokowski, "Glass lasers," in *Handbook of Laser Science and Technology, Vol. I. Lasers and Masers*, ed. M. J. Weber, Boca Raton, FL: CRC Press, 1982.

[12] Laser Program Annual Report, M. L. Rufer and P. W. Murphy, eds., Lawrence Livermore National Laboratory, UCRL-50021-85, 1986.

[13] J. McKenna, "The focusing of light by a dielectric rod," *Appl. Opt.* **2**, 303–310, 1963.

[14] C. H. Cooke, J. McKenna, and J. G. Skinner, "Distribution of absorbed power in a side-pumped ruby rod," *Appl. Opt.* **3**, 957–961, 1964.

[15] G. E. Devlin, J. McKenna, A. D. May, and A. L. Schawlor, "Composite rod optical masers," *Appl. Opt.* **1**, 11–15, 1962.

[16] M. J. Lubin, J. M. Soures, and L. M. Gelman, "A larger aperture Nd-glass face pumped laser amplifier for high peak power applications," General Electric Report No. 72CRD143, May 1972.

[17] T. Y. Fan and R. C. Byer, "Diode laser-pumped solid state lasers," *IEEE J. Quant. Electron.* **24**, 895–912, 1988.

[18] T. J. Kane and R. C. Byer, "Monolithic, unidirectional single-mode Nd:YAG ring laser," *Opt. Lett.* **10**, 65–67, 1985.

[19] Y. Jeong, A. J. Boyland, J. K. Sahu, S. Chung, J. Nilsson, and D. N. Payne, "Multi-kilowatt single-mode ytterbium-doped large-core fiber laser," *J. Opt. Soc. Korea* **13**, 416–422, 2009.

[20] M. J. F. Digonnet, ed., *Rare Earth Doped Fiber Lasers And Amplifiers*, 2nd revised edn., Basingstoke: Taylor & Francis Ltd., 2001.

[21] F. J. Duarte, ed., *Tunable Laser Applications*, 2nd edn., New York: CRC Press, 2009.

[22] H. Statz and G. deMars, "Transients and oscillation pulses in masers," in *Quantum Electronics*, Cambridge, MA: Columbia University Press, 1960, pp. 530–537.

[23] L. F. Johnson, "Optically pumped pulsed crystal lasers other than ruby," in *Lasers: A Series of Advances*, Vol. 1, ed. A. K. Levine, New York: Marcel Dekker, 1966.

[24] R. W. Hellworth, "Control of fluorescent pulsations," in *Advances in Quantum Electronics*, ed. J. R. Singer, New York: Columbia University Press, 1961, pp. 334–341.

[25] F. J. McClung and R. W. Hellworth, "Giant optical pulsations from ruby", *J. Appl. Phys.* **33**, 828–829, 1962.

[26] R. W. Hellworth, "Q modulation of lasers," in *Lasers: A Series of Advances*, Vol. 1, ed. A. K. Levine, New York: Marcel Dekker, 1966, pp. 253–294.

[27] R. J. Collins and P. Kisliuk, "Control of population inversion in pulsed optical masers by feedback modulation," *J. Appl. Phys.* **33**, 2009–2011, 1962.

[28] G. Bret and F. Gines, "Giant-pulse laser and light amplifier using variable transmission coefficient glasses as light switches," *Appl. Phys. Lett.* **4**, 175–176, 1964.

[29] A. Szabo and R. A. Stern, "Theory of laser giant pulsing by a saturable absorber," *J. Appl. Phys.* **36**, 1562–1566, 1965.

[30] W. G. Wagner and B. A. Lengyel, "Evolution of the giant pulse in a laser," *J. Appl. Phys.* **34**, 2042–2046, 1963.

[31] W. H. Press, S. A. Teukolsky, W. T. Vetterling, and B. P. Flannery, *Numerical Recipes, The Art of Scientific Computing*, 3rd edn., Cambridge: Cambridge University Press, 2007.

# 8 Gas lasers

## 8.1 Introduction

Gas lasers operate using a large number of different atomic and molecular gases and gas mixtures. On a macroscopic scale gases are intrinsically uniform, so the properties of the laser medium are immune to the defects and structural issues that affect lasers using solid materials. In addition, continuous variations of the composition and pressure of a gas medium offer a degree of flexibility in laser design. On the other hand, gases must be held in a container and lack the robustness offered by solid materials. Just as vacuum and gas-filled electronic tubes have been largely supplanted by solid-state devices, except in niche applications, gas lasers, especially atomic-gas lasers, have been largely replaced by lasers using solid active media. Even so, some gas lasers remain important in various applications, and the renewed interest in optically pumped gas lasers, which until relatively recently had seemed only of historical interest, suggests that no laser should necessarily be written off as "out-dated" prematurely. In this chapter we shall consider some of the fundamental processes which are used to produce, and maintain, population inversion in atomic gases. We shall see that the technological features of gas lasers, and the efficiency with which they can be made to operate, are intimately connected with the particular mechanism used to excite the upper laser level. Our attention will be concentrated on a consideration of gas lasers in which the laser action involves energy levels of a neutral or ionized atom. Consideration of most of the features of molecular-gas lasers will be delayed until the next chapter.

## 8.2 Excitation schemes

Most gas lasers are excited by electron collisions in various types of gas-discharge and electron-beam apparatus. (In the context of "gas lasers" at this point we exclude those gaseous lasers for which the pumping occurs as a result of chemical or photochemical processes.) However, there are several gas laser systems, particularly operating in the far infrared, in which optical pumping (with auxiliary lasers) has been successful. A special case of optical pumping involves the production of plasmas in which a population inversion results. These lasers are excited with another intense short-pulse laser, which focused on a solid target creates the plasma. Laser action in these laser-produced plasmas can provide emission into the soft X-ray region.

### 8.2.1 Optical pumping

Because gaseous systems do not in general have broad absorption bands, optical pumping of laser action in such systems with non-laser sources is not generally feasible. An early laser system that was an exception to this general rule was the pure-cesium laser first operated by Jacobs, Rabinowitz, and Gould [1, 2], and illustrated in Fig. 8.1. This system is only of academic interest, but it does provide a good example of where conventional optical pumping of gases can be successful. Cesium vapor has a strong absorption on the

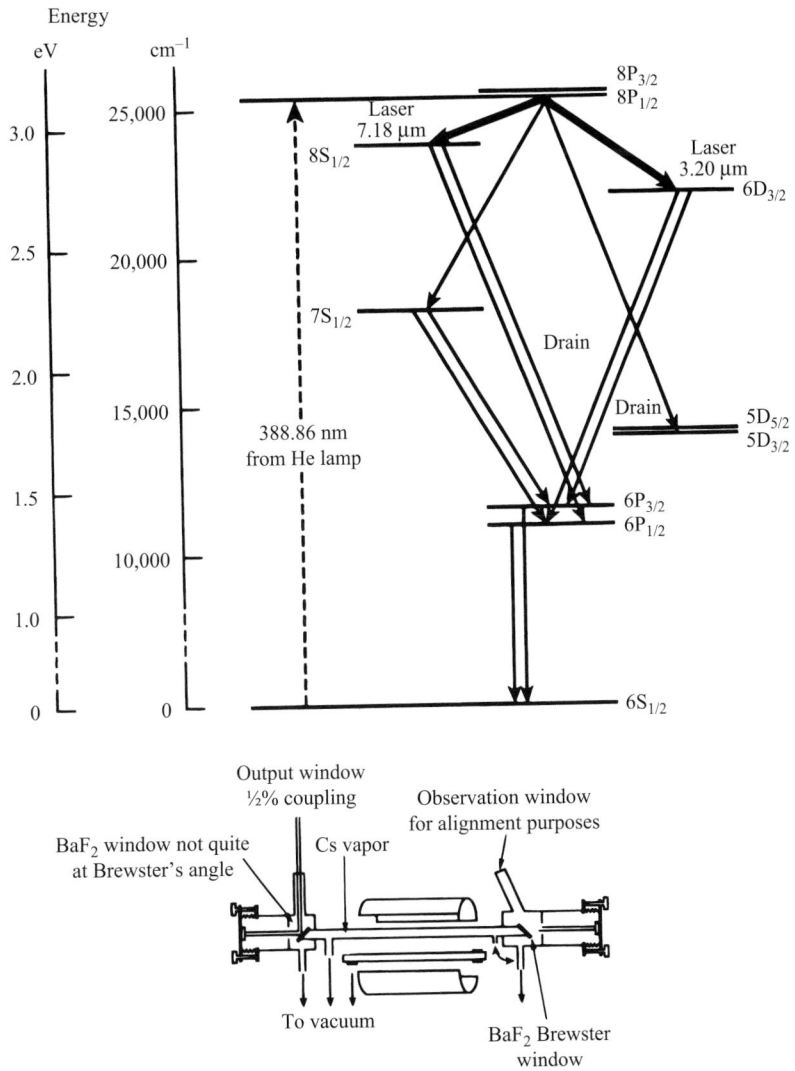


**Fig. 8.1** The optically pumped cesium laser of Jacobs *et al.* [1], showing the construction of the laser system and the coincidence between the helium pump line and cesium absorption. Laser radiation is coupled from the cavity of one of the laser tube end windows, which is set not quite at Brewster's angle and provides about 0.5% output coupling.

transition $6S_{1/2} \rightarrow 8P^0_{1/2}$ at 388.8616 nm, which is in very good coincidence with the strong helium line at 388.864 nm from the transition $3p\ ^3P_{2,1} \rightarrow 2s\,^3S_1$, as shown in Fig. 8.1. The line-broadening of the Cs atoms is sufficiently large that significant absorption of the helium transition occurs. Laser action from the Cs atoms was obtained at 7.18 μm, 3.20 μm, 1.38 μm, and 0.8919 μm. Population inversion results in these transitions because their lower levels decay rapidly with lifetimes shorter than that of their upper level. The power output at 7.18 μm was about 25 μW for an input of 800 W to the helium lamp. Rather more power output and efficiency has been obtained from the optically pumped mercury laser, which was first developed by Djeu and Burnham [3], where pumping of the mercury upper laser level is accomplished by successive absorption of two photons from a mercury pumping lamp, as illustrated in Fig. 8.2. Artusy, Holmes, and Siegman obtained ~3 mW output at 5461 Å with 300 W input to the pumping lamp [4]. The development of powerful semiconductor diode lasers, whose output wavelength can be controlled, as well as the linewidth of their laser emission, has led to a resurgence in interest in optically pumped gas lasers. This is especially true for the diode-pumped alkali-vapor gas lasers (DPALs). These DPALS were first proposed by Krupke in 2001 [5]. They use an alkali-metal vapor in a carrier-gas environment, which may involve a noble gas, usually helium, with possible addition of an additional quenching gas such as ethane or methane [6]. These lasers use a semiconductor diode laser (or lasers) to pump the $D_2$

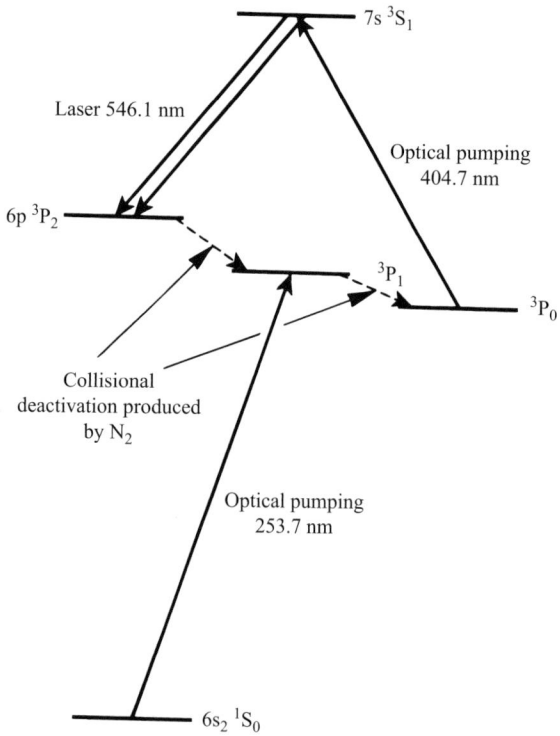


**Fig. 8.2** The energy-level scheme and excitation pathways in the optically pumped mercury laser.

**Table 8.1** Pump and lasing wavelengths of DPALS and their $\Delta E$ energy spacings

| Atom | $\lambda_{pump}(D_2)$ (nm) | $\lambda_{laser}(D_1)$ (nm) | $\Delta E$ (cm$^{-1}$) |
|---|---|---|---|
| Li | 670.96 | 670.98 | 0.34 |
| Na | 589.16 | 589.76 | 17.20 |
| K | 766.70 | 770.11 | 57.7 |
| Rb | 780.25 | 794.98 | 237.5 |
| Cs | 852.35 | 894.59 | 554.1 |

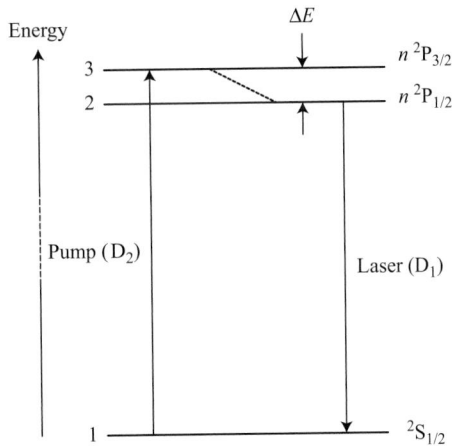


**Fig. 8.3** The energy-level scheme and pumping processes in a DPAL.

transition of the alkali metal. There is then an internal energy conversion to the upper level of the $D_1$ transition, which lases. The energy lost in this process ranges from 0.34 cm$^{-1}$ in lithium to 554.1 cm$^{-1}$ in cesium. In sodium it is 17.2 cm$^{-1}$. The excitation scheme is shown in Fig. 8.3. In sodium the D lines are the familiar lines emitted by sodium-vapor street lamps. Because these are three-level lasers the pump must be sufficiently intense to deplete the ground $^2S_{1/2}$ state. Because the lasing wavelength is only slightly longer than the pump wavelength, as shown in Table 8.1, these lasers have very high quantum efficiency, approaching 100%. The availability of increasingly powerful diode-laser pumps using GaAlAs and InGaAs that match the absorption of the alkali-metal vapors K, Rb, and Cs has suggested that DPALs *may* be scaled to MW powers [6].

Optical pumping of laser transitions in molecular gases by auxiliary lasers has had much more widespread success than optical pumping with conventional light sources. The availability of powerful pump lasers with either discrete tunability, for example the many vibration–rotation transitions in the $CO_2$ laser, or continuous tunability, for example from titanium–sapphire lasers, allows a laser to be operated in very close coincidence with an absorption of a molecular vapor, particularly since molecules possess many closely spaced absorption lines. The resultant selectively pumped level serves as the upper level of the laser transition. In this way visible and infrared laser oscillation (5,440 Å to 1.355 μm) has been obtained from iodine vapor pumped with a frequency-doubled (0.53 μm) Nd:YAG laser [7]. However, of more importance are the very many far-infrared laser transitions, spanning the far infrared right up into the millimeter-wave region, which can be excited by optically pumping various molecules such as $CD_3OH$, $C_2H_5OH$, $CH_3F$, $C_2H_5F$, $C_2H_2F_2$, $C_2H_4F_4$, HCOOH, $NH_3$, $CH_3NH_2$, $CH_3CN$, $CH_2CHCN$, $CH_3CCH$, $CH_3Cl$, $C_2H_5Cl$, $CH_2Cl_2$, $CH_2CHCl$, $CH_3Br$, and $CH_3I$, with $CO_2$ lasers [8–11]. The excitation scheme of a typical far-infrared laser of this type is shown schematically in Fig. 8.4. Further discussion of some of the technological aspects of these lasers appears later in Section 10.8.

### 8.2.2 Electron-impact excitation

Most successful excitation of laser action in gases is accomplished by electron-impact processes, for example directly, as in the pure neutral-argon laser,

$$\mathrm{Ar} + \mathrm{e} \rightarrow \mathrm{Ar}_{\mathrm{u}}^{*} + \mathrm{e},$$

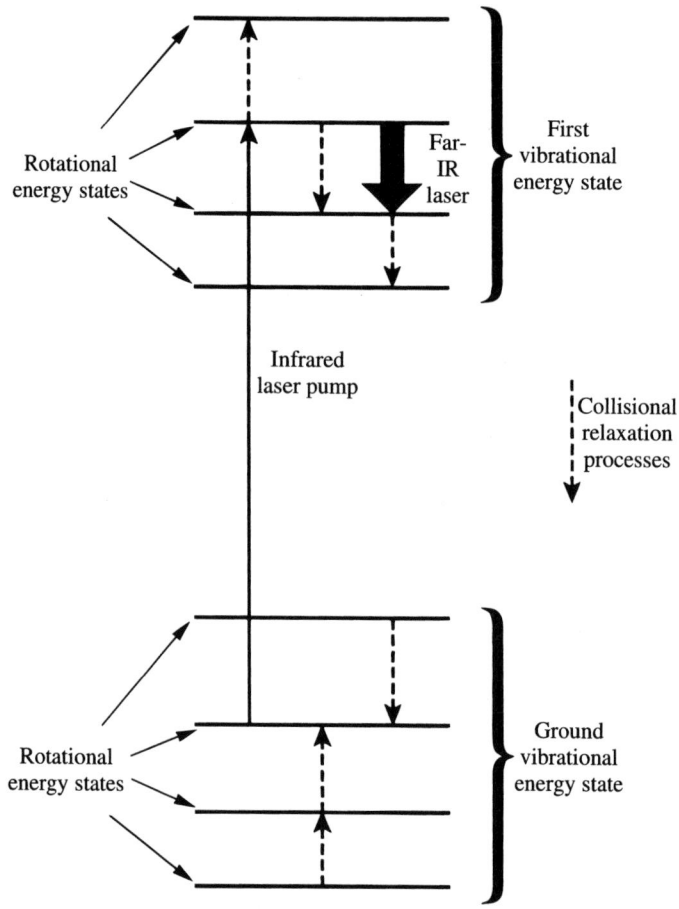


Fig. 8.4 A schematic energy-level diagram and excitation scheme in a laser-pumped far-infrared laser.

where $Ar_u^*$ denotes the upper laser level; by electron impact followed by radiative cascade,

$$\begin{aligned} Ar + e &\rightarrow Ar^* + e, \\ Ar^* &\rightarrow Ar_u^* + h\nu; \end{aligned}$$

or by excitation transfer from some species that is itself excited by electron impact, as for example in the He–Ne laser:

$$\begin{aligned} He + e &\rightarrow He_M^* + e, \\ He_M^* + Ne &\rightarrow He + Ne^*, \end{aligned}$$

where $He_M^*$ is a metastable helium atom. This process requires a close energy coincidence unless one of the colliding species is molecular and can take up excess energy, as for example in the Ne–$O_2$ system

$$\begin{aligned} Ne + e &\rightarrow Ne^*, \\ Ne^* + O_2 &\rightarrow O_u^* + O + Ne, \end{aligned}$$

or via a repulsive state of a molecule,

$$\begin{aligned} M^* + AB &\rightarrow AB^* + M, \\ AB^* &\rightarrow A_u^* + B, \end{aligned}$$

or

$$\begin{aligned} AB + e &\rightarrow AB^* + e, \\ AB^* &\rightarrow A_u^* + B. \end{aligned}$$

## 8.3 The argon-ion laser

An important gas-laser system where the excitation is by electron impact is the argon-ion laser [12]. In this system the excitation follows a number of paths, illustrated in Fig. 8.5, depending for their relative importance on the exact conditions in the system:

$$\begin{aligned} Ar + e &\rightarrow Ar_u^* + 2e && \text{one-step process} \\ Ar + e &\rightarrow Ar^+ + 2e \\ Ar^+ + e &\rightarrow Ar_u^{+*} + e && \text{two-step process} \\ Ar + e &\rightarrow Ar^+ + 2e \\ Ar^+ + e &\rightarrow Ar^{+*} \\ Ar^{+*} &\rightarrow Ar_u^{+*} + h\nu && \text{cascade process} \end{aligned}$$

and other similar processes involving metastable argon species. Similar processes prevail in the excitation of other noble-gas and similar ion lasers. The efficiency of direct

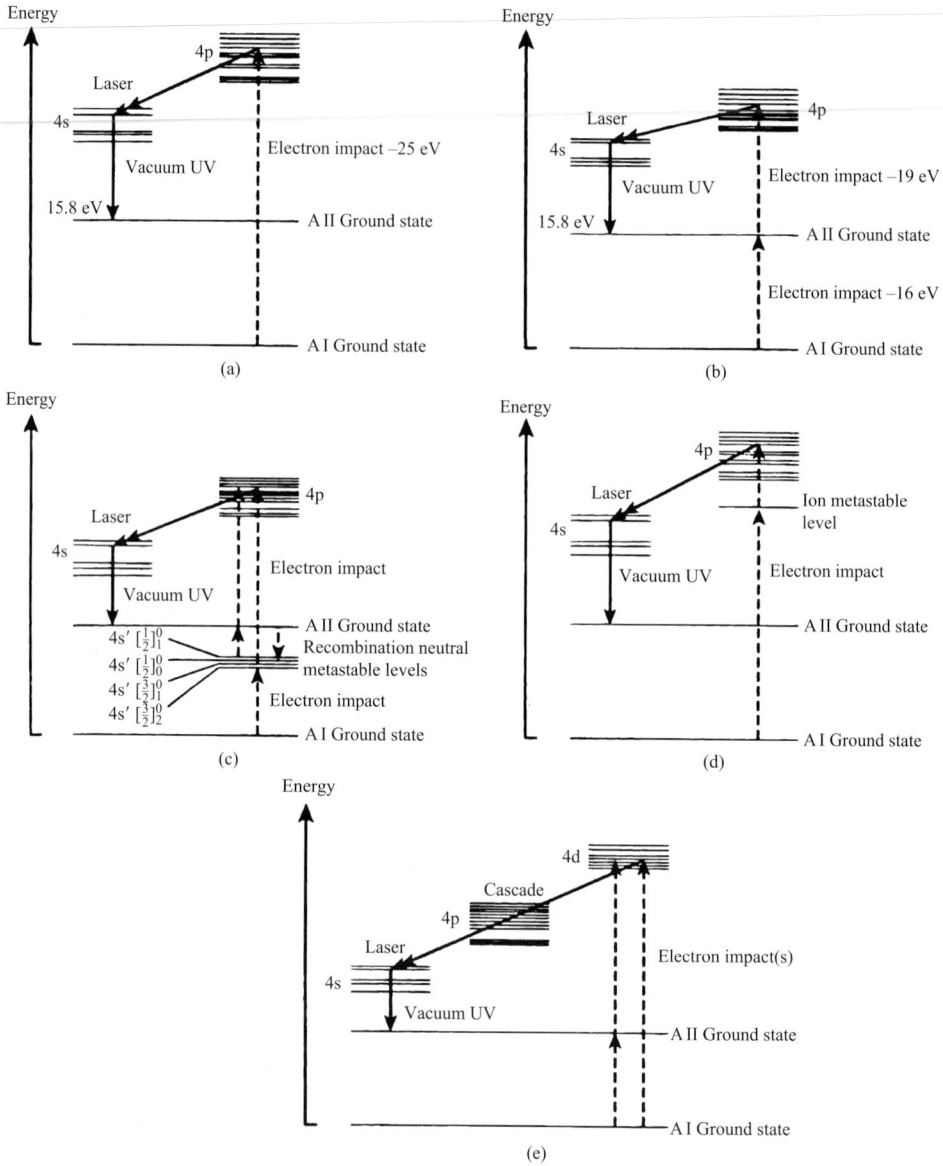


**Fig. 8.5** Excitation processes that occur in argon-ion lasers: (a) single-step excitation from a neutral ground-state atom; (b) two-step excitation involving the ion ground state; (c) one- and two-step excitation processes from neutral metastable levels; (d) two-step excitation involving ion metastable levels; and (e) one- and two-step excitation involving cascades.

electron-pumped lasers of this type depends on the relative cross-sections for the excitation of various laser levels. In a gas discharge, where the excitation is generally performed, the electrons have a range of energies. The *cross-section*, $\sigma_j(E)$, for excitation of level $j$ from the ground state by collision with an electron of energy $E$ is essentially a probability factor indicating how likely it is that this excitation will take place. We often assume that the distribution of electron energies $f(E)$ follows a Maxwellian distribution appropriate to some electron temperature $T_e$,

$$f(E) = 2\left(\frac{E}{\pi k_B^3 T_e^3}\right)^{1/2} e^{-E/(k_B T_e)}, \tag{8.1}$$

where $f(E)dE$ is the fraction of electrons with energy between $E$ and $E + dE$. The cross-section $\sigma_j(E)$ for a collisional excitation to a state $j$ when the electron has energy $E$ is defined from the rate of such processes:

$$R_j(E) = NN_e\sigma_j(E)v, \tag{8.2}$$

where $N$ is the density of the particles being excited, $N_e$ is the electron density, and $v$ is the electron velocity; $N_e v$ is the flux of colliding electrons. The total excitation rate is

$$R_j = NN_e \int_{E_j}^{\infty} f(E)\sigma_j(E)\left(\frac{2\epsilon}{m}\right)^{1/2} dE. \tag{8.3}$$

Electron-impact excitation of the laser system is effective when the rate of excitation of the upper level sufficiently exceeds that of the lower in relation to the ratio of their effective lifetimes.

The argon-ion laser and a number of other gaseous ion lasers fall into a category of gas laser that offers a significant advantage over gas lasers of the same general type as the helium–neon laser. This advantage is associated with the saturation of the pumping process.

## 8.4 Pumping saturation in gas-laser systems

In gas lasers such as the helium–neon laser, and many other neutral-gas lasers, such as the helium–xenon, argon, and xenon lasers, the excitation of the upper laser level involves either energy-transfer processes involving the collision of neutral species, or electron collision with neutral species leading to the direct excitation of the upper laser level. When the excitation involves energy transfer from excited neutral species the efficiency of excitation of the upper laser level depends on the concentration of these excited neutral species. When the concentration of these excited neutral species is maximized the rate of excitation of the upper laser level is also maximized. In general the rate of excitation of these neutral species in a gas discharge is a saturating function of the current density: a point at which the population of the neutral species attains a value essentially independent of the current density is reached. This situation arises because the neutral species are both created and destroyed by electron collisions. Thus in either an excitation-transfer or a direct-electron-impact-excited gas laser the population of the upper laser level cannot be increased beyond a certain point by further increasing the discharge current. In such systems the same saturation also occurs for the lower laser level but may, and frequently does, occur only when the lower level has already attained a population that quenches laser action, as illustrated in Fig. 8.6. The discharge current at which this saturation occurs is relatively low in neutral-gas laser systems (and in some ion laser system where the excitation involves

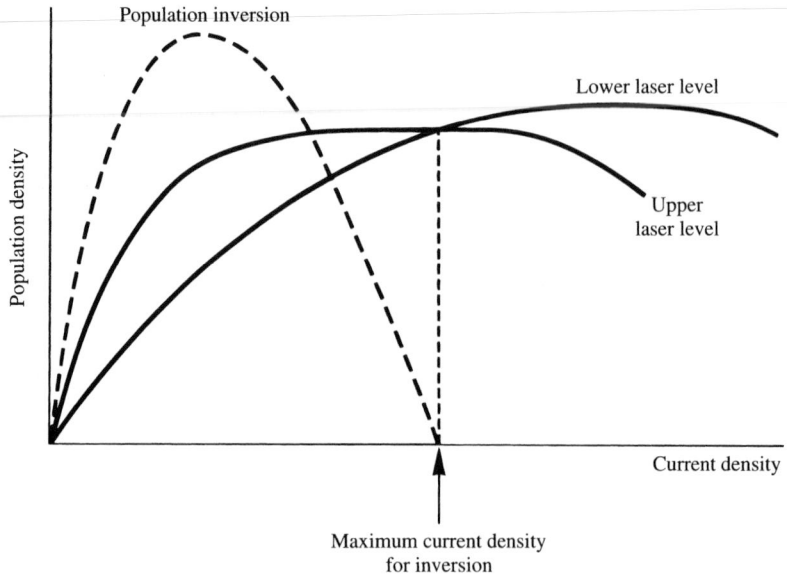


**Fig. 8.6** Schematic variation of upper and lower laser population densities and population inversion in a laser where pumping saturation of the upper laser level occurs before pumping saturation of the lower.

energy transfer from neutral species). For example, saturation of the upper laser level population (and hence the power output) occurs in helium–neon lasers at a current density $\lesssim 1$ A cm$^{-2}$. The same is true for CW helium–xenon and xenon lasers operating in the near infrared. However, if these lasers are excited in high repetition-rate short-pulse discharges at high pressure, average power outputs of several watts can be obtained [13]. The different electron energies that result in this mode of excitation allow high gas pressures and pulsed-discharge currents to be used. The gas in these lasers is excited in a transverse geometry with gas flow, an excitation technique that has proved very valuable in exciting various high-pressure gas lasers that will be descibed in Chapter 10. However, in laser systems where the excited levels are ionic, this saturation does not occur until very much higher current densities are reached; for example, in the pure-noble-gas ion lasers not until the current density reaches millions of amps m$^{-2}$. Since the power output is a function of the current density (typically it is $\propto I^n$, where $n$ is between 1 and 6) the saturated power output of ion lasers can be much larger than from neutral lasers. For example, an argon-ion laser 1 m long can under appropriate conditions, deliver a power output of tens of watts [14], whereas a helium–neon laser of equivalent length might deliver 100 mW at 632.8 nm at best.

## 8.5 Pulsed gas lasers

Because ion lasers, particularly the noble-gas ion lasers, depend for their efficient operation on the presence of significant numbers of ions in the discharge plasma, they operate best under discharge conditions of current and pressure very different from those found in

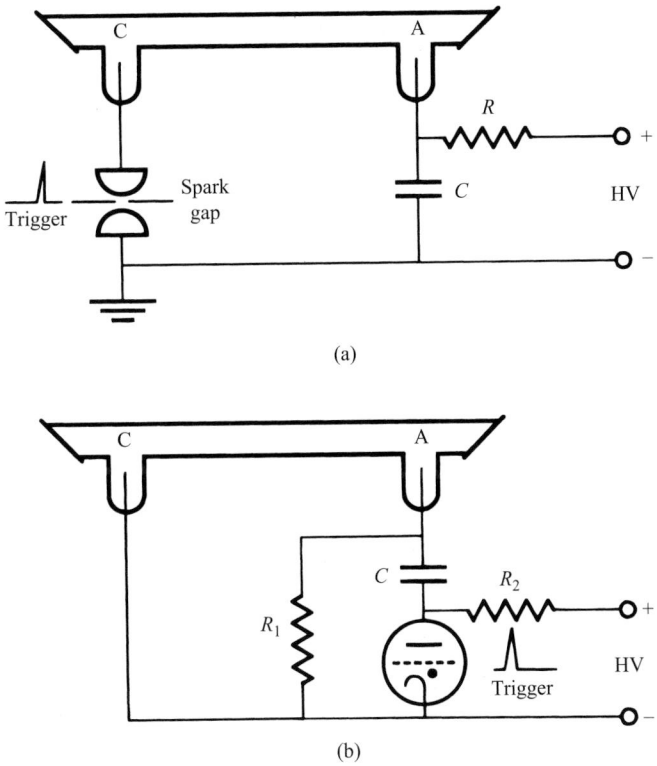


**Fig. 8.7** (a) A schematic representation of the arrangement for pulsed excitation of a gas laser tube in which the switching element is a triggered spark gap (C is the cathode; A is the anode). The charging time constant $RC$ is chosen to be less than the inverse of the operating pulse-repetition frequency (PRF). (b) As (a) but with a thyratron as the switching element. The charging time constant $R_2C$ is chosen to be less than 1/PRF. $R_1$ provides virtual grounding of the cathode during each charging cycle: $R_1 < CR_2$. However, $R_1$ is chosen to be larger than the impedance of the operating discharge. In high-PRF operation the charging resistors can be replaced by chokes (inductors).

neutral-gas lasers. By operating under conditions of extreme ionization, laser action even between levels of nickel-like gold $Au^{51+}$ has been obtained. Laser action can be obtained in such systems most easily by pulsed-discharge techniques, whereby a capacitor charged to high voltage is switched through the laser tube with a thyratron or spark gap as illustrated in Fig. 8.7. In this mode of operation, because the overall duty cycle of the discharge tube is small, thermal dissipation problems are small and the technology of the laser tube is simple, requiring only pyrex glass, or perhaps more refractive quartz tubulation, with ambient cooling. Currents of several tens of thousands of amps are easily handled. There are also some important pulsed gas lasers that operate on energy levels of neutral species such as copper and xenon. Some interesting laser transitions that can be excited in this way are listed in Table 8.2. These lasers are generally excited in tubes of small diameter (5–10 mm) with current densities up to $10^8$ A m$^{-2}$.

**Table 8.2** Some interesting pulsed laser transitions. Most of these data were obtained using a 7-mm-bore by 1.5-m discharge tube [15].

| Lasing species | Wavelength (nm) | Lasing species | Wavelength (nm) |
|---|---|---|---|
| Ne III | 247.34 | Xe III or IV | 231.54 |
| Ne II | 332.37 | Xe IV | 333.09 |
| Ne II | 337.35 | Xe IV | 364.54 |
| Ar IV | 211.4 | Xe IV | 430.58 |
| Ar III | 351.11 | Xe IV | 495.42 |
| Ar III | 363.79 | Xe IV | 500.78 |
| Ar II | 457.94 | Xe IV | 515.91 |
| Ar II | 476.49 | Xe IV | 526.02 |
| Ar II | 487.99 | Xe IV | 535.29 |
| Ar II | 496.51 | Xe IV | 539.46 |
| Ar II | 501.72 | O II | 374.95 |
| Ar II | 514.53 | O III | 375.99 |
| Kr IV | 219.19 | O III | 559.24 |
| Kr IV | 225.46 | N IV | 347.88 |
| Kr III | 350.74 | N III | 409.73 |
| Kr II | 743.58 | Cu I vapor | 510.54 |
| Cu I vapor | 570 | Cu I vapor | 578.2 |
| Xe I | 1,733 | Xe I | 2,027 |
| Xe I | 3,508 | Xe I | 3,652 |

If an intense pulsed gas discharge is confined to a small capillary tube then very highly ionized species can be produced and inversion that leads to soft-X-ray emission obtained. For example, Rocca *et al.* [16] were the first to demonstrate that this means of excitation could produce a compact soft-X-ray laser. They used 60-ns-long pulses with an amplitude of 40 kA in a capillary of diameter 4 mm to produce inversion in neon-like eight-times-ionized argon ($Ar^{8+}$) and emission near 47 nm. In this type of laser there is generally no cavity but directional, line-narrowed, laser emission results from the high gain produced and amplified spontaneous emission (ASE). Soft-X-ray emission in an ASE mode has been observed in plasmas created by focusing intense short-pulse lasers onto solid targets, for example [17], at 18.9 nm in nickel-like molybdenum, $Mo^{14+}$. By this means laser emission has been observed from 3.56 nm in Ni-like gold to 46.9 nm in neon-like argon [18].

Soft-X-ray emission has been achieved at wavelengths down to 1.47 nm in singly ionized neon, where the inversion was produced on an inner-shell $K\alpha$ transition [19]. This laser was excited by short pulses from an X-ray free-electron laser (see Chapter 11) focused into neon gas to produce an energetic plasma. Table 8.3 lists some of the XUV laser lines that have been observed from intense capillary discharges and laser-induced plasmas.

**Table 8.3** Pulsed XUV lasers [18]

| Ion | Wavelength(s) (nm) | Core | Ion | Wavelength(s) (nm) | Core |
|---|---|---|---|---|---|
| $_{79}$Au | 3.56 | Ni-like | $_{39}$Y 2 | 4.01 | Ni-like |
| $_{74}$W | 4.32 | Ni-like | $_{36}$Kr | 32.8 | Ni-like |
| $_{73}$Ta | 4.48 | Ni-like | $_{47}$Ag | 9.9365, 10.0377 | Ne-like |
| $_{72}$Hf | 4.65 | Ni-like | $_{42}$Mo | 10.64, 13.10, 13.27 | Ne-like |
| $_{70}$Yb | 5.609, 5.026 | Ni-like | $_{41}$Nb | 13.86, 14.04, 14.59 | Ne-like |
| $_{67}$Ho | 5.63, 6.20 | Ni-like | $_{40}$Zr | 15.04 | Ne-like |
| $_{66}$Dy | 5.85, 6.41 | Ni-like | $_{39}$Y 1 | 5.5 | Ne-like |
| $_{65}$Tb | 5.9, 6.7 | Ni-like | $_{38}$Sr | 15.98, 16.41, 16.65 | Ne-like |
| $_{64}$Gd | 6.33, 6.86 | Ni-like | $_{37}$Rb | 16.50, 17.35, 17.61 | Ne-like |
| $_{63}$Eu | 6.583, 7.100 | Ni-like | $_{34}$Se | 18.2, 20.6, 20.9 | Ne-like |
| $_{62}$Sm | 7.36, 6.85 | Ni-like | $_{33}$As | 21.884, 22.256 | Ne-like |
| $_{60}$Nd | 7.92 | Ni-like | $_{32}$Ge | 19.6, 23.2, 23.6 | Ne-like |
| $_{59}$Pr | 8.2 | Ni-like | $_{31}$Ga | 24.670, 25.111 | Ne-like |
| $_{58}$Ce | 8.6 | Ni-like | $_{30}$Zn | 21.2, 26.2, 26.7 | Ne-like |
| $_{57}$La | 8.9 | Ni-like | $_{29}$Cu | 22.11, 27.93, 28.47 | Ne-like |
| $_{54}$Xe | 9.64, 9.98 | Ni-like | $_{28}$Ni | 23.1 | Ne-like |
| $_{52}$Te | 11.1 | Ni-like | $_{26}$Fe | 25.49 | Ne-like |
| $_{50}$Sn | 11.97 | Ni-like | $_{25}$Mn | 22.1, 26.9 | Ne-like |
| $_{49}$In | 12.58 | Ni-like | $_{24}$Cr | 28.55, 40.22 | Ne-like |
| $_{48}$Cd | 13.17 | Ni-like | $_{23}$V 2 | 6.1, 30.4 | Ne-like |
| $_{47}$Ag | 13.89 | Ni-like | $_{22}$Ti | 32.63 | Ne-like |
| $_{46}$Pd | 14.68 | Ni-like | $_{21}$Sc | 31.2, 35.2 | Ne-like |
| $_{42}$Mo | 18.90 | Ni-like | $_{20}$Ca | 38.3 | Ne-like |
| $_{41}$Nb | 20.33 | Ni-like | $_{19}$K | 42.1 | Ne-like |
| $_{40}$Zr | 22.02 | Ni-like | $_{18}$Ar | 46.875 | Ne-like |

## 8.6 Continuous-wave ion lasers

The most important lasers in this category are the argon-, krypton-, xenon-, and neon-ion lasers, whose most important lines and typical relative power outputs are listed in Table 8.4. Because these lasers operate best at very high current densities, usually in excess of 100 A cm$^{-2}$, and have very low overall electrical efficiency, $\lesssim$ 0.1%, the amount of heat which is dissipated in the laser tube is very high and requires the selection of special refractory materials for the laser tube, as well as careful attention to thermal-stress and cooling problems. Because of the long-term reliability problems that result, high-current CW ion lasers are being replaced in many applications with diode-pumped solid-state lasers (see Chapter 11). In the first CW ion lasers, water-cooled fused quartz tubes were found adequate, but subsequently demand for higher laser power outputs, and hence power inputs, showed that other materials had greater strength and thermal shock resistance, could

**Table 8.4** Continuous-wave gaseous ion lasers [12, 20]

| Species | Wavelength (nm) | Species | Wavelength (nm) |
|---|---|---|---|
| Ne II | 332.4 | Kr II | 337.5 |
| Ne II | 337.8 | Kr II | 350.7 |
| Ne II | 339.3 | Kr II | 356.4 |
| Ar II | 333.4 | Kr II | 406.7 |
| Ar II | 333.6 | Kr II | 413.1 |
| Ar II | 351.1[a] | Kr II | 461.9 |
| Ar II | 363.8[a] | Kr II | 468.0 |
| Ar II | 454.5 | Kr II | 476.2 |
| Ar II | 457.9 | Kr II | 482.5 |
| Ar II | 465.8 | Kr II | 530.9[a] |
| Ar II | 476.5 | Kr II | 568.1[a] |
| Ar II | 488.0[a] | Kr II | 647.0 |
| Ar II | 496.5 | Kr II | 676.4 |
| Ar II | 501.17 | Kr II | 752.5 |
| Ar II | 514.5[a] | Kr II | 793.1 |
| Ar II | 528.7 | Kr II | 799.3 |
| Ar II | 1,092.3 | Xe III | 374.6 |
| Xe III | 378.1 | | |

[a] Strongest lines.

withstand greater thermal loading and the internal erosion caused by the gaseous discharge plasma itself, and were easier to cool. The various materials that have been used in modern ion lasers include beryllium oxide (BeO), graphite, discharge-confining structures made up of separate spaced metal disks, and, to a lesser degree, familiar refractory ceramics based on alumina ($Al_2O_3$). Important design features are illustrated by Fig. 8.8, which shows details of a metal-disk laser system. This figure indicates a number of the solutions to the problems of hot low-pressure plasma confinement, which were necessitated by the development of high-power ion lasers. Some interesting features are worthy of note.

(i) The Brewster windows, familiar in almost all gas-laser systems, can be replaced by antireflection-coated perpendicular windows. Such windows do not impose a prefered polarization direction on the laser output, which as a consequence tends to be unpolarized, or sometimes circularly polarized. However, the power output of the laser is not generally increased by the use of such windows instead of Brewster windows.

(ii) The population inversion usually occurs in a fairly dense gas-discharge plasma produced by passing a high current through a small-diameter tube or the central holes in a line of spaced insulating or metal disks.

(iii) The current density in this capillary region is increased by the application of an axial magnetic field whose magnitude does not generally exceed 1,000 gauss.

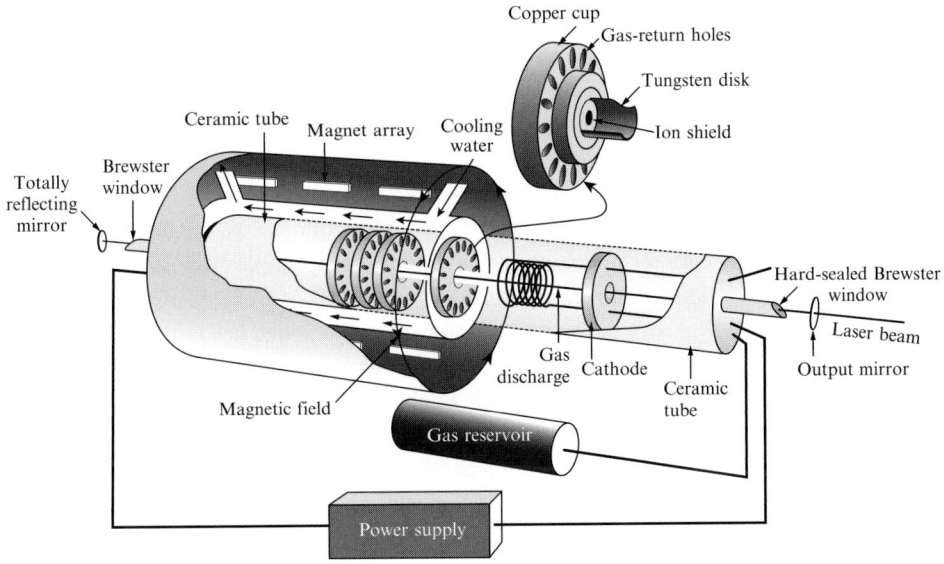


**Fig. 8.8** Construction of a modern ion laser.

The optimum field strength is governed by other factors such as the gas pressure, the discharge current, the tube, diameter, and the magnitude of Zeeman splitting of the laser transition induced by the field.

(iv) The high current flowing through the gas heats it considerably, so that it expands from the capillary region into the cooler parts of the system, leaving an active particle density in the capillary region much lower than the static filling pressure of the system would indicate.

(v) The high discharge current drives neutral particles towards the anode and creates a pressure differential across the tube, which must be prevented as much as possible by the use of an external pressure-equalizing flow path (through which the main discharge current should show no tendency to go), usually formed by internal bypass holes built into the disk structure of a segmented laser tube.

(vi) The high discharge currents ($>10$ A) for operation are generally supplied by a heated thermionic emitting cathode, of either oxide-coated or dispenser type (a cathode with a porous tungsten matrix that is impregnated with emitting material, which continually diffuses to and replenishes the emitting surface).

(vii) The main discharge capillary is water-cooled, either directly or by radiational cooling of the hot confining disks through an outer water-cooled fused-quartz envelope. Lower-power devices can operate with forced air cooling.

(viii) The hot discharge region should be thermally isolated from the resonator system, otherwise thermal instability of the resonator length, particularly during the warm-up period just after the laser has been turned on, will produce unstable operation.

(ix) Efficient laser action, at the 100-W level in the blue–green region, from tubes a little over a meter long, can be obtained from argon lasers operated in relatively wide-bore water-cooled quartz tubes at very high currents (>100 A) [14]. Krypton offers slightly less powerful operation in the yellow and red. A mixture of Kr and Ar can be used to provide a "white"-light laser output, which has been useful for TV and wide-screen display purposes.

(x) Continuous-wave ultraviolet laser action can be obtained at higher current densities than for operation in the visible, for example from excited singly ionized neon or doubly ionized argon and krypton.

(xi) Reactive gases are most conveniently excited in a radio-frequency external-electrode fused-quartz system. Internal thermionic cathodes cannot withstand the chemical reactivity of gases such as chlorine. It should be pointed out that there are significant physical differences between the low-current-density glow-discharge plasmas used to excite helium–neon, neutral-noble-gas, carbon dioxide, carbon monoxide, and other similar neutral-species lasers on the one hand, and the high-current, low-pressure, quasi-arc discharge used to excite noble-gas and other similar ion lasers on the other. The properties of these gas-laser-discharge plasmas are discussed in Section 8.8.

## 8.7 "Metal"-vapor ion lasers

These are ion lasers in which the laser transitions occur between singly ionized states (usually of low ionization potential) of metallic or semimetallic elements. They are excited by collisional processes involving helium or neon metastable atoms and ions in a discharge consisting largely of helium (or neon) with admixture of a small amount of the metal vapor. The excitation processes of greatest importance are, for example, in the case of helium–metal-vapor discharges

$$\mathrm{M} + \mathrm{He}_{\mathrm{M}}^{*} \rightarrow \mathrm{M}^{+*} + \mathrm{He} + \mathrm{e},$$

where $\mathrm{He}_{\mathrm{M}}$ is a metastable helium atom and $\mathrm{M}^{+*}$ is the upper laser level. This process is called Penning ionization [21]. Charge transfer is also an important excitation mechanism:

$$\mathrm{M} + \mathrm{He}^{+} \rightarrow \mathrm{M}^{+*} + \mathrm{He}.$$

Large cross-sections exist for these processes when the energy of the final state $\mathrm{M}^{+*}$ is close to (within $k_{\mathrm{B}}T$ of) the energy of the exciting particle $\mathrm{He}_{\mathrm{M}}^{*}$ or $\mathrm{He}^{+}$ (or the corresponding neon species) [22, 23].

Materials that have been excited into laser action in this way include cadmium, selenium, tellurium, iodine, arsenic, zinc, lead, tin, mercury, copper, silver, and gold [24]. The charge-transfer process is the more common of the two above excitation mechanisms. As an excitation process, it offers the significant advantage over Penning ionization of having a much higher current-saturation level since the exciting species are ions, rather than

**Table 8.5** Continuous-wave metal-vapor ion lasers and their excitation mechanism [12]

| Species | Wavelength (nm) | Excitation mechanism |
|---|---|---|
| Cd II | 325.0 | Penning ionization |
| Cd II | 441.6 | Penning ionization |
| Zn II | 491.2 | Charge transfer |
| Zn II | 492.4 | Charge transfer |
| Zn II | 589.4 | Penning ionization |
| Zn II | 747.9 | Penning ionization |
| Se II | 497.6 | Charge transfer |
| Se II | 499.3 | Charge transfer |
| Se II | 517.6 | Charge transfer |
| Se II | 522.5 | Charge transfer |
| Se II | 522.9 | Charge transfer |
| Se II | 530.7 | Charge transfer |

neutral metastable species. The latter, for reasons discussed previously, have populations that saturate at low current levels.

Some of the more important metal-vapor laser transitions are listed in Table 8.5 together with their dominant excitation mechanism. In many respects these lasers are comparable to the helium–neon laser, insofar as they operate in relatively low-current-density glow discharges (of at most a few A $cm^{-2}$) and offer comparable power outputs and efficiencies. The He–Cd laser does, however, offer the two very attractive short wavelengths 441.6 nm ($\sim$100 mW from a 1-m tube) and 325 nm ($\sim$3 mW from a 1-m tube), whilst the He–Se laser offers slightly greater efficiency for the production of blue–green laser radiation than that of the argon-ion laser at low power levels ($\lesssim$100 mW).

The discharge structures in which laser action is excited in metal-vapor lasers are of two main types: (a) cataphoretic flow and (b) hollow cathode [25], which are shown in Fig. 8.9. In the cataphoretic-flow discharge, metal atoms are evaporated from an oven at the anode end of a helium or neon gas discharge at a few torr pressure and are swept along the discharge towards the cathode by cataphoresis (a process whereby heavy atoms in a glow discharge at relatively low current experience a net cathode-directed force). If the conditions of operation are optimum, a uniform distribution of a few mtorr of metal atoms occurs throughout the positive column of the discharge. When they reach the cathode end of the positive column the metal atoms largely recondense, so it is possible to use a thermionic cathode, although a large cold cathode (usually aluminum) is adequate to provide the currents necessary for laser oscillation. In the hollow-cathode structure the excitation region consists largely of the extended negative glow inside the cathode. This is a region containing large numbers of energetic electrons, which generally makes it more suitable for the operation of charge-transfer-excited lasers in which the production of helium ions is necessary.

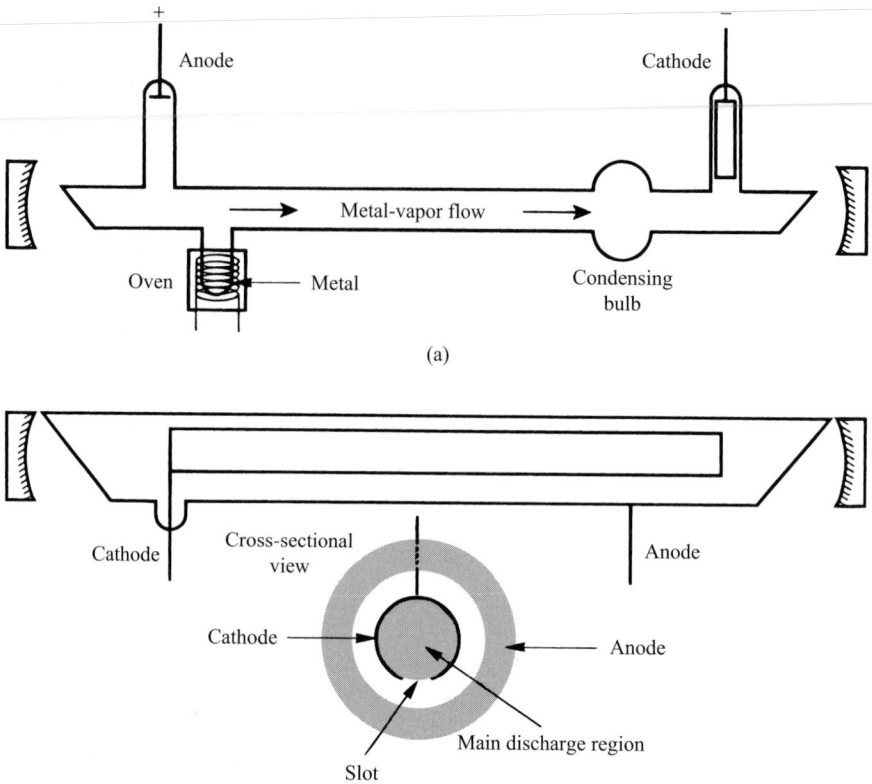


**Fig. 8.9** Metal-vapor laser structures: (a) a cataphoretic flow metal-vapor laser and (b) a slotted hollow cathode metal-vapor laser.

## 8.8 Gas discharges for exciting gas lasers

If one steadily increases the current through a gas discharge and plots the resulting $V/I$ characteristic of the tube, the schematic behavior illustrated in Fig. 8.10 is observed. The general arrangement for conducting such an experiment is also shown in the figure. In the normal glow and arc regions the $V/I$ characteristic is almost flat, or has a negative slope, so the series ballast resistor is very important in order to prevent current runaway. The various regions of operation of the discharge illustrated in Fig. 8.10 are characterized by particular physical phenomena. A $\rightarrow$ B: the Townsend dark discharge occurs when the number of electrons flowing through the gas is insufficient, and their mean energies are too low, for significant secondary ionization and subsequent cumulative current growth to occur. The current flow will consist mainly of the drift of residual free charges in the gas, for example those produced by cosmic radiation and radioactivity of the discharge tube and its gas fill. B $\rightarrow$ C: cumulative secondary ionization grows and increases the current flow and the necessary operating voltage falls. C $\rightarrow$ D: the normal-glow region. The discharge is now self-sustaining by secondary ionization; the current is liberated from the cathode

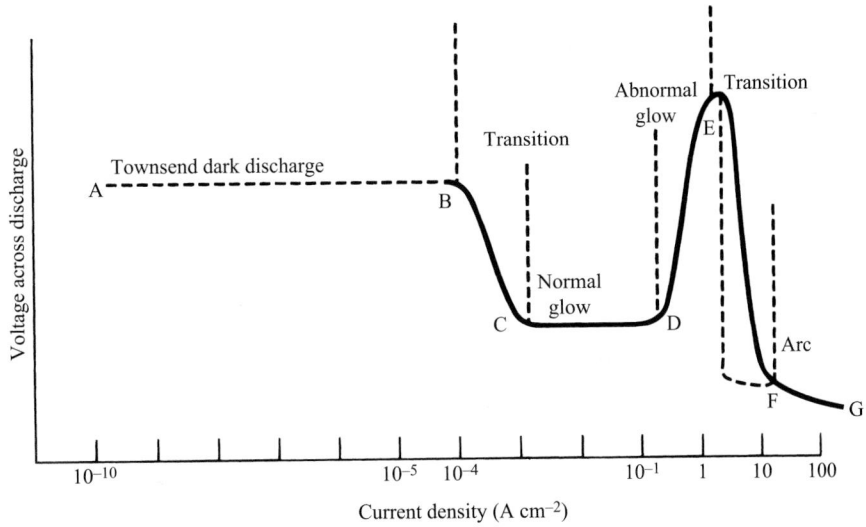


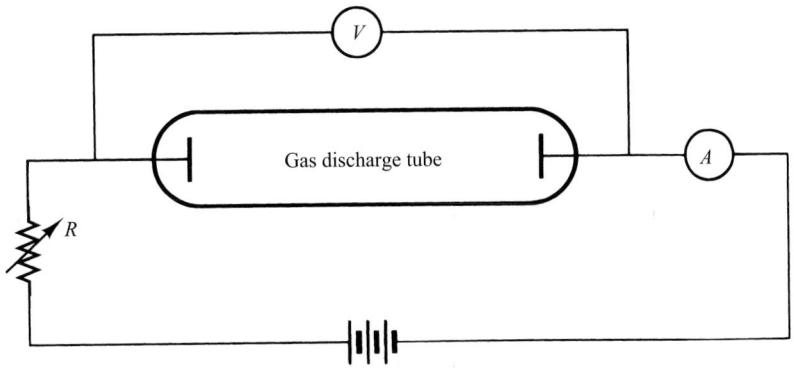


**Fig. 8.10** A schematic $I$–$V$ characteristic of a gas discharge over several orders of current density.

by ion bombardment (plus thermionic emission if the discharge is being operated with a thermionic cathode). The normal glow is characterized by a constant current density at the cathode surface (and in the discharge tube itself if this allows expansion of the column of conducting gas with increasing current). Because the normal glow has such a flat $V/I$ characteristic it is essential to stabilize the current with a ballast resistor. D → E: when the whole cathode surface has been covered in the operation of a glow discharge, the sustaining voltage must now be increased to provide additional current flow. E → F: the transition to arc operation can occur abruptly, and this region might not be observed very readily in practice. F → G: the arc is characterized by high current flow and low sustaining voltage. Emission from the cathode, if this is not provided by external heating of thermionic materials, can occur because of the production, by ion bombardment, of local thermionically emitting hot spots. The gas-discharge plasma of the arc is becoming highly ionized.

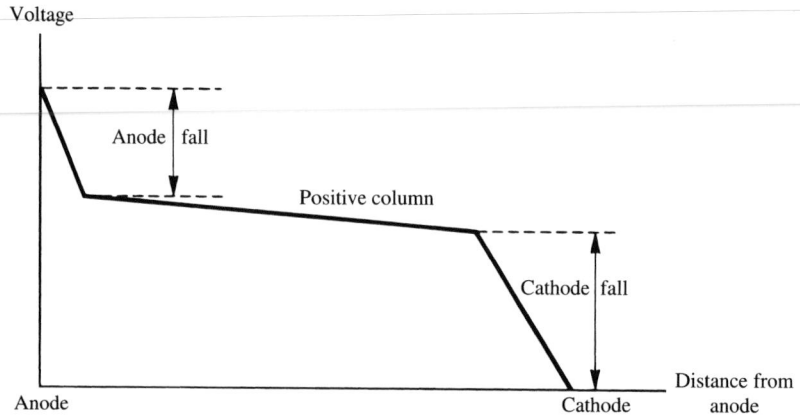


**Fig. 8.11** A schematic voltage drop with position along a gas discharge tube in the glow or arc region.

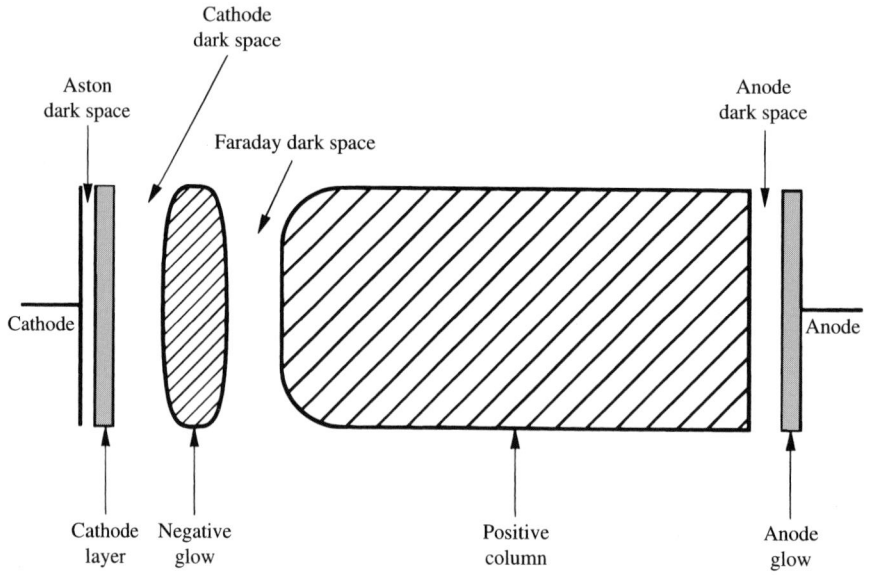


**Fig. 8.12** A schematic diagram of the appearance of a glow discharge in a cylindrical tube, showing the various regions of the discharge.

In the region of glow and arc-discharge operation, the voltage/length characteristic of the gas discharge is schematically of the form shown in Fig. 8.11. The bulk of the region between the anode and the cathode is filled by the luminous positive column, which is a quasi-neutral plasma. In glow-discharge-excited lasers the charged-particle density in this region is typically $10^{22}$–$10^{24}$ $m^{-3}$. In the arc region the charged-particle density is typically $>10^{25}$ $m^{-3}$. Although the arc is a more highly ionized plasma than the glow-discharge, the electron temperature in the plasma is not necessarily higher.

In the region of glow-discharge operation the discharge is characterized by a number of distinct luminous and non-luminous regions, as illustrated in Fig. 8.12. Electrons that leave the cathode pick up energy in the cathode fall until they acquire sufficient energy to excite visible radiation from the particles of the gas; this produces the negative glow. The region near the cathode where they have insufficient energy to excite such radiation is the cathode dark space. In the negative glow the electrons lose energy by exciting gas particles until they no longer have sufficient energy to excite visible radiation, so another dark space, the Faraday dark space, results before the electrons have picked up sufficient energy to excite visible radiation again at the head of the positive column. This successive acceleration and deceleration of the discharge electrons can cause the positive column to break up into bright and dark segments called *striations*.

## 8.9 Rate equations for gas-discharge lasers

The discharge plasma of an electrically excited gas laser is characterized by a number of parameters that vary from one type of laser to another and crucially affect the way in which such different lasers operate. Since the electrons of the discharge are fundamentally responsible, directly or indirectly, for the excitation both of the upper and of the lower laser levels, the energy distribution of these electrons and their density are crucial in determining whether a population inversion can be produced in the gas discharge. It is possible to formulate simple rate equations that describe the excitation of the upper and lower levels of a gaseous laser system. Let us consider the simple system shown in Fig. 8.13, where both the upper and the lower levels of the laser transition are populated by electron collision. This will allow us to see how the operation of the system is affected by changes in the electron energy distribution and density. Let the ground-state density in the system be $N_0$, let the densities of the upper and lower laser levels be $N_2$ and $N_1$, respectively, and let the density of a general higher lying level be $N_i$. The electron density is $N_e$. The rate of change of population of the upper laser level is

$$\frac{dN_2}{dt} = N_0 N_e \langle \sigma_{01} v \rangle + \sum_i N_i A_{i1} - \frac{N_1}{\tau_1}, \tag{8.4}$$

where the summation over $i$ here includes the upper laser level and $\langle \sigma_{01} v \rangle$ is the average of $\sigma_{01} v$ over the electron velocity distribution. In equilibrium

$$\frac{dN_2}{dt}, \frac{dN_1}{dt} = 0.$$

The condition for population inversion is that

$$\frac{N_2}{g_2} - \frac{N_1}{g_1} > 0,$$

i.e.,

$$\frac{\tau_2}{g_2}\left(N_0 N_e \langle \sigma_{02} v \rangle + \sum_i N_i A_{i2}\right) > \frac{\tau_1}{g_1}\left(N_0 N_e \langle \sigma_{01} v \rangle + \sum_i N_i A_{i1}\right). \tag{8.5}$$

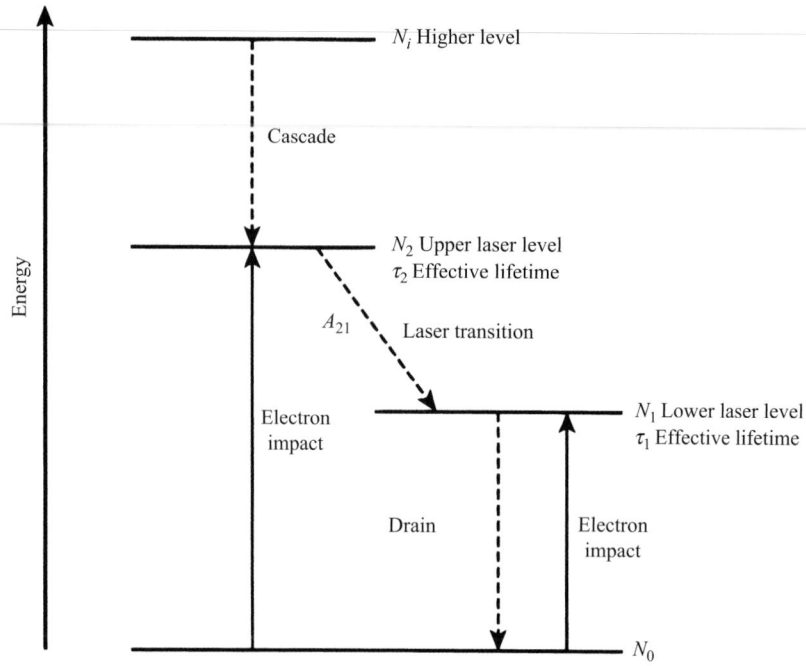


**Fig. 8.13** A schematic energy-level diagram of a gas laser system in which levels are excited by electron collisions.

If we neglect the radiative-cascade processes, this reduces to

$$\langle \sigma_{02} v \rangle > \frac{\tau_1}{\tau_2} \frac{g_2}{g_1} \langle \sigma_{01} v \rangle. \tag{8.6}$$

Thus the obtaining of population inversion is assisted by (i) a favorable cross-section ratio and (ii) favorably small values of the lifetime ratio, $\tau_1/\tau_2$, and of the ratio of degeneracy factors, $g_2/g_1$. The magnitude of the achieved population inversion is

$$N_0 N_e \left( \langle \sigma_{02} v \rangle \frac{\tau_2}{g_2} - \langle \sigma_{01} v \rangle \frac{\tau_1}{g_1} \right),$$

which increases with the electron density. Let us examine the cross-section ratio in a little more detail. The cross-section for the upper and lower laser levels may be of the form shown schematically in Fig. 8.14, in which case a favorable cross-section ratio exists for electron energies above $E_m$ and an optimum electron energy distribution exists for the velocity-averaged cross-section difference given by $\langle \sigma_{02} v \rangle - \langle \sigma_{01} v \rangle$ to be as large as possible. If the electron energy distribution is Maxwellian and can be characterized by an electron temperature $T_e$, then, in such a system as the above, an increase of the electron temperature will not automatically destroy the inversion, but it will eventually reduce its magnitude. An optimum value of the electron temperature exists for the production of the maximum inversion. In some laser systems the cross-sections may instead be of the form shown schematically in Fig. 8.15. A situation such as that shown in Fig. 8.15 can arise when the electron-impact process exciting the upper laser level corresponds to an "optically

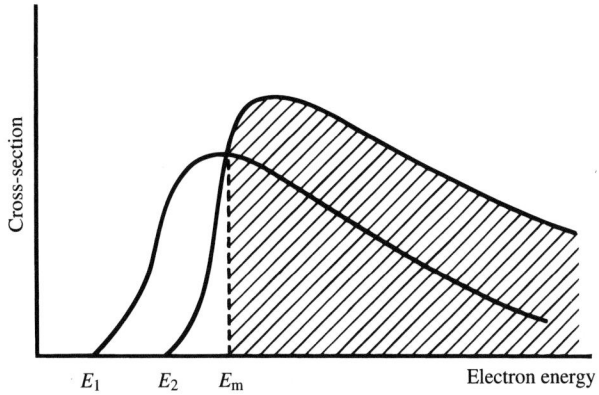


**Fig. 8.14** The schematic variation of the collision cross-section for excitation of upper and lower laser levels as a function of electron energy. Electrons with energy above $E_m$ are more efficient at exciting the upper laser level than the lower. In this case inversion is not reduced by an increase in electron energy.

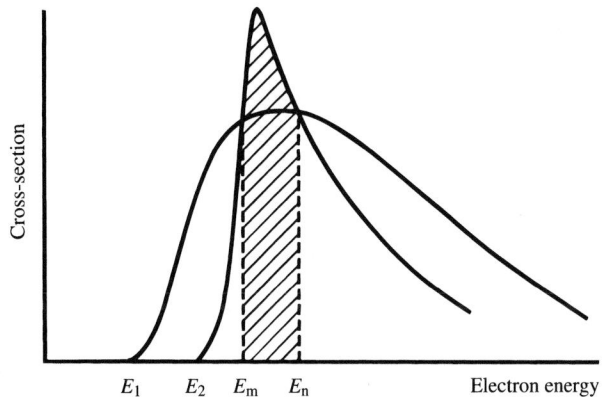


**Fig. 8.15** The schematic variation of excitation cross-section with electron energy for upper and lower laser levels where inversion can be achieved only over a specific range of electron energies.

forbidden" transition and the cross-section for the lower laser level corresponds to an "optically allowed" transition. For example, in the two-step excitation of the singly ionized noble-gas-ion laser levels from the ion ground state, the excitation of the upper laser levels is "optically forbidden," since these levels do not make electric dipole transitions to the ground state. The excitation of the lower laser levels from the ion ground state is allowed, as one would expect, since the lower laser levels make very strong, short-lifetime optical transitions to the ground state. In systems of this kind an inversion is possible only (in our simplified model) for a small range of electron temperatures, and a continued increase of the electron temperature will ultimately destroy the inversion. The electron temperature which prevails in a gas discharge is mainly a function of the electric field gradient $E$ and the pressure $p$. The electric field influences the rate at which electrons pick up energy between collisions, while the pressure controls the mean free path between these collisions. It is clear that

$$T_e = f\left(\frac{E}{p_0}\right),$$

where $p_0$ is the pressure that corresponds to the (hot) particle density in the tube reduced to S.T.P. In CW argon-ion lasers the optimum electron temperature is in the range 2–6 eV; typical operating conditions involve axial fields $\sim$1 kV m$^{-1}$ and pressures $p_0 \sim 50$ mtorr, i.e., $E/p_0 \sim 20$ kV m$^{-1}$ torr$^{-1}$ (the actual filling pressure of the laser is much higher than $p_0$, but gas heating drives gas from the hot discharge capillary).

In $CO_2$–$N_2$–He lasers, which are discussed in detail in the next chapter, the excitation process involves energy transfer from vibrationally excited nitrogen molecules. The optimum electron temperature for direct excitation of the $N_2(v = 1)$ level is very low (since the threshold energy is $\sim$0.25 eV). Most coaxial gas-discharge $CO_2$ lasers operate with higher electron temperatures than this, and excite higher levels of the $N_2$ molecule that subsequently reach the $v = 1$ level by cascade. Typical axial fields and pressures are $\sim$10 kV m$^{-1}$, and $p_0 \simeq 10$ torr, i.e., $E/p_0 \simeq 1$ kV m$^{-1}$ torr$^{-1}$. Gas discharges used to excite lasers are quasi-neutral, the mean electron and ion densities at any point in the plasma are equal, and, unless the gas is highly ionized, the density of neutral atoms exceeds that of the charged particles. Some of the detailed aspects of laser operation are controlled by the temperature (and consequent Doppler broadening) of these particles in the plasma and their spatial distribution and velocities. Although some gas lasers are not excited in cylindrical structures, most gas lasers utilize discharges in cylindrical tubes for their operation. Generally speaking, the longer the wavelength of the laser transition the wider the bore of the discharge tube which will be used to excite it. This is particularly a consequence of the Fresnel diffraction condition for low-loss transverse-mode operation, i.e., the fact that

$$\frac{D}{\lambda \ell} \gtrsim 1,$$

where $D$ is the diameter of the discharge tube and $\ell$ is its length. The tube bore is also determined by the generally lower electron densities needed to achieve population inversion in lasers operating at longer wavelengths. This allows wider-bore tubes to be used at the same overall current, and means that a greater volume of gas can be excited.

## 8.10 Problem

8.1 In the helium–neon laser helium metastable states ($2^1$S) transfer their excitation to neon atoms, producing a population of neon atoms in the $5s'\left[\frac{1}{2}\right]_1^0$ state.[1] Two important laser transitions originate from this state:

[1] In the notation for the neon states the subscript is the total angular momentum $J$ of the state, so $g = 2J + 1$.

$$5s'\left[\tfrac{1}{2}\right]_1^0 \rightarrow 3p'\left[\tfrac{3}{2}\right]_2 \text{ at } 632.8 \text{ nm},$$

$$5s'\left[\tfrac{1}{2}\right]_1^0 \rightarrow 4p'\left[\tfrac{3}{2}\right]_2 \text{ at } 3.39\ \mu\text{m}.$$

(a) Calculate the minimum length of laser tube required in order to produce laser oscillation at each of these two wavelengths. (b) What would happen if a small quantity of methane that produced an absorption of 4 $m^{-1}$ at 3.39 μm were added to the helium–neon mixture? Consider the longitudinal mode at the line center only. You may assume that helium metastable states are destroyed only by transfer of their energy to neon atoms and by collisions with the walls of the discharge tube. The only important collisional process involving neon atoms is the energy-transfer process from helium $2\,^1S$ metastable states. The lower levels of both laser transitions are unpopulated. The following parameters apply.

Discharge-tube diameter 7 mm.
Helium pressure 1 torr.
Gas temperature 400 K.
Mirror reflectance 98% at 3.39 μm and 28% at 3.39 μm.
Rate of production of helium $2^1S$ metastable states $10^{23}$ $m^{-3}s^{-1}$.
Rate coefficient for transfer process $4.8 \times 10^{-15}$ $s^{-1}$: He($2\,^1S$)+Ne (ground state) → Ne($5s'[\frac{1}{2}]_1^0$) + He(ground state).
Lifetime of Ne $5s'[\frac{1}{2}]_1^0$ 62.5 ns.
Einstein $A$ coefficient for the transitions at 632.8 nm $A = 1.4 \times 10^6$ $s^{-1}$, 3.39 μm $9.96 \times 10^5$ $s^{-1}$.

Hint: use kinetic theory to estimate the average column destruction rate of helium metastable states at the discharge-tube wall.

The rate coefficient $k_T$ for the collisional transfer reaction (a) is defined from rate of production of Ne $5s'[\frac{1}{2}]_1^0$ states per volume $= k_T$[He $2^1S$][Ne(ground state)], where [He $2^1S$] and [Ne(ground state)] are the population densities (atoms per unit volume) in the helium metastable and neon ground states, respectively.

## References

[1] S. Jacobs, P. Rabinowitz, and G. Gould, "Coherent light amplification in optically pumped Cs vapor," *Phys. Rev. Lett.* **7**, 415–417, 1961.

[2] P. Rabinowitz, S. Jacobs, and G. Gould, "Continuously optically pumped Cs laser," *Appl. Opt.* **1**, 511–516, 1962.

[3] N. Djeu and R. Burnham, "Optically pumped CW Hg laser at 546.1 nm," *Appl. Phys. Lett.* **25**, 350–351, 1974.

[4] M. Artusy, N. Holmes, and A. E. Siegman, "D.C.-excited and sealed off operation of the optically pumped 546.1 nm Hg laser," *Appl. Phys. Lett.* **28**, 1331–1334, 1976.

[5] W. F. Krupke, "Diode pumped alkali laser," U. S. Patent No. 6,643,311, 2001.

[6] W. F. Krupke, "Diode pumped alkali lasers (DPALs) – A review," *Progr. Quant. Electron.* **36**, 4–28, 2012.

[7] R. L. Byer, R. L. Herbst, H. Kildal, and M. D. Levenson, "Optically pumped molecular iodine vapor-phase laser," *Appl. Phys. Lett.* **20**, 463–466, 1972.

[8] T. Y. Chang and T. Bridges, "Laser action at 452, 496, and 541 μm in optically pumped $CH_3$," *Opt. Commun.* **1**, 1117–1118, 1970.

[9] T. Y. Chang, "Optically pumped submillimeter sources," *IEEE Trans. Microwave Theory Tech.* **22**, 983–988, 1974.

[10] D. T. Hodges, "A review of advances in optically pumped far IR lasers," *Infrared Phys.* **18**, 375–384, 1978.

[11] D. Pereira, J. C. S. Moraes, E. M. Telles, A. Scalabrin, F. Strumia, A. Moretti, G. Carelli, and C. A. Massa, "A review of optically pumped far-infrared laser lines from methanol isotopes," *Int. J. Infrared Millimeter Waves* **15**, 1–44, 1994.

[12] C. C. Davis and T. A. King, "Gaseous ion lasers," in *Advances in Quantum Electronics*, Vol. 3, ed. D.W. Goodwin, London: Academic Press, 1975.

[13] S. E. Schwarz, T. A. DeTemple, and R. Targ, "High-pressure pulsed xenon laser," *Appl. Phys. Lett.* **17**, 305–306, 1970.

[14] C. P. Wang and ShaoChi Lin, "Performance of a large-bore high-power argon ion laser," *J. Appl. Phys.* **44**, 4681–4682, 1973.

[15] J. B. Marling, "Ultraviolet ion laser performance and spectroscopy–part 1: new strong noble gas transitions below 2500 Å," *IEEE J. Quant. Electron.* **11**, 822–834, 1975.

[16] J. J. Rocca, V. Shlyaptsev, F. G. Tomasel, O. D. Cortazar, D. Hartshorn, and J. L. A. Chilla, "Demonstration of a discharge pumped table-top soft-X-ray laser," *Phys. Rev. Lett.* **73**, 2192–2195, 1994.

[17] B. M. Luther, Y. Wang, M. A. Larotonda, D. Alessi, M. Berrill, M. C. Marconi, J. J. Rocca, and V. N. Shlyaptsev. "Saturated high-repetition-rate 18.9-nm tabletop laser in nickellike molybdenum," *Opt. Lett.* **30**, 165–167, 2005.

[18] H. Daido, "Review of soft X-ray laser researches and development," *Rep. Prog. Phys.* **65**, 1513–1576, 2002.

[19] N. Rohringer, D. Ryan, R. A. London, M. Purvis, F. Albert, J. Dunn, J. D. Bozek, C. Bostedt, A. Graf, R. Hill, S. P. Hau-Riege, and J. J. Rocca. "Atomic inner-shell X-ray laser at 1.46 nanometres pumped by an X-ray free-electron laser," *Nature* **481**, 488–491, 2012.

[20] T. K. Tio, H. H. Luo, and S. C. Lin, "High CW power ultraviolet generation from wall-confined noble-gas ion lasers," *Appl. Phys. Lett.* **29**, 795–797, 1976.

[21] C. E. Webb, A. R. Turner-Smith, and J. M. Green, "Optical excitation in charge transfer and Penning ionization," *J. Phys.* **B3**, L134–L138, 1970.

[22] D. S. Duffendack and K. T. Thomson, "Some factors affecting action cross section for collisions of the second kind between atoms and ions," *Phys. Rev.* **43**, 106–111, 1933.

[23] O. S. Duffendack and W. H. Gran, "Regularity along a series in the variation of the action cross section with energy discrepancy in impacts of the second kind," *Phys. Rev.* **51**, 804–809, 1937.

[24] W. B. Bridges, "Ionized gas lasers," in *Handbook of Laser Science and Technology, Vol. II. Gas Lasers*, ed. M. J. Weber, FL: CRC Press, Boca Raton, 1982.

[25] J. A. Piper and C. E. Webb, "A hollow cathode device for CW helium–metal vapour laser systems," *J. Phys. D: Appl. Phys.* **6**, 400–407, 1973.

# 9 Molecular gas lasers I

## 9.1 Introduction

Many important laser systems operate using molecular species. The laser transition occurs between energy levels of the molecule, which may be in the gaseous, liquid, or solid state. To understand more about molecular lasers it is important to consider the additional complexity of the energy-level structure of a molecule compared with that of an atom. In this chapter we will explain the three different kinds of energy level – electronic, vibrational, and rotational – which occur for molecular species, and then go on to explain how such a complex energy-level structure allows the possibility of laser oscillation over a very broad wavelength range.

## 9.2 The energy levels of molecules

### 9.2.1 Electronic energy states

In an atom, the orbiting electrons move in the spherically symmetric potential of the nucleus, and the various energy levels of the system correspond, in a simple sense, to different orbital arrangements of these electrons. For example, an excited electron frequently moves into an orbit that takes it further from the nucleus. In a molecule the electrons travel in orbits that surround all the nuclei of the molecule, although quite often there will be considerable localization of some electrons near a particular nucleus. The *electronic* energy states of the molecule result from different arrangements of the orbiting electrons about the nuclei. Electrons that move from one electronic energy level of a molecule to another experience changes in energy that are broadly comparable to such jumps in atoms. The emitted radiation is typically in the ultraviolet–near-infrared region; that is, the energy changes involved are on the order of 10,000 $cm^{-1}$. However, the nuclei of a molecule can also move in ways that give rise to *vibrational* and *rotational* energy states.

### 9.2.2 Vibrational energy states

The nuclei of a stable molecule are held together by forces that make it more desirable for the molecule to stay together rather than having the atoms fly apart. The molecule

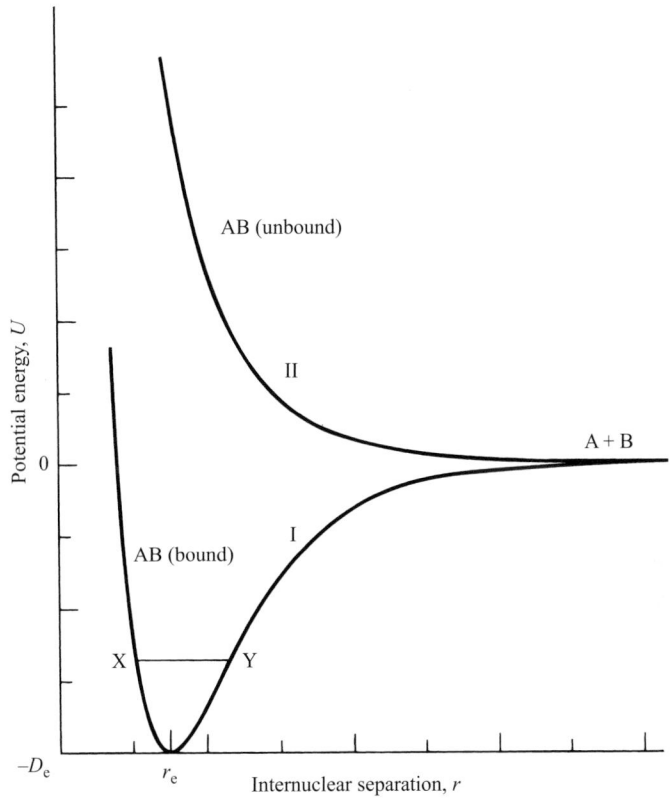


**Fig. 9.1** A schematic potential-energy-level diagram of a diatomic molecule AB.

minimizes its energy by remaining as a molecule. To illustrate this, let us consider the simple situation in which two atoms A and B are brought together. Figure 9.1 shows the potential energy of the composite (A + B) molecule as the internuclear separation varies. At large internuclear separation the atoms A and B are essentially independent. As A and B approach, two alternatives exist. If the reduction in overall potential energy that results from electrons initially on A and B now orbiting both nuclei offsets the Coulomb repulsion between the A and B nuclei, a *bound* molecule results. This is shown as the potential curve I with the energy minimum at internuclear separation $r_e$. If the nuclei A and B are brought closer still, the potential energy rises again as Coulombic repulsion becomes important. If the depth of the potential dip or *well*, $D_e$, on curve I is great enough, roughly $> k_B T$, the molecule will be stable enough to resist a tendency to fly apart because of thermal excitation.

If, as A and B are brought closer from large internuclear separations, the potential energy rises continually, as in curve II, then no *bound* state of the molecule occurs. The nuclei A and B will find it advantageous, in an energy sense, to fly apart. Curve II is referred to an *unbound* state of the molecule AB.

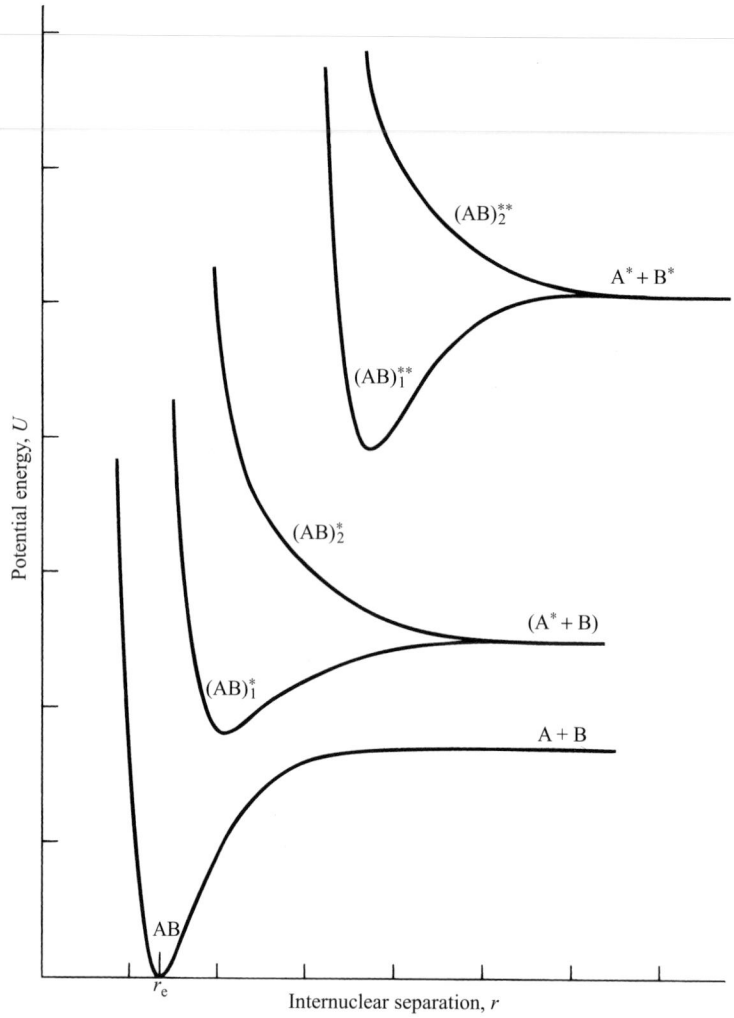


**Fig. 9.2** A potential-energy-level diagram showing the ground and excited electronic states.

The molecule AB can also be formed by joining together atoms that are themselves in excited states. For example,

$$A + B^* \rightarrow (AB)^*,$$
$$A^* + B^* \rightarrow (AB)^{**}.$$

In this way we can explain the existence of excited electronic states, which may be bound or unbound, as shown in Fig. 9.2.

When a bound molecule, such as AB, sits in the potential minimum on curve I in Fig. 9.1 its nuclei do not remain stationary – they vibrate with respect to each other. We can think of any molecule as a collection of masses $M_i$ held together by springs. Each spring has an equilibrium length $r_{ij}$ and a spring constant $k_{ij}$. Any displacement of any of the masses from

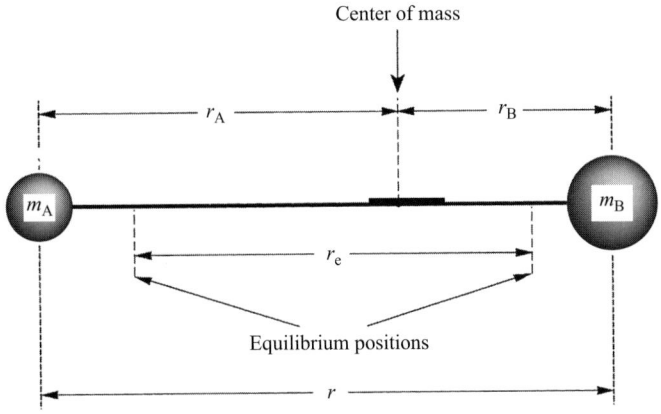


**Fig. 9.3** A simple model for a vibrating diatomic molecule.

their equilibrium positions will cause the structure to vibrate. The form of these vibrations can be complex for molecules with many nuclei. However, for a simple diatomic molecule AB the only possible vibration of the molecule involves a change in the internuclear separation. So, for example, a molecular vibration would correspond to motion along a line such as XY inside the potential well in Fig. 9.1. The line is horizontal because as the molecule vibrates its energy does not change. We refer to a line such as XY as a *vibrational* energy level. We can analyze such vibrational motion with reference to Fig. 9.3.

In equilibrium the spacing of the two nuclei is $r_e$. At any instant during vibration the spacing is $r(t)$. In the simplest model of the force between the two nuclei it is assumed that the force acting on each nucleus is proportional to the change in internuclear separation. During the vibration the center of mass of the molecule does not move, so we measure the position of each nucleus relative to the center of mass. For nucleus A the equation of motion is

$$m_A \frac{d^2 r_A}{dt^2} = -k[r(t) - r_e], \tag{9.1}$$

where $k$ is the *force constant* of the bond between the two nuclei (analogous to the spring constant of a spring connecting two vibrating masses). For nucleus B

$$m_B \frac{d^2 r_B}{dt^2} = -k[r(t) - r_e]. \tag{9.2}$$

The distances to the center of mass must satisfy the equations

$$r_A + r_B = r, \tag{9.3}$$

$$m_A r_A = m_B r_B, \tag{9.4}$$

which gives

$$r_A = \left(\frac{m_B}{m_A + m_B}\right) r,$$
$$r_B = \left(\frac{m_A}{m_A + m_B}\right) r. \tag{9.5}$$

From either Eq. (9.1) or Eq. (9.2),

$$\left(\frac{m_A m_B}{m_A + m_B}\right) \frac{d^2 r}{dt^2} = -k[r(t) - r_e], \tag{9.6}$$

which is generally written in the form

$$\mu \frac{d^2 r}{dt^2} = -k[r(t) - r_e], \tag{9.7}$$

where $\mu = m_A m_B/(m_A + m_B)$ is called the *reduced mass*.

Equation (9.7) is the equation of a simple harmonic oscillator. The solution to the equation is

$$r = r_0 \sin(\nu_{vib} t + \phi) + r_e, \tag{9.8}$$

where the vibrational frequency $\nu_{vib} = \sqrt{k/\mu}$. This motion is quantized, just as were the oscillations of the radiation field considered in Chapter 1. The only permitted energies of the vibrational motion are

$$E_{vib} = h\nu_{vib}(v + \tfrac{1}{2}), \tag{9.9}$$

where $v$ is an integer called the *vibrational quantum number*. These energy levels are equally spaced and are shown inside the simple-harmonic potential well in Fig. 9.4. In real diatomic molecules the potential well shown in Fig. 9.1 is not of the symmetrical shape shown in Fig. 9.4. In particular, the well has only finite height in the direction of increasing nuclear separation. This reflects the fact that if the molecule vibrates energetically enough it will fly apart. In a real potential well of the shape shown in Fig. 9.5 the discrete vibrational levels are almost equally spaced near the bottom of the well, but become closer together in energy as the vibrational quantum number $v$ increases. A molecule can jump from one vibrational energy level in the well to another. Usually this process is relatively unlikely except for $\Delta v = \pm 1$. If $\Delta v = +1$ the molecule absorbs a vibrational quantum; if $\Delta v = -1$ it emits a vibrational quantum. The energies of such vibrational quanta are on the order of 1,000 $\text{cm}^{-1}$, so the emitted or absorbed radiation is in the infrared region.

## 9.3 Vibrations of a polyatomic molecule

When a molecule with more than two nuclei vibrates, more than one kind of vibration can occur. However, it is always possible to treat any complex vibrational motion of such a polyatomic molecule as a superposition of motions in a finite number of characteristic

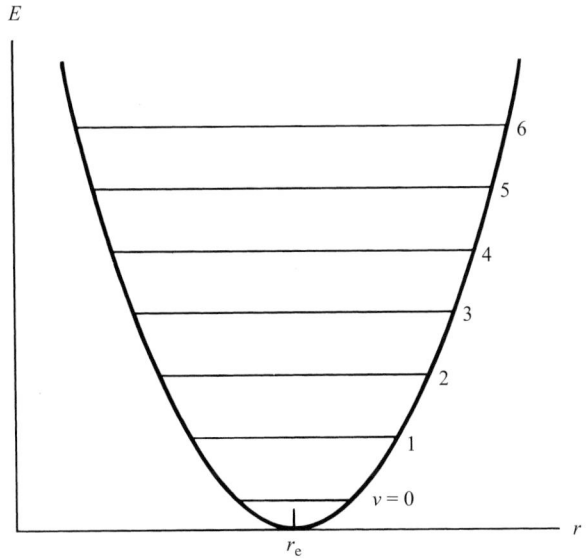


Fig. 9.4 Equally spaced vibrational energy levels in a simple harmonic potential well.

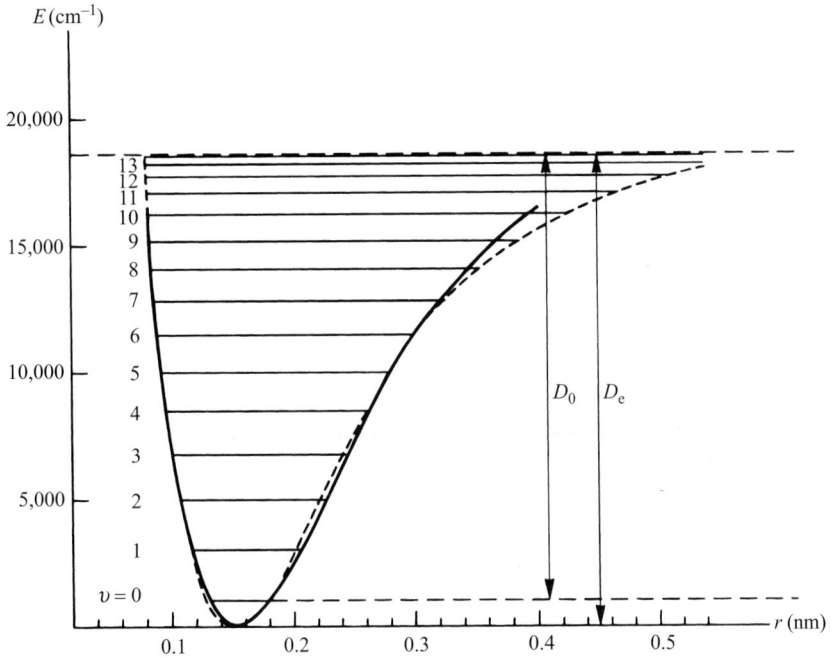


Fig. 9.5 Vibrational energy levels in an anharmonic potential well.

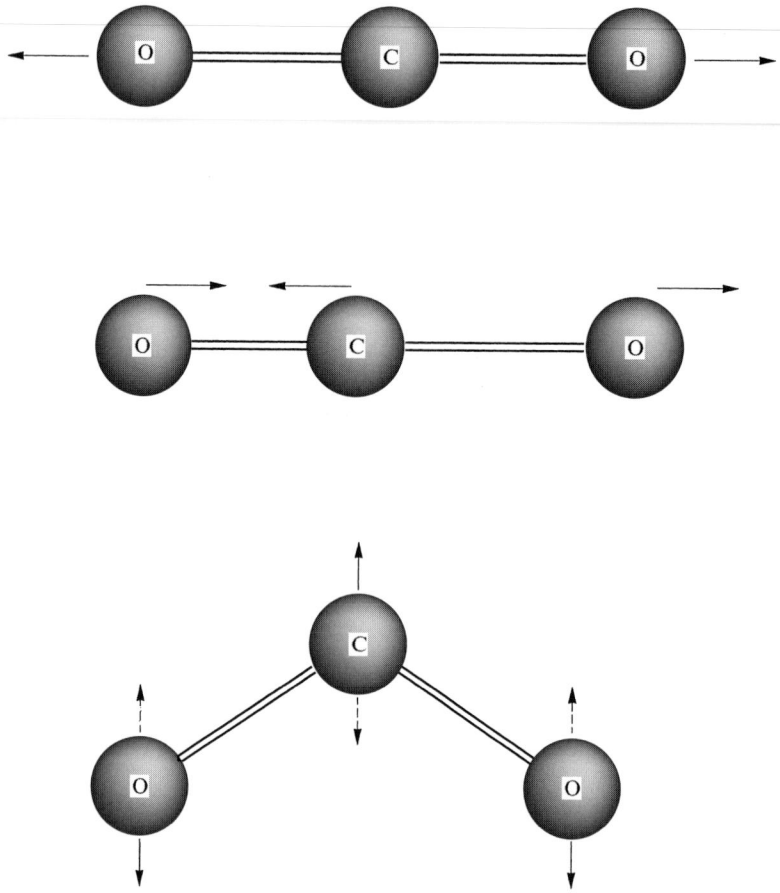


Fig. 9.6 Normal modes of vibration of the $CO_2$ molecule.

vibrations called *normal modes*. For example, in the molecule $CO_2$, shown in Fig. 9.6, where the oxygen, carbon, and oxygen nuclei are in a straight line in the equilibrium position, there are three such normal modes of vibration, which are also shown in Fig. 9.6. The mode with the highest vibrational frequency is called the *antisymmetric stretch*. In this characteristic motion the length of one C—O bond increases as the other decreases: the molecule remains linear. In the mode of next-highest vibrational frequency, the *symmetric stretch*, the vibration is symmetric in the sense that the two oxygen atoms vibrate in antiphase and the carbon nucleus remains stationary. The lowest vibrational frequency mode is the *bending* vibration in which the molecule deviates from linearity. To describe the energy levels of the $CO_2$ molecule, it is necessary to specify the number of vibrational quanta present in each mode. For example, a level designated $(v_1, v_2, v_3)$ has $v_1$ quanta in the symmetric stretching vibration, $v_2$ in the bending vibration, and $v_3$ in the asymmetric stretch. Its total vibrational energy is

$$E_{\text{vib}}(v_1, v_2, v_3) = h\nu_1(v_1 + \tfrac{1}{2}) + h\nu_2(v_2 + \tfrac{1}{2}) + h\nu_3(v_3 + \tfrac{1}{2}). \tag{9.10}$$

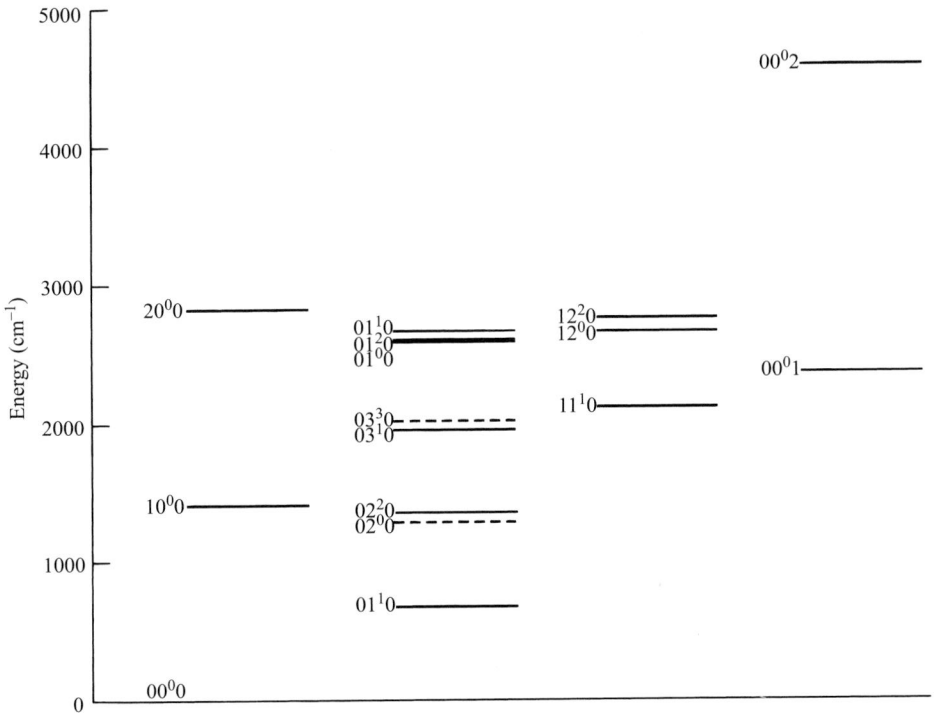


**Fig. 9.7** The vibrational energy-level diagram of the carbon dioxide molecule in its lowest electronic state.

Figure 9.7 shows the actual vibrational energy levels of the $CO_2$ molecule in its lowest electronic state. The superscript ° in the designation (10°0), for example, indicates whether the molecule has also acquired angular momentum about its axis as a result of the whirling motion that results if two bending vibrations that are not in phase occur at right angles to each other. A very important series of laser transitions occurs between the (00°1) → (02°0) and (00°1) → (10°0) vibrational levels, as we shall discuss in more detail shortly.

In general, a molecule containing $N$ nuclei has $3N - 6$ vibrational normal modes, unless the molecule is linear, in which case there are $3N-5$: in $CO_2$, $3N-5 = 4$: one antisymmetric stretch, one symmetric stretch, and two bending modes. When a polyatomic molecule jumps from one vibrational level to another, large simultaneous changes in $v_1, v_2, v_3$, etc., are usually unlikely.

## 9.4 Rotational energy states

Just to complicate the situation further, molecules can also possess rotational energy – imagine twirling a bar-bell about its center of mass! This rotational motion is simplest to understand for the rigid bar-bell diatomic molecule AB shown in Fig. 9.8. If the angular

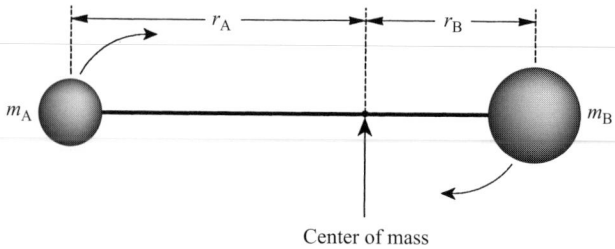


**Fig. 9.8** A diatomic molecule AB used to illustrate the concept of rotational energy states.

velocity of the molecule about its center of mass is $\omega$ then the rotational energy is

$$E_{\text{rot}} = \tfrac{1}{2} I\omega^2, \tag{9.11}$$

where $I$ is the moment of inertia of the molecule about the axis of rotation. Equation (9.11) can also be written in the form

$$E_{\text{rot}} = \frac{L^2}{2\omega}, \tag{9.12}$$

where $L = I\omega$ is the angular momentum.

The angular momentum is

$$\begin{aligned} L &= m_A r_A^2 \omega + m_B r_B^2 \omega \\ &= \mu r^2 \omega. \end{aligned} \tag{9.13}$$

The rotational energy is also quantized, because for rotational motion quantum theory shows that angular momentum can take only certain discrete values. The resultant quantized rotational energy levels are

$$E_{\text{rot}} = BJ(J+1) - DJ^2(J+1)^2 + \cdots, \tag{9.14}$$

where $J$ is an integer and usually $D \ll B$. These energies are usually quoted in $\text{cm}^{-1}$ units, so $B$ is usually also measured in $\text{cm}^{-1}$ units. Therefore, a typical set of rotational levels increases in spacing as $J$ increases, as shown in Fig. 9.9.

A molecule can change its rotational state with the subsequent absorption or emission of radiation. Because typical rotational energy levels are spaced by just a few $\text{cm}^{-1}$ these transitions occur in the far-infrared region, roughly 50–1,000 μm.

## 9.5 Rotational populations

When a molecular gas is in thermal equilibrium, the number of molecules in a given rotational level per unit volume depends statistically on the total number of molecules and the temperature. If $N_v$ is the population density of vibrational level $v$ then we can write

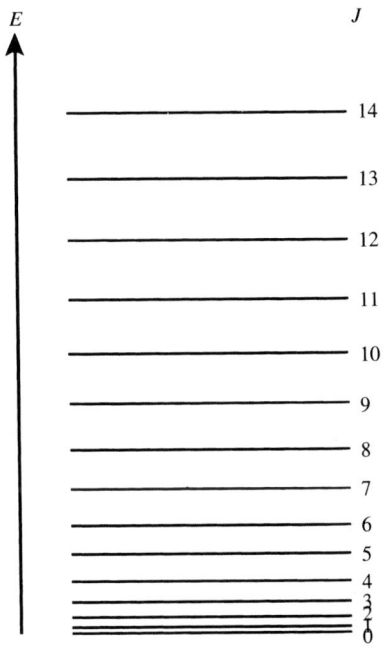


Fig. 9.9 The rotational energy-level diagram associated with a particular vibrational energy state.

$$N_v = \sum_{J=0}^{\infty} N_J(v), \tag{9.15}$$

where $N_J(v)$ is the population density of the $J$th rotational level of vibrational level $v$. In thermal equilibrium the $N_J(v)$ obey Maxwell–Boltzmann statistics, so that

$$N_J(v) = N_0(v)g(J)e^{E_J/(k_B T_{rot})}, \tag{9.16}$$

where $N_0(v)$ is the population density of the lowest rotational level ($J = 0$) of vibrational level $v$. The parameter $T_{rot}$ is called the rotational temperature, which is normally the same as the macroscopic temperature of the gas.[1] $g(J)$ is the degeneracy factor of rotational level $J$, where

$$g(J) = 2J + 1. \tag{9.17}$$

So, from Eqs. (9.14) and (9.16),

$$N_J(v) = N_0(v)(2J + 1)e^{-hcBJ(J+1)/(k_B T_{rot})}. \tag{9.18}$$

The factor of $hc$ ($c = 3 \times 10^8$ m s$^{-1}$) in the exponential of Eq. (9.18) is included in order to correct the $B$ constant from cm$^{-1}$ to joules. Figure 9.10 shows schematically the variation of the rotational population of the different rotational levels of a vibrational level. Note that the population peaks for a certain value of $J$: $N_J(v)$ increases because of the degeneracy factor, but eventually decreases because of the exponential factor.

[1] The macroscopic temperature, more properly called the *translational* temperature, is a measure of the average energy and velocity distribution of the molecules.

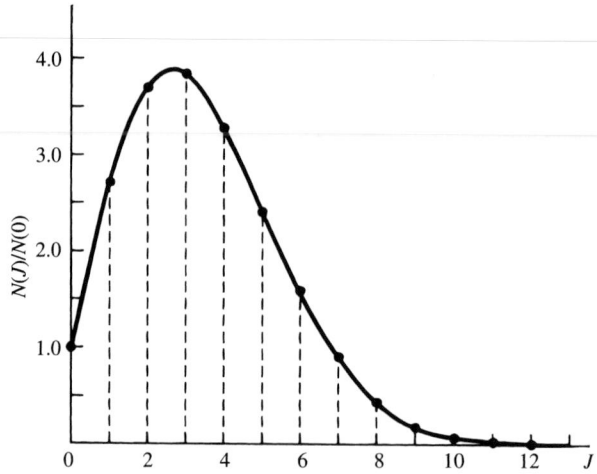


**Fig. 9.10** The variation of population in thermal equilibrium among the various rotational levels of a particular vibrational level.

If we examine the possibility of population inversion on a transition between rotational levels of two different vibrational levels, we can write the gain as

$$\begin{aligned}\gamma &\propto N_2 - \frac{g_2}{g_1}N_1 \\ &\propto N_0(v')(2J'+1)e^{-AJ'(J'+1)} - \left(\frac{2J'+1}{2J''+1}\right)N_0(v'')(2J''+1)e^{-AJ''(J''+1)} \\ &\propto N_0(v')(2J'+1)e^{-AJ'(J'+1)}\left[1 - \frac{N_0(v'')}{N_0(v')}e^{-A[J''(J''+1)-J'(J'+1)]}\right], \qquad (9.19)\end{aligned}$$

where $A$ is a constant.[2] If Eq. (9.19) is examined carefully, it can be seen that, even if $N_0(v') \leq N_0(v'')$, if $J'' > J'$, i.e., for a P-branch transition, it is still possible for the transition to exhibit gain. This is called a *partial* population inversion. It is quite common for molecular lasers to operate in this way. The vibrational levels as a whole do not have inverted populations, but an individual pair of rotational levels does.

## 9.6 The overall energy state of a molecule

A molecule simultaneously possesses rotational, vibrational, and electronic energy. A jump from one energy level can involve a change in one, two, or all three of these types of excitation. For example, Fig. 9.11 shows schematically the vibrational–rotational energy levels of a molecule in a particular electronic state. An emission would correspond to a

[2] In these molecular systems it is customary to use single primes to denote the upper level and double primes to denote the lower.

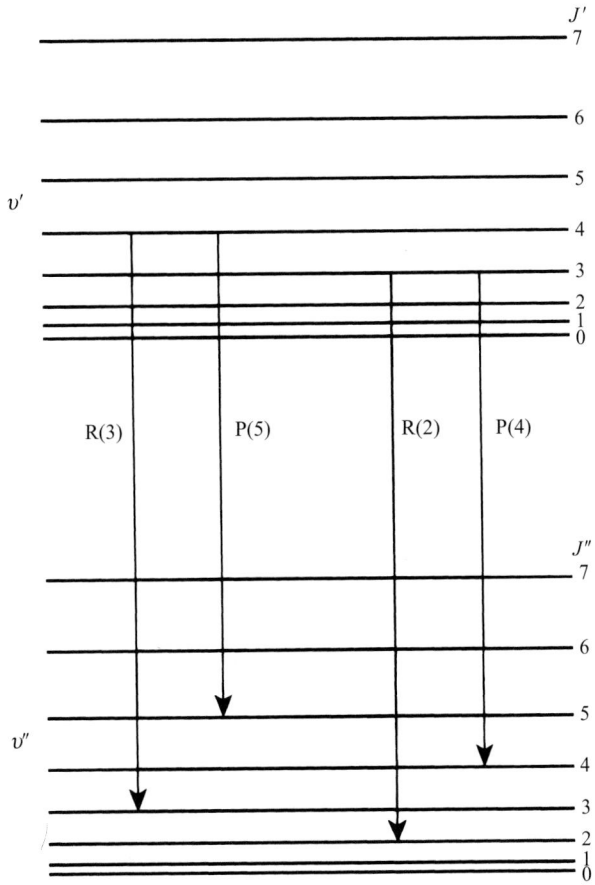


**Fig. 9.11** A schematic vibrational–rotational energy-level diagram of a molecule in a particular electronic state.

change from upper level $(v', J')$ to lower level $(v'', J'')$. Transitions for which $J' - J'' = +1$ are called R-branch transitions, and are labeled by the rotational quantum number of the lower level. Transitions for which $J' - J'' = -1$ are called P-branch transitions and are also labeled by the rotational quantum number of the lower level. Transitions for which $J' - J'' = 0$ (Q-branch transitions) also sometimes occur. Changes of $J$ greater than 1 are sometimes seen, but are usually much less likely. Figure 9.12 shows a transition from a vibrational–rotational level of one electronic state to a vibrational–rotational level of a lower electronic state. Since the emission of the photon is essentially instantaneous on the time scale of molecular vibration, the internuclear separation does not change during the transition, so the jump is always a vertical line. This is very important in determining whether given changes of electronic state are likely or not.

Armed with this brief introduction to the energy levels of molecules we are now in a position to discuss the fascinating variety of molecular lasers and the methods used to excite them.

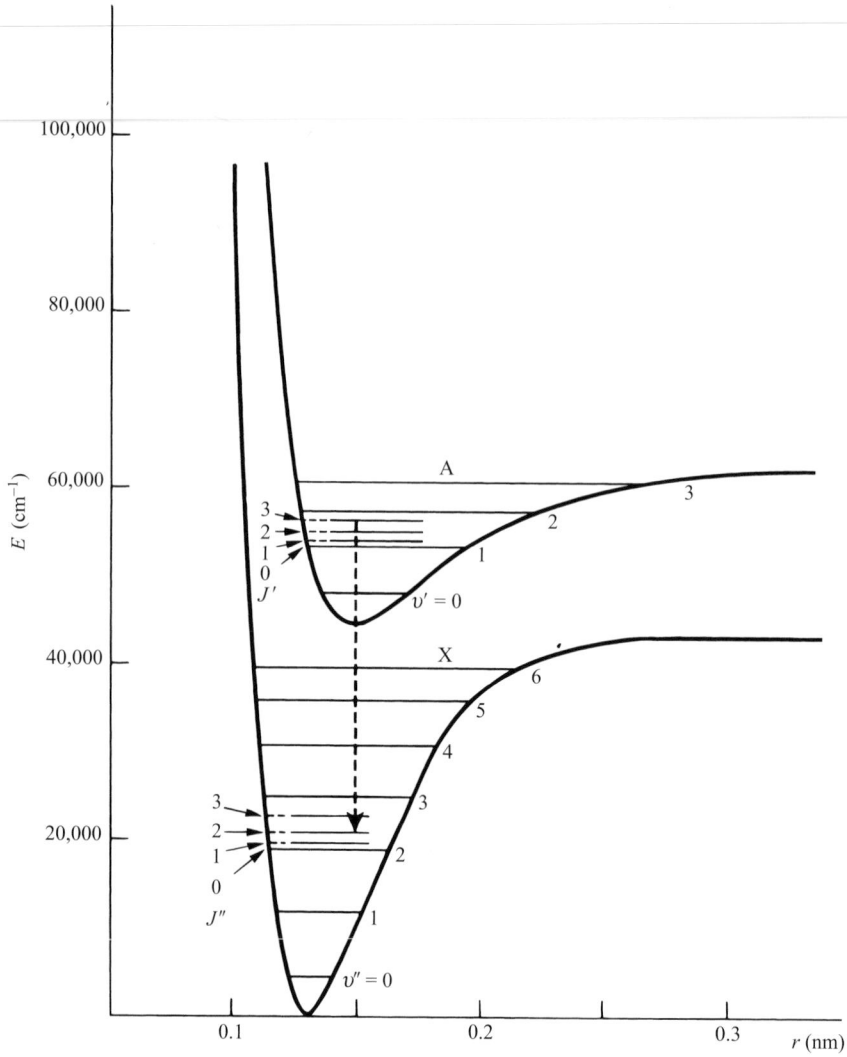


**Fig. 9.12** A schematic diagram showing a transition between vibrational–rotational levels of two different electronic states.

## 9.7 The carbon dioxide laser

Carbon dioxide lasers can be excited in a number of ways: they can be operated in a CW mode using both low- and high-pressure gas mixtures containing $CO_2$. For CW operation one can use both direct-current and radio-frequency excitation. They can be operated in a pulsed mode, expecially at pressures on the order of atmospheric pressure, in a pulsed mode when excited by high-energy electron beams, and they can be excited as a result of chemical reactions. We shall touch briefly on all these methods of excitation, but in the present chapter will concern ourselves principally with low-pressure gas-discharge operation.

A typical low-pressure CW $CO_2$ laser operates using a gas mixture containing $CO_2$, nitrogen, and helium in the approximate ratio $CO_2$: $N_2$:He = 1:1:8. The optimum total gas pressure varies depending on the diameter of the discharge tube but is typically on the order of 20 torr. Operating currents are low, for example 50 mA in a tube of diameter 10 mm, so these lasers are easy to construct. If the laser has Brewster windows, they must be made of a material that is transparent in the 10-μm range: sodium chloride, potassium chloride, zinc selenide, and germanium are commonly used. The laser tube is usually water-cooled. Figure 9.13 shows some of the essential features of a modern laser of this kind. The $CO_2$ laser was originally reported in 1964 by C. K. N. Patel at Bell Telephone Laboratories [1]. He initially reported low-power operation using a discharge through $CO_2$ alone, but soon discovered that the addition of nitrogen, in particular, enhanced the power output enormously [2]. The effect of nitrogen is nothing short of spectacular: a laser 1 m long with an internal diameter of 10 mm will deliver a power output of several milliwatts when operated with pure $CO_2$, but can easily provide more than 10 W when nitrogen is added and the mixture, pressure, and current are optimized. The role of nitrogen in the excitation of the laser can be illustrated with reference to Fig. 9.14, which shows some of the low-lying vibrational energy levels of the $CO_2$ molecule in its ground electronic state. Electron collisions excite ground-state nitrogen molecules to their first vibrational level ($v = 1$). Once in this level, the nitrogen molecules are metastable; they cannot readily lose energy radiatively, and can return to the ground state only as a result of collisions. Because $N_2$

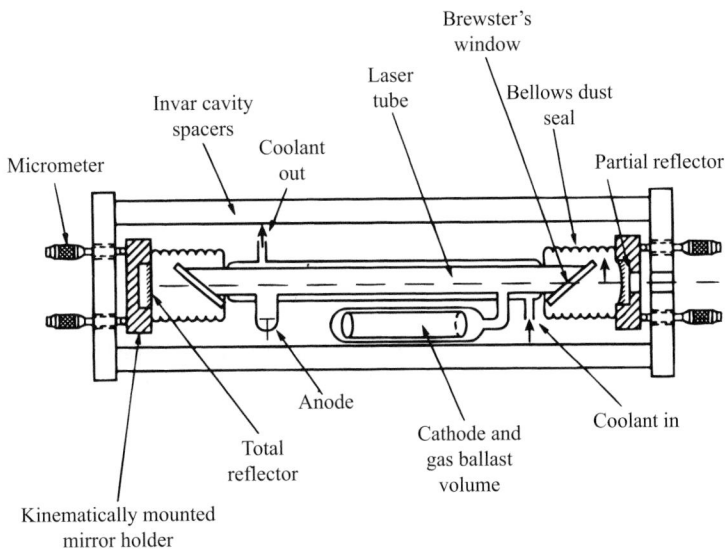


Fig. 9.13 The essential features of a modern CW low-pressure $CO_2$ laser.

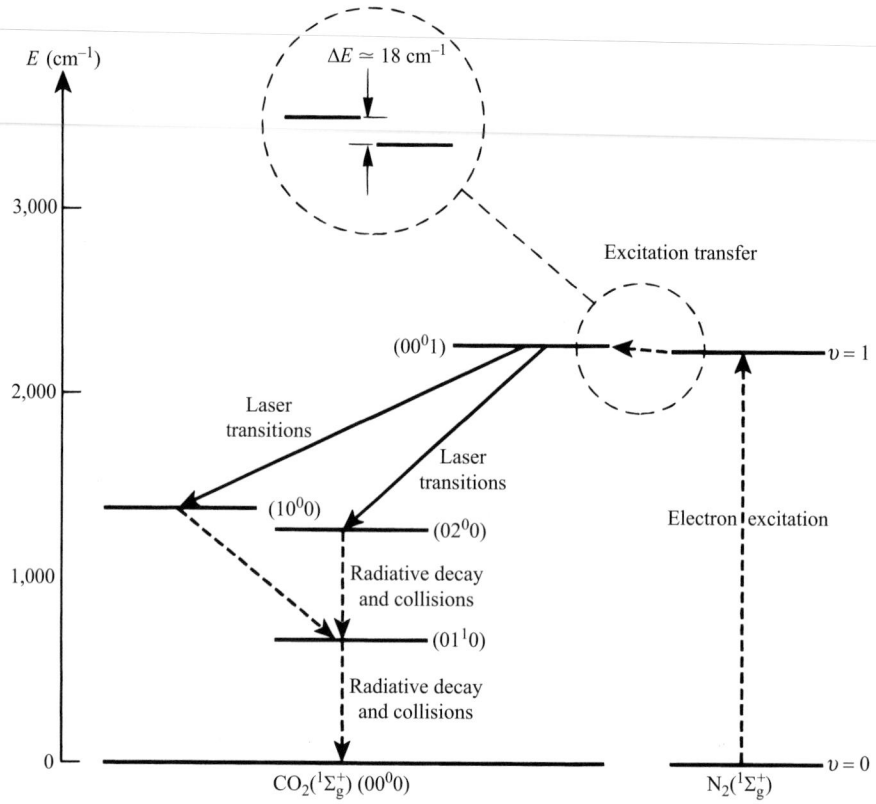


**Fig. 9.14** Excitation processes and vibrational energy levels important in the $CO_2$ laser, where vibrational energy is transferred from nitrogen molecules.

is what is called a *homonuclear* diatomic,[3] as it vibrates it has no dipole moment, and a dipole moment is needed for spontaneous emission to occur. The population of $N_2(v = 1)$ molecules builds up, and these molecules transfer their excitation to $CO_2$ molecules, thereby populating the upper $CO_2$ laser level $(00°1)$:

$$N_2(v = 1, E = 2{,}330.7\ \text{cm}^{-1}) + CO_2(00°0)$$

$$\rightarrow N_2(v = 0) + CO_2(00°1)(E = 2{,}349.2\ \text{cm}^{-1}) - 18.5\ \text{cm}^{-1}.$$

This excitation-transfer reaction is efficient because of a close energy coincidence between $N_2(v = 1)$ and $CO_2$ in the $(00°1)$ vibrational level (which has one quantum of excitation in the antisymmetric vibration). The energy discrepancy of the excitation-transfer reaction ($18.5\ \text{cm}^{-1}$), is much less than $k_B T$ at ambient temperature ($k_B T = 208\ \text{cm}^{-1}$ at 300 K). The highly selective excitation of the $(00°1)$ level produces population inversion between this level and the $(10°0)$ and $(02°0)$ levels. The upper laser level has a radiative lifetime

[3] Its nuclei are identical, except for the possibility of their being different isotopes of nitrogen.

of 4 ms. The lower laser levels have long radiative lifetimes ($\simeq$4.5 s) but are depopulated efficiently by collisions with nitrogen molecules, helium atoms, and, to a lesser extent, other $CO_2$ molecules. Collisional relaxation of the lower laser level is the principal role of the helium, although its presence also makes operation of a uniform gas discharge much easier. The (10°0) and (02°0) lower laser levels are close enough in energy that they, too, constantly exchange energy through collisions.

The optimum electron temperature for direct excitation of the $N_2(v = 1)$ level is very low (since the threshold energy is $\sim\frac{1}{4}$ eV). Most coaxial gas-discharge $CO_2$ lasers operate with higher electron temperatures than this, and excite higher levels of the $N_2$ molecule that subsequently reach the $v = 1$ level of the $N_2$ molecule by cascade. Typical axial fields and pressures are $\sim$10 kV m$^{-1}$ and $p_0 \simeq 10$ torr, i.e., $E/p_0 \simeq 1$ kV m$^{-1}$ torr$^{-1}$.

Because the upper and lower laser levels are vibrational levels, they each have associated rotational levels. Consequently a very large number of laser transitions between different rotational levels of the upper and lower vibrational levels will be possible, as can be seen from Table 9.1. The laser transitions between (00°1) and (10°0) are centered around 10.6 μm, while those between (00°1) and (02°0) fall around 9.6 μm. The individual laser transitions are labeled as P- or R-branch as described previously. The strongest transition is the 10.6-μm P(20) transition of the (00°1) → (10°0) *band*,[4]

$$CO_2(00°1)(J' = 19) \rightarrow CO_2(10°0)(J'' = 20) + h\nu(\lambda = 10.591\ \mu\text{m}).$$

Because the lower laser level of the $CO_2$ laser is $\simeq$1,390 cm$^{-1}$ above the ground state, this is a four-level laser. However, if the laser gas is allowed to become too hot, this ceases to be the case. This is why cooling of the laser discharge is necessary. The gain of the laser will actually decrease as the temperature of the discharge gas rises. With water cooling to keep the walls of the discharge tube at ambient temperature (300 K) the gain of a $CO_2$ laser is much higher than, for example, that of a helium–neon laser. This is reflected in the optimum mirror reflectance for maximum power output, which for a $CO_2$ laser roughly depends on its length/diameter ratio, $L/d$, according to the formula [4]

$$R \simeq 1 - L/(500d). \tag{9.20}$$

For $L/d = 100$ the optimum reflectance is 80%: for a helium–neon laser of comparable dimensions the optimum reflectance is $\simeq$99%. The $CO_2$ laser is a very efficient laser; its quantum efficiency is $\simeq$40%, and overall electrical efficiencies close to this value have been achieved in practice. The maximum achievable power output from these CW low-pressure lasers is about 300 W m$^{-1}$ of discharge length, although much higher output powers can be achieved by using different excitation schemes, as we shall see in the next chapter.

[4] A *band* is a group of transitions between the rotational levels of two vibrational levels.

**Table 9.1** Laser transitions from $^{12}C^{16}O_2$ lasers in the 9–11-μm region. Wavelengths and energies are calculated from accurate values of the molecular constants [3].

| Wavelength (μm) | Energy ($cm^{-1}$) | Transition | Wavelength (μm) | Energy ($cm^{-1}$) | Transition |
|---|---|---|---|---|---|
| 9.09349 | 1,099.6872 | R(62) | 9.41471 | 1,062.165965 | P(2) |
| 9.09976 | 1,098.9301 | R(60) | 9.42889 | 1,060.570666 | P(4) |
| 9.10623 | 1,098.14940 | R(58) | 9.444333 | 1,058.948714 | P(6) |
| 9.11291 | 1,097.344886 | R(56) | 9.455805 | 1,057.300161 | P(8) |
| 9.11979 | 1,096.516356 | R(54) | 9.47306 | 1,055.625068 | P(10) |
| 9.12689 | 1,095.663612 | R(52) | 9.48835 | 1,053.923503 | P(12) |
| 9.13420 | 1,094.786462 | R(50) | 9.50394 | 1,052.195545 | P(14) |
| 9.14173 | 1,093.884721 | R(48) | 9.51981 | 1,050.441282 | P(16) |
| 9.14948 | 1,092.958211 | R(46) | 9.53597 | 1,048.660810 | P(18) |
| 9.15745 | 1,092.006758 | R(44) | 9.55243 | 1,046.854234 | P(20) |
| 9.16565 | 1,091.030196 | R(42) | 9.56918 | 1,045.021670 | P(22) |
| 9.17407 | 1,090.028367 | R(40) | 9.58623 | 1,043.163239 | P(24) |
| 9.18273 | 1,089.001119 | R(38) | 9.60357 | 1,041.279074 | P(26) |
| 9.19161 | 1,087.948306 | R(36) | 9.62122 | 1,039.36915 | P(28) |
| 9.20073 | 1,086.869791 | R(34) | 9.63917 | 1,037.434110 | P(30) |
| 9.21009 | 1,085.765445 | R(32) | 9.65742 | 1,035.473616 | P(32) |
| 9.21969 | 1,084.635145 | R(30) | 9.67597 | 1,033.487000 | P(34) |
| 9.22953 | 1,083.478778 | R(28) | 9.69483 | 1,031.477430 | P(36) |
| 9.23961 | 1,082.296237 | R(26) | 9.71400 | 1,029.442092 | P(38) |
| 9.24995 | 1,081.087426 | R(24) | 9.73348 | 1,027.382171 | P(40) |
| 9.26053 | 1,079.852255 | R(22) | 9.75326 | 1,025.297865 | P(42) |
| 9.27136 | 1,078.590644 | R(20) | 9.77336 | 1,023.189375 | P(44) |
| 9.28244 | 1,077.302520 | R(18) | 9.79377 | 1,021.056912 | P(46) |
| 9.29379 | 1,075.987820 | R(16) | 9.81450 | 1,018.900693 | P(48) |
| 9.30539 | 1,074.646490 | R(14) | 9.83554 | 1,016.720942 | P(50) |
| 9.31725 | 1,073.278484 | R(12) | 9.85690 | 1,014.517888 | P(52) |
| 9.32937 | 1,071.883766 | R(10) | 9.87858 | 1,012.291767 | P(54) |
| 9.34176 | 1,070.462308 | R(8) | 9.90057 | 1,010.042823 | P(56) |
| 9.35441 | 1,069.014093 | R(6) | 9.92289 | 1,007.771302 | P(58) |
| 9.36734 | 1,067.539110 | R(4) | 9.94552 | 1,005.47746 | P(60) |
| 9.38053 | 1,066.037360 | R(2) | 9.96849 | 1,003.1615 | P(62) |
| 9.39400 | 1,064.508853 | R(0) | 9.99177 | 1,000.8238 | P(64) |
| | | | 10.01538 | 998.4646 | P(66) |
| 10.02591 | 997.41550 | R(62) | 10.44059 | 957.800537 | P(4) |
| 10.03347 | 996.66441 | R(60) | 10.45823 | 956.184982 | P(6) |
| 10.04132 | 995.884686 | R(58) | 10.47619 | 954.545087 | P(8) |
| 10.04948 | 995.076610 | R(56) | 10.49449 | 952.880850 | P(10) |
| 10.05793 | 994.240442 | R(54) | 10.51312 | 951.192264 | P(12) |
| 10.06668 | 993.376427 | R(52) | 10.53209 | 949.479314 | P(14) |
| 10.07572 | 992.484803 | R(50) | 10.55140 | 947.741979 | P(16) |

**Table 9.1** (Cont.)

| Wavelength (μm) | Energy ($cm^{-1}$) | Transition | Wavelength (μm) | Energy ($cm^{-1}$) | Transition |
|---|---|---|---|---|---|
| 10.08506 | 991.565748 | R(48) | 10.57105 | 945.980230 | P(18) |
| 10.09469 | 990.619630 | R(46) | 10.59104 | 944.194030 | P(20) |
| 10.10462 | 989.646506 | R(44) | 10.61139 | 942.383336 | P(22) |
| 10.11484 | 988.646626 | R(42) | 10.63210 | 940.548098 | P(24) |
| 10.12535 | 987.620181 | R(40) | 10.65316 | 938.688257 | P(26) |
| 10.13616 | 986.567352 | R(38) | 10.67459 | 936.803747 | P(28) |
| 10.14725 | 985.488312 | R(36) | 10.69639 | 934.894496 | P(30) |
| 10.15865 | 984.383226 | R(34) | 10.71857 | 932.960421 | P(32) |
| 10.17033 | 983.252249 | R(32) | 10.74112 | 931.001434 | P(34) |
| 10.18231 | 982.095531 | R(30) | 10.76406 | 929.017437 | P(36) |
| 10.19458 | 980.913211 | R(28) | 10.78739 | 927.008325 | P(38) |
| 10.20725 | 979.705421 | R(26) | 10.81111 | 924.973985 | P(40) |
| 10.22001 | 978.472286 | R(24) | 10.83524 | 922.914294 | P(42) |
| 10.23317 | 977.213922 | R(22) | 10.85978 | 920.829123 | P(44) |
| 10.24663 | 975.930439 | R(20) | 10.88473 | 918.718331 | P(46) |
| 10.26039 | 974.621939 | R(18) | 10.91010 | 916.581770 | P(48) |
| 10.27445 | 973.288517 | R(16) | 10.93590 | 914.419283 | P(50) |
| 10.28880 | 971.930258 | R(14) | 10.96214 | 912.230703 | P(52) |
| 10.30347 | 970.547244 | R(12) | 10.98882 | 910.015853 | P(54) |
| 10.31843 | 969.139547 | R(10) | 11.01595 | 907.774549 | P(56) |
| 10.33370 | 967.707233 | R(8) | 11.04354 | 905.50659 | P(58) |
| 10.34928 | 966.250361 | R(6) | 11.07160 | 903.21177 | P(60) |
| 10.36518 | 964.768982 | R(4) | 11.10014 | 900.88992 | P(62) |
| 10.38138 | 963.263140 | R(2) | 11.12915 | 898.54082 | P(64) |
| 10.39790 | 961.732874 | R(0) | 11.15867 | 896.1643 | P(66) |
| | | | 11.18868 | 893.7602 | P(68) |
| 10.42327 | 959.391745 | P(2) | | | |

## 9.8 The carbon monoxide laser

Carbon monoxide, CO, is a diatomic molecule with the simple vibrational energy-level structure shown in Fig. 9.15. Because the vibration of the molecule is anharmonic, the vibrational energy levels become slightly closer together as the vibrational quantum number increases. If pure CO is excited in a gas discharge, laser oscillation can occur between the rotational levels of adjacent vibrational levels [5], as shown in Fig. 9.15. Usually the laser transition is to the next lowest vibrational level, but it can also be to a level two levels lower. In theory the laser process should be 100% efficient. The lower level of one laser transition can serve as the upper level of another laser transition. Further, a process called

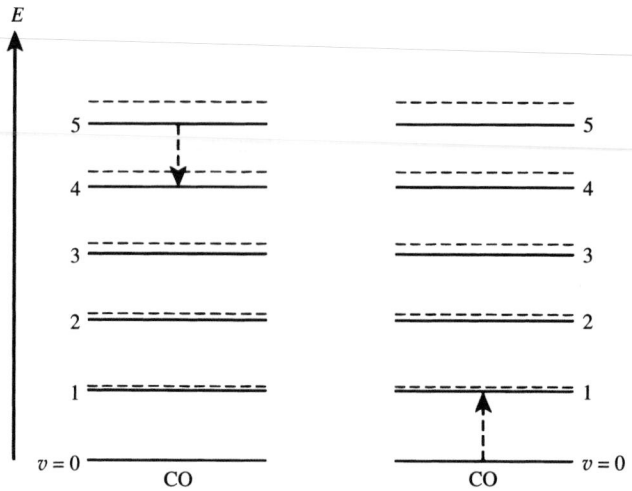


**Fig. 9.15** The vibrational energy-level diagram of CO in the ground electronic state.

*laddering* or *Treanor* pumping redistributes vibrational energy, so none is lost except by laser emission [6]. For example, if laser oscillation occurs from level $v$ to level $v-1$, the following sort of process can redistribute the energy back into level $v$:

$$\mathrm{CO}(v) \rightarrow \mathrm{CO}(v-1) + h\nu,$$
$$\mathrm{CO}(v-1) + \mathrm{CO}(v=0) \rightleftarrows \mathrm{CO}(v-2) + \mathrm{CO}(v=1) - \Delta E_{v-1,0},$$
$$\mathrm{CO}(v-1) + \mathrm{CO}(v=1) \rightleftarrows \mathrm{CO}(v) + \mathrm{CO}(v=0) + \Delta E_{v-1,1}.$$

For every CO molecule that loses a quantum of vibrational energy in a collision, another CO molecule gains one. The small energy discrepancy $\Delta E$ arises because the vibrational energies that are exchanged are not quite equal. The larger the value of $v$, the greater the energy discrepancy, so the exchange process becomes more efficient in pumping higher-$v$ molecules up and lower-$v$ molecules down. This process is conducive to maintaining population in the higher vibrational levels. The strongest laser transitions occur in the 5-μm region in vibrational jumps $7 \rightarrow 6$, $6 \rightarrow 5$, and $5 \rightarrow 4$, although laser action all the way from $37 \rightarrow 36$ down to $1 \rightarrow 0$ has been observed.

The CO laser is technologically important since it is a very powerful laser. The construction of a typical low-pressure device would be very similar to that shown in Fig. 9.12, although for maximum power output the laser tube must be kept colder than is the case for a $CO_2$ laser. Water cooling can be used but liquid-nitrogen cooling enhances power output considerably. It has been determined that the addition of other gases also improves performance. The $N_2$ stores vibrational energy that it can transfer to CO molecules, and it also helps maintain the electron energy distribution in the discharge within a favorable range for laser pumping. The addition of a small quantity of xenon also helps: it lowers the $E/p_0$ ratio[5] in the discharge because of the low ionization potential of the xenon. The addition of some oxygen helps to prevent the decomposition of CO and the formation of carbon deposits. Helium is added as an inert buffer to help maintain a smooth, uniform gas

[5] See Chapter 8.

**Table 9.2** Some molecular gas-discharge lasers

| Molecule | Laser wavelength range (μm) |
|---|---|
| $N_2O$ | 10–11 |
| $H_2O$ | 119 |
| HCN | 337 |
| DCN | 195 |

discharge. Typical operating parameters described for a laser 1.25 m long with a 21-mm bore are CO, 0.5 torr; He,16 torr; $N_2$, 1.5 torr; $O_2$, 0.01 torr; Xe, 0.3 torr [7]. The discharge required 12.4 kV at 12 mA. This laser provided a CW power of 70 W with an electrical efficiency of 47%.

## 9.9 Other gas-discharge molecular lasers

Many small molecules will lase when excited in low-pressure gas discharges. Table 9.2 lists some molecules that have been found useful in special applications. The three far-infrared lasers listed in Table 9.2 fall into a category that will be discussed further in the next chapter. These lasers use transitions between adjacent rotational levels of the *same* vibrational level.

## References

[1] C. K. N. Patel, "Continuous-wave laser action on vibrational–rotational transitions of $CO_2$," *Phys. Rev.* **136A**, 1187–1193, 1964.

[2] C. K. N. Patel, "CW high power $N_2$–$CO_2$ laser," *Appl. Phys. Lett.* **7**, 15–17, 1965.

[3] T. Y. Chang, "Vibrational transition lasers," in *Handbook of Laser Science and Technology*, ed. M. J. Weber, Boca Raton, FL: CRC Press, 1982.

[4] D. C. Tyte, "Carbon dioxide laser," in *Advances in Quantum Electronics*, Vol. 1, ed. D. W. Goodwin, London: Academic Press, 1970.

[5] C. K. N. Patel, "Vibrational–rotational laser action in carbon monoxide." *Phys. Rev.* **141**, 71–83, 1966.

[6] C. E. Treanor, J. W. Rich, and R. G. Rehm. "Vibrational relaxation of anharmonic oscillators with exchange-dominated collisions," *J. Chem. Phys.* **48**, 1798–1807, 1968.

[7] M. L. Bhaumik, W. B. Lacina, and M. M. Mann, "Enhancement of CO laser efficiency by addition of xenon," *IEEE J. Quant. Electron.* **6**, 575–576, 1970.

# 10 Molecular gas lasers II

## 10.1 Introduction

In this chapter we shall continue our discussion of molecular gas lasers. Many of the lasers to be discussed here provide substantial CW and pulsed power output, or have unusual and innovative technical features. Some of the lasers to be discussed have already been encountered in another context in earlier chapters. For many lasers a change in the method of excitation enhances some important aspects of laser performance, for example providing higher power output or operation in a new wavelength range. Radical departures from traditional methods of gas-discharge excitation have been particularly important in allowing the development of many of the laser systems to be described in the present chapter.

## 10.2 Gas transport lasers

In many laser systems a fundamental limit to the average output power is set by the build-up of waste heat that results from inefficient laser operation. Even in the relatively high-efficiency CW $CO_2$ and CO lasers, collisions that destroy vibrationally excited molecules, rather than just leading to energy exchange from one molecule to another, cause the temperature of the laser medium to rise. In these lasers the temperature rise reduces the population inversion through thermal excitation of the lower laser level. The rise in temperature also reduces the gain through an increase in the Doppler width. Waste heat can also produce changes in the optical properties of the laser medium in a spatially inhomogeneous way. This leads to a phenomenon called *thermal lensing*. We have already seen in Chapter 7 that this can be a serious problem in optically pumped solid-state lasers. The change from rod-shaped amplifiers to large disks placed at Brewster's angle was driven by the need to provide a better surface-area-to-volume ratio to enhance heat removal.

In a cylindrical gas-discharge laser, heat must be removed by transport of hot particles to the cooled tube walls. Except at very low pressures, diffusion is the process that controls the way hot particles reach the tube walls. The characteristic time for this process is the *thermal conduction time*,

$$T_c \simeq \frac{D^2}{\bar{c}\lambda}, \tag{10.1}$$

where $D$ is the diameter of the tube, $\bar{c}$ is the mean velocity of the gas particles, and $\lambda$ is the *mean free path* of these particles.[1] When diffusive thermal conduction is the only means for removing waste heat, there is an upper limit of about 1 kW $m^{-1}$ for the input power, regardless of the tube diameter. Consequently, CW gas-discharge $CO_2$ lasers operating at 30% efficiency have a fundamental power output limit of around 300 W $m^{-2}$. A further limit to the laser power is set by the saturation intensity, which, as we saw in Chapter 2 is related to the rate of removal of particles from the lower laser level. For the $CO_2$ lasers we have discussed so far, which are *static* lasers, the saturation intensity is ~60 W $cm^{-2}$.

In a *gas transport* laser, by flowing gas through the system we can improve laser operation in a number of ways.

(a) The gas flow provides convective cooling. Waste heat is removed at a rate determined by the residence time of particles in the system, which is a function of the flow velocity $v$ and the dimension over which particles flow through the system. The least-efficient flow geometry is shown in Fig. 10.1(a). In this arrangement the laser axis is coincident with the flow direction. The characteristic time for heat removal is $\tau_{\text{flow}} = L/v$. For a transverse flow geometry, such as that shown in Fig. 10.1(b), $\tau_{\text{flow}} = D/v$ and the removal of waste heat is much more efficient.

(b) The gas flow removes particles in the lower laser level and other undesirable metastables from the amplifying region. This can enhance the power output capability even of atomic gas lasers in which build-up of metastables can destroy the inversion or lead to a small saturation intensity. Examples that fall within this category are the copper-vapor and helium–xenon lasers [2] in which high-pulse-repetition-frequency operation can produce average power outputs of tens of watts.

(c) The gas flow removes decomposition products of the discharge. In chemical lasers, which we shall discuss later in this chapter, the flow removes undesirable reaction products from the laser.

(d) The gas flow can be used as a vehicle for mixing the laser gases, or, in the case of a chemical laser, the reacting species. For example, in a $CO_2$ laser, nitrogen molecules can be excited in a discharge region downstream and $CO_2$ molecules injected upstream, where laser action can then occur. This is an example of a *flow-mixing* laser, as shown in Fig. 10.1(c).

Gas transport lasers can be classified according to the relative directions of the laser axis ($L$), current flow ($I$), and gas velocity ($v$). For example Figure 10.1(a) is $(LIv)_x$. Figure 10.1(b) could be $(LI)_x(v)_z$, $(L)_x(Iv)_z$, or $(L)_x(I)_y(v)_z$ depending on whether the electrodes were placed for longitudinal current flow along the laser axis, for current flow transverse to the laser axis in the direction of gas flow, or for current flow transverse both to the laser axis and to the gas flow. Figure 10.2 gives an example of an $(L)_x(Iv)_z$ system. Geometries involving transverse current flow are generally to be preferred, because lower operating voltages are required to excite a shorter distance between electrodes.

When gas transport techniques are used in a laser oscillator the output transverse-mode pattern can become unsymmetrical because the system lacks cylindrical symmetry and

[1] The mean free path is the average distance traveled by a gas particle between collisions [1].

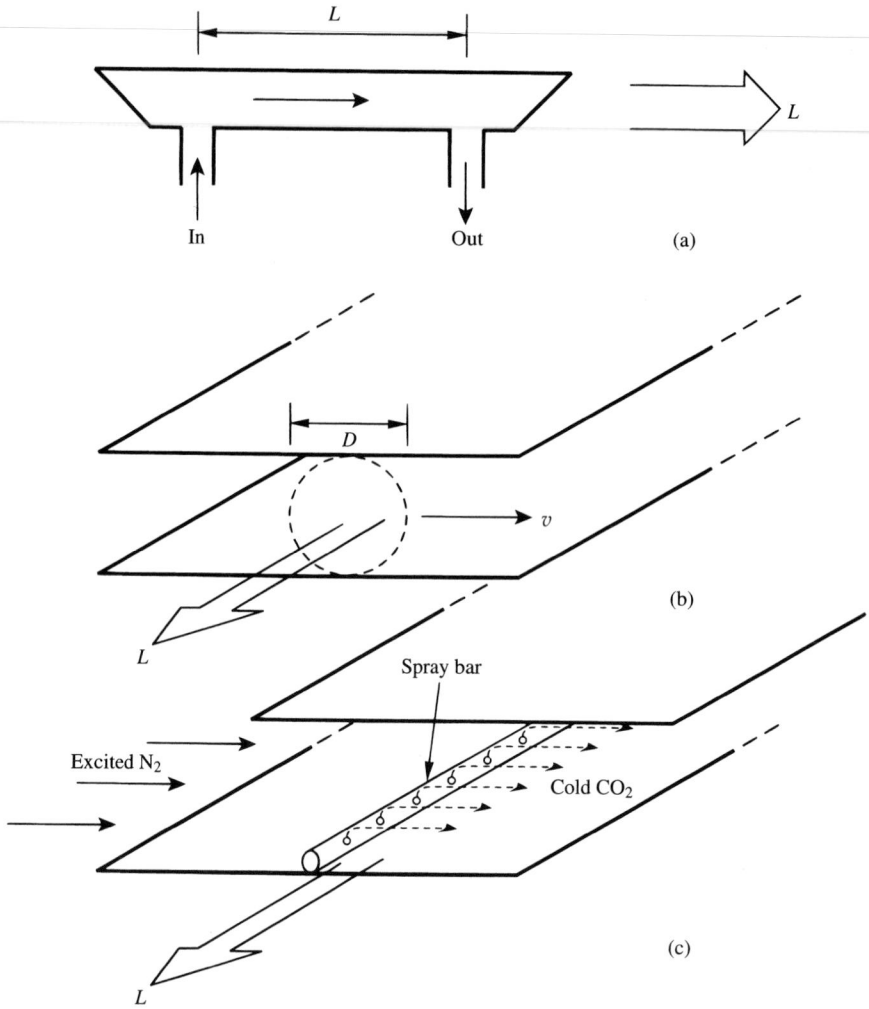


**Fig. 10.1** Some examples of gas-transport-laser geometries: (a) the axial flow laser $(L/v)_x$ geometry, in which the resonator axis, current direction, and gas velocity are in a common axial direction; (b) a transverse-flow laser; and (c) a flow-mixing laser.

can be influenced by flow distortion. For some industrial applications, such as cutting and welding, the resultant laser beam can still be focused sufficiently well to be useful. However, if improved beam geometry is desired, the gas transport device can be used as an amplifier for the beam from an oscillator, giving a controlled beam profile. To enhance its efficiency when used in this mode the beam to be amplified can be multipassed as shown in Fig. 10.3.

Gas transport $CO_2$ lasers provide large power output from relatively small packages. Figure 10.4 shows schematically how such a device could be operated in practice. The hot gases emerging from the gain region are cooled in a heat exchanger and then recirculated.

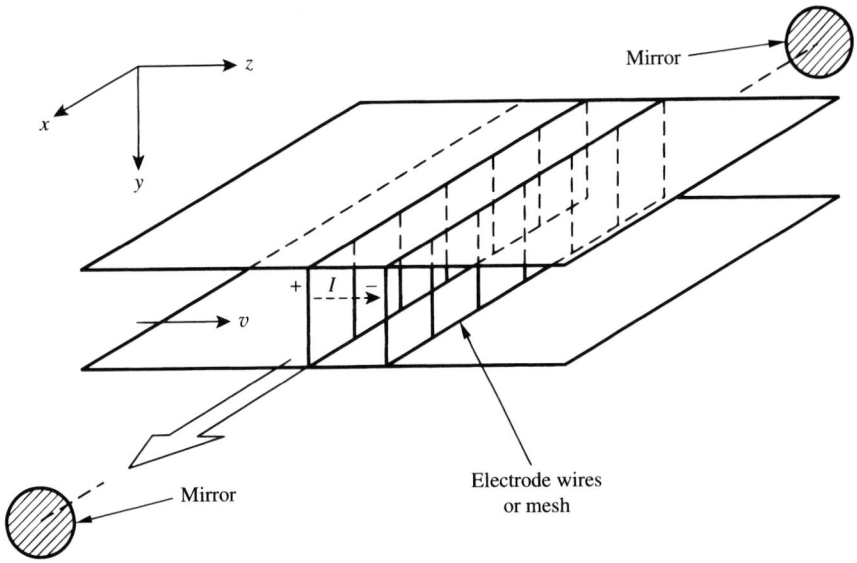


**Fig. 10.2** The $(L)_x(Iv)_z$ gas-transport-laser geometry, where $L$ is the laser-beam direction, $I$ the current direction, and $v$ the gas-flow-velocity direction. The gas flow passes through the grid or mesh electrodes.

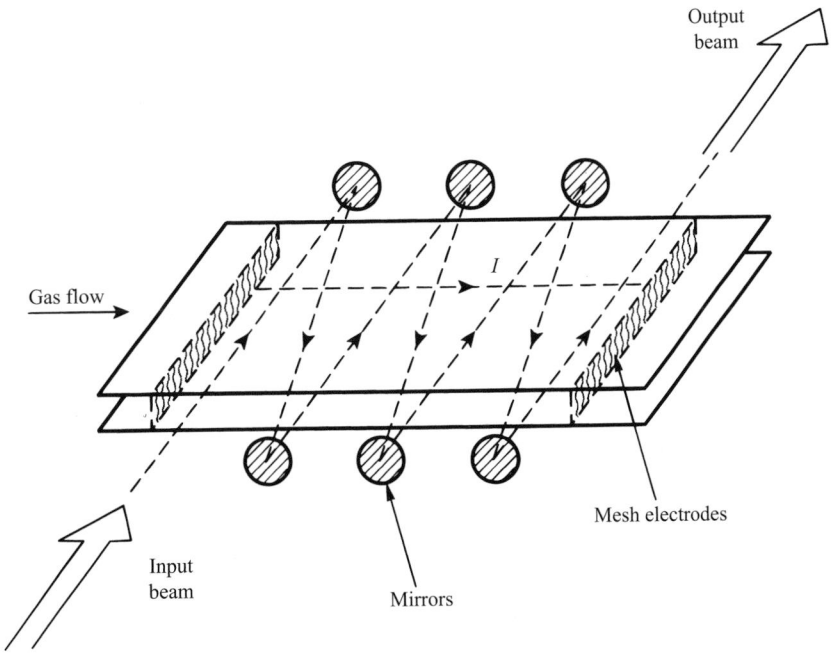


**Fig. 10.3** A schematic diagram showing the arrangement of a gas-transport-laser amplifier.

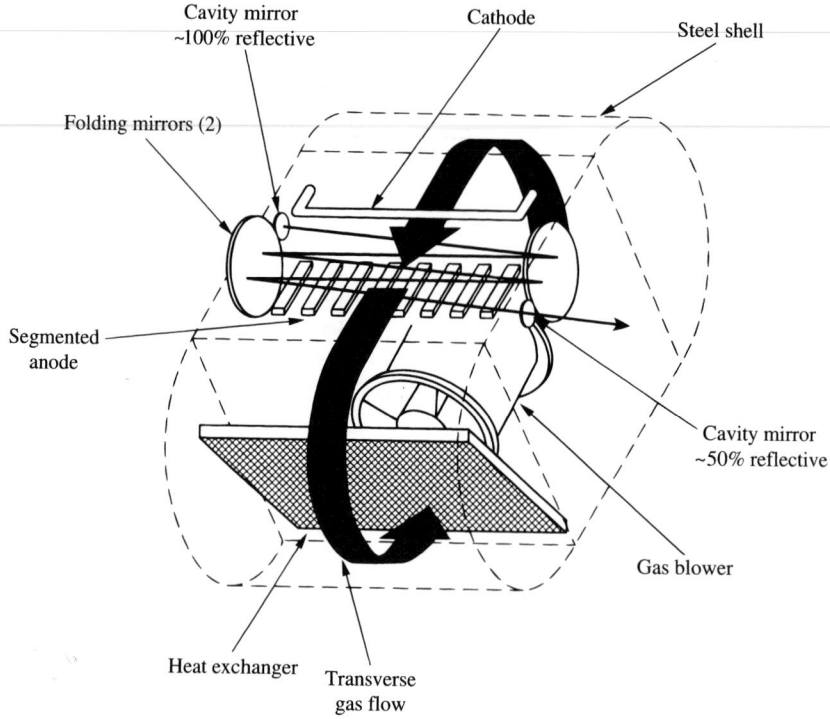


**Fig. 10.4** A schematic diagram of the GTE/Sylvania gas-transport laser [3].

The power output increases with the rate of flow of gas through the laser cavity, as shown in Fig. 10.5. A power output of 1 kW CW can be obtained from a $CO_2$ laser of this kind only about 0.5 m in length. Larger devices can provide CW power outputs in excess of 10 kW. High-power CW $CO_2$ lasers are technologically important in industrial applications such as cutting, drilling, welding, and the marking of parts.

## 10.3 Gas dynamic lasers

In the gas transport lasers described above, the main purpose of the gas flow is to remove waste heat generated from the primary electrical (or chemical) excitation. In gas dynamic lasers the population inversion itself is produced as a result of the flow of gas from a high- to a low-pressure region. The inversion is produced "gas dynamically" by the rapid expansion of the gas. The subsequent gas flow also removes waste heat and reaction products from the gain region.

All gases cool when they expand, even slowly, from high to low pressure, provided that they are above a characteristic temperature called the *Joule–Thomson* temperature, which for most gases is at cryogenic temperatures. If the expansion of the gas is rapid, through a nozzle, cooling always occurs, and is very efficient. By "cooling" in this context we mean

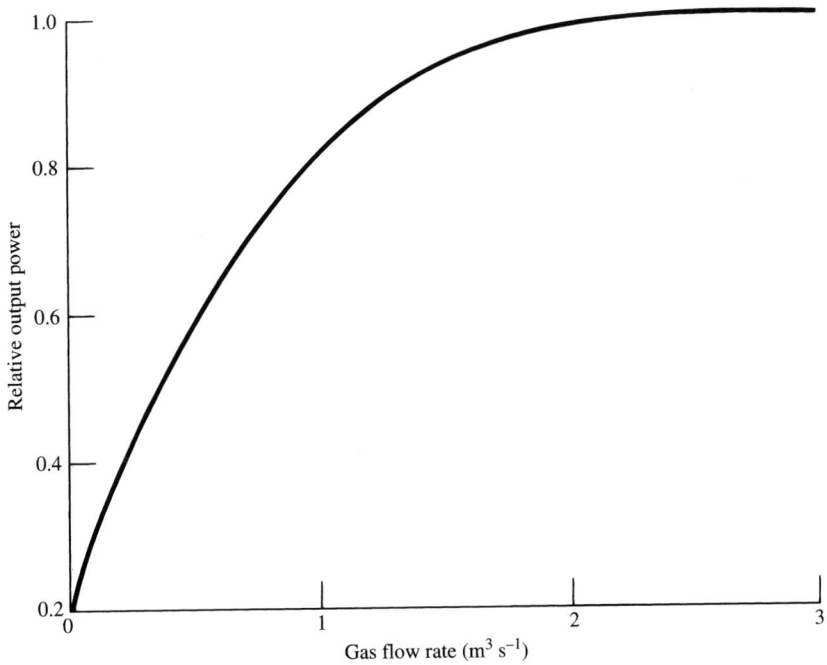


**Fig. 10.5** Relative output power versus gas flow rate for a GTE/Sylvania gas-transport laser.

a decrease in the translational temperature of the molecules. Expansion of a hot molecular gas through a supersonic nozzle channels the random kinetic motion of the hot molecules into directed motion in the direction of the flow. Since energy must be conserved, the random component of molecular velocity must decrease, and the gas cools translationally. The rotational temperature remains in equilibrium with the translational temperature, so it also falls. However, the gas need not equilibrate insofar as its vibrational populations are concerned. A population inversion can result if excitation remains *frozen* in higher vibrational levels while at the same time lower vibrational levels lose population through collisions. For example, in the case of $CO_2$, in thermal equilibrium at temperature $T$, the population densities of the upper (00°1) and lower (10°0, 02°0) levels, $N_2$ and $N_1$, respectively, satisfy the relation

$$\frac{N_2}{N_1} = \frac{g_2}{g_1} e^{-h\nu/(k_B T_{\text{trans}})}. \tag{10.2}$$

After expansion of the $CO_2$ gas through a supersonic nozzle, the population density $N_1$ falls rapidly through collision. The lower laser level comes rapidly into thermal equilibrium with the ground state. However, the upper laser level is not as efficiently relaxed by collisions, and downstream of the nozzle a population inversion can result. Figure 10.6 shows a schematic diagram of a gas dynamic $CO_2$ laser nozzle. The gas flows from left to right. The high-pressure, high-temperature region to the left of the nozzle is called the *stagnation region*. A typical operating mixture might be 7.5% $CO_2$, 1.2% $H_2O$, and 91.3% $N_2$ at 17 atm and 1,400 K. After expansion through the nozzle the directed gas velocity

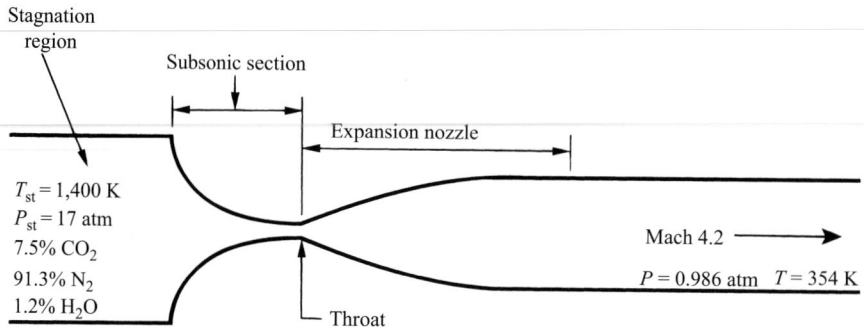


**Fig. 10.6** A simple gas-dynamic-laser expansion nozzle.

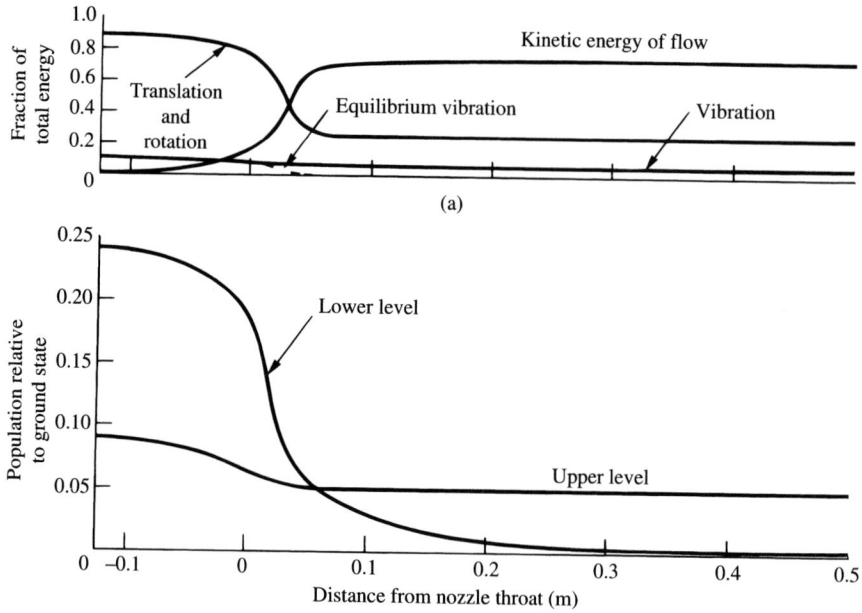


**Fig. 10.7** (a) The partition of energy in the gas-dynamic-laser arrangement shown in Fig. 10.6. (b) The relative population of upper and lower $CO_2$ laser levels.

is Mach 4.2,[2] the pressure has fallen to 0.986 atm and the translational temperature has fallen to 354 K. The water vapor is added to the gas mix because it is particularly efficient at relaxing the $CO_2$ lower laser levels. Vibrationally excited nitrogen also persists downstream of the nozzle and assists in population of the $CO_2$ upper laser level. The reason why population inversion results downstream of the nozzle is best illustrated with the aid of Fig. 10.7, which shows the conversion of energy initially stored in translation and rotation in the stagnation region into directed motion downstream of the nozzle. Figure 10.7 also shows the reduction of the lower laser level population downstream and

[2] The Mach number is the ratio of the gas-flow velocity to the velocity of sound in the gas.

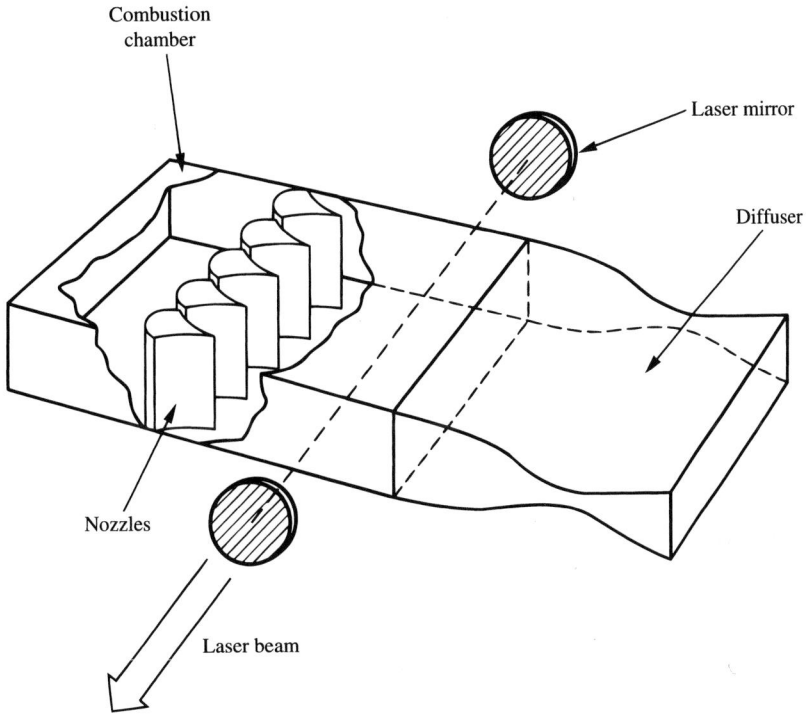


Fig. 10.8 A schematic diagram of the arrangement of a high-power, multinozzle gas-dynamic laser [4].

the *freezing-in* of the upper level population. To extract laser power from the system the laser axis is set up to run orthogonal to the gas flow a few centimeters down from the nozzle, where the maximum inversion exists. In practice an arrangement of many nozzles in parallel would be used, as shown in Fig. 10.8. The energy theoretically available from such a laser is essentially that energy stored in the vibrationally excited hot nitrogen and in the asymmetric stretching mode of $CO_2$. This energy is 35 kJ per kg of gas with a composition of 10% $CO_2$ and 90% $N_2$ at a stagnation temperature of 1,400 K. To achieve the high temperatures and pressures needed in the stagnation region, electrical heating can be used. However, it is fortuitous that the burning of appropriate hydrocarbon fuels with air will produce just the hot high-pressure gases needed. Continuous burning of fuel will provide CW laser operation. Pulsed ignition of the fuel will give explosive production of gas in the stagnation region followed by pulsed laser operation. Some of the fuel mixtures that have been used in gas dynamic $CO_2$ lasers include CO (or $C_2N_2$), air, $CH_4$, and $N_2$ (the methane provides the hydrogen needed for water production); and $CO_2$, $N_2$, $O_2$, $H_2$ (for pulsed operation). Other hydrocarbons such as acetylene and propane can also be used, or more exotic nitrogen-containing hydrocarbons, which when burnt provide $CO_2$, $H_2O$, and $N_2$ in the correct proportions. Gas dynamic laser action in carbon monoxide can be obtained in similar ways, but the available power outputs are much smaller.

The gas dynamic $CO_2$ laser is the highest-power CW laser currently operating. This laser falls into the category of potential "Star Wars" weaponry. Pilotless drone aircraft have been shot down with lasers of this kind. Continuous-wave power outputs of several megawatts have been reported. It is worth briefly commenting on some of the technological problems associated with such high-power lasers. The windows and mirrors of the laser system must withstand very great energy fluxes without damage. This requires windows of the lowest absorptivity possible and mirrors of solid, water-cooled metal, such as copper or copper alloys, whose intrinsic infrared reflectance is high. Because of the violent excitation conditions in these lasers it is difficult to obtain good beam quality. This is not surprising when one considers the fact that the optical cavity runs right through the exhaust of an arrangement that is not much different from a rocket or jet engine. However, oscillator/amplifier configurations, unstable resonators,[3] and special beam-profiling techniques are available to tackle the problem of output beam quality.

## 10.4 High-pressure pulsed gas lasers

These lasers, of which the pulsed high-pressure $CO_2$ laser is the best example, are frequently called TEA lasers – *transversely excited atmospheric-pressure lasers*. Their development arose from a desire to obtain larger power outputs from the $CO_2$ laser. Conventional discharge-excited lasers generally operate at relatively low pressure (<50 torr), since it is only at such pressures that a stable, uniform glow discharge can be excited in a longitudinally excited discharge. An increase of pressure in the discharge leads to the formation of narrow current-bearing filaments (called sparks, arcs, or lightning streamers). Consequently, the laser medium is not uniformly excited. Furthermore, at high pressures the $E/p_0$ value of the discharge may become too low for efficient excitation of the laser levels (see Chapter 8). Transverse excitation overcomes this problem since the applied voltage acts across a smaller dimension of the laser structure. If uniform transverse excitation of a high-pressure discharge can be achieved, the possibility of large laser powers exists since there are many more laser molecules available in the gas – if they can be excited. The problem of filament formation can be overcome in the case of transverse pulsed excitation. The power input requirements would be prohibitive for CW operation in this way – even if such excitation were achieved. Several methods have been used to prevent filament formation in pulsed excitation, but they all have one feature in common: they excite the gas in such a way as to encourage uniform glow-discharge excitation and discourage the formation of filaments.

(i) *Rogowski-profile electrodes*. If a transversely excited laser structure is constructed like a large plane-parallel capacitor, sparks will form at the edges of the electrodes because edge effects cause the electric field to be higher there. To prevent this, the electrodes must be curved away at the edges in such a way as to ensure that the field is never higher at the edges than in the center of the structure. This is done with

[3] See Chapter 15.

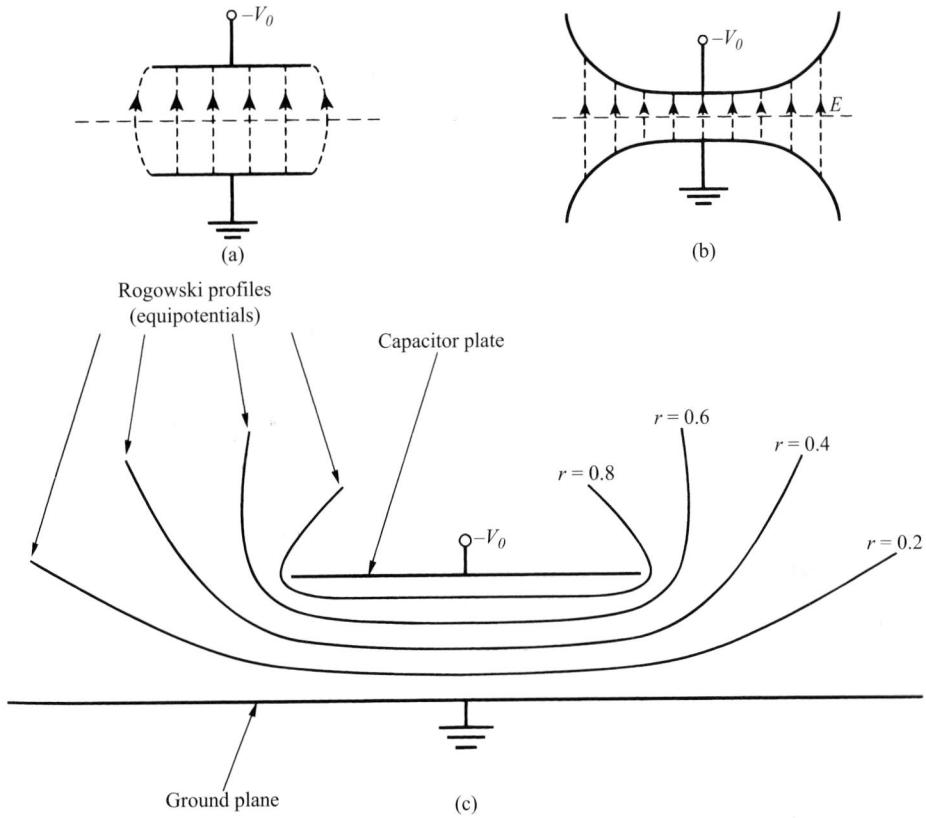


**Fig. 10.9** (a) Plane-parallel electrodes have a high field at their edges. (b) A double Rogowski profile has a uniform field between the electrodes. (c) Rogowski profiles – the equipotentials of a two-dimensional parallel-plate capacitor. The potential of each profile satisfies $V = rV_0$.

electrodes formed in a *Rogowski* profile. Two such electrodes can be used, as shown in Fig. 10.9(b), or one such electrode and a much larger plane electrode placed where the equipotential of the double Rogowski structure would be. The Rogowski profiles follow the equipotential surfaces of the parallel-plate capacitor structure shown in Fig. 10.9(c). The parametric equations of the Rogowski profile are [5]

$$x = \frac{a}{\pi}[\phi + e^{\phi}\cos(r\pi)],$$
$$y = \frac{a}{\pi}[r\pi + e^{\phi}\sin(r\pi)].$$

For electrodes fabricated so that $r\pi = \pi/2$ or $r\pi = 2\pi/3$, it has been found experimentally that sparking does not occur at the edge of the electrodes. However, even in structures with $\pi/2$ or $2\pi/3$ electrodes, fine adjustment of the operating voltage and gas mixture is still necessary in order to prevent filamentary behavior. These devices use large operating voltages – typically 20–50 kV.

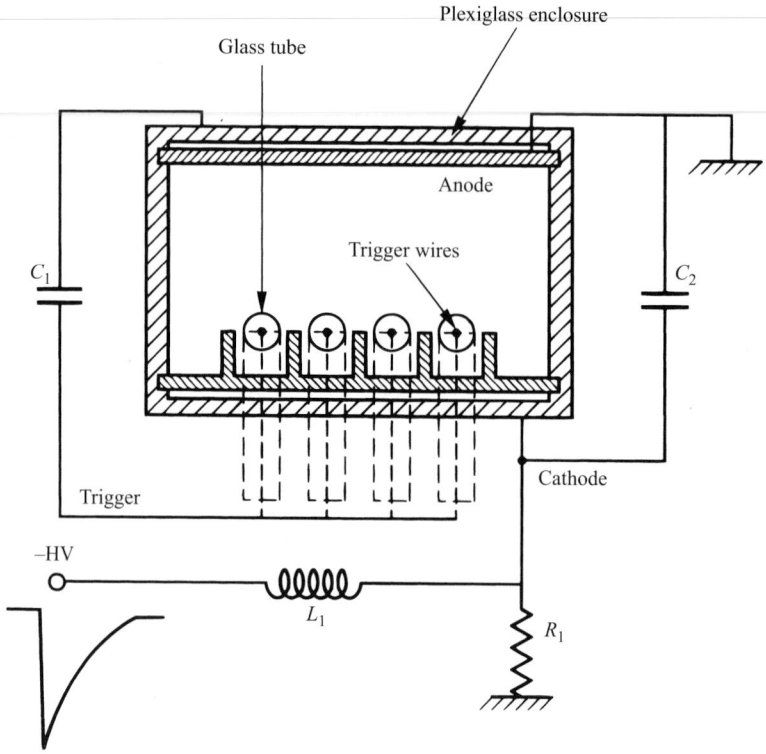


**Fig. 10.10** A head-on view of a double-discharge TEA laser structure. Preionization is created by striking a discharge with the trigger wires.

(ii) *Preionization techniques*. In a transversely excited structure, even one with Rogowski electrodes, a uniform discharge can be encouraged to form by the use of preionization techniques. Some charge carriers are injected into the space between the main electrodes just prior to application of the main excitation voltage pulse. Preionization can be provided by arrays of insulated trigger wires placed near the cathode as shown in Fig. 10.10, which are excited by a high-voltage pulse and lead to production of a corona discharge near the surface of the electrode. Alternatively, as shown in Fig. 10.11, preionization can be provided by firing an array of sparks below a mesh anode or at the side of the structure. These sparks feed energetic ultraviolet photons into the laser gas, causing photoionization. Pulsed $CO_2$ lasers of this sort have been constructed with volumetric power outputs of 35 $l^{-1}$. The typical laser pulse from these devices lasts about 100 μs, so peak power outputs are on the order of 10 MW $J^{-1}$. Shorter pulses can also be obtained. Outputs of many kilojoules in 1 ns can be produced by large oscillator/amplifier configurations.

(iii) *Electron-beam preionization and excitation*. High-energy electron beams (e-beams) can be used both as a means for preionizing a large volume of high-pressure gas and as the primary excitation source in a high-pressure gas laser. In its simplest form such an e-beam is generated by applying a large voltage pulse, typically >100 kV,

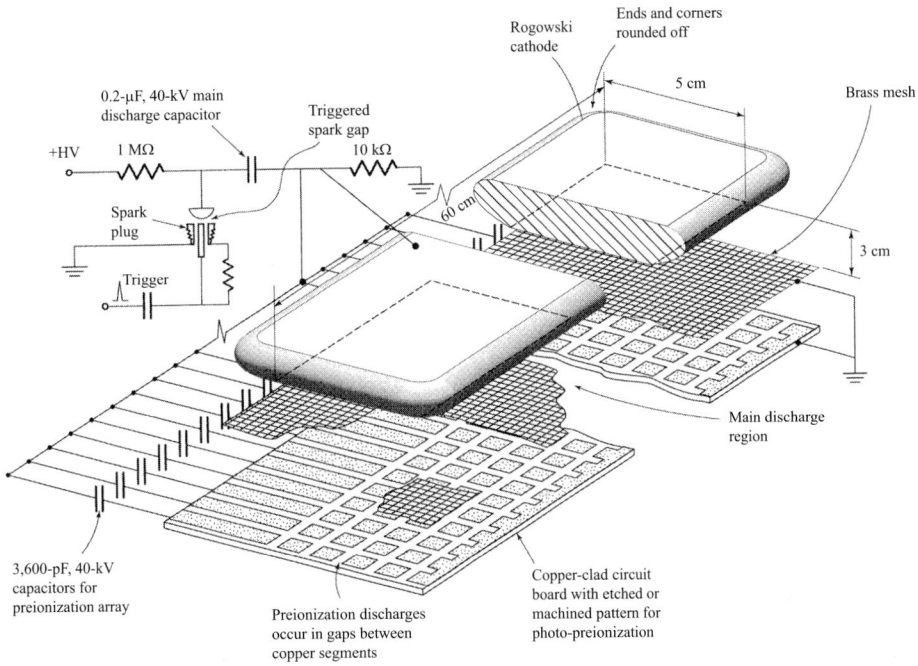


**Fig. 10.11** The TEA $CO_2$ laser structure using ultraviolet photo-preionization with a spark array. Typical dimensions are shown.

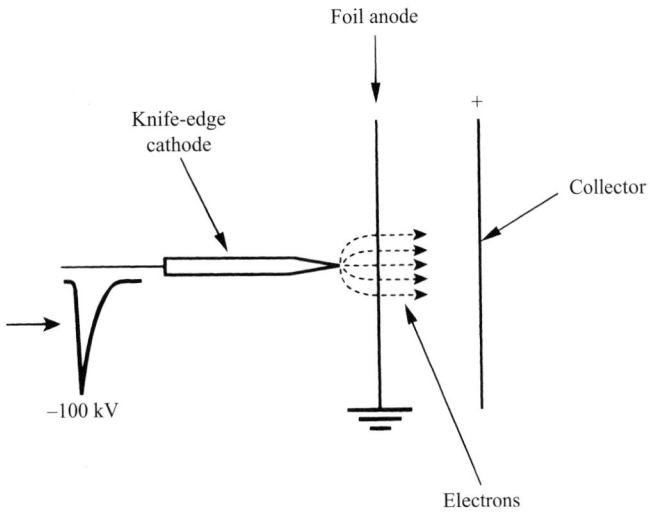


**Fig. 10.12** A schematic diagram of a simple e-beam ionizer for a TEA laser.

between a knife-edge cathode and a thin metal foil anode in a vacuum. Electrons are drawn from the cathode by field emission and accelerate towards the anode. Most of the high-energy electrons penetrate the foil and are then available to excite, or preionize, a volume of high-pressure gas on the other side of the foil, as shown in Fig. 10.12.

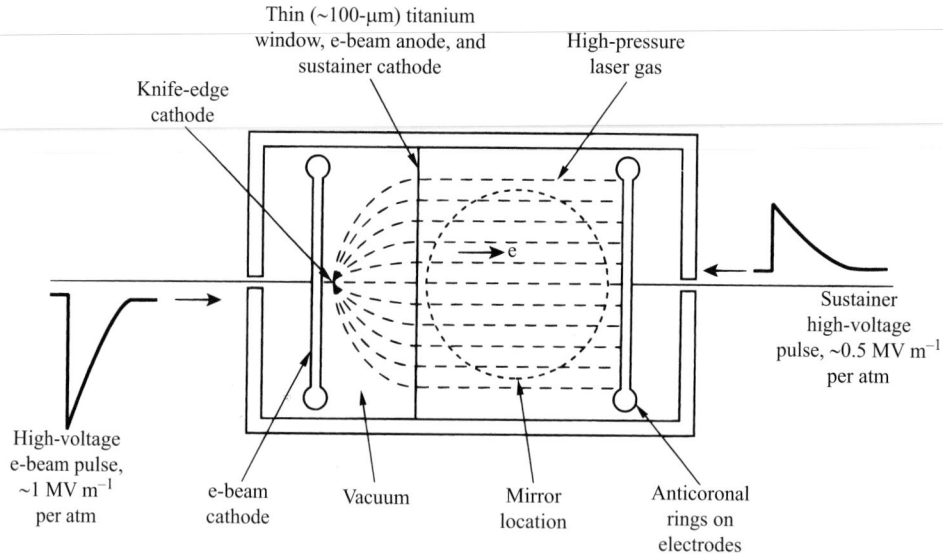


**Fig. 10.13** The transverse cross-section of an e-beam pulser/sustainer laser.

In TEA lasers it is uncommon for the voltage and pressure conditions necessary for uniform, and therefore efficient, discharge operation to be the same as the voltage and pressure conditions that provide the optimum $E/p_0$ ratio for excitation of the upper laser level. To solve this problem an e-beam pulser/sustainer arrangement can be used, as shown schematically in Fig. 10.13. The e-beam energy is chosen to provide uniform penetration of electrons throughout the volume to be excited. The current needed from the e-beam for preionization is relatively low. Once preionization has occurred, the main discharge capacitor is able to provide current to the discharge. The voltage, $V$, to which this capacitor is charged is usually insufficient to cause gas breakdown without preionization, but is chosen to provide the desired $E/p_0$ ratio for laser excitation. The bulk of the excitation energy is provided by this *sustainer* capacitor, whose capacitance can be chosen to provide the length of excitation pulse required.

## 10.5 Ultraviolet molecular gas lasers

Several important molecular gas lasers use fast-pulse electrical discharge or e-beam pumping to excite laser action on electronic transitions in the ultraviolet. Fast-pulse, high-power excitation is required because (a) the upper laser level is short-lived; (b) the lower laser level may be longer lived than the upper, so the system is self-terminating; and (c) a large upper laser level population is required in order to achieve threshold gain at short wavelengths. Figure 10.14 shows an energy-level diagram for the nitrogen molecule showing the electronic states that are important in this laser. Electrons excite $N_2$ molecules from the lowest vibrational level of the $^1\Sigma_g^+$ ground electron state to various vibrational levels of

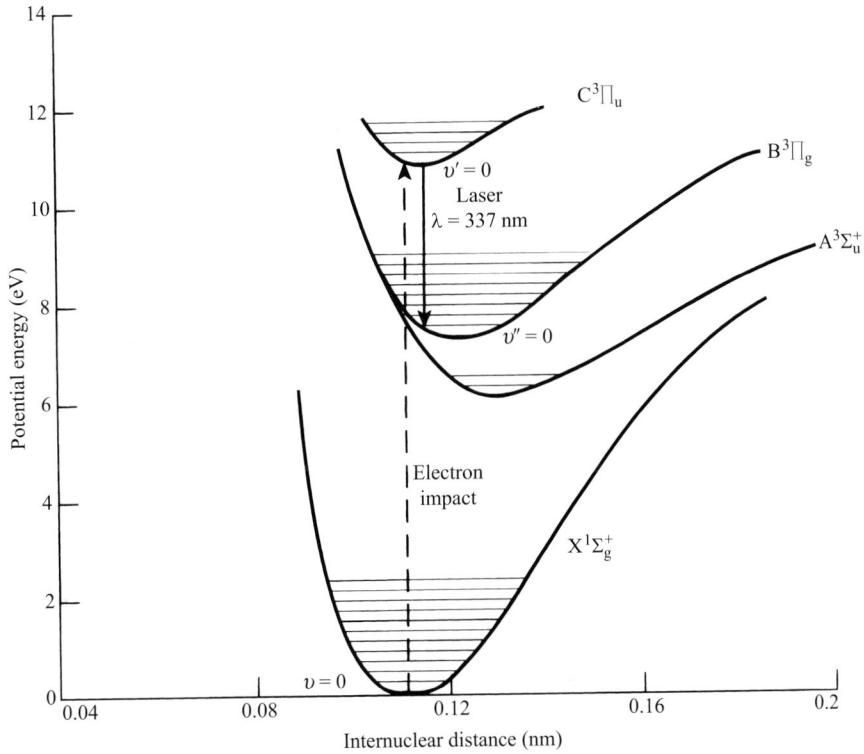


**Fig. 10.14** The energy-level diagram for the $N_2$ molecule, showing an excitation pathway and a laser transition from the upper level ($C^3\Pi_u(v' = 0)$) to the lower level ($B^3\Pi_g(v'' = 0)$) at 337 nm.

the $C^3\Pi_u$ state. Laser emission then occurs to various vibrational levels of the $B^3\Pi_g$ state. This state cannot lose energy readily, so the laser is self-terminating. The most important laser transitions are around 337.1 nm. Lasers of this kind can generate megawatt peak powers in pulses 10–20 ns long. Broadly similar laser action can be obtained using CO ($\lambda \sim 190$ nm) and hydrogen ($\lambda \sim 160$ and 116 nm) [6].

There exists a much more important series of lasers in which the laser transition occurs between two electronic states, the lower of which is unbound. The lower laser level produced by stimulated emission flies apart on the time scale of a molecular vibration ($\sim 10^{-13}$ s). There could not be a more efficient way to depopulate the laser level and provide a perfect population inversion! Electronic transitions of this sort have been known for many years to occur in high-pressure, rare-gas discharges, and do indeed serve as practical ultraviolet high-intensity sources. Although the rare gases do not form stable molecules such as $He_2, Ne_2, \ldots$ etc. (called *dimers*) in the ground state, these molecules are known to exist in excited electronic states. These molecules are called *excimers*, a contraction from *excited dimer*. The $Xe_2^*$ excimer laser was the first of this type to be operated; its energy-level diagram is Shown in Fig. 10.15. This laser is excited by a high-energy e-beam using high-pressure xenon (1–30 atm) [7]. Although there are very many different

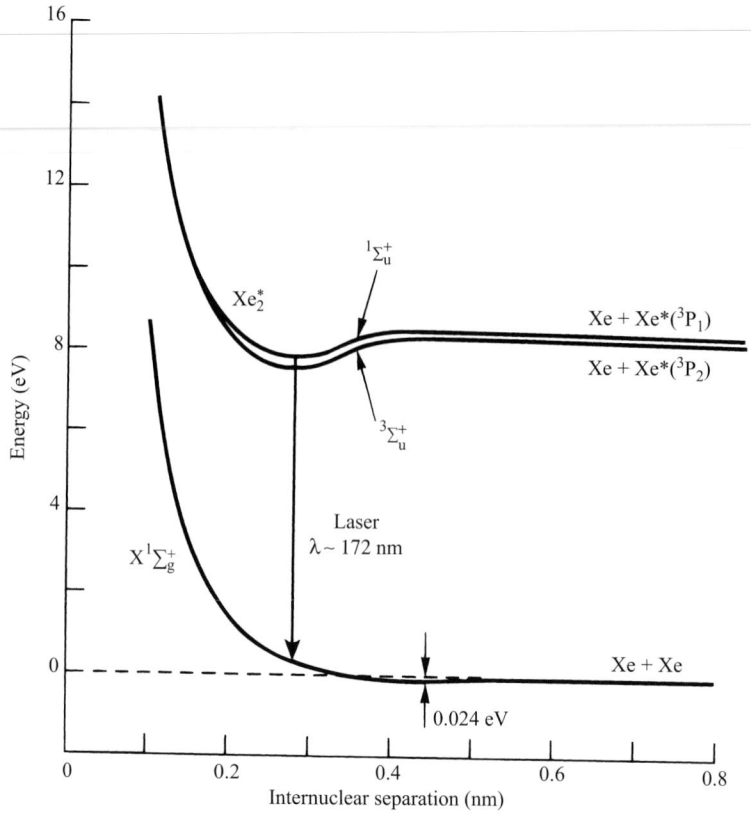


**Fig. 10.15** Partial energy-level diagrams of the $Xe_2$ molecule showing some stable bound states of $Xe_2^*$ and the laser transition. There is a very weak van der Waals binding force between the two Xe atoms in the ground state.

collisional reactions operative in the excitation process, it can be represented in a simple way as

$$\begin{aligned} Xe + e &\rightarrow Xe^*, \\ Xe^* + Xe + Xe &\rightarrow Xe_2^* + Xe, \\ Xe_2^* &\rightarrow Xe_2 + h\nu, \\ Xe_2 &\rightarrow Xe + Xe. \end{aligned} \tag{10.3}$$

It is clear from Fig. 10.15 that the laser transition can occur over a range of energies, so the laser is tunable over a small range near 171.6 nm.

If spontaneous emission on the ($^1\Sigma_u^+ \rightarrow {}^1\Sigma_g^+$) transition occurs, the observed line profile is very broad. In consequence of this and the short operating wavelength of the laser, the population inversion needed for oscillation is very large. This is the principal reason why high-power pulsed excitation is needed to reach threshold. Such required pumping powers are too great to be realistic on a CW basis.

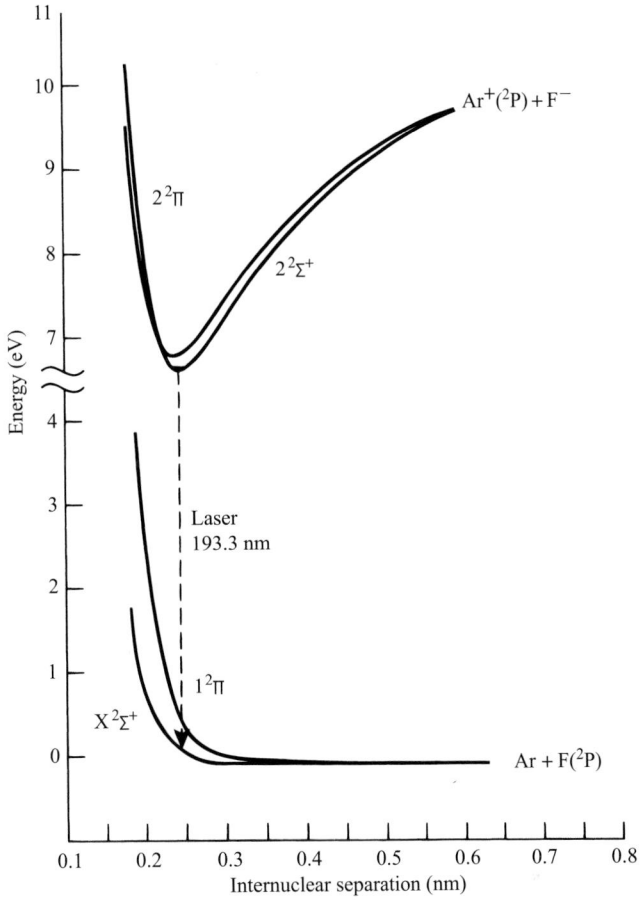


**Fig. 10.16** A partial energy-level diagram of the Ar–F system showing stable states of ArF* and the important laser transition at 193 nm.

Most ultraviolet molecular gas lasers are the so-called *excimer* lasers. These lasers use electronic transitions between an excited bound state of a molecule formed from two different atoms, frequently a rare gas[4] and a halogen, for example ArF*, and the unbound (repulsive) ground state of the molecule, as shown in Fig. 10.16. Since these molecular lasers utilize molecules that are not strictly dimers, they should really be called *exciplex* lasers. This word was introduced as a contraction of *excited complex*. There are several important rare-gas halide and other halide lasers, listed in Table 10.1. These lasers provide not only ultraviolet radiation but also intense visible radiation, for example from HgBr. These lasers are attractive because of their large energy outputs and potentially great electrical efficiency. For example, e-beam excitation of a mixture containing 5 torr of $F_2$ in 2 atm of argon has generated 55-ns pulses of energy 92 J and a peak power of 1.6 GW [8].

[4] Not all the *rare* gases are particularly rare; however, the alternative name *noble* gas, which was coined to describe their lack of reactivity, is not particularly appropriate either, since some of the rare-gas halides are actually stable in the ground state, for example $XeF_6$.

**Table 10.1** Exciplex lasers

| Exciplex molecule | Principal laser wavelengths (nm) |
|---|---|
| ArCl | 169, 175 |
| ArF | 193.3 |
| ArO | 558 |
| HgBr | 499–504.6 |
| HgCl | 551.6–559 |
| KrCl | 222 |
| KrF | 248.4–249.1 |
| KrO | 557.81 |
| XeBr | 281.8 |
| XeCl | 307–308.43 |
| XeF | 348.8–354.0, 483, 486 |
| XeO | 537.6, 544.2 |
| $Kr_2F$ | 430 |
| $Xe_2Cl$ | 518 |

Most of these lasers provide greatest power and efficiency when e-beam excited. However, most of them can also be excited in TEA laser structures, provided that care is taken to minimize the inductance of the discharge circuit so that fast-rise-time electrical discharges can be obtained. Typical operating conditions of the rare-gas halide lasers require a large pressure of the rare gas with a much smaller amount of the halogen, or a halogen-bearing molecule. For example, the ArF* laser operates well with an Ar–$NF_3$ mixture. All these lasers have broad spontaneous emission profiles and need large population inversions in order to reach threshold. However, they store a great deal of energy in the excited molecules, and, once laser action can be achieved, very large output energies are possible. These lasers have been seriously investigated as drivers for laser-induced thermonuclear fusion.

## 10.6 Photodissociation lasers

These are lasers in which an ultraviolet flashlamp or the Sun is used to decompose (photolyse) a molecule into fragments, one of which is produced in an excited state. The best example is the atomic-iodine photodissociation laser. Several fluoro-organic iodides when photolysed produce atomic iodine in an excited atomic state. One of the best molecules for this is i-$C_3F_7I$ (iso-perfluoropropyl iodide), which is a liquid at room temperature but has a vapor pressure of more than 200 torr. An ultraviolet photon at ~270 nm decomposes this molecule as follows:

$$\text{i-C}_3\text{F}_7\text{I} + h\nu(\lambda \sim 270\ \text{nm}) \rightarrow \text{i-C}_3\text{F}_7 + \text{I}^*(^2\text{P}_{1/2}). \tag{10.4}$$

A rare gas is usually added to the laser mixture. Laser oscillation occurs at 1.315 μm:

$$I^*(^2P_{1/2}) \rightarrow I(^2P_{3/2}, \text{ ground state}) + h\nu(\lambda = 1.315\ \mu\text{m}). \tag{10.5}$$

This might appear to be a three-level laser level, since the lower laser is the atomic ground state; however, the ground-state atomic iodine is quite rapidly removed to re-form the parent molecule (and, unfortunately, some molecular iodine):

$$I + \text{i-}C_3F_7 + M \rightarrow \text{i-}C_3F_7I + M \tag{10.6}$$

(M is any other molecule, or rare-gas atom, present),

$$I + I + M \rightarrow I_2 + M. \tag{10.7}$$

Molecular iodine makes subsequent laser action less efficient, so it must be removed.

When operated in a pulsed mode, large oscillator/amplifier(s) iodine laser systems have yielded output energies of 1 kJ in 1 ns – a peak power of $10^{12}$ W [9]. These systems are similar to optically pumped solid-state lasers, except that a quartz tube full of the laser gas mixture replaces the crystalline laser rod. These lasers can also be operated in CW mode with a continuous ultraviolet lamp as the pump, provided that the laser gas is recirculated and the iodine removed [10].

## 10.7 Chemical lasers

A chemical laser system is one in which the reaction of two or more molecules leads to the generation of reaction products in excited states. An important laser of this kind is the hydrogen fluoride laser. In this laser, electrical initiation of a reaction between hydrogen and fluorine leads to the following reactions:

$$H_2 + e \rightarrow 2H + e, \tag{10.8a}$$

$$H + F_2 \rightarrow HF(v^{**}) + F + 4.24\,\text{eV}, \tag{10.8b}$$

$$F + H_2 \rightarrow HF(v^*) + H + 1.38\,\text{eV}, \tag{10.8c}$$

where the asterisks indicate vibrationally excited molecules. Reactions (10.8b) and (10.8c) cyclically repeat, and constitute what is called a *chain reaction*. Each of these reactions liberates energy, which can emerge in part as vibrational energy of the hydrogen fluoride molecule. Sufficient energy is liberated by reaction (10.8b) to excite the HF to vibrational levels up to $v = 6$. Laser oscillation is possible on vibrational transitions $6 \rightarrow 5$, $5 \rightarrow 4$, $4 \rightarrow 3$, $3 \rightarrow 2$, $2 \rightarrow 1$, and $1 \rightarrow 0$. The wavelength range runs from 2.41–3.38 μm. Reaction (10.8c) will excite HF molecules up to $v = 3$. The DF laser operates in a similar fashion but has laser transitions $4 \rightarrow 3$, $3 \rightarrow 2$, $2 \rightarrow 1$, and $1 \rightarrow 0$ in the 3.5–4.06-μm range. We can illustrate schematically the progress of a reaction such as (10.8c) with reference to Fig. 10.17. The energy of the various molecules involved is plotted as a function of a variable called the *reaction coordinate*, which is a parameter that indicates in a qualitative manner the progress of the reaction from starting material (reactants)

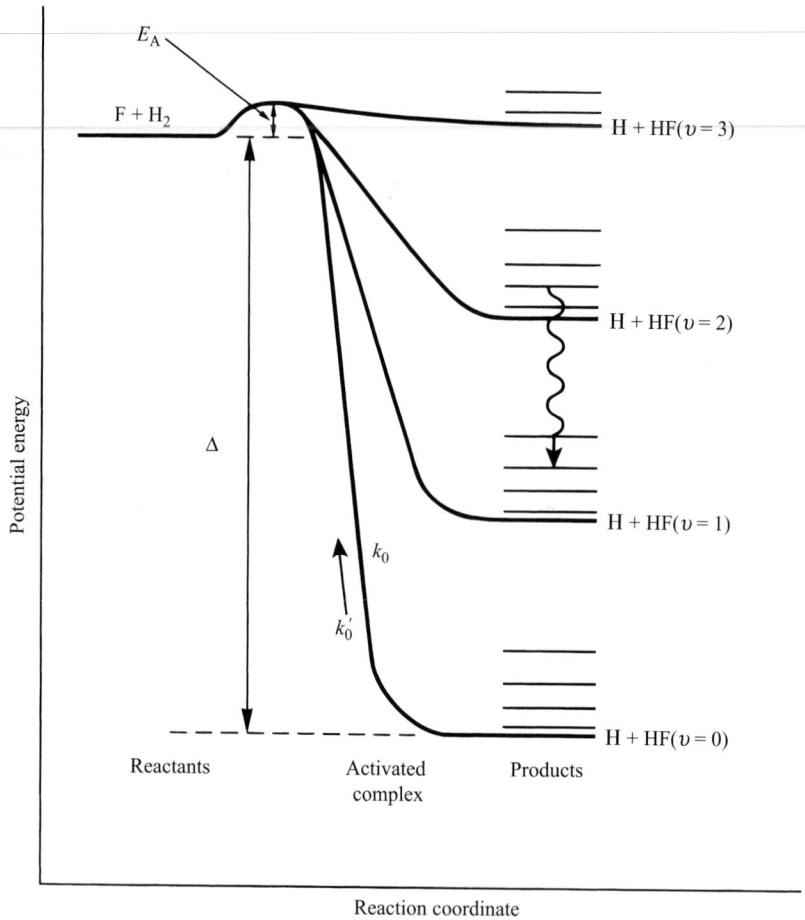


**Fig. 10.17** A schematic representation of the progress of the reaction between fluorine atoms and molecular hydrogen to generate vibrationally excited HF. $E_A$ is the activation energy of the reaction [11].

to reaction products. The activation energy needed to cause the reaction to start is generally supplied by the kinetic energy of a colliding hydrogen atom and fluorine molecule. The energy $\Delta E$ liberated by the reaction allows the possibility of the excitation of several vibrational levels of the product.

In the case of the HF or DF chemical laser, the starting material can be a mixture of $H_2$ ($D_2$) and $F_2$, but often is a mixture of $H_2$ ($D_2$) and some other fluorinated molecule such as $NF_3$, $SF_6$, $IF_5$, or $SO_2F_2$. The reaction between $H_2$ and $F_2$ is very violent, and the laser can constitute a virtual "bomb." Very-high-energy systems of this kind use high-pressure starting mixtures and are initiated by e-beam excitation to generate a uniform reaction throughout the laser volume. Pulses in excess of 4 kJ in 25 ns have been obtained. Very large CW powers are also possible. Figure 10.18 shows an interesting system in which a mixture of ethylene ($C_2H_4$), nitrogen trifluoride ($NF_3$), and helium is burnt to generate fluorine atoms that then react with deuterium to generate vibrationally excited

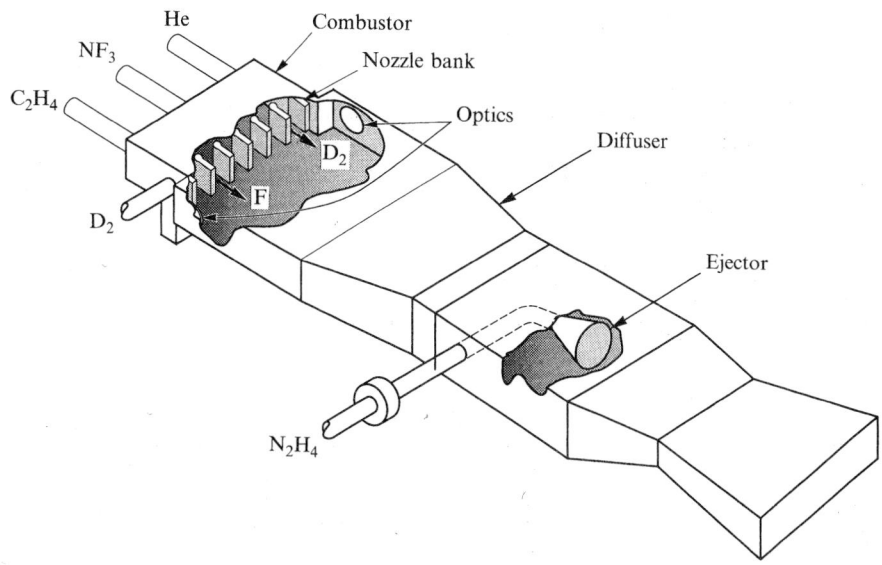


Fig. 10.18 A schematic diagram of a high-power CW DF chemical laser.

DF in a supersonic flow from a bank of nozzles. The removal of gases downstream of the nozzles is assisted by the injection at high pressure of hydrazine ($N_2H_4$) to provide momentum to the gas flow. This scheme is the basis for the highest-power CW chemical laser currently in operation, known as the Middle Infrared Advanced Chemical Laser (MIRACL), at White Sands, New Mexico, which can generate output powers in excess of 2 MW from vibrationally excited DF. This laser has proved capable of shooting down small missiles.

There are several other important chemical lasers. The CO chemical laser generally uses the following reaction sequence:

$$CS_2 + O \rightarrow CS + SO, \tag{10.9}$$

$$CS + O \rightarrow CO(v^*) + S + 3.6\ \text{eV}. \tag{10.10}$$

This reaction is a "flame" reaction, and continues as long as oxygen and carbon disulfide are supplied. Continuous powers of 700 W have been obtained from lasers of this kind [12].

The hydrogen chloride laser uses a reaction scheme

$$H + Cl_2 \rightarrow HCl(v^{**}) + Cl,$$

$$Cl + H_2 \rightarrow HCl(v^*) + H,$$

and lases in the 3.57–4.11-μm region.

One final interesting system is the chemical iodine laser, which involves the generation of *electronically* excited iodine atoms as a result of the reactions

$$O_2(^1\Delta) + O_2(^1\Delta) \rightarrow O_2(^1\Sigma) + O_2(^3\Sigma, \text{ ground state}),$$
$$O_2(^1\Sigma) + I_2 \rightarrow O_2(^3\Sigma) + 2I(^2P_{3/2}, \text{ ground state}),$$
$$O_2(^1\Delta) + I(^2P_{3/2}, \text{ ground state}) \rightarrow O_2(^3\Sigma) + I^*(^2P_{1/2}).$$

Laser emission occurs at 1.315 μm, the same wavelength as in the previously described iodine photodissociation laser. The electronically excited molecular oxygen $O_2(^1\Delta)$ is generated by a chemical reaction between gaseous chlorine and liquid hydrogen peroxide:

$$Cl_2 + H_2O_2 + 2KOH \rightarrow O_2(^1\Delta) + 2KCl + 2H_2O.$$

This laser system is both powerful and efficient [13]. Continuous-wave powers of tens of kilowatts have been reported [14] and this laser is a useful industrial high-power laser. The United States Air Force Airborne Laser (ABL) system uses a COIL laser mounted on a Boeing 747 and has been reported to generate hundreds of kilowatts.[5]

## 10.8 Far-infrared lasers

Far-infrared (FIR) lasers are generally molecular gas lasers in which the laser transition is between two rotational levels associated with the same vibrational level. As we have already mentioned briefly in Chapters 8 and 9, these lasers can be optically pumped by an auxiliary laser or excited by an electrical discharge. A few technological features of these devices make them different from the gas lasers we have discussed so far. Because of their long operating wavelengths, these lasers have narrow gain linewidths. Consequently, only one axial model of a resonator lies under the gain profile. The length, $L$, of the laser resonator must be physically adjustable so that $mc/(2L) = \nu_0$, where $\nu_0$ is the center frequency of the gain profile. The variation of the power output as the resonator length is varied is shown schematically in Fig. 10.19. In order to minimize diffraction losses, characterized by the Fresnel number $D^2/(\lambda L)$, where $D$ is the diameter of the tube, laser tubes of relatively large diameter are generally used. This is particularly true in the design shown in Fig. 10.20(a). This is a traditional laser resonator in which the optical pumping is done by a $CO_2$ laser beam that enters through one of the resonator mirrors. A simple way for the $CO_2$ laser beam to enter is for it to be focused through a small hole in the center of one laser mirror. The FIR laser output can be extracted through a hole in the other resonator mirror, or through a *dichroic* or *hybrid* laser mirror.[6] It is more common these days, however, to

[5] The ABL project was canceled in 2011.

[6] A dichroic mirror has different reflectance characteristics in two wavelength regions. In this case it would have high reflectance in the 10 μm region and the desired output coupling reflectance at the FIR wavelength. A hybrid mirror performs a dichroic function, but uses a metal grid deposited on to a dielectric material to provide the required reflectance characteristics. Such metal grids can provide a linearly polarized output beam. If the grid consists of a series of parallel wires, the **E**-field polarization parallel to the wires induces a current in the wires and suffers high loss. The polarization perpendicular to the wire suffers much less loss and is transmitted.

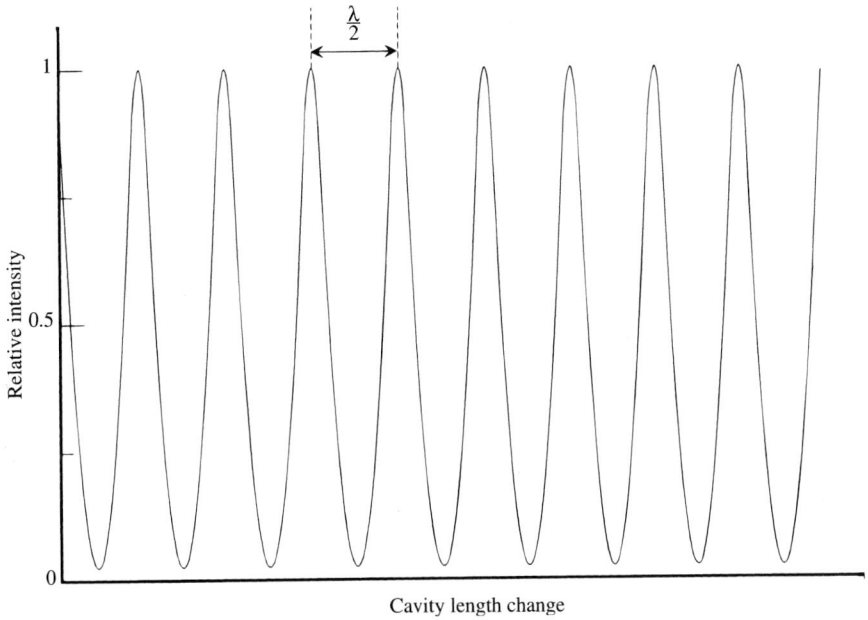


**Fig. 10.19** The schematic variation of the power output of a far-infrared laser as the length of the resonator is adjusted.

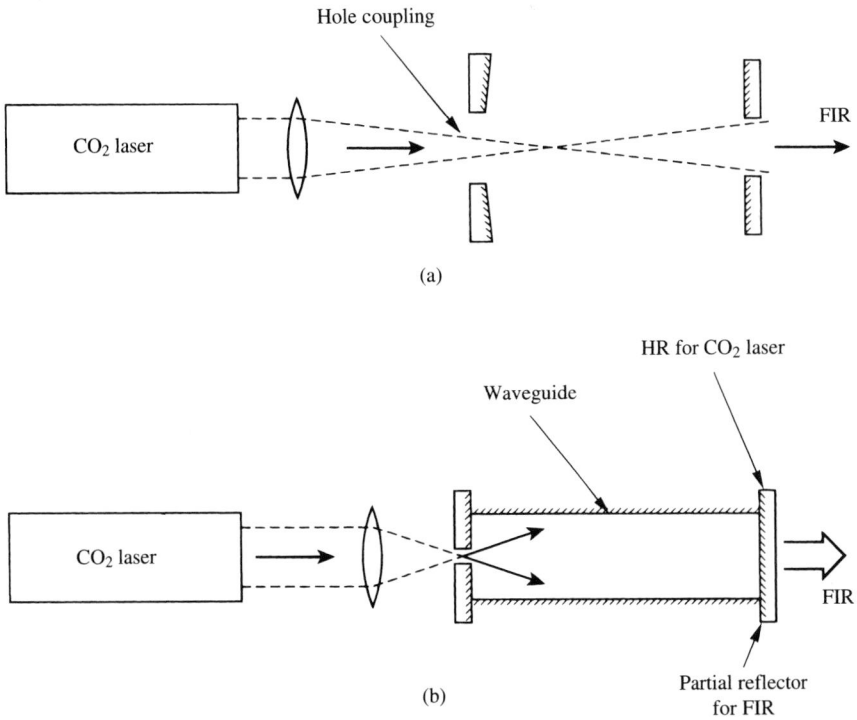


**Fig. 10.20** (a) A far-infrared laser pumped by a $CO_2$ laser that enters through a hole in one of the resonator mirrors.
(b) A far-infrared waveguide laser design.

**Table 10.2** Important optically pumped far-infrared lasers

| Molecule | Wavelength (μm) | Typical power output (mW) | Pump laser and wavelength (μm) |
|---|---|---|---|
| $NH_3$ | 81.5 | 50 | $N_2O$, 10.78 |
| $^{15}NH_3$ | 152.9 | 200 | $^{13}CO_2$, 10.78 |
| $^{10}BCl_3$ | 19.4 | 50 | $CO_2$ |
| | 19.1 | 50 | $CO_2$ |
| | 18.3 | 50 | $CO_2$ |
| $^{11}BCl_3$ | 20.6 | 100 | $CO_2$ |
| | 20.2 | 100 | $CO_2$ |
| $CH_3F$ (methyl fluoride) | 372.68 | 10 | $CO_2$, 10.16 |
| | 192.78 | 10 | $CO_2$, 10.17 |
| HCOOH (formic acid) | 513.0157 | 5 | $CO_2$, 9.23 |
| | 302.2781 | 5 | $CO_2$, 9.27 |
| HCOOD | 926.2087 | 10 | $CO_2$, 10.29 |
| | 919.9355 | 10 | $CO_2$, 10.17 |
| DCOOH | 380.5654 | 10 | $CO_2$, 10.30 |
| $CH_3Cl$ (methyl chloride) | 334.0 | 50 | $CO_2$, 9.75 |
| $CH_2F_2$ (difluoromethane) | 214.5791 | 500 | $CO_2$, 9.20 |
| | 184.30590 | 500 | $CO_2$, 9.21 |
| $CH_3{}^{79}Br$ | 414.98 | 20 | $CO_2$, 10.38 |
| $CH_3I$ (methyl iodide) | 1,253.738 | 10 | $CO_2$, 10.72 |
| $CH_3OH$ (methanol) | 170.57637 | 10 | $CO_2$, 9.69 |
| | 164.7832 | 10 | $CO_2$, 9.33 |
| | 118.8349 | 20 | $CO_2$, 9.69 |
| | 96.52239 | 20 | $CO_2$, 9.33 |

build the laser tube as an overmoded waveguide. The laser tube can be made of polished copper or of glass coated with gold on its interior, as shown schematically in Fig. 10.20(b). The $CO_2$ beam reflects off the interior walls, and makes multiple passes inside the resonator so as to pump the absorbing molecules efficiently. The FIR laser oscillation occurs in a transverse mode of the waveguide structure. Diffraction losses are reduced because the oscillating wavelength cannot leak sideways from the resonator: it is guided by the tube walls.

Some important FIR lasers are listed in Table 10.2.

## 10.9 Problem

10.1 From the data in Fig. 10.7, calculate the population inversion and gain as a function of distance from the nozzle.

# References


[1] Sir J.H. Jeans, *An Introduction to the Kinetic Theory of Gases*, Cambridge: Cambridge University Press, 1962.

[2] C.C. Davis, "Neutral gas lasers," in *Handbook of Laser Science and Technology, Vol. I, Lasers and Masers*, ed. M.J. Weber, Boca Raton, FL: CRC Press, 1982, pp. 3–168.

[3] W.B. Tiffany, R. Targ, and J.D. Foster, "Kilowatt $CO_2$ gas-transport laser," *Appl. Phys. Lett.* **15**, 91–93, 1969.

[4] J.D. Anderson, Jr., *Gasdynamic Lasers: An Introduction*, New York: Academic Press, 1976.

[5] J.D. Cobine, *Gaseous Conductors*, New York: Dover, 1958.

[6] R.S. Davis and C.K. Rhodes, "Electronic transition lasers," in *Handbook of Laser Science and Technology, Vol. I, Lasers and Masers,* ed. M.J. Weber, Boca Raton, FL: CRC Press, 1982, pp. 273–312.

[7] M.H.R. Hutchinson, "Excimer lasers," in *Tunable Lasers*, eds. L.F. Mollenhauer and J.C. White, Berlin: Springer-Verlag, 1987, pp. 19–56.

[8] J.M. Hoffman, A.K. Hays, and G.C. Tisone, "High-power noble-gas-halide lasers," *Appl. Phys. Lett.* **28**, 538–539, 1976.

[9] G. Brederlow, E. Fill, and K.J. White, *The High-Power Iodine Laser*, Berlin: Springer-Verlag, 1983.

[10] L.A. Schlie and R.D. Rathge, "Long operating time CW atomic iodine probe laser at 1.315 μm," *IEEE J. Quant. Electron.* **20**, 1187–1196, 1984.

[11] K.L. Kompa, *Chemical Lasers*, Berlin: Springer-Verlag, 1973.

[12] R.J. Richardson, H.Y. Agero, H.V. Lilenfield, T.J. Menne, J.A. Smith, and C.E. Wiswall, "CO chemical laser utilizing combustor-generated reactants," *J. Appl. Phys.* **50**, 7939–7947, 1979.

[13] G.N. Hays and G.A. Fisk, "Chemically pumped iodine laser as a fusion driver," *IEEE J. Quant. Electron.* **17**, 1823–1827, 1981.

[14] D.L. Carroll, D.M. King, L. Fockler, D. Stromberg, W.C. Solomon, L.H. Sentman, and C.H. Fisher, "High-performance chemical oxygen–iodine laser using nitrogen diluent for commercial applications," *IEEE J. Quant. Electron.* **36**, 40–51, 2000.

# 11 Tunable lasers

## 11.1 Introduction

All lasers are to some extent *tunable*. Their output frequency can be varied continuously without discontinuous changes in output power by moving the position of the oscillating modes under the gain profile. However, if the gain profile is not very wide then this range of tunability is limited. For any atomic gas laser, for example, where gain profiles typically have Doppler widths on the order of 1 GHz, tunability of a single axial mode over about a 1 GHz range can be accomplished by changing the optical length $n\ell$ of the cavity. This can be done by moving one of the mirrors with a piezoelectric transducer[1] and thereby varying the geometric length $\ell$, or by adjusting the laser pressure so as to adjust the index $n$. Although a tunable frequency range of 1 GHz might seem large in absolute terms, it represents a very small fraction of the operating frequency of the laser, $10^{14}-10^{15}$ Hz, say. Discontinuous tunability in the infrared can be obtained by using a molecular gas laser, for which several vibrational–rotational transitions have gain. Systems, such as the $CO_2$ laser, can offer many lines over a relatively broad wavelength region, but a graph of output power vs. frequency is not continuous.

To achieve continuous tunable laser operation over a broad wavelength region we must use an amplifying medium with a broad gain profile. The gain of such a laser is inversely proportional to the width of this gain profile, so, all else being equal, we must pump the system harder, and more efficiently, to reach threshold. We must also provide wavelength-selective feedback in the laser cavity to encourage laser oscillation at the desired wavelength.

## 11.2 The titanium–sapphire laser

Titanium ($Ti^{3+}$)-doped sapphire has a very broad absorption and emission spectrum, as seen previously in Chapter 7. It has a high fluorescence quantum efficiency of 80%, which has made it an attractive medium for tunable CW solid-state laser operation [1–3]. It can be tuned from approximately 650 nm to 1,100 nm, and is generally optically pumped with another laser. The sapphire host medium in this laser has excellent thermal conductivity, which helps to avoid the thermal problems that can occur in other host materials such as

[1] See Chapter 18.

glass. Because it has a relatively short upper-state lifetime (3.2 μs) and a high saturation intensity it is generally pumped with a high-power focused pump laser, with powers that can range up to 20 W. Flashlamp-pumped operation is also possible [4]. At one time argon-ion lasers were widely used for this purpose, but they have been largely replaced by frequency-doubled Nd:YAG or Nd:$YVO_4$ lasers (see Chapter 7), themselves optically pumped with GaAs semiconductor lasers (see Chapter 12). Titanium–sapphire lasers themselves have mostly replaced dye lasers as tunable sources in the visible and near infrared. Their radiation can also be doubled to provide blue and ultraviolet output [5]. The broad gain profile that can be achieved in these lasers makes them the most popular medium for the generation of femtosecond mode-locked laser pulses (see Chapter 6).

## 11.3 Organic dye lasers

Although they were not, chronologically, the first tunable lasers, lasers using a liquid gain medium consisting of organic dye molecules in solution provide an interesting case study of the characteristics of tunable lasers. Although they are not as widely used as they once were, they are inexpensive to build and operate. They also retain importance as the best representives of the class of liquid lasers. *Dye* molecules are so called because they impart characteristic and generally bright colors to otherwise colorless media. They do this by selectively absorbing radiation, so that, for example, a dye looks red because it absorbs incident blue and green light. The behavior of fluorescent and phosphorescent materials in lighting and displays can also be conveniently described in many ways by describing these phenomena in dye media.

### 11.3.1 Energy-level structure

From a structural standpoint organic dyes are hydrocarbon molecules that contain carbon–carbon double bonds and, in particular, sequences of alternate single and double bonds. Such structures are said to be *conjugated*. Figure 11.1 shows three examples. All these molecules contain the hexagonal carbon-ring structures (benzene rings) that characterize *aromatic* hydrocarbons. It is the relatively loosely bound electrons associated with these benzene rings that lead to the interesting electronic energy-level structure of organic dyes. These loosely bound electrons actually travel freely from nucleus to nucleus within the plane of the ring. This orbital motion in the ground state involves an even number of electrons that can be grouped together in pairs. Each pair consists of one electron with the direction of its spin momentum vector pointing up and one electron with its spin momentum vector pointing down, so the overall angular momentum is zero. In a single benzene ring we can group these electrons into three pairs above the plane of the ring and three pairs below. A single electron from one of these pairs can be excited into a range of states, as shown in Fig. 11.2. The states labeled $S_1$ and $S_2$ are called *singlet* states; $S_0$ is the ground state. The states $T_1$ and $T_2$ are called *triplet* states. The distinction between these two kinds of states is that in a singlet state the excited electron has moved to a higher energy but

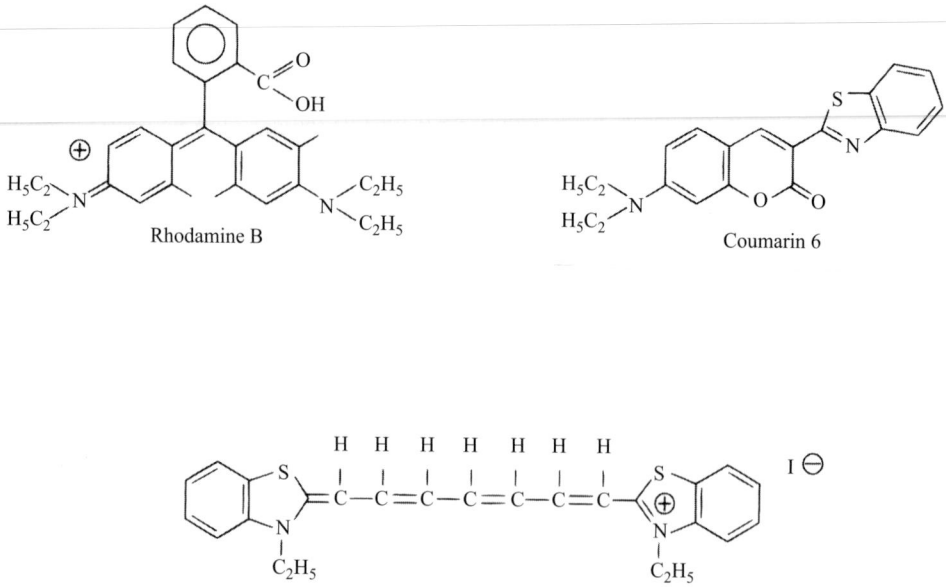


**Fig. 11.1** Some aromatic organic dye molecules.

retained its original spin orientation, so the overall angular momentum of the electrons remains zero, whereas in a triplet state the excited electron has moved to a higher energy state but its spin direction has flipped over, say from up to down, so the electrons now have net angular momentum. Each singlet and triplet state has associated with it the vibrational and rotational energy states we have discussed previously. At ambient temperature most dye molecules will be in the lowest vibrational level of $S_0$. Incident light of appropriate energy will excite these molecules to higher singlet states. Excitation of the vibrational levels of $S_1$ usually involves visible or ultraviolet photons. Thus we obtain a characteristic absorption spectrum, which *may* show vibrational structure, although in solution this structure will likely be smeared out. Once a vibrational level of $S_1$ has been excited, the molecule, generally on a time scale of about 1 ps, *internally converts* to the lowest vibrational level of $S_1$. *Fluorescence* is spontaneous emission from this lowest vibrational level of $S_1$ to the various vibrational levels of $S_0$, as shown in Fig. 11.2.[2]

In a simple model the fluorescence and absorption spectra should be mirror images, as shown in Fig. 11.3(a). This is not quite the case in reality, as the real spectra in Fig. 11.3(b) show. Molecules can also move non-radiatively from $S_1 \rightarrow T_1$ (or $S_2 \rightarrow T_2$ etc.). This process, called *intersystem crossing*, is relatively slow compared with internal conversion, but can occur as fast as within about 10 ns. Spontaneous emission from $T_1 \rightarrow S_0$ is called *phosphorescence*. Because this process involves the electron losing energy *and* flipping its spin, it is a relatively unlikely process and can have a lifetime from 1 μs to many

[2] Only one aromatic molecule, azulene (and its derivatives), fluoresces readily from $S_2 \rightarrow S_0$ [6].

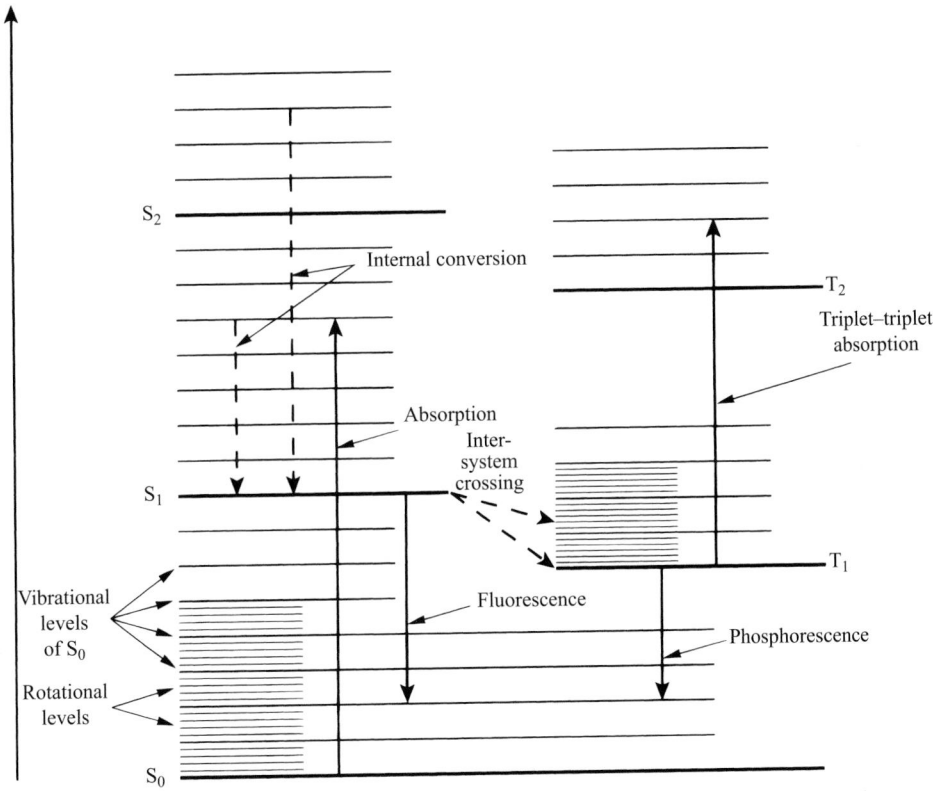


**Fig. 11.2** A schematic energy-level diagram for an aromatic organic dye molecule, showing various important radiative and non-radiative processes.

milliseconds. If molecules reach $T_0$, then absorption to higher triplet states $T_1, T_2, \ldots$ is an efficient process. All these processes are summarized in Fig. 11.2.

### 11.3.2 Pulsed laser excitation

The earliest dye lasers were excited by the pulses from $Q$-switched ruby lasers. Dye solutions of concentration varying from $10^{-6}$ to $10^{-3}$ mol $\ell^{-1}$ were placed in small optical cuvettes inside a laser cavity and the pump laser radiation was directed into the cuvette either from the side of the cuvette, or through one of the dichroic laser mirrors, as shown in Fig. 11.4. The pumping process involves the following steps: (i) rapid excitation of a vibrational–rotational level of $S_1$; (ii) internal conversion to the lowest vibrational level of $S_1$; and (iii) laser emission to those vibrational–rotational levels of $S_0$ sufficiently far above the ground state that their populations are low. A large population density in $S_1$ is required in order to achieve threshold because of the large $\Delta\nu$ of the fluorescence spectrum. The threshold will increase even further if the distributed loss, $\alpha$, within the resonator is not minimized. Distributed loss occurs for various reasons: in particular, intersystem crossing

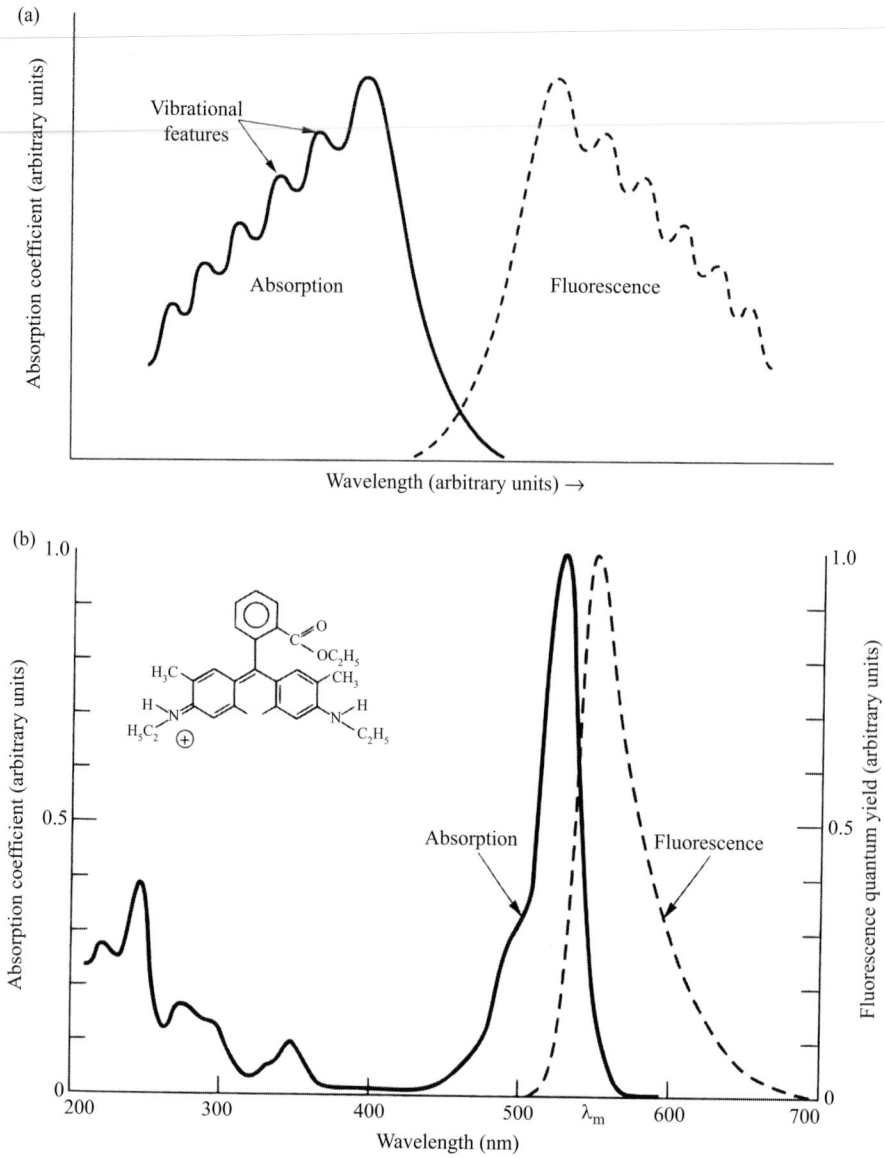


**Fig. 11.3** (a) Idealized absorption and fluorescence spectra of an aromatic organic dye. The vibrational structure is usually smeared out in solution. (b) Absorption and emission spectra of the organic dye Rhodamine 6G dissolved in ethanol.

populates $T_1$ and triplet–triplet absorption occurs in the same wavelength region as fluorescence. It has been discovered, however, that the $T_1$ population can be kept small by several means. These include using dyes where intersystem crossing is inefficient, and adding oxygen or detergent to the dye solution, both of which rapidly relax $T_1$ molecules to $S_0$. Absorption from $S_1 \rightarrow S_2$ can also occur at the laser wavelength, which is an important loss mechanism in some dyes. Regions of unpumped dye can contribute to $\alpha$

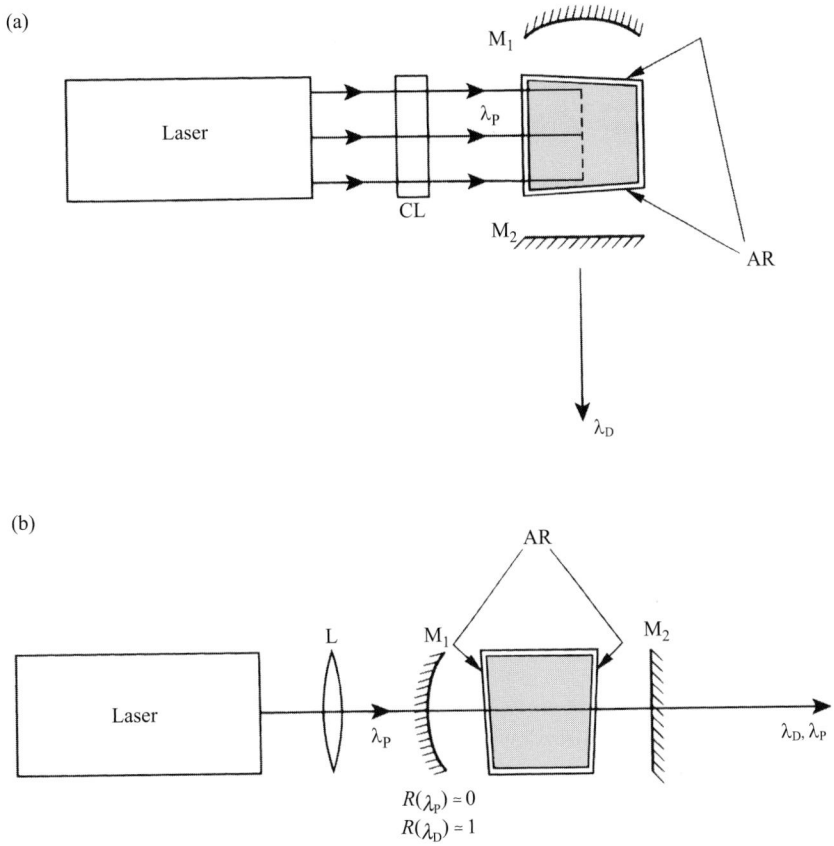


**Fig. 11.4** Simple arrangements for (a) side pumping and (b) end pumping of simple pulsed dye lasers. In each case the dye solution is contained in an optical cuvette with wedged, antireflection-coated (AR) faces. In (a) the laser beam is focused with a cylindrical lens (CL) to a line image in the cuvette. In (b) the pump laser is focused with a spherical lens (L). $M_1$ and $M_2$ are the dye-laser cavity mirrors.

through absorption of the laser wavelength, particularly if this is near the wavelength $\lambda_m$ in Fig. 11.3(b).

Pulsed dye-laser operation can also be achieved with fast flashlamp excitation in geometrical arrangements similar to those used to excite solid-state crystalline lasers. Figure 11.5 shows schematically a coaxial flashlamp arrangement which has been used to obtain dye-laser pulses up to 400 J in energy [7].

Although a dye laser is predominantly homogeneously broadened, this does not preclude the simultaneous oscillation of several simultaneous modes. With such a broad gain profile, many modes within a relatively broad wavelength region will have almost identical gain. There is insufficient time for the strongest mode to suppress its neighbors when the laser emits only short pulses. Furthermore, spatial inhomogeneity of the optical pumping and thermal and acoustic fluctuations of the liquid gain medium add an effective inhomogeneous contribution to the broadening. The output of a pulsed dye laser can be

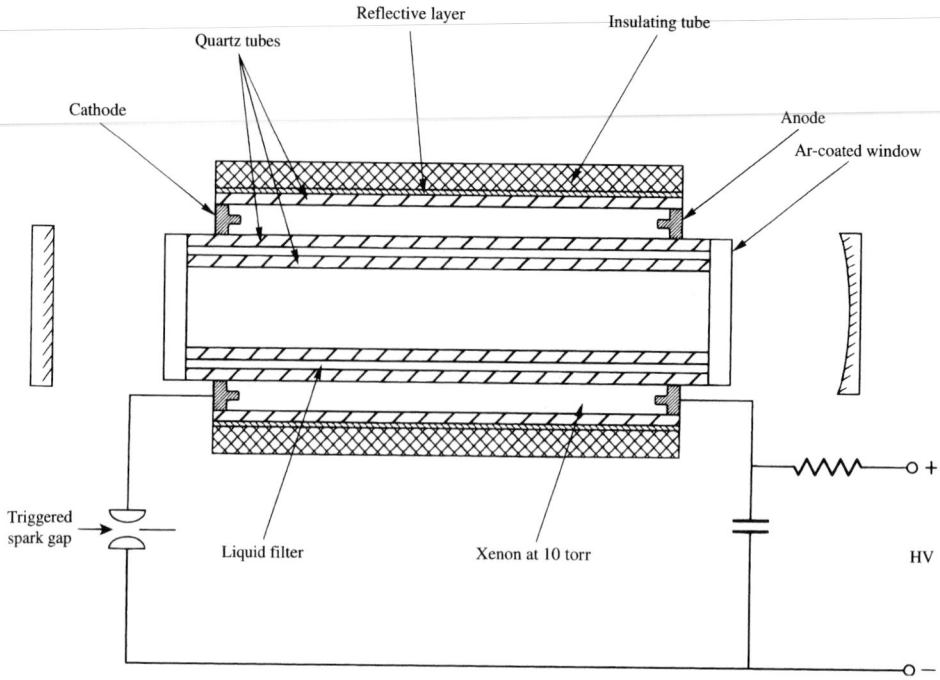


**Fig. 11.5** A schematic diagram of a high-energy pulsed dye laser using a coaxial flashlamp. The liquid filter absorbs short-wavelength ultraviolet light that can photodissociate dye molecules.

several nanometers wide unless additional frequency selectivity is incorporated into the resonant cavity. If one of the cavity mirrors is replaced by a reflective diffraction grating, as shown in Fig. 11.6, only the wavelength that satisfies the diffraction condition will experience reflection back along the resonator axis. It is common to include an intracavity beam-expanding telescope so that a large area of the diffraction grating is illuminated. For a reflective diffraction grating with groove spacing $d$ the condition for retroreflection is

$$m\lambda = 2d\sin\theta, \tag{11.1}$$

where $m$ is an integer, and $\theta$ is the angle between the resonator axis and the normal to the grating. The output wavelength of the laser can be tuned by altering the angle $\theta$. The output linewidth will also be reduced because only a small range of wavelengths will diffract within an angular range $\Delta\theta$ that keeps them traveling back and forth within the resonator. A reduction in the output linewidth by about a factor of 100 can be achieved with a grating having a small groove spacing, say $d = 0.5$ μm. The diffraction gratings used for this purpose are generally fabricated with a groove profile to maximize their efficiency for diffracting at the angle for which $m$ in Eq. (11.1) is equal to unity. For additional reduction in the output linewidth intracavity Fabry–Perot etalons, prisms, or birefringent-crystal waveplates[3] can be used. Further details are available elsewhere [9, 10]. One popular simple

[3] See Chapter 17 for a detailed discussion of waveplates – components that affect the polarization state of a light beam in a way that depends on the wavelength.

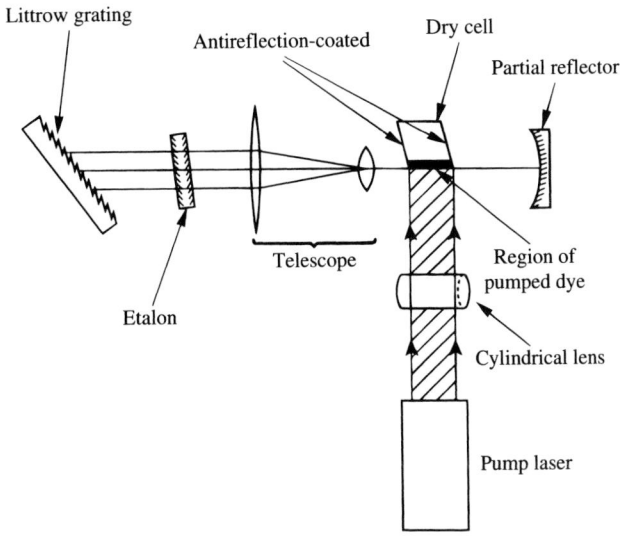


**Fig. 11.6** A pulsed-dye-laser design that uses a reflective diffraction grating for wavelength tuning and an intracavity etalon for line-narrowing. This is referred to as a Hänsch-type dye laser [8].

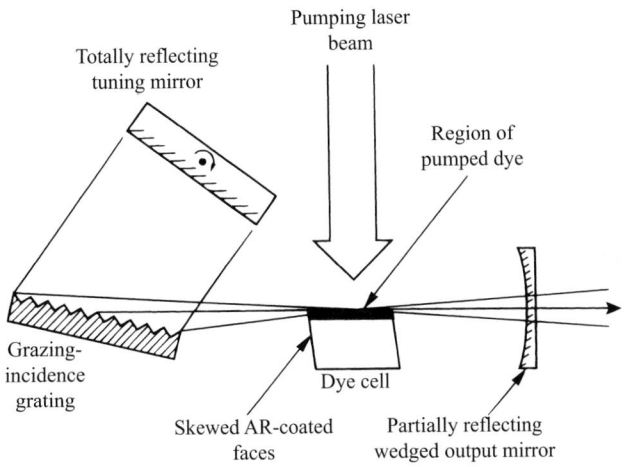


**Fig. 11.7** A Littman-type pulsed dye laser [11, 12] that uses a grazing-incidence grating for simultaneous tuning and line-narrowing.

arrangement is shown in Fig. 11.7. Line-narrowing of the output radiation in this case is accomplished by using a diffraction grating at a large angle of incidence (so-called *grazing* incidence). The pump laser is focused into the dye cuvette, often with a cylindrical lens so as to produce a line image. For high-power operation the dye would be continuously recirculated through a filter from a reservoir of dye solution.

### 11.3.3 Continuous-wave dye-laser operation

The pump intensity required to reach threshold in a CW dye laser is very high, typically 50 kW $cm^{-2}$. It is achieved by focusing a high-power visible CW laser into a small volume of dye solution (~50 μl). The dye solution is recirculated by spraying it from a slit nozzle in the form of a thin, planar jet, collecting the jet, and returning it to the dye reservoir. Frequently, the plane of the dye jet is set at Brewster's angle. Because of its desirable solvent and viscosity properties, ethylene glycol is frequently used as the solvent for the dye. Tuning can be accomplished with either an intracavity prism or a reflective grating. Figure 11.8 shows such an arrangement in which the dispersion of the tuning prism allows the pump laser beam to enter without having to pass through one of the dye-laser mirrors. Because this three-mirror resonator involves oblique incidence on one of the cavity mirrors, it can introduce astigmatism into the resultant laser beam. However, it is possible to place a Brewster-angle-oriented dye cell or dye jet so that its astigmatism compensates for that of the resonator [13]. The most commonly used pump lasers are argon- and krypton-ion lasers, which operate in the ultraviolet–red part of the spectrum, frequency-doubled Nd:YAG lasers operating at 530 nm, frequency-tripled Nd:YAG lasers operating at 355 nm, or ultraviolet excimer lasers. Figure 11.9 shows the tuning capability of various dyes pumped by various argon- and krypton-ion laser lines. The use of different dye–solvent combinations provides tunability over a broad region.

Because the pump laser is focused to a small spot (10–20 μm in diameter) in the dye solution, it is essential to design a dye-laser cavity with a small spot size, and to locate its beam waist at the focus of the pump beam.

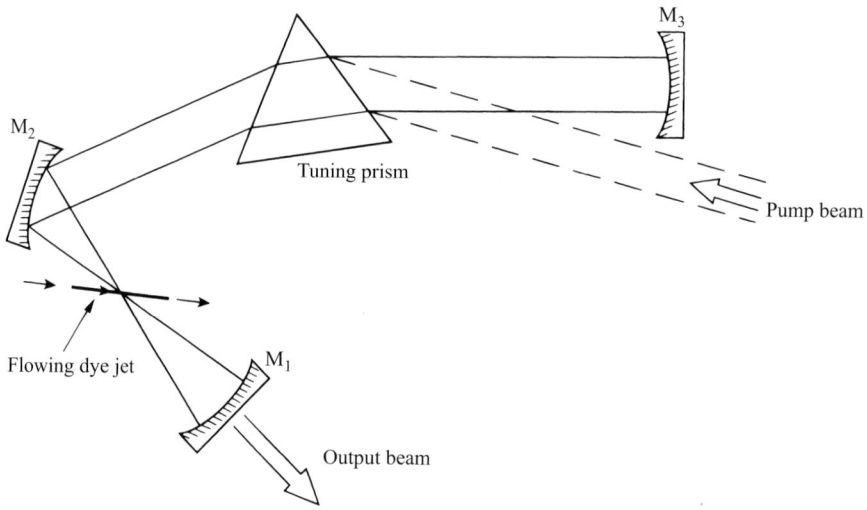


**Fig. 11.8** A schematic diagram of a CW dye-laser design incorporating an astigmatically compensated laser cavity and an intracavity flowing dye jet at Brewster's angle.

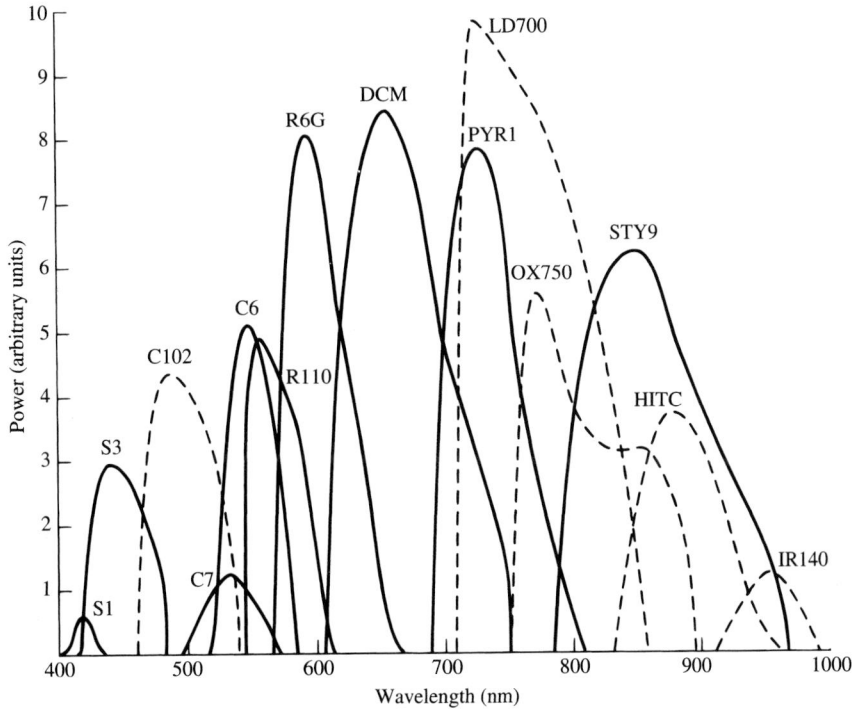


**Fig. 11.9** Typical tuning curves of various dyes pumped by argon- and krypton-ion laser lines. The dyes and pump wavelength are as follows: S1 (Stilbene 1, $Ar^+$ 351–364 nm); S3 (Stilbene 3, $Ar^+$ 351–364 nm); C102 (Coumarin 102, $Kr^+$ 407–423 nm); C7 (Coumarin 7, $Ar^+$ 476 nm); C6 (Coumarin 6, $Ar^+$ 488 nm); R110 (Rhodamine 110; $Ar^+$ 458–514 nm); R6G (Rhodamine 6G, $Ar^+$ 458–514 nm); DCM ($Ar^+$ 458–514 nm); PYR1 (Pyridene 1, $Ar^+$ 458–514 nm); LD700 ($Kr^+$ 647 and 676 nm); Ox750 (Oxazine 750, $Kr^+$ 647 and 676 nm); STY (Styryl 9, $Ar^+$ 458–514 nm); HITC (HITC-P, $Kr^+$ 647 and 676 nm); and IR140 ($Kr^+$ 752 and 799 nm) (tuning curves courtesy of Spectra-Physics, Inc.).

## 11.4 Calculation of threshold pump power in dye lasers

The calculation of the threshold pump power in a dye laser follows similar lines to the calculation performed for optically pumped solid-state lasers in Chapter 7. However, we must allow for the possible build-up of triplet-state molecules, particularly in the $T_1$ state, since these constitute a pump-dependent distributed absorption loss. The various important pathways in the system are shown schematically in Fig. 11.10.

Pump radiation populates a vibrational level of $S_1$ but very rapid internal conversion transfers these molecules to the lowest vibrational level of $S_1$. If the ground-state dye concentration is $N$, then the absorption coefficient for the pump radiation can be written as

$$\alpha_p = \sigma_p N, \tag{11.2}$$

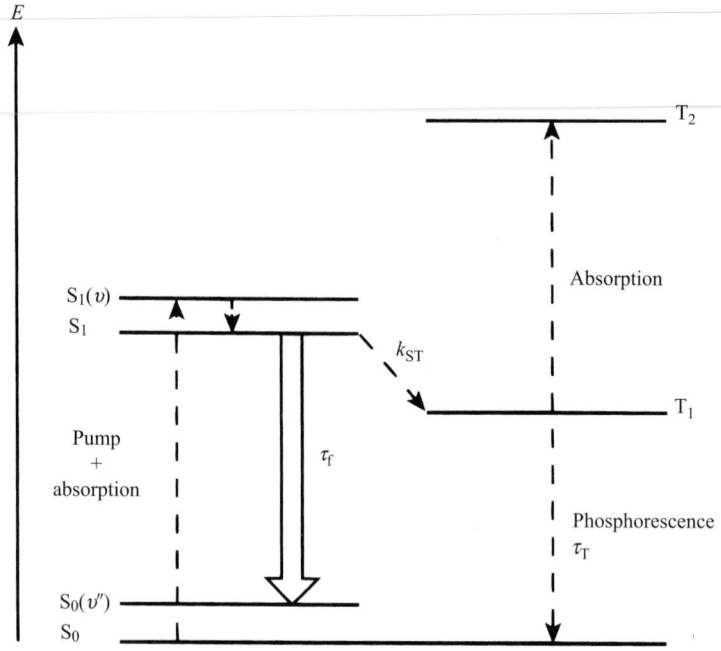


Fig. 11.10 Excitation and relaxation pathways in a dye laser.

where $\sigma_p$ is the absorption cross-section at the pump wavelength. If all the pump radiation of intensity $I_p$ is absorbed by the dye solution of thickness $d$ then we can write the pumping rate as

$$R = \frac{I_p}{h\nu_p d}. \tag{11.3}$$

Molecules leave $S_0$ by spontaneous emission (fluorescence) with lifetime $\tau_f$ and by transferring to $T_1$ with rate constant $k_{ST}$. Clearly the upper-laser-level population density $N_2$ satisfies the equation

$$\frac{d}{dt}N_2 = R - \frac{N_2}{\tau_f} - k_{ST}N_2. \tag{11.4}$$

### 11.4.1 Pulsed operation

In short-pulse operation with an excitation pulse faster than $1/k_{ST}$ (typically $\simeq$ 100 ns) we can neglect transfer to the triplet state. The pump intensity needed to achieve an upper-level population density then satisfies

$$I_p = N_2 h\nu_p d/\tau_f. \tag{11.5}$$

The lower-laser-level population is close to zero, so to reach threshold $N_2 \simeq N_t$. If we neglect distributed loss, which is reasonable in pulsed operation, then from Eq. (5.17) the

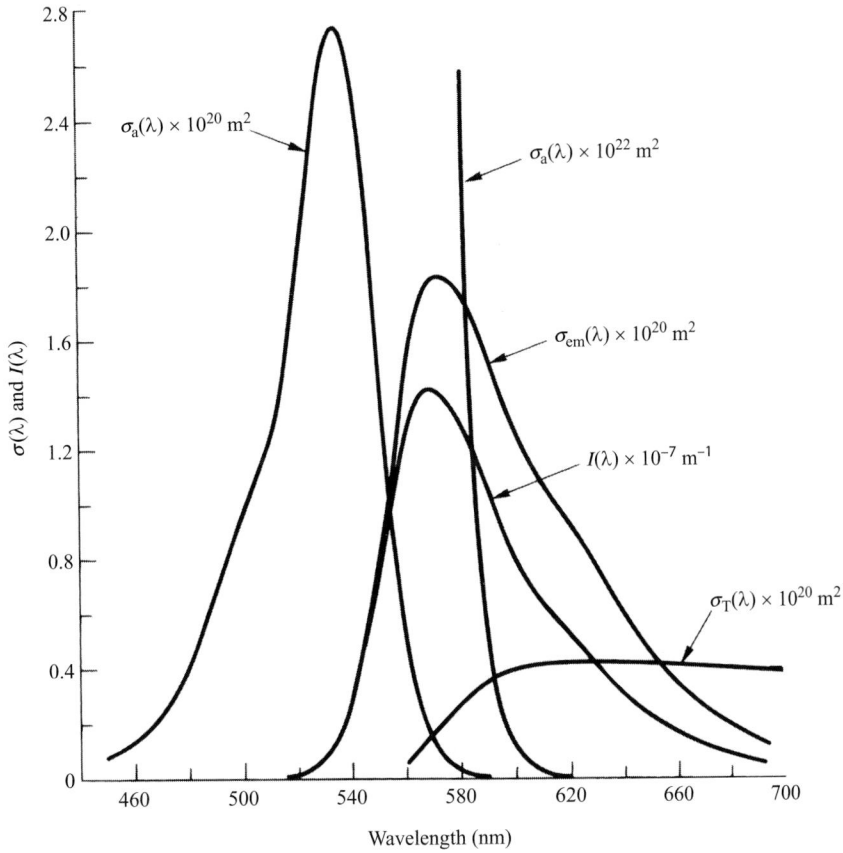


**Fig. 11.11** Absorption and emission cross-sections for the dye Rhodamine 6G. Data from Snaveley [14]. The emission spectrum has been normalized so that $\int I(\lambda)d\lambda = 0.92$, the measured fluorescence quantum efficiency.

threshold inversion density can be written as

$$N_t = -\frac{8\pi\,\tau_f\,\ell n(r_1 r_2)}{\lambda^2 g(\nu_0, \nu)\ell}. \tag{11.6}$$

For a homogeneously broadened fluorescence spectrum $g(\nu_0, \nu) = 2/(\pi\ \Delta\nu)$. If we design a laser cavity such that $R_1 = 1$ and $R_2 = 1 - T$, then $-\ell n(r_1 r_2) = T/2$. Combining Eqs. (11.5) and (11.6) gives

$$I_p = 2\pi^2 h\nu_p\,\Delta\nu\,Td/(\lambda^2\ell). \tag{11.7}$$

In a transversely pumped geometry, such as that shown in Fig. 11.4, some reasonable values for the parameters are $d = 1$ mm, $\ell = 10$ mm, and $T = 0.01$. For Rhodamine 6G in aqueous solution a frequency-doubled[4] Nd:YAG laser operating at 530 nm is a suitable pump. The peak of the fluorescence spectrum is $\lambda = 580$ nm with an FWHM of about 60 nm, equivalent to $5.4 \times 10^{13}$ Hz, as shown in Fig. 11.11. The calculated threshold pump

[4] See Chapter 20.

intensity is $2.1 \times 10^4$ W cm$^{-2}$. Such intensities are easily obtained with focused pump beams of modest energy. For example, with a line image of the pump beam in the dye solution with a width of 1 mm, the total pump peak is only 2.1 kW. With a $Q$-switched laser pulse 10 ns in length, the threshold pump energy is only 21 μJ.

### 11.4.2 Continuous-wave operation

In a CW dye laser, we can no longer neglect the distributed loss that results from transfer of molecules to the triplet state. If the phosphorescence lifetime in Fig. 11.10 is $\tau_T$ the population of the triplet state obeys the equation

$$\frac{d}{dt}N_T = N_2 k_{ST} - \frac{N_T}{\tau_T}. \tag{11.8}$$

The S → T transfer rate can be estimated from the fluorescence quantum efficiency $\phi$ by assuming that any molecule that does not fluoresce to the ground state crosses to $T_1$. Clearly,

$$\phi = \frac{1/\tau_f}{1/\tau_f + k_{ST}}, \tag{11.9}$$

so

$$k_{ST} = \frac{1}{\tau_f}(1 - \phi). \tag{11.10}$$

In equilibrium, from Eq. (11.8),

$$N_T = N_2 k_{ST} \tau_T. \tag{11.11}$$

The threshold gain is

$$\gamma_t(\lambda) = \alpha(\lambda) - \frac{1}{2\ell}\,\ell n(R_1(\lambda)R_2(\lambda)), \tag{11.12}$$

where the explicit dependence of all the quantities on wavelength has been noted, particularly because laser oscillation potentially can occur over a broad wavelength region. The distributed loss comes from $S_0 \rightarrow S_1$ and $T_1 \rightarrow T_2$ absorption, so

$$\alpha(\lambda) = N_0\sigma_a(\lambda) + N_T\sigma_T(\lambda), \tag{11.13}$$

where $\sigma_a(\lambda)$ and $\sigma_T(\lambda)$ are the absorption cross-sections for absorption (at the laser wavelength, $\lambda$) from the ground state and from $T_1$, respectively. If we write

$$\gamma_t(\lambda) = N_2\sigma_L(\lambda), \tag{11.14}$$

where $\sigma_L$ is the stimulated emission cross-section, then Eq. (11.12) becomes

$$N_2\sigma_L(\lambda) = N_0\sigma_a(\lambda) + N_T\sigma_T(\lambda) - \frac{1}{2\ell}\,\ell n(R_1(\lambda)R_2(\lambda)). \tag{11.15}$$

Now the total dye concentration $N$ must satisfy

$$N = N_0 + N_2 + N_T, \tag{11.16}$$

so, from Eq. (11.15),

$$N_2(\lambda) = \frac{N\sigma_2(\lambda) - (1/(2\ell))\ell\mathrm{n}(R_1(\lambda)R_2(\lambda))}{\sigma_\mathrm{L}(\lambda) + \sigma_\mathrm{a}(\lambda) - k_\mathrm{ST}\tau_\mathrm{T}(\sigma_\mathrm{T} - \sigma_\mathrm{a})}. \tag{11.17}$$

Values for the various cross-sections used in Eq. (11.17) are shown for the dye Rhodamine 6G in Fig. 11.11. If the laser cavity does not provide wavelength-selective feedback, the laser will oscillate at the wavelength for which $N_2(\lambda)$ is a minimum – the *intrinsic* wavelength. The threshold inversion increases as the laser wavelength is tuned to values below the intrinsic value because $\sigma_\mathrm{a}(\lambda)$ increases. The ideal dye is one for which $\sigma_\mathrm{a}(\lambda) \simeq 0$ and $k_\mathrm{ST}\tau_\mathrm{T} = 0$. The latter condition can be obtained, approximately, by adding a triplet-state quencher such as a detergent to the solution. In this case

$$N_2(\lambda) = \frac{-\frac{1}{2}\,\ell\mathrm{n}(R_1(\lambda)R_2(\lambda))}{\sigma_\mathrm{L}(\lambda)}. \tag{11.18}$$

This inversion density is equivalent to the one considered previously in pulsed operation. The threshold pump intensities are comparable. In practice, desirable CW pump intensities exceed 100 kW cm$^{-2}$. A 10-W argon-ion laser focused to a Gaussian spot of spot size $w_0 = 5$ μm has a peak intensity of 25 MW cm$^{-2}$.

Dye lasers have largely been replaced with tunable solid-state devices. The dyes used photodegrade, and liquid spills are common. The laboratories of most dye-laser users are stained with the dyes that they use! There has been some progress in developing solid-state dye lasers, in which the dye is encapsulated in a polymer matrix [15], but photodegradation of the dye requires the use of a rotating dye disk, which constantly moves the pump beam to another region of the solid-state dye medium to avoid photodegradation and the build-up of triplets. There is also ongoing work on organic semiconductor lasers [16], which share some of the characteristics of both dye lasers and semiconductor lasers. Semiconducting organic polymers can have broad absorption and emission spectra, and can be tuned over a significant range. To date lasers using these materials have generally been optically pumped, although electric pumping has also been successful in a microcavity structure [17]. Light-emitting diodes using organic polymers are much more established.

## 11.5 Inorganic liquid lasers

Several rare-earth ions will exhibit laser action when a molecule containing the ion is dissolved in an appropriate solvent. Some examples are given in Table 11.1. These lasers have characteristics that are in some ways intermediate between those of crystalline solid-state lasers and those of glass lasers containing the active ion. For example, Fig. 11.12 shows the fluorescence spectra of neodymium ions in two different inorganic solvents, selenium oxychloride and phosphorous oxychloride,[5] compared with their fluorescence spectrum in glass. The emission of the ions is broader in the glass, which can be interpreted

[5] The laser medium is usually made by dissolving $Nd_2O_3$ in the solvent; typical $Nd^{3+}$-ion concentrations are in the range $6 \times 10^{24}$–$6 \times 10^{26}$ m$^{-3}$.

**Table 11.1** Some liquid lasers

| Ion | Solvent | Laser wavelength (μm) |
|---|---|---|
| $Nd^{3+}$ | $SeOCl_2$–$SnCl_4$ | 1.056, 1.33 |
| | $POCl_3$–$SnCl_4$ | 1.052–1.06 |
| | $POCl_3$–$ZrCl_4$ | 1.056 |
| $Eu^{3+}$ (in a complex with phosphorus ions and benzoylacetonate) | Ethanol | 0.6129 |
| $Eu^{+}$ (in a complex with dimethyl ammonium ions and trifluoro-acetylacetonate) | Acetonitrile | 0.6119 |
| $Tb^{3+}$ (in a complex with phosphorus ions and thenoyltrifluoracetonate) | Polymethylmethacrylate (plexiglas) | 0.545 |

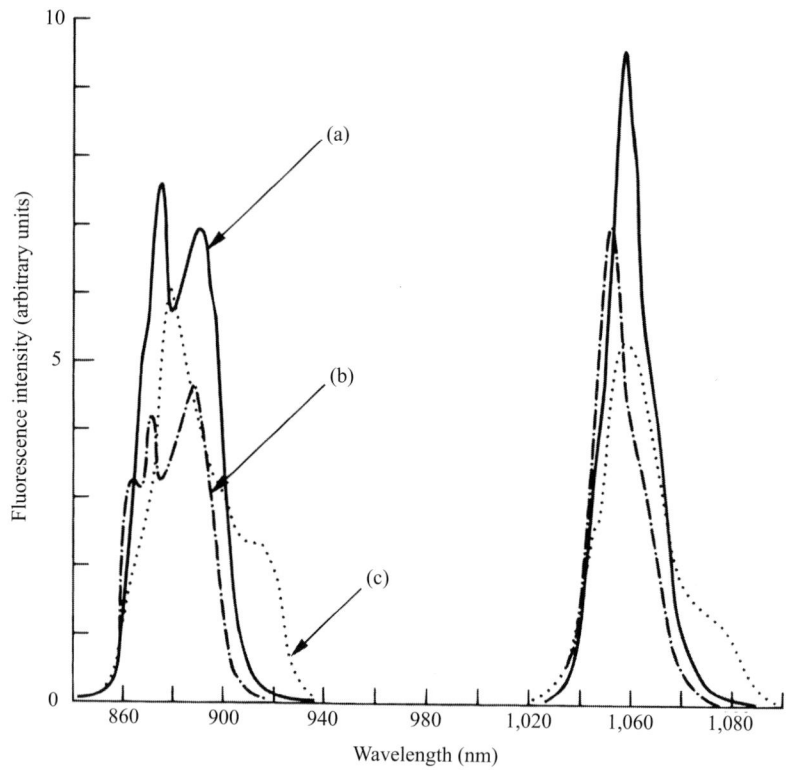


**Fig. 11.12** Fluorescence spectra of $Nd^{3+}$ ions in (a) $SeOCl_2$ solvent; (b) $POCl_3$ solvent; and (c) glass.

as evidence of the broad range of local environments experienced by $Nd^{3+}$ ions in the glass. In the liquid system some local structure exists insofar as each $Nd^{3+}$ ion bonds to molecules of the inorganic solvent in a specific way. $Nd^{3+}$ liquid lasers can provide substantial power outputs, pulsed outputs of several hundred joules, and peak powers of several gigawatts have been obtained. Because the laser emission wavelength of $Nd^{3+}$ ions in liquid solution is still 1.06 μm, hybrid high-energy oscillator/amplifier systems in which Nd:YAG and Nd:glass components are also used can be constructed. One important advantage and one important disadvantage of the liquid system in this regard are worth mentioning. A liquid gain medium is self-repairing. If optical damage occurs and the liquid is circulated, a homogeneous medium is presented for the next firing of the laser. However, the index of refraction of a liquid changes $10^2$–$10^3$ times faster with temperature than occurs in a solid, so liquid gain media excited by flashlamps can show very strong thermal lensing effects.

## 11.6 Free-electron lasers

An electron traveling undisturbed will not spontaneously emit energy. However, if the electron accelerates it must absorb energy, whereas if it decelerates it must radiate it. For example, if an electron increases its velocity in a particle accelerator, it does so by absorbing energy from the electromagnetic field supplied by the accelerating electrode. If an electron is constrained to move in a circle by a combination of electric and magnetic fields then it emits *synchrotron* radiation. Since the electron is experiencing a central acceleration, it should absorb energy. However, if the electron is constrained to a circular path, it cannot increase its kinetic energy, so the energy absorbed must be reradiated. The spontaneous radiation emitted by an electron as it decelerates is called *Bremsstrahlung*. A special kind of radiation called *Čerenkov* radiation is emitted when an electron with velocity $v$, where $v > c_0/n$, enters a medium of refractive index $n$. This radiation is emitted because the electron cannot have a velocity in the medium that is greater than the velocity of light in the medium.

The general form of the electromagnetic field produced by an accelerating or decelerating electron is complex, but a couple of special cases will serve to illustrate some features of the field. If the electron is non-relativistic (its velocity is much less than the velocity of light) then the electric and magnetic field vectors of the radiation produced by the electrons are [18]

$$\mathbf{E} = \frac{e}{4\pi\epsilon_0 c^2 r^3}\mathbf{r} \times (\mathbf{r} \times \dot{\mathbf{v}}), \tag{11.19}$$

and

$$\mathbf{B} = \frac{e}{4\pi\epsilon_0 c^3 r^2}\dot{\mathbf{v}} \times \mathbf{r}, \tag{11.20}$$

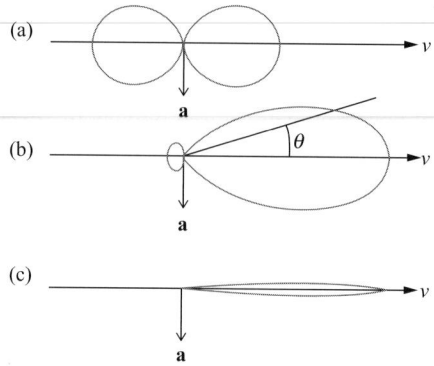


**Fig. 11.13** (a) Angular distribution of the electromagnetic field produced by a low-velocity electron. The field pattern is symmetrical about the direction of electron motion [9]. (b) The radiation pattern becomes more strongly directed in the forward direction as the speed of the electron increases relative to the velocity of light. (c) If the acceleration is perpendicular to the velocity then as the speed of the accelerated electron approaches the velocity of light the radiation is strongly directed into the forward direction [18].

where **v** is the velocity vector of the electron and **r** is the vector from the electron to the point of observation.[6] Equations (11.19) and (11.20) are similar to the field of a small radiating electric dipole, and have the form shown in Fig. 11.13(a).

If an electron accelerates or oscillates perpendicular to its direction of motion, then it experiences a lateral acceleration/deceleration $\mathbf{a} = \dot{\mathbf{v}}$. The radiation pattern of the electron is now of the form [18]

$$I(\theta) \propto \frac{(1 - f\cos\theta)^2 - (1 - f^2)\sin^2\theta\cos^2\phi}{(1 - f\cos\theta)^5}, \tag{11.21}$$

where $\theta$ and $\phi$ are the polar and azimuthal angles, and $f = |\mathbf{v}|/c$. At non-relativistic velocities the emission pattern of the accelerated electron is like a dipole antenna as shown in Fig. 11.13(a). The radiation pattern becomes more strongly directed in the forward direction as the speed of the electron increases relative to the velocity of light, as shown in Fig. 11.13(b). If the acceleration is perpendicular to the velocity then, as the speed of the accelerated electron approaches the velocity of light, the radiation is strongly directed into the forward direction, as shown in Fig. 11.13(c). For relativistic electrons constrained to move in a circle this gives rise to *synchrotron* radiation [19].

The above line of discussion is intended to show that for an accelerated or decelerated electron there is a spontaneous emission process. In the free-electron laser the spontaneous emission of the electron is channeled into a particular mode (or set of modes) of the radiation field by selecting the velocity of the beam in such a way that the radiation emitted by each electron adds coherently to the radiation from the other electrons in the beam. In such *free-electron* lasers (FELs) [20–22] an electron beam is perturbed in such a way as

[6] Strictly, **r** is the *retarded* position vector of the electron–the position at which the electron was when it generated the effect that propagates with the velocity of light to the point where the field is observed. For velocities much smaller than the velocity of light (*non-relativistic* velocities) this distinction is unimportant.

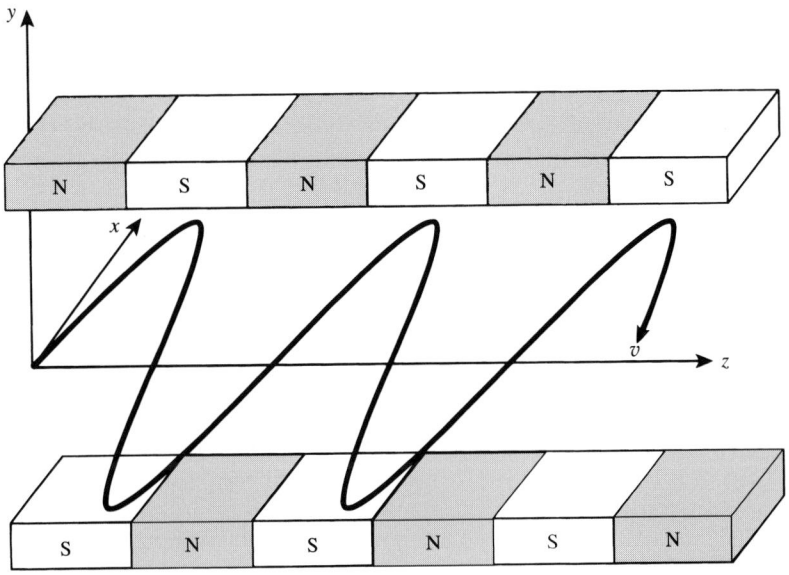


Fig. 11.14 Geometry of a planar *Bremsstrahlung* FEL in which an electron beam passes through the alternating north–south magnetic fields of a "wiggler" [22].

to encourage its interaction with a photon beam and stimulated emission into a particular direction and frequency band. Free electrons are not restricted to specific energies as are bound electrons; they can have any energy. The electrons make what are called *free–free* transitions – they jump between two energy states in the continuum of free-electron states.

In a *Bremsstrahlung* FEL a relativistic electron beam is fired through a periodic static magnetic field (an *undulator* or *wiggler*), which forces the electrons to oscillate and radiate energy in a narrow frequency band. There are two principal undulator configurations: planar and helical. In the planar configuration the electrons pass through a periodic magnetic field oriented perpendicular to the main axis of electron propagation. For example, in Fig. 11.14 an electron beam traveling in the $z$ direction has its velocity in the $y$ direction modulated by the Lorentz force, $\mathbf{F} = -e(\mathbf{v} \times \mathbf{B})$, of the alternating magnetic flux $\mathbf{B}$ provided by the north and south poles of the undulator magnet. In a helical undulator the magnetic field perpendicuar to the beam axis rotates by 360° for one period of the undulator. Figure 11.15(a) shows an arrangement of permanent magnets that can produce this field variation [23]. The magnets in this case are arranged around the beam axis with their magnetic field directions. The magnets are staggered in the $z$ direction so that the undulator period $\ell_0$ is determined by twice the length of the individual magnets along the axis direction, as shown in Fig. 11.15(b).

In an FEL the electrons are generally traveling at close to the velocity of light, so relativistic effects are important. The electron energy is $E = \gamma mc^2$, where $\gamma$ is the relativistic factor $(1 - v^2/c^2)^{-1/2}$. If the undulator field has a period $\ell_0$ then the electrons oscillate about the beam axis with this spatial period. If the electron velocity along the beam axis ($z$ axis) is $v_z$ then the electron motion is synchronous with the light wave traveling

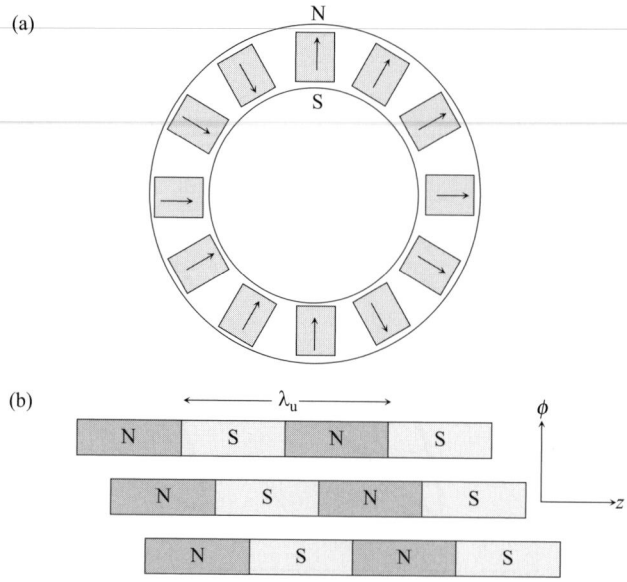


**Fig. 11.15** (a) The arrangement of magnets around the electron-beam propagation axis in an FEL with a helical wiggler. (b) The axial staggering of the magnets in a helical wiggler.

along the $z$ axis if it maintains its phase with respect to the light of wavelength $\lambda$. This requires that

$$v_z = c\ell_0/(\ell_0 + \lambda). \tag{11.22}$$

The average electron velocity in the axial direction is

$$v_z = c\left(1 - \frac{1}{2\gamma^2}\right)\left(1 + \frac{K^2}{2}\right), \tag{11.23}$$

where $K = \ell_0 e B_u/(2\pi m_e c)$ is called the undulator parameter [24]. $B_u$ is the RMS value of the undulator magnetic flux density. For a planar undulator $B_u = B_p/2$ and for a helical undulator $B_u = B_p$, where $B_p$ is the peak magnetic flux density on the beam axis.

The electrons in the beam see the undulator field as a pump electromagnetic wave. These FELs can operate in two regimes, depending on whether the electron beam is sufficiently dense for particle interactions within the beam to be important or not. When the electrons in the beam do not interact the laser operates in the Compton regime, which was first proposed by Madey in 1971 [25]. If interactions are important, they operate in the Raman regime [26–28]. The FELs operating in the Compton regime use very-high-energy, low-density beams, typically with beam energies in the range 20 MeV to 3 GeV and currents of tens of mA, and provide stimulated emission at short wavelengths (in the near-infrared, visible, and ultraviolet ranges). The FELs operating in the Raman regime utilize the interactions between particles in the beam to enhance the gain at specific frequencies. These devices typically use pulse-line accelerators with energies up to about 10 MeV and currents up to 3 kA, and provide substantial gain and power output in the middle-infrared-to-millimeter-wave range. The undulator fields usually have periods on the order of a few centimeters and use fields of about 0.1 T.

We can represent the axial variation of the undulator magnetic field in Fig. 11.15 as

$$B_x(z) = B_0 \sin(k_0 z), \tag{11.24}$$

where $k_0 = 2\pi/\ell_0$, with $\ell_0$ the undulator period. The motion of an electron is governed by the Lorentz force acting on it. Because an electron must be assumed to be traveling at near the velocity of light, its mass, measured in the rest frame of the magnet, has increased to $\gamma m$. Consequently, it obeys the equation of motion

$$\frac{d}{dt}(\gamma m \mathbf{v}) = \frac{e}{c}(\mathbf{v} \times \mathbf{B}). \tag{11.25}$$

If we assume that $\gamma$ does not change, which is true only if we neglect radiation, then

$$\frac{d\mathbf{v}}{dt} = \frac{e}{\gamma m}(\mathbf{v} \times \mathbf{B}). \tag{11.26}$$

Equation (11.26) can be written in the form

$$\frac{dv_x}{dt} = -\frac{\Omega v_z}{\gamma} \sin(k_0 z), \tag{11.27}$$

$$\frac{dv_z}{dt} = \frac{\Omega v_x}{\gamma} \sin(k_0 z), \tag{11.28}$$

where $\Omega = eB_0/(mc)$. If $v_x \ll v_z$ we can put $z = v_z t$ and solve Eqs. (11.27) and (11.28) to give

$$v_x = \frac{v_0 \Omega}{\gamma \omega_0} \cos(\omega_0 t), \tag{11.29}$$

$$v_z = v_0 - \frac{2 v_0 \Omega^2}{4\gamma^2 \omega_0^2} \cos(2\omega_0 t), \tag{11.30}$$

where $v_0$ is the velocity of the electron beam before it enters the undulator field and $\omega_0 = k_0 v_z$. The position of the electron satisfies the equations

$$x = \frac{v_0 \Omega}{\gamma \omega_0^2} \sin(\omega_0 t), \tag{11.31}$$

$$z = v_0 t - \frac{v_0 \Omega^2}{4\gamma^2 \omega_0^3} \sin(2\omega_0 t). \tag{11.32}$$

Therefore, the electron undergoes a lateral simple-harmonic motion as it traverses the undulator field. The oscillating electron emits radiation at a wavelength that must be calculated by considering the Doppler shift resulting from the electron motion.

Insofar as the electron is concerned, the undulator field appears as a time-varying magnetic (and electric) field; that is, as an electromagnetic wave. In this field the electron oscillates with a frequency

$$\omega_0' = \gamma k_0 v_z = \gamma k_0 v_0; \tag{11.33}$$

the relativistic factor $\gamma$ appears because the electrons experience many more undulations of the field within their unit of time. To the observer at rest in the laboratory this corresponds to the relativistic phenomenon of a moving clock appearing to run slow. The electron can interact with radiation that has frequency $\omega_0'$ (within the frame of reference traveling with the electron). This radiation is observed in the laboratory at frequency $\omega_s$, where

$$\omega_0' = \gamma(\omega_s - \mathbf{k}_s \cdot \mathbf{v}), \tag{11.34}$$

and $\omega_s$ is greater than $\omega_0'$ if the radiating electron is traveling towards the observer. In Eq. (11.34) $k_s = \omega_s/c$ is the wave vector observed in the laboratory frame. If $\mathbf{k}$ is nearly parallel to $\mathbf{v}$ then

$$\omega_0' = \gamma\omega_s\left(1 - \frac{v}{c}\cos\theta\right), \tag{11.35}$$

where $\theta$ is the angle between $\mathbf{k}$ and $\mathbf{v}$ ($\simeq v_0$). For $\theta$ equal to a small angle, $\cos\theta = 1-\theta^2/2$, and Eq. (11.35) gives

$$\gamma k_0 v_0 = \gamma\omega_s\left(1 - \frac{v_0}{c}\right). \tag{11.36}$$

Since $1 - v_0/c \simeq 1/(2\gamma^2)$, $k_0 = 2\pi/\ell_0$, and $\omega_s = 2\pi c/\lambda_s$ we get

$$\lambda_s \simeq \frac{\ell_0}{2\gamma^2}. \tag{11.37}$$

This radiation is emitted in the forward direction, parallel to the electron velocity.

A more precise treament of the oscillation frequency gives [25]

$$\lambda_s = \frac{\ell_0}{2\gamma^2}(1 + K^2), \tag{11.38}$$

where $\gamma$ is the relativistic factor $(1 - v_z^2/c^2)^{-1/2}$. For relativistic electrons $v_z \simeq c$ and

$$\lambda = \frac{\ell_0}{2\gamma^2}. \tag{11.39}$$

Clearly, if $\lambda_0$ is on the order of centimeters, large values of $\gamma$ are needed in order to achieve short-wavelength operation.

Doppler shifting of the emitted radiation wavelength is quite significant. For example, if $\ell_0 \sim 10$ mm and the electron-beam energy is 2 MeV then $\lambda_s$ is in the sub-millimeter region. If the beam energy increases to 50 MeV the emitted radiation moves into the near-infrared.

Figure 11.16 shows schematically the construction and two modes of operation of a modern FEL. Electron pulses (bunches) are created at the start by illuminating the cathode (photocathode) with pulses from the drive laser. This is a much more efficient way of generating a bright electron beam than relying on thermionic emission from a cathode. If short laser pulses are used, then the electron pulses emerging from the photocathode are also short. These electron pulses leave the electron gun and booster before they are injected into the path through the superconducting linear accelerator (*linac*). In Fig. 11.16 all the curved electron paths are created by deflection magnets and stabilized with focusing magnets along the beam path. On entering the linac they have energy $E_i$. The RF drive phase is adjusted to accelerate these electrons. After emerging from the linac they have energy $E_0$.

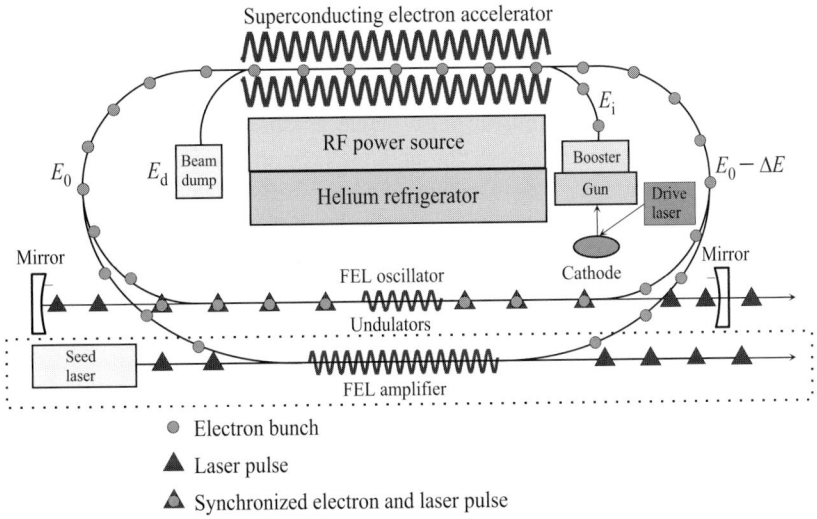


**Fig. 11.16** A schematic diagram showing the arrangement of a modern FEL.

In the operation of an FEL oscillator system they are passed along the FEL oscillator path, which lies between the mirrors of an optical resonator. The electrons passing the undulator give up some of their energy as light. The stream of electron pulses is synchronized with the resulting optical pulses that bounce back and forth between the laser mirrors. After exciting the undulator region, the electrons have energy $E_0 - \Delta E$. These electrons then re-enter the linac, where they join the new electron pulses from the electron gun. The RF phase for these lower-energy electrons is set so that they give up further energy to the RF driving field and thereby reduce the RF drive power required to operate the system. After emerging from the linac at reduced energy $E_d$ they are sent to a beam dump by a magnetic deflector that deflects them more than it deflects the high-energy electrons that have been accelerated in the linac. In the FEL amplifier arrangement, shown inside the dotted lines in Fig. 11.16, a seed laser injects pulses that are synchronized with the electron pulses passing the undulator so that they experience amplification by extracting energy from the electrons.

In a self-amplification spontaneous-emission (SASE) FEL light pulses grow in intensity on a single pass through the undulator without the need for a seed laser. The theory of operation of FELs is not quite so straightforward as the simple discussion here suggests [21, 29], and additional effects must be considered when the electron beam interacts with itself so as to form "bunches" or when the electron beam has a high particle density so that plasma-like phenomena must be considered. In an FEL oscillator system the interaction of the cavity modes of the resonator with the electron beam must be considered. In such a system the light pulses "bouncing" back and forth between the cavity mirrors must be synchronous with the electron pulses passing through the undulator. The round-trip time of light pulses in the cavity is longer than the vacuum velocity of light because the light pulses are slowed by interaction with the electron beam. The light pulses interact strongly with the electron beam only when they are co-propagating with the electrons.

By controlling the electron-beam energy and undulator period, FELs can be operated over a broad range of wavelengths, from the X-ray region to the millimeter-wave region. They can generate pulses in the sub-picosecond regime and, for X-ray FELs, in the attosecond range. The optical pulse length is closely correlated with the length of the electron pulses passing through the undulator. Short-wavelength FELs use larger structures than those used for long-wavelength FELs. Peak pulse powers for FELs can be higher than 1 GW and average powers are currently about 15 kW, but considerable research is being carried out in order to move this to much higher values for laser-weapon applications, especially on board warships [30]. Because of their flexibility, FELs are important research tools and are the only means for generating very high powers in some parts of the spectrum, especially in the far-IR and millimeter-wave region [31].

## 11.7 Problems

11.1 Calculate the spectral width of the dye-laser output when wavelength selection is provided by a diffraction grating if the laser beam divergence is $\theta_{\text{beam}}$. Hint: consider the *angular dispersion* of the grating $d\theta/d\lambda$ calculated from Eq. (11.1).

11.2 (Computer project) Use the data in Fig. 11.11 to calculate the intrinsic wavelength of a Rhodamine 6G dye laser as a function of concentration $N$.

## References

[1] P. Albers, E. Stark, and G. Hiber, "Continuous-wave laser operation and quantum efficiency of titanium-doped sapphire," *J. Opt. Soc. Am.* **B3**, 134–139, 1986.

[2] A. Sanchez, R. E. Fahey, A. J. Strauss, and R. L. Aggarwal. "Room-temperature continuous-wave operation of a Ti:$Al_2O_3$ laser," *Opt. Lett.* **11**, 363–364, 1986.

[3] W. R. Rapoport and C. P. Khattak, "Titanium sapphire laser characteristics," *Appl. Opt.* **27**, 2677–2684, 1988.

[4] A. Hoffstadt, "Design and performance of a high-average-power flashlamp-pumped Ti:sapphire laser and amplifier," *IEEE J. Quant. Electron.* **33**, 1850–1863, 1997.

[5] S. Bourzeix, M. D. Plimmer, F. Nez, L. Julien, and F. Biraben, "Efficient frequency doubling of a continuous-wave titanium–sapphire laser in an external enhancement cavity," *Opt. Commun.* **99**, 89–94, 1993.

[6] J. B. Birks, *Photophysics of Aromatic Molecules*, New York: Wiley-Interscience, 1970.

[7] F. N. Baltakov, B. A. Barikhin, and L. V. Sukhanov. "400-J pulsed laser using a solution of rhodamine-6G in ethanol," *JETP Lett.* **19**, 174–175, 1974.

[8] T. W. Hänsch, "Repetitively pulsed tunable dye laser for high resolution spectroscopy," *Appl. Opt.* **11**, 895–898, 1972.

[9] F. J. Duarte and L. W. Hillman, eds., *Dye Laser Principles*, New York: Academic, 1990.

[10] F. P. Schäfer, ed., *Dye Lasers*, Vol. 1, 2nd revised edn., Berlin: Springer-Verlag, 1977.

[11] M. G. Littman and H. J. Metcalf, "Spectrally narrow dye laser without beam expander," *Appl. Opt.* **17**, 2224–2227, 1978.

[12] M. Littman and J. Montgomery, "Grazing-incidence designs improve pulsed dye lasers," *Laser Focus* **24**, 70–86, 1988.

[13] H. W. Kogelnik, E. P. Ippen, A. Dienes, and C. V. Shank. "Astigmatically compensated cavities for CW dye lasers," *IEEE J. Quant. Electron.* **8**, 373–379, 1972.

[14] B. B. Snavely, "Continuous-wave dye lasers," in *Dye Lasers*, ed. F. P. Schäfer, Vol. 1, 2nd revised edn., Berlin: Springer-Verlag, 1977.

[15] R. Bornemann, U. Lemmer, and E. Thiel, "Continuous-wave solid-state dye laser," *Opt. Lett.* **31**, 1669–1671, 2006.

[16] I. D. W. Samuel and G. A. Turnbull, "Organic semiconductor lasers," *Chem. Rev.* **107**, 1272–1295, 2007.

[17] Xingyuan Liu, Huibin Li, Chunyan Song, Yaqin Liao, and Miaomiao Tian, "Microcavity organic laser device under electrical pumping," *Opt. Lett.* **34**, 503–505, 2009.

[18] W. K. H. Panofsky and M. Phillips, *Classical Electricity and Magnetism*, Reading, MA: Addison-Wesley, 1955.

[19] A. Hofman, *The Physics of Synchrotron Radiation*, Cambridge: Cambridge University Press, 2004.

[20] H. P. Freund and T. M. Antonsen, *Principles of Free-Electron Lasers*," 2nd edn., London: Chapman and Hall, 1996.

[21] T. C. Marshall, *Free Electron Lasers*, New York: Macmillan, 1985.

[22] D. Prosnitz, "Free electron lasers," in *Handbook of Laser Science and Technology, Vol. I, Lasers and Masers*, ed. M. J. Weber, Boca Raton, FL: CRC Press, 1982.

[23] G. Bekefi and J. Ashkenazy, "Permanent magnet helical wiggler for free electron laser and cyclotron maser applications," *Appl. Phys. Lett.* **51**, 700–701, 1987.

[24] S. Reiche, "GENESIS 1.3: a fully 3D time-dependent FEL simulation code," *Nucl. Instrum. Methods Phys. Res. A: Accelerators, Spectrometers, Detectors Associated Equipment* **429**, 243–248, 1999.

[25] J. M. J. Madey, "Stimulated emissions of bremsstrahlung in a periodic magnetic field," *J. Appl. Phys.* **42**, 1906–1913, 1971.

[26] V. L. Granatstein, S. P. Schlesinger, M. Herndon, R. K. Parker, and J. A. Pasour, "Production of megawatt submillimeter pulses by stimulated magneto-Raman scattering," *Appl. Phys. Lett.* **30**, 384–386, 1977.

[27] T. J. Orzechowski, B. Anderson, W. M. Fawley, D. Prosnitz, E. T. Scharlemann, S. Yarema, D. Hopkins, A. C. Paul, A. M. Sessler, and J. Wurtele, "Microwave-radiation from a high-gain free-electron laser-amplifier," *Phys. Rev. Lett.* **54**, 889–892, 1985.

[28] D. A. Kirkpatrick, G. Bekefi, A. C. Dirienzo, H. P. Freund, and A. K. Ganguly. "A millimeter and submillimeter wavelength free-electron laser," *Phys. Fluids B Plasma Phys.* **1**, 1511–1518, 1989.

[29] E. L. Saldin, E. A. Schneidmiller, and M. V. Yurkov, "The physics of free electron lasers. An introduction," *Phys. Rep.* **260**, 187–327, 1995.

[30] *Scientific Assessment of High Power Free Electron Laser Technology*, National Academy of Sciences Report, Committee on a Scientific Assessment of Free-Electron Laser Technology for Naval Applications, 2008.

[31] P. G. O'Shea and H. P. Freund, "Free-electron lasers: status and applications," *Science* **292**, 1853–1858, 2001.

# 12 Semiconductor lasers

## 12.1 Introduction

The semiconductor laser, in various forms, is the most widely used of all lasers, is manufactured in the largest quantities, and is of the greatest practical importance. Every CD (compact disk), DVD (digital versatile disk or digital video disk), and Blu-ray player contains one. Most of the world's long-, and medium-, distance communication takes place over optical fibers along which propagate the beams from semiconductor lasers. High-power semiconductor lasers are increasingly part of laser systems for engraving, cutting, welding, and medical applications. Semiconductor lasers operate by using the jumps in energy that can occur when electrons travel between semiconductors containing different types and levels of controlled impurities (called *dopants*). In this chapter we will discuss the basic semiconductor physics that is necessary to understand how these lasers work, and how various aspects of their operation can be controlled and improved. Central to this discussion will be what goes on at the junction between p- and n-type semiconductors. The ability to grow precisely doped single- and multi-layer semiconductor materials and fabricate devices of various forms – at a level that could be called molecular engineering – has allowed the development of many types of structure with which one can make efficient semiconductor lasers. In some respects the radiation from semiconductor lasers is far from ideal, since its coherence properties are far from perfect, being intermediate between those of a low-pressure gas laser and an incoherent line source. We will discuss how these coherence properties can be controlled in semiconductor lasers.

## 12.2 Semiconductor physics background

When atoms combine to form a solid the sharply defined energy states that we associate with free atoms, such as in a gas, broaden considerably because of the large number of interactions between an atom and its neighbors. The result is a series of broad energy states, called *bands*. The energy states of the outermost electrons of the atoms constituting the solid broaden and combine to produce a band of filled energy states called the *valence band*. Excited energy states of the atoms broaden and combine to give rise to a series of excited energy bands, which at absolute zero are empty. The lowest, empty energy band is called the *conduction band*. The relative disposition of the valence and conduction bands determines whether the solid will be an insulator, a metal, or a semiconductor.

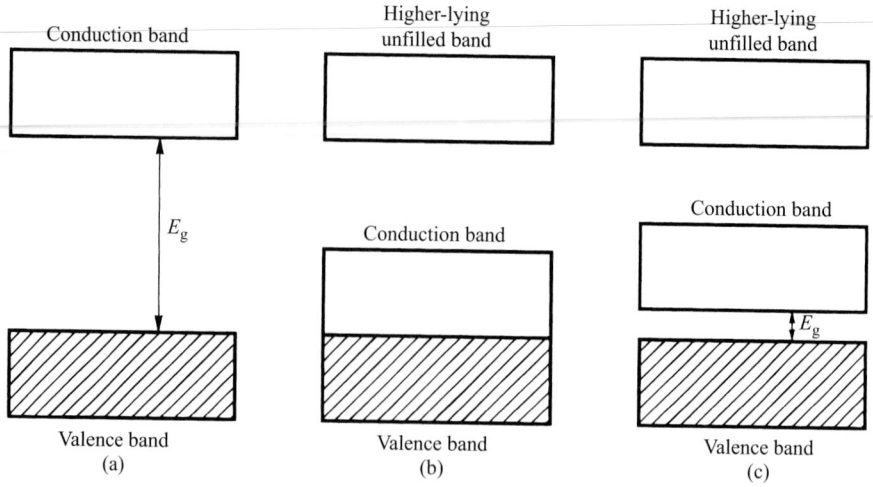


**Fig. 12.1** Simple band diagrams for (a) an insulator, (b) a metal, and (c) a semiconductor, where $E_g \sim k_B T$.

The band structure can be drawn in a one-dimensional energy representation as shown in Fig. 12.1. The energy spacing, $E_g$, between the top of the highest filled energy band and the bottom of the lowest unfilled energy band is called the *band gap*. In an insulator, $E_g \gg k_B T$, so hardly any electrons can be thermally excited into the conduction band to give rise to electrical conductivity. Electrical conductivity can result only if there are vacant unfilled states within an energy band. In a semiconductor $E_g$ is a few times larger than $k_B T$; for a very small energy gap the material can be referred to as a semimetal – such as bismuth.

The most important semiconductor material in electronics is silicon (in the early days it was germanium). However, the most important semiconductor materials in photonics are binary materials, such as the III–V and II–VI[1] semiconductors GaAs, GaN, GaP, and InSb, ternary materials such as $Al_xGa_{1-x}As$ (where $x$ is a fractional factor, $x \leq 1$, that determines the precise stochiometry of the material), and quaternary materials such as $In_xGa_{1-x}As_yP_{1-y}$. The ternary and quaternary materials can be viewed as binary materials in which one or both principal constituents have been replaced to some extent by atoms from the same group in the periodic table. Although a broad range of multicomponent crystals can be grown in this way, arbitrary compositions are not all possible [1]. The most important parameter that changes from one semiconductor system to the next is its band gap, as can be seen in Table 12.1. We shall see that the primary process that leads to light generation in semiconductors involves an electron falling from the conduction band to the valence band and releasing a photon. The characteristic wavelength associated with the band-gap energy is $\lambda = ch/E_g$.

When an electric field is applied to a semiconductor conduction results because electrons in the conduction band acquire directed velocities. These velocities are in the opposite

[1] The roman numerals indicate the group in the periodic table to which each atomic constituent belongs.

**Table 12.1** Band-gap characteristics of important semiconductor materials

| Material | Energy gap at 300 K (eV) | $\lambda = ch/E_g$ (μm) |
|---|---|---|
| Diamond[a] | 5.4 | 0.23 |
| Germanium | 0.66 | 1.88 |
| Silicon | 0.66 | 1.88 |
| InSb | 0.17 | 7.3 |
| InAs | 0.36 | 3.5 |
| InP | 1.35 | 0.92 |
| GaN | 3.4 | 0.389 |
| GaP | 2.25 | 0.55 |
| GaP | 2.25 | 0.55 |
| GaAs | 1.42 | 0.87 |
| $In_{0.8}Ga_{0.2}As_{0.34}P_{0.65}$ | 1.1 | 1.13 |
| $In_{0.5}Ga_{0.5}P$ | 2.0 | 0.62 |
| GaSb | 0.68 | 0.73 |
| AlSb | 1.6 | 1.58 |
| $Al_xGa_{1-x}As$ $(0 \leq x < 0.37)$ | 1.42–1.92 | 0.65–0.87 |
| $In_{0.53}Ga_{0.47}As$ | 0.74 | 1.67 |

[a] Although diamond is an insulator, it can be made conductive by irradiation with ultraviolet light.

direction to the applied field because the electrons are negatively charged.[2] Each electron moving in a material has momentum $p = m^* v$. The mass is written as $m^*$ to indicate that this is the *effective* mass of an electron moving in the lattice, which because of the dual wave/particle character of the electron and its interaction with the lattice is not the same as the free-electron mass, and may be negative. The momentum can be related to the wavelength of the electron, viewed as a wave, by the famous de Broglie relation

$$p = h/\lambda, \tag{12.1}$$

where $h$ is Planck's constant. Since the wave vector $\mathbf{k}$ of the electron has magnitude $|\mathbf{k}| = 2\pi/\lambda$, we can relate the momentum, kinetic energy, $E$, and wave vector by

$$E = \frac{1}{2}m^* v^2 = \frac{p^2}{2m^*}, \tag{12.2}$$

with

$$p = \hbar k, \tag{12.3}$$

[2] It sometimes happens that because of the interaction between the electrons and the lattice the electrons do not move in this intuitive way; instead they behave as if their mass had becomed *negative*. For further information on the interesting way in which charge carriers move in semiconductors consult Kittel [2].

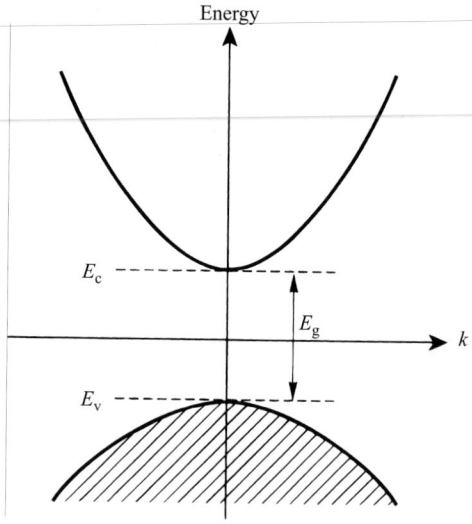


**Fig. 12.2** A parabolic energy-band diagram for an insulator or semiconductor crystal where no carriers have been thermally excited into the conduction band. Cross-hatching indicates filled states at 0 K.

where $\hbar = h/(2\pi)$, and

$$E = \frac{\hbar^2 k^2}{2m^*}. \tag{12.4}$$

In this simple picture the energy bands can be drawn as parabolas on an $E$–$k$ diagram. Figure 12.2 gives an example.

For the conduction band, $E$ in Eq. (12.2) would be measured from the bottom of the band at energy $E_c$; in the valence band it would be measured from the top of the band at energy $E_v$. Clearly in Fig. 12.2 the effective mass of electrons in the valence band is negative. To explain how electrons move within a band we must allow for the existence of at least one vacant state in the band. We must also remember that, because electrons behave like waves, they are quantized just like the modes of the black-body radiation field. Consequently, in a cubical crystal of side $L$ the allowed wave vectors of the electron can only take values

$$k_i = \frac{2\pi n}{L} \quad (i = x, y, z), \tag{12.5}$$

where $n$ is a positive or negative integer, or zero. When this feature of the energy bands is included, there are discrete electron states available in the valence and conduction bands as shown in Fig. 12.3.

For a filled energy band, such as the valence band shown in Fig. 12.3, the vector sum of the momenta of all the electrons is zero, that is

$$\sum_i \mathbf{k}_i = 0. \tag{12.6}$$

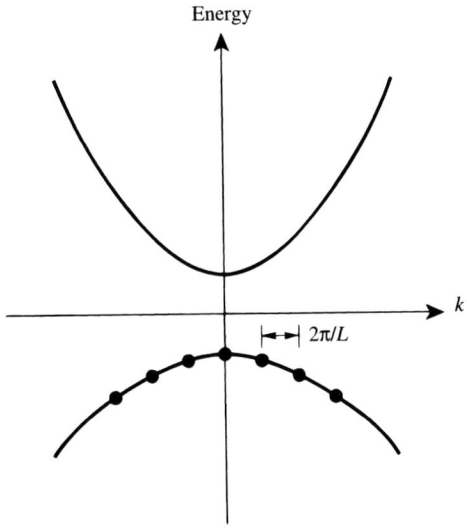


Fig. 12.3 A parabolic energy-band diagram showing the filled discrete states in the valence band at 0 K.

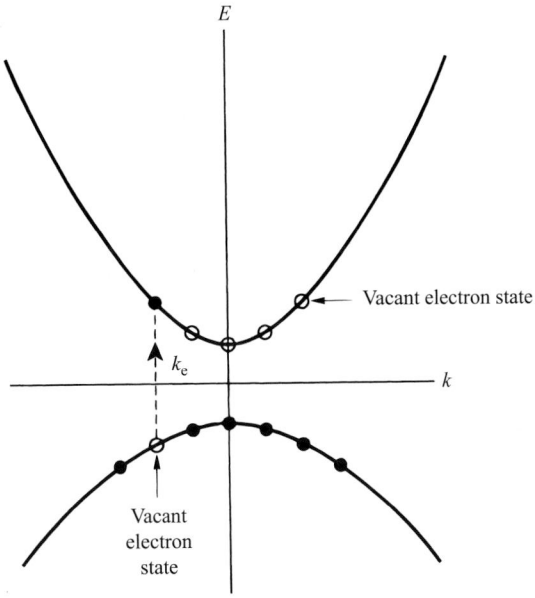


Fig. 12.4 A parabolic energy-band diagram of a direct-band-gap semiconductor in which an electron has been excited from the valence band to a vacant state in the conduction band, leaving a vacancy in the valence band.

If an electron is excited from the valence band to the conduction band by absorbing a photon, then for a *direct*-band-gap transition the electron does not change its wave vector significantly[3] in crossing the gap, as shown in Fig. 12.4. If the energy minimum of the conduction band is not at the same $k$ value as the energy maximum, then we have an

[3] This is a very small change in momentum because of the momentum of the absorbed photon.

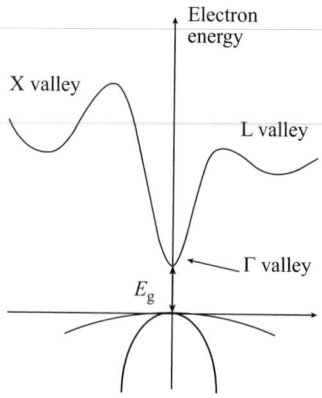


**Fig. 12.5** The multiple valleys in the conduction-band energy versus **k** diagram with their minima at values of **k** of zero in the Γ valley and at higher values in the L and X valleys.

*indirect*-band-gap semiconductor. In many semiconductors there are multiple valleys in the conduction-band energy versus **k** diagram with their minima at values of **k** of zero (the Γ valley), L and X, as shown in Fig. 12.5. If the L valley is the lowest in energy then the transition from this valley to the valence band will be an indirect L-gap transition. If the lowest energy valley is the X valley then the transition from this valley to the valence band will be an indirect X-gap transition. In GaAs the Γ valley has the lowest energy, and GaAs is a direct-band-gap material. There can also be more than one band for the holes, as shown in Fig. 12.5, where the shallower band is the heavy-hole band. In AlAs the X valley has the lowest energy, and this is an indirect-band-gap semiconductor with $E_g = 2.16$ eV.

The vacancy left in the valence band behaves very much like a positive charge, called a *hole*. The wave vector of the electron is $\mathbf{k}_e$, so, to conserve the total momentum of all the electrons (both in the valence band and in the conduction band), the total momentum of the electron in the valence band is

$$\mathbf{k}_h = -\mathbf{k}_e. \tag{12.7}$$

Therefore, the hole is not really located on the $E$–$k$ diagram at the location indicated by the electron vacancy. If an electric field is applied to a semiconductor with a conduction-band electron and a valence-band hole, then electrical conduction occurs through motion of the conduction-band electron in the opposite direction to the applied field, and of the positively charged valence-band hole in the same direction as the field. The effective mass of the hole is positive, opposite in sign to that of the valence-band electrons, but equal to them in magnitude:

$$m_h^* = -m_e^* \text{ (valence band)}. \tag{12.8}$$

Under the action of an applied field the vacancy in the valence band does not move in the simple fashion that would be expected from Fig. 12.4. Motion of electrons into and out of the vacancy leads to motion of the vacancy in the *opposite* direction to the applied field because the equivalent hole is moving in the *same* direction as the field.

Fortunately, in describing the basic properties of semiconductors that allow them to be both detectors and generators of light, we do not need to worry about the somewhat complex behavior of vacancies, but can treat the material as containing negatively charged electrons and positively charged holes.

## 12.3 Carrier concentrations

In a very pure (*intrinsic*) semiconductor, except at absolute zero, there are some electrons in the conduction band, and, of course, corresponding holes in the valence band. To determine their concentrations at temperature $T$ we must find the probability that an electron is thermally excited across the band gap. This probability is not $\sim e^{-E_g/(k_B T)}$, as might be expected from Maxwell–Boltzmann statistics, because electrons distribute themselves among available energy states according to Fermi–Dirac statistics [3]. Electrons avoid each other to the extent that no two electrons can occupy exactly the same energy state. Consequently, the probability that an electron will occupy an energy state at energy $E_k$ is given by the Fermi–Dirac distribution

$$f(E_k) = \frac{1}{e^{(E_k-\mu)/(k_B T)} + 1}. \tag{12.9}$$

The reference energy $\mu$ is called the Fermi level.[4] For $E_k - \mu \gg k_B T$ this can be approximated as

$$f(E_k) = e^{(\mu-E_k)/(k_B T)}. \tag{12.10}$$

Figure 12.6 shows the shape of this distribution relative to the band gap in an intrinsic semiconductor both at absolute zero and at a finite temperature.

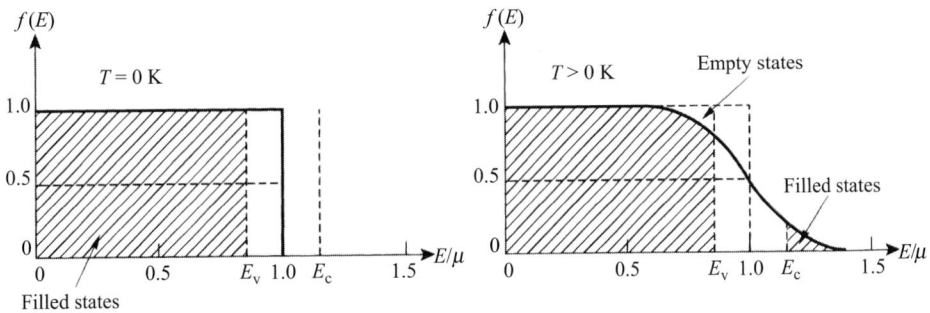


**Fig. 12.6** The Fermi–Dirac distribution shown relative to the band gap in an intrinsic semiconductor.

[4] For those readers with an interest in chemical thermodynamics, $\mu$ is the *chemical potential* of the electrons in the semiconductor.

A few significant features of Fig. 12.6 should be noted. At absolute zero, or if the effective masses of electrons and holes are equal, the Fermi level is situated in the center of the band gap. At the Fermi level $f(E) = \frac{1}{2}$. This is the probability of a state *at that energy* being occupied. Since there are no energy states in the band gap, there will be no electrons in a state at energy $\mu$. All states up to $E_v$, the top of the valence band, are filled.

At $T > 0$ K some states above $E_c$ will become filled, while vacancies will exist in the top of the valence band.

To calculate the electron concentration we need, in addition to the probability of occupation of each mode, the actual number of states involved. This calculation is very similar to the one we performed earlier for black-body radiation. The number of states per unit volume for which the electron waves have wave vectors within a small range $dk$ at $k = |\mathbf{k}|$ is (cf. Eq. (1.34))

$$\rho(k)dk = \frac{k^2}{\pi^2}\,dk. \tag{12.11}$$

The electrons also have two different kinds of state that are analogous to the two polarization states of electromagnetic waves. These two states are called the *spin* states of the electrons.

For an electron in the conduction band the total energy is

$$E_k = E_c + \frac{\hbar k^2}{2m_e^*}, \tag{12.12}$$

where $m_e^*$ is the electron effective mass in the conduction band. Therefore, from Eq. (12.12),

$$k = \sqrt{\frac{2m_e^*}{\hbar^2}}\sqrt{E_k - E_c} \tag{12.13}$$

and

$$dk = \frac{1}{2}\sqrt{\frac{2m_e^*}{\hbar^2}}\frac{dE}{\sqrt{E_k - E_c}}. \tag{12.14}$$

The density of states $\rho(E_k)$, which tells us how many electron states exist in a range of energies between $E_k$ and $E_k + dE_k$, is found from

$$\rho(E_k)dE_k = \rho(k)dk. \tag{12.15}$$

It follows that

$$\rho(E_k)dE_k = \frac{k^2}{\pi^2}dk \tag{12.16}$$

and

$$\rho(E_k)dE_k = \frac{1}{2\pi^2}\left(\frac{2m_e^*}{\hbar^2}\right)^{3/2}\sqrt{E_k - E_c}\,dE_k. \tag{12.17}$$

The total number of electrons in the conduction band is

$$n = \int_{E_c}^{\infty} \rho(E_k) f(E_k) dE_k. \tag{12.18}$$

Substitution from Eq. (12.10) and Eq. (12.17) gives

$$n = \frac{1}{2\pi^2} \left( \frac{2m_e^*}{\hbar^2} \right)^{3/2} e^{\mu/(k_B T)} \int_{E_c}^{\infty} \sqrt{E_k - E_c} e^{-E_k/(k_B T)} \, dE_k, \tag{12.19}$$

which can be integrated to give

$$n = 2 \left( \frac{m_e^* k_B T}{2\pi \hbar^2} \right)^{3/2} e^{(\mu - E_c)/(k_B T)}. \tag{12.20}$$

We can derive a similar result for the concentration of holes in the valence band if we note that the Fermi–Dirac distribution for the holes is

$$f_h(E) = 1 - f(E), \tag{12.21}$$

since a hole is equivalent to a "missing" electron. Therefore,

$$f_h(E) = \frac{1}{e^{(\mu - E_k)/(k_B T)} + 1}. \tag{12.22}$$

The corresponding hole concentration is

$$p = 2 \left( \frac{m_h^* k_B T}{2\pi \hbar^2} \right)^{3/2} e^{(E_v - \mu)/(k_B T)}, \tag{12.23}$$

and, noting that $E_c - E_v = E_g$,

$$np = 4 \left( \frac{k_B T}{2\pi \hbar^2} \right)^3 \left( m_e^* m_h^* \right)^{3/2} e^{-E_g/(k_B T)} = K^2 e^{-E_g/(k_B T)}. \tag{12.24}$$

## 12.4 Intrinsic and extrinsic semiconductors

The conductivity of a semiconductor is determined by the concentration of conduction-band electrons and valence-band holes, and by the ability of these charges to move through the solid under the action of an electric field, characterized by their mobilities $\mu_e$ and $\mu_h$. The conductivity is

$$\sigma = ne\mu_e + pe\mu_h, \tag{12.25}$$

where $e$ is the magnitude of the electronic charge. The conductivity of semiconductors can be controlled by doping. It generally lies between the values for metals and insulators, at ambient temperatures typically in the range from $10^{-7}$ S m$^{-1}$ to $10^4$ S m$^{-1}$,[5] but depends strongly on temperature. A true insulator would have $\sigma < 10^{-9}$ S m$^{-1}$, while a good conductor typically has $\sigma > 10^6$ S m$^{-1}$. Very pure semiconductors (in which $n = p$ in

[5] The siemens (S) is the SI unit of conductance, equivalent to the unit $\Omega^{-1}$.

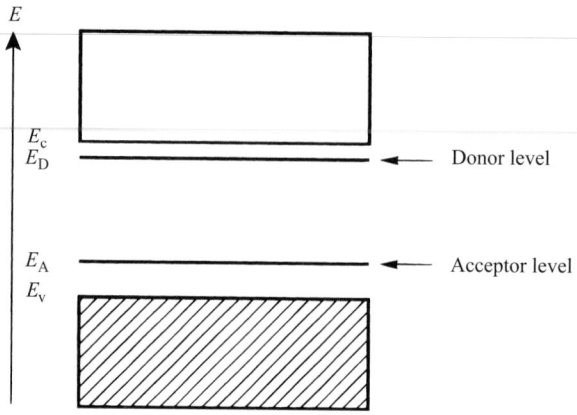


**Fig. 12.7** A schematic band diagram showing the positions of donor and acceptor levels in the band gap of a doped semiconductor.

Eq. (12.24)) generally have lower conductivity than *impurity* semiconductors, in which impurity dopants are added to provide *extrinsic* conductivity. In an n-type semiconductor an impurity is added that can release electrons. For example, GaAs can be made n-type by doping with silicon or germanium in gallium sites. The impurity group-IV atom has one more electron than the group-III atom it replaces, which it can *donate* to the conduction band. GaAs can be made p-type by doping with zinc or cadmium (group-II elements) in gallium sites, where the group-II atoms have one too few electrons and would like to *accept* another. Conversely, GaAs can also be made p-type by replacing arsenic (group V) with a group-IV atom (Si or Ge). In the large-band-gap semiconductor material GaN the semiconductor can be made n-type by doping with silicon or oxygen, or p-type by doping with magnesium. In an extrinsic semiconductor the impurity atoms occupy energy levels within the band gap, as shown in Fig. 12.7.

When an impurity atom is added to a pure semiconductor to make it n- or p-type the growth of the crystal is affected and can lead to defects in the lattice structure. These defects result because the differing sizes of the impurity atoms cause compressive or tensile stress in the structure. These stresses increase with doping density. The practical importance of various doped semiconductors depends on the appropriate impurities and growth techniques being developed.

In n-type material the energy gap that electrons must cross to reach the conduction band is lower than in intrinsic material, so $n > p$. However, if the doping level is not too great Eq. (12.24) still holds. In p-type material $p > n$.

When an intrinsic semiconductor is doped the Fermi level moves from the center of the band gap. It moves up towards the conduction band with increasing n doping and down towards the valence band with p doping. Its precise location is a function of the doping level. At high enough doping levels the Fermi level moves into the conduction band in n-type material and into the valence band in p-type material. When this happens, the material is said to be *degenerately* doped. This happens, because at high doping levels, $\sim 10^{18}$ m$^{-3}$ in GaAs, the impurity atoms begin to interact with each other and their once

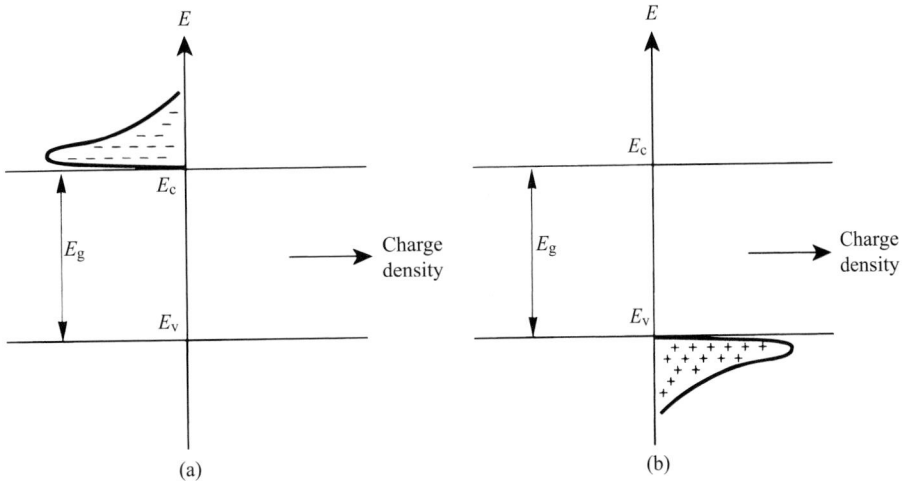


**Fig. 12.8** The schematic variation of carrier concentration in degenerately doped (a) n- and (b) p-type semiconductors.

sharp energy levels start to spread and merge with the neighboring band. The schematic charge distribution with energy in degenerately doped n- and p-type materials is shown in Fig. 12.8.

There are a few holes in the valence band in the n-type material and a few electrons in the conduction band in the p-type material, but their numbers are small, and they are not shown in Fig. 12.8. If we could somehow superimpose the two halves of Fig. 12.8 we would have a condition of population inversion – filled electron energy states above empty ones. In essence, we do this by forward biasing a p–n junction. However, before describing precisely how this is done and how a semiconductor laser is constructed, we must first describe a little of the physics that goes on at the junction between two differently doped semiconductors.

## 12.5 The p–n junction

In its simplest form the p–n junction is the interface region between two identical semiconductors, one of which is n-doped, the other p-doped. This is called a p–n *homojunction*.

The Fermi level in a semiconductor corresponds to its chemical potential, so that when a p–n junction is formed the Fermi levels line up with each other, leading to the electron-energy-versus-position diagram shown in Fig. 12.9. The low-resistance conduction direction of the diode is shown running from p $\rightarrow$ n. In Fig. 12.9 the n-type material is more heavily doped than the p-type material, so its Fermi level is closer to the conduction band. The materials are designated p and $n^+$ to indicate this. Typical doping levels might result from a donor concentration of $n_D \sim 10^{24}$ m$^{-3}$ and an acceptor concentration of $n_A \sim 10^{22}$ m$^{-3}$. The vacuum level is the energy to which an electron would need to be raised in order for it to escape completely from the semiconductor. The electrons in

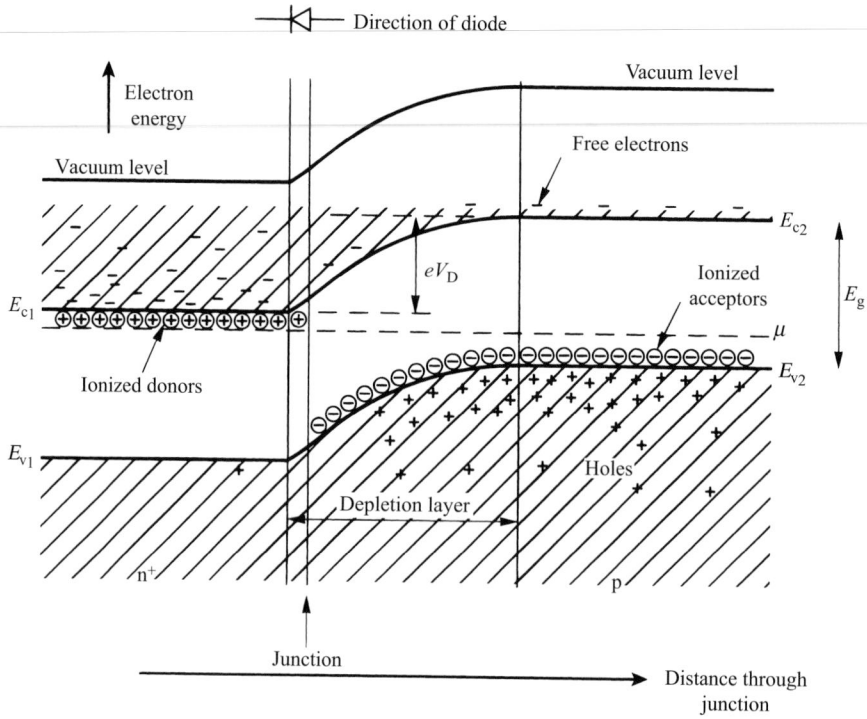


**Fig. 12.9** The band diagram of a p–n junction in equilibrium, showing the relative positions of electrons and holes.

the n-type material come largely from ionized donors whose energy is near the conduction-band edge $E_{c_1}$. The holes in the p-type material come largely from ionized acceptors whose energy is near the valence-band edge $E_{v_2}$. In thermal equilibrium an electron that travels from $n^+$ to p must climb an energy hill, shown in Fig. 12.9 as $eV_D$, where $V_D$ is called the diffusion potential. Note that

$$eV_D = E_g - (E_{c_1} - \mu) - (\mu - E_{v_2}). \tag{12.26}$$

In a region called the *depletion layer*, which spans the junction, electrons fall down the potential hill, while holes "fall" up the hill (because they are positively charged their energy is lower on the p-side of the junction). However, across the junction the product $pn$ remains constant, according to Eq. (12.24). The hill constitutes an energy barrier to the passage of electrons from $n^+ \rightarrow$ p or of holes from p $\rightarrow n^+$. The variation of the electron and hole concentrations across the junction is shown schematically in Fig. 12.10.

The concentrations of electrons and holes far from the junction are, respectively, $n_{10}$ and $p_{10}$ in the $n^+$ material and $n_{20}$ and $p_{20}$ in the p-type material. From Eq. (12.24)

$$n_{10}p_{10} = n_{20}p_{20} = n_i^2, \tag{12.27}$$

where $n_i$ is the equilibrium electron (and hole) concentration in undoped (intrinsic) material.

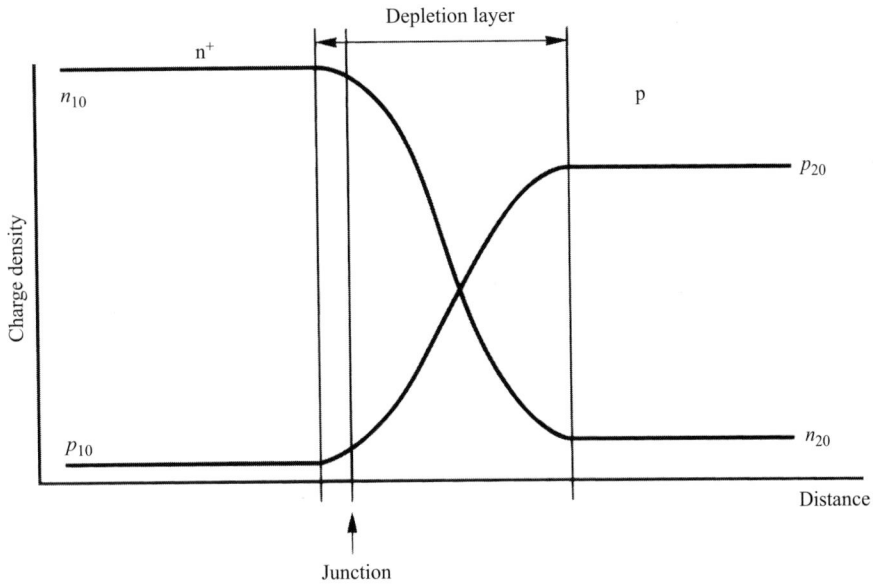


**Fig. 12.10** The schematic charge-density variation of electrons and holes across a p–n junction in equilibrium.

The probability factor that describes the likelihood of a charge carrier being able to "climb" the hill is $e^{-e(V_D-V)/(k_BT)}$. It should not be too surprising therefore that the current–voltage relation for the diode is [2]

$$I = I_S(e^{eV/(k_BT)} - 1). \tag{12.28}$$

When the p–n junction is forward biased the charge concentration across the junction changes to the schematic form shown in Fig. 12.11.

Current flow injects extra electrons into the p-type material, and a relatively smaller number of extra holes into the n-type material (because the p-type material is less heavily doped). Note that, to preserve electrical neutrality across the junction, the concentration of free electrons in the $n^+$ material and that of holes in the p-type material[6] rise to balance the injected excess carriers. In each material these excess carriers must disappear within a short distance of the junction since they constitute a deviation from the carrier distribution at thermal equilibrium.[7]

The excess carriers disappear by recombination as electrons and holes recombine. Of course, all this means is that an electron at a given energy falls into an empty electron state of lower energy. If this recombination involves a band-to-band radiative transition as shown schematically in Fig. 12.12, then a photon is produced, which may escape from the semiconductor and be observed as *injection luminescence*. This process, in semiconductors, is the spontaneous emission process with which we are already familiar. The efficiency

[6] The *majority* carriers in each material, respectively.

[7] The characteristic distance over which these excess carriers disappear is closely related to the so-called *diffusion length*, which for electrons in p-type material is $L_p = \sqrt{D_e\tau_p}$, where $D_e$ is the electron diffusion coefficient (in units of $m^2\ s^{-1}$) and $\tau_p$ is the lifetime of free electrons in p-type material.

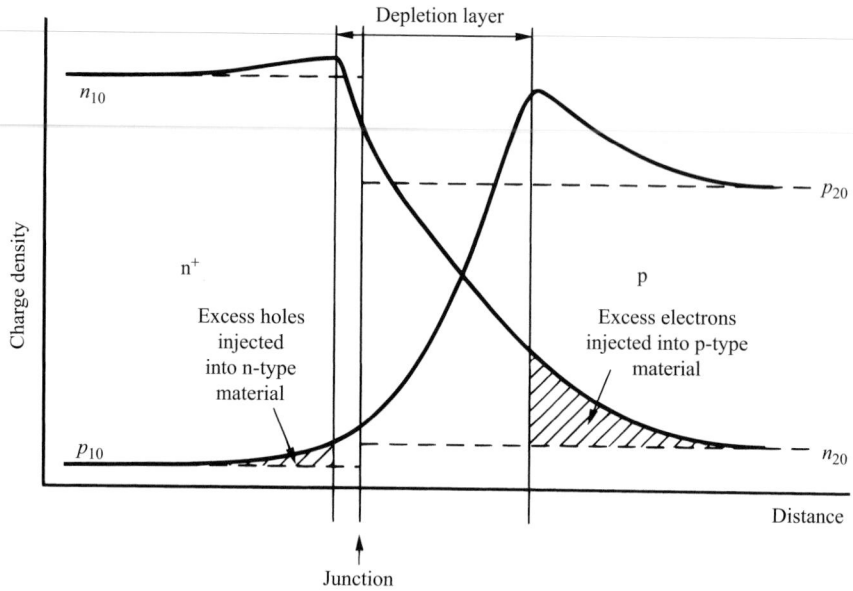


**Fig. 12.11** The charge-density variation of electrons and holes across a forward-biased p–n junction.

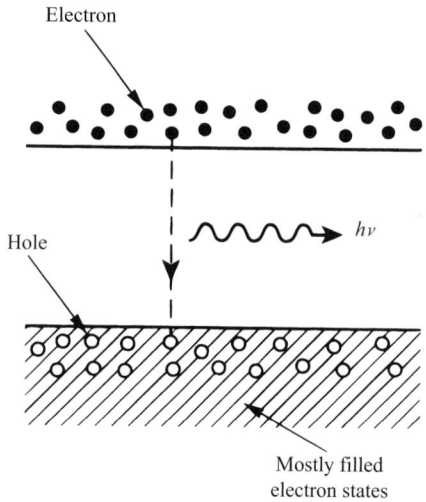


**Fig. 12.12** Band-to-band radiative recombination.

with which excess carriers recombine radiatively strongly determines the utility of the junction as a *light-emitting diode* (LED). Early semiconductor diode lasers were constructed by using a p–n junction using degenerately doped materials as shown in Fig. 12.13. Under forward bias the energy-band diagram would distort near the junction, as shown in Fig. 12.14, to provide a region where filled electron states in the valence band find themselves directly above empty electron states (holes) in the valence band. The resulting population inversion

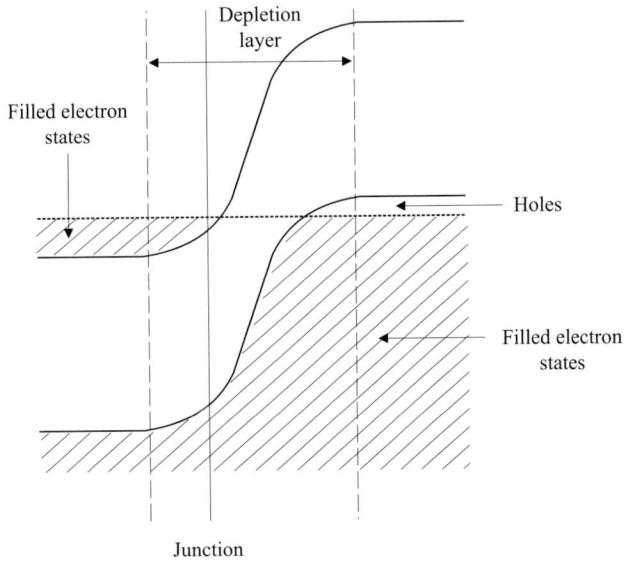


**Fig. 12.13** The schematic band structure at the junction between two degenerately doped semiconductors at low temperature.

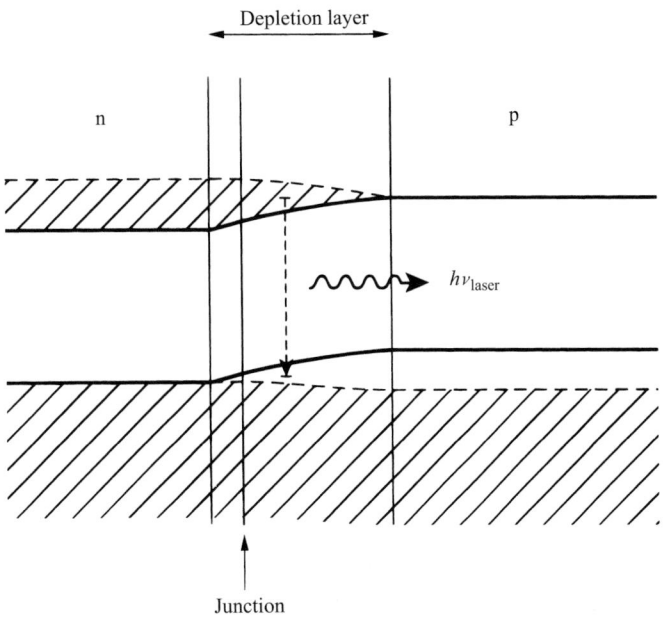


**Fig. 12.14** Forward-biased p–n junction between degenerately doped semiconductors, showing schematically how a population inversion can be obtained.

could be large enough to permit laser oscillation at high bias currents and/or low temperatures. Further discussion of these devices can be found elsewhere [4–7].

## 12.6 Recombination and luminescence

The rate at which electrons and holes recombine near a p–n junction is determined by the excess volume density of these carriers[8] and their lifetimes. For example, for excess electrons recombining in p-type material[9]

$$\frac{d(\Delta n)}{dt} = -\frac{\Delta n}{\tau_p}, \tag{12.29}$$

where $\Delta n$ is the excess electron volume density injected into the p-region and $\tau_p$ is the lifetime of these electrons in the p-type material. A similar relation holds for holes injected into the n-region.

Equation (12.29) is analogous to Eq. (1.26), the characteristic equation describing population loss from an excited state by spontaneous emission. The actual rate of recombination depends on the local electron, $n$ and $p$, hole concentrations, $n$ and $p$, respectively. If both radiative and non-radiative recombination processes occur then the recombination rate can be written as

$$\frac{d(\Delta n)}{dt} = -\Delta n\left(\frac{1}{\tau_{rr}} + \frac{1}{\tau_{nr}}\right), \tag{12.30}$$

where $\tau_{rr}$ and $\tau_{nr}$ are lifetimes for radiative and non-radiative recombination, respectively. The fraction of recombinations that occurs radiatively is called the *internal quantum efficiency*, $\eta_{int}$, where

$$\eta_{int} = \frac{1}{1 + \tau_{rr}/\tau_{nr}}. \tag{12.31}$$

The overall recombination rate $R_{rr}$ is

$$R_{rr} = rnp, \tag{12.32}$$

where $r$ is the recombination rate constant (in units of $m^3\ s^{-1}$) which is characteristic of the material. Consequently, the generated optical power per unit volume is

$$\phi = rnp\eta_{int}\overline{h\nu}, \tag{12.33}$$

where $\overline{h\nu}$ is the average photon energy of radiative recombinations.

In an efficient LED (or semiconductor laser) most of the carriers that recombine do so radiatively, and other recombination processes in which electron energy is lost in producing phonons, or in recombination with ionized impurities, are undesirable. In most of these devices much of the current through the junction is carried by electrons, which generally

[8] Namely densities above their equilibrium values with no forward bias applied.

[9] Structures in which excess holes recombine in n-type material behave similarly, and their analysis is similar to the analysis for structures in which electrons carry most of the current.

have higher mobility than the holes: for example in GaAs, $\mu_e = 0.8\ \mathrm{m^2\ V^{-1}\ s^{-1}}$ and $\mu_h = 0.03\ \mathrm{m^2\ V^{-1}\ s^{-1}}$. We define the *injection efficiency*, $\eta_i$, as

$$\eta_i = \frac{\text{electron current}}{\text{total current}}. \tag{12.34}$$

The optical power generated by an injected current $I$ is therefore

$$\phi = \eta_{\text{int}} \frac{I}{e} \overline{h\nu}. \tag{12.35}$$

In a good LED using a direct-band-gap material $\eta_{\text{int}} \simeq 0.5$. Most of the radiation is generated within a few diffusion lengths of the junction; in typical devices this is a few tens of micrometers, but depends on the temperature and doping level.

The probability of band-to-band recombination, which is the most desirable process for generating high-energy photons, is reduced in indirect-band-gap semiconductors such as silicon, germanium, and gallium phosphide (GaP). Since these materials have relatively large band gaps, this is unfortunate if short-wavelength LEDs are to be constructed. Fortunately, the probability of radiative recombination can be increased by appropriate doping, for example by doping GaP with nitrogen or replacing a GaP pair with ZnO (called "isoelectronic" doping).

### 12.6.1 The spectrum of recombination radiation

We expect the photons generated by band-to-band recombination to contain a range of energies rising from a minimum value $\lambda_g = ch/E_g$. To calculate the shape of the recombination radiation we must consider the electron and hole population densities as functions of energy. For the electrons the population density as a function of energy is, cf. Eq. (12.14),

$$n(E_2) = \rho_c(E_2) f(E_2), \tag{12.36}$$

where $E_2$ is the energy within the conduction band and $\rho_c(E_2)$ is the density of states within the band.

For the holes in the valence band the population density as a function of the energy $E_1$ within the band[10] is

$$p(E_1) = \rho_v(E_1)[1 - f(E_1)]. \tag{12.37}$$

From Eq. (12.17)

$$\rho_c(E_2) = 4\pi \left(\frac{2m_e^*}{h^2}\right)^{3/2} \sqrt{E_2 - E_c}, \tag{12.38}$$

and similarly

$$\rho_v(E_1) = 4\pi \left(\frac{2m_h^*}{h^2}\right)^{3/2} \sqrt{E_v - E_1}. \tag{12.39}$$

The rate of spontaneous emission resulting from recombination will depend on several factors: (i) the number of electron states within each small range of energy, $dE_2$, in the

[10] $E_1$ is the energy of the vacant electron state in the valence band.

conduction band; (ii) the probability that these states are occupied, $f(E_2)$; (iii) the number of electron states within each small range of energies, $dE_1$, in the valence band; and (iv) the probability that these states are empty, $1 - f(E_1)$. Therefore, we write the total rate per unit volume at which recombinations occur as

$$R_{\text{spont}} = \int\int A\rho_c(E_2)f_2(E_2)\rho_v(E_1)[1 - f_1(E_1)]dE_1\, dE_2, \tag{12.40}$$

where $A$ is an Einstein coefficient for spontaneous emission and the integrals run over the conduction and valence bands, respectively. If $A$ does not depend on the energies $E_2$ and $E_1$, then

$$\begin{aligned} R_{\text{spont}} &= A\int \rho_c(E_2)f_2(E_2)dE_2 \int \rho_v(E_1)[1 - f_1(E_1)]dE_1 \\ &= Anp, \end{aligned} \tag{12.41}$$

where $n$ and $p$ are the total electron and hole densities, respectively. A typical value for $A$ in the case of GaAs is $10^{-16}$ m$^3$ s$^{-1}$.

Unfortunately, in calculating the spectral distribution of recombination radiation we need to use Eqs. (12.36) and (12.37), which are complicated. Consequently, we make the simplifying assumption that the density of charged particles decreases exponentially from the band edge; for example

$$n(E_2) = n_c e^{-(E_2 - E_c)/(k_B T)}, \tag{12.42}$$

and

$$p(E_1) = p_v e^{-(E_v - E_1)/(k_B T)}. \tag{12.43}$$

The luminescence spectrum will have a lineshape

$$g(\nu) \propto \int_{E_c}^{E_v + h\nu} n(E_2)p(E_1)dE_2, \tag{12.44}$$

subject to the constraint $E_2 - E_1 = h\nu$. Substitution from Eqs. (12.42) and (12.43) and integration gives

$$g(\nu) \propto (h\nu - E_g)e^{-(h\nu - E_g)/(k_B T)}. \tag{12.45}$$

This lineshape function is shown in Fig. 12.15.

Actual recombination spectra are more symmetric than Fig. 12.15, as shown in Fig. 12.16, because of energy-band smearing that occurs near the band edge of real semiconductors.[11]

### 12.6.2 External quantum efficiency

It is not sufficient to build an LED by using an active region where $\eta_{\text{int}}$ is high. This will guarantee efficient photon generation, but, unless these photons escape from the device, useful light will not be generated. There are four principal ways in which generated photons

[11] Called band-tailing.

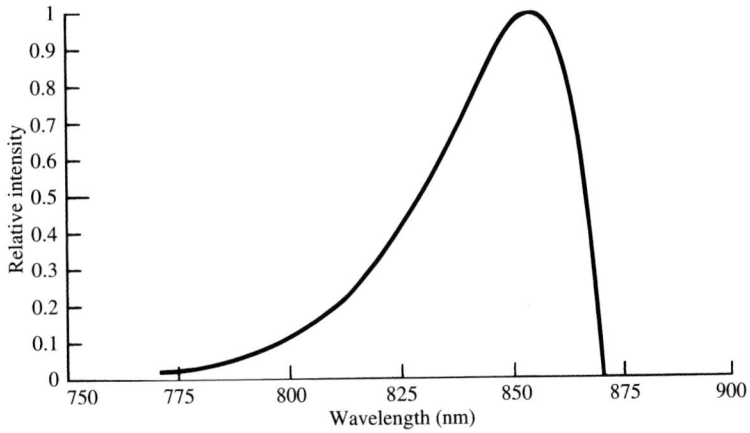


Fig. 12.15 Recombination spectrum of GaAs calculated from the simple model given by Eq. (12.45).

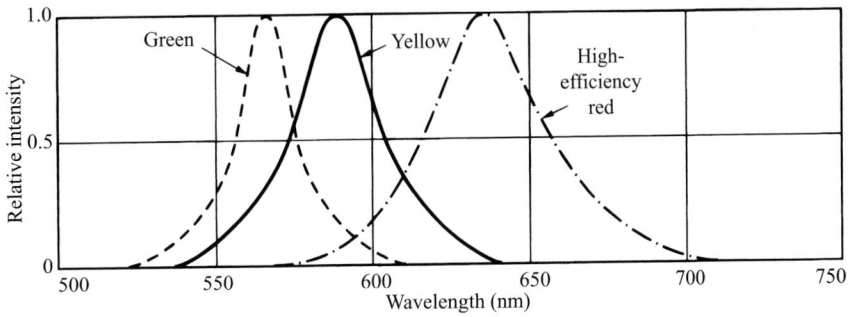


Fig. 12.16 Spectra of Hewlett-Packard red, yellow, and green LEDs at 25°C.

fail to escape from the active region, which are shown in Fig. 12.17. Of potentially greatest importance is the re-absorption of photons in bulk material between the active layer and the surface. Reflectance losses at the outer semiconductor/air interface can be reduced by coating this interface appropriately. The half of the radiation that is not emitted towards the exit surface can be re-directed with a reflective layer.

The *external quantum efficiency* $\eta_{\text{ext}}$ is the ratio

$$\eta_{\text{ext}} = \frac{\text{photons leaving exit interface}}{\text{number of carriers crossing junction}}. \tag{12.46}$$

The most efficient way to maximize $\eta_{\text{ext}}$ is to build a planar structure in which the active layer is very close to the surface. This is exemplified by the Burrus-type surface-emitting LED [8] shown in Fig. 12.18 with an attached optical fiber.

A structure of this kind is fabricated in a series of steps involving the deposition of one layer of material on another with selective etching to produce, for example, the "etched well" which allows the end of the optical fiber to be placed close to the active region.

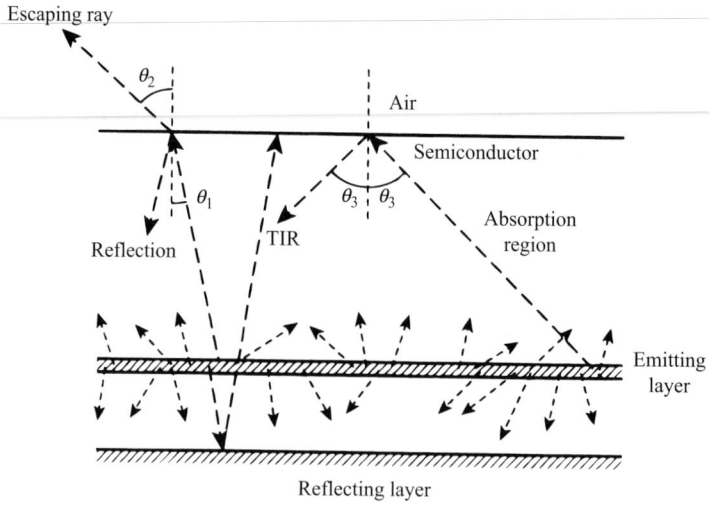


**Fig. 12.17** A schematic illustration of factors that affect the external quantum efficiency. The totally internally reflected (TIR) ray is incident on the interface at an angle $\theta_3 > \theta_c$.

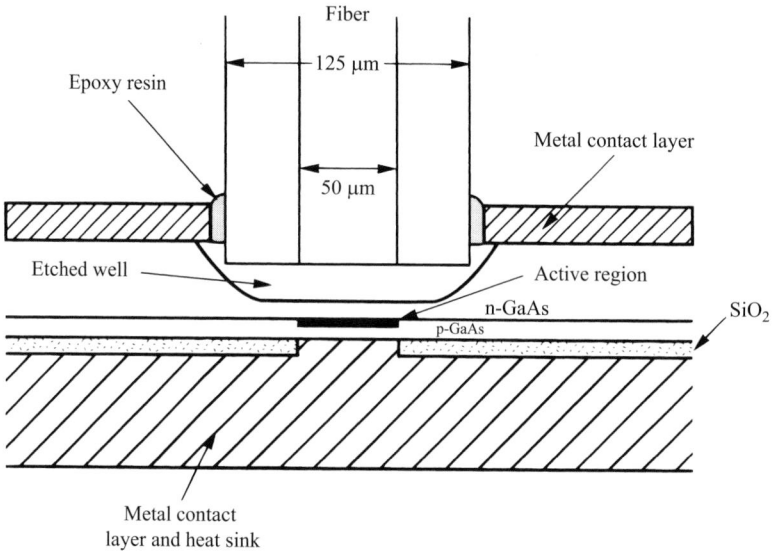


**Fig. 12.18** A Burrus-type LED with attached optical fiber.

Figure 12.19 shows a simple structure, widely used in simple solid-state lamps, in which the hemispherical (or ellipsoidal) shape permits a greater fraction of the emitted light to escape through the outer surface.

In all these structures a potentially important loss mechanism is the re-absorption of emitted photons within the unexcited regions between the active layer and the outer surface. This occurs because the emitted photons have $h\nu > E_g$, so they can easily be

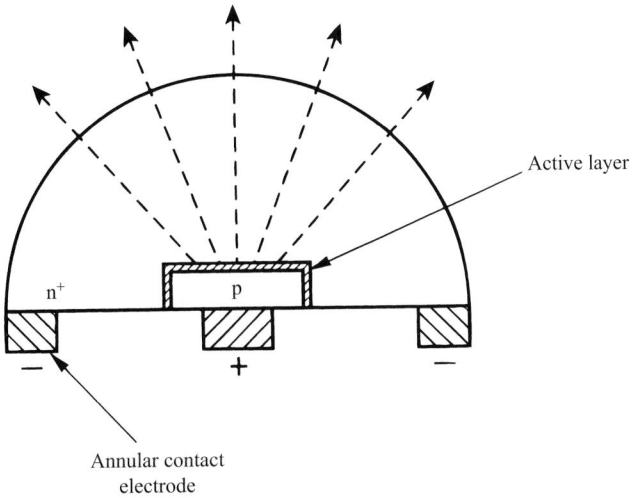


Fig. 12.19 A simple hemispherical LED.

re-absorbed. However, it is possible to build composite structures in which the band gaps of the semiconductor regions near the active layer are larger so that re-absorption cannot occur. This is made possible through the incorporation of *heterojunctions*.[12]

## 12.7 Heterojunctions

Although it is possible to build LEDs and obtain semiconductor laser oscillation by using a forward-biased homojunction, as was the case for early devices, such devices are not optimal and are no longer in practical use. Their efficiency can be characterized by the use of the internal quantum efficiency, $\eta_{\text{int}}$, considered as

$$\eta_{\text{int}} = \frac{\text{number of band-to-band radiative recombinations}}{\text{number of carriers crossing junction}}. \tag{12.47}$$

If both radiative and non-radiative recombinations occur, then, for electrons injected into p-type material,

$$\frac{1}{\tau_{\text{p}}} = A + X, \tag{12.48}$$

where $A$ is an Einstein coefficient describing radiative recombination and $X$ is a coefficient for non-radiative recombination.

Clearly,

$$\eta_{\text{int}} = \frac{1}{1 + X/A}, \tag{12.49}$$

[12] A heterojunction is a junction between two different semiconductors, for example GaAs and GaAlAs.

so we want $A \gg X$ for efficient photon generation. Unfortunately, this is not the whole story. If we want a bright, efficient, photon emitter we need to guarantee that as many carriers as possible recombine soon after crossing the junction, and do not escape to travel into the bulk material far from the junction. In an ideal device electrons and holes would be "trapped" in the region where recombination is desired. This trapping can be effected by the use of layered semiconductor structures that use heterojunctions. A heterojunction results when one semiconductor layer is grown on a different semiconductor material.

### 12.7.1 Growth of multi-layer semiconductor structures

Most semiconductor light-emitting structures, irrespective of whether they are LEDs or lasers, use multiple layers of different semiconductor materials. Depending on the application, the various layers involved can be n- or p-doped, or can be intrinsic (undoped). A complete structure will also involve metallization layers for electrical contact, and there may be a substrate layer such as silicon, or sapphire, on which the semiconducting layers are grown. Many of the advances in device fabrication have resulted from improved growth techniques that allow disparate materials to be grown on each other. A most important consideration in growing one material on another is the matching of their *lattice constants*. The lattice constant is the spacing between the constituent unit cells, or elemental units, within the crystal structure.[13] These spacings are typically referred to as $a$, $b$, and $c$ in the $x$, $y$, and $z$ directions. In the most general case, the lattice constants differ in three othogonal directions in the crystal structure. In a cubic crystal, such as GaAs, all three lattice constants are equal. When one material is grown on another, unless the lattice constants of the materials are closely matched, strain will result at the interface. This strain can lead to defects and even cracking of the whole grown structure. It is possible to avoid such effects by gradually grading the composition of the interface. For example, the tertiary material InGaP can be grown on GaAs by gradually changing the stoichiometry across the junction. The lattice contacts and band gaps of various important materials are shown in Fig. 12.20. This figure shows that GaAs and AlAs have almost the same lattice constant.

Sophisticated growth techniques, especially molecular beam epitaxy (MBE)[14] have allowed the development of complex multi-layered semiconductor structures and improved semiconductor lasers, LEDs, and also optical detectors. The compositions of individual layers can be adjusted continuously over broad ranges, provided that the value of a stoichiometry parameter $x$, in for example $In_xGa_{1-x}As$, allows a layer to be grown without excessive defects. This ability to control stoichiometry also changes the band gap, so the process can be referred to as *band-gap engineering*. If different materials are grown on each other the lattice mismatch results in either tensile or compressive strain at the interface. However, in thin layers, and by the use of graded composition, strain can be

[13] See Appendix 8.

[14] In MBE atomic elements to be deposited during film growth, such as Ga and As, are heated in ultra-high vacuum. The sublimating elements travel without collisions to a substrate, where they are deposited and grow a crystal structure. The growth rates achieved are quite slow (~100 nm/h), but this allows the depositing atoms to form a crystal structure epitaxially on a lattice-matched substrate. MBE was invented by Cho and Arthur [9].

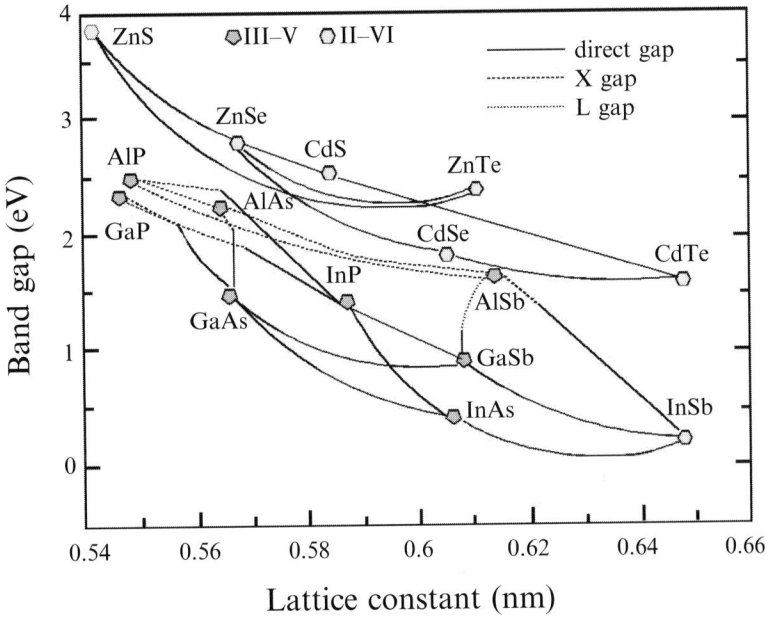


Fig. 12.20 Lattice constants and band gaps of various binary semiconductors.

compensated, and extremely variable layered structures can be fabricated. This ability to change stoichiometry, either continuously or in steps, can allow the fabrication of *superlattices*: structures in which a macroscopic periodically repeated layered strucure is a close analog to the periodicity of a natural crystal, but in an engineered way on a larger scale.

### 12.7.2 Ternary and quaternary lattice-matched materials

Because GaAs and AlAs have almost the same lattice constant, as shown in Fig. 12.20 it is possible to replace Ga atoms in GaAs with Al to make the ternary material $Ga_{1-x}Al_xAs$ without any significant change in the atomic arrangement of the crystal lattice. For compositions containing up to 37% Al, the ternary material remains a direct-band-gap semiconductor. For higher Al compositions the ternary material becomes an indirect-X-gap material. At the same time the band gap changes from 1.42 eV (GaAs) to 1.92 eV ($Ga_{0.63}Al_{0.37}As$). Because lattice spacing of the GaAlAs is very close to that of GaAs, it is possible to grow a layer of GaAlAs on GaAs, or vice versa, without introducing strain at the interface between the two materials. The two materials are said to be *lattice-matched*. The junction between the GaAs and the GaAlAs is called a heterojunction: either material can additionally be n- or p-doped. So, for example, we could have n–N, n–P, P–N, or p–P heterojunctions.[15] Other important lattice-matched heterojunctions are $GaAs/In_xGa_{1-x}As_{1-y}P_y$, $InP/In_rGa_{1-r}As_{1-s}P_s$, and $GaSb/In_tGa_{1-t}As_uSb_{1-u}$. Not all

[15] The upper-case letters refer to the semiconductor with the larger band gap, in this case GaAlAs.

compositional ranges (values of $x$, $y$, $r$, $s$, $t$, and $u$) lead to lattice-matched, direct-band-gap semiconductors, and some compositions cannot be grown [2]. The band gap depends on composition, for example in $Ga_{1-x}Al_xAs$ it is $E_g = 1.424 + 1.155x + 0.37x^2$ eV and in $In_xGa_{1-x}As_{1-y}P_y$ lattice matched to InP it is $E_g = 0.75 + 0.462y + 0.138y^2$ eV [10]. Some specific examples of important ternary and quaternary materials are listed in Table 12.1 in Section 12.2.

### 12.7.3 Energy barriers and rectification

We will use the system GaAl/GaAlAs as the archetype for explaining the properties of heterojunctions that make them desirable for the fabrication of LEDs and lasers. Because these two materials have different energy gaps, when a junction between them is formed, energy barriers to the flow of electrons and holes result. The Fermi levels in the two materials line up across the junction so that deformation of the band edge occurs near the junction as shown schematically in Fig. 12.21.

The n–P heterojunction has rectification properties like a traditional diode. Moreover, even the n–N and p–P heterojunctions can show diode-like behavior because of the energy barriers in Figs. 12.21(b) and (c) [11, 12]. For example, there is a significant probability that an electron in the conduction band moving from n $\rightarrow$ N in Fig. 12.21(b) or from p $\rightarrow$ P in Fig. 12.21(c) will be reflected because of the energy discontinuity at the junction.

### 12.7.4 The double heterostructure

To maximize the recombination efficiency, and therefore potential light output, at a p–n junction it is desirable to place a heterojunction on each side of the region where electrons and holes recombine. A schematic diagram of such a double heterostructure is given in Fig. 12.22.

The active GaAs layer is sandwiched between two layers of GaAlAs whose larger energy gaps reflect both electrons and holes back into the active layer, as shown schematically in Fig. 12.23. The contact layers are highly doped n- and p-type layers of GaAs. The + superscript is used to denote a high doping level. The metal contact layer can be deposited most easily on a small-band-gap material with a relatively large conductivity. Otherwise, a rectifying metal/semiconductor junction will result.[16]

One additional advantage of this type of structure has already been mentioned, namely the transparency of the confining layers to photons emitted from the active layer because of the larger band gap of the material in these layers. But, there is a second advantage: the confining layers have a lower refractive index than that of the active layer. Consequently, photons emitted parallel, or nearly so, to the junctions can be guided by total internal reflection and this leads to enhanced efficiency of edge-emitting LEDs and lasers. Figure 12.24 shows an edge-emitting LED design. The insulating $SiO_2$ layer confines the current to a specific lateral region. The emerging optical beam has an approximately elliptical shape

[16] Called a Schottky diode.

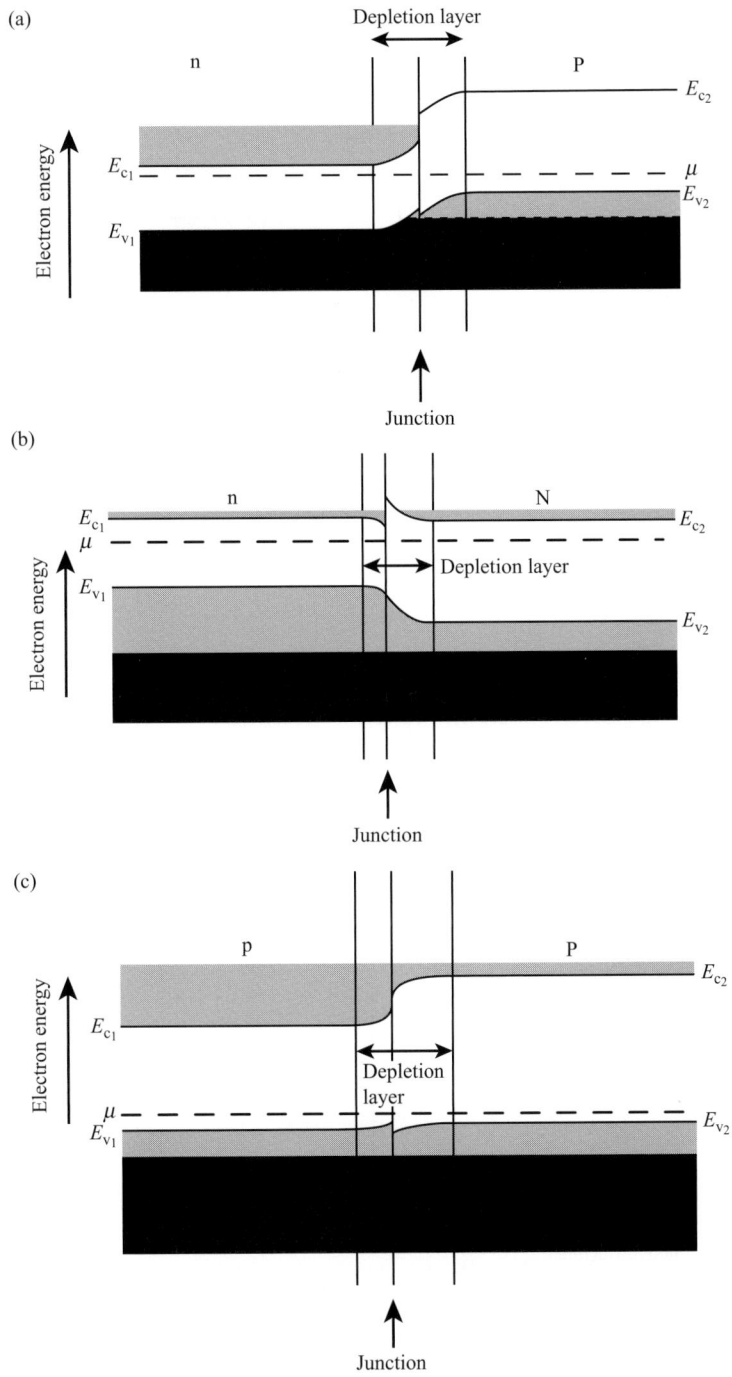


**Fig. 12.21** (a) The band-energy diagram of an n–P heterojunction in thermal equilibrium. (b) The band-energy diagram of an n–N heterojunction in thermal equilibrium. (c) The band-energy diagram of a p–P heterojunction in thermal equilibrium.

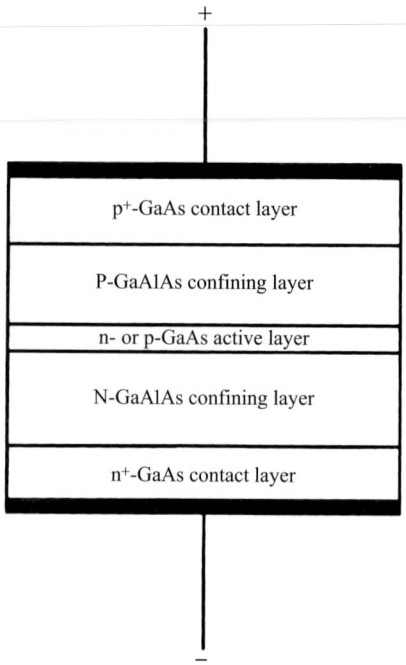


**Fig. 12.22** A schematic diagram of a layered double heterostructure.

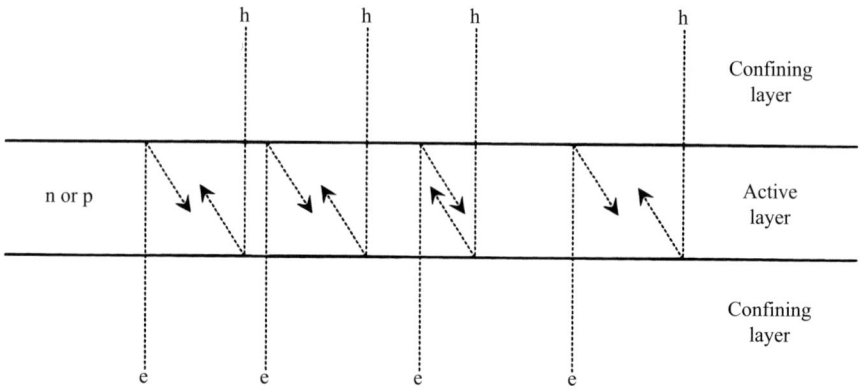


**Fig. 12.23** A schematic illustration of how electrons (e) and holes (h) reflect at the edges of the confining layer in a double heterostructure.

whose semimajor axis is parallel to the junction. However, the small thickness of the active layer leads to diffraction such that this elliptical beam becomes elliptical with its semimajor axis perpendicular to the junction as the beam travels away from the device.

The double heterostructure shown in Fig. 12.24 is biased so that the p–N (or n–P) junction at the edge of the active layer is conducting.

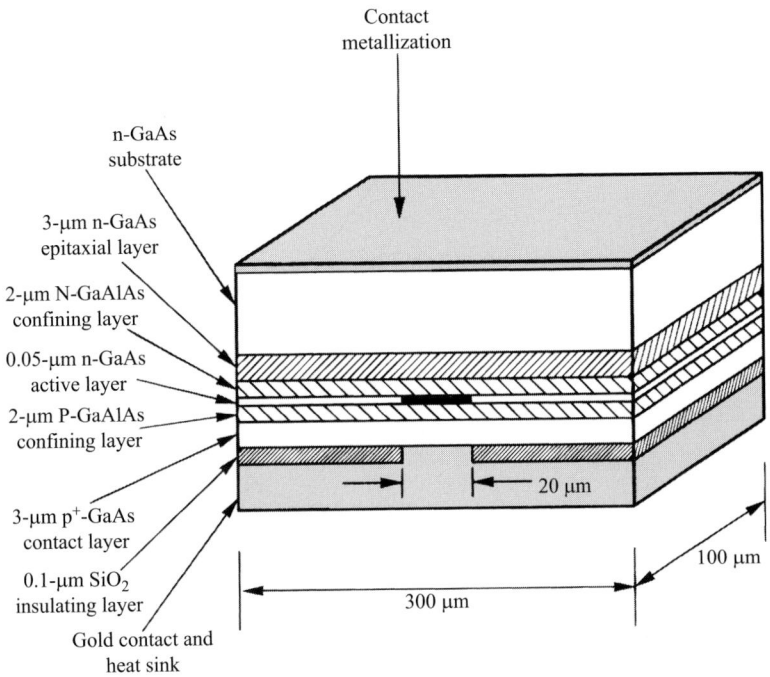


Fig. 12.24 An edge-emitting LED design.

If the active layer is sufficiently thin it can be assumed that the electron concentration $n$ does not vary much within it. The recombination rate per unit volume within this layer is

$$-\frac{dn}{dt} = \frac{n}{\tau_{rr}} + \frac{n}{\tau_{nr}}, \tag{12.50}$$

where $\tau_{rr}$ and $\tau_{nr}$ are the recombination times for radiative and non-radiative recombination, respectively, within the active layer.

However, an additional term should be added to Eq. (12.50) in order to include *surface-recombination* effects that occur at the heterojunction boundaries of the layer. This surface recombination occurs at a rate per unit area of

$$-\frac{dn}{dt} = nv_S, \tag{12.51}$$

where $v_S$ is called the *recombination velocity*. Surface recombination is enhanced by any lattice mismatch at the heterojunction: it can also occur at the outside bounding faces of the active layer, but this effect is not significant except in a device of very small area. Therefore, for an active layer of thickness $d$ the overall recombination time $\tau$ satisfies

$$\frac{1}{\tau} = \frac{1}{\tau_{rr}} + \frac{1}{\tau_{nr}} + \frac{2v_S}{d}. \tag{12.52}$$

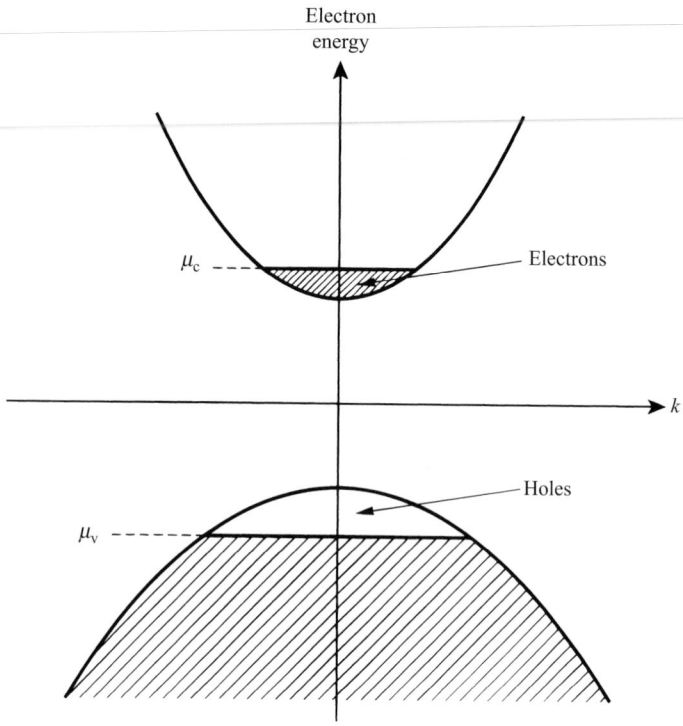


Fig. 12.25 Population inversion shown in an $E$–$k$ diagram.

The factor of 2 comes from the two interfaces of the active layer, and we have assumed that $v_S$ is the same for both interfaces. A typical value for $v_S$ at a GaAs/GaAlAs heterojunction is 10 m s$^{-1}$. The internal quantum efficiency within the active layer is clearly

$$\eta_{\text{int}} = \frac{\tau}{\tau_{\text{rr}}}. \tag{12.53}$$

The radiative recombination time in the active layer will vary depending on the doping level and depends also on the injected current, but values in the range 1–100 ns are typical. The non-radiative lifetime $\tau_{\text{nr}}$ is decreased by the presence of impurities and lattice defects, but a value of 100 ns might be typical. Consequently, for an active layer of thickness 0.05 μm and for $\tau = 1$ ns, $\eta_{\text{int}} \simeq 0.7$.

## 12.8 Semiconductor lasers

The edge-emitting LED design shown in Fig. 12.24 can, in the right circumstances, become an edge-emitting semiconductor laser. All that is required is a condition of population inversion within the active region that provides sufficient gain to overcome cavity losses.

In many such lasers the cavity is formed by the plane-parallel facets at opposite ends of the active region. However, before discussing these and other semiconductor laser designs in more detail we need to return to some fundamental considerations.

In common with all lasers, gain results in the active region if the rate of stimulated emission exceeds that of absorption. The probability of band-to-band stimulated emission depends on four factors: the Einstein coefficient, $B_{21}$, for the transition; the energy density of the stimulating radiation, $\rho(\nu_{21})$; the probability that the upper energy level is occupied by an electron, $f_2$; *and* the probability $1 - f_1$ that the lower energy level is *not* occupied by an electron; namely that it is occupied by a hole. Formally, we can write

$$\text{gain} \propto [B_{21}f_2(1 - f_1) - B_{12}f_1(1 - f_2)]\rho(\nu_{21}), \tag{12.54}$$

which, if we assume that $B_{12} = B_{21}$, gives

$$\text{gain} \propto f_2 - f_1. \tag{12.55}$$

The existence of gain is a non-equilibrium situation that can be represented schematically as the existence of a region of filled electron states above empty electron states, as shown in Fig. 12.25.

Even though the system as a whole is not in thermal equilibrium, the distribution of electrons among available energy states can still be derived from a Fermi–Dirac distribution function, but using a different Fermi level for the conduction and valence bands, respectively. We can represent the energies that bound the filled electron states in the conduction and valence bands as $\mu_c$ and $\mu_v$, respectively, which are referred to as quasi-Fermi levels. From Eq. (12.9)

$$f_2(E_2) = \frac{1}{e^{(E_2-\mu_c)/(k_B T)} + 1}, \tag{12.56}$$

$$f_1(E_1) = \frac{1}{e^{(E_1-\mu_v)/(k_B T)} + 1}. \tag{12.57}$$

The condition $f_1 > f_2$ results if

$$(\mu_c - \mu_v) > h\nu_{21}, \tag{12.58}$$

a condition first derived by Bernard and Duraffourg [13], and, of course, $h\nu_{21} > E_g$.

In a heavily doped p-type semiconductor the Fermi level can be forced down into the valence band. If such a material is used as the active region in an N–p–P double heterostructure then the energy levels in thermal equilibrium appear as shown in Fig. 12.26. However, under sufficient forward bias the situation shown in Fig. 12.27 will result. Within the active layer the situation shown in Fig. 12.25 has been created: there are filled electron states above quasi-Fermi levels $\mu_N$ in the conduction band and $\mu_P$ in the valence band.

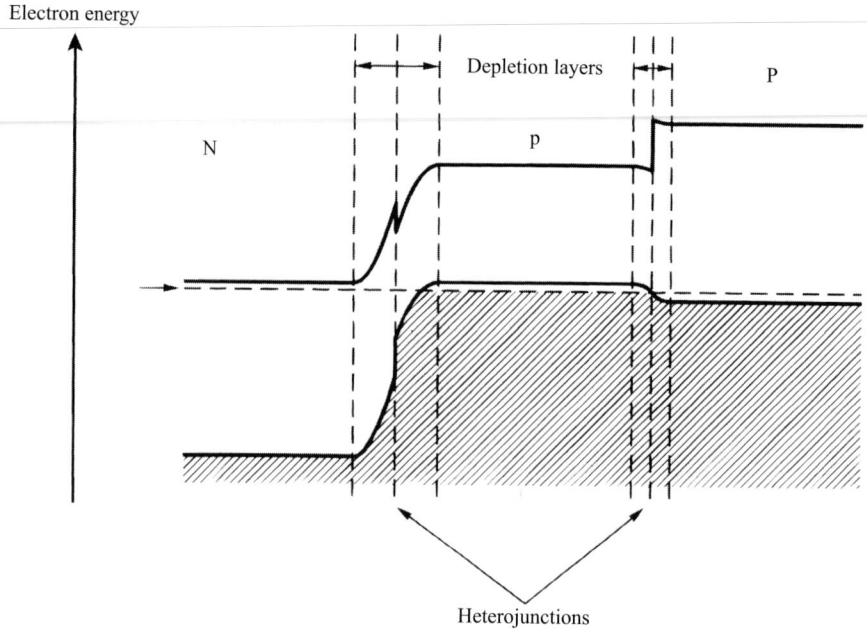


**Fig. 12.26** A schematic band diagram in thermal equilibrium of an N–p–P double heterostructure in which the p region is heavily doped. Note that the Fermi level in the p region has moved down into the valence band.

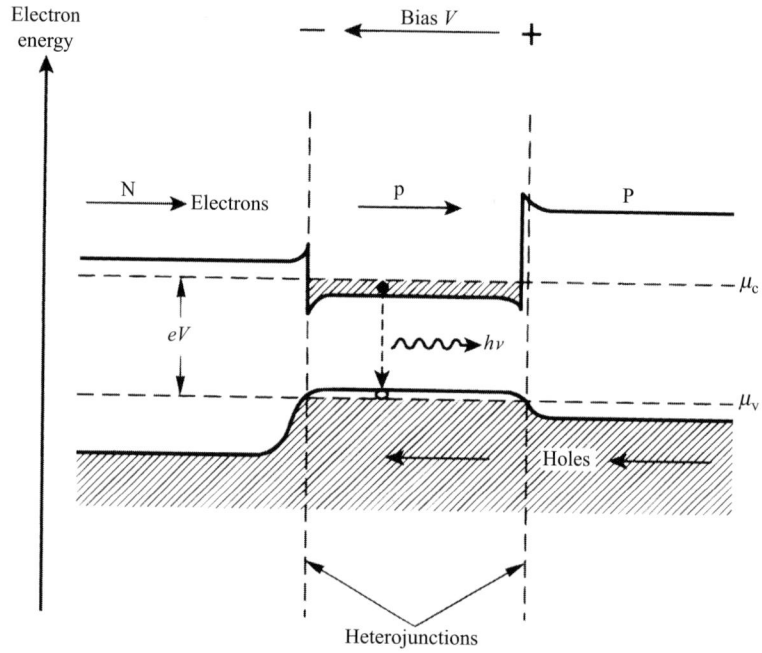


**Fig. 12.27** A schematic band diagram of a forward-biased N–p–P double heterostructure in which the p region is heavily doped. Electrons flowing from the N region and holes flowing from the P region are trapped in the p-region potential well, creating a population inversion.

## 12.9 The gain coefficient of a semiconductor laser

To derive an expression for the gain coefficient of a semiconductor laser it is best to return to the fundamentals discussed in Chapter 2. In that chapter we saw that the increase in intensity, $dI_\nu$ (W m$^{-2}$) of a small band of frequencies near $\nu$ on passing a distance $dz$ through a medium could be written as

$$\begin{aligned} dI_\nu &= (\text{no. of stimulated emissions/volume} - \text{no. of absorptions/volume}) \\ &\quad \times h\nu\, dz, \\ &= \left[ B_{21} N_2 g(\nu', \nu) \frac{I_\nu}{c} - B_{12} N_1 g(\nu', \nu) \frac{I_\nu}{c} \right] h\nu\, dz, \end{aligned} \tag{12.59}$$

where $N_2 g(\nu', \nu)$ is the density of filled upper level states per frequency interval, $N_1 g(\nu', \nu)$ is the density of filled lower level states per frequency interval, and $I_\nu$ (W m$^{-2}$) is the intensity of input monochromatic radiation.

In a semiconductor laser the probability of stimulated emission depends on several factors: (i) an Einstein coefficient $B$, which we shall assume applies also to absorption processes; (ii) the density of states in the upper level, $\rho_c(E_2)$, and the probability that these states are occupied, $f_2(E_2)$; and (iii) the density of states in the lower level, $\rho_v(E_1)$, and the probability that these states are empty, $1 - f_1(E_1)$. Consequently, Eq. (12.59) becomes

$$\begin{aligned} dI_\nu = \Bigg\{ & \rho_c(E_2)\rho_v(E_1) f_2(E_2)[1 - f_1(E_1)] B h \frac{I_\nu}{c} \\ & - \rho_c(E_2)\rho_v(E_1) f_1(E_1)[1 - f_2(E_2)] B h \frac{I_\nu}{c} \Bigg\} h\nu\, dz. \end{aligned} \tag{12.60}$$

The additional factor $h$ in Eq. (12.60) enters because we will be integrating over energy rather than frequency.

If the dispersive nature of the medium is taken into account then[17] the relation between the Einstein $A$ and $B$ coefficients in this case is

$$\frac{A}{B} = \frac{8\pi h \nu^3 n^2 n_g}{c_0^3}. \tag{12.61}$$

Therefore,

$$dI_\nu = \frac{h c_0^2 A}{8\pi n n_g \nu^2} \rho_c(E_2)\rho_v(E_1)[f_2(E_2) - f_1(E_1)] I_\nu\, dz, \tag{12.62}$$

where $dI_\nu$ now has dimensions W m$^{-2}$ J$^{-1}$. The coefficient, $B$, for stimulated emission for states distributed over an energy range $dE$ has units J$^{-1}$ m$^6$ s$^{-2}$. Bearing in mind that the gain coefficient is defined as

$$\gamma(\nu) \frac{1}{I_\nu} \frac{dI_\nu}{dz},$$

[17] See Problem (1.1).

Eq. (12.62) can be transformed into an integral over energy $dE_2$,

$$\gamma(\nu) = \frac{hc_0^2}{8\pi n n_g \nu^2} \int_{E_c}^{E_v+h\nu} A\rho_c(E_2)\rho_v(E_2 - h\nu)[f_2(E_2) - f_1(E_2 - h\nu)]dE_2. \quad (12.63)$$

The limits of integration are the same as applied in our previous calculation of the spectrum of recombination radiation.

How does the gain $\gamma(\nu)$ depend on the current flowing through the junction? For a current $I$ crossing a p–n junction of width $w$ and height $h$ the injected current density is

$$J = \frac{I}{hw}. \quad (12.64)$$

In a double heterojunction laser, ideally, all the injected carriers recombine in the region between the two heterojunctions, of spacing $d$. Therefore, the total emitted spontaneous power is

$$R_{\text{spont}} = \frac{\eta_{\text{int}} J}{ed}. \quad (12.65)$$

So, from Eq. (12.41),

$$Anp = \frac{\eta_{\text{int}} J}{ed}. \quad (12.66)$$

It is not straightforward to calculate the gain from Eq. (12.63) because the Fermi–Dirac functions $f_2(E_2)$ and $f_1(E_1)$ depend on the locations of the quasi-Fermi levels $\mu_c$ and $\mu_v$ in the conduction and valence bands, respectively, in the active region. But, we can estimate typical values that might be expected.

### 12.9.1 Estimation of semiconductor laser gain

We consider the case of an n-doped GaAs active region with a donor concentration of $n_D = 10^{24}\ \text{m}^{-3}$, an internal quantum efficiency of 0.6 and an injected current of 10 mA that enters a 10-μm-wide by 10-μm-high active region whose thickness between heterojunctions is 0.5 μm. From Eqs. (12.66) and (12.64), taking $A = 10^{-16}\ \text{m}^3\ \text{s}^{-1}$,

$$np = \frac{0.6 \times 0.01}{1.6 \times 10^{-19} \times 100 \times 10 \times 0.5 \times 10^{-18} \times 10^{-16}} = 7.5 \times 10^{47}\ \text{m}^{-6}. \quad (12.67)$$

Furthermore, in the active region $n - p = n_D = 10^{24}\ \text{m}^{-3}$, which gives $n = 1.5 \times 10^{24}\ \text{m}^{-3}$ and $p = 0.5 \times 10^{24}\ \text{m}^{-3}$.

The electron density satisfies

$$n = \int_{E_c}^{\infty} \rho_c(E_2) f_2(E_2) dE_2. \quad (12.68)$$

In a degenerately doped semiconductor the quasi-Fermi levels are well inside the conduction and valence bands, respectively. Consequently, if the temperature is sufficiently low the Fermi–Dirac distribution can be represented as a step function. We can approximate

$f_2(E_2) = 1$ for $E_2 \leq \mu_c$ and $f_2(E_2) = 0$ for $E_2 > \mu_c$, with similar relations for the holes. Consequently, using Eq. (12.38)

$$n = \int_{E_c}^{\mu_c} \rho_c(E_2) dE_2 = \frac{1}{2\pi^2} \left( \frac{2m_e^*}{\hbar^2} \right)^{3/2} \int_{E_c}^{\mu_c} \sqrt{E_2 - E_c}\, dE_2 \tag{12.69}$$

and

$$p = \int_{\mu_v}^{E_v} \rho_v(E_1) dE_1 = \frac{1}{2\pi^2} \left( \frac{2m_h^*}{\hbar^2} \right)^{3/2} \int_{\mu_v}^{E_v} \sqrt{E_v - E_1}\, dE_1. \tag{12.70}$$

In GaAs $m_e^* = 0.067 m_e$ and $m_h^* = 0.48 m_e$, so evaluating the constants and performing the integration gives

$$n = 1.23 \times 10^{54} (\mu_c - E_c)^{3/2}, \tag{12.71}$$

$$p = 2.35 \times 10^{55} (E_v - \mu_v)^{3/2}. \tag{12.72}$$

With the electron and hole concentrations calculated previously, we obtain

$$\mu_c - E_c = 1.14 \times 10^{-20}\ \text{J} = 0.071\ \text{eV} \tag{12.73}$$

and

$$E_v - \mu_v = 7.88 \times 10^{-22}\ \text{J} = 0.0048\ \text{eV}. \tag{12.74}$$

The band-gap energy in GaAs is 1.42 eV, so this calculation predicts that gain will result for photon energies from $\mu_c - \mu_v$ to $E_g$, i.e., wavelengths from 832 nm to 873 nm. With the parameters calculated so far, we can calculate the gain at a wavelength within this range, at, say, 838 nm $\equiv$ 1.48 eV. In this case

$$\gamma(\nu) = \frac{hc_0^2}{8\pi n n_g \nu^2} \int_{E_a}^{E_b} A \times 6.5 \times 10^{109} \times \sqrt{E_2 - E_c}\sqrt{E_v - E_2 + h\nu}\, dE_2. \tag{12.75}$$

The limits on the integral are set by the condition $f_2(E_2) > f_1(E_1)$ and $E_2 - E_1 = h\nu$. Clearly,

$$E_a = E_c + 0.055\ (\text{eV}), \tag{12.76}$$

$$E_b = E_c + 0.06\ (\text{eV}). \tag{12.77}$$

If we assume that $A$ is independent of energy and has a value of $10^{-16}$ m$^3$ s$^{-1}$, and use the values $n = 3.6$ and $n_g = 4$ appropriate to GaAs, Eq. (12.75) becomes

$$\gamma(\nu) = 2.52 \times 10^7 \int_{E_c+0.055}^{E_c+0.06} \sqrt{E_2 - E_c}\sqrt{E_v - E_2 + 1.48}\, dE_2, \tag{12.78}$$

where the intergand is expressed in units of eV. Now $E_v - E_2 + 1.48 = 0.06 - (E_2 - E_c)$, so, on writing $E_2 - E_c = 0.06 \cos^2\theta$, the integral is transformed to give

$$\gamma(\nu) = 1.62 \times 10^5 \int_0^{0.293} \sin^2\theta \cos^2\theta\, d\theta = 1.06 \times 10^4\ \text{m}^{-1} = 10.6\ \text{mm}^{-1}. \tag{12.79}$$

Such a high gain figure is very typical for modern semiconductor lasers. The effective gain in the laser is, of course, reduced by the distributed loss coefficient $\alpha$, which in semiconductor lasers results from scattering. This scattering increases with doping level, but will be in the range $10^3$–$10^4$ m$^{-1}$. If $\alpha = 10^3$ m$^{-1}$, the effective $\gamma$ is 9.6 mm$^{-1}$. In a laser 500 μm long the single-pass intensity increase would be $e^{4.8} = 122$.

## 12.10 Threshold current and power–voltage characteristics

The active region in a semiconductor laser is typically very thin, ranging from a few micrometers down to a few nanometers in so-called single-quantum-well (SQW) lasers, which we will discuss briefly later. The laser cavity is typically a few hundred micrometers long, although shorter devices are also made. Semiconductor lasers with longer cavities are also fabricated. In these lasers additional structures are incorporated into the cavity region to provide single-frequency operation or to allow intracavity modulation. Because the gain can be very high, the natural Fresnel reflectance at the cleaved facet ends of the laser provides more than sufficient feedback to produce laser oscillation. In fact, in applications in which a semiconductor laser structure is to be used purely as an amplifier, extraordinary steps must be taken to reduce the facet reflectance, perhaps to below $10^{-4}$ [14], to prevent spurious oscillation.

The performance of a semiconductor laser is defined by several parameters: its threshold current density $J_{\text{th}}$, its power-versus-voltage ($P$–$I$) characteristic, and its effective current density $J_{\text{eff}}$, which is defined as (cf. Eq. (12.66))

$$J_{\text{eff}} = \frac{\eta_{\text{int}} J}{d}. \tag{12.80}$$

The threshold current density $J_{\text{th}}$ is the value at which the gain exceeds the cavity losses, which we expect to be

$$\gamma_{\text{th}} = \alpha - \frac{1}{\ell} \ln\sqrt{R_1 R_2}, \tag{12.81}$$

where $R_1$ and $R_2$ are the end-facet reflectances. However, an additional factor must be included in Eq. (12.81). As has already been mentioned, there is a refractive-index discontinuity on both sides of the active layer in a double-heterostructure laser, as shown schematically in Fig. 12.28.

This gives rise to a *waveguiding* effect or confinement of the transverse mode of the laser to a region in, and on both sides of, the active region. The confinement factor, $\Gamma$, is a measure of the fraction of the oscillating field distribution that experiences gain within the active layer. The threshold gain is therefore

$$\gamma_{\text{th}} = \frac{1}{\Gamma}\left(\alpha - \frac{1}{\ell} \ln\sqrt{R_1 R_2}\right). \tag{12.82}$$

In many semiconductor lasers, so long as they oscillate in a single transverse mode [15] the gain is directly proportional to the excess current density above threshold, namely,

$$\gamma = \beta[J_{\text{eff}} - (J_{\text{eff}})_{\text{th}}]. \tag{12.83}$$

One representative $P - I$ characteristic is of the general form shown in Fig. 12.29, although very-different-looking curves are observed experimentally.

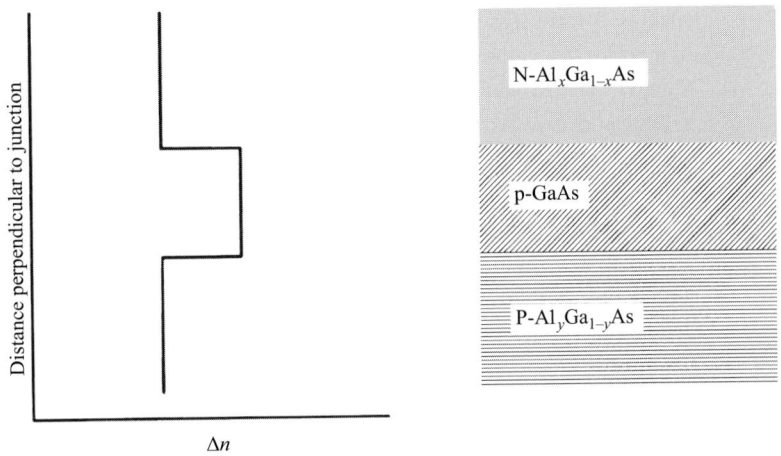


**Fig. 12.28** The schematic variation of refractive index across a N–p–P double heterojunction.

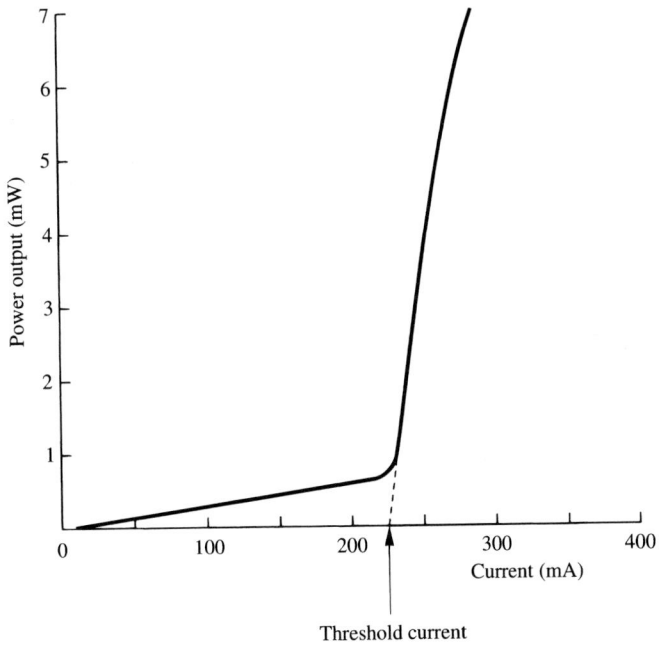


**Fig. 12.29** The *P–I* characteristic of an InGaAsP/InP double-heterojunction laser [18].

## 12.11 Longitudinal and transverse modes

Semiconductor lasers do not generate laser radiation whose linewidth is narrow compared with that from low-pressure gas lasers. They can often be encouraged to oscillate in a single transverse mode, but will generally produce several longitudinal modes, of typical spacing 0.1 nm (for a 200-μm-long GaAs laser). The gain profile is generally Lorentzian, of a width determined by the phonon collisions within the semiconductor. The calculated Schawlow–Townes linewidths of small narrow-stripe lasers are typically a few MHz, but values as narrow as this are quite difficult to achieve in practice. Unless special measures are taken, typical linewidths are tens of MHz. Lasers with extended cavities, either external or incorporated into the semiconductor structure, have generated laser linewidths as narrow as 10 kHz.

The active region is generally a rectangular parallelepiped with an end aperture of dimension $wd$ with waveguiding refractive-index discontinuities both in the direction perpendicular to the junction and, sometimes, additionally in the direction parallel to the junction – in so-called *index-guided* stripe-geometry lasers. The dimension $wd$ and the extent to which the laser is above threshold lead to the possibility of different transverse emitted field distributions as the current is increased, as shown in Fig. 12.30.

If the stripe width $w$ is maintained sufficiently small, then a single laser field distribution can be maintained to relatively high power levels. The conditions under which this will occur depend on the index discontinuities involved, as will be discussed later in Chapter 16.

Below threshold a semiconductor laser emits a broad, continuous spectrum like an LED. However, as the current density rises above threshold distinct longitudinal modes will be observed to oscillate. Particularly in index-guided lasers, one or two modes may begin to dominate as the current rises far above threshold, as shown in Fig. 12.31.

Because the index of the active region, and the confining layer, varies with wavelength, the longitudinal modes of a semiconductor laser are not equally spaced. Schematically, we can write

$$n(\nu)\nu = \frac{mc_0}{2\ell}, \tag{12.84}$$

where $n(\nu)$ is the refractive index that applies at the frequency of the cavity resonance of index $m$, and the mode spacing is $\Delta\nu = c_0/n_g$, where $n_g$ is the group refractive index defined by

$$n_g = n(\nu) + n(\nu)\frac{dn(\nu)}{d\nu}. \tag{12.85}$$

In most semiconductor lasers the gain region is not cylindrically symmetric about the axial direction. As a result, even if the laser operates in a single transverse mode, the output beam will not be cylindrically symmetric, except in the case of vertical-cavity surface-emitting lasers (VCSELs). The non-circular output beam can be corrected with external optical components called *anamorphic* optics, and in some designs the laser structure can be designed to minimize the non-circularity.

## 12.12 Quantum-well lasers

Quantum-well semiconductor lasers use a very thin active region, for example a 20-nm-thick layer of GaAs bounded on each side by GaAlAs. If the laser has only one such

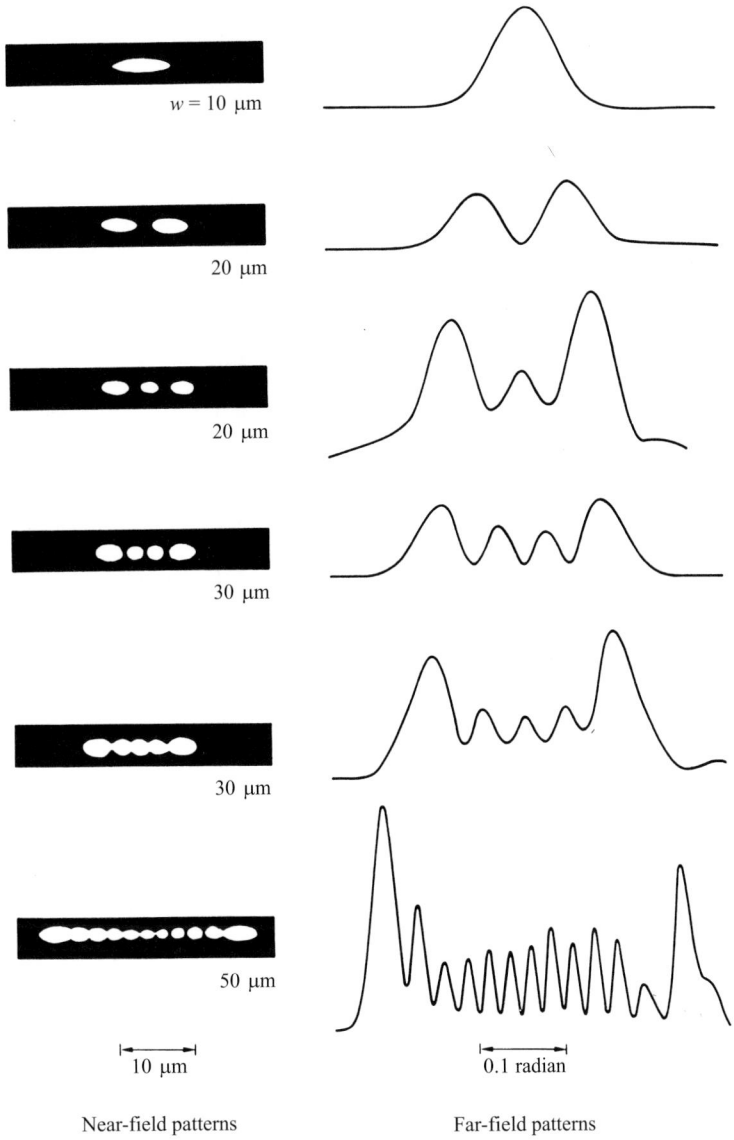


**Fig. 12.30** Near-field and far-field intensity patterns for different stripe widths and currents in a double-heterojunction planar stripe laser [19] (reproduced with the permission of the publisher).

thin layer it is called a single-quantum-well (SQW) laser. On the other hand, if the device is fabricated as a multi-layer sandwich of thin layers of alternating GaAs and GaAlAs, as shown schematically in Fig. 12.32, then it is called a multiple-quantum-well (MQW) laser. In these lasers the total number of injected carriers necessary to achieve inversion is dramatically reduced compared with the number in conventional heterostructure devices.

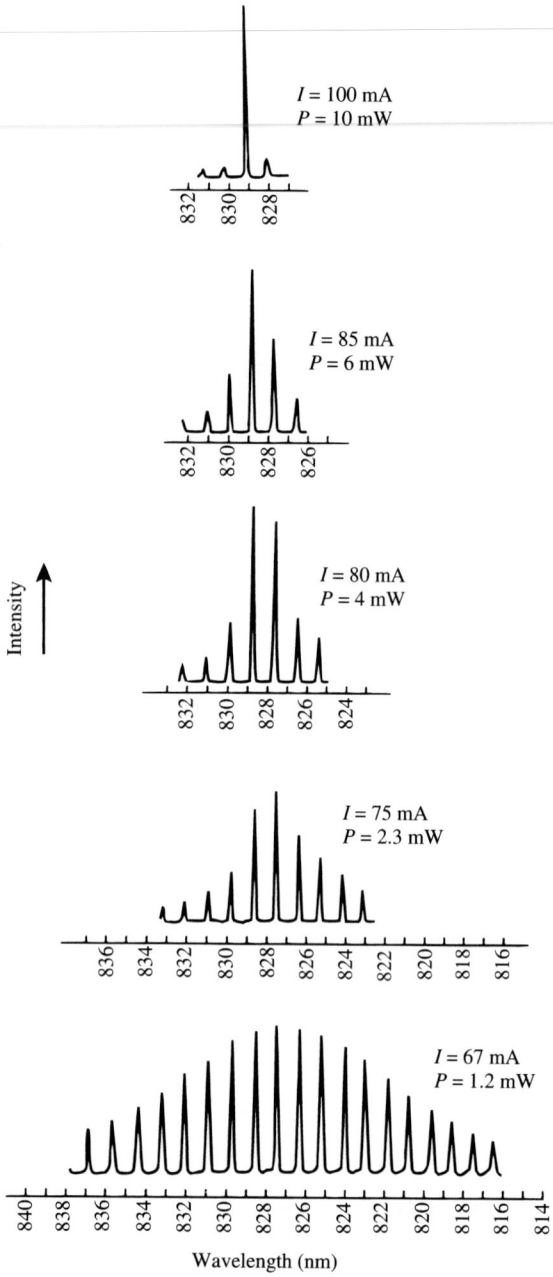


**Fig. 12.31** Output frequency spectra of a fairly "well-behaved" CW laser diode at different currents [20].

A detailed treatment of the operation of these devices is beyond our scope here, but it is still possible to explain semiquantitatively why they have attractive properties. In the energy-band diagram of an SQW laser shown in Fig. 12.33 the quantum well is a region of minimum potential energy for electrons in the conduction band and holes in the valence band. When the quantum well is thin the electrons cannot have arbitrary wave

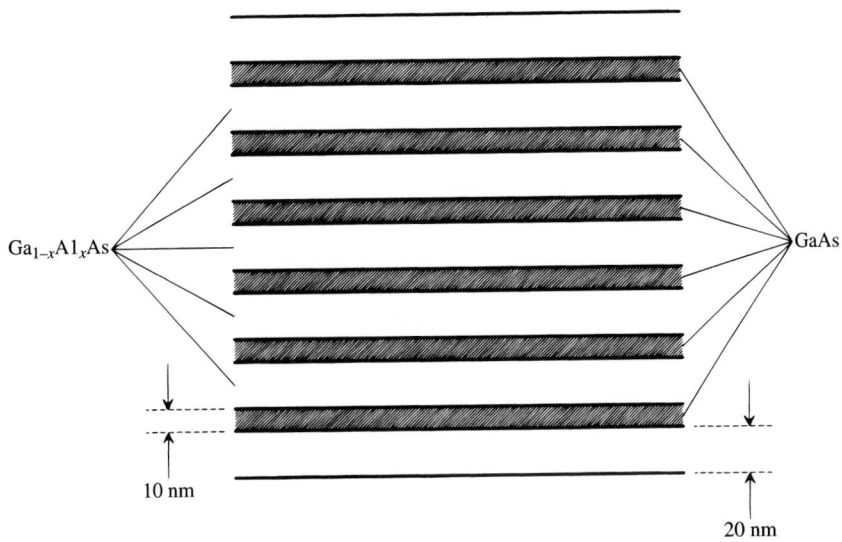


**Fig. 12.32** The layered structure of the active region of a multiple-quantum-well semiconductor laser.

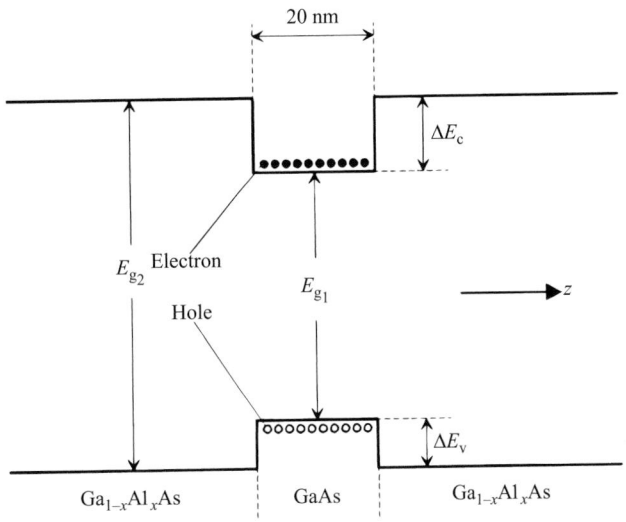


**Fig. 12.33** The energy-band diagram of an SQW laser.

vectors in the direction perpendicular to the walls of the potential well. These particles have wave-like properties and must satisfy periodic boundary conditions within a box (just like black-body radiation). A quantum-mechanical analysis allows only **k** values given by $k_\perp = n\pi/w$, where $w$ is the width of the well and $n$ is a positive integer.[18] Consequently, from

[18] This result is exact only for an infinitely deep potential well. For a well of finite depth it is approximately correct for energy states near the bottom of the well.

Eq. (12.4) the allowed energies of electrons associated with their motion perpendicular to the wall are

$$E_n = \frac{\pi^2 \hbar^2}{2m_e w^2} n^2. \tag{12.86}$$

There are similar discrete energy states for the holes in their potential-energy minimum in the conduction band.

The reduced threshold injection current in a quantum-well device results largely because the very thin quantum well modifies the density of states available to electrons and holes. The number of modes that exist in the two-dimensional region parallel to the layers is easily calculated as

$$N_{2D}(k_\parallel) = \frac{k_\parallel^2 L_x L_y}{2\pi} = \frac{k_\parallel^2 A}{2\pi}, \tag{12.87}$$

where $A = L_x L_y$ is the area of the junction region.

Therefore, the two-dimensional density of states is

$$\rho(k_\parallel) dk_\parallel = \frac{dN_{2D}(k_\parallel)}{dk_\parallel} dk_\parallel = \frac{A k_\parallel}{\pi} dk_\parallel, \tag{12.88}$$

or, in terms of energy per unit area,

$$\rho_{QW}(E) = \frac{1}{A}\frac{dN_{2D}}{dE} = \frac{k_\parallel}{\pi}\frac{dk_\parallel}{dE}. \tag{12.89}$$

The number of modes that exist in the $z$ (perpendicular) direction is severely reduced because of the small size in this direction. Qualitatively, there may be only one such mode, corresponding roughly to a half wavelength of the electron wave fitting in the energy minimum in the conduction band in Fig. 12.33. The total energy of an electron in the quantum well can be written as

$$E = \frac{\hbar^2 k_\parallel^2}{2m_e} + E_Q, \tag{12.90}$$

where $E_Q$ is the energy associated with the restricted motion of the electrons perpendicular to the junction.

From Eq. (12.90)

$$k_\parallel = \sqrt{\frac{2m_e}{\hbar^2}(E - E_Q)}. \tag{12.91}$$

Therefore,

$$\frac{dk_\parallel}{dE} = \frac{m_e/\hbar^2}{k_\parallel} \tag{12.92}$$

and

$$\rho_{QW}(E) = \frac{m_e}{\pi \hbar^2}. \tag{12.93}$$

This result holds when only a single mode exists perpendicular to the thin layer. At critical electron energies the electron wavelength becomes short enough that additional modes perpendicular to the layers become possible. These are the $n = 1, 2, 3$, etc. modes that

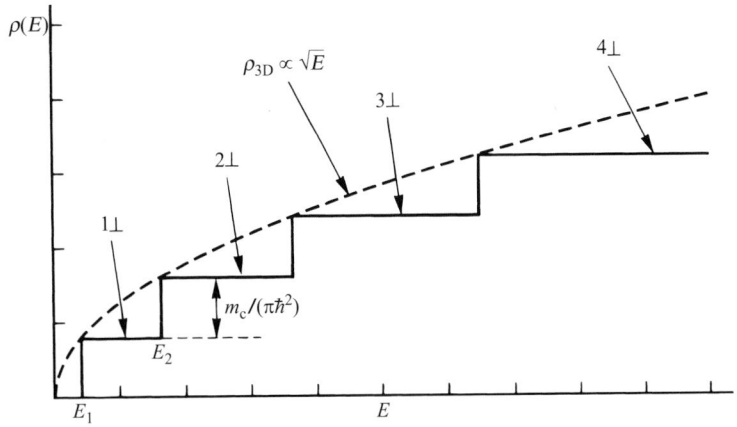


**Fig. 12.34** The density of states in an SQW compared with the full three-dimensional density-of-states function: $E_1$ is the minimum electron energy for one mode perpendicular to the layers, $E_2$ is that for two modes, and so on. The label 2⊥ on the curve, for example, indicates that there are two permitted modes in the perpendicular direction.

were just discussed. The consequence of this is a step-like density-of-states function, as shown in Fig. 12.34. Each step corresponds to an increase in the quantum-well density of states $m_e/(\pi\hbar^2)$.

Quantum-well lasers have lower thresholds and higher gains and generate narrower-linewidth laser radiation than do more conventional heterostructure lasers. An SQW laser generates the narrowest linewidth and has the lowest threshold current, but at current densities above threshold the effective gain of an MQW laser is higher. This is a consequence of the ratio $\Gamma$ between the width of the active region and the width of the oscillating intracavity mode. To provide modal confinement, the region on each side of the quantum well is frequently a region of graded index, as shown in Fig. 12.35.

## 12.13 Semiconductor laser structures

There are three principal heterostructure arrangements used in edge-emitting semiconductor lasers. These are shown schematically in Fig. 12.36. The simple design shown in Fig. 12.36(a) has largely been replaced with designs like those in Figs. 12.36(b) and 12.36(c). The separate confinement heterostructures (SCHs) shown in Figs. 12.36(b) and (c) incorporate additional layers on each side of the active single-quantum-well or multiple-quantum-well active region. The energy gap of the confinement layer is larger than the energy gap in the active region and consequently this layer has a lower refractive index. This provides better confinement of the laser mode than is the case in the simple double heterostructure shown in Fig. 12.36(a). In most semiconductor lasers the structure is fabricated so as to confine the current to a specific region within the device. Such *stripe-geometry* lasers can be fabricated in several ways, as shown schematically in Fig. 12.37.

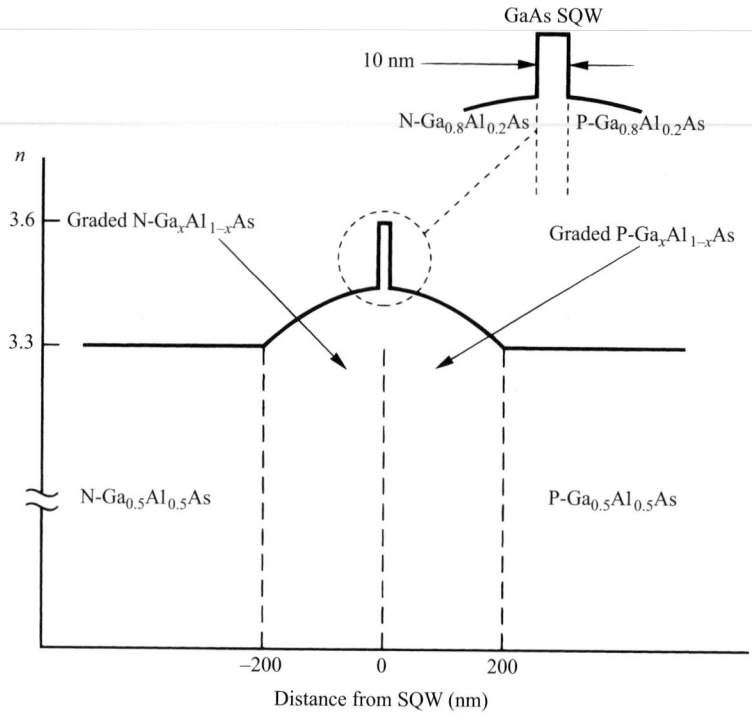


**Fig. 12.35** The schematic variation of the refractive-index change across a QW laser with a graded layer of $Ga_{0.5}Al_{0.5}As \rightarrow Ga_{0.8}Al_{0.2}As$ on each side of the QW (GaAs) region. The Ga concentration increases towards the QW region in the graded layers on either side [16].

If the region of current flow is defined by a conductive stripe on one of the electrodes, or by making the confining layer resistive by proton bombardment or oxygen implantation (which creates defects in the lattice) then the active region will be laterally restricted to the region where current flows as shown in Figs. 12.37(b), (c), and (f). If there is no additional modification of the active region in the lateral direction, then the laser is described as *gain-guided* as in Figs. 12.37(a) and (f). On the other hand, if the fabrication process involves surrounding an active region of GaAs by GaAlAs on all four sides as in Fig. 12.37, then the laser is described as *index-guided*. Index-guided structures provide the greatest control of the emitted transverse mode, at the expense of fabrication complexity. There are many other structures that incorporate the essential fabrication schemes shown in Fig. 12.37. An important semiconductor laser design is the graded-index separate-confinement heterostructure (GRINSCH), one example of which is shown in Fig. 12.38. The active layer in this laser is very thin – a single quantum well (SQW)[19] with graded layers of GaAlAs on each side to confine the mode optically. The active region could also be a multiple quantum well of, for example, four 0.013-μm-thick layers of $Ga_{0.94}Al_{0.06}As$ separated

[19] See Section 12.12.

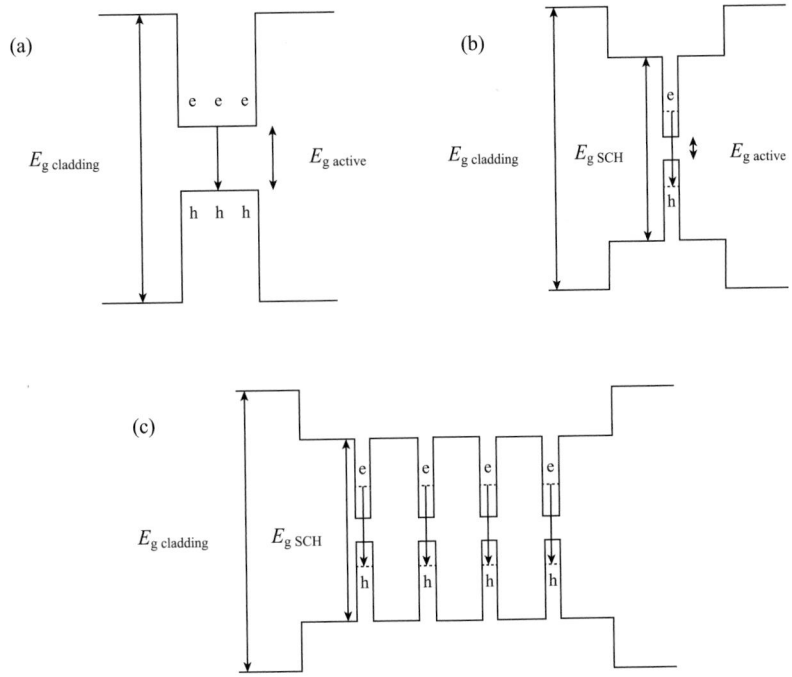


**Fig. 12.36** The three principal heterostructure arrangements used in edge-emitting semiconductor lasers.

by three 0.009-μm-thick layers of $Ga_{0.8}Al_{0.2}As$ [17]. For additional technical details the reader should consult the specialized literature [18–24].

### 12.13.1 Distributed-feedback and distributed-Bragg-reflection lasers

Most edge-emitting semiconductor lasers that we have discussed so far have a tendency to oscillate simultaneously on several cavity modes. Generally, special features must be incorporated into the laser structure to encourage single-mode operation. This can be done by including a periodic structure or *grating* in the device near the active layer where the oscillating transverse mode will interact with it. Figure 12.39 shows a planar waveguiding structure that includes a surface corrugation of periodicity $d$. Light that satisfies the Bragg condition will be reflected efficiently back on itself. This condition is set by waves scattering off successive corrugations reflecting and constructively interfering. If the index of the guiding layer is $n$, the Bragg condition can be written as

$$2d \sin\theta = m\lambda_0/n. \tag{12.94}$$

The angle $\theta$ is generally close to 90°; its effect can be included in an *effective* refractive index, $n_{\text{eff}}$, for the guided wave so that the grating period $d$ required for operation at wavelength $\lambda_0$ is

$$d = \frac{m\lambda_0}{2n_{\text{eff}}}. \tag{12.95}$$

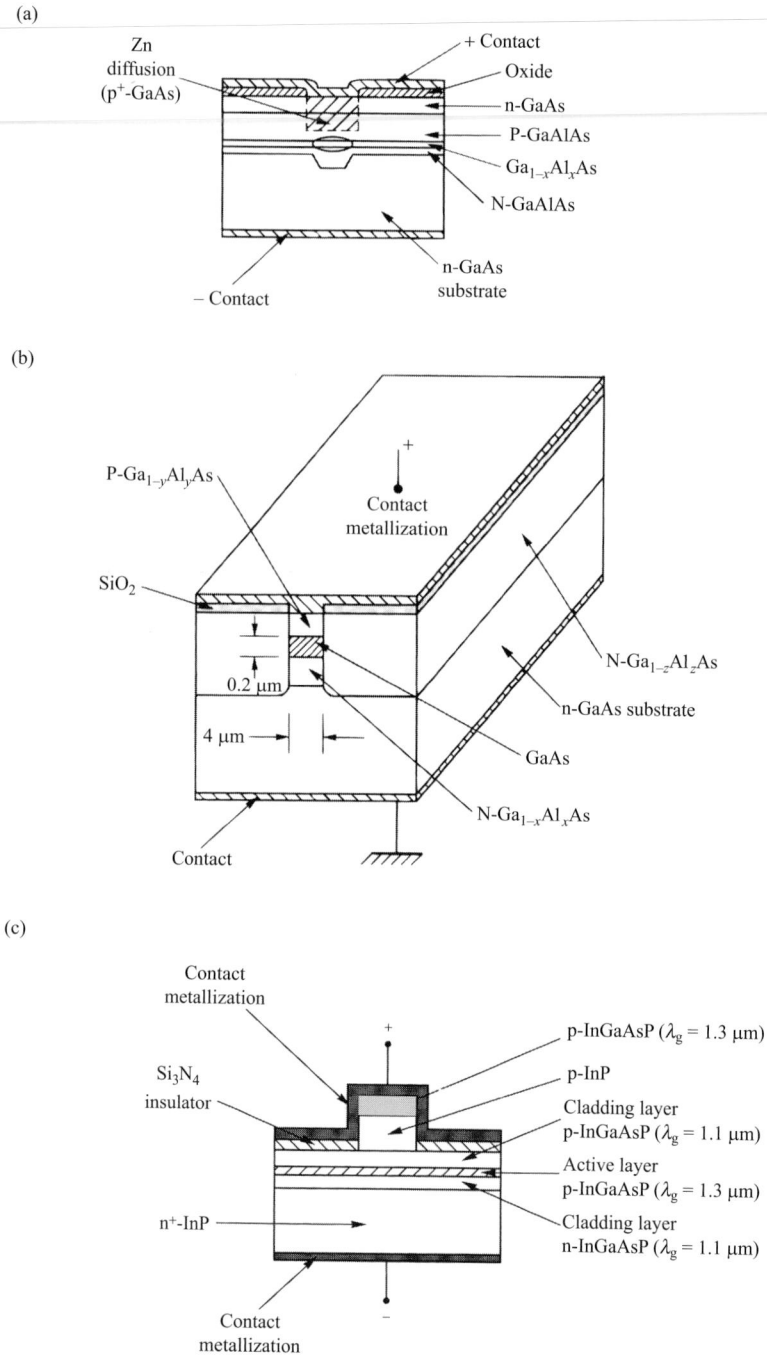


**Fig. 12.37** The schematic construction of various types of stripe-geometry, double-heterostructure lasers: (a) stripe confinement by oxide layer and deep zinc diffusion; (b) a buried stripe; (c) a cladded ridge [21]; (d) an etched mesa; (e) a constricted double heterostructure (CDH) (will give single-transverse-mode operation) [22]; and (f) stripe definition by oxygen implantation or proton bombardment.

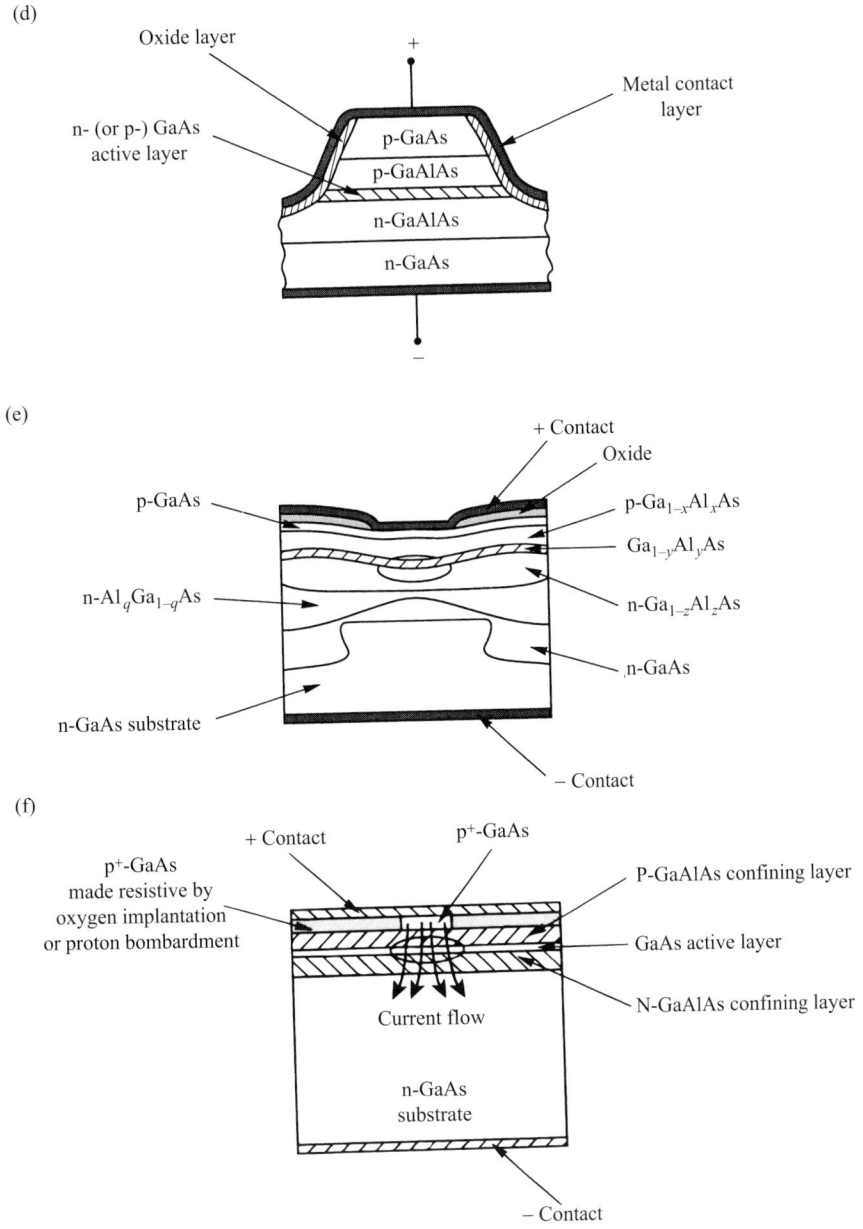


Fig. 12.37 (*cont.*)

If the corrugations are symmetrical, either sinusoidal or rectangular, then the even-order diffractions, for example $m = 2$, are suppressed.

A laser in which a periodic structure is placed in close proximity to the active region over a substantial length of the gain region is called a *distributed-feedback* (DFB) laser. It oscillates at a wavelength determined by the Bragg reflection condition given in Eq. (12.94)

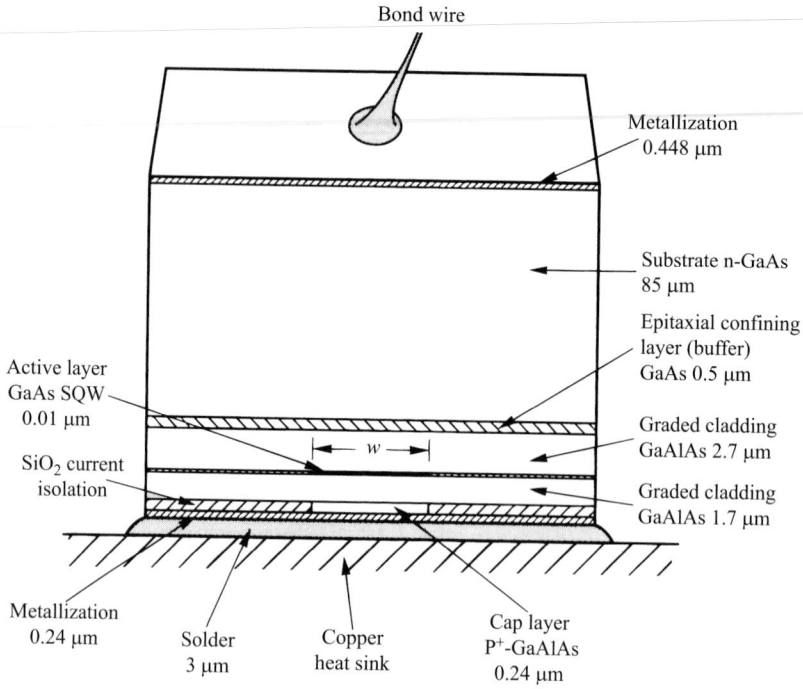


**Fig. 12.38** The GRINSCH SQW design used by McDonnell–Douglas Electronics Systems Co. Typical thicknesses of the various layers are shown. For a narrow-stripe laser $w \sim 5$ μm; for a broad-stripe laser $w \sim 60$ μm.

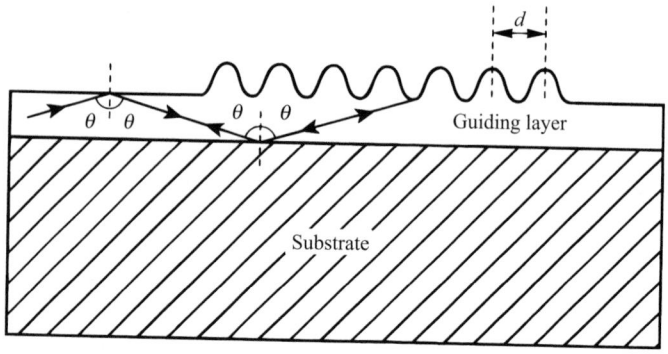


**Fig. 12.39** Bragg reflection in a periodic waveguide structure.

above. Figure 12.40(a) shows how such a DFB laser is fabricated. The corrugations are separated from the active region and are formed between an $Al_{0.15}Ga_{0.85}As$ layer and an $Al_{0.05}Ga_{0.95}As$ layer. The thin p-type $Al_{0.15}Ga_{0.85}As$ layer and an n-type $Al_{0.3}Ga_{0.7}As$ layer confine the optical field to the active region. However, sufficient field amplitude leaks through the p-type $Al_{0.15}Ga_{0.85}As$ to allow interaction with the corrugation.

The corrugations on the confining layer are produced by photolithography. The periodic structure can be produced by holographic techniques in which a periodic interference

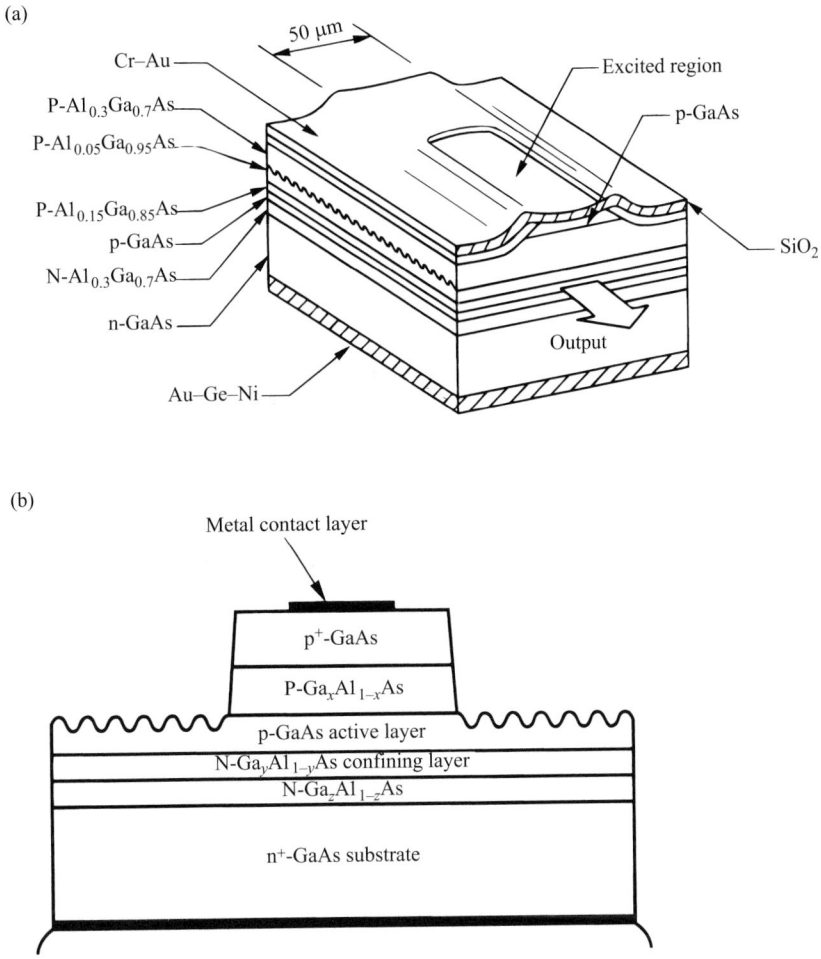


**Fig. 12.40** A GaAlAs double-heterojunction DFB laser structure [25].

pattern is projected onto the surface to be etched. If a laser beam of wavelength $\lambda$ is split into two parts and then the two beams are made to reinterfere at an angle of $2\theta$ to each other, with each at an angle of $\theta$ to the surface normal, a grating of period $d = \lambda/(2\sin\theta)$ results. Alternatively, a phase mask is used to project the desired pattern onto the structure during fabrication. In general, a layer of photoresist is periodically exposed in preparation for etching. For a laser incorporating a periodic structure a detailed analysis can provide specific information about the threshold gain, and will predict the precise resonance frequencies at which laser oscillation will occur [4, 23, 24].

Selective cavity reflectance can also be provided in a semiconductor laser structure in which a grating is used at each end of the gain region to provide feedback at a wavelength that will satisfy the Bragg reflection condition. Such lasers are called *distributed-Bragg-reflection* (DBR) lasers. Figure 12.40(b) shows an example of such a laser. In the figure the direction of light propagation is in the plane of the figure.

A grating structure, which can be fabricated at one or both ends of the cavity, provides selective cavity reflection to keep the left and right traveling waves in phase. DBR lasers can provide stable, single-frequency operation and tunability. The gain and Bragg-reflection regions are separate, which allows independent control of the carrier density in the grating region. The active region may be quite short, in which case $m = 1$ may correspond to a wavelength that is too long, and $m = 3$ Bragg reflection is used. Changes in carrier density in this region change the effective index of the cavity mode and tune the laser frequency. DBR can also provide a narrower oscillating linewidth (<1 nm) than is obtained with conventional double-heterojunction lasers. An additional advantage of both DFB and DBR lasers is that their lasing frequency is less dependent on temperature than is that of a Fabry–Perot cavity semiconductor laser. The emission from the latter follows the temperature dependence of the energy gap, whereas the emission from the DFB and DBR structures follows the smaller temperature dependence of the refractive index and grating spacing.

## 12.14 Surface-emitting lasers

In a surface-emitting laser the laser beam is emitted perpendicular to the wafer. These new structures can be fabricated by conventional planar integrated-circuit techniques and can provide emission over a broad area. In contrast to the edge-emitting devices we have considered so far, they also can be scaled down to *very* small size. Literally, many of these miniature lasers can be fabricated in an area smaller than the head of a pin.

One version of a surface-emitting laser is shown in Fig. 12.41. In this *grating-coupled* surface-emitting laser a periodic structure is etched onto the surface of the confining layer on one side of the junction. If the grating period, $d$, is chosen appropriately, then a diffraction maximum exists in the vertical direction for waves traveling in the active layer. This results if

$$m\lambda = d. \tag{12.96}$$

In principle, a laser of this kind can emit over a large surface area, and provide a beam of high power and low beam divergence. Pulsed powers in excess of 4 W have been achieved [26] and high CW powers [27].

However, the most technologically important surface-emitting lasers are *vertical-cavity surface-emitting lasers* (VCSELs). These lasers are schematically of the form shown in Fig. 12.42. The active layer is bounded on each side in the vertical direction with alternate high- and low-index layers, each of which is designed to be a quarter wavelength thick at the laser wavelength. Each of these stacks of $\lambda/4$ layers acts as a narrow-band, high-reflectivity mirror, a *distributed Bragg reflector* (DBR), thereby forming a vertical cavity. For operation in the 650–1300-nm region these lasers are fabricated using heterostructures based on GaAs and $Ga_xAl_{1-x}As$ with an active region using multiple quantum wells of InGaAs separated by barriers of GaAs with confinement layers of GaAlAs. VCSELs for operation in the important 1,550-nm region for use in fiber-optic communication systems

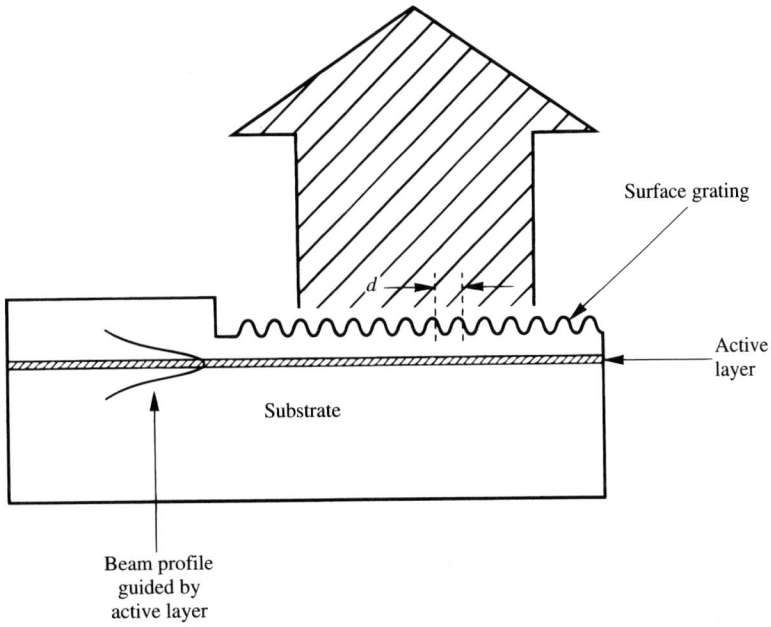


Fig. 12.41 A grating-coupled surface-emitting laser.

use InAlGaAs–InP/AlGaAs–GaAs heterostructures with AlGaAs/GaAs DBRs. A modern structure is shown in Fig. 12.43 [28]. There are several advantages to VCSEL laser structures.

(a) A very large number of small lasers can be fabricated to occupy a given area. A collaboration between AT&T Bell Laboratories and Bellcore has produced VCSEL microlasers at densities of over 20,000 per square millimeter [29]. These "microlasers" are only about 5.5 μm high and have diameters in the region of 1–5 μm. The active layer is only 24 nm thick. The threshold currents for lasing are only about 1 mA.
(b) Because these lasers are very short their cavity mode spacing is very large, ~180 nm for a 5.5-μm-long laser. This ensures that they operate in a single longitudinal mode.
(c) The small diameter and cylindrical shape of these lasers provide a circular output beam, provided that higher-order modes are suppressed by fabrication techniques [30].
(d) Large-area VCSELs can be fabricated to give milliwatt output powers [28].
(e) An array of VCSELs can be fabricated to provide large output powers, or to provide two-dimensional optical processing and connectivity capability.

Closely related to VCSELs are the vertical-external-cavity surface-emitting lasers (VECSELs). These are vertically integrated lasers in which only one cavity reflector is incorporated into the laser structure, and the second cavity reflector is external to the semiconductor structure. VECSELs can be optically pumped with another semiconductor laser and can provide a single-transverse-mode output at high power levels [31]. Figure 12.44 shows the schematic construction of a VECSEL [12, 32].

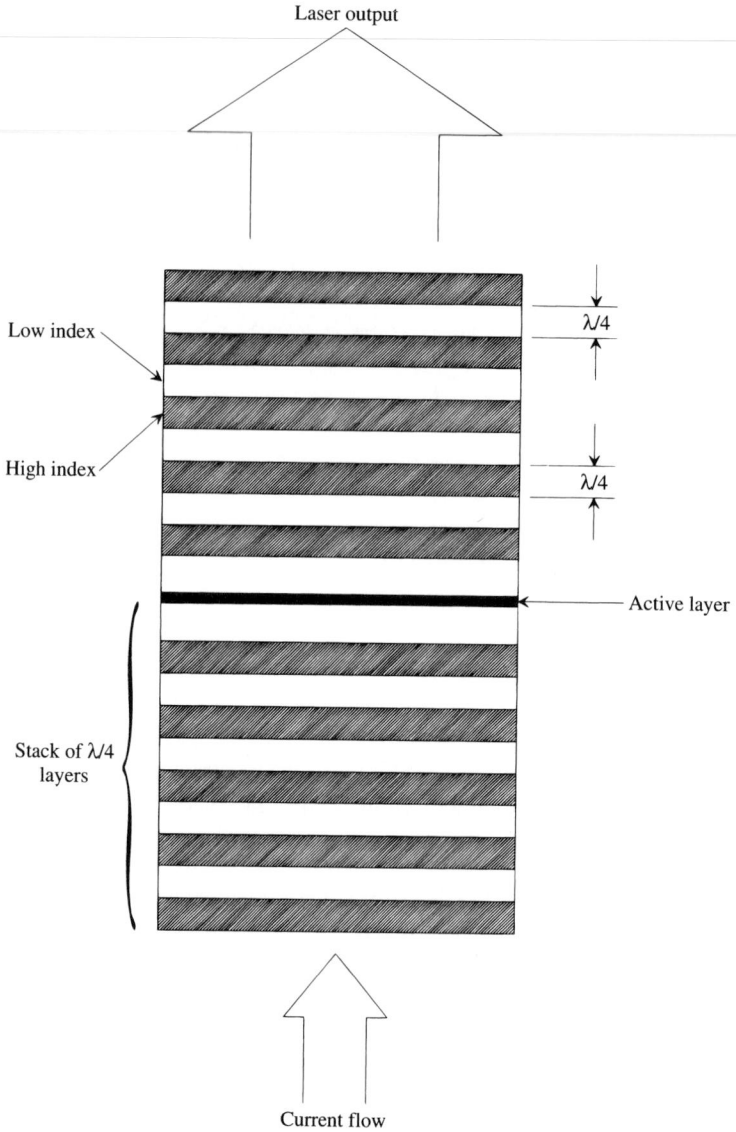


**Fig. 12.42** The schematic structure of a vertical-cavity surface-emitting laser.

## 12.15 Laser diode arrays and broad-area lasers

There is a limit to the maximum optical power that can be extracted from an edge-emitting, stripe-geometry laser. A primary limit is set by the intracavity power density, which at high enough levels will lead to optical damage to the laser facets. An oscillating laser mode that is 4 μm wide by 500 nm high with an output power of 100 mW will be close to the optical

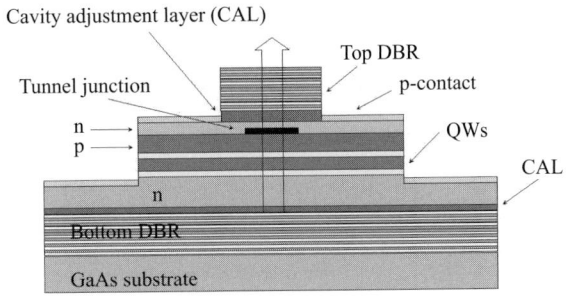


**Fig. 12.43** The design of a 1,550-nm VCSEL using InAlGaAs–InP/AlGaAs–GaAs heterostructures with AlGaAs/GaAs DBRs.

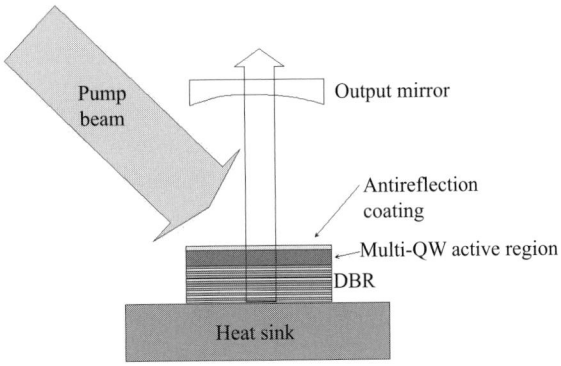


**Fig. 12.44** A vertical-external-cavity surface-emitting laser (VECSEL).

damage threshold, since its intracavity peak power density is on the order of 1 MW mm$^{-2}$. A successful approach that has been employed to produce higher output power than this has involved the use of laser diode *arrays*. In these devices, many stripe lasers are fabricated in close proximity to each other, as shown in Fig. 12.45.

The current is confined to a series of stripe regions by producing regions of semi-insulating semiconductor through proton implantation, which produces defects in the semiconductor lattice. Alternatively, stripe metal contacts on the top surface can be used to delineate the regions of current flow. Typically, ten stripe lasers would be fabricated with 4-μm-wide active regions on 10-μm-wide centers. The proximity of the individual stripes to each other causes overlap of adjacent oscillating intracavity field distributions. A phase-locking of the fields in the separate stripes occurs, leading to a combined, coherent wave front being emitted from the whole facet. The optical quality of this wave front is not generally high, since it has several maxima and minima – very much like the diffraction pattern that would result from a series of regularly spaced small apertures. The total power output from an array of ten elements can be as high as 1 W. Arrays of up to 40 elements that produce 2.5 W have been fabricated. If a larger number of stripe lasing elements is included in a single line then parasitic laser oscillation in the transverse direction can occur. Even higher powers can be produced by two-dimensional arrays. For example a CW output

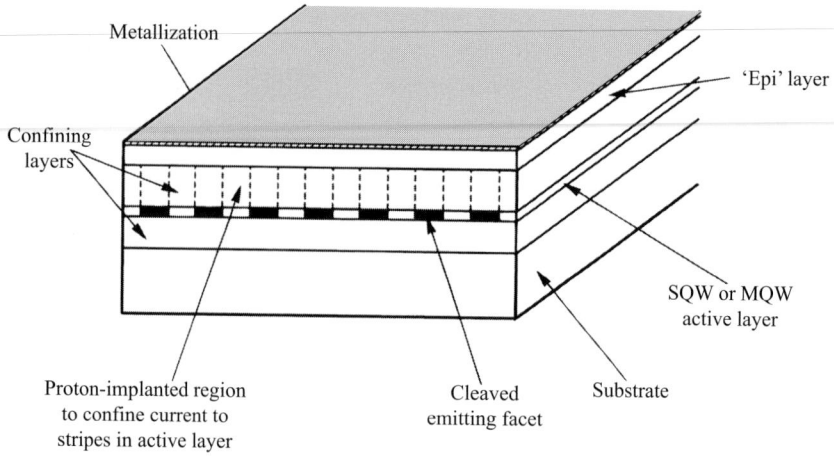


Fig. 12.45 An edge-emitting laser diode array.

of 38 W has been produced from 20 ten-stripe phased arrays spaced on 500-μm-wide centers.

A broad-area laser (BAL) is an alternative approach to generating a high output power. In these devices the active region is made large, but is not divided into stripes. For example, a 100-μm-wide multiple-quantum-well structure has produced a CW power of more than 10 W at a current of 20 A [33]. The most important application of these high-power arrays and BALs is in the pumping of solid-state lasers, which have already been described in Chapter 7.

## 12.16 Blue–green lasers

To build a blue semiconductor laser, one requires an active region with a large band gap. For operation in the wavelength range 360–480 nm the band gap required varies from 2.58 to 3.44 eV. GaN when grown in its wurtzite crystal form has a band gap of 3.4 eV, which varies with temperature. AlN in its wurtzite crystal form has a bandgap of 6.2 eV. GaN and AlN have almost the same lattice constant, 0.319 and 0.311 nm, respectively. $Ga_{1-x}Al_xN$ is a direct-band-gap material, so tunable stoichiometries can be used to build heterostructures. With the addition of indium, (almost) lattice-matched structures using InGaAlN can be fabricated. The biggest obstacles to using these materials to build practical LEDs and lasers were the lack of good lattice-matched substrates for growth and the difficulties of producing low-defect-density p- and n-doped materials. In 1991 Shuji Nakamura of Nichia Chemical Industries Ltd., Tokushima, Japan, developed the first practical blue LED, using silicon to achieve n-doping and magnesium to achieve p-doping [34]. These devices were made possible by the growth of GaN on a sapphire substrate. These early devices had too many defects to make a laser, but, after improvements in

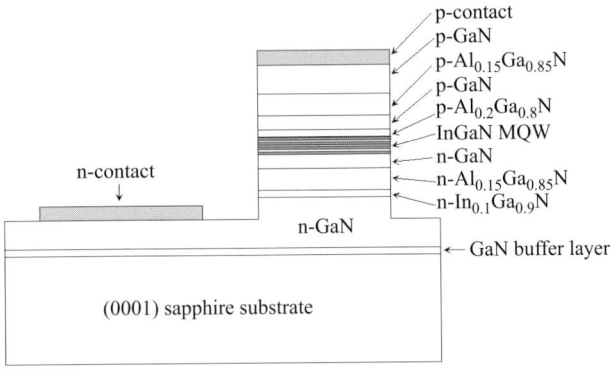


**Fig. 12.46** The structure of the first blue semiconductor laser based on InGaAsP.

growth techniques, in 1996 Nakamura and his coworkers produced the first GaN-based blue laser [35].[20] The structure of the first blue semiconductor laser based on InGaAsP is shown in Fig. 12.46. This blue laser technology has matured since this first report and devices providing hundreds of mW at bias currents of hundreds of mA operating at room temperature are available. Blue–green semiconductor lasers based on the II-VI material ZnSe [36–38] have also been fabricated, but to date the lifetime of these lasers has been relatively short, on the order of 500 h, and they operate best at cryogenic temperatures. BluRay and HD-DVD technologies offer higher storage densities than earlier technologies because they use blue lasers. Green semiconductor lasers have proved more difficult to fabricate than blue semiconductors lasers because the material with a band gap in the green is InGaP, which is more difficult to dope and grow in defect-free form [39]. To date, most green lasers use frequency doubling of diode-pumped Nd lasers at 1.06 μm (see Chapter 7).

## 12.17 Quantum cascade lasers

All the semiconductor lasers discussed so far have involved electron–hole transitions from the conduction band to the valence band in a heterostructure involving p–n junctions. Quantum cascade lasers (QCLs), on the other hand, use multiple stacked heterostructures where all the layers are either undoped or n-doped, perhaps with different doping densities, but p–n junctions are not involved. In a conventional QCL the laser transitions occur between energy levels that are all in the conduction band: these are *inter-sub-band* lasers. These lasers were first proposed by Kazarinov and Suris [40], and the first practical QCL was demonstrated by Faist *et al.* [41]. If the layered heterostructure is sufficiently forward-biased, typically with a field $\sim 10\,\mathrm{V\,m^{-1}}$, then the conduction-band diagram of the alternating layers in one active region becomes a staircase-like structure as shown in

[20] For his pioneering work in developing the necessary growth techniques to allow the fabrication of blue LEDs and semiconductor lasers, Shuji Nakamura was awarded the 2006 Millennium Technology Prize.

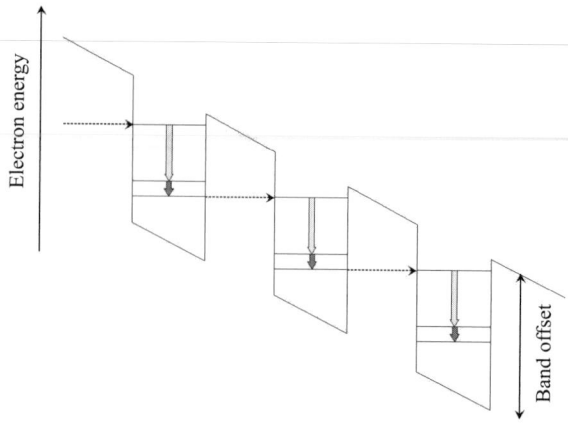


Fig. 12.47 The conduction-band diagram of the alternating layers in one active region of a quantum cascade laser becomes a staircase-like structure.

Fig. 12.47. The band offset is the difference in energy for electrons in the conduction band between the bottom of the wells and the tops of the intervening barriers. In this scheme the position of the energy states in the quantum wells can be tailored by adjusting their thicknesses. So, in this example, electrons are injected into the level with $n = 3$, and the laser transition is to level $n = 2$. Levels 2 and 3 relax quickly and electrons then tunnel into the next well to populate the $n = 3$ level. In the first QCL the quantum wells were undoped $Ga_{0.47}In_{0.53}As$ wells separated by undoped $Al_{0.48}In_{0.52}As$ barriers. These material stoichiometries are lattice-matched to InP, so this can serve as the substrate on which the QW layers are grown. The term cascade laser is used because the output of such a stage serves as the input to the succeeding stage. Typically QCLs have about 30–50 stages and operate within the 4–12-μm range. Electrons are injected into these undoped alternating layers from the adjacent n-doped region. The injector regions are designed as a superlattice with a low density of states at the same energy as the $n = 3$ level in the active region (a minigap) and a high density of states (a miniband) in the range of energies corresponding to levels $n = 2$ and $n = 1$ to facilitate extraction of electrons from the lower laser level. The active region of a QCL is usually four or five quantum wells thick, the first well next to the injector structure is very narrow, and there are typically 20–35 or so active regions separated by injection regions. Individual wells are only a few nanometers thick. Figure 12.48 shows one stage of a multiple-stage QCL with adjacent injection and electron-extraction regions [42]. In this laser the laser transition is from level $n = 4$ to level $n = 3$ in the QWs. The transitions $3 \rightarrow 2$ and $2 \rightarrow 1$ correspond to the energy of one phonon in the structure, so electrons relax quickly from these levels and a high level of population inversion is achieved. There has been substantial interest in the use of quantum cascade lasers to generate far-infrared radiation in the terahertz region [43–45]. There are also QCLs that use multiple-quantum-well heterostructures in which the laser transitions are between QW states in the conduction and valence bands. Thus they are known as interband cascade lasers. They were first developed by Yang [46]. They can achieve tunable operation at shorter wavelength than can conventional QCLs.

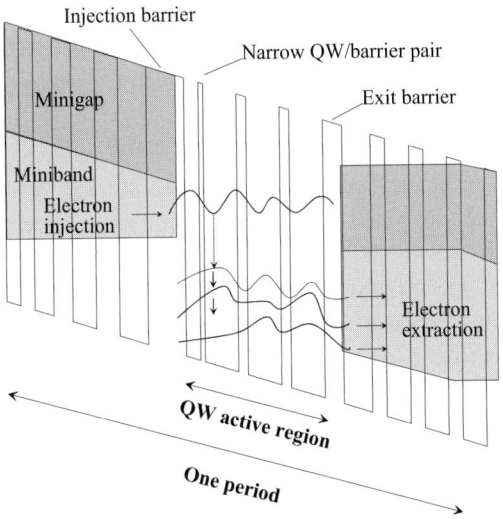


Fig. 12.48 The energy-level diagram of one stage of a multiple-stage QCL with adjacent injection and electron-extraction regions. Each stage (injector plus active region) is typically tens of nanometers thick. The probability distributions for electrons and the energy levels are calculated from Schrödinger's equation. In the work of Capasso and coworkers [41] the energy well and barriers are made of AlInAs and GaInAs alloys, respectively, lattice-matched to InP.

## 12.18 Silicon lasers

We have concentrated in this chapter on semiconductor lasers that are driven by electric currents passing through a semiconductor structure. Laser operation in semiconductors produced by optical pumping or electron-beam excitation [47] is also currently of interest, especially as approaches to developing semiconductor lasers based on silicon. Silicon semiconductor devices dominate the industry except as optical emitters. This is because of their indirect band gap. Consequently, the integration of current optical devices with conventional silicon electronics involves the fabrication of hybrid structures. There is continuing interest in developing optical emitters based on silicon that can more easily be integrated with conventional silicon electronics, and progress towards this goal is being made [48]. Because of its indirect band gap, like that of germanium, radiative emission in silicon is inefficient. Electrons in the conduction band reside largely in the X valley, which is not aligned with the hole valley in the valence band. In radiative emission a third particle must carry away the excess momentum. By using nanostructure silicon materials such as nano-particles or porous silicon increased emission efficiency has been achieved. To date optically pumped laser emission based on stimulated Raman scattering has been observed [49]. Improvements in the growth of other materials on silicon have also allowed the demonstration of an optically pumped Ge-on-Si laser [50]. There is certain to be continuing progress in the development of integrated silicon photonic systems.

## 12.19 Modulation of semiconductor lasers

One of the most important applications of semiconductor lasers is as sources in optical fiber communication systems (see Chapter 23). In this context the laser output must be amplitude- or phase-modulated in some way to encode the information to be transmitted. It is also desirable for the laser to produce a circularly symmetric output or the efficiency of the injection of the laser beam into a cylindrical fiber will be reduced. An in-depth discussion of fiber-optic communication systems is beyond our scope here. The interested reader should consult the specialized literature [51–55]. A few points are worth noting. The output of a semiconductor laser can be intensity-modulated by modulating the drive current. For small-sized laser structures with low capacitance this can be carried out to very high data rates, at least to 35 Gb $s^{-1}$ [56]. When a laser is intensity-modulated by changing the drive current there are, however, other consequences. Changing the drive current changes the electron density in the cavity, and the refractive index of the active region depends on the electron density. This causes the laser frequencey to change (*chirp*) during the modulation. These chirp effects must be taken into account in optical communication sytems using directly modulated lasers. Controlled amplitude and phase modulation is best achieved using an external modulator such as a Mach–Zehnder electro-optic modulator or a separate semiconductor modulating structure such as an electro-absorption modulator [57] (see Chapter 18).

## 12.20 Problems

12.1 Prove that the maximum fraction of the emitted radiation from the active layer in Fig. 12.17 that might escape from the semiconductor/air interface if the reflectance $R$ at the outer semiconductor/air surface is assumed to be independent of angle is $f = \frac{1}{2}n_s^2$, where $n_s$ is the refractive index of the semiconductor.

12.2 An edge-emitting LED with a center wavelength of 810 nm has an active region 100 μm wide by 1 μm thick. Estimate the distance from the output facet at which the output beam will have become circular.

12.3 The recombination time in a GaAs/GaAlAs laser is ~4 ns, the cavity length between end facets of the diode is 500 μm, the reflectance at each end is $R = 0.31$, the distributed loss is 1,000 $m^{-1}$, and the oscillating beam width is 4 μm. If all the electrons injected into the junction region combine there, calculate the threshold current density for oscillation. Take the gain ($m^{-1}$) of the laser to be $N/10^{23}$, where $N$ is the carrier density ($m^{-3}$) in the junction region.

12.4 A GaAs laser has a cavity length of 0.3 mm and a refractive index of 4. Its threshold gain coefficient is 6 $mm^{-1}$. If the small-signal gain is increased to 12 $mm^{-1}$, how many longitudinal modes will oscillate if the gain curve has a Gaussian frequency variation with a full width at half maximum height of 0.1 μm at 1.3 μm?

12.5 Calculate the wavelength of peak intensity for the recombination spectrum given by Eq. (12.45). Plot the variation of the wavelength of peak emission for GaAs from 0 to 500 K.

12.6 The transverse laser mode in the active region of a semiconductor laser is wider than the active region so the effective gain is reduced by a confinement factor $\Gamma$. At the same time the real $n'$ and imaginary $K$ parts of the refractive index of the active region depend on the current. Henry [58] intoduced an $\alpha_h$ factor, which relates the changes in the real and imaginary parts of the refractive index $n = n' - jK$, according to $\alpha_h = n'/K$. If the real part of the refractive index changes by $\Delta n'$, then the change in material loss in the active region is

$$\Delta\alpha_m = 2\frac{\omega}{c_0}\frac{\Delta n'}{\alpha_h}.$$

The effective change in intracavity loss is reduced by the confinement factor to $\Delta\alpha = \Gamma\alpha_m$. Derive expressions for the minimum cavity reflectances (assumed the same at each end of the cavity) needed for laser oscillation, and the change in laser oscillation wavelength resulting from current injection.

## Further reading

W. W. Chow and S. W. Koch, *Semiconductor-Laser Fundamentals: Physics of the Gain Materials*, New York: Springer, 2010.

E. Kapon, *Semiconductor Lasers I: Fundamentals (Optics and Photonics) (Pt. 1)*, San Diego, CA: Academic Press, 1999.

E. Kapon, *Semiconductor Lasers II: Materials and Structures (Optics and Photonics)* San Diego CA: Academic Press, 1999.

T. Numai, *Fundamentals of Semiconductor Lasers*, New York: Springer, 2010.

S. Nakamura, S. Pearton, and G. Fasol, *The Blue Laser Diode: The Complete Story*, New York: Springer, 2010.

S. F. Yu, *Analysis and Design of Vertical Cavity Surface Emitting Lasers*, Hoboken, NJ: Wiley, 2003.

## References

[1] J. Gowar, *Optical Communication Systems*, 2nd edn., London: Prentice Hall, 1984.

[2] C. Kittel, *Introduction to Solid State Physics*, 8th edn., New York: Wiley, 2004.

[3] F. Reif, *Fundamentals of Statistical and Thermal Physics*, New York: McGraw Hill, 1965.

[4] A. Yariv, *Introduction to Optical Electronics*, 1st edn., New York: Holt, Rinehart and Winston Inc., 1971.

[5] M. I. Nathan, "Semiconductor lasers," *Appl. Opt.* **5**, 1514–1528, 1966.

[6] H. C. Casey, Jr., and M. B. Panish, *Heterostructure Lasers*, Parts A and B, New York: Academic Press, 1978.

[7] M. B. Panish, "Heterostructure injection lasers," *Proc. IEEE* **64**, 1512–1540, 1976.

[8] C. A. Burrus and B. I. Miller, "Small-area, double-heterostructure, aluminum–gallium arsenide electroluminescent diode source for optical-fiber transmission lines," *Opt. Commun.* **4**, 307–309, 1971.

[9] A. Y. Cho and J. R. Arthur, Jr., "Molecular beam epitaxy," *Prog. Solid State Chem.* **10**, 157–192, 1975.

[10] Y. Yamazoe, T. Nishino, Y. Hamakawa, and T. Kariya, "Bandgap energy of InGaAsP quaternary alloy," *Jap. J. Appl. Phys.* **19**, 1473–1479, 1980.

[11] A. Chandra and L. F. Eastman, "Rectification at n–n GaAs:(Ga, Al)As heterojunctions," *Electron. Lett.* **15**, 90–91, 1979.

[12] J. F. Womac and R. H. Rediker, "The graded-gap $Al_xGa_{1-x}As$–GaAs heterojunction," *J. Appl. Phys.* **43**, 4129–4133, 1972.

[13] M. G. Bernard and G. Duraffourg, "Laser conditions in semiconductors," *Phys. Stat. Solidi* **1**, 699–703, 1961.

[14] I. F. Wu, I. Riant, J.-M. Verdiell, and M. Dagenais, "Real-time in situ monitoring of antireflection coatings for semiconductor laser amplifiers by ellipsometry," *IEEE Photonics Technol. Lett.* **4**, 991–993, 1992.

[15] R. L. Hartmann and R. W. Dixon, "Reliability of DH GaAs lasers at elevated temperatures," *Appl. Phys. Lett.* **26**, 239–240, 1975.

[16] A. Yariv, *Quantum Electronics*, 3rd edn., New York: Wiley, 1989.

[17] D. R. Scifres, R. D. Burnham, and W. Streifer, "High power coupled multiple stripe quantum well injection lasers," *Appl. Phys. Lett.* **41**, 118–120, 1982.

[18] G. H. Olsen, C. J. Nuese, and M. Ettenberg, "Low-threshold 1.25 μm vapor-grown InGaAsP CW lasers," *Appl. Phys. Lett.* **34**, 262–264, 1979.

[19] H. Yonezu, I. Sakuma, K. Kobayashi, T. Kamejima, M. Unno, and Y. Nannichi, "A GaAs–$Al_xGa_{1-x}As$ double heterostructure planar stripe laser," *Jap. J. Appl. Phys.* **12**, 1585–1592, 1973.

[20] H. Kressel, M. Effenberg, J. P. Wittke, and I. Ladany, "Laser diodes and LEDs for optical fiber communication," in *Semiconductor Devices for Optical Communication*, ed. H. Kressel, New York: Springer-Verlag, 1982, pp. 9–62.

[21] I. P. Kaminow, L. W. Stulz, J. S. Ko, A. G. Dentai, R. E. Nahory, J. C. DeWinter, and R. L. Hartman, "Low threshold InGaAsP ridge waveguide lasers at 1.3 μm," *IEEE J. Quant. Electron.* **19**, 1312–1319, 1983.

[22] D. Botez, "CW high-pressure single-mode operation of constricted double-heterojunction AlGaAs lasers with a large optical cavity," *Appl. Phys. Lett.* **36**, 190–192, 1980.

[23] P. K. Cheo, *Fiber Optics and Optoelectronics*, 2nd edn., Englewood Cliffs, NJ: Prentice Hall, 1990.

[24] A. Yariv, *Optical Electronics*, 3rd edn., New York: Holt, Rinehart and Winston, 1985.

[25] K. Aiki, M. Nakamura, J. Umeda, A. Yariv, A. Katziv, and H. W. Yen, "GaAs–GaAlAs distributed feedback laser with separate optical and carrier confinement," *Appl. Phys. Lett.* **27**, 145–146, 1975.

[26] D. F. Welch, R. Parke, A. Hardy, R. Waarts, W. Striefer, and D. R. Scifres. "High-power, 4 W pulsed, grating-coupled surface emitting laser," *Electron. Lett.* **25**, 1038–1039, 1989.

[27] J. S. Mott and S. H. Macomber, "Two-dimensional surface emitting distributed feedback laser array," *IEEE Photon. Technol. Lett.* **1**, 202–204, 1989.

[28] A. Caliman, A. Mereuta, G. Suruceanu, V. Iakovlev, A. Sirbu, and E. Kapon, "8 mW fundamental mode output of wafer-fused VCSELs emitting in the 1550-nm band," *Opt. Express* **19**, 16996–17001, 2011.

[29] J. L. Jewel, "Microlasers," *Sci. Am.* Nov. 1991, 86–94.

[30] E. Söderberg, J. S. Gustavsson, P. Modh, A. Larsson, Z. Zhang, J. Berggren, and M. Hammar, "Suppression of higher order transverse and oxide modes in 1.3-μm InGaAs VCSELs by an inverted surface relief," *IEEE Photon. Technol. Lett.* **19**, 327–329, 2007.

[31] A. Laurain, M. Myara, G. Beaudoin, I. Sagnes, and A. Garnache, "Multiwatt-power highly coherent compact single-frequency tunable vertical-external-cavity surface-emitting semiconductor laser," *Opt. Express* **18**, 14627–14636, 2010.

[32] M. Kuznetsov, F. Hakimi, R. Sprague, and A. Mooradian, "High-power (>0.5-W CW) diode-pumped vertical-external-cavity surface-emitting semiconductor lasers with circular $TEM_{00}$ beams," *IEEE Photon. Technol. Lett.* **9**, 1063–1065, 1997.

[33] H. Wenzel, P. Crump, A. Pietrzak, X. Wang, G. Erbert, and G. Tränkle, "Theoretical and experimental investigations of the limits to the maximum output power of laser diodes," *New. J. Phys.* **12**, 085007, 2010.

[34] S. Nakamura, T. Mukai, and M. Senoh, "High-power GaN p–n-junction blue-light-emitting diodes," *Jap. J. Appl. Phys.* **30**, L1998–L2001, 1991.

[35] S. Nakamura, M. Senoh, S. Nagahama, N. Iwasa, T. Yamada, T. Matsushita, H. Kiyoku, and Y. Sugimoto, "InGaN-based multi-quantum-well-structure laser diodes," *Jap. J. Appl. Phys.* **35**, L74–L76, 1996.

[36] M. A. Haase, J. Qiu, J. M. DePuydt, and H. Cheng, "Blue–green laser diodes," *Appl. Phys. Lett.* **59**, 1272–1274, 1991.

[37] K. Katayama, H. Yao, F. Nakanishi, H. Doi, A. Saegusa, N. Okuda, T. Yamada, H. Matsubara, M. Irikura, T. Matsuoka, T. Takebe, S. Nishine, and T. Shirakawa, "Lasing characteristics of low threshold ZnSe-based blue/green laser diodes grown on conductive ZnSe substrates," *Appl. Phys. Lett.* **73**, 102–104, 1998.

[38] J.-P. Faurie and E. Tournié, "ZnSe-based heterostructures for blue–green lasers," *C. R. Acad. Sci. Paris* Série IV, **1**, 23–33, 2000.

[39] H. Ohta, S. P. DenBaars, and S. Nakamura, "Future of group-III nitride semiconductor green laser diodes," *J. Opt. Soc. Am.* **B27**, 45–49, 2010.

[40] R. F. Kazarinov and R. A. Suris, "Possibility of the amplification of electromagnetic waves in a semiconductor with a superlattice," *Sov. Phys. Semicond.* **5**, 707–709, 1971.

[41] J. Faist, F. Capasso, D. L. Sivco, C. Sirtori, A. L. Hutchinson, and A. Y. Cho, "Quantum cascade laser," *Science* **264**, 553–556, 1994.

[42] M. Beck, D. Hofstetter, T. Aellen, J. Faist, U. Oesterle, M. Ilegems, E. Gini, and H. Melchior, "Continuous wave operation of a mid-infrared semiconductor laser at room temperature," *Science* **295**, 301–305, 2002.

[43] S. Kumar, "Recent progress in terahertz quantum cascade lasers," *IEEE J. Selected Topics Quant. Electron.* **17**, 38–47, 2011.

[44] R. Kohler, A. Tredicucci, F. Beltram, H. E. Beere, E. H. Linfield, A. G. Davies, D. A. Ritchie, R. C. Iotti, and F. Rossi, "Terahertz semiconductor-heterostructure laser," *Nature* **417**, 156–159, 2002.

[45] A. Tredicucci, R. Köhler, L. Mahler, H. E. Beere, E. H. Linfield, and D. A Ritchie, "Terahertz quantum cascade lasers–first demonstration and novel concepts," *Semicond. Sci. Technol.* **20**, S222–S227, 2005.

[46] R. Q. Yang, "Infra-red laser based on intersubband transitions in quantum-wells," *Superlattices Microstruct.* **17**, 77–83, 1995.

[47] S. V. Gronin, S. V. Sorokin, I. V. Sedova, P. S. Kop'ev, S. V. Ivanov, M. M. Zverev, N. A. Gamov, D. V. Peregoudov, and V. B. Studionov, "Effective green semiconductor lasers with multiple CdSe/ZnSe QD active region for electron beam pumping," *Acta Phys. Polon. A* **114**, 1115–1122, 2008.

[48] D. Liang and J. E. Bowers, "Recent progress in lasers on silicon," *Nature Photonics* **5**, 511–517, 2010.

[49] H. S. Rong, R. Jones, A. S. Liu, O. Cohen, D. Hak, A. Fang, and M. Paniccia, "A continuous-wave Raman silicon laser," *Nature* **433**, 725–728, 2005,

[50] J. Liu, X. Sun, R. Camacho-Aguilera, L. C. Kimerling, and J. Michel, "Ge-on-Si laser operating at room temperature," *Opt. Lett.* **35**, 679–681, 2010.

[51] L. G. Kazovsky, S. Benedetto, and A. E. Willner, *Optical Fiber Communication Systems*, Norwood, MA: Artech House, 2012.

[52] G. P. Agrawal, *Fiber-Optic Communication Systems,* 4th edn., Hoboken, NJ: Wiley, 2010.

[53] G. Keiser, *Optical Fiber Communications*, 3rd edn., New York: McGraw-Hill, 1983.

[54] M. Bass and E. W. Van Stryland, *Fiber Optics Handbook*, New York: McGraw-Hill, 2002.

[55] J. C. Palais, *Fiber Optic Communications*, 5th edn., Upper Saddle River, NJ: Pearson/Prentice Hall, 2005.

[56] W. Hofmann, T. Grundl, M. Horn, P. Wolf, R. D. Nagel, E. Ronneberg, G. Bohm, D. Bimberg, and M. C. Amann, "1550-nm high-speed short-cavity VCSELs," *IEEE J. Selected. Topics Quant. Electron.* **17**, 1158–1166, 2011.

[57] M. Chaciński, U. Westergren, B. Stoltz, R. Driad, R. E. Makon, V. Hurm, and A. G. Steffan, "100 Gb/s ETDM transmitter module," *IEEE J. Selected Topics in Quantum Electronics* **16,** 1321–1327, 2010.

[58] C. H. Henry, "Theory of the linewidth of semiconductor laser," *IEEE J. Quant. Electron.* **18**, 259–264, 1982.

# 13 Passive optical systems

## 13.1 Introduction

Practical photonic systems can conveniently be divided into four distinct parts: (a) the optical source (or sources), (b) a passive optical structure, (c) an active optical structure, and (d) a detection system. A good example of such a system is the human eye, which is shown schematically in Fig. 13.1. In this system the source of light is the Sun, or artificial lighting that renders objects visible to the eye. The passive optical structure includes the cornea, the intraocular fluid, and other fixed structures in the eye. The active optical structure includes the deformable eye lens, whose shape is controlled by the ciliary muscles, and the iris, whose diameter is adjusted to control the amount of light entering the eye. The detector in this system is the retina, where photons are absorbed by special molecules, leading to chemical reactions that produce charges and subsequent electrical signals to the brain along the optic nerve.

We have already explored in some detail the fundamental physics, constructional details, and properties of lasers. We have also seen the connection between the amplifying medium and the optical structures that turn the laser amplifier into an oscillator. In this and the following chapters we will further examine the characteristics of various passive and active optical structures that are important in photonic systems. Initially we will discuss the characteristics of important passive optical components and systems. These will include optical materials, lenses, mirrors, prisms, diffraction gratings, interferometers, crystals, polarizers, and optical fibers. We will begin by discussing some very useful, and quite simple, techniques for analyzing such optical systems. Most of our attention will be directed towards optical systems that have axial symmetry. In our analysis we will be confronted with the problem of how the components of the system affect the passage of light. To describe the behavior of the system in detail we need to know how a wave passing through is modified in terms of intensity, phase shift, direction of propagation, and polarization state. Our discussion will be simplified by assuming that we are dealing with light that is monochromatic, or at least contains only a small range of wavelengths. This allows us to neglect the effects of *dispersion*, namely the variation of the refractive index with wavelength. Dispersion, defined quantitatively as $dn/d\lambda$, causes the properties of an optical system containing transmissive components, such as lenses and optical fibers, to vary with wavelength.

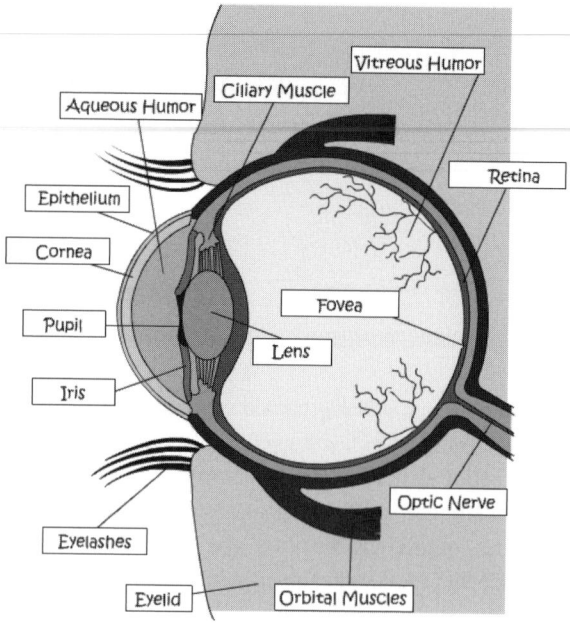


**Fig. 13.1** The structure of the human eye (from sciencewithme.com).

## 13.2 The propagation of rays and waves through isotropic media

An isotropic optical material is one whose properties are independent of the direction of propagation or polarization state of an electromagnetic wave passing through the material. Several simple relations hold for such a material.

(i) The polarization **P** induced by the electric field **E** of the wave is parallel to **E**, so

$$\mathbf{P} = \epsilon_0 \chi \mathbf{E}, \tag{13.1}$$

where $\chi$, the electric susceptibility, is a scalar.

(ii) It follows that the electric displacement vector **D** of the wave is parallel to **E**,

$$\mathbf{D} = \epsilon_0 \mathbf{E} + \mathbf{P}, \tag{13.2}$$

which, from Eq. (13.1), gives

$$\mathbf{D} = \epsilon_0 (1 + \chi) \mathbf{E} \tag{13.3}$$

or

$$\mathbf{D} = \epsilon_0 \epsilon_r \mathbf{E}, \tag{13.4}$$

where $\epsilon_r$ is called the dielectric constant of the material. For an absorbing (or amplifying) material $\epsilon_r$ is complex, $\epsilon_r = \epsilon' - j\epsilon''$. The refractive index of the material is $n = \sqrt{\epsilon_r}$, or in a lossy material $n = \sqrt{\epsilon'}$. The dielectric constant, $\epsilon_r$, and consequently

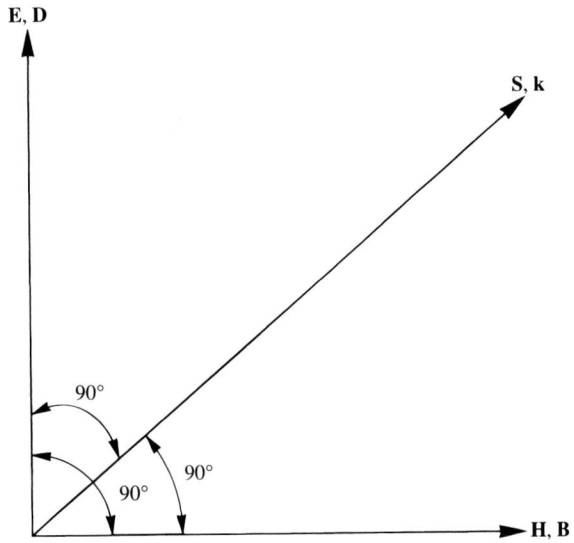


Fig. 13.2 Angular relationships among the vectors D, E, B, H, k, and S in an isotropic medium.

the refractive index, $n$, depend on the frequency of the wave. This phenomenon is called dispersion, and is the reason why a prism separates white light into its constituent colors.

(iii) The Poynting vector of the wave, $\mathbf{S} = \mathbf{E} \times \mathbf{H}$, which represents the magnitude and direction of energy flow, is parallel to $\mathbf{k}$, the wave vector. The wave vector $\mathbf{k}$ is perpendicular to the surfaces of constant phase (phase fronts), and is by definition perpendicular to the vectors $\mathbf{D}$ and $\mathbf{B}$ as shown in Fig. 13.2.

The time-averaged magnitude of $\mathbf{S}$ is called the intensity of the wave:

$$I = \langle |\mathbf{S}| \rangle_{\text{av}}. \tag{13.5}$$

The local direction of $\mathbf{S}$ is called the *ray* direction. Only for plane waves is the ray direction the same at all points on the phase front. Many properties of an optical system can be best understood by considering what happens to a ray of light entering the system. How does its direction change as it encounters mirrors and lenses, and what happens to the intensity of the wave to which the ray belongs as it crosses interfaces between different components of the system?

## 13.3 Simple reflection and refraction analysis

The phenomena of reflection and refraction are most easily understood in terms of plane electromagnetic waves – those sorts of waves where there is a unique direction of energy flow (the ray direction). Other types of wave, such as spherical waves and Gaussian beams,

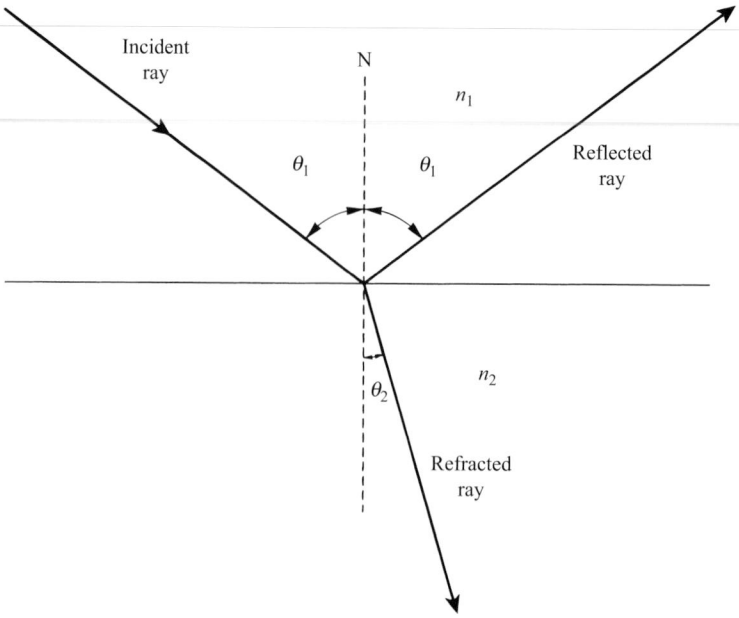


**Fig. 13.3** Incident-, reflected-, and refracted-ray directions at the planar boundary between two media.

are also important; however, the part of their wave front which strikes an optical component can frequently be approximated as a plane wave, so plane-wave considerations of reflection and refraction still hold true. The ray direction is the same as the direction of the Poynting vector $\mathbf{S} = \mathbf{E} \times \mathbf{H}$. In isotropic media (gases, liquids, glasses, and crystals of cubic symmetry), the ray direction and the direction of $\mathbf{k}$, the propagation vector, are the same. Except for plane waves, the direction of $\mathbf{k}$ varies from point to point on the phase front.

When light strikes a plane mirror, or the planar boundary between two media with different refractive indices, the *angle of incidence* is always equal to the *angle of reflection*, as shown in Fig. 13.3. The directions of the incident ray, the reflected ray, and the local surface normal lie in the same plane. This is the fundamental *law of reflection*.

When a light ray crosses the boundary between two media with different refractive indices, the *angle of refraction* $\theta_2$, shown in Fig. 13.3, is related to the angle of incidence $\theta_1$, by *Snell's law*

$$\frac{\sin\theta_1}{\sin\theta_2} = \frac{n_2}{n_1}. \tag{13.6}$$

This result does not hold true in general unless both media are isotropic. The directions of the incident ray, the refracted ray, and the local surface normal lie in the same plane.

Since $\sin\theta_2$ cannot be greater than unity for a *real* angle, if $n_2 < n_1$ the maximum angle of incidence for which there can be a refracted wave is called the *critical angle* $\theta_c$, where

$$\sin\theta_c = n_2/n_1, \tag{13.7}$$

as illustrated in Fig. 13.4(a).

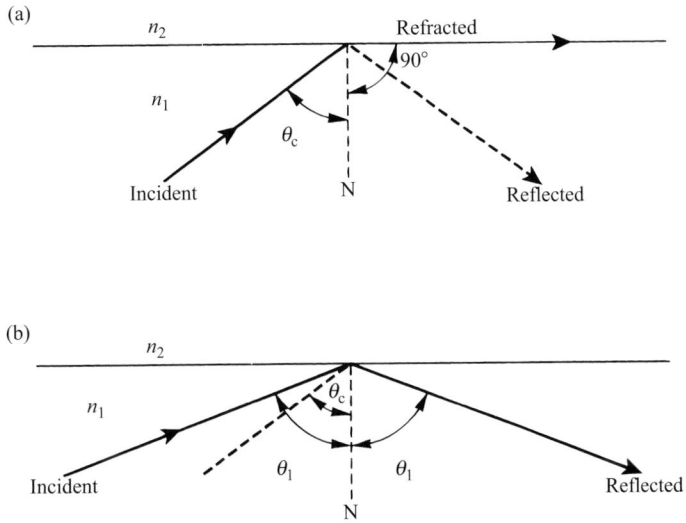


**Fig. 13.4** (a) A ray incident at the critical angle striking the boundary between a dense and a less-dense medium. (b) Total internal reflection at the boundary between a dense and a less-dense medium.

If $\theta_1$ exceeds $\theta_c$ the boundary acts as a very good mirror, as illustrated in Fig. 13.4(b). This phenomenon is called *total internal reflection* (TIR). Several types of reflecting prisms operate this way. When total internal reflection occurs, there is no transmission of energy through the boundary. However, the fields of the wave do not abruptly go to zero at the boundary. There is said to be an *evanescent wave* on the other side of the boundary, the field amplitudes of which decay exponentially with distance. Because of the existence of this evanescent wave, other optical components should not be brought too close to a totally reflecting surface, or energy will be coupled to them via the evanescent wave and the efficiency of total internal reflection will be reduced. With extreme care this effect can be used to produce a variable-reflectivity, totally internally reflecting surface, as shown in Fig. 13.5. If one or both of the media in Fig. 13.2 are made from an anisotropic material, such as calcite, crystalline quartz, ammonium dihydrogen phosphate (ADP), potassium dihydrogen phosphate (KDP), or β-barium borate (BBO) then in general the incident wave will split into two components, one of which obeys the simple form of Snell's law (the *ordinary wave*) and one of which does not (the *extraordinary wave*). This phenomenon is called *double refraction*, and is discussed in detail in Chapter 17.

If an optical system contains only planar interfaces, the path of a ray of light through the system can be easily calculated using only the law of reflection and Snell's law. This simple approach is called *geometrical optics*. It ignores diffraction effects, which can be neglected provided that the lateral dimensions (*apertures*) of the system are *all* much larger than the wavelength (say $10^4$ times larger). Analysis of the behavior of light rays in more complex systems containing non-planar components, but where diffraction effects are still negligible, can still be carried out by geometrical optics. This can become computationally difficult, but can often be simplified by *paraxial-ray* analysis [1]. Transmitted and reflected

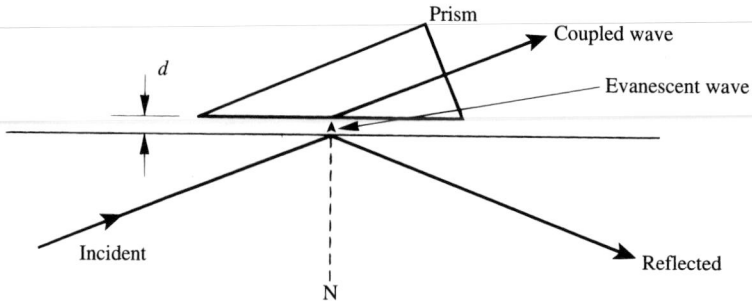


**Fig. 13.5** A schematic arrangement with which evanescent coupling could be observed. The spacing between the boundary and the coupling prism must be very small ($d$ on the order of the wavelength) for significant coupling to occur.

intensities and polarization states cannot be determined by the above methods, and are most easily determined by the *method of impedances*.

## 13.4 Paraxial-ray analysis

A plane wave is characterized by a unique propagation direction given by the wave vector **k**. All fields associated with the wave are, at a given time, equal at all points in infinite planes orthogonal to the propagation direction. In real optical systems such plane waves do not exist, since the finite size of the elements of the system restricts the lateral extent of the waves. Non-planar optical components will cause further deviations of the wave from planarity. Consequently, the wave acquires a ray direction that varies from point to point on the phase front. The behavior of the optical system must be characterized in terms of the deviations its elements cause to the bundle of rays that comprise the propagating, laterally restricted wave. This is most easily done in terms of paraxial rays. In a cylindrically symmetric optical system, for example, a coaxial system of spherical lenses or mirrors, the *paraxial rays* are those rays whose directions of propagation occur at sufficiently small angles $\theta$ to the symmetry axis of the system that it is possible to replace $\sin\theta$ or $\tan\theta$ by $\theta$ – in other words paraxial rays obey the small-angle approximation.

### 13.4.1 Matrix formulation

In an optical system whose symmetry axis is in the $z$ direction, a paraxial ray in a given cross-section ($z$ = constant) is characterized by its distance $r$ from the $z$ axis and the angle $r'$ it makes with that axis. If the values of these parameters in two planes of the system (an *input* and an *output* plane) are $r_1 r'_1$ and $r_2 r'_2$, respectively, as shown in Fig. 13.6(a), then in the paraxial-ray approximation there is a linear relation between them, of the form

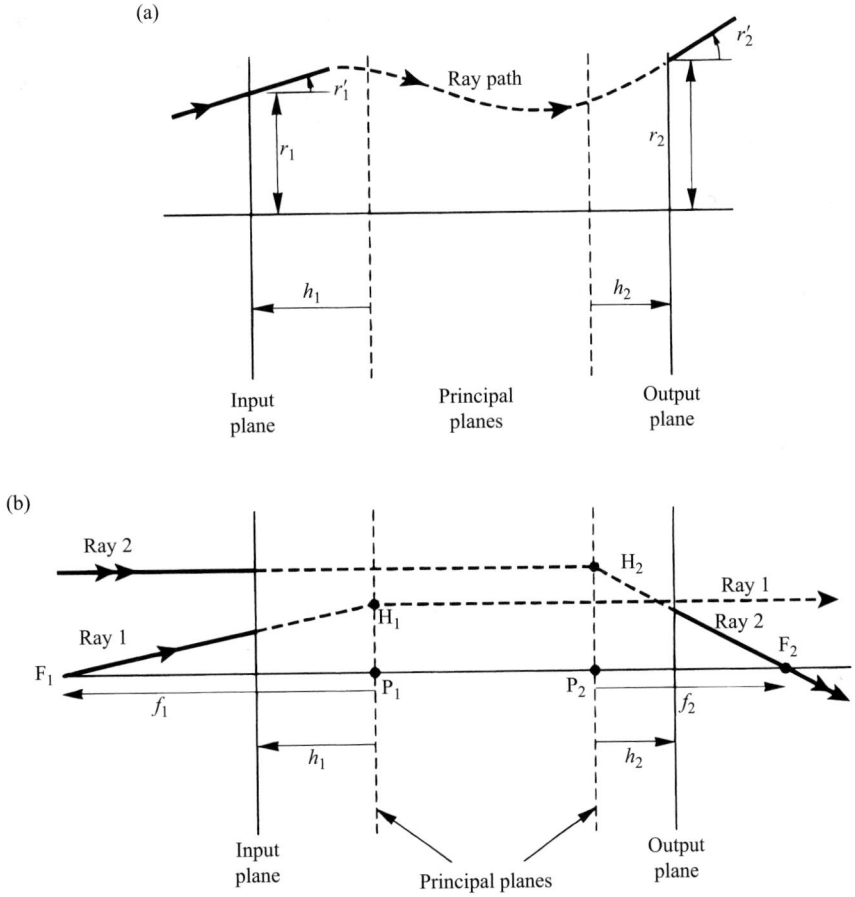


**Fig. 13.6** (a) The paraxial-ray path from the input plane to the output plane. (b) Principal planes, points, and rays of a paraxial optical system.

$$
\begin{aligned}
r_2 &= Ar_1 + Br_1', \\
r_2' &= Cr_1 + Dr_1',
\end{aligned}
\tag{13.8}
$$

or, in matrix notation,

$$
\begin{pmatrix} r_2 \\ r_2' \end{pmatrix} = \begin{pmatrix} A & B \\ C & D \end{pmatrix} \begin{pmatrix} r_1 \\ r_1' \end{pmatrix}. \tag{13.9}
$$

The matrix

$$
\begin{pmatrix} A & B \\ C & D \end{pmatrix}
$$

is called the ray-transfer matrix, **M**; its determinant is usually unity,[1] i.e., $AD - BC = 1$.

Optical systems made of isotropic material are generally reversible – a ray that travels from right to left with input parameters $r_2$ and $r_2'$ will leave the system with parameters $r_1$ and $r_1'$ – thus

$$\begin{pmatrix} r_1 \\ r_1' \end{pmatrix} = \begin{pmatrix} A' & B' \\ C' & D' \end{pmatrix} \begin{pmatrix} r_2 \\ r_2' \end{pmatrix}, \tag{13.10}$$

where the reverse ray-transfer matrix satisfies

$$\begin{pmatrix} A' & B' \\ C' & D' \end{pmatrix} = \begin{pmatrix} A & B \\ C & D \end{pmatrix}^{-1}.$$

The ray-transfer matrix allows the properties of an optical system to be described in general terms by the location of its *focal points* and *principal planes*, whose locations are determined from the elements of the matrix. The significance of these features of the system can be illustrated with the aid of Fig. 13.6(b). An input ray that passes through the *first focal point*, $F_1$ (or would pass through this point if it did not first enter the system) emerges traveling parallel to the axis. The intersection point of the extended input and output rays, point $H_1$ in Fig. 13.6(b), defines the location of the *first principal plane*. Conversely, an input ray traveling parallel to the axis will emerge at the output plane and pass through the second focal point, $F_2$ (or appear to have come from this point). The intersection of the extension of these rays, point $H_2$, defines the location of the *second principal plane*. Rays 1 and 2 in Fig. 13.6(b) are called the *principal rays* of the system. The location of the principal planes allows the corresponding emergent ray paths to be determined as shown in Fig. 13.6(b). The dashed lines in this figure, which permit geometrical construction of the location of output rays 1 and 2, are called *virtual ray paths*. Both $F_1$ and $F_2$ lie on the axis of the system. The axis of the system intersects the principal planes at the *principal points*, $P_1$ and $P_2$, in Fig. 13.6(b).

The distance $f_1$ from the first principal plane to the first focal point is called the *first focal length*; $f_2$ is the *second focal length*.

In most practical situations, the refractive indices of the media to the left of the input plane (the *object space*) and to the right of the output plane (the *image space*) are equal. In this case we can derive simple relations between the focal lengths $f_1$ and $f_2$ and $h_1$ and $h_2$, the distances of the input and output planes from the principal planes, measured in the sense shown in Fig. 13.6(b).

We can break up the system shown in Fig. 13.6(b) into three parts, namely the region from the input plane to the first principal plane, the region between the two principal planes, and the region from the second principal plane to the output plane.

If we write the transfer matrix from the left to the right principal planes as

$$\begin{pmatrix} A'' & B'' \\ C'' & D'' \end{pmatrix},$$

[1] When the media to the left of the input plane and to the right of the output plane have the same refractive index.

then the overall transfer matrix is

$$\begin{pmatrix} A & B \\ C & D \end{pmatrix} = \begin{pmatrix} 1 & h_2 \\ 0 & 1 \end{pmatrix} \begin{pmatrix} A'' & B'' \\ C'' & D'' \end{pmatrix} \begin{pmatrix} 1 & h_1 \\ 0 & 1 \end{pmatrix}, \tag{13.11}$$

which gives

$$\begin{pmatrix} A'' & B'' \\ C'' & D'' \end{pmatrix} = \begin{pmatrix} 1 & -h_2 \\ 0 & 1 \end{pmatrix} \begin{pmatrix} A & B \\ C & D \end{pmatrix} \begin{pmatrix} 1 & -h_1 \\ 0 & 1 \end{pmatrix}, \tag{13.12}$$

and therefore

$$\begin{pmatrix} A'' & B'' \\ C'' & D'' \end{pmatrix} = \begin{pmatrix} A - h_2 C & B - h_1 A - h_2(D - h_1 C) \\ C & D - h_1 C \end{pmatrix}. \tag{13.13}$$

Clearly, $C'' = C$ for the second principal ray in Fig. 13.6(b), and the distance of the ray from the axis, $r$, does not change, therefore $A'' = 1$, giving

$$r = (A - h_2 C)r, \tag{13.14}$$

so

$$h_2 = \frac{A - 1}{C}. \tag{13.15}$$

Furthermore, for the first principal ray, whose input angle is $r'$,

$$r = A''r + B''r', \tag{13.16}$$

so $B'' = 0$.

Consequently, if the media to the left of the input plane and the right of the output plane are the same, using $\det(M) = 1$ gives

$$h_1 = \frac{D - 1}{C}. \tag{13.17}$$

Note that both $A''$ and $D''$ are equal to unity.

For the second principal ray in Fig. 13.6(b)

$$\begin{pmatrix} r_2 \\ r_2' \end{pmatrix} = \begin{pmatrix} A'' & B'' \\ C'' & D'' \end{pmatrix} \begin{pmatrix} r_1 \\ 0 \end{pmatrix}, \tag{13.18}$$

so

$$r_2' = C'' r_1. \tag{13.19}$$

Furthermore, from Fig. 13.6(b), it is easy to see that

$$-r_2' = \frac{r_1}{f_2}. \tag{13.20}$$

Combining Eqs. (13.19) and (13.20) gives

$$C'' = -\frac{1}{f_2}. \tag{13.21}$$

By a similar procedure using the first principal ray it can be shown that

$$C'' = -\frac{1}{f_1}. \tag{13.22}$$

We have already seen from Eq. (13.13) that $C'' = C$, therefore

$$f_1 = f_2 = f. \tag{13.23}$$

If the media to the left and right of the input plane are the same, then both principal focal lengths are equal. If the elements of the transfer matrix are known, the locations of the focal points and principal planes are determined. Graphical construction of ray paths through the system using the methods of *ray tracing* is then straightforward (see Section 13.10).

In using matrix methods for optical analysis, a consistent sign convention must be employed. In the present discussion, a ray is assumed to travel in the positive-$z$ direction from left to right through the system. The distance from the first principal plane to an object is measured positive from right to left – in the negative-$z$ direction. The distance from the second principal plane to an image is measured positive from left to right – in the positive-$z$ direction. The lateral distance of the ray from the axis is positive in the upward direction, negative in the downward direction. The acute angle between the system axis direction and the ray, say $r_1'$ in Fig. 13.6(a), is positive if a counterclockwise motion is necessary to go from the positive-$z$ direction to the ray direction. When the ray crosses a spherical interface the radius of curvature is positive if the interface is convex to the input ray. The use of ray-transfer matrices in optical system analysis can be illustrated with some specific examples.

(i) *A uniform optical medium.* In a uniform optical medium of length $d$ no change in ray angle occurs, as illustrated in Fig. 13.7(a), so

$$\begin{aligned} r_2' &= r_1', \\ r_2 &= r_1 + d r_1'. \end{aligned} \tag{13.24}$$

Therefore,

$$\mathbf{M} = \begin{pmatrix} 1 & d \\ 0 & 1 \end{pmatrix}. \tag{13.25}$$

The focal length of this system is infinite and it has no specific principal planes.

(ii) *A planar interface between two different media.* At the interface, as shown in Fig. 13.7(b), $r_1 = r_2$, and, from Snell's law, using the approximation $\sin\theta = \theta$,

$$r_2' = \frac{n_1}{n_2} r_1'. \tag{13.26}$$

Therefore

$$\mathbf{M} = \begin{pmatrix} 1 & 0 \\ 0 & n_1/n_2 \end{pmatrix}. \tag{13.27}$$

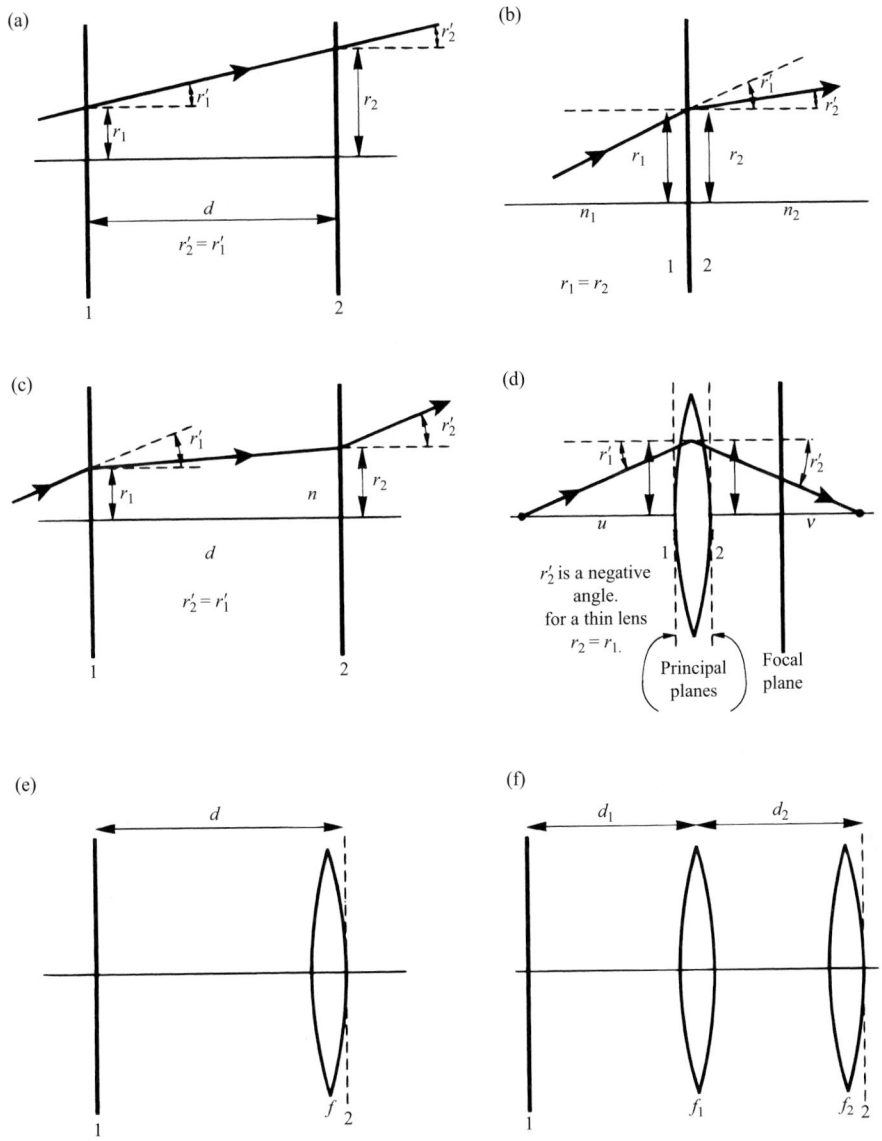


**Fig. 13.7** Simple optical systems: (a) a length of uniform medium; (b) the planar interface between two media; (c) a parallel-sided slab; (d) a thin lens; (e) a length of uniform medium plus a thin lens; and (f) two spaced thin lenses.

(iii) *A parallel-sided slab of refractive index n bounded on both sides with media of refractive index 1* (Fig. 13.7(c)). In this case,

$$\mathbf{M} = \begin{pmatrix} 1 & d/n \\ 0 & 1 \end{pmatrix}. \tag{13.28}$$

The principal planes of this system are the boundary faces of the optically dense slab.

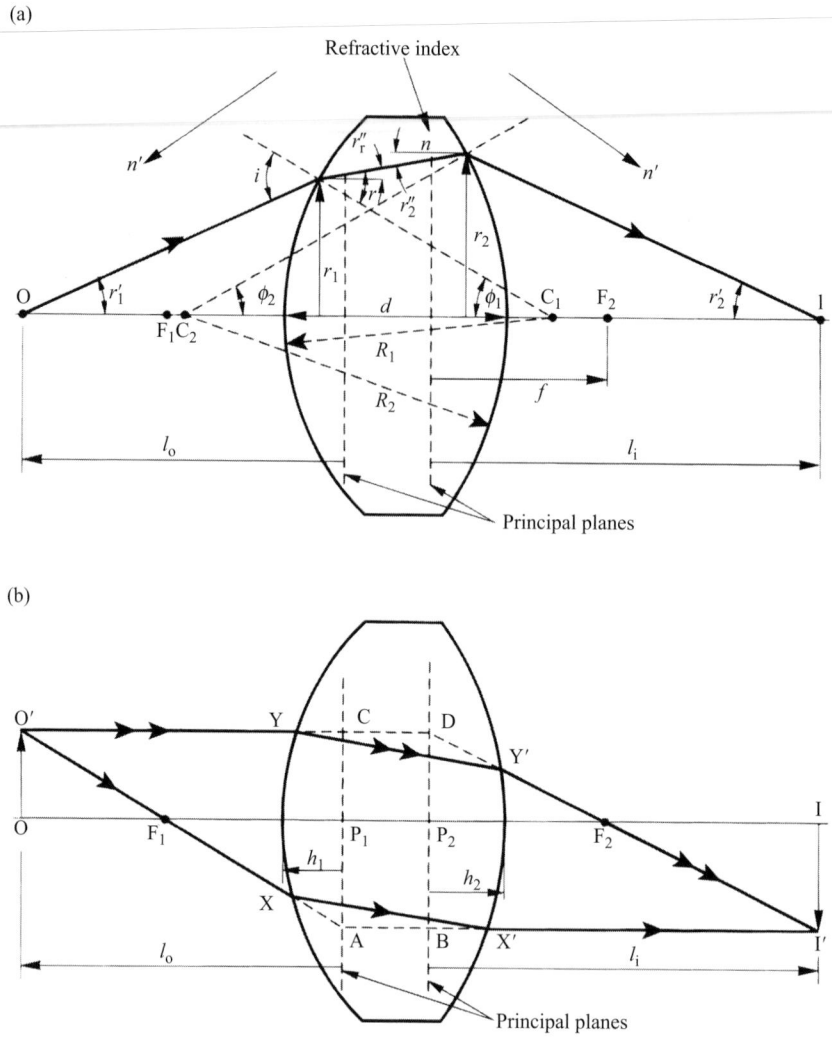


**Fig. 13.8** (a) The ray path through a thick lens. $C_1$ and $C_2$ are the centers of curvature of the two spherical surfaces of which the faces of the lens are a part. (b) Principal rays and planes for a thick lens. The dashed lines show the use of the principal planes for determining the exit trajectories of the input principal rays. The actual ray paths through the lens are shown by the solid lines.

(iv) *A thick lens.* The ray-transfer matrix of the thick lens shown in Fig. 13.8(a) is the product of the three transfer matrices:

$$\mathbf{M}_3\mathbf{M}_2\mathbf{M}_1 = \begin{pmatrix} \text{matrix for} \\ \text{second spherical} \\ \text{interface} \end{pmatrix} \begin{pmatrix} \text{matrix for} \\ \text{medium of} \\ \text{length } d \end{pmatrix} \begin{pmatrix} \text{matrix for} \\ \text{first spherical} \\ \text{interface} \end{pmatrix}. \tag{13.29}$$

Note the order of these three matrices; $\mathbf{M}_1$ comes on the right because it operates first on the column vector that describes the input ray.

At the first spherical surface

$$n'(r_1' + \phi_1) = nr = n(r_2'' + \phi_1), \tag{13.30}$$

which, since $\phi_1 = r_1/R_1$, can be written as

$$r_2'' = \frac{n'r_1'}{n} + \frac{(n' - n)r_1}{nR_1}. \tag{13.31}$$

The transfer matrix at the first spherical surface is

$$\mathbf{M}_1 = \begin{pmatrix} 1 & 0 \\ (n' - n)/(nR_1) & n'/n \end{pmatrix} = \begin{pmatrix} 1 & 0 \\ -D_1/n & n'/n \end{pmatrix}, \tag{13.32}$$

where $D_1 = (n - n')/R_1$ is called the power of the surface. If $R_1$ is measured in meters the units of $D_1$ are *diopters*.

In the paraxial approximation all the rays passing through the lens travel the same distance $d$ in the lens. Thus,

$$\mathbf{M}_2 = \begin{pmatrix} 1 & d \\ 0 & 1 \end{pmatrix}. \tag{13.33}$$

The ray-transfer matrix at the second interface is

$$\mathbf{M}_3 = \begin{pmatrix} 1 & 0 \\ (n - n')/(n'R_2) & n/n' \end{pmatrix} = \begin{pmatrix} 1 & 0 \\ -D_2/n' & n/n' \end{pmatrix}, \tag{13.34}$$

which is identical in form to $\mathbf{M}_1$. Note that in this case both $r_2'$ and $R_2$ are negative. The overall transfer matrix of the thick lens is

$$\mathbf{M} = \mathbf{M}_3\mathbf{M}_2\mathbf{M}_1 = \begin{pmatrix} 1 - dD_1/n & dn'/n \\ dD_1D_2/(nn') - D_1/n' - D_2/n' & 1 - dD_2/n \end{pmatrix}. \tag{13.35}$$

If Eq. (13.35) is compared with Eqs. (13.15) and (13.17), it is clear that the locations of the principal planes of the thick lens are

$$h_1 = \frac{d}{(n/n')(1 + D_1/D_2 - dD_1/n)}, \tag{13.36}$$

$$h_2 = \frac{d}{(n/n')(1 + D_2/D_1 - dD_2/n)}. \tag{13.37}$$

A numerical example will best illustrate the location of the principal planes for a biconvex thick lens.

Suppose that $n' = 1$ (air), $n = 1.5$ (glass), $R_1 = -R_2 = 50$ mm, and $d = 10$ mm. In this case $D_1 = D_2 = 0.01\ \text{mm}^{-1}$ and, from Eqs. (13.36) and (13.37), $h_1 = h_2 = 3.448$ mm. These principal planes are symmetrically placed inside the lens. Figure 13.8(b) shows how the principal planes can be used to trace the principal ray paths through a thick lens.

From Eqs. (13.18) and (13.11)

$$r_2' = \frac{-r_1}{f} + \left(1 - \frac{dD_2}{n}\right) r_1', \tag{13.38}$$

where the focal length is

$$f = \left( \frac{D_1 + D_2}{n'} - \frac{dD_1D_2}{nn'} \right)^{-1}. \tag{13.39}$$

If $\ell_o$ is the distance from the object O to the first principal plane, and $\ell_i$ is the distance from the second principal plane to the image I in Fig. 13.8(b), then, from the similar triangles $OO'F_1$ and $P_1AF_1$ and $P_2DF_2$ and $II'F_2$,

$$\frac{OO'}{II'} = \frac{OF_1}{F_1P_1} = \frac{\ell_o - f_1}{f_1},$$
$$\frac{OO'}{II'} = \frac{P_2F_2}{F_2I} = \frac{f_2}{\ell_i - f_2}. \tag{13.40}$$

If the media on both sides of the lens are the same, then $f_1 = f_2 = f$, and it immediately follows that

$$\frac{1}{\ell_o} + \frac{1}{\ell_i} = \frac{1}{f}. \tag{13.41}$$

This is the fundamental imaging equation.

(v) *A thin lens.* If a lens is sufficiently thin that to a good approximation $d = 0$, the transfer matrix is

$$\mathbf{M} = \begin{pmatrix} 1 & 0 \\ -(D_1 + D_2)/n' & 1 \end{pmatrix}. \tag{13.42}$$

As shown in Fig. 13.7(d), the principal planes of such a thin lens are at the lens. The focal length of the thin lens is $f$, where

$$\frac{1}{f} = \frac{D_1 + D_2}{n'} = \left( \frac{n}{n'} - 1 \right) \left( \frac{1}{R_1} - \frac{1}{R_2} \right), \tag{13.43}$$

so the transfer matrix can be written very simply as

$$\mathbf{M} = \begin{pmatrix} 1 & 0 \\ -1/f & 1 \end{pmatrix}. \tag{13.44}$$

This result can be derived in a very direct simple way. Since the height of a ray passing through a thin lens does not change, and is independent of the angle of incidence, it is clear that $A = 1$ and $B = 0$. Furthermore, by definition $C = 1/f$. Since the determinant of the ray-transfer matrix is unity for a thin lens immersed in a uniform medium, the final element of the ray-transfer matrix is $D = 1$.

The focal length of the lens depends on the refractive index of the lens material and the refractive index of the medium within which it is immersed. In air

$$\frac{1}{f} = (n - 1) \left( \frac{1}{R_1} - \frac{1}{R_2} \right). \tag{13.45}$$

For a biconvex lens $R_2$ is negative and

$$\frac{1}{f} = (n-1)\left(\frac{1}{|R_1|} + \frac{1}{|R_2|}\right). \tag{13.46}$$

For a biconcave lens

$$\frac{1}{f} = -(n-1)\left(\frac{1}{|R_1|} + \frac{1}{|R_2|}\right). \tag{13.47}$$

The focal length of any diverging lens is negative. For a thin lens, object and image distances are measured to a common point. It is common practice to rename the distances $\ell_o$ and $\ell_i$ in this case so that the imaging equation, Eq. (13.41), reduces to its familiar form

$$\frac{1}{v} + \frac{1}{u} = \frac{1}{f}, \tag{13.48}$$

where $u$ is called the *object distance* and $v$ the *image distance*.

(vi) *A length of uniform medium plus a thin lens* (Fig. 13.7(e)). This is a combination of systems (i) and (v); its overall transfer matrix is found from Eqs. (13.25) and (13.44) as

$$\mathbf{M} = \begin{pmatrix} 1 & 0 \\ -1/f & 1 \end{pmatrix}\begin{pmatrix} 1 & d \\ 0 & 1 \end{pmatrix} = \begin{pmatrix} 1 & d \\ -1/f & 1-d/f \end{pmatrix}. \tag{13.49}$$

(vii) *Two thin lenses.* As a final example of the use of ray-transfer matrices, consider the combination of two thin lenses shown in Fig. 13.7(f). The transfer matrix of this combination is

$$\mathbf{M} = \begin{pmatrix} \text{matrix for} \\ \text{second lens, } f_2 \end{pmatrix}\begin{pmatrix} \text{matrix for} \\ \text{uniform medium, } d_2 \end{pmatrix}\begin{pmatrix} \text{matrix for} \\ \text{first lens, } f_1 \end{pmatrix} \times \begin{pmatrix} \text{matrix for} \\ \text{uniform medium, } d_1 \end{pmatrix}, \tag{13.50}$$

which can be shown to be

$$\mathbf{M} = \begin{pmatrix} 1 - d_2/f_1 & d_1 + d_2 - d_1 d_2/f_1 \\ -1/f_1 - 1/f_2 + d_2/(f_1 f_2) & 1 - d_1/f_1 - d_1/f_2 - d_2/f_2 + d_1 d_2/(f_1 f_2) \end{pmatrix}. \tag{13.51}$$

The focal length of the combination is

$$f = \frac{f_1 f_2}{d_2 - (f_1 + f_2)}. \tag{13.52}$$

The optical system consisting of two thin lenses is the archetypal system used in analyses of the stability of lens waveguides and optical resonators [1].

(viii) *Spherical mirrors.* The object and image distances from a spherical mirror also obey Eq. (13.48), where the focal length of the mirror is $R/2$; $f$ is positive for a concave mirror, negative for a convex mirror. Positive object and image distances for a mirror are measured positive in the normal direction away from its surface. If a negative image distance results from Eq. (13.48) this implies a *virtual image* (behind the mirror).

### 13.4.2 Paraxial ray tracing

Practical implementation of paraxial-ray analysis in optical system design can be very conveniently carried out graphically by paraxial *ray tracing*. In paraxial ray tracing a few simple rules allow geometrical construction of the principal ray paths from an object point, although these constructions do not take into account the non-ideal behavior, or *aberrations*, of real lenses. The first principal ray from a point on the object passes through (or its projection passes through) the first focal point. From the point where this ray, or its projection, intersects the first principal plane the output ray is drawn parallel to the axis. The actual ray path between input and output planes can be found in simple cases, for example, the path XX′ in the thick lens shown in Fig. 13.8(b). The second principal ray is directed parallel to the axis; from the intersection of this ray, or its projection, with the second principal plane the output ray passes through (or appears to have come from) the second focal point. The actual ray path between input and output planes can be found in simple cases, for example, the path YY′ in the thick lens shown in Fig. 13.8(b). The intersection of the two principal rays in the image space produces the image point which corresponds to the same point on the object. If only the back-projections of the output principal rays appear to intersect, this intersection point lies on a *virtual* image.

In the majority of applications of ray tracing a quick analysis of the system is desired. In this case, if all lenses in the system are treated as thin lenses, the positions of the principal planes need not be calculated beforehand, and ray tracing becomes particularly easy. For a thin lens a third principal ray is useful for determining image location. The ray from a point on the object which passes through the center of the lens is not deviated by the lens. The use of ray tracing to determine the size and position of the real image produced by a convex lens and the virtual image produced by a concave lens is shown in Fig. 13.9. Figure 13.9(a) shows a converging lens where the input principal ray parallel to the axis actually passes through the focal point. Figure 13.9(b) shows a diverging lens where this ray emerges from the lens, appearing to have come from the focal point. More complex systems of lenses can be analyzed in the same way. Once the image location has been determined by use of the principal rays, the path of any group of rays, a ray *pencil*, can be found; see, for example, the cross-hatched ray pencil shown in Fig. 13.9(a).

The use of ray-tracing rules to analyze spherical mirror systems is similar to that described above, although the ray striking the center of the mirror in this case reflects so that the angle of incidence equals the angle of reflection. Figures 13.9(c) and 13.9(d) illustrate the use of ray tracing to locate the images produced by concave and convex mirrors. In the case of a convex mirror the ray-tracing construction is modified in the same way as it was for a defocusing lens. A ray parallel to the axis that would pass through the focal point *if it did not first encounter the mirror surface* reflects parallel to the axis. A ray parallel to the axis reflects at the mirror surface so as to appear in projection to have come from the focal point. Figure 13.10 shows how a concave mirror can produce a virtual image if the object is closer to the mirror than its focal point.

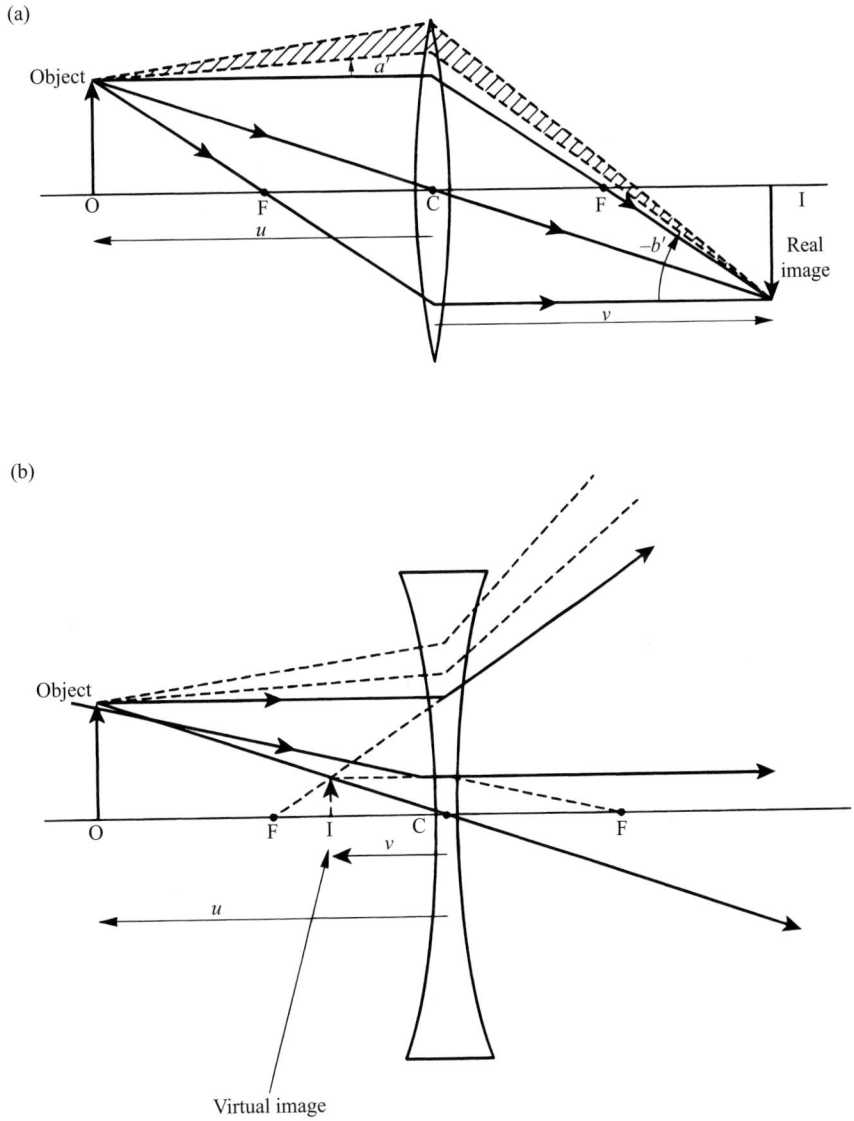


**Fig. 13.9** Ray-tracing diagrams: (a) real-image production by a converging lens; (b) virtual-image production by a diverging lens; (c) real-image production by a concave mirror; and (d) virtual-image production by a convex mirror.

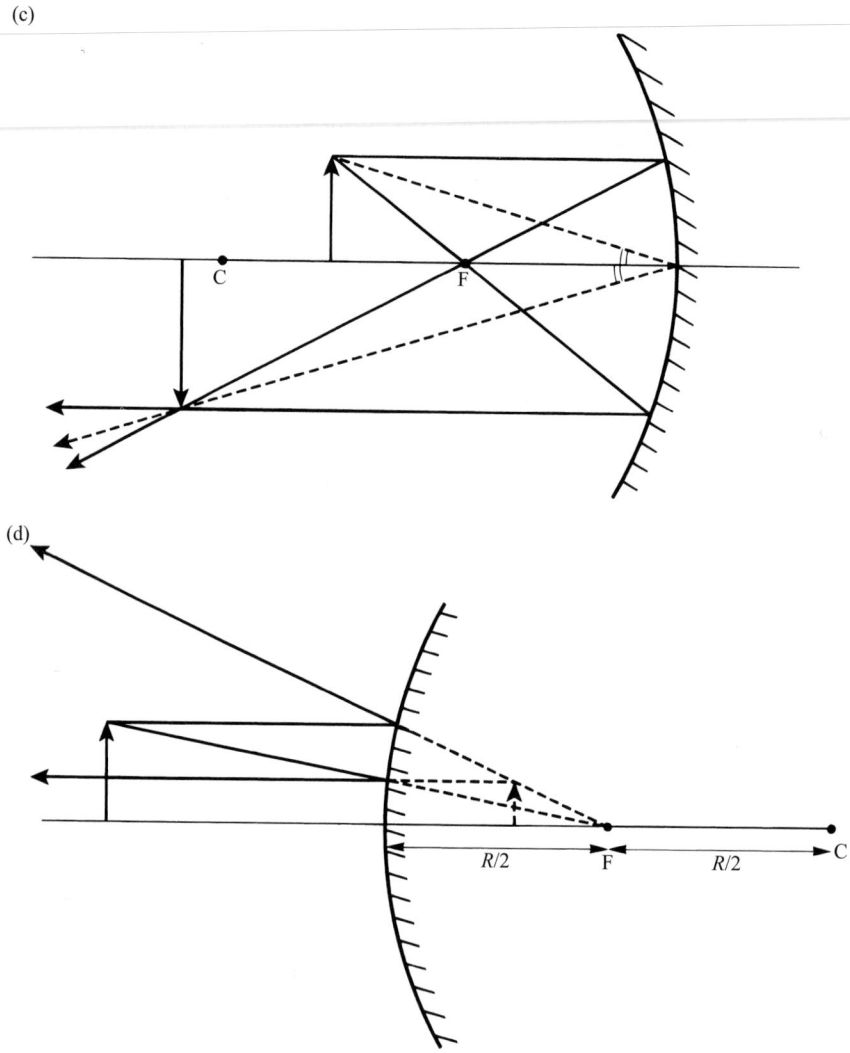


Fig. 13.9 (*cont.*)

### 13.4.3 Imaging and magnification

In Fig. 13.9(a) the ratio of the height of the image to the height of the object is called the *magnification*, $m$. If the distance from the axis of a ray through the top of the object is $a$ and the distance through the top of the image is $b$, then

$$m = \frac{b}{a}, \tag{13.53}$$

so in Fig. 13.9(a), since $b$ is negative, the magnification is negative, because the image is inverted.

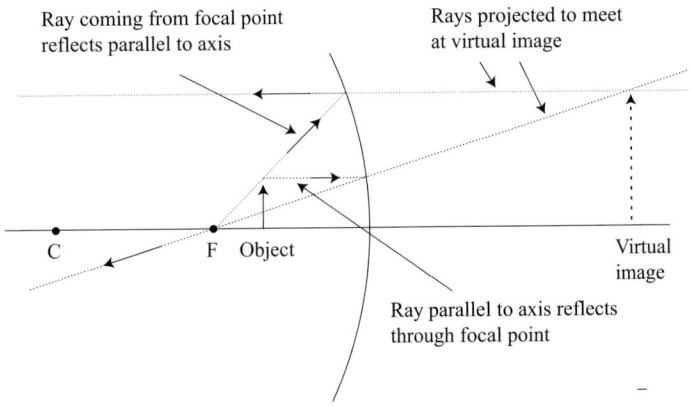


**Fig. 13.10** How a concave mirror forms a virtual image if the object is closer to the mirror than its focal point.

For a more general system the magnification can be obtained from the ray-transfer matrix equation

$$\begin{pmatrix} b \\ b' \end{pmatrix} = \begin{pmatrix} A & B \\ C & D \end{pmatrix} \begin{pmatrix} a \\ a' \end{pmatrix} = \mathbf{M} \begin{pmatrix} a \\ a' \end{pmatrix}, \tag{13.54}$$

where $a'$ is the angle of a ray through a point on the object and $b'$ is the angle of a ray through the corresponding point on the image. The matrix $\mathbf{M}$ in this case includes the entire system from object O to image I. So, in Fig. 13.9(a),

$$\mathbf{M} = \begin{pmatrix} \text{matrix for} \\ \text{uniform medium} \\ \text{of length } v \end{pmatrix} \begin{pmatrix} \text{matrix for} \\ \text{lens} \end{pmatrix} \begin{pmatrix} \text{matrix for} \\ \text{uniform medium} \\ \text{of length } u \end{pmatrix}. \tag{13.55}$$

For imaging, $b$ must be independent of $a'$, so $B = 0$.

The magnification is

$$m = \frac{b}{a} = A, \tag{13.56}$$

so the ray-transfer matrix of the imaging system can be written

$$\mathbf{M} = \begin{pmatrix} m & 0 \\ -1/f & 1/m \end{pmatrix}, \tag{13.57}$$

where it should be noted that the result $\det(M) = 1$ has been used.

The *angular magnification* of the system is defined as

$$m' = \left(\frac{b'}{a'}\right)_{a=0}, \tag{13.58}$$

which gives $m' = 1/m$.

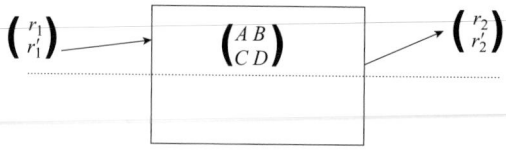


**Fig. 13.11** Input and output rays of a general optical system described by a ray-transfer matrix.

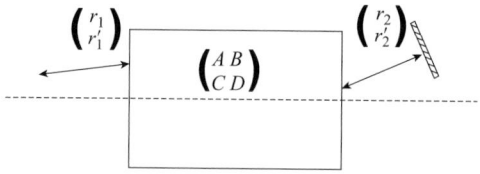


**Fig. 13.12** Input and output rays if a plane mirror is placed perpendicular to the exit ray.

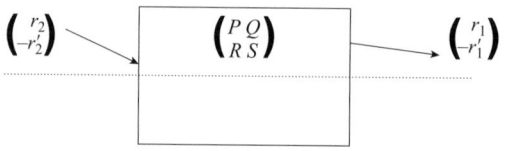


**Fig. 13.13** Reversed rays traveling though an optical system.

Note that $mm' = 1$. This result demonstrates a very important fact about imaging systems: a large linear magnification must be accompanied by a small angular magnification, or vice versa. So, for example, a microscope that has a large linear magnification has a small angular magnification. It produces an image whose linear size is really larger than that of the object. In contrast, a telescope makes distant objects appear larger because such an instrument has a large angular magnification. If the Moon is examined through a telescope, the image is not larger than the Moon itself, but the Moon appears bigger because the telescope provides the illusion that the Moon is closer. Our eyes interpret the size of a distant object on the basis of the angle it subtends. A distant planet appears smaller than the Moon, not because its actual size is less, but because it subtends a much smaller angle at our eyes.

### 13.4.4 Mirrors and the use of ray-transfer matrices

If an optical system is represented by a ray-transfer matrix then the input and output ray parameters are related by Eq. (13.9), as shown in Fig. 13.11. If we place a plane mirror perpendicular to the output ray, then the ray retraces its trajectory and re-emerges at the input plane traveling in the opposite direction along its input trajectory, as shown in Fig. 13.12.

However, if we represent the reverse passage through the system as shown in (Fig. 13.13), then clearly the angles involved obey

$$\begin{pmatrix} r_1 \\ -r_1' \end{pmatrix} = \begin{pmatrix} P & Q \\ R & S \end{pmatrix} \begin{pmatrix} r_2 \\ -r_2' \end{pmatrix}, \tag{13.59}$$

where the signs of the angles have been changed to correspond to the rays traveling in the reverse direction.

If Eq. (13.59) is compared with Eq. (13.10), it is clear that the matrix

$$\begin{pmatrix} P & Q \\ R & S \end{pmatrix}$$

is not the same as

$$\begin{pmatrix} A & B \\ C & D \end{pmatrix}^{-1}.$$

However, if we write

$$\begin{pmatrix} r_1 \\ -r_1' \end{pmatrix} = \begin{pmatrix} 1 & 0 \\ 0 & -1 \end{pmatrix} \begin{pmatrix} r_1 \\ r_1' \end{pmatrix} \tag{13.60}$$

or

$$\begin{pmatrix} r_1 \\ r_1' \end{pmatrix} = \begin{pmatrix} 1 & 0 \\ 0 & -1 \end{pmatrix} \begin{pmatrix} r_1 \\ -r_1' \end{pmatrix}, \tag{13.61}$$

then

$$\begin{pmatrix} 1 & 0 \\ 0 & -1 \end{pmatrix} \begin{pmatrix} r_1 \\ -r_1' \end{pmatrix} = \begin{pmatrix} 1 & 0 \\ 0 & -1 \end{pmatrix} \begin{pmatrix} P & Q \\ R & S \end{pmatrix} \begin{pmatrix} 1 & 0 \\ 0 & -1 \end{pmatrix} \begin{pmatrix} r_2 \\ r_2' \end{pmatrix}, \tag{13.62}$$

which gives

$$\begin{pmatrix} r_1 \\ r_1' \end{pmatrix} = \begin{pmatrix} 1 & 0 \\ 0 & -1 \end{pmatrix} \begin{pmatrix} P & Q \\ R & S \end{pmatrix} \begin{pmatrix} 1 & 0 \\ 0 & -1 \end{pmatrix} \begin{pmatrix} r_2 \\ r_2' \end{pmatrix}. \tag{13.63}$$

Therefore,

$$\begin{pmatrix} 1 & 0 \\ 0 & -1 \end{pmatrix} \begin{pmatrix} P & Q \\ R & S \end{pmatrix} \begin{pmatrix} 1 & 0 \\ 0 & -1 \end{pmatrix} = \begin{pmatrix} A & B \\ C & D \end{pmatrix}^{-1}, \tag{13.64}$$

and

$$\begin{pmatrix} P & Q \\ R & S \end{pmatrix} = \begin{pmatrix} 1 & 0 \\ 0 & -1 \end{pmatrix} \begin{pmatrix} A & B \\ C & D \end{pmatrix}^{-1} \begin{pmatrix} 1 & 0 \\ 0 & -1 \end{pmatrix}. \tag{13.65}$$

If an optical system includes a mirror placed perpendicular to the system axis as shown in Fig. 13.14, then the mirror causes rays to return through the system. One convenient way to visualize this is to look at the "unfolded system." If a mirror is placed perpendicular to the exit ray, then the ray returns through the system, but in unfolded format the ray angles and distances from the axis are as shown schematically in Fig. 13.15.

The output ray parameters in this "unfolded" system obey

$$\begin{pmatrix} r_3 \\ r_3' \end{pmatrix} = \begin{pmatrix} P & Q \\ R & S \end{pmatrix} \begin{pmatrix} 1 & 0 \\ 0 & -1 \end{pmatrix} \begin{pmatrix} A & B \\ C & D \end{pmatrix} \begin{pmatrix} r_1 \\ r_1' \end{pmatrix}, \tag{13.66}$$

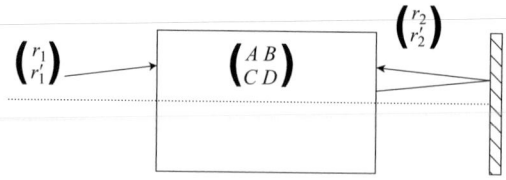


**Fig. 13.14** An optical system with a plane mirror placed perpendicular to the system axis to reflect output rays back through the system.

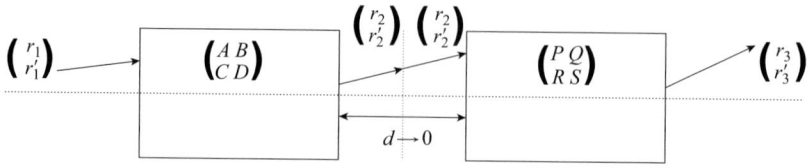


**Fig. 13.15** The "unfolded" version of Fig. 13.14.

which can be written as

$$\begin{pmatrix} r_3 \\ r_3' \end{pmatrix} = \begin{pmatrix} 1 & 0 \\ 0 & -1 \end{pmatrix} \begin{pmatrix} A & B \\ C & D \end{pmatrix}^{-1} \begin{pmatrix} 1 & 0 \\ 0 & -1 \end{pmatrix} \begin{pmatrix} 1 & 0 \\ 0 & -1 \end{pmatrix} \begin{pmatrix} A & B \\ C & D \end{pmatrix} \times \begin{pmatrix} r_1 \\ r_1' \end{pmatrix}, \tag{13.67}$$

and, therefore,

$$\begin{pmatrix} r_3 \\ r_3' \end{pmatrix} = \begin{pmatrix} 1 & 0 \\ 0 & -1 \end{pmatrix} \begin{pmatrix} r_1 \\ r_1' \end{pmatrix}. \tag{13.68}$$

Note that in this "unfolded" version of the problem the sign of the emerging ray has been changed. Now consider what happens when a plane mirror is placed perpendicular to the system axis as shown in Fig. 13.14.

The reflected ray is shown. Now "unfold" the system, which results in Fig. 13.15.

In this unfolded system the plane mirror has no effect. Multiply the various ray-transfer matrices involved, which gives

$$\begin{pmatrix} r_3 \\ r_3' \end{pmatrix} = \begin{pmatrix} P & Q \\ R & S \end{pmatrix} \begin{pmatrix} A & B \\ C & D \end{pmatrix} \begin{pmatrix} r_1 \\ r_1' \end{pmatrix} \tag{13.69}$$

and

$$\begin{pmatrix} r_3 \\ r_3' \end{pmatrix} = \begin{pmatrix} 1 & 0 \\ 0 & -1 \end{pmatrix} \begin{pmatrix} A & B \\ C & D \end{pmatrix}^{-1} \begin{pmatrix} 1 & 0 \\ 0 & -1 \end{pmatrix} \begin{pmatrix} A & B \\ C & D \end{pmatrix} \begin{pmatrix} r_1 \\ r_1' \end{pmatrix}. \tag{13.70}$$

In Eq. (13.70) the effect of the plane mirror is contained in the second

$$\begin{pmatrix} 1 & 0 \\ 0 & -1 \end{pmatrix}$$

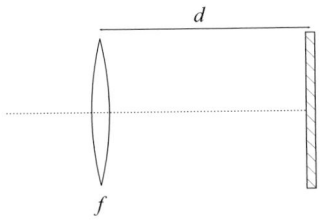


Fig. 13.16 A thin lens followed by a plane mirror a distance $d$ away.

matrix, while the first matrix of this form merely changes the sign convention for the ray emerging at the original input plane after its passage back through the system.

### 13.4.5 Example

Consider the case of a thin lens followed by a length $d$ of uniform medium, as shown in Fig. 13.16.

The ray transfer from left to right is

$$\begin{pmatrix} A & B \\ C & D \end{pmatrix} = \begin{pmatrix} 1 & d \\ 0 & 1 \end{pmatrix} \begin{pmatrix} 1 & 0 \\ -1/f & 1 \end{pmatrix} = \begin{pmatrix} 1 - d/f & d \\ -1/f & 1 \end{pmatrix}. \tag{13.71}$$

The ray-transfer matrix from right to left is

$$\begin{pmatrix} P & Q \\ R & S \end{pmatrix} = \begin{pmatrix} 1 & 0 \\ -1/f & 1 \end{pmatrix} \begin{pmatrix} 1 & d \\ 0 & 1 \end{pmatrix} = \begin{pmatrix} 1 & d \\ -1/f & 1 - d/f \end{pmatrix}. \tag{13.72}$$

It is easy to check that for these matrices Eq. (13.70) is correct. If we place the plane mirror perpendicular to the axis at the focus of the thin lens then the ray parameters on return to the input plane are

$$\begin{pmatrix} r_3 \\ r_3' \end{pmatrix} = \begin{pmatrix} 1 & 0 \\ 0 & -1 \end{pmatrix} \begin{pmatrix} 1 & -f \\ -1/f & 0 \end{pmatrix} \begin{pmatrix} 1 & 0 \\ 0 & -1 \end{pmatrix} \begin{pmatrix} 0 & f \\ -1/f & 1 \end{pmatrix} \begin{pmatrix} r_1 \\ r_1' \end{pmatrix}, \tag{13.73}$$

which gives finally

$$\begin{pmatrix} r_3 \\ r_3' \end{pmatrix} = \begin{pmatrix} -1 & 2f \\ 0 & -1 \end{pmatrix} \begin{pmatrix} r_1 \\ r_1' \end{pmatrix}. \tag{13.74}$$

### 13.4.6 Imaging and non-imaging optical systems

The techniques of optical system analysis that we have discussed so far will provide an approximate description of simple systems. However, in many situations a more detailed analysis is needed. Additional parameters are introduced in order to characterize the system. More sophisticated methods of analysis can be illustrated through a more detailed discussion of *imaging* systems and an introduction to *non-imaging* optical systems. In any experimental setup that involves the collection of light from a source and its delivery to a light-detection system, the properties of the optical system between source and detector

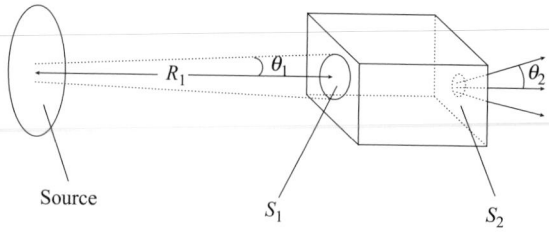


**Fig. 13.17** A generalized diagram of a non-imaging system. Light from the source enters the system through the aperture $S_1$ over an angular range defined by $\theta_1$, and leaves through an aperture $S_2$ over a range of angles defined by $\theta_2$.

should be optimized. This optimization will take different forms depending on the application. For example, in an experiment in which a weak fluorescent signal from a liquid in a cuvette is to be detected, it is generally the aim to maximize the amount of fluorescence collected and deliver this light to the active surface of the detector. This application does not require an *imaging* optical system, although an imaging system is often quite satisfactory. On the other hand, if an image of some object is being delivered to a CCD camera, then an imaging optical system is required.

## 13.5 Non-imaging light collectors

The properties of a *non-imaging* optical system can be described schematically with the aid of Fig. 13.17.

Light that enters the front aperture of the system within some angular range will leave the system through the exit aperture. In such an arrangement, the *brightness* of the light leaving the exit aperture cannot exceed the brightness of the light entering the entrance aperture. The brightness of an emitting object is measured in units of $\mathrm{W\,m^{-2}\,sr^{-1}}$. In Fig. 13.17, if the brightness of the radiation entering the aperture is $B_1$, then the power entering $S_1$ from the source is

$$P_1 = \pi B_1 S_1 \theta_1^2, \tag{13.75}$$

where the angular range of the entering rays is $\theta_1$. Assuming that $R_1^2 \gg S_1$, this angular range is $\theta_1 \simeq (1/R)\sqrt{S_1/\pi}$, or $\theta_1^2 \simeq S_1/(\pi R_1^2)$. If all the power entering $S_1$ leaves through $S_2$, then the brightness of the emerging radiation is

$$B_2 \simeq \frac{P_1}{\pi S_2 \theta_2^2} = \frac{\pi B_1 S_1 \theta_1^2}{\pi S_2 \theta_2^2}, \tag{13.76}$$

where $\theta_2$ is the angular spread of the emerging rays. So

$$B_2 = \frac{B_1 S_1 \theta_1^2}{S_2 \theta_2^2}. \tag{13.77}$$

Hence, if $B_2 \leq B_1$,

$$\theta_2 \geq \sqrt{\frac{S_2}{S_1}}\theta_1. \tag{13.78}$$

Non-imaging systems are very important in passive solar energy conversion and also in mirror systems used to optically pump solid-state lasers, as discussed briefly in Chapter 7. For further details of this interesting branch of optics the interested reader should consult the books by Winston and co-authors [2, 3].

## 13.6 Generalized imaging systems

In a generalized imaging system, light from an object, located in the *object plane*, passes through the system until it reaches the *image plane*. In an ideal imaging system all the light rays that leave a point on the object arrive at a single point in the imaging plane. Furthermore, the location of points in the image plane is such as to preserve the scaling of the image relative to the object. The image may be magnified or demagnified, but it should not be distorted. A type of imaging called *aplanatic imaging* occurs when rays parallel to the axis produce a sharp image irrespective of their distance from the axis. More specifically, if the angle of an input ray with the axis is $\theta_1$ and the output angle with the axis is $\theta_2$, then an aplanatic system satisfies the *sine condition*, namely

$$\frac{\sin\theta_1}{\sin\theta_2} = \text{constant}, \tag{13.79}$$

for rays at all distances from the axis that can pass through the system. *Stigmatic imaging* is sharp imaging of a point object to a point image. An object point and the corresponding image point are together called *conjugate points*.

## 13.7 The numerical aperture

If an optical system is illuminated by a point source on its axis, then the amount of light collected by the system depends on the angle that an effective aperture, called the *entrance pupil*, subtends at the source. This concept will be discussed further in the next section. In Fig. 13.18 this angle is defined by

$$\sin\theta = \frac{D}{2d}. \tag{13.80}$$

The *numerical* aperture of the system is given by

$$\text{NA} = n\sin\theta, \tag{13.81}$$

where $n$ is the refractive index of the object space (which is generally 1, but could be higher – for example, in microscopy when an oil-immersion objective lens is used). Now, a

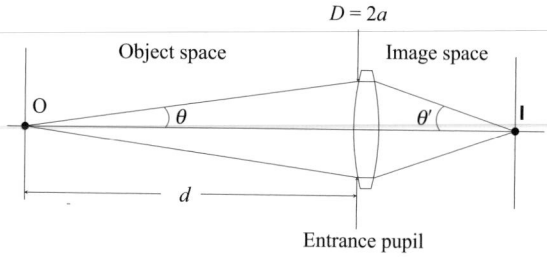


**Fig. 13.18** A simple lens system to illustrate the concepts of the entrance pupil and numerical aperture.

good imaging system obeys the sine condition, which relates the angles subtended by the exit and entrance pupils and the magnification $m$ by

$$m = \frac{n \sin\theta}{n' \sin\theta'}, \tag{13.82}$$

where the primed quantities refer to the image space. The numerical aperture in the image space is

$$(\mathrm{NA})' = n' \sin\theta'. \tag{13.83}$$

The entrance pupil plays an important geometrical role in determining the amount of light from an object (or light source) that can enter and pass through an optical system. For example, for an entrance-pupil diameter (EPD) of $D_1$ located a distance $d_1$ from a point source of light, the fraction of the light emitted by the source that will pass through the system is

$$\mathcal{F} = \frac{1}{2}(1 - \cos\theta) = \frac{1}{2}\left[1 - \left(1 + \frac{D^2}{4d^2}\right)^{-1/2}\right], \tag{13.84}$$

where $\theta$ is the angle shown in Fig. 13.18. For $D \ll d$, Eq. (13.84) reduces to

$$\mathcal{F} = \frac{D^2}{16d^2}. \tag{13.85}$$

For a lens the ratio of the focal length to the EPD is called the *f*/number ($f/\#$) of the lens. For an $f/1$ lens the ratio of its focal length to the aperture (the open-lens diameter) is equal to unity. For a point source at the focal point of such a lens, $\mathcal{F} = 0.067$.

## 13.8 Apertures, stops, and pupils

Apertures control the ray trajectories that can pass through an optical system. The *aperture stop* is the aperture within the system that limits the angular spread, or diameter, of a cone of rays from an axial point on the object, which can pass through the system. Another aperture in the system may restrict the size or angular extent of an object that the system can image. Awareness of the position and size of these stops in an optical design is crucial

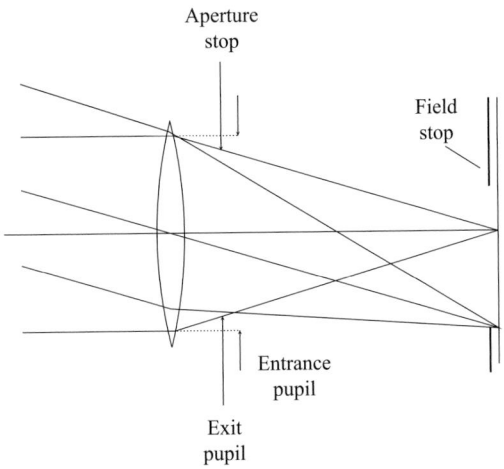


**Fig. 13.19** A single-lens imaging system illustrating the concepts of the aperture stop, entrance pupil, and exit pupil.

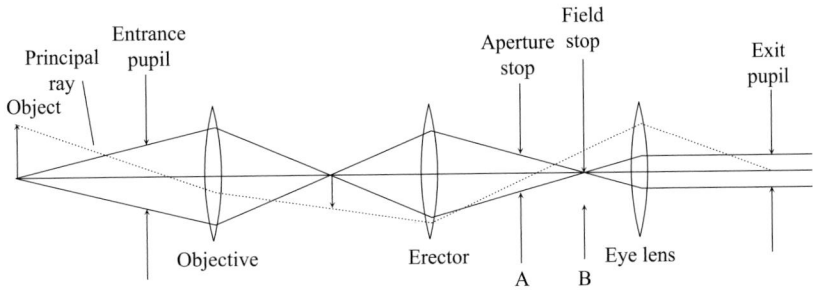


**Fig. 13.20** A three-lens imaging system showing an aperture stop, a field stop, and the location of the entrance and exit pupils.

in determining the light-gathering power of the system, and its ability to deliver light to an image or detector. The role of the aperture and field stop can be illustrated with Fig. 13.19, which shows a single lens being used to image light from a distant object. The circular aperture behind the lens clearly limits the diameter of a bundle of rays from an axial point on the object and is clearly the aperture stop. The second stop, placed in the image plane, clearly controls the maximum angle at which rays from off-axis points on the object can pass through the system. This is the *field stop* in this case.

Whether a given aperture in an optical system is an aperture stop or a field stop is not always so easy to determine as was the case in Fig. 13.19. Figure 13.20 shows a three-element optical system, which actually constitutes what is called a Gullstrand opthalmoscope.[2] The first lens, closest to the object, called the *objective* lens, produces a real inverted image, which is then re-inverted by the *erector* lens, to form an image at the focal point of the *eyepiece* lens. Aperture A is the aperture stop, and clearly restricts the angular range of rays from an axial point on the object that can pass through the system.

[2] This device is used by physicians for examining the retina of a patient's eye [4].

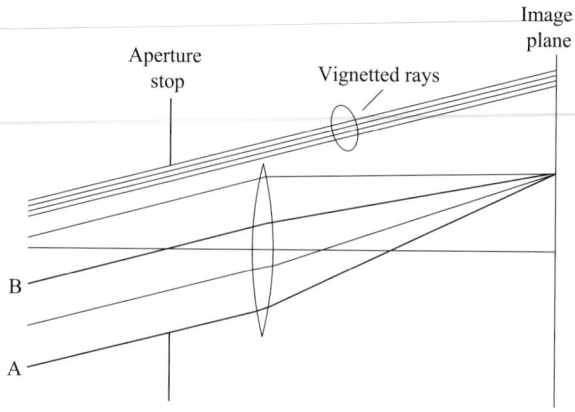


**Fig. 13.21** An aperture stop in front of a single lens showing the marginal ray A, the chief ray B, and vignetting.

Aperture B restricts from what height on the object rays can originate and still pass through the system. This aperture is acting as a field stop.

The *entrance pupil* is the image of the aperture stop produced by all the elements of the system between the aperture stop and the object. In Fig. 13.20 the two elements producing the image are the erector lens and the objective lens. The *exit* pupil is the image of the aperture stop produced by all the elements between the aperture stop and the image, which in Fig. 13.20 is just the eye lens. The size and location of the entrance and exit pupils determine the light-gathering and light-delivery properties of the system. For example, if the entrance pupil has diameter $D$, and lies a distance $L$ from a point source of radiant power $P$ (in watts) then the power entering the entrance pupil is

$$P_1 = \frac{P}{2}(1 - \cos\theta), \tag{13.86}$$

where $\tan\theta = D/(2L)$.

For a point source that is far from the entrance pupil this result becomes

$$P_1 = \frac{PD_1^2}{16L^2}. \tag{13.87}$$

The solid angle subtended by the entrance pupil in this case is $\Delta\omega = D_1^2/(4L^2)$. The light that leaves the system can be characterized by the total power that leaves the exit pupil, and the angular range of the rays that leave the exit pupil. Somewhere in an imaging system there is always an *aperture stop*. This may be an actual aperture in an opaque thin screen placed on the system axis, or it may be the clear aperture of a lens or mirror in the system. In either case the aperture stop limits the range of rays that can pass through the system. This concept is illustrated in Fig. 13.21. A ray of light such as A that enters the system and just passes the edge of the aperture stop is called a *marginal ray*. A ray such as B that enters the system and passes through the center of the aperture stop is called a *chief* or *principal ray*. An aperture stop will always allow some light from every point on an object to reach the corresponding point on the image.

The ray that enters the system half-way between the highest and lowest rays of an oblique beam is called the *chief ray* of the beam. In the absence of aberrations the principal ray and chief ray are the same, and both pass through the center of the entrance pupil and the aperture stop.

## 13.9 Vignetting

In our previous discussion we have simplified the actual performance of a real optical system, because for oblique rays passing through the system the roles of aperture and field stops become intertwined. This can be illustrated with Fig. 13.21. A family of oblique rays that fills the aperture stop might not all reach the image plane because the following lens is acting as a stop. This is called *vignetting*. An imaging system does not actually remove points of the object from the image, but parts of the image that depend for their illumination on oblique rays (especially at large angles) become less bright. Vignetting is often deliberately allowed to occur in an imaging system so as to remove from the system certain rays that would otherwise suffer severe aberration.

## 13.10 Exact ray tracing and aberrations

If an imaging system is built with a series of lenses and/or mirrors then there are inevitably various aberrations. For refraction at a spherical surface this results because, for angles of incidence and refraction $\theta_1$ and $\theta_2$, respectively, when light travels from a medium with refractive index $n_1$ to one of refractive index $n_2$,

$$n_1 \sin\theta_1 = n_2 \sin\theta_2. \tag{13.88}$$

The paraxial approximation, which lacks aberrations, fails because $\theta_1 \neq \sin\theta_1$; in reality $\sin\theta = \theta - \theta^3/3! + \theta^5/5! - \theta^7/7!$. The overall effect of aberrations can be greatly reduced, and some types of aberrations can be eliminated by using several spherical surfaces, or, in some situations, by using an appropriate *aspherical* surface, a surface that is not part of a sphere but is generally describable in terms of some other conic function or power series of the coordinates. Analysis of imaging systems is most conveniently carried out using exact ray-tracing techniques. In exact ray tracing the small-angle approximation is not made, and Snell's law is solved exactly for each ray, at each interface. By injecting many light rays into the system over some defining aperture or range of angles the performance of the system in imaging, focusing, light collection, or light delivery can be evaluated precisely. These days this procedure is carried out with optical design software. The most notable software packages in this regard are Code V[3] and Zemax.[4] Examples of their use are

[3] Available from Optical Research Associates.
[4] Available from Zemax Development Corporation.

**Table 13.1** Some important wavelengths used in optical design

| Wavelength (nm) | Designation | Source |
|---|---|---|
| 312.59 | | Mercury |
| 334.15 | | Mercury |
| 365.01 | $\ell$ | Mercury |
| 404.66 | h | Mercury |
| 435.83 | g | Mercury |
| 479.99 | F′ | Mercury |
| 486.13 | F | Hydrogen |
| 546.07 | e | Mercury |
| 587.56 | d | Helium |
| 589.29 | D | Sodium (doublet) |
| 632.80 | | He–Ne laser |
| 643.85 | c′ | Cadmium |
| 656.27 | c | Hydrogen |
| 706.52 | r | Helium |
| 852.11 | s | Cesium |
| 1,013.98 | t | Mercury |
| 1,060.00 | | Neodymium laser |
| 1,529.58 | | Mercury |
| 1,970.09 | | Mercury |
| 2,325.42 | | Mercury |

provided in Moore, Davis, and Coplan [5]. The aberrations in an optical system can be greatly reduced in multi-element systems, such as modern camera lenses, where the radii and spacings of multiple spherical (or aspherical) elements, as well as the optical materials used, are jointly optimized. Since most common imaging systems use various kinds of glass, it is worthwhile describing the variable parameters for a glass, such as its refractive index $n$ and its $dn/d\lambda$, both of which vary with wavelength. The dispersion is frequently characterized by the Abbé $V$ number, which provides a relative measure of dispersion:

$$V = \frac{n_{\mathrm{d}} - 1}{n_{\mathrm{F}} - n_{\mathrm{C}}}, \tag{13.89}$$

where the refractive index $n$ is specified at the three wavelengths d (the helium d line at 587.6 nm), F (the hydrogen F line at 486.1 nm), and C (the hydrogen C line at 656.3 nm). Some of the standard wavelengths that are used in optical design are listed in Table 13.1. The different glasses that are available from a manufacturer are generally plotted on a glass chart, which plots $(n_{\mathrm{d}} - 1)$ against $V$. Figure 13.22 shows such a chart for glasses available from Schott. Other major suppliers of optical glass are Corning, Hoya, and Ohara.

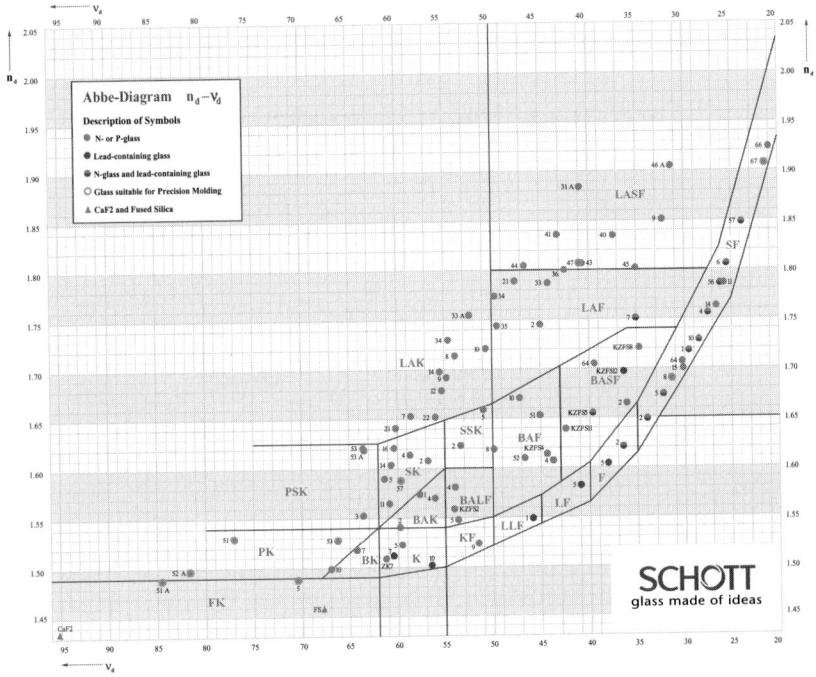


**Fig. 13.22** The Schott glass chart showing how the refractive index $n_d$ and Abbé number $V$ vary for different glasses from www.us.schott.com/advanced_optics.

Two old classifications of glass into *crown* and *flint* glasses can be related to the glass chart. Crown glasses are glasses with a $V$ value greater than 55 if $n_d < 1.6$ and $V > 50$ if $n_d > 1.6$. The flint glasses have $V$ values below these limits. The rare-earth glasses contain rare-earth oxides instead of $SiO_2$, which is the primary constituent of crown and flint glasses.

## 13.11 Chromatic aberrations

Because, in general, the index of refraction increases with decreasing wavelength, the focal length of a singlet lens[5] will, for example, be shorter for blue light than it is for red light. This effect leads to imperfect imaging so that in white-light illumination both axial and off-axis object points will be imaged not as bright points but as regions that show blurring from red to blue. For axial object points this blurring is spherically symmetric and results from the image being closer to the lens for blue light than for red light. This is simple *chromatic aberration*. For off-axis object points, the magnification of the system varies with wavelength, leading to what is called *lateral color* or *transverse chromatic aberration*. Chromatic aberration can be corrected for two colors (generally red and blue) and

[5] A simple lens with just two spherical surfaces.

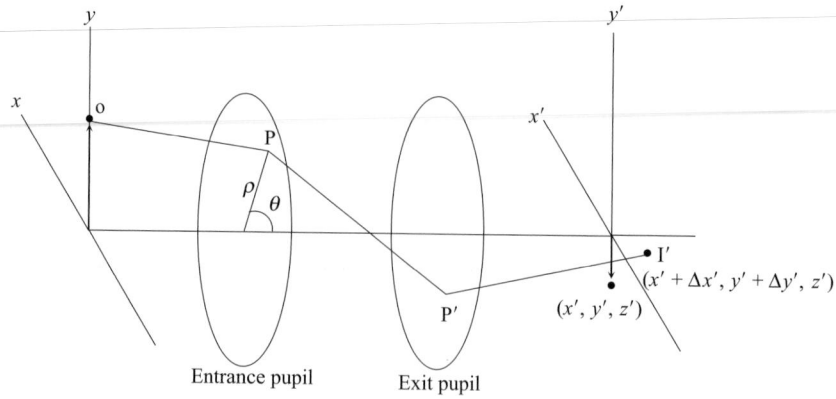


**Fig. 13.23** A diagram of a generalized imaging system used to describe aberrations.

the imaging of axial points. Such lenses take the form of an achromatic doublet where a doublet lens is generally fabricated from a positive (converging) crown-glass element and a negative (diverging) flint-glass element. If a compound lens is designed to remove chromatic aberration at three wavelengths it is called an *apochromat*, whereas if it is corrected for four wavelengths it is called a *superachromat*. Achromatic doublets, because they have at least three spherical surfaces (or four in an air-spaced achromat) can also be better corrected for other aberrations than a singlet lens.

## 13.12 Geometrical aberrations

In a perfect imaging system all rays from a point on the object would pass through an identical point on the image and there would be a linear relation between the coordinates of points in the image and corresponding points on the object. For an axially symmetric system we can arbitrarily choose the object point 0 in Fig. 13.23 to be at the $(x, y, z)$ point $(0, h, 0)$ in the object. We take the polar coordinates of a ray from this point to a point P in the entrance pupil as $(\rho, \theta)$. In the general case this ray will intersect the image plane at point I′ with coordinates in the image plane $(x', y')$. The aberrations $\Delta x'$ and $\Delta y'$ are the displacements of $(x', y')$ from the paraxial image point $\mathrm{I}\,(x'_0, y'_0, z')$ for which there is a linear transformation from the coordinates of the point on the object. The lowest-order aberration terms contain terms in $h^3$, $h^2\rho$, or $h\rho^2$, and are referred to as *third-order aberrations*. Although there are differences in terminology concerning these terms in the literature, we can write (following Born and Wolf [6])

$$\begin{aligned}\Delta y' = {} & B\rho^3 \cos\theta - Fh\rho^2(2 + \cos(2\theta)) \\ & + (2C + D)h^2\rho\cos\theta - Eh^3\end{aligned} \tag{13.90}$$

and

$$\Delta x' = B\rho^3 \sin\theta - Fh\rho^2 \sin(2\theta) + Dh^2\rho \sin\theta. \tag{13.91}$$

The coefficients $B, C, D, E$, and $F$ characterize the five *primary* or *Seidel* aberrations. Their magnitude in any given optical design can be calculated and visualized with ray-tracing optical software.

(i) *Spherical aberration*, $B \neq 0$. When all the coefficients in Eqs. (13.90) and (13.91) are zero except for $B$, the remaining effect is spherical aberration. This aberration is strikingly observed in the imaging of an axial point source, which will be imaged to a circular bright region whose radius reveals the extent of the aberration. For an axial point $h = 0$ and from Eqs. (13.90) and (13.91) $\Delta y' = B\rho^3 \cos\theta$ and $\Delta x' = B\rho^3 \sin\theta$, which can be written as

$$\Delta r = B\rho^3. \tag{13.92}$$

For focusing of parallel light the "best-form" singlet lens for minimum spherical aberration is close to a convex plane (with the convex surface towards the incoming parallel light).

(ii) *Coma*, $F \neq 0$. Coma is the aberration in which the image of an off-axis point varies for rays passing through different regions of the entrance pupil. It is difficult to produce a lens for which coma alone can be observed. The spot diagram associated with coma shows the characteristic "flaring" of the spot pattern, like a comet tail, which gives the aberration its name. Coma can be controlled by varying the curvatures of surfaces in the system.

(iii) *Astigmatism*, $C \neq 0$. All the aberrations discussed so far are defects in the imaging of *meridional rays*, that is, rays in the plane containing the axis of the lens and the line from the object point through the center of the lens. The imaging is different for *sagittal rays*, that is, rays in the sagittal plane, which is perpendicular to the meridional plane, as illustrated in Fig. 13.24. The resulting aberration is called *astigmatism*. It can be controlled by lens curvature and refractive-index variations, and by the use of apertures to restrict the range of angles and off-axis distances at which rays can traverse the lens.

(iv) *Distortion*, $E \neq 0$. In distortion the magnification varies across the image plane. The height of the image above the axis relative to the height of the object varies across the image. Each object point appears as a point in the image, and the images of object points on a plane orthogonal to the axis are also on such a plane. However, the magnification of an object line segment may vary with its distance from the axis. This is called *distortion*. It takes two common forms: *pincushion* and *barrel* distortion, as illustrated in Fig. 13.25. Distortion is sensitive to the lens shape and spacing, and to the size and position of apertures in the system – called *stops*.

For further details of aberrations and how to deal with them, we refer the reader to specialized texts on optics (especially lens design), such as those by Born and Wolf [6], Ditchburn [7], Levi [8], Smith [9, 10], Shannon [11], and Laikin [12].

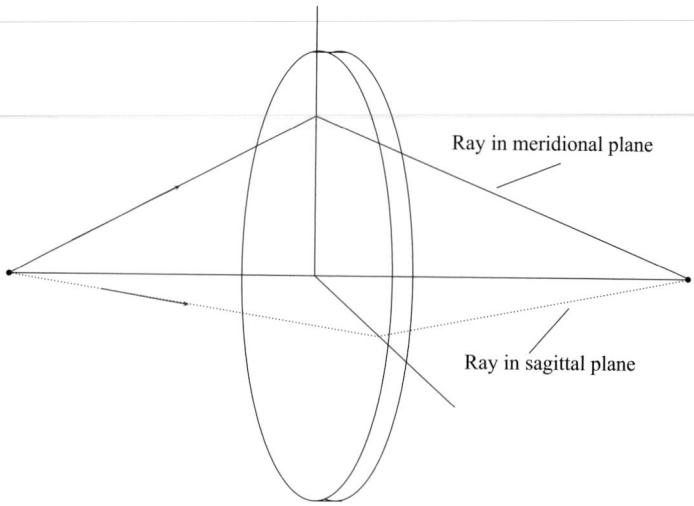


**Fig. 13.24** The meridional and sagittal planes for rays passing through a simple lens.

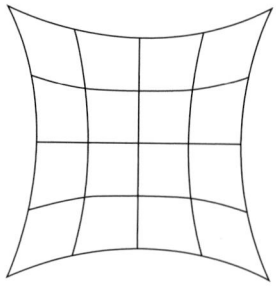

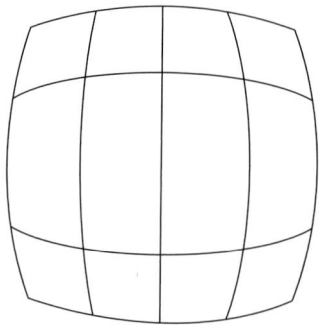

**Fig. 13.25** Pincushion-and-barrel distortion.

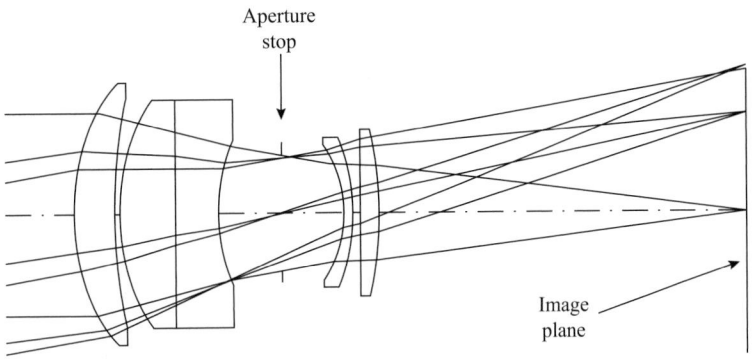


Fig. 13.26 The layout of a telephoto lens, U.S. Patent No. 2,906,173.

(v) *Curvature of field*, $D \neq 0$. In this aberration the image of a plane object perpendicular to the axis is sharp over a curved surface in the image space.

## 13.13 Spot diagram

If very many rays are launched from a point on the object so as to cover the extrance pupil of the system, the resulting pattern of ray intersections with the image plane is called the *spot diagram*. A tight spot diagram indicates that the effects of several aberrations, notably spherical aberration, astigmatism, coma, and curvature of field have been markedly reduced. It must be remembered that this does not imply that the aberrations themselves have necessarily been eliminated, but rather that their effects have been reduced in the situation being examined. A tighter spot diagram can always be produced by reducing the size of the entrance pupil.[6] Figure 13.26 shows the layout of a telephoto lens, and Fig. 13.27 shows its corresponding spot diagram for simultaneous inclusion of three wavelengths, a blue, a green, and a red, entering parallel to the axis and at angles of 7° and 10° to the axis. The 0° spot is dominated by spherical aberration, while the 7° and 10° rays show strong coma.

## 13.14 The modulation transfer function

A common way of characterizing the performance of an imaging system is through its *modulation transfer function* (MTF). This refers to the ability of the system to replicate, in the image, periodic features in the object. An optical test pattern used to show this consists of a series of white and black bands. These bands will be smeared out in the

[6] A pinhole camera is free of all aberrations in a geometrical-optics description. In practice the phenomenon of diffraction, which becomes important for small apertures, will cause diffraction effects to blur the image.

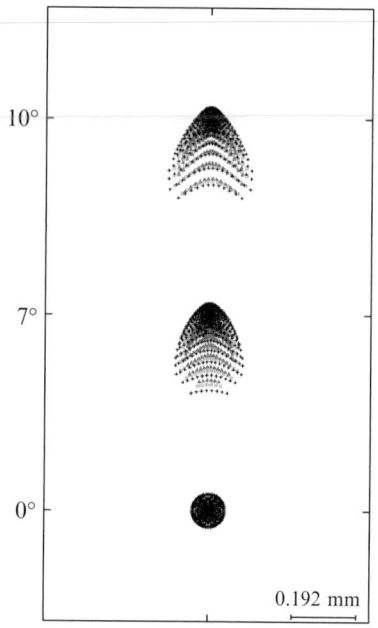


Fig. 13.27 A spot diagram of the telephoto lens shown in Fig. 13.26.

image to a greater or lesser extent because of aberrations, and ultimately, if geometrical aberrations are absent, by diffraction. The relative brightness of the object and image will appear schematically as shown in Fig. 13.28. We can characterize the *visibility*, *contrast*, or *modulation* of the image as

$$\text{visibility} = \frac{\max - \min}{\max + \min}. \tag{13.93}$$

If the visibility is plotted versus the number of lines per millimeter in the object, in this case this shows the modulation transfer function. Figure 13.29 shows an example for a good camera lens. The MTF is ultimately limited by diffraction. In this case, for a pattern with $\nu$ lines per millimeter,

$$\text{MTF}(\nu) = \frac{1}{\pi}(2\phi - \sin(2\phi)), \tag{13.94}$$

where

$$\phi = \cos^{-1}\left(\frac{\lambda_0 \nu}{2\text{NA}}\right). \tag{13.95}$$

In Eq. (13.95) $\lambda_0$ is measured in millimeters and NA is the numerical aperture of the system. A quantity closely related to the MTF that is used for characterizing the MTF and also the focusing of a light beam, or laser beam, is the *Strehl ratio*. The Strehl ratio is the ratio of the illumination at the center of a focused or circular image to the illumination of an equivalent unaberrated image. The interested reader can see a more detailed discussion in Smith [9].

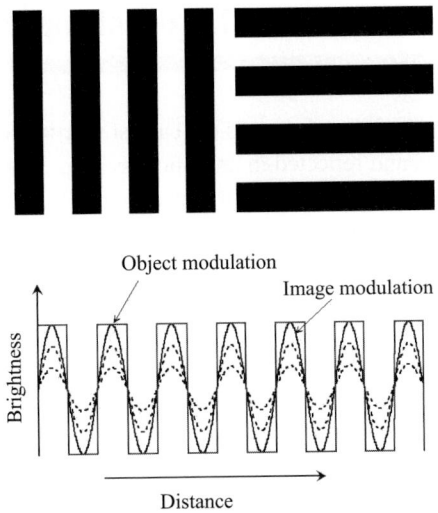


Fig. 13.28 The modulation transfer function.

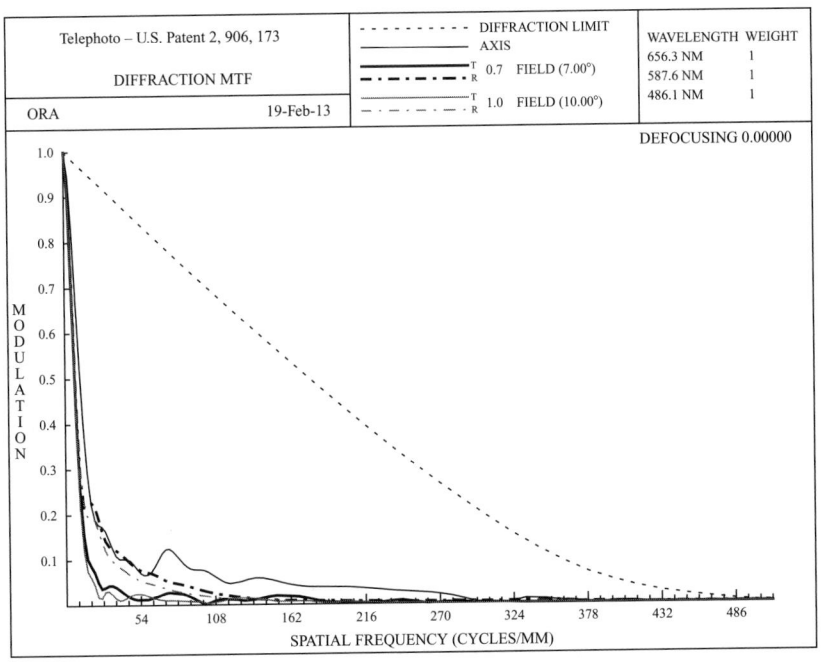


Fig. 13.29 The modulation transfer function of the telephoto lens shown in Fig. 13.26.

## 13.15 The use of impedances in optics

The method of impedances is the easiest way to calculate the fraction of incident intensity transmitted and reflected in an optical system. It is also a good way to follow the changes in polarization state that result when light passes through an optical system.

The impedance of a plane wave traveling in a medium of relative permeability $\mu_r$ and dielectric constant $\epsilon_r$ is

$$Z = \sqrt{\frac{\mu_r \mu_0}{\epsilon_r \epsilon_0}} = Z_0 \sqrt{\frac{\mu_r}{\epsilon_r}}. \tag{13.96}$$

If $\mu_r = 1$, as is usually the case for optical media, the impedance can be written

$$Z = Z_0/n. \tag{13.97}$$

This impedance relates the transverse **E** and **H** fields of the wave:

$$Z = \frac{E_{tr}}{H_{tr}}. \tag{13.98}$$

When a plane wave crosses a planar boundary between two different media the components of both **E** and **H** parallel to the boundary have to be continuous across that boundary. Figure 13.30(a) illustrates a plane wave polarized in the plane of incidence striking a planar boundary between two media of refractive indices $n_1$ and $n_2$, respectively. In terms of the magnitudes of the vectors involved

$$E_i \cos\theta_1 + E_r \cos\theta_1 = E_t \cos\theta_2 \tag{13.99}$$

and

$$H_i - H_r = H_t. \tag{13.100}$$

Equation (13.100) can be written

$$\frac{E_i}{Z_1} - \frac{E_r}{Z_1} = \frac{E_t}{Z_2}. \tag{13.101}$$

It is easy to eliminate $E_t$ between Eqs. (13.99) and (13.101) to give

$$\rho = \frac{E_r}{E_i} = \frac{Z_2 \cos\theta_2 - Z_1 \cos\theta_1}{Z_2 \cos\theta_2 + Z_1 \cos\theta_1}, \tag{13.102}$$

where $\rho$ is the *reflection coefficient* of the surface. The fraction of the incident energy reflected from the surface is called the reflectance, $R$, where

$$R = \rho^2. \tag{13.103}$$

Similarly, the *transmission coefficient* of the boundary is

$$\tau = \frac{E_t \cos\theta_2}{E_i \cos\theta_1} = \frac{2Z_2 \cos\theta_2}{Z_2 \cos\theta_2 + Z_1 \cos\theta_1}. \tag{13.104}$$

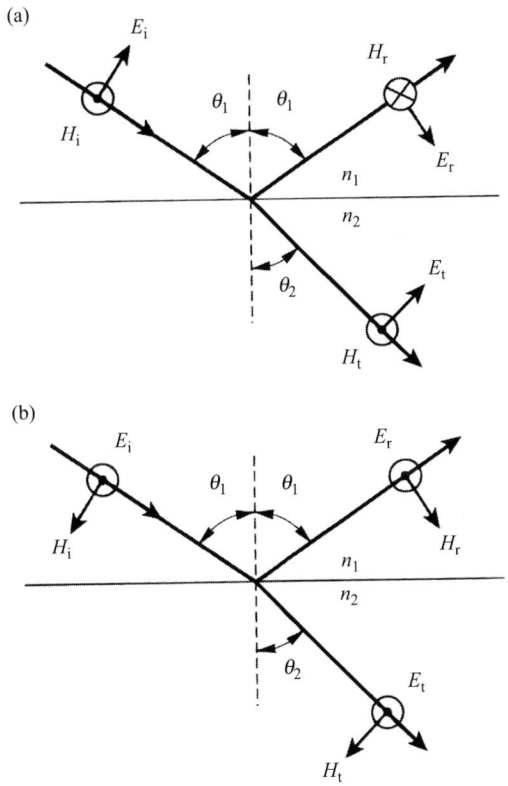


**Fig. 13.30** Field vectors of a plane wave striking a planar boundary: (a) an electric field in the plane of incidence – a P wave or TM wave; and (b) an electric field perpendicular to the plane of incidence – an *S* wave or TE wave.

By application of a similar treatment to the geometry shown in Fig. 13.30(b) it can be shown that for a plane wave polarized perpendicular to the plane of incidence

$$\rho = \frac{Z_2 \sec\theta_2 - Z_1 \sec\theta_1}{Z_2 \sec\theta_2 + Z_1 \sec\theta_1}, \tag{13.105}$$

$$\tau = \frac{2Z_2 \sec\theta_2}{Z_2 \sec\theta_2 + Z_1 \sec\theta_1}. \tag{13.106}$$

If the effective impedance for a plane wave polarized in the plane of incidence (*P polarization*) and incident on a boundary at angle $\theta$ is defined as

$$Z' = Z\cos\theta, \tag{13.107}$$

and that for a wave polarized perpendicular to the plane of incidence (*S polarization*) is defined as

$$Z' = Z\sec\theta, \tag{13.108}$$

then a universal pair of formulas for $\rho$ and $\tau$ results:

$$\rho = \frac{Z_2' - Z_1'}{Z_1' + Z_2'}, \tag{13.109}$$

$$\tau = \frac{2Z_2'}{Z_1' + Z_2'}. \tag{13.110}$$

It will be apparent from an inspection of Fig. 13.30 that $Z'$ is just the ratio of the electric field component parallel to the boundary and the magnetic field component parallel to the boundary. For reflection from an ideal mirror $Z_2' = 0$. Note that, from Eqs. (13.91) and (13.92),

$$\tau = 1 + \rho. \tag{13.111}$$

In normal incidence Eqs. (13.84) and (13.87) become identical and can be written as

$$\rho = \frac{Z_2 - Z_1}{Z_2 + Z_1} = \frac{n_1 - n_2}{n_1 + n_2}. \tag{13.112}$$

Note that there is a change of phase of $\pi$ in the reflected field relative to the incident field when $n_1 < n_2$.

Since intensity $\propto$ (electric field)$^2$, the fraction of the incident energy that is reflected is

$$R = \rho^2 = \left(\frac{n_1 - n_2}{n_1 + n_2}\right)^2. \tag{13.113}$$

$R$ increases with the index mismatch between the two media, and with the number of interfaces, as shown in Fig. 13.31. However, it should be noted that if a series of boundaries are parallel and the incident light is fairly monochromatic, then Fabry–Perot effects occur and the transmittance of the structure must be calculated differently – see Section 13.15.3.

It is important to note that, although the energy reflectance at a boundary is $|\rho|^2$, the energy transmittance is *not* $|\tau|^2$. This is because the impedances on the two sides of the boundary are different. For example, if a wave is incident from medium 1 to medium 2 its intensity is $I_1 = |E_i|^2/(2Z_1)$, and the transmitted intensity is $I_2 = |E_t|^2/(2Z_2)$. The transmittance of the boundary is

$$T = \frac{I_2}{I_1} = \tau^2 \frac{Z_1}{Z_2}. \tag{13.114}$$

If the wave were traveling in the reverse direction across the boundary then its transmission coefficient, as in Eq. (13.110), would be

$$\tau' = \frac{2Z_1'}{Z_1' + Z_2'} \tag{13.115}$$

and

$$\tau\tau' = \frac{4Z_1'Z_2'}{(Z_1' + Z_2')^2} = \frac{\tau^2 Z_1}{Z_2} = T. \tag{13.116}$$

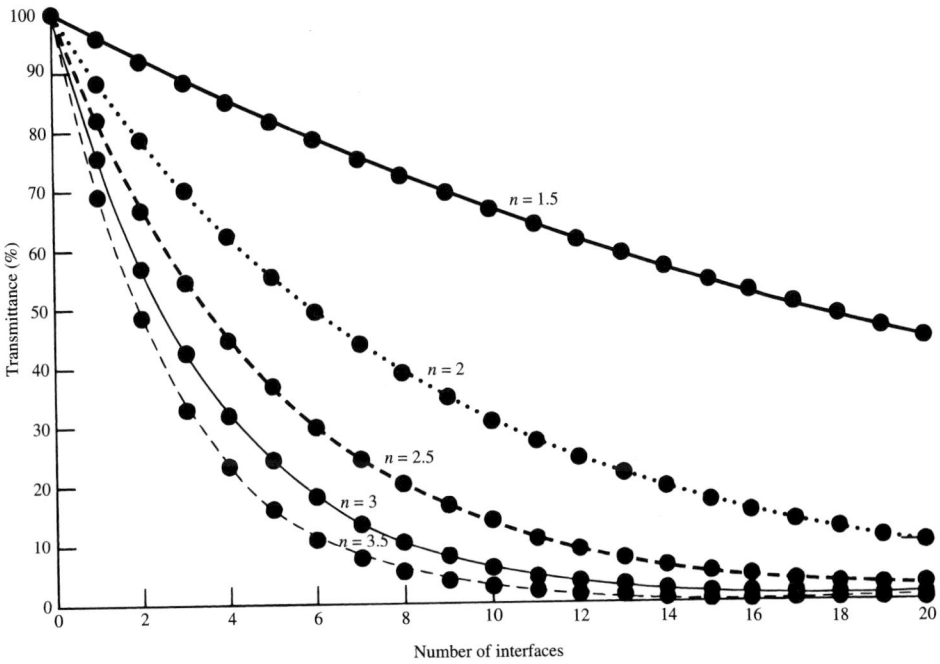


**Fig. 13.31** The change in the reflectance at the boundary between two media as the index mismatch increases and as the number of independent boundaries crossed increases.

If there is no absorption of energy at the boundary the fraction of energy transmitted, called the *transmittance*, is

$$T = 1 - R = \frac{4n_1 n_2}{(n_1 + n_2)^2}. \tag{13.117}$$

### 13.15.1 Reflectance for waves incident on an interface at oblique angles

If the wave is not incident normally, it must be decomposed into two linearly polarized components, namely one polarized in the plane of incidence and one polarized perpendicular to the plane of incidence.

For example, consider a plane-polarized wave incident on an air/glass ($n = 1.5$) interface at an angle of incidence of 30° with a polarization state exactly intermediate between the S and P polarizations.

The angle of refraction at the boundary is found from Snell's law,

$$\sin\theta_2 = \sin 30°/1.5,$$

so $\theta_2 = 19.47°$.

The effective impedance of the P component in the air is

$$Z'_{P1} = 376.7\cos\theta_1 = 326.23\,\Omega,$$

and that in the glass is

$$Z'_{P2} = \frac{376.7}{1.5} \cos\theta_2 = 236.77\,\Omega.$$

Thus, from Eq. (13.112), the reflection coefficient for the P component is

$$\rho_P = \frac{236.77 - 326.23}{236.77 + 326.23} = -0.159.$$

The fraction of the intensity associated with the P component that is reflected is $\rho_P^2 = 0.0253$.

For the S component of the input wave

$$Z'_{S1} = 376.7/\cos\theta_1 = 434.98\,\Omega,$$
$$Z'_{S2} = 376.7/(1.5\cos\theta_2) = 266.37\,\Omega,$$

and

$$\rho_S = \frac{266.37 - 434.98}{266.37 + 434.98} = -0.240.$$

The fraction of the intensity associated with the S-polarization component that is reflected is $\rho_S^2 = 0.0578$. Since the input wave contains equal amounts of S and P polarization, the overall reflectance in this case is

$$R = \langle \rho^2 \rangle_{avg} = 0.0416 \simeq 4\%.$$

Note that the reflected wave now contains more S polarization than P, so the polarization state of the reflected wave has been changed.

### 13.15.2 Brewster's angle

Returning to Eq. (13.112), it might be asked whether the reflectance is ever zero. It is clear that $\rho$ will be zero if

$$n_1 \cos\theta_2 = n_2 \cos\theta_1, \tag{13.118}$$

which, from Snell's law, Eq. (13.6), gives

$$\cos\theta_1 = \frac{n_1}{n_2}\sqrt{1 - \left(\frac{n_1^2}{n_2^2}\right)\sin^2\theta_1}, \tag{13.119}$$

giving the solution

$$\theta_1 = \theta_B = \arcsin\sqrt{\frac{n_2^2}{n_1^2 + n_2^2}} = \arctan\left(\frac{n_2}{n_1}\right). \tag{13.120}$$

The angle $\theta_B$ is called *Brewster's angle*; a wave polarized in the plane of incidence and incident on a boundary at this angle is totally transmitted. This fact is put to good use in the design of low-reflection-loss windows in laser systems, as we have seen previously.

If Eq. (13.105) is inspected carefully, it will be seen that there is no angle of incidence that yields zero reflection for a wave polarized perpendicular to the plane of incidence.

### 13.15.3 Transformation of impedance through multi-layer optical systems

The concept of impedance allows the reflection and transmission characteristics of multi-layer optical systems to be evaluated very simply. If the incident light is incoherent then the overall transmission of a multi-layer structure is just the product of the transmittances of its various interfaces. For example, an air/glass interface transmits about 96% of the light in normal incidence. The transmittance of a parallel-sided slab is $0.96 \times 0.96 \equiv 92\%$. This simple result ignores the possibility of interference effects between reflected and transmitted waves at the two faces of the slab. If the faces of the slab are very flat and parallel, and the light is coherent, such effects cannot be ignored. In this case, the method of transformed impedances is useful.

Consider the three-layer structure shown in Fig. 13.32(a). The path of a ray of light through the structure is shown. The angles $\theta_1, \theta_2$, and $\theta_3$ can be calculated from Snell's law. As an example consider a wave polarized in the plane of incidence. The effective impedances of media 1, 2, and 3 are

$$\begin{aligned} Z_1' &= Z_1 \cos\theta_1 = Z_0 \cos\theta_1/n_1, \\ Z_2' &= Z_2 \cos\theta_2 = Z_0 \cos\theta_2/n_2, \\ Z_3' &= Z_3 \cos\theta_3 = Z_0 \cos\theta_3/n_3. \end{aligned} \tag{13.121}$$

It can be shown that the reflection and transmission coefficients of the structure are exactly the same as the equivalent structure in Fig. 13.32(b) for normal incidence, where the effective thickness of layer 2 is now

$$d' = d\cos\theta_2. \tag{13.122}$$

The reflection coefficient of the structure can be calculated from its equivalent structure using the concept of transformed impedance [13].

The transformed impedance of medium 3 at the boundary between media 1 and 2 is

$$Z_3'' = Z_2' \left( \frac{Z_3' \cos(k_2 d') + jZ_2' \sin(k_2 d')}{Z_2' \cos(k_2 d') + jZ_3' \sin(k_2 d')} \right), \tag{13.123}$$

where $k_2 = 2\pi/\lambda_2 = 2\pi n_2/\lambda_0$.

The reflection coefficient of the whole structure is now just

$$\rho = \frac{Z_3'' - Z_1'}{Z_3'' + Z_1'}, \tag{13.124}$$

and its transmittance is

$$T = 1 - |\rho|^2. \tag{13.125}$$

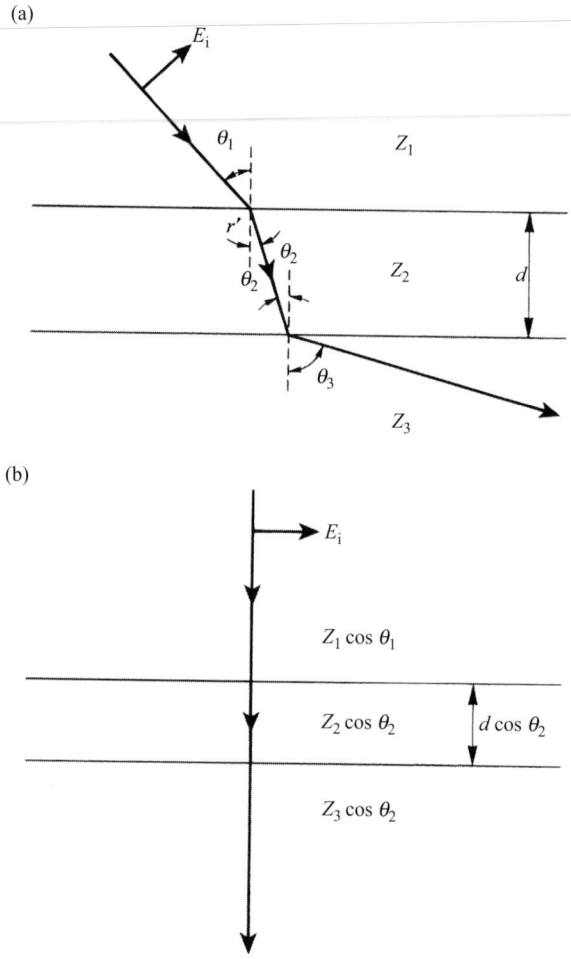


**Fig. 13.32** Waves passing through a dielectric slab: (a) waves incident at an angle being refracted and transmitted; and (b) the normal-incidence structure equivalent to (a).

In a structure with more layers, the transformed-impedance formula Eq. (13.123) can be used sequentially, starting at the last optical surface and working back to the first.

### 13.15.4 The effect of an oblique slab on the reflected polarization state

If a linearly polarized wave is a mixture of P and S waves then its resultant polarization direction makes an angle $\alpha$ with the P direction where $\tan\alpha = E_S/E_P$, where $E_P$ and $E_S$ are the components of the input electric vector in the P and S directions. After reflection from the slab, the two orthogonal components of the electric field are $\rho_P E_P$ and $\rho_S E_S$, respectively, where $\rho_P$ and $\rho_S$ are the corresponding reflection coefficients. If the input wave is at Brewster's angle, then the reflected light is pure S-polarized, and the transmitted light is more P-polarized. This serves as the basis for the *stacked-plate* polarizer, which

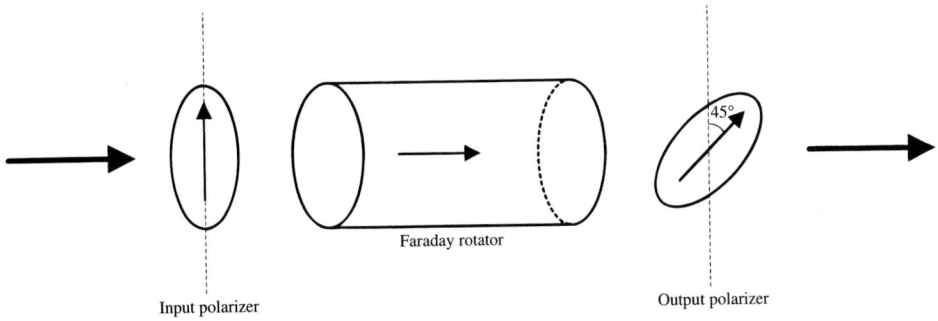


Fig. 13.33 Schematic construction of a Faraday isolator. A section of Faraday material producing a 45° rotation of the plane of linear polarization is placed between two polarizers oriented at 45° with respect to each other.

consists of a series of slabs all placed at Brewster's angle. The larger the number of slabs the more closely the transmitted light approaches pure P polarization.

### 13.15.5 Example

A linearly polarized wave at $\lambda_0 = 500\,\text{nm}$ consisting of 70% P and 30% S polarization (by intensity) strikes a parallel-sided slab of material with ($n = 3.5$) that is 1.025 mm thick, at an angle of incidence of 60°. The incident polarization makes an angle $\alpha = \arctan\sqrt{0.3/0.7} = 33.211°$ with the P direction. Use of Eqs. (13.86) and (13.87) gives $\rho_P = -0.265 + j0.265$, $\rho_S = -0.858 + j0.293$. The reflected light is now elliptically polarized because the amplitudes of the reflected P and S components are neither equal nor in phase. The trajectory traced out by the resultant reflected electric field vector can be calculated using

$$E_P(t) = \text{Re}\left(\sqrt{0.7}\rho_P e^{j\omega t}\right) \tag{13.126}$$

and

$$E_S(t) = \text{Re}\left(\sqrt{0.3}\rho_S e^{j\omega t}\right). \tag{13.127}$$

The polarization state of the transmitted light cannot be determined directly using impedance techniques, but must be determined using transfer matrices (see Section 14.5).

### 13.15.6 Polarization changes

In summary, a change in polarization state can occur if (a) a wave reflects off an interface for which the reflection coefficient is different for light polarized in the plane of incidence (a *P wave*) and light polarized perpendicular to the plane of incidence (an *S wave*), or is transmitted through a slab structure for which the transmission coefficients for the P and S components of the wave are different. In addition, the wave can be transmitted through an anisotropic, or optically active, medium that alters the polarization state of the wave. These effects are considered in further detail in Chapters 17 and 20. Optically active media rotate the plane of polarization of a linearly polarized wave passing through them. This can occur in solids or liquids in which the constituent particles are *chiral* in character, that is, they have distinct left- or right-handed symmetry. A good example would be an aqueous sugar solution. A solution of the sugar dextrose, for example, rotates the plane of polarization of a linearly polarized wave in a clockwise direction viewing in the direction of propagation, while its *enantiomorph*,[7] laevulose, rotates the plane of polarization, in a counterclockwise direction. One way to account for the phenomenon of optical activity is for left- and right-hand circularly polarized light to have different phase velocities in the medium. Optical activity can also be induced by the action of an external field. For example, in the liquid nitrobenzene, application of a transverse electric field causes the liquid to become optically active. This is the Kerr effect, which can be used to make shutters for $Q$-switching. Many materials become optically active if they are placed in an axial magnetic field. This is the Faraday effect. The actual amount of rotation is described by the *Verdet* constant $\alpha$ for the material. For a slab of material of length $\ell$ and an applied axial flux density $\mathbf{B}$

$$\theta_{\text{rot}} = \alpha \ell |\mathbf{B}|. \tag{13.128}$$

If $\alpha$ is positive the direction of rotation is clockwise if the light propagates in the same direction as $\mathbf{B}$, and it is counterclockwise if the direction of propagation is antiparallel to $\mathbf{B}$. Thus, a to-and-fro pass of light through the field doubles the angle of rotation.[8] The Faraday effect can be used to make an optical isolator, the operation of which is shown schematically in Fig. 13.33.

A plane-polarized wave is rotated by 45° on passing through the Faraday-active medium from left to right. It then passes through a second polarizer oriented at 45° to the input polarizer. A wave passing from right to left also experiences a 45° rotation, but cannot then pass the entrance linear polarizer. Thus, the source of radiation is protected from undesirable feedback from optical components further down the chain. The use of isolators is extremely important in high-power oscillator/amplifier(s) systems. Without their use a reflected wave of initially weak intensity could be amplified on its return passage through the system, with disastrous consequences for the oscillator.

## 13.16 Problems

13.1 A slab of glass with $n = 1.55$ is exactly 3.5 mm thick and is bounded on both sides by air, $n = 1$. Coherent light of wavelength 1 μm is incident on the slab. Calculate (a) the fractional intensity transmission for P waves incident at an angle of 30°, (b) the fractional intensity transmission for S waves incident at an angle of 30°, (c) the rotation of the plane of polarization of the transmitted and reflected waves for a wave initially composed of 50% S and 50% P polarization, and (d) the rotation of the plane of polarization of the transmitted and reflected waves for a wave initially composed of 30% S and 70% P polarization.

[7] The molecule with the same atomic constituents, but the mirror-image structure.

[8] This is in contrast to natural optical activity, for which a to-and-fro pass leads to no net rotation.

13.2 How would the answers to parts (a) and (b) of Problem 13.1 be modified if the light were incoherent?

13.3 A thick lens with $n = 1.55$ has $R_1 = 60$ mm and $R_2 = -40$ mm, with $d = 15$ mm. Calculate the positions of the principal planes and the focal length of the lens. Draw a sketch to illustrate your answer.

13.4 An object is placed 500 mm from the $R_1$ face of the thick lens in Problem 13.3. Calculate where the image is.

13.5 For a thin lens with $u = 50$ mm and $f = 20$ mm draw a ray-tracing diagram to illustrate the location of the image and its magnification. State the image distance, magnification, and angular magnification found from the diagram. Repeat for $u = 20$ mm, $f = 50$ mm and $u = 50$ mm, $f = -20$ mm. Mention any important differences among these three situations.

13.6 Develop a matrix approach for dealing with flat and spherical mirrors in paraxial-ray analysis. Hint: a mirror folds a ray path back on itself.

13.7 What is the overall transmittance through a glass parallel-sided slab of thickness 3 mm and refractive index 1.55, when light of wavelength 632.8 nm is incident at an angle of 40° if the incident light is coherent and is an equal mixture of S and P polarizations?

13.8 (Computer project) Write a computer program to plot the variation in reflectance for P and S waves incident on an air/glass boundary at angles from 0 to $\pi/2$. Take $n_1 = 1$ and $n_2 = 1.5$. Calculate the S-wave reflectance at Brewster's angle.

13.9 (Computer project) Write a computer program to calculate the reflectance as a function of wavelength in normal incidence for a single $\lambda_s/4$ antireflection layer, where $\lambda_s$ is the wavelength in the layer. Plot $R$ from $\lambda_s/2$ to $1.5\lambda_s$.

13.10 (Computer project) Write a computer program to calculate and plot the reflectance of a single $\lambda_s/4$ antireflection layer designed for normal incidence as the angle of incidence varies from 0 to 90°.

13.11 For a spherical mirror of aperture diameter $D$ and focal length $f$ calculate the blurring along the axis, caused by spherical aberration, of the image of an object at a distance $u$ from the mirror. What is the magnitude of the blurring if $u = 3$ m, $D = 3$ m, and $f = 2.5$ m? Hint: calculate the path of the extreme ray – one that just hits the edge of the mirror – by exact trigonometry.

13.12 (Computer project) A triangular prism has index $n = 1.5$ and apex angle $\theta = 30°$. Plot the angular deviation of a ray of light passing through the prism (and undergoing two refractions) as a function of the angle of incidence. For what angle of incidence is this angle a minimum? Use this result as a clue in deriving the minimum deviation angle theoretically.

## References

[1] H. Kogelnik and T. Li, "Laser beams and resonators," *Proc. IEEE* **54**, 1312–1329, 1966.

[2] R. Winston, J. C. Miñano, and P. Benitz, *Nonimaging Optics*, Burlington, MA: Elsevier, 2005.

[3] W. T. Wellford and R. Winston, *High Collection Nonimaging Optics*, Waltham, MA: Academic Press, 1989.

[4] R. Kingslake, *Optical System Design,* Orlando, FL: Academic Press, 1983.

[5] J. H. Moore, C. C. Davis, and M. A. Coplan, *Building Scientific Apparatus*, 4th edn., Cambridge: Cambridge University Press, 2009.

[6] M. Born and E. Wolf, *Principles of Optics*, 7th edn., Cambridge: Cambridge University Press, 1999.

[7] R. W. Ditchburn, *Light*, 3rd edn., New York: Academic Press, 1976.

[8] L. Levi, *Applied Optics*, Vol. 1, New York: Wiley, 1968.

[9] W. J. Smith, *Modern Optical Engineering*, 3rd edn., Bellingham, WA: SPIE Press, 2000.

[10] W. J. Smith, *Modern Lens Design: A Resource Manual*, New York: McGraw-Hill, 1992.

[11] R. R. Shannon, *The Art and Science of Optical Design*, Cambridge: Cambridge University Press, 1997.

[12] M. Laikin, *Lens Design,* 3rd edn., New York: Marcel Dekker, 2001.

[13] S. Ramo, J. R. Whinnery, and R. Van Duzer, *Fields and Waves in Communication Electronics*, 3rd edn., New York: Wiley, 1993.

# 14 Periodic optical systems, resonators, and inhomogeneous media

## 14.1 Introduction

In this chapter we shall use a variety of analytic techniques to analyze periodic optical systems and understand their behavior. These are optical systems in which a ray crosses a repeating pattern of interfaces, a so-called *stratified medium*, or passes through a sequence of optical components that repeat in a periodic fashion along the axis.

These periodic systems include multi-layer structures that have special reflective or transmissive properties and structures whose dielectic properties vary symmetrically in two or three dimensions. Such periodic structures are often called *photonic crystals*. *Periodic lens sequences* are axisymmetric arrangements of multiple lenses arranged along an axial direction. They can be described as stable or unstable. Stable lens sequences have the ability to confine a propagating bundle of rays near to the axis in such a way that the rays never deviate by more than a certain finite distance from the axis and remain confined. Periodic lens sequences can also be used to study the paths of light rays bouncing back and forth between pairs of mirrors – *optical resonators*. This will allow us to deduce the *stability* condition for such resonators.

## 14.2 Plane waves in media with complex refractive indices

The relative permittivity and the relative permeability characterize a medium in terms of its difference from a vacuum. Most materials that are important in a discussion of lasers and optical devices are not strongly magnetic, and it is generally legitimate to assume that for such materials $\mu_r = 1$. To include the properties of real, rather than ideal, matter in our discussion, as well as negative-refractive-index materials, we allow the possibility that both the dielectric constant and the relative permeability can be represented by complex numbers. The frequency-dependent dielectric constant can be written as

$$\epsilon_r(\omega) = \epsilon'(\omega) - j\epsilon''(\omega). \tag{14.1}$$

There is a corresponding definition for the complex susceptibility:

$$\chi(\omega) = \chi'(\omega) - j\chi''(\omega). \tag{14.2}$$

The susceptibility and dielectric constant are closely related: $\chi'(\omega) = \epsilon'(\omega) - 1$ and $\chi''(\omega) = \epsilon''(\omega)$. The minus sign in Eqs. (14.1) and (14.2) is a sign convention. With this sign convention, when $\chi''(\omega)$ is positive the medium absorbs energy from the wave.

When an electromagnetic wave propagates in the $z$ direction through a medium with complex dielectric constant and/or relative permeability the variation of the fields of the wave is, taking the electric field as an example,

$$\mathbf{E} = \mathbf{E}_0 e^{j\omega t - \gamma z}, \tag{14.3}$$

where the new propagation constant is

$$\gamma = j\omega\sqrt{\mu_0 \mu_r \epsilon_r \epsilon_0}. \tag{14.4}$$

Both $\epsilon'$ and $\epsilon''$ vary with frequency. The phase velocity of light in the medium is

$$c = \frac{1}{\sqrt{\mu_r \mu_0 \epsilon_r \epsilon_0}} = \frac{c_0}{n}, \tag{14.5}$$

where the refractive index is defined as $n = \sqrt{\mu_r \epsilon_r}$. Most optical media are generally not strongly magnetic, which allows us to take $\mu_r = 1$ and write

$$\gamma = \frac{j\omega}{c_0}\sqrt{\epsilon' - j\epsilon''}. \tag{14.6}$$

If the medium does not have large absorption or gain, we can make the assumption that $\epsilon'' \ll \epsilon'$, and Eq. (14.6) becomes

$$\gamma = \frac{j\omega}{c}\left(1 - j\frac{\epsilon''}{\epsilon'}\right), \tag{14.7}$$

where $c = c_0/\sqrt{\epsilon'}$ is the modified phase velocity of light in the medium. The variation of the electric field amplitude of the wave as it propagates is now

$$E = E_0 e^{-\frac{\alpha}{2}z} e^{j(\omega t - kz)}, \tag{14.8}$$

where $k = \omega/c$ and $\alpha = k\epsilon''/\epsilon'$. In the general case we can write $\gamma$ in terms of its real and imaginary parts as

$$\gamma = \alpha + j\beta. \tag{14.9}$$

The phase variation of the wave as it propagates depends on $\beta$ and its attenuation on $\alpha$. The phase velocity is $c = \omega/\beta$. In a medium with a complex dielectric constant the *refractive index* is also complex, and is written as

$$n - jK = \sqrt{\epsilon_r}. \tag{14.10}$$

In this case $\epsilon_r = (n - jK)^2$. For example, at a wavelength of 633 nm for gold the refractive index is $0.122 - j3.968$ and $\epsilon_r = -15.73 - j0.968$.

## 14.3 Negative-refractive-index materials

There has been much recent interest in metamaterials, whose optical behavior depends on the mesoscopic structure of the material more than on the properties of individual atoms in the structure [1, 2]. For such materials both $\epsilon_r$ and $\mu_r$ can have negative real parts. We write

$$\mu_r = \mu' - j\mu''. \tag{14.11}$$

In this case the propagation constant can be written as

$$\gamma = jk\sqrt{(\mu' - j\mu'')(\epsilon' - j\epsilon'')}, \tag{14.12}$$

which can be written as

$$\gamma = jk\sqrt{(\mu'\epsilon')\left(1 - j\frac{\mu''}{\mu'}\right)\left(1 - j\frac{\epsilon''}{\epsilon'}\right)}. \tag{14.13}$$

If both $\mu'$ and $\epsilon'$ are negative, this equation is of the form

$$\gamma = jk\sqrt{(\mu'\epsilon')(1 - jX)(1 - jY)}, \tag{14.14}$$

where both $X$ and $Y$ are absolutely positive for non-amplifying materials. The location of $\gamma$ in the complex plane depends on the choice of the sign of the square root $\sqrt{\mu'\epsilon'}$. For non-amplifying materials $\gamma$ can only lie in the second or third quadrant of the complex plane, which requires the negative sign of the square root $\sqrt{\mu'\epsilon'}$. Thus for such materials the refractive index is $n = -|\sqrt{mu'\epsilon'}|$, which is negative. For a negative-refractive-index material the impedance $Z = Z_0/n$ is also negative.

## 14.4 Structures with periodically varying dielectric properties

The behavior of an electromagnetic wave that enters a medium consisting of a series of parallel interfaces between media with different dielectric properties is very interesting. The overall reflectance and transmittance of the structure depend on the wavelength, the spacing of the interfaces, the dielectric constants involved, the angle of incidence, and the wavelength of the light. If the structure consists of many alternating layers of two different materials with thicknesses $d_1$ and $d_2$, respectively, then the structure can be described as a *Bragg grating* or a one-dimensional *photonic crystal*. Various hybrid structures or metamaterials that possess a so-called *negative refractive index* have recently been fabricated, or proposed. Stratified media containing such materials can be analyzed in a unified way using techniques that are commonly used for analyzing conventional stratified structures.

A classic periodic structure consists of a sequence of slabs of dielectric material with alternating different refractive indices, $n_1$ and $n_2$, respectively, mounted on a substrate material of refractive index $n_3$ with a wave entering from a medium (generally air) of refractive index $n_0$, as shown in Fig. 14.1. Such structures can be fabricated to act as

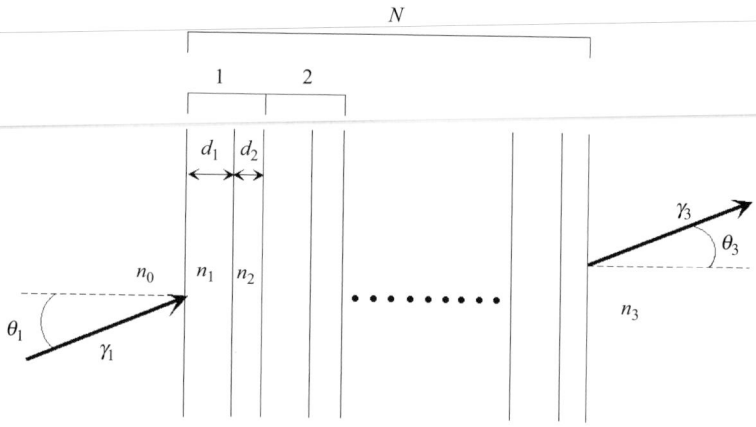


**Fig. 14.1** A periodic slab structure of alternating different materials of refractive indices $n_1$ and $n_2$, respectively, mounted on a substrate of refractive index $n_3$ with a wave entering from a medium of refractive index $n_0$.

high-reflectance mirrors in some specific wavelength region; simultaneously such a structure acts as a *band-stop* filter. If the periodicity of refractive index occurs along the axis inside an optical fiber (see Chapter 16) then the resulting structure is referred to as a *fiber Bragg grating*.[1] It is interesting to examine the analysis of such multi-layer structures by several techniques.

## 14.5 The transfer matrix for a single slab

We can begin to demonstrate the properties of stratified media by using the slab geometry shown in Fig. 14.2(a). A plane wave at angular frequency $\omega$ enters the first boundary of the structure at an angle of incidence $\theta_1$, refracts at the first boundary according to Snell's law to be traveling at angle $\theta_2$ to the normal, and refracts at the next boundary to be propagating at angle $\theta_3$ in medium 3. The propagation constants of the waves in each region are $\gamma_1, \gamma_2$, and $\gamma_3$, respectively. We can analyze this structure using the impedance ideas first introduced in Chapter 13. The problem reduces to the slab problem shown in Fig. 14.2(b), where all waves are traveling normally to the slab, whose effective thickness is $d' = d\cos\theta_1$. For a geometry appropriate to a P wave (or TM wave), as shown in Fig. 14.3, the transverse electric and magnetic fields of the incident wave are related by

$$\frac{E_x}{H_y} = Z_1 \cos\theta_1 = Z_{1\mathrm{P}}, \tag{14.15}$$

where $Z_{1\mathrm{P}}$ is the effective impedance for this P wave. In medium 2, which is bounded at $z = 0$ and $z = d$, there are waves traveling to both the right and the left as shown in Fig. 14.3.

[1] So called because of a close connection with the Bragg diffraction discussed in Chapter 5.

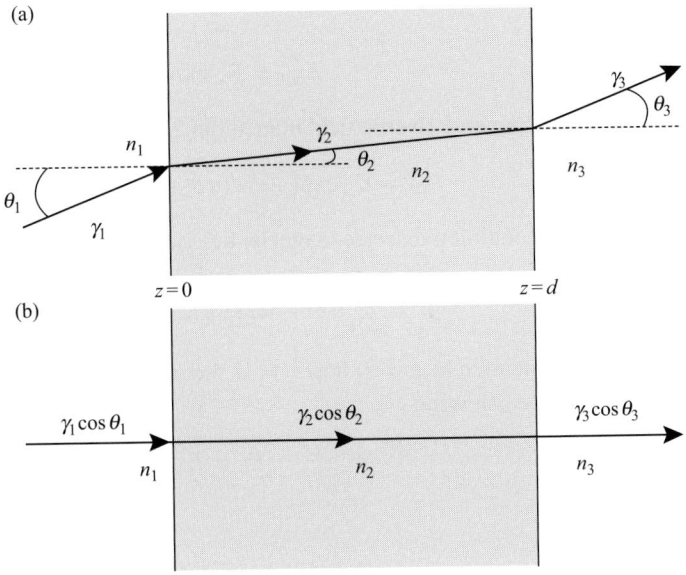


**Fig. 14.2** (a) Wave propagation through a slab at an angle of incidence of $\theta_1$. (b) The equivalent problem reduced to normal incidence.

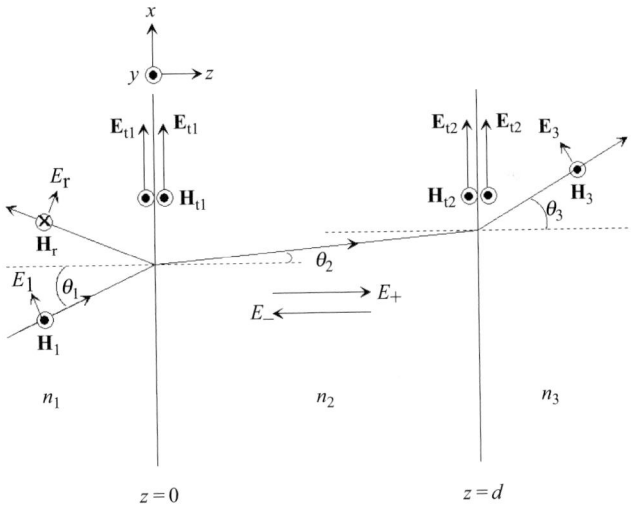


**Fig. 14.3** The field compoments of a P wave entering, bouncing inside a slab, and leaving the slab.

The transverse electric fields of these waves can be represented as having amplitudes $E_+$ traveling to the right and $E_-$ traveling to the left. Specifically, the $z$ variation of these fields is

$$E_{\text{right}} = E_+ \exp(-\gamma_2 \cos\theta_2 \, z) \tag{14.16}$$

and

$$E_{\text{left}} = E_- \exp(\gamma_2 \cos\theta_2 z). \tag{14.17}$$

The total transverse electric field in medium 2 is

$$E_{2x} = E_+ \exp(-\gamma_2 \cos\theta_2 z) + E_- \exp(\gamma_2 \cos\theta_2 z). \tag{14.18}$$

The corresponding transverse magnetic field is

$$H_{2y} = \frac{E_+}{Z_{2\text{P}}} \exp(-\gamma_2 \cos\theta_2 z) - \frac{E_-}{Z_{2\text{P}}} \exp(\gamma_2 \cos\theta_2 z). \tag{14.19}$$

At the boundaries at $z = 0$ and $z = d$ the transverse fields are continuous. At the first boundary we can write

$$E_0 = E_+ + E_- \tag{14.20}$$

and

$$H_0 = \frac{E_+ - E_-}{Z_{2\text{P}}}. \tag{14.21}$$

At the boundary at $z = d$

$$E_d = E_+ \exp(-\gamma_2 \cos\theta_2 d) + E_- \exp(\gamma_2 \cos\theta_2 d) \tag{14.22}$$

and

$$H_d = \frac{E_+}{Z_{2\text{P}}} \exp(-\gamma_2 \cos\theta_2 d) - \frac{E_-}{Z_{2\text{P}}} \exp(\gamma_2 \cos\theta_2 d). \tag{14.23}$$

Equations (14.22) and (14.23) can be rearranged to give

$$\begin{pmatrix} E_d \\ H_d \end{pmatrix} = \begin{pmatrix} \cosh(\gamma_{2z} d) & -Z_{2\text{P}} \sinh(\gamma_{2z} d) \\ -(1/Z_{2\text{P}})\sinh(\gamma_{2z} d) & \cosh(\gamma_{2z} d) \end{pmatrix} \begin{pmatrix} E_0 \\ H_0 \end{pmatrix}. \tag{14.24}$$

This analysis provides a transfer matrix $M_{\text{P}}$ for the slab that can be written as

$$M_{2\text{P}} = \begin{pmatrix} \cosh(\gamma_{2z} d) & -Z_{2\text{P}} \sinh(\gamma_{2z} d) \\ -(1/Z_{2\text{P}})\sinh(\gamma_{2z} d) & \cosh(\gamma_{2z} d) \end{pmatrix}. \tag{14.25}$$

Note that $M_{2\text{P}}$ is a unimodular matrix, meaning that its determinant is equal to unity. By a similar analysis the transfer matrix for S waves is

$$M_{2\text{S}} = \begin{pmatrix} \cosh(\gamma_{2z} d) & Z_{2\text{S}} \sinh(\gamma_{2z} d) \\ (1/Z_{2\text{S}})\sinh(\gamma_{2z} d) & \cosh(\gamma_{2z} d) \end{pmatrix}. \tag{14.26}$$

The matrices provide a relationship between the lateral fields at a boundary in a stratified medium and at the next boundary in the medium. In a lossless medium the ray-transfer matrix for the P wave can be written as

$$M_{2\text{P}} = \begin{pmatrix} \cos(k_2 \cos\theta_2 d) & -jZ_{2\text{P}} \sin(k_2 \cos\theta_2 d) \\ -j(1/Z_{2\text{P}})\sin(k_2 \cos\theta_2 d) & \cos(k_2 \cos\theta_2 d) \end{pmatrix} \tag{14.27}$$

and for the S-wave

$$M_{2S} = \begin{pmatrix} \cos(k_2 \cos\theta_2\, d) & jZ_{2S} \sin(k_2 \cos\theta_2\, d) \\ j(1/Z_{2S})\sin(k_2 \cos\theta_2\, d) & \cos(k_2 \cos\theta_2\, d) \end{pmatrix}. \tag{14.28}$$

These results can be summarized as

$$\begin{pmatrix} E_{t_1} \\ H_{t_1} \end{pmatrix} = M \begin{pmatrix} E_{t_2} \\ H_{t_2} \end{pmatrix}, \tag{14.29}$$

where $E_{t_1}$ and $H_{t_1}$ are the total transverse electric and magnetic field amplitudes at the first boundary, respectively, and $E_{t_2}$ and $H_{t_2}$ are the total transverse electric and magnetic fields at the second boundary, respectively. In terms of reflection and transmission coefficients

$$E_{t_1} = E_{t_{\text{incident}}} + rE_{t_{\text{incident}}} \tag{14.30}$$

and

$$H_{t_1} = H_{t_{\text{incident}}} - rH_{t_{\text{incident}}}, \tag{14.31}$$

where all the fields are values parallel to the boundaries.

## 14.6 The etalon

There are two interesting cases of a simple slab structure that are of special interest. The first of these is of a parallel-sided lossless slab (an *etalon*) of refractive index $n$ bounded on both sides by air. In this case the transfer matrix for P waves can be written as

$$M_{2P} = \begin{pmatrix} \cos(k_2 \cos\theta_2\, d) & -j(Z_0/n)\cos\theta_2 \sin(k_2 \cos\theta_2\, d) \\ -j[n/(Z_0 \cos\theta_2)]\sin(k_2 \cos\theta_2\, d) & \cos(k_2 \cos\theta_2\, d) \end{pmatrix}. \tag{14.32}$$

Figure 14.4 shows how the transmittance of such a slab varies with the angle of incidence both for the P and S waves. Figure 14.5 shows how the transmittance of the slab varies with wavelength when used in normal incidence. The transmittance of the slab is unity whenever the effective thickness of the slab $d\cos\theta_2$ is a multiple of half wavelengths for the wavelength $\lambda_2 = \lambda/n$ in the material of the slab. In this case the transfer matrix becomes an identity matrix. Note that Fig. 14.4 also shows the transmittance maximum at Brewster's angle.

## 14.7 A thin metal film

The transmittance of a metal film on a substrate is of special interest when the incident light passes from a high-index material through the film and then into a low-index material. Figure 14.6 shows the reflectance of a 50-nm-thick gold film at a wavelength of 632.8 nm as the angle of incidence is varied. At this wavelength the refractive index of the gold is complex, $n_{\text{gold}} = 0.122 - j3.968$. The dramatic dip in the reflectance occurs when

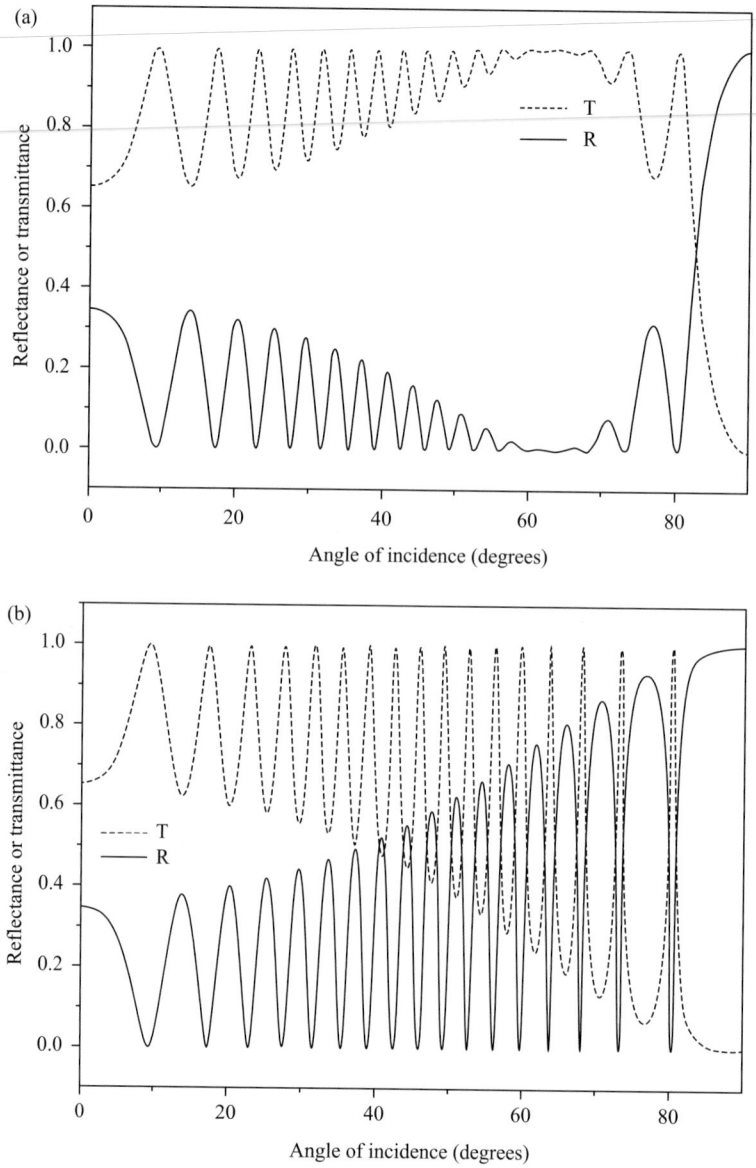


**Fig. 14.4** The transmittance and reflectance of (a) a P wave and (b) an S wave passing through a single slab (etalon) in air as a function of the angle of incidence with $n = 2$, $d = 20\ \mu\text{m}$, and $\lambda = 633$ nm (T, transmittance; R, reflectance).

the incident wave can excite a surface electromagnetic wave called a *surface plasmon polariton* [3–5] at the interface between the gold and the low-index material. The angle at which this occurs is called the *surface plasmon resonance* angle. This angle is extremely sensitive to the presence of any material on the surface of the gold film and can serve as an extremely sensitive sensor for material that sticks to the gold surface.

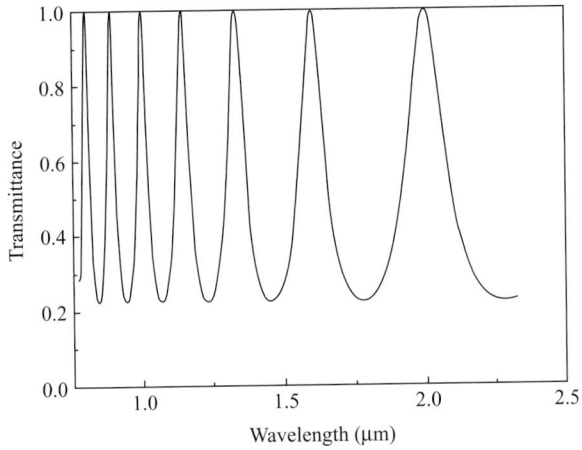


**Fig. 14.5** The transmittance of an etalon in air at normal incidence as the incident wavelength is varied, with $d = 1\mu\text{m}$ and $n = 1$.

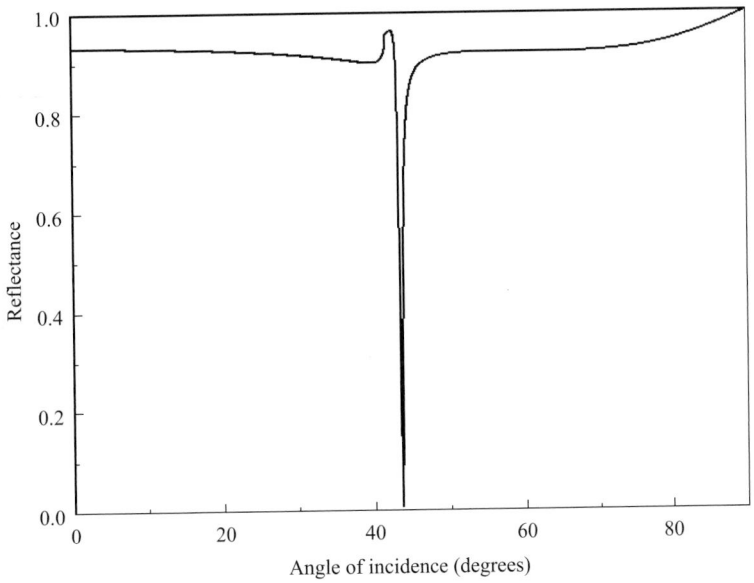


**Fig. 14.6** The reflectance of a thin gold film for a wave passing through the film from the high-index to the low-index side, showing the dramatic dip in reflectance at the surface plasmon resonance angle, with $n_s = 1.5$, $d = 50$ nm, $n_{\text{gold}} = 0.122 - j^{3.968}$, and $\lambda = 633$ nm.

## 14.8 A stratified medium of multiple slabs

Consider a stratified medium made up of a series of $2N$ slabs, which alternate between media of thickness $d_1$ and refractive index $n_1$ and thickness $d_2$ with refractive index $n_2$, as shown in Fig. 14.1. The stratified medium is bounded on the left with a medium of refractive index $n_0$ and on the right with a medium of refractive index $n_3$. The wave is incident at angle $\theta_0$ and leaves in medium 3 at angle $\theta_3$. The transverse fields are continuous at the first and final boundaries. The incident, reflected, and final transmitted electric amplitudes parallel to the boundaries are $E_i$, $E_r$, and $E_t$. The overall reflection and transmission coefficients of the structure are $r$ and $t$, where

$$r = \frac{E_r}{E_i} \tag{14.33}$$

and

$$t = \frac{E_t}{E_i}. \tag{14.34}$$

The overall transfer matrix of the slabs is

$$M = (M_2 M_1)^N, \tag{14.35}$$

where $M_1$ and $M_2$ are found from Eq. (14.25) or Eq. (14.26), depending on whether a P- or S-wave geometry is being considered. $M$ has the form

$$M = \begin{pmatrix} m_{11} & m_{12} \\ m_{21} & m_{22} \end{pmatrix}. \tag{14.36}$$

Using the continuity of the transverse fields at the first and final boundaries gives the following equations:

$$t = m_{11}(1 + r) + \frac{m_{12}}{Z_0'}(1 - r) \tag{14.37}$$

and

$$t = m_{21} Z_3'(1 + r) + \frac{m_{22} Z_3'}{Z_0'}(1 - r), \tag{14.38}$$

where $Z_0'$ and $Z_3'$ are the effective impedances of media 0 and 3, which for P waves are $Z_0' = Z_0 \cos\theta_0$ and $Z_3' = Z_3 \cos\theta_3$. Equations (4.34) and (4.35) can be solved to give

$$r = \frac{(m_{22} + m_{21} Z_0')Z_3' - (m_{12} + m_{11} Z_0')}{(m_{22} - m_{21} Z_0')Z_3' - (m_{12} - m_{11} Z_0')} \tag{14.39}$$

and

$$t = \frac{2Z_3'}{(m_{22} - m_{21} Z_0')Z_3' - (m_{12} - m_{11} Z_0')}. \tag{14.40}$$

Figure 14.7 shows the transmittance properties of a slab structure with 20 pairs of alternating layers of refractive indices 1.3 and 1.5 on a substrate of refractive index 1.5. Each layer is a quarter wavelength thick. This structure is a good example of a high-reflectance multi-layer dielectric mirror.

As the number of alternating layers increases the structure can be referred to as a one-dimensional (1D) *photonic crystal*. The transmittance of the structure approaches zero over a range of wavelengths as the number of layers increases, as shown in Fig. 14.8 for a range of wavelengths near the important fiber-optic communication wavelength of

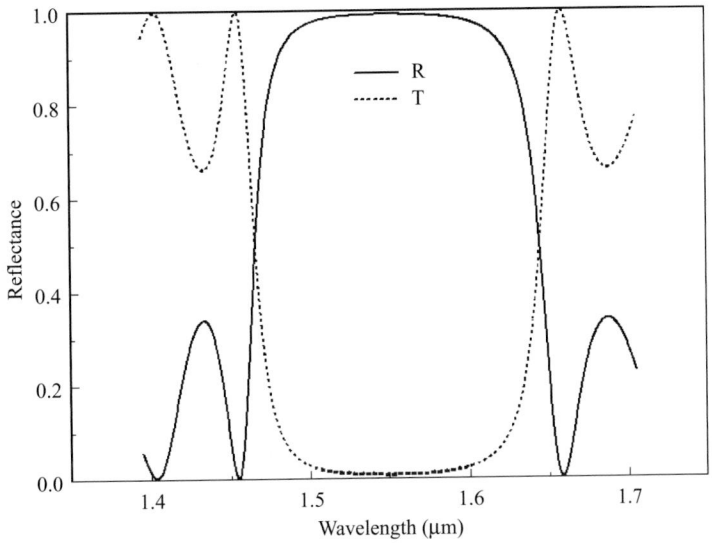


**Fig. 14.7** The reflectance (R) and transmittance (T) properties of a slab structure with 20 pairs of alternating layers of refractive indices 1.3 and 1.5 on a substrate of refractive index 1.5. Each layer is a quarter wavelength thick.

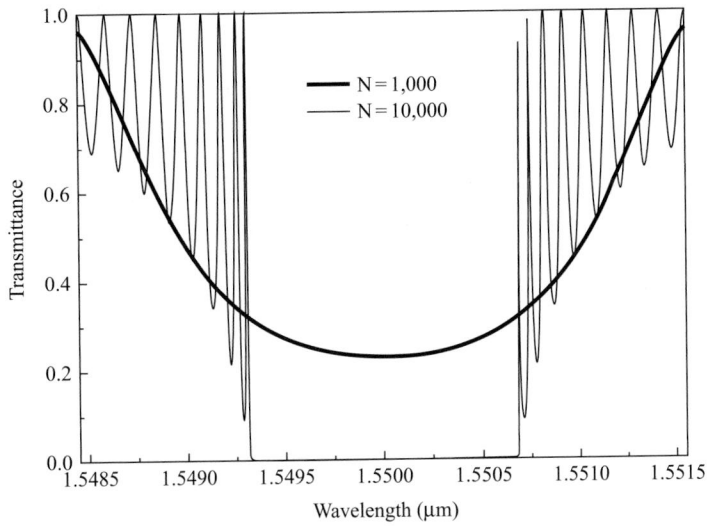


**Fig. 14.8** The transmittance of a one-dimensional photonic crystal with a large number, $N$ of alternating layers of refractive indices $n_1 = 1.46$ and $n_2 = 1.458$.

1.55 μm. The range of wavelengths that cannot propagate through the structure, because all the light is reflected, is called the *stop band*. The high reflectance of the structure results because the multiple reflections that occur as light propagates through the structure are all in phase. The length of one (high-index-low-index) period is $\lambda/2$, and a reflected wave returns to the start of the period after traveling one wavelength, and comes back in phase. This result can be stated mathematically as $2d_{\text{period}} = \lambda$. There is a very good analogy between the properties of this periodic structure and the phenomenon of X-ray Bragg diffraction in crystals. The wavelength of X-rays can be comparable to the spacings of atoms in a crystal. There are many imaginary planes within the crystal lattice on which atoms are regularly arranged. If X-rays reflect off successive planes so that the additional path difference is a multiple of whole wavelengths, then a Bragg reflection maximum occurs. The angular pattern of reflected X-rays reveals the internal atomic arrangement within the crystal. Two- (2D) and three-dimensional (3D) photonic crystals exploit similar ideas and involve periodic refractive-index patterns in two or three dimensions. A light wave traveling in a particular direction in such a structure can be very strongly reflected if the periodicity in that direction satisfies an in-phase reflection condition. There will then be stop bands for light of a particular range of wavelengths and/or directions. In such a photonic crystal no waves can propagate in these forbidden regions.

## 14.9 Fiber Bragg gratings

An important device called a *fiber Bragg grating* uses a stratified structure with many periods of high–low index change. If a silica optical fiber is illuminated with intense ultraviolet light then the refractive index of the core of the fiber can be modified. This occurs because of the germanium doping of the core, which is the most common fiber fabrication dopant used to elevate the core index above the cladding. If the fiber is impregnated with hydrogen gas then the photosensitivity of the fiber is further enhanced. Ultraviolet wavelengths near 255 nm are best for producing the index changes in the core [6]. If the regions of high and low index are each $\lambda_0/(4n)$ long, then the fiber becomes strongly reflective at this wavelength, as is shown in Fig. 14.9. The wavelength range that is reflected strongly is a stop band in transmission, so this is another example of a 1D photonic crystal.

## 14.10 Analysis of multi-layer structures by impedance techniques

As discussed in Chapter 13, if we replace each layer in a periodic structure by its equivalent impedance and equivalent thickness, the structure can be analyzed by the transformed impedance approach. In normal incidence the effective impedances are $Z_1 = Z_0/n_1$ and $Z_2 = Z_0/n_2$, respectively, and the effective thicknesses are the actual thicknesses, $d_1$ and $d_2$, respectively. We start at the last boundary between medium 1 and air. The transformed impedance at the next to last boundary is

$$Z_1^{\mathrm{T}} = Z_1\left(\frac{Z_0\cos(k_1d_1) + jZ_2\sin(k_1d_1)}{Z_1\cos(k_1d_1) + jZ_0\sin(k_1d_1)}\right), \tag{14.41}$$

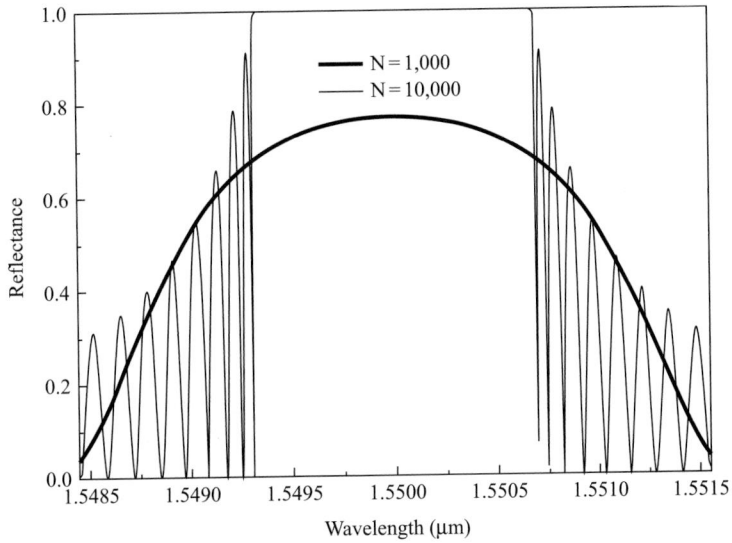


**Fig. 14.9** The high reflectance of a Bragg grating fiber-optic reflector, which is also an example of a one-dimensional photonic crystal.

as in Eq. (13.123). If we move one more boundary to the left the next transformed impedance can be written as

$$Z_2^{\mathrm{T}} = Z_2\left(\frac{Z_1^{\mathrm{T}}\cos(k_2 d_2) + jZ_1\sin(k_2 d_2)}{Z_2\cos(k_2 d_2) + jZ_1^{\mathrm{T}}\sin(k_2 d_2)}\right), \tag{14.42}$$

and so on, until we reach the last boundary where the final transformed impedance for a structure with $n$ layers is $Z_n^{\mathrm{T}}$. The reflection coefficient of the entire structure is then

$$\rho = \frac{Z_n^{\mathrm{T}} - Z_0}{Z_n^{\mathrm{T}} + Z_0}. \tag{14.43}$$

## 14.11 Photonic crystals

A structure that has periodicity in one, two, or three dimensions has many similarities, apart from scale, to the periodic atomic arrangements that occur in crystals. For this reason such structures are generally called photonic crystals, and schematic pictures of how 1D, 2D, and 3D versions might be arranged are shown in Fig. 14.10. Research in this field was pioneered by Eli Yablonovitch [7, 8]. When electromagnetic waves or electrons (which are also waves) propagate through a material they scatter off atoms or periodic structures. In atomic or molecular crystals X-rays scatter substantially in specific directions. These directions of strong scattering result because the waves scattered from different atoms

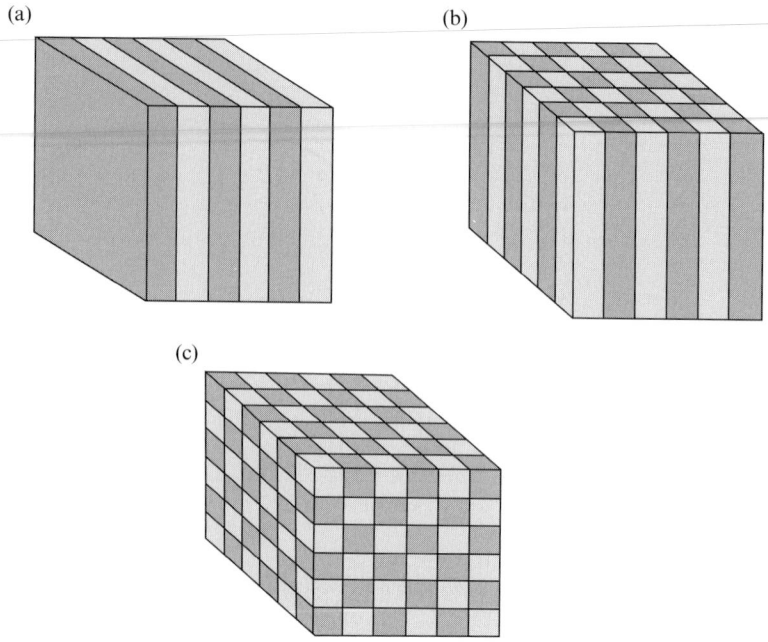


Fig. 14.10 Schematic arrangement of refractive-index variations inside (a) one-dimensional, (b) two-dimensional, and (c) three-dimensional photonic crystals.

add constructively in these directions of constructive *coherent* scattering. This X-ray diffraction bears great similarity to the diffraction of light waves from diffraction gratings, which are used to separate light into its wavelength components. The patterns of scattered X-rays in these cases depend on the structural arrangement of atoms in the crystal, and these patterns can be used to interpret the structure of the crystal. Perhaps the most famous example of this comes from the work of Watson and Crick [9], who analyzed the X-ray-diffraction results of Rosemary Franklin [10, 11] to deduce the double-helical structure of DNA.

As a guide to further reading, bear in mind the following: central to an understanding of the optical properties of a photonic crystal is the concept of its *reciprocal lattice* [12]. In a 2D or 3D periodic structure one can visualize lattice lines or planes that correspond to arrangements that repeat throughout the structure, as shown schematically for a 2D square lattice in Fig. 14.11(a). A vector to a point in the reciprocal lattice corresponds to the normals to these various lattice lines or planes, and the magnitude of this vector is the reciprocal of the spacing between these real lines or planes. The connection between the reciprocal lattice and constructive scattering comes from the directions of a wave and the spacing of the lattice lines or planes that are perpendicular to the propagation direction. The reciprocal lattice also represents the Fourier components of a spatial Fourier transform of the structure. Just as a real crystal has a *unit cell*, which represents the basic repeating structural unit that is repeated to form the whole structure, so the reciprocal lattice also has a unit cell. This unit cell in the reciprocal lattice is called the *Brillouin zone*. In a periodic

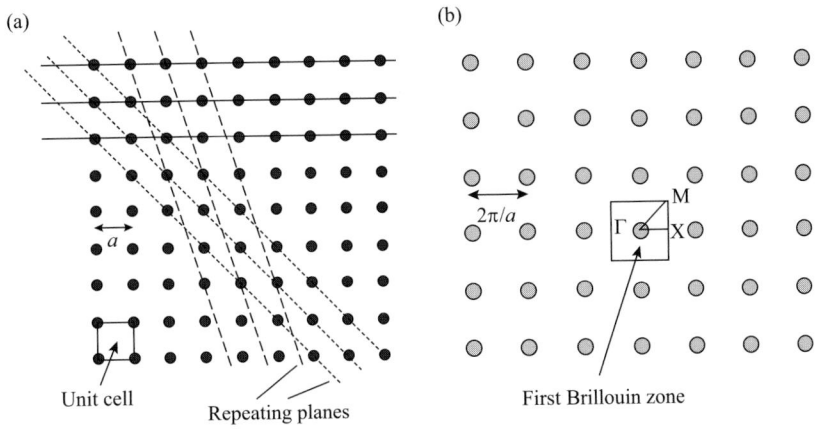


**Fig. 14.11** (a) A two-dimensional lattice, with various repeating planes shown, and (b) its reciprocal lattice, showing the first Brillouin zone and characteristic locations within the zone called Γ, M, and X.

1D structure where the repeating unit cell has physical dimension $a$, the reciprocal lattice is also a set of points on a line but spaced by $2\pi/a$. The first Brillouin zone is the region between $-\pi/a$ and $\pi/a$. For a 2D square lattice with spacing $a$ the reciprocal lattice is also a square lattice of spacing $2\pi/a$, as shown in Fig. 14.11(b). In this square reciprocal lattice the center of the lattice $\mathbf{k} = 0$ is called the Γ point. A corner of the first Brillouin zone is called the M point, and the center of an edge of the first Brillouin zone is called the X point. In more complex structures different points in the first Brillouin zone have their own letter designations.

In a photonic crystal, because light scatters constructively only in specific directions, at a given wavelength there will be directions in which light does not propagate, because in these directions the structure reflects light very efficiently. The inability of light to propagate in these specific directions is a function of the wave vector $\mathbf{k}$ of the light, which contains information about the wavelength and direction of the light. Conversely, in directions where light propagation occurs without loss (or with negligible loss) the scattering that still occurs cancels out coherently. The overall optical properties of a photonic crystal are described by a dispersion diagram, which shows the allowed and forbidden transmission regions through the structure on an $\omega$–$k$ diagram. An example of such a dispersion diagram for TM waves in a 2D periodic structure is given in Fig. 14.12. The frequency ranges where, irrespective of light direction, no propagation occurs are called photonic band gaps. We have already seen how these are calculated for a 1D photonic crystal. The detailed calculation of dispersion diagrams in 2D and 3D structures is beyond our scope, but further details can be obtained elsewhere [13]. We saw in Chapter 12 that for electron waves in a semiconductor these band gaps also occurred and corresponded to electron energies at which no electron states exist. If defects are introduced into a photonic crystal, for example as shown schematically in Figs. 14.13(a) and (b), then it is possible to create microcavities where light is trapped, and this is becoming an increasingly popular way to control light on a microscale or nanoscale [14]. If a series of cylindrical channels is arranged about a

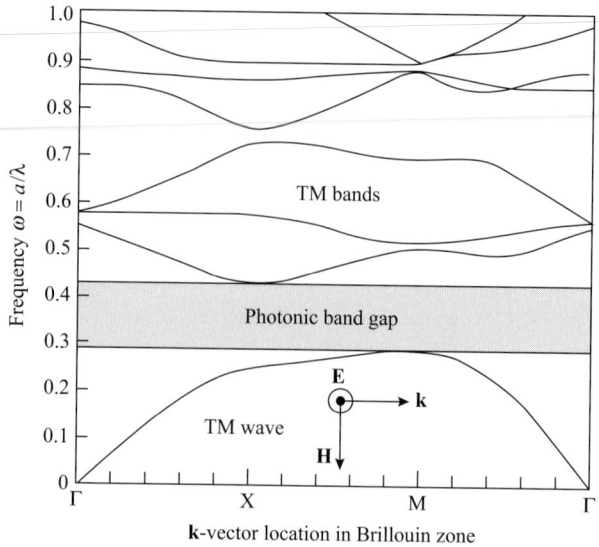


**Fig. 14.12** The dispersion diagram for TM waves passing through a two-dimensional photonic crystal made of circular rods embedded in a different dielectric in which the dielectric constant of the rod structures is 12 times larger than that of the gap regions.

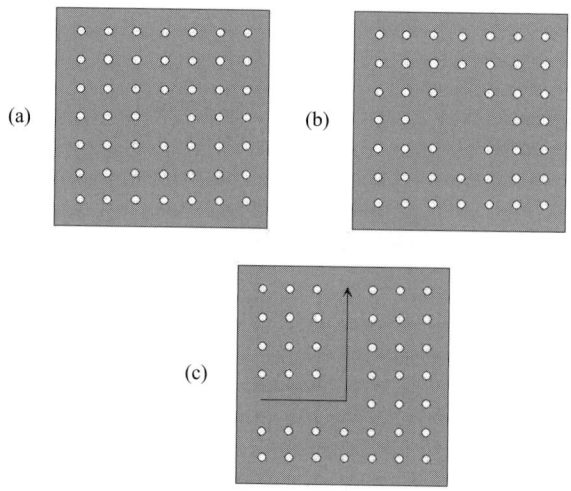


**Fig. 14.13** Two-dimensional photonic crystal lattices with defects that can creat microcavities to trap light, (a) and (b), or (c) create a lossless pathway for light to make a 90° bend.

center solid core, as shown in Fig. 14.13(b), then there results a so-called "holey" fiber structure that guides light efficiently without loss along its axis (see Chapter 16). By introducing channels into the photonic crystal where, for example, there are no holes in a 2D lattice of holes, light can even be guided without loss around a 90° bend [15], as shown in Fig. 14.13(c).

## 14.12 Periodic lens sequences

If we increase the number of lenses in an optical system we can produce arrangements that periodically refocus a ray of light propagating paraxially through the system. When this periodic refocusing prevents the ray being lost from the lens sequence, then we have constructed a stable lens waveguide. As an example let us analyze the biperiodic lens sequence shown in Fig. 14.14. The "unit cell" of this periodic structure consists of the two lenses of focal lengths $f_1$ and $f_2$. Now, in Fig. 14.14,

$$\begin{aligned} r_2 &= Ar_1 + Br_1', \quad & r_3 &= Ar_2 + Br_2', \\ r_2' &= Cr_1 + Dr_1', \quad & r_3' &= Cr_2 + Dr_2', \end{aligned} \tag{14.44}$$

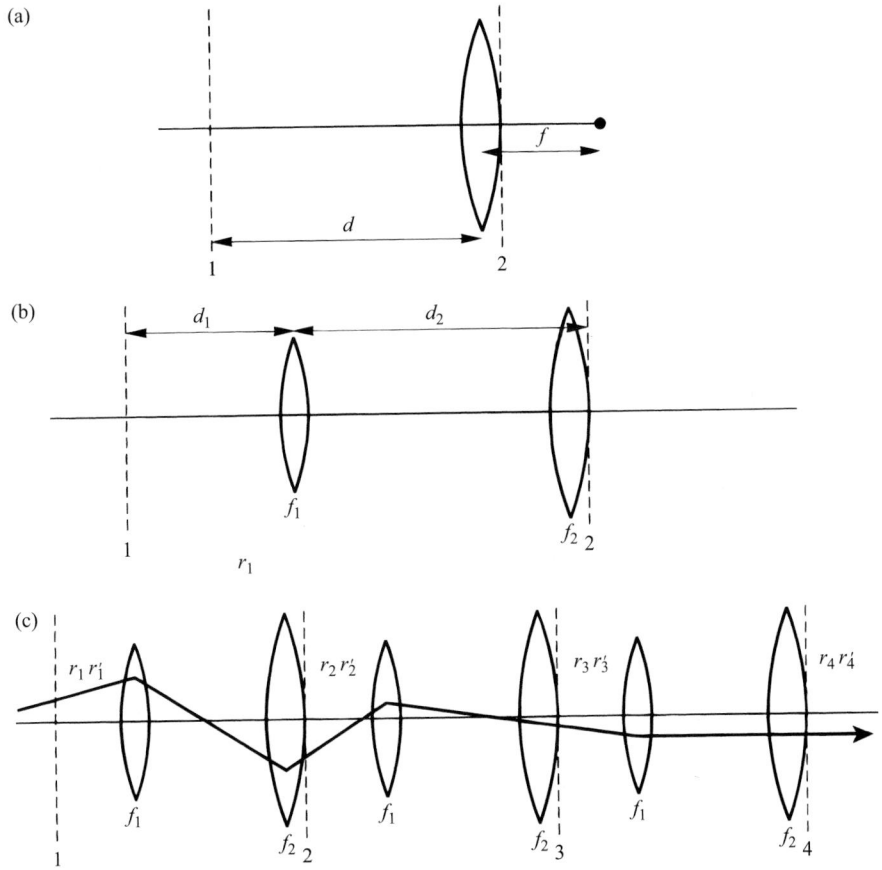


**Fig. 14.14** A biperiodic lens sequence: (a) the basic unit on which the sequence is based, namely a length of uniform medium plus a lens; (b) a unit cell of the sequence consisting of two lenses of different focal lengths; and (c) three unit cells of the sequence (the paraxial-ray parameters are indicated at the input and output planes of each unit cell).

or in general, for transmission through $n$ sections (each consisting in this case of two lenses),

$$\begin{pmatrix} r_{n+1} \\ r'_{n+1} \end{pmatrix} = \begin{pmatrix} A & B \\ C & D \end{pmatrix}^n \begin{pmatrix} r_1 \\ r'_1 \end{pmatrix}. \tag{14.45}$$

We can evaluate such a matrix product by application of Sylvester's theorem,

$$\begin{pmatrix} A & B \\ C & D \end{pmatrix}^n = \frac{1}{\sin\phi} \begin{pmatrix} A\sin(n\phi) - \sin[(n-1)\phi] & B\sin(n\phi) \\ C\sin(n\phi) & D\sin(n\phi) - \sin[(n-1)\phi] \end{pmatrix}, \tag{14.46}$$

where

$$\cos\phi = \frac{1}{2}(A + D). \tag{14.47}$$

At first glance there might appear to be a potential problem with Eq. (14.46) if $\sin\phi \to 0$. However, in this case all the sine terms approach zero, and the matrix becomes a null matrix. No real optical system is described by such a matrix.

Clearly,

$$r_{n+1} = \{[A\sin(n\phi) - \sin[(n-1)\phi]]r_1 + B(\sin(n\phi))r'_1\}/\sin\phi. \tag{14.48}$$

If the angular factors in (14.45) remain real, then $r_{n+1}$ will oscillate as a function of $n$ (the number of pairs of lenses traversed), and the structure has periodic refocusing properties. However, if $\phi$ becomes imaginary, then we can write

$$\sin\phi = \sin(j\psi) = \frac{1}{2j}(e^{-\psi} - e^{\psi}) = j\sinh(\psi). \tag{14.49}$$

Now the dependence of $r_{n+1}$ on $n$ has become a hyperbolic function, and the ray diverges more and more from the axis as it passes through the system.

The condition for $\phi$ to be real and $\sin\phi$ to remain oscillatory is

$$|\cos\phi| \leq 1, \tag{14.50}$$

which from (14.47) gives

$$\left|\frac{1}{2}(A + D)\right| \leq 1. \tag{14.51}$$

If this condition is satisfied the lens or other optical sequence is said to be stable.

## 14.13 The identical-thin-lens waveguide

If $f_1 = f_2 = f$ and $d_1 = d_2 = d$ for thin lenses in the lens sequence considered above, then the unit cell of the structure becomes a single lens, whose transfer matrix is

$$\begin{pmatrix} A & B \\ C & D \end{pmatrix} = \begin{pmatrix} 1 & d \\ -1/f & 1 - d/f \end{pmatrix}, \tag{14.52}$$

and the stability condition becomes

$$\left|1 - \frac{d}{2f}\right| \leq 1, \tag{14.53}$$

which is satisfied if

$$0 \leq d \leq 4f. \tag{14.54}$$

So a sequence of diverging lenses ($f < 0$) is, of course, unstable. If Eq. (14.53) is not satisfied, a ray rapidly escapes from the lens waveguide.

If a complete cone of optical rays is propagating through the system (a family of rays that lie inside a cone of semivertical angle $r_1'$ at the input plane of the system) then the family of rays constitutes a beam whose radius at the $n$th lens is

$$r_{n+1} = \frac{1}{\sin\phi}\{[A\sin(n\phi) - \sin[(n-1)\phi]]r_1 + B(\sin(n\phi))r_1'\}, \tag{14.55}$$

where $A = 1$ and $B = d$.

In this case

$$\cos\phi = 1 - \frac{d}{2f}, \tag{14.56}$$

and

$$\sin\phi = \sqrt{\frac{d}{f} - \frac{d^2}{4f^2}} = \frac{d}{2f}\sqrt{\frac{4f}{d} - 1}, \tag{14.57}$$

so

$$\begin{aligned} r_{n+1} &= \frac{1}{\sin\phi}[(\sin(n\phi) - \sin(n\phi)\cos\phi + \cos(n\phi)\sin\phi)r_1 + d(\sin(n\phi))r_1'] \\ &= \left(\frac{d}{2f}r_1 + dr_1'\right)\frac{\sin(n\phi)}{\sin\phi} + r_1\cos(n\phi). \end{aligned} \tag{14.58}$$

We can write this in the form

$$r_{n+1} = r_{\max}\sin(n\phi + \alpha), \tag{14.59}$$

where

$$r_{\max}\cos\alpha = \frac{dr_1/(2f) + dr_1'}{\sin\phi}; \quad r_{\max}\sin\alpha = r_1. \tag{14.60}$$

Therefore,

$$r_{\max}^2 = \frac{(dr_1/(2f) + dr_1')^2}{\sin^2\phi} + r_1^2, \tag{14.61}$$

which from (14.57) gives

$$r_{\max}^2 = \left(\frac{4f}{4f - d}\right)(r_1^2 + dr_1r_1' + dfr_1'^2), \tag{14.62}$$

and

$$\tan\alpha = \frac{r_1 \sin\phi}{dr_1/(2f) + dr_1'} = \frac{\sqrt{4f/d - 1}}{1 + 2fr_1'/r_1}. \tag{14.63}$$

Note that the stability condition is effectively restated by Eqs. (14.59) and (14.60): if $d > 4f$ then $r_{\max}$ and $\tan\alpha$ become imaginary.

## 14.14 The propagation of rays in mirror resonators

When a ray of light strikes a spherical mirror of radius of curvature $R$, the path of the reflected ray is the same, apart from being in the mirror-image direction, as that of the same ray traversing a thin lens of focal length $R/2$, i.e., the spherical mirror is equivalent to a superimposed "one-way" thin lens and plane mirror *when the reflected ray is considered to be unaffected by the lens* as illustrated in Fig. 14.15. So a two-mirror system with mirror radii $R_1$ and $R_2$ spaced $d$ apart, is equivalent to a biperiodic lens sequence with lenses of focal length $f_1 = R_1/2$ and $f_2 = R_2/2$ spaced the same distance $d$ apart. The only difference between the two structures is that the ray path in the mirror system is folded back on itself. By considering this duality between curved mirrors and biperiodic lens sequences we can arrive at some important considerations regarding the stability of such curved-mirror resonators.

The curved-mirror resonator shown in Fig. 14.16 is equivalent to the biperiodic lens sequence shown in Fig. 14.16. For a symmetrical mirror resonator the equivalent biperiodic lens sequence becomes an identical lens waveguide as shown in Fig. 14.17. We can resolve the paraxial propagation of a ray in such a resonator into independent motions in the $x$ and $y$ planes. Following Eq. (14.59), after $n$ reflections we write the coordinates of the ray as

$$x_n = x_{\max}\sin(n\phi + \alpha_x), \quad y_n = y_{\max}\sin(n\phi + \alpha_y), \tag{14.64}$$

where $\alpha_x$ and $\alpha_y$ are constants.

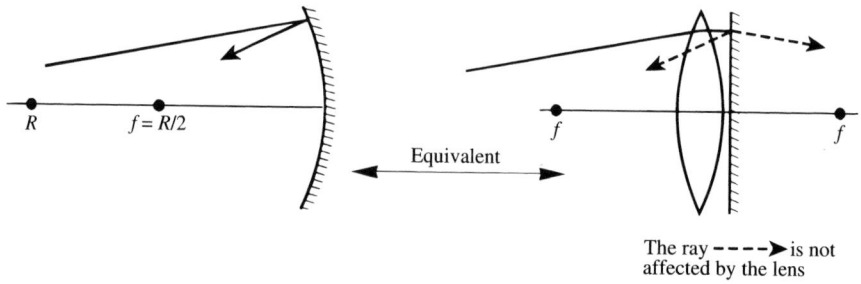


**Fig. 14.15** A ray striking a concave mirror is equivalent to a thin-lens–flat-mirror combination in which the ray interacts with the lens just once.

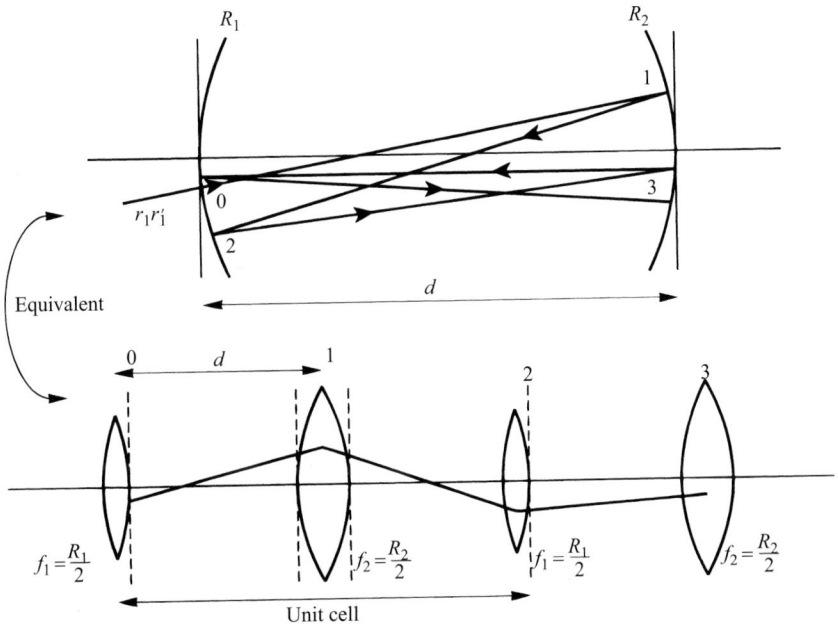


**Fig. 14.16** A curved-mirror resonator and the equivalent biperiodic lens sequence.

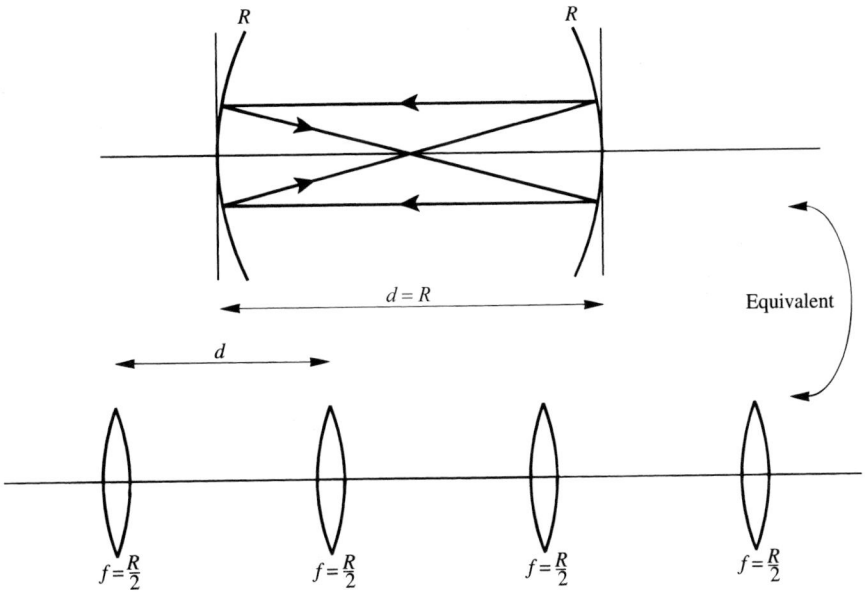


**Fig. 14.17** A confocal symmetric resonator and the equivalent lens sequence.

The locus of points $x_n$ and $y_n$ on the surface of one of the mirrors is an ellipse.

It must be remembered that these considerations apply only to rays propagating sufficiently nearly parallel to the axis of the system that the paraxial-ray approximation still holds.

If in (14.47) $\phi$ satisfies

$$2p\phi = 2q\pi, \tag{14.65}$$

where $p$ and $q$ are any two integers, then after $p$ round trips the ray will retrace its pattern between the mirrors.

For example if $q = 1$ and $p = 2$, so that $\phi = \pi/2$, then $d = 2f = R$ (from $\cos\phi = 1 - d/(2f)$) and the ray retraces its path after two round trips as illustrated in Fig. 14.17. When this situation exists the resonator is said to be *symmetrical confocal* since the two mirrors have a common focal point.

For a general resonator with mirrors of radii $R_1$ and $R_2$ spaced a distance $d$ apart, the stability condition, i.e., the condition that a paraxial ray remains so and eventually retraces its pattern between the two mirrors, is the same as the stability condition for the equivalent biperiodic lens sequence. Now, for the equivalent biperiodic lens sequence, the unit-cell transfer matrix is, from (14.51),

$$\begin{pmatrix} A & B \\ C & D \end{pmatrix} \equiv \begin{pmatrix} 1 - d/f_1 & 2d - d^2/f_1 \\ -1/f_1 - 1/f_2 + d/(f_1 f_2) & 1 - d/f_1 - 2d/f_2 + d^2/(f_1 f_2) \end{pmatrix}, \tag{14.66}$$

and the stability condition is $|\frac{1}{2}(A + D)| \leq 1$, which gives

$$\left| \frac{1}{2}\left(2 - \frac{2d}{f_1} - \frac{2d}{f_2} + \frac{d^2}{f_1 f_2}\right) \right| \leq 1. \tag{14.67}$$

This can be simplified to the stability condition

$$0 \leq 1 - \frac{d}{2f_2} - \frac{d}{2f_1} + \frac{d^2}{4f_1 f_2} \leq 1. \tag{14.68}$$

This stability condition for the biperiodic lens sequence can be rewritten as

$$0 \leq \left(1 - \frac{d}{2f_2}\right)\left(1 - \frac{d}{2f_1}\right) \leq 1. \tag{14.69}$$

For the equivalent spherical mirror resonator, the stability condition is

$$0 \leq \left(1 - \frac{d}{R_2}\right)\left(1 - \frac{d}{R_1}\right) \leq 1, \tag{14.70}$$

where $d$ is the spacing between the mirrors. The mirror curvatures and spacings that satisfy this equation can be represented on a resonator stability diagram, as shown in Fig. 14.18. It can be seen from Fig. 14.18 that the plane-parallel, confocal, and concentric cavities are right on the margin of stability – in unstable equilibrium. In these situations a slight perturbation of the cavity, for example, either by increasing the mirror spacing in the concentric case or by any convex distortion of one of the plane mirrors in the plane-parallel case, will move the resonator into the unstable region.

Although we have considered the stability properties of these resonators by employing a geometrical-optics approach, by considering the paths of rays between the reflectors we shall see in the next chapter that the same stability conditions apply to the real types of wave that propagate between the reflectors of such resonators. However, before considering these waves in a little more detail, we shall deal with a ray-propagation problem that

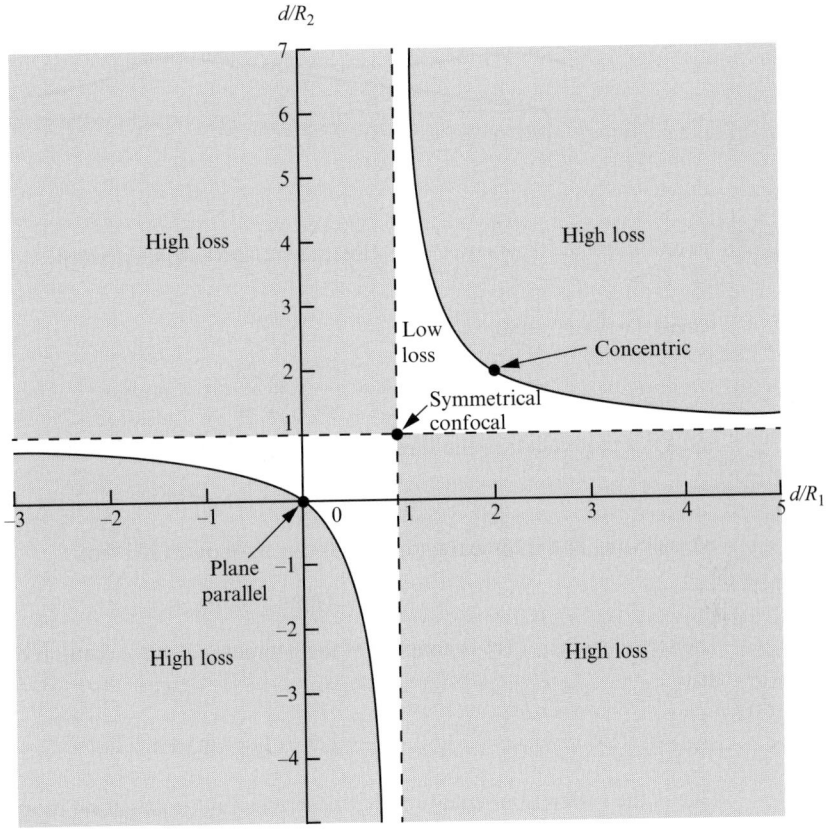


Fig. 14.18 The stability diagram for optical resonators [16].

arises frequently in laser systems and in the propagation of waves through thin-film and fiber-optic waveguides.

## 14.15 The propagation of rays in isotropic media with refractive-index gradients

In this case by "isotropic" we continue to imply that in the medium **P** and **E** are parallel, and that the direction of the wave normal is the same as the direction of energy propagation. Energy propagation occurs in the direction of the Poynting vector **S**, which is the direction of the geometrical-optical light ray. Insofar as the (scalar) refractive index is concerned, the medium is not isotropic.

The light wave in the medium, at any point, will have a field dependence of the general form

$$U_0 = U(x, y, z)e^{j(\omega t - \mathbf{k}\cdot\mathbf{r})}, \tag{14.71}$$

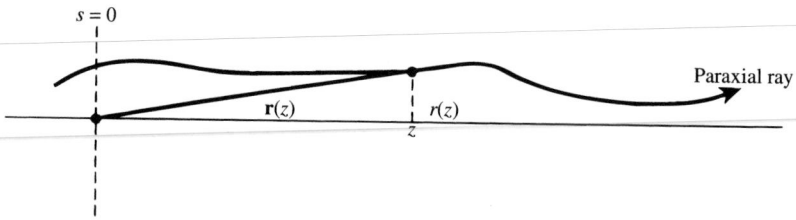


Fig. 14.19 The path of a paraxial ray in a system where the refractive index varies with position.

where the wave vector is

$$\mathbf{k} = \frac{2\pi}{\lambda}\hat{\mathbf{s}} = \frac{2\pi}{\lambda_0}n\hat{\mathbf{s}} = k_0 n\hat{\mathbf{s}}, \tag{14.72}$$

and $\hat{\mathbf{s}}$ is a unit vector in the direction of energy flow. So

$$U_0 = U(x, y, z, t)e^{jk_0 n(\hat{\mathbf{s}}\cdot\mathbf{r})}. \tag{14.73}$$

We call $n(\hat{\mathbf{s}} \cdot \mathbf{r})$ the optical path $\zeta(r)$. It can be shown [17] that

$$\text{grad}\ \zeta = n\hat{\mathbf{s}}. \tag{14.74}$$

The surface where $\zeta(\mathbf{r})$ is constant is the geometrical wave front. It can also be shown [17] that[2]

$$\frac{d}{ds}\left(n\frac{d\mathbf{r}}{ds}\right) = \text{grad}\ n. \tag{14.75}$$

This is the differential equation that describes ray propagation in an optically inhomogeneous medium. For paraxial rays $\mathbf{r}(z) = r(z)\hat{\boldsymbol{\rho}} + \mathbf{z}$, where $\hat{\boldsymbol{\rho}}$ is a unit vector in the direction of $r$ and $d/ds = d/dz$ as shown in Fig. 14.19, so (14.75) becomes

$$\frac{d}{dz}\left\{n\frac{d}{dz}\left[r(z)\hat{\boldsymbol{\rho}} + \mathbf{z}\right]\right\} = \text{grad}\ n, \tag{14.76}$$

which, if $n$ does not vary in the $z$ direction, gives

$$n\frac{d}{dz}\left[\frac{dr(z)}{dz}[\hat{\boldsymbol{\rho}} + \hat{\mathbf{k}}]\right] = \text{grad}\ n, \tag{14.77}$$

where $\hat{\mathbf{k}}$ is a unit vector in the $z$ direction, thus

$$\frac{d^2 r(z)}{dz^2}\hat{\boldsymbol{\rho}} = \frac{1}{n}\ \text{grad}\ n. \tag{14.78}$$

It is not surprising that Eq. (14.78) shows that grad $n$, the direction of maximum rate of change of the refractive index, is in the radial direction.

If $n$ is independent of position,

$$r = az + b, \tag{14.79}$$

where $a$ and $b$ are constants of integration. The path of propagation of the ray is a straight line, as expected.

[2] See Appendix 7.

## 14.16 Media with a Gaussian radial index variation

A commonly encountered axisymmetric index variation is Gaussian, or approximately so. The refractive index as a function of distance $r$ from the axis of symmetry is

$$n = n_0 e^{-r^2/(2\sigma^2)}. \tag{14.80}$$

For distances from the axis that satisfy $r \ll \sigma$, Eq. (14.80) can be written as

$$n = n_0 \left(1 - \frac{r^2}{2\sigma^2}\right) = n_0 - \frac{1}{2} n_2 r^2, \tag{14.81}$$

where $n_2 = n_0/\sigma^2$. This is a quadratic index profile. Such a Gaussian or quadratic index profile can arise in a number of ways.

(i) It can arise from propagation of a laser beam with a transverse Gaussian-like intensity profile in a slightly absorbing medium. The heating caused by absorption of the beam is largest on axis, and, if $dn/dT < 0$, as is the case for most materials, then the index is smallest on axis. We will see that this leads to spreading and defocusing of the beam. If, however, $dn/dT > 0$, as is the case in certain glasses, then thermal self-focusing of the beam can occur.

(ii) In cylindrical laser crystals the pumping radiation heats up the crystal most on its outer surface, which generally leads to a refractive index that increases towards the center of the rod (for $dn/dT < 0$).

(iii) Gas lenses with such index profiles can be made for focusing and guiding laser beams.

(iv) It can arise in optical waveguides made from composite materials having different refractive indices: either as a planar sandwich, as in a heterostructure laser, or in a cylindrical configuration, such as a fiber-optic light guide.

(v) It can arise in small-gradient-index lenses, which take the form of small cylinders with plane ends, which are now in widespread use in optical systems. These lenses are fabricated by introducing a radial concentration of dopant into the glass. They are frequently referred to as GRIN or SELFOC lenses.

For a Gaussian refractive-index variation given by Eq. (14.80),

$$\text{grad}\, n = \frac{dn}{dr} = -\frac{r}{\sigma^2} n \hat{\boldsymbol{\rho}}, \tag{14.82}$$

and the equation for light rays becomes

$$\frac{d^2 r}{dz^2} + \frac{r}{\sigma^2} = 0. \tag{14.83}$$

If the initial parameters of the ray are $r_0$ and $r_0'$ at the plane $z = 0$, then we know that $r(0) = r_0$ and $dr/dz(r = 0) = r_0'$, and the solution of (14.80) can be easily shown to be

$$r(z) = \cos\left(\frac{z}{\sigma}\right) r_0 + \sigma \sin\left(\frac{z}{n\sigma}\right) r_0'. \tag{14.84}$$

The small angle of the ray with respect to the axis is

$$r'(z) = \frac{dr}{dz} = -\frac{1}{\sigma}\sin\left(\frac{z}{\sigma}\right) r_0 + \cos\left(\frac{z}{\sigma}\right) r'_0. \tag{14.85}$$

It can be seen from Eq. (14.84) that the ray oscillates back and forth around the axis, and the medium acts like a lens, provided that $\sigma > 0$. If the medium has $\sigma < 0$, then the solution of (14.80) is

$$r(z) = \cosh\left(\frac{z}{|\sigma|}\right) r_0 + |\sigma|\sinh\left(\frac{z}{n|\sigma|}\right) r'_0, \tag{14.86}$$

and

$$r'(z) = \frac{dr}{dz} = \frac{1}{\sigma}\sinh\left(\frac{z}{|\sigma|}\right) r_0 + \cosh\left(\frac{z}{|\sigma|}\right) r'_0. \tag{14.87}$$

In this case a propagating ray will diverge from the axis and not experience periodic focusing within the medium.

The ray-transfer matrix for a Gaussian medium that extends from $z = 0$ to $z = d$ from Eqs. (14.84) and (14.85) is

$$\begin{pmatrix} A & B \\ C & D \end{pmatrix} = \begin{pmatrix} \cos(d/\sigma) & \sigma\sin[d/(n\sigma)] \\ -(1/\sigma)\sin(d/\sigma) & \cos(d/\sigma) \end{pmatrix}. \tag{14.88}$$

Note that this matrix does not include the matrices for entering and leaving this graded-index medium from air. Provided that the index variation near the axis is small, these matrices are given by Eq. (13.27). The overall ray-transfer matrix from air, through a length $d$ of the graded medium and back into air, is

$$\begin{aligned} \begin{pmatrix} A & B \\ C & D \end{pmatrix} &= \begin{pmatrix} 1 & 0 \\ 0 & n_0 \end{pmatrix} \begin{pmatrix} \cos(d/\sigma) & \sigma\sin[d/(n_0\sigma)] \\ -(1/\sigma)\sin(d/\sigma) & \cos(d/\sigma) \end{pmatrix} \begin{pmatrix} 1 & 0 \\ 0 & 1/n_0 \end{pmatrix} \\ &= \begin{pmatrix} \cos(d/\sigma) & (\sigma/n_0)\sin[d/(n\sigma)] \\ -(n_0/\sigma)\sin(d/\sigma) & \cos(d/\sigma) \end{pmatrix}. \end{aligned} \tag{14.89}$$

## 14.17 The propagation of spherical waves

Although the transverse modes that propagate inside a laser cavity are not plane waves, in many cases we can derive useful information about them by considering them as such, for example, in the calculation of longitudinal-mode frequencies. Also, although these transverse modes are not spherical waves, they frequently exhibit properties similar to those of spherical waves, and it is worthwhile reviewing briefly some of the propagation characteristics of spherical waves.

Spherical waves are waves that originate from a point source in an isotropic medium. For these waves **P** is parallel to **E**, and $n$ must be uniform or spherically symmetric. When they propagate their radius of curvature increases as shown in Fig. 14.20, according to

$$R_2(z_2) = R_1(z_1) + z_2 - z_1. \tag{14.90}$$

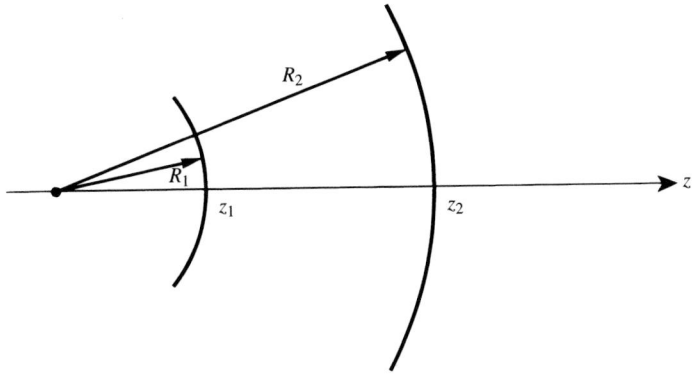


**Fig. 14.20** The change in radius of curvature of a spherical wave as it propagates away from its source.

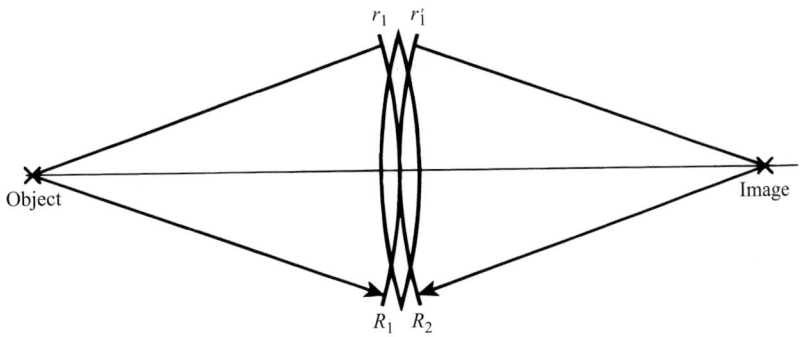


**Fig. 14.21** Focusing of a spherical wave by a lens.

We use the sign convention that the spherical wave has positive radius of curvature when it is convex in the positive-$z$ direction (the direction of propagation). When such a wave encounters a thin lens it is focused (if the lens has a positive focal length) as shown in Fig. 14.21. In this case

$$\frac{1}{R_1} - \frac{1}{R_2} = \frac{1}{f},$$

so

$$\frac{1}{R_2} = \frac{1}{R_1} - \frac{1}{f} \tag{14.91}$$

(note that $R_2$ is negative). If instead we consider one ray OX that helps to make up this spherical wave, then, if $r_1$ and $r_1'$ are the input parameters of this ray,

$$R_1 = r_1/r_1',$$

or, in general, at any plane

$$R(z) = \frac{r(z)}{r'(z)}. \tag{14.92}$$

For propagation of this ray through any optical system with transfer matrix

$$\begin{pmatrix} A & B \\ C & D \end{pmatrix}$$

we know that

$$\begin{pmatrix} r_2 \\ r_2' \end{pmatrix} = \begin{pmatrix} A & B \\ C & D \end{pmatrix} \begin{pmatrix} r_1 \\ r_1' \end{pmatrix},$$

so

$$\begin{aligned} r_2 &= Ar_1 + Br_1', \\ r_2' &= Cr_1 + Dr_1'. \end{aligned} \tag{14.93}$$

The radius of curvature of the outgoing spherical wave (of which the ray is a part) is

$$R_2 = \frac{r_2}{r_2'} = \frac{Ar_1 + Br_1'}{Cr_1 + Dr_1'} = \frac{A(r_1/r_1') + B}{C(r_1/r_1') + D}, \tag{14.94}$$

which gives the simple final result

$$R_2 = \frac{AR_1 + B}{CR_1 + D}. \tag{14.95}$$

Therefore, we can calculate the change in the radius of curvature of the spherical wave as it propagates by considering the transfer matrix of the system. As a spherical wave comes closer to being a plane wave, its radius approaches infinity.

## 14.18 Problems

14.1 An input ray enters the following sequence with ray parameters $r_{\text{in}} = 0.1$ m and $r_{\text{in}}' = 0.01$ rad: a thin lens of focal length 3 m, an air path of length 0.5 m, and a rectangular glass slab of length 1 m and refractive index 1.5. The final flat face of the glass slab is coated to reflect light. What are the ray parameters when the ray returns to the input plane?

14.2 A Galilean telescope, consisting of a converging lens and a diverging lens, is being used to expand a laser beam. If the two lenses of the telescope have focal lengths $f_1$ and $f_2$ and are placed a distance $d$ apart, use ray-transfer matrices to show what configuration should be adopted in order to minimize the beam divergence of the output beam.

14.3 Find a proof for Sylvester's theorem.

14.4 A spherical wave of radius of curvature 100 mm enters a planar, parallel-sided slab with a graded index of the form $n(r) = 1.5 - \frac{1}{2}n_2 r^2$, where $n_2 = 10^2$ m$^{-2}$. How thick must the slab be for the central part of the spherical wavefront to emerge as a plane wave?

# References


[1] W. Cai and V. Shalaev, *Optical Metamaterials: Fundamentals and Applications*, New York: Springer, 2009.

[2] A. A. Maradudin, ed., *Structured Surfaces as Optical Metamaterials,* Cambridge: Cambridge University Press, 2011.

[3] V. M. Agranovich and D. L. Mills, *Surface Polaritons – Electromagnetic Waves at Surfaces and Interfaces*, Amsterdam: North-Holland, 1982.

[4] H. Raether, *Surface Plasmons on Smooth and Rough Surfaces and on Gratings*, New York: Springer, 1988.

[5] D. Sarid and W. Challener, *Modern Introduction to Surface Plasmons*, Cambridge: Cambridge University Press, 2010.

[6] E. K. Illy and H. J. Booth, "Comparison of fiber Bragg grating writing at multiple UV wavelength," in *Proceedings of ECOC*, Copenhagen, 2002.

[7] E. Yablonovitch, "Inhibited spontaneous emission in solid-state physics and electronics," *Phys. Rev. Lett.* **58**, 2059–2062, 1987.

[8] E. Yablonovitch, "Photonic crystals," *J. Mod. Opt.* **41**, 173–194, 1994.

[9] J. D. Watson and F. H. C. Crick, "Molecular structure of nucleic acids: a structure for deoxyribose nucleic acid," *Nature* **171**, 737–738, 1953.

[10] R. E. Franklin and R. G. Gosling. "Molecular configuration in sodium thymonucleate," *Nature* **171**, 740–741, 1953.

[11] R. Franklin and R. G. Gosling, "Evidence for 2-chain helix in crystalline structure of sodium deoxyribonucleate," *Nature* **172**, 156–157, 1953.

[12] C. Kittel, *Introduction to Solid State Physics,* 8th edn., New York: Wiley, 2005.

[13] J. D. Joannopoulos, S. G. Johnson, J. N. Winn, and R. D. Meade, *Photonic Crystals: Molding the Flow of Light*, 2nd edn., Princeton, MA: Princeton University Press, 2008 (available online at http://ab-initio.mit.edu/book/photonic-crystals-book.pdf).

[14] J. Vučković, M. Lončar, H. Mabuchi, and A. Scherer, "Design of photonic crystal microcavities for cavity QE," *Phys. Rev. E.* **65**, 01660811, 2001.

[15] A. Mekis, J. C. Chen, I. Kurland, S. Fan, P. R. Villeneuve, and J. D. Joannopoulos, "High transmission through sharp bends in photonic crystal waveguides," *Phys. Rev. Lett.* **77**, 3788–3790, 1996.

[16] H. Kogelnik and T. Li, "Laser beams and resonators," *Proc. IEEE* **54**, 1312–1329, 1966.

[17] M. Born and E. Wolf, *Principles of Optics*, 7th edn., Cambridge: Cambridge University Press, 1999.

# 15 The optics of Gaussian beams

## 15.1 Introduction

In this chapter we shall examine solutions to Maxwell's equations for the electromagnetic field that take the form of *beam waves*. Of particular interest are beam waves called *Gaussian beams*. These beams of light have a characteristic radial intensity profile whose width varies in the direction of propagation of the beam. These Gaussian beams have almost spherical phase fronts, like spherical waves, so we can match their phase fronts to the curvatures of the mirrors of an optical resonator to determine what form of beam will result from a particular resonator geometry.

## 15.2 Beam-wave solutions of the wave equation

We expect intuitively that the transverse modes of a laser system will take the form of narrow beams of light that propagate between the mirrors of the laser resonator and maintain a field distribution that remains distributed around and near the axis of the system. We shall therefore need to find solutions of the wave equation that take the form of narrow beams and then see how we can make these solutions compatible with a given laser cavity.

The wave equation for any field or potential component $U$ of an electromagnetic wave is

$$\nabla^2 U - \mu_r \mu_0 \epsilon_r \epsilon_0 \frac{\partial^2 U}{\partial t^2} = 0, \tag{15.1}$$

where $\mu_r$ in optical media is generally 1, and $\epsilon_r$ is the dielectric constant, which may be a function of position. The time-harmonic beam-wave solutions that we seek are of the form

$$U(x, y, z, t) = U(x, y, z) e^{j[\omega t - \mathbf{k}(\mathbf{r}) \cdot \mathbf{r}]}. \tag{15.2}$$

We allow the wave vector $\mathbf{k}(\mathbf{r})$ to be a function of $\mathbf{r}$ to include situations where the medium has a non-uniform refractive index. From Eqs. (15.1) and (15.2)

$$\nabla^2 U + \mu_r \mu_0 \epsilon_r \epsilon_0 \omega^2 U = 0, \tag{15.3}$$

where $\epsilon_r$ and $\mu_r$ may be functions of $\mathbf{r}$. We have seen previously that the propagation constant in the medium is $k = \omega \sqrt{\mu_r \mu_0 \epsilon_r \epsilon_0}$, so

$$\nabla^2 U + k(\mathbf{r})^2 U = 0. \tag{15.4}$$

This is the time-independent form of the wave equation, frequently referred to as the *Helmholtz* equation.

In general, if the medium is absorbing, or exhibits gain, then its dielectric constant $\epsilon_r$ has real and imaginary parts.

$$\epsilon_r = 1 + \chi(\omega) = 1 + \chi'(\omega) - j\chi''(\omega), \tag{15.5}$$

and

$$k = k_0\sqrt{\epsilon_r}, \tag{15.6}$$

where $k_0 = \omega\sqrt{\mu_r\mu_0\epsilon_0}$

If the medium were conducting, with conductivity $\sigma$, then its complex propagation vector would obey

$$k^2 = \omega^2\mu\epsilon_r\epsilon_0\left(1 - j\frac{\sigma}{\omega\epsilon_r\epsilon_0}\right), \tag{15.7}$$

so conductivity can be included as a contribution to the imaginary part of the dielectric constant [1].

The simplest solutions of the time-independent wave equation above are transverse plane waves. These simple solutions are not, however, adequate to describe the field distributions of transverse modes in laser systems. Let us look for solutions to the wave equation whose amplitude varies in some way transverse to their direction of propagation. Such solutions will be of the form $U(x,y,z) = \psi(x,y,z)e^{-jkz}$ for waves propagating in the positive-$z$ direction. For functions $\psi(x,y,z)$ that are localized near the $z$ axis the propagating wave takes the form of a narrow beam. Further, because $\psi(x,y,z)$ is not uniform, the surfaces of constant phase in the wave will no longer necessarily be plane. If we can find solutions of the wave equation where $\psi(x,y,z)$ gives phase fronts that are spherical (or approximately so over a small region) then we can make the propagating-beam solution $U(x,y,z) = \psi(x,y,z)e^{-jkz}$ satisfy the boundary conditions in a resonator with spherical reflectors, provided that the mirrors are placed at the positions of phase fronts whose curvature equals that of the mirror. Thus the propagating-beam solution becomes a satisfactory transverse mode of the resonator. For example, in Fig. 15.1, if the propagating beam is to be a satisfactory transverse mode, then a spherical mirror of radius $R_1$ must be

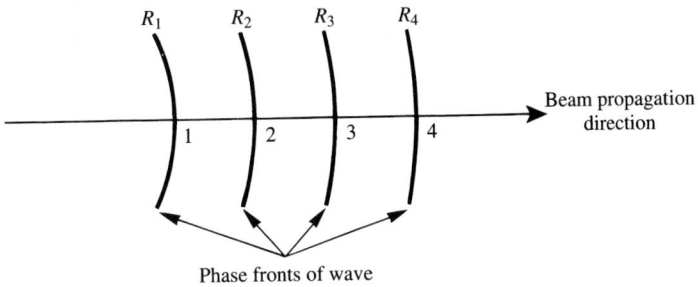


**Fig. 15.1** The changing curvature of a spherical wave as it propagates in the $z$ direction.

placed at position 1 or one of radius $R_2$ at position 2, etc. Mirrors placed in this way lead to reflection of the wave back on itself.

On substituting $U(x,y,z) = \psi(x,y,z)e^{-jkz}$ into Eq. (15.4) we get

$$\left(\frac{\partial^2\psi}{\partial x^2}\right)e^{-jkz} + \left(\frac{\partial^2\psi}{\partial y^2}\right)e^{-jkz} + \left(\frac{\partial^2\psi}{\partial z^2}\right)e^{-jkz} - 2jk\frac{\partial\psi}{\partial z}e^{-jkz}$$
$$- k^2\psi(x,y,z)e^{-jkz} + k^2\psi(x,y,z)e^{-jkz} = 0, \tag{15.8}$$

which reduces to

$$\frac{\partial^2\psi}{\partial x^2} + \frac{\partial^2\psi}{\partial y^2} - 2jk\frac{\partial\psi}{\partial z} + \frac{\partial^2\psi}{\partial z^2} = 0. \tag{15.9}$$

If the beam-like solution we are looking for remains paraxial then $\psi$ will vary only slowly with $z$, so we can neglect $\partial^2\psi/\partial z^2$ and get

$$\frac{\partial^2\psi}{\partial x^2} + \frac{\partial^2\psi}{\partial y^2} - 2jk\frac{\partial\psi}{\partial z} = 0. \tag{15.10}$$

We try as a solution

$$\psi(x,y,z) = \exp\left\{-j\left[P(z) + \frac{k}{2q(z)}r^2\right]\right\}, \tag{15.11}$$

where $r^2 = x^2 + y^2$ is the square of the distance of the point $x, y$ from the axis of propagation. Here $P(z)$ represents a phase-shift factor and $q(z)$ is called the beam parameter. We shall see the significance of these parameters shortly.

On substituting into Eq. (15.10) and using the relations below that follow from Eq. (15.11),

$$\frac{\partial\psi}{\partial x} = -\frac{jk}{2q(z)}\exp\left\{-j\left[P(z) + \frac{k}{2q(z)}r^2\right]\right\}\cdot 2x, \tag{15.12}$$

$$\frac{\partial^2\psi}{\partial x^2} = -\frac{jk}{q(z)}\exp\left\{-j\left[P(z) + \frac{k}{2q(z)}r^2\right]\right\}$$
$$- \frac{k^2}{4q^2(z)}\exp\left\{-j\left[P(z) + \frac{k}{2q(z)}r^2\right]\right\}\cdot 4x^2, \tag{15.13}$$

$$\frac{\partial\psi}{\partial z} = \exp\left\{-j\left[P(z) + \frac{k}{2q(z)}r^2\right]\right\}\left[-j\left(\frac{dP}{dz} - \frac{k}{2q^2}\frac{dq}{dz}r^2\right)\right], \tag{15.14}$$

we get

$$-2k\left[\frac{dP}{dz} + \frac{j}{q(z)}\right] - \left[\frac{k^2}{q^2(z)} - \frac{k^2}{q^2(z)}\frac{dq}{dz}\right](r^2) = 0. \tag{15.15}$$

Since this equation must be true for all values of $r$, the coefficients of different powers of $r$ must be independently equal to zero, so

$$\frac{dq}{dz} = 1 \tag{15.16}$$

and

$$\frac{dP}{dz} = -\frac{j}{q(z)}. \tag{15.17}$$

The solution $\psi(x,y,z) = \exp\{-j\left[P(z) + kr^2/(2q(z))\right]\}$ is called the *fundamental* Gaussian-beam solution of the time-independent wave equation, since its "intensity" as a function of $x$ and $y$ is

$$UU^* = \psi\psi^* = \exp\left\{-j\left[P(z) + \frac{k}{2q(z)}r^2\right]\right\}\exp\left\{j\left[P^*(z) + \frac{k}{2q^*(z)}r^2\right]\right\}, \tag{15.18}$$

where $P^*(z)$ and $q^*(z)$ are the complex conjugates of $P(z)$ and $q(z)$, respectively, so

$$UU^* = \exp\{-j\left[P(z) - P^*(z)\right]\}\exp\left\{-\frac{jkr^2}{2}\left[\frac{1}{q(z)} - \frac{1}{q^*(z)}\right]\right\}. \tag{15.19}$$

For convenience we introduce two real beam parameters, $R(z)$ and $w(z)$, which are related to $q(z)$ by

$$\frac{1}{q} = \frac{1}{R} - \frac{j\lambda}{\pi w^2}, \tag{15.20}$$

where both $R$ and $w$ depend on $z$. It is important to note that $\lambda = \lambda_0/n$ is the wavelength *in the medium*. From Eq. (15.16) above we can see that

$$\begin{aligned} UU^* &\propto \exp\left[-\frac{jkr^2}{2}\left(-\frac{j\lambda}{\pi w^2} - \frac{j\lambda}{\pi w^2}\right)\right] \\ &\propto \exp\left(-\frac{2r^2}{w^2}\right). \end{aligned} \tag{15.21}$$

Thus, the beam intensity shows a Gaussian dependence on $r$; the physical significance of $w(z)$ is that it is the distance from the axis at point $z$ where the intensity of the beam has fallen to $1/e^2$ of its peak value on axis and its amplitude to $1/e$ of its axial value: $w(z)$ is called the *spot size*. With these parameters

$$U(x,y,z) = \exp\left\{-j\left[kz + P(z) + \frac{kr^2}{2}\left(\frac{1}{R} - \frac{j\lambda}{\pi w^2}\right)\right]\right\}. \tag{15.22}$$

We can integrate Eqs. (15.16) and (15.17). From Eq. (15.16)

$$q = q_0 + z, \tag{15.23}$$

where $q_0$ is a constant of integration that gives the value of the beam parameter at plane $z = 0$. From Eq. (15.17)

$$\frac{dP(z)}{dz} = -\frac{j}{q(z)} = -\frac{j}{q_0 + z}, \tag{15.24}$$

so

$$P(z) = -j[\ell\mathrm{n}(z + q_0)] + (\theta + j\,\ell\mathrm{n}\,q_0), \tag{15.25}$$

where the constant of integration, $(\theta + j\,\ell n\, q_0)$, is written in this way so that, on substituting from Eqs. (15.23) and (15.25) into Eq. (15.22), we get

$$U = \exp\left\{-j\left[kz - j\,\ell n\left(1 + \frac{z}{q_0}\right) + \theta + \frac{kr^2}{2}\left(\frac{1}{R} - \frac{j\lambda}{\pi w^2}\right)\right]\right\}. \tag{15.26}$$

The factor $e^{-j\theta}$ is only a constant phase factor, which we can arbitrarily set to zero, and then we get

$$U(x, y, z) = \exp\left\{-j\left[kz - j\,\ell n\left(1 + \frac{z}{q_0}\right) + \frac{kr^2}{2}\left(\frac{1}{R} - \frac{j\lambda}{\pi w^2}\right)\right]\right\}. \tag{15.27}$$

The radial variation in phase of this field for a particular value of $z$ is

$$\phi(z) = kz + \frac{kr^2}{2R} - \mathcal{R}\left[j\,\ell n\left(1 + \frac{z}{q_0}\right)\right]. \tag{15.28}$$

There is no radial phase variation if $R \to \infty$. We choose the value of $z$ where this occurs as our origin $z = 0$. We call this location the *beam waist*; the complex beam parameter here has the value

$$\frac{1}{q_0} = -j\frac{\lambda}{\pi w_0^2}, \tag{15.29}$$

where $w_0 = w(0)$ is called the *minimum spot size*. At an arbitrary point $z$

$$q = \frac{j\pi w_0^2}{\lambda} + z, \tag{15.30}$$

and, using $1/q = 1/R - j\lambda/(\pi w^2)$, we have the relationship for the *spot size* $w(z)$ at position $z$:

$$w^2(z) = w_0^2\left[1 + \left(\frac{\lambda z}{\pi w_0^2}\right)^2\right]. \tag{15.31}$$

So $w_0$ is clearly the *minimum* spot size. The value of $R$ at position $z$ is

$$R(z) = z\left[1 + \left(\frac{\pi w_0^2}{\lambda z}\right)^2\right]. \tag{15.32}$$

We shall see shortly that $R(z)$ can be identified as the radius of curvature of the phase front at $z$. From Eq. (15.31) we can see that $w(z)$ expands with distance $z$ along the axis along a hyperbola that is inclined to the axis at an angle $\theta_{\text{beam}} = \tan^{-1}(\lambda/(\pi w_0))$. This is a small angle if the beam is to have small divergence, in which case

$$\theta_{\text{beam}} \simeq \lambda/(\pi w_0), \tag{15.33}$$

where $\theta_{\text{beam}}$ is the half angle of the diverging beam shown in Fig. 15.2. From Eq. (15.27) it is clear that the surfaces of constant phase are those surfaces satisfying

$$kz + \mathcal{R}\left[-j\,\ell n\left(1 + \frac{z}{q_0}\right)\right] + \frac{kr^2}{2}\left(\frac{1}{R}\right) = \text{constant}, \tag{15.34}$$

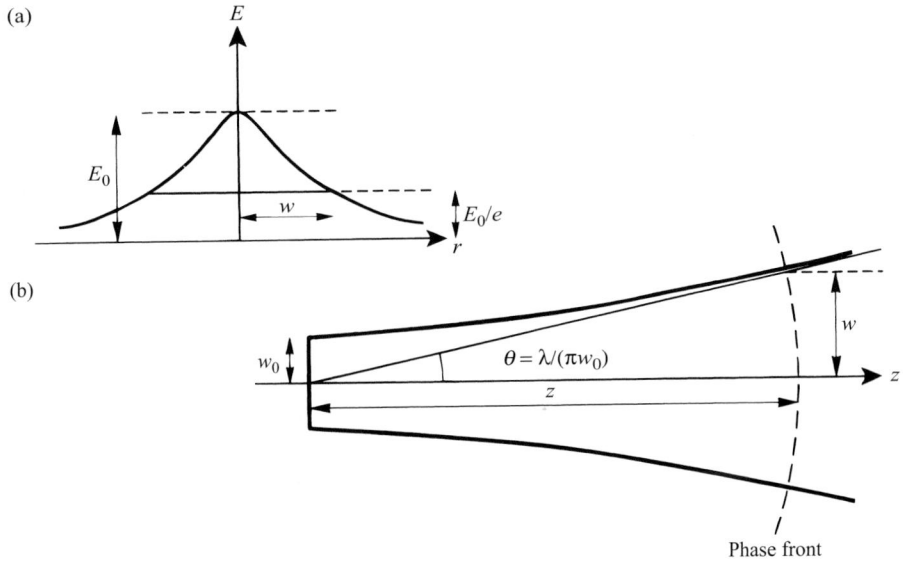


**Fig. 15.2** Characteristics of a Gaussian beam: (a) the Gaussian field distribution along a phase front; and (b) a beam contour showing the hyperbolic envelope of the beam-expansion contour and the asymptotic angle $\theta$.

which gives

$$k\left(z + \frac{r^2}{2R}\right) = \text{constant} - \mathcal{R}\left[-j\,\ell\text{n}\left(1 + \frac{z}{q_0}\right)\right]. \tag{15.35}$$

Now,

$$\ell\text{n}\left(1 + \frac{z}{q_0}\right) = \ell\text{n}\left(1 - \frac{jz\lambda}{\pi w_0^2}\right) = \ell\text{n}\sqrt{1 + \left(\frac{\lambda z}{\pi w_0^2}\right)^2} - j\tan^{-1}\left(\frac{\lambda z}{\pi w_0^2}\right), \tag{15.36}$$

where we have used the relation

$$\ell\text{n}(x + jy) = \ell\text{n}\sqrt{x^2 + y^2} + j\tan^{-1}(y/x). \tag{15.37}$$

So, from Eq. (15.37),

$$\mathcal{R}\left[-j\,\ell\text{n}\left(1 + \frac{z}{q_0}\right)\right] = \tan^{-1}\left(\frac{\lambda z}{\pi w_0^2}\right). \tag{15.38}$$

The term $kz = 2\pi z/\lambda$ in Eq. (15.35) is very large; except very close to the beam waist and for long wavelengths, it generally dominates the $z$ dependence of Eq. (15.35). Also, $\tan^{-1}[\lambda z/(\pi w_0^2)]$ at most approaches a value of $\pi/2$, so we can write

$$k\left(z + \frac{r^2}{2R}\right) = \text{constant} \tag{15.39}$$

as the equation of the surface of constant phase at $z$, which can be rewritten

$$z = \frac{\text{constant}}{k} - \frac{r^2}{2R}. \tag{15.40}$$

This is a parabola, which for $r^2 \ll z^2$ is a very close approximation to a spherical surface of which $R(z)$ is the radius of curvature.

The complex phase shift a distance $z$ from the point where $q$ is purely imaginary, the point called the beam waist, is found from Eq. (15.17),

$$\frac{dP(z)}{dz} = -\frac{j}{q} = -\frac{j}{z + j\pi w_0^2/\lambda}. \tag{15.41}$$

Thus,

$$P(z) = -j\,\ell\text{n}\left[1 - j\left(\frac{\lambda z}{\pi w_0^2}\right)\right] = -j\,\ell\text{n}\sqrt{1 + \left(\frac{\lambda z}{\pi w_0^2}\right)^2} - \tan^{-1}\left(\frac{\lambda z}{\pi w_0^2}\right), \tag{15.42}$$

where the constant of integration has been chosen so that $P(z) = 0$ at $z = 0$.

The real part of this phase shift can be written as

$$\Phi = \tan^{-1}\left(\frac{\lambda z}{\pi w_0^2}\right), \tag{15.43}$$

which marks a distinction between this Gaussian beam and a plane wave. The imaginary part is, from Eqs. (15.31) and (15.42),

$$\mathcal{I}\,[P(z)] = -j\,\ell\text{n}\,[w(z)/w_0]. \tag{15.44}$$

This imaginary part gives a real intensity dependence of the form $w_0^2/w^2(z)$ on axis, which we would expect because of the expansion of the beam. For example, part of the "phase factor" in Eq. (15.11) gives

$$e^{-j\mathcal{I}[P(z)]} = e^{-\ell\text{n}(w/w_0)} = \frac{w_0}{w}. \tag{15.45}$$

Thus, from Eqs. (15.11), (15.20), and (15.44) we can write the spatial dependence of our Gaussian beam

$$U = \psi(x, y, z)e^{-jkz} \tag{15.46}$$

as

$$U(r, z) = \frac{w_0}{w}\exp\left[-j(kz - \Phi) - r^2\left(\frac{1}{w^2} + \frac{jk}{2R}\right)\right]. \tag{15.47}$$

This field distribution is called the *fundamental* or $TEM_{00}$ Gaussian mode. Its amplitude distribution and beam contour in planes of constant $z$ and planes lying in the $z$ axis, respectively, are shown in Fig. 15.2.

In any plane the radial intensity distribution of the beam can be written as

$$I(r) = I_0 e^{-2r^2/w^2}, \tag{15.48}$$

where $I_0$ is the axial intensity.

## 15.3 Higher-order modes

In the preceding discussion, only one possible solution for $\psi(x,y,z)$ has been considered, namely the simplest solution where $\psi(x,y,z)$ gives a beam with a Gaussian dependence of intensity as a function of its distance from the axis: the width of this Gaussian beam changes as the beam propagates along its axis. However, higher-order beam solutions of Eq. (15.10) are possible.

### 15.3.1 Beam modes with Cartesian symmetry

For a system with Cartesian symmetry we can try for a solution of the form

$$\psi(x,y,z) = g\left(\frac{x}{w}\right) h\left(\frac{y}{w}\right) \exp\left\{-j\left[P + \frac{k}{2q}(x^2+y^2)\right]\right\}, \tag{15.49}$$

where $g$ and $h$ are functions of $x$ and $z$ and of $y$ and $z$, respectively, $w(z)$ is a Gaussian beam parameter, and $P(z)$ and $q(z)$ are the beam parameters used previously.

For real functions $g$ and $h$, $\psi(x,y,z)$ represents a beam whose intensity patterns scale according to a Gaussian beam parameter $w(z)$. If we substitute this higher-order solution into Eq. (15.10) then we find that the solutions have $g$ and $h$ obeying the following relation:

$$gh = H_m\left(\sqrt{2}\frac{x}{w}\right) H_n\left(\sqrt{2}\frac{y}{w}\right), \tag{15.50}$$

where $H_m$ and $H_n$ are solutions of the differential equation

$$\frac{d^2H_m}{dx^2} - 2x\frac{dH_m}{dx} + 2mH_m = 0. \tag{15.51}$$

The solution to Eq. (15.51) is the Hermite polynomial $H_m$ of order $m$: $m$ and $n$ are called the transverse mode numbers. Some of the low-order Hermite polynomials are

$$H_0(x) = 1,\ H_1(x) = x,\ H_2(x) = 4x^2 - 2,\ H_3(x) = 8x^3 - 12x. \tag{15.52}$$

The overall amplitude behavior of this higher-order beam is

$$U_{m,n}(r,z) = \frac{w_0}{w} H_m\left(\sqrt{2}\frac{x}{w}\right) H_m\left(\sqrt{2}\frac{x}{w}\right) \exp\left[-j(kz - \Phi) - r^2\left(\frac{1}{w^2} + \frac{jk}{2R}\right)\right]. \tag{15.53}$$

The parameters $R(z)$ and $w(z)$ are the same for all these solutions. However, $\Phi$ depends on $m$, $n$, and $z$ as

$$\Phi(m,n,z) = (m+n+1)\tan^{-1}\left(\frac{\lambda z}{\pi w_0^2}\right). \tag{15.54}$$

Since this Gaussian beam involves a product of Hermite and Gaussian functions it is called a *Hermite–Gaussian* (HG) mode, and has the familiar intensity pattern observed in the output of many lasers, as shown in Fig. 15.3. The mode is designated a $\mathrm{TEM}_{mn}$ HG mode.

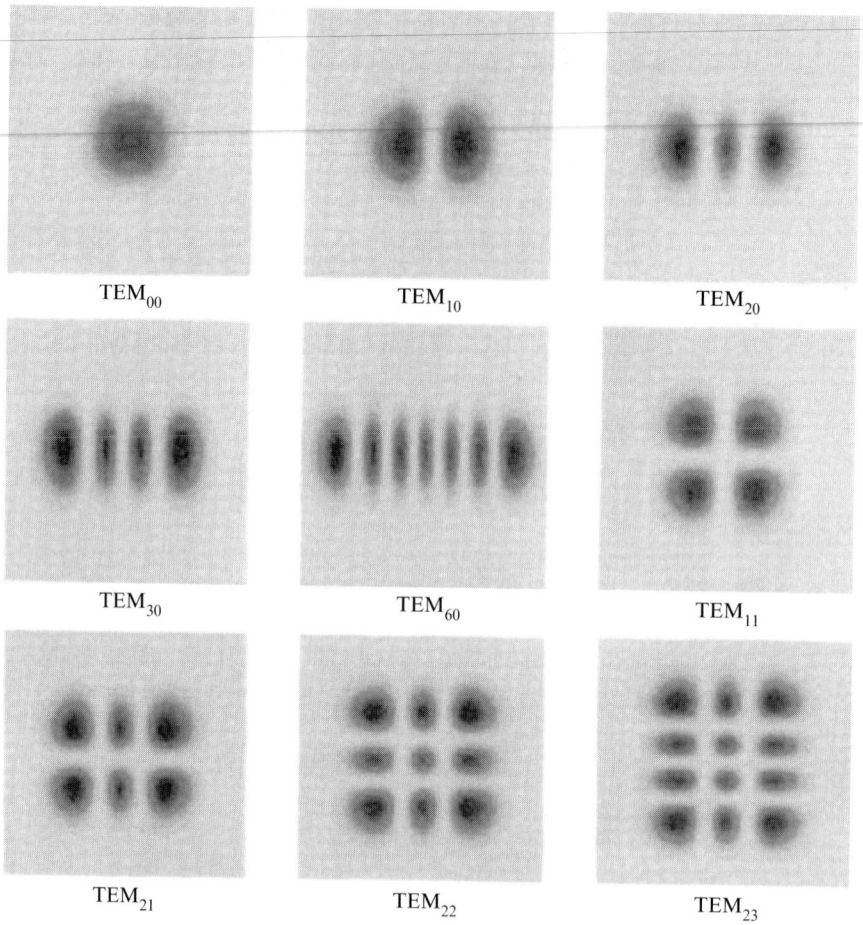


**Fig. 15.3** Hermite–Gaussian laser mode-intensity patterns.

For example, in the plane $z = 0$ the electric field distribution of the TEM$_{mn}$ mode is, if the wave is polarized in the $x$ direction,

$$E_{m,n}(x,z) = E_0 H_m\left(\sqrt{2}\frac{x}{w_0}\right) H_n\left(\sqrt{2}\frac{y}{w_0}\right) \exp\left(-\frac{x^2+y^2}{w_0^2}\right). \tag{15.55}$$

The lateral intensity variation of the mode is

$$I(x,y) \propto |E_{m,n}(x,z)|^2, \tag{15.56}$$

and, in surfaces of constant phase,

$$E_{m,n}(x,z) = E_0 \frac{w_0}{w} H_m\left(\sqrt{2}\frac{x}{w}\right) H_n\left(\sqrt{2}\frac{y}{w}\right) \exp\left(-\frac{x^2+y^2}{w^2}\right), \tag{15.57}$$

where it is important to remember that $w$ is a function of $z$. Some examples of the field and intensity variations described by Eqs. (15.55) and (15.56) are shown in Figs. 15.4 and 15.5.

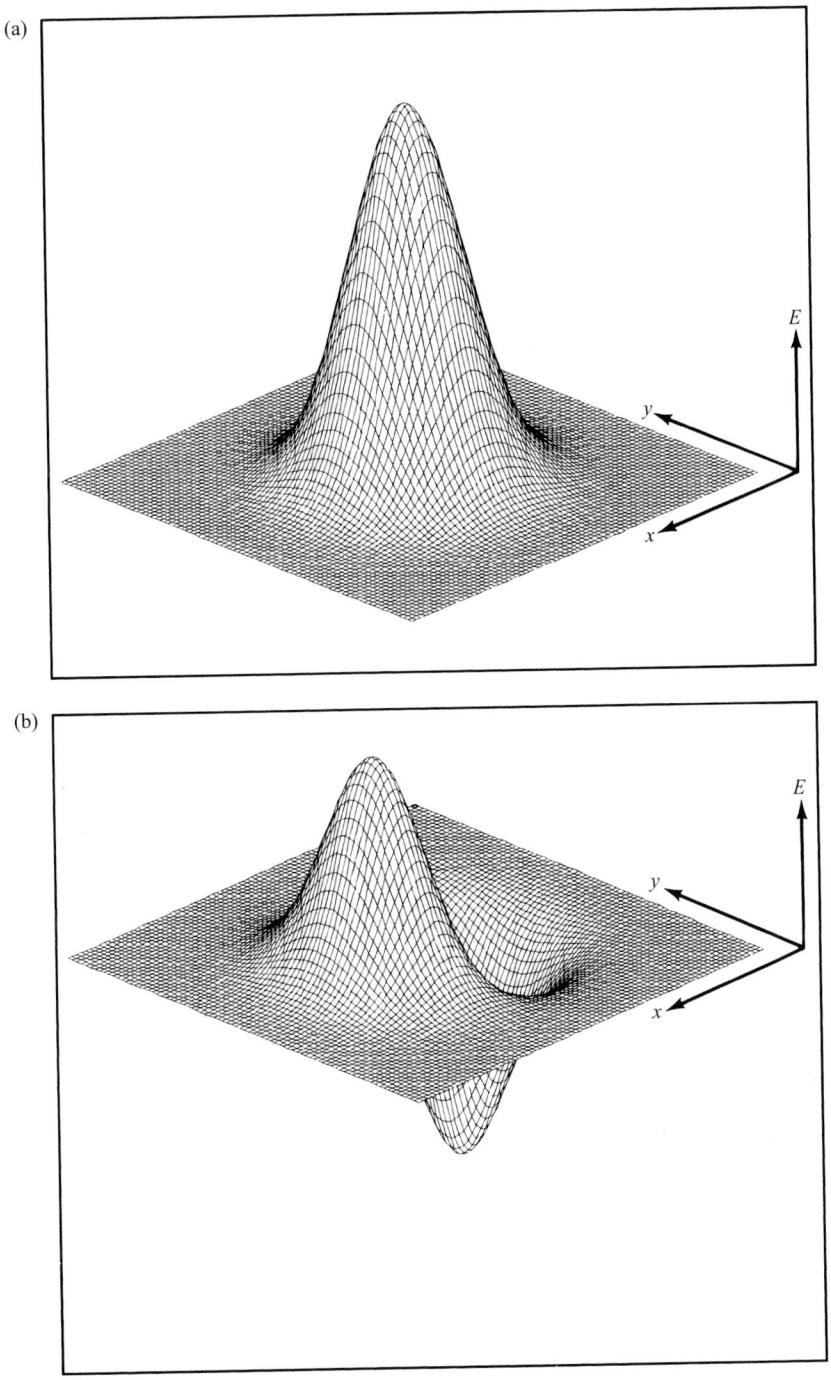


**Fig. 15.4** Three-dimensional plots of the field distribution of some Hermite–Gaussian laser modes.

(c)

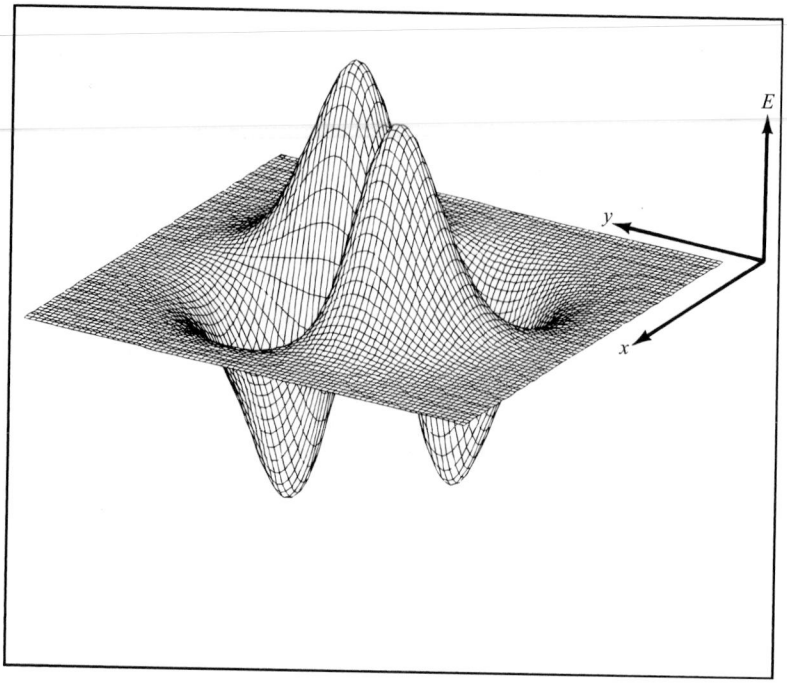


(d)

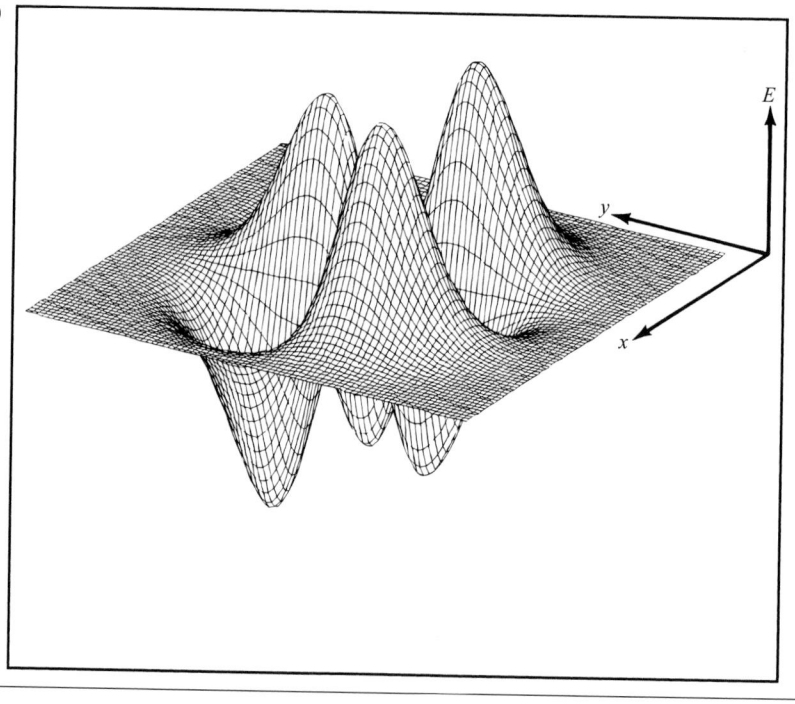


**Fig. 15.4** *(cont.)*

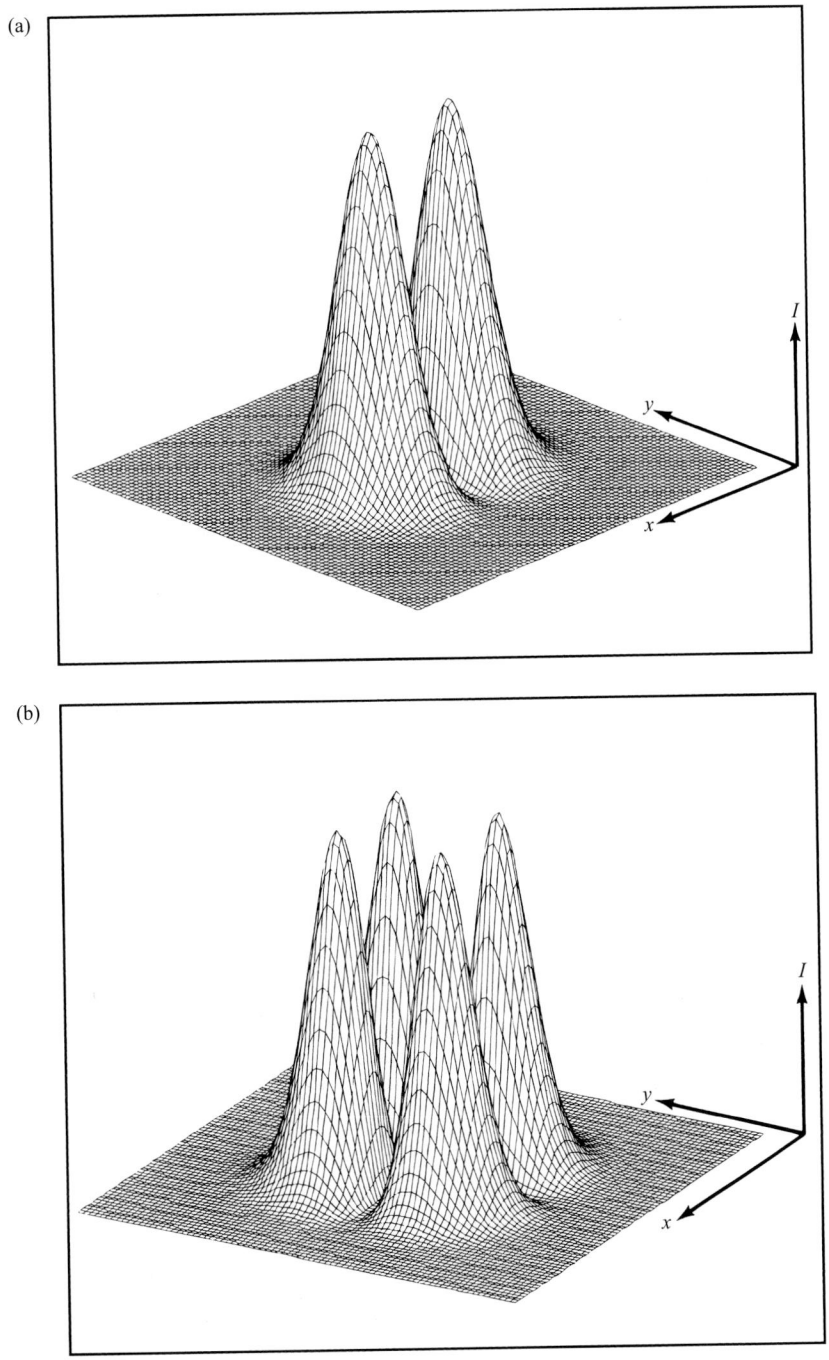


**Fig. 15.5** Three-dimensional plots of the intensity distribution of some Hermite–Gaussian laser modes.

(c)

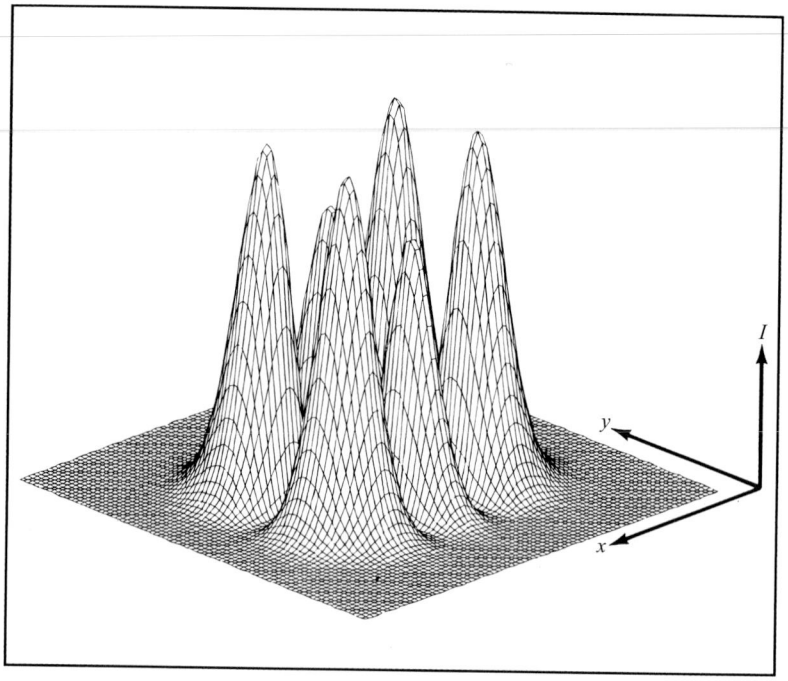


(d)

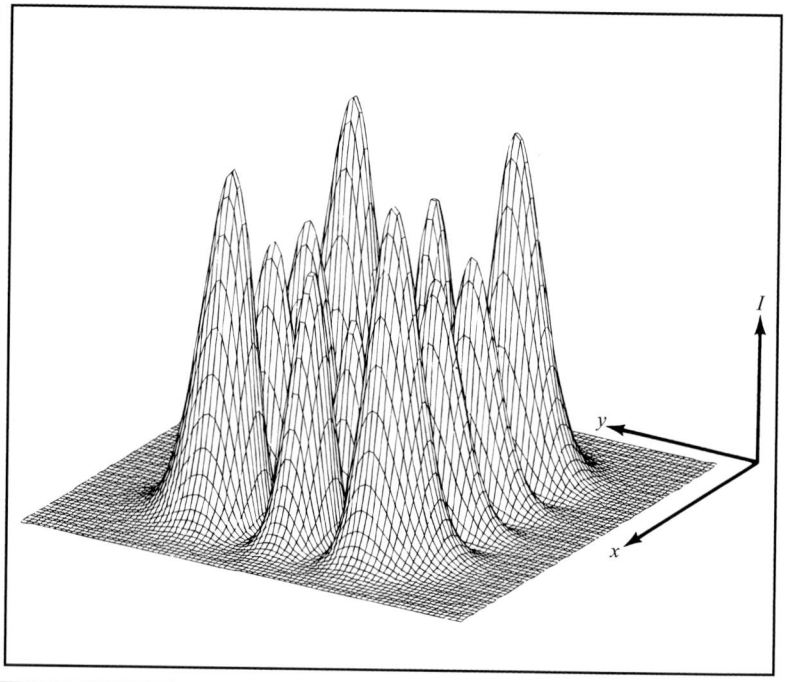


**Fig. 15.5** (*cont.*)

The phase variation

$$\Phi(m,n,z) = (m+n+1)\tan^{-1}\left(\frac{\lambda z}{\pi w_0^2}\right) \tag{15.58}$$

means that the phase velocity increases with mode number so that in a resonator different transverse modes have different resonance frequencies.

### 15.3.2 Cylindrically symmetric higher-order beams

In this case we try for a solution of Eq. (15.10) in the form

$$\psi = g\left(\frac{r}{w}\right)\exp\left\{-j\left(P(z) + \frac{k}{2q(z)}r^2 + \ell\phi\right)\right\}, \tag{15.59}$$

and find that

$$g = \left(\sqrt{2}\frac{r}{w}\right)^{\ell} L_p^{\ell}\left(2\frac{r^2}{w^2}\right), \tag{15.60}$$

where $L_p^{\ell}$ is an associated Laguerre polynomial, $p$ and $\ell$ are the radial and angular mode numbers, and $L_p^{\ell}$ obeys the differential equation

$$x\frac{d^2L_p^{\ell}}{dx^2} + (\ell+1-x)\frac{dL_p^{\ell}}{dx} + pL_p^{\ell} = 0. \tag{15.61}$$

Some of the associated Laguerre polynomials of low order are

$$L_0^{\ell}(x) = 1, \tag{15.62}$$

$$L_1^{\ell}(x) = \ell + 1 - x, \tag{15.63}$$

$$L_2^{\ell}(x) = \frac{1}{2}(\ell+1)(\ell+2) - (\ell+2)x + \frac{1}{2}x^2. \tag{15.64}$$

These are called Laguerre–Gaussian (LG) modes. Some examples of their intensity-mode patterns are given in Fig. 15.6. These "axisymmetric" modes are designated $\mathrm{TEM}_{pl}$. Figures 15.7 and 15.8 give some three-dimensional representations of their field and intensity distributions. The angular variation of these patterns is shown explicitly by looking at the magnitude of the electric field, which from Eq. (15.59) is

$$|E(r,\phi)| = E_0\left(\sqrt{2}\frac{r}{w}\right)^{\ell} L_p^{\ell}\left(2\frac{r^2}{w^2}\right)e^{-r^2/w^2}\cos(\ell\phi). \tag{15.65}$$

A true axisymmetric intensity distribution frequently results in practice because the choice of origin for $\phi$ is arbitrary. The modes with $\phi = \phi_0$ and those with $\phi = \phi_0 + \pi/2$ give identical patterns, which, superimposed, have cylindrical symmetry. The best example of this is the $\mathrm{TEM}_{01}^{*}$ mode, called the *donut* mode. This mode can be regarded as the superposition of two LG modes, or the superposition of the two HG modes $\mathrm{TEM}_{01}$ and $\mathrm{TEM}_{10}$. As is the case for beams with Cartesian symmetry, the beam parameters $w(z)$ and $R(z)$ are

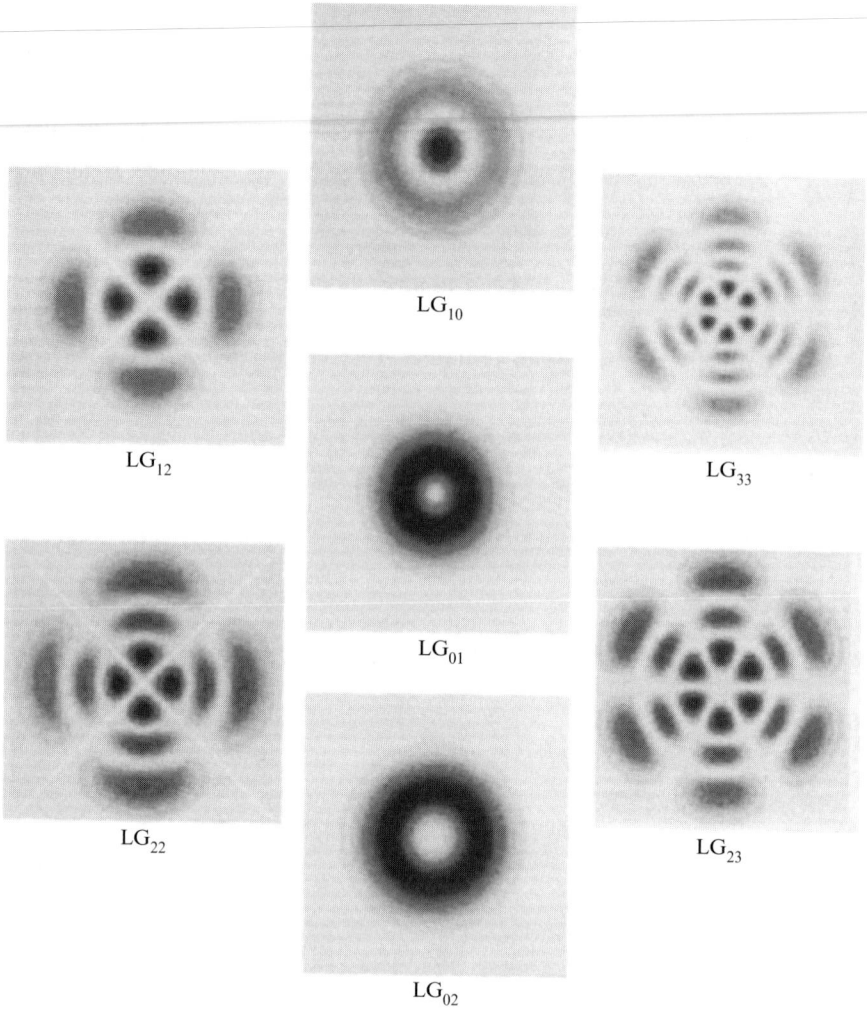


Fig. 15.6 Laguerre–Gaussian laser mode intensity patterns.

the same for all cylindrical modes. The phase factor $\Phi$ for these cylindrically symmetric modes is

$$\Phi(p, \ell, z) = (2p + \ell + 1)\tan^{-1}\left(\frac{\lambda z}{\pi w_0^2}\right). \tag{15.66}$$

The following question could be asked – why should lasers that frequently have apparently cylindrical symmetry both in construction and excitation geometry generate transverse modes with Cartesian rather than radial symmetry? The answer is that this usually results because there is indeed some feature of the laser's construction or method of excitation that removes an apparent equivalence of all radial directions. For example, in a laser with Brewster windows the preferred polarization orientation imposes a directional

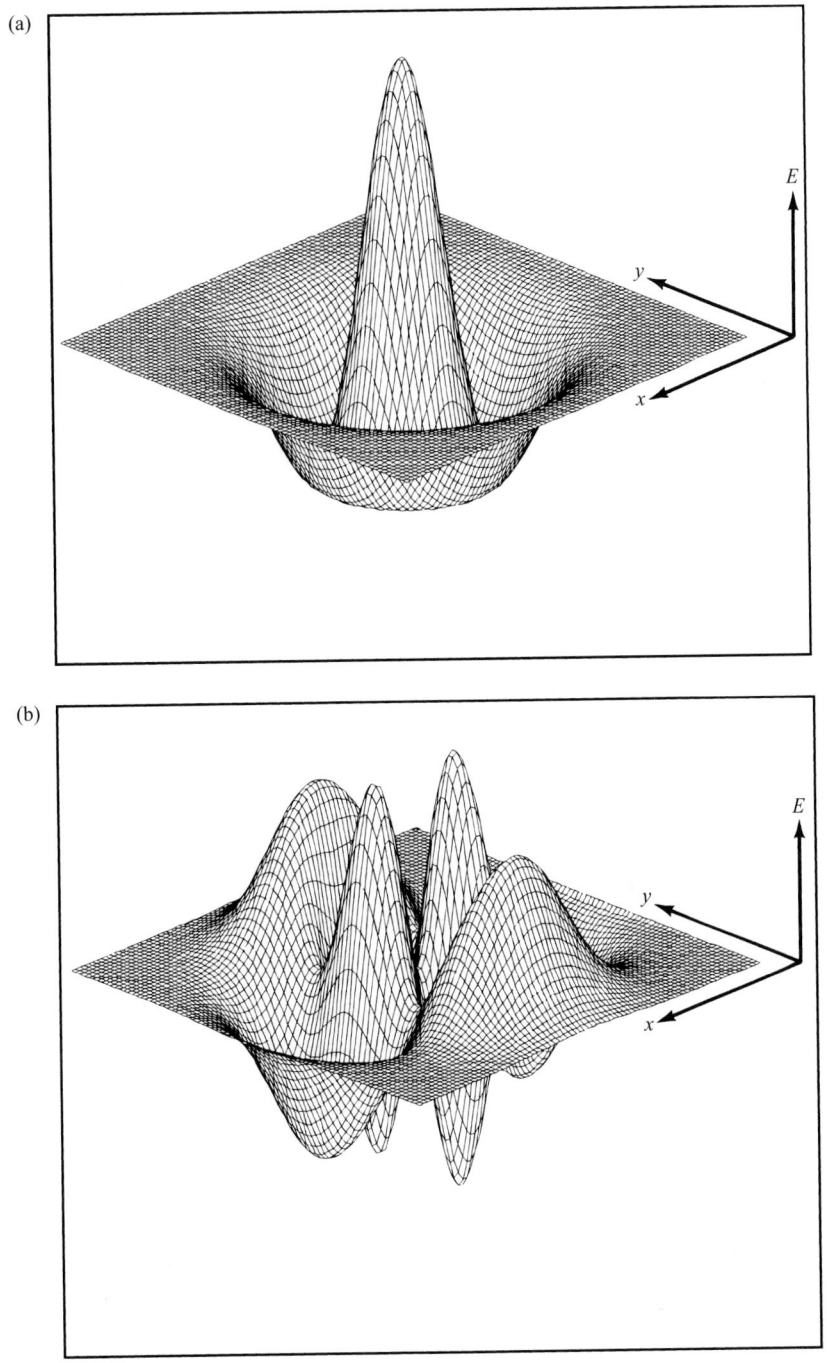


**Fig. 15.7** Three-dimensional plots of the field distribution of some Laguerre–Gaussian laser modes.

(c)

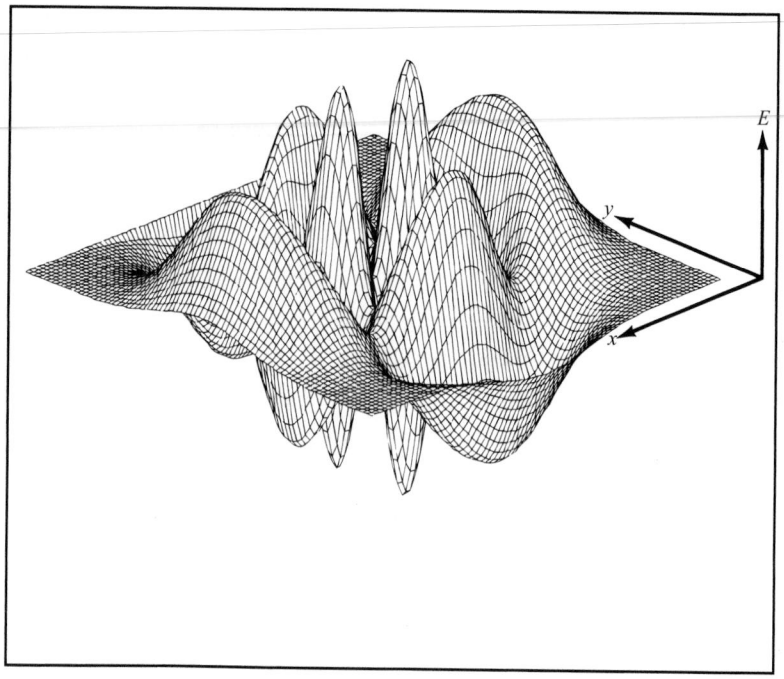


Fig. 15.7 (*cont.*)

constraint that forms modes of Cartesian (HG) rather than radial (LG) symmetry. The simplest way to observe the HG modes in the laboratory is to place a thin wire inside the laser cavity. The laser will then choose to operate in a transverse mode that has a nodal line in the location of the intracavity obstruction. Adjustment of the laser mirrors to slightly different orientations will usually change the output transverse mode, although if the mirrors are adjusted too far from optimum the laser will go out.

## 15.4 Transformation of Gaussian beams by general optical systems

We have seen that a Gaussian laser beam of whatever order is characterized by a complex beam parameter $q(z)$, which changes as the beam propagates in an isotropic, homogeneous material according to $q(z) = q_0 + z$, where

$$q_0 = \frac{j\pi w_0^2}{\lambda} \qquad \text{and} \qquad \frac{1}{q(z)} = \frac{1}{R(z)} - \frac{j\lambda}{\pi w^2}. \tag{15.67}$$

Here $R(z)$ is the radius of curvature of the (approximately) spherical phase front at $z$, and is given by

$$R(z) = z\left[1 + \left(\frac{\pi w_0^2}{\lambda z}\right)^2\right]. \tag{15.68}$$

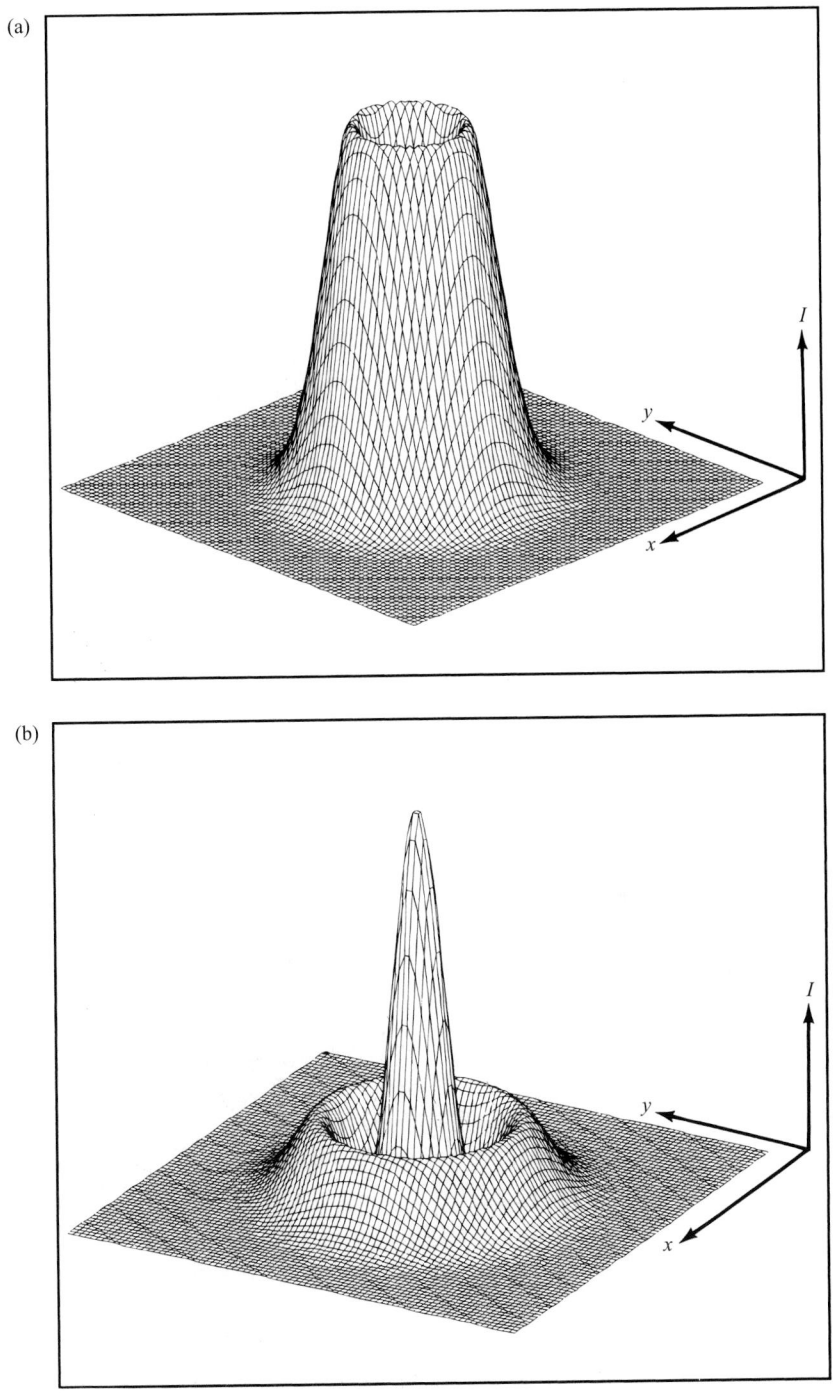


**Fig. 15.8** Three-dimensional plots of the intensity distribution of some Laguerre–Gaussian laser modes.

(c)

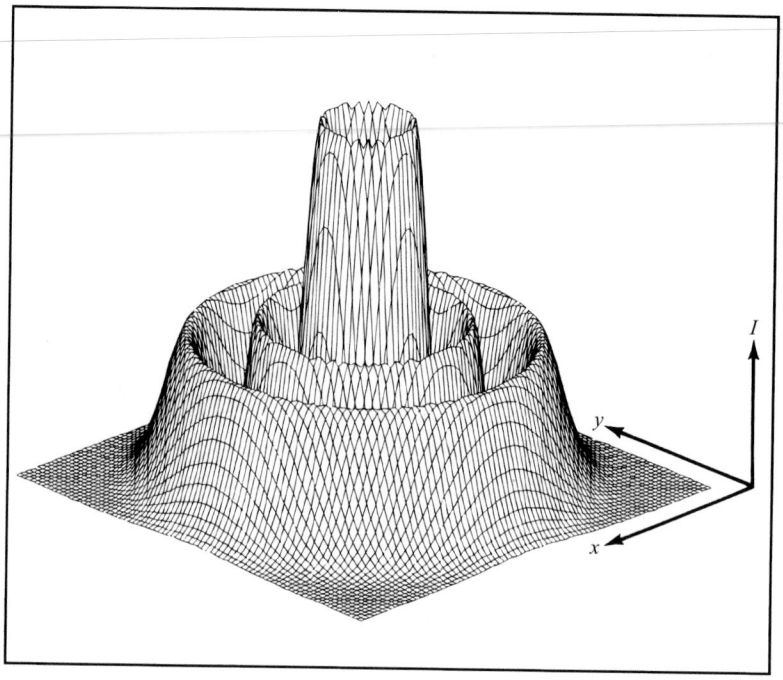


(d)

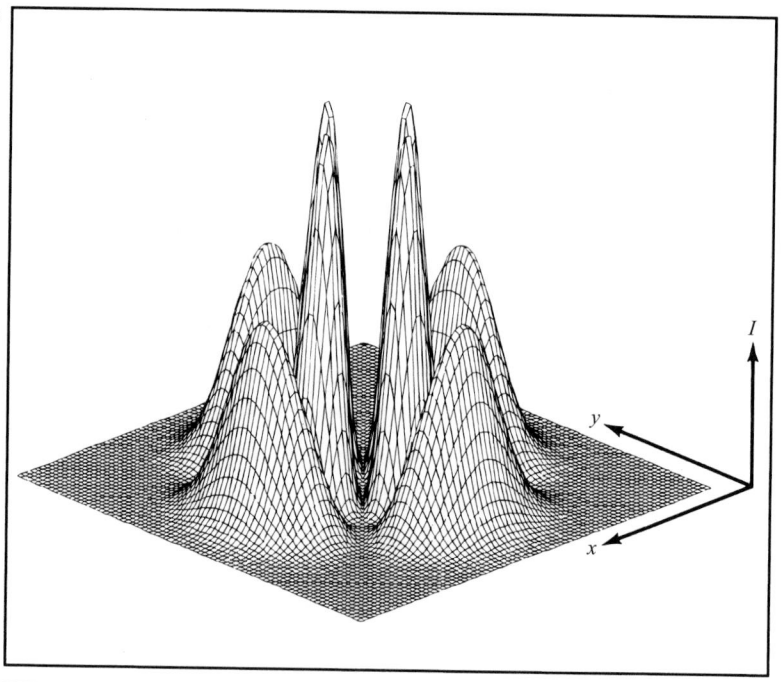


**Fig. 15.8** (*cont.*)

(e)

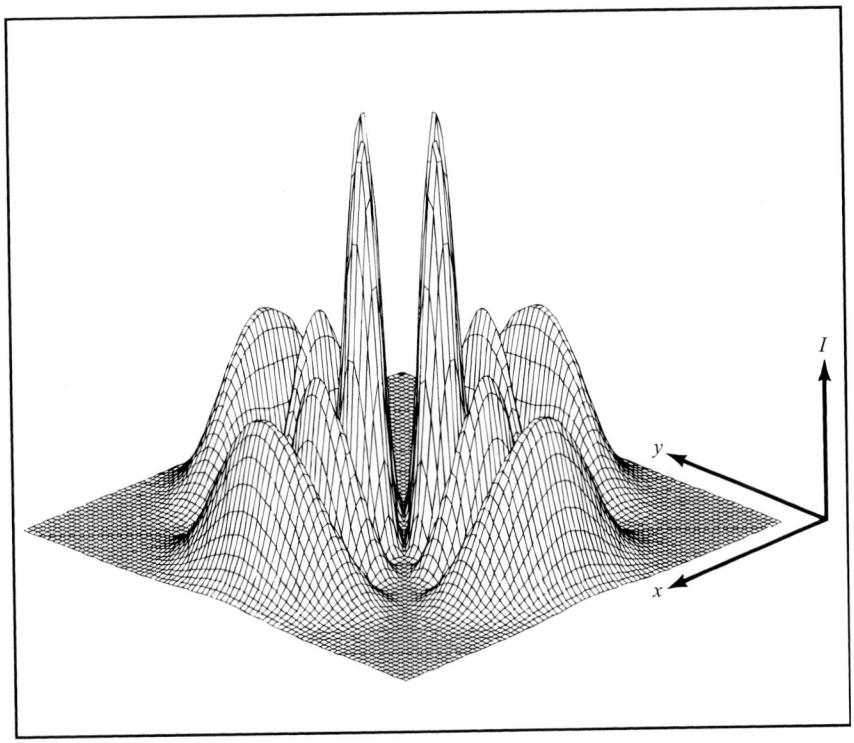


Fig. 15.8 (*cont.*)

This Gaussian beam becomes a true spherical wave as $w \to \infty$. Now, a spherical wave changes its radius of curvature as it propagates according to $R(z) = R_0 + z$, where $R_0$ is its radius of curvature at $z = 0$. So the complex beam parameter of a Gaussian wave changes in just the same way as it propagates as does the radius of curvature of a spherical wave.

When a Gaussian beam strikes a thin lens the spot size, which measures the transverse width of the beam intensity distribution, is unchanged at the lens. However, the radius of curvature of its wave front is altered in just the same way as a spherical wave. If $R_1$ and $R_2$ are the radii of curvature of the incoming and outgoing waves measured at the lens, as shown in Fig. 15.9, then, as in the case of a true spherical wave,

$$\frac{1}{R_2} = \frac{1}{R_1} - \frac{1}{f}. \tag{15.69}$$

So, for the change of the overall beam parameter, since $w$ is unchanged at the lens, we have the following relationship among the beam parameters, *measured at the lens*:

$$\frac{1}{q_2} = \frac{1}{q_1} - \frac{1}{f}. \tag{15.70}$$

As we have seen, the complex beam parameter of a Gaussian beam is transformed by a thin lens in just the same way as the radius of curvature of a spherical wave. Now, the

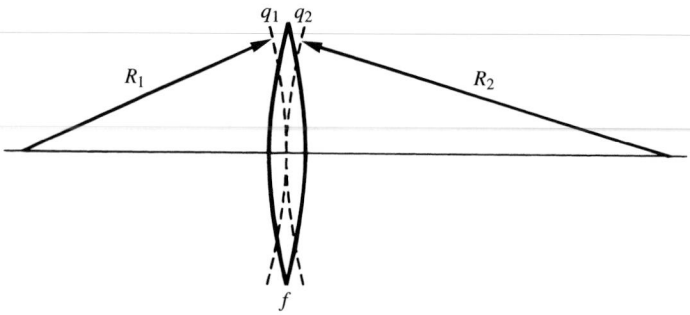


Fig. 15.9 Transformation of a Gaussian beam by a lens. The input and output phase-front curvatures and beam parameters are indicated.

transformation of the radius of curvature of a spherical wave by an optical system with transfer matrix

$$\begin{pmatrix} A & B \\ C & D \end{pmatrix}$$

obeys

$$R_2 = \frac{AR_1 + B}{CR_1 + D}. \tag{15.71}$$

Thus, by continuing to draw a parallel between the $q$ of a Gaussian beam and the $R$ of a spherical wave, we can postulate that the transformation of the complex beam parameter obeys a similar relation, i.e.,

$$q_2 = \frac{Aq_1 + B}{Cq_1 + D}, \tag{15.72}$$

where

$$\begin{pmatrix} A & B \\ C & D \end{pmatrix}$$

is the transfer matrix for paraxial rays. This is demonstrably true for propagation over a distance $d$ through an isotropic medium, since it gives

$$q_2 = q_1 + d. \tag{15.73}$$

It is also true for a Gaussian beam crossing a thin lens, since Eq. (15.70) also follows from Eq. (15.72) and the ray-transfer matrix for a thin lens.

Equation (15.72) can be justified for a general optical system by considering that such a system could, in principle, be replaced by an arrangement of spaced thin lenses. Each thin lens, or length of uniform medium, obeys Eq. (15.68).

The analysis of Gaussian beam propagation in more complex optical systens can be extended to tilted surfaces and non-spherical surfaces [2–4], but this analysis is beyond our scope here.

## 15.5 The transformation of a Gaussian beam by a lens

A lens can be used to focus a laser beam to a small spot, or systems of lenses may be used to expand the beam and recollimate it. An ideal thin lens in such an application will not change the transverse-mode intensity pattern measured at the lens, but it will alter the radius of curvature of the phase fronts of the beam.

If $q_1$ and $q_2$ are measured at distances $d_1$ and $d_2$, respectively, from the lens, as shown in Fig. 15.10, then at the lens

$$\frac{1}{q_2 - d_2} = \frac{1}{q_1 + d_1} - \frac{1}{f}, \tag{15.74}$$

which gives

$$q_2 = \frac{(1 - d_2/f)q_1 + (d_1 + d_2 - d_1 d_2/f)}{(-q_1/f) + (1 - d_1/f)}. \tag{15.75}$$

If the lens is placed at the beam waist of the input beam, then

$$q_1 = \frac{j\pi w_0^2}{\lambda}, \tag{15.76}$$

where $w_0$ is the spot size of the input beam. The beam parameter immediately after the lens is, using Eq. (15.72)

$$q_2 = \frac{j\pi w_0^2 f}{\lambda f - j\pi w_0^2}, \tag{15.77}$$

which can be rewritten as

$$q_2 = \frac{j\pi w_0^2}{\lambda[1 + \pi w_0^2/(\lambda f)]^2} - \frac{f}{1 + [\lambda f/(\pi w_0^2)]^2}. \tag{15.78}$$

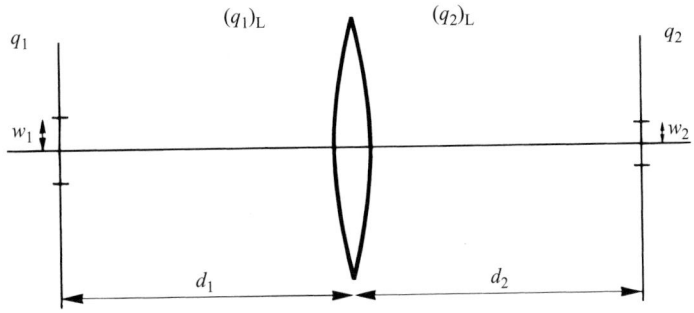


**Fig. 15.10** The optical system diagram for transformation of a Gaussian beam from an input to an output plane.

If we move a distance $d_f$ to the right of the lens, then at this new location $q_2(d_f) = q_2 + d_f$. If this point corresponds to the new beam waist of the focused beam, then $\mathcal{R}(1/q_2(d_f)) = 0$. Clearly, for $d_f$ to be the distance to the focus we must have

$$d_f = \frac{f}{1 + \left[\lambda f/(\pi w_0^2)\right]^2}. \tag{15.79}$$

Almost always $\lambda f/(\pi w_0^2) \ll 1$, so the new beam waist is very close to the focal point of the lens.

Examination of the imaginary part of the right-hand side of Eq. (15.78) reveals that the spot size of the focused beam is

$$w_2 = \frac{f\lambda/(\pi w_0)}{\sqrt{1 + [\lambda f/(\pi w_0^2)]^2}}, \tag{15.80}$$

which, provided that $\lambda f/(\pi w_0^2) \gg 1$, as is frequently the case, gives

$$w_2 \simeq f\theta, \tag{15.81}$$

where $\theta = \lambda/(\pi w_0)$ is the half angle of divergence of the input beam.

It is straightforward to show that in general the minimum spot size of a $\mathrm{TEM}_{00}$ Gaussian beam focused by a lens is

$$w_f = \frac{f\lambda}{\pi w_1}\left[\left(1 - \frac{f}{R_1}\right)^2 + \left(\frac{\lambda f}{\pi w_1^2}\right)^2\right]^{-1/2}, \tag{15.82}$$

where $w_1$ and $R_1$ are the laser-beam spot size and radius of curvature at the input face of the lens. If the lens is placed very close to or very far from the beam waist of the beam being focused, Eq. (15.82) reduces to

$$w_f = f\theta_B, \tag{15.83}$$

where $\theta_B$ is the beam divergence at the input face of the lens.

Thus, if the focusing lens is placed a great distance from, or very close to, the input beam waist then the size of the focused spot is always close to $f\theta$. If, in fact, the beam incident on the lens were a plane wave, then the finite size of the lens (radius $r$) would be the dominant factor in determining the size of the focused spot. We can take the "spot size" of the plane wave as approximately the radius of the lens, and from Eq. (15.67), setting $w_0 = r$, the radius of the lens, we get

$$w_2 = \frac{\lambda f}{\pi r}. \tag{15.84}$$

This focused spot cannot be smaller than a certain size since for any lens the value of $r$ clearly has to satisfy the condition $r \leq f$.

Thus, the minimum focal spot size that can result when a plane wave is focused by a lens is

$$(w_2)_{\min} \sim \frac{\lambda}{\pi}. \tag{15.85}$$

We do not have an equality sign in Eq. (15.85) because a lens for which $r = f$ does not qualify as a thin lens, so Eq. (15.85) does not really hold exactly.

We can see from Eq. (15.80) that in order to focus a laser beam to a small spot we must use either a lens of very short focal length or a beam of small beam divergence. We cannot, however, reduce the focal length of the focusing lens indefinitely, since, when the focal length does not satisfy $f_1 \ll w_0, w_1$, the lens ceases to satisfy our definition of it as a thin lens ($f \ll r$). To obtain a laser beam of small divergence, we must expand and recollimate the beam. There are two simple ways of doing this.

(i) With a Galilean telecope, as shown in Fig. 15.11. The expansion ratio for this arrangement is $-f_2/f_1$, where it should be noted that the focal length $f_1$ of the diverging input lens is negative. This type of arrangement has the advantage that the laser beam is not brought to a focus within the telescope, so the arrangement is very suitable for the expansion of high-power laser beams. Very-high-power beams can cause air breakdown if brought to a focus, which considerably reduces the energy transmission through the system.

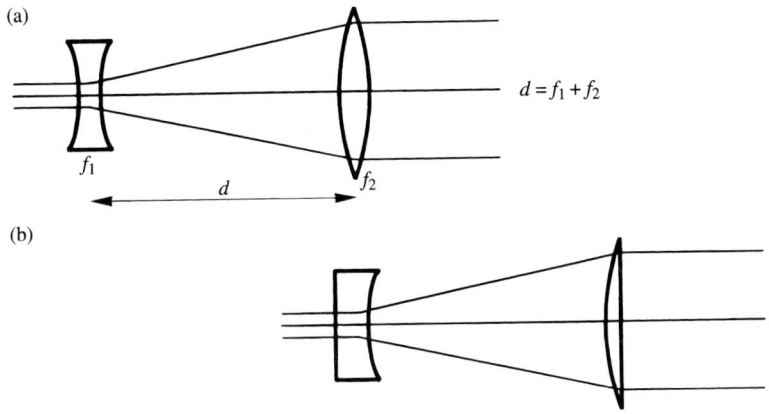


Fig. 15.11 Galilean telescope laser-beam expanders: (a) using bispherical lenses; and (b) using plano-spherical lenses.

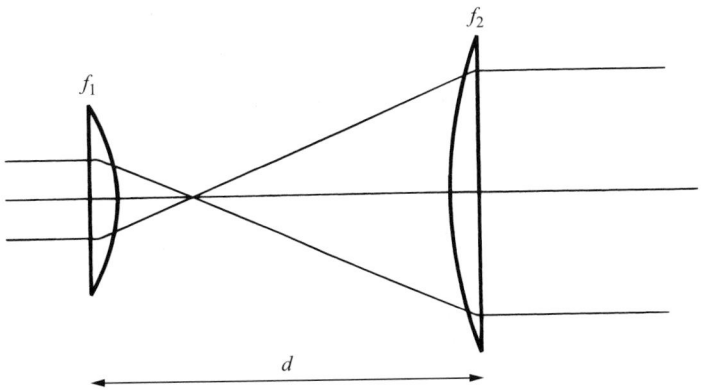


Fig. 15.12 An astronomical-telescope laser-beam expander in which the beam is brought to an internal focus.

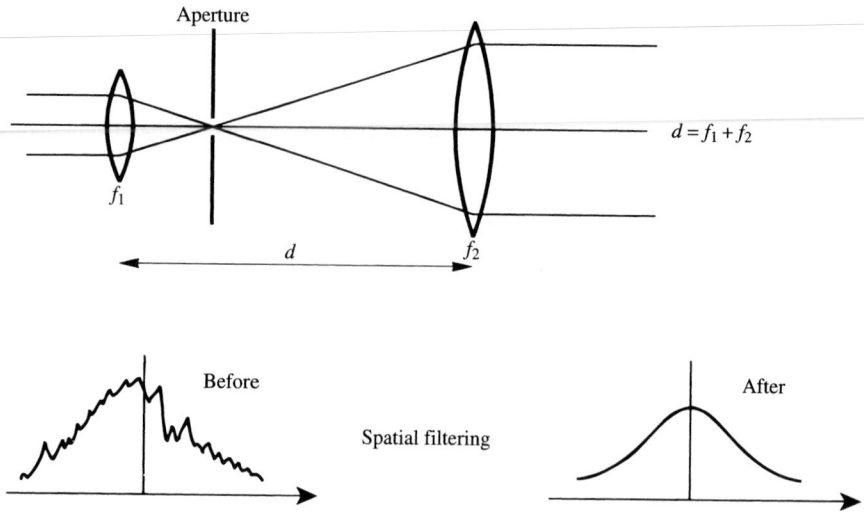


**Fig. 15.13** A beam expander incorporating a spatial filter for smoothing the intensity profile.

(ii) With an astronomical telecope, as shown in Fig. 15.12. The expansion ratio for this arrangement is $f_2/f_1$. The beam is brought to a focus within the telescope, which can be a disadvantage when expanding high-intensity laser beams because air breakdown at the common focal point can occur. The telescope can be evacuated or filled to high pressure to help prevent such breakdown occurring. An advantage of this system is that by placing a small circular aperture at the common focal point it is possible to obtain an output beam with a smoother radial intensity profile than the input beam. The aperture should be chosen to have a radius about the same size as, or slightly larger than, the spot size of the focused Gaussian beam at the focal point. This process is called *spatial filtering* and is illustrated in Fig. 15.13. Both in astronomical and in Galilean telescopes, spherical aberration[1] is reduced by the use of bispherical lenses. This distributes the focusing power over the maximum number of surfaces.

## 15.6 Gaussian beams in lens waveguides

In a biperiodic lens sequence containing equally spaced lenses of focal lengths $f_1$ and $f_2$ the transfer matrix for $n$ unit cells of the sequence is, from Eq. (14.46),

$$\begin{pmatrix} A & B \\ C & D \end{pmatrix}^n = \frac{1}{\sin\phi}\begin{pmatrix} A\sin(n\phi) - \sin[(n-1)\phi] & B\sin(n\phi) \\ C\sin(n\phi) & D\sin(n\phi) - \sin[(n-1)\phi] \end{pmatrix}, \quad (15.86)$$

[1] The type of lens aberration in which rays of light traveling parallel to the axis that strike a lens at different distances from the axis are brought to different focal points.

so, from Eq. (15.86) and writing $q_2 = q_{n+1}$, since we are interested in propagation through $n$ unit cells of the sequence,

$$q_{n+1} = \left(\frac{1}{\sin\phi}\right)\frac{[A\sin(n\phi) - \sin[(n-1)\phi]]q_1 + B\sin(n\phi)}{C\sin(n\phi)q_1 + D\sin(n\phi) - \sin[(n-1)\phi]}. \tag{15.87}$$

The condition for stable confinement of the Gaussian beam by the lens sequence is the same as in the case of paraxial rays. This is the condition that $\phi$ remains real, i.e., $|\cos\phi| \leq 1$, where

$$\cos\phi = \tfrac{1}{2}(A + D) = \tfrac{1}{2}\left(1 - \frac{d}{f_1} - \frac{d}{f_2} + \frac{d^2}{2f_1f_2}\right), \tag{15.88}$$

which gives

$$0 \leq \left(1 - \frac{d}{2f_1}\right)\left(1 - \frac{d}{2f_2}\right) \leq 1. \tag{15.89}$$

## 15.7 The propagation of a Gaussian beam in a medium with a quadratic refractive-index profile

We can most simply deduce what happens to a Gaussian beam in such a medium, whose refractive-index variation is given by $n(r) = n_0 - \frac{1}{2}n_2r^2$ by using Eq. (14.88) noting that $\sigma = \sqrt{n_0/n_2}$. The transfer matrix of the medium is

$$\begin{pmatrix} \cos\left(z\sqrt{n_2/n_0}\right) & \sqrt{n_0/n_2}\sin\left(z\sqrt{n_2/n_0}\right) \\ -\sqrt{n_2/n_0}\sin\left(z\sqrt{n_2/n_0}\right) & \cos\left(z\sqrt{n_2/n_0}\right) \end{pmatrix}. \tag{15.90}$$

Thus, if the input beam parameter is $q_0$ we have

$$q_{\text{out}}(z) = \frac{\cos\left(z\sqrt{n_2/n_0}\right)q_{\text{in}} + \sqrt{n_0/n_2}\sin\left(z\sqrt{n_2/n_0}\right)}{-\sqrt{n_2/n_0}\sin\left(z\sqrt{n_2/n_0}\right)q_{\text{in}} + \cos\left(z\sqrt{n_2/n_0}\right)}, \tag{15.91}$$

where $1/q_{\text{in}} = 1/R_0 - j\lambda/(\pi w_0^2)$. The condition for stable propagation of this beam is that $q_{\text{out}}(z) = q_{\text{in}}$, which, from Eq. (15.91), gives

$$-\sqrt{\frac{n_2}{n_0}}\sin\left(z\sqrt{n_2/n_0}\right)q^2 + \cos\left(z\sqrt{n_2/n_0}\right)q = \cos\left(z\sqrt{n_2/n_0}\right)q + \sqrt{n_0/n_2}\sin\left(z\sqrt{n_2/n_0}\right), \tag{15.92}$$

which has the solution

$$q^2 = -\frac{n_0}{n_2} \quad \text{and} \quad q = j\sqrt{\frac{n_0}{n_2}}. \tag{15.93}$$

This implies the propagation of a Gaussian beam with planar phase fronts and constant spot size

$$w = \left(\frac{n_0}{n_2}\right)^{1/4} \sqrt{\frac{\lambda}{\pi}}. \tag{15.94}$$

Once again we stress that $\lambda = \lambda_0/n$ is the wavelength in the medium.[2]

Propagation of such a Gaussian beam without spreading is clearly very desirable in the transmission of laser beams over long distances. In most modern optical communication systems laser beams are guided inside optical fibers that have a maximum refractive index on their axis. These fibers, although they might not have an exactly parabolic radial index profile, achieve the same continuous refocusing effect. We shall consider the properties of such fibers in greater detail in Chapter 16.

## 15.8 The propagation of Gaussian beams in media with spatial gain or absorption variations

We can describe the effect of gain or absorption on the field amplitudes of a wave propagating through an amplifying or lossy medium by using a complex propagation constant $k'$, where

$$k' = k + j\gamma/2. \tag{15.95}$$

This result is exact at line center; otherwise the value of $k$ would be written as $k + \Delta k$ as in Eq. (4.3).

For a wave traveling in the positive-$z$ direction the variation of the field amplitude with distance is $e^{-jk'z}$.

Consequently, $\gamma$ can be identified as the intensity gain coefficient of the medium and $k$ as the effective magnitude of the wave vector, satisfying

$$k = \frac{2\pi}{\lambda}, \tag{15.96}$$

where $\lambda$ is the wavelength in the medium.

If we replace $k$ by $k'$ in Eq. (15.47) we can represent the field of the Gaussian beam as

$$U(r,z) = \frac{w_0}{w} \exp\left[-j(k'z - \Phi) - r^2\left(\frac{1}{w^2} + \frac{jk'}{2R}\right)\right], \tag{15.97}$$

which gives

$$U(r,z) = \frac{w_0}{w} e^{\gamma z/z} e^{-j(kz-\phi)} e^{-r^2(1/w^2 - \gamma/(4R))}, \tag{15.98}$$

[2] In the relationship between the beam parameter $q$ and the radius and spot size of the Gaussian beam a uniform refactive index is assumed. In taking $n = n_0$ in a medium with a quadratic index profile we are assuming only a small variation of the index over the width of the beam. So, for example, $n_2 w^2 \ll n_0$.

with intensity variation

$$I(r,z) = \left(\frac{w_0}{w}\right)^2 e^{\gamma z} e^{-2r^2(1/w^2 - \gamma/(4R))}. \tag{15.99}$$

The intensity grows with distance as $e^{\gamma z}$, as usual, but there is a modification of the "apparent" spot size to

$$w' = \frac{w}{\sqrt{1 - \gamma w^2/(4R)}}. \tag{15.100}$$

This increase in spot size is negligible even for an extremely high-gain laser. For example, the xenon laser operating at 3.5 μm has shown [5] a small signal gain of 400 dB $\mathrm{m}^{-1}$. In this case $10\log_{10}(I/I_0) = 400$ and $I/I_0 = e^{\gamma}$, giving $\gamma = 92.1\ \mathrm{m}^{-1}$.

The smallest phase-front curvature that this laser beam could exhibit is, from Eq. (15.32),

$$R_{\min} = \frac{2\pi w_0^2}{\lambda}, \tag{15.101}$$

where the spot size is $\sqrt{2}w_0$. Therefore,

$$\frac{\gamma w^2}{4R} = \frac{\gamma\lambda}{4\pi} = 2.6 \times 10^{-5}, \tag{15.102}$$

so there is a negligible change in the spot size.

In practice, the occurrence of gain saturation leads to much greater modification of a propagating Gaussian beam than the subtle effect described above. This is an example of a situation in which the gain medium affects the Gaussian beam because the gain is spatially non-uniform. To illustrate this let us consider a medium where the gain varies quadratically with distance from the axis.

## 15.9 Propagation in a medium with a parabolic gain profile

It is convenient in this case to include the gain as part of a complex refractive index. If the modified propagation constant is $k'$, we can write

$$k' = n'k_0, \tag{15.103}$$

where $k_0$ is the propagation constant *in vacuo*.

Therefore,

$$n' = \frac{k}{k_0} + \frac{j\gamma}{2k_0}, \tag{15.104}$$

where $k$ is the real part of $k'$.

Consequently, if the gain varies parabolically with distance from the axis,

$$\gamma(r) = \gamma_0 - \gamma_2 r^2. \tag{15.105}$$

The modified propagation constant is

$$k'(r) = k + \frac{1}{2}j(\gamma_0 - \gamma_2 r^2), \tag{15.106}$$

and the complex refractive index is

$$n(r) = \frac{k + \frac{1}{2}j(\gamma_0 - \gamma_2 r^2)}{k_0}, \tag{15.107}$$

giving

$$n(r) = n_0 + \frac{j\gamma_0}{2k_0} - \frac{j\gamma_2}{2k_0}r^2. \tag{15.108}$$

The transfer matrix **M** for a length $z$ of this medium is similar to Eq. (15.90), namely

$$\mathbf{M} = \begin{pmatrix} \cos\left(z\sqrt{n_2/n'}\right) & \sqrt{n'/n_2}\sin\left(z\sqrt{n_2/n'}\right) \\ -\sqrt{n_2/n'}\sin\left(z\sqrt{n_2/n'}\right) & \cos\left(z\sqrt{n_2/n'}\right) \end{pmatrix}, \tag{15.109}$$

where $n_2 = j\gamma_2/k_0$ and $n' = n_0 + j\gamma_0/(2k_0)$.

The beam parameter of a Gaussian beam propagating through such a medium will be constant, as was the case for Eq. (15.92), if

$$q^2 = -\frac{n'}{n_2} \quad \text{and} \quad q = j\sqrt{\frac{n'}{n_2}}. \tag{15.110}$$

Thus,

$$q = \sqrt{\frac{\gamma_2}{2n_0k_0}}\left[\left(1 - \frac{\gamma_0}{4k_0}\right) - j\left(1 + \frac{\gamma_0}{4k_0}\right)\right]; \tag{15.111}$$

$k_0 = 2\pi/\lambda_0$ is much larger than achievable values of $\gamma_0$, even for millimeter-wave lasers, so

$$q = \sqrt{\frac{\gamma_2}{2n_0k_0}}(1 - j), \tag{15.112}$$

which, from Eq. (15.20), gives

$$R = \sqrt{\frac{2n_0k_0}{\gamma_2}} = 2\sqrt{\frac{\pi}{\lambda\gamma_2}} \tag{15.113}$$

and

$$w^2 = 2\sqrt{\frac{\lambda}{\pi\gamma_2}}. \tag{15.114}$$

Remember that $\lambda$ is the wavelength in the medium of refractive index $n_0$. Thus, we have a Gaussian beam that does not spread yet has spherical wave fronts.

We can understand this phenomenon qualitatively, as follows. The Gaussian beam, in the absence of gain, expands as it propagates. However, in the presence of gain that is highest on axis there is a compensatory effect where by the amplification of the beam makes it wish to decrease in width as it propagates: the peak grows more than do the wings of the profile.

A radial distribution of gain occurs in most gas lasers because of radial variations in the electron density or in the density of metastable states. This affects radially the rate of excitation of laser levels, either those excited directly by electrons or those excited by transfer of energy from metastable states. The radial variation is close to quadratic in lasers where the excitation is by single-collision direct electron impact, but where the radial electron density is parabolic as a function of radius. The radial electron density is nearly parabolic in gas discharges at moderate to high pressures where the flow of charged particles to the walls occurs by ambipolar diffusion.[3] Quasi-parabolic gain profiles can also result in optically pumped cylindrical solid-state lasers.

## 15.10 Gaussian beams in plane and spherical mirror resonators

We have already mentioned that a beam-like solution of Maxwell's equations will be a satisfactory transverse mode of a plane or spherical mirror resonator, provided that we place the resonator mirrors at points where their radii of curvature match the radii of curvature of the phase fronts of the beam. So for a Gaussian beam a double-concave mirror resonator matches the phase fronts as shown in Fig. 15.14, and plano-spherical and concave–convex resonators match as shown in Figs. 15.15 and 15.16.

We can consider these resonators in terms of their equivalent biperiodic lens sequences as shown in Fig. 15.17. Propagation of a Gaussian beam from plane 1 to plane 3 in the biperiodic lens sequence is equivalent to one complete round trip inside the equivalent

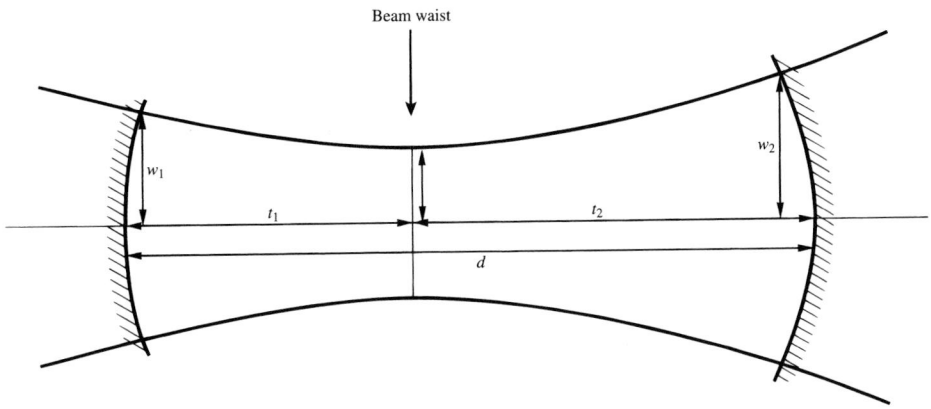


Fig. 15.14 A resonator using two concave mirrors, showing the location of the beam waist and the contour of the Gaussian beam.

[3] A situation in which positive ions and electrons diffuse relative to each other at a rate set by the concentration gradient of each species. The discharge usually remains electrically neutral because the charge densities of positive and negative charge are spatially equal.

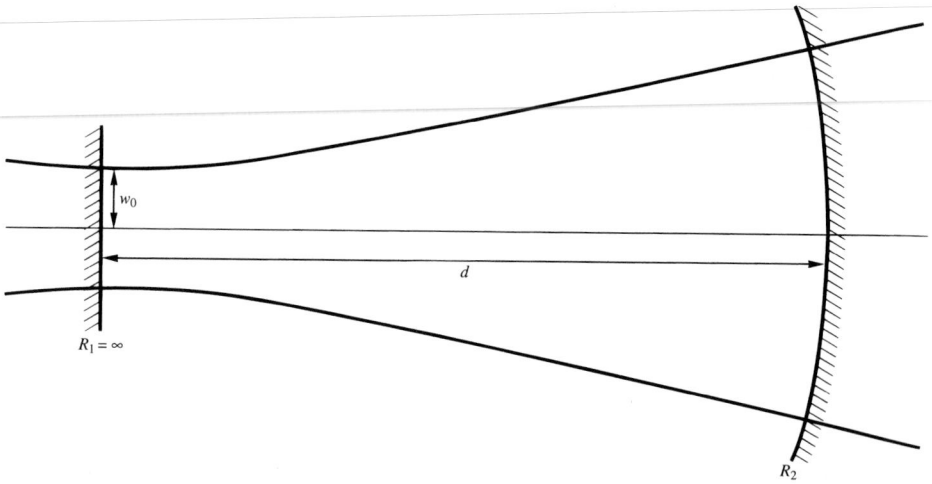


**Fig. 15.15** A laser resonator that uses a plane mirror and a concave mirror.

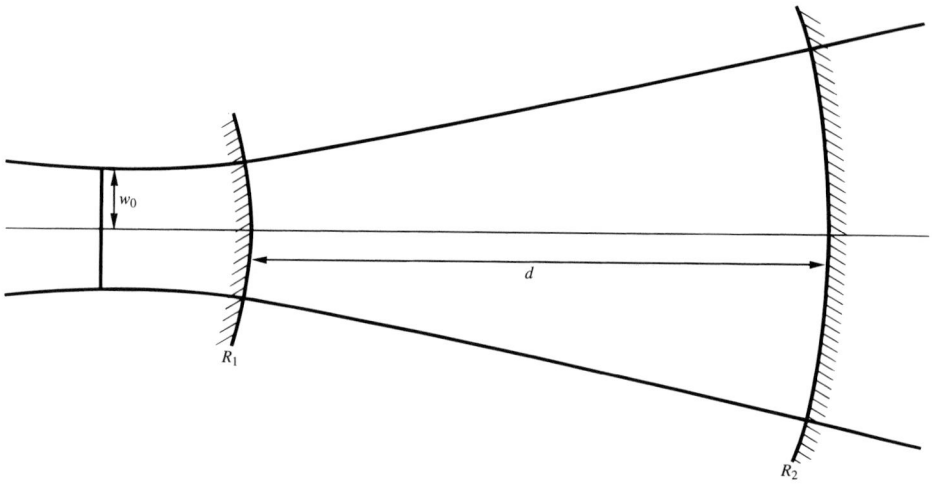


**Fig. 15.16** Convex–concave laser resonator.

spherical mirror resonator. If the complex beam parameters at planes 1, 2, and 3 are $q_1$, $q_2$, and $q_3$, then

$$\frac{1}{q_1 + d} - \frac{1}{f_1} = \frac{1}{q_2}, \tag{15.115}$$

which gives

$$q_2 = \frac{(q_1 + d)f_1}{f_1 - q_1 - d} \tag{15.116}$$

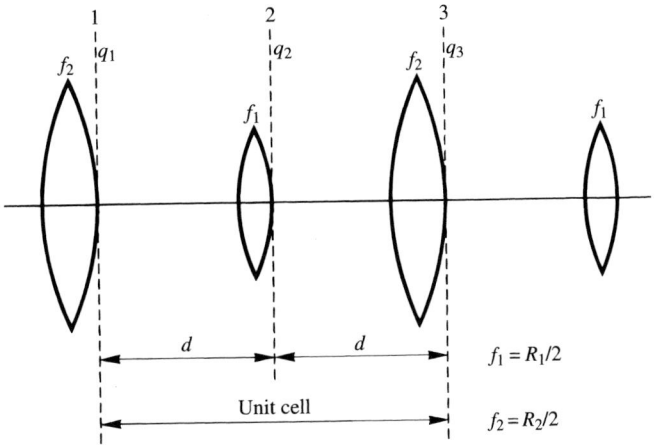


**Fig. 15.17** The biperiodic lens sequence equivalent to the resonator in Fig. 15.14.

and

$$\frac{1}{q_2 + d} - \frac{1}{f_2} = \frac{1}{q_3}. \tag{15.117}$$

If the Gaussian beam is to be a real transverse mode of the cavity, then we want it to repeat itself after a complete round trip – at least insofar as spot size and radius of curvature are concerned – that is, we want

$$q_3 = q_1 = q. \tag{15.118}$$

## 15.11 Symmetrical resonators

If the two mirrors of the resonator have equal radii of curvature, $R = 2f$, the condition that the beam be a transverse mode is

$$q_3 = q_2 = q_1 = q. \tag{15.119}$$

From Eq. (15.119) above, we have

$$fq - (q + d)q = (q + d)f, \tag{15.120}$$

which gives

$$\frac{1}{q^2} + \frac{1}{df} + \frac{1}{fq} = 0, \tag{15.121}$$

which has roots

$$\frac{1}{q} = -\frac{1}{2f} \pm \frac{1}{2}\sqrt{\frac{1}{f^2} - \frac{4}{fd}} = -\frac{1}{2f} \pm j\sqrt{\frac{1}{fd} - \frac{1}{4f^2}}. \tag{15.122}$$

Now, $1/q = 1/R - j\lambda/(\pi w^2)$, so, for a real spot size, i.e., $w^2$ positive, we must take the negative sign above. We do not get a real spot size if the resonator is geometrically unstable ($d > 4f$).

The spot size at either of the two mirrors (of equal curvature) is found from

$$\frac{\lambda}{\pi w^2} = \sqrt{\frac{1}{fd} - \frac{1}{4f^2}},$$

$$\frac{\lambda}{\pi w^2} = \sqrt{\frac{2}{Rd} - \frac{1}{R^2}},$$

which gives

$$w^2 = \frac{\lambda R/\pi}{\sqrt{2R/d - 1}}. \tag{15.123}$$

The radius of curvature of the phase fronts at the mirrors is the same as the radii of curvature of the mirrors.

Because this resonator is symmetrical, we expect the beam waist to be in the center of the resonator, at $z = d/2$ if we measure from one of the mirrors. So, since $q_0 = q + z$ and $q_0 = j\pi w_0^2/\lambda$, from Eq. (15.122) we get

$$w_0^2 = \frac{\lambda}{2\pi}\sqrt{d(2R - d)}. \tag{15.124}$$

Note that the spot size gets larger as the mirrors become closer and closer to being planar.

Now we know that a TEM mode, the $\text{TEM}_{mn}$ Cartesian mode, for example, propagates as

$$U_{m,n}(r,z) = \frac{w_0}{w} H_m\left(\sqrt{2}\frac{x}{w} H_m\right)\left(\sqrt{2}\frac{y}{w}\right) \exp\left[-j(kz - \Phi) - r^2\left(\frac{1}{w^2} + \frac{jk}{2R}\right)\right], \tag{15.125}$$

where the phase factor associated with the mode is

$$\Phi(m,n,z) = (m + n + 1)\tan^{-1}\left(\frac{\lambda z}{\pi w_0^2}\right). \tag{15.126}$$

On axis, $r = 0$, the overall phase of the mode is $kz - \Phi$.

In order for standing-wave resonance to occur in the cavity, this phase shift from one mirror to the other must correspond to an integral number of half wavelengths, i.e.,

$$kd - 2(m + n + 1)\tan^{-1}\left(\frac{\lambda d}{2\pi w_0^2}\right) = (q + 1)\pi, \tag{15.127}$$

where the cavity in this case holds $q + 1$ half wavelengths. For cylindrical $\text{TEM}_{p\ell}$ modes the resonance condition would be

$$kd - 2(2p + \ell + 1)\tan^{-1}\left(\frac{\lambda d}{2\pi w_0^2}\right) = (q + 1)\pi. \tag{15.128}$$

If we write $\Delta\nu = c/(2d)$ (the frequency spacing between modes), then, since $k = 2\pi\nu/c$,

$$kd = 2\pi\nu d/c = \pi\nu/\Delta\nu, \tag{15.129}$$

we get

$$\frac{\nu}{\Delta\nu} = (q+1) + \frac{2}{\pi}(m+n+1)\tan^{-1}\left(\frac{1}{\sqrt{2R/d-1}}\right). \tag{15.130}$$

By writing $2\tan^{-1}\left(1/\sqrt{2R/d-1}\right) = x$ and then employing $\sin^2(x/2) = \frac{1}{2}(1-\cos x)$ and $\cos^2(x/2) = \frac{1}{2}(1+\cos x)$, we get the formula

$$\frac{\nu}{\Delta\nu} = (q+1) + \frac{1}{\pi}(m+n+1)\cos^{-1}(1-d/R) \tag{15.131}$$

for the resonance frequencies of the longitudinal modes of the $\mathrm{TEM}_{mn}$ Cartesian mode in a symmetrical mirror cavity.

From inspection of the above formula we can see that the resonance frequencies of longitudinal modes of the same order (the same integer $q$) depend on $m$ and $n$. When the cavity is plane–plane, in which case $R = \infty$ and $\nu/\Delta\nu = q+1$, the familiar result for a plane-parallel Fabry–Perot cavity results.

When $d = R$ the cavity is said to be confocal, and we have the special relations

$$w^2 = \frac{\lambda d}{\pi}, \qquad w_0^2 = \frac{\lambda d}{2\pi}, \tag{15.132}$$

and

$$\nu/\Delta\nu = (q+1) + \frac{1}{2}(m+n+1). \tag{15.133}$$

The beam spot size increases by a factor of $\sqrt{2}$ between the center and the mirrors.

## 15.12 Examples of resonator design

We consider the problem of designing a symmetrical (double-concave) resonator with $R_1 = R_2 = R$ and mirror spacing $d = 1$ m for $\lambda = 488$ nm.

If the system were made confocal, $R_1 = R_2 = R = d$, then the minimum spot size would be

$$(w_0)_{\text{confocal}} = \sqrt{\frac{\lambda d}{2\pi}} = 0.28 \text{ mm}.$$

At the mirrors $(w_{1,2})_{\text{confocal}} = w_0\sqrt{2} \simeq 0.39$ mm.

If we wanted to increase the mirror spot size to 2 mm, say, we would need to use mirrors of larger radius $R \gg d$ such that

$$w^2 = \frac{\lambda R}{\pi}\Big/\sqrt{\frac{2R}{d}-1} = 4\times 10^{-6},$$

which, using

$$\frac{w}{\sqrt{\lambda d/2\pi}} \simeq \sqrt{\frac{2R}{d}},$$

gives $R \simeq 346d = 346$ m (so $R \gg d$).

Thus, it can be seen that, to increase the spot size even by the small factor above, we need to go to mirrors of very large radius. Even in this large-radius spherical-mirror resonator the $\mathrm{TEM}_{00}$ transverse mode is still a very narrow beam.

In the general case the resonator mirrors do not have equal radii of curvature, so we need to solve Eqs. (15.115) and (15.117) for $q_1 = q_3 = q$. To find the position of the mirrors relative to the position of the beam waist, it is easiest to proceed directly from the equation describing the variation of the curvature of the wave front with distance inside the resonator, i.e.,

$$R(z) = z\left[1 + \left(\frac{\pi w_0^2}{\lambda z}\right)^2\right], \tag{15.134}$$

where $z$ is measured from the beam waist. The radius of the Gaussian beam is negative for $z < 0$, and positive for $z > 0$. It is easier, however, to treat the distances of the mirrors from the beam waist in the general resonator shown in Fig. 15.14 both as positive, so that for two concave mirrors the radius of curvature of the Gaussian beam and that of the mirror match and are both positive. With this approach the curvature of the Gaussian beam at mirror 1 is

$$R_1 = t_1[1 + (z_\mathrm{R} t_1)^2], \tag{15.135}$$

where we have introduced a parameter called the *Rayleigh range*, $z_\mathrm{R} = \pi w_0^2/\lambda$, which is the distance over which a propagating Gaussian beam increases its area by a factor of two. From Eq. (15.135) we have

$$R_1 t_1 - t_1^2 - z_\mathrm{R}^2 = 0. \tag{15.136}$$

Similarly, the curvature of the Gaussian beam at mirror 2 is

$$R_2 = t_2[1 + (z_\mathrm{R} t_2)^2], \tag{15.137}$$

which gives

$$R_2 t_2 - t_2^2 - z_\mathrm{R}^2 = 0. \tag{15.138}$$

Therefore, we get

$$t_1 = \frac{R_1}{2} \pm \frac{1}{2}\sqrt{R_1^2 - 4z_\mathrm{R}^2} \tag{15.139}$$

and

$$t_2 = \frac{R_2}{2} \pm \frac{1}{2}\sqrt{R_2^2 - 4z_\mathrm{R}^2}. \tag{15.140}$$

If we choose the minimum spot size that we want in the resonator the above two equations will tell us where to place our mirrors.

If we write $d = t_2 + t_1$ then we can solve Eqs. (15.139) and (15.140) for $w_0$.

If we choose the positive sign in Eqs. (15.139) and (15.140), then

$$t_1 = -\frac{R_1}{2} + \frac{1}{2}\sqrt{R_1^2 - 4z_R^2}, \tag{15.141}$$

and

$$t_2 = \frac{R_2}{2} + \frac{1}{2}\sqrt{R_2^2 - 4z_R^2}. \tag{15.142}$$

Therefore

$$d = \frac{R_2 + R_1}{2} + \frac{1}{2}\sqrt{R_2^2 - 4z_R^2} + \frac{1}{2}\sqrt{R_1^2 - 4z_R^2}, \tag{15.143}$$

so

$$(2d - R_2 + R_1)^2 = R_2^2 - 4z_R^2 + R_1^2 - 4z_R^2 + 2\sqrt{\left[R_1^2 - 4z_R^2\right]\left[R_2^2 - 4z_R^2\right]}. \tag{15.144}$$

Equation (15.144) can be solved for $w_0$ to give

$$w_0^4 = \left(\frac{\lambda}{\pi}\right)^2 \frac{d(R_1 - d)(R_2 - d)(R_2 + R_1 - d)}{(R_2 + R_1 - 2d)^2}. \tag{15.145}$$

The spot sizes at the two mirrors are found from

$$w_1^4 = \left(\frac{\lambda R_1}{\pi}\right)^2 \frac{R_2 - d}{R_1 - d}\frac{d}{R_2 + R_1 - d} \tag{15.146}$$

and

$$w_2^4 = \left(\frac{\lambda R_2}{\pi}\right)^2 \frac{R_1 - d}{R_2 - d}\frac{d}{R_2 + R_1 - d}, \tag{15.147}$$

where we must continue to remember that, for the double-concave mirror resonator we are considering, $R_1$, the radius of curvature of the Gaussian beam at mirror 1, is being treated as positive.

The distances $t_1$ and $t_2$ between the beam waist and the mirrors are

$$t_1 = \frac{d(d - R_2)}{2d - R_1 - R_2} \tag{15.148}$$

and

$$t_2 = \frac{d(d - R_1)}{2d - R_1 - R_2}. \tag{15.149}$$

For this resonator the resonance condition for the longitudinal modes of the $TEM_{mn}$ Cartesian mode is

$$\frac{\nu}{\Delta\nu} = (q+1) + \frac{1}{\pi}(m+n+1)\cos^{-1}\sqrt{\left(1+\frac{d}{R_1}\right)\left(1-\frac{d}{R_2}\right)}, \tag{15.150}$$

where the sign of the square root is taken to be the same as the sign of $1 - d/R_1$, which is the same as the sign of $1 - d/R_2$ for a stable resonator.

## 15.13 Diffraction losses

In our discussions of transverse modes we have not taken into account the finite size of the resonator mirrors, so the discussion is strictly valid only for mirrors whose physical size makes them much larger than the spot size of the mode at the mirror in question. If this condition is not satisfied, some of the energy in the Gaussian beam leaks around the edge of the mirrors. This causes modes of large spot size to suffer high diffraction losses. So, since plane-parallel resonators in theory have infinite spot size, these resonators suffer enormous diffraction losses.

Because higher-order $TEM_{mn}$ modes have larger spot sizes than low-order modes, we can prevent them from operating in a laser cavity by increasing their diffraction loss. This is done by placing between the mirrors an aperture whose size is less than the spot size of the high-order mode at the position of the aperture, yet greater than the spot size of a lower-order mode (usually $TEM_{00}$) that it is desired to have operate in the cavity.

The diffraction loss of a resonator is usually described in terms of its Fresnel number $a_1 a_2/(\lambda d)$, where $a_1$ and $a_2$ are the radii of the mirrors, and $d$ is their spacing. Very roughly, the fractional diffraction loss at a mirror is $\approx w/a$. When the diffraction losses of a resonator are significant then the real field distribution of the modes can be found using Fresnel–Kirchhoff diffraction theory. If the field distribution at the first mirror of the resonator is $U_{11}$, then the field $U_{21}$ on a second mirror is found from the sum of all the Huygens secondary wavelets originating on the first mirror as [6]

$$U_{21}(x,y) = \frac{jk}{4\pi}\int_1 U_{11}(x,y)\frac{e^{-jkr}}{r}(1+\cos\theta)dS. \tag{15.151}$$

The integral is taken over the surface of mirror 1: $r$ is the distance from a point on mirror 1 to the point of interest on mirror 2; $\theta$ is the angle which this direction makes with the axis of the system; and $dS$ is an element of area. Having found $U_{21}(x,y)$ by the above method, we can recalculate the field on the first mirror as

$$U_{12}(x,y) = \frac{jk}{4\pi}\int_2 U_{21}(x,y)\frac{e^{jkr}}{r}(1+\cos\theta)dS, \tag{15.152}$$

and, if we require $U_{12}(x,y)$ to be equivalent to $U_{11}(x,y)$, apart from an arbitrary phase factor, then our transverse field distribution becomes self-consistent and is a solution for the transverse-mode distribution of a resonator with finite aperture mirrors. This is the approach that was first used by Fox and Li [7] to analyze the transverse field distributions that would result in optical resonators.

## 15.14 Unstable resonators

Many high-power lasers have sufficiently large gain that they do not require the amount of feedback provided by a stable optical resonator with high-reflectance mirrors. To achieve maximum power output from such lasers, one resonator mirror must have relatively low reflectance and the laser beam should have a large enough spot size so as to extract energy from a large volume of the medium. However, as the spot size of the oscillating transverse mode becomes larger, an increasing portion of the energy lost from the cavity "leaks" out past the edge of the mirrors. As the configuration of the cavity approaches the unstable regions in Fig. 14.18 the spot size grows infinitely large. At the same time, the divergence of the output beam falls, making focusing and control of the output easier. Operation of the resonator in the unstable region provides efficient energy extraction from a high-gain laser, and provides a low-divergence beam of generally quite high transverse-mode quality.

Schawlow first suggested the use of an unstable resonator, with its associated intracavity beam-defocusing properties, to counteract the tendency of some lasers to focus internally into high-intensity filaments. This property of unstable resonators is valuable in reducing the occurrence of "hot spots" in the output beam. These can result if a high-gain laser with spatial variations in its refractive-index profile (produced, for example, by non-homogeneous pumping) is operated in a stable resonator arrangement. A simple unstable resonator configuration that accomplishes this is shown in Fig. 15.18. The output laser beam will have a characteristic "donut"-shaped intensity pattern in the near field.

The loss per pass of an unstable resonator can be determined approximately by a simple geometrical optics analysis first presented by Siegman [8]. The geometry used in the analysis is shown in Fig. 15.19. The wave originating from mirror $M_1$ is assumed to be a

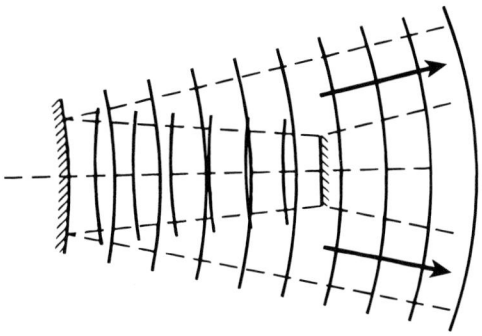

Fig. 15.18 A Cassegrain unstable resonator.

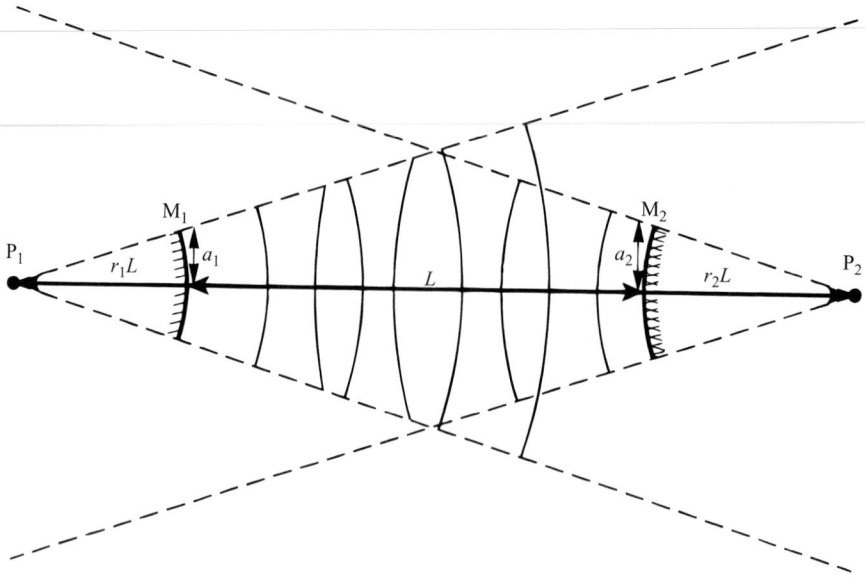


Fig. 15.19 The geometry for unstable resonator calculation [8].

spherical wave whose center of curvature is at point $P_1$ (which is not in general the same point as the center of curvature of mirror $M_1$). The right-traveling wave is partially reflected at mirror $M_2$, and is now assumed to be a spherical wave that has apparently originated at point $P_2$. This wave now reflects from $M_1$, again, as if it had come from point $P_1$, and another round-trip cycle commences. The positions of $P_1$ and $P_2$ must be self-consistent in order for this to occur; each must be the virtual image of the other in the appropriate spherical mirror. The distances of $P_1$ and $P_2$ from $M_1$ and $M_2$ are written as $r_1L$ and $r_2L$, respectively, where $L$ is the resonator length, and $r_1$ and $r_2$ are measured in the positive sense in the direction of the arrows in Fig. 15.19. If, for example, $P_2$ were to the left of $M_2$, then $r_2$ would be negative. We also use our previous sign convention that the radii of curvature of $M_1$ and $M_2$, $R_1$ and $R_2$, respectively, are both negative.

From the fundamental lens equation we have

$$\frac{1}{r_1} - \frac{1}{r_2 + 1} = -\frac{2L}{R_1} = 2(g_1 - 1) \tag{15.153}$$

and

$$\frac{1}{r_2} - \frac{1}{r_1 + 1} = -\frac{2L}{R_2} = 2(g_2 - 1), \tag{15.154}$$

where $g_1$ and $g_2$ are dimensionless parameters.

So

$$r_1 = \pm\frac{\sqrt{1 - 1/(g_1 g_2)} - 1 + 1/g_1}{2 - 1/g_1 - 1/g_2} \tag{15.155}$$

and

$$r_2 = \pm\frac{\sqrt{1 - 1/(g_1 g_2)} - 1 + 1/g_2}{2 - 1/g_1 - 1/g_2}. \tag{15.156}$$

These equations determine the locations of the virtual centers of the *spherical* waves in the resonator.

A wave of unit intensity that leaves mirror $M_1$ will partially spill around the edge of mirror $M_2$ – this is sometimes loosely referred to as a *diffraction loss*, but it is not. A fraction $\Gamma_1$ of the energy is reflected at $M_1$, so a fraction $1 - \Gamma_1$ is lost from the resonator. A fraction $\Gamma_2$ of this reflected wave reflects from mirror $M_2$. Thus, after one complete round trip, our initial intensity has been reduced to $\Gamma^2 \equiv \Gamma_1\Gamma_2$.

We can get some feeling for how the loss per round trip actually depends on the geometry of the resonator by considering the resonator in Fig. 15.19 as a purely two-dimensional one. This is equivalent to having infinitely long strip-shaped mirrors that curve only in the lateral direction.

From the geometry of Fig. 15.19 it is easy to see that

$$\Gamma_1 = \frac{r_1 a_2}{(r_1 + 1)a_1} \tag{15.157}$$

and

$$\Gamma_2 = \frac{r_2 a_1}{(r_2 + 1)a_2}. \tag{15.158}$$

Thus,

$$\Gamma_{2D}^2 = \Gamma_1\Gamma_2 = \pm\frac{r_1 r_2}{(r_1 + 1)(r_2 + 1)}. \tag{15.159}$$

Surprisingly, the round-trip energy lost is *independent* of the dimensions of the two mirrors.

## 15.15 Other beam waves

All beam waves of finite width expand eventually as they propagate. However, specific non-Gaussian beam waves have interesting diffraction properties and can be useful in certain applications, [9–11]. A beam wave with the spatial field distribution at $z = 0$

$$\psi(r, z) = \psi_0 e^{-r^2/w^2} J_0(\alpha r), \tag{15.160}$$

where $J_0(r)$ is the zeroth-order Bessel function of the first kind, is called a *Bessel–Gauss* field. For $w_0 \to \infty$ this is a Bessel beam, and the parameter $\alpha$ is a measure of the localization of the central peak of the Bessel function. For $\alpha \to \infty$ we have a regular Gaussian beam with minimum spot size $w_0$. For $\alpha = 0$ and $w_0 \to \infty$ we have a plane wave. The plane wave $\psi_0 e^{-jkz}$ and Bessel beam $\psi_0 J_0 e^{-j\beta z}$ are both exact solutions of the Helmholtz equation. In the case of the Bessel beam this requires $k^2 = \alpha^2 + \beta^2$. Both the plane wave and the Bessel beam do not spread as they propagate – they do not diffract. But such waves cannot be realized in practice because they must be of infinite lateral extent and carry infinite energy. The Gaussian and Bessel–Gauss beams are solutions to the Helmholtz equation only in the paraxial approximation. Bessel–Gauss beams can in certain circumstances diffract less than a Gaussian beam over some distance beyond an aperture even if partially occluded over a small distance [12], so they have become important in optical-tweezer applications.[4] A Bessel–Gauss beam can be generated by focusing a Gaussian lens with an *axicon* lens, which is a lens with a conical surface profile. Just as there exist elliptical Gaussian beams, which are non-circular and have different minimum spot sizes in orthogonal planes, so there are elliptical non-diffracting beams related to Bessel–Gauss waves that are called Mathieu beams [13].

## 15.16 Problems

15.1 Compute and plot the fraction of laser power in a $TEM_{00}$ mode of spot size $w_0$ that passes through an aperture of radius $a$ as a function of the ratio $a/w_0$.

15.2 A Gaussian beam has a minimum spot size $w_0 = 10$ μm and a wavelength of 880 nm. Calculate (a) the radius of curvature 1 m from the beam waist and (b) the spot size 1 m from the beam waist. Where does the beam have its minimum radius of curvature?

15.3 The beam in Problem 15.2 is focused by a lens with focal length 10 mm placed at the beam waist. Calculate the position of the new beam waist, and the new minimum spot size.

15.4 Design a GRIN lens system of reasonable parameters to collimate the laser beam in Problem 15.2.

15.5 The laser beam in Problem 15.2 is allowed to enter a medium with a quadratic index profile, with $n_0 = 1.5$ and $n_2 = 0.1\ \mathrm{m}^{-2}$. The beam enters a section of this medium 2 m long with its beam waist at the front face. The beam emerges from the medium and travels 2 m further. What are its spot size and radius of curvature at this point?

15.6 Prove that, when a laser beam of spot size $w_0$ is focused by a thin lens placed at the beam waist, there is a maximum possible distance at which a focus can be achieved. What is this maximum distance if $w_0 = 30$ mm? What happens to the Gaussian beam when it is not brought to a focus by the lens? Where has the beam waist moved to?

15.7 Prove that, to produce the smallest spot size at range $d$ with a thin lens placed at the beam waist, the optimum thin lens has $f = d$.

15.8 A semiconductor laser operating at wavelength 1.55 μm emits an elliptical Gaussian beam with minimum spot sizes in two orthogonal directions at the output facet of 100 μm and 10 μm, respectively. At what distance from the facet has the output beam become circular? What is the spot size at this point?

15.9 A cylindrical lens (one with two different focal lengths in orthogonal planes) is placed at the circular beam location in Priblem 15.8. What two focal lengths are required to bring the semiconductor laser to a circular focus 20 mm from this point?

[4] An application of focused laser beams where the forces exerted by the laser field on a dielectric particle can hold it and manipulate it in space.

15.10 For a resonator with two concave mirrors with $R_1 = 1$ m and $R_2 = 2$ m, for what value of $d$ and specific ray parameters does the intracavity ray retrace its path after two round trips? Draw ray-tracing diagrams to illustrate these situations.

15.11 A Gaussian beam with a minimum spot size of 20 μm is focused by a thin lens with a focal length of 5 mm that is placed 10 mm from the beam waist. Calculate the location and size of the new beam waist. Take the wavelength as 1.55 μm.

15.12 A laser operating at 1.3 μm has a minimum spot size of 0.5 mm. Design a GRIN lens, choosing reasonable values for its length, diameter, and $n_2$ parameter, for focusing this beam to a spot size of 10 μm. Place the GRIN lens 20 mm from the original beam waist.

15.13 A laser operating at 633 nm has a resonator with two concave mirrors of radii 10 m and 5 m, respectively, placed 0.5 m apart. Calculate (a) the spot size of the resultant Gaussian beam and (b) the location of its beam waist.

15.14 Design a symmetrical biconcave resonator ($R_1 = R_2$) with a minimum spot size of 1 mm for operating at 1.06 μm and a spacing of 100 mm.

15.15 Prove Eq. (15.82).

15.16 Explain the different resonator arrangements that correspond to the different choices of sign in Eqs. (15.139) and (15.140). Do the different choices affect the spot size?

## References

[1] S. Ramo, J. R. Whinnery, and T. Van Duzer, *Fields and Waves in Communication Electronics*, 3rd edn., New York: Wiley, 1993.

[2] D. Kessler and R. V. Shack, "$\bar{y}y$ diagram, a powerful optical design method for laser systems," *Appl. Opt.* **31**, 2692–2707, 1992.

[3] A. A. Tovar and L. W. Casperson, "Generalized beam matrices: Gaussian beam propagation in misaligned complex optical systems," *J. Opt. Soc. Am.* **A12**, 1522–1533, 1995.

[4] L. C. Andrews, W. B. Miller, and J. C. Ricklin, "Geometrical representation of Gaussian beams propagating through complex paraxial optical systems," *Appl. Opt.* **32**, 5918–5929, 1993.

[5] J. W. Kluver, "Laser amplifier noise at 3.5 microns in helium–xenon," *J. Appl. Phys.* **37**, 2987–2999, 1966.

[6] M. Born and E. Wolf, *Principles of Optics*, 7th edn., Cambridge: Cambridge University Press, 1999.

[7] A. G. Fox and T. Li, "Resonant modes in a maser interferometer," *Bell Sys. Tech. J.* **40**, 453–488, 1961.

[8] A. E. Siegman, "Unstable optical resonators," *Appl. Opt.* **13**, 353–367, 1974.

[9] J. Durnin, J. J. Miceli, Jr., and J. H. Eberly, "Comparison of Bessel and Gaussian beams," *Opt. Lett.* **13**, 79–80, 1988.

[10] P. L. Overfelt and C. S. Kenney, "Comparison of the propagation characteristics of

Bessel, Bessel–Gauss, and Gaussian beams diffracted by a circular aperture," *J. Opt. Soc. Am.* **A8**, 572–745, 1991.

[11] D. McGloin and K. Dholaki, "Bessel beams: Diffraction in a new light," *Contemp. Phys.* **46**, 15–28, 2010.

[12] V. Garcés-Chávez, D. McGloin, H. Melville, W. Sibbett, and K. Dholakia, "Simultaneous micromanipulation in multiple planes using a self-reconstructing light beam," *Nature* **410**, 415–417, 2002.

[13] J. C. Gutiérrez-Vega, M. D. Iturbe-Castillo, and S. Chávez-Cerda,"Alternative formulation for invariant optical fields: Mathieu beams," *Opt. Lett.* **25**, 1493–1496, 2000.

# 16 Optical fibers and waveguides

In this chapter we shall discuss both from a ray and from a wave standpoint how light can be guided along by planar and cylindrical dielectric waveguides. We shall explain why optical fibers are important and useful in optical communication systems and discuss briefly how these fibers are manufactured. Some practical details of how fibers are used and how they are integrated with other important optical components will conclude the chapter.

## 16.1 Introduction

We saw in the previous chapter that a Gaussian beam can propagate without beam expansion in an optical medium whose refractive index varies in an appropriate manner in the radial direction. This is a rather specific example of how an optical medium can guide light energy. However, we can discuss this phenomenon in more general terms. By specifying the spatial variation in the refractive index, and through the use of the wave equation with appropriate boundary conditions, we can show that dielectric waveguides will support certain "modes" of propagation. However, it is helpful initially to see what can be learned about this phenomenon from ray optics.

## 16.2 Ray theory of cylindrical optical fibers

### 16.2.1 Meridional rays in a step-index fiber

A step-index fiber has a central core of index $n_1$ surrounded by cladding of index $n_2$, where $n_2 < n_1$. When a ray of light enters such an optical fiber, as shown in Fig. 16.1, it will be guided along inside the core of the fiber if the angle of incidence between core and cladding is greater than the critical angle. Two distinct types of ray can travel along inside the fiber in this way: *meridional* rays travel in a plane that contains the fiber axis, whereas *skew* rays travel in a non-planar zig-zag path and may never cross the fiber axis, as illustrated in Fig. 16.2.

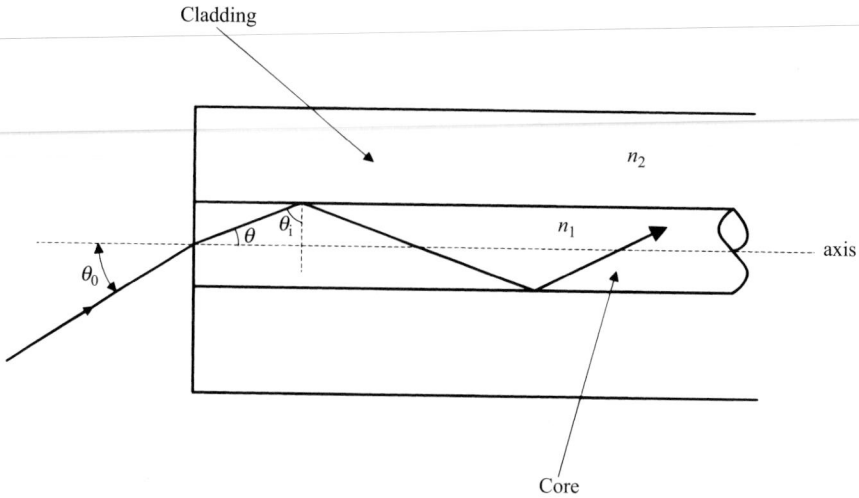


**Fig. 16.1** A meridional ray entering a fiber and being guided by total internal reflection.

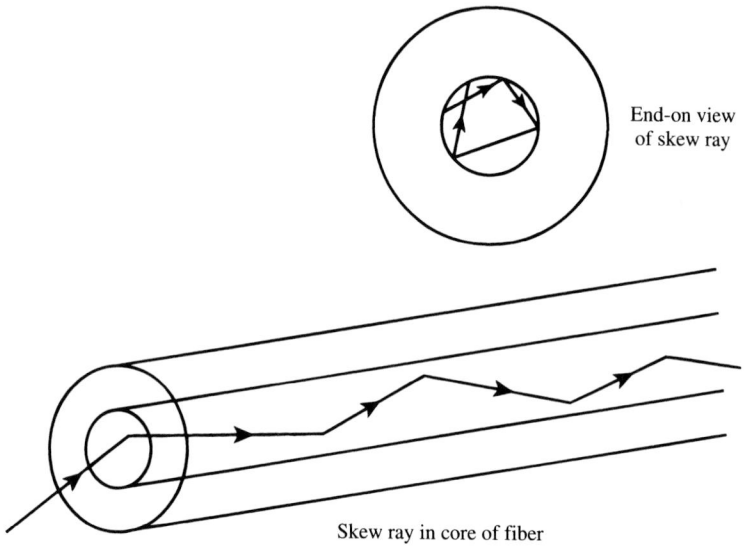


**Fig. 16.2** Skew rays.

For the meridional ray in Fig. 16.1, total internal reflection (TIR) occurs within the core if $\theta_i > \theta_c$, or $\sin\theta_i > n_2/n_1$. From Snell's law, applied to the ray entering the fiber,

$$\sin\theta = \sin\theta_0/n_1. \tag{16.1}$$

Since $\theta + \theta_i = 90°$, the condition for TIR is

$$\sin\theta_0 < \sqrt{n_1^2 - n_2^2}. \tag{16.2}$$

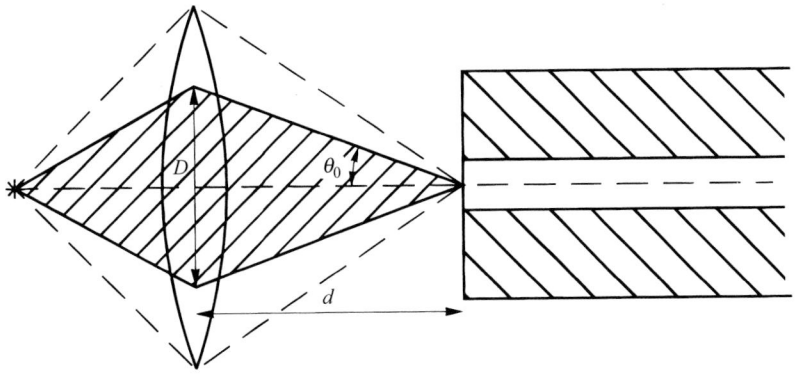


Fig. 16.3 The geometry for focusing light into a fiber.

For most optical fibers the relative difference between the index of the core and that of the cladding, $\Delta = (n_1 - n_2)/n_1$, is small, so Eq. (16.2) can be written

$$\sin\theta_0 < \sqrt{(n_1 - n_2)(n_1 + n_2)}, \tag{16.3}$$

which, since $n_1 \simeq n_2$, can be written

$$\sin\theta_0 < n_1\sqrt{2\Delta}. \tag{16.4}$$

If a lens is used to focus light from a point source into a fiber, as shown in Fig. 16.3, then there is a maximum aperture size $D$ that can be used. When the end of the fiber is a distance $d$ from the lens, light rays outside the cross-hatched region enter the fiber at angles too great to allow TIR. The quantity $\sin\theta_0 \simeq D/(2d)$ is called the numerical aperture (NA) of the system. So, from Eq. (16.4),

$$\mathrm{NA} = \sin\theta_0 = \sqrt{n_1^2 - n_2^2} = n_1\sqrt{2\Delta}. \tag{16.5}$$

If the point source in Fig. 16.3 is placed a great distance from the lens, then $d = f$. In this case

$$\mathrm{NA} = \frac{D}{2f}. \tag{16.6}$$

If the lens is chosen to be no larger than necessary, then the lens diameter will be $D$. The ratio $f/D$ is a measure of the focusing/light-collecting properties of the lens, called the $f$/number. So to match a distant source to the fiber $2\mathrm{NA} = 1/(f/\mathrm{number})$.

This analysis of a step-index cylindrical optical fiber by means of simple ray theory has so far considered only the paths of meridional rays, namely those rays whose trajectories lie within the plane containing the fiber axis. We have also tacitly assumed that Snell's law strictly delineates between those rays that totally internally reflect, and are *bound* within the fiber, and those that escape or *refract* into the cladding. However, if we consider the trajectories of skew rays we shall see that a third class of rays, so-called *tunneling* or *leaky* rays, exists. These are rays that appear at first glance to satisfy the requirement from Snell's law for TIR. However, because the refraction occurs at a curved surface they can in certain circumstances leak propagating energy into the cladding. In our description of the phenomenon we follow the essential arguments and notation given by Snyder and Love [1].

Figure 16.4 shows the paths of skew rays in step-index and graded-index fibers projected onto a plane perpendicular to the fiber axis. The path of a particular skew ray is characterized by the angle $\theta_z$ it makes with the $z$ axis and the angle $\theta_\phi$ that it makes with the azimuthal direction. In a step-index fiber these angles are the same at all points along the ray path, whereas in a graded-index fiber they vary with the radial distance of the ray from the fiber axis. The projected ray paths of skew rays are bounded, for a particular family of skew rays, by the inner and outer *caustic* surfaces.

### 16.2.2 Step-index fibers

In a step-index fiber the outer caustic surface is the boundary between core and cladding. Figure 16.4(a) shows clearly that the radius of the inner caustic in a step-index fiber is

$$r_{\mathrm{ic}} = \rho \cos\theta_\phi. \tag{16.7}$$

For meridional rays $\theta_\phi = \pi/2$ and $r_{\mathrm{ic}} = 0$.

When a skew ray strikes the boundary between core and cladding it makes an angle $\alpha$ with the surface normal, where

$$\cos\alpha = \sin\theta_z \sin\theta_\phi. \tag{16.8}$$

Simple application of Snell's law predicts that TIR should occur if

$$\sin\alpha > \frac{n_{\mathrm{cladding}}}{n_{\mathrm{core}}}. \tag{16.9}$$

We introduce a new angle $\theta_{\mathrm{c}}$, which is related to the critical angle $\alpha_{\mathrm{c}}$ by

$$\theta_{\mathrm{c}} + \alpha_{\mathrm{c}} = 90^\circ.$$

Rays can be classified according to the values of these various angles:

$\theta \le \theta_z < \theta_{\mathrm{c}}$, bound rays;
$0 \le \alpha \le \alpha_{\mathrm{c}}$, refracting rays;
$\theta_{\mathrm{c}} \le \theta_z \le 90^\circ$ and $\alpha_{\mathrm{c}} \le \alpha \le 90^\circ$, tunneling rays.

It is the existence of this third class of rays that is somewhat surprising [2]. In order to explain the existence of such rays we need to consider various qualities that remain constant, or *invariant*, for rays in a step-index fiber. These invariants follow for simple geometrical reasons and are

$$\overline{\beta} = n_{\mathrm{core}} \cos\theta_z, \tag{16.10}$$

$$\overline{\ell} = n_{\mathrm{core}} \sin\theta_z \cos\theta_\phi. \tag{16.11}$$

It follows from Eq. (16.8) that

$$\overline{\beta}^2 + \overline{\ell}^2 = n_{\mathrm{core}}^2 \sin^2\alpha. \tag{16.12}$$

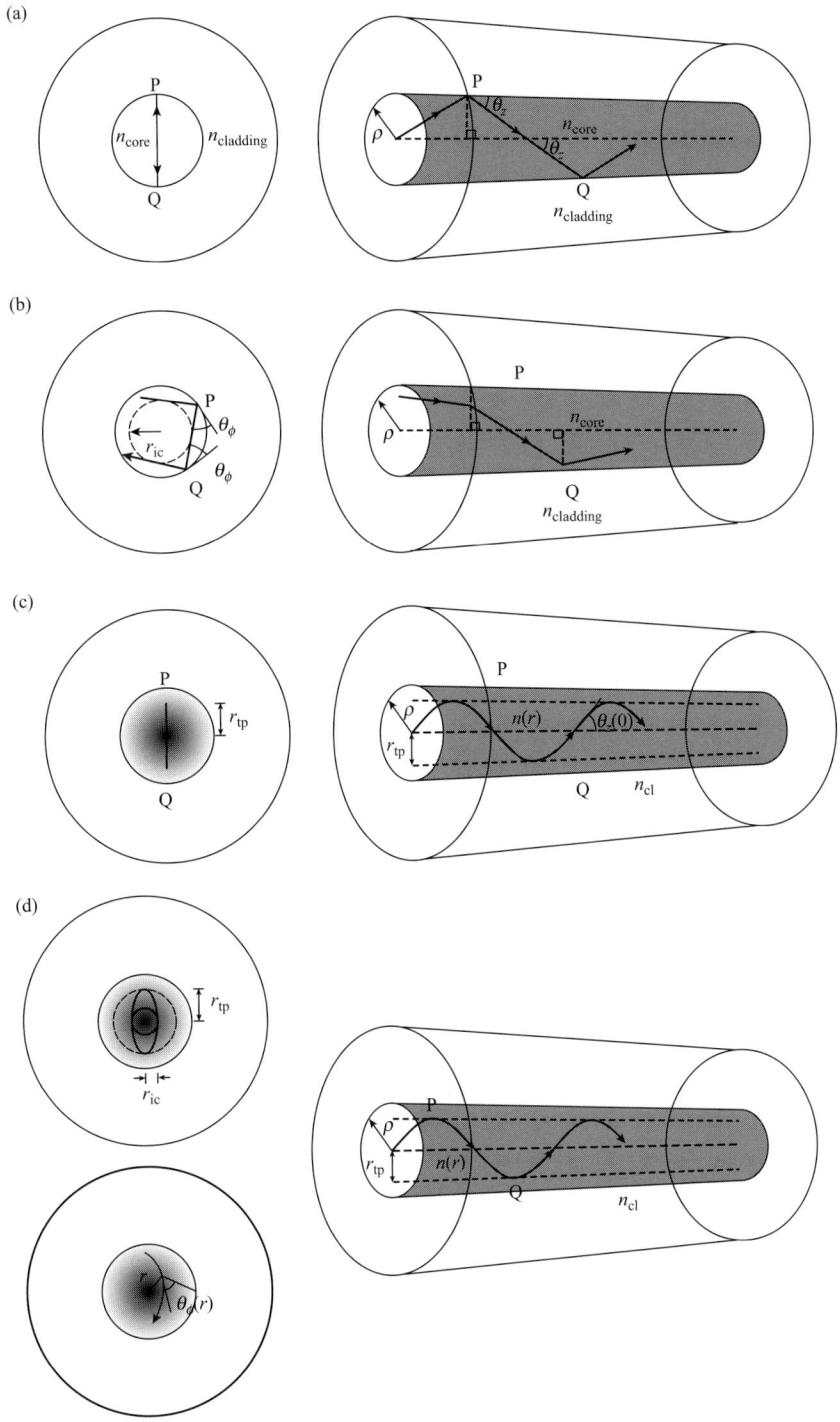


**Fig. 16.4** (a) The projected path of a meridional ray in a step-index cylindrical fiber; (b) the projected path of a skew ray in a step-index cylindrical fiber; (c) the projected path of a meridional ray in a graded-index cylindrical fiber; and (d) the projected path of a skew ray in a step-index fiber, showing the radial dependence of the azimuthal angle $\theta_\phi(r)$.

For meridional rays $\overline{\ell} = 0$, whereas for skew rays $\overline{\ell} > 0$. In terms of these ray invariants the classification of rays in a step-index fiber becomes

$$n_{\text{cladding}} < \overline{\beta} \leq n_{\text{core}}, \quad \text{bound rays;} \tag{16.13}$$

$$0 \leq \overline{\beta}^2 + \overline{\ell}^2 < n_{\text{cladding}}^2, \quad \text{refracting rays;} \tag{16.14}$$

$$n_{\text{cladding}}^2 < \overline{\beta}^2 + \overline{\ell}^2 \leq n_{\text{core}}^2 \text{ and } 0 \leq \overline{\beta} < n_{\text{cladding}}, \quad \text{tunneling rays.} \tag{16.15}$$

An important consequence of the ray invariants is that the length of time it takes light to travel a distance $L$ along the fiber is independent of the skewness. From Fig. 16.4 it is easy to see that the distance from P to Q is

$$L_{\text{p}} = 2\rho \frac{\sin\theta_\phi}{\sin\theta_z} = \frac{2\rho N_{\text{core}}\sqrt{n_{\text{core}}^2 - \overline{\beta}^2 - \overline{\ell}^2}}{n_{\text{core}}^2 - \overline{\beta}^2}. \tag{16.16}$$

The corresponding distance traveled in the axial direction is

$$L = \frac{L_{\text{p}}}{\cos\theta_z}, \tag{16.17}$$

and the time taken is

$$\tau = \frac{L_{\text{p}} n_{\text{core}}}{c_0 \cos\theta_z}. \tag{16.18}$$

The time taken per unit distance traveled is

$$\frac{\tau}{L} = \frac{n_{\text{core}}}{c_0 \cos\theta_z}, \tag{16.19}$$

which clearly depends only on the axial angle $\theta_z$, and does not depend on the skew angle $\theta_\phi$. This has important consequences for optical pulse propagation along a fiber. In this ray model an optical pulse injected at angle $\theta_z$ will take a single fixed time to travel along the fiber, irrespective of whether it travels as a meridional or skew ray.

### 16.2.3 Graded-index fibers

To discover the important ray invariants in a cylindrically symmetric graded-index fiber, we use Eq. (15.31):

$$\frac{d}{ds}\left[n(\mathbf{r})\frac{d\mathbf{r}}{ds}\right] = \text{grad}\, n(\mathbf{r}). \tag{16.20}$$

If we transform to cylindrical coordinates, for the case in which $n$ varies only radially

$$\frac{d}{ds}\left[n(r)\frac{dr}{ds}\right] - rn(r)\left(\frac{d\phi}{ds}\right)^2 = \frac{dn(r)}{dr}, \tag{16.21}$$

$$\frac{d}{ds}\left[n(r)\frac{d\phi}{ds}\right] + \frac{2n(r)}{r}\frac{d\phi}{ds}\frac{dr}{ds} = 0, \tag{16.22}$$

$$\frac{d}{ds}\left[n(r)\frac{dz}{ds}\right] = 0. \tag{16.23}$$

The path of a ray in projection in this case is shown in Fig. 16.4. At each point along the ray path the tangent to the ray path makes an angle $\theta_z(r)$ with the fiber axis.

In the projected path shown in Fig. 16.4 the projected tangent to the ray path makes an angle (in degrees) with the radius of $90 - \theta_\phi(r)$.

As before, from Eq. (16.8),

$$\cos\alpha(r) = \sin\theta_z(r)\sin\theta_\phi(r). \tag{16.24}$$

Various other relations follow:

$$\frac{dr}{ds} = \cos\alpha(r) \tag{16.25}$$

and

$$\frac{dz}{ds} = \cos\theta_z(r), \tag{16.26}$$

and, because

$$ds^2 = dr^2 + r^2\,d\phi^2 + dz^2, \tag{16.27}$$

it also follows that

$$\frac{d\phi}{ds} = \frac{1}{r}\sin\theta_z(r)\cos\theta_\phi(r). \tag{16.28}$$

Therefore, we can integrate Eq. (16.23) to give

$$\overline{\beta} = n(r)\frac{dz}{ds} = n(r)\cos\theta_z(r), \tag{16.29}$$

where $\overline{\beta}$ is a ray invariant.

Integration of Eq. (16.22) gives

$$\overline{\ell} = \frac{r^2}{\rho}n(r)\frac{d\phi}{ds} = \frac{r}{\rho}n(r)\sin\theta_z(r)\cos\theta_\phi(r), \tag{16.30}$$

where we have introduced a normalizing radius $\rho$ so that $\overline{\ell}$ remains dimensionless. This could be a reference radius, or, more usually, the boundary between core and cladding when only the core has a graded index.

If we eliminate $\theta_z$ from Eqs. (16.29) and (16.30), we get

$$\cos\theta_\phi(r) = \frac{\rho}{r}\frac{\overline{\ell}}{\sqrt{n^2(r) - \overline{\beta}^2}}. \tag{16.31}$$

Now, as shown schematically in Fig. 16.4, skew rays take a helical path in the core, but not one whose cross-section is necessarily circular. A particular family of skew rays will not come closer to the axis than the inner caustic and will not go further from the axis than the outer caustic. Inspection of Fig. 16.4 reveals that on each caustic $\theta_\phi(r) = 0$. Consequently, the radii of the inner and outer caustic, $r_{\text{ic}}$ and $r_{\text{tp}}$, respectively, are the roots of

$$g(r) = n^2(r) - \overline{\beta}^2 - \overline{\ell}^2\frac{\rho^2}{r^2} = 0. \tag{16.32}$$

### 16.2.4 Bound, refracting, and tunneling rays

We are now in a position to explain mathematically the distinction between these three classes of rays. From Eq. (16.29) we can write

$$\frac{dz}{ds} = \frac{\overline{\beta}}{n(r)}, \tag{16.33}$$

and from Eq. (16.30)

$$\frac{d\phi}{ds} = \frac{\rho\overline{\ell}}{r^2 n(r)}. \tag{16.34}$$

Substitution from Eq. (16.34) into Eq. (16.21), noting that

$$\frac{d}{ds} \equiv \frac{dz}{ds}\frac{d}{dz}, \tag{16.35}$$

gives

$$\overline{\beta}^2 \frac{d^2 r}{dz^2} - \overline{\ell}^2 \frac{\rho^2}{r^3} = \frac{1}{2}\frac{dn^2(r)}{dr}. \tag{16.36}$$

Writing $d^2r/dz^2 = r'\,dr'/dr$, where $r' = dr/dz$, gives

$$\overline{\beta}^2 r' \frac{dr'}{dr} - \overline{\ell}^2 \frac{\rho^2}{r^3} = \frac{1}{2}\frac{dn^2(r)}{dr}, \tag{16.37}$$

which can be integrated to give

$$\frac{1}{2}\overline{\beta}^2 (r')^2 + \frac{1}{2}\frac{\overline{\ell}^2 \rho^2}{r^2} = \frac{1}{2}n^2(r) + \text{constant}. \tag{16.38}$$

Using Eq. (16.32) allows Eq. (16.38) to be written as

$$\overline{\beta}^2 \left(\frac{dr}{dz}\right)^2 = g(r) + \overline{\beta}^2 + \text{constant}. \tag{16.39}$$

At the outer caustic surface $(dr/dz)_{r_{\text{tp}}} = 0$, and also $g(r_{\text{tp}}) = 0$, so the constant of integration in Eq. (16.39) is $-\overline{\beta}^2$. Finally, therefore, we have

$$\overline{\beta}^2 \left(\frac{dr}{dz}\right)^2 = g(r) = n^2(r) - \overline{\beta}^2 - \overline{\ell}^2 \frac{\rho^2}{r^2}. \tag{16.40}$$

Ray paths can exist only if $g(r)$ is non-negative.

The roots of $g(r)$ determine the locations of the inner and outer caustic surfaces. For a refracting ray there is no outer caustic surface: the ray reaches the core boundary (or the outer surface of the fiber). If we assume for simplicity that $n(r)$ increases monotonically from the axis, this requires $g(\rho) > 0$. From Eq. (16.40), this requires

$$n^2_{\text{cladding}} > \overline{\beta}^2 + \overline{\ell}^2. \tag{16.41}$$

There is only an inner caustic in this case, as shown in Fig. 16.5(c).

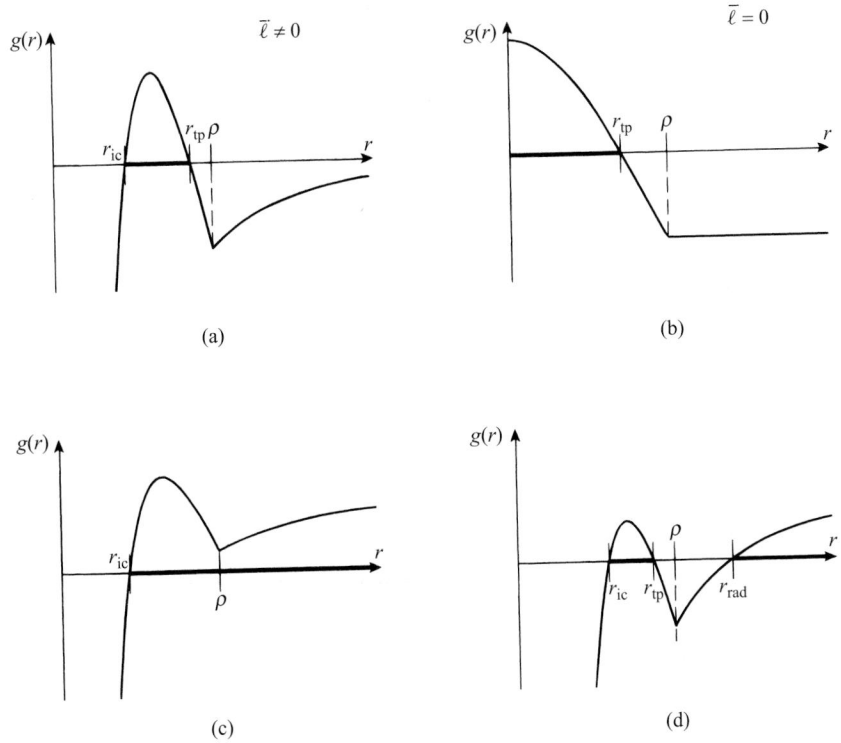


**Fig. 16.5** The schematic form of the function $g(r)$, Eq. (16.40) for a graded-index fiber for different classes of ray: (a) a skew bound ray; (b) a meridional bound ray; (c) a refracting ray; and (d) a tunneling ray.

For a ray to be bound, $g(r) < 0$ everywhere in the cladding, one requires

$$n_{\text{cladding}} < \overline{\beta}, \tag{16.42}$$

which additionally requires, not surprisingly, that

$$n_{\text{cladding}} < \overline{\beta} \leq n_{\text{core}}. \tag{16.43}$$

Both meridional and skew bound rays exist. For the meridional rays $\overline{\ell} = 0$ and $g(r)$ has only a single root, corresponding to $r_{\text{tp}}$, where

$$n(r_{\text{tp}}) = \overline{\beta}, \tag{16.44}$$

as shown in Fig. 16.5(b). For the skew bound rays there are two roots of $g(r)$, corresponding to the inner and outer caustic surfaces, as shown in Fig. 16.5(a).

Tunneling rays fall into a different category, where

$$\overline{\beta} < n_{\text{cladding}} \text{ and } n^2_{\text{cladding}} < \overline{\beta}^2 + \overline{\ell}^2. \tag{16.45}$$

The behavior of $g(r)$ in this case is such that, between $r_{\text{tp}}$ and an outer radius $r_{\text{rad}}$, $g(r)$ is negative and no ray paths exist. However, for $r > r_{\text{rad}}$, $g(r)$ is non-negative, as shown in

Fig. 16.5(d). The ray "tunnels" from radius $r_{tp}$ to $r_{rad}$, where it becomes a free ray. Since $g(r_{rad}) = 0$, it follows that

$$r_{rad} = \frac{\rho\bar{\ell}}{\sqrt{n_{cladding}^2 - \bar{\beta}^2}}. \tag{16.46}$$

The mechanism by which energy leaks from the core to the outer part of the fiber is similar to the evanescent coupling of energy between two closely spaced surfaces, at one of which TIR occurs, as was discussed in Chapter 13. Depending on the circumstances, tunneling or "leaky" rays may correspond to a very gradual loss of energy from the fiber, and the fiber can guide energy in such rays over relatively long distances.

## 16.3 Ray theory of a dielectric-slab guide

Before considering the wave theory of a cylindrical fiber in detail, we can gain further insight into the propagation characteristics of meridional rays by analyzing the two-dimensional problem of a dielectric-slab waveguide.

A dielectric slab of dielectric constant $\epsilon_1$ will guide rays of light by TIR, provided that the medium in which it is immersed has dielectric constant $\epsilon_2 < \epsilon_1$. For light rays making angle $\theta$ with the axis as shown in Fig. 16.6, TIR will occur, provided that $\sin\theta_i > \sqrt{\epsilon_2/\epsilon_1}$.

Although there might appear to be an infinite number of such rays, this is not so. As the ray makes its zig-zag path down the guide, the associated wave fronts must remain in phase or the wave amplitude will decay because of destructive interference. We can analyze the slab as a Fabry–Perot resonator. Only those rays which satisfy the condition for maximum stored energy in the slab will propagate. This corresponds to the upward and downward components of the traveling wave constructively interfering. The component of the propagation constant perpendicular to the fiber axis for the ray shown in Fig. 16.1 is $k_1 \cos\theta_i$, so, for constructive interference,

$$4dk_1 \cos\theta_i = 2m\pi, \tag{16.47}$$

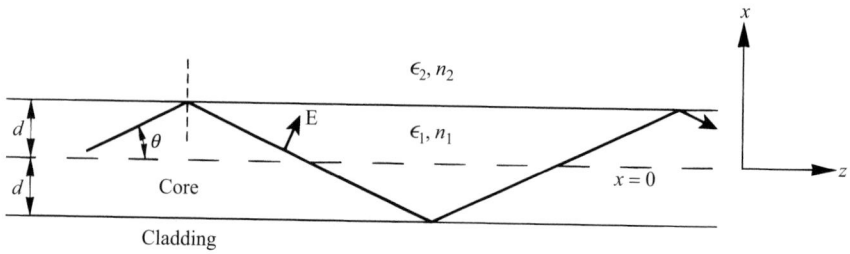


Fig. 16.6 The geometry describing meridional rays in a two-dimensional slab waveguide.

which gives the condition

$$\cos\theta_i = \frac{m\lambda_0}{4n_1 d}. \tag{16.48}$$

For simplicity we have neglected the phase shift that occurs when the wave totally internally reflects at the upper and lower core/cladding boundaries.

A given value of $m$ in Eq. (16.47) corresponds to a phase shift of $2m\pi$ per round trip between the upper and lower boundaries of the slab. Such a *mode* will not propagate if $\theta$ is greater than the critical angle. In other words, the cutoff condition for the $m$th mode is

$$\frac{d}{\lambda_0} = \frac{m}{4\sqrt{n_1^2 - n_2^2}}. \tag{16.49}$$

This condition is identical for waves polarized in the $xz$ plane (P waves) or polarized in the $y$ direction (S waves). Only the lowest mode, $m = 0$, has no cutoff frequency. A dielectric waveguide that is designed to allow the propagation of just the lowest mode is called a *single-mode* waveguide. The thickness of the guiding layer needs to be quite small to accomplish this. For example, at an operating wavelength of 1.55 μm and with indices $n_1 = 1.5$ and $n_2 = 1.485$, the maximum guide width that will support single-mode operation is 3.66 μm. Such waveguides are important because, in a simple sense, there is only one possible ray path in the guide. A short pulse of light injected into such a guide will emerge as a single pulse at its far end. In a multimode guide, a single pulse can travel along more than one path, and can emerge as multiple pulses. This is undesirable in a digital optical communication link. We can estimate when this would become a problem by using Eq. (16.48).

For the $m = 0$ mode in a guide of total length $L$ the path length is $L$. For the $m = 1$ mode, the total path length along the guide is $L + \Delta L$, where

$$L + \Delta L = L \Big/ \sqrt{1 - [\lambda_0/(4n_1 d)]^2}. \tag{16.50}$$

A single, short optical pulse injected simultaneously into these two modes will emerge from the fiber as two pulses separated in time by an interval $\Delta\tau$, where

$$\Delta\tau = \frac{n_1\,\Delta L}{c_0} = \frac{n_1 L}{c_0\sqrt{1 - [\lambda_0/(4n_1 d)]^2}} - \frac{n_1 L}{c_0}. \tag{16.51}$$

This effect is called *group delay*. For example, if $L = 1$ km, $n_1 = 1.5$, and $d/\lambda_0 = 5$, then $\Delta\tau = 2.78$ ns. Clearly, communication rates in excess of about 200 MHz would be impossible in this case. As the number of modes increases, the bandwidth of the system decreases further.

## 16.4 The Goos–Hänchen shift

We can examine the discussion of the previous section in a little more detail by including the phase shift that occurs when a wave undergoes total internal reflection. For an S wave striking the core/cladding boundary the reflection coefficient is

$$\rho = \frac{n_1 \cos\theta_1 - n_2 \cos\theta_2}{n_1 \cos\theta_1 + n_2 \cos\theta_2}, \tag{16.52}$$

where we have made use of the effective impedance for S waves at angle of incidence, $\theta_1$, or refraction, $\theta_2$, at the boundary, namely

$$Z'_{\text{core}} = \frac{Z_0}{n_1 \cos\theta_1}, \tag{16.53}$$

$$Z'_{\text{cladding}} = \frac{Z_0}{n_2 \cos\theta_2}. \tag{16.54}$$

For angles of incidence greater than the critical angle, $\cos\theta_2$ becomes imaginary and can be written as

$$\cos\theta_2 = j\sqrt{\left(\frac{n_1}{n_2}\right)^2 \sin^2\theta_1 - 1}. \tag{16.55}$$

In terms of the critical angle, $\theta_c$,

$$\begin{aligned}\cos\theta_2 &= j\frac{n_1}{n_2}\sqrt{\sin^2\theta_1 - \sin^2\theta_c} \\ &= j\frac{n_1}{n_2}X.\end{aligned} \tag{16.56}$$

If we write the phase shift on reflection for the S wave as $\phi_s$, then, from Eqs. (16.52) and (16.56),

$$\rho = |\rho| e^{j\phi_s} = \frac{\cos\theta_1 - jX}{\cos\theta_1 + jX}. \tag{16.57}$$

It is easy to show that $|\rho| = 1$: all the wave energy is reflected as a result of TIR. Therefore, if we write

$$\cos\theta_1 - jX = e^{j\phi_s/2},$$

then

$$\tan(\phi_s/2) = -\frac{\sqrt{\sin^2\theta_1 - \sin^2\theta_c}}{\cos\theta_1}. \tag{16.58}$$

For P waves the phase shift on reflection, $\phi_p$, satisfies

$$\tan(\phi_p/2) = -\frac{\sqrt{\sin^2\theta_1 - \sin^2\theta_c}}{(n_2/n_1)^2 \cos\theta_1}. \tag{16.59}$$

In practical optical waveguides usually $n_1 \simeq n_2$, so $\phi_s \simeq \phi_p = \phi$. We can put Eq. (16.58) in a more convenient form by using angles measured with respect to the fiber axis. We define

$$\theta_z = \pi/2 - \theta_1,$$
$$\theta_a = \pi/2 - \theta_c.$$

In a fiber with $n_1 \simeq n_2$, both $\theta_z$ and $\theta_a$ will be small angles. With these definitions

$$\phi = -2\frac{\sqrt{\theta_a^2 - \theta_z^2}}{\theta_z}. \tag{16.60}$$

If we represent the electric field of a guided wave incident on the slab boundary as

$$E_i(z) = E_1 e^{j\beta_1 z} \tag{16.61}$$

then, after TIR, the wave becomes

$$E_r(z) = E_1 e^{j[\beta_1 z + \phi(\beta_1)]}, \tag{16.62}$$

where $\beta_1 = k_1 \sin\theta_1$ is the propagation constant parallel to the $z$ axis of the slab. Because this wave is laterally confined within the core we can include the effects of diffraction by considering the wave as a group of rays near the value $\theta_1$. The phase shift $\phi(\beta_1)$ can be expanded as a Taylor series for propagation constants near $\beta_1$:

$$\phi(\beta) = \phi(\beta_1) + (\beta - \beta_1)\left(\frac{\partial\phi}{\partial\beta}\right)_{\beta_1}. \tag{16.63}$$

Therefore, the reflected wave can be written as

$$\begin{aligned} E_r(z) &= E_1 \exp\left[j(\beta z + \phi(\beta_1) + (\beta - \beta_1))\left(\frac{\partial\phi}{\partial\beta}\right)_{\beta_1}\right] \\ &= E_0 \exp\left[j(\beta z + \beta)\left(\frac{\partial\phi}{\partial\beta}\right)_{\beta_1}\right], \end{aligned} \tag{16.64}$$

where all the terms independent of angle have been incorporated into the new complex amplitude $E_0$. The additional phase factor $\beta\,(\partial\phi/\partial\beta)_{\beta_1}$ can be interpreted as a shift, $Z_s$, in the axial position of the wave as it reflects at the boundary, as shown in Fig. 16.7.

This shift in axial position $Z_s$ is called the Goos–Hänchen shift [3],

$$Z_s = -\left(\frac{\partial\phi}{\partial\beta}\right)_{\beta_1}, \tag{16.65}$$

which can be evaluated as

$$Z_s = -\left(\frac{\partial\phi}{\partial\theta_z}\frac{\partial\theta_z}{\partial\beta}\right)_{\beta_1}, \tag{16.66}$$

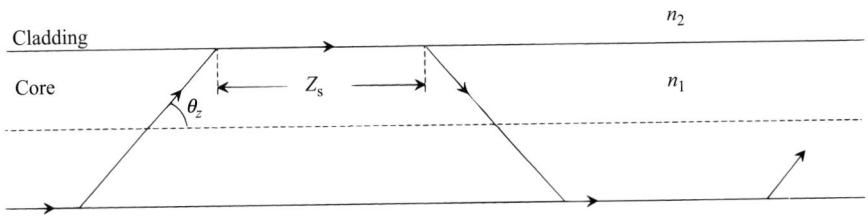


Fig. 16.7 The Goos–Hänchen shift.

where

$$\frac{\partial \theta_z}{\partial \beta} = -\frac{1}{k \sin \theta_z}. \tag{16.67}$$

By differentiating Eq. (16.60) and assuming that the angles $\theta_z$ and $\theta_a$ are small, we get

$$Z_s \simeq \frac{\lambda}{\pi n_1 \theta_z} \frac{1}{\sqrt{\theta_a^2 - \theta_z^2}}. \tag{16.68}$$

This shift in axial position makes the distance traveled by the ray in propagating a distance $\ell$ along the fiber shorter than it would be without the shift. However, the shift is very small unless the ray angle is close to the critical angle, in which case the evanescent portion of the associated wave penetrates very far into the cladding.

## 16.5 Wave theory of the dielectric-slab guide

When a given mode propagates in a slab, light energy is guided along in the slab medium or core. There are, however, finite field amplitudes that decay exponentially into the external medium. The propagating waves in this case, because they are guided along by the slab, but are not totally confined within it, are often called *surface waves*. This distinguishes them from the types of wave that propagate inside, and are totally confined by, hollow conducting structures – such as rectangular and cylindrical microwave waveguides. We can find the electromagnetic field profiles in the two-dimensional slab guide shown in Fig. 16.6 by using Maxwell's equations. We look for a propagating solution with fields that vary like $e^{j\omega t - \gamma z}$, where $\gamma$ is the propagation constant.

There are both P- and S-wave solutions to the wave equation. Clearly, as can be seen from Fig. 16.6, the P wave has an $E_z$ component of its electric field, so it could also be called a transverse-magnetic-field (TM) wave. Similarly, the S wave is a transverse-electric-field (TE) wave. From the curl equations, written in the form

$$\begin{aligned} \text{curl } \mathbf{E} &= -j\omega\mu\mathbf{H}, \\ \text{curl } \mathbf{H} &= j\omega\epsilon\mathbf{E}, \end{aligned} \tag{16.69}$$

the following equations result:

$$\frac{\partial E_z}{\partial y} + \gamma E_y = -j\omega\mu H_x, \tag{16.70}$$

$$-\gamma H_x - \frac{\partial H_z}{\partial x} = j\omega\epsilon E_y, \tag{16.71}$$

$$-\gamma E_x - \frac{\partial E_z}{\partial x} = -j\omega\mu H_y, \tag{16.72}$$

$$\frac{\partial H_z}{\partial y} + \gamma H_y = j\omega\epsilon E_x, \tag{16.73}$$

$$\frac{\partial E_y}{\partial x} - \frac{\partial E_x}{\partial y} = -j\omega\mu H_z, \tag{16.74}$$

$$\frac{\partial H_y}{\partial x} - \frac{\partial H_x}{\partial y} = j\omega\epsilon E_z. \tag{16.75}$$

An expression for $H_x$ can be found by eliminating $E_y$ from Eqs. (16.70) and (16.71), or vice versa. $E_x$ and $H_y$ can be found from Eqs. (16.72) and (16.73):

$$E_x = -\frac{1}{\gamma^2 + k^2}\left(\gamma\frac{\partial E_z}{\partial x} + j\omega\mu\frac{\partial H_z}{\partial y}\right), \tag{16.76}$$

$$H_y = -\frac{1}{\gamma^2 + k^2}\left(j\omega\epsilon\frac{\partial E_z}{\partial x} + \gamma\frac{\partial H_z}{\partial y}\right), \tag{16.77}$$

$$E_y = \frac{1}{\gamma^2 + k^2}\left(-\gamma\frac{\partial E_z}{\partial y} + j\omega\mu\frac{\partial H_z}{\partial x}\right), \tag{16.78}$$

$$H_x = \frac{1}{\gamma^2 + k^2}\left(j\omega\epsilon\frac{\partial E_z}{\partial y} + \gamma\frac{\partial H_z}{\partial x}\right), \tag{16.79}$$

where $k = \omega\sqrt{\mu\epsilon}$.

From the Helmholtz equation (Eq. (16.4)),

$$\nabla^2\mathbf{E} + k^2\mathbf{E} = 0; \quad \nabla^2\mathbf{H} + k^2\mathbf{H} = 0. \tag{16.80}$$

Since $\partial^2/\partial z^2$ is equivalent to multiplication by $-\gamma^2$, Eq. (16.80) can be written as

$$\frac{\partial^2\mathbf{E}}{\partial x^2} + \frac{\partial^2\mathbf{E}}{\partial y^2} = -(\gamma^2 + k^2)\mathbf{E}, \tag{16.81}$$

$$\frac{\partial^2\mathbf{H}}{\partial x^2} + \frac{\partial^2\mathbf{H}}{\partial y^2} = -(\gamma^2 + k^2)\mathbf{H}. \tag{16.82}$$

## 16.6 P waves in the slab guide

For P waves in the slab guide $E_y = 0$ and $\partial/\partial y \equiv 0$, so, from Eq. (16.77),

$$H_y = -\frac{j\omega\epsilon}{\gamma^2 + k^2}\frac{\partial E_z}{\partial x}, \tag{16.83}$$

and, from Eq. (16.81),

$$\frac{\partial^2 E_z}{\partial x^2} = -(\gamma^2 + k^2)E_z. \tag{16.84}$$

The solutions to Eq. (16.84) are sine or cosine functions, or exponentials. If we wish the waves to be confined to the slab we must choose the solution in medium 2 where the field amplitudes decay exponentially in the positive (or negative)-$x$ direction. So, for example, using Eqs. (16.83) and (16.84), the $x$ dependences of $E_z$ and $H_y$ in the cladding will be

$$E_{z_2} = Be^{-\alpha_x x}; \quad H_{y_2} = -\frac{j\omega\epsilon_2}{\alpha_x}Be^{-\alpha_x x}, \quad x \geq d, \tag{16.85}$$

and

$$E_{z_2} = Be^{\alpha_x x}; \quad H_{y_2} = \frac{j\omega\epsilon_2}{\alpha_x} Be^{\alpha_x x}, \quad x \leq -d. \tag{16.86}$$

Inside the slab the fields are

$$E_{z_1} = A\sin(k_x x); \quad H_{y_1} = -\frac{j\omega\epsilon_1}{k_x} A\cos(k_x x) \tag{16.87}$$

or

$$E_{z_1} = A\cos(k_x x); \quad H_{y_1} = \frac{j\omega\epsilon_1}{k_x} A\sin(k_x x), \tag{16.88}$$

where we have taken note that in medium 2

$$\alpha_x^2 = -(\gamma^2 + k_2^2), \tag{16.89}$$

where $\alpha_x > 0$, and in medium 1

$$k_x^2 = \gamma^2 + k_1^2, \tag{16.90}$$

where $k_x > 0$.

A propagating wave exists in the slab, provided that $\gamma$ is imaginary. If $\gamma$ were to be real, then the $e^{-\gamma z}$ variation of the fields would represent an attenuated, non-propagating *evanescent* wave. For $\gamma$ to be imaginary, from Eq. (16.90), $k_1^2 > k_x^2$; therefore, $k_x^2 < \omega^2\mu_0\epsilon_1$.

To find the relation between $\alpha_x$ and $k_x$ we use the boundary condition that at $x = d$ the $E_z$ and $H_y$ fields (which are the tangential fields at the boundary between the two media) must be continuous. Therefore,

$$\left(E_{z_1}/H_{y_1}\right)_{x=d} = \left(E_{z_2}/H_{y_2}\right)_{x=d}. \tag{16.91}$$

For the modes with odd $E_z$ field symmetry in the slab (the sine solution above),

$$-\frac{k_x}{j\omega\epsilon_1}\tan(k_x d) = -\frac{\alpha_x}{j\omega\epsilon_2}, \tag{16.92}$$

so

$$\alpha_x = \frac{\epsilon_2}{\epsilon_1} k_x \tan(k_x d). \tag{16.93}$$

It is straightforward to show that, for the even-symmetry $E_z$ field solution in the slab (the cosine solution above),

$$\alpha_x = -\frac{\epsilon_2}{\epsilon_1} k_x \cot(k_x d). \tag{16.94}$$

To determine the propagation constant $\gamma$ for the odd-symmetry modes we must solve simultaneously Eqs. (16.89), (16.90), and (16.93). Elimination of $\gamma$ from Eqs. (16.89) and (16.90) gives

$$\alpha_x = \sqrt{k_1^2 - k_2^2 - k_x^2} = \sqrt{\omega^2\mu_0(\epsilon_1 - \epsilon_2) - k_x^2}. \tag{16.95}$$

We have assumed that $\mu_1 = \mu_2 = \mu_0$, since most optically transparent materials have relative permeabilities very close to unity.

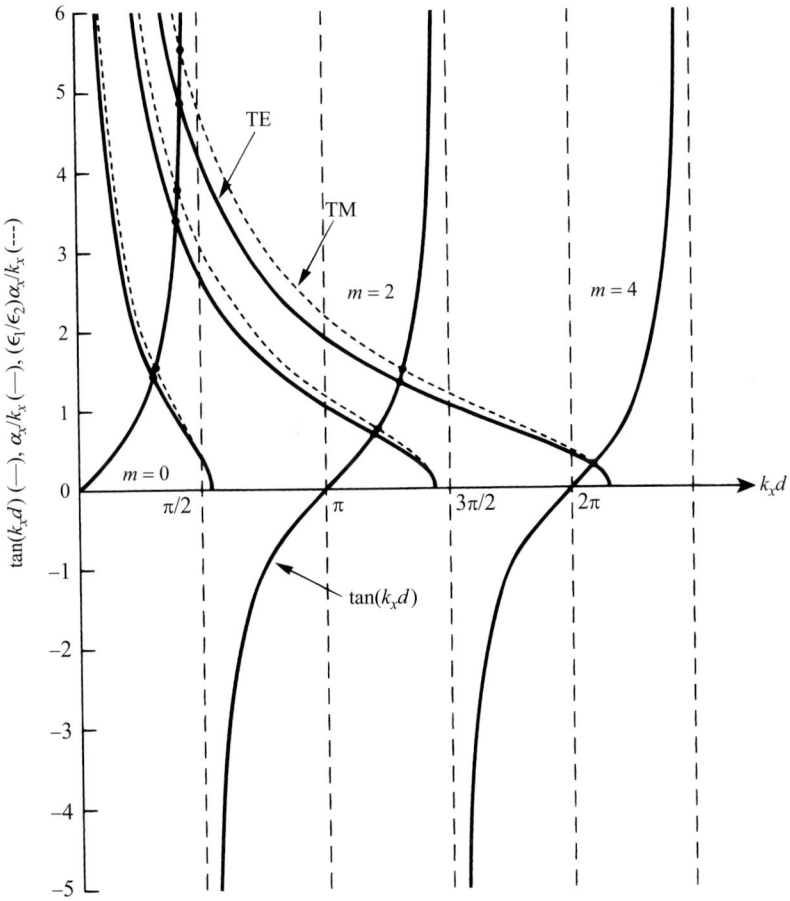


**Fig. 16.8** Graphical solution of Eqs. (16.96) and (16.121) for the odd-symmetry TM (P-wave) and TE (S-wave) modes in a slab waveguide. For these curves $n_1 = 1.5$ and $n_2 = 1.4$. The three pairs of curves from left to right correspond to $\lambda_0 = 2\ \mu\text{m}$, 0.75 μm, and 0.5 μm, respectively.

From Eqs. (16.93) and (16.95),

$$\tan(k_x d) = \frac{\epsilon_1}{\epsilon_2}\sqrt{\frac{\omega^2\mu_0(\epsilon_1 - \epsilon_2)}{k_x^2} - 1}. \tag{16.96}$$

The solution to this equation can be illustrated graphically, as shown in Fig. 16.8. Whatever the value of $\omega$, there is always at least one solution to this equation. For small values of $k_x d$, Eq. (16.93) gives

$$\alpha_x = \frac{\epsilon_2}{\epsilon_1}k_x^2 d, \tag{16.97}$$

and from Eq. (16.95)

$$k_x^4 d^4 + \left(\frac{\epsilon_1}{\epsilon_2}\right)^2 k_x^2 d^2 = \left(\frac{\epsilon_1}{\epsilon_2}\right)^2 \omega^2 \mu_0 (\epsilon_1 - \epsilon_2) d^2. \tag{16.98}$$

Neglecting the term in $k_x^4 d^4$ gives

$$k_x \simeq \omega^2 \mu_0 (\epsilon_1 - \epsilon_2) = k_1 - k_2, \tag{16.99}$$

and from Eq. (16.90), if $k_x \to 0$,

$$\gamma \to \pm j k_1. \tag{16.100}$$

Therefore as either $\epsilon_1 \to \epsilon_2$ or $d$ becomes very small, $\alpha_x$ also decreases, and energy extends far into medium 2. The propagation constant of the wave approaches the value it would have in medium 2 alone.

The solution of Eq. (16.95) with the smallest value of $k_x d$ is called the *fundamental* P-wave mode of the guide. This mode can propagate at any frequency – it has no cutoff frequency. However, this is not true for the even-symmetry $E_z$ field solution. In this case, from Eqs. (16.94) and (16.95),

$$\cot(k_x d) = -\frac{\epsilon_1}{\epsilon_2}\sqrt{\frac{\omega^2 \mu_0 (\epsilon_1 - \epsilon_2)}{k_x^2} - 1}. \tag{16.101}$$

The graphical solution of this equation is given in Fig. 16.9. In order for this equation to have a solution, $k_x d$ must be greater than $\pi/2$. This cannot be so while at the same time $\omega \to 0$.

## 16.7 Dispersion curves and field distributions in a slab waveguide

For a propagating mode to exist, from either Eq. (16.96) or (16.101),

$$\omega^2 \mu_0 (\epsilon_1 - \epsilon_2) \geq k_x^2. \tag{16.102}$$

In this case Eq. (16.90) implies

$$\omega^2 \mu_0 (\epsilon_1 - \epsilon_2) \geq \gamma^2 + \omega^2 \mu_0 \epsilon_1, \tag{16.103}$$

which, on writing $\gamma = \pm j\beta$, implies $\beta \geq \omega^2 \mu_0 \epsilon_2$. By solving Eq. (16.96) or Eq. (16.101) for various values of $\omega$ we can plot the variation of $\beta$ with $\omega$ for the S- and P-wave modes. These curves are called dispersion curves.

From Eq. (16.90),

$$\beta^2 = k_1^2 - k_x^2 = \omega^2 \mu_0 \epsilon_1 - k_x^2. \tag{16.104}$$

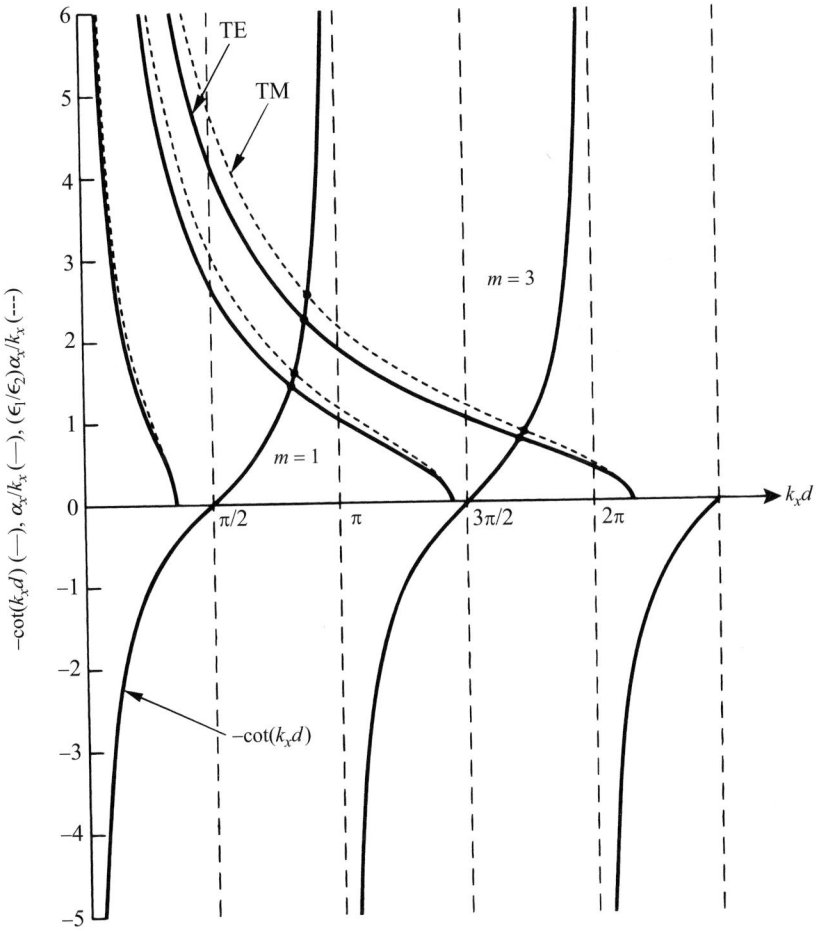


**Fig. 16.9** Graphical solution of Eqs. (16.101) and (16.128) for even-symmetry TM (P-wave) and TE (S-wave) modes in a slab waveguide. For these curves $n_1 = 1.5$ and $n_2 = 1.4$. The three pairs of curves from left to right correspond to $\lambda_0 = 3\ \mu\text{m}$, $0.75\ \mu\text{m}$, and $0.5\ \mu\text{m}$, respectively. For the wavelength of $3\ \mu\text{m}$ the curves do not intersect, and Eq. (16.101), for example, will have no solution.

By solving for $k_x$ from either Eq. (16.96) for the odd-symmetry $E_z$ solution or Eq. (16.101) for the even-symmetry solution, we can produce the dispersion curves shown in Fig. 16.10. The curve labeled $m = 0$ is the fundamental mode, the curves labeled $m = 1, 2, 3$ etc. are alternate even-, odd-, even-symmetry $E_z$ field higher modes. The $E_x$ and $H_y$ components of the P-wave solution are easily found from Eqs. (16.76) and (16.77). Some of their $E_z$ field distributions within the slab are shown in Fig. 16.11. For the fundamental mode,

$$\begin{aligned} E_x(\text{in medium 1}) &= -\frac{\gamma}{\gamma^2 + k_1^2}\frac{\partial E_z}{\partial x} = -\frac{A\gamma}{k_x}\cos(k_x x), \\ E_x(\text{in medium 2}) &= -\frac{\gamma}{\gamma^2 + k_2^2}\frac{\partial E_z}{\partial x} = -\frac{B\gamma}{\alpha_x}e^{-\alpha_x x}. \end{aligned} \tag{16.105}$$

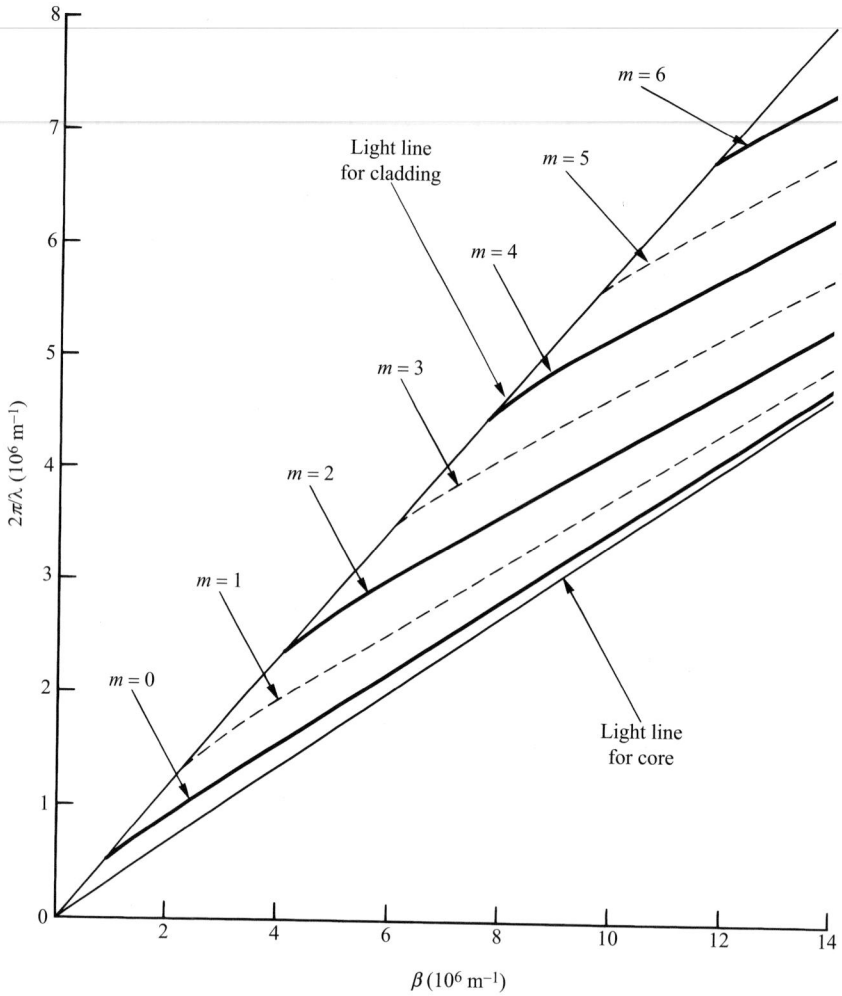


**Fig. 16.10** Dispersion curves for P-wave modes in a slab waveguide. The solid dispersion curves are for the odd-symmetry $E_z$ modes and the dashed curves are for even-symmetry $E_z$ modes. The curve has been calculated for $n_1 = 3, n_2 = \sqrt{3}$, and $d = 0.6$ μm, which are unrealistic values for a practical waveguide, but chosen to illustrate the dispersion curves clearly. The curves marked $m = 0, 2, 4$, and 6 have been calculated using Eq. (16.96); those for $m = 1, 3$, and 5 have been calculated from Eq. (16.101).

Note that for each mode as the frequency increases the wave is confined more and more to the core and its propagation constant $\beta \to k_1$:

$$\begin{aligned} H_y(\text{in medium 1}) &= -\frac{j\omega\epsilon_1}{\gamma^2 + k_1^2}\frac{\partial E_z}{\partial x} = -\frac{jA\omega\epsilon_1}{k_x}\cos(k_x x), \\ H_y(\text{in medium 2}) &= -\frac{j\omega\epsilon_2}{\gamma^2 + k_2^2}\frac{\partial E_z}{\partial x} = -\frac{jB\omega\epsilon_2}{\alpha_x}e^{-\alpha_x x}. \end{aligned} \tag{16.106}$$

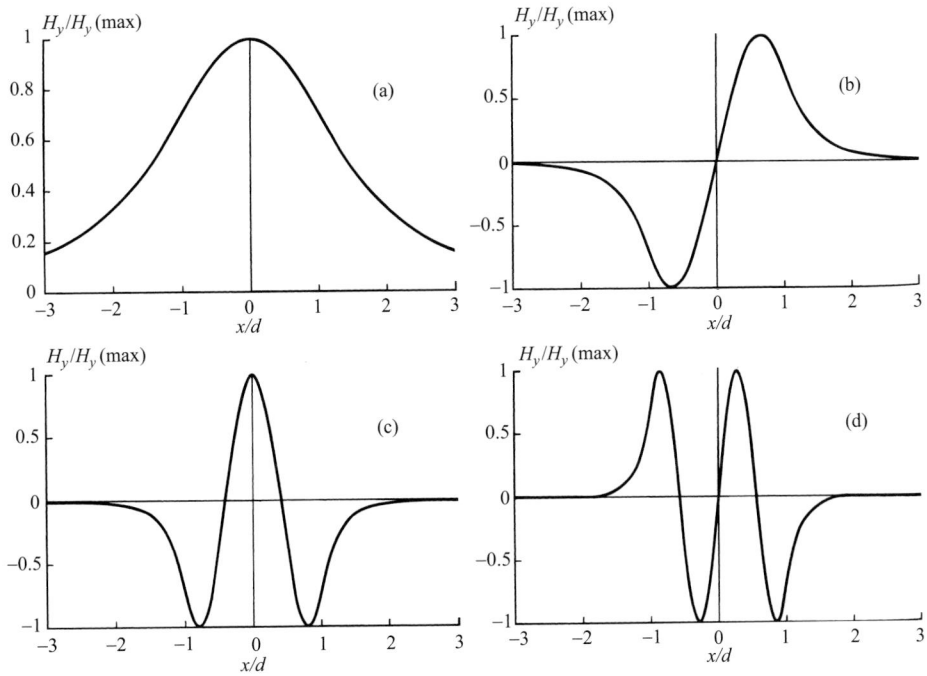


**Fig. 16.11** $H_y$ field variation for P-wave (TM-wave) odd- and even-symmetry modes in a slab waveguide. The curves have been calculated for $n_1 = 1.5$ and $n_2 = 1.485$: (a) $k_x d = \pi/4$, odd $E_z$ symmetry; (b) $k_x d = 3\pi/4$, even $E_z$ symmetry; (c) $k_x d = 5\pi/4$, odd $E_z$ symmetry; and (d) $k_x d = 7\pi/4$, even $E_z$ symmetry.

The tangential component of the magnetic field must be continuous at, say, $x = d$. So, from Eq. (16.106),

$$\frac{A}{B} = \frac{\epsilon_2 k_x e^{-\alpha_x d}}{\epsilon_1 \alpha_x \cos(k_x d)}, \tag{16.107}$$

which, from Eq. (16.93), gives

$$\frac{A}{B} = \frac{e^{-\alpha_x d}}{\sin(k_x d)}. \tag{16.108}$$

The actual value of $A$ (or $B$) is determined from the energy flux in the dielectric slab. The transverse intensity distribution in the guide is

$$I = E_x H_y, \tag{16.109}$$

which, from Eqs. (16.105) and (16.106), gives

$$I_1(\text{in medium 1}) = \frac{j\gamma\omega\epsilon_1 A^2}{k_x^2}\cos^2(k_x x), \tag{16.110}$$

$$I_2(\text{in medium 2}) = \frac{j\gamma\omega\epsilon_2 \beta^2}{\alpha_x^2} e^{-2\alpha_x x}. \tag{16.111}$$

The total energy flux per unit width is

$$\phi = 2\int_0^d I_1\,dx + 2\int_d^\infty I_2\,dx. \tag{16.112}$$

## 16.8 S waves in the slab guide

For S waves in the slab guide the electric field points only in the $y$ direction, $\partial/\partial y \equiv 0$, and there are $H_x$ and $H_z$ components of the magnetic field. In this case we start from Eq. (16.82) and write

$$\frac{\partial^2 H_z}{\partial x^2} = -(\gamma^2 + k^2)H_z. \tag{16.113}$$

For the modes with odd symmetry of the $H_z$ field

$$H_{z_1} = C\sin(k_x x), \qquad |\,x\,| \leq d, \tag{16.114}$$

$$H_{x_1} = \frac{C\gamma k_x}{\gamma^2 + k_1^2}\cos(k_x x) = \frac{C\gamma}{k_x}\cos(k_x x), \tag{16.115}$$

$$E_{y_1} = \frac{jC\omega\mu_1}{\gamma^2 + k_1^2}k_x\cos(k_x x) = \frac{jC\omega\mu_1}{k_x}\cos(k_x x), \tag{16.116}$$

$$H_{z_2} = De^{-\alpha_x x}, \quad x \geq d, \tag{16.117}$$

$$H_{x_2} = -\frac{C\gamma\alpha_x}{\gamma^2 + k_2^2}e^{-\alpha_x x} = \frac{C\gamma}{\alpha_x}e^{-\alpha_x x}, \tag{16.118}$$

$$E_{y_2} = -\frac{jD\omega\mu_2\alpha_x}{\gamma^2 + k_2^2}e^{-\alpha_x x} = \frac{jD\omega\mu_2}{\alpha_x}e^{-\alpha_x x}. \tag{16.119}$$

It is straightforward to show, in a similar manner to before, that

$$\alpha_x = \frac{\mu_2}{\mu_1}k_x\tan(k_x d) \tag{16.120}$$

and

$$\tan(k_x d) = \frac{\mu_1}{\mu_2}\sqrt{\frac{\omega^2\mu_0(\epsilon_1 - \epsilon_2)}{k_x^2} - 1}. \tag{16.121}$$

Even as $\omega \to 0$ there is always at least one solution to the equation. This is the *fundamental*, $m = 0$, S-wave mode of the guide. The graphical solution of Eq. (16.121) in Fig. 16.8 illustrates this.

Modes that have even symmetry of the $H_z$ field have field components

$$H_{z_1} = C\cos(k_x x), \quad |x| \leq d, \tag{16.122}$$

$$H_{x_1} = -\frac{C\gamma}{k_x}\sin(k_x x), \tag{16.123}$$

$$E_{y_1} = -\frac{jC\omega\mu_1}{k_x}\sin(k_x x), \tag{16.124}$$

$$H_{z_2} = De^{-\alpha_x x}, \quad x \geq d, \tag{16.125}$$

$$H_{x_2} = \frac{C\gamma}{\alpha_x}e^{-\alpha_x}, \tag{16.126}$$

$$E_{y_2} = \frac{jD\omega\mu_2}{\alpha_x}e^{-\alpha_x x}. \tag{16.127}$$

The defining equation for the modes is

$$\alpha_x = -\frac{\mu_2}{\mu_1}k_x\cot(k_x d). \tag{16.128}$$

Because most optical materials have $\mu \simeq \mu_0$, the cutoff frequencies of the S-wave modes above the fundamental are slightly lower, by a factor of $\epsilon_2/\epsilon_1$, than those for the corresponding P-wave modes, as shown in Fig. 16.9.

## 16.9 Practical slab guide geometries

Slab guides that are infinite in one dimension are convenient mathematically but do not exist in practice. Practical slab guides utilize index changes or discontinuities in both the $x$ and the $y$ direction (for a guide intended to propagate waves in the $z$ direction). Some examples are shown in Fig. 16.12. Such structures can be fabricated by methods similar to those used in integrated-circuit manufacture. In each case the fields outside the guiding layer decay exponentially. The fields are broadly similar to those discussed previously for the infinite slab guide. However, the modes that exist are no longer pure P- or S-wave modes; rather, they possess all components of **E** and **H**, although they can still be predominantly P or S in character in some circumstances.

Slab guide geometries exist in most semiconductor diode lasers and in many LEDs as discussed in Chapter 12. Consequently it is possible to integrate the light source with a slab waveguide and then incorporate active components such as modulators, couplers, and switches into the structure. The resultant *integrated optics* package can serve as a complete transmitting module with full potential for amplitude or phase modulation, or multiplexing. We shall meet examples of these devices later, in Chapter 18. For a guided-wave optical communication link over long distances, slab waveguides are not practical. The information-carrying light beam must be transmitted in an optical fiber – a cylindrical dielectric waveguide.

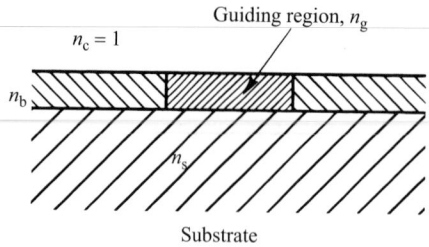


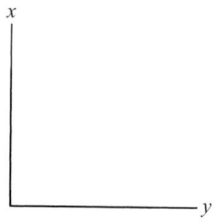


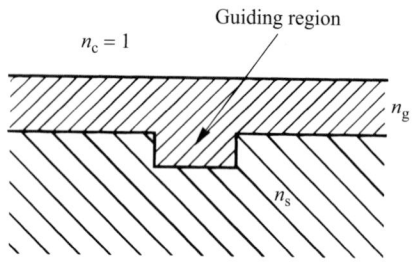


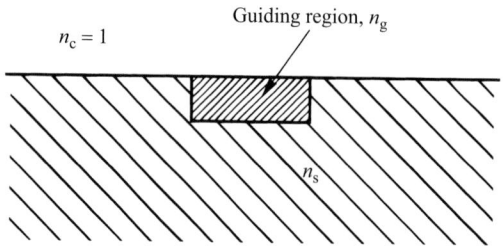


**Fig. 16.12** Some practical slab waveguide geometries.

## 16.10 Cylindrical dielectric waveguides

The ray description of the slab waveguide is useful for describing the path of a meridional ray down a fiber. However, to obtain a more detailed understanding of the propagation characteristics of the fiber we again use Maxwell's equations. In cylindrical coordinates the curl equation (Eq. (16.56)) is

$$\begin{aligned}\text{curl}\,\mathbf{E} &= \left(\frac{1}{r}\frac{\partial E_z}{\partial \phi} - \frac{\partial E_\phi}{\partial z}\right)\hat{\mathbf{r}} + \left(\frac{\partial E_r}{\partial z} - \frac{\partial E_z}{\partial r}\right)\hat{\boldsymbol{\phi}} + \left[\frac{1}{r}\frac{\partial (rE_\phi)}{\partial r} - \frac{1}{r}\frac{\partial E_r}{\partial \phi}\right]\hat{\mathbf{z}} \\ &= j\omega\mu\mathbf{H},\end{aligned} \tag{16.129}$$

with a similar equation for curl $\mathbf{H}$. Here $E_r, E_\phi$, and $E_z$, respectively, are the radial, azimuthal, and axial components of the electric field, with associated unit vectors

$\hat{\mathbf{r}}, \hat{\boldsymbol{\phi}}$, and $\hat{\mathbf{z}}$. The Helmholtz equations Eqs. (16.81) and (16.82) become

$$\begin{aligned} \frac{\partial^2 \mathbf{E}}{\partial r^2} + \frac{1}{r}\frac{\partial \mathbf{E}}{\partial r} + \frac{1}{r^2}\frac{\partial^2 \mathbf{E}}{\partial \phi^2} &= -(\gamma^2 + k^2)\mathbf{E}, \\ \frac{\partial^2 \mathbf{H}}{\partial r^2} + \frac{1}{r}\frac{\partial \mathbf{H}}{\partial r} + \frac{1}{r^2}\frac{\partial^2 \mathbf{H}}{\partial \phi^2} &= -(\gamma^2 + k^2)\mathbf{H}, \end{aligned} \tag{16.130}$$

where, as before, $k^2 = \omega^2 \mu \epsilon$.

We start by examining the solution of Eq. (16.130) for the axial component of the field. We look for a separable solution of the form

$$E_z(r, \phi) = R(r)\Phi(\phi). \tag{16.131}$$

The $z$ dependence of $E_z$ is $e^{j\omega t - \gamma z}$, as before, but this factor will be omitted for simplicity. Substituting from (16.131) into (16.130) gives

$$r^2 \frac{R''}{R} + \frac{rR'}{R} + k_c^2 r^2 = -\frac{\Phi''}{\Phi}, \tag{16.132}$$

where primes denote differentiation, and we have written $k_c^2 = \gamma^2 + k^2$.

The only way for Eq. (16.132) to be satisfied, since the left-hand side is only a function of $r$, and the right-hand side is only a function of $\phi$, is for each side to be independently equal to a constant. Therefore,

$$r^2 \frac{R''}{R} + \frac{rR'}{R} + k_c^2 r^2 = \nu^2, \tag{16.133}$$

which is usually written as

$$R'' + \frac{1}{r}\frac{\partial R}{\partial r} + \left(k_c^2 - \frac{\nu^2}{r^2}\right) R = 0, \tag{16.134}$$

and

$$-\frac{\Phi''}{\Phi} = \nu^2. \tag{16.135}$$

The general solution of Eq. (16.135) is

$$\Phi(\phi) = A \sin(\nu\phi) + B \cos(\nu\phi), \tag{16.136}$$

where $\nu$ must be a positive or negative integer, or zero, so that the function $\Phi(\phi)$ is single-valued: i.e., $\Phi(\phi + 2\pi) = \Phi(\phi)$.

Equation (16.134) is called Bessel's equation. Its solutions are Bessel functions, which are the cylindrical geometry equivalents of the real or complex exponential solutions we encountered in the slab guide.

If $k_c^2 > 0$ the general solution of Eq. (16.134) is

$$R(r) = CJ_\nu(k_c r) + DY_\nu(k_c r), \tag{16.137}$$

where $J_\nu(k_c r)$ is the Bessel function of the first kind and $Y_\nu(k_c \nu)$ is the Bessel function of the second kind.[1] These are oscillatory functions of $r$, as shown in Fig. 16.13.

[1] The $Y_\nu$ Bessel function of the second kind is also written as $N_\nu$ by some authors.

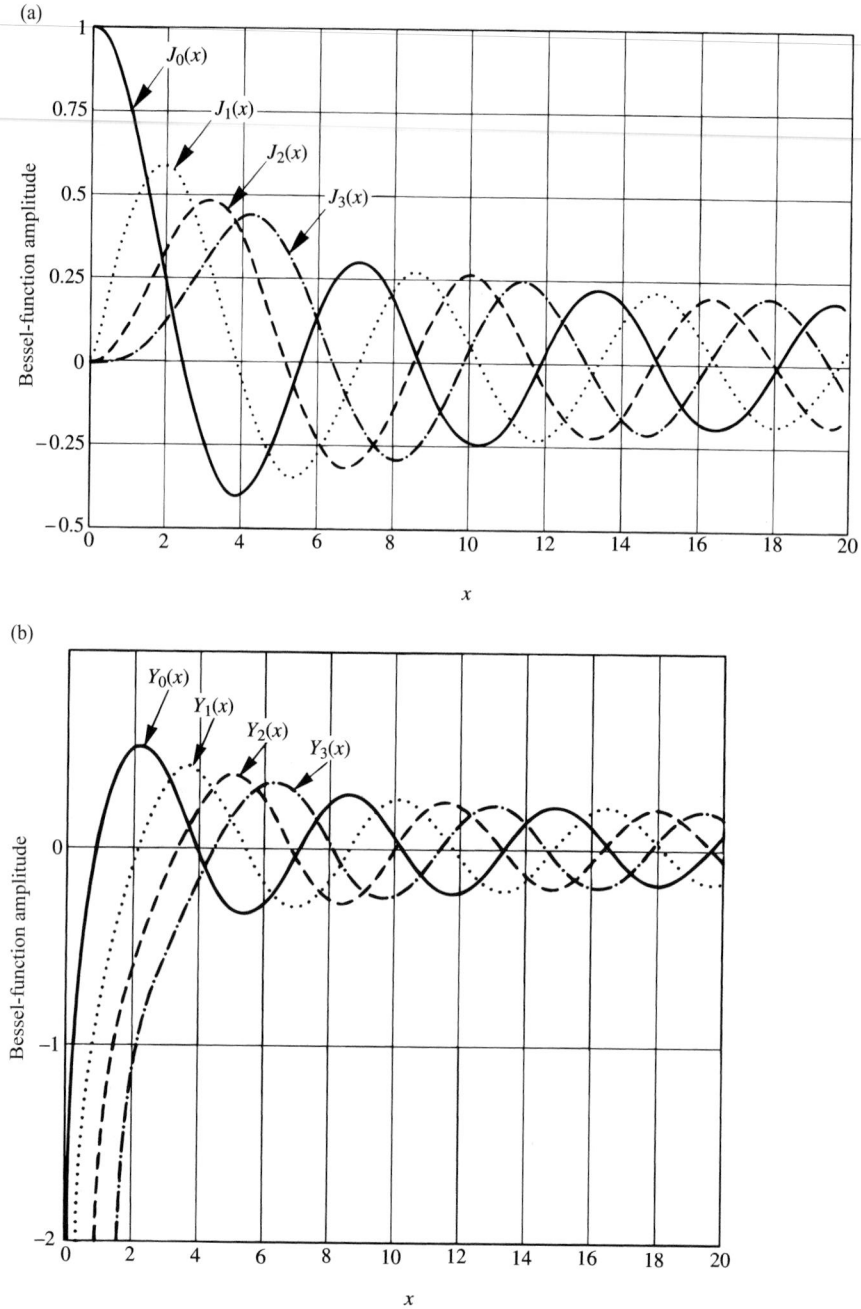


**Fig. 16.13** Bessel functions of (a) the first kind and (b) the second kind.

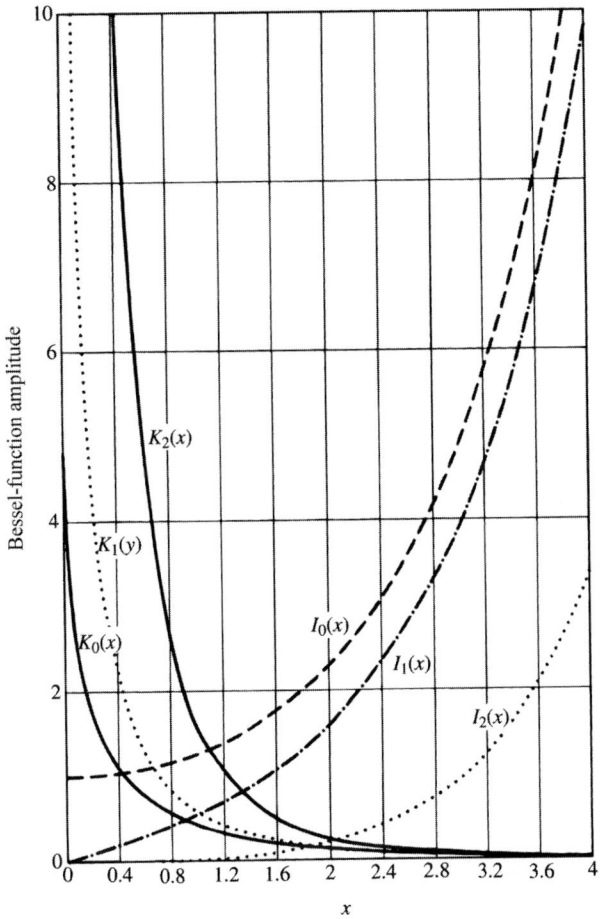


**Fig. 16.14** Modified Bessel functions of the first and second kinds.

If $k_c^2 < 0$ the general solution of Eq. (16.134) is

$$R(r) = C' I_\nu(|k_c|r) + D' K_\nu(|k_c|r), \tag{16.138}$$

where $I_\nu(|k_c|r)$ and $K_\nu(|k_c|r)$ are the modified Bessel functions of the first and second kinds, respectively. These are, respectively, monotonically increasing and decreasing functions of $r$ as shown in Fig. 16.14.

Before proceeding further we can make several physical observations to help simplify our analysis.

(i) We cannot use the equations above to solve for wave propagation in a graded-index fiber – a fiber in which the dielectric constant varies smoothly in the radial direction. In such fibers $\epsilon$ is a function of $r$ (the quadratic index fiber is an example), and the wave-equation analysis is much more complex. References [1] and [4] can be consulted for details.

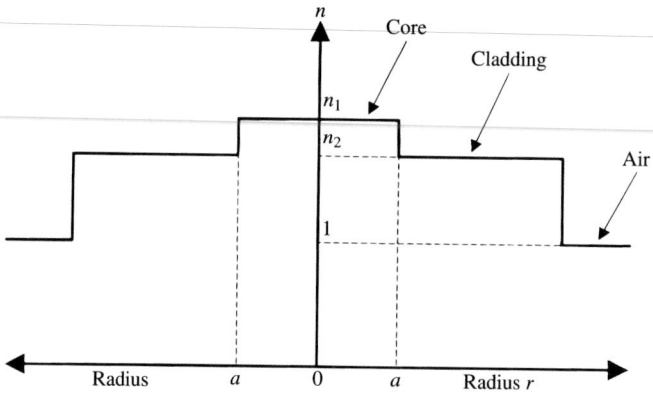


**Fig. 16.15** The radial refractive-index profile of a step-index cylindrical fiber.

(ii) We can use these equations to find the field components in a step-index fiber, whose radial refractive-index profile is as shown in Fig. 16.15. We find a solution for the core, where the dielectric constant $\epsilon = n_1^2$, and match this solution to an appropriate solution for the cladding, where $\epsilon_2 = n_2^2$.

(iii) In the core we expect an oscillatory form of the axial field to be appropriate (as it was for the slab guide). Since the function $Y_\nu(k_c r)$ blows up as $r \to 0$, we reject it as physically unreasonable. Only the $J_\nu(k_c r)$ variation need be retained.

(iv) We are interested in fibers with $n_1 > n_2$ that guide energy. The fields in the cladding must decay as we go further away from the core. Since $I_\nu(k_c r)$ blows up as $r \to \infty$, we reject it as physically unreasonable. Only the $K_\nu(k_c r)$ variation need be retained.

(v) In the core $k_{c_1}^2 > 0$ implies

$$\gamma^2 + k_1^2 > 0, \tag{16.139}$$

and for a propagating solution $\gamma \equiv j\beta$; therefore,

$$k_1^2 > \beta^2. \tag{16.140}$$

In the cladding $k_{c_2}^2 < 0$, and therefore $\gamma^2 + k_2^2 < 0$, which implies $k_2^2 < \beta^2$. In summary,

$$k_2 < \beta < k_1. \tag{16.141}$$

The closer $\beta$ approaches $k_1$, the more fields are confined to the core of the fiber, while, for $\beta$ approaching $k_2$, they penetrate far into the cladding. When $\beta$ becomes equal to $k_2$ the fields are no longer guided by the core: they penetrate completely into the cladding.

(vi) If $k_c^2 > 0$, then the axial fields in the cladding would no longer be described by the $K_\nu$ Bessel function; instead they would be described by the $J_\nu$ Bessel function. These oscillatory fields no longer decay evanescently in the cladding, and the wave is no longer guided by the core. We can take $k_c^2 = 0$ as a demarcation point between a wave being guided and a wave leaking from the core.

If we take the $\sin(\nu\phi)$ azimuthal variation for the axial electric field, then in the core we have

$$E_{z_1} = AJ_\nu(k_{c_1}r)\sin(\nu\phi) \qquad (r \le a), \tag{16.142}$$

and in the cladding

$$E_{z_2} = CK_\nu(|k_{c_2}|r)\sin(\nu\phi) \qquad (r \ge a), \tag{16.143}$$

where the subscripts 1 and 2 indicate core and cladding fields, respectively. These fields must be continuous at the boundary between the core and the cladding, so

$$E_{z_1}(r = a) = E_{z_2}(r = a). \tag{16.144}$$

Far into the cladding the evanescent nature of the fields becomes apparent, since for large values of $r$

$$K_\nu(|k_{c_2}|r) \propto \frac{e^{-|k_{c_2}|r}}{\sqrt{|k_{c_2}|r}}. \tag{16.145}$$

The corresponding magnetic fields are

$$H_z = BJ_\nu(k_{c_1})\cos(\nu\phi) \qquad (r \le a), \tag{16.146}$$

$$H_z = DK_\nu(k_{c_2})\cos(\nu\phi) \qquad (r \ge a). \tag{16.147}$$

In this case, where we have assumed the $\sin(\nu\phi)$ for the electric field, we must have the $\cos(\nu\phi)$ variation for the $H_z$ field to allow matching of the tangential fields (which include both $z$ and $\phi$ components) at the core/cladding boundary. We must allow for the possibility of both axial electric and magnetic field components in the fiber: so-called *hybrid* modes. However, if $\nu = 0$, then Eq. (16.143) shows that $E_z = 0$: this is a TE mode. If we had chosen the $\cos(\nu\phi)$ variation for the axial electric field, then with $\nu = 0$ we would have a TM mode. Thus, TE and TM modes exist only for $\nu = 0$. To find the relationship between the field components orthogonal to the fiber axis and the $E_z$ and $H_z$ components we work from Eq. (16.129) and the corresponding equation for curl **H**. Some tedious algebra will yield the results:

$$E_r = -\frac{j}{k_c^2}\left(\frac{\gamma}{j}\frac{E_z}{\partial r} + \omega\mu\frac{1}{r}\frac{\partial H}{\partial\phi}\right), \tag{16.148}$$

$$E_\phi = -\frac{j}{k_c^2}\left(\frac{\gamma}{j}\frac{1}{r}\frac{\partial E_z}{\partial\phi} - \omega\mu\frac{\partial H_z}{\partial r}\right), \tag{16.149}$$

$$H_r = -\frac{j}{k_c^2}\left(\frac{\gamma}{j}\frac{\partial H_z}{\partial r} - \omega\epsilon\frac{1}{r}\frac{\partial E_z}{\partial\phi}\right), \tag{16.150}$$

$$H_\phi = -\frac{j}{k_c^2}\left(\frac{\gamma}{j}\frac{1}{r}\frac{\partial H_z}{\partial\phi} + \omega\epsilon\frac{\partial E_z}{\partial r}\right). \tag{16.151}$$

Note that, for a TE mode ($E_z = 0$), $E_r \sim \partial H_z/\partial\phi$, so that, if $H_z$ has a variation $\sim\cos(\nu\phi)$, where $\nu$ is a constant, then $E_r$ must vary as $\sin(\nu\phi)$. For a TM mode ($H_z = 0$), $H_r \sim \partial E_z/\partial\phi$, with similar consequences. Bearing these facts in mind, let us write down the full variation of the fields in the core and cladding as

$$E_z = AJ_\nu\left(\frac{ur}{a}\right)\sin(\nu\phi), \tag{16.152}$$

$$E_r = \left[-\frac{A\gamma}{u/a}J'_\nu\left(\frac{ur}{a}\right) + B\frac{j\omega\mu_0}{(u/a)^2}\frac{\nu}{r}J_\nu\left(\frac{ur}{a}\right)\right]\sin(\nu\phi), \tag{16.153}$$

$$E_\phi = \left[-\frac{A\gamma}{(u/a)^2}\frac{\nu}{r}J_\nu\left(\frac{ur}{a}\right) + B\frac{j\omega\mu_0}{u/a}J'_\nu\left(\frac{ur}{a}\right)\right]\cos(\nu\phi), \tag{16.154}$$

$$H_z = BJ_\nu\left(\frac{ur}{a}\right)\cos(\nu\phi), \tag{16.155}$$

$$H_r = \left[A\frac{j\omega\epsilon_1}{(u/a)^2}\frac{\nu}{r}J_\nu\left(\frac{ur}{a}\right) - \frac{B\gamma}{u/a}J'_\nu\left(\frac{ur}{a}\right)\right]\cos(\nu\phi), \tag{16.156}$$

$$H_\phi = \left[-A\frac{j\omega\epsilon_1}{u/a}J'_\nu\left(\frac{ur}{a}\right) + \frac{B\gamma}{(u/a)^2}\frac{\nu}{r}J_\nu\left(\frac{ur}{a}\right)\right]\sin(\nu\phi), \tag{16.157}$$

for the core, where $u = k_{c_1}a = a\sqrt{k_1^2 + \gamma^2}$ is called the *normalized transverse phase constant* in the core region, and

$$E_z = CK_\nu\left(\frac{wr}{a}\right)\sin(\nu\phi), \tag{16.158}$$

$$E_r = \left[\frac{C\gamma}{w/a}K'_\nu\left(\frac{wr}{a}\right) - D\frac{j\omega\mu_0}{(w/a)^2}\frac{\nu}{r}K_\nu\left(\frac{wr}{a}\right)\right]\sin(\nu\phi), \tag{16.159}$$

$$E_\phi = \left[\frac{C\gamma}{(w/a)^2}\frac{\nu}{r}K_\nu\left(\frac{wr}{a}\right) - D\frac{j\omega\mu_0}{w/a}K'_\nu\left(\frac{wr}{a}\right)\right]\cos(\nu\phi), \tag{16.160}$$

$$H_z = DK_\nu\left(\frac{wr}{a}\right)\cos(\nu\phi), \tag{16.161}$$

$$H_r = \left[-C\frac{j\omega\epsilon_2}{(w/a)^2}\frac{\nu}{r}K_\nu\left(\frac{wr}{a}\right) + \frac{D\gamma}{w/a}K'_\nu\left(\frac{wr}{a}\right)\right]\cos(\nu\phi), \tag{16.162}$$

$$H_\phi = \left[C\frac{j\omega\epsilon_2}{w/a}K'_\nu\left(\frac{wr}{a}\right) - \frac{D\gamma}{(w/a)^2}\frac{\nu}{r}K_\nu\left(\frac{wr}{a}\right)\right]\sin(\nu\phi), \tag{16.163}$$

for the cladding, where $w = |k_c|a = -a\sqrt{k_2^2 + \gamma^2}$ is called the *normalized transverse attenuation coefficient in the cladding*. If $w = 0$, this determines the *cutoff* condition for the particular field distribution.

### 16.10.1 Boundary conditions

To determine the relationship among the constants $A$, $B$, $C$, and $D$ in Eqs. (16.152)–(16.163) we need to match appropriate tangential and radial components of the fields at the boundary. If we neglect magnetic behavior and put $\mu_1 = \mu_2 = \mu_0$, for the tangential fields

$$\begin{aligned} E_{z_1} &= E_{z_2}, \\ E_{\phi_1} &= E_{\phi_2}, \\ H_{z_1} &= H_{z_2}, \\ H_{\phi_1} &= H_{\phi_2}, \end{aligned} \tag{16.164}$$

and for the radial fields

$$\epsilon_1 E_{r_1} = \epsilon_2 E_{r_2}, \qquad H_{r_1} = H_{r_2}, \tag{16.165}$$

where all fields are evaluated at $r = a$, and the subscripts 1 and 2 refer to core and cladding fields, respectively. It can be shown that use of these relationships in conjunction with Eqs. (16.152)–(16.163) leads to the condition

$$\left[\frac{J'_\nu(u)}{uJ_\nu(u)} + \frac{K'_\nu(w)}{wK_\nu(w)}\right]\left[\frac{\epsilon_1}{\epsilon_2}\frac{J'_\nu(u)}{uJ_\nu(u)} + \frac{K'_\nu(w)}{wK_\nu(w)}\right] = \nu^2\left(\frac{1}{u^2} + \frac{1}{w^2}\right)\left(\frac{\epsilon_2}{\epsilon_2}\frac{1}{u^2} + \frac{1}{w^2}\right). \tag{16.166}$$

Furthermore,

$$u^2 + w^2 = (k_1^2 + \gamma^2)a^2 - (k_2^2 + \gamma^2)a^2 = (k_1^2 - k_2^2)a^2. \tag{16.167}$$

Since $k_1 = 2\pi n_1/\lambda_0$ and $k_2 = 2\pi n_2/\lambda_0$, Eq. (16.167) gives

$$u^2 + w^2 = 2k_1^2 a^2 \Delta, \tag{16.168}$$

where $\Delta$ is the relative refractive-index difference between core and cladding:

$$\Delta = \frac{n_1^2 - n_2^2}{2n_1^2} \simeq \frac{n_1 - n_2}{n_1}. \tag{16.169}$$

A fiber is frequently characterized by its $V$ number, where

$$V = \sqrt{u^2 + w^2}, \tag{16.170}$$

which can be written as

$$V = \frac{2\pi a}{\lambda_0}\sqrt{n_1^2 - n_2^2} = \frac{2\pi a}{\lambda_0}\mathrm{NA} = kn_1 a\sqrt{2\Delta}. \tag{16.171}$$

$V$ is also called the normalized frequency.

Equations (16.166) and (16.168) can be solved for $u$ and $w$ to determine the specific field variations in the fiber.

## 16.11 Modes and field patterns

The modes in a cylindrical dielectric waveguide are described as TE, TM, or hybrid. The hybrid modes fall into two categories, namely HE modes for which the axial electric field $E_z$ is significant compared with $E_r$ or $E_\phi$, and EH modes for which the axial magnetic field $H_z$ is significant compared with $H_r$ or $H_\phi$. These modes are further characterized by two integers, $\nu$ and $\ell$. A mode characterized as $\mathrm{HE}_{12}$, for example, has azimuthal field variations like $\sin\phi$ or $\cos\phi$, and, since $\ell = 2$, has two radial oscillations of the axial electric field within the core. For the TE and TM modes $\nu = 0$. For the $E_z$ field variation given by Eq. (16.152),

$$E_z = AJ_\nu\left(\frac{ur}{a}\right)\sin(\nu\phi), \tag{16.172}$$

the number of radial oscillations of the axial electric field is determined by how many times $J_\nu(x)$ has crossed the axis as $x$ varies from 0 to $ur/a$. Each time the Bessel function oscillates from positive to negative, or vice versa, there is a new value of $ur/a$ where the oscillatory function can merge smoothly with the radially decaying field distribution in the cladding.

The field distributions are easiest to write down for the $\mathrm{TE}_{0\ell}$ and $\mathrm{TM}_{0\ell}$ modes. On setting $\nu = 0$ in Eqs. (16.152)–(16.157) we get the fields of the $\mathrm{TE}_{0\ell}$ mode, for example, in the core:

$$E_z = 0, \tag{16.173}$$

$$E_r = 0, \tag{16.174}$$

$$E_\phi = -B\frac{j\omega\mu_0}{u/a}J_1\left(\frac{ur}{a}\right), \tag{16.175}$$

$$H_z = BJ_0\left(\frac{ur}{a}\right), \tag{16.176}$$

$$H_r = \frac{B\gamma}{u/a}J_1\left(\frac{ur}{a}\right). \tag{16.177}$$

We have made use of the relation $J_0'(x) = -J_1(x)$. The $\mathrm{TE}_{0\ell}$ modes correspond to choosing the constants $A = C = 0$ in Eqs. (16.152)–(16.163).

We could have chosen the $\cos(\nu\phi)$ variation for the $E_z$ field in which we would have got the fields of the $\mathrm{TM}_{0\ell}$ mode by setting $\nu = 0$, which in this case implies $B = D = 0$.

The transverse fields of the mode give rise to a characteristic Poynting vector or intensity distribution, which can be called the *mode* pattern. For the $\mathrm{TE}_{0\ell}$ mode this is

$$S(r) = E_\phi H_r \propto J_1^2\left(\frac{ur}{a}\right). \tag{16.178}$$

Because $J_1(0) = 0$, this is a donut-shaped intensity distribution, as shown in Fig. 16.16. The intensity distribution for the $\mathrm{TM}_{01}$ mode looks the same. We shall see shortly that these are *not* the *fundamental* modes of the step-index fiber. These modes cannot propagate at a given frequency for an arbitrarily small core diameter – they have a *cutoff* frequency. The only mode that has no cutoff frequency is the $\mathrm{HE}_{11}$ mode. To discuss this further, and in order to find a way of grouping modes with similar mode patterns together, we need to work in the *weakly guiding approximation* [4].

## 16.12 The weakly guiding approximation

In the weakly guiding approximation we neglect the difference in refractive index between core and cladding and set $\epsilon_1 = \epsilon_2$. In this case the boundary-condition relation, Eq. (16.166), becomes

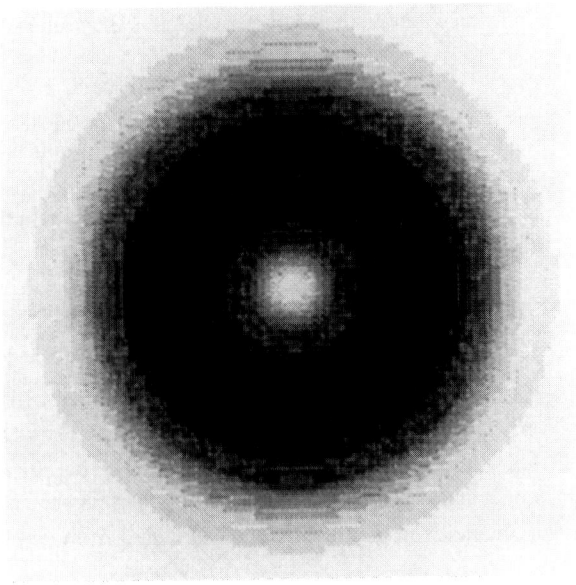

Fig. 16.16 The intensity pattern of the axisymmetric $TE_{01}$ mode.

$$\frac{J'_\nu(u)}{uJ_\nu(u)} + \frac{K'_\nu(w)}{wK_\nu(w)} = \pm\nu\left(\frac{1}{u^2} + \frac{1}{w^2}\right). \tag{16.179}$$

For the $TE_{0\ell}$ and $TM_{0\ell}$ mode this gives

$$\frac{J_1(u)}{uJ_0(u)} + \frac{K_1(w)}{wK_0(w)} = 0, \tag{16.180}$$

where we have made use of the relation $K'_0(w) = -K_1(w)$.

By the use of a series of relationships between Bessel functions, given in Appendix 10, when the positive sign is taken on the right-hand side of Eq. (16.179), the following equation results:

$$\frac{J_{\nu+1}(u)}{uJ_\nu(u)} + \frac{K_{\nu+1}(w)}{wK_\nu(w)} = 0. \tag{16.181}$$

This equation describes the EH modes. When the negative sign is taken in Eq. (16.179), the following equation for the HE modes is obtained:

$$\frac{J_{\nu-1}(u)}{uJ_\nu(u)} - \frac{K_{\nu-1}(w)}{wK_\nu(w)} = 0. \tag{16.182}$$

By the use of the Bessel-function relations given in Appendix 10, Eq. (16.182) can be rearranged into the form

$$\frac{J_{\nu-1}(u)}{uJ_{\nu-2}(u)} + \frac{K_{\nu-1}(w)}{wK_{\nu-2}(w)} = 0. \tag{16.183}$$

If we introduce a new integer $m$, which takes the values

$$m = \begin{cases} 1 & \text{for TE and TM modes,} \\ \nu + 1 & \text{for EH modes,} \\ \nu - 1 & \text{for HE modes,} \end{cases} \tag{16.184}$$

then a unified boundary condition relation that includes Eqs. (16.180), (16.181), and (16.183) results. It is

$$\frac{uJ_{m-1}(u)}{J_m(u)} = -\frac{wK_{m-1}(w)}{K_m(w)}. \tag{16.185}$$

If $m - 1 < 0$, as would be the case for $\mathrm{HE}_{\nu\ell}$ modes with $\nu = 1$, then the Bessel-function relations $J_{-\nu} = (-1)^{\nu} J_{\nu}$ and $K_{-\nu} = K_{\nu}$ would be useful in evaluating Eq. (16.185).

If the $V$ number of the fiber, given by Eqs. (16.170) and (16.171), is known, then the values of $u$ and $w$ that satisfy Eq. (16.106) can be determined. Once these parameters are known, the fields in the fiber can be calculated from Eqs. (16.152)–(16.163).

## 16.13 Mode patterns

In the weakly guiding approximation the Poynting-vector variation within the fiber is again calculated from

$$S = E_{\mathrm{t}} H_{\mathrm{t}}, \tag{16.186}$$

where the transverse fields in the fiber are

$$\begin{aligned} \mathbf{E}_{\mathrm{t}} &= E_r \hat{\mathbf{r}} + E_{\phi} \hat{\boldsymbol{\phi}}, \\ \mathbf{H}_{\mathrm{t}} &= H_r \hat{\mathbf{r}} + H_{\phi} \hat{\boldsymbol{\phi}}. \end{aligned} \tag{16.187}$$

The transverse intensity distribution has a characteristic mode pattern. Similarities between these mode patterns allow groups of TE, TM, and hybrid modes to be grouped together as LP (linearly polarized) modes, as shown in Fig. 16.17. Each LP mode is characterized by the two integers $m$ and $\ell$: $\ell$ describes the number of radial oscillations of the field that occur in the core; and $m$ describes the number of variations of $E_x$ or $E_y$ that occur as $\phi$ varies from 0 to $2\pi$.

In the weakly guiding approximation the transverse field distributions of these LP modes can be written in the form [4]

$$E_x = E_m \frac{J_m(ur/a)}{J_m(u)} \cos(m\phi), \tag{16.188}$$

$$H_y = \frac{E_m}{Z_{\mathrm{core}}}, \tag{16.189}$$

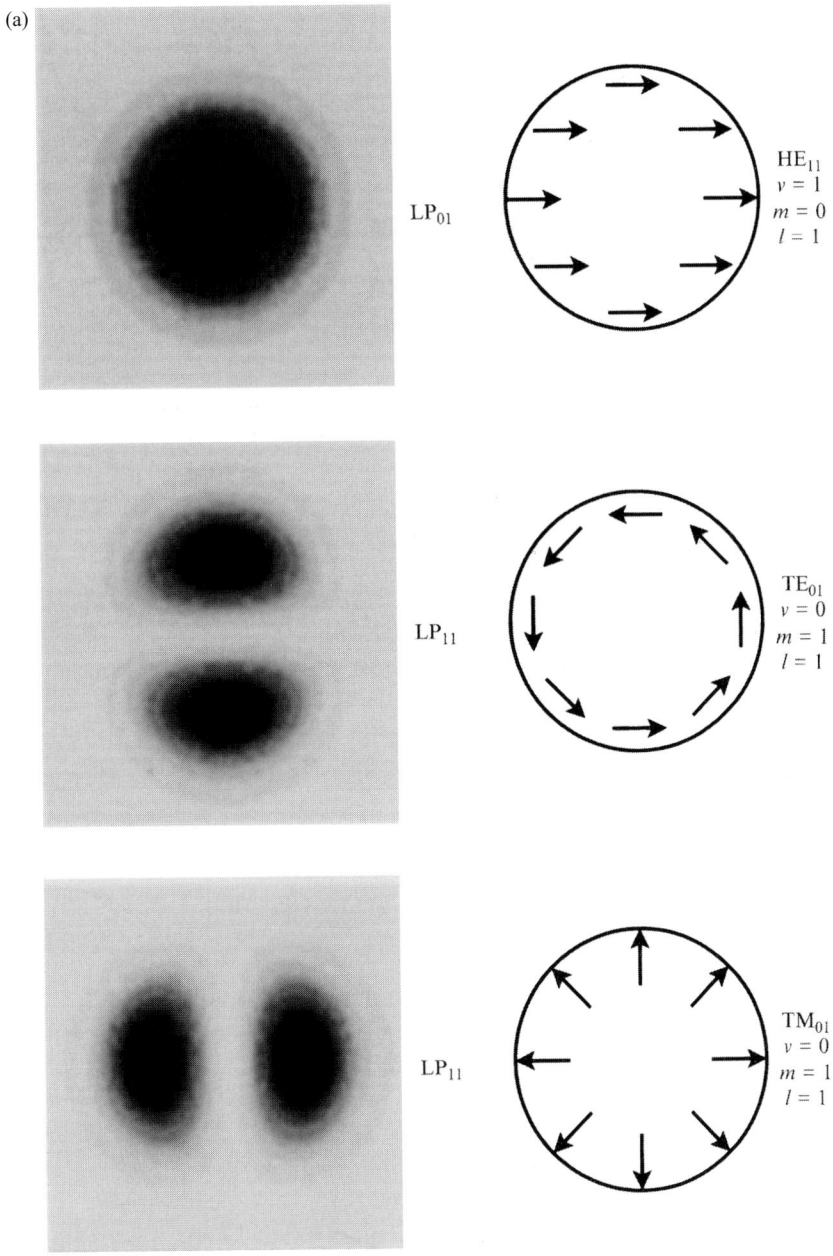


**Fig. 16.17** (a) Characteristic field and intensity patterns of low-order linearly polarized modes. Although the transverse field distributions are different for different TE, TM, and hybrid modes that constitute a given LP mode, their $x$-polarized intensity distributions are qualitatively similar. (b) Intensity patterns of the higher-order LP modes $LP_{22}$ and $LP_{12}$ calculated for $u = 6$.

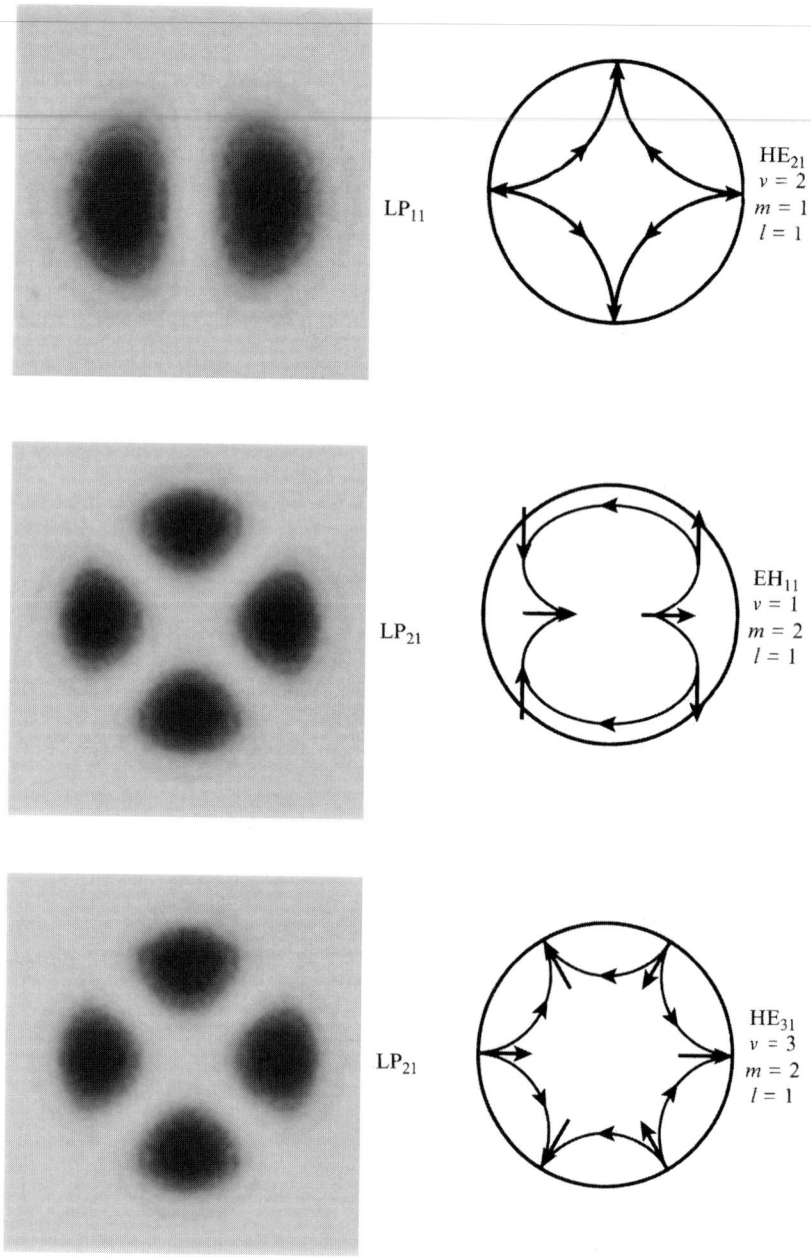


**Fig. 16.17** *(cont.)*

for the core; and

$$E_x = E_m \frac{K_m(ur/a)}{K_m(u)} \cos(m\phi), \tag{16.190}$$

$$H_y = \frac{E_x}{Z_{\text{cladding}}}, \tag{16.191}$$

(b)

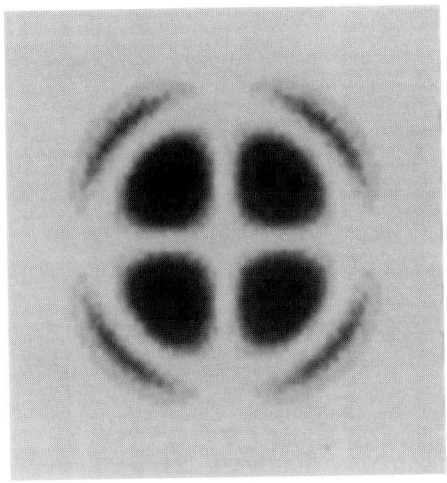

$LP_{12}$

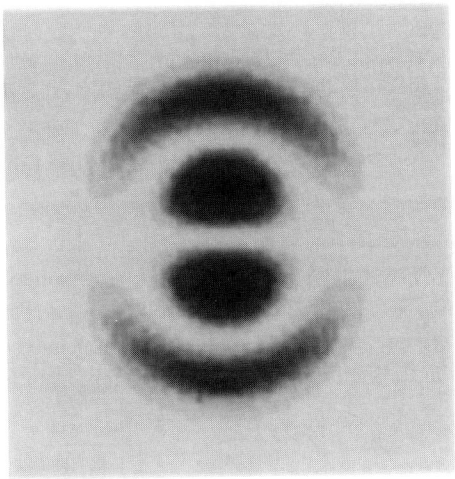

$LP_{22}$

**Fig. 16.17** *(cont.)*

for the cladding; where $E_x$ is a constant, and $Z_{\text{core}}$ and $Z_{\text{cladding}}$ are the impedances of core and cladding, respectively. The designation of these modes as "linearly polarized" does not imply that the transverse field points in a specific unique direction everywhere within the fiber. We have already seen that this was not the case for the $TE_{0\ell}$ mode; see Eqs. (16.173)–(16.177). However, if the Poynting vector obtained from $E_xH_y$ or $-E_yH_x$ is examined, LP modes with the same $m$ will have similar transverse mode patterns, as shown in Fig. 16.17. The field patterns associated with the various modes belonging to a particular $LP_{m\ell}$ mode are actually quite different [5, 6], as is also shown schematically in Fig. 16.17.

The schematic Poynting-vector distributions for the LP modes shown in Fig. 16.17 are the patterns appropriate to either $E_x H_y$ or $-E_y H_x$. For example, for the $TE_{0\ell}$ mode discussed earlier

$$E_x = -E_\phi \sin\phi, \tag{16.192}$$

$$E_y = E_\phi \cos\phi, \tag{16.193}$$

$$H_x = H_r \cos\phi, \tag{16.194}$$

$$H_y = H_r \sin\phi, \tag{16.195}$$

and the $x$-polarized intensity distribution in the core can be written as

$$I_x \propto J_1^2\left(\frac{ur}{a}\right)\sin^2\phi. \tag{16.196}$$

For $LP_{m\ell}$ modes with $\ell \geq 1$ there are four constituent transverse or hybrid modes. This results because each hybrid mode consists of two degenerate modes, corresponding to the choice of the $\cos(\nu\phi)$ or $\sin(\nu\phi)$ azimuthal variation of the fields. The TM or TE modes (for which $\nu = 0$) are singly degenerate and axisymmetric. The $LP_{01}$ mode ($HE_{11}$) is doubly degenerate, corresponding to the choice of the origin for $\phi$. For example, if $\phi = 0$ corresponds to the $x$ direction, the mode will be linearly polarized in the $x$ direction. If $\phi = 0$ corresponds to the $y$ direction, the mode will be linearly polarized in the $y$ direction.

## 16.14 Cutoff frequencies

The field in the cladding decays with increasing distance from the core so long as $w > 0$. However, as $w \to 0$, the propagation constant $\beta$ approaches the value $k_2$ appropriate to the cladding, and the wave is no longer confined. The condition $w = 0$ determines the cutoff condition for the mode in question, which, from Eq. (16.185), becomes

$$\frac{uJ_{m-1}(u)}{J_m(u)} = 0. \tag{16.197}$$

The unique property of the $HE_{11}$ mode can be illustrated with this equation. This is the $LP_{01}$ mode with $m = 0$. Its Poynting-vector distribution is axisymmetric.

Only the $J_0(u)$ Bessel function is non-zero at $u = 0$, so Eq. (16.197) has the solution $u = 0$ in this case. Therefore, the $HE_{11}$ mode has no cutoff frequency – or, viewed another way, it can propagate in a fiber of small core radius $a$. Since

$$u = a\sqrt{k_1^2 + \gamma^2}, \tag{16.198}$$

and at cutoff

$$u = V = k_0 n_1 a\sqrt{2\Delta}, \tag{16.199}$$

$u = 0$ implies that the wave frequency is $\omega = 0$. All other modes in the fiber have a cutoff frequency below which they cannot propagate. The cutoff frequencies are determined by

**Table 16.1** Low-order zeros of Bessel functions

| $j_0$ | $j_1$ | $j_2$ | $j_3$ |
|---|---|---|---|
| $j_{01} = 2.405$ | $j_{11} = 3.832$ | $j_{21} = 5.136$ | $j_{31} = 6.380$ |
| $j_{02} = 5.520$ | $j_{12} = 7.106$ | $j_{22} = 8.417$ | $j_{32} = 9.761$ |
| $j_{03} = 8.654$ | $j_{13} = 10.173$ | $j_{23} = 11.620$ | $j_{33} = 13.015$ |

the zeros $J_{m-1,\ell}$ of the Bessel function $J_{m-1}$, a few values of which are given in Table 16.1. For the $\mathrm{TM}_{0\ell}$ and $\mathrm{TE}_{0\ell}$ modes the cutoff condition is

$$\frac{J_0(u)}{J_1(u)} = 0, \tag{16.200}$$

so $u = j_{0\ell}$ at cutoff.

For the $\mathrm{EH}_{\nu\ell}$ hybrid mode with $\nu \geq 2$ the cutoff condition is

$$\frac{J_\nu(u)}{J_{\nu+1}(u)} = 0, \tag{16.201}$$

so $u = j_{\nu,\ell}$ at cutoff.

For the $\mathrm{HE}_{\nu\ell}$ hybrid modes the cutoff condition is

$$\frac{J_{\nu-2}(u)}{J_{\nu-1}(u)} = 0, \tag{16.202}$$

so $u = j_{\nu-2,\ell}$ at cutoff.

The smallest $j_{m-1,\ell}$ is $j_{01}$, so the $\mathrm{LP}_{11}$ mode has the lowest non-zero cutoff frequency. The cutoff wavelength for this mode is found from

$$k_0 n_1 a\sqrt{2\Delta} = 2.405, \tag{16.203}$$

so the cutoff wavelength is

$$\lambda_c = \frac{2\pi n_1 a\sqrt{2\Delta}}{2.405}. \tag{16.204}$$

If a given mode in the fiber is not below cutoff then its normalized phase constant $u$ has to take values that are bounded by values at which the $J_{m-1}(u)$ and $J_m(u)$ Bessel functions in Eq. (16.106) are equal to zero. This is shown in Fig. 16.18 [4].

The value of $u$ must be between the zeros of $J_{m-1}(u)$ and $J_m(u)$; that is, for modes with $\ell = 0$ or 1,

$$J_{m-1,\ell} < u < J_{m,\ell}. \tag{16.205}$$

If this condition is not met then it is not possible to match the $J$ Bessel function to the decaying $K$ Bessel function at the core boundary. At the lower limit of the permitted values of $u$ in Fig. 16.18, the cutoff frequency is $V = u$, while at the upper limit $V \to \infty$.

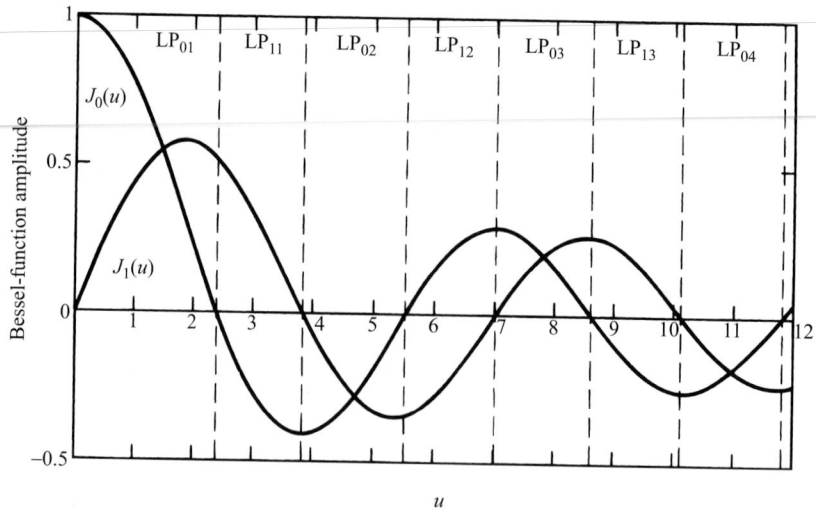


**Fig. 16.18** Superimposed variation of the Bessel functions $J_0(u)$ and $J_1(u)$, showing how the zeros of the Bessel functions bound the values of $u$ where different LP modes operate [4].

### 16.14.1 Example

For a fiber with $a = 3$ μm, $n_1 = 1.5$, and $\Delta = 0.002$ the cutoff wavelength is $\lambda_c = 744$ nm. This fiber would support only a single mode for $\lambda > \lambda_c$.

Alternatively, to design a fiber for single-frequency operation at $\lambda_0 = 1.55$ μm, we want

$$a < \frac{2.405\lambda_0}{2\pi n_1\sqrt{2\Delta}}. \tag{16.206}$$

For $n_1 = 1.5$ and $\Delta = 0.002$, this requires $a < 6.254$ μm.

A fiber satisfying this condition would be a single-mode fiber for 1.55 μm; it would not necessarily be a single-mode fiber for shorter wavelengths. Single-mode fibers for use in optical communication applications at the popular wavelengths of 810 nm, 1.3 μm, and 1.55 μm all have small core diameters, although for mechanical strength and ease of handling their overall cladding diameter is typically 125 μm. The fiber will have additional protective layers of polymer, fiber, and even metal armoring (for rugged applications such as underground or overhead communications links, or for laying under the ocean). Figure 16.19 shows the construction of various multifiber cables of this kind.

## 16.15 Multimode fibers

If the core of a step-index fiber has a larger radius relative to the operating wavelength than the value given by Eq. (16.206), then more than one mode can propagate in the fiber. The

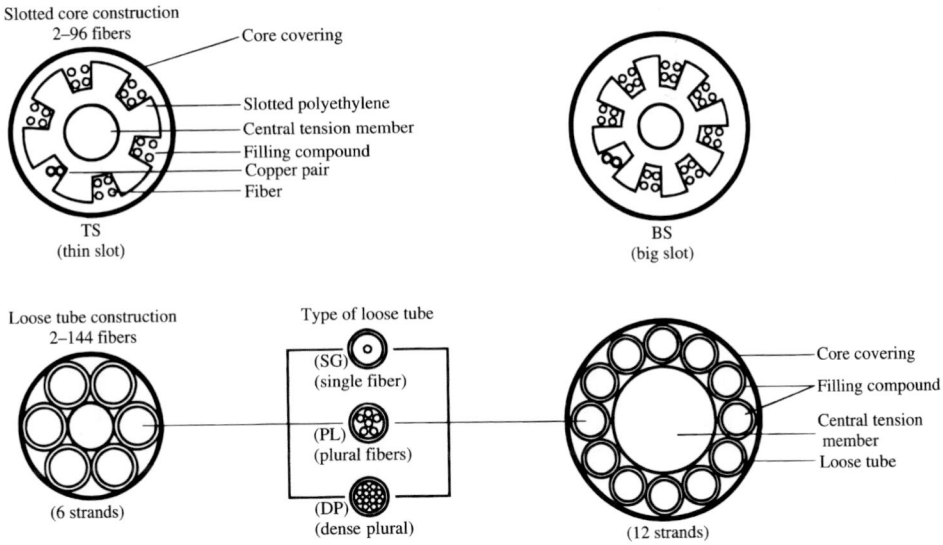


**Fig. 16.19** Construction of multifiber optical cables.

number of modes that can propagate in a given fiber can be determined from its $V$ number. For $V < 2.405$ the fiber will allow only the $HE_{11}$ mode to propagate. For relatively small $V$ numbers the number of modes permitted to propagate can be determined by counting the number of Bessel-function zeros $j_{\nu\ell} < V$. For large $V$ numbers a good estimate of the number $M$ of permitted modes is $M = V^2/2$.

In a multimode fiber most of the modes will be well above cutoff and well confined to the core. The fraction of the mode power that propagates in the cladding decreases markedly for these modes, as is shown in Fig. 16.20.

## 16.16 Fabrication of optical fibers

There are two principal groups of fabrication techniques for the manufacture of low-attenuation optical fibers: methods involving the direct drawing of fiber from molten glass in a crucible; and various vapor-deposition methods in which a "preform" with appropriate graded-index properties is produced and then drawn into a fiber. All these fabrication procedures rely on the use of highly pure starting material, whether this be a borosilicate glass or silica, with added dopants. The brief discussion here can be supplemented with information from more specialized texts [6–9].

Figure 16.21 shows schematically the double-crucible method for the continuous production of clad fibers. The starting material can be fed in either in powder form or as preformed glass rods. This technique cannot be used for high-melting-point glasses, such

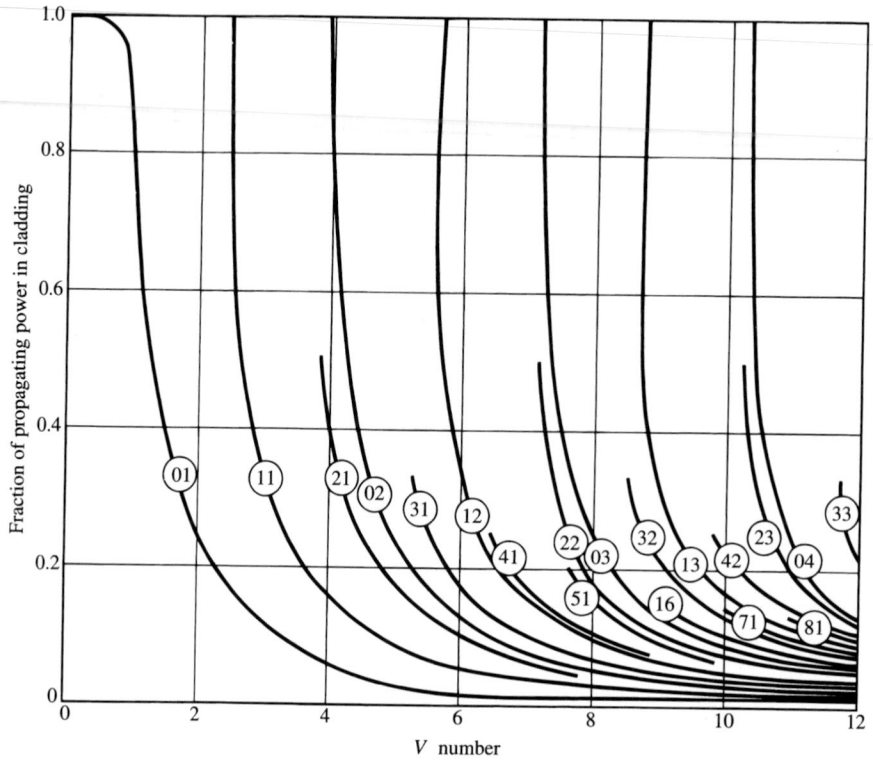


**Fig. 16.20** The fraction of the power that propagates in the cladding for various LP modes as a function of the *V* number [4].

as fused silica. The core and cladding materials can be doped to give the core a higher index than the cladding, for example, by doping with tantalum and sodium, respectively. By varying the drawing temperature, the degree of interdiffusion between core and cladding can be controlled to give a weakly guiding "step-index" or "graded-index" fiber. The addition of thallium to the core, followed by its diffusion into the cladding during the drawing process, also gives a graded-index fiber. The double-crucible method allows the continuous production of fiber with attenuation as low as 5 dB $km^{-1}$.

The production of single-mode versus multimode fiber is determined by the size of the drawing aperture from the inner crucible, the temperature of the molten glass, and the rate at which the fiber is drawn.

For the production of the lowest-attenuation fibers based on silica doped with germanium, fluorine, or phosphorus, particularly for the manufacture of single-mode fiber, a doped preform is fabricated by some form of vapor deposition. The three principal vapor-deposition processes are shown schematically in Fig. 16.22.

In the process of outside chemical vapor deposition (OCVD) materials such as $SiCl_4$, $GeCl_4$, $POCl_3$, $BBr_3$, and $BCl_3$ are oxidized in a burner and deposited on a rotating silica mandrel as a "soot." By varying the vapor to be decomposed, layers of soot particles can

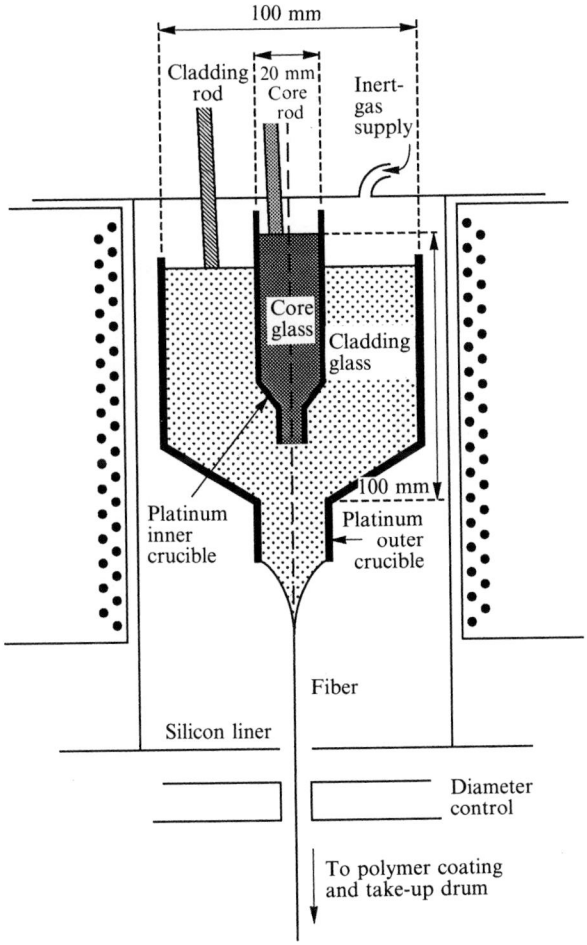


Fig. 16.21 A schematic illustration of the double-crucible method for fabricating optical fibers.

be built up, layer upon layer, to generate the final refractive-index profile desired. The mandrel, with soot over-layers, is then melted into a preform in a furnace. The preform can then be heated and drawn into a fiber.

In the process of inside chemical vapor deposition (ICVD) the chemical vapors are decomposed inside a silica tube by heating them externally or with an internal RF-generated plasma. The tube with its inner layers of soot can then be collapsed and inserted into a second silica (cladding) tube to serve as the core of the final drawn fiber.

Vapor-phase axial deposition (VAD) is a variant on the OCVD process in which the core and cladding glasses are deposited simultaneously onto the end of the preform, which is continuously drawn up into an electric furnace. The perform is rotated as deposition proceeds, and as it is drawn up into the furnace its diameter decreases. The final preform is then taken away and heated for fiber drawing [6].

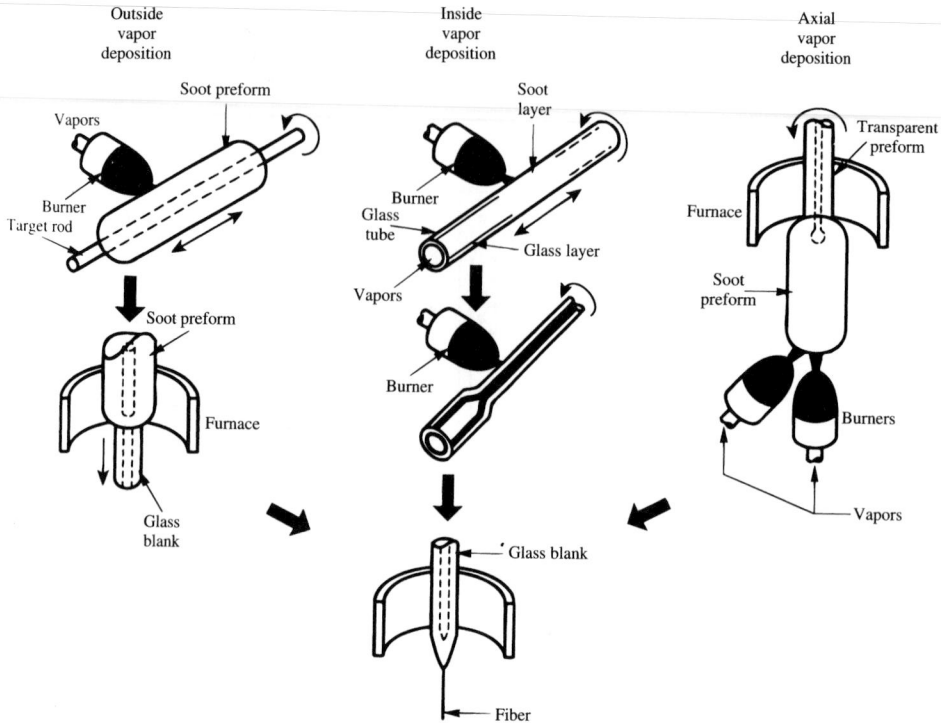


Fig. 16.22 Vapor-deposition methods for fabricating doped fiber "preforms" [10].

## 16.17 Dispersion in optical fibers

The use of optical fibers to transmit data at ever increasing rates, which in the laboratory have already exceeded 10 Gbit $s^{-1}$ over kilometer-long fibers, has required careful consideration of the effects of dispersion in the fiber. In its simplest sense dispersion is the spreading in width of a narrow optical pulse as it travels along the fiber. This spreading is greatest in multimode fibers, where in a ray description each different mode corresponds to a ray trajectory of different length inside the fiber. The different pulse transit times along these different ray paths lead to broadening or splitting of a narrow optical pulse. This is called *modal* dispersion. For example, for a step-index fiber the difference in transit time between a ray that propagates along the axis and one that enters just within the NA is easily shown to be

$$\begin{aligned}\Delta T &= \frac{n_1^2}{n_2}\frac{\ell}{c_0}\Delta \\ &\simeq \frac{n_1 \ell \Delta}{c_0}.\end{aligned} \tag{16.207}$$

For $\Delta = 0.002$ and $n_1 = 1.5$, this modal dispersion is 10 ns $km^{-1}$.

In single-mode fibers modal dispersion does not exist, but dispersion still occurs. This residual dispersion comes from the *material* dispersion $dn/d\lambda$, the variation of the refractive index of core and cladding with wavelength (or frequency), and *waveguide* dispersion, the variation of the waveguide propagation constant $\beta$ with frequency. A detailed analysis of these effects has been given by Gowar [11], but a few general observations can be made.

### 16.17.1 Material dispersion

The transit time of a pulse in a single-mode fiber depends, among other factors, on its group velocity,[2] defined as

$$v_g = \frac{d\omega}{d\beta}. \tag{16.208}$$

In the simplest case of a pulse totally confined to the core,

$$\beta = k_1 = \frac{\omega n_1}{c_0}. \tag{16.209}$$

It is easy to show that

$$v_g = \frac{c_0}{n_1 - \lambda\, dn_1/d\lambda}. \tag{16.210}$$

We can define a *group refractive index*, $n_{g_1}$, such that

$$v_g = \frac{c_0}{n_{g_1}}. \tag{16.211}$$

If a narrow optical pulse is traveling in the core, then its different spectral components will travel with different velocities. The spectral components cover a wavelength range $\Delta\lambda$ that depends on the monochromaticity of the source. For a light-emitting diode (LED) typically $\Delta\lambda \simeq 40$ nm at 800 nm; for a semiconductor laser the equivalent spectral frequency width will be in the range 1–100 MHz. The transit time of the pulse will be

$$T = \frac{\ell}{v_{g_1}} = \left(n_1 - \lambda\frac{dn_1}{d\lambda}\right)\frac{\ell}{c_0}, \tag{16.212}$$

and a narrow pulse (impulse) will spread to a width

$$\Delta T = -\frac{\ell}{c_0}\lambda^2\frac{d^2n}{d\lambda^2}\left(\frac{\Delta\lambda}{\lambda}\right). \tag{16.213}$$

This pulse broadening is minimized for a source with specific spectral purity $\Delta\lambda/\lambda$ when the quantity

$$|Y_m| = \left|\lambda^2\frac{d^2n}{d\lambda^2}\right| \tag{16.214}$$

[2] The group velocity is a measure of the velocity with which the centroid of a pulse, which itself contains a range of frequencies, will travel. It is, therefore, an important parameter in characterizing the performance of an optical communication system that transmits information as a train of light pulses.

is minimized. For doped silica fibers $Y_m = 0$ for wavelengths in the range 1.22 μm < λ < 1.37 μm, the minimum being at $\lambda_{md} = 1.276$ μm for a pure silica fiber. The spreading of the pulse will not actually be zero, even at this operating wavelength, because of the spectral width of the source. The residual pulse broadening that occurs for a source with $\lambda_0 = \lambda_{md}$ can be shown to be [11]

$$\frac{\Delta T}{\ell} = (-)\frac{\lambda_0^3}{8c_0}\left(\frac{d^3n}{d\lambda^3}\right)_{\lambda_0}\left(\frac{\Delta\lambda}{\lambda}\right)^2. \tag{16.215}$$

For pure silica at $\lambda_0 = 1.276$ μm, $\Delta T/\ell = 2 \times 10^{-11}(\Delta\lambda/\lambda)^2$ s m$^{-1}$.

For a high-data-rate optical communication link using a single-mode semiconductor laser running at 10 Gbits s$^{-1}$ the effective laser linewidth is broadened to ≃20 GHz, so $\Delta\lambda/\lambda = 0.00009$ and $\Delta T/\ell \simeq 1.5 \times 10^{-16}$ s km$^{-1}$.

Operation at a wavelength of 1.55 μm, at which the attenuation of silica is a minimum, but $Y_m = -0.01$, leads to more broadening. Even so, this broadening is extremely small. For the optical link described above running at 1.55 μm, $\Delta T/\ell \simeq 3.5$ ps km$^{-1}$. In so-called *dispersion-shifted* fibers, doping, usually of germania, $GeO_2$, can move the minimum dispersion wavelength out to the minimum attenuation wavelength.

### 16.17.2 Waveguide dispersion

When a mode propagates, its propagation constant is $\gamma = j\beta$, where $\beta$ simultaneously satisfies

$$\beta = \sqrt{k_1^2 - (u/a)^2} \tag{16.216}$$

and

$$\beta = \sqrt{(w/a)^2 + k_2^2}. \tag{16.217}$$

At cutoff $w = 0$ and $\beta = k_2$, and the wave propagates completely in the cladding. As the frequency $\omega$ of the wave rises far above the cutoff value $\beta \to k_1$. The variation of $\beta$ with $\omega$ is called a waveguide dispersion curve. At any point on a dispersion curve the *phase* velocity of the mode is $\omega/\beta$, while the *group* velocity is $d\omega/d\beta$. The dispersion curve for a particular mode is bounded by the $\beta$ values appropriate for core and cladding, as shown in Fig. 16.23. The so-called *light lines* have $\omega/\beta$ slopes $c_1$ and $c_2$, where $c_1$ and $c_2$ are the velocities of light in core and cladding, respectively. Waveguide dispersion can be minimized by varying the radial refractive-index distribution, for example, as in the so-called W fiber, whose radial refractive-index variation is shown schematically in Fig. 16.24.

It is also possible to design a single-mode fiber so that the waveguide dispersion and residual material dispersion cancel each other out at wavelengths longer than 1.276 μm; for example, to 1.6 μm in a fiber with core diameter of 4 μm [11].

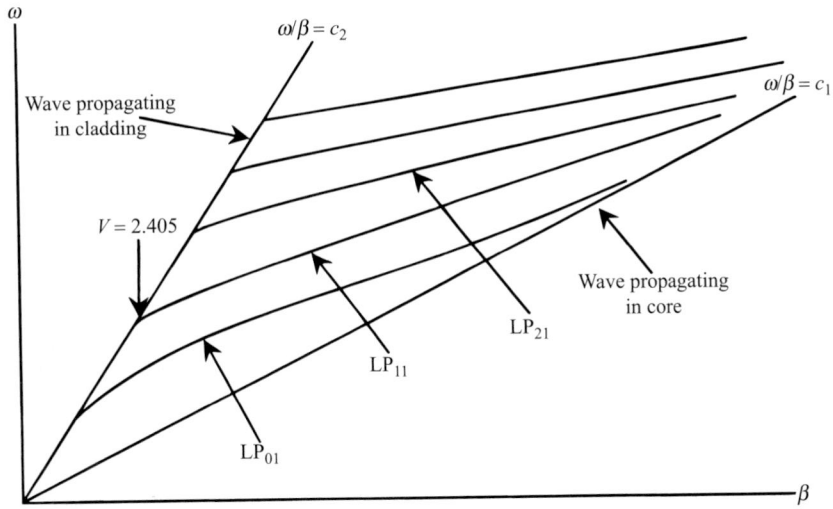


**Fig. 16.23** Schematic variation of dispersion curves of different LP modes in a step-index fiber, neglecting the variation of index with frequency and greatly exaggerating the difference between $n_1$ and $n_2$.

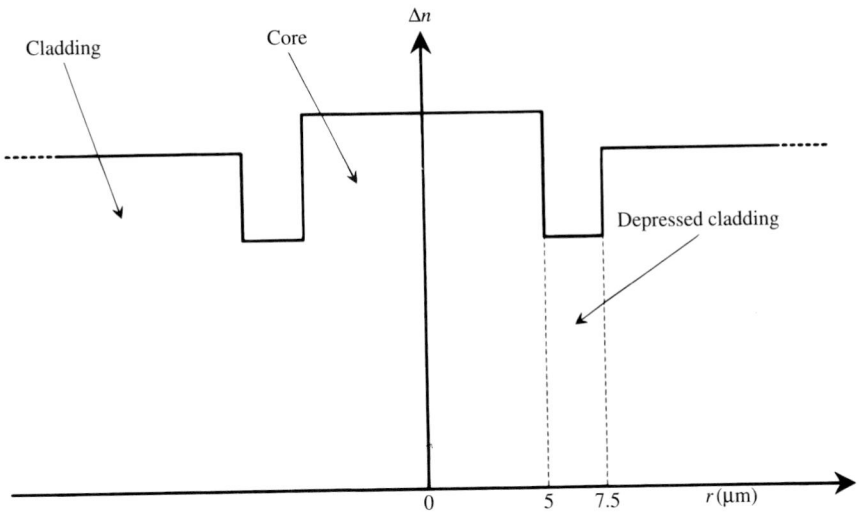


**Fig. 16.24** The schematic radial refractive-index variation of a W fiber.

## 16.18 Holey fibers

Optical fibers can be drawn so that they have a pattern of cylindrical holes running parallel to the axis of the fiber, as shown in Fig. 16.25. The center of the fiber may be solid or may also be an air-filled cylindrical channel. These fibers can be called *photonic crystal fibers*

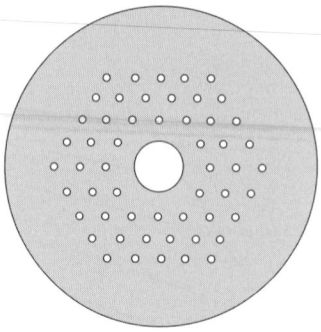

Fig. 16.25 The cylindrical channels parallel to the fiber axis in a photonic crystal or "holey" fiber.

and bear some similarity to the photonic crystals discussed in Chapter 14 [12]. They can be fabricated by stacking together silica rods and tubes to make a preform and drawing the composite arrangement of tubes in a fiber draw tower. If the size, arrangement, and spacing of the cylindrical channels (holes) are chosen appropriately then photonic band gaps can develop, which for example prevent any propagation of light perpendicular to the fiber axis [13, 14]. To achieve light guiding in these fibers the air-filling fraction in the fiber must be high and the spacing between air channels must be small. Even if photonic band gaps are not responsible for the light guiding, as in a *holey* fiber, the propagation characteristics of the fiber can be very interesting. In general, holey fibers guide light efficiently because of the index mismatch between the material of the fiber and the air holes. In these "holey" fibers there is no photonic band gap for propagation in any direction. Simple holey fibers with a simple hexagonal pattern of holes around a solid core can be fabricated with losses comparable to those in conventional single-mode fibers, and have very small bending loss, which can be less than 1/500 that of conventional single-mode fiber [15]. They can also be fabricated to have near zero dispersion at visible and near-infrared wavelengths. Holey fibers fabricated in materials other than silica, such as the chalcogenides, like gallium lanthanum sulfide [16] and tellurium arsenic selenide [17], can extend light guiding to infrared wavelengths in the 5–10-μm region.

## 16.19 Solitons

The material of an optical fiber is not only dispersive, but also nonlinear. For sufficiently intense optical pulses propagating in the fiber the variation of the refractive index with intensity,

$$n = n_0 + n_2 I, \tag{16.218}$$

where $I$ is the intensity within the fiber, affects the dispersion characteristics of the fiber. It can be shown [18] that, if the fiber is being operated at a wavelength for which $dv_g/d\lambda < 0$

(or $d^2\beta/d\omega^2 < 0$), a so-called region of *anomalous*[3] dispersion, effects of nonlinearity and waveguide dispersion can cancel each other out. The consequence of this is the development of a stable pulse, called a *soliton*, that propagates along the fiber without broadening.

If the intensity profile of the pulse with time is written as

$$I(z,t) \propto |E(z,t)e^{j\omega_0 t}|^2, \tag{16.219}$$

where $E(z,t)$ is the electric field envelope, then the behavior of the pulse is dscribed by the *nonlinear Schrödinger* equation, which in normalized form can be written as

$$-j\frac{\partial u}{\partial q} + \frac{1}{2}\frac{\partial^2 u}{\partial s^2} + |u|^2 u = 0, \tag{16.220}$$

where the relationship between the actual and normalized variables is given by

$$u = \tau\sqrt{\frac{n_2\omega}{2Zc_0}\left|\frac{d^2\beta}{d\omega^2}\right|^2}\,E(z,t), \tag{16.221}$$

$$s = (t - z/v_g)/\tau, \tag{16.222}$$

$$q = \left|\frac{d^2\beta}{d\omega^2}\right|\frac{z}{\tau^2}. \tag{16.223}$$

Here $s$ is a dimensionless variable that measures the relative position in time of points of the pulse envelope relative to the location of the point that travels with the group velocity $v_g$.

One solution of Eq. (16.220) is the so-called *fundamental* soliton, which is described by

$$u(q,\tau) = \operatorname{sech}\tau\, e^{-jq/2}. \tag{16.224}$$

Apart from a varying phase factor, this pulse propagates without changing its shape: the effects of dispersion and nonlinearity have cancelled each other out.

## 16.20 Erbium-doped fiber amplifiers

The development of dispersion-shifted fiber and the use of soliton transmission offer solutions to the problem of very-high-data-rate transmission along very long optical fibers. In dispersion-shifted fiber the combined effects of material and waveguide dispersion can be reduced to a very low value [19],

$$D_s = 0.1 \text{ ps km}^{-1} \text{ nm}^{-1}, \tag{16.225}$$

[3] Anomalous dispersion occurs on the long-wavelength side of the minimum dispersion (material dispersion plus waveguide dispersion) in the fiber, which in silica implies $\lambda > 1.3$ μm.

where the actual pulse spreading will be calculated as

$$\frac{\Delta T}{\ell} = D_s \, \Delta\lambda. \tag{16.226}$$

For example, with a bit rate of 2.5 Gbit $s^{-1}$ the maximum transmission distance is about 40,000 km. Unfortunately, dispersion-shifted fibers operating at 1.55 μm have roughly 0.03 dB $km^{-1}$ greater loss than single-mode fibers operating at the attenuation minimum near 1.3 μm in pure silica fibers. On the other hand, the dispersion in the low-attenuation fibers operating at 1.3 μm is greater

$$D_s \simeq 17 \text{ ps km}^{-1} \text{ nm}^{-1}, \tag{16.227}$$

so the maximum transmission distance at 2.5 Gbit $s^{-1}$ is 235 km. Many of the currently deployed long-haul optical fiber links were designed for operation at 1.3 μm; however, as the demand for higher data rates increases the use of dispersion-shifted fiber operating at 1.55 μm becomes very attractive. Because of the higher attenuation in these fibers the optical signal intensity must be periodically regenerated. This has been done traditionally with a *repeater*. This is a module incorporating an optical detector, pulse-reshaping electronics, and a laser to retransmit the bit stream with renewed intensity. Such repeaters placed at intervals along a long fiber link periodically correct for the effects of dispersion, since pulse reshaping prevents these effects being cumulative. This is unnecessary at 1.55 μm because dispersion is negligible – only pulse amplification is necessary. It is very attractive to perform this amplification without a traditional repeater. A semiconductor laser amplifier (a semiconductor laser structure with very low reflectivity or angled facets to prevent self-oscillation) can do the job, but the development of erbium-doped fiber amplifiers (EDFAs) presents a better solution.

Very early in the development of solid-state lasers it was observed that erbium, $Er^{3+}$, doped into various hosts, including glass [20], would operate as an optically pumped laser, and it was demonstrated that laser oscillation could be obtained using a doped glass fiber [21]. It was some time later, in 1987, before Payne and his coworkers demonstrated the first erbium-doped fiber with amplification properties that made it useful in telecommunication applications [22]. The attractive feature of EDFAs is that they can be pumped with the light from semiconductor lasers operating at 800 nm, 980 nm, or 1,480 nm. The energy-level scheme and pumping process responsible for creating gain around 1.5 μm with a 980-nm pump is shown in Fig. 16.26. Because the upper level of the amplifying transition $^4I_{13/2}$ is metastable, with a lifetime of around 11.4 ms, it is possible to create significant gain with modest CW pump powers (tens of milliwatts). The fluorescence peak at 1.54 μm dominates the spectrum of $Er^{3+}$-doped silica glass when the material is pumped with shorter-wavelength radiation [23]. The width of the emission is about 30 nm (FWHM) so semiconductor lasers operating at 1.55 μm are well within the gain curve. Gains in excess of 30 dB are readily obtained. Erbium concentrations typically range up to 1,000 ppm: fibers with higher doping concentration will have a larger absorption coefficient for the pump laser, so a shorter length of fiber is needed in order to give equivalent

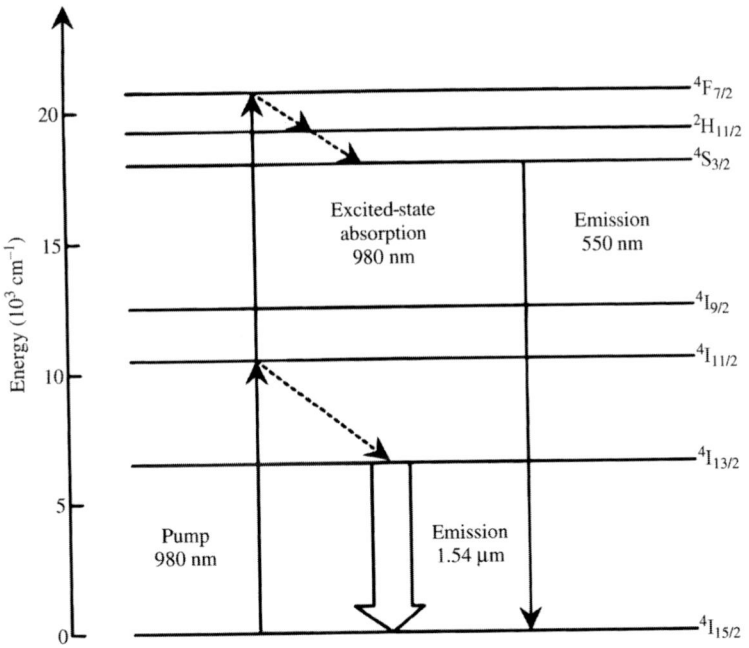


**Fig. 16.26** A partial energy-level diagram of $Er^{3+}$ ions in silica glass showing the 980-nm pump process and the amplifying transition at 1.54 μm. Excited-state absorption can also occur, leading to green fluorescence, and may reduce amplifier efficiency.

gain. Typical lengths of erbium-doped fiber range from meters to hundreds of meters. For further details concerning the optimization of amplifier performance with respect to gain, bandwidth, fiber doping concentration, length, and noise behavior the interested reader should consult the specialized literature [19, 24].

Figure 16.27 shows schematically four configurations in which EFDAs can be operated. The use of isolators can enhance amplifier performance by discouraging feedback-induced oscillation from the couplers, or in between lengths of erbium-doped fiber to reduce the possibility of oscillation or amplified spontaneous emission (ASE). If 980-nm pumps are used then provision must be made for the pump light not to pass through the isolator, since an in-fiber isolator optimized for 1.55 μm will pass 1.48-μm pump light efficiently but not 980 nm.

In summary, the advantages of EDFAs in advanced optical communication networks are many: high gain, low noise, high power-conversion efficiency, high saturation intensity, insensitivity to polarization states, and wide spectral bandwidth. Because of their wide spectral bandwidth, they are particularly attractive in schemes involving *wavelength-division multiplexing* (WDM). In such a scheme multiple channels of information are transmitted along the same fiber at different wavelengths. For example, with a bandwidth of 30 nm at 1.55 μm the frequency bandwidth of an EDFA is 3.74 THz. Therefore, for a series of lasers operating at wavelengths $\lambda_1$, $\lambda_2$, etc., near 1.55 μm, 375

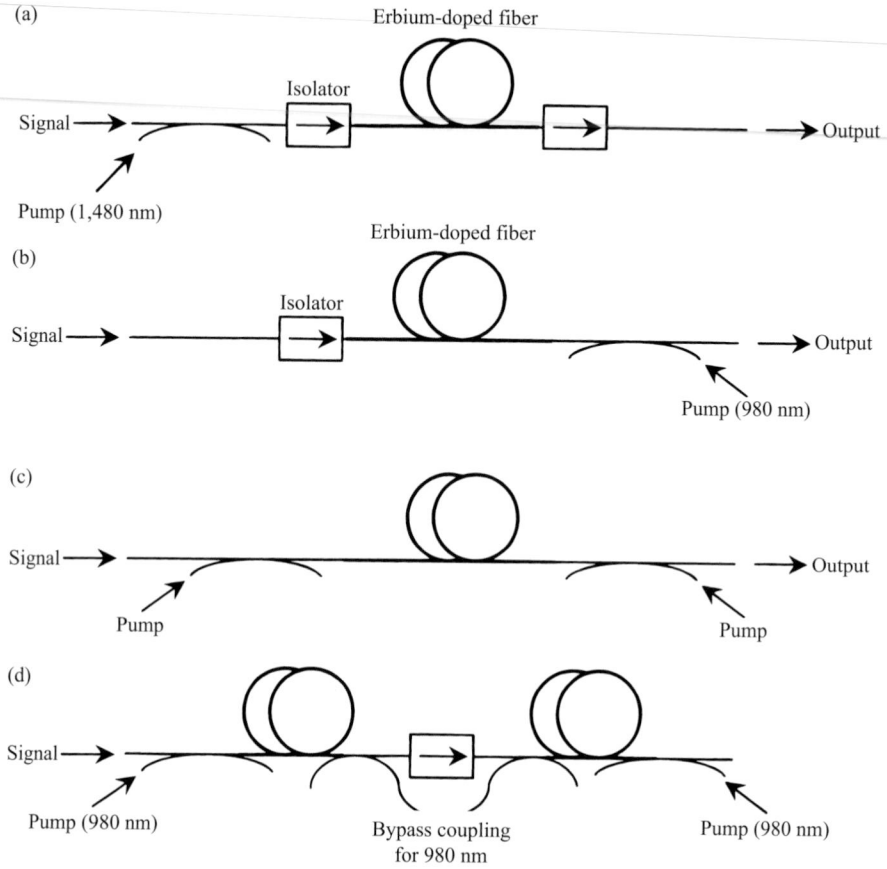


**Fig. 16.27** Four configurations for operating erbium-doped fiber amplifiers: (a) forward pumping, (b) backward pumping, (c) bidirectional pumping, and (d) bidirectional pumping with a bypass coupler so that 980-nm pump radiation can pass the isolator.

2.5 Gbit $s^{-1}$ channels spaced by 10 GHz could be amplified simultaneously. Furthermore, EDFAs can do this with low cross-talk.[4]

Because many fibers that are already in service were designed for use at 1.3 μm, intensive research efforts to produce doped fiber amplifiers with gain at this wavelength are continuing. Praseodymium-, $Pr^{3+}$-, and neodymium-, $Nd^{3+}$-, doped fluorozirconate (ZBLAN) glass has received a great deal of attention in this regard [19]. Amplifers for the 1.3-μm region have not achieved the commercial success of EDFAs [25]. They have lower efficiencies and require higher pump powers.

[4] The effect whereby modulation at one wavelength $\lambda_i$ is transferred to another amplified laser beam at wavelength $\lambda_j$.

## 16.21 Coupling optical sources and detectors to fibers

In order to couple light from a source into a fiber efficiently the light from the source must, in the first instance, be injected within the numerical aperture. In the case of multimode fibers this is generally sufficient to ensure efficient collection and efficient guiding of the injected source light. However, when step-index single-mode fibers are used with a laser source, the injected light must additionally be matched as closely as possible to the lateral intensity profile of the mode that will propagate in the fiber. This cannot be done exactly because the injected light will typically have a Gaussian radial intensity profile, whereas the single-mode lateral intensity profile is a zeroth-order Bessel function with exponentially decaying wings. For optimum coupling a circularly symmetric Gaussian laser beam should be focused so that its beam waist is on, and parallel to, the cleaved end face of the fiber, as shown in Fig. 16.28.

Optimum coupling of the Gaussian beam into the $HE_{11}$ mode occurs when $w_0 = 0.61a$ and the $V$ number of the fiber is $V = 3.8$ [1], in other words close to the value at which the next highest mode, $HE_{12}$, will propagate. In this situation, if Fresnel losses at the fiber entrance face are ignored, the maximum coupling efficiency is 85%. The coupling efficiency from a Gaussian beam to a single-mode or multimode fiber is somewhat complex to calculate in the general case [1]. To optimize coupling the input Gaussian must be matched to the equivalent Gaussian profile of the propagating mode. A useful approximation that can be used for a step-index fiber of core radius $a$ is

$$w_{\text{mode}} = \frac{a}{\sqrt{2\,\ell\text{n}\,V}}. \tag{16.228}$$

If the end of the fiber is illuminated with too large a focused spot, then significant energy enters higher-order modes, which leak out of the cladding. This is readily demonstrated in the laboratory. If a visible laser is focused into a fiber, the first few centimeters of the fiber

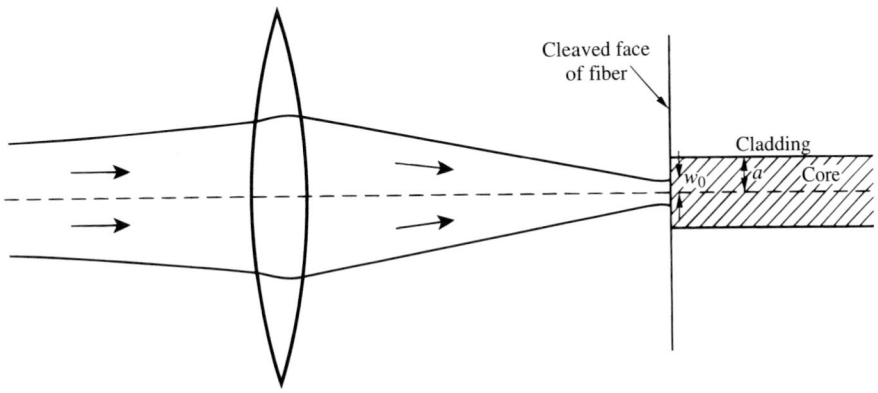


**Fig. 16.28** The geometry for focusing a $TEM_{00}$ Gaussian beam into a single-mode fiber.

will glow brightly as non-propagating modes are excited and then energy leaks from the cladding.

In the arrangement shown in Fig. 16.28 the lens can very conveniently be replaced by a GRIN lens, which allows butt coupling of source to GRIN. The parameters of the GRIN lens must be chosen commensurate with the input laser spot size in order to give an appropriate beam waist at the output face of the GRIN.

If the laser source to be coupled to the fiber is not generating a circularly symmetric Gaussian beam, then anamorphic focusing optics must be used to render the beam circular and focus it onto the end face of the fiber. A typical semiconductor laser may generate an elliptical Gaussian beam, one in which the beam is characterized by different spot sizes in two orthogonal directions. In this case a combination of a cylindrical and spherical lens can be used to give improved focusing. If the emitting facet of the semiconductor laser is sufficiently small, a microfabricated spherical lens can be used to couple source and fiber.

The problem of coupling an optical detector to a fiber efficiently is an easy one. Most optical detectors are of larger area than the modal spot leaving the fiber, so direct butt coupling from fiber to detector is straightforward.

### 16.21.1 Fiber connectors

If two identical fibers are to be joined together this can be done either with mechanical coaxial connectors or through a *fusion* splicer. In either case the flat, orthogonally cleaved ends of the two fibers must be brought together so that their two faces are in very close

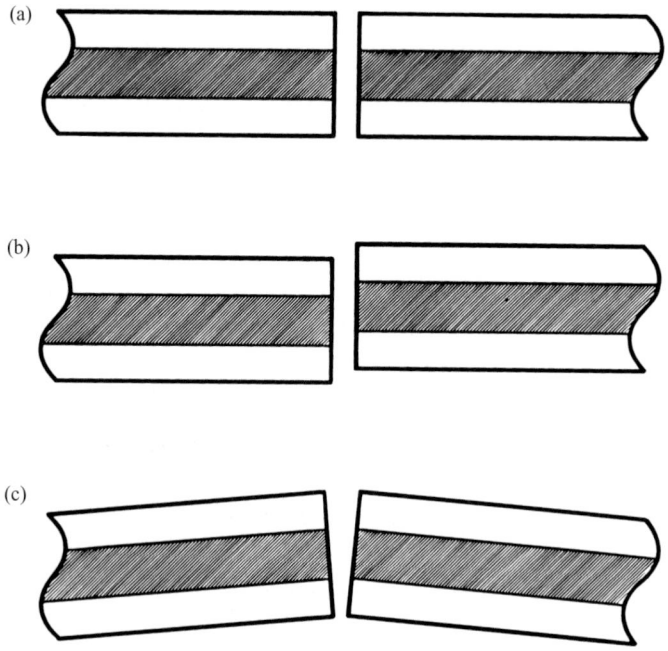


Fig. 16.29 Alignment of optical fibers for connection: (a) optimum, (b) offset, and (c) tilted.

proximity and parallel, and the fibers are coaxial, as shown in Fig. 16.29(a). Any tilt or offset of the fibers will reduce the coupling efficiency [26].

Precision, mechanical connectors are able to couple two fibers with typical losses of 0.1–0.2 dB. Generally, the connector will contain a small quantity of index-matching fluid that is squeezed in the small interface gap between the two fibers.

In a fusion splicer the two fiber ends in Fig. 16.29(a) are brought into contact and external heat quickly applied to melt and fuse the two fibers together. Such splices will have a typical insertion loss of 0.1 dB.

## 16.22 Problems

16.1 A lens of focal length 10 mm is being used to focus light from a point source into a multimode fiber with $n_{\text{core}} = 1.5$ and $n_{\text{cladding}} = 1.48$. What is the practical maximum fraction of the light that can be injected into the fiber and be guided? How are source, lens, and fiber positioned to accomplish this? Use a simple ray model to obtain your answer.

16.2 The curves in Fig. 16.5 have been calculated for a graded-core fiber with $n(r) = 3.5 - \frac{1}{2}n_2 r^2$ and $n_{\text{cladding}} = 1.625$. The core radius is 25 μm. From the curves (a)–(d) estimate the ray invariants $\overline{\beta}$ and $\overline{\ell}$ in each case.

16.3 The Goos–Hänchen shift can be regarded as equivalent to the penetration of the totally internally reflected ray into the cladding, so the "effective" thickness of the slabs is $2d_{\text{eff}}$. Derive an expression for $d_{\text{eff}}$ using the parameters in Eq. (16.60).

16.4 Prove that for an optical fiber being operated at a center wavelength $\lambda_0$ with a source of spectral width $\Delta\lambda$ the minimum pulse broadening as a function of distance caused by material dispersion is

$$\frac{\Delta T}{\ell} = (-)\frac{\lambda_0^3}{8c_0}\left(\frac{d^3 n}{d\lambda^3}\right)_{\lambda_0}\left(\frac{\Delta\lambda}{\lambda}\right)^2.$$

16.5 A fiber with a parabolic index profile has

$$n(r) = n_0 - \tfrac{1}{2}n_2 r^2$$

with $n_0 = 1.5$ and $n_2 = 10^9\ \text{m}^{-2}$. Design a single-lens focusing arrangement that will optimally couple a laser with $\lambda_0 = 180$ nm and $w_0 = 2$ mm into the fiber. How would you modify the focusing arrangement if the laser emitted a Gaussian beam with $w_0 = 5$ μm from the laser facet?

16.6 Repeat Problem 16.5 but design a focusing system using a parabolic-index GRIN lens that is 20 mm long. What parabolic profile parameter $n_2$ for the GRIN lens is needed?

16.7 Use Eqs. (16.105) and (16.106) to plot transverse-mode patterns of the P-wave and S-wave modes in a slab guide. Choose reasonable values for the wavelength and slab dimensions. Use $n_2 = 3$ and $n_2 = 1$.

16.8 Calculate the cutoff wavelengths for the modes $LP_{01}$, $LP_{11}$, and $LP_{21}$ in a cylindrical fiber with $n_{core} = 1.5$, $\Delta = 0.005$, and $r = 3$ μm. What is the $V$ number of this fiber for the fundamental mode at a wavelength half its minimum wavelength for single-mode operation?

16.9 A planar slab guide has $n_2 = 1.5$, $n_1 = 1.498$, and $2d = 10$ μm. Calculate the maximum electric field in the guiding layer for a $TM_{01}$ mode if the total guided flux is 1 W mm$^{-1}$. The wavelength of the injected radiation is 1.3 μm.

16.10 Plot the transverse intensity distribution in a cylindrical guide with parameters similar to those of the slab guide described in Problem 16.9 for the $HE_{22}$ and $EH_{12}$ modes. Choose an appropriate wavelength so that these modes will propagate.

16.11 Evaluate the number of modes, $M$, that will propagate in fibers of core diameters 10 μm and 50 μm, when they are excited by sources of wavelength 1.55 μm and 0.85 μm, respectively. Take $n_{core} = 1.465$ and $n_{cladding} = 1.460$.

16.12 What fraction of the power travels outside the cladding in a cylindrical single-mode step-index guide with $2a = 5$ μm, $\lambda_0 = 1.55$ μm, $n_{core} = 1.48$, and $\Delta = 0.005$?

16.13 Prove Eq. (16.210).

16.14 Prove that Eq. (16.166) is the necessary condition for the field boundary conditions at the core/cladding boundary to be satisfied.

16.15 Plot the normalized propagation constant $\beta\lambda/(2\pi)$ for the modes $HE_{11}$, ($HE_{21}$, $TE_{01}$, and $TM_{01}$), and ($HE_{12}$, $HE_{11}$, and $HE_{31}$) in a step-index fiber with $n_1 = 1.5$, $n_2 = 1.485$, and $a = 2$ μm for $V$ numbers from 0 to 10.

16.16 Repeat Problem 16.15 in the weakly guiding approximation and show that the seven dispersion curves in Problem 16.15 coalesce into three dispersion curves characteristic of the $LP_{01}$, $LP_{11}$, and $LP_{21}$ modes [11].

16.17 Plot dispersion curves for the S-wave modes in a slab waveguide with the same parameters as those in Fig. 16.10.

16.18 If a single-mode fiber is tapered gradually (adiabatically), then it makes a transition from being a silica-core/silica-cladding fiber to being effectively a silica-core/air-cladding fiber. In the process the taper can go from single mode to multimode, and back to single mode again [27]. Model this process for a step-index fiber with $n_1 = 1.47$ and $n_2 = 1.458$ as it is tapered from an initial core diameter of 5 μm and outside diameter of 125 μm down to a total diameter of 100 nm. Use a wavelength of 1.55 μm.

## References

[1] A. W. Snyder and J. D. Love, *Optical Waveguide Theory*, London: Chapman and Hall, 1983.

[2] A. W. Snyder, "Leaky-ray theory of optical waveguides of circular cross section," *Appl. Phys.* **4**, 273–298, 1974.

[3] A. W. Snyder and J. D. Love, "The Goos–Hänchen shift," *Appl. Opt.* **15**, 236–238, 1976.

[4] D. Gloge, "Weakly guiding fibers," *Appl. Opt.* **10**, 2252–2258, 1971.

[5] T. Okoshi, *Optical Fibers*, New York: Academic Press, 1982.

[6] P. K. Cheo, *Fiber Optics and Optoelectronics*, 2nd edn., Englewood Cliffs, NJ: Prentice Hall, 1990.

[7] J. E. Midwinter, *Optical Fibers for Transmission*, New York: Wiley, 1979.

[8] S. E. Miller and A. G. Chynoweth, eds., *Optical Fiber Telecommunications*, New York: Academic, 1979.

[9] P. C. Schultz, "Progress in optical waveguide process and materials," *Appl. Opt.* **18**, 3684–3693, 1979.

[10] D. B. Keck, "Single-mode fibers out perform multimode cables," *IEEE Spectrum* March, 30–37, 1983.

[11] J. Gowar, *Optical Communication Systems*, Englewood Cliffs, NJ: Prentice Hall, 1984.

[12] T. A Birks, P. J. Roberts, P. St. J. Russell, D. M. Atkin, and T. J. Shepherd, "Full 2-D photonic band gaps in silica/air structures," *Electron. Lett.* **31**, 1941, 1995.

[13] J. C. Knight, J. Broeng, T. A. Birks, and P. St. J. Russell, "Photonic band gap guidance in optical fibers," *Science* **282**, 1476–1478, 1998.

[14] R. F. Cregan, B. J. Mangan, J. C. Knight, T. A. Birks, P. St. J. Russell, P. J. Roberts, and D. C. Allan, "Single-mode photonic band gap guidance of light in air," *Science* **285**, 1537–1539, 1999.

[15] Y. Bing, K. Ohsono, Y. Kurosawa, T. Kumagai, and M. Tachikura, "Low-loss holey fiber," *Hitachi Cable Rev.* **24**, 1–4, 2005.

[16] T. M. Monro, Y. D. West, D. W. Hewak, N. G. R. Broderick, and D. J. Richardson, "Chalcogenide holey fibres," *Electron. Lett.* **36**, 1998–2000, 2000.

[17] F. Désévédavy, G. Renversez, J. Troles, P. Houizot, L. Brilland, I. Vasilief, Q. Coulombier, N. Traynor, F. Smektala, and J.–L. Adam, "Chalcogenide glass hollow core photonic crystal fibers," *Opt. Mater.* **32**, 1532–1539, 2012.

[18] H. A. Haus, *Waves and Fields in Optoelectronics*, Englewood Cliffs, NJ: Prentice-Hall, 1984.

[19] A. Bjarklev, *Optical Fiber Amplifiers: Design and System Applications*, Boston, MA: Artech House, 1993.

[20] E. Snitzer and R. Woodcock, "$Yb^{3+}$–$Er^{3+}$ glass laser," *Appl. Phys. Lett.* **6**, 45–46, 1965.

[21] C. J. Koester and E. Snitzer, "Amplification in a fiber laser," *Appl. Opt.* **3**, 1182–1186, 1964.

[22] R. J. Means, L. Reekie, I. M. Jauncy, and D. N. Payne, "High-gain rare-earth-doped fiber amplifier operating at 1.54 μm," *Proceedings of the Conference on Optical Fiber Communications, OFC '87*, Vol. 3, Washington, DC, 1987, p. 187.

[23] B. J Ainslie, "A review of the fabrication and properties of erbium-doped fibers for optical amplifiers," *J. Lightwave Technol.* **9**, 220–227, 1991.

[24] *Journal of Lightwave Technology, Special Issue on Optical Amplifiers*, Vol. 9, No. 2, 1991.

[25] P. C. Becker, N. A. Olsson, and J. R. Simpson, *Erbium-Doped Fiber Amplifiers: Fundamentals and Technology*. Amsterdam: Academic Press/Elsevier, 1999.

[26] D. Marcuse, "Loss analysis of single-mode fiber splices," *Bell. Sys. Tech. J.* **56**, 703–718, 1977.

[27] A. J. Fielding, K. Edinger, and C. C. Davis, "Experimental observation of mode evolution in single-mode tapered optical fibers," *J. Lightwave Technol.* **17**, 1649–1656, 1999.

# 17 The optics of anisotropic media

## 17.1 Introduction

In this chapter we discuss wave propagation in anisotropic media. We shall see that in such media the electric vector of a propagating wave is not in general parallel to its polarization direction, which is defined by the direction of its electric displacement vector. Further, for propagation of plane waves in a particular direction through an anisotropic medium two distinct possible polarization directions exist, and waves having these polarization directions propagate with different velocities. We shall discuss an ellipsoidal surface called the indicatrix and show how with its aid the allowed polarization directions and their corresponding refractive indices can be determined for wave propagation in a given direction. Other three-dimensional surfaces related to the indicatrix and their use in describing different optical properties of anisotropic media are also discussed. We shall concentrate our attention primarily on uniaxial crystals, which have optical properties that can be referred to an indicatrix with two equal axes, and will discuss how such crystals can be used to control the polarization characteristics of light.

Important anisotropic optical media are generally crystalline, and their optical properties are closely related to various symmetry properties possessed by crystals. To assist the reader who is not familiar with basic ideas of crystal symmetry, Appendix 8 summarizes a number of aspects of this subject that should be helpful in reading this chapter and some of those succeeding it.

## 17.2 The dielectric tensor

In an isotropic medium the propagation characteristics of electromagnetic waves are independent of their propagation direction. This generally implies that there is no direction within such a medium that is any different from any other. Clearly then, we can class gases and liquids, but not liquid crystals, as isotropic media, provided that there are no externally applied fields present. Such a field would, of course, imply the existence of a unique direction in the medium – that of the field. As an example of a situation in which an isotropic medium becomes anisotropic in an external field, we mention the case of a gas in a magnetic field, where the gas changes the polarization characteristics of a wave that propagates in the field direction. This phenomenon is called the Faraday effect. In most circumstances, cubic crystals of the highest symmetry, namely crystal classes m3m and 432, behave as isotropic media.

In an isotropic medium the electric displacement vector **D** and its associated electric field **E** are parallel; in other words, we write

$$\mathbf{D} = \epsilon_{\mathrm{r}}\epsilon_0\mathbf{E}, \tag{17.1}$$

where $\epsilon_{\mathrm{r}}$ is the scalar dielectric constant, which in the general case is a function of frequency. This is equivalent to saying that the polarization induced by the field and the field itself are parallel:

$$\mathbf{P} = \epsilon_0\chi\mathbf{E}, \tag{17.2}$$

where $\chi$ is the scalar susceptibility. We will restrict ourselves in what follows to materials that are neither absorbing nor amplifying, so that both $\epsilon_{\mathrm{r}}$ and $\chi$ are real.

In an anisotropic medium **D** and **E** are no longer necessarily parallel, and we write

$$\mathbf{D} = \overline{\overline{\epsilon}}_{\mathrm{r}}\epsilon_0\mathbf{E}. \tag{17.3}$$

where $\overline{\overline{\epsilon}}_{\mathrm{r}}$ is the dielectric tensor,[1] which in matrix form referred to three arbitrary orthogonal axes is

$$\overline{\overline{\epsilon}}_{\mathrm{r}} \equiv \begin{pmatrix} \epsilon_{xx} & \epsilon_{xy} & \epsilon_{xz} \\ \epsilon_{yx} & \epsilon_{yy} & \epsilon_{yz} \\ \epsilon_{zx} & \epsilon_{zy} & \epsilon_{zz} \end{pmatrix}. \tag{17.4}$$

So, for example,

$$\begin{aligned} D_x &= \epsilon_0(\epsilon_{xx}E_x + \epsilon_{xy}E_y + \epsilon_{xz}E_z), \\ D_y &= \epsilon_0(\epsilon_{yx}E_x + \epsilon_{yy}E_y + \epsilon_{yz}E_z), \\ D_z &= \epsilon_0(\epsilon_{zx}E_x + \epsilon_{zy}E_y + \epsilon_{zz}E_z). \end{aligned} \tag{17.5}$$

By making the appropriate choice of axes the dielectric tensor can be diagonalized. With this choice of axes, called the *principal axes* of the material, Eq. (17.5) in matrix form becomes

$$\begin{pmatrix} D_x \\ D_y \\ D_z \end{pmatrix} = \epsilon_0 \begin{pmatrix} \epsilon_x & 0 & 0 \\ 0 & \epsilon_y & 0 \\ 0 & 0 & \epsilon_z \end{pmatrix} \begin{pmatrix} E_x \\ E_y \\ E_z \end{pmatrix}, \tag{17.6}$$

where $\epsilon_x$, $\epsilon_y$, and $\epsilon_z$ are called the principal dielectric constants.

Alternatively, we can describe the anisotropic character of the medium with the aid of the susceptibility tensor $\overline{\overline{\chi}}$,

$$\mathbf{P} = \epsilon_0\overline{\overline{\chi}}\,\mathbf{E}, \tag{17.7}$$

[1] For a brief review of some aspects of tensor algebra that should be of value, see Appendix 9.

where $\overline{\overline{\chi}}$ has the following matrix form when referred to three arbitrary orthogonal axes:

$$\overline{\overline{\chi}} \equiv \begin{pmatrix} \chi_{xx} & \chi_{xy} & \chi_{xz} \\ \chi_{yx} & \chi_{yy} & \chi_{yz} \\ \chi_{zx} & \chi_{zy} & \chi_{zz} \end{pmatrix}. \tag{17.8}$$

In the principal coordinate system

$$\overline{\overline{\chi}} \equiv \begin{pmatrix} \chi_x & 0 & 0 \\ 0 & \chi_y & 0 \\ 0 & 0 & \chi_z \end{pmatrix}. \tag{17.9}$$

Since $\mathbf{D} = \epsilon_0 \mathbf{E} + \mathbf{P}$, it is clear that in a principal coordinate system

$$\epsilon_x = 1 + \chi_x, \text{ etc.} \tag{17.10}$$

We can understand why crystals with less than cubic symmetry have susceptibilities that depend on the direction of the applied field by considering the physical relationship between the polarization and the applied field. When a field $\mathbf{E}$ is applied to a crystal, it displaces both electrons and nuclei from their equilibrium positions in the lattice and induces a net dipole movement per unit volume (polarization), which we can write in the form

$$\mathbf{P} = \sum_j N_j e_j \, \Delta \mathbf{r}_j, \tag{17.11}$$

where $N_j$ is the density of species $j$ with charge $e_j$ in the crystal and $\Delta \mathbf{r}_j$ is the displacement of this charged species from its equilibrium position. We should note that, when the applied field is at optical frequencies, only electrons make any significant contribution to this polarization, particularly the most loosely bound outer valence electrons of the ions within the lattice. The ions themselves are too heavy to follow the rapidly oscillating applied field. If the applied electric field has components, $E_x, E_y$, and $E_z$, then in equilibrium we can write

$$E_x = -k_{jx} (\Delta \mathbf{r}_j)_x \tag{17.12}$$

and two other similar equations, where $k_{jx}$ is a restoring force constant appropriate to the $x$ component of the displacement of the charge $j$ from its equilibrium position; thus

$$\Delta \mathbf{r}_j = \left( \frac{E_x}{k_{jx}} \hat{\mathbf{i}} + \frac{E_y}{k_{jy}} \hat{\mathbf{j}} + \frac{E_z}{k_{jz}} \hat{\mathbf{k}} \right), \tag{17.13}$$

which is a vector parallel to $\mathbf{E}$ only if all three force constants are equal. The equality of these force constants for displacement along three orthogonal axes would imply an arrangement with cubic symmetry of the ions in the lattice about the charged particle being considered.

## 17.3 Stored electromagnetic energy in anisotropic media

If we wish the stored energy density in an electromagnetic field to be the same in an anisotropic medium as it is in an isotropic one then we require

$$U = \frac{1}{2}(\mathbf{E}\cdot\mathbf{D} + \mathbf{B}\cdot\mathbf{H}), \tag{17.14}$$

which gives

$$U = \frac{1}{2}(\overline{\overline{\epsilon}}_{\mathrm{r}}\epsilon_0\mathbf{E}\cdot\mathbf{E} + \mathbf{B}\cdot\mathbf{H}). \tag{17.15}$$

Now, in any medium the net power flow into unit volume is[2]

$$\begin{aligned}\frac{\partial U}{\partial t} &= -\nabla\cdot(\mathbf{E}\times\mathbf{H})\\ &= \mathbf{E}\cdot\frac{\partial\mathbf{D}}{\partial t} + \mathbf{H}\cdot\frac{\partial\mathbf{B}}{\partial t}\\ &= \epsilon_0\mathbf{E}\cdot\overline{\overline{\epsilon}}_{\mathrm{r}}\frac{\partial\mathbf{E}}{\partial t} + \mathbf{H}\cdot\frac{\partial\mathbf{B}}{\partial t},\end{aligned} \tag{17.16}$$

where we have assumed that the medium is non-conductive so that $\mathbf{j} = 0$. Equation (17.16) gives the rate of change of stored energy within unit volume, which must also be given by the time derivative of Eq. (17.15):

$$\frac{\partial U}{\partial t} = \frac{1}{2}\left[\epsilon_0\overline{\overline{\epsilon}}_{\mathrm{r}}\frac{\partial}{\partial t}(\mathbf{E}\cdot\mathbf{E}) + \frac{\partial}{\partial t}(\mathbf{B}\cdot\mathbf{H})\right]. \tag{17.17}$$

Most optical crystals are not, or are only very slightly, magnetic, so we can assume that $\mathbf{B} = \mu_0\mathbf{H}$ since $\mu_{\mathrm{r}} \simeq 1$. In this case

$$\frac{1}{2}\frac{\partial}{\partial t}\mathbf{B}\cdot\mathbf{H} = \mathbf{H}\cdot\frac{\partial\mathbf{B}}{\partial t}. \tag{17.18}$$

On comparing the first term on the right-hand side of Eq. (17.16) with the first two terms on the right-hand side of Eq. (17.17) we have

$$\epsilon_0\mathbf{E}\cdot\overline{\overline{\epsilon}}_{\mathrm{r}}\frac{\partial\mathbf{E}}{\partial t} = \frac{1}{2}\left(\epsilon_0\overline{\overline{\epsilon}}_{\mathrm{r}}\mathbf{E}\cdot\frac{\partial\mathbf{E}}{\partial t} + \epsilon_0\overline{\overline{\epsilon}}_{\mathrm{r}}\frac{\partial\mathbf{E}}{\partial t}\cdot\mathbf{E}\right), \tag{17.19}$$

which implies that

$$\overline{\overline{\epsilon}}_{\mathrm{r}}\mathbf{E}\cdot\frac{\partial\mathbf{E}}{\partial t} = \overline{\overline{\epsilon}}_{\mathrm{r}}\frac{\partial\mathbf{E}}{\partial t}\cdot\mathbf{E}. \tag{17.20}$$

[2] See Appendix 5.

Written out in full, Eq. (17.20) is

$$\begin{aligned}
&\epsilon_{xx}E_x\dot{E}_x + \epsilon_{xy}E_y\dot{E}_x + \epsilon_{xz}E_z\dot{E}_x + \epsilon_{yx}E_x\dot{E}_y + \epsilon_{yy}E_y\dot{E}_y \\
&\quad + \epsilon_{yz}E_z\dot{E}_y + \epsilon_{zx}E_x\dot{E}_z + \epsilon_{zy}E_y\dot{E}_z + \epsilon_{zz}E_z\dot{E}_z \\
&= \epsilon_{xx}\dot{E}_xE_x + \epsilon_{xy}\dot{E}_yE_x + \epsilon_{xz}\dot{E}_zE_x + \epsilon_{yx}\dot{E}_xE_y + \epsilon_{yy}\dot{E}_yE_y \\
&\quad + \epsilon_{yz}\dot{E}_zE_y + \epsilon_{zx}\dot{E}_xE_z + \epsilon_{zy}\dot{E}_yE_z + \epsilon_{zz}\dot{E}_zE_z.
\end{aligned} \tag{17.21}$$

Clearly $\epsilon_{xy} = \epsilon_{yx}, \epsilon_{xz} = \epsilon_{zx}$, etc., so the dielectric tensor has only six independent terms. Working in the principal coordinate system means that all the off-diagonal terms of the dielectric tensor become zero, which greatly simplifies consideration of the wave-propagation characteristics of anisotropic crystals. In this case we can write

$$D_x = \epsilon_0\epsilon_xE_x, \quad D_y = \epsilon_0\epsilon_yE_y, \quad D_z = \epsilon_0\epsilon_zE_z, \tag{17.22}$$

where we are writing $\epsilon_x = \epsilon_{xx}$ etc. for simplicity, and the electrical energy density in the crystal becomes

$$U_{\mathrm{E}} = \frac{1}{2}\mathbf{E}\cdot\mathbf{D} = \frac{1}{2\epsilon_0}\left(\frac{D_x^2}{\epsilon_x} + \frac{D_y^2}{\epsilon_y} + \frac{D_z^2}{\epsilon_z}\right). \tag{17.23}$$

Equation (17.23) shows that the electric displacement vectors from a given point that correspond to a constant stored electrical energy describe an ellipsoid.

## 17.4 Propagation of monochromatic plane waves in anisotropic media

Let us assume that a monochromatic plane wave of the form

$$\mathbf{D} = \mathbf{D}_0e^{j(\omega t - \mathbf{k}\cdot\mathbf{r})} = \mathbf{D}_0e^{j\phi} \tag{17.24}$$

can propagate through an anisotropic medium. The direction of $\mathbf{D}$ specifies the direction of polarization of this wave. The wave vector $\mathbf{k}$ of this plane wave is normal to the wave front (the plane where the phase of the wave is everywhere equal) and has magnitude $|\mathbf{k}| = \omega/c$. The phase velocity of the wave is $c$, which is related to the velocity of light *in vacuo*, $c_0$, by the refractive index $n$ experienced by the wave according to $c = c_0/n$. The phase velocity of the wave in a particular wave-vector direction in the crystal can also be written as $c = \omega/|\mathbf{k}|$.

We assume that Maxwell's equations still hold, so that in the absence of currents or free charges

$$\text{div}\,\mathbf{D} = 0, \tag{17.25}$$

$$\text{div}\,\mathbf{B} = 0, \tag{17.26}$$

$$\text{curl}\,\mathbf{E} = -\frac{\partial \mathbf{B}}{\partial t}, \tag{17.27}$$

$$\text{curl}\,\mathbf{H} = \frac{\partial \mathbf{D}}{\partial t}. \tag{17.28}$$

We stress that, because **D** and **E** are now related by a tensor operation, div **E** is no longer zero.

Taking the curl of both sides of Eq. (17.27),

$$\text{curl curl}\,\mathbf{E} = -\frac{\partial}{\partial t}(\text{curl}(\mu_0\mathbf{H})) = -\mu_0\frac{\partial}{\partial t}\left(\frac{\partial \mathbf{D}}{\partial t}\right) = -\mu_0\frac{\partial^2}{\partial t^2}(\epsilon_0\,\overline{\overline{\epsilon}}_{\mathrm{r}}\mathbf{E}), \tag{17.29}$$

and using the vector identity curl curl $\mathbf{E} = \text{grad}(\text{div}\,\mathbf{E}) - \nabla^2\mathbf{E}$ gives

$$\nabla^2\mathbf{E} - \text{grad}(\text{div}\,\mathbf{E}) = \epsilon_0\mu_0\,\overline{\overline{\epsilon}}_{\mathrm{r}}\frac{\partial^2\mathbf{E}}{\partial t^2}. \tag{17.30}$$

In the principal coordinate system the $x$ component of Eq. (17.30) is, for example,

$$\frac{\partial^2 E_x}{\partial x^2} + \frac{\partial^2 E_x}{\partial y^2} + \frac{\partial^2 E_x}{\partial z^2} - \frac{\partial}{\partial x}\left(\frac{\partial E_x}{\partial x} + \frac{\partial E_y}{\partial y} + \frac{\partial E_z}{\partial z}\right) = \epsilon_0\mu_0\epsilon_x\frac{\partial^2 E_x}{\partial t^2}. \tag{17.31}$$

For a plane wave of the form above, again in the principal coordinate system,

$$\epsilon_0\epsilon_x E_x = D_x = (\mathbf{D}_0)_x e^{j[\omega t - (k_x x + k_y y + k_z z)]}, \tag{17.32}$$

where $k_x$, $k_y$, and $k_z$ are the three orthogonal components of the wave vector and $\epsilon_x$ depends on the angular frequency $\omega$. Similar equations apply for $E_y$ and $E_z$. Substituting for $E_x = D_x/(\epsilon_0\epsilon_x)$ in Eq. (17.32) gives

$$\frac{D_x}{\epsilon_x}(k_x^2 + k_y^2 + k_z^2) - k_x\left(\frac{k_x D_x}{\epsilon_x} + \frac{k_y D_y}{\epsilon_y} + \frac{k_z D_z}{\epsilon_z}\right) = \epsilon_0\mu_0\omega^2 D_x \tag{17.33}$$

and two other similar equations. On recognizing that

$$\epsilon_0\mu_0 = \frac{1}{c_0^2} \tag{17.34}$$

and

$$(k_x^2 + k_y^2 + k_z^2) = |\mathbf{k}|^2, \tag{17.35}$$

and writing

$$\frac{\epsilon_0 c_0^2}{|\mathbf{k}|^2}(k_x E_x + k_y E_y + k_z E_z) = P^2, \tag{17.36}$$

Eq. (17.33) becomes

$$D_x = -\frac{k_x P^2}{c^2 - c_x^2}. \tag{17.37}$$

Similarly,

$$D_y = -\frac{k_y P^2}{c^2 - c_y^2}, \tag{17.38}$$

$$D_z = -\frac{k_z P^2}{c^2 - c_z^2}, \tag{17.39}$$

where $c_x = c_0/\epsilon_x^{1/2}$, $c_y = c_0/\epsilon_y^{1/2}$, and $c_z = c_0/\epsilon_z^{1/2}$ are called the principal phase velocities of the crystal. Now, since div $\mathbf{D} = -j(k_x D_x + k_y D_y + k_z D_z) = 0$, we have

$$\frac{k_x^2}{c^2 - c_x^2} + \frac{k_y^2}{c^2 - c_y^2} + \frac{k_z^2}{c^2 - c_z^2} = 0. \tag{17.40}$$

Multiplying both sides of Eq. (17.41) by $c^2$ and rearranging gives

$$\frac{k_x^2 n_x^2}{n^2 - n_x^2} + \frac{k_y^2 n_y^2}{n^2 - n_y^2} + \frac{k_z^2 n_z^2}{n^2 - n_z^2} = 0, \tag{17.41}$$

where we have put $n_x = \sqrt{\epsilon_x}$, $n_y = \sqrt{\epsilon_y}$, and $n_z = \sqrt{\epsilon_z}$; $n_x, n_y$, and $n_z$ are called the principal refractive indices of the crystal. Equations (17.40) and (17.41), which is called Fresnel's equation, are quadratic in $c^2$ and $n^2$, respectively. Thus in general there are two possible solutions $c_1, c_2$ and $n_1, n_2$, respectively, for the phase velocity and refractive index of a monochromatic wave propagating through an anisotropic medium with wave vector $\mathbf{k}$. However, when $\mathbf{k}$ lies in certain specific directions the roots of Eqs. (17.40) and (17.41) become equal. These special directions within the crystal are called *optic axes*. The fact that, for example, $\sqrt{n_1^2}$ has two roots $+n_1$ and $-n_1$ merely corresponds to each solution of Eq. (17.41) allowing a wave to propagate in a given direction with wave vector either $\mathbf{k}$ or $-\mathbf{k}$.

## 17.5 The two possible directions of D for a given wave vector are orthogonal

We can show that the two solutions of Eqs. (17.40) and (17.41) correspond to two different possible linear polarizations of a wave propagating with wave vector $\mathbf{k}$ and that these two solutions have mutually orthogonal polarization. Let us first illustrate this by considering some special cases.

If we send a monochromatic wave traveling in the $z$ direction, but linearly polarized in the $x$ direction into an anisotropic crystal, then, from Maxwell's equations, Eqs. (17.27) and (17.28),

$$\frac{\partial E_x}{\partial z} = -\mu_r \mu_0 \frac{\partial H_y}{\partial t}, \tag{17.42}$$

$$\frac{\partial H_y}{\partial z} = -\epsilon_x \epsilon_0 \frac{\partial E_x}{\partial t}, \tag{17.43}$$

where we have taken the axes $x, y$, and $z$ to correspond to the principal coordinate system of the crystal. Taking the $z$ derivative of Eq. (17.43) and substituting for $\partial H_y/\partial z$ from (17.43) gives

$$\frac{\partial^2 E_x}{\partial z^2} = \mu_r \mu_0 \epsilon_x \epsilon_0 \frac{\partial^2 E_x}{\partial t^2}, \tag{17.44}$$

which is a one-dimensional wave equation with a solution of the form

$$E_x = |\mathbf{E}_0|_x e^{j(\omega t - k_x z)}, \tag{17.45}$$

where $k_x = \omega\sqrt{\mu_r \mu_0 \epsilon_x \epsilon_0}$.

Thus the wave propagates along the $z$ axis with a phase velocity $c_1 = c_0/n_x$, where $n_x^2 = \epsilon_x$.

If we repeat this derivation, but for a wave propagating in the $z$ direction and polarized in the $y$ direction, we find that the wave now propagates with a phase velocity $c_2 = c_0/n_y$, where $n_y^2 = \epsilon_y$. This illustrates, that, at least in this special case, a wave propagating in the $z$ direction of the principal coordinate system has two possible orthogonal allowed linear polarizations, $E_x$ and $E_y$, that travel with respective phase velocities $c_1 = c_0/n_x$ and $c_2 = c_0/n_y$. It is straightforward to show that for propagation in the $x$ direction there are two possible orthogonal linear polarizations, $E_y$ and $E_z$, with corresponding phase velocities $c_0/n_y$ and $c_0/n_z$, with similar behavior for a wave propagating in the $y$ direction. Clearly, in these special cases in which a wave propagates along one principal axis and is polarized along a second, both **D** and **E** are parallel, and the two possible orientations of **D** for a given wave vector are orthogonal. However, for propagation in an arbitrary direction **D** and **E** are no longer parallel. To find the angular relationship between the two directions of **D** which are allowed for a particular wave vector we use Eqs. (17.37), (17.38), and (17.39). If we designate quantities that refer to the two solutions by the subscripts 1 and 2, then

$$\begin{aligned} D_{j_1} &= -\frac{k_j P_1^2}{c_1^2 - c_j^2}, \\ D_{j_2} &= -\frac{k_j P_2^2}{c_2^2 - c_j^2}, \end{aligned} \tag{17.46}$$

with $j = x, y, z$, and

$$\begin{aligned} \mathbf{D}_1 \cdot \mathbf{D}_2 &= \sum_{j=x,y,z} \frac{k_j^2 P_1^2 P_2^2}{(c_1^2 - c_j^2)(c_2^2 - c_j^2)} \\ &= \frac{P_1^2 P_2^2}{c_2^2 - c_1^2} \sum_{j=x,y,z} \frac{k_j^2}{c_1^2 - c_j^2} - \frac{k_j^2}{c_2^2 - c_j^2}, \end{aligned} \tag{17.47}$$

which is zero by virtue of Eq. (17.40). Thus, the two possible directions of **D** for a given wave vector are orthogonal.

## 17.6 Angular relationships involving D, E, H, k, and the Poynting vector S

The electric displacement vector **D** and wave vector **k** are, by definition and by virtue of Eq. (17.25), mutually perpendicular. Further, since the values of **D**, **E**, and **H** are constant over the phase front of a plane wave, the phase factor $\phi$ must be the same for all these three vectors. We can show that this is so, provided that certain angular relationships hold. For **D**, **E**, and **H** with a phase dependence of the form $e^{j(\omega t - \mathbf{k}\cdot\mathbf{r})}$ we can replace the operation $\partial/\partial t$ by multiplication by $j\omega$ and the operation $\partial/\partial x$ by multiplication by $-jk_x$. So, for example,

$$\frac{\partial \mathbf{D}}{\partial t} = j\omega\mathbf{D}, \tag{17.48}$$

$$\text{curl}\,\mathbf{E} = j(\mathbf{E}\times\mathbf{k}), \tag{17.49}$$

which can be verified by writing out both sides in Cartesian coordinates, and

$$\text{div}\,\mathbf{E} = -j(\mathbf{E}\cdot\mathbf{k}), \tag{17.50}$$

with similar relations for curl **H**, etc.

Thus, from Eqs. (17.49) and (17.27),

$$\mathbf{E}\times\mathbf{k} = \omega\mu\mu_0\mathbf{H}, \tag{17.51}$$

and similarly

$$\mathbf{H}\times\mathbf{k} = \omega\mathbf{D}. \tag{17.52}$$

Thus, **H** is normal to **D**, **E**, and **k**, and the three are coplanar. **D** and **E** make an angle $\alpha$ with one another, where

$$\alpha = \arccos\left(\frac{\mathbf{E}\cdot\mathbf{D}}{|\mathbf{E}|\,|\mathbf{D}|}\right) = \arccos\left(\frac{\epsilon_0(\overline{\overline{\epsilon}}_r\mathbf{E})\cdot\mathbf{E}}{|\mathbf{E}|\,|\mathbf{D}|}\right). \tag{17.53}$$

These angular relationships are illustrated in Fig. 17.1. The direction of the Poynting vector $\mathbf{S} = \mathbf{E}\times\mathbf{H}$ is, by definition, perpendicular to both the **E** and the **H** vector. **S** defines the direction of energy flow within the medium, and we identify it as the direction of the *ray* of familiar geometrical optics. Whereas in isotropic media the ray is always parallel to the wave vector, and is therefore perpendicular to the wave front, this is no longer so in anisotropic media, except for propagation along one of the principal axes.

To summarize, we have shown that transverse electromagnetic plane waves can propagate through anisotropic media, but for propagation in a general direction two distinct allowed linear polarizations specified by the direction of **D** can exist for the wave. We have shown that these two allowed polarizations are orthogonal and that the wave propagates with a phase velocity (the velocity of the surface of constant phase – the wave front) that depends on which of these two polarizations it has. Clearly, a wave of arbitrary polarization that enters such an anisotropic medium will not in general correspond to one of the

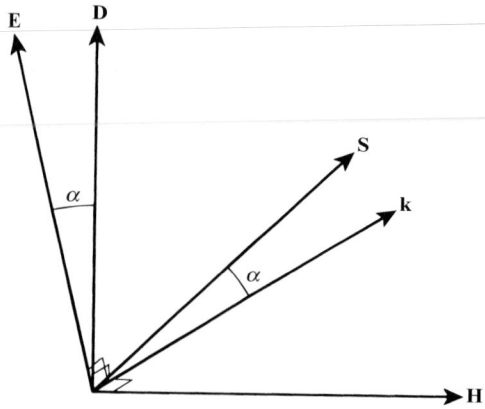


Fig. 17.1 Angular relationship between D, E, H, k, and S vectors in an anisotropic medium.

allowed polarizations, and will therefore be resolved into two linearly polarized components polarized along the allowed directions. Each component propagates with a different phase velocity. All that remains for us to do to characterize fully these allowed polarization directions is to specify their orientation with respect to the principal axes of the medium. This is done with the aid of a geometrical figure called the *index ellipsoid* or *indicatrix*.

## 17.7 The indicatrix

The indicatrix, wave-normal, or index ellipsoid is an ellipsoid with the equation

$$\frac{x^2}{n_x^2} + \frac{y^2}{n_y^2} + \frac{z^2}{n_z^2} = 1, \tag{17.54}$$

which allows us to determine the refractive index for monochromatic plane waves as a function of their direction of polarization. It is, apart from a scale factor, equivalent to the surface mapped out by the **D** vectors corresponding to a constant energy density at a given frequency. This ellipsoid can be visualized as oriented inside a crystal consistent with the symmetry axes of the crystal. For example, in any crystal with perpendicular symmetry axes, such as those belonging to the cubic, tetragonal, hexagonal, trigonal, and orthorhombic crystal systems, the axes of the ellipsoid, which are the principal axes of the crystal, are parallel to the three axes of symmetry of the crystal. For the orientation of the indicatrix to be consistent with the symmetry of the crystal, planes of mirror symmetry within the crystal must coincide with planes of symmetry of the indicatrix, namely the $xy$, $yz$, and $zx$ planes. In the monoclinic system, crystal symmetry is referred to three axes, two of which, namely the $a$ and $c$ axes of crystallographic terminology, intersect at acute and obtuse angles, and a third, the $b$ axis, which is perpendicular to the $a$ and $c$ axes. In such crystals, one of the axes of the indicatrix must coincide with the $b$ axis, but the other two axes can have any orientation, although this is fixed for a given crystal and wavelength.

In triclinic crystals whose symmetry is referred to three unequal-length, non-orthogonal axes, the indicatrix can take any orientation, although this is fixed for a given crystal and wavelength.

We use the geometrical properties of the indicatrix to determine the refractive indices and polarizations of the two monochromatic plane waves that can propagate through the crystal with a given wave vector. This is illustrated in Fig. 17.2, which shows the direction of the wave vector **k** for a monochromatic wave propagating through a crystal, drawn relative to the orientation of the indicatrix in the crystal. The plane surface that is orthogonal to the wave vector, and that passes through the center of the indicatrix, intersects this ellipsoid in an ellipse called the *intersection ellipse*. The semiaxes of this ellipse define the directions of the two allowed **D** polarizations which can propagate through the crystal with the given wave vector **k**, and the lengths of these semiaxes give the refractive indices experienced by these two polarizations. That this interpretation is consistent follows for a number of reasons. For a wave of given intensity propagating through the crystal, the stored energy density must be independent of the polarization direction, otherwise we could either absorb

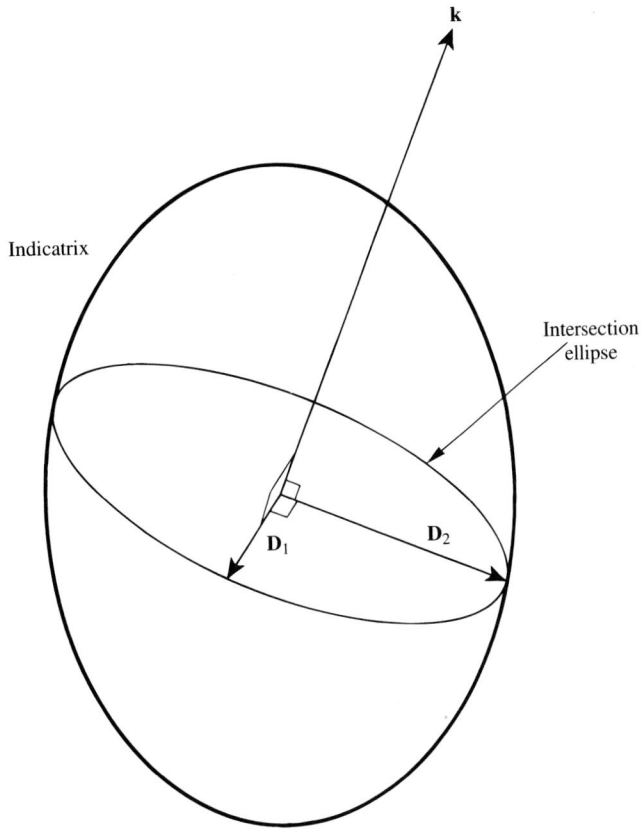


**Fig. 17.2** The ellipsoidal indicatrix showing the wave vector direction **k** and the intersection ellipse that is perpendicular to **k**. The two permitted linear polarization directions $\mathbf{D}_1$ and $\mathbf{D}_2$ lie along the semiaxes of the intersection ellipse.

**Table 17.1** Uniaxial crystal classes

| Hexagonal | Trigonal | Tetragonal |
|---|---|---|
| $\bar{6}2m$ | 3m | $\bar{4}2m$ |
| 6mm | 32 | 4mm |
| 622 | 3 | 422 |
| $\bar{6}$ | | $\bar{4}$ |
| 6 | | 4 |

or extract energy from the wave merely by rotating the crystal. This cannot be so, since we have already specified that the crystal is transparent. Consequently, for a given wave vector the many possible **D** polarizations trace out an ellipse, equivalent to the intersection ellipse. Further, it can be shown that the lengths of the semiaxes of the intersection ellipse corresponding to a given wave vector $\mathbf{k} = k_x\hat{\mathbf{i}} + k_y\hat{\mathbf{j}} + k_z\hat{\mathbf{k}}$ are the two roots $n_1$ and $n_2$ of Fresnel's equation, Eq. (17.41). Thus, only the two **D** vectors parallel to these semiaxes simultaneously satisfy the condition of both being on the ellipse and having the appropriate refractive indices to satisfy Fresnel's equations.

In the general case there are two **k** vector directions through the center of the indicatrix for which the intersection ellipse is a circle. This is a fundamental geometrical property of ellipsoids. These two directions are called the *principal optic axes*, and are fixed for a given crystal and frequency of light. Waves can propagate along these optic axes, with any arbitrary polarization, since in these directions the refractive index is not a function of polarization.

In cubic crystals the indicatrix is a sphere called the isotropic indicatrix; there are no specific optic axes since the indicatrix is anaxial and the propagation of waves is independent of the directions of both **k** and **D**.

In crystals belonging to the tetragonal, hexagonal, and trigonal crystal systems the crystal symmetry requires that $n_x = n_y$, and the indicatrix reduces to an ellipsoid of revolution. In this case there is only one optic axis, oriented along the axis of highest symmetry of the crystal, namely the $z$ axis (or $c$ axis). These crystal classes, listed in Table 17.1, are said to be uniaxial: a discussion of their properties is considerably simpler than for crystals belonging to the less-symmetric orthorhombic, monoclinic, and triclinic crystal systems, which are *biaxial.*

## 17.8 Uniaxial crystals

The equation of the uniaxial indicatrix is

$$\frac{x^2 + y^2}{n_o^2} + \frac{z^2}{n_e^2} = 1, \tag{17.55}$$

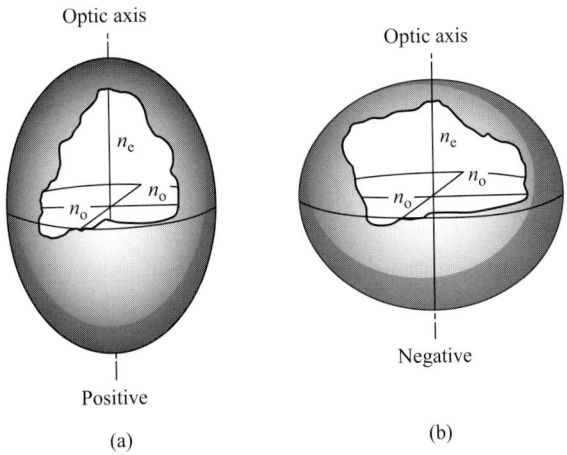


**Fig. 17.3** (a) Indicatrix in a positive crystal, (b) indicatrix in a negative crystal.

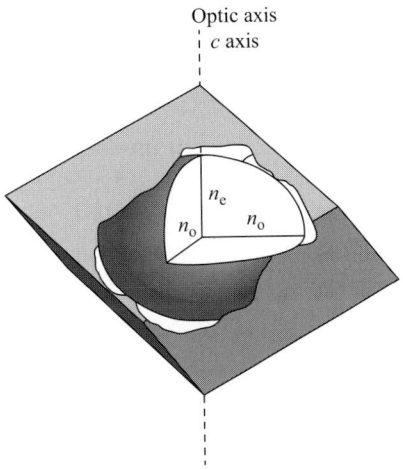


**Fig. 17.4** Rhomboidal calcite crystal showing the orientation of the indicatrix.

where $n_o$ is the index of refraction experienced by waves polarized perpendicular to the optic axis, called *ordinary* or O waves; and $n_e$ is the index of refraction experienced by waves polarized parallel to the optic axis, called *extraordinary* or E waves. If $n_e > n_o$ the indicatrix is a prolate ellipsoid of revolution, as shown in Fig. 17.3(a), and such a crystal is said to be *positive uniaxial*.

If $n_e < n_o$ the indicatrix is an oblate ellipsoid of revolution as shown in Fig. 17.3(b), and the crystal is said to be *negative uniaxial*. Figure 17.4 shows the orientation of this ellipsoid inside a negative uniaxial crystal of calcite.

Because uniaxial crystals have indicatrices that are circularly symmetric about the $z$ (optic) axis, their optical properties depend only on the polar angle $\theta$ which the wave vector $\mathbf{k}$ makes with the optic axis and not on the azimuthal orientation of $\mathbf{k}$ relative to

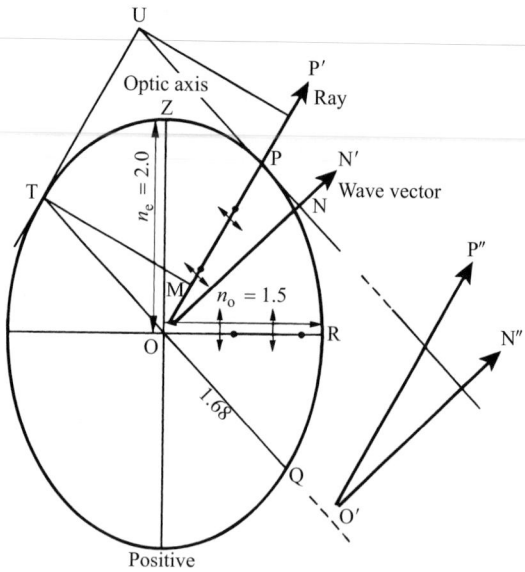


**Fig. 17.5** Section of positive uniaxial indicatrix containing optic axis showing wave-normal and ray directions.

the $x$ and $y$ axes. Thus, we can illustrate all their optical characteristics by considering propagation in any plane containing the optic axis, as shown in Fig. 17.5. For propagation with the wave vector in the direction ON, the two allowed polarizations are as indicated: perpendicular to the optic axis, and in the plane containing ON and the optic axis OZ. The major semiaxis of the intersection ellipse is OQ in this case. The length OQ gives the refractive index for waves polarized parallel to OQ (E waves). The refractive index of the O wave is given by the minor semiaxis of the intersection ellipse, which in a uniaxial crystal is independent of direction and is equal to the length OR. In Fig. 17.5 the O wave with wave vector in the direction ON propagates with phase velocity $c_0/n_o = c_0/1.5$, while the E wave propagates with phase velocity $c_0/\mathrm{OQ} = c_0/1.68$. Thus, in this positive crystal the O wave travels faster than the E wave.

In a uniaxial crystal the O waves have no component of **D** along the $z$ axis; thus, from Eq. (17.7),

$$D_x = \epsilon_0 \epsilon_x E_x, \tag{17.56}$$

$$D_y = \epsilon_0 \epsilon_y E_y. \tag{17.57}$$

In uniaxial crystals, since $\epsilon_x = \epsilon_y$, **D** and **E** are parallel for the O wave, and the O ray is parallel to the O wave vector. However, the ray and wave vector of the E wave are not parallel, except for propagation along, or perpendicular to, the optic axis. We can find the direction of the E ray by a simple geometrical construction shown in Fig. 17.5. The wave vector is in the direction ON, and TQ is the major axis of the intersection ellipse. The tangent to the ellipse at the point T is parallel to the ray direction OP. Note that the tangent to the ellipse at point P is orthogonal to the wave vector. The wave index of refraction of the extraordinary wave with wave vector **k** is given by the length OT. The length TM gives the ray index of refraction, which is a measure of the velocity with which energy flows along the ray OP.

Light propagating along the ray OP as an E wave is not polarized perpendicular to the ray direction; however, the **E** vector of this wave is perpendicular to the ray, parallel to the direction MT in Fig. 17.5. Since light propagating along the ray OP with wave vector **k** travels a distance OP in the time during which the wave front (which is parallel to the direction PN) travels a distance ON, it is clear that the velocity of light along the ray is greater than the wave-normal velocity (the phase velocity). If the wave vector is in the direction OR in Fig. 17.5 then both wave-normal and ray are parallel for the E and O waves. As indicated by the dots, the O wave propagates more quickly in this positive crystal.

## 17.9 Index surfaces

The refractive index $n_e(\theta)$ of an E wave propagating at angle $\theta$ to the optic axis in a uniaxial crystal can be calculated with reference to Fig. 17.6. For this wave the value of $n_e(\theta)$ is given by the length OX in the figure, the major semiaxis of the intersection ellipse. The Cartesian coordinates of the point X relative to the origin O are

$$\begin{aligned} x &= -n_e(\theta)\cos\theta, \\ z &= n_e(\theta)\sin\theta, \\ y &= 0. \end{aligned} \tag{17.58}$$

This point lies on the indicatrix, so

$$\frac{n_e^2(\theta)\cos^2\theta}{n_o^2} + \frac{n_e^2(\theta)\sin^2\theta}{n_e^2} = 1, \tag{17.59}$$

which gives

$$n_e(\theta) = \frac{n_o n_e}{\sqrt{n_e^2\cos^2\theta + n_o^2\sin^2\theta}}. \tag{17.60}$$

We can use this relationship to specify a surface called the extraordinary index surface, which shows geometrically the index of refraction of extraordinary waves in a uniaxial crystal as a function of their direction of propagation. The Cartesian coordinates on this surface must satisfy

$$n_e^2(\theta) = x^2 + y^2 + z^2 \tag{17.61}$$

and also

$$\sin^2\theta = \frac{x^2 + y^2}{x^2 + y^2 + z^2}, \tag{17.62}$$

$$\cos^2\theta = \frac{z^2}{x^2 + y^2 + z^2}. \tag{17.63}$$

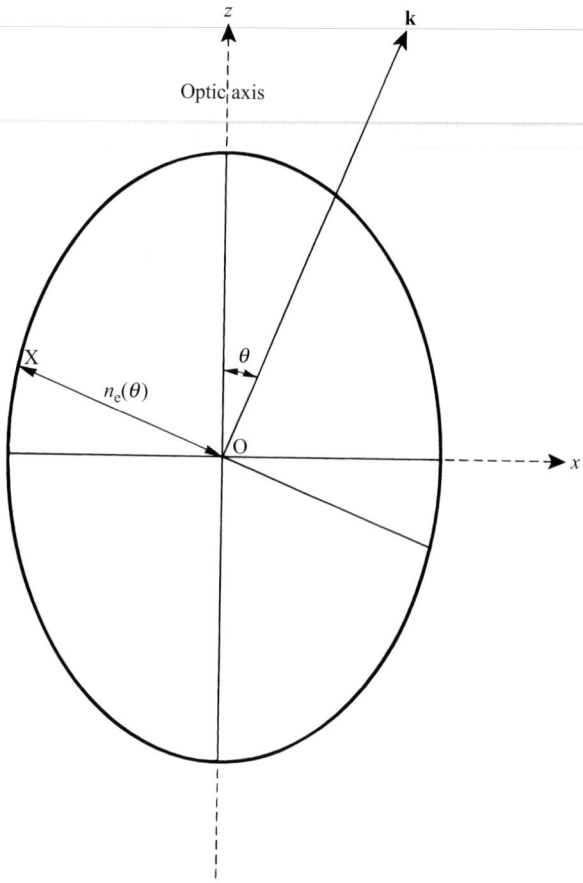


Fig. 17.6 Section of uniaxial indicatrix containing the optic axis.

Substituting from Eqs. (17.61), (17.62), and (17.63) into Eq. (17.60) gives

$$\frac{x^2 + y^2}{n_e^2} + \frac{z^2}{n_o^2} = 1. \tag{17.64}$$

We might have expected this, since for an E wave propagating perpendicular to the optic axis $z = 0$ and $\sqrt{x^2 + y^2} = n_e(0) = n_e$. For an E wave propagating along the $z$ axis (which in this case is actually also an O wave) $x^2 + y^2 = 0$ and $z = n_e(90) = n_o$.

The index surface for O waves is, of course, a sphere, since the index of refraction of such waves is independent of their propagation direction. The equation of this surface is

$$x^2 + y^2 + z^2 = n_o^2. \tag{17.65}$$

Sections of the ordinary- and extraordinary-index surfaces for both positive and negative uniaxial crystals which contain the optic axis are shown in Fig. 17.7. Such sections are called principal sections.

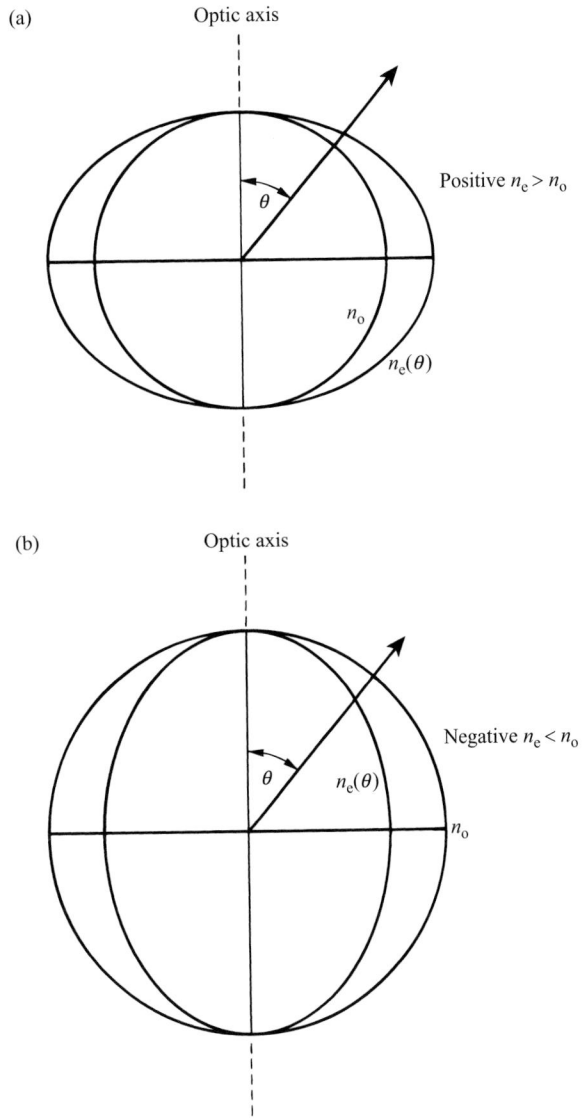


**Fig. 17.7** Ordinary and extraordinatry index surfaces: (a) positive crystal; (b) negative crystal.

## 17.10 Other surfaces related to the uniaxial indicatrix

There are several other surfaces related to the uniaxial indicatrix whose geometrical properties describe various aspects of the propagation characteristics of uniaxial crystals.

The wave-velocity surface describes the velocity of waves in their direction of propagation; this surface is, like the index surface, a two-shelled surface. The wave-velocity surface for the O waves is clearly

$$n_o^2(x^2 + y^2 + z^2) = c_0^2, \tag{17.66}$$

while for the E waves

$$n_e^2(\theta)(x^2 + y^2 + z^2) = c_0^2, \tag{17.67}$$

which, since

$$\sin\theta = \frac{x^2 + y^2}{x^2 + y^2 + z^2}, \quad \cos^2\theta = \frac{z^2}{x^2 + y^2 + z^2},$$

from Eq. (17.60) gives

$$\frac{x^2 + y^2}{n_e^2} + \frac{z^2}{n_o^2} = \frac{(x^2 + y^2 + z^2)^2}{c_0^2}. \tag{17.68}$$

This surface is not an ellipsoid but an ovaloid of revolution.

The ray-velocity surface describes the velocity of rays in their direction of propagation, which for E rays is, as we have mentioned already, not the same as the direction of their wave vectors except for propagation along a principal axis. Along the $x$ and $y$ principal axes the extraordinary ray and wave velocities are both $c_0/n_e$, while along the $z$ axis they both have the value $c_0/n_o$. It is left as an exercise to the reader to show that the ray-velocity surface for extraordinary rays is an ellipsoid of revolution with semiaxes $c_0/n_o$ and $c_0/n_e$ satisfying the equation

$$\frac{x^2 + y^2}{1/n_e^2} + \frac{z^2}{1/n_o^2} = c_0^2. \tag{17.69}$$

The ray-velocity surface for O rays is clearly a sphere with equation

$$x^2 + y^2 + z^2 = \frac{c_0^2}{n_o^2}. \tag{17.70}$$

Figure 17.8 shows principal sections of ray-velocity surfaces in both positive and negative crystals.

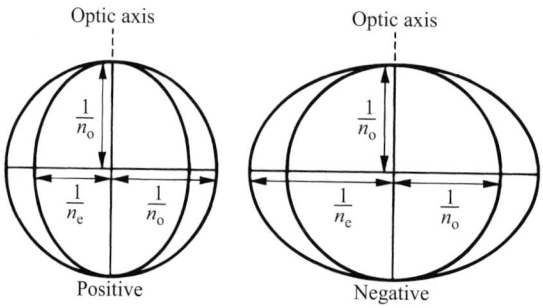


**Fig. 17.8** Ray-velocity surfaces for positive and negative uniaxial crystals.

## 17.11 Huygenian constructions

When a linearly polarized wave of arbitrary polarization direction enters an anisotropic medium, it will be resolved into two components polarized along the two allowed polarization directions determined by the direction of the wave vector relative to the axes of the indicatrix. In a uniaxial crystal these two components propagate as O and E waves respectively. At the entry surface of the anisotropic medium refraction of these waves, and of their corresponding rays, occurs. For normal incidence at the boundary of the anisotropic medium no refraction of either O or E wave vectors occurs, so within the medium the wave vectors of both O and E waves remain parallel. However, except for propagation along a principal axis, the E ray deviates from the common direction of the O and E wave vectors and the O ray. Thus at the exit surface of the medium, where refraction renders O and E rays parallel once again, the E ray will have been laterally displaced from the O ray. In simple terms we can say that the E ray is not refracted at the surface of the medium according to Snell's law. For other than normal incidence the O and E wave vectors refract separately according to Snell's law. For the O wave, calculation of the wave-vector direction inside the medium is straightforward. For angle of incidence $\theta_i$ the angle of refraction obeys

$$\sin\theta_i = n_o \sin\theta_r. \tag{17.71}$$

However, for the E wave the refraction of the wave vector obeys

$$\sin\theta_i = n_e(\theta)\sin\theta_r, \tag{17.72}$$

where $\theta$ is the angle that the refracted wave vector makes with the optic axis. To illustrate these geometrical optical properties of anisotropic media it is instructive to employ *Huygenian constructions* using the ray-velocity surfaces for the O and E rays. In Huygenian constructions we treat each point on the entrance boundary of the anisotropic media as a secondary emitter of electromagnetic waves. In a given time the distance traveled by all the O or E rays leaving this point as a function of direction traces out the appropriate ray-velocity surface. The geometrical paths of the O and E rays are perpendicular to the envelope of the ray-velocity surfaces that arise from these secondary emitters. To illustrate this, consider first the simple case shown in Fig. 17.9, where light enters a negative uniaxial crystal normally traveling along the optic axis. Secondary emitters at A and B give rise to the ray-velocity surfaces shown and contribute to the envelope PP′, which in this case is both the wave front and the ray front (the surface normal to the ray direction). Figure 17.10 illustrates the situation that results when light is incident normally on a positive uniaxial crystal traveling in a plane perpendicular to the optic axis. In this case, the ray-velocity surfaces which arise from secondary emitters at A and B define two envelopes, the surfaces OO′ and EE′, which are, respectively, the ordinary wave and ray front and the extraordinary wave and ray front. The O ray travels faster than the E ray, as indicated by the dots and arrows. Both in Fig. 17.9 and in Fig. 17.10 the wave vector and ray, which are perpendicular to their respective fronts, remain parallel. When the incident radiation does

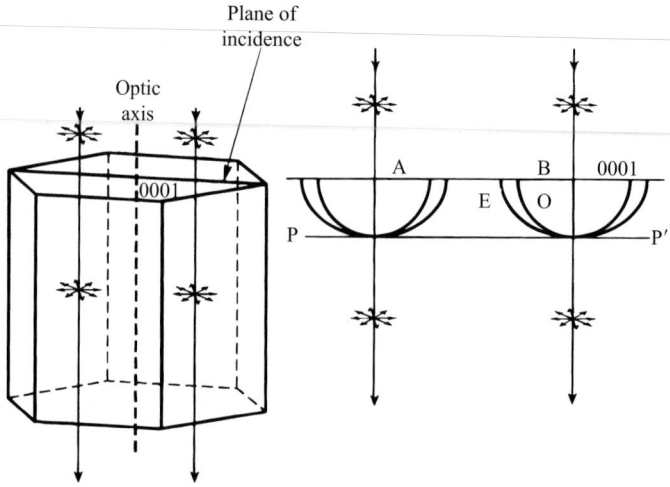


**Fig. 17.9** Huygenian construction for light entering a negative uniaxial crystal normally and traveling along the optic axis [1]. The number on the crystal face is an index that gives the orientation of the face [2].

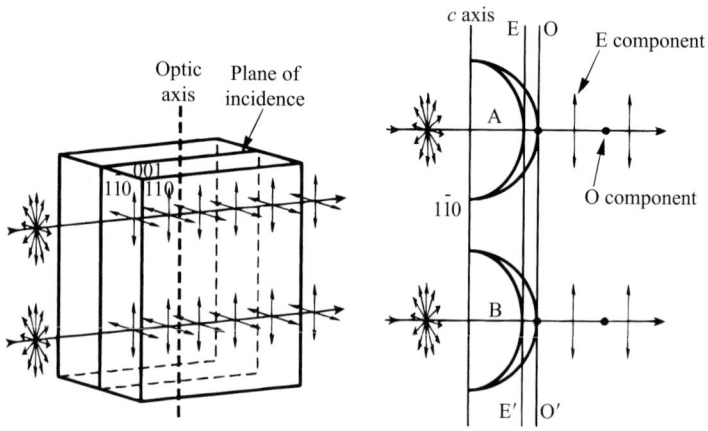


**Fig. 17.10** Huygenian construction for light entering a positive uniaxial crystal normally and traveling perpendicular to the optic axis [1].

not travel along a principal axis in the crystal the O and E rays are no longer coincident, as shown in Fig. 17.11, which shows light traveling at an angle to the optic axis striking a planar uniaxial crystal slab cut normal to the optic axis. In this case the utility of the ray-velocity surface in determining the ray directions within the crystal is clear. The incident wave fronts are parallel to PP′; by the time the wave front reaches point A the secondary emitter at B has given rise to the O and E ray-velocity surfaces shown. The O and E ray fronts are the tangents from A to the respective ray-velocity surfaces. The ray directions are perpendicular to these ray fronts. Once again, in this positive crystal the O ray travels more quickly than the E ray. On leaving the crystal the O and E rays become parallel to each other once more, but the E ray has been laterally displaced. This phenomenon is called

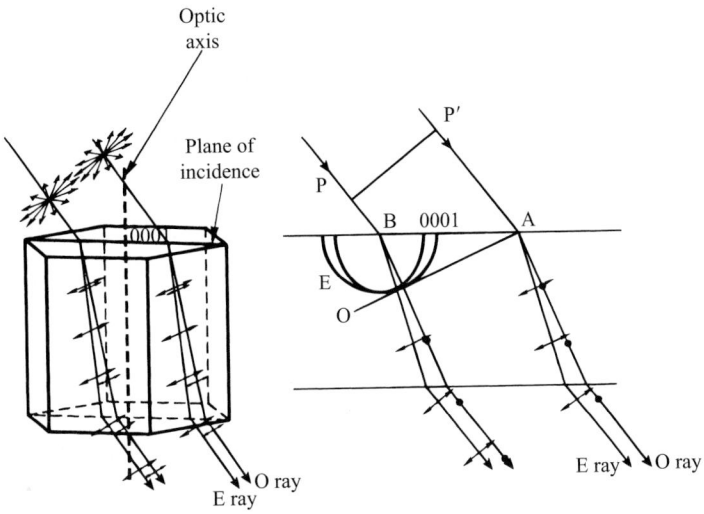


**Fig. 17.11** Huygenian construction for light that does not enter the crystal normally and does not propagate along a principal axis direction [1].

*double refraction*. If the input to the crystal were a narrow laser beam, linearly polarized at some angle between the two allowed polarization directions within the crystal, it would be resolved into an O ray and an E ray, and would emerge from the crystal as two separate orthogonally polarized beams.

Further examples of double refraction are shown in Figs. 17.12 and 17.13, with the paths of the O and E rays determined by Huygenian constructions. Figure 17.13 shows that, even in normal incidence, if the input wave vector is not along a principal axis the E ray is displaced laterally from the O ray. The refraction of the O ray in every case follows the

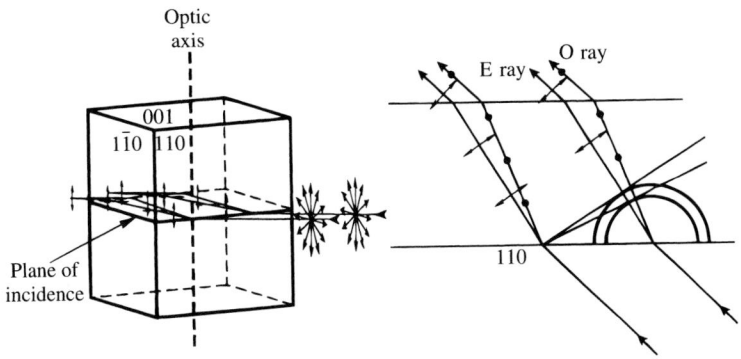


**Fig. 17.12** Huygenian construction for light traveling perpendicular to optic axis that refracts on entering the crystal [1]. O and E waves (and rays) refract at different angles. The indices on the crystal faces specify their orientation [2].

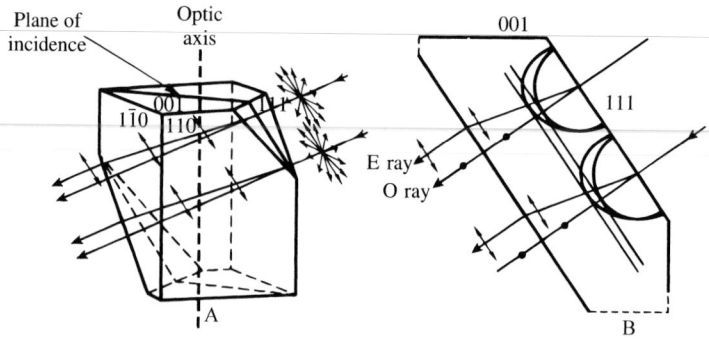


**Fig. 17.13** Huygenian construction for a rather general case [1].

familiar refraction laws of geometrical optics, with the angles of incidence $\theta_i$ and refraction $\theta_r$ at the crystal boundary obeying Snell's law:

$$n_o = \frac{\sin\theta_i}{\sin\theta_r}. \tag{17.73}$$

The path of the E ray is controlled by the anisotropic character of the medium and must be determined from a non-spherical ray-velocity surface.

The separation of the O and E rays within an anisotropic medium is the basic phenomenon which is operative in the various polarizing devices used for controlling the direction of polarization of light beams, particularly laser beams. The construction and mode of operation of some common types of laser polarizer are illustrated in Fig. 17.14.

## 17.12 Retardation

Unless a light beam is traveling in the direction of a principal axis, and is polarized parallel to a principal axis, when it is incident normally on a planar uniaxial crystal slab, it will be resolved within this material into O and E waves. If the direction of the wave vector is at an angle $\theta$ to the optic axis then the wave velocities of these waves will be $c_0/n_o$ and $c_0/n_e(\theta)$, respectively. In a positive crystal the O wave will travel faster than the E wave, and in a negative crystal vice versa.

The wave vector of the ordinary wave is

$$|\mathbf{k}_o| = k_o = \frac{\omega n_o}{c_0} = \frac{2\pi n_o}{\lambda_0} \tag{17.74}$$

and that of the extraordinary wave is

$$|\mathbf{k}_e| = k_e = \frac{\omega n_e(\theta)}{c_0} = \frac{2\pi n_e(\theta)}{\lambda_0}, \tag{17.75}$$

where $\omega$ and $\lambda_0$ are the angular frequency and wavelength *in vacuo*, respectively, of the incident wave.

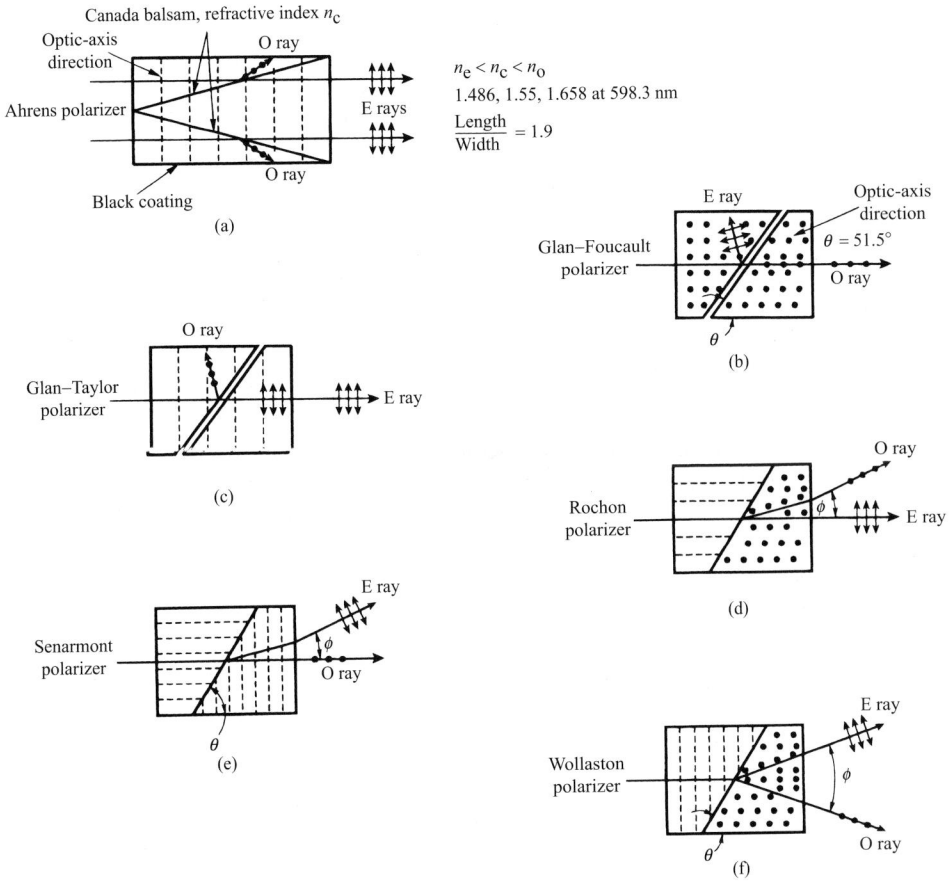


**Fig. 17.14** Construction of various kinds of polarizer. (For (d) – (f) the separation of the O and E rays is much exaggerated.)

On passing through a crystal of thickness $L$, the phase changes for the O and E waves, respectively, are

$$\phi_o = k_o L = \frac{2\pi n_o L}{\lambda_0} \tag{17.76}$$

and

$$\phi_e = k_e L = \frac{2\pi n_e(\theta) L}{\lambda_0}. \tag{17.77}$$

The phase difference (*retardation*) introduced by the crystal is

$$\Delta\phi = \phi_e - \phi_o = \frac{2\pi L}{\lambda_0}[n_e(\theta) - n_o]. \tag{17.78}$$

For an incident wave of the form

$$D = A\cos(\omega t - kr) \tag{17.79}$$

linearly polarized at an angle $\beta$ to the ordinary polarization direction, the ordinary and extraordinary waves will be

$$D_o = A\cos\beta\cos(\omega t - k_o r), \tag{17.80}$$

$$D_e = A\sin\beta\cos(\omega t - k_e r), \tag{17.81}$$

and at the exit face of the crystal

$$D_o = A\cos\beta\cos(\omega t - \phi_o), \tag{17.82}$$

$$D_e = A\sin\beta\cos(\omega t - \phi_e). \tag{17.83}$$

If the input to the crystal is a narrow beam of light, these two orthogonally polarized output beams will in general be displaced laterally from one another and will not recombine to form a single output beam. However, if the input to the crystal is a plane wave, or a narrow beam traveling perpendicular to the optic axis, these two electric vectors recombine to form a resultant single displacement vector with magnitude

$$D_{out} = \sqrt{D_o^2 + D_e^2}, \tag{17.84}$$

which makes an angle $\alpha$ with the ordinary polarization direction, such that

$$\tan\alpha = \frac{D_e}{D_o} = \tan\beta\,\frac{\cos(\omega t - \phi_e)}{\cos(\omega t - \phi_o)} \tag{17.85}$$

as illustrated in Fig. 17.15.

In the simplest case, in which $\phi_e - \phi_o = 2m\pi$, with $m$ being any positive or negative integer or zero, $\tan\alpha = \tan\beta$ and the output wave is linearly polarized in the same direction as the input.

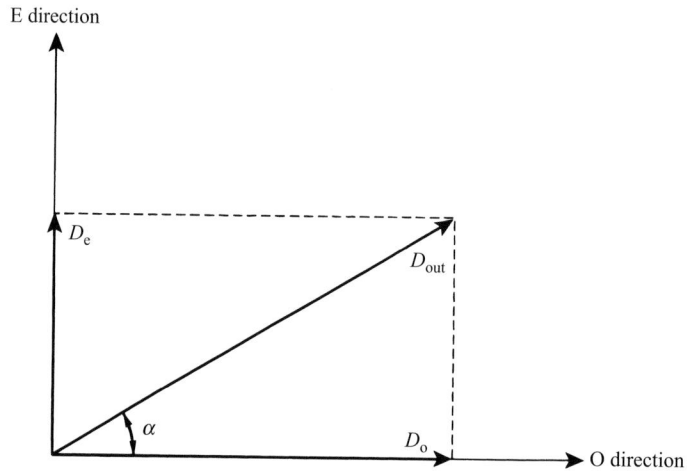


**Fig. 17.15** Ordinary and extraordinary D vector directions at the output face of a crystal slab.

If $\phi_e - \phi_o = (2m+1)\pi$, then $\tan\alpha = -\tan\beta$; thus $\alpha = -\beta$ and the output wave is linearly polarized but rotated by $2\beta$ from its original polarization direction. This occurs through a rotation by angle $\beta$ towards the O polarization direction followed by a further rotation through angle $\beta$.[3] Since a retardation $\phi = (2m+1)\pi$ is equivalent to a path difference of $(2m+1)\lambda/2$, a crystal that rotates the plane of linear polarization by $2\beta$ is called a ($(2m+1)$th-order) half-wave (retardation) plate. It is most usual to cut such a crystal so that its faces are parallel to the optic axis and to polarize the input at 45° to the optic axis, in which case the output is linearly polarized and rotated by 90° from the input. When the input is polarized in this manner and $\phi = \phi_e - \phi_o = (2m+1)\pi/2$ then $\tan\alpha = \pm\tan(\omega t - \phi_o)$.

For $m$ even, $\tan\alpha = \omega t - \phi_o$. For $m$ odd $\tan\alpha = -(\omega t - \phi_o)$. In both these cases the resultant displacement vector has, from Eqs. (17.82), (17.83), and (17.84), with $\beta = 45°$, magnitude

$$\begin{aligned} D_{\text{out}} &= \frac{A}{\sqrt{2}}\sqrt{\cos^2(\omega t + \phi_o) + \cos^2(\omega t - \phi_o - \phi)} \\ &= \frac{A}{\sqrt{2}}, \end{aligned} \tag{17.86}$$

so the output displacement vector has a constant magnitude but rotates about the direction of propagation with constant angular velocity $\omega$. This is circularly polarized light. If the electric vector rotates in a clockwise direction when viewed in the direction of propagation, as when $m$ is odd above, the light is said to be *left-hand circularly polarized*, as illustrated in Fig. 17.16. When the electric vector rotates counterclockwise the light is *right-hand circularly polarized*. A crystal that introduces a retardation $\phi$ of $(2m+1)\pi/2$ is called a *quarter-wave plate* (strictly a $(2m+1)$th-order quarter-wave plate).

In the general case when $\phi$ is not an integral number of half wavelengths, or if the input to a quarter-wave plate is not polarized at 45° to the optic axis, the resultant output will be elliptically polarized, as illustrated also in Fig. 17.16. In this case the displacement vector and the electric vector trace out an ellipse as they rotate in time as the wave propagates. To prove this we rewrite Eqs. (17.82) and (17.83) as

$$D_o = p = a\cos z, \tag{17.87}$$

$$D_e = q = b\cos(z - \phi) = b\cos z\cos\phi + b\sin z\sin\phi, \tag{17.88}$$

where $a = A\cos\beta$, $b = A\sin\beta$, and $z = (\omega t - \phi_o)$.

From Eq. (17.88)

$$\frac{q^2}{b^2} - \frac{2q}{b}\cos z\cos\phi + \cos^2 z\cos^2\phi = \sin^2 z\sin^2\phi, \tag{17.89}$$

[3] For this reason a waveplate alone cannot be used to make an optical isolator as can a Faraday rotator.

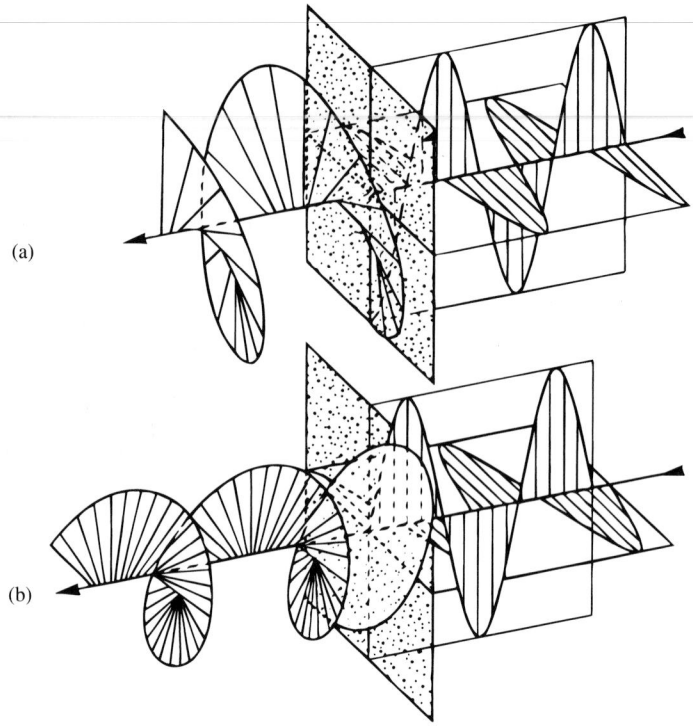


Fig. 17.16 Left-hand circularly polarized light (a) and (b) left-hand elliptically polarized light [1].

and, on substituting from Eq. (17.87),

$$\frac{q^2}{b^2} - \frac{2pq}{ab}\cos\phi + \frac{p^2}{a^2}\cos^2\phi = \left(1 - \frac{p^2}{a^2}\right)\sin^2\phi, \tag{17.90}$$

giving finally

$$\frac{p^2}{a^2} + \frac{q^2}{b^2} - \frac{2pq}{ab}\cos\phi - \sin^2\phi = 0. \tag{17.91}$$

This is the equation of an ellipse, one axis of which makes an angle $\psi$ with the O polarization direction, where

$$\tan(2\psi) = \frac{2ab\cos\phi}{a^2 - b^2}. \tag{17.92}$$

This ellipse has its axes coincident with the O and E vibration directions if $\psi = (2m + 1)\pi/2$. Thus, a quarter-wave plate produces elliptically polarized light if the incident light is not polarized exactly half-way between the O and E allowed polarization directions.

A first-order quarter-wave plate is very thin; as can be seen from Eq. (17.78), its thickness $L$ is

$$L = \frac{\lambda_0}{4(n_e - n_o)}. \tag{17.93}$$

For calcite (Iceland spar), a mineral form of calcium carbonate that is commonly used to make polarizing optics, $n_o = 1.658$ and $n_e = 1.486$, so for a wavelength of 500 nm the thickness of a first-order quarter-wave plate is only 726.7 nm $\sim$ 0.0007 mm. It is not practical to cut a crystalline slab as thin as this, except for a birefringent material such as mica that cleaves readily into very thin slices. Also, $(2m + 1)$th-order quarter-wave plates can be of practical thickness for large values of $m$ but suffer severely from the effects of temperature: the plate only has to expand or contract very slightly and it will cease to be a quarter-wave plate at the wavelength for which it was designed. To overcome this drawback, temperature-compensated plates can be made. These consist of one $(2m + 1)$th-order quarter-wave plate stacked on top of a plate that produces a retardation of $m\pi$ but whose optic axis is perpendicular to the optic axis (E-wave polarization direction) of the first quarter-wave plate. The total retardation of the combination is

$$\phi = \phi_1 - \phi_2 = (2n + 1)\pi/2 - n\pi = \pi/2, \tag{17.94}$$

so it is equivalent to a first-order quarter-wave plate but is not sensitive to temperature changes.

## 17.13 Biaxial crystals

The optical properties of biaxial crystals can be related to an ellipsoid with three unequal axes, the biaxial indicatrix, whose equation is

$$\frac{x^2}{n_x^2} + \frac{y^2}{n_y^2} + \frac{z^2}{n_z^2} = 1, \tag{17.95}$$

where $n_x$, $n_y$, and $n_z$ are the three principal refractive indices of the material. It is the normal convention to label the axes so that $n_x < n_y < n_z$. When this is done the two optic axes, namely those directions through the crystal along which the direction of propagation of waves is independent of their polarization direction, lie in the $xz$ plane. If the two optic axes are closer to the $z$ axis than they are to the $x$ axis, the crystal is said to be *positive*, otherwise it is *negative*. Some of the important features of positive and negative biaxial indicatrices are illustrated in Fig. 17.17. The acute angle between the two optic axes is labeled $2V$ – the *optic angle*. Whichever axis bisects this acute angle is called the *acute bisectrix*, this is the $z$ axis in positive crystals, or the $x$ axis in negative crystals. The other axis in each case is called the *obtuse bisectrix*.

The angle between one of the optic axes and the $z$ axis is $V_z$, and is given by the expression

$$V_z = \tan^{-1}\sqrt{\frac{1/n_x^2 - 1/n_y^2}{1/n_y^2 - 1/n_z^2}}, \tag{17.96}$$

which is an angle $<45°$ in a positive crystal.

Because of the lack of rotational symmetry of the biaxial indicatrix, the optical properties of biaxial crystals are more complicated than for uniaxial ones. For wave propagation along

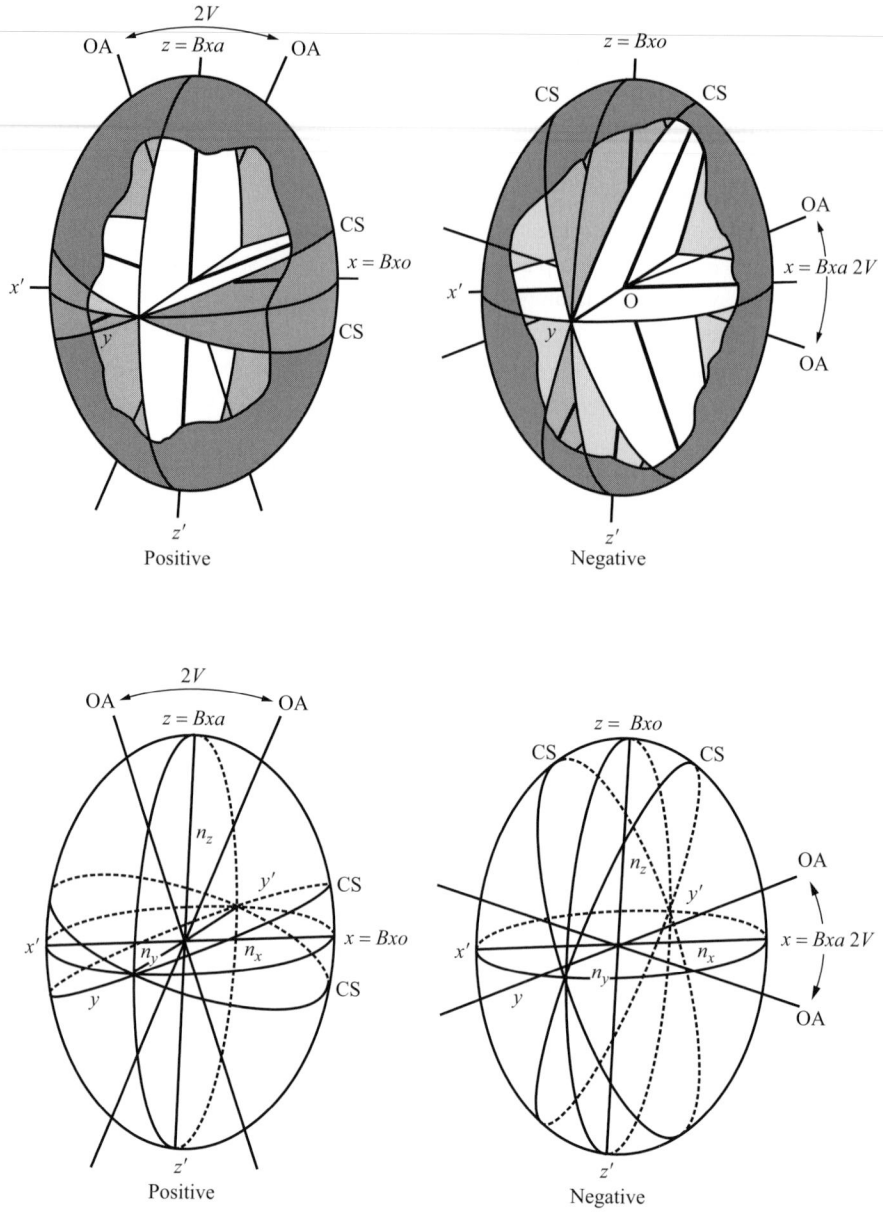


**Fig. 17.17** Biaxial indicatrices.

the three principal axes the wave-normals and rays of both of the two allowed polarization directions are parallel. For propagation in a symmetry plane of the biaxial indicatrix, as shown in Fig. 17.18, the two allowed polarization directions are perpendicular to, and in, the plane of symmetry. The component polarized perpendicular to the symmetry plane has its polarization direction parallel to a principal axis, and therefore has its **E** direction parallel to **D**. Thus, for this polarization the wave-normal direction ON and the ray $OS_1$

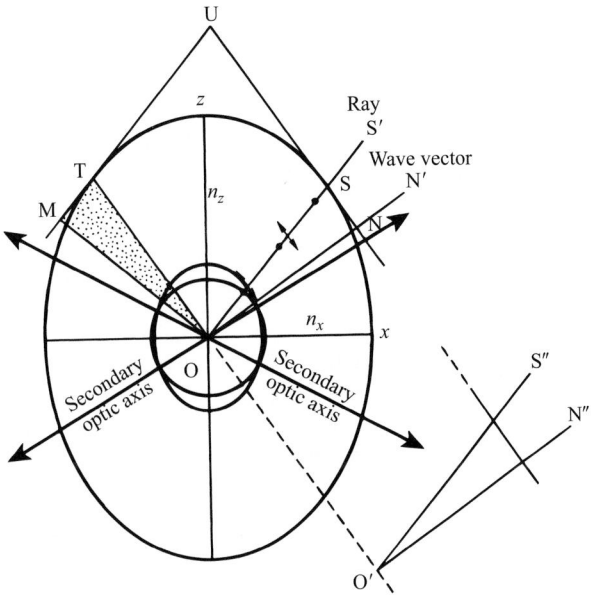


**Fig. 17.18** A wave and ray in a symmetry plane of a biaxial indicatrix [1].

are parallel. Both the wave and the ray refractive indices for this polarization have the value $n_y$. The ray direction for the wave polarized in the symmetry plane is the direction OS, which is found by a construction similar to the one described for finding the direction of the E ray in a uniaxial crystal. The wave refractive index for this polarization is given by the length OT and its ray refractive index by the length OM; OM is less than OT since the ray travels further than the wave in the same period of time. There are two directions for which OM $= n_y$ in Fig. 17.18. These are ray directions for which the ray velocity is the same for both polarization directions; of course, the wave vectors of these two waves are not collinear with their rays. These directions of equal ray velocity are called the secondary optic axes, and are generally within two degrees of the primary optic axes, which are directions of equal wave velocity. For propagation in an arbitrary direction not lying in a symmetry plane, both allowed polarization directions have rays which are not parallel to their wave-normals. This is illustrated in Fig. 17.19, which shows the two ray directions $OS_1$ and $OS_2$ that correspond to the two allowed polarization directions $OP_1$ and $OP_2$ of a wave propagating in the general direction ON.

The surfaces which are related to the indicatrix of a biaxial crystal are more complicated than for uniaxial crystals, although for propagation in symmetry planes of the indicatrix they are somewhat like, but not quite the same as, sections of uniaxial surfaces. Figure 17.20 illustrates the ray-velocity surface of a biaxial crystal. This is a two-shelled surface, with one shell for each polarization direction corresponding to a given ray direction. The four dimples where one shell cuts through the other (two are visible in Fig. 17.20) are the directions of the secondary optic axes.

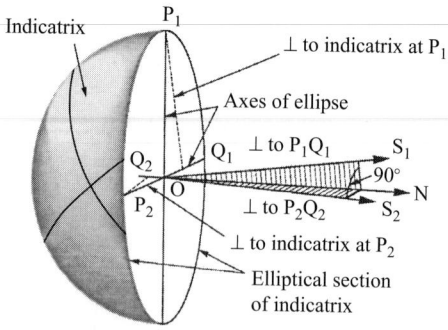


**Fig. 17.19** Construction that indicates how wave vector and ray directions are related for the two permitted polarization directions in a biaxial crystal [1].

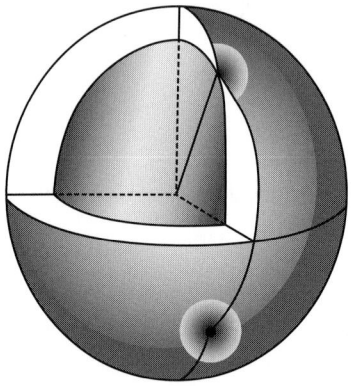

**Fig. 17.20** Cut away view of the ray velocity surface of a biaxial crystal [1].

## 17.14 Intensity transmission through polarizer/waveplate/polarizer combinations

A wave is passed through a linear polarizer (P) whose preferred direction is at angle $\beta$ to the O direction of a succeeding waveplate W, as shown in Fig. 17.21, and is then transmitted through a second linear polarizer (A) whose preferred axis makes an angle $\psi$ with the O direction of the waveplate. The second polarizer, usually called the analyzer, will not in general permit all the radiation emerging from the waveplate to pass. This PWA combination serves as an adjustable attenuator of a light beam.

From Eqs. (17.82) and (17.83), it is easy to see that the **D** vector transmitted through the analyzer has magnitude

$$D = A\cos\beta\cos(\omega t - \phi_o)\cos\psi + A\sin\beta\cos(\omega t - \phi_e)\sin\psi, \tag{17.97}$$

which can be rearranged to give

$$D = D_0\cos(\omega t - \phi_e + \chi), \tag{17.98}$$

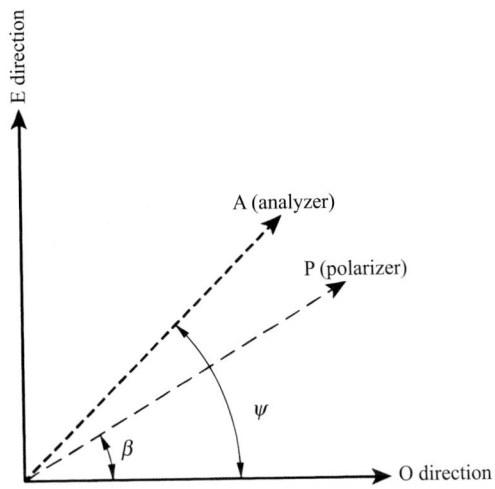


**Fig. 17.21** Orientation directions for analysis of polarizer/waveplate/analyzer combinations.

where

$$D_0 = A\sqrt{(\cos\beta\cos\psi\cos\Delta\phi + \sin\beta\sin\psi)^2 + (\cos\beta\cos\psi\sin\Delta\phi)^2} \tag{17.99}$$

and

$$\tan\chi = \frac{\cos\beta\cos\psi\sin\Delta\phi}{\cos\beta\cos\psi\cos\Delta\phi + \sin\beta\sin\psi}, \tag{17.100}$$

where $\Delta\phi = \phi_e - \phi_o$ is the retardation produced by the waveplate.

### 17.14.1 Examples

To illustrate the value of Eq. (17.99), some specific examples are in order.

(1) The input and output polarizers are parallel: $\psi = \beta$. In this case

$$D_0 = A\sqrt{1 - \tfrac{1}{2}\sin^2(2\beta)(1 - \cos\Delta\phi)}. \tag{17.101}$$

If $\Delta\phi = 0$ (or if $\Delta\phi$ is any multiple of $2\pi$) then $D_0 = A$: all the light is transmitted, as we would have expected.

If $\beta = 45°$ then

$$D_0 = A\sqrt{1 - \tfrac{1}{2}(1 - \cos\Delta\phi)}. \tag{17.102}$$

If $\Delta\phi = \pi$ (or any odd multiple of $\pi$) then $D_0 = 0$. If $\Delta\phi = \pi/2$ (or any odd multiple of $\pi/2$) then $D_0 = A/\sqrt{2}$. Since the transmitted intensity is proportional to $D_0^2$, half of the incident light is transmitted.

(2) The input and output polarizers are crossed: $\psi = \beta \pm \pi/2$. In this case

$$D_0 = \frac{1}{\sqrt{2}}\sin(2\beta)\sqrt{1 - \cos\Delta\phi}. \tag{17.103}$$

**Table 17.2** Transmittance of PWA combinations

| P/A | Type of W | Transmittance |
|---|---|---|
| ‖ | 0 | 1 |
| ‖ | $\lambda/4$ | 1/2 |
| ‖ | $\lambda/2$ | 0 |
| ⊥ | 0 | 0 |
| ⊥ | $\lambda/4$ | 1/2 |
| ⊥ | $\lambda/2$ | 1 |

If $\beta = 45°$ then

$$D_0 = \frac{1}{\sqrt{2}}\sqrt{1 - \cos\Delta\phi}. \tag{17.104}$$

If $\Delta\phi = 0$ then $D_0 = 0$.

If $\Delta\phi = (2m+1)\pi$, the waveplate is a half-wave plate, and $D_0 = A$. All the incident light is transmitted.

If $\Delta\phi = (2m+1)\pi/2$, the waveplate is a quarter-wave plate and $D_0 = A/\sqrt{2}$. Half the incident light is transmitted.

These findings are summarized in Table 17.2.

## 17.15 The Jones calculus

Over 50 years ago R. Clark Jones introduced a very useful technique for describing the change in polarization state of a light wave as it passed through an optical system containing various interfaces and polarizing elements [3] (see also [4–9]). The *Jones calculus* treats the optical system as a linear system characterized by an appropriate Jones matrix that transforms vectors describing the polarization state of the wave. In this sense the Jones calculus is analogous to paraxial-ray analysis.

### 17.15.1 The Jones vector

The electric field of a light wave that is linearly polarized along the $x$ axis can be written as

$$\mathbf{E}_x = A_x \sin(\omega t + \phi_x)\hat{\mathbf{i}}, \tag{17.105}$$

which, in complex exponential notation, is

$$\mathbf{E}_x = A_x e^{j\phi_x} e^{j\omega t}\hat{\mathbf{i}}. \tag{17.106}$$

On the other hand a light wave that is linearly polarized along the $y$ axis can be written as

$$\mathbf{E}_y = A_y e^{j\phi_y} e^{j\omega t}\hat{\mathbf{j}}. \tag{17.107}$$

The superposition of the electric fields $\mathbf{E}_x$ and $\mathbf{E}_y$ leads in general to elliptically polarized light with electric field

$$\mathbf{E} = \left(A_x e^{j\phi_x}\hat{\mathbf{i}} + A_y e^{j\phi_y}\hat{\mathbf{j}}\right) e^{j\omega t}. \tag{17.108}$$

The complex amplitudes of the $x$ and $y$ components of this wave form the two elements of the Jones vector $\mathbf{J}$, where

$$\mathbf{J} = \begin{pmatrix} A_x e^{j\phi_x} \\ A_y e^{j\phi_y} \end{pmatrix}. \tag{17.109}$$

Clearly, light that is linearly polarized along the $x$ axis has

$$\mathbf{J} = \begin{pmatrix} A_x e^{j\phi_x} \\ 0 \end{pmatrix}, \tag{17.110}$$

with a similar result for light that is linearly polarized along $y$.

Light that is linearly polarized at angle $\beta$ to the axis has

$$\mathbf{J}_\beta = \begin{pmatrix} E_0 \cos\beta \\ E_0 \sin\beta \end{pmatrix}. \tag{17.111}$$

Right-hand circularly polarized light has

$$\mathbf{J}_{\text{rcp}} = \begin{pmatrix} E_0 e^{j\phi_x} \\ E_0 e^{j(\phi_x + \pi/2)} \end{pmatrix} \tag{17.112}$$

and left-hand circularly polarized light has

$$\mathbf{J}_{\text{lcp}} = \begin{pmatrix} E_0 e^{j\phi_x} \\ E_0 e^{j(\phi_x - \pi/2)} \end{pmatrix}. \tag{17.113}$$

In an isotropic medium the intensity of the wave is

$$I \propto \mathbf{E}^* \cdot \mathbf{E}, \tag{17.114}$$

which, in terms of the Jones vector, gives

$$I \propto \mathbf{J}^* \cdot \mathbf{J} = A_x^2 + A_y^2. \tag{17.115}$$

It is generally simple and convenient to use the Jones vector in its normalized form, in which

$$\mathbf{J}^* \cdot \mathbf{J} = 1. \tag{17.116}$$

In this case the three vectors in Eqs. (17.111), (17.112), and (17.113) would become

$$\mathbf{J}_\beta = \begin{pmatrix} \cos\beta \\ \sin\beta \end{pmatrix}, \tag{17.117}$$

$$\mathbf{J}_{\text{rcp}} = \frac{1}{\sqrt{2}}\begin{pmatrix} 1 \\ e^{j\pi/2} \end{pmatrix} \text{ or } \frac{1}{\sqrt{2}}\begin{pmatrix} e^{-j\pi/4} \\ e^{j\pi/4} \end{pmatrix}, \tag{17.118}$$

$$\mathbf{J}_{\text{lcp}} = \frac{1}{\sqrt{2}}\begin{pmatrix} 1 \\ e^{-j\pi/2} \end{pmatrix} \text{ or } \frac{1}{\sqrt{2}}\begin{pmatrix} e^{j\pi/4} \\ e^{-j\pi/4} \end{pmatrix}. \tag{17.119}$$

The second description for circularly polarized light is an alternative symmetrical way of writing the column vector, since only the phase difference between the $x$ and $y$ components is significant.

In this notation a general elliptically polarized beam has a Jones vector that can be written as

$$\mathbf{J} = \begin{pmatrix} \cos\beta\, e^{-j\Delta/2} \\ \sin\beta\, e^{j\Delta/2} \end{pmatrix}. \tag{17.120}$$

### 17.15.2 The Jones matrix

In a linear system description the output Jones vector of a light wave after it has interacted with an optical system has components that are linearly related to its input components.

If

$$\mathbf{J}_{\text{in}} = E_x\hat{\mathbf{i}} + E_y\hat{\mathbf{j}} \tag{17.121}$$

and

$$\mathbf{J}_{\text{out}} = E'_x\hat{\mathbf{i}} + E'_y\hat{\mathbf{j}}, \tag{17.122}$$

where phase factors are included in the complex amplitudes, we can now write

$$\begin{aligned} E'_x &= m_{11}E_x + m_{12}E_y, \\ E'_y &= m_{21}E_x + m_{22}E_y, \end{aligned} \tag{17.123}$$

or, in matrix form,

$$\begin{pmatrix} E'_x \\ E'_y \end{pmatrix} = \begin{pmatrix} m_{11} & m_{12} \\ m_{21} & m_{22} \end{pmatrix}\begin{pmatrix} E_x \\ E_y \end{pmatrix}. \tag{17.124}$$

Equation (17.124) introduces the Jones matrix $\mathbf{M}$ with elements $m_{ij}$; it can be rewritten as

$$\mathbf{J}_{\text{out}} = \mathbf{M}\mathbf{J}_{\text{in}}. \tag{17.125}$$

The determination of the Jones matrix for common optical elements is straightforward. This can be demonstrated with a few examples.

(a) For an isotropic element: since $\mathbf{J}_{\text{out}} = \mathbf{J}_{\text{in}}$, clearly

$$\mathbf{M} = \begin{pmatrix} 1 & 0 \\ 0 & 1 \end{pmatrix}. \tag{17.126}$$

(b) For a linear polarizer oriented along the $x$ axis:

$$\mathbf{M} = \begin{pmatrix} 1 & 0 \\ 0 & 0 \end{pmatrix}. \tag{17.127}$$

(c) For a linear polarizer oriented at an angle $\theta$ to the $x$ axis:

$$\mathbf{M} = \begin{pmatrix} \cos^2\theta & \sin\theta\cos\theta \\ \sin\theta\cos\theta & \sin^2\theta \end{pmatrix}. \tag{17.128}$$

(d) The case of a waveplate that produces a retardation $\phi_x - \phi_y = \Gamma$ is worthy of closer consideration. If the waveplate has its principal axes parallel to the $x$ and $y$ axes then the refractive indices seen by the $x$ and $y$ components of the wave are $n_x$ and $n_y$, respectively. In this case

$$\begin{aligned} E'_x &= E_x e^{-jk_0\ell n_x}, \\ E'_y &= E_y e^{-jk_0\ell n_y}, \end{aligned} \tag{17.129}$$

where $k_0 = 2\pi/\lambda_0$ and $\ell$ is the thickness of the retarder. Since $\Gamma = k_0\ell(n_y - n_x)$, Eq. (17.129) can be written as

$$\begin{aligned} E'_x &= E_x e^{-jk_0\ell n_y} e^{j\Gamma}, \\ E'_y &= E_y e^{-jk_0\ell n_y}, \end{aligned} \tag{17.130}$$

or, in symmetrical form,

$$\begin{aligned} E'_x &= E_x e^{-j\phi} e^{j\Gamma/2}, \\ E'_y &= E_y e^{-j\phi} e^{-j\Gamma/2}, \end{aligned} \tag{17.131}$$

where $\phi = k_0\ell(n_x + n_y)/2$. Therefore, the Jones matrix of the waveplate is

$$\mathbf{M} = e^{-j\phi} \begin{pmatrix} e^{j\Gamma/2} & 0 \\ 0 & e^{-j\Gamma/2} \end{pmatrix}. \tag{17.132}$$

The phase factor $e^{-j\phi}$ in Eq. (17.129) can be omitted for most practical purposes.

A quarter-wave plate has a Jones matrix

$$\mathbf{M} = \begin{pmatrix} e^{j\pi/4} & 0 \\ 0 & e^{-j\pi/4} \end{pmatrix}. \tag{17.133}$$

A linearly polarized wave with

$$\mathbf{J} = \begin{pmatrix} 1/\sqrt{2} \\ 1/\sqrt{2} \end{pmatrix} \tag{17.134}$$

is linearly polarized at 45° to the fast (and slow) axis of the waveplate. Its electric field is in the $\mathbf{E}_1$ direction in Fig. 17.22. The output Jones vector is

$$\begin{aligned}\mathbf{J}_{\text{out}} &= \begin{pmatrix} e^{j\pi/4} & 0 \\ 0 & e^{-j\pi/4} \end{pmatrix} \begin{pmatrix} 1/\sqrt{2} \\ 1/\sqrt{2} \end{pmatrix} \\ &= \frac{1}{\sqrt{2}} \begin{pmatrix} e^{j\pi/4} \\ e^{-j\pi/4} \end{pmatrix},\end{aligned} \tag{17.135}$$

which has transformed the linearly polarized input light to left-hand circular polarization.

If the input linearly polarized light is rotated 90° with respect to the axis of the quarter-wave plate to the $\mathbf{E}_2$ direction in Fig. 17.22, then the output Jones vector will be

$$\begin{aligned}\mathbf{J}_{\text{out}} &= \begin{pmatrix} e^{j\pi/4} & 0 \\ 0 & e^{-j\pi/4} \end{pmatrix} \begin{pmatrix} 1/\sqrt{2} \\ -1/\sqrt{2} \end{pmatrix} \\ &= \frac{1}{\sqrt{2}} \begin{pmatrix} e^{j\pi/4} \\ e^{j3\pi/4} \end{pmatrix} \\ &= \frac{1}{\sqrt{2}} e^{j\pi/2} \begin{pmatrix} e^{-j\pi/4} \\ e^{j\pi/4} \end{pmatrix}.\end{aligned} \tag{17.136}$$

This is right-hand circularly polarized light, so a quarter-wave plate will convert linearly polarized light into left- or right-hand circularly polarized light depending on its orientation.

It is of value to view the action of the quarter-wave plate in a coordinate system that is rotated by 45°. What is now the Jones matrix for the plate for an input linearly

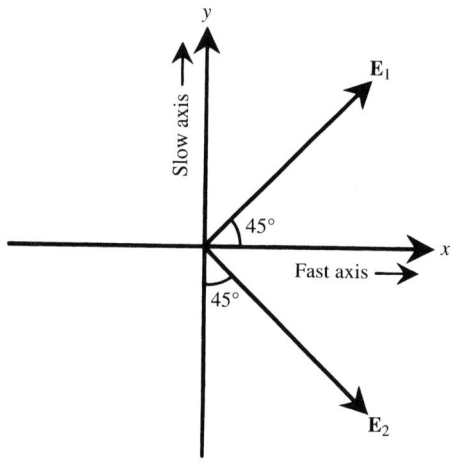


Fig. 17.22 Input electric field directions for producing left- and right-hand circularly polarized light from linearly polarized light oriented at 45° to the fast and slow axis of a quarter-wave plate.

polarized wave with Jones vector

$$\mathbf{J}_{\text{in}} = \begin{pmatrix} 1 \\ 0 \end{pmatrix}?$$

Such a wave is polarized in the $x'$ direction. Now the transformation between the $xy$ and $x'y'$ coordinate systems is

$$\begin{pmatrix} x' \\ y' \end{pmatrix} = \frac{1}{\sqrt{2}} \begin{pmatrix} 1 & -1 \\ 1 & 1 \end{pmatrix} \begin{pmatrix} x \\ y \end{pmatrix} = \mathbf{S} \begin{pmatrix} x \\ y \end{pmatrix} \tag{17.137}$$

and

$$\begin{pmatrix} x \\ y \end{pmatrix} = \frac{1}{\sqrt{2}} \begin{pmatrix} 1 & 1 \\ -1 & 1 \end{pmatrix} \begin{pmatrix} x' \\ y' \end{pmatrix} = \mathbf{S}^{-1} \begin{pmatrix} x \\ y \end{pmatrix}. \tag{17.138}$$

In terms of the rotation matrix $\mathbf{S}$, in the $x'y'$ coordinate system

$$\mathbf{J}'_{\text{out}} = \mathbf{S}\mathbf{J}_{\text{out}}. \tag{17.139}$$

In the $xy$ coordinate system

$$\mathbf{J}_{\text{out}} = \mathbf{M}\mathbf{J}_{\text{in}}, \tag{17.140}$$

where

$$\mathbf{M} = \begin{pmatrix} e^{j\pi/4} & 0 \\ 0 & e^{-j\pi/4} \end{pmatrix} \tag{17.141}$$

and

$$\mathbf{J}_{\text{in}} = \mathbf{S}^{-1} \begin{pmatrix} 1 \\ 0 \end{pmatrix} = \mathbf{S}^{-1}\mathbf{J}'_{\text{in}}. \tag{17.142}$$

Combining Eqs. (17.139), (17.140), and (17.142) gives

$$\mathbf{J}'_{\text{out}} = \mathbf{S}\mathbf{M}\mathbf{S}^{-1}\mathbf{J}'_{\text{in}}, \tag{17.143}$$

so the new Jones matrix in the $x'y'$ coordinate system is

$$\mathbf{M}' = \mathbf{S}\mathbf{M}\mathbf{S}^{-1}. \tag{17.144}$$

In this case

$$\begin{aligned} \mathbf{M}' &= \frac{1}{2} \begin{pmatrix} 1 & -1 \\ 1 & 1 \end{pmatrix} \begin{pmatrix} e^{j\pi/4} & 0 \\ 0 & e^{-j\pi/4} \end{pmatrix} \begin{pmatrix} 1 & 1 \\ -1 & 1 \end{pmatrix} \\ &= \frac{1}{\sqrt{2}} \begin{pmatrix} 1 & j \\ j & j \end{pmatrix}. \end{aligned} \tag{17.145}$$

We can generalize from this example: if coordinate system $x'y'$ is obtained by a counterclockwise rotation by angle $\alpha$ from coordinate system $xy$, then the transformation matrix is

$$\mathbf{S} = \begin{pmatrix} \cos\alpha & \sin\alpha \\ -\sin\alpha & \cos\alpha \end{pmatrix}. \tag{17.146}$$

The transformation of the Jones matrix from $xy$ to $x'y'$ obeys Eq. (17.144).

(e) A Faraday rotator that rotates the plane of linear polarization in a counterclockwise direction by an angle $\theta$ has a Jones matrix

$$\mathbf{M}_{\mathrm{Fr}} = \begin{pmatrix} \cos\theta & -\sin\theta \\ \sin\theta & \cos\theta \end{pmatrix}. \tag{17.147}$$

## 17.16 Mueller calculus

The Jones calculus analyzes the passage of light through optical systems using field amplitudes and phases, so it works best for coherent light. In 1943 Hans Mueller devised an alternative calculus that decribes the changes in the Stokes parameters (introduced in Chapter 1) as light propagates through an optical system. A Mueller matrix describes the change in the Stokes parameters as they pass through a system. If the input Stokes parameters are written as a vector $\mathbf{S}_{\mathrm{i}} = (I_{\mathrm{i}}, Q_{\mathrm{i}}, U_{\mathrm{i}}, V_{\mathrm{i}})^{\mathrm{T}} = (S_0, S_1, S_2, S_3)^{\mathrm{T}}$,[4] then the output Stokes vector is $\mathbf{S}_{\mathrm{o}}$, where

$$\mathbf{S}_{\mathrm{o}} = \mathbf{M}\mathbf{S}_{\mathrm{i}}. \tag{17.148}$$

Here $\mathbf{M}$ is the Mueller matrix, which for some simple systems has the following forms. For a horizontal linear polarizer

$$\mathbf{M} = \frac{1}{2}\begin{pmatrix} 1 & 1 & 0 & 0 \\ 1 & 1 & 0 & 0 \\ 0 & 0 & 0 & 0 \\ 0 & 0 & 0 & 0 \end{pmatrix}. \tag{17.149a}$$

The Stokes vector for unpolarized light is $(1, 0, 0, 0)^{\mathrm{T}}$. The Stokes vector for horizontally polarized light is $(1, 1, 0, 0)^{\mathrm{T}}$. For a vertical linear polarizer (parallel to the $y$ axis)

$$\mathbf{M} = \frac{1}{2}\begin{pmatrix} 1 & -1 & 0 & 0 \\ -1 & 1 & 0 & 0 \\ 0 & 0 & 0 & 0 \\ 0 & 0 & 0 & 0 \end{pmatrix}. \tag{17.149b}$$

The Stokes vector for vertically polarized light is $(1, -1, 0, 0)^{\mathrm{T}}$. For a linear polarizer at an angle of $\phi$ to the $x$ axis

$$\mathbf{M} = \frac{1}{2}\begin{pmatrix} 1 & \cos(2\phi) & \sin(2\phi) & 0 \\ \cos(2\phi) & \cos^2(2\phi) & \sin(2\phi)\cos(2\phi) & 0 \\ \sin(2\phi) & \sin(2\phi)\cos(2\phi) & \sin^2(2\phi) & 0 \\ 0 & 0 & 0 & 0 \end{pmatrix}. \tag{17.149c}$$

[4] These are two commonly used notations for the components of the Stokes vector.

For a $\lambda/4$ plate with its fast axis horizontal

$$\mathbf{M} = \begin{pmatrix} 1 & 0 & 0 & 0 \\ 0 & 1 & 0 & 0 \\ 0 & 0 & 0 & 1 \\ 0 & 0 & -1 & 0 \end{pmatrix}. \tag{17.149d}$$

For a $\lambda/4$ plate with its fast axis vertical

$$\mathbf{M} = \begin{pmatrix} 1 & 0 & 0 & 0 \\ 0 & 1 & 0 & 0 \\ 0 & 0 & 0 & -1 \\ 0 & 0 & 1 & 0 \end{pmatrix}. \tag{17.149e}$$

We can find the Mueller matrix for a $\lambda/2$ plate from the matrix for a $\lambda/4$ plate. For a half-wave plate with its fast axis vertical

$$\mathbf{M} = \begin{pmatrix} 1 & 0 & 0 & 0 \\ 0 & 1 & 0 & 0 \\ 0 & 0 & 0 & -1 \\ 0 & 0 & 1 & 0 \end{pmatrix}^2 = \begin{pmatrix} 1 & 0 & 0 & 0 \\ 0 & 1 & 0 & 0 \\ 0 & 0 & -1 & 0 \\ 0 & 0 & 0 & -1 \end{pmatrix}. \tag{17.149f}$$

For a right-hand circular polarizer

$$\mathbf{M} = \frac{1}{2} \begin{pmatrix} 1 & 0 & 0 & 1 \\ 0 & 0 & 0 & 0 \\ 0 & 0 & 0 & 0 \\ 1 & 0 & 0 & 1 \end{pmatrix}, \tag{17.149g}$$

and for a left-hand circular polarizer

$$\mathbf{M} = \frac{1}{2} \begin{pmatrix} 1 & 0 & 0 & -1 \\ 0 & 0 & 0 & 0 \\ 0 & 0 & 0 & 0 \\ -1 & 0 & 0 & 1 \end{pmatrix}. \tag{17.149h}$$

The matrix that introduces a linear birefringence of $\delta$ rad with a fast axis in the horizontal direction is

$$\mathbf{M} = \begin{pmatrix} 1 & 0 & 0 & 0 \\ 0 & 1 & 0 & 0 \\ 0 & 0 & \cos\delta & \sin\delta \\ 0 & 0 & -\sin\delta & \cos\delta \end{pmatrix}. \tag{17.149i}$$

This brief survey provides the essentials needed in order to describe the polarization change that occurs in a multi-element optical system. For more details the interested reader is referred to the specialized literature [10–16].

## 17.17 Problems

17.1 Prove that in a uniaxial crystal the maximum angular separation of O and E rays occurs when the wave vector makes an angle $\theta$ with the optic axis that satisfies

$$\theta = \arctan(n_e/n_o).$$

17.2 A uniaxial crystal has $n_o = 1.5$ and $n_e = 2.0$. Calculate the index of refraction for E waves traveling at 30° and 45° to the optic axis.

17.3 What is the wave velocity for the waves in Problem 17.2 traveling at 30° and 45° to the optic axis?

17.4 What are the ray velocities in Problem 17.2 for rays traveling at 30° and 45° to the optic axis?

17.5 What is the angle between the optic axis and the ray in Problem 17.2 when the wave vector is at 45° to the optic axis?

17.6 The indicatrix of a particular crystal is

$$ax^2 + by^2 + cz^2 + dyz = 1.$$

Where are the principal axes?

17.7 A laser beam traveling at an angle of incidence of 45° strikes a uniaxial crystal slab that is 20 mm thick. The faces of the slab are parallel to each other and are perpendicular to the optic axis; $n_o = 3$ and $n_e = 2$. Calculate the lateral separation of the exiting O and E rays.

17.8 In a crystal with $n_e = 1.5$ and $n_o = 2$, a *ray* is traveling at an angle of 45° to the optic axis. What is the direction of the wave vector and what value of $n_e(\theta)$ does this wave see?

17.9 A wave enters the plane surface of a very birefringent uniaxial crystal with $n_o = 1.5$, $n_e = 4$ from air at angle of incidence $\theta_1$. The wave refracts according to Snells law, where $\sin\theta_1 = n_e(\theta)\sin\theta_2$, where $\theta$ is the angle that the refracted $|\mathbf{k}|$ vector makes with the optic axis in the crystal. Show that in this situation for a range of angles of incidence the ray direction in the crystal actually appears to show *negative refraction* where the ray angle inside the crystal is on the same side of the surface normal as the incident ray in air.

## References

[1] E. Wahlstrom, *Optical Crystallography*, 4th edn., London: Wiley, 1969.

[2] C. Kittel, *Introduction to Solid State Physics*, 8th edn., New York: Wiley, 2004.

[3] R. Clark Jones, "A new calculus for the treatment of optical systems, I. Description and discussion of the calculus," *J. Opt. Soc. Am.* **31**, 488–493, 1941.

[4] H. Hurwitz and R. Clark Jones, "A new calculus for the treatment of optical systems, II. Proof of three general equivalence theorems," *J. Opt. Soc. Am.* **31**, 493–499, 1941.

[5] R. Clark Jones, "A new calculus for the treatment of optical systems, III. The Sohncke theory of optical activity," *J. Opt. Soc. Am.* **31**, 500–503, 1941.

[6] R. Clark Jones, "A new calculus for the treatment of optical systems, IV," *J. Opt. Soc. Am.* **32**, 486–493, 1942.

[7] R. Clark Jones, "A new calculus for the treatment of optical systems, V. A more general formulation and description of another calcular," *J. Opt. Soc. Am.* **37**, 107–110, 1947.

[8] R. Clark Jones, "A new calculus for the treatment of optical systems, VII. Properties of the *N*-matrices," *J. Opt. Soc. Am.* **38**, 671–683, 1948.

[9] R. Clark Jones, "A new calculus for the treatment of optical systems, VIII. Electromagnetic theory," *J. Opt. Soc. Am.* **46**, 126–131, 1956.

[10] D. S. Kliger, J. W. Lewis, and C. E. Randall, *Polarized Light in Optics and Spectroscopy*, San Diego, CA: Academic Press, 1990.

[11] W. A. Shercliff, *Polarized Light: Production and Use*, Cambridge, MA: Harvard University Press, 1962.

[12] R. M. A. Azzam and N. M. Bashara, *Ellipsometry and Polarized Light*, Amsterdam: North-Holland, 1977.

[13] A. Yariv and P. Yeh, *Optical Waves in Crystals*, New York: Wiley, 2003.

[14] E. Collett. *Field Guide to Polarization*, Bellingham, WA: SPIE, 2005.

[15] J. R. Priebe. "Operational form of the Mueller matrices," *J. Opt. Soc. Am.* **59**, 176–180, 1968,

[16] D. H. Goldstein, *Polarized Light*, 3rd edn., Boca Raton, FL: CRC Press, 2010.

# 18 The electro-optic and acousto-optic effects and modulation of light beams

In this chapter we shall explain how the distortion produced in a crystal lattice by the application of an electric field or by the passage of a sound wave affects the propagation of light through the crystal. These effects – the electro-optic and acousto-optic effects, and related effects such as field-induced changes in the absorption of a material – are of considerable practical importance since they can be used to amplitude- and phase-modulate light beams, shift their frequencies, and alter the direction in which they travel.

## 18.1 Introduction to the electro-optic effect

When an electric field is applied to a crystal, the ionic constituents move to new locations determined by the field strength, the charge on the ions, and the restoring force. As we saw in Chapter 17, unequal restoring forces along three mutually perpendicular axes in the crystal lead to anisotropy in the optical properties of the medium. When an electric field is applied to such a crystal, in general, it causes a change in the anisotropy. These changes can be described in terms of the modification of the indicatrix by the field – both in terms of the principal refractive indices of the medium and in terms of the orientation of the indicatrix. If these effects can be described, to first order, as being linearly proportional to the applied field then the crystal exhibits the *linear* electro-optic effect. We shall see that this results only if the crystal lattice lacks a center of symmetry.[1] So, some cubic crystals can exhibit the linear electro-optic effect. If the crystal possesses a center of symmetry (or is even an isotropic material such as a gas or liquid) a change in optical properties can result that depends, to first order, on the square of the applied field. This is the *quadratic* electro-optic or Kerr effect. Both the linear and quadratic electro-optic effects can be used effectively in various optical devices.

[1] A crystal possesses a center of symmetry if there are identical particles in the lattice at vectors $\mathbf{r}$ and $-\mathbf{r}$, where $\mathbf{r}$ is a position vector measured from an appropriate origin.

## 18.2 The linear electro-optic effect

For any Cartesian coordinate system the equation of the indicatrix has the general form

$$\left(\frac{1}{n^2}\right)_1 x^2 + \left(\frac{1}{n^2}\right)_2 y^2 + \left(\frac{1}{n^2}\right)_3 z^2 + 2\left(\frac{1}{n^2}\right)_4 yz + 2\left(\frac{1}{n^2}\right)_5 xz + 2\left(\frac{1}{n^2}\right)_6 xy = 1. \tag{18.1}$$

If $x$, $y$, and $z$ are chosen to be principal axes, the cross-terms in $yz$, $xz$, and $xy$ disappear, and, in general, the equation of a biaxial indicatrix (Eq. (17.95)) results. For the linear electro-optic effect, an applied electric field changes each of the coefficients $(1/n^2)_i$ above in a way that is linearly dependent on the $E_x$, $E_y$, and $E_z$ components of the field. This change in these coefficients is described by the electro-optic tensor $\overline{\overline{r}}$ through the relationship

$$\Delta\left(\frac{1}{n^2}\right)_i = \sum_{j=1}^{3} r_{ij}E_j, \tag{18.2}$$

where $E_1 = E_x$, $E_2 = E_y$, and $E_3 = E_z$. So, for example,

$$\Delta\left(\frac{1}{n^2}\right)_1 = r_{11}E_x + r_{12}E_y + r_{13}E_z \tag{18.3}$$

and

$$\begin{pmatrix} \Delta(1/n^2)_1 \\ \Delta(1/n^2)_2 \\ \Delta(1/n^2)_3 \\ \Delta(1/n^2)_4 \\ \Delta(1/n^2)_5 \\ \Delta(1/n^2)_6 \end{pmatrix} = \begin{pmatrix} r_{11} & r_{12} & r_{13} \\ r_{21} & r_{22} & r_{23} \\ r_{31} & r_{32} & r_{33} \\ r_{41} & r_{42} & r_{43} \\ r_{51} & r_{52} & r_{53} \\ r_{61} & r_{62} & r_{63} \end{pmatrix} \begin{pmatrix} E_x \\ E_y \\ E_z \end{pmatrix}. \tag{18.4}$$

Typical magnitudes of the $r_{ij}$ coefficients are $10^{-12}$ m V$^{-1}$. The form of the matrix describing $\overline{\overline{r}}$ depends on the symmetry of the crystal and is closely related to the symmetry of the piezoelectric[2] tensor $\overline{\overline{d}}$, which relates polarization produced in a medium to stress $\sigma$.[3] The

[2] *Piezoelectricity* is the phenomenon in which application of a force to a crystal causes a voltage to develop between the faces of the crystal. The voltage appears because the force has caused polarization – separation of positive and negative charges within the medium.

[3] *Stress* is force per unit area and can take the form of compressive or shear stress. For a more detailed discussion see Nye [1].

matrix describing $\overline{\overline{r}}$ is the transpose of the matrix describing $\overline{\overline{d}}$. Nye has tabulated in a convenient way the form of $\overline{\overline{d}}$. Table 18.1 is a slight variant on the format given by Nye.[4]

With an electric field applied the equation of the indicatrix changes from Eq. (18.1) to

$$\begin{aligned}&\left[\frac{1}{n_1^2}+\Delta\left(\frac{1}{n^2}\right)_1\right]x^2+\left[\frac{1}{n_2^2}+\Delta\left(\frac{1}{n^2}\right)_2\right]y^2+\left[\frac{1}{n_3^2}+\Delta\left(\frac{1}{n^2}\right)_3\right]z^2\\&\quad+2\left[\frac{1}{n_4^2}+\Delta\left(\frac{1}{n^2}\right)_4\right]yz+2\left[\frac{1}{n_5^2}+\Delta\left(\frac{1}{n^2}\right)_5\right]xz\\&\quad+2\left[\frac{1}{n_6^2}+\Delta\left(\frac{1}{n^2}\right)_6\right]xy=1.\end{aligned}\tag{18.5}$$

The use of the index $i$ on $\Delta(1/n^2)_i$ is an example of what is called *contracted* notation: $i$ takes the values 1, 2, 3, 4, 5, and 6, where $1 \equiv xx$, $2 \equiv yy$, $3 \equiv zz$, $4 \equiv yz$, $5 \equiv xz$, and $6 \equiv xy$. Therefore the index $i$ represents the specific quadratic term with which it is associated in the equation of the indicatrix. We shall have cause to use this notation further.

So, if our original axes were principal axes, the indicatrix becomes

$$\begin{aligned}&\left[\frac{1}{n_x^2}+\Delta\left(\frac{1}{n^2}\right)_1\right]x^2+\left[\frac{1}{n_y^2}+\Delta\left(\frac{1}{n^2}\right)_2\right]y^2\\&\quad+\left[\frac{1}{n_z^2}+\Delta\left(\frac{1}{n^2}\right)_3\right]z^2+2\Delta\left(\frac{1}{n^2}\right)_4yz+2\Delta\left(\frac{1}{n^2}\right)_5xz\\&\quad+2\Delta\left(\frac{1}{n^2}\right)_6xy=1.\end{aligned}\tag{18.6}$$

The application of the electric field has introduced cross-terms, so $x$, $y$, and $z$ are no longer appropriate principal axes – the indicatrix has been changed in shape and rotated in space.

[4] His piezoelectric coefficients $d_{ij}$ describe polarization produced by stress in the form

$$\begin{pmatrix}P_x\\P_y\\P_z\end{pmatrix}=\begin{pmatrix}d_{11}&d_{12}&d_{13}&d_{14}&d_{15}&d_{16}\\d_{21}&d_{22}&d_{23}&d_{24}&d_{25}&d_{26}\\d_{31}&d_{32}&d_{33}&d_{34}&d_{35}&d_{36}\end{pmatrix}\begin{pmatrix}\sigma_{11}\\\sigma_{22}\\\sigma_{33}\\\sigma_{23}\\\sigma_{13}\\\sigma_{12}\end{pmatrix},$$

so, for example,

$$P_x=d_{11}\sigma_{11}+d_{12}\sigma_{22}+d_{13}\sigma_{33}+d_{14}\sigma_{23}+d_{15}\sigma_{13}+d_{16}\sigma_{12}.$$

Here $\sigma_{12}$ is a shear stress – it is the stress produced in the $y$ direction by a force acting on the face of a volume perpendicular to $x$. To be consistent with our notation for $\overline{\overline{r}}$, where factors of 2 appear in the cross-terms in Eq. (18.1) we need to write

$$P_x=d_{11}\sigma_{11}+d_{12}\sigma_{22}+d_{13}\sigma_{33}+\left(\tfrac{1}{2}d_{14}\right)2\sigma_{23}+\left(\tfrac{1}{2}d_{15}\right)2\sigma_{13}+\left(\tfrac{1}{2}d_{16}\right)2\sigma_{12}.$$

So all Nye's $d_{ij}$ coefficients where $i \neq j$ are double the value they have in our definition, where in symmetry terms $d_{ij} \equiv r_{ji}$.

**Table 18.1** The electro-optic $\bar{\bar{r}}$ for the 32 crystal symmetry classes. · Zero element, ● nonzero element, ●——● equal nonzero elements, ●- - - - - - ○ equal nonzero elements of opposite sign. Centrosymmetric classes – all elements of $\bar{\bar{r}}$ are zero.

Triclinic

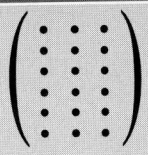

Example: Calcium thiosulphate ($CaS_2O_3.6H_2O$) strontium tartrate ($SrH_2(C_4H_4O_6)_2.4H_2O$)

Monoclinic

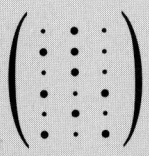

2 (symmetry axis parallel to $y$)

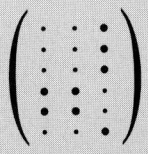

2 (symmetry axis parallel to $z$)

Example: lithium sulphate ($LiSO_4.H_2O$), tartaric acid triglycine sulphate

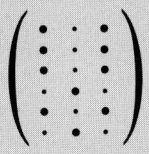

m (perpendicular to $y$)

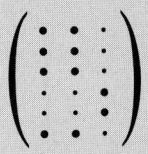

m (perpendicular to $z$)

Example: potassium nitrite ($KNO_2$)

Orthorhombic

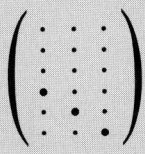

222

Example: α–iodic acid (α–$HIO_3$), magnesium sulphate ($MgSO_4.7H_2O$), Rochelle salt ($KNaC_4H_4O_6.4H_2O$)

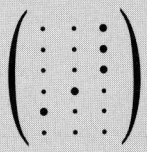

mm2

Example: Barium sodium niobate ($Ba_2NaNb_5O_{15}$), polyvinylidine fluoride (PVF), $(CH_2CF_2)_n$

Tetragonal

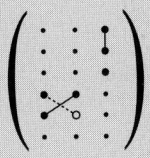

4

Example: iodo succinimide $(CH_2CO)_2NI$

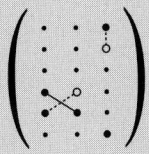

$\bar{4}$

Example: pentaerythritol ($C(CH_2OH)_4$), cahnite ($Ca_4B_2As_2O_{12}.4H_2O$)

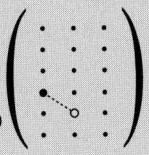

422

Example: nickel sulphate $NiSO_4.6H_2O$

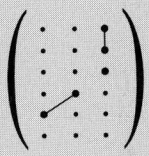

4 mm

Example: $BaTiO_3$

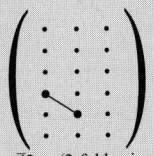

$\bar{4}2$ m (2-fold axis parallel to $x$)

Example: KDP, ADP, $CdGeAs_2$, $AgGaSe_2$, $AgGaS_2$

**Table 18.1** (cont.)

Cubic

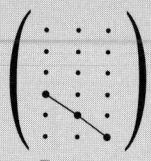

$\bar{4}3$ mm, 23

Examples: ($\bar{4}3$ m): GaAs, InAs, GaP, ZnSe, CdTe, InSb. (23): Sodium chlorate ($NaClO_3$), sodium bromate ($NaBrO_3$)

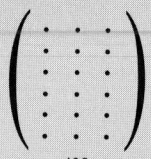

432 (lacks a center of symmetry, but other symmetry elements make all elements zero)

Trigonal

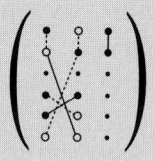

Example: sodium periodate $NaIO_4.3H_2O$

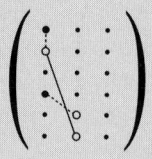

Examples: quartz, Te, mercury sulphide ($\alpha$–HgS)

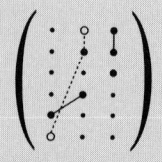

3m (m perpendicular to $x$) standard orientation

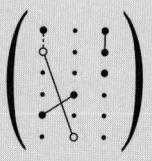

3m (m perpendicular to $y$)

Examples: $LiNbO_3$, $LiTaO_3$, proustite ($Ag_3AsS_3$), pyragyrite ($Ag_3SbS_3$)

Hexagonal

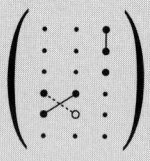

6

Examples: lithium iodate ($LiIO_3$), lodoform ($CHI_3$)

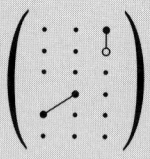

6 mm

Examples: CdS, CdSe, ZnS, ZnO

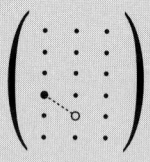

622

Examples: barium aluminate ($BaAl_2O_4$) kalsilite ($KAlSiO_4$)

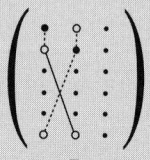

$\bar{6}$

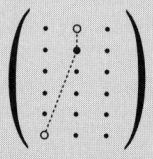

$\bar{6}$ m2 (m perpendicular to $x$) standard orientation

Examples: benitoite ($BaTiSi_3O_9$)

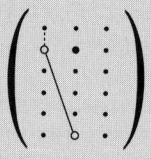

$\bar{6}$ m2 (m perpendicular to $y$)

In a centrosymmetric crystal the change in shape of the indicatrix produced by a field **E** must be the same as for field –**E**, since these two fields are identical insofar as the lattice is concerned. Therefore

$$\Delta\left(\frac{1}{n^2}\right)_i = \sum_{j=1}^{3} r_{ij}E_j = -\sum_{j=1}^{3} r_{ij}E_j, \tag{18.7}$$

which can be true only if $r_{ij} = 0$. Consequently, there is no *linear* electro-optic effect (or any piezoelectric effect) in a centrosymmetric crystal.

The linear electro-optic effect is closely related to the inverse piezoelectric effect – where application of an electric field to a non-centrosymmetric crystal leads to an actual change in crystal shape [1]. The change in crystal shape leads to strain and the shape and orientation of the indicatrix are altered. With this additional contribution to the change in index coefficients we can write

$$\Delta\left(\frac{1}{n^2}\right)_i = \sum_{k=1}^{6} p_{ik}S_j + \sum_{j=1}^{3} r_{ij}E_j; \tag{18.8}$$

here $j$ takes the values 1, 2, and 3; $i$ and $k$ take the values $1 \rightarrow 6$ (in contracted notation); $S_j$ is a component of the strain;[5] and $p_{ik}$ is the *elasto-optic* or *photoelastic* tensor. At high frequencies the inertia of the crystal prevents it straining macroscopically, so the first term on the right-hand side of Eq. (18.8) is zero. At high frequencies the residual change in indices is called the *Pockels* effect. At low frequencies the elasto-optic effects cannot be ignored. However, since the deformation leading to strain is generally caused by the inverse piezoelectric effect[6]– and is therefore also related to the applied field – it is possible to incorporate all changes in index into a single low-frequency linear electro-optic tensor $\bar{\bar{r}}^{\mathrm{dc}}$ and write

$$\Delta\left(\frac{1}{n^2}\right)_i = \sum_{j=1}^{3} r_{ij}^{\mathrm{dc}}E_j. \tag{18.9}$$

We shall not deal explicitly with this point further, but it should be noted that, whenever a coefficient $r_{ij}$ is written in what follows, this coefficient may be significantly different in magnitude for low and high frequencies. Table 18.2 gives some examples for important linear electro-optic materials.

One additional effect is worth noting – the photoelastic effect. This is the change in index coefficients produced directly by applied stress (no applied electric field is involved). For this effect, from Eq. (18.8) with no applied electric field,

[5] Each strain component is a measure of the distortion of the crystal shape in a particular axial direction or of the rotation of the structure about a particular axial direction. For further details see Nye [1].

[6] If strain results solely from the inverse piezoelectric effect then

$$r_{ij}^{\mathrm{dc}} = r_{ij} + \sum_{k=1}^{6} p_{ik}d_{jk}.$$

**Table 18.2** Low- and high-frequency values of elements of the electro-optic tensor for various crystals

| Material | Symmetry | Electro-optic coefficients ($10^{-12}$ m $V^{-1}$) at specific wavelength spanner (μm) | |
|---|---|---|---|
| | | Low frequency | High frequency |
| KDP | $\bar{4}2m$ | $r_{63} = -10.5$<br>$r_{41} = 8.8$ | $r_{63} = 8.15$ |
| ADP | $\bar{4}2m$ | $r_{63} = -8.5$<br>$r_{41} = 8.6$ | $r_{63} = 5.5$ |
| CDA ($CsH_2AsO_4$) | $\bar{4}2m$ | $r_{63} = 8.6$ | |
| $LiNbO_3$ | 3m | $r_{33} = 32.2$ ($\lambda = 0.633$)<br>$r_{13} = 10$ ($\lambda = 0.633$)<br>$r_{22} = 7$ ($\lambda = 0.633$)<br>$r_{51} = 32$ ($\lambda = 0.633$) | $r_{33} = 30.8$ ($\lambda = 0.633$)<br>$r_{13} = 8.6$ ($\lambda = 0.633$)<br>$r_{22} = 3.4$ ($\lambda = 0.633$)<br>$r_{51} = 28$ ($\lambda = 0.633$) |
| $LiTaO_3$ | 3m | | $r_{33} = 35.8$ ($\lambda = 0.633$)<br>$r_{13} = 7$ ($\lambda = 0.633$)<br>$r_{22} \simeq 1$ ($\lambda = 0.633$)<br>$r_{51} = 20$ ($\lambda = 0.633$) |
| $BaTiO_3$ | 4mm | $r_{51} = 1{,}640$ ($\lambda = 0.546$) | $r_{51} = 820$ ($\lambda = 0.546$)<br>$r_{33} = 28$ ($\lambda = 0.633$)<br>$r_{13} = 8$ ($\lambda = 0.633$) |
| $PbTiO_3$ | 4mm | | $r_{33} = 5.9$ ($\lambda = 0.633$)<br>$r_{13} = 13.8$ ($\lambda = 0.633$) |
| $Ba_2NaNb_5O_{15}$ | mm2 | $r_{33} = 48$ ($\lambda = 0.633$)<br>$r_{13} = 15$ ($\lambda = 0.633$)<br>$r_{23} = 13$ ($\lambda = 0.633$)<br>$r_{42} = 92$ ($\lambda = 0.633$)<br>$r_{51} = 90$ ($\lambda = 0.633$) | $r_{33} = 29$ ($\lambda = 0.633$)<br>$r_{13} = 7$ ($\lambda = 0.633$)<br>$r_{23} = 8$ ($\lambda = 0.633$)<br>$r_{42} = 75$ ($\lambda = 0.633$)<br>$r_{51} = 88$ ($\lambda = 0.633$) |
| GaAs | $\bar{4}3m$ | $r_{41} = 1.6$ ($\lambda = 10.6$) | $r_{41} = 1.5$ ($\lambda = 10.6$) |
| ZnSe | $\bar{4}3m$ | $r_{41} = 2.0$ ($\lambda = 0.546$) | $r_{41} = 2.0$ ($\lambda = 0.633$) |
| ZnS | 6mm | | $r_{33} = 1.8$ ($\lambda = 0.633$)<br>$r_{13} = 0.9$ ($\lambda = 0.633$) |
| $SiO_2$(quartz) | 32 | $r_{11} = -0.47$ ($\lambda = 0.5$)<br>$r_{41} = 0.2$ | $r_{11} = 0.29$ ($\lambda = 0.633$) |
| CuCl | $\bar{4}3m$ | $r_{41} = 3.6$ ($\lambda = 0.633$)<br>$r_{41} = 3.2$ ($\lambda = 10.6$) | $r_{41} = 2.35$ ($\lambda = 0.633$)<br>$r_{41} = 2.2$ ($\lambda = 3.39$) |

$$\Delta\left(\frac{1}{n^2}\right)_i = \sum_{k=1}^{6} p_{ik} S_j. \tag{18.10}$$

In a medium in which the photoelastic effect is isotropic, Eq. (18.10) can be written in the simple form

$$\frac{1}{n^2} - \frac{1}{n_0^2} = pS, \tag{18.11}$$

which gives

$$\Delta n = -\frac{n_0^3}{2} pS, \tag{18.12}$$

where $p$ is the photoelastic or elasto-optic constant and $S$ is the strain (deformation/length). Equation (18.12) is particularly useful in describing the change in refractive index that occurs when a sound wave passes through an isotropic medium.

The photoelastic effect can also be described by the equation

$$\Delta\left(\frac{1}{n^2}\right)_i = \sum_{k=1}^{6} \pi_{ik}\sigma_k. \tag{18.13}$$

The $\pi_{ik}$ are piezo-optical coefficients whose typical magnitude is $10^{-12}$ m$^2$ N$^{-1}$; the $\sigma_k$ are the components of the stress (for further discussion see Nye [1]).

## 18.3 The quadratic electro-optic effect

In a crystal possessing a center of symmetry, or in an isotropic medium, there is no linear electro-optic effect. However, there is a change in the constants of the indicatrix that is related to second-order products of applied electric field components.

For an isotropic medium the index of refraction of the medium changes according to

$$\frac{1}{n^2} = \frac{1}{n_0^2} + RE^2, \tag{18.14}$$

where $R$ is a coefficient for the quadratic electro-optic or *Kerr* effect. Usually, the effect is characterized by the Kerr constant

$$B = \frac{\Delta n}{\lambda_0 E^2} = \frac{n_0^3 R}{2\lambda_0}. \tag{18.15}$$

In a centrosymmetric crystal whose indicatrix can be written as

$$\sum_{i=1}^{3} \frac{x_i^2}{n_i^2} = 1, \tag{18.16}$$

the quadratic electro-optic effect changes the indicatrix to [2]

$$\sum_{i=1}^{3}\sum_{j=1}^{3}\sum_{k=1}^{3}\sum_{\ell=1}^{3}\left(\frac{1}{n_{ij}^2} + R_{ijk\ell}E_kE_\ell\right)x_ix_j = 1, \tag{18.17}$$

where $n_{ij} = n_i$ for $i = j$, and $n_{ij} = 0$ for $i \neq j$.

In contracted notation the changes in coefficients of the indicatrix are

$$\Delta\left(\frac{1}{n^2}\right)_i = \sum_{k=1}^{3}\sum_{\ell=1}^{3} R_{ik\ell}E_kE_\ell. \tag{18.18}$$

## 18.4 Longitudinal electro-optic modulation

To illustrate how an applied electric field affects an electro-optic crystal it is instructive to consider the specific important material potassium dihydrogen phosphate (KDP, $KH_2PO_4$). The related materials KD*P ($KD_2PO_4$), ADP (ammonium dihydrogen phosphate, $(NH_4)H_2PO_4$), and AD*P ($(NH_4)D_2PO_4$) behave similarly. For these materials, which are of symmetry $\bar{4}2m$, the electro-optic tensor has the form

$$\overline{\overline{r}} = \begin{pmatrix} 0 & 0 & 0 \\ 0 & 0 & 0 \\ 0 & 0 & 0 \\ r_{41} & 0 & 0 \\ 0 & r_{41} & 0 \\ 0 & 0 & r_{63} \end{pmatrix}. \tag{18.19}$$

In the presence of an applied field $\mathbf{E} = E_x\hat{\mathbf{i}} + E_y\hat{\mathbf{j}} + E_z\hat{\mathbf{k}}$ the indicatrix, written in its original principal-axes coordinates, is

$$\frac{x^2+y^2}{n_o^2} + \frac{z^2}{n_e^2} + 2r_{41}E_x yz + 2r_{41}E_y xz + 2r_{63}E_z xy = 1. \tag{18.20}$$

For a field applied in the $z$ direction only, Eq. (18.20) becomes

$$\frac{x^2+y^2}{n_o^2} + \frac{z^2}{n_e^2} + 2r_{63}E_z xy = 1. \tag{18.21}$$

To find the new principal axes we note that Eq. (18.21) is symmetric in $x$ and $y$, so the new $x$ and $y$ axes must be rotated by 45° from the original axes, as shown in Fig. 18.1. The $z$ axis is unchanged.

The old axes are related to the new according to

$$\begin{aligned} x &= x'\cos 45^\circ + y'\sin 45^\circ, \\ y &= -x'\sin 45^\circ + y'\cos 45^\circ. \end{aligned} \tag{18.22}$$

Substitution into Eq. (18.21) gives

$$\left(\frac{1}{n_o^2} - r_{63}E_z\right)x'^2 + \left(\frac{1}{n_o^2} + r_{63}E_z\right)y'^2 + \frac{z^2}{n_e^2} = 1. \tag{18.23}$$

The indicatrix now no longer has any cross-terms, so our choice of new principal axes was correct. The crystal has now become biaxial (but only slightly so) with

$$n_{x'}^2 = \frac{n_o^2}{1 - n_o^2 r_{63}E_z}, \qquad n_{y'}^2 = \frac{n_o^2}{1 + n_o^2 r_{63}E_z}, \qquad n_{z'}^2 = n_e^2. \tag{18.24}$$

If $n_o^2 r_{63}E_z \ll 1$, which is true for any reasonable electric field, these equations become

$$n_{x'} = n_o\left(1 + \tfrac{1}{2}n_o^2 r_{63}E_z\right), \qquad n_{y'} = n_o\left(1 - \tfrac{1}{2}n_o^2 r_{63}E_z\right), \qquad n_{z'} = n_e. \tag{18.25}$$

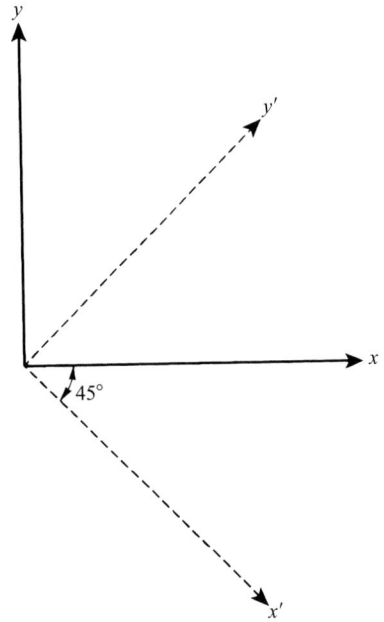


Fig. 18.1 Rotation of principal axes in a crystal of $\bar{4}2m$ symmetry when an electric field is applied in the $z$ direction.

If a wave propagates in the $z$ direction through such a crystal the retardation is

$$\Delta\phi = \frac{2\pi L}{\lambda_0}(n'_x - n'_y) = \frac{2\pi L n_o^3 r_{63} E_z}{\lambda_0}. \tag{18.26}$$

The crystal has become a waveplate whose retardation is linearly proportional to the field. It is common practice to characterize such a crystal by the voltage, $V_\pi$, necessary to produce a retardation of $\pi$. Since this applied voltage acts across a length $L$ of crystal, $E_z = V_\pi/L$, so, from Eq. (18.26),

$$V_\pi = \frac{\lambda_0}{2n_o^3 r_{63}}. \tag{18.27}$$

For a crystal of KDP that has $r_{63} = -10.5 \times 10^{-12}$ m V$^{-1}$ and $n_o = 1.51$, and for an operating wavelength of 632.8 nm, Eq. (18.27) gives $V_\pi = 8{,}752$ V. A crystal of KDP with this longitudinally applied voltage has been transformed into a half-wave plate and will, therefore, transform a linearly polarized input wave into a linearly polarized output beam whose polarization direction has been rotated from its original direction. If the input wave is polarized at 45° to the $x'$ and $y'$ directions, as shown in Fig. 18.2, that is, polarized parallel to the original $x$ or $y$ directions of the undisturbed indicatrix, then the half-wave plate rotates the plane of linear polarization by 90°. When the crystal is operated in this way, as the applied voltage increases from 0 to $V_\pi$ the output makes a transition from linear polarization to elliptical polarization with the major axis aligned along the original polarization direction, to circular polarization, to elliptical polarization with the minor axis aligned along the original polarization direction, and to 90° rotated linear polarization, as

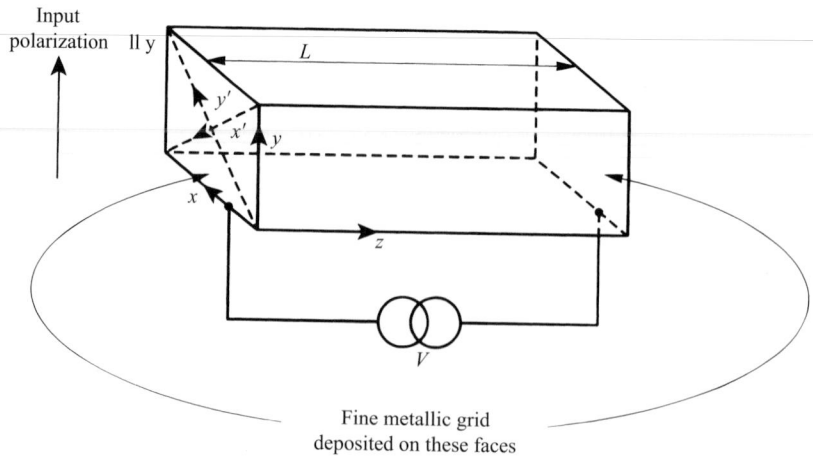


**Fig. 18.2** KDP crystal orientation for longitudinal electro-optic modulation.

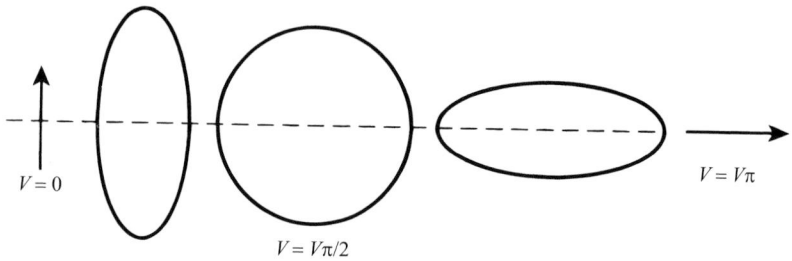


**Fig. 18.3** Changes in polarization state resulting from different amounts of electro-optic retardation for a crystal operated as in Fig. 18.2.

shown in Fig. 18.3. We shall see shortly how this modulation of the polarization state can be used to make an electro-optic amplitude modulator.

Longitudinal electro-optic modulators of this sort, in which the field is applied along the direction of propagation of light, are somewhat inconvenient insofar as the light must pass through the electrodes used to apply the field. To minimize loss, the electrodes are usually deposited as a fine metallic grid on the crystal surfaces shown in Fig. 18.2.

## 18.5 Transverse electro-optic modulation

It is more convenient to apply the electric field transversely between electrodes placed on side faces of the crystal as shown in Fig. 18.4. As an example of how this is done, we can consider the important electro-optic material lithium niobate $LiNbO_3$.[7]

[7] This material is also strongly piezoelectric and is used to fabricate acousto-optic modulators.

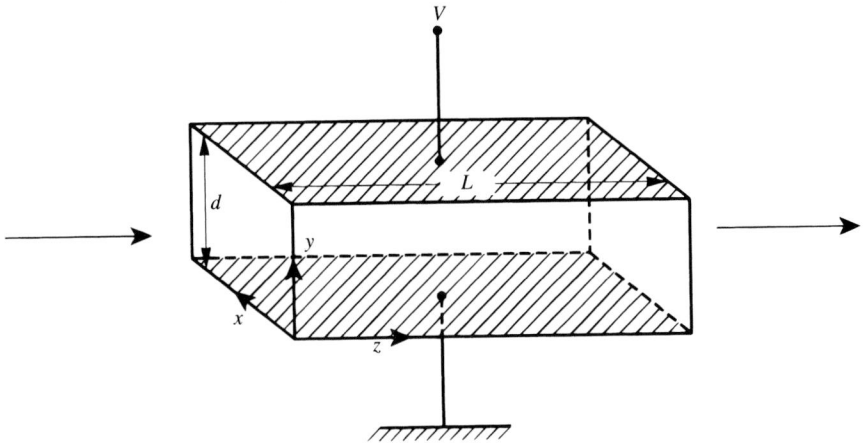


Fig. 18.4 A $LiNbO_3$ crystal setup for use as a transverse electro-optic amplitude modulator.

Its crystal symmetry is 3m, so the form of its linear electro-optic tensor is

$$\bar{\bar{r}} = \begin{pmatrix} 0 & -r_{22} & r_{13} \\ 0 & r_{22} & r_{13} \\ 0 & 0 & r_{33} \\ 0 & r_{51} & 0 \\ r_{51} & 0 & 0 \\ -r_{22} & 0 & 0 \end{pmatrix}. \tag{18.28}$$

Without an applied electric field the crystal has uniaxial symmetry with an indicatrix of the form

$$\frac{x^2}{n_o^2} + \frac{y^2}{n_o^2} + \frac{z^2}{n_e^2} = 1. \tag{18.29}$$

With a field applied in the $y$ principal-axis direction the indicatrix becomes

$$\left(\frac{1}{n_o^2} - r_{22}E_y\right)x^2 + \left(\frac{1}{n_o^2} + r_{22}E_y\right)y^2 + \frac{z^2}{n_e^2} + 2r_{51}E_y yz = 1. \tag{18.30}$$

Clearly, because of the introduction of the cross-term in $yz$ the original principal areas are no longer appropriate. The effect of the field has been to render the crystal biaxial and rotate the indicatrix in the $yz$ plane. Let us assume that the indicatrix has been rotated so that its new principal axes $y'$ and $z'$ are at an angle $\theta$ to the original principal axes, as shown in Fig. 18.5. The $x$ direction remains a principal axis with the field $E_y$ applied, as no cross-term involving $x$ has been introduced into the equation of the indicatrix.

We can write the old coordinates in terms of the new as

$$\begin{aligned} x &= x', \\ y &= y'\cos\theta - z'\sin\theta, \\ z &= z'\cos\theta + y'\sin\theta. \end{aligned} \tag{18.31}$$

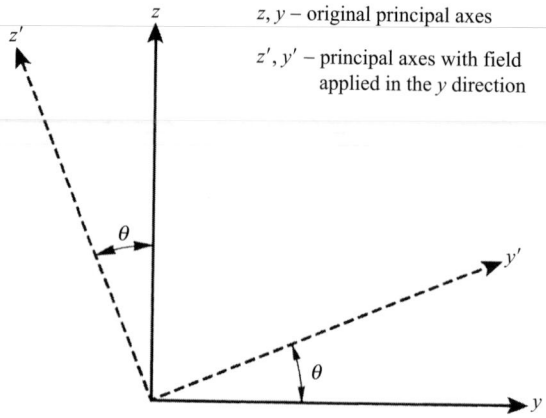


**Fig. 18.5** Rotation of the principal axes in $LiNbO_3$ when a field is applied in the $y$ principal-axis direction.

If we substitute from Eq. (18.31) into Eq. (18.30), the condition that $x'$, $y'$, and $z'$ are the new principal axes is that the cross-term in $z'$ and $y'$ must vanish, that is,

$$\left(\frac{1}{n_e^2} - \frac{1}{n_o^2} - r_{22}E_y\right)\sin\theta\cos\theta + r_{51}E_y(2\cos^2\theta - 1) = 0. \tag{18.32}$$

We can assume that $r_{51}E_y$ is small enough that $\theta$ is a small angle and write $\sin\theta \simeq \theta$ and $\cos\theta \simeq 1$. Equation (18.32) then gives

$$\theta = \frac{-r_{51}E_y}{1/n_e^2 - 1/n_o^2 - r_{22}E_y}. \tag{18.33}$$

With $r_{51} = 28 \times 10^{-12}$ m V$^{-1}$, $r_{22} = 3.4 \times 10^{-12}$ m V$^{-1}$, $n_e = 2.21$, $n_o = 2.3$, and a voltage of 1 kV applied across a 1-mm-thick crystal,

$$\theta = \frac{-28 \times 10^{-12} \times 10^6}{(1/2.21)^2 - (1/2.3)^2 - 3.4 \times 10^{-12} \times 10^6} = 1.78 \text{ mrad} = 0.1^\circ. \tag{18.34}$$

Indeed, $\theta$ is a small angle. Since, in $LiNbO_3$, $n_o > n_e$, $\theta$ is a negative angle, so in reality the rotation of the axes is clockwise, not counterclockwise as shown in Fig. 18.5. With respect to the new principal axes, the equation of the indicatrix is now

$$\left(\frac{1}{n_o^2} - r_{22}E_y\right)x'^2 + \left[\left(\frac{1}{n_o^2} + r_{22}E_y\right)\cos^2\theta + \frac{\sin^2\theta}{n_e^2} + r_{51}E_y\sin(2\theta)\right]y'^2$$
$$+ \left[\left(\frac{1}{n_o^2} + r_{22}E_y\right)\sin^2\theta + \frac{\cos^2\theta}{n_e^2} - r_{51}E_y\sin(2\theta)\right]z'^2 = 1. \tag{18.35}$$

With $\cos\theta \simeq 1$ and $\sin\theta \simeq \theta$,

$$\left(\frac{1}{n_o^2} - r_{22}E_y\right)x'^2 + \left(\frac{1}{n_o^2} + r_{22}E_y + 2r_{51}E_y\theta + \frac{\theta^2}{n_e^2}\right)y'^2$$
$$+ \left[\left(\frac{1}{n_o^2} + r_{22}E_y\right)\theta^2 - 2r_{51}E_y\theta + \frac{1}{n_e^2}\right]z'^2 = 1. \tag{18.36}$$

The new principal refractive indices satisfy the equations

$$\begin{aligned}
\frac{1}{n_{x'}^2} &= \frac{1}{n_o^2} - r_{22}E_y, \\
\frac{1}{n_{y'}^2} &= \frac{1}{n_o^2} + r_{22}E_y + 2r_{51}E_y\theta + \frac{\theta^2}{n_o^2}, \\
\frac{1}{n_{z'}^2} &= \left(\frac{1}{n_o^2} + r_{22}E_y\right)\theta^2 - 2r_{51}E_y\theta + \frac{1}{n_e^2}.
\end{aligned} \tag{18.37}$$

Since $\theta$ is very small, we can actually neglect the rotation of the indicatrix and write its equation as

$$\frac{x^2}{n_x^2} + \frac{y^2}{n_y^2} + \frac{z^2}{n_z^2} = 1, \tag{18.38}$$

where

$$\begin{aligned}
n_x &= \frac{n_o}{\sqrt{1 - n_o^2 r_{22}E_y)}} = n_o\left(1 + \tfrac{1}{2}n_o^2 r_{22}E_y\right), \\
n_y &= \frac{n_o}{\sqrt{1 + n_o^2 r_{22}E_y)}} = n_o\left(1 - \tfrac{1}{2}n_o^2 r_{22}E_y\right), \\
n_z &= n_e.
\end{aligned} \tag{18.39}$$

For waves traveling in the $z$ direction the indices of refraction experienced for polarizations along $x$ and $y$ are $n_x$ and $n_y$, respectively. For the transversely operated electro-optic modulator shown in Fig. 18.4 the retardation is

$$\Delta\phi = \frac{2\pi L}{\lambda_0}(n_x - n_y) = \frac{4\pi L n_o^3 r_{22} V}{\lambda_0 d}. \tag{18.40}$$

The half-wave voltage $V_\pi$ of the modulator is the voltage required to make it act as a half-wave plate.

From Eq. (18.40), if $\phi = \pi$ then

$$V_\pi = \frac{\lambda_0 d}{4L n_o^3 r_{22}}. \tag{18.41}$$

For example, with $d = 5$ mm, $L = 10$ mm, $n_o = 2.3$, and $r_{22} = 3.4 \times 10^{-12}$ m V$^{-1}$ at an operating wavelength of 530 nm, $V_\pi = 1{,}600$ V.

To illustrate further how a change in the orientation of the applied field and the light-propagation direction affects the way in which an electro-optic crystal behaves, let us now consider what happens when light propagates through an $LiNbO_3$ crystal in the $y$ direction, but with the field applied along the $z$ direction.

In this case the indicatrix becomes

$$\left(\frac{1}{n_o^2} + r_{13}E_z\right)x^2 + \left(\frac{1}{n_o^2} + r_{13}E_z\right)y^2 + \left(\frac{1}{n_e^2} + r_{33}E_z\right)z^2 = 1. \tag{18.42}$$

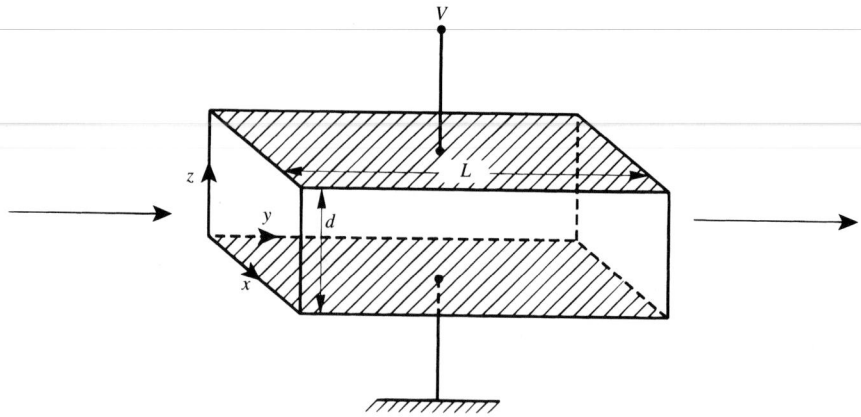


Fig. 18.6 $LiNbO_3$ crystal orientation for transverse electro-optic modulation.

The application of the field has left the crystal uniaxial, but the new ordinary and extraordinary refractive indices satisfy

$$\frac{1}{n_o'^2} = \left(\frac{1}{n_o^2} + r_{13}E_z\right),$$
$$\frac{1}{n_e'^2} = \left(\frac{1}{n_e^2} + r_{33}E_z\right). \tag{18.43}$$

Since $r_{13}E_z \ll 1/n_o^2$ and $r_{33}E_z \ll 1/n_e^2$ for any reasonable applied field strengths,

$$n_o' = \frac{n_o}{\sqrt{1 + r_{13}n_o^2E_z}} = n_o\left(1 - \tfrac{1}{2}r_{13}n_o^2E_z\right),$$
$$n_e' = \frac{n_e}{\sqrt{1 + r_{33}n_e^2E_z}} = n_e\left(1 - \tfrac{1}{2}r_{33}n_e^2E_z\right). \tag{18.44}$$

If a wave passes through such a crystal in the $y$ direction it will, in general, split into an O wave polarized in the $x$ direction, and an E wave polarized along the $z$ direction. For the geometry shown in Fig. 18.6 the retardation is

$$\Delta\phi = \frac{2\pi L}{\lambda_0}(n_e' - n_o), \tag{18.45}$$

which gives

$$\Delta\phi = \frac{2\pi L}{\lambda_0}\left[n_e - n_o + \tfrac{1}{2}\left(r_{33}n_e^3 - r_{13}n_o^3\right)E_z\right]. \tag{18.46}$$

The half-wave voltage of the crystal used in this way is

$$V_\pi = \frac{\lambda_0 d}{L\left(r_{33}n_e^3 - r_{13}n_o^3\right)}. \tag{18.47}$$

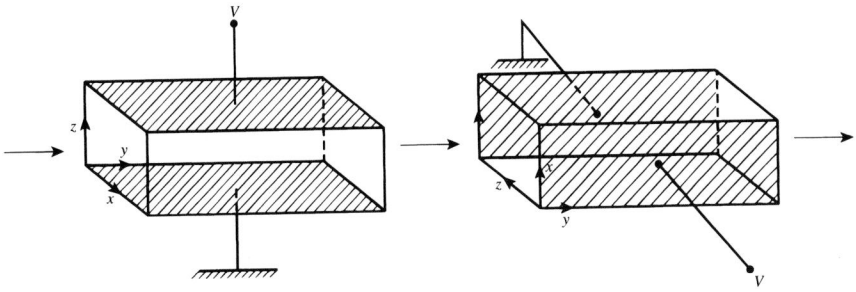


Fig. 18.7 A transverse electro-optic modulator using two $LiNbO_3$ crystals to cancel out zero-field birefringence.

With $d = 5$ mm, $L = 10$ mm, and $r_{33}n_e^3 - r_{13}n_o^3 = 224 \times 10^{-12}$ m V$^{-1}$, and an operating wavelength of 530 nm, $V_\pi = 1{,}183$ V. Although this is a slightly lower half-wave voltage than that in the configuration corresponding to Eq. (18.41) discussed previously, there is a potential drawback to using the crystal in this way. Equation (18.46) shows that the crystal operated in this orientation acts as a waveplate – because the wave is not propagating along the optic axis. The field-independent birefringent retardation $2\pi L(n_e - n_o)/\lambda_0$ will be temperature-sensitive and can lead to practical stability problems. However, in practice this difficulty can be overcome by building an electro-optic device using pairs of crystals in series as shown in Fig. 18.7. The orientations of the $x$ and $z$ axes in the two crystals are arranged to be orthogonal so that the constant term in Eq. (18.46) cancels out. In order that the field-dependent retardation in Eq. (18.46) does not also cancel out, the field polarity in the second crystal must be reversed.

## 18.6 Electro-optic amplitude modulation

We have seen through the above example that an electro-optic crystal can be used to produce a retardation that is directly proportional to an applied voltage. To use this effect to make a useful amplitude modulator we must place the electro-optic retardation modulator in an experimental arrangement that incorporates a polarizer, waveplate, and analyzer, as shown in Fig. 18.8. This figure shows an $LiNbO_3$ crystal operated in the transverse field geometry in which the light propagates in the $z$ direction and the field is applied in the $y$ direction. We can analyze the performance of such an arrangement by using Eq. (17.100). A specific case will serve as a good example. If the input and output polarizers are crossed then the fractional transmitted intensity (neglecting reflection losses at any interfaces involved) is, without the electro-optic crystal,

$$\frac{I_t}{I_0} = \tfrac{1}{2}\sin^2(2\beta)(1 - \cos\Delta\phi_{WP}), \tag{18.48}$$

where $\Delta\phi_{WP}$ is the retardation produced by the waveplate and $\beta$ is the angle between the O direction of the waveplate and the input polarization direction. It is usual to set $\beta = 45°$ and to insert the electro-optic crystal so that one of its preferred polarization directions

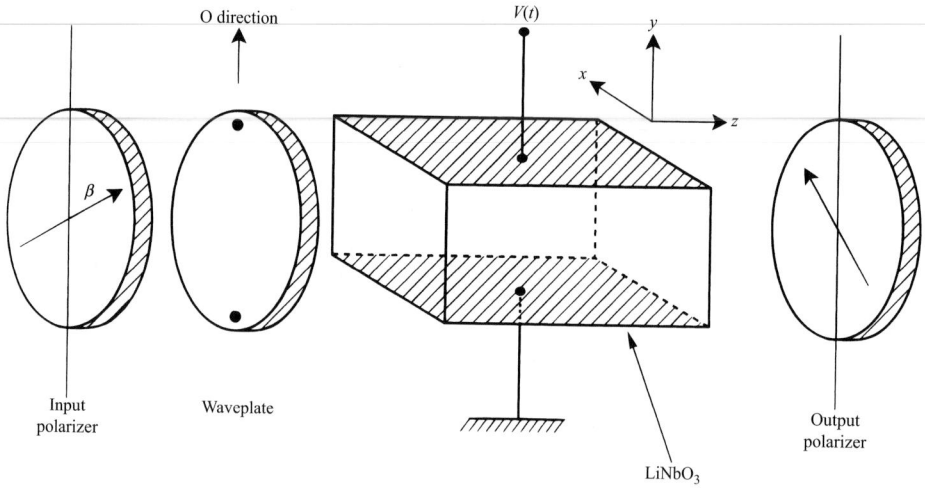


**Fig. 18.8** A transverse electro-optic amplitude modulator using $LiNbO_3$.

is parallel to the O direction of the waveplate. In this case the overall retardation can be written as

$$\Delta\phi = \Delta\phi_{\mathrm{WP}} + aV(t), \tag{18.49}$$

where $\Delta\phi_{\mathrm{WP}}$ is the retardation produced by the waveplate and $a$ is the constant of proportionality between the applied voltage $V(t)$ and the retardation in the electro-optic crystal. The constant $a$ may be positive or negative depending on the orientation of the preferred directions in the crystal. In this case from Eq. (18.48) we can write

$$\begin{aligned}\frac{I_{\mathrm{T}}}{I_0} &= \tfrac{1}{2}\{1 - \cos[\Delta\phi_{\mathrm{WP}} + aV(t)]\} \\ &= \tfrac{1}{2}\{1 - \cos\Delta\phi_{\mathrm{WP}}\cos[aV(t)] + \sin\Delta\phi_{\mathrm{WP}}\sin[aV(t)]\}\,.\end{aligned} \tag{18.50}$$

The amplitude modulation will be linear only if the time-dependent part of $I_{\mathrm{T}}/I_0$ is proportional to $V(t)$. This will be true only if (i) $\Delta\phi_{\mathrm{WP}} = \pi/2$, making the crystal a quarter-wave plate, and (ii) $aV(t) \ll 1$, so that $\sin[aV(t)] = aV(t)$. If these conditions are satisfied,

$$\frac{I_{\mathrm{T}}}{I_0} = \tfrac{1}{2}[1 + aV(t)]. \tag{18.51}$$

If the input modulating signal is

$$V(t) = V_0 \sin(\omega_{\mathrm{m}} t) \tag{18.52}$$

then the intensity is

$$\frac{I_{\mathrm{T}}}{I_0} = \tfrac{1}{2}(1 + aV_0 \sin\omega_{\mathrm{int}}). \tag{18.53}$$

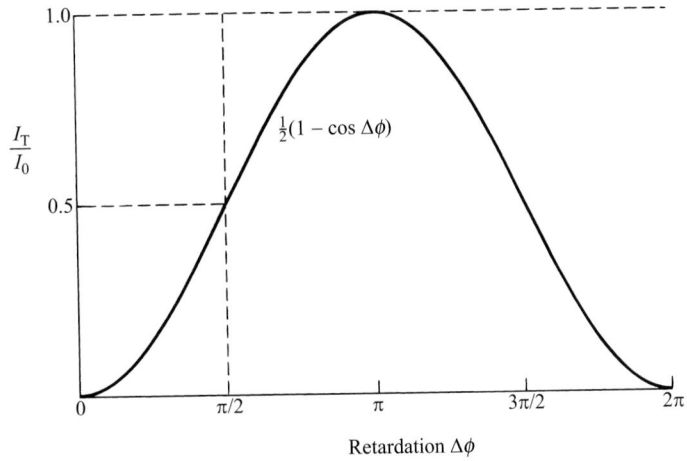


**Fig. 18.9** The transmittance characteristic of the electro-optic modulator shown in Fig. 18.8 as a function of the total retardation.

The modulated output is a faithful representation of the original modulation as shown in Fig. 18.9, which plots Eq. (18.50) as a function of the total retardation. Modulation about the "bias" point $\Delta\phi_{WP} = \pi/2$ produces linear modulation. For the geometry shown in Fig. 18.8 the retardation produced by the $LiNbO_3$ is given by Eq. (18.40) and

$$a = \frac{4\pi L n_o^3 r_{22}}{\lambda_0 d}. \tag{18.54}$$

Without the waveplate, the modulation would be nonlinear since, from Eq. (18.48),

$$\frac{I_T}{I_0} = \tfrac{1}{2}[1 - \cos(aV(t))], \tag{18.55}$$

which for $aV(t) \ll 1$ gives

$$\frac{I_T}{I_0} = \tfrac{1}{2}a^2 V^2(t). \tag{18.56}$$

This is clearly nonlinear modulation.

## 18.7 Electro-optic phase modulation

If the input polarizer in Fig. 18.8 is rotated so that the light is linearly polarized parallel to the $y$ direction, and the waveplate and output polarizer are removed, the arrangement of linear polarizer and crystal becomes an electro-optic phase modulator. In this situation the electric field of the light wave emerging from the crystal can be written as

$$E_y^{\omega}(t) = E_0 \cos[\omega t + \phi(t)], \tag{18.57}$$

where $\phi(t)$ is the phase modulation, which, from Eq. (18.39), is

$$\phi(t) = \left(\frac{n_o^3 r_{22}}{2d}\right) V(t). \tag{18.58}$$

If $V(t) = V_0 \sin(\omega_m t)$, then Eq. (18.57) can be written as

$$\begin{aligned} E_y^{\omega}(t) &= E_0 \cos[\omega t + m \sin(\omega_m t)] \\ &= E_0\{\cos(\omega t)\cos[m \sin(\omega_m t)] - \sin(\omega t)\sin[m \sin(\omega_m t)]\}, \end{aligned} \tag{18.59}$$

where

$$m = \left(\frac{n_o^3 r_{22}}{2d}\right) V_0 \tag{18.60}$$

is called the *depth of modulation*.

The frequency spectrum of the phase-modulated wave is interesting. We use the following identities [3]:

$$\cos[m \sin(\omega_m t)] = J_0(m) + 2\sum_{k=1}^{\infty} J_{2k}(m)\cos(2k\omega_m t), \tag{18.61}$$

$$\sin[m \sin(\omega_m t)] = 2\sum_{k=0}^{\infty} J_{2k+1}(m)\sin[(2k+1)\omega_m t], \tag{18.62}$$

$$2 \sin A \sin B = \cos(A - B) - \cos(A + B), \tag{18.63}$$

$$2 \cos A \cos B = \cos(A - B) + \cos(A + B), \tag{18.64}$$

and write Eq. (18.59) as

$$\begin{aligned} E_y^{\omega}(t) = E_0\{&J_0(m)\cos(\omega t) + J_1(m)\cos[(\omega + \omega_m)t] - J_1(m)\cos[(\omega - \omega_m)t] \\ &+ J_2(m)\cos[(\omega + 2\omega_m)t] + J_2(m)\cos[(\omega - 2\omega_m)t] \\ &+ J_3(m)\cos[(\omega + 3\omega_m)t] - J_3(m)\cos[(\omega - 3\omega_m)t] \\ &+ \cdots\}. \end{aligned} \tag{18.65}$$

The frequency spectrum of the phase-modulated light has acquired all of the frequencies $\omega \pm n\omega_m$, where $n$ is an integer, with amplitudes given by the Bessel-function amplitudes $J_n(m)$. The center frequency $\omega$ is called the *carrier* frequency. Figure 18.10 shows such a spectrum, calculated for $m = 1$.

Phase modulation has important communication applications, since the term $\phi(t)$ in Eq. (18.57) can be modulated in an arbitrary way to include information on the light beam. This is the optical analog of radiowave phase modulation (PM) or frequency modulation (FM)[8] [4–6].

[8] The concepts of PM and FM are closely related since the overall phase of the wave $\Phi(t) = \omega t + \phi(t)$ obeys $d\Phi/dt = \omega + d\phi/dt$, so $\omega + d\phi/dt$ can be regarded as the *instantaneous* frequency of the wave. Loosely speaking, we could say that FM $= d/dt$(PM).

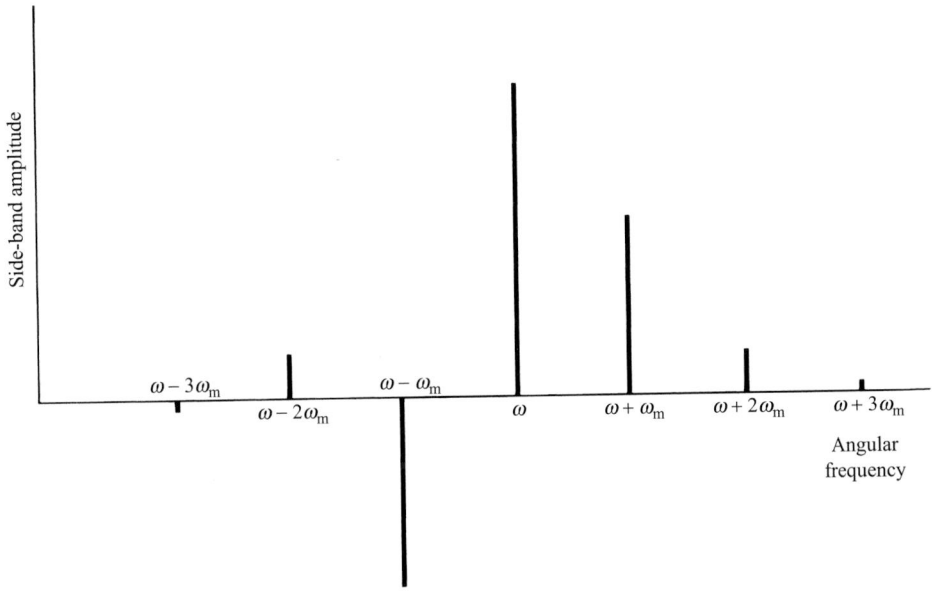


**Fig. 18.10** The phase-modulated spectrum for a modulation depth $m = 1$.

## 18.8 High-frequency waveguide electro-optic modulators

In a high-frequency electro-optic modulator the electrical drive to the modulator must be applied using some form of stripline configuration so that the induced phase or polarization modulation propagates at, or close to, the phase velocity of the light signal traveling through the modulator. If this is not done, the light propagating through the crystal sees an electric field of varying phase as it propagates, and the effects of the field on the phase or polarization state of the wave can cancel out. An excellent review of electro-optic modulators has been given by Alferness [7].

### 18.8.1 Straight-electrode modulators

In a stripline electro-optic modulator of the straight type shown in Fig. 18.11, the high-frequency response is limited by traveling-wave-mismatch effects. We can represent the drive voltage at position $z$ along this structure as

$$V(z,t) = V_0 \sin\left(\omega_m t - \frac{\omega_m n_m}{c_0} z\right), \tag{18.66}$$

where $c_0$ is the velocity of light *in vacuo*, $\omega_m$ is the drive frequency, and $n_m$ is the effective refractive index for the drive field (such that the phase velocity of the drive field is $c_0/n_m$).

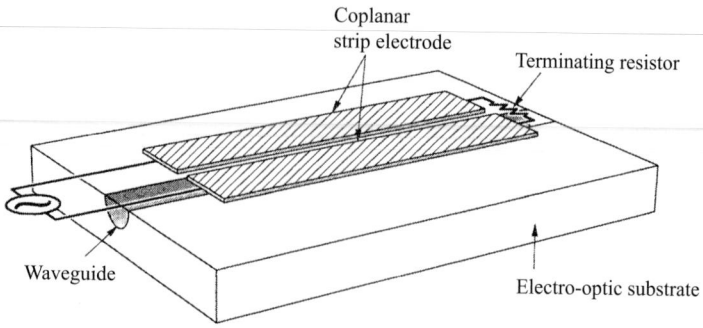


**Fig. 18.11** A traveling wave electro-optic modulator (courtesy of Suzanne Hiser).

The optical field propagates with phase velocity

$$c = \frac{c_0}{n}, \tag{18.67}$$

where $n$ is the effective refractive index for the waveguide mode.

If the optical field enters the crystal at $z = 0$ at time zero, then the voltage that is seen as it propagates can be found from Eq. (18.66) by writing $t = zn/c_0$, which is the time it takes the optical wave to reach location $z$:

$$V(z) = V_0 \sin\left[\frac{\omega_m z}{c_0}(n - n_m)\right]. \tag{18.68}$$

The electro-optically induced change in the index $n$ can be written as

$$\Delta n(z) = aV(z), \tag{18.69}$$

where $a$ is a constant that depends on the specific geometry and crystal being used.

The total phase shift experienced in propagating a length $\ell$ through the crystal is

$$\begin{aligned}\Delta\phi &= \int_0^\ell \frac{\omega\,\Delta n(z)}{c_0}\,dz \\ &= \frac{aV_0\omega}{c_0}\int_0^\ell \sin\left[\frac{\omega_m z}{c_0}(n - n_m)\right]dz,\end{aligned} \tag{18.70}$$

which gives

$$\Delta\phi = -\frac{aV_0}{n - n_m}\left\{\cos\left[\frac{\omega_m \ell}{c_0}(n - n_m)\right] - 1\right\}. \tag{18.71}$$

We make the substitution

$$\theta = \pi\omega_m/\omega_c, \tag{18.72}$$

where

$$\omega_c = \frac{\pi c_0}{(n_m - n)\ell} \tag{18.73}$$

is called the cutoff frequency.

The total phase shift can be written as

$$\begin{aligned}\Delta\phi &= \frac{aV_0\ell}{\pi c_0}\omega_c(\cos\theta - 1)\\ &= \frac{aV_0\ell\omega_m}{2c_0(\theta/2)}\left[\cos^2(\theta/2) - \sin^2(\theta/2) - 1\right]\\ &= \frac{aV_0\ell\omega_m}{2c_0(\theta/2)}\left[-2\sin^2(\theta/2)\right]\\ &= -\frac{aV_0\omega_m\ell}{c_0}\left[\frac{\sin(\theta/2)}{\theta/2}\right]\sin(\theta/2).\end{aligned} \tag{18.74}$$

Equation (18.74) can be written as

$$\begin{aligned}\Delta\phi &= -\frac{aV_0\omega_m^2\pi\ell}{2\omega_c c_0[\pi\omega_m/(2\omega_c)]}\frac{\sin^2(\theta/2)}{\theta/2}\\ &= -\frac{a\pi V_0\omega_m^2\ell}{2c_0\omega_c}\operatorname{sinc}^2(\theta/2),\end{aligned} \tag{18.75}$$

giving finally[9]

$$\Delta\phi = -\frac{aV_0\omega_m^2\ell^2}{2c_0^2}(n_m - n)\operatorname{sinc}^2(\theta/2). \tag{18.76}$$

The optimum length for the modulator satisfies

$$\frac{\omega_m\ell(n - n_m)}{c_0} = \pi;\ \text{or}\ \omega_m = \omega_c, \theta = \pi. \tag{18.77}$$

For $LiNbO_3$, $n_m = 4.2$, and $n = 2.146$; so, for example, with $\ell = 10$ mm the cutoff frequency is 7.36 GHz.

The electric field acts on the waveguide region of the electro-optic crystal as shown in Fig. 18.12. In $LiNbO_3$, for example, the waveguiding region is made by diffusing titanium into the crystal. The buffer layer is usually $Al_2O_3$ or $SiO_2$, and prevents the propagating wave, in this geometry a TM wave, from interacting with the metal electrodes. A practical device might have an indiffused channel 7 μm wide, a buffer layer 0.26 μm thick, and a 3-μm-thick Cr/Au electrode on the top. The electrode thickness is chosen to be larger than the skin depth at the design frequency. The index difference between guiding region and substrate is typically 0.01–0.02.

In the *Z*-cut, TM mode of operation, the electric field of the mode has $x$ and $z$ components.

[9] $\operatorname{sinc}(\theta/2) = \sin(\theta/2)/(\theta/2)$.

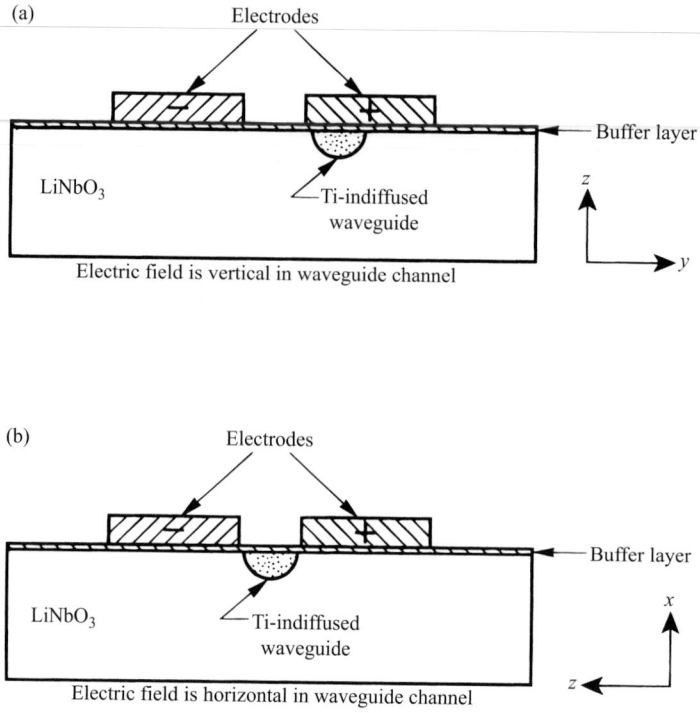


**Fig. 18.12** Geometries of high-frequency $LiNbO_3$ waveguide electro-optic modulators (courtesy of Suzanne Hiser): (a) *Z*-cut, TM mode; and (b) *X*-cut, TE mode.

For $LiNbO_3$, the linear electro-optic tensor is

$$\bar{\bar{r}} = \begin{pmatrix} 0 & -3.4 & 8.6 \\ 0 & 3.4 & 8.6 \\ 0 & 0 & 30.8 \\ 0 & 28 & 0 \\ 28 & 0 & 0 \\ -3.4 & 0 & 0 \end{pmatrix} \times 10^{-12}\ \mathrm{m\ V^{-1}}. \tag{18.78}$$

For this material typical optical losses are $\sim$1 dB cm$^{-1}$ at $\lambda = 633$ nm and $\sim$0.3 dB cm$^{-1}$ at $\lambda = 1.32$ μm.

The column vector of index changes once again obeys

$$\Delta\left(\frac{1}{n^2}\right) = \bar{\bar{r}} \begin{pmatrix} E_x \\ E_y \\ E_z \end{pmatrix}, \tag{18.79}$$

where $E_x$, $E_y$, and $E_z$ are the components of the externally applied electric field.

For the $Z$-cut the applied field has only a $z$ component and therefore the only change in the indicatrix is

$$\Delta\left(\frac{1}{n^2}\right)_3 = r_{33}E_z. \tag{18.80}$$

For the $X$-cut, the field is also applied in the $z$ direction, and the wave propagates in the $y$ direction. The only electric field component of the mode is in the $z$ direction, which is a TE mode – it has $x$- and $y$-directed magnetic field components. With the $X$-cut, $\Delta(1/n^2)_3 = r_{33}E_z$ is still the only change in the indicatrix.

## 18.9 Other high-frequency electro-optic devices

Various kinds of waveguide structure can be fabricated in $LiNbO_3$, as shown in Fig. 18.13.

In the Mach–Zehnder interferometer configuration the waves propagating in the two arms can experience different phase shifts, so they can emerge in or out of phase. The device can be used as an amplitude modulator or switch. If the phase shift between the two arms is $\phi$ the output intensity $I_{\text{out}}$ will be related to the input intensity $I_{\text{in}}$ according to

$$\frac{I_{\text{out}}}{I_{\text{in}}} = \tfrac{1}{2}(1 + \cos\phi) = \cos^2(\phi/2). \tag{18.81}$$

Variation of $\phi$ in such a device will not produce linear modulation. However, if the device is biased by $\pi/2$, for example, by having one interferometer arm slightly longer than the other (as shown in Fig. 18.13(d)) then we can write $\phi = \pi/2 + \Delta\phi$. In this case

$$\begin{aligned}\frac{I_{\text{out}}}{I_{\text{in}}} &= \tfrac{1}{2}[1 + \cos[\pi/2 + \Delta\phi]] \\ &= \tfrac{1}{2}[1 + \cos(\pi/2)\cos\Delta\phi - \sin(\pi/2)\sin\Delta\phi],\end{aligned} \tag{18.82}$$

which gives

$$\frac{I_{\text{out}}}{I_{\text{in}}} = \tfrac{1}{2} - \tfrac{1}{2}\Delta\phi \tag{18.83}$$

for small phase modulation $\Delta\phi$.

This provides linear modulation of the intensity about a mean value $I_{\text{out}} = \frac{1}{2}I_{\text{in}}$. To achieve passive biasing of $LiNbO_3$, the path difference is $\Delta L = 152$ nm; this provides $\phi = \pi/2$. Typical devices are 10–20 mm long, with the guiding regions 20 μm apart.

Because it is not possible to match the velocity of the excitation field and optical field over a long distance at high modulation frequencies, a phase-reversed scheme can be used to extend the bandwidth. After each length of the device over which the optical and drive fields get out of phase by $\pi$ the polarity of the drive field is reversed, as shown in Fig. 18.14. Such a device will in theory have a pass-band response. A broad frequency response

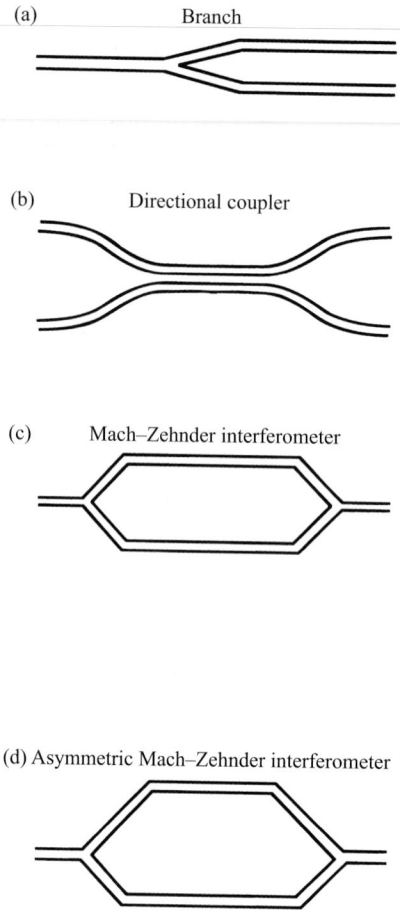


**Fig. 18.13** Integrated waveguide structures.

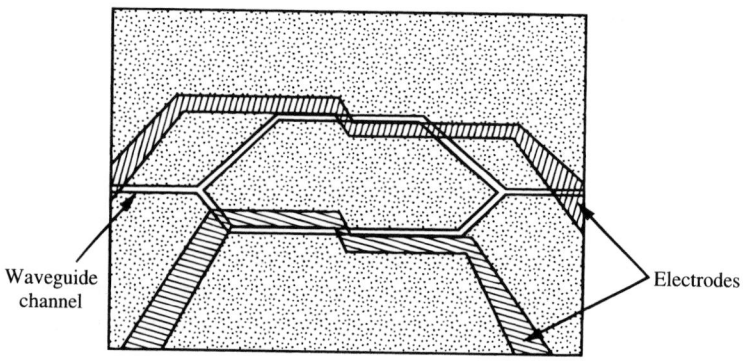


**Fig. 18.14** A phase-reversed traveling-wave electro-optic modulator with periodic phase-reversal electrodes, $N = 2$ (courtesy of Suzanne Hiser).

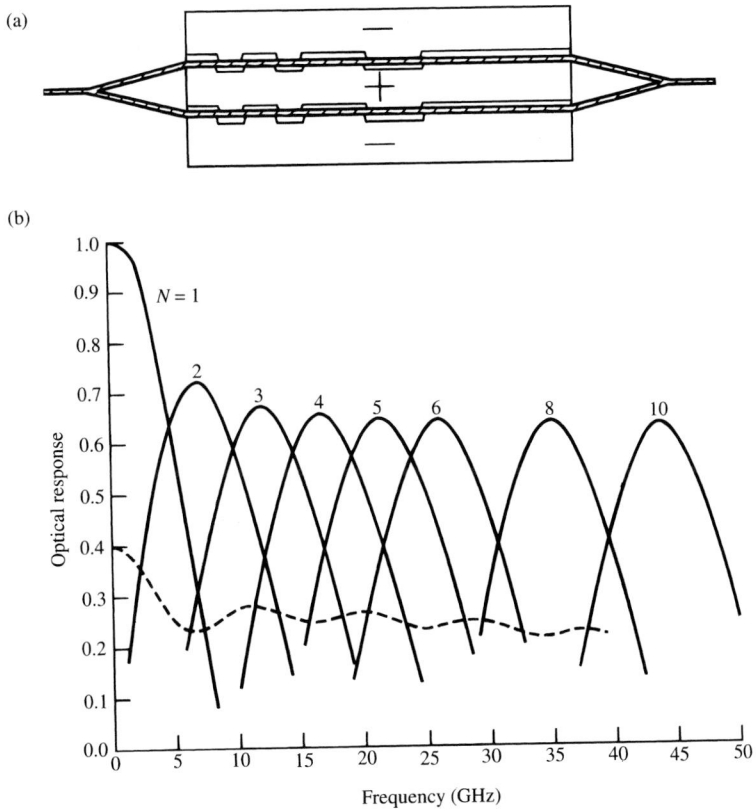


**Fig. 18.15** (a) A traveling-wave electro-optic modulator with 13-bit Barker code, aperiodic, phase-reversal electrodes. (b) The frequency response of various periodic and aperiodic traveling-wave electro-optic modulators. The dashed line is for a 13-bit Barker code aperiodic, phase-reversal electrode configuration (courtesy of Suzanne Hiser).

can be obtained by aperiodically varying the length between phase reversals of the drive field, as shown in Fig. 18.15(a). By increasing the number of aperiodic phase reversal electrodes a much broader response than that with a periodic phase-reversed structure with different numbers of sections $N$ can be obtained, as shown in Fig. 18.15(b). The placement of electrodes in these traveling-wave modulators can be optimized for broadband, high-frequency performance [8].

Low-voltage electro-optic modulators have also been built using microring resonators. A microring resonator is a circular waveguide fabricated on a substrate by micro-lithography. Such resonators have been fabricated using semiconductors, lithium niobate [9], and polymers [10]. The operation of such a ring resonator involves the multiple total internal reflection of light rays around the perhipheral outer boundary of a circular region in which the ring has a higher refactive index than that of the surrounding substrate. These are called *whispering-gallery* modes.[10] Because a curved dielectric waveguide of this kind is

[10] So called because in a circular gallery or building speech whispered at one point can be heard all round the circumference. This phenomenon was first described by Lord Rayleigh in about 1878 when he observed it in St. Paul's Cathedral in London.

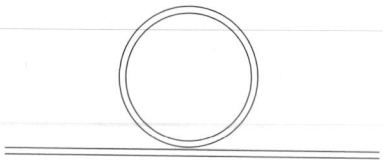

Fig. 18.16 A microring resonator shown closely coupled to a straight dielectric waveguide, which can be used to fabricate an electro-absorption modulator.

always leaky, the index mismatch between core and cladding, and the ring radius, must be carefully chosen to provide a high $Q$ for the ring. For high-index materials such as silicon, ring radii can be made as small as a few micrometers. Since large index differences are not readily available for lithium niobate or polymer devices; their ring radii are in the range of tens to hundreds of micrometers. If a microring resonator is placed adjacent to a straight dielectric waveguide as shown schematically in Fig. 18.16, then electromagnetic energy can couple strongly between the straight waveguide and the microring. The condition for this to happen is that an integer multiple of the wavelength of the light in the straight waveguide matches the optical path length around the microring. If the effective refractive indices of light in the straight waveguide and microring are $n_1$ and $n_2$, respectively, and the circumference of the ring is $L$, this condition is

$$m\lambda_0/n_1 = n_2 L, \tag{18.84}$$

where $m$ is an integer and $\lambda_0$ is the vacuum wavelength of the light. Light traveling around the ring has a propagation constant $\beta$, with $\beta L = 2m\pi$, so light making a round trip in the ring comes back to its starting point in phase. If the ring is made of electro-optic material and electrodes are placed around the ring, then the coupling between straight waveguide and ring can be modulated. Microring resonators have received great attention in recent years and have a host of applications in integrated photonic systems [11–13].

## 18.10 Electro-absorption modulators

The absorption of light by a semiconductor can be modified by the application of an electric field [14]. In a bulk semiconductor this results from the Franz–Keldysh effect, which can be regarded as a shift in the band-gap energy with the applied electric field. In a quantum well (QW) the change in absorption results from the *quantum confined Stark effect*. In the smaller-band-gap material sandwiched between two larger-band-gap materials electron–hole pairs are bound together to form an exciton. An applied electric field pulls the electron and hole in opposite directions, but they are trapped within the quantum well. The electric field moves the discrete electron states in the QW to lower energies and the allowed hole states to higher energies. This changes the absorption spectrum of the QW. Both the Franz–Keldysh effect [15] and quantum confined Stark effect [16] can be used to build

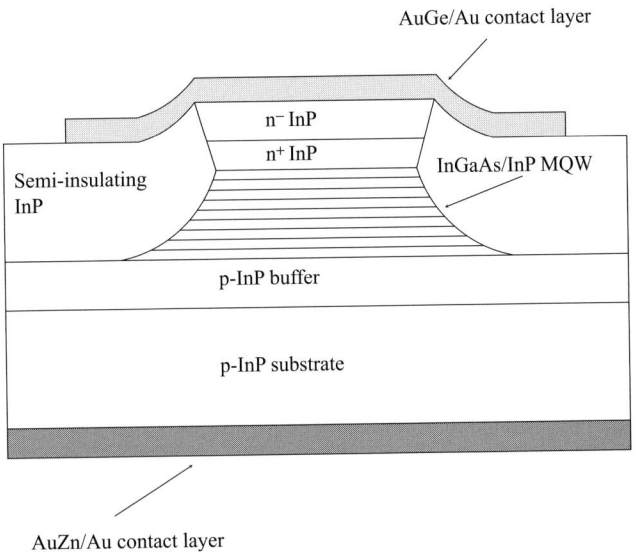


**Fig. 18.17** The structure of a multiple-quantum-well (MQW) waveguide electro-absorption modulator that uses the quantum-confined Stark effect [14].

electro-absorption modulators (EAMs). These EAMs, especially those using QWs, can be operated at low voltages and at high speeds. Figure 18.17 shows an example of the cross-section of a waveguide EAM based on InGaAs/InP [16]. Such EAMs have become very important as amplitude modulators in optical communication systems.

## 18.11 Electro-optic beam deflectors

The electro-optic effect can also be used to deflect or scan a laser beam. The simplest way to do this is to use a prism of electro-optic material in which an applied field varies the bulk index of an ordinary or extraordinary wave passing through the material, as shown in Fig. 18.18.

Figure 18.19 shows a double-prism electro-optic beam deflector that uses two KDP prisms with their optic axes reversed with respect to each other. By applying a voltage as shown, the index of refraction along the path cd can be made larger than that along the path ab. This causes the phase fronts passing through the crystal to tilt, and the output beam deflects.[11] Electro-optic beam deflectors have not found as widespread use as have acousto-optic beam deflectors, which are our next topic in this chapter.

[11] Problem 18.1 involves a quantitative analysis of this device.

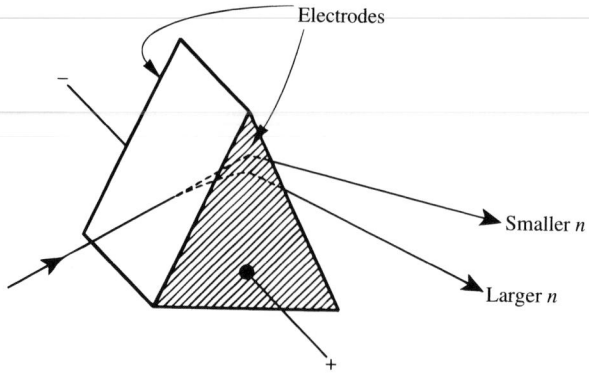


**Fig. 18.18** A simple single-prism electro-optic laser-beam deflector.

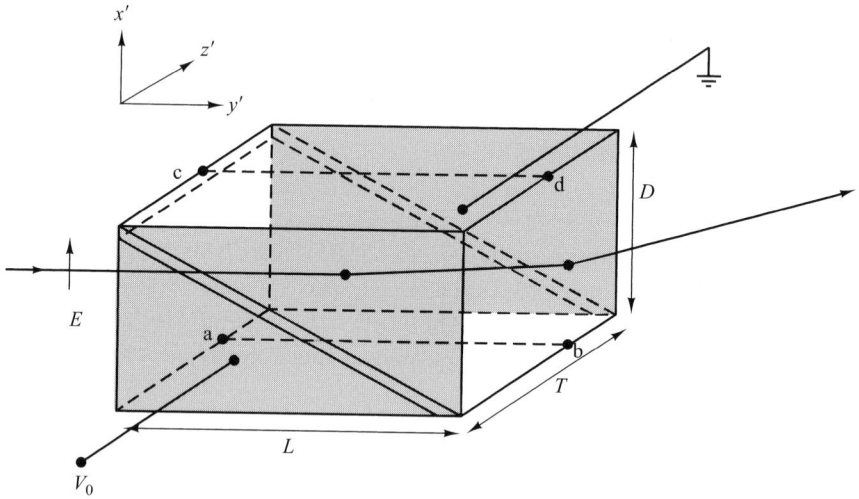


**Fig. 18.19** A double-prism KDP electro-optic beam deflector. The upper and lower prisms have their $z$ (optic) axes reversed with respect to each other. The prisms are cut so that their orthogonal edges are parallel to the $x'$, $y'$, and $z'$ directions – the principal-axis directions in KDP when an external electric field is applied in the $z$ direction.

## 18.12 Acousto-optic modulators

When a sound wave propagates through a medium it produces regions of compression and rarefaction. The medium may support both longitudinal and shear sound waves, depending on the sound-wave-propagation direction relative to the crystal symmetry axes. For a plane sound wave at frequency $\omega_s$ traveling in the $z$ direction, the change in density as a function of position can be written as

$$\Delta\rho = \Delta\rho_0 \cos(\omega_s t - k_s z), \tag{18.85}$$

where $k_s = 2\pi/\lambda_s$, with $\lambda_s$ being the wavelength of the sound wave. For small relative changes in density we expect the change in refractive index to be $\Delta n \propto \Delta\rho$. From Eqs. (18.12) and (18.85), it is clear that

$$\Delta n = -\frac{n_0^3}{2} p S_0 \cos(\omega_s t - k_s z), \tag{18.86}$$

where $S_0$ is the peak value of the strain produced by the passage of the sound wave. Equation (18.86) is most useful in describing the photoelastic effect in an isotropic medium, where the behavior is described by a single photoelastic coefficient $p$. In anisotropic media the change in index will depend on the direction both of the light wave and of the sound wave relative to the symmetry axes. We will not deal with the details of acousto-optic effects in such anisotropic materials here; detailed information is available in the specialized literature [17–20].

If a light wave passes through a medium with a spatial variation in refractive index like Eq. (18.67), it is diffracted. This phenomenon is closely related to Brillouin scattering, the interaction of light with naturally occurring density fluctuations in media, which was first predicted in 1922 [21].

Practical application of the diffraction of light by sound waves is accomplished in one of two geometries, shown in Figs. 18.20 and 18.21, in which an acoustical transducer (a piezoelectric transducer) is used to drive a sound wave through the acousto-optic medium. In the geometry shown in Fig. 18.20, where the interaction is called the *Debye–Sears* effect [22], the interaction between the light wave and the sound wave occurs over a relatively short length – in the so-called *Raman–Nath* regime [23]. In the geometry of Fig. 18.21 the light wave interacts with essentially plane sound waves and the phenomenon is described as acoustic *Bragg* diffraction: it is as if the light reflects and diffracts from the planes of

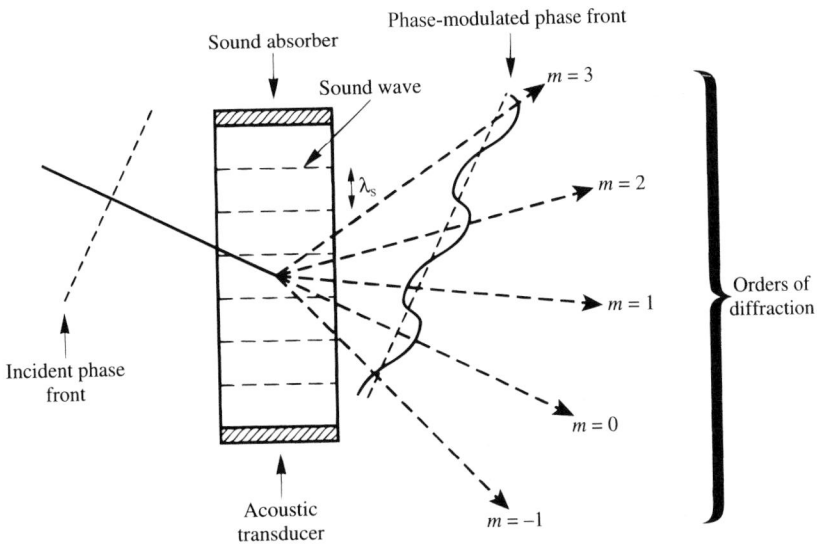


Fig. 18.20 An acousto-optic device using the Debye–Sears effect.

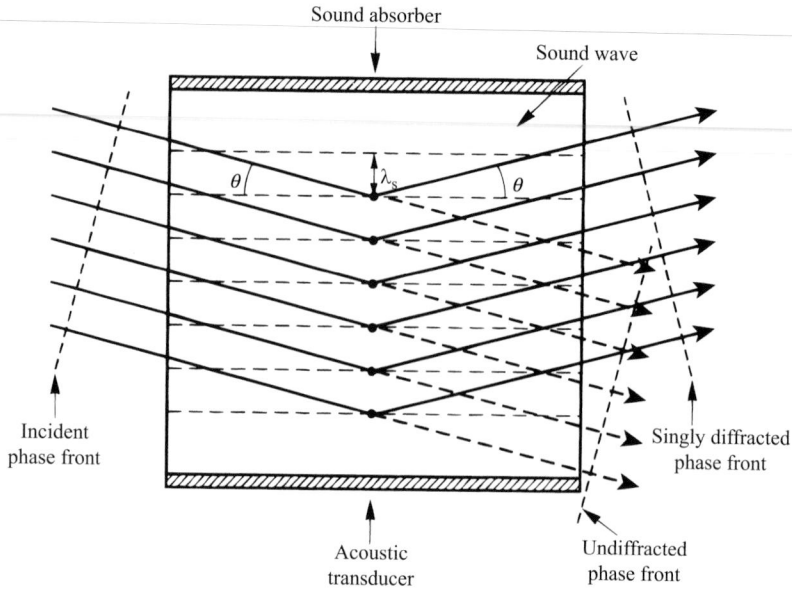


**Fig. 18.21** An acousto-optic device using Bragg diffraction.

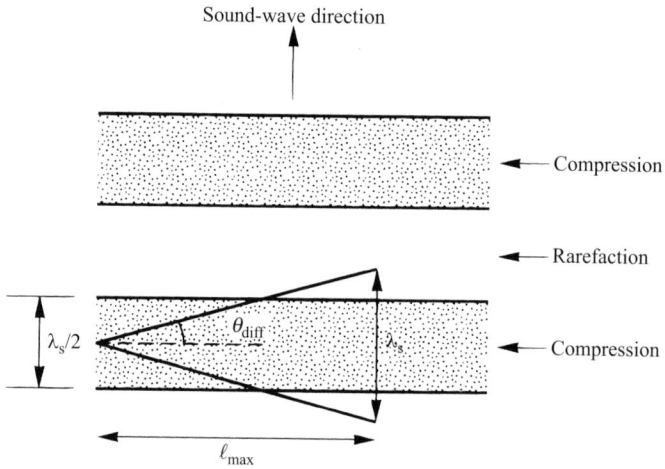


**Fig. 18.22** The schematic distribution of density changes produced by passage of a sound wave through a medium, showing how the maximum interaction length in the Raman–Nath regime is calculated.

density variation in the medium. We can delineate the difference between Raman–Nath and Bragg diffraction with the aid of Fig. 18.22. The regions of compression or rarefaction in the medium have width $\sim\lambda_s/2$. Light traveling in one of these layers spreads by diffraction at an angle of

$$\theta_{diff} \sim \frac{2\lambda}{\lambda_s}. \tag{18.87}$$

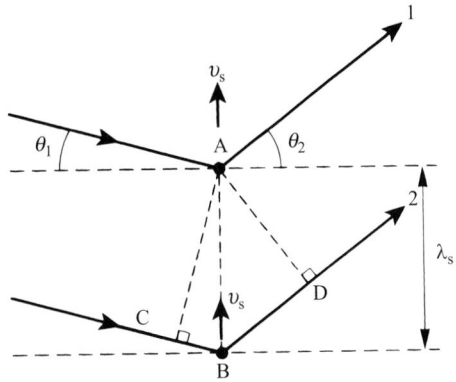


**Fig. 18.23** A simple diagram for calculating diffraction from sound waves.

The maximum useful interaction length, $\ell_{max}$ over which the wave remains within a compressed (or rarefied layer) satisfies

$$\tan\theta_{diff} = \frac{\lambda_s}{2\ell_{max}}, \tag{18.88}$$

which for small angles gives

$$\ell_{max} = \frac{\lambda_s^2}{4\lambda}. \tag{18.89}$$

For example, with 1.3-μm laser radiation interacting with 40-MHz sound waves in $TeO_2$ (an important acousto-optic material) in a crystal orientation where $v_s = 4.2$ km s$^{-1}$ and $\lambda_s = 0.105$ mm, $\ell_{max} = 2$ mm.

An alternative parameter that has been introduced to delineate the Raman–Nath and Bragg diffraction regimes is [24]

$$Q = \frac{2\pi\lambda_0 L}{n\lambda_s^2}. \tag{18.90}$$

For $Q < 1$, Raman–Nath diffraction is dominant, while for $Q > 7$ the device operates in the Bragg regime.

The light emerging from a Debye–Sears acousto-optic modulator contains several angular diffraction orders, each of which is Doppler shifted by virtue of the fact that the light has interacted with a *moving* sound wave. We can illustrate this with the simple model shown in Fig. 18.23. We model the moving sound wave as a series of small scatterers traveling with velocity $v_s$. Light reflected from A and B will constructively interfere if

$$\sin\theta_1 + \sin\theta_2 = \frac{m\lambda}{\lambda_s}, \tag{18.91}$$

where $m$ is an integer.[12]

[12] Equation (18.91) should be slightly modified in an anisotropic material to account for the different refractive indices experienced by waves traveling in the $\theta_1$ and $\theta_2$ directions. However, this difference in index is generally small, and we shall not worry about it.

We can calculate the Doppler shift of the light diffracted in Fig. 18.23 by considering the shortening of the ray path that occurs as the sound wave moves from point B to point A. In moving from B to A, the sound wave causes the distance traveled by a light ray to change from the path length of ray 1 to that of ray 2. This reduction in path length is $m\lambda$, and it occurs in a time $1/f_s$, where $f_s$ is the sound frequency. Therefore, light observed along path 1 appears to come from a source whose velocity is

$$v_{\text{source}} = m\lambda f_s \tag{18.92}$$

towards the observer. The resultant Doppler shift is

$$\Delta\nu = v_{\text{source}}\frac{\nu_0}{c}, \tag{18.93}$$

where $\nu_0$ is the light frequency. Therefore,

$$\Delta\nu = mf_s. \tag{18.94}$$

Each diffracted order is Doppler shifted by the appropriate multiple of the sound frequency. The Doppler shift is negative for waves diffracted in the downward direction shown in Fig. 18.20.

When an acousto-optic device works in the Bragg regime, as shown in Fig. 18.21, the incident light wave interacts with broad sound-wave phase fronts. This imposes a more restrictive condition on the permitted directions for diffraction. Not only must the incident and diffracted angles satisfy Eq. (18.91), but also the angles $\theta_1$ and $\theta_2$ must be equal – as if the acoustic phase fronts act as mirrors for the incident light. This situation is illustrated in Fig. 18.24. The Bragg diffraction condition is[13]

$$\sin\theta = \frac{m\lambda}{2\lambda_s}, \tag{18.95}$$

where $m = 1$. Brillouin was the first to point out that, because the sound wave produces a sinusoidal disturbance in the medium, higher-order diffraction effects, for which $m > 1$, do not occur.

If the sound wave is moving in the upward direction in Fig. 18.24, with velocity $v_s^+$, then the diffracted light is up-shifted to the frequency $\nu + f_s$. If the sound wave is traveling downwards with velocity $v_s^-$, then the diffracted wave is down-shifted to the frequency $\nu - f_s$.

Acousto-optic interactions can be viewed in two additional ways. In the particle-interaction model of the process each incident light photon collides with a quantized particle of acoustic energy (a phonon), creating a photon with a different energy and momentum. This is shown schematically in Fig. 18.25. An incident photon with a wave vector $\mathbf{k}_1$ collides with a phonon with wave vector $\mathbf{k}_s$ to create a photon with wave vector $\mathbf{k}_2$. Since the momentum of each of these particles is $\hbar\mathbf{k}$, momentum is conserved through the vector relationship

$$\mathbf{k}_2 = \mathbf{k}_1 \pm \mathbf{k}_s, \tag{18.96}$$

[13] Once again we are neglecting the different refractive indices that would be experienced by the incident and diffracted waves in an anisotropic medium.

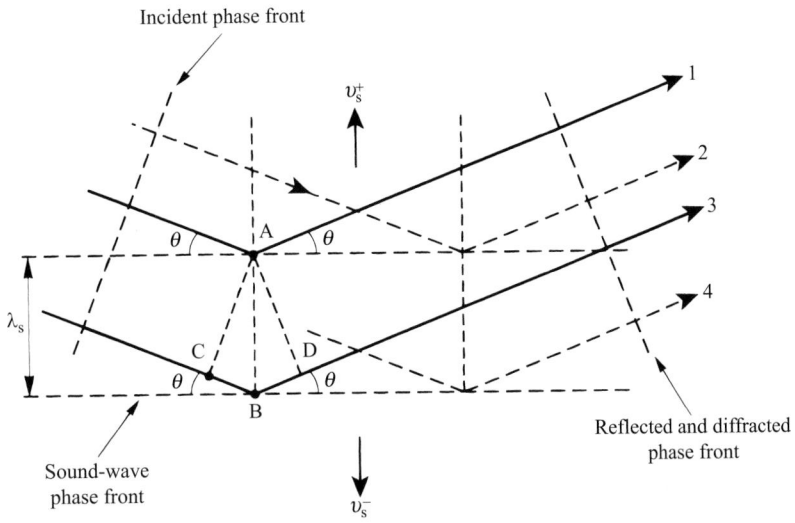


**Fig. 18.24** An illustration of the Bragg condition for diffraction from a sound wave. If CB + BD $= m\lambda$ then $\sin\theta = m\lambda/(2\lambda_s)$, and the diffracted rays from all sound-wave phase fronts separated by $\lambda_s$ remain in phase.

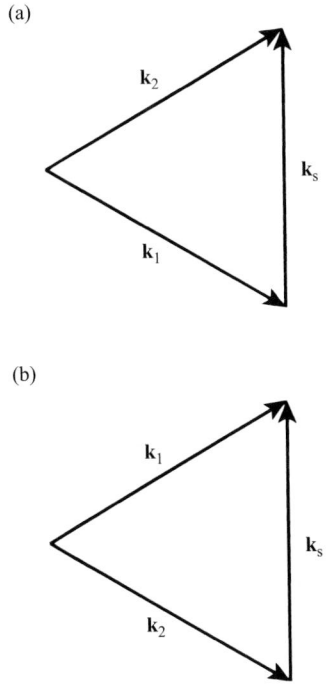


**Fig. 18.25** Vector addition of wave vectors in the particle-interaction model of an acousto-optic interaction: (a) Doppler frequency up-shift; and (b) Doppler frequency down-shift.

and energy is conserved through the relationship

$$\omega_2 = \omega_1 \pm \omega_s. \tag{18.97}$$

The $\pm$ sign in Eqs. (18.96) and (18.97) corresponds to the two geometries shown in Fig. 18.25.

The interchange of energy among the three waves involved in Eq. (18.97) is also viewable as a general class of interaction called a *parametric* interaction. These are nonlinear interactions, and will be discussed in greater detail in Chapters 19, and 20.

## 18.13 Applications of acousto-optic modulators

Acousto-optic modulators can be used as frequency shifters, frequency modulators, optical beam deflectors and scanners, and optical switches. Used in one of these specific ways they find application in communication systems, sensors, optical information storage and retrieval, xerography, laser control, and spectrum analysis. They are generally used in the Bragg regime. Their performance efficiency is characterized by their diffraction efficiency $\eta$, which is [18]

$$\eta = \sin^2\left[\frac{\pi L}{\lambda \cos\theta}\sqrt{\frac{M_2 I_s}{2}}\right]. \tag{18.98}$$

In this equation $L$ is the interaction width of the light beam with the sound wave, $I_s$ is the intensity of the sound wave, $\theta$ is the Bragg angle and $M_2$ is a *figure of merit* (FOM) for the acousto-optic material [25].[14] The FOM is a material constant that allows the relative efficiencies of different acousto-optic materials to be compared. It is given by

$$M_2 = \frac{n_0^6 p^2}{\rho v_s^3}, \tag{18.99}$$

where $n_0$ is the refractive index, $p$ is the photoelastic constant, $\rho$ is the density, and $v_s$ is the sound velocity.

### 18.13.1 The diffraction efficiency of $TeO_2$

For a longitudinal sound wave the $M_2$ FOM for $TeO_2$ is $34.5 \times 10^{-15}$ s$^3$ kg$^{-1}$ and the sound velocity is 4.2 km s$^{-1}$. Consequently, at 40 MHz the sound wavelength is 0.105 mm. For a laser wavelength of 1.06 μm with a refractive index of 2.26 the Bragg angle is

$$\theta = \arcsin[\lambda/(2\lambda_s)] = 0.128°.$$

[14] Other figures of merit, $M_1$ and $M_3$, have also been proposed to include factors such as bandwidth [26] and an independence of beam dimensions [27], respectively.

For an acousto-optic modulator with a cross-section of 1 mm × 1 mm and an injected sound-wave power of 1 W, from Eq. (18.98) the diffraction efficiency is

$$\eta = \sin^2\left[\frac{\pi \times 10^{-3} \times 2.26}{1.06 \times 10^{-6}}\sqrt{\frac{34.5 \times 10^{-15} \times 1}{2 \times 10^{-6}}}\right],$$

which gives $\eta = 59\%$.

### 18.13.2 Acousto-optic modulators

The Doppler shift of the Bragg-diffracted light in Fig. 18.21 is the same as the frequency applied to the acoustic transducer. If the drive signal is of the form

$$V(t) = V_0 \cos\{[\omega_s + \Delta\omega(t)]t\}, \tag{18.100}$$

which is a frequency-modulated carrier wave at frequency $\omega_s$, then, provided that $\Delta\omega \ll \omega_s$, the beam deflection angle is essentially constant but the frequency of the light within this beam will also be frequency modulated. If $\Delta\omega$ is not small with respect to $\omega_s$ then the deflection angle will be modulated. Insofar as the diffracted light in a particular direction is concerned, this corresponds to intensity-modulation. However, the easiest way to achieve intensity modulation is to utilize the undiffracted beam and modulate its intensity by varying the drive signal so as to transfer incident light intensity into the diffracted beam. Intracavity acousto-optic modulators can be used in this way to both $Q$-switch and mode-lock lasers. The maximum intensity-modulation frequency that can be achieved in this way depends on the transit time of the sound wave through the diameter of the incident light beams, so for high modulation frequencies the beam must be focused into the acousto-optic medium.

### 18.13.3 Acousto-optic beam deflectors and scanners

If an acousto-optic device is operated at the Bragg angle corresponding to a drive frequency $\omega_s$, then variation of $\omega_s$ will lead to variation in the diffraction angle, thereby serving to deflect or scan the diffracted beam. Because the Bragg condition is strictly satisfied only at frequency $\omega_s$, deviations from this value deflect the beam, but with a concomitant reduction in diffracted intensity as the frequency moves further from $\omega_s$. In Fig. 18.26 a variation in $\omega_s$ to $\omega_s + \Delta\omega_s$ changes $\theta$ to $\theta + \Delta\theta$. From Eq. (18.91) we can write

$$\sin\theta + \sin(\theta + \Delta\theta) = \frac{k_s + \Delta k_s}{k}, \tag{18.101}$$

where $k_s = 2\pi f_s/v_s$ and $k = 2\pi/\lambda$. Incorporating the Bragg condition (Eq. (18.95)) into Eq. (18.101) for small deflection angles gives

$$\Delta\theta = \frac{\Delta k_s}{k}. \tag{18.102}$$

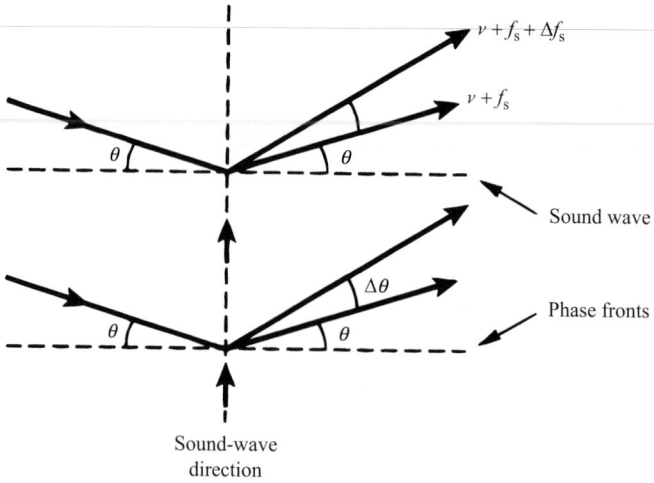


**Fig. 18.26** The variation in diffraction angle that results from a change in sound-wave frequency.

The utility of a beam scanner is frequently determined by the number of resolvable deflected spots $N$ that it can produce. For a Gaussian beam of minimum spot size $\omega_0$ focused at the Bragg angle

$$N = \frac{\Delta\theta_{\max}}{\theta_{\text{beam}}}, \tag{18.103}$$

which gives

$$N = \frac{\pi\omega_0\,\Delta\theta_{\max}}{\lambda}. \tag{18.104}$$

The maximum deflection angle is

$$\Delta\theta_{\max} = \frac{\lambda\,\Delta f_{\max}}{v_s}, \tag{18.105}$$

and, since the sound-wave transit time through the beam is

$$\tau = \frac{2\omega_0}{v_s}, \tag{18.106}$$

we get

$$N = \frac{\pi}{2}\tau\,\Delta f_{\max}. \tag{18.107}$$

A maximum reasonable sound-wave frequency variation is $\Delta f_{\max} = f_s/2$, where $f_s$ is the sound-wave frequency corresponding to the Bragg angle. In this case

$$\tau = \frac{2\omega_0}{f_s\lambda_s} \tag{18.108}$$

and

$$N = \frac{\pi\omega_0}{2\lambda_s}. \tag{18.109}$$

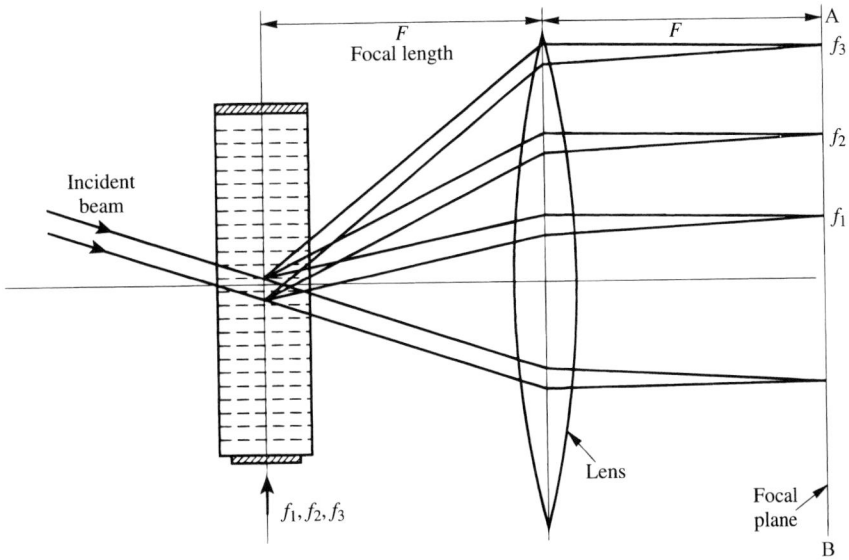


**Fig. 18.27** Acousto-optic Bragg diffraction for radio-frequency spectrum analysis.

The maximum number of resolvable spots is really the maximum number of distinguishable angles by which the incident beam is deflected. Each diffracted beam could be focused to a distinct spot with an appropriate lens.

### 18.13.4 Radio-frequency spectrum analysis

If the radio-frequency (RF) drive signal applied to the transducer in Fig. 18.21 has a frequency spectrum $F(\omega_s)$, then each frequency component in this spectrum produces an angular deflection whose angle depends on $\omega_s$, and whose intensity depends on the amplitude $F(\omega_s)$ of the drive signal applied to the transducer at frequency $\omega_s$. If the overall diffracted beam is focused by a lens, the different angular components of this beam will be focused to spots in different lateral locations in the focal plane of the lens, as shown in Fig. 18.27. The overall arrangement acts like a Fourier-transform device, whereby input RF frequencies $f_1, f_2$, and $f_3$ produce different angles of diffraction, which are focused by the lens at different spatial locations. If an array of light detectors is placed along the line AB then a rapid parallel computation of the input RF spectrum occurs – this is true *optical computation*.

## 18.14 Construction and materials for acousto-optic modulators

An acousto-optic modulator is fabricated from a material with a high acousto-optic figure of merit, such as tellurium dioxide ($TeO_2$) or lead molybdate ($PbMoO_4$), to which has

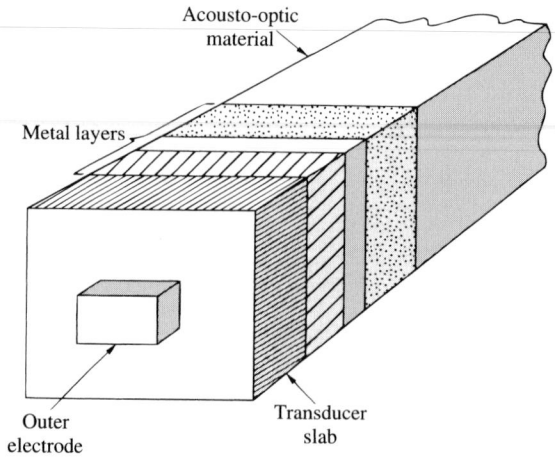


Fig. 18.28 An acousto-optic device, showing the transducer and metal layers for acoustic matching.

been bonded a piezoelectric transducer material such as lithium niobate ($LiNbO_3$). The transducer is generally driven at RF: 40 MHz and 80 MHz are common values for fixed-frequency devices. Consequently, it is also important that the acousto-optic material have low sound-wave attenuation at the sound-wave frequency being used. Since the acoustic attenuation in these materials varies roughly as $f_s^2$ the attenuation coefficient is quoted in units of dB $m^{-1}$ $GHz^{-2}$. Table 18.3 lists several important acousto-optic materials and their performance characteristics.

The transducer is a thin ceramic slab bonded to one or more metallized layers on the acousto-optic material. A second metallized layer is placed on the other side of the slab to provide electrical contact, as shown schematically in Fig. 18.28. Thus, the transducer presents an essentially capacitive load to the drive circuit. For high-frequency operation the thickness of the transducer is reduced to a few micrometers, and in this way acousto-optic modulators operating at several gigahertz have been fabricated. The performance of the device is optimized by matching the transducer electrically to the device circuit and acoustically to the acousto-optic material [29]. The maximum frequency, $f_{max}$, at which an acousto-optic modulator can modulate a light beam is limited by the transit time of the sound wave across the beam diameter $D$

$$f_{max} = \frac{v_s}{D}. \tag{18.110}$$

However, a beam focused to a small beam waist will not be able to interact coherently along a long path through the sound wave. Generally speaking, except in mode-locking applications, the sound wave travels unidirectionally in the acousto-optic material, and a reflected sound wave can be suppressed by an acoustic absorber or matched acoustic load on the end of the material remote from the transducer. In high-frequency devices the sound wave attenuates significantly as it leaves the transducer, so in a sufficiently long device there is negligible reflected sound wave.

**Table 18.3** Characteristics of acousto-optic materials (additional values can be found in [18] and [20])

| Material | Optical transmission (μm) | Wavelength (μm) | Density (kg m$^{-3}$) | Mode and propagation direction | $v_s$ (km s$^{-1}$) | Polarization direction | $n$ | $M_2$ (FOM) ($10^{-15}$ s$^{-3}$ kg$^{-1}$) |
|---|---|---|---|---|---|---|---|---|
| Fused silica | 0.2–4.5 | 0.633 | 2,650 | L | 5.96 | ⊥ | 1.46 | 1.51 |
| Diamond | 0.2–5 | 0.589 | 3,520 | L[100] | 17.5 | ‖ | 2.417 | 1.02 |
| $TeO_2$ | 0.35–5 | 0.633 | 6,210 | L[001] | 4.2 | ⊥ | 2.26 | 34.5 |
| $PbMoO_4$ | 0.42–5.5 | 0.633 | 6,950 | L[001] | 3.63 | ‖ | 2.262 | 36.3 |
| | | | | | | ⊥ | 2.386 | 36.1 |
| $LiNbO_3$ | 0.4–4.5 | 0.633 | 4,640 | L[100] | 6.57 | 35° to $y$ | 2.2 | 7.0 |
| | | | | S[001] | 3.59 | ⊥ | 2.29 | 2.92 |
| $Pb_5(GeO_4)(VO_4)_2$ | 0.52–5.5 | 0.633 | 7,150 | L[001] | 3.45 | ‖ | 2.275 | 50.6 |
| | | | | L[010] | 3.11 | ‖ | 2.281 | 35.6 |
| α-ZnS | 0.4–12 | 0.633 | 4,090 | L[001] | 5.82 | ‖ | 2.351 | 3.4 |
| | | 0.633 | | S[001] | 2.63 | See [18] | 2.348 | 8.4 |
| β-ZnS | 0.4–12 | 0.633 | 4,100 | L[110] | 5.51 | ‖ | 2.35 | 3.41 |
| α-HgS | 0.62–16 | 0.633 | 8,100 | L[001] | 2.45 | ⊥ | 2.887 | 960 |
| | | 1.153 | | | | | | |
| $Tl_3AsS_4$ | 0.6–12 | 0.633 | 6,200 | L[001] | 2.15 | ‖ | 2.825 | 800 |
| | | 1.153 | | | | | | |
| Ge | 2–20 | 10.6 | 5,320 | L[111] | 5.50 | ‖ | 4.00 | 840 |
| | | 10.6 | | S[100] | 3.51 | Arbitrary | 4.00 | 290 |
| GaAs | 1–11 | 1.153 | 5,340 | L[110] | 5.15 | ‖ | 3.37 | 104 |
| | | 1.153 | | S[100] | 3.32 | Arbitrary | 3.37 | 46.3 |
| KDP | 0.25–1.7 | 0.633 | 2,340 | L[100] | 5.50 | ‖ | 1.51 | 1.91 |
| ZnTe | 0.55–20 | 0.633 | 5,670 | L[110] | 3.87 | ‖ | 2.983 | 45.1 |
| | | 1.153 | | | | | | |
| $LiIO_3$ | 0.3–6 | 0.633 | 4,500 | L[001] | 4.13 | ⊥ | 1.88 | 8.0 |
| | | 0.633 | | L[100] | 4.3 | [001] | 1.74 | 13.0 |
| GaP | 0.6–10 | 0.633 | 4,130 | L[110] | 6.32 | ‖ | 3.31 | 44.6 |
| | | | | S[100] | 4.13 | Arbitrary | 3.31 | 24.1 |
| Si | 1.5–10 | 10.6 | 2,330 | L[111] | 9.85 | ‖ | 3.42 | 6.2 |

L, longitudinal mode; S, shear mode. Directions are given in crystallographic notation: for example [100] ≡ $x$ direction, [001] ≡ $y$ direction; [110] ≡ in the $xy$ plane at 45° to $x$. For further details on this notation consult Kittel [28]. ‖ and ⊥, parallel and perpendicular to the acoustic-wave vector, respectively.

## 18.15 Problem

18.1 Figure 18.19 shows a double-prism electro-optic beam deflector described by Fowler and Schlafer [30]. Prove that the deflection angle of an input beam polarized in the extraordinary direction is, for small deflection angles,

$$\theta = \frac{L}{A} n_0^3 r_{63} V_0,$$

where $A$ is the area of the input face, and $V_0$ is the applied field.

## References

[1] J. F. Nye, *Physical Properties of Crystals*, Oxford: Oxford University Press, 1985.

[2] G. W. C. Kaye and T. H. Laby, *Tables of Physical and Chemical Constants*, 15th edn., New York: Longman, 1985.

[3] M. Abramowitz and I. A. Stegun, *Handbook of Mathematical Functions*, New York: Dover, 1965.

[4] H. Taub and D. L. Schilling, *Principles of Communication Systems*, 2nd edn., New York: McGraw-Hill, 1996.

[5] S. Haykin, *Communication Systems*, 5th edn., New York: Wiley, 2009.

[6] B. P. Lathi and Zhi Ding, *Modern Digital and Analog Communication Systems*, 4th edn., Oxford: Oxford University Press, 2009.

[7] R. C. Alferness, "Waveguide electrooptic modulators," *IEEE Trans. Microwave Theory Technol.* **30**, 1121–1137, 1982.

[8] K. W. Hui, K. S. Chiang, B. Wu, and Z. H. Zhang, "Electrode optimization for high-speed traveling-wave integrated optic modulators," *J. Lightwave Technol.* **16**, 232–238, 1998.

[9] A. Guarino, G. Poberaj, D. Rezzonico, R. Degl'Innocenti, and P. Günter, "Electro-optically tunable microring resonators in lithium niobate," *Nature Photonics* **1**, 407–410, 2007.

[10] P. Rabiei, W. H. Steier, C. Zhang, and L. R. Dalton, "Polymer micro-ring filters and modulators," *J. Lightwave Technol.* **20**, 1968–1975, 2002.

[11] B. E. Little, S. T. Chu, W. Pan, and Y. Kokubun, "Microring resonator arrays for VLSI photonics," *IEEE Photon. Technol. Lett.* **12**, 323–325, 2000.

[12] A. B. Matsko and V. S. Ilchenko, "Optical resonators with whispering-gallery modes – part I: basics," *IEEE J. Selected Topics Quant. Electron.* **12**, 3–14, 2006.

[13] V. S. Ilchenko and A. B. Matsko, "Optical resonators with whispering-gallery modes – part II: applications," *IEEE J. Selected Topics Quant. Electron.* **12**, 15–32, 2006.

[14] J. I. Pankove, *Optical Processes in Semiconductors*, New York: Dover Publications Inc., 1971.

[15] G. E. Stillman, C. M. Wolfe, C. O. Bozler, and J. A. Rossi, "Electroabsorption in GaAs and its application to waveguide detectors and modulators," *Appl. Phys. Lett.* **28**, 544–546, 1976.

[16] U. Koren, B. I. Miller, T. L. Koch, G. Eisenstein, R. S. Tucker, I. Bar-Joseph, and D. S. Chemia, "Low-loss InGaAs/InP multiple quantum well optical electroabsorption waveguide modulator," *Appl. Phys. Lett.* **51**, 1132–1134, 1987.

[17] E. G. Spencer, P. V. Lenzo, and A. A. Ballman, "Dielectric materials for electro-optic, elastooptic and ultrasonic device applications," *Proc. IEEE* **55**, 2074–2108, 1967.

[18] N. Uchida and N. Niizeki, "Acousto-optic deflection materials and techniques," *Proc. IEEE* **61**, 1073–1092, 1973.

[19] A. Korpel, *Acousto-Optics*, New York: Marcel Dekker, 1988.

[20] I. C. Chang, "Acousto-optic devices and applications," *IEEE Trans. Sonics Ultrasonics* **23**, 2–22, 1976.

[21] L. Brillouin, "Diffusion de la lumière et des rayons X par un corps transparent homogène: influence de l'agitation thermique," *Ann. Phys. (Paris)*, 9th series, **17**, 88–122, 1922.

[22] P. Debye and F. W. Sears, "On the scattering of light by supersonic waves," *Proc. Natl. Acad. Sci*, **18**, 409–414, 1932.

[23] C. V. Raman and N. S. N. Nath, "The diffraction of light by sound waves of high frequency, part II," *Proc. Ind. Acad. Sci.* **A2**, 413–420, 1936.

[24] W. R. Klein and B. D. Cook, "Unified approach to ultrasonic light diffraction," *IEEE Trans. Sonics Ultrasonics* **14**, 123–124, 1967.

[25] T. M. Smith and A. Korpel, "Measurement of light–sound interaction efficiency in solids," *IEEE J. Quant. Electron.* **1**, 283–284, 1965.

[26] E. Gordon, "A review of acousto-optical deflection and modulation devices," *Proc. IEEE* **54**, 1391–1401, 1966.

[27] R. Dixon, "Photoelastic properties of selected materials and their relevance for application to acoustic light modulations and scanners," *J. Appl. Phys.* **38**, 5149–5153, 1967.

[28] C. Kittel, *Introduction to Solid State Physics*, 8th edn., New York: Wiley, 2005.

[29] E. H. Young, Jr., and S.-K. Yao, "Design considerations for acousto-optic devices," *Proc. IEEE* **69**, 54–64, 1981.

[30] V. J. Fowler and J. Schlafer, "A survey of laser beam deflection techniques," *Proc. IEEE* **54**, 1437–1444, 1966.

# 19 Introduction to nonlinear processes

## 19.1 Introduction

In this chapter we shall begin our discussion of nonlinear phenomena that are important in optics. When one or more electromagnetic waves propagate through any medium they produce polarizations in the medium that, in principle, oscillate at all the possible sum and difference frequencies that can be generated from the incoming waves. These polarizations, which oscillate at these new frequencies, give rise to corresponding electromagnetic waves. Thus, we get phenomena such as harmonic generation, for example, when infrared light is converted into visible or ultraviolet light, and various other frequency-mixing processes. These nonlinear processes can be described by a series of nonlinear susceptibilities or mixing coefficients. These coefficients will be defined and their origin traced to the anharmonic character of the potential that describes the interaction of particles in the medium.

## 19.2 Anharmonic potentials and nonlinear polarization

When an electromagnetic wave propagates through a medium a *total* electric field acts on each particle of the medium. This total field contains components at all the frequencies contained in the input wave or waves. Each particle of the medium will be displaced from its equilibrium position by the action of this field. Positive ions and nuclei will be displaced in the direction of the field, while negative ions and electrons will be displaced in the opposite direction to that of the field. The resultant separation of centers of positive and negative charge creates dipoles in the medium. The resultant dipole moment per unit volume (polarization) can be written in S.I. units as

$$\mathbf{P} = \epsilon_0(\chi^{(1)}\mathbf{E} + \chi^{(2)}\mathbf{E}^2 + \chi^{(3)}\mathbf{E}^3 + \cdots), \tag{19.1}$$

where $\chi^{(1)}$ is the already familiar linear susceptibility, $\chi^{(2)}$ is the second-order nonlinear polarizability, and so on. Various authors have used older systems of units to describe nonlinear phenomena, and as a consequence Eq. (19.1) can look different, as will the units used to describe the various susceptibilities [1]. In the S.I. system used here $\chi^{(1)}$ is dimensionless, $\chi^{(2)}$ has units of $\mathrm{m\,V^{-1}}$, and $\chi^{(3)}$ has units of $\mathrm{m^2}\,V^{-2}$. In the older system in which the susceptibilities are given in electrostatic units (esu), $\chi^{(1)}$ (esu) $= [1/(4\pi)]\chi^{(1)}$ (S.I.), $\chi^{(2)}$ (esu) $= [3 \times 10^4/(4\pi)]\chi^{(2)}$ (S.I.), and $\chi^{(3)}$ (esu) $= [9 \times 10^8/(4\pi)]\chi^{(1)}$ (S.I.).

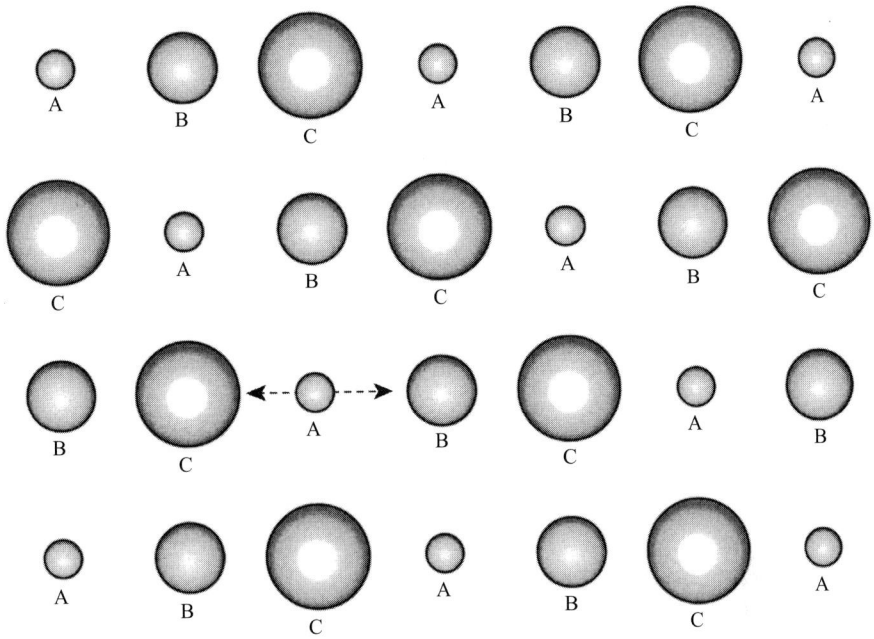


**Fig. 19.1** A rectangular lattice structure that lacks a center of symmetry.

We can explain the dependence of the polarization on the square and higher powers of the field in terms of the restoring force that acts on a particle of the medium when it is displaced from its equilibrium position. Consider the simple rectangular lattice structure shown in Fig. 19.1. If particle A is displaced towards particle B it will be pushed back to equilibrium by the repulsion of particle B and the attraction of particle C. We can describe the size and direction of the restoring force on particle A in terms of its potential energy as it moves along the line between B and C. This potential will be of the schematic form shown in Fig. 19.2.

It is clear that this potential is not symmetric about $x = 0$. The restoring force acting on particle A as it moves a distance $\Delta$ towards particle B is clearly less than that when it moves the same distance towards C. This occurs because the lattice structure lacks a center of symmetry. In this case $V(-x) \neq V(x)$.[1]

We can always write the potential $V(x)$ as a power series

$$V(x) = \frac{1}{2}kx^2 + \frac{1}{3}bx^3 + \frac{1}{4}cx^4 + \frac{1}{5}dx^5 + \cdots, \tag{19.2}$$

where in general $b, c, d, \ldots \ll k$.

If $b, c, d, \ldots = 0$ the potential is a *harmonic* potential; otherwise it is called an *anharmonic* potential. If the structure is centrosymmetric so that $V(-x) = V(x)$ then $b, c, d, \ldots = 0$.

[1] A structure has a center of symmetry if a particle placed at position vector $\mathbf{r}$ sees exactly the same location and distribution of surrounding particles as a particle placed at $-\mathbf{r}$. Crystalline structures can lack a center of symmetry, but gases and liquids have a centrosymmetric structure.

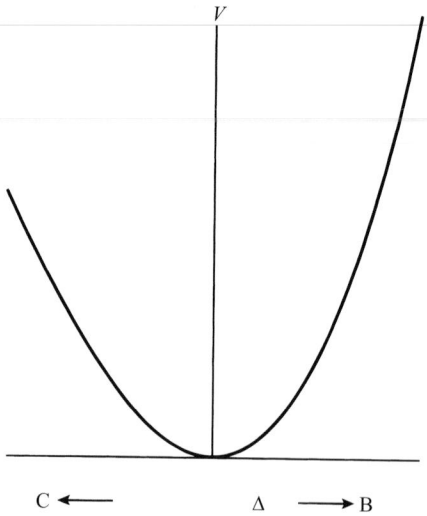


Fig. 19.2 The non-symmetrical potential that results for displacement within a lattice that lacks a center of symmetry.

The restoring force that acts on the particle is

$$F(x) = -\frac{\partial V}{\partial x} = -kx - bx^2 - \cdots, \tag{19.3}$$

so, for example,

$$\begin{aligned} F(\Delta) &= -k\Delta - b\Delta^2, \\ F(-\Delta) &= k\Delta - b\Delta^2. \end{aligned} \tag{19.4}$$

These are unequal forces for displacements of equal magnitude in opposite directions.

For very small displacements $\Delta$ the forces become equal and a particle mass $m$ displaced from its equilibrium position will execute a simple harmonic oscillation with frequency $\omega_0 = \sqrt{k/m}$. When an electric field[2] $\frac{1}{2}[E(\omega)e^{j\omega t} + \text{c.c.}]$ is applied to the particles making up the medium, it is generally the outermost electrons on the atoms of the medium that move the most. This is particularly so at optical frequencies, at which the inertia of the much heavier nuclei and positive ions as a whole prevents them following the rapid oscillation of the applied field to any significant extent.

We can write down an equation of motion for one of the electrons in a very similar form to the one we used in the electron oscillator model for radiative transitions (see Eq. (2.110)):

$$\frac{d^2x}{dt^2} + 2\Gamma\frac{dx}{dt} + \frac{k}{m}x + \frac{b}{m}x^2 = -\frac{e}{2m}E(\omega)e^{j\omega t} + E^*(\omega)e^{-j\omega t}. \tag{19.5}$$

[2] This field is, strictly speaking, the *local* field within the medium that is seen by individual atoms or ions. We shall discuss at the conclusion of this chapter the distinction between this local field and the *externally applied field*.

For simplicity, we are ignoring the higher-order nonlinear terms involving $x^3, x^4$, etc. We try as a solution to Eq. (19.5)

$$x(t) = \frac{x_1}{2}(e^{j\omega t} + e^{-j\omega t}) + \frac{x_2}{2}(e^{2j\omega t} + e^{-2j\omega t}). \tag{19.6}$$

Substituting, and comparing the coefficients of the terms containing $e^{j\omega t}$, gives

$$x_1\left[(\omega_0^2 - \omega^2) + 2j\omega\Gamma + \frac{b}{m}x_2\right] = -\frac{e}{m}E(\omega), \tag{19.7}$$

where, as before, $\omega_0 = \sqrt{k/m}$. If we assume that $bx_2/m \ll (\omega_0^2 - \omega^2) + 2j\omega\Gamma$, then we have

$$x_1 = -\frac{e}{m}E(\omega)\frac{1}{(\omega_0^2 - \omega^2) + 2j\omega\Gamma}. \tag{19.8}$$

For $N$ identical electrons per unit volume the resultant polarization that oscillates at angular frequency $\omega$ is

$$P^{\omega}(t) = -\frac{Ne}{2}(x_1 e^{j\omega t} + \text{c.c.}), \tag{19.9}$$

which can be written in terms of the linear susceptibility as

$$P^{\omega}(t) = \frac{\epsilon_0 \chi^{(1)}(\omega)}{2}[E(\omega)e^{j\omega t} + \text{c.c.}]. \tag{19.10}$$

So

$$\chi^{(1)}(\omega) = \frac{Ne^2}{m\epsilon_0}\frac{1}{(\omega_0^2 - \omega^2) + 2j\omega\Gamma}. \tag{19.11}$$

Near resonance, $\omega \simeq \omega_0$, and Eq. (19.11) would reduce to Eq. (2.119). If we substitute from Eq. (19.6) into Eq. (19.5) and compare the coefficients of the terms containing $e^{2j\omega t}$, we get

$$x_2 = -\frac{be^2E^2(\omega)}{2m^3}\frac{1}{(\omega_0^2 - 4\omega^2 + 4j\omega\Gamma)[(\omega_0^2 - \omega^2) + 2j\omega\Gamma]^2}. \tag{19.12}$$

The resultant polarization at the second harmonic is

$$P^{2\omega}(t) = -\frac{Ne}{2}(x_2 e^{2j\omega t} + \text{c.c.}), \tag{19.13}$$

which can be written in terms of the Fourier amplitude $P(2\omega)$ as

$$P^{2\omega}(t) = \frac{1}{2}[P(2\omega)e^{2j\omega t} + \text{c.c.}]. \tag{19.14}$$

Clearly, $P(2\omega) = -Nex_2$. If we define a second-harmonic-generation (SHG) coefficient $d(2\omega)$ by the relationship

$$P(2\omega) = \epsilon_0 d(2\omega)E(\omega), \tag{19.15}$$

then

$$d(2\omega) = \frac{bNe^3}{2\epsilon_0 m^3} \frac{1}{(\omega_0^2 - 4\omega^2 + 4j\omega\Gamma)[(\omega_0^2 - \omega^2) + 2j\omega\Gamma]^2}, \tag{19.16}$$

which, with the use of Eq. (19.11) for the linear susceptibility, gives

$$d(2\omega) = \frac{b\epsilon_0^2}{2N^2e^3}[\chi^{(1)}(2\omega)][\chi^{(1)}(\omega)]^2, \tag{19.17}$$

where $d(2\omega)$ has units of m $V^{-1}$. Equation (19.17) shows that, within the framework of this simple model, the SHG coefficient can be determined from the *linear* susceptibilities at the fundamental and second-harmonic frequencies. Estimates of SHG properties made in this way have met with some success [2, 3].

The SHG coefficient defined in Eq. (19.15) is a scalar quantity. In reality, in anisotropic media the polarization of the second-harmonic frequency will depend on the direction of propagation of the fundamental wave, and its state of polarization, so a SHG tensor $\overline{\overline{d}}$ must be used. This will be discussed in more detail in the next section.

If the fundamental wave can be written as

$$E(t) = E(\omega)\cos(\omega t - \mathbf{k_1} \cdot \mathbf{r}) \tag{19.18}$$

then

$$P^{2\omega}(t) = \epsilon_0 d(2\omega)E^2(\omega)\cos(2\omega t - 2\mathbf{k_1} \cdot \mathbf{r}), \tag{19.19}$$

so the $d(2\omega)$ coefficient is a *practical* coefficient that determines the amplitude of the *real* polarization at the second-harmonic frequency.

We could continue this form of analysis to derive coefficients for third-harmonic generation and higher terms. However, it is better in many ways to generalize our discussion to include all possible nonlinear interactions that can occur when one or more waves at different frequencies interact within a medium. This is done by the introduction of a series of nonlinear susceptibilities.

## 19.3 Nonlinear susceptibilities and mixing coefficients

If two waves, with angular frequencies $\omega_1$ and $\omega_2$, respectively, interact within a nonlinear medium, then the second-order polarization $\epsilon_0\chi^{(2)}\mathbf{E}^2$ will clearly contain the frequencies $|\omega_1 \pm \omega_2|, 2\omega_1, 2\omega_2$, and DC. The magnitudes of the polarization produced at all these sum and difference frequencies are formally related to the electric field amplitudes involved by nonlinear susceptibility tensors. These tensors reveal the relationship between the Cartesian components of the induced nonlinear polarization and the Cartesian components of the interacting fields. Unfortunately, the way in which these nonlinear susceptibilities have been defined in the literature has not always been consistent, and has been the source of much confusion for many years. We will follow the approach of Byer [4] and Shen [5], and relate *real* polarizations to *real* fields. We take each electric field as

$$\mathbf{E}^{\omega_i}(t) = \mathbf{E}(\omega_i)\cos(\omega_i t - \mathbf{k}_i \cdot \mathbf{r}), \tag{19.20}$$

which can be written in terms of its Fourier amplitudes as

$$\mathbf{E}^{\omega_i}(t) = \frac{1}{2}[\mathbf{E}(\omega_i)e^{j(\omega_i t - \mathbf{k}_i \cdot \mathbf{r})} + \text{c.c.}]. \tag{19.21}$$

For a second-order process involving the interaction of two fields at frequencies $\omega_1$ and $\omega_2$ the nonlinear susceptibility tensor for sum-frequency generation is defined by the relation

$$\mathbf{P}^{\omega_3}(t) = \epsilon_0 \bar{\bar{\chi}}^{(2)}(-\omega_3, \omega_1, \omega_2)\mathbf{E}^{\omega_1}(t)\mathrm{E}^{\omega_2}(t). \tag{19.22}$$

The sum frequency is $\omega_3 = \omega_1 + \omega_2$, which is reflected in the formal notation $\bar{\bar{\chi}}^{(2)}(-\omega_3, \omega_1, \omega_2)$. In this specification of the nonlinear susceptibility, the algebraic sum of the frequencies contained within parentheses is zero. This reflects the conservation of energy during the nonlinear process. Specifically, in terms of the components of the vectors involved,

$$P_i^{\omega_3}(t) = \epsilon_0 \sum_{jk} \chi_{ijk}^{(2)}(-\omega_3, \omega_1, \omega_2)E_j^{\omega_1}(t)E_k^{\omega_2}(t), \tag{19.23}$$

where $i, j$, and $k$ can take the values $x, y$, or $z$.

The polarization at the frequency $\omega_3$ is a *real* polarization, which can be written in terms of its Fourier amplitudes $P(\omega_3)$ as

$$\mathbf{P}^{\omega_3}(t) = \frac{1}{2}\left[\mathbf{P}(\omega_3)e^{j(\omega_3 t - \mathbf{k}_3 \cdot \mathbf{r})} + \text{c.c.}\right]. \tag{19.24}$$

Substitution of Eqs. (19.24) and (19.21) into Eq. (19.23) reveals the relationship between the Fourier amplitudes:

$$\mathbf{P}(\omega_3) = \epsilon_0 \frac{\bar{\bar{\chi}}^{(2)}}{2}(-\omega_3, \omega_1, \omega_2)\mathbf{E}(\omega_1)\mathbf{E}(\omega_2)e^{-j[(\mathbf{k}_3 - \mathbf{k}_1 - \mathbf{k}_2) \cdot \mathbf{r}]}, \tag{19.25}$$

or, in terms of the components of the vectors,

$$P_i(\omega_3) = \frac{\epsilon_0}{2} \sum_{jk} \chi_{ijk}^{(2)}(-\omega_3, \omega_1, \omega_2)E_j(\omega_1)E_k(\omega_2)e^{-j[(\mathbf{k}_3 - \mathbf{k}_1 - \mathbf{k}_2) \cdot \mathbf{r}]}. \tag{19.26}$$

Equation (19.26) can be written in the repeated-subscript shorthand notation of tensor algebra as

$$P_i(\omega_3) = \frac{\epsilon_0}{2}\chi_{ijk}E_j(\omega_1)E_k(\omega_2). \tag{19.27}$$

The repeated subscripts $j$ and $k$ on the right-hand side of this equation imply the summation over $j$ and $k$. From now on, for simplicity, we will omit writing the exponential factor in Eq. (19.26) explicitly.[3] An alternative way of writing Eqs. (19.21) and (19.26) that is based

[3] We shall see later that a most desirable situation for nonlinear mixing results when $\mathbf{k}_3 - \mathbf{k}_1 - \mathbf{k}_2 = 0$, namely the so-called *phase-matched* condition.

on a nonlinear mixing tensor $\overline{\overline{d}}$ is also used. In terms of this tensor the relationship between the Fourier amplitudes is

$$\mathbf{P}(\omega_3) = \epsilon_0 \cdot \overline{\overline{d}}^{(2)}(-\omega_3, \omega_1, \omega_2)E(\omega_1)E(\omega_2). \tag{19.28}$$

The relationship between $\overline{\overline{\chi}}^{(2)}$ and $\overline{\overline{d}}^{(2)}$ is such that

$$\chi^{(2)}_{ijk}(-\omega_3, \omega_1, \omega_2) = 2d_{ijk}(-\omega_3, \omega_1, \omega_2). \tag{19.29}$$

It is instructive to examine in full the dependence of one of the polarization components in Eq. (19.26) on the various tensor components and frequencies involved. The summation covers all field components and frequencies that are *distinguishable* from each other. For example:

$$\begin{aligned} P_x(\omega_3) = \frac{\epsilon_0}{2}[&\chi_{xxx}E_x(\omega_1)E_x(\omega_2) + \chi_{xyy}E_y(\omega_1)E_y(\omega_2) \\ &+ \chi_{xzz}E_z(\omega_1)E_z(\omega_2) \\ &+ \chi_{xyz}E_y(\omega_1)E_z(\omega_2) + \chi_{xzy}E_z(\omega_1)E_y(\omega_2) \\ &+ \chi_{xzx}E_z(\omega_1)E_x(\omega_2) + \chi_{xxz}E_x(\omega_1)E_z(\omega_2) \\ &+ \chi_{xxy}E_x(\omega_1)E_y(\omega_2) + \chi_{xyx}E_y(\omega_1)E_x(\omega_2)], \end{aligned} \tag{19.30}$$

where we have omitted the full frequency description of the $\chi$s. Note that $\chi_{xyz}(-\omega_3, \omega_1, \omega_2)$ $E_y(\omega_1)E_z(\omega_2)$ and $\chi_{xzy}(-\omega_3, \omega_1, \omega_2)E_z(\omega_1)E_y(\omega_2)$ are physically distinguishable. A unique frequency is associated with each field component.

The form of the tensor $\overline{\overline{\chi}}^{(2)}$ has to be consistent with the symmetry of the crystal. In particular, $\overline{\overline{\chi}}^{(2)}$ vanishes if the medium possesses inversion symmetry. To show why this must be so, consider the polarization component given by Eq. (19.26), which in repeated-subscript notation is

$$P_i(\omega_3) = \frac{\epsilon_0}{2}\chi_{ijk}E_j(\omega_1)E_k(\omega_2). \tag{19.31}$$

If we reverse the direction of both electric fields then the polarization must also reverse direction because the fields see an identical arrangement of atoms in the medium when they are reversed. Consequently,

$$-P_i(\omega_3) = \frac{\epsilon_0}{2}\chi_{ijk}E_j(\omega_1)E_k(\omega_2). \tag{19.32}$$

Equations (19.31) and (19.32) are inconsistent unless $\chi_{ijk} = 0$.

For the 21 of the 32 crystal classes that lack inversion symmetry the number of non-zero, independent elements of $\chi_{ijk}$ depends on the symmetry. Only for the triclinic class 1 are all 27 elements of $\overline{\overline{\chi}}^{(2)}$ non-zero and independent.

Equation (19.23) can be generalized to the mixing of $n$ different waves so that

$$\begin{aligned} P_\ell^{\omega=\omega_1+\omega_2+\cdots+\omega_n}(t) = \epsilon_0 \sum_{\ell_1,\ell_2,\ldots,\ell_n} &\chi^{(n)}_{\ell\ell_1\ell_2,\ldots,\ell_n}(-\omega, \omega_1, \omega_2, \ldots, \omega_n) \\ &\times E^{\omega_1}_{\ell_1}(t)E^{\omega_2}_{\ell_2}(t), \ldots, E^{\omega_n}_{\ell_n}(t). \end{aligned} \tag{19.33}$$

In terms of the Fourier amplitudes involved

$$P_\ell(\omega) = \frac{\epsilon_0}{2^{(n-1)}} \sum_{\ell_1,\ell_2,\ldots,\ell_n} \chi^{(-\omega,\omega_1,\omega_2,\ldots,\omega_n)}_{\ell\ell_1,\ell_2,\ldots,\ell_n} E_{\ell_1}(\omega_1)E_{\ell_2}(\omega_2)\ldots E_{\ell_n}(\omega_n), \tag{19.34}$$

and in terms of a nonlinear mixing tensor

$$\mathbf{P}(\omega) = \epsilon_0 \overline{\overline{d}}^{(n)}(-\omega,\omega_1,\omega_2,\ldots,\omega_n)\mathbf{E}(\omega_1)\ldots\mathbf{E}(\omega_n). \tag{19.35}$$

Comparison of Eqs. (19.34) and (19.35) shows that

$$\overline{\overline{d}}^{(n)} = \overline{\overline{\chi}}^{(n)}/2^{(n-1)}. \tag{19.36}$$

The above results must be modified if one or more of the frequencies $\omega, \omega_1, \omega_2, \ldots, \omega_n$ is DC.

For a DC field $E_i(t) = E(0)$. There is no factor of $\frac{1}{2}$ in this equation as there is in Eq. (19.21). Consequently, if $p$ of the interacting fields is at DC Eq. (19.36) becomes

$$\overline{\overline{d}}^{(n)} = \overline{\overline{\chi}}^{(n)}/2^{n-1-p}. \tag{19.37}$$

## 19.4 Second-harmonic generation

An interesting, and potentially confusing, situation results if two or more of the frequencies in Eq. (19.34) become equal to each other. To show here that the definition of nonlinear susceptibilities remains consistent in this case, let us consider the specific situation of second-harmonic generation (SHG). If the frequencies $\omega_1$ and $\omega_2$ approach each other in Eq. (2.30), then when $\omega_1 = \omega_2$ we get

$$\begin{aligned} P_x(2\omega) = \frac{\epsilon_0}{2}\Big[&\chi_{xxx}E_x(\omega)E_x(\omega) + \chi_{xyy}E_y(\omega)E_y(\omega) + \chi_{xzz}E_z(\omega)E_z(\omega) \\ &+ \chi_{xyz}E_y(\omega)E_z(\omega) + \chi_{xzy}E_z(\omega)E_y(\omega) \\ &+ \chi_{xzx}E_z(\omega)E_x(\omega) + \chi_{xxz}E_x(\omega)E_z(\omega) \\ &+ \chi_{xxy}E_x(\omega)E_y(\omega) + \chi_{xyx}E_y(\omega)E_x(\omega)\Big]. \end{aligned} \tag{19.38}$$

Each $\chi_{ijk}$ written above is shorthand for $\chi_{ijk}(-2\omega,\omega,\omega)$. Note that each of the nine terms is still physically distinguishable from the others. However, a fundamental question must be asked. Is the second harmonic being generated by the mixing together of two *distinct* waves, each at the same frequency $\omega$, or are the fields of a single wave interacting nonlinearly with themselves to generate the second harmonic? The actual values of the $\chi_{ijk}$ coefficients must differ in these two circumstances, otherwise we would have an apparent discontinuity in the amount of second harmonic generated. This distinction can be clearly illustrated for the situation in which the input electric fields have only an $x$-directed component. For two distinct waves, each of intensity $I$, the electric field of each wave is

$E_x(\omega) = \sqrt{2ZI}$, where $Z$ is the impedance of the medium at frequency $\omega$. Therefore, from Eq. (19.38),

$$P_x(2\omega) = \frac{\epsilon_0}{2}\chi_{xxx}^{\text{mix}}(-2\omega, \omega, \omega)(2ZI), \tag{19.39}$$

where the superscript mix stresses the independence of the two waves.

For SHG from a single input wave of intensity $2I$, whose field amplitude is $E_x(\omega) = \sqrt{4ZI}$, Eq. (19.38) gives

$$P_x(2\omega) = \frac{\epsilon_0}{2}\chi_{xxx}^{\text{SHG}}(-2\omega, \omega, \omega)(4ZI). \tag{19.40}$$

Equations (19.39) and (19.40) must correspond to the same induced polarization amplitude, therefore

$$\chi_{xxx}^{\text{SHG}}(-2\omega, \omega, \omega) = \tfrac{1}{2}\chi_{xxx}^{\text{mix}}(-2\omega, \omega, \omega). \tag{19.41}$$

SHG is such an important and useful nonlinear process that the above fine distinction is avoided through the use of the SHG tensor $\overline{\overline{d}}^{2\omega}$ defined by

$$P_i(2\omega) = \epsilon_0 \sum_{jk} d_{ijk}E_j(\omega)E_k(\omega)e^{-j[(\mathbf{k}_{2\omega}-2\mathbf{k}_\omega)\cdot\mathbf{r}]}. \tag{19.42}$$

In the use of the $d_{ijk}$ coefficients no distinction is made between terms like $d_{xyz}E_y(\omega)E_z(\omega)$ and $d_{xzy}E_z(\omega)E_y(\omega)$, so in matrix notation

$$\begin{pmatrix} P_x(2\omega) \\ P_y(2\omega) \\ P_z(2\omega) \end{pmatrix} = \epsilon_0 \begin{pmatrix} d_{xxx} & d_{xyy} & d_{xzz} & d_{xyz} & d_{xzx} & d_{xxy} \\ d_{yxx} & d_{yyy} & d_{yzz} & d_{yyz} & d_{yzx} & d_{yxy} \\ d_{zxx} & d_{zyy} & d_{zzz} & d_{zyz} & d_{zzx} & d_{zxy} \end{pmatrix} \times \begin{pmatrix} E_x^2(\omega) \\ E_y^2(\omega) \\ E_z^2(\omega) \\ 2E_y(\omega)E_z(\omega) \\ 2E_z(\omega)E_x(\omega) \\ 2E_x(\omega)E_y(\omega) \end{pmatrix}. \tag{19.43}$$

Equation (19.43) can be simplified further through the use of contracted notation, in which

$$\begin{aligned} x &= 1, & y &= 2, & z &= 3, \\ xx &= 1, & yy &= 2, & zz &= 3, \\ yz &= 4, & zx &= 5, & xy &= 6, \end{aligned} \tag{19.44}$$

and $\overline{\overline{d}}$ can be written as

$$\overline{\overline{d}} = \begin{pmatrix} d_{11} & d_{12} & d_{13} & d_{14} & d_{15} & d_{16} \\ d_{21} & d_{22} & d_{23} & d_{24} & d_{25} & d_{26} \\ d_{31} & d_{32} & d_{33} & d_{34} & d_{35} & d_{36} \end{pmatrix}. \tag{19.45}$$

### 19.4.1 Symmetries and Kleinman's conjecture

The formally defined nonlinear susceptibility coefficients, for example, $\chi_{ijk}(-\omega_3, \omega_1, \omega_2)$, are related to each other in various ways. Common sense dictates that any permutation of the frequencies $\omega_1, \omega_2$, and $\omega_3$ cannot affect the value of the coefficient, provided the axial indices $i, j, k$ are also permuted in the same way. Therefore

$$\chi_{ijk}(-\omega_3, \omega_1, \omega_2) = \chi_{ikj}(-\omega_3, \omega_2, \omega_1) = \chi_{kij}(\omega_2, -\omega_3, \omega_1) \tag{19.46}$$

etc. Furthermore, if the nonlinear polarization is to be a real quantity, then $[P_i(\omega_3)]^* = P_i(-\omega_3)$, which provides the additional relationship

$$[\chi_{ijk}(-\omega_3, \omega_1, \omega_2)]^* = \chi_{ijk}(\omega_3, -\omega_1, -\omega_2). \tag{19.47}$$

If the nonlinear material is transparent at all three frequencies $\omega_1, \omega_2$, and $\omega_3$, then $\overline{\overline{\chi}}$ will be real, and if $\overline{\overline{\chi}}$ is independent of frequency all permutations of axial index and associated frequency become irrelevant:

$$\chi_{ijk}(-\omega_3, \omega_1, \omega_2) = \chi_{ijk} = \chi_{kij} = \chi_{jki} \tag{19.48}$$

etc. This is Kleinman's conjecture [6]. Because all materials are to some extent dispersive, it cannot be absolutely true, but it has been verified experimentally to be a very good approximation in most practical nonlinear media. If Kleinman's conjecture is valid we have relationships such as $d_{14} = d_{25} = d_{36}, d_{15} = d_{31}, d_{23} = d_{34}$, etc. Independently of Kleinman's conjecture the elements of the SHG tensor are also in general not all independent, but are constrained by the symmetry of the crystal. For triclinic crystals of class 1 all elements of $\overline{\overline{d}}$ are non-zero and independent, whereas for crystal classes with inversion symmetry all the elements of $\overline{\overline{d}}$ are zero. In other cases many elements of $\overline{\overline{d}}$ will either be zero or related to the elements of the matrix. In most nonlinear crystals that are widely used, such as ADP, KDP, and $LiNbO_3$, there are only one or two important elements of the $\overline{\overline{d}}$ tensor, as can be seen from the values tabulated for these and other crystals in Table 19.1.

## 19.5 The linear electro-optic effect

If one of the input waves, say $\omega_2$, is at DC then Eq. (19.35) becomes

$$\mathbf{P}(\omega_1) = \epsilon_0 \overline{\overline{d}}^{\,\mathrm{EO}}(-\omega_1, \omega_1, 0)\mathbf{E}(\omega_1)\mathbf{E}(0), \tag{19.49}$$

where the tensor $\overline{\overline{d}}^{\,\mathrm{EO}}$ describes the linear electro-optic effect, which we have already discussed, but in a phenomenologically different way. Symmetry requires that $d_{ijk}(-\omega, \omega, 0) = d_{kij}(0, -\omega, \omega)$, where $d_{kij}(0, -\omega, \omega)$ is a component of the optical rectification (OR) tensor $\overline{\overline{d}}^{\,\mathrm{OR}}$, so the symmetries of the linear electro-optic mixing tensor and the tensor for optical rectification should be the same. Further, since SHG and OR occur simultaneously when a wave interacts nonlinearly with a medium, we expect the symmetry but not the values of the individual elements $\overline{\overline{d}}^{\,2\omega}$, $\overline{\overline{d}}^{\,\mathrm{EO}}$, and $\overline{\overline{d}}^{\,\mathrm{OR}}$ to be the same. The specific

**Table 19.1** Important elements of the $\overline{\overline{d}}$ second-harmonic-generation tensor for various nonlinear materials

| Crystal | Symmetry group | $\lambda(\mu m)$ | $d_{ij}(10^{-12}\ m\ V^{-1})$ |
|---|---|---|---|
| $Ba_2NaNb_5O_{15}$ (banana) | mm2 | 1.06 | $d_{15} = 7.4 \pm 2.3$, $d_{24} = 6.5 \pm 0.7$ |
| | 3m | 1.15 | $d_{33} = 9 \pm 0.7$ |
| $Ag_3AsS_3$ (proustite) | 3m | 1.5 | $d_{15} = 15.3$, $d_{22} = 25.5$ |
| $LiNbO_3$ | 3m | 1.15 | $d_{31} = -5.95 \pm 0.5$, $d_{22} = 2.1 \pm 2$ |
| | | | $d_{33} = -34.4 \pm 9$ |
| HgS | 32 | 10.6 | $d_{11} = 77 \pm 26$ |
| $SiO_2$ | 32 | 1.06 | $d_{11} = 0.32 \pm 0.02$, $d_{14} = 0.003$ |
| Te | 32 | 10.6 | $d_{11} = 922 \pm 293$ |
| $BaTiO_3$ | 4mm | 1.06 | $d_{15} = 17.7 \pm 1.5$, $d_{31} = -18.8 \pm 1.5$ |
| | | | $d_{33} = -7.1 \pm 0.5$ |
| $(NH_4)H_2PO_4$ (ADP) | $\bar{4}2m$ | 1.06 | $d_{36} = 0.57 \pm 0.07$, $d_{14} = 0.63 \pm 0.03$ |
| $(NH_4)D_2PO_4$ (AD*P) | $\bar{4}2m$ | 0.694 | $d_{36} = 0.56$ |
| $KD_2PO_4$ (KD*P, DKDP) | $\bar{4}2m$ | 1.06 | $d_{36} = 0.46 \pm 0.02$ |
| $KH_2PO_4$ (KDP) | $\bar{4}2m$ | 1.06 | $d_{36} = 0.63 \pm 0.06$ |
| $KTiOPO_4$ (KTP) | $\bar{4}mm2$ | 1.06 | $d_{31} = 6.5$, $d_{32} = 5.0$ |
| | | | $d_{33} = 13.7$, $d_{24} = 6.6$, $d_{15} = 6.1$ |
| $\beta$-$BaB_2O_4$(BBD) | 3m | 1.06 | $d_{22} = 2.22 \pm 0.09$ |
| $AgGaS_2$ | $\bar{4}2m$ | 10.06 | $d_{36} = 13 \pm 2$ |
| $AgGaSe_2$ | $\bar{4}2m$ | 10.06 | $d_{36} = 33 \pm 3$ |
| $CdGeAs_2$ | $\bar{4}2m$ | 10.6 | $d_{36} = 877 \pm 175$ |
| $LiIO_3$ | 6 | 1.06 | $d_{31} = -7.11 \pm 0.51$, $d_{33} = -7.22 \pm 0.51$ |
| | | | $d_{15} = -5.5$, $d_{14} = 0.31$ |
| $LiB_3O_5$ (LBO) | mm2 | 1.06 | $d_{31} = 1.09 \pm 0.09$, $d_{32} = 1.17 \pm 0.14$ |
| CdTe | $\bar{4}3m$ | 10.6 | $d_{14} = 168 \pm 67$ |
| GaAs | $\bar{4}3m$ | 1.06 | $d_{36} = 137 \pm 27$ |
| | | 10.6 | $d_{36} = 134 \pm 45$ |
| GaP | $\bar{4}3m$ | 10.6 | $d_{14} = 100 \pm 18$ |
| InAs | $\bar{4}3m$ | 10.6 | $d_{14} = 419 \pm 126$ |
| ZnS | $\bar{4}3m$ | 10.6 | $d_{14} = 30.6 \pm 8.4$ |
| CdS | 6mm | 1.06 | $d_{15} = 17.7 \pm 1.5$, $d_{31} = -16.2 \pm 1$ |
| | | | $d_{33} = 32 \pm 2$ |
| | | 10.6 | $d_{15} = 29 \pm 7$, $d_{31} = 26.4 \pm 6.3$ |
| | | | $d_{33} = 44 \pm 13$ |

The $d_{ij}$ coefficients are given in contracted notation; if Kleinman's conjecture were correct $d_{14} \equiv d_{xyz} \equiv d_{zxy} \equiv d_{36}$, for example. The values given represent a compilation of values from many sources, including manufacturers' data sheets, and values listed by Zernike and Midwinter [7], Kurtz [8], Sutherland [9], and Boyd and Fischer [10]. There is some disagreement in the literature concerning the value of many of these coefficients [11]; the errors listed come from the same sources, but are probably optimistic values.

form of $\bar{\bar{d}}^{2\omega}$ for each of the 32 crystal classes has been tabulated by Byer [4], for example, for class 3m

$$\bar{\bar{d}} = \begin{pmatrix} 0 & 0 & 0 & 0 & d_{15} & -d_{22} \\ -d_{22} & d_{22} & 0 & d_{15} & 0 & 0 \\ d_{31} & d_{31} & d_{33} & 0 & 0 & 0 \end{pmatrix}. \tag{19.50}$$

However, we do not need to tabulate all the forms of $\bar{\bar{d}}$ for the different crystal classes here. The symmetry of $\bar{\bar{d}}$ is the same as the symmetry of the piezoelectric tensor discussed in Chapter 18. There it is pointed out that the symmetry of $\bar{\bar{d}}$ is the same as the symmetry of the transpose of the linear electro-optic tensor $\bar{\bar{r}}$. Therefore, the symmetry of $\bar{\bar{d}}$ can be determined from the tabulated symmetries given in Chapter 18. This is, perhaps, not too surprising in view of the fact that $\bar{\bar{d}}^{\mathrm{EO}}$ and $\bar{\bar{r}}$ describe the same phenomenon.

## 19.6 Parametric and other nonlinear processes

Parametric amplification is a process in which a weak electric field at frequency $\omega_s$, called the *signal*, interacts nonlinearly with a strong electric field at frequency $\omega_p$, called the *pump*, in such a way that the amplitude of the signal at frequency $\omega_s$ is increased. To conserve energy in this process, as photons are converted from frequency $\omega_p$ to frequency $\omega_s$, additional photons at frequency $\omega_i$, called the *idler* frequency, must be generated simultaneously in such a way that

$$\omega_s + \omega_i = \omega_p. \tag{19.51}$$

The process is described by the nonlinear susceptibility tensor $\bar{\bar{\chi}}(\omega_p - \omega_s, -\omega_p, \omega_s)$. Parametric amplification has been studied in conventional nonlinear electric circuits for many years, for example, in resonant circuits containing voltage-dependent capacitors. The interested reader can find brief discussions of such circuits in references [12] and [13].

Higher-order nonlinear processes described by the tensor $\bar{\bar{\chi}}^{(3)}$ are of some importance because they are permitted to occur in media that possess a center of inversion symmetry, such as gases and liquids. We will not discuss any of these phenomena in detail here but will list some of the important ones. Third-harmonic generation, which is described by the tensor $\bar{\bar{\chi}}(-3\omega, \omega, \omega, \omega)$, has been studied quite widely as a means for generating shorter wavelengths from powerful laser sources [14]. The quadratic electro-optic effect is described by the tensor $\chi(-\omega, \omega, 0, 0)$. Coherent anti-Stokes Raman scattering, which is described by the tensor $\chi(-\omega_2, \omega_1, \omega_1, -\omega_2)$, is a useful spectroscopic technique for studying species in difficult situations – such as during combustion. If the nonlinear medium is dispersive, then the nonlinear tensors become complex and the real and imaginary parts of the appropriate tensor may describe different processes; for example, $\mathcal{R}\bar{\bar{\chi}}(-\omega, \omega, \omega - \omega)$ describes intensity-induced birefringence, and $\mathcal{I}(-\omega, \omega, \omega_1 - \omega)$ describes two-photon absorption.

There are also important nonlinear interactions that depend on *magnetic* field components of the nonlinearly interacting waves. For example the polarization effect described by

$$\mathbf{P}(\omega) = -j\epsilon_0 \bar{\bar{\chi}}^{\mathrm{B}}(-\omega, \omega, 0)\mathbf{E}(\omega)\mathbf{B}(0), \tag{19.52}$$

which depends linearly on a DC magnetic flux density $\mathbf{B}(0)$, describes the Faraday effect – the rotation of the plane of polarization of linearly polarized light as it propagates through a Faraday-active medium along an axially directed magnetic field. The tensor $\bar{\bar{\chi}}^{\mathrm{BB}}(-\omega, \omega, 0, 0)$ describes the birefringence induced by a DC magnetic flux density – the Cotton–Mouton effect [15]. For a fuller discussion of these various phenomena the interested reader should consult Rabin and Tang [16].

An interesting, unrelated, nonlinear effect in which a light wave changes the refractive index of a material is the photorefractive effect, which cannot be simply described by a nonlinear susceptibility [1, 17]. The photorefractive effect results from the field-induced separation of charge carriers, such as electrons and holes, in a material. It is generally a slow process, and the refractive-index change can remain for long periods after the illuminating light has been removed. If two light waves interfere inside a photorefractive material then regions of high and low local light intensity result. In the bright regions electrons can be excited into the conduction band and then diffuse into the dark regions. The resulting electric fields produced by the separated electrons and holes change the refractive index by the electro-optic effect. The resultant patterns of changed refractive index can be used to write a hologram (see Chapter 23) into a material such as $BaTiO_3$, $LiNbO_3$, $As_2S_3$, and certain organic polymers.

## 19.7 Macroscopic and microscopic susceptibilities

In our general discussion of nonlinear susceptibilities we have been dealing with applied fields $\mathbf{E}$ and resultant *macroscopic* polarizations $\mathbf{P}$, which are related schematically by Eq. (19.1). However, when we attempt to relate these macroscopically measured quantities to an atomic model of the nonlinear process, such as in Eq. (19.5), we must be careful to make the distinction between the applied *macroscopic* field and the field actually experienced by the individual particles of the medium. This *local* field is different from the macroscopic field because of the polarization of the medium. We follow the arguments presented by Kittel [18] in relating the macroscopic and local fields.

The macroscopic field $\mathbf{E}$ generates an induced linear polarization in the medium according to

$$\mathbf{P} = \epsilon_0 \chi^{(1)} \mathbf{E}. \tag{19.53}$$

We will neglect the tensor character of $\chi^{(1)}$ to make our analysis a little more straightforward. Each particle in the medium sees a local electric field $\mathbf{E}_{\mathrm{loc}}$, where

$$\mathbf{E}_{\mathrm{loc}} = \mathbf{E} + \mathbf{E}_1 + \mathbf{E}_2. \tag{19.54}$$

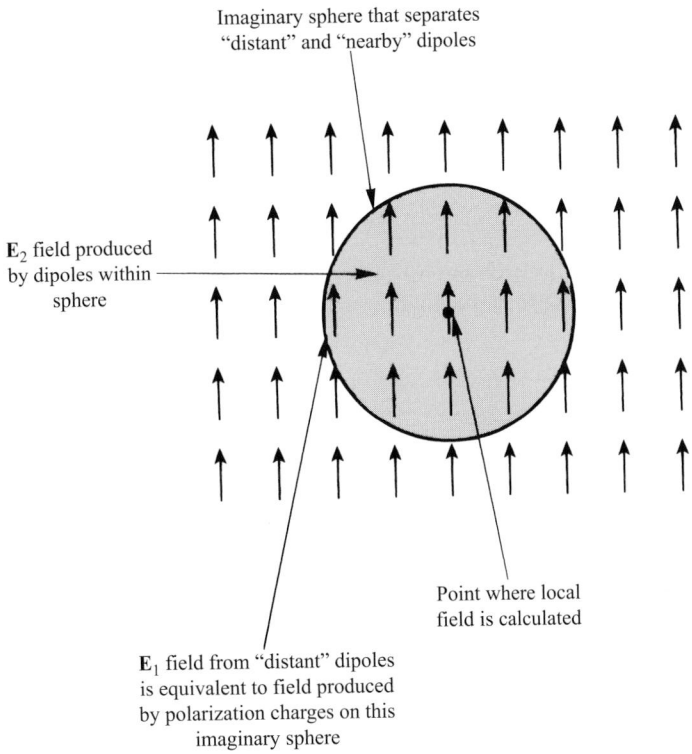


**Fig. 19.3** The imaginary spherical region around a particle in a polarized medium that delineates nearby and distant dipoles.

The induced dipoles, which give rise to the macroscopic polarization $\mathbf{P}$, macroscopically contribute to the fields $\mathbf{E}_1$ and $\mathbf{E}_2$. The field $\mathbf{E}_1$ is produced, at the point of interest, by all the dipoles that are so far away as to form a continuous polarization. $\mathbf{E}_2$ is the field of dipoles sufficiently close to the observation point that their discrete nature is apparent. The field $\mathbf{E}_1 + \mathbf{E}_2$ is the total field acting at the observation point that results from all the dipoles in the medium:[4]

$$\mathbf{E}_1 + \mathbf{E}_2 = \frac{1}{4\pi\epsilon_0} \sum_i \frac{3(\boldsymbol{\mu}_i \cdot \mathbf{r}_i)}{r_i^5} \mathbf{r}_i - \frac{\boldsymbol{\mu}_i}{r_i^3}, \tag{19.55}$$

where $\boldsymbol{\mu}_i$ is the dipole moment (vector) of the $i$th particle of the medium. We can separate the dipoles near the observation point from the distant dipoles by an imaginary sphere centered at the observation point as shown in Fig. 19.3. The field $\mathbf{E}_1$, which comes from the distant dipole can be shown to be [18]

$$\mathbf{E}_1 = \frac{\mathbf{P}}{3\epsilon_0}, \tag{19.56}$$

[4] We are not including in our analysis the so-called depolarization field, which results from dipoles on the outside surface of a finite object of specific shape [18]. This field is zero for long thin crystals.

where $\mathbf{P}$ is the total macroscopic polarization defined in Eq. (19.53), which is related to the $\mathbf{D}$ vector that appears in Maxwell's equations by

$$\mathbf{D} = \epsilon_0 \mathbf{E} + \mathbf{P}. \tag{19.57}$$

In an isotropic non-metallic medium it can be shown that $\mathbf{E}_2 = 0$. Therefore in an isotropic medium

$$\mathbf{E}_{\text{loc}} = \mathbf{E} + \frac{\mathbf{P}}{3\epsilon_0}. \tag{19.58}$$

In non-isotropic media $\mathbf{E}_{\text{loc}}$ is determined by a Lorentz correction factor $\mathcal{L}$ such that

$$\mathbf{E}_{\text{loc}} = \mathbf{E} + \frac{\mathcal{L}}{\epsilon_0}\mathrm{P}. \tag{19.59}$$

On a microscopic scale, each individual dipole moment $\boldsymbol{\mu}_i$ is related to the local field by

$$\boldsymbol{\mu}_i = \overline{\alpha}_i \mathbf{E}_{\text{loc}}, \tag{19.60}$$

where $\overline{\alpha}_i$ is the spatially averaged polarizability of the $i$th particle. Clearly,

$$\mathbf{P} = \sum_i N_i \overline{\alpha}_i \mathbf{E}_{\text{loc}}, \tag{19.61}$$

where $N_i$ is the number of particles of type $i$ per unit volume of the medium. Combining Eqs. (19.58) and (19.61) gives

$$\mathbf{P} = \sum_i N_i \overline{\alpha}_i \left( \mathbf{E} + \frac{\mathbf{P}}{3\epsilon_0} \right), \tag{19.62}$$

and, by rearranging, we obtain

$$\mathbf{P} = \frac{\sum_i N_i \overline{\alpha}_i \mathbf{E}}{1 - \sum_i N_i \overline{\alpha}_i / (3\epsilon_0)}. \tag{19.63}$$

Also,

$$\mathbf{D} = \left( \epsilon_0 + \frac{\sum_i N_i \overline{\alpha}_i}{1 - \sum_i N_i \overline{\alpha}_i / 3(\epsilon_0)} \right) \mathbf{E} = \epsilon_0 \epsilon_{\text{r}} \mathbf{E}, \tag{19.64}$$

where $\epsilon_{\text{r}}$ is the dielectric constant. Remembering that $n = \sqrt{\epsilon}$, Eq. (19.64) gives

$$\frac{\epsilon_{\text{r}} - 1}{\epsilon_{\text{r}} + 2} = \frac{n^2 - 1}{n^2 + 2} = \frac{1}{3\epsilon_0} \sum_i N_i \overline{\alpha}_i, \tag{19.65}$$

which is called the Clausius–Mossotti equation. It relates a macroscopic quantity $\epsilon_{\text{r}} = 1 + \chi^{(1)}$ to a microscopic one $\alpha$. For linear phenomena, $\mathbf{P}$ is related to $\mathbf{E}$ by Eq. (19.53). Combining Eqs. (19.58) and (19.53) gives

$$\mathbf{E}_{\text{loc}} = \frac{\epsilon_{\text{r}} + 2}{3} \mathbf{E} = f\mathbf{E}, \tag{19.66}$$

where $\epsilon_r = 1 + \chi$ is the dielectric constant of the medium. It is easy to show that

$$f = 1 + \chi^{(1)}\mathcal{L}. \tag{19.67}$$

The value of $f$ predicted by Eq. (19.67) is satisfactory for ionic crystals of cubic symmetry, but in crystals of other symmetry its exact calculation can be complicated. We will treat it as an empirical factor that relates the local field to the externally applied field.

Equation (19.61) can be written in the form

$$\mathbf{P} = \epsilon_0 \chi_m^{(1)} \mathbf{E}_{loc}, \tag{19.68}$$

where the linear microscopic susceptibility $\chi_m^{(1)}$ is clearly

$$\chi_m^{(1)} = \frac{1}{\epsilon_0} \sum_i N_i \overline{\alpha}_i. \tag{19.69}$$

Combining Eqs. (19.68) and (19.66) gives

$$\mathbf{P} = \epsilon_0 f \chi_m^{(1)} \mathbf{E}_{loc}. \tag{19.70}$$

Clearly,

$$\chi^{(1)} = f \chi_m^{(1)}. \tag{19.71}$$

When nonlinear phenomena occur in a medium the connection between microscopic and macroscopic susceptibilities becomes more subtle. Ducuing [19] has presented a simple analysis of this situation. Consider the case of SHG in which a macroscopic nonlinear polarization $P_{NLS}(2\omega)$ is generated according to

$$P_{NLS}(2\omega) = \frac{\epsilon_0}{2} \chi^{(2)}(2\omega) E^2(\omega), \tag{19.72}$$

where $\chi^{(2)}(2\omega)$ is a shorthand notation for $\chi^{(2)}(-2\omega, \omega, \omega)$. For simplicity we will ignore the vector or tensor qualities of the variables in Eq. (19.72). The subscript NLS, which stands for "nonlinear source," is included to stress that the origin of the polarization is distinctly nonlinear, and that this polarization can serve as a source of waves in Maxwell's equations. We shall examine this point in detail in the next chapter. The subtlety that arises is that the electric field $E(2\omega)$ generated as a result of $P_{NLS}(2\omega)$ makes an additional contribution to the *total* polarization at frequency $2\omega, P(2\omega)$, through the linear susceptibility at frequency $2\omega$. On a microscopic scale each particle of the medium experiences a local second-harmonic field $E_{loc}(2\omega)$, and the microscopic nonlinear polarization is

$$P_m(2\omega) = \frac{\epsilon_0}{2} \chi_m^{(2)}(2\omega)[E_{loc}(\omega)]^2 = \frac{\epsilon_0}{2} \chi_m^{(2)}(2\omega) f^2(\omega) E^2(\omega). \tag{19.73}$$

Equation (19.73) effectively defines the microscopic nonlinear susceptibility $\chi_m^{(2)}(2\omega)$.

The total polarization at frequency $2\omega$ is

$$P(2\omega) = \epsilon_0 \chi_m^{(1)}(2\omega) E_{loc}(2\omega) + \frac{\epsilon_0}{2} \chi_m^{(2)}(2\omega) f^2(\omega) E^2(\omega). \tag{19.74}$$

The local field at frequency $2\omega$ is

$$E_{loc}(2\omega) = E(2\omega) + \frac{\mathcal{L}(2\omega)}{\epsilon_0} P(2\omega). \tag{19.75}$$

Combining Eqs. (19.74) and (19.75) gives

$$P(2\omega) = \frac{\epsilon_0 \chi_{\mathrm{m}}^{(1)}(2\omega) E_{\mathrm{loc}}(2\omega) + (\epsilon_0/2)\chi_{\mathrm{m}}^{(2)}(2\omega) f^2(\omega) E^2(\omega)}{1 - \chi_{\mathrm{m}}^{(1)}(2\omega)\mathcal{L}(2\omega)}, \tag{19.76}$$

and, from Eqs. (19.67) and (19.71),

$$\mathcal{L}(2\omega) = \frac{f(2\omega) - 1}{\chi^{(1)}(2\omega)} = \frac{f(2\omega) - 1}{f(2\omega)\chi_{\mathrm{m}}^{(1)}(2\omega)}. \tag{19.77}$$

Combining Eqs. (19.76) and (19.77) gives

$$P(2\omega) = \epsilon_0 \chi_{\mathrm{m}}^{(1)}(2\omega) f(2\omega) E(2\omega) + \frac{\epsilon_0}{2} f(2\omega) f^2(\omega) \chi_{\mathrm{m}}^{(2)}(2\omega) E^2(\omega). \tag{19.78}$$

On close examination, the first term on the right-hand side of Eq. (19.78) should be obvious: this linear term would be expected to occur if any second-harmonic field is present. The second term in Eq. (19.78) clearly must correspond to the nonlinear source polarization defined in Eq. (19.72), since it is the only term present that depends on $E^2(\omega)$. Therefore, comparing Eqs. (19.72) and (19.78) gives

$$\chi^{(2)}(2\omega) = f(2\omega) f^2(\omega) \chi_{\mathrm{m}}^{(2)}(2\omega). \tag{19.79}$$

This concludes our rather formal discussion of nonlinear processes and how they are quantified. In the next chapter we will pursue this discussion further, but on a more practical basis.

## 19.8 Problem

19.1 By substituting from Eq. (19.6) into Eq. (19.5) and examining the DC terms, derive an expression for the coefficient for optical rectification $d(0)$ defined by the relationship

$$P(\omega = 0) = \epsilon_0 d(0) E^*(\omega) E(\omega).$$

## References

[1] R. W. Boyd, *Nonlinear Optics*, 3rd edn., Burlington, MA: Academic Press, 2008.

[2] R. C. Miller, "Optical second harmonic generation in piezoelectric crystals," *Appl. Phys. Lett.* **5**, 17–19, 1964.

[3] C. G. B. Garrett and F. N. H. Robinson, "Miller's phenomenological rule for computing nonlinear susceptibilities," *IEEE J. Quant. Electron.* **2**, 328–329, 1966.

[4] R. L. Byer, "Parametric oscillation and nonlinear materials," in *Nonlinear Optics*, eds. P. G. Harper and B. S. Wherrett, London: Academic Press, 1977, pp. 47–160.

[5] Y. R. Shen, *The Principles of Nonlinear Optics*, New York: Wiley, 1984.

[6] D. A. Kleinman, "Nonlinear dielectric polarization in optical media," *Phys. Rev.* **126**, 1977–1979, 1962.

[7] F. Zernike and J. E. Midwinter, *Applied Nonlinear Optics*, New York: Wiley, 1973.

[8] S. K. Kurtz, "Measurement of nonlinear optical susceptibilities," in *Quantum Electronics*, Vol. 1A, eds. H. Rabin and C. L. Tang, New York: Academic Press, 1975, pp. 209–281.

[9] R. L. Sutherland, D. G. McLean, and S. Kirkpatrick, *Handbook of Nonlinear Optics*, Boca Raton, FL: CRC Press, 2003.

[10] R. W. Boyd and G. L. Fischer, "Nonlinear optical materials," in *Encyclopedia of Materials: Science and Technology,* Amsterdam: Elsevier, 2001, pp. 6237–6244.

[11] S. Singh, "Nonlinear optical materials," in *Handbook of Lasers*, ed. R. J. Pressley, Cleveland, OH: CRC Press, 1971, pp. 489–525.

[12] A. Yariv, *Optical Electronics*, 3rd edn., New York: Holt, Rinehart & Winston, 1985.

[13] S. Y. Liao, *Microwave Devices and Circuits*, 2nd edn., Englewood Cliffs, NJ: Prentice-Hall, 1985.

[14] C. R. Vidal, "Four-wave frequency mixing in gases," in *Tunable Lasers*, eds. L. R. Mollenauer and J. C. White, Berlin: Springer-Verlag, 1987, pp. 57–113.

[15] R. W. Ditchburn, *Light*, 3rd edn., London: Academic Press, 1976.

[16] H. Rabin and C.L. Tang, eds., *Quantum Electronics*, Vols. 1A and 1B, New York: Academic Press, 1975.

[17] P. Yeh, *Introduction to Photorefractive Nonlinear Optics*, New York: Wiley, 1993.

[18] C. Kittel, *Introduction to Solid State Physics*, 8th edn., New York: Wiley, 2005.

[19] J. Ducuing, "Microscopic and macroscopic nonlinear optics," in *Nonlinear Optics*, eds. P. G. Harper and B. S. Wherrett, London: Academic Press, 1977, pp. 11–45.

# 20 Wave propagation in nonlinear media

## 20.1 Introduction

In this chapter we will follow up our rather formal discussion of nonlinear phenomena begun in the last chapter. We will describe through the use of Maxwell's equations how nonlinear interactions actually lead to the production of waves at new frequencies in a medium. Specifically, we will examine the practically useful phenomena of second-harmonic generation (SHG) and parametric amplification and oscillation. For each phenomenon, specific examples of how these processes are optimized in practice will be given. Practical applications of nonlinear optical phenomena are widespread.

## 20.2 Electromagnetic waves and nonlinear polarization

To show how a nonlinearly generated polarization leads to the production of waves at new frequencies, the following equations of electromagnetic theory will be used: Ampère's law,

$$\text{curl}\,\mathbf{H} = \mathbf{j} + \frac{\partial \mathbf{D}}{\partial t}, \tag{20.1}$$

where $\mathbf{j}$(A m$^{-2}$) is the current density, Faraday's law,

$$\text{curl}\,\mathbf{E} = -\mu \frac{\partial \mathbf{H}}{\partial t}, \tag{20.2}$$

and Ohm's law,

$$\mathbf{j} = \sigma \mathbf{E}, \tag{20.3}$$

where $\sigma$ (S m$^{-1}$) is the conductivity of the medium, and

$$\mathbf{D} = \epsilon_0 \mathbf{E} + \mathbf{P}. \tag{20.4}$$

When a second-order nonlinear process occurs in the medium the total polarization can be written as

$$\mathbf{P} = \epsilon_0 \overline{\overline{\chi}}^{(1)} \mathbf{E} + \mathbf{P}_{\text{NLS}}, \tag{20.5}$$

where the nonlinear source (NLS) polarization has Cartesian components

$$(P_{\text{NLS}})_i = \epsilon_0 \sum_{jk} \chi_{ijk} E_j E_k. \tag{20.6}$$

From Eqs. (20.1) and (20.4) we have

$$\begin{aligned}\text{curl}\,\mathbf{H} &= \sigma\mathbf{E} + \epsilon_0\frac{\partial\mathbf{E}}{\partial t} + \frac{\partial\mathbf{P}}{\partial t}\\ &= \sigma\mathbf{E} + \epsilon_0\overline{\overline{\epsilon}}_{\mathrm{r}}\frac{\partial\mathbf{E}}{\partial t} + \frac{\partial}{\partial t}\mathbf{P}_{\mathrm{NLS}}.\end{aligned} \tag{20.7}$$

In Eq. (20.7) we have included all the *linear* polarization effects in the term containing the dielectric tensor $\overline{\overline{\epsilon}}_{\mathrm{r}}$. Use of the vector identity curl curl $\mathbf{E}$ = grad div $\mathbf{E} - \nabla^2\mathbf{E}$ in conjunction with Eqs. (20.2) and (20.7) gives

$$\nabla^2\mathbf{E} = \mu\sigma\frac{\partial\mathbf{E}}{\partial t} + \mu\epsilon_0\epsilon_{\mathrm{r}}\frac{\partial^2\mathbf{E}}{\partial t^2} + \mu\frac{\partial^2}{\partial t^2}\mathbf{P}_{\mathrm{NLS}}. \tag{20.8}$$

For simplicity we are neglecting the anisotropy of the medium: in other words we are assuming that $\epsilon_{\mathrm{r}}$ is a scalar and div $\mathbf{E} = 0$. Equation (20.8) is a modified version of the wave equation with the addition of a loss term (because of non-zero conductivity $\sigma$) and a new source term.

If we are dealing with plane waves (and laser beams come close to being treatable in this way unless we are interested in fine detail), then without loss of generality we can analyze a situation in which all the waves propagate in the $z$ direction and the only field components present lie in the $x$ or $y$ directions. In this case Eq. (20.8) becomes

$$\frac{\partial^2 E_k}{\partial z^2} = \mu\sigma_k\frac{\partial E_k}{\partial t} + \mu\epsilon_0\epsilon_{\mathrm{r}}\frac{\partial^2 E_k}{\partial t^2} + \mu\epsilon_0\frac{\partial^2}{\partial t^2}\chi_{kij}E_iE_j, \tag{20.9}$$

where $\epsilon_{\mathrm{r}}$ is the dielectric constant appropriate to the frequency of wave $E_k$, and $\sigma_k$ is the appropriate conductivity. The subscripts $i,j$, and $k$ can take only the values $x$ and $y$.[1] Equation (20.9) shows clearly the consistency relationship among the various field components present. If the field components $E_i$ and $E_j$ are at frequencies $\omega_1$ and $\omega_2$, respectively, then clearly $E_k$ must contain a component of frequency $\omega_3 = \omega_1 + \omega_2$ in order for both sides of Eq. (20.9) to balance.

We can write the plane waves present as

$$\begin{aligned}E_i^{\omega_1}(z,t) &= \frac{1}{2}\left[E_{1i}(z)e^{j(\omega_1 t - k_1 z)} + \text{c.c.}\right],\\ E_j^{\omega_2}(z,t) &= \frac{1}{2}\left[E_{2j}(z)e^{j(\omega_2 t - k_2 z)} + \text{c.c.}\right],\\ E_k^{\omega_3}(z,t) &= \frac{1}{2}\left[E_{3k}(z)e^{j(\omega_3 t - k_3 z)} + \text{c.c.}\right].\end{aligned} \tag{20.10}$$

The NLS polarization at frequency $\omega_3$, which drives the nonlinearly generated wave at frequency $\omega_3$, is, cf. Eq. (19.23),

$$\left[P_{\mathrm{NLS}}^{\omega_3}(z,t)\right]_k = \frac{\epsilon_0\chi_{kij}}{4}E_{1i}(z)E_{2j}(z)e^{j[(\omega_1+\omega_2)t-(k_1+k_2)z]} + \text{c.c.} \tag{20.11}$$

[1] The repeated indices in the product $\chi_{kij}E_iE_j$ in Eq. (20.9) and in the following equations imply the summation $\sum_{ij}\chi_{kij}E_iE_j$.

The left-hand side of Eq. (20.9) can be written as

$$\begin{aligned}\frac{\partial^2 E_k^{\omega_3}}{\partial z^2}(z,t) &= \frac{1}{2}\frac{\partial^2}{\partial z^2}\left[E_{3k}(z)e^{j(\omega_3 t - k_3 z)} + \text{c.c.}\right] \\ &= -\frac{1}{2}\left[k_3^2 E_{3k}(z)e^{j(\omega_3 t - k_3 z)} + 2jk_3\frac{dE_{3k}(z)}{dz}e^{j(\omega_3 t - k_3 z)}\right. \\ &\quad \left. - \frac{d^2 E_{3k}(z)}{dz^2}e^{j(\omega_3 t - k_3 z)}\right] + \text{c.c.} \end{aligned} \tag{20.12}$$

If $E_{3k}$ is sufficiently slowly varying in the $z$ direction, then we can assume that $2k_3\, dE_{3k}/dz \gg d^2E_{3k}/dz^2$, and neglect the term containing $d^2E_{3k}/dz^2$. With this approximation we have

$$\frac{\partial^2 E_k^{\omega_3}}{\partial z^2}(z,t) = -\frac{1}{2}\left[k_3^2 E_{3k}(z)e^{j(\omega_3 t - k_3 z)} + 2jk_3\frac{dE_{3k}(z)}{dz}e^{j(\omega_3 t - k_3 z)}\right] + \text{c.c.} \tag{20.13}$$

Similar relations exist for $\partial^2 E_i^{\omega_1}/\partial z^2(z,t)$ and $\partial^2 E_j^{\omega_2}/\partial z^2(z,t)$.

If we substitute into Eq. (20.9) and compare the terms that oscillate at frequency $\omega_3 = \omega_1 + \omega_2$, we have

$$\begin{aligned}&\left[-\frac{k_3^2}{2}E_{3k}(z) - jk_3\frac{dE_{3k}(z)}{dz}\right]e^{j(\omega_3 t - k_3 z)} \\ &\quad = \frac{1}{2}(j\omega_3\mu\sigma_3 + \omega_3^2\mu\epsilon_0\epsilon_3)E_{3k}e^{j(\omega_3 t - k_3 z)} \\ &\qquad + \frac{\omega_3^2\mu\epsilon_0\chi_{kij}}{4}E_k E_j e^{j[(\omega_1+\omega_2)t - (k_1+k_2)z]}. \end{aligned} \tag{20.14}$$

Noting that $\omega_3^2\mu\epsilon_0 = k_3^2$ in this case, where $\epsilon_3$ is the dielectric constant at frequency $\omega_3$, Eq. (20.14) can be simplified as

$$jk_3\frac{dE_{3k}}{dz}e^{-jk_3 z} = -\frac{j\omega_3\mu\sigma_3}{2}E_{3k}e^{-jk_3 z} + \frac{\omega_3^2\mu\epsilon_0\chi_{kij}}{4}E_{1i}E_{2j}e^{-j(k_3-k_2)z}, \tag{20.15}$$

giving

$$\frac{dE_{3k}}{dz} = -\frac{\sigma_3}{2}\sqrt{\frac{\mu}{\epsilon_3\epsilon_0}}E_{3k} - \frac{j\omega_3}{4}\sqrt{\frac{\mu}{\epsilon_3\epsilon_0}}\epsilon_0\chi_{kij}E_{1i}E_{2j}e^{-j(k_1+k_2-k_3)z}. \tag{20.16}$$

Bearing in mind that NLS polarization like Eq. (20.11) can also be written for nonlinear generation of frequencies $\omega_1$ and $\omega_2$, for example,

$$\left[P_{\text{NLS}}^{\omega_2}(z,t)\right]_j = \frac{\epsilon_0\chi_{jki}}{4}E_{3k}(z)E_{1i}^*(z)e^{j[(\omega_3-\omega_1)t-(k_3-k_1)z]} + \text{c.c.}, \tag{20.17}$$

it is straightforward to show that

$$\frac{dE_{1i}}{dz} = -\frac{\sigma_1}{2}\sqrt{\frac{\mu}{\epsilon_1\epsilon_0}}E_{1i} - \frac{j\omega_1}{4}\sqrt{\frac{\mu}{\epsilon_1\epsilon_0}}\epsilon_0\chi_{ikj}E_{3k}E_{2j}^*e^{-j(k_3-k_2-k_1)z} \tag{20.18}$$

and

$$\frac{dE_{2j}}{dz} = -\frac{\sigma_2}{2}\sqrt{\frac{\mu}{\epsilon_2\epsilon_0}}E_{2j} - \frac{j\omega_2}{4}\sqrt{\frac{\mu}{\epsilon_2\epsilon_0}}\epsilon_0\chi_{jki}E_{3k}E_{1i}^*e^{-j(k_3-k_1-k_2)z}. \tag{20.19}$$

We have omitted the precise frequency dependence of, for example, $\chi_{kij} = \chi_{kij}(-\omega_3, \omega_1, \omega_2)$, but this is important and must not be completely ignored – as we shall see later. If all three frequencies $\omega_1, \omega_2$, and $\omega_3$ are different, and Kleinman's conjecture is valid, we can ignore index permutations on $\chi_{ijk}$ and write them all as $\chi$.

Equations (20.16), (20.18), and (20.19) are the fundamental equations that couple three waves through a second-order nonlinearity. We can use them to describe various important nonlinear processes.

Equations (20.16), (20.18), and (20.19) can be written in a simple, compact form as

$$\frac{dE_{1i}}{dz} = -\alpha_1 E_{1i} - j\kappa_1 E_{3k} E_{2j}^* e^{-j\Delta kz}, \tag{20.20}$$

$$\frac{dE_{2j}}{dz} = -\alpha_2 E_{2j} - j\kappa_2 E_{3k} E_{1i}^* e^{-j\Delta kz}, \tag{20.21}$$

$$\frac{dE_{3k}}{dz} = -\alpha_3 E_{3k} - j\kappa_3 E_{1i} E_{2j} e^{j\Delta kz}, \tag{20.22}$$

where

$$\Delta k = k_3 - k_2 - k_1, \tag{20.23}$$

$$\alpha_\ell = \frac{\sigma_\ell}{2}\sqrt{\frac{\mu}{\epsilon_\ell \epsilon_0}}, \quad \ell = 1, 2, 3, \tag{20.24}$$

and

$$\kappa_\ell = \frac{\omega_\ell}{4}\sqrt{\frac{\mu}{\epsilon_\ell \epsilon_0}}\epsilon_0 \chi, \quad \ell = 1, 2, 3, \tag{20.25}$$

which can be written as

$$\kappa_\ell = \frac{\omega_\ell \chi}{4 n_\ell c_0}, \quad \ell = 1, 2, 3. \tag{20.26}$$

Note that $\alpha_\ell$ is the electric field absorption coefficient for each wave. In the absence of any nonlinearity $\kappa_i = 0$ and each of the Eqs. (20.20)–(20.22) has the general solution

$$E_i = E_0 e^{-\alpha_\ell z}. \tag{20.27}$$

If the nonlinear crystal is transparent at all three frequencies, $\omega_1, \omega_2$, and $\omega_3$, then the loss terms in Eqs. (20.20)–(20.22) can be neglected. Furthermore, if $k_3 = k_1 + k_2$, the so-called *phase-matched* condition, then we have three simple coupled equations:

$$\frac{dE_{1i}}{dz} = -j\kappa_1 E_{2j}^* E_{3k}, \tag{20.28}$$

$$\frac{dE_{2j}}{dz} = -j\kappa_2 E_{1i}^* E_{3k}, \tag{20.29}$$

and

$$\frac{dE_{3k}}{dz} = -j\kappa_3 E_{1i} E_{2j}. \tag{20.30}$$

Because the crystal is transparent at all three frequencies, $\kappa_\ell = \kappa_\ell^*$.

The intensity of each wave is

$$I_\ell = \frac{E_\ell E_\ell^*}{2Z_\ell}, \ \ell = 1, 2, 3, \tag{20.31}$$

where $Z_\ell$ is the characteristic impedance for the frequency and polarization direction involved. The rate of intensity change of each of the waves is

$$\frac{dI_\ell}{dz} = \frac{1}{2Z_\ell}\left(E_\ell \frac{dE_\ell^*}{dz} + E_\ell^* \frac{dE_\ell}{dz}\right). \tag{20.32}$$

So, for example, for the wave at $\omega_1$, from Eq. (20.28),

$$\frac{dI_1}{dz} = \frac{j\kappa_1}{2Z_1}\left(E_{1i}E_{2j}E_{3k}^* - E_{1i}^*E_{2j}^*E_{3k}\right). \tag{20.33}$$

Similarly,

$$\frac{dI_2}{dz} = \frac{j\kappa_2}{2Z_2}\left(E_{1i}E_{2j}E_{3k}^* - E_{1i}^*E_{2j}^*E_{3k}\right) \tag{20.34}$$

and

$$\frac{dI_3}{dz} = -\frac{j\kappa_3}{2Z_3}\left(E_{1i}E_{2j}E_{3k}^* - E_{1i}^*E_{2j}^*E_{3k}\right). \tag{20.35}$$

Each impedance factor can be written as

$$Z_\ell = \frac{1}{n_\ell}\sqrt{\frac{\mu}{\epsilon_0}} = \frac{\mu}{n_\ell c_0}, \tag{20.36}$$

so

$$\frac{\kappa_\ell}{Z_\ell} = \frac{\omega_\ell \chi}{4\mu}. \tag{20.37}$$

Therefore, Eqs. (20.33)–(20.35) can be combined to read

$$\frac{1}{\omega_1}\frac{dI_1}{dz} = \frac{1}{\omega_2}\frac{dI_2}{dz} = -\frac{1}{\omega_3}\frac{dI_3}{dz}, \tag{20.38}$$

or, in terms of the photon fluxes $N_\ell$ at each of the frequencies,

$$\frac{dN_1}{dz} = \frac{dN_2}{dz} = -\frac{dN_3}{dz}. \tag{20.39}$$

Equations (20.38) and (20.39) are alternative statements of the *Manley–Rowe* relations [1]. They are essentially statements of the principle of energy conservation. They hold for any nonlinear interaction involving three frequencies. For a sum-frequency-generation experiment giving $\omega_1 + \omega_2 = \omega_3$ each photon generated at $\omega_3$ is accompanied by the disappearance of one photon at $\omega_1$ and one at $\omega_2$. In a difference-frequency-generation experiment giving $\omega_3 - \omega_1 = \omega_2$ each photon generated at $\omega_2$ is accompanied by the disappearance of one photon at $\omega_3$ and the *appearance* of one photon at $\omega_1$. For a parametric process of the form $\omega_3 = \omega_1 + \omega_2$, each photon at $\omega_3$ that disappears leads to the simultaneous production of two photons – one at $\omega_1$ and one at $\omega_2$.

## 20.3 Second-harmonic generation

The field of modern nonlinear optics was heralded by the observation of radiation at the second-harmonic wavelength of 347 nm when a ruby laser, with a fundamental wavelength of 694 nm, was focused into a quartz crystal [2]. This process of SHG is described by setting $\omega_1 = \omega_2 = \omega$ and $\omega_3 = 2\omega$, in Eqs. (20.17)–(20.19).

If we assume that very little fundamental radiation at frequency $\omega$ is converted into the second harmonic then we can set $dE_{1i}/dz = 0$ and $E_{1i} = E_{1i}(0)$. In other words we ignore *depletion*. If, furthermore, we neglect losses by setting $\alpha_3 = 0$, then Eq. (20.16) becomes

$$\frac{dE_k(2\omega)}{dz} = -\frac{j\omega}{2}\sqrt{\frac{\mu}{\epsilon_3\epsilon_0}}\epsilon_0\chi_{kij}E_{1i}(0)E_{1j}(0)e^{j\Delta kz}, \tag{20.40}$$

where $\Delta k = k_3 - 2k_1$ is the wave-vector mismatch.

In an anisotropic material the magnitude of $\Delta k$ will depend on the direction of propagation in relation to the principal axes of the material (see Chapter 17), as will the value of $\epsilon_3$. Introduction of the experimental SHG coefficient $d_{kij}^{2\omega}$ allows Eq. (20.40) to be written as

$$\frac{dE_k(2\omega)}{dz} = -j\omega\sqrt{\frac{\mu}{\epsilon_3\epsilon_0}}\epsilon_0 d_{kij}^{2\omega}E_{1i}(0)E_{1j}(0)e^{j\Delta kz}. \tag{20.41}$$

Strictly speaking, Eq. (20.41) involves a summation over the components $E_{1i}$ and $E_{1j}$ of the input beam. However, in any given situation of propagation direction and input beam polarization we can use the *effective* SHG coefficient $d_{\text{eff}}$ that describes the situation. The value of $d_{\text{eff}}$ for some specific crystals and geometries is discussed in the next section.

If SHG takes place in a crystal of length $L$ and there is no input at $2\omega$, then the solution of Eq. (20.41) with the boundary condition $E_k(2\omega)(z = 0) = 0$ is

$$E_k(2\omega)(z = L) = -\omega\sqrt{\frac{\mu}{\epsilon_3\epsilon_0}}\epsilon_0 d_{\text{eff}}E_{1i}(0)E_{1j}(0)\frac{e^{j\Delta kL} - 1}{\Delta k}. \tag{20.42}$$

In an anisotropic medium with a propagation direction $\mathbf{k}$, chosen as the $z$ direction in this case, $E_{1i}$ and $E_{1j}$ will correspond to the permitted polarization directions $\mathbf{D}_1$ and $\mathbf{D}_2$ in the crystal (see Section 17.7). Therefore, it is logical to choose these two directions as our $x$ and $y$ axes. The fundamental will have an ordinary and/or extraordinary component, and, depending on the direction of $E_k(2\omega)$ (the $x$ or $y$ direction), the second harmonic will also have ordinary and extraordinary components. The simplest case arises when the fundamental is *either* an ordinary or an extraordinary wave, in which case the second-harmonic intensity is

$$I(2\omega) = \frac{E_k^*(2\omega)E_k(2\omega)}{2Z(2\omega)}, \tag{20.43}$$

where $Z(2\omega)$ is the impedance of the medium at the second-harmonic frequency

$$Z(2\omega) = \sqrt{\frac{\mu}{\epsilon_3\epsilon_0}}, \tag{20.44}$$

and the value of $\epsilon_3$ will depend on the polarization direction of the second harmonic.

From Eqs. (20.42), (20.43), and (20.44)

$$I(2\omega) = \frac{1}{2}\sqrt{\frac{\mu}{\epsilon_3\epsilon_0}}\omega^2\epsilon_0^2 d_{\text{eff}}^2 |E_{1i}|^4 L^2 \frac{\sin^2(\Delta k L/2)}{(\Delta k L/2)^2}. \tag{20.45}$$

Introducing the intensity of the fundamental beam

$$I(\omega) = \frac{|E_{1i}|^2}{2Z(\omega)} \tag{20.46}$$

gives

$$I(2\omega) = \frac{2\omega^2 d_{\text{eff}}^2}{c_0^3 n^{2\omega}(n^{\omega})^2} I^2(\omega) L^2 \frac{\sin^2(\Delta k L/2)}{(\Delta k L/2)^2}, \tag{20.47}$$

where $c_0 = 1/\sqrt{\mu_0\epsilon_0}$ is the velocity of light *in vacuo*, $n^{2\omega} = \sqrt{\epsilon_3}$ is the refractive index at the second-harmonic frequency, and $n^{\omega} = \sqrt{\epsilon_1}$ is the refractive index at the fundamental frequency. Equation (20.47) can be written as[2]

$$I(2\omega) = A I^2(\omega) L^2 \operatorname{sinc}^2(\Delta k L/2), \tag{20.48}$$

where

$$A = \frac{2\omega^2 d_{\text{eff}}^2}{c_0^3 n^{2\omega}(n^{\omega})^2}. \tag{20.49}$$

Equation (20.48) shows clearly that the second-harmonic intensity is proportional to the square of the input intensity and is maximized if $\Delta k = 0$. If $\Delta k = 0$, the process is phase-matched. The relative variation of the second-harmonic conversion efficiency described by the function $\sin^2(\Delta k L/2)/(\Delta k L/2)^2$ is shown in Fig. 20.1. There is a rapid reduction in SHG with detuning from $\Delta k = 0$. Because of dispersion, the variation of refractive index with wavelength, the refractive indices $n^{2\omega}$ and $n^{\omega}$ will be different if both correspond to waves polarized in the same direction. In this case the second-harmonic radiation generated at each point within the medium will get out of step with second-harmonic radiation generated at later points, leading to alternating constructive and destructive interference of the second-harmonic waves. In the phase-matched condition this does not happen. To achieve the condition $\Delta k = 0$ requires $n^{2\omega} = n^{\omega}$, which can be accomplished by using the birefringence of an anisotropic material and obtaining SHG that is polarized orthogonally to the fundamental.

## 20.4 The effective nonlinear coefficient $d_{\text{eff}}$

For a given crystal there will be specific propagation directions and polarizations for optimizing a particular nonlinear process. The generated polarization that results in a particular geometry involves summations over input field components and elements of the

[2] $\operatorname{sinc}(x) = \sin(x)/x$.

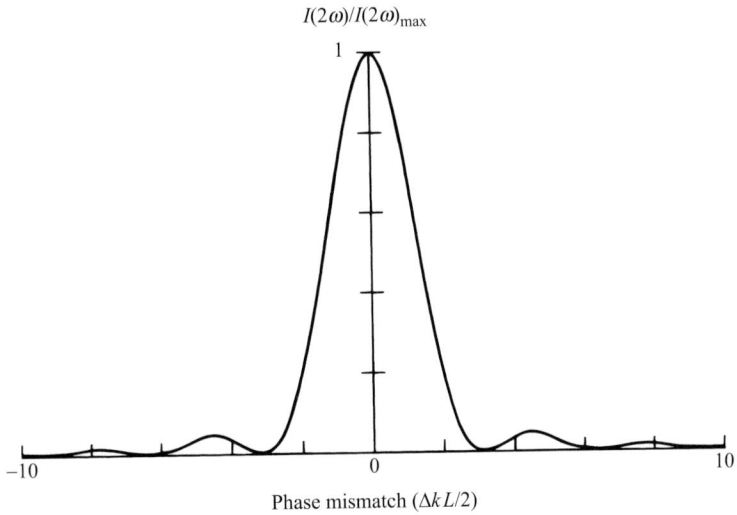


**Fig. 20.1** The variation of the second-harmonic conversion efficiency as a function of wave-vector mismatch $\Delta k$ and crystal length $L$ given by the relation $\text{sinc}^2(\Delta k L/2)$.

appropriate nonlinear tensor $\bar{\bar{d}}$, for example, as given by Eq. (19.27). However, it is possible to write a simple equation for the various amplitudes involved. For example, for a second-order process

$$P(\omega_1 + \omega_2) = d_{\text{eff}} E(\omega_1) E(\omega_2), \tag{20.50}$$

where $d_{\text{eff}}$ is an effective nonlinear coefficient that includes all the summations that apply in the particular geometry in question. This concept is best illustrated with an example. Consider the case of SHG in tellurium, which belongs to point group 32 and is positive uniaxial ($n_{\text{e}} > n_{\text{o}}$). For this crystal, if we assume Kleinman's conjecture,

$$\begin{aligned} P_x(2\omega) &= d_{11} E_x^2(\omega) - d_{11} E_y^2(\omega), \\ P_y(2\omega) &= -2 d_{11} E_x(\omega) E_y(\omega), \\ P_z(2\omega) &= 0. \end{aligned} \tag{20.51}$$

For type-I phase matching in this crystal we have a fundamental E wave and a second-harmonic O wave. The geometry for phase matching in this case is shown in Fig. 20.2. The fundamental ray direction **S** is shown traveling at an angle $\psi$ to the optic axis: this is not exactly equal to the phase-matching angle because of the angle between **S** and **k**. For a ray at the azimuthal angle shown the input electric field components are

$$\begin{aligned} E_x &= \cos\psi \cos\phi |\mathbf{E}(\omega)|, \\ E_y &= \cos\psi \sin\phi |\mathbf{E}(\omega)|, \\ E_z &= -\sin\psi |\mathbf{E}(\omega)|. \end{aligned} \tag{20.52}$$

From Eq. (20.51) the components of the second-harmonic polarization are

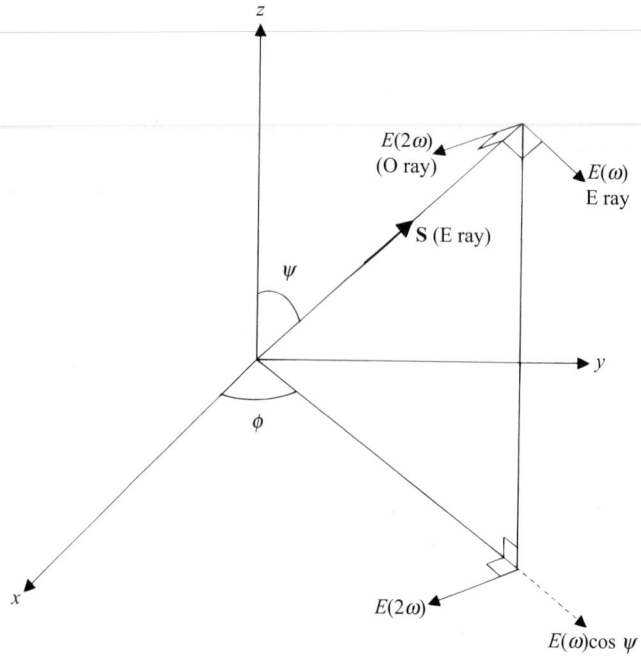


**Fig. 20.2** The geometry for type-I phase matching in a positive uniaxial crystal showing fundamental field and second-harmonic polarization directions.

$$P_x(2\omega) = d_{11}\cos^2\psi\cos(2\phi)|\mathbf{E}(\omega)|^2,$$
$$P_y(2\omega) = -d_{11}\cos^2\psi\sin(2\phi)|\mathbf{E}(\omega)|^2. \tag{20.53}$$

The *effective* polarization that points in the correct direction to generate a second-harmonic O wave must point in the direction shown in Fig. 20.2. This effective polarization is

$$\begin{aligned} P_{\text{eff}}(2\omega) &= d_{11}\cos^2\psi\cos(2\phi)\sin\phi|\mathbf{E}(\omega)|^2 + d_{11}\cos^2\psi\sin(2\phi)\cos\phi|\mathbf{E}(\omega)|^2 \\ &= d_{11}\cos^2\psi\sin(3\phi)|\mathbf{E}(\omega)|^2. \end{aligned} \tag{20.54}$$

Comparison with Eq. (20.50) shows that

$$d_{\text{eff}} = d_{11}\cos^2\psi\sin(3\phi). \tag{20.55}$$

Therefore, for maximum second-harmonic conversion efficiency in this case the fundamental should propagate at the phase-matching angle, and at an azimuthal angle of either 30° or 90°. Table 20.1 lists the $d_{\text{eff}}$ coefficients for the 13 uniaxial crystal classes that lack inversion symmetry, assuming the validity of Kleinman's conjecture. These coefficients can be used to describe both SHG and frequency mixing.

**Table 20.1** The $d_{\text{eff}}$ coefficients when Kleinman's conjecture is valid

| Crystal class | Two O rays and one E ray | Two E rays and one O ray |
|---|---|---|
| 622 | 0 | 0 |
| 6mm | $-d_{15}\sin\psi$ | 0 |
| $\bar{6}$m2 | $-d_{22}\cos\psi\sin(3\phi)$ | $d_{22}\cos^2\psi\cos(3\phi)$ |
| $\bar{6}$ | $\cos\psi[d_{11}\cos(3\phi)$ $-d_{22}\sin(3\phi)]$ | $\cos^2\psi[d_{11}\sin(3\phi)$ $+d_{22}\cos(3\phi)]$ |
| 6 | $d_{15}\sin\psi$ | 0 |
| $\bar{4}$2m | $-d_{14}\sin\psi\sin(2\phi)$ | $d_{14}\sin(2\psi)\cos(2\phi)$ |
| $\bar{4}$ | $-\sin\psi[d_{14}\sin(2\phi)$ $+d_{15}\cos(2\phi)]$ | $\sin(2\psi)[d_{14}\cos(2\phi)$ $-d_{15}\sin(2\phi)]$ |
| 422 | 0 | 0 |
| 4mm | $d_{15}\sin\psi$ | 0 |
| 4 | $d_{15}\sin\psi$ | 0 |
| 32 | $d_{11}\cos\psi\cos(3\phi)$ | $d_{11}\cos^2\psi\sin(3\phi)$ |
| 3m | $d_{15}\sin\psi - d_{22}\cos\psi\sin(3\phi)$ | $d_{22}\cos^2\psi\cos(3\phi)$ |
| 3 | $d_{15}\sin\psi + \cos\psi[d_{11}\cos(3\phi)$ $-d_{22}\sin(3\phi)]$ | $\cos^2\psi[d_{11}\sin(3\phi)$ $+d_{22}\cos(3\phi)]$ |

## 20.5 Phase matching

Collinear phase matching occurs when the waves involved have a common wave-vector direction. In this case, the vector condition

$$\Delta\mathbf{k} = \mathbf{k}_3 - \mathbf{k}_1 - \mathbf{k}_2 = 0 \tag{20.56}$$

reduces, in the case of SHG, to

$$\Delta k = 2\pi\left[\frac{n(2\omega)}{\lambda(2\omega)} - \frac{2n(\omega)}{\lambda(\omega)}\right] = 0, \tag{20.57}$$

so $n(2\omega) = n(\omega)$.

In the most general case involving non-collinear waves and many frequencies through a susceptibility tensor $\chi(-\omega, \omega_1, \omega_2, \ldots, \omega_n)$ the phase-matching condition is

$$\mathbf{k}(\omega) - \mathbf{k}(\omega_1) - \mathbf{k}(\omega_2) - \cdots - \mathbf{k}(\omega_n) = 0. \tag{20.58}$$

### 20.5.1 Second-harmonic generation

For SHG we can achieve phase matching in a sufficiently birefringent crystal by having a fundamental *ordinary* wave and a second-harmonic *extraordinary wave*, or vice versa. So, for example in a positive uniaxial crystal we seek the condition $n_{\text{o}}^{2\omega}(\theta) = n_{\text{e}}^{\omega}$. How this is done is illustrated by the phase-matching diagram shown in Fig. 20.3(a). This diagram

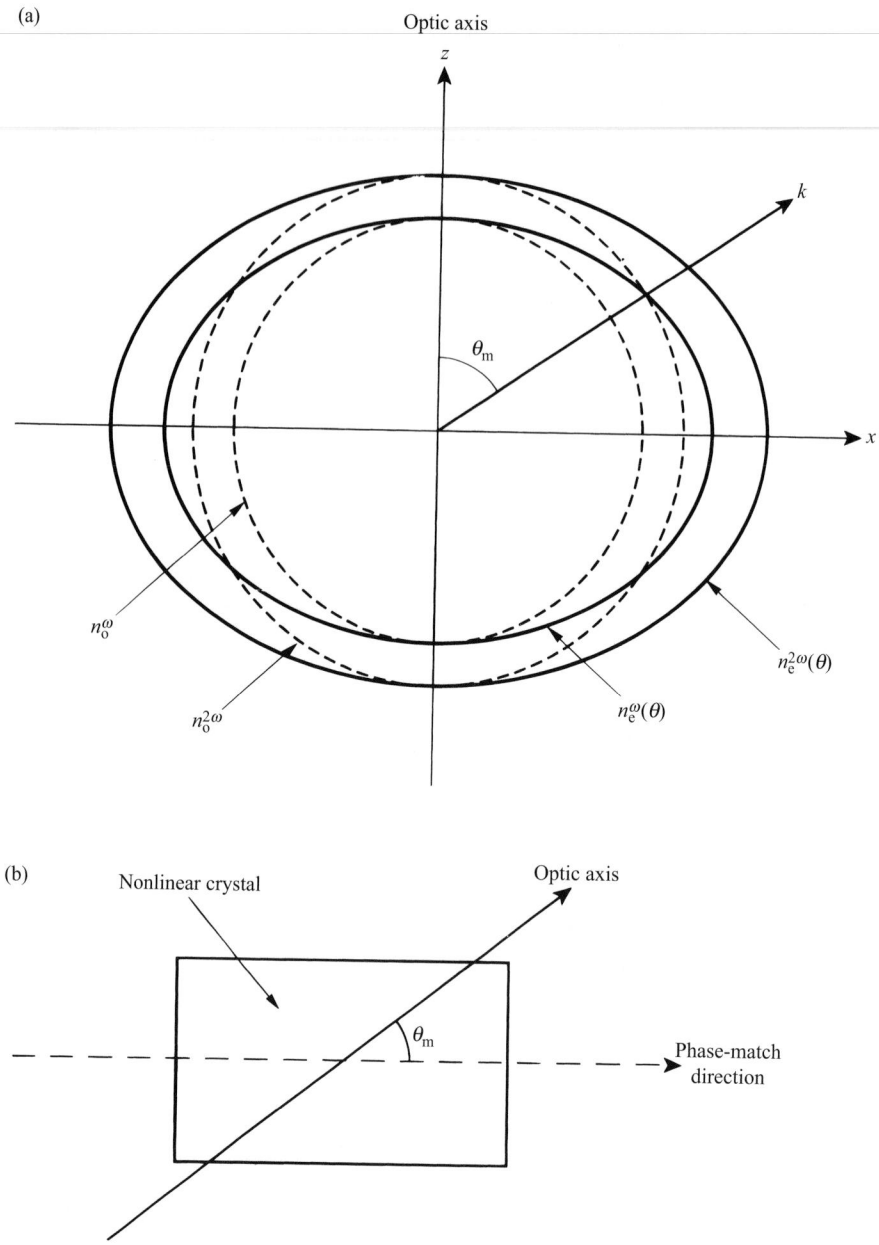


**Fig. 20.3** The geometry for type-I phase matching in a positive uniaxial crystal: (a) indicatrices and (b) the crystal cut with faces perpendicular to the phase-matching direction.

shows principal sections of the ordinary and extraordinary index surfaces of a positive uniaxial crystal for both the fundamental and second-harmonic frequencies. Normal dispersion in such a crystal leads to $n_o^{2\omega} > n_o^{\omega}$ and $n_e^{2\omega}(\theta) > n_e^{\omega}(\theta)$, as shown.

If the birefringence in a negative crystal is large enough, then the ordinary index surface at the fundamental frequency will intersect the extraordinary index surface at the second harmonic at the *phase-matching angle*, $\theta_m$. From Eq. (18.58) the condition for phase matching for a negative uniaxial crystal becomes

$$n_e^{2\omega}(\theta_m) = \frac{n_o^{2\omega} n_e^{2\omega}}{\sqrt{(n_e^{2\omega})^2 \cos^2\theta_m + (n_o^{2\omega})^2 \sin^2\theta_m}} = n_o^{\omega}, \tag{20.59}$$

which can be rearranged to give

$$\sin^2\theta_m = \frac{(n_o^{2\omega})^{-2} - (n_o^{\omega})^{-2}}{(n_o^{2\omega})^{-2} - (n_e^{2\omega})^{-2}}. \tag{20.60}$$

In a positive crystal phase matching is achieved with an extraordinary wave at the fundamental and an ordinary wave at the second harmonic. It is easy to show in this case that

$$\sin^2\theta_m = \frac{(n_o^{2\omega})^{-2} - (n_o^{\omega})^{-2}}{(n_e^{\omega})^{-2} - (n_o^{\omega})^{-2}}. \tag{20.61}$$

In practical applications of phase matching a crystal would be cut with end faces perpendicular to the phase-matching direction as shown in Fig. 20.3(b).

### 20.5.2 Example

Tellurium, a positive uniaxial crystal belonging to class 32, has been used for SHG with a $CO_2$ laser. Interpolation from the refractive-index data of Caldwell and Fan [3] gives, at 10.6 μm and 5.3 μm,

$$n_o^{\omega} = 4.794, \qquad n_o^{2\omega} = 4.856, \qquad n_e^{\omega} = 6.243.$$

Substitution into Eq. (20.61) gives $\theta_m = 14.4°$.

Phase-matching angles are frequently calculated through the use of *Sellmaier* equations, which are semi-empirical formulas that give the variation of the ordinary and extraordinary refractive indices of a material. A typical pair of Sellmaier equations would be of the form [4]

$$n_o^2(\lambda) = a_o + \sum_k \frac{b_{ok}}{\lambda^2 - \lambda_k^2}, \tag{20.62}$$

$$n_e^2(\lambda) = a_e + \sum_k \frac{b_{ek}}{\lambda^2 - \lambda_k^2}, \tag{20.63}$$

where the summation covers the various absorption resonances, $\lambda_k$, that contribute to dispersion. Usually only one or two such resonances are important, so Eqs. (20.62) and (20.63) often have only one or two terms in the summation. For example, for the nonlinear material proustite $Ag_3AsS_3$ [5],

$$n_o^2(\lambda) = 9.220 + \frac{0.4454}{\lambda^2 - 0.1264} - \frac{1{,}733}{1{,}000 - \lambda^2}, \tag{20.64}$$

$$n_e^2(\lambda) = 7.007 + \frac{0.3230}{\lambda^2 - 0.1192} - \frac{660}{100 - \lambda^2}, \tag{20.65}$$

where $\lambda$ is measured in micrometers.

### 20.5.3 Phase matching in sum-frequency generation

In the examples of phase-matched SHG we have considered so far the fundamental waves are polarized in a unique direction, while the second-harmonic waves are polarized in an orthogonal direction. This is a special case of *type-I* phase matching. Generalized to the mixing of two waves at frequencies $\omega_1$ and $\omega_2$ to generate a sum frequency $\omega_3 = \omega_1 + \omega_2$, type-I phase matching has the waves at $\omega_1$ and $\omega_2$ polarized in the same direction.

*Type-II* phase matching occurs in this case when the waves at $\omega_1$ and $\omega_2$ are orthogonally polarized. For collinear phase matching we want

$$\mathbf{k}_3 = \mathbf{k}_1 + \mathbf{k}_2; \tag{20.66}$$

that is,

$$\frac{n_3}{\lambda_3} = \frac{n_1}{\lambda_1} + \frac{n_2}{\lambda_2}. \tag{20.67}$$

Therefore, we can summarize the phase-matching conditions for collinear sum-frequency generation as follows.

For a *positive uniaxial crystal*,

type I: $\omega_1 \equiv$ E wave, $\omega_2 \equiv$ E wave, and $\omega_3 \equiv$ O wave,

$$n_o^{\omega_3} = \frac{\lambda_3}{\lambda_1} n_e^{\omega_1} + \frac{\lambda_3}{\lambda_2} n_e^{\omega_2}; \tag{20.68}$$

type II: $\omega_1 \equiv$ O wave, $\omega_2 \equiv$ E wave, and $\omega_3 \equiv$ O wave,

$$n_o^{\omega_3} = \frac{\lambda_3}{\lambda_1} n_o^{\omega_1} + \frac{\lambda_3}{\lambda_2} n_e^{\omega_2}. \tag{20.69}$$

For a *negative uniaxial crystal*,

type I: $\omega_1 \equiv$ O wave, $\omega_2 \equiv$ O wave, and $\omega_3 \equiv$ E wave,

$$n_e^{\omega_3} = \frac{\lambda_3}{\lambda_1} n_o^{\omega_1} + \frac{\lambda_3}{\lambda_2} n_o^{\omega_2}; \tag{20.70}$$

type II: $\omega_1 \equiv$ E wave, $\omega_2 \equiv$ O wave, and $\omega_3 \equiv$ E wave,

$$n_e^{\omega_3} = \frac{\lambda_3}{\lambda_1} n_e^{\omega_1} + \frac{\lambda_3}{\lambda_2} n_o^{\omega_2}. \tag{20.71}$$

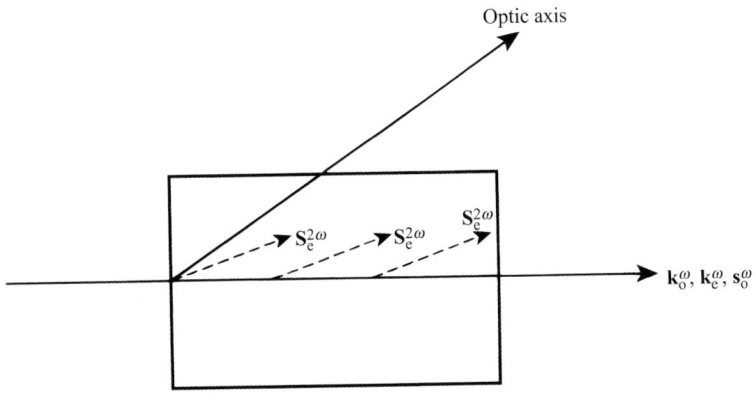


**Fig. 20.4** Beam walk-off in second-harmonic generation in a negative uniaxial crystal that results when the phase-matching direction is not a principal-axis direction. The wave vectors both of the fundamental and of the second harmonic and of the ray direction of the fundamental all propagate collinearly through the crystal.

## 20.6 Beam walk-off and 90° phase matching

The refractive indices of crystals depend on temperature so the phase-matching angle can be temperature-tuned. A highly desirable geometry results if the phase-matching angle can be tuned to 90° in a uniaxial crystal. If this can be accomplished, and it is not feasible in every case, then beam walk-off caused by double refraction can be eliminated. This beam walk-off results because in either type-I or type-II phase matching both ordinary and extraordinary waves are participating in the nonlinear interactions. For collinear phase matching all the wave-normals involved point in the same direction, but the ray directions (Poynting vectors) do not. Nonlinear generation usually involves narrowly focused laser beams. Such focused beams generate large electric fields and lead to efficient nonlinear conversion. However, along the path of such a beam in the crystal, which will follow the ray direction, a nonlinearly generated wave of orthogonal polarization will travel at a different angle. The input and nonlinearly generated rays do not remain collinear along the path. For example, for SHG in a negative uniaxial crystal, the second-harmonic ray continually "walks-off" from the fundamental as shown in Fig. 20.4. The problem of beam walk-off is more serious for mixing processes using type-II phase matching. In this case the input waves are orthogonally polarized and walk-off from each other until no nonlinear mixing occurs. To first order, if phase matching occurs at 90° to the optic axis then no beam walk-off occurs. Table 20.2 lists some of the crystals, wavelengths, and temperatures for which this can be accomplished.

When a focused Gaussian beam is used to maximize the second-harmonic or mixing efficiency in a nonlinear crystal there is a second-order contribution to beam walk-off because the focused beam contains a range of effective wave-vector directions [6].

**Table 20.2** Conditions for 90° phase matching (PM) in selected crystals

| Crystal | Wavelength (μm) | 90° PM temperature (°C) |
|---|---|---|
| KDP | 0.5145 | −11.0 |
| ADP | 0.5145 | ∼ −9.2 |
| $LiNbO_3$ | 1.1523 | 210 |
| $Ag_3AsS_3$ | 1.152 | 12 |
| CD*A ($CsH_2AsO_4$) | 1.06 | 103 |

## 20.7 Second-harmonic generation with Gaussian beams

To maximize the conversion efficiency in SHG it is usual to focus a fundamental Gaussian-mode laser beam into the nonlinear crystal. For high-energy pulsed lasers this is generally done by placing the nonlinear crystal outside the fundamental laser cavity and focusing the fundamental beam with a lens. The second-harmonic pulse will be of shorter duration than the fundamental because its time profile will be

$$I(2\omega, t) \propto I^2(\omega, t). \tag{20.72}$$

For CW SHG an efficient way to achieve maximum fundamental intensity in the crystal is to place the nonlinear crystal inside the laser resonator. If the nonlinear crystal is not too long, we can treat the phase fronts of the input Gaussian beam within the crystal as plane. The SHG results therefore from a plane wave with a Gaussian transverse intensity distribution

$$I(\omega, r) = I_0 e^{-2r^2/\omega^2}, \tag{20.73}$$

with total power

$$P(\omega) = \frac{1}{2}\pi\omega_0^2 I_0. \tag{20.74}$$

If we assume phase matching then the radial variation of the second-harmonic intensity is, from Eq. (20.48),

$$I(2\omega, r) = AL^2 I_0^2 e^{-4r^2/\omega^2}, \tag{20.75}$$

so the spot size of the second-harmonic beam in the crystal is $\omega/\sqrt{2}$.

The total second-harmonic power is

$$P(2\omega) = 2\pi AL^2 I_0^2 \int_0^\infty r e^{-4r^2/\omega^2}\, dr, \tag{20.76}$$

which gives

$$P(2\omega) = \frac{AL^2}{\pi\omega^2} P^2(\omega). \tag{20.77}$$

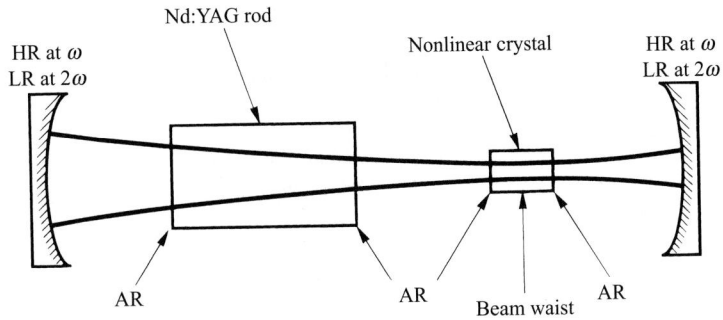


Fig. 20.5 A schematic diagram of the arrangement for intracavity SHG of the Nd:YAG laser. HR, high reflectance; LR, low reflectance.

## 20.7.1 Intracavity SHG

The crystal is placed at the beam waist of the intracavity mode as shown in Fig. 20.5. To minimize undesirable losses, which would prevent the threshold for oscillation at the fundamental being reached, the nonlinear crystal should have negligible loss at the fundamental wavelength. In addition, to prevent intracavity reflection losses or alignment problems, all the intracavity faces should be antireflection-coated at the fundamental wavelength. Intracavity SHG provides a high fundamental intensity, but there may be instability of the generated second-harmonic light if the laser operates on more than one axial mode.

The use of second-harmonic generation to produce new wavelengths has even spread into the low-cost consumer marketplace. Green laser pointers use a diode-pumped solid-state laser with an intracavity crystal to generate green light from the output of a neodymium laser. For example, Fig. 20.6 shows the internal construction of a green laser pointer where a crystal of neodymium-doped yttrium vanadate ($Nd:YVO_4$) pumped by a laser diode is frequency-doubled with a KTP crystal. These powerful small lasers are not generally eye-safe [7].

## 20.7.2 External SHG

The efficiency of SHG outside the cavity can be enhanced by placing the nonlinear crystal in an external resonator, which has mirrors that are highly reflective at the fundamental wavelength. This can be done in two ways, as shown in Fig. 20.7.

The reflectors can be separate from the nonlinear crystal, but it is better to incorporate them directly on the polished crystal faces. In the geometry of Fig. 20.7(a) the second harmonic is phase-matched in both directions, so light emerges through both mirrors. Unidirectional SHG can be obtained through the use of a ring resonator, as shown in Fig. 20.7(b). The path of the second harmonic is only phase-matched in the direction shown, and it emerges predominantly through the right-hand mirror. Although it is possible to make the left-hand mirror in Fig. 20.7(a) highly reflective at the second harmonic, there is then the

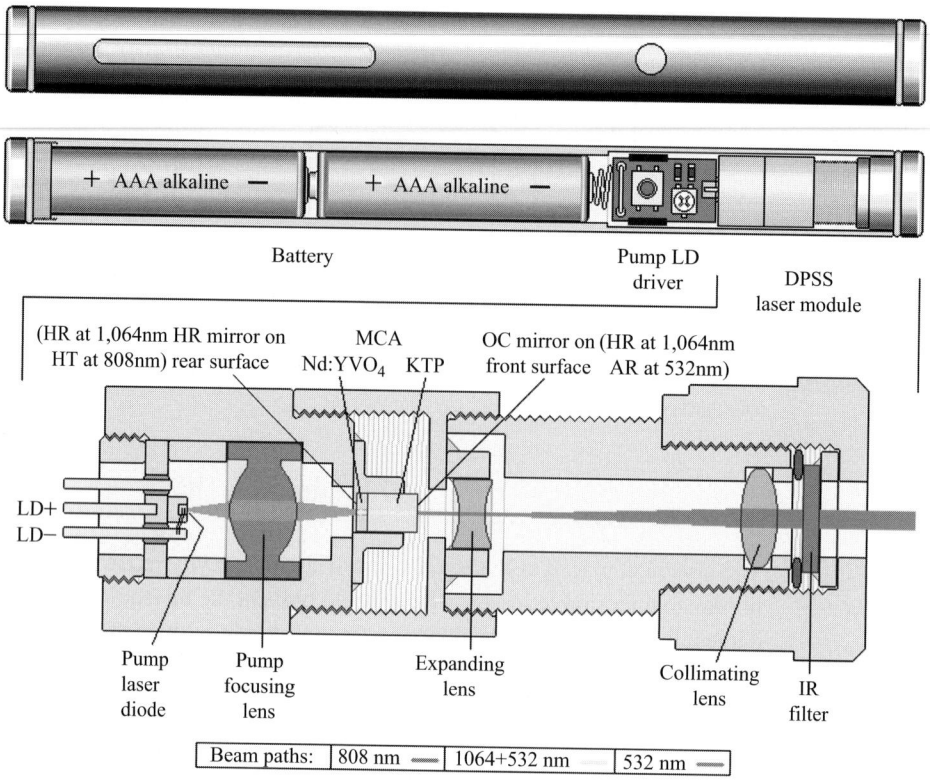


**Fig. 20.6** Construction of a green laser pointer using a multiple crystal assembly (MCA) that uses frequency doubling of a diode-pumped Nd:$YVO_4$ laser with a KTP crystal.

possibility of destructive interference between the counterpropagating second-harmonic beams in the crystal.

### 20.7.3 Periodic poling for phase matching

If the crystal orientation in a birefringent material is periodically reversed, as shown schematically in Fig. 20.8, then in each section of the composite material the phase mismatch that develops in one section can be reversed in alternate sections. The orientation of alternating sections can be related to the orientation of the principal crystal symmetry $c$ axis in the sections (see Chapter 17). This technique was first proposed independently by Armstrong *et al.* [8] and by Franken and Ward [9]. If the sign of the nonlinear susceptibility is flipped in alternating sections, then the input beam continues to be converted efficiently to the second harmonic or new frequency being generated. Otherwise the mismatch $\Delta k\, L/2$ shown in Eq. (20.48) causes the conversion efficiency to oscillate rather than grow as the length of the crystal is increased. The periodicity of the alternating section is generally a multiple of the wavelength that is driving the nonlinear process. The overall process can be described as *quasi-phase-matched*. These periodically poled (PP) materials are much

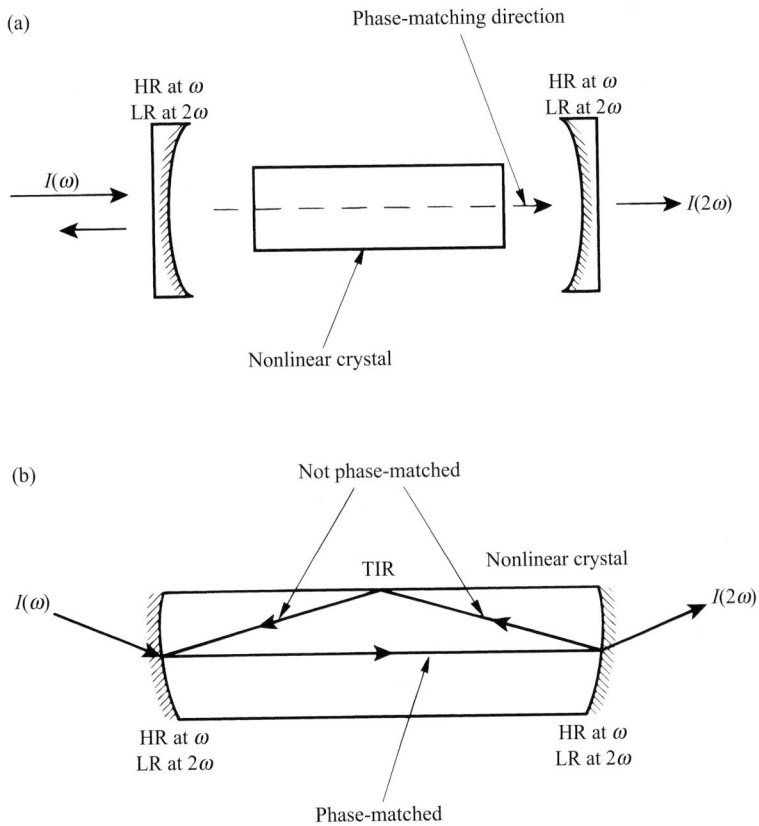


**Fig. 20.7** Arrangements for second-harmonic generation outside the laser cavity: (a) the conventional standing-wave arrangement and (b) a monolithic ring resonator. TIR, total internal reflection.

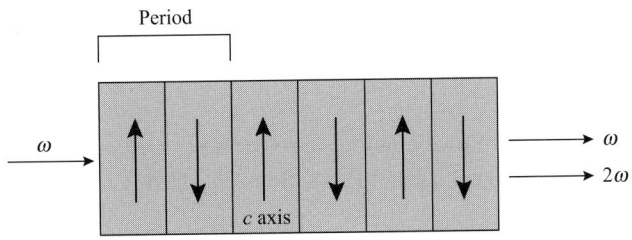


**Fig. 20.8** Periodic poling of a crystal for quasi-phase matching.

more efficent in generating new wavelengths than unpoled non-phase-matched materials. Although phase matching can often be achieved by choosing different directions in a crystal, this may require non-collinear phase matching and can lead to walk-off effects. In periodically poled lithium niobate (PPLN) the orientation of ferroelectric domains in the crystal can be reversed by the application of a high electric field. By applying short high-voltage pulses to an electrode structure on the crystal the built-in electric dipoles in domains can be flipped, leading to the periodically poled material. PPLN is more than 20 times more efficient at SHG than unpoled non-phase-matched material. The details of periodic poling and its optimization have been discussed in detail by Fejer *et al.* [10].

### 20.7.4 The effects of depletion on SHG

If a high conversion efficiency to the second harmonic is obtained, then it is no longer reasonable to ignore depletion. If type-I phase matching is being used, we can rewrite Eqs. (20.28) and (20.30) as

$$\frac{dE(\omega)}{dz} = -j\kappa_1 E^*(\omega)E(2\omega), \tag{20.78}$$

$$\frac{dE(2\omega)}{dz} = -j\kappa_3 E(\omega)E(\omega). \tag{20.79}$$

However, some care must be taken at this point in determining the values of the $\kappa$s. Written out in full,

$$\kappa_1 = \frac{\omega}{4n_1c_0}\chi(-\omega, -\omega, 2\omega), \tag{20.80}$$

$$\kappa_3 = \frac{2\omega}{4n_3c_0}\chi(-2\omega, \omega, \omega), \tag{20.81}$$

where $\chi(-\omega, -\omega, 2\omega)$ is a coefficient that describes difference-frequency generation – it enters because when depletion occurs second-harmonic radiation can be reconverted back to the fundamental by the process, whereas $\chi(-2\omega, \omega, \omega)$ describes second-harmonic generation. Therefore, as discussed in Chapter 19,

$$\chi(-2\omega, \omega, \omega) = \frac{1}{2}\chi(-\omega, -\omega, 2\omega). \tag{20.82}$$

Therefore, in terms of the $d_{\text{eff}}$ coefficient for SHG and with $n_1 = n_3 = n$,

$$\kappa_1 = \kappa = \frac{\omega d_{\text{eff}}}{n_2c_0}, \tag{20.83}$$

$$\kappa_3 = \kappa = \frac{\omega d_{\text{eff}}}{nc_0}. \tag{20.84}$$

The Manley–Rowe relations require that energy be conserved in the conversion of fundamental to second harmonic, so, from Eq. (20.38),

$$\frac{1}{\omega}\frac{d(I(\omega)/2)}{dz} = -\frac{1}{2\omega}\frac{dI(2\omega)}{dz}, \tag{20.85}$$

where we are imagining the SHG process to be equivalent to the mixing together of two beams each of intensity $I(\omega)/2$.

Therefore, following the treatment given by Byer [11],

$$\frac{d}{dz}[I(\omega) + I(2\omega)] = 0 \tag{20.86}$$

and

$$I(\omega) + I(2\omega) = I_0, \tag{20.87}$$

where $I_0$ is a constant, which can be written in terms of the electric fields as

$$|E(\omega)^2| + |E(2\omega)|^2 = 2ZI_0, \tag{20.88}$$

where $Z_0$ is the impedance of the medium, which is identical for the two frequencies $\omega$ and $2\omega$ under phase-matched conditions. If we write the complex amplitudes $E(\omega)$ and $E(2\omega)$ in terms of amplitude and phase factors we have

$$\begin{aligned} E(\omega) &= \mathcal{E}(\omega)e^{j\phi_\omega}, \\ E(2\omega) &= \mathcal{E}(2\omega)e^{j\phi_{2\omega}}, \end{aligned} \tag{20.89}$$

and Eqs. (20.78) and (20.79) become

$$\frac{d\mathcal{E}(\omega)}{dz} = -j\kappa\mathcal{E}(\omega)\mathcal{E}(2\omega)e^{j(2\phi_\omega - \phi_{2\omega})}, \tag{20.90}$$

$$\frac{d\mathcal{E}(2\omega)}{dz} = -j\kappa\mathcal{E}(\omega)\mathcal{E}(\omega)e^{j(2\phi_\omega - \phi_{2\omega})}. \tag{20.91}$$

For the amplitudes $\mathcal{E}(\omega)$ and $\mathcal{E}(2\omega)$ to be real it is clearly necessary for the factor $je^{-j(2\phi_\omega - \phi_{2\omega})}$ to be real. We choose the negative sign for Eq. (20.90) because we expect the fundamental to decrease in amplitude as it is converted to the second harmonic. Therefore, we have

$$\frac{d\mathcal{E}(\omega)}{dz} = -\kappa\mathcal{E}(\omega)\mathcal{E}(2\omega), \tag{20.92}$$

$$\frac{d\mathcal{E}(2\omega)}{dz} = \kappa\mathcal{E}(\omega)\mathcal{E}(\omega). \tag{20.93}$$

Substituting from Eq. (20.88) into Eq. (20.93) gives

$$\frac{d\mathcal{E}(2\omega)}{dz} = \kappa[2ZI_0 - \mathcal{E}^2(2\omega)], \tag{20.94}$$

and, on integrating,

$$\int_0^{\mathcal{E}(2\omega,L)} \frac{d\mathcal{E}(2\omega)}{2ZI_0 - \mathcal{E}^2(2\omega)} = \kappa \int_0^L dz, \tag{20.95}$$

which gives

$$\frac{1}{\sqrt{2ZI_0}} \tanh^{-1}\left[\frac{\mathcal{E}(2\omega,L)}{\sqrt{2ZI_0}}\right] = \kappa L_0, \tag{20.96}$$

so

$$\mathcal{E}(2\omega,L) = \sqrt{2ZI_0}\tanh(\kappa L\sqrt{2ZI_0}), \tag{20.97}$$

and finally,

$$I(2\omega,L) = I_0 \tanh^2(\Gamma L), \tag{20.98}$$

where

$$\Gamma = \kappa\sqrt{2ZI_0}. \tag{20.99}$$

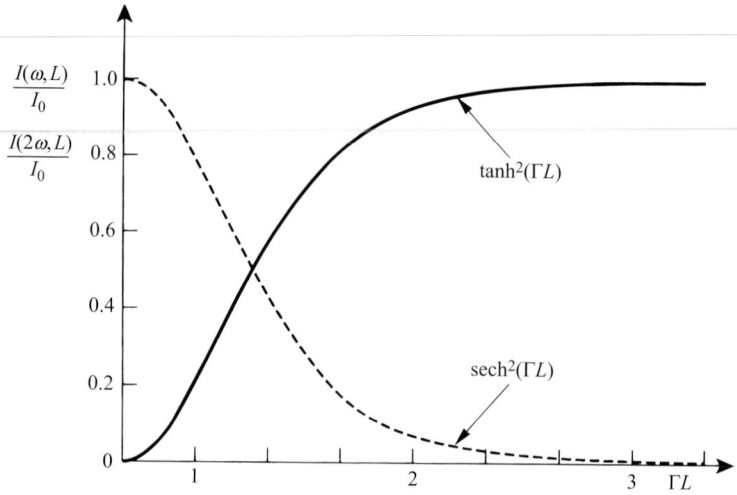


**Fig. 20.9** Growth of the second harmonic and decay of the fundamental as a function of the crystal length during efficient second-harmonic generation.

From Eq. (20.88)

$$\begin{aligned} I(\omega, L) &= I_0\left[1 - \tanh^2(\Gamma L)\right] \\ &= I_0 \operatorname{sech}^2(\Gamma L). \end{aligned} \tag{20.100}$$

Figure 20.9 shows the growth of the second harmonic towards 100% conversion efficiency, and the decay of the fundamental to zero predicted by Eqs. (20.98) and (20.100) for plane waves. For Gaussian beams the conversion efficiency is actually limited to about 60%.

## 20.8 Up-conversion and difference-frequency generation

These are two processes that can be analyzed in a similar way to SHG. In up-conversion, or sum-frequency generation, two lower-energy photons, $\omega_1$ and $\omega_2$, combine in a nonlinear medium to generate a higher-energy photon, $\omega_3$. One application of the process $\omega_1 + \omega_2 = \omega_3$ is for a strong laser beam, at frequency $\omega_1$, in the visible to be mixed with a weak infrared signal, $\omega_2$. The up-converted photons, $\omega_3$, will then be capable of detection with very sensitive photodetectors. This provides a means for the detection of very weak infrared light signals [12]. The up-conversion process can also be used to provide visual imaging of an infrared-illuminated scene [13]. The efficiency and practical utility of these approaches depend critically on phase matching. Incoming radiation will be up-converted most efficiently at angles that vary with wavelength [6].

Difference-frequency generation can be used to obtain coherent infrared radiation at wavelengths that are difficult to generate by other means [14, 15]. The process $\omega_3 = \omega_1 - \omega_2$ will work efficiently only if the nonlinear medium is transparent at all three

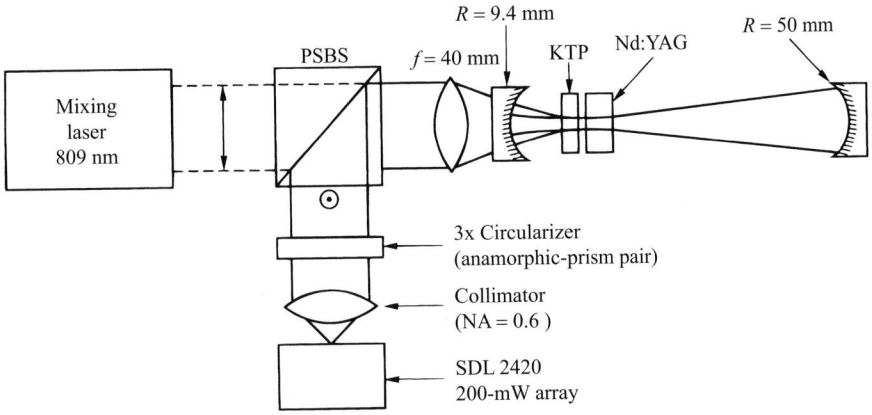


**Fig. 20.10** The arrangement for generating blue–green light by mixing the output of a semiconductor laser with that from a diode-pumped Nd:YAG laser operating at 1.06 μm [22].

wavelengths involved. GaAs has been used successfully for the generation of tunable far-infrared radiation by the mixing together of different lines from two $CO_2$ lasers [16, 17]. $LiNbO_3$ has been used to generate far-infrared radiation with a wavelength as long as 1.23 mm by the mixing of two ruby-laser beams. However, discretely or continuously tunable infrared radiation can also be generated by other means, such as stimulated Raman scattering, use of line-tunable $CO_2$ lasers [18–20], or with a spin-flip Raman laser [21], a discussion of which is outside the scope of this book.

An important application of sum-frequency generation is in the generation of blue–green light by mixing the output of a semiconductor diode laser with the output of a diode-pumped Nd:YAG laser. An 810-nm diode laser mixed with a 1.06-μm Nd:YAG laser provides radiation at 459 nm, which is transmitted efficiently through water. Figure 20.10 shows a configuration that has been used experimentally to do this.

## 20.9 Optical parametric amplification

This is a process in which a weak laser beam at frequency $\omega_1$, the *signal*, is amplified by a nonlinear interaction involving a strong laser beam at frequency $\omega_3$, the *pump*. To conserve energy, additional photons at a frequency $\omega_2$, the *idler*, must be simultaneously generated. The process is represented as

$$\omega_3 = \omega_1 + \omega_2.$$

In a sense this process is the reverse of sum-frequency generation. Its attractiveness comes from its tunability, since in theory the pump photons at $\omega_3$ can be divided in infinitely many ways into pairs of photons at frequencies $\omega_1$ and $\omega_2$. To describe this process mathematically we again use Eqs. (20.20)–(20.22).

If we neglect depletion of the pump we can assign the pump field a constant value $E_3 = E_P$. Further, by assuming a particular set of polarizations for the fields involved we can drop the subscripts on the electric field components and write these equations as

$$\frac{dE_1}{dz} + \alpha_1 E_1 = -j\kappa_1 E_P E_2^* e^{-j\Delta kz},$$
$$\frac{dE_2^*}{dz} + \alpha_2 E_2^* = j\kappa_2 E_P^* E_1 e^{j\Delta kz}. \tag{20.101}$$

We look for a solution in which both the signal field, $E_1$, and the idler field, $E_2$, grow exponentially, and try as our solution

$$E_1(z) = E_S e^{(\Gamma' - j\Delta k/2)z},$$
$$E_2(z) = E_I e^{(\Gamma' - j\Delta k/2)z}. \tag{20.102}$$

Substitution into Eq. (20.98) gives

$$E_S(\Gamma' + \alpha_1 - j\,\Delta k/2) + E_I^*(j\kappa_1 E_P) = 0,$$
$$E_S(-j\kappa_2 E_P^*) + E_I^*(\Gamma' + \alpha_2 + j\,\Delta k/2) = 0. \tag{20.103}$$

In order for Eq. (20.103) to represent the solution for arbitrary initial values of the signal and idler fields, the determinant of the coefficients must be zero; therefore

$$(\Gamma' + \alpha_1 - j\,\Delta k/2)(\Gamma' + \alpha_2 + j\,\Delta k/2) - \kappa_1\kappa_2|E_P|^2 = 0. \tag{20.104}$$

Solving for $\Gamma'$ gives

$$\Gamma' = -\left(\frac{\alpha + \alpha_2}{2}\right) \pm \frac{1}{2}\left[(\alpha_1 - \alpha_2)^2 - 2j\,\Delta k(\alpha_1 - \alpha_2) - 4\left(\frac{\Delta k}{2}\right)^2 + 4\kappa_1\kappa_2|E_P|^2\right]^{1/2}. \tag{20.105}$$

Equation (20.105) can be put into a simpler form if we assume that $\alpha_1 = \alpha_2 = \alpha$ and introduce a parametric gain coefficient $\Gamma$ defined by

$$\Gamma^2 = \kappa_1\kappa_2|E_P|^2 = \frac{\omega_1\omega_2 d_{\text{eff}}^2|E_P|^2}{n_1 n_2 c_0^2}, \tag{20.106}$$

so that

$$\Gamma' = -\alpha \pm \left[\Gamma^2 - \left(\frac{\Delta k}{2}\right)^2\right]^{1/2}. \tag{20.107}$$

If we introduce the effective gain coefficient $g$ defined by

$$g = \left[\Gamma^2 - \left(\frac{\Delta k}{2}\right)^2\right]^{1/2}, \tag{20.108}$$

then we can write the general solution to Eq. (20.102) as

$$E_1(z) = \left(E_{1A}e^{gz} + E_{1B}e^{-gz}\right)e^{-\alpha z}e^{-j\Delta kz/2},$$
$$E_2(z) = \left(E_{2A}e^{gz} + E_{2B}e^{-gz}\right)e^{-\alpha z}e^{-j\Delta kz/2}. \tag{20.109}$$

If the parametric-amplification process occurs in a nonlinear medium that stretches from $z = 0$ to $z = \ell$ and the input signal and idler amplitudes satisfy $E_1(z = 0) = E_S(0)$ and $E_2(z = 0) = E_I(0)$, then Eqs. (20.109) and (20.101) can be solved to determine the constants $E_{1A}, E_{1B}, E_{2A}$, and $E_{2B}$. The solution is

$$E_S(\ell)e^{\alpha\ell} = E_S(0)e^{-j\frac{\Delta k\ell}{2}}\left[\cosh(g\ell) + \frac{j\,\Delta k}{2g}\sinh(g\ell)\right] - \frac{j\kappa_1}{g}E_P E_I^*(0)e^{-j\frac{\Delta k\ell}{2}}\sinh(g\ell), \tag{20.110}$$

$$E_I(\ell)e^{\alpha\ell} = E_I(0)e^{-j\frac{\Delta k\ell}{2}}\left[\cosh(g\ell) + \frac{j\,\Delta k}{2g}\sinh(g\ell)\right] - \frac{j\kappa_2}{g}E_P E_S^*(0)e^{-j\frac{\Delta k\ell}{2}}\sinh(g\ell), \tag{20.111}$$

where $\kappa_i = \omega_i d_{\text{eff}}/(n_i c_0)$, as before.

Note that Eqs. (20.110) and (20.111) are symmetric under interchange of the designation S (signal) and I (idler), as expected.

A few interesting features of the above analysis can be pointed out. Unless $\Gamma > \Delta k/2$ no continuous growth of signal or idler occurs. In this case $g$ becomes purely imaginary and the exponential growth predicted by Eqs. (20.110) and (20.111) becomes only oscillatory since $\cosh(ix) = \cos x$ and $\sinh(ix) = j\sin x$.

Optimum signal and idler generation results if $\Delta k = 0$. This is the phase-matching condition that has already been discussed, and the same relationships between the refractive indices and wavelengths involved hold true as were given in Eqs. (20.68)–(20.71).[3] If no idler input is supplied, then the parameter amplifier can still generate an idler wave by amplifying its own noise. In this case with phase matching Eqs. (20.110) and (20.111) would be

$$E_S(\ell)e^{\alpha\ell} = E_S(0)\cosh(g\ell) \tag{20.112}$$

and

$$E_I(\ell)e^{\alpha\ell} = -\frac{j\kappa_2}{g}E_P E_S^*(0)\sinh(g\ell). \tag{20.113}$$

The fractional single-pass increase in intensity corresponding to Eq. (20.112) is

$$\frac{I_S(\ell)}{I_S(0)} = e^{-2\alpha\ell}\cosh^2(g\ell), \tag{20.114}$$

which, for $g\ell \gg 1$, can be written

$$I_S(\ell) = I_S(0)e^{2(g-\alpha)\ell}. \tag{20.115}$$

In a phase-matched situation

$$g^2 = \frac{\omega_1\omega_2 d_{\text{eff}}^2|E_P|^2}{n_1 n_2 c_0^2}. \tag{20.116}$$

[3] The additional complexities introduced by the use of Gaussian beams and the effects of beam walk-off are not dealt with here. For further details the interested reader should consult the specialized literature [23–26].

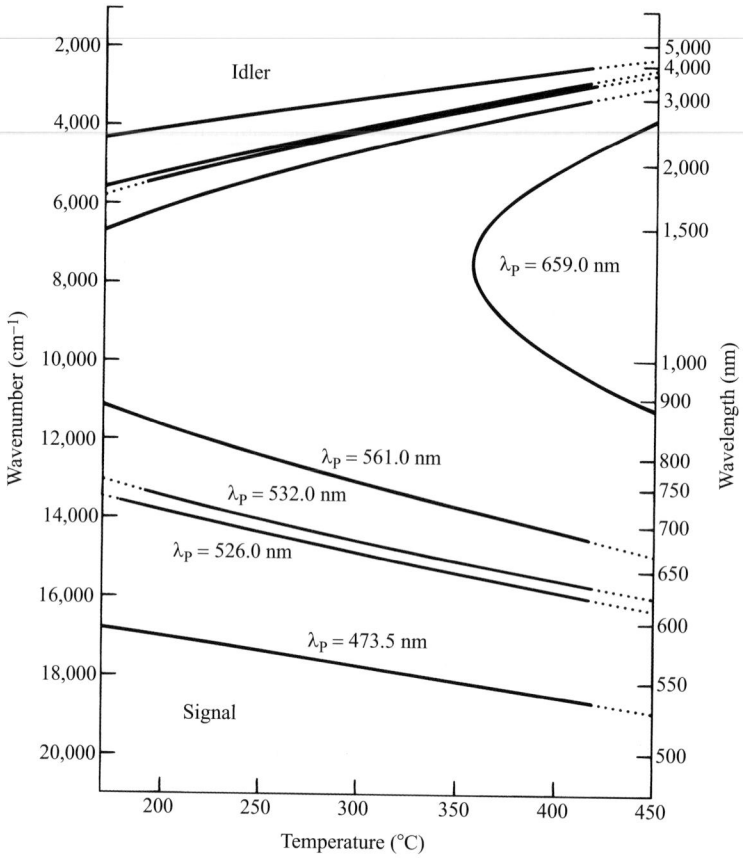


**Fig. 20.11** The temperature tuning curve for $LiNbO_3$.

## 20.9.1 Example

$LiNbO_3$ is a popular nonlinear material for use in parametric amplifiers and oscillators. For a pump wavelength of 532 nm (obtained from a frequency-doubled Nd:YAG laser) and for signal and idler wavelengths of 647 nm and 3 μm, we can use $n_1 = n_2 = n_o = 2.24$. This is a negative uniaxial crystal with $n_e < n_o$, and phase matching can be achieved with signal and idler waves propagating as ordinary waves. The pump at $\omega_3$ propagates as an extraordinary wave. The phase-matching condition is given by Eq. (20.70). This crystal can be temperature-tuned to achieve 90° phase matching, as shown in Fig. 20.11 [13]. We can use $n_3 = n_e = 2.16$ and $d_{\text{eff}} = d_{31} = 6.25 \times 10^{-12}$ m V$^{-1}$.

For a pump intensity of 10 GW m$^{-2}$, which is well below the crystal damage threshold of 500–1,400 GW m$^{-2}$, the pump field is

$$|E_P|^2 = \frac{2Z_0 I_P}{n_e}. \tag{20.117}$$

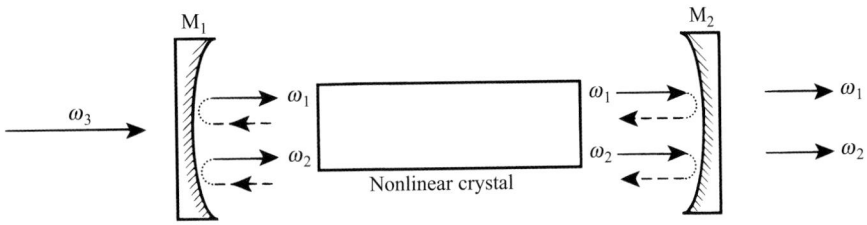


Fig. 20.12 The arrangement of a parametric oscillator. Mirror $M_1$ has high transmittance for the pump wavelength.

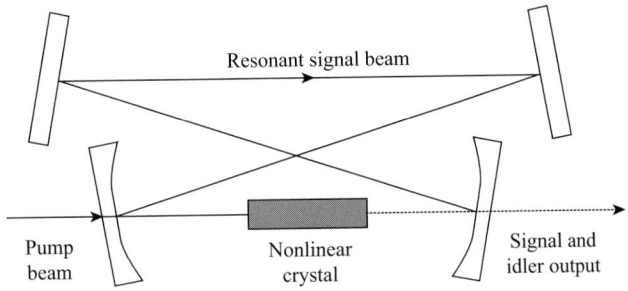


Fig. 20.13 A ring optical parametric oscillator.

Substitution of numerical values into Eq. (20.116) gives $g = 23.5\ \mathrm{m}^{-1}$. In practice the crystal in such a system might be 10 mm long, so $g\ell \simeq 0.235$.

It is clear from this example that substantial gains in such a parametric amplifier are not obtained without large pump intensities. However, by placing the pumped crystal in a resonant cavity to provide feedback at the signal and/or idler wavelengths, the device can be converted into a parametric oscillator.

## 20.10 Parametric oscillators

Although Eq. (20.115) suggests that for the possibility of oscillation $g > \alpha$, namely that gain must exceed loss, we can better examine the conditions necessary for the onset of parametric oscillation by using Eqs. (20.110) and (20.111).

In the schematic parametric-oscillator arrangements shown in Figs. 20.12 and 20.13 the pump wave generally makes only a single pass through the crystal, so the signal and idler waves are amplified only when they travel in the same direction as the pump. A more stable, lower-loss device can be fabricated if the mirrors are deposited directly onto the ends of the crystal.

The resonant cavity formed by the two mirrors $M_1$ and $M_2$ may provide significant feedback at both signal and idler wavelengths, in which case it is referred to as a doubly resonant optical parametric oscillator (DRO). If the cavity provides feedback at only signal or idler wavelengths it is a singly resonant optical parametric oscillator (SRO). We can incorporate both these situations into our analysis by enlarging our definition of the loss coefficients $\alpha_1$ and $\alpha_2$. So far, these coefficients have referred to intracavity losses within the nonlinear medium. However, the leakage of an oscillating field through $M_1$ or $M_2$ is equivalent to additional intracavity loss.

As we have seen in Chapter 4, this loss is related to the passive cavity lifetime $\tau$. If the reflectances of the cavity mirrors at the signal and idler wavelengths are $R_1(\omega_1)$ and $R_2(\omega_2)$ then

$$\tau_1 \simeq \frac{L_1}{c_0(1-R_1)},$$
$$\tau_2 \simeq \frac{L_2}{c_0(1-R_2)}, \tag{20.118}$$

where $L_1$ and $L_2$ are the optical lengths of the cavity at the signal and idler wavelengths, respectively.[4] In a passive cavity a field at frequency $\omega$ would decay according to

$$E_1(t) = E_1(0)e^{-t/2\tau_1}. \tag{20.119}$$

It is straightforward to show that, if the effect of the intracavity loss coefficients $\alpha_i$ is included, then

$$\tau_i \simeq \frac{L_i}{c_0(1+2\alpha_i\ell - R_i)}, \tag{20.120}$$

provided that both $\alpha_i\ell, R_i \ll 1$.

In the phase-matched case Eqs. (20.110) and (20.111) become

$$E_S(\ell) = e^{-\alpha\ell}\left[E_S(0)\cosh(\Gamma\ell) - \frac{j\kappa_1}{\Gamma}E_P E_I^*(0)\sinh(\Gamma\ell)\right], \tag{20.121}$$

$$E_I^*(\ell) = e^{-\alpha\ell}\left[E_I^*(0)\cosh(\Gamma\ell) + \frac{j\kappa_2}{\Gamma}E_P^* E_S(0)\sinh(\Gamma\ell)\right]. \tag{20.122}$$

With an additional pass through the nonlinear medium and reflections at the cavity mirrors, Eqs. (20.121) and (20.122) become, at the threshold of oscillation when changes in field amplitudes occur on a round trip in the cavity,

$$E_S(0) = R_1 e^{-2\alpha\ell}\left[E_S(0)\cosh(\Gamma\ell) - \frac{j\kappa_1}{\Gamma}E_P E_I^*(0)\sinh(\Gamma\ell)\right], \tag{20.123}$$

$$E_I^*(0) = R_2 e^{-2\alpha\ell}\left[E_I^*(0)\cosh(\Gamma\ell) + \frac{j\kappa_2}{\Gamma}E_P^* E_S(0)\sinh(\Gamma\ell)\right]. \tag{20.124}$$

If we incorporate mirror reflectance and intracavity losses into a single loss coefficient we can write

$$R_1 e^{-2\alpha\ell} = e^{-\alpha_S\ell}, \tag{20.125}$$
$$R_2 e^{-2\alpha\ell} = e^{-\alpha_I\ell}. \tag{20.126}$$

Note that, in terms of $\alpha_S$ and $\alpha_I$,

$$\tau_1 \simeq \frac{L_1}{c_0\alpha_S}, \quad \tau_2 \simeq \frac{L_2}{c_0\alpha_I}. \tag{20.127}$$

[4] For example, for a cavity of geometric length $\ell$ and crystal length $L$, $L_1 = (\ell - L) + n_1 L$.

A non-trivial solution of the simultaneous equations requires once again that the determinant of coefficients be zero. Noting that $\Gamma^2 = \kappa_1\kappa_2|E_P|^2$, it is straightforward to show that

$$\cosh(\Gamma\ell) = 1 + \frac{\alpha_S\alpha_I\ell^2}{2 + \alpha_S\ell + \alpha_I\ell}. \tag{20.128}$$

In the DRO case $\alpha_S\ell$ and $\alpha_I\ell$ will both be small, so the threshold condition can be written as

$$\Gamma^2\ell^2 \simeq \alpha_S\alpha_I\ell^2. \tag{20.129}$$

For the SRO threshold with $\alpha_I\ell \gg \alpha_S\ell$

$$\Gamma^2\ell^2 \sim 2\alpha_S\ell. \tag{20.130}$$

Thus, the SRO threshold is $2/\alpha_I$ greater than the DRO. We can express Eq. (20.129) in terms of the threshold pump intensity (in the crystal) by using Eqs. (20.106), (20.117), (20.125), and (20.126). If the mirror reflectance $R_1$ (signal) and $R_2$ (idler) are $\simeq 1$ and intracavity losses are negligible, then we can write Eqs. (20.125) and (20.126) as

$$\alpha_S\ell = 1 - R_1, \tag{20.131}$$

$$\alpha_I\ell = 1 - R_2. \tag{20.132}$$

The threshold pump intensity is

$$(I_P)_{\text{threshold}} = \frac{n_1n_2n_3c_0^2(1-R_1)(1-R_2)}{2\omega_1\omega_2d_{\text{eff}}^2Z_0\ell^2}. \tag{20.133}$$

### 20.10.1 Example

Let us consider an $LiNbO_3$ crystal with $\ell = 50$ mm, $n_1 = n_2 = 2.24$, $n_3 = 2.16$, $\lambda_1 = 647$ nm, $\lambda_2 = 3$ μm, and $\lambda_3 = 532$ nm. We consider a DRO with 99% reflectance mirrors, so $(1-R_1)$ and $(1-R_2)$ are both 0.01. Substitution of these values into Eq. (20.133) gives $(I_P)_{\text{threshold}} \simeq 7 \times 10^5$ W m$^{-2}$ or 0.7 W mm$^{-2}$; this is a modest pump intensity.

This illustrates the relative ease (*in principle*) with which a DRO can be operated. Indeed, many investigators have operated such devices using crystals such as $LiNbO_3$ and $Ba_2NaNbO_{15}$[5] with pump powers of only a few milliwatts focused into the crystal. However, the practical difficulties of actually operating parametric oscillators should not be minimized. In order to achieve stable DRO operation much care must be devoted to the phase-matching condition (90° phase matching prevents beam walk-off) and the actual resonance conditions for signal and idler wavelengths. In practice this requires a stable single-frequency pump laser and careful cavity-length control to achieve resonance for both signal and idler waves. These stringent requirements generally mean that pulsed parametric oscillators are generally operated in an SRO configuration, whereas CW systems may be operated in a DRO configuration. The signal chooses an appropriate cavity resonance, and then the idler adjusts itself to satisfy energy conservation. Byer and coworkers

[5] Colloquially referred to as "banana."

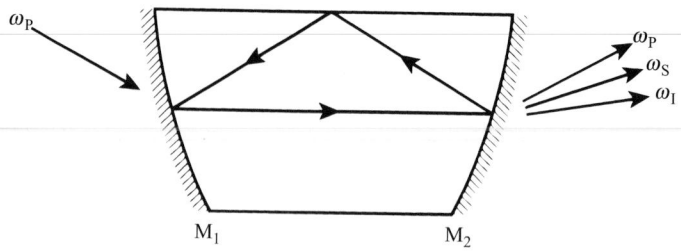


Fig. 20.14 A monolithic ring doubly resonant optical parametric oscillator. Mirror $M_1$ is transparent to the pump wavelength, while both mirrors $M_1$ and mirror $M_2$ are reflective at the signal and idler wavelengths. The angles between the emerging (depleted) pump, signal, and idler beams are exaggerated, but result from the different refractive indices of the nonlinear crystal at the three frequencies $\omega_P$, $\omega_S$, and $\omega_I$.

[24] have achieved considerable success in the operation of CW $LiNbO_3$ DROs that use a monolithic ring configuration, like that shown in Fig. 20.14. In this ring geometry the parametric-amplification process occurs unidirectionally along the single-pass pump direction.

## 20.11 Parametric-oscillator tuning

Although non-collinear phase matching has been used in parametric oscillators, collinear phase matching is generally used. For example, the pump may be an extraordinary wave, traveling at angle $\theta$ to the optic axis, and the signal and idler are ordinary waves. The collinear phase-matching condition can be restated as

$$\omega_3 n_e^{\omega_3}(\theta) = \omega_1 n_o^{\omega_1} + \omega_2 n_o^{\omega_2}, \tag{20.134}$$

and, for energy conservation,

$$\omega_3 = \omega_1 + \omega_2. \tag{20.135}$$

Equations (20.134) and (20.135) can be solved to determine $\omega_1$ and $\omega_2$, for fixed $\omega_3$ (say), provided that the variation of $n_e(\theta)$ with the tuning parameter is known.

The use of angle, temperature, and electro-optic tuning is discussed in much more detail than here by Byer and coworkers [24] and by Shen [18].

In temperature tuning, which allows the use of 90° phase matching, we can write for the type-I phase matching previously mentioned

$$\omega_3 n_e^{\omega_3}(T) = \omega_1 n_o^{\omega_1}(T) + \omega_2 n_o^{\omega_2}(T). \tag{20.136}$$

At temperature $T + \Delta T$, the output frequencies change to $\omega_1 + \Delta\omega$ and $\omega_2 - \Delta\omega$. Provided that we are not too close to the *degenerate* operating point,[6] where $\omega_1 = \omega_2$, we can write

[6] At the degenerate operating points where $\omega_1 = \omega_2 = \omega_3/2$, the properties of the emitted degenerate radiation can become very interesting. The device becomes a subharmonic generator – providing two identical photons for each pump photon. In the ideal case this is called "squeezed light" [27–29].

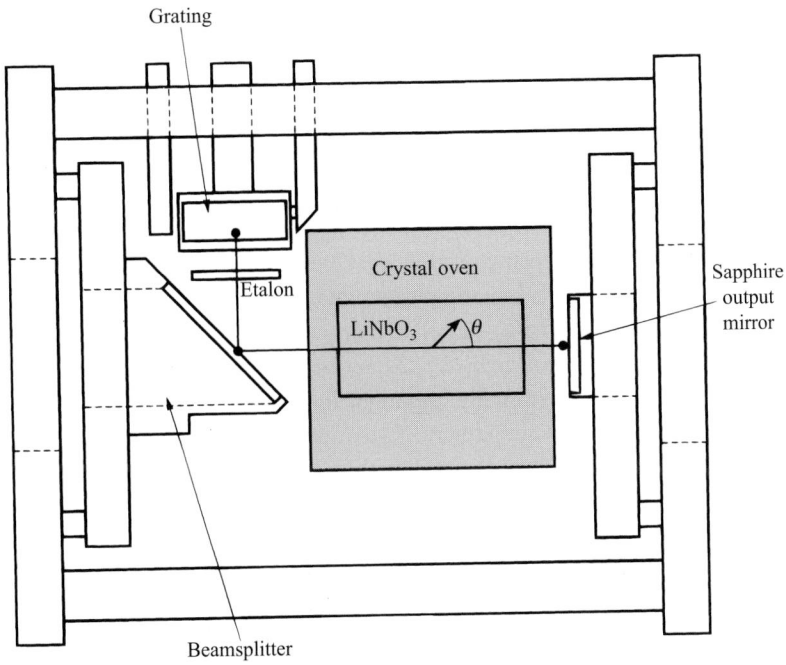


**Fig. 20.15** An angle-tuned $LiNbO_3$ singly resonant [11] parametric oscillator.

$$\omega_3\left[n_e^{\omega_3}(T)\frac{\partial n_e^{\omega_3}}{\partial T}\Delta T\right] = (\omega_1+\Delta\omega)\left[n_o^{\omega_1}(T)+\frac{\partial n_o^{\omega_1}}{\partial T}\Delta T+\frac{\partial n_o^{\omega_1}}{\partial \omega_1}\Delta\omega\right] + (\omega_2-\Delta\omega)\left[n_o^{\omega_2}(T)+\frac{\partial n_o^{\omega_2}}{\partial T}\Delta T+\frac{\partial n_o^{\omega_2}}{\partial \omega_2}\Delta\omega\right], \quad (20.137)$$

which gives

$$\Delta\omega = \frac{\omega_3(\partial n_e^{\omega_3}/\partial T)-\omega_1(\partial n_o^{\omega_1}/\partial T)-\omega_2(\partial n_o^{\omega_2}/\partial \omega_2)}{n_o^{\omega_1}-n_o^{\omega_2}+\omega_1(\partial n_o^{\omega_1}/\partial\omega)-\omega_2(\partial n_o^{\omega_2}/\partial\omega)}. \quad (20.138)$$

Near the degenerate operating condition $T = T_0$ and $\omega_1 = \omega_3/2$. In this case the calculation of the tuning equation, Eq. (20.138), cannot neglect higher-order terms in the Taylor-sense expansions of the variation of $n_o^{\omega_1}$ and $n_o^{\omega_2}$ with frequency. In this case

$$\Delta\omega = \frac{\omega_3\left[\partial n_e^{\omega_3}/\partial T-\partial n_e^{\omega_1}/\partial T\right]_{T=T_0}}{\left[2\,\partial n_o^{\omega_1}/\partial\omega+\omega_3(\partial^2 n_o^{\omega}/\partial\omega^2)/2\right]_{\omega=\omega_3/2}}(\Delta T)^{1/2}. \quad (20.139)$$

Figure 20.11 shows temperature tuning curves calculated in this way for $LiNbO_3$ parametric oscillators.

In angle tuning, which can be accomplished in an arrangement shown in Fig. 20.15, a change in angle from $\theta$ to $\theta + \Delta\theta$ changes the operating frequencies to $\omega_1 + \Delta\omega$ and $\omega_2 - \Delta\omega$. Not too close to the degeneracy point

$$\Delta\omega \simeq \omega_3 \frac{\partial n_e^{\omega_3}}{\partial\theta}\Delta\theta \Bigg/ \left[\left(\omega_1 \frac{\partial n_o^{\omega_1}}{\partial\omega_1} - \omega_2 \frac{\partial n_o^{\omega_2}}{\partial\omega_2}\right) + n_o^{\omega_1} - n_o^{\omega_1}\right]. \tag{20.140}$$

Near the degenerate operating point

$$\Delta\omega = \left[\omega_3 \left(\frac{\partial n_e^{\omega_3}}{\partial\theta}\right)_{\theta_0} \Bigg/ \left(2\frac{\partial n_o}{\partial\omega} + \frac{\omega_3}{2}\frac{\partial^2 n_o}{\partial\omega^2}\right)_{\omega=\omega_3/2}\right]^{1/2} (\Delta\theta)^{1/2}. \tag{20.141}$$

Optical parametric oscillators (OPOs) operated at the degenerate operating point essentially convert each pump photon into two identical photons. These "two-photon" sources have very interesting correlation properties. In a quantum-mechanical description they can generate *squeezed* light [27–29] and *entangled* quantum states [30, 31]. Consequently they have become extremely important for studying fundamental quantum phenomena, which are beyond our scope here. Even operated away from the degeneracy point, OPOs still generate correlated photons at the signal and idler frequencies. The parametric process actually occurs at the single photon level. One pump photon is "split" into correlated photons at different frequencies. A comprehensive discussion has been provided by Mandel and Wolf [28].

## 20.12 Phase conjugation

Before completing this chapter two additional nonlinear optical phenomena are worthy of brief mention. The first of these, optical phase conjugation, uses a nonlinear mixing process to generate the perfect counterpropagating wave, both in direction and in phase, of an incident wave.

For example, an incident wave of the form

$$E_1 = E_0 e^{j(\omega t - kz)} \tag{20.142}$$

will, after reflection at a phase-conjugate mirror, become

$$E_2 = aE_0^* e^{j(\omega t + kz)}, \tag{20.143}$$

where $a$ is a constant that indicates the reflection efficiency of the "mirror." The reflected wave retraces the path of the incident wave, and its phase variation in space will replicate exactly the phase variation of the input wave. If one could observe the variation in time of the moving phase fronts of the reflected wave, it would appear exactly as if an equivalent "movie" of the incident phase fronts were being run backwards. The interesting consequence of this is that, if a wave travels through a medium that distorts its optical phase fronts, such as an aberrated optical system or an optical fiber, or along an atmospheric path, then, after a reflection at a phase-conjugate mirror, the wave will retrace its path and the phase distortions produced on its outward path to the phase-conjugate mirror will be exactly unfolded on the return path, as shown schematically in Fig. 20.16. The wave returns to its starting point undistorted.

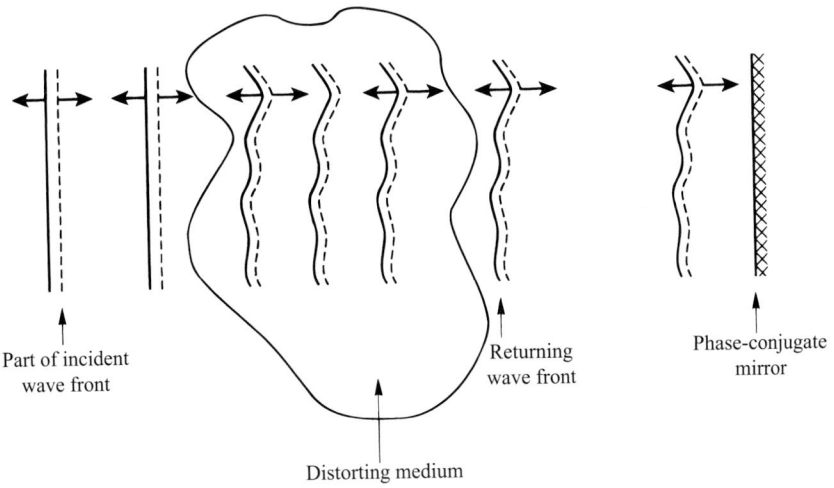


**Fig. 20.16** A schematic illustration of wave-front correction of a distorted wave front by reflection from a conjugate mirror.

Of particular interest is the ability of phase conjugation to compensate for the random fluctuations imposed on the phase front of an optical wave as it traverses a path through the atmosphere. This is of interest in connection with line-of-sight coherent optical communication through the atmosphere, in optical astronomy, and in imaging through the turbulent atmosphere. The atmosphere behaves as a random medium in the sense that its refractive index varies randomly in time according to

$$n(\mathbf{r}, t) = n_0 + n_1(\mathbf{r}, t).$$

The fluctuating contribution to the index, $n_1(\mathbf{r}, t)$, contributes to the haze seen over a hot highway and to the twinkling of distant lights or stars at night. These fluctuations occur on a time scale of up to about 1 kHz.

To illustrate the attractiveness of phase conjugation in correcting for optical properties of the atmosphere, a futuristic example can be given. Suppose that it is desired to focus an intense laser beam onto a target through the turbulent atmosphere, for example, in a "Star Wars" application. It is possible, in principle, to generate the phase conjugate of a weak signal that has traveled back from target to source location and then amplify this weak phase-conjugate signal and project it back to a small focal spot on the target. There continues to be great interest in the generation of "beacons" or "guide stars" in this application. In astronomy, in principle, a distant star acts as a point source, so the blurred image of the star received at the ground can be corrected by the use of deformable or multi-segmented mirrors. This is the field of *adaptive optics* [32], in which a distorted image is corrected by correcting the blurred image of a point source within the field of view by reflecting the distorted image off a deformable mirror to produce a point image. For the focusing of a laser beam on a distant target there generally exists no guide star to provide feedback to an adaptive optical system; however, in principle, phase conjugation can provide a point "beacon" on the target [33].

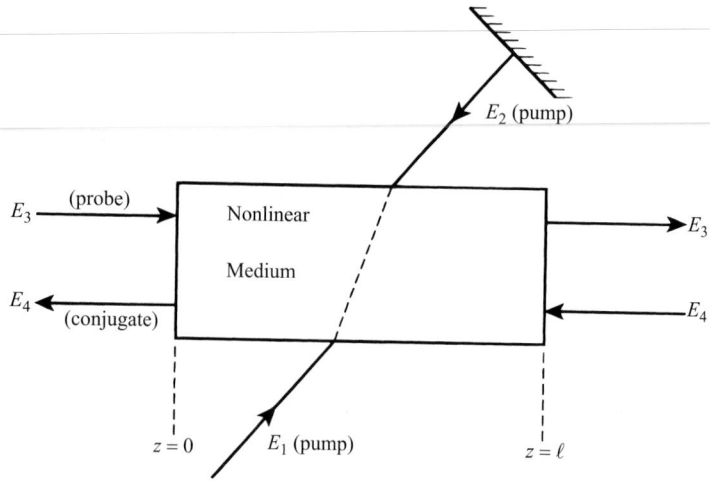


**Fig. 20.17** The arrangement for phase conjugation by four-wave mixing.

How do we make a phase-conjugate mirror? We use a nonlinear mixing process involving a third-order nonlinear-susceptibility tensor $\chi^{(3)}(-\omega_4, \omega_1, \omega_2, -\omega_3)$. Because four frequencies are involved, this process is called four-wave mixing. In practice all of the frequencies $\omega_1, \omega_2, \omega_3$, and $\omega_4$ may be the same. A common geometry for doing this is shown in Fig. 20.17. In this arrangement a pump laser beam passes through the nonlinear medium and is then reflected back on itself. The probe beam is incident on the nonlinear medium at an arbitrary angle. In a phase-matched situation where all waves have the same frequency, if $\mathbf{k}_1 = -\mathbf{k}_2$, then automatically $\mathbf{k}_4 = -\mathbf{k}_3$. In and of itself, this explains how the phase-conjugate wave is generated in the counterpropagating direction to the probe.

If we write the electric fields of the waves involved in the form

$$\mathbf{E}_i(t) = \tfrac{1}{2} E_i e^{j(\omega_i t - \mathbf{k}_i \cdot \mathbf{r})} + \text{c.c.}, \tag{20.144}$$

then the polarization that results at frequency $\omega_4 = \omega_1 + \omega_2 - \omega_3$ can be written (see Chapter 19) as

$$P_i^{\omega_4} = 6\epsilon_0 \chi_{ijk\ell}(-\omega_4, \omega_1, \omega_2, -\omega_3) E_{1j} E_{2k} E_{3\ell}^* e^{-j(\mathbf{k}_1 + \mathbf{k}_2 - \mathbf{k}_3)\cdot\mathbf{r}}. \tag{20.145}$$

If specific polarizations are chosen for all the waves involved, then a simplified version of Eq. (20.145) is

$$P^{\omega_4} = \epsilon_0 \chi^{(3)} E_1 E_2 E_3^* e^{-j(\mathbf{k}_1 + \mathbf{k}_2 - \mathbf{k}_3)\cdot\mathbf{r}}. \tag{20.146}$$

If all four waves have the same frequency $\omega$, then the polarization $P(\omega)$ can arise in five ways – from the combinations $E_1 E_2 E_3^*$, $E_1 E_1^* E_3$, $E_2 E_2^* E_3$, $E_3 E_3^* E_3$, and $E_4 E_4^* E_4$. Other combinations would not generate frequency $\omega$, or be phase-matched. For example, the combination $E_1 E_2 E_3$ would generate a nonlinear polarization at frequency $3\omega$. If we substitute the nonlinear polarization $P^{\omega_4}$ into the wave equation, in an analogous manner to obtaining Eqs. (20.18) and (20.19), we get in a lossless medium

$$\frac{dE_4}{dz} = \frac{j\omega}{2}\sqrt{\frac{\mu}{\epsilon}}\epsilon_0\chi^{(3)}\left(|E_1|^2 + |E_2|^2 + |E_3|^2 + |E_4|^2\right)E_4 + \frac{j\omega}{2}\sqrt{\frac{\mu}{\epsilon}}\epsilon_0\chi^{(3)}E_1E_2E_3^*, \tag{20.147}$$

where we have assumed that the process is phase-matched with $\mathbf{k}_1 = -\mathbf{k}_2$, and $\mathbf{k}_4 = -\mathbf{k}_3$, and the + signs on the right-hand side of Eq. (20.147) result because $E_4$ is propagating in the $-z$ direction. In a similar way

$$\frac{dE_3}{dz} = -\frac{j\omega}{2}\sqrt{\frac{\mu}{\epsilon}}\epsilon_0\chi^{(3)}\left(|E_1|^2 + |E_2|^2 + |E_3|^2 + |E_4|^2\right)E_3 - \frac{j\omega}{2}\sqrt{\frac{\mu}{\epsilon}}\epsilon_0\chi^{(3)}E_1E_2E_4^*. \tag{20.148}$$

If the pump beams are intense relative to the probe and conjugate beams, then we can neglect terms in $|E_3|^2$ and $|E_4|^2$ to get

$$\frac{dE_4}{dz} = \frac{j\omega}{2}\sqrt{\frac{\mu}{\epsilon}}\epsilon_0\chi^{(3)}\left(|E_1|^2 + |E_2|^2\right)E_4 + \frac{j\omega}{2}\sqrt{\frac{\mu}{\epsilon}}\epsilon_0\chi^{(3)}E_1E_2E_3^* \tag{20.149}$$

and

$$\frac{dE_3}{dz} = -\frac{j\omega}{2}\sqrt{\frac{\mu}{\epsilon}}\epsilon_0\chi^{(3)}\left(|E_1|^2 + |E_2|^2\right)E_3 - \frac{j\omega}{2}\sqrt{\frac{\mu}{\epsilon}}\epsilon_0\chi^{(3)}E_1E_2E_4^*. \tag{20.150}$$

If we treat $|E_1|^2$ and $|E_2|^2$ as constants (so we are neglecting any depletion effects) then we can redefine $E_3$ and $E_4$ as

$$E_3' = E_3 \exp\left[-j\omega\sqrt{\frac{\mu}{\epsilon}}\epsilon_0\chi^{(3)}(|E_1|^2 + |E_2|^2)\right] \tag{20.151}$$

and

$$E_4' = E_4 \exp\left[j\omega\sqrt{\frac{\mu}{\epsilon}}\epsilon_0\chi^{(3)}(|E_1|^2 + |E_2|^2)\right]. \tag{20.152}$$

All this really means is that we have multiplied the fields $E_3$ and $E_4$ by a constant phase factor. Equations (20.149) and (20.150) now reduce to the simple forms

$$\frac{dE_4'}{dz} = \frac{j\omega}{2}\sqrt{\frac{\mu}{\epsilon}}\epsilon_0\chi^{(3)}E_1E_2E_3'^* \tag{20.153}$$

and

$$\frac{dE_3'^*}{dz} = -\frac{j\omega}{2}\sqrt{\frac{\mu}{\epsilon}}\epsilon_0\chi^{(3)}E_1^*E_2^*E_4'. \tag{20.154}$$

Writing $\kappa^* = (\omega/2)\sqrt{\mu/\epsilon}\epsilon_0\chi^{(3)}E_1E_2$ gives us the final coupled equations describing phase conjugation:

$$\frac{dE_3}{dz} = -j\kappa^*E_4^* \tag{20.155}$$

and

$$\frac{dE_4}{dz} = j\kappa^* E_3. \tag{20.156}$$

If we take $E_3(z=0) = E_P$ and $E_4(z=\ell) = 0$, which represents the practical situation of a finite pump entering the nonlinear medium and zero initial conjugate wave entering the nonlinear medium at $z = \ell$, then the solution of Eqs. (20.155) and (20.156) is [34, 35]

$$E_4(z=0) = -j\left[\frac{\kappa}{|\kappa|}\tan(|\kappa|\ell)\right]E_P^* \tag{20.157}$$

and

$$E_3(z=\ell) = E_P/\cos(|\kappa|\ell). \tag{20.158}$$

Equation (20.157) shows that the nonlinearly reflected wave at $z = 0$ is the conjugate of the pump multiplied by a "gain" factor $-j(\kappa^*/|\kappa|)\tan(|\kappa|\ell)$. This "gain" factor will not be significant in general without sufficient pump intensity. We can illustrate this by considering the use of carbon disulfide, $CS_2$, a material with a large $\chi^{(3)}$.

### 20.12.1 Phase conjugation in $CS_2$

For this medium $\epsilon_0\chi^{(3)} \equiv 6\epsilon_0\chi_{yxxy}(-\omega,\omega,\omega,-\omega) \simeq 2.025\times10^{31}$ C m V$^{-3}$ and $n = 1.62$. For this element of the nonlinear susceptibility tensor the pump beams $\omega_1$ and $\omega_2$ are polarized orthogonally to the probe and conjugate beams. If $E_1 = E_2 = E$, then $E^2 = 2ZI_P$ and

$$|\kappa| = \left(\frac{\pi c_0}{\lambda n^2}\right)(\epsilon_0\chi^{(3)})(2Z_0^2 I_P). \tag{20.159}$$

At a pump intensity of 1 MW mm$^{-2}$, $|\kappa| = 38.8$ m$^{-1}$, so for a "gain" of 1 we need a length of $\simeq$26 mm. Intensities of this magnitude require the use of pulsed lasers, but media with much larger effective $\epsilon_0\chi^{(3)}$ that allow the observation of phase conjugation with CW lasers are available.

The practical ability of a phase-conjugate mirror to remove distortions from a laser beam is illustrated in Fig. 20.18.

## 20.13 Optical bistability

Optical bistability is a phenomenon in which the output intensity from an optical system can have two potential values for the same value of input intensity for some range of value of these parameters. In other words its $I_{out}$–$I_{in}$ curve appears as shown schematically in Fig. 20.19. This is just like the hysteresis curve familiar in an analysis of the magnetization of ferromagnetic materials [37].

Just as the magnetic recording medium on a disk or tape has the ability to store information because the magnetization exhibits hysteresis, so optical bistable devices can perform

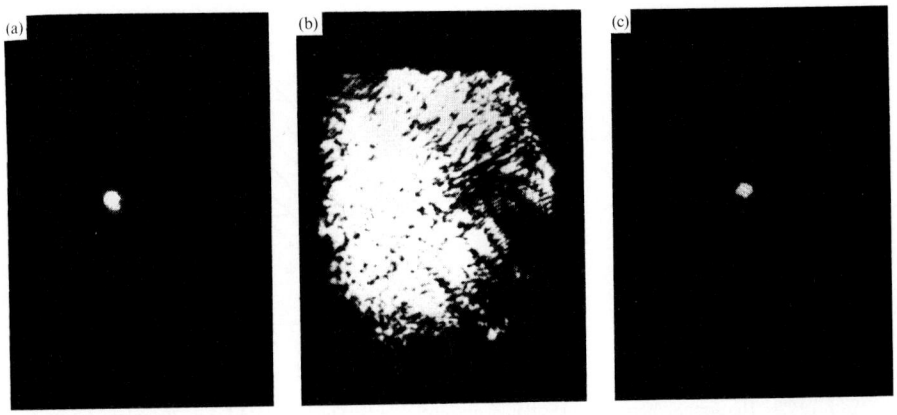


**Fig. 20.18** An illustration of how a phase-conjugate mirror can recreate an undistorted wave front from one that has been distorted in propagating through a medium (or aberrated optical system). Photograph (a) shows an unperturbed focused laser beam; (b) shows this beam after it has been phase aberrated on passage through an etched glass plate; and (c) shows the beam after it has reflected from a plane-conjugate mirror and passed again through the same etched glass plate. From reference [36]. Reproduced with the permission of the publisher.

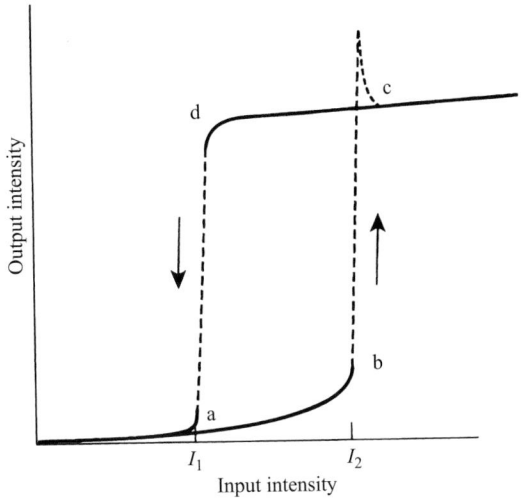


**Fig. 20.19** Input intensity–output intensity characteristic curves for a bistable medium.

a memory function. For example, if the input intensity is increased from zero to $I_2$ out to point b in Fig. 20.19 then the output intensity makes a rapid transition to point c. The output intensity could at this point be regarded as representing a "1," and this value will be sustained as long as the input intensity is maintained above value $I_1$ – point d on the curve. The input intensity must be reduced below $I_1$ (after it has been above $I_2$) in order for the output intensity to return to a low-value point a, which can be regarded as representing a "0."

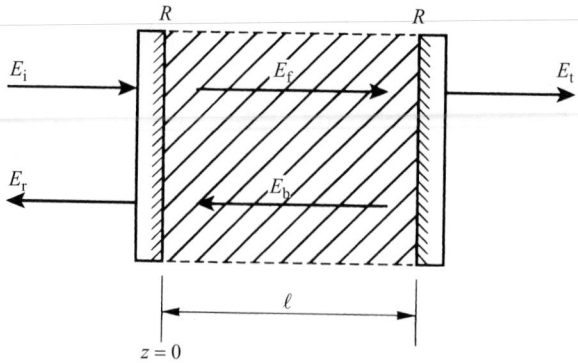


**Fig. 20.20** The arrangement for illustration of absorptive optical bistability. A Fabry–Perot interferometer having mirrors of reflectance $R$ contains an absorbing medium. The various traveling-wave directions involved are shown.

Many bistable systems consist of a nonlinear medium placed between the mirrors of an optical cavity, and we will use such a system to present a brief introduction to the phenomenon. The model we will discuss is the one first described by Szöke, Danen, Goldhar, and Kurnit [38]. A Fabry–Perot interferometer containing an absorbing medium, shown schematically in Fig. 20.20, is illuminated at the peak of one of its transmission resonances. The input wavelength therefore satisfies $m\lambda/2 = \ell$, where $\ell$ is the spacing of the two mirrors, each assumed to have reflectance $R$. The intensity absorption coefficient of the medium is $\alpha$.

We can model the system in a fairly simple way. If the input electric field amplitude is $E_i$, then the transmitted electric field amplitude is (see Section 4.5)

$$E_t = E_i T e^{-\alpha\ell/2}\left(1 + Re^{-\alpha\ell}e^{-j\delta} + R^{2-2\alpha\ell}e^{-2j\delta} + \cdots\right), \tag{20.160}$$

where $T$ and $R$ are the transmittance and reflectance, respectively, of the mirrors, which are assumed to be identical, and the parameter $\delta$ is the phase shift experienced on a single pass from one mirror to the other. Summing the geometric series in Eq. (20.160) gives

$$E_t = \frac{TE_i e^{-\alpha\ell/2}}{1 - Re^{-j\delta}e^{-\alpha\ell}}. \tag{20.161}$$

At resonance $\delta = 0$, and, since intensity $\propto$ |electric field|$^2$, Eq. (20.161) gives

$$\frac{I_t}{I_i} \simeq \frac{e^{-\alpha\ell}}{\left[\left(1 - Re^{-\alpha\ell}\right)/(1 - R)\right]^2}. \tag{20.162}$$

If $\alpha\ell \ll 1$, we can expand the exponential terms in Eq. (20.162) to give

$$\frac{I_t}{I_i} \simeq \frac{1}{(1 + \alpha\ell)\left[1 + R\alpha\ell/(1 - R)\right]^2} \simeq \frac{1}{(1 + k)^2}, \tag{20.163}$$

provided that $1 - R \gg \alpha\ell$, and where

$$k = \frac{R\alpha\ell}{1 - R}. \tag{20.164}$$

Interesting things happen if $\alpha$ is intensity-dependent (as it actually always is). If the absorbing medium is treated as a simple two-level homogeneously broadened system, then its absorption coefficient will be

$$\alpha = \frac{\alpha_0}{1 + I_e/I_s}, \tag{20.165}$$

where $\alpha_0$ is its absorption coefficient at zero intensity, $I_e$ is the effective intracavity intensity, and $I_s$ is the saturation intensity. When this saturation of absorption is included we can write

$$k = \frac{k_0}{1 + I_e/I_s}, \tag{20.166}$$

where $k_0 = R\alpha_0\ell/(1 - R)$. We need to relate the effective intracavity intensity to the incident intensity, bearing in mind that the intracavity field $E_{\text{cavity}}$ is a standing wave, which for $\alpha\ell, T \ll 1$, will be

$$E_{\text{cavity}} = E_f \cos(\omega t - kz) + E_b \cos(\omega t + kz), \tag{20.167}$$

where $E_f$ and $E_b$ are the forward and backward traveling-wave amplitudes, respectively, and in this case $E_f \simeq E_b$. Rearrangement of Eq. (20.167) gives

$$E_{\text{cavity}} = 2E_b \cos(\omega t)\cos(kz). \tag{20.168}$$

Since $E_b = E_t/\sqrt{T}$, Eq. (20.168) can be rewritten as

$$\frac{I_{\text{cavity}}}{I_t} = \frac{2}{T}\cos^2(\omega t)\cos^2(kz). \tag{20.169}$$

The average of $\cos^2(kz)$ over the whole cavity is $\frac{1}{2}$, so

$$\overline{I_{\text{cavity}}} = \frac{I_t}{T}\cos^2(\omega t). \tag{20.170}$$

Therefore, we can replace the effective intracavity intensity by $I_e = I_t/T$, and, from Eq. (20.163),

$$\frac{I_i}{I_t} = \left[1 + \frac{k_0}{1 + I_t/(TI_s)}\right]^2. \tag{20.171}$$

If we write $Y = I_i/(TI_s)$ and $X = I_t/(TI_s)$, where $Y$ and $X$ now represent normalized input and output intensities, respectively, then

$$Y = X\left(1 + \frac{k_0}{1 - X}\right)^2. \tag{20.172}$$

Equation (20.172) is a fundamental equation of bistability. For a given value of $Y$, there are two possible values for $X$. However, for true bistable behavior the curve of $Y$ versus $X$ must pass through a minimum between the two values of $X$. Figure 20.21 shows curves calculated from Eq. (20.172) for various values of $k_0$. Note, for example, that the curve for $k_0 = 15$ will permit bistability, whereas that for $k_0 = 5$ will not. We can identify when bistable behavior can occur by noting that it requires $dY/dX < 0$.

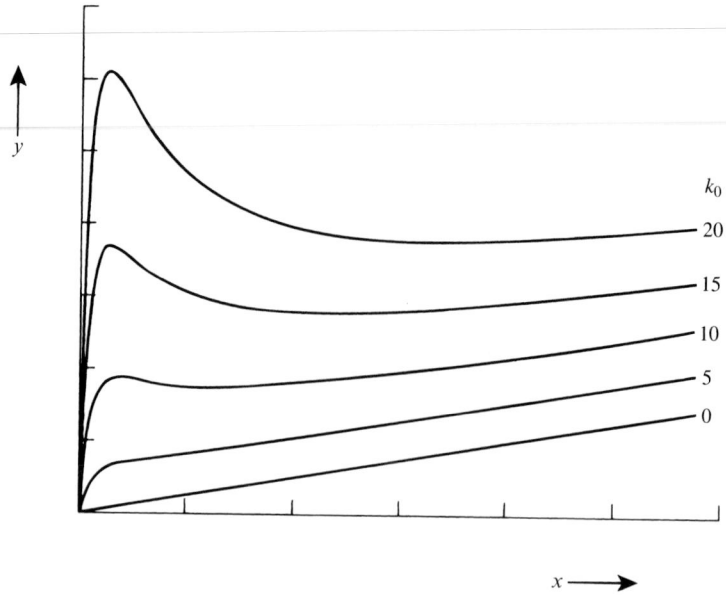


Fig. 20.21 Bistability curves calculated from Eq. (20.172).

The condition $dY/dX = 0$ gives, from Eq. (20.172),

$$X = \frac{k_0 - 2 \pm \sqrt{k_0(k_0 - 8)}}{2}. \tag{20.173}$$

For a physically meaningful value of $X$ this requires $k_0 \geq 8$; that is,

$$\frac{\alpha_0 \ell R}{1 - R} > 8, \tag{20.174}$$

which, for $1 - R \simeq 0$, gives

$$\frac{\alpha_0 \ell}{T} > 4. \tag{20.175}$$

The actual behavior of a real bistable system will deviate from the predictions of this simple steady-state absorptive optical bistable model because it neglects simple features of real systems such as linear background absorption, transverse intensity profiles, and the effect of standing waves in the cavity. Bistability can also result not only from absorptive effects but from dispersive effects in the nonlinear medium. A full review of the whole subject has been given by Gibbs [39]. In potentially important applications of bistability the state of the bistable system is switched from one of its two stable states to the other by the action of a light signal. Thus, the transient behavior of the system is important.

The ability of "light to control light" in an optical bistable system opens up the possibility of using such devices in all-optical computers and, as all-optical switches, in optical communication systems. The challenge is to make these devices smaller, faster, and economically practical.

## 20.14 Practical details of the use of crystals for nonlinear applications

One of the commonest methods for growing nonlinear crystals is the Czochralski method, which was discussed in Chapter 7. The resultant boule will be oriented according to the orientation of the seed crystal – quite often along the $c$ axis of the crystal, which is identical with the optic axis in uniaxial crystals. After growth, the orientation of the boule can be confirmed by X-ray crystallographic techniques. For angle-tuned phase matching a section must be cut from the boule so that its faces are perpendicular to the phase-matching direction, as shown in Fig. 20.3(b). To minimize Fresnel reflection losses it may be desirable to antireflection coat the input and output faces of the nonlinear crystal. Some crystals, such as ADP and KDP, are hygroscopic and must be protected from atmospheric moisture. Such crystals can be placed in a hermetically sealed container with antireflection-coated input and output windows. If the windows have a refractive index close to that of the crystal, and an appropriate *index-matching* fluid is placed around the crystal, then Fresnel losses can be minimized. An excellent crystal for generation of higher harmonics from Nd:YAG lasers is beta-barium borate ($\beta$-$BaB_2O_4$), commonly referred to as BBO, which has successfully been used to generate the fifth harmonic of 1.06 μm at 213 nm.

Efficient nonlinear generation frequently requires the use of very intense pulsed lasers. Therefore, of particular importance in the selection of materials for nonlinear applications is the resistance of those materials to optical damage. Damage can result from thermal heating, multiphoton absorption, self-focusing and dielectric breakdown, among other mechanisms [40]. The intensities necessary to induce damage depend on the material, wavelength, pulse duration of the intense source, and whether the nonlinear material is being operated in a phase-matched arrangement [41]. Semiconductor materials damage in the 0.5–1 TW $m^{-2}$ range[7] for 100-ns pulses. Sapphire damages at around 250 TW $m^{-2}$ (250 MW $mm^{-2}$) and BBO at 46 TW $m^{-2}$ at 1.06 μm for pulses ~10 ns long. Most nonlinear materials have damage thresholds in the range 0.1–10 TW $mm^{-2}$, which for a $Q$-switched laser pulse 10 ns long limits maximum pulse energies to 0.1 J $mm^{-2}$. It is easy to see from these numbers how laser pulses of relatively small energy focused onto small spot sizes can lead to damage. A 1-J laser pulse, 10 ns long focused to a Gaussian spot size of 1 μm has a peak intensity of 6.4 TW $m^{-2}$, and generates a peak electric field at the focus of almost 70 MV $m^{-1}$. This is well in excess of the value necessary to produce dielectric breakdown, which is ~3 MV $m^{-1}$.

The phenomenon of self-focusing, which can lead to optical damage, is an interesting one in its own right. This is a nonlinear phenomenon in which the refractive index becomes a function of the laser intensity, namely

$$n = n_0 + n_1 I. \tag{20.176}$$

[7] 1 TW (terawatt) is $10^{12}$ W.

If a fundamental Gaussian-mode laser beam enters such a medium it is focused because the radial intensity variation in the Gaussian beam leads to the production of a positive lens. The radial index profile in the medium will be

$$n(r) = n_0 + n_1 I_0 e^{-r^2/(2\omega^2)}, \tag{20.177}$$

which leads to focusing. However, the effect can be negated by diffraction, which makes the Gaussian beam want to diverge more rapidly as its spot size gets smaller. A more detailed discussion can be found elsewhere [18, 25, 26]. An interesting consequence of self-focusing is for any spatial non-uniformity of a high-intensity laser beam to be amplified so that the beam may break up into a collection of tiny filaments [42]. In very-high-energy pulsed laser systems, such as are used for laser-induced fusion studies (see Chapter 23), the laser profile must be continuously smoothed, for example, by spatial filtering (see Chapter 15), to prevent self-focusing and internal damage.

If high-energy, very short (sub-picosecond) laser pulses are focused into a gas then self-focusing leads to a plasma filament, on the order of 100 μm wide, which can guide the laser light as a continuous narrow filament for many meters [43–47]. The ionized filament is maintained by the balance among self-focusing in the gas, defocusing caused by the radial variation of refractive index created by the plasma, and diffraction.

## 20.15 Problems

20.1 Use Eqs. (20.64) and (20.65) to calculate the phase-matching conditions for SHG of a 10.6-μm laser.

20.2 Prove Eq. (20.139).

20.3 Prove that in a uniaxial crystal

$$\frac{dn_e^{\omega_3}(\theta)}{d\theta} = -\frac{n_e^{\omega_3}(\theta)^3}{2}\sin(2\theta)\left[\left(\frac{1}{n_e^{\omega_3}}\right)^2 - \left(\frac{1}{n_o^{\omega_3}}\right)^3\right],$$

where $\theta$ is the angle of wave propagation with respect to the optic axis.

20.4 Prove Eq. (20.140).

20.5 Prove Eq. (20.141).

20.6 Prove that, in a phase-matched up-conversion process, the conversion efficiency of the detected frequency $\omega_2$ to the up-converted frequency $\omega_3$ is

$$\frac{I_3(z)}{I_2(0)} = \frac{\omega_3}{\omega_2}\sin^2\Gamma(z),$$

where

$$\Gamma = \sqrt{\kappa_3^2|E_1|^2},$$

where $E_1$ is the field amplitude of the pump beam at frequency $\omega$ and $\kappa = \omega d_{\text{eff}}/(nc)$. Neglect any depletion of the strong pump beam. Explain why the $z$

dependence of the conversion is different from the SHG conversion efficiency given by Eq. (20.98).

20.7 $BeSO_4 \cdot 4H_2O$ is a crystal used for generation of ultraviolet light by SHG. Its refractive indices can be described by the Sellmaier equations

$$n_o^2 = 2.1545 + \frac{0.00835}{\lambda^2 - 0.01606} - 0.03573\lambda^2$$

and

$$n_e^2 = 2.0335 + \frac{0.00806}{\lambda^2 - 0.01354} - 0.01970\lambda^2,$$

with $\lambda$ given in micrometers.

Calculate the phase-matching angles for both type-I and type-II SHG with a fundamental input at 530 nm.

## References


[1] J. M. Manley and H. E. Rowe, "General energy in nonlinear resistances," *Proc. IRE* **47**, 2115–2116, 1959.

[2] P. A. Franken, A. E. Hill, C. L. Peters, and G. Weinrich, "Generation of optical harmonics," *Phys. Rev. Lett.* **7**, 118–119, 1961.

[3] R. W. Caldwell and H. Y. Fan, "Optical properties of tellurium and selenium," *Phys. Rev.* **114**, 664–675, 1959.

[4] M. Born and E. Wolf, *Principles of Optics*, 7th Edn., Cambridge: Cambridge University Press, 1999.

[5] M. V. Hobden, "The dispersion of the refractive indices of proustite ($Ag_3AsS_3$)," *Opto-Electronics* **1**, 159, 1969.

[6] F. Zernike and J. E. Midwinter, *Applied Nonlinear Optics*, New York: Wiley, 1973.

[7] J. Galang, A. Restelli, E. W. Hagley, and C. W. Clark, "A green laser pointer hazard," National Institute of Standards Technology Note 1668, 9 pages, 2012. Available at http://arxiv.org/ftp/arxiv/papers/1008/1008.1452.pdf.

[8] J. A. Armstrong, N. Bloembergen, J. Ducuing, and P. S. Pershan, "Interactions between light waves in a nonlinear dielectric," *Phys. Rev.* **127**, 1918–1939, 1962.

[9] P. A. Franken and J. F. Ward, "Optical harmonics and nonlinear phenomena," *Rev. Mod. Phys.* **35**, 23–39, 1963.

[10] M. M. Fejer, G. A. Magel, D. H. Jundt, and R. L. Byer, "Quasi-phase-matched second harmonic generation: tuning and tolerances," *IEEE J. Quant. Electron.* **28**, 2631–2654, 1992.

[11] R. L. Byer, "Parametric oscillators and nonlinear materials," in *Nonlinear Optics*, eds. P. F. Harper and B. S. Wherrett, London: Academic Press, 1977.

[12] H. A. Smith and H. Mahr, "An infrared detector for astronomy using up-conversion techniques," paper presented at the International Quantum Electronics Conference, Kyoto, Japan, September 1970.

[13] J. E. Midwinter, "Image conversion from 10.6 μm to the visible in lithium niobate," *Appl. Phys. Lett.* **12**, 68–70, 1968.

[14] R. L. Byer, "Parametric oscillation and mixing," in *Nonlinear Infrared Generation*, Y. R. Shen, Berlin: Springer-Verlag, 1977, pp. 81–137.

[15] F. Zernike, in *Methods of Experimental Physics, Vol. XV, Quantum Electronics*, Part B, ed. C. L. Tang, New York: Acadamic Press, 1979, pp. 143–183.

[16] B. Lax, R. L. Aggarwal, and G. Favrot, "Far-infrared step-tunable coherent radiation source: 70 μm to 2 mm," *Appl. Phys. Lett.* **23**, 679–681, 1973.

[17] B. Lax and R. L. Aggarwal, "Far-infrared generation and applications," *J. Opt. Soc. Am.* **64**, 533, 1974.

[18] Y. R. Shen, *The Principles of Nonlinear Optics*, New York: Wiley, 1984.

[19] R. Frey and F. Pradere, "High-efficiency narrow-linewidth Raman amplification and spectral compression," *Opt. Lett.* **5**, 374–376, 1980.

[20] J. J. Tiee and C. Wittig, "$CF_4$ and NOCl molecular lasers operating in the 16-μm region," *Appl. Phys. Lett.* **30**, 420–422, 1977.

[21] M. J. Colles and C. R. Pidgeon, "Tunable lasers," *Rep. Prog. Phys.* **38**, 329–460, 1975.

[22] W. Lenth, R. M. Macfarlane, and W. P. Risk, "Generation of blue–green laser radiation using nonlinear frequency up conversion," paper presented at the 1988 meeting of the Lasers and Electro-optics Society, Santa Clara, CA, November, 1988.

[23] R. L. Byer, "Optical parametric oscillators," in *Quantum Electronics*, Vol. 1B, eds. H. Rabin and C. C. Tang, New York: Academic Press, 1975, pp. 587–702.

[24] R. C. Eckardt, C. D. Nabors, W. J. Kozlovsky, and R. L. Byer, "Optical parametric oscillator frequency tuning and control," *J. Opt. Soc. Am.* **B8**, 646–667, 1991.

[25] A. Yariv, *Quantum Electronics*, 3rd edn., New York: Wiley, 1989.

[26] R. W. Boyd, *Nonlinear Optics*, 3rd edn., Burlington, MA: Academic Press, 2008.

[27] D. F. Walls, "Squeezed states of light," *Nature* **305**, 141–146, 1983.

[28] L. Mandel and E. Wolf, *Optical Coherence and Quantum Optics*, Cambridge: Cambridge University Press, 1995.

[29] R. London, *The Quantum Theory of Light*, 3rd edn., Oxford: Oxford University Press, 2000.

[30] H. P. Yuen, "Two-photon coherent states of the radiation field," *Phys. Rev.* **A13**, 2226–2243, 1976.

[31] H. P. Yuen and J. H. Shapiro, "Generation and detection of two-photon coherent states in degenerate four-wave mixing," *Opt. Lett.* **3**, 334–336, 1979.

[32] R. K. Tyson, *Principles of Adaptive Optics*, Boca Raton, FL: CRC Press, 2010.

[33] V. Markov, A. Khizhnyak, P. Sprangle, A. Ting, J. R. Penano, B. Hafizi, C. C. Davis, and L. DeSandre, "Nonlinear OPC technique for laser beam control in turbulent atmosphere," paper present at the SPIE Conference on Optics in Atmospheric Propagation and Adaptive Systems, Edinburgh, September 2012.

[34] A. Yariv and D. M. Pepper, "Amplified reflection, phase conjugation and oscillation in degenerate four-wave mixing," *Opt. Lett.* **1**, 16–18, 1977.

[35] G. C. Bjorklund, "Conjugate wave-front generation and image reconstruction by four-wave mixing," *Appl. Phys. Lett.* **31**, 592–594, 1977.

[36] C. R. Guiliano, "Applications of optical phase conjugation," *Phys. Today* April, p. 27, 1981.

[37] I. D. Mayergoyz, *Mathematical Models of Hysteresis*, New York: Springer-Verlag, 1991.

[38] A. Szöke, V. Danen, J. Goldhar, and N. A. Kurnit, "Bistable optical element and its applications," *Appl. Phys. Lett.* **15**, 376–379, 1969.

[39] H. M. Gibbs, *Optical Bistability: Controlling Light with Light*, Orlando, FL: Academic Press, 1985.

[40] J. F. Ready, *The Effects of High Power Laser Radiation*, New York: Academic Press, 1971.

[41] A. G. Glass and A. H. Guenther, "Damage in laser materials," *Appl. Opt.* **11**, 832–840, 1972.

[42] S. C. Abbi and H. Malin, "Correlation of filaments in nitrobenzene with laser spikes," *Phys. Rev. Lett.* **26**, 604–606, 1971.

[43] A. Braun, G. Korn, X. Liu, D. Du, J. Squier, and G. Mourou, "Self channeling of high-peak-power femtosecond laser pulses in air," *Opt. Lett.* **20**, 73–75, 1995.

[44] E. T. J. Nibbering, P. F. Curley, G. Grillon, B. S. Prade, M. A. Franco, F. Salin, and A. Mysyrowicz, "Conical emission from self-guided femtosecond pulses in air," *Opt. Lett.* **21**, 62–64, 1996.

[45] H. Schillinger and R. Sauerbrey, "Electrical conductivity of long plasma channels in air generated by self-guided femtosecond laser pulses," *Appl. Phys.* **B68**, 753–756, 1999.

[46] S. Tzortzakis, M. A. Franco, Y.-B. André, A. Chiron, B. Lamouroux, B. S. Prade, and A. Mysyrowicz, "Formation of a conducting channel in air by self-guided femtosecond laser pulses," *Phys. Rev. E* **60**, R3505–R3507, 1999.

[47] B. La Fontaine, F. Vidal, Z. Jiang, C. Y. Chien, D. Comtois, A. Desparois, T. W. Johnson, J.-C. Kieffer, and H. Pépin, "Filamentation of ultrashort pulse laser beams resulting from their propagation over long distances in air," *Phys. Plasmas* **6**, 1615–1621, 1999.

# 21 Detection of optical radiation

## 21.1 Introduction

In this chapter we will review some of the fundamentals of the optical detection process. This will include a discussion of the randomly fluctuating signals, known as *noise*, that appear at the output of any detector. We will then examine some of the practical characteristics of various types of optical detector. The chapter will conclude with a discussion of the limiting detection sensitivities of important detectors used in various ways.

Photon detectors operate by absorbing the photons coming from a source and using the absorbed energy to produce a change in the electrical characteristics of their active element(s). This can occur in many ways. In a photomultiplier or vacuum photodiode the incoming photons are absorbed in a photoemissive surface, and free electrons are produced by the photoelectric effect. These electrons can be accelerated and detected as an electric current. In a semiconductor photodiode or photovoltaic detector, absorption of a photon at a p–n or p–i–n junction creates an electron–hole pair. The electron and hole separate because of the energy barrier at the junction – each carrier moves to the region where it can reduce its potential energy, as shown in Fig. 21.1.

Thermal radiation detectors use the heating effect produced by absorbed photons to change some characteristic of the detector element. In a bolometer, the heating of carriers changes their mobility and consequently the resistivity of the detector element. In a thermopile, the heating effect is used to generate a voltage through the thermoelectric (Seebeck) effect.

Pyroelectric detectors utilize the change in surface charge that results when certain crystals (ones that can possess an internal electric dipole moment) are heated.

## 21.2 Noise

### 21.2.1 Shot noise

The ability of a photodetector to detect an incoming light signal is limited by the intrinsic fluctuations, or noise, both of the incoming light itself and of the background electric current fluctuations generated by the detector. In the photon description of a monochromatic light beam an incoming beam of intensity $\langle I \rangle$ has an average photon flux associated with it of $\langle N \rangle$ photons $m^{-2}$, where

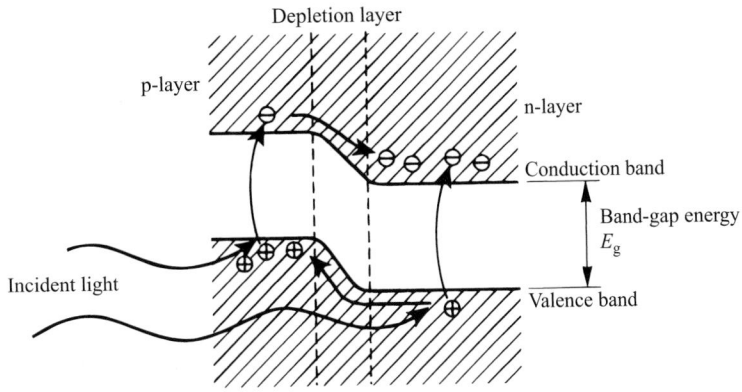


**Fig. 21.1** Photoexcitation at a p–n junction.

$$\langle N \rangle = \frac{\langle I \rangle}{h\nu}. \tag{21.1}$$

If the rate of arrival of photons at the detector is examined closely, it will be observed to fluctuate about this average value. Strictly speaking, all we can ever really observe is the rate of appearance of photoelectrons or carriers produced by the disappearance of photons at the active surface of the detector. We assume that these events are directly related by a quantum efficiency factor $\eta$, where

$$\eta = \frac{\text{number of carriers produced}}{\text{number of absorbed photons}}. \tag{21.2}$$

The fluctuations in photon flux, $N$, give rise to a fluctuation in the rate of production of photoproduced carriers. These fluctuations appear as a randomly varying current superimposed on the average current. The average current from the photodetector is

$$\bar{i} = \frac{\eta e}{h\nu} \langle I \rangle A, \tag{21.3}$$

which can be written as

$$\bar{i} = \mathcal{R} \langle I \rangle A, \tag{21.4}$$

where $A$ is the detector area, and $\mathcal{R} = e\eta/(h\nu)$ is the responsivity, which has units of A $W^{-1}$. The fluctuations in current resulting from the fluctuating appearance of photoproduced carriers are called *shot*, or *photon*, noise. The frequency spectrum of these current fluctuations can be calculated by considering the frequency components in the current produced by the elemental current of a single photoproduced carrier. If we take one of these elemental current pulses to be of a Gaussian shape, we can write

$$i(t) = \frac{e}{\sigma\sqrt{2\pi}} \exp\left(\frac{-t^2}{2\sigma^2}\right), \tag{21.5}$$

where the shape of the pulse is normalized so that

$$\int_{-\infty}^{\infty} i(t)dt = e. \tag{21.6}$$

The charge on an electron is $-e$.

The frequency spectrum of this current is

$$F(\omega) = \frac{1}{2\pi} \int_{-\infty}^{\infty} i(t) \exp(-j\omega t) dt, \tag{21.7}$$

which gives

$$F(\omega) = \frac{e}{2\pi} \exp\left(-\frac{\sigma^2 \omega^2}{2}\right). \tag{21.8}$$

For sufficiently low frequencies, for example, such that $\sigma^2\omega^2/2 < 0.01$, the exponential factor is very close to unity, and the Fourier transform is flat. Simply stated, this result says that randomly occurring narrow pulses contribute broadband "white" noise.

In terms of its Fourier transform the current can be written as

$$i(t) = \int_{-\infty}^{\infty} F(\omega) \exp(j\omega t) d\omega. \tag{21.9}$$

If this current flows into a resistance $R$, then the average power dissipated over time $T$, where $T$ is greater than the pulse duration, is

$$P = \frac{1}{T} \int_{-T/2}^{T/2} i^2(t) R \, dt = \frac{R}{T} \int_{-T/2}^{T/2} i(t) dt \int_{-\infty}^{\infty} F(\omega) \exp(j\omega t) d\omega, \tag{21.10}$$

which can be rearranged as

$$P = \frac{R}{T} \int_{-\infty}^{\infty} F(\omega) \left[ \int_{-\infty}^{\infty} i(t) \exp(j\omega t) dt \right] d\omega. \tag{21.11}$$

The limits on the inner integral have been set to $\pm\infty$ without significant error since the pulse is localized within the time interval $T$. Since $i(t)$ is real, it follows from Eq. (21.7) that

$$F(\omega) = \frac{1}{2\pi} \int_{-\infty}^{\infty} i(t) \exp(j\omega t) dt = F(-\omega). \tag{21.12}$$

Substitution into Eq. (21.9) gives

$$P = \frac{2\pi R}{T} \int_{-\infty}^{\infty} |F(\omega)|^2 \, d\omega, \tag{21.13}$$

which can be written as

$$P = \frac{4\pi R}{T} \int_{0}^{\infty} |F(\omega)|^2 \, d\omega. \tag{21.14}$$

The fraction of the power spectrum between $\omega$ and $\omega + d\omega$ is $P(\omega)d\omega$, where the power spectral density function is clearly

$$P(\omega) = \frac{4\pi |F(\omega)|^2 R}{T}. \tag{21.15}$$

For a flow of $\eta\langle N\rangle$ charge carriers per second, which are assumed to be uncorrelated, the total power spectral density is $\eta\langle N\rangle P(\omega)$. The spectral energy density supplied per second is

$$U(\omega) = \eta\langle N\rangle P(\omega)T, \tag{21.16}$$

where $T = 1$ s, which can be written as

$$U(\omega) = 4\pi\,\eta\langle N\rangle |F(\omega)|^2 R. \tag{21.17}$$

This energy can be considered to result from a noise-current generator whose mean-squared magnitude is

$$\langle i_{\mathrm{N}}^2(\omega)\rangle = 4\pi\,\eta\langle N\rangle |F(\omega)|^2\, d\omega. \tag{21.18}$$

For frequencies where $F(\omega)$ is flat

$$\langle i_{\mathrm{N}}^2(\omega)\rangle = \frac{\eta\langle N\rangle e^2}{\pi}\, d\omega. \tag{21.19}$$

Since $\eta\langle N\rangle = \bar{i}/e$ and $d\omega = 2\pi\ df$, we can write the shot-noise spectrum in terms of conventional frequency as

$$\langle i_{\mathrm{N}}^2(f)\rangle = 2e\bar{i}\ df. \tag{21.20}$$

Even if all other sources of noise in the detector and its following electronics can be eliminated, shot noise is inescapable. It is produced by the signal. The best signal-to-noise (power) ratio to be expected is the shot-noise-limited (SNL) value, for a bandwidth $\Delta f$,

$$\left(\frac{S}{N}\right)_{\mathrm{SNL}} = \frac{\bar{i}^2 R}{i_{\mathrm{N}}^2 R} = \frac{[e\eta\langle I\rangle A/(h\nu)]^2}{[2e^2\eta\langle I\rangle A/(h\nu)]\Delta f}, \tag{21.21}$$

giving

$$\left(\frac{S}{N}\right)_{\mathrm{SNL}} = \frac{\eta\langle I\rangle A}{2h\nu\ \Delta f} = \frac{\eta P}{2h\nu\ \Delta f}, \tag{21.22}$$

where $P = \langle I\rangle A$ is the total power (in watts) falling on the detector.

This signal-to-noise ratio is frequently called the limiting *video* or *direct* detection limit.[1] It is the best that can be done with a detector directly receiving only the incoming light it is desired to detect. For a detector with unity quantum efficiency the minimum detectable light signal ($S/N = 1$) is

$$P_{\min} = 2h\nu\ \Delta f. \tag{21.23}$$

The sampling time for digital data, which is determined by Nyquist's theorem, is generally set so that data are sampled at twice the maximum frequency being observed, i.e., $\Delta T = 1/(2\ \Delta f)$. The minimum detectable signal therefore corresponds to one detected photon per sampling interval. In practice, additional sources of noise are present in a detection system involving a detector and other electronic devices. Principal among these are $1/f$ noise, Johnson noise (also called thermal, Nyquist, or resistance noise), and, in semiconductor detectors, generation–recombination noise.

[1] Also called the photon-noise or quantum-limited signal-to-noise ratio.

### 21.2.2 Johnson noise

At temperatures above absolute zero the thermal energy of the charge carriers in any resistor leads to fluctuations in local charge density. These fluctuating charges cause local voltage gradients that can drive a corresponding current into the rest of the circuit. The quantitative treatment of this thermal noise can be carried out in several different ways [1]. Nyquist [2] proposed a theorem that stated that the noise power generated by a circuit element did not depend specifically on the nature of the circuit element, but depended only on the temperature of the component and the frequency band being examined. He proved this result thermodynamically and derived a quantitative expression for the noise power by considering the energy flow between two resistors of equal value connected together by a cable with characteristic impedance $R$. The noise power can also be calculated by considering the voltages and currents associated with an $RLC$ tuned circuit. However, we choose here to calculate the noise power by relating the current and voltage fluctuations to the available radiation energy at temperature $T$ described by black-body radiation. We do this by considering the circuit shown in Fig. 21.2(a) in which an antenna is connected to a resistance $R$. The antenna can be modeled as a voltage source with an associated series resistance $R_r$ – called its radiation resistance. If the antenna is a wire of length $\ell$, then the electric field of the black-body radiation will induce a voltage in the antenna. For radio and microwave frequencies, we can use the Rayleigh–Jeans approximation to Planck's radiation formula

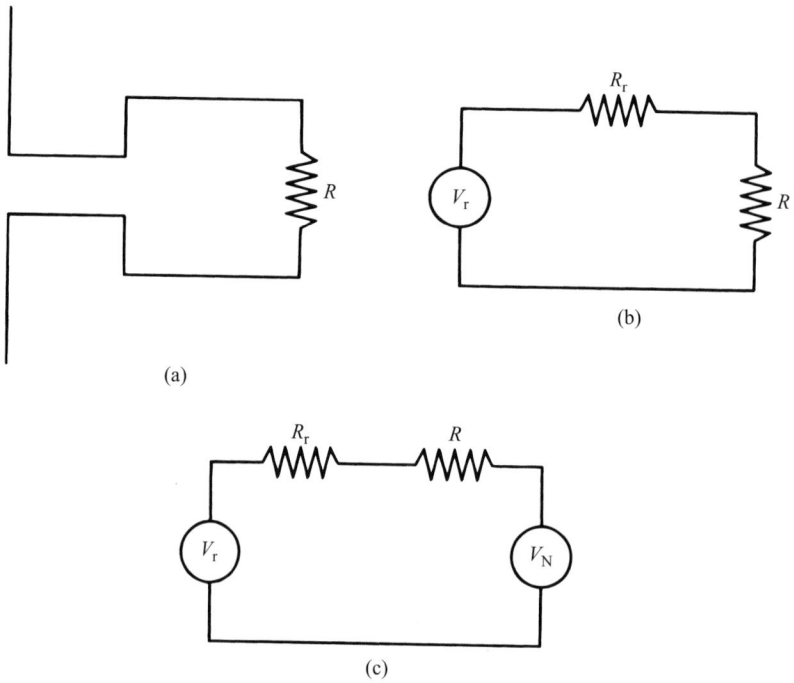


**Fig. 21.2** Circuits used in an analysis of Johnson noise: (a) an antenna connected to a resistor; (b) including the radiation resistance of the antenna; and (c) showing the noise source associated with the resistor.

to calculate this induced voltage. The energy density of the radiation interacting with the antenna within a frequency band of width $\Delta f$ at frequency $f$ is

$$\Delta\rho = \frac{8\pi f^2}{c^3} k_B T \, \Delta f. \tag{21.24}$$

On average, because the black-body radiation is unpolarized, only one-third of this energy actually interacts with a linear antenna oriented (say) in the $x$ direction. The mean-squared electric field in the antenna direction within the frequency band $\Delta f$ is[2]

$$\Delta\langle E_x^2\rangle = \frac{cZ_0 \, \Delta\rho}{3}. \tag{21.25}$$

The mean-square voltage induced in the antenna within the band $\Delta f$ is

$$\Delta\langle V_r^2\rangle = \ell^2 \Delta(E_x)^2 = \frac{8\pi f^2}{3c^2} \ell^2 k_B T Z_0 \, \Delta f. \tag{21.26}$$

This voltage induces a current $i_a$ in the antenna, which can flow into an external short-circuit.

Since an ideal antenna cannot dissipate any energy, it must reradiate the energy it receives. We can model this reradiation process by associating a radiation resistance $R_r$ with the antenna as shown in Fig. 21.2(b) such that the *apparent* dissipation in $R_r$ balances the received power. For this to be so

$$\frac{\Delta(V_r^2)}{R_r} = i_r^2 R_r. \tag{21.27}$$

Now the power radiated by an antenna of length $\ell$ carrying an oscillating current of magnitude $i_r$ is [3, 4]

$$W = \frac{\pi Z_0 i_r^2 \ell^2 f^2}{3c^2}. \tag{21.28}$$

Equating $W$ with $\frac{1}{2} i_r^2 R_r$ gives

$$R_r = \frac{2\pi f^2 \ell^2 Z_0}{3c^2}. \tag{21.29}$$

If the antenna is connected to a resistor $R$, as shown in Fig. 21.2(b), then the fluctuating current in the antenna will dissipate a power $V_r^2 R/(R_r + R)^2$ in the resistor $R$. If the entire circuit shown in Fig. 21.2(b) is in thermal equilibrium, then for the resistor $R$ to heat would violate the second law of thermodynamics. Consequently, the resistor $R$ must drive power back into the antenna to balance the power it receives. Consequently, the mean-square voltage generated by $R$ within the band $\Delta f$ must satisfy

$$\frac{\langle V_N^2\rangle}{R} = \frac{\Delta\langle V_r^2\rangle}{R_r} = \frac{8\pi f^2 \ell^2}{3c^2} k_B T Z_0 \, \Delta f \times \frac{3c^2}{2\pi f^2 \ell^2 Z_0} = 4 k_B T \, \Delta f, \tag{21.30}$$

[2] For a plane linearly polarized wave the intensity within $\Delta f$ can be written as $\Delta\langle I\rangle = \Delta\langle E_x^2\rangle/\langle 2Z_0\rangle = c\,\Delta\rho/6$. The factor of 6 comes from the three orthogonal propagation directions in space and the two orthogonal polarization directions associated with each of them.

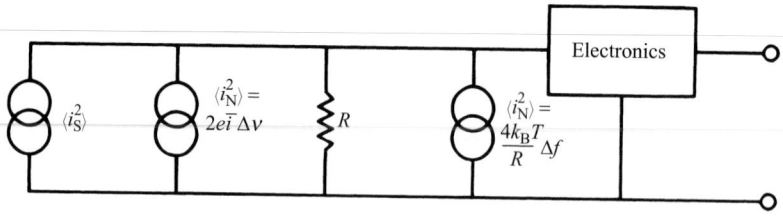


**Fig. 21.3** The equivalent circuit of a shot-noise-limited detector driving a resistive load.

which gives

$$\langle V_N^2 \rangle = 4k_B TR\,\Delta f. \tag{21.31}$$

The overall situation can be represented as shown in Fig. 21.2(c). Alternatively, the noisy resistor can be treated as a resistance $R$ with a parallel current source whose mean-square value within the band $\Delta f$ is

$$\langle i_N^2 \rangle = \frac{4k_B T}{R}\,\Delta f. \tag{21.32}$$

Thus, this source of noise can be reduced by cooling the offending component to a low temperature.

If a shot-noise-limited detector drives a circuit with an input resistance $R$, the equivalent current involves both the shot-noise current source and the Johnson-noise current source, as shown in Fig. 21.3.

### 21.2.3 Generation–recombination noise and 1/*f* noise

These two types of noise are important in semiconductor detectors. $1/f$ noise, or current noise, has a power spectrum that depends inversely on the frequency. It can be described by a noise-current generator

$$\langle i_N^2 \rangle = \frac{K\langle i \rangle^{\alpha}\,\Delta f}{f^{\beta}}, \tag{21.33}$$

where $\langle i \rangle$ is the average current through the detector, $K$ is a constant, and typically $\alpha = 2$ and $\beta = 1$. The noise, which depends inversely on the frequency, the $1/f$ noise, originates from many causes, such as the diffusion of charge carriers, the presence of impurity atoms and lattice defects in the material, and interaction of charge carriers with the surface of the semiconductor. To achieve high $S/N$ ratios low-frequency detection should be avoided because of $1/f$ noise – which may increase in magnitude down to frequencies of $10^{-4}$ Hz. The trend towards higher and higher data rates in optical communication systems moves system operation well away from $1/f$ noise, although at high and intermediate frequencies generation–recombination (GR) noise will still be a factor in determining the $S/N$ ratio. Generation–recombination noise arises from statistical fluctuations in the number of carriers in the detector. In this sense it is closely related to photon noise, the fluctuation in the number of generated carriers, but the GR noise results from the secondary-carrier-density fluctuations arising from random electron–hole recombinations. This recombination process has a characteristic lifetime $\tau_0$, which can be very short, 1 ns or less, so we expect this source of noise to contribute up to frequencies $\sim 1/\tau_0$. This is indeed, the case, and the GR noise spectrum is flat up to a frequency $\sim 1/\tau_0$ and can be shown to correspond to an equivalent noise power generator [5]

$$\langle i_N^2 \rangle = \frac{4\langle i \rangle^2 \tau_0}{\langle N \rangle (1 + 4\pi^2 f^2 \tau_0^2)}, \tag{21.34}$$

where $\langle N \rangle$ is the average number of charge carriers. If the time it takes a charge carrier to travel through the detector into the external current is $\tau_d$ then we can write

$$\frac{\langle N \rangle}{\tau_d} = \frac{\langle i \rangle}{e}, \tag{21.35}$$

and the GR noise current generator can be written as

$$\langle i_N^2 \rangle = \frac{4e\langle i \rangle (\tau_0/\tau_d)}{1 + 4\pi^2 f^2 \tau_0^2}. \tag{21.36}$$

Thus, we expect the GR noise to be less in a detector with a rapid recombination time and also less in heavily doped semiconductors where the number of charge carriers will be greater at a given average current $\langle i \rangle$ than in an intrinsic material.

Thus, in a semiconductor detector there are contributions from $1/f$ noise, GR noise, and thermal noise. The relative contributions of these noise sources vary with frequency in the schematic way shown in Fig. 21.4.

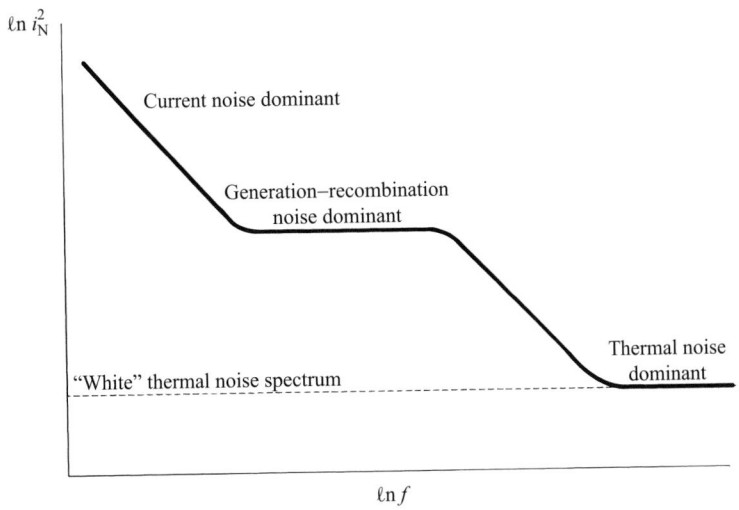


Fig. 21.4 The variation of noise with frequency in a semiconductor detector.

## 21.3 Detector performance parameters

### 21.3.1 Noise equivalent power

It is important to consider under what conditions of operation the performance of a detector will be primarily limited by GR noise, since for all but low frequencies GR noise generally dominates over $1/f$ and Johnson noise. This can be carried out by considering the *noise equivalent power* (NEP) of the detector, which for a good detector is a practical measure of the magnitude of the GR noise. The NEP is the RMS value of a sinusoidally modulated light signal falling on the detector that gives rise to an RMS electrical signal equal to the RMS noise voltage. The NEP is usually specified in terms of a black-body source, a reference bandwidth, usually 1 Hz, and the modulation frequency of the radiation.

For example NEP (500 K, 900, 1) implies a black-body illuminator whose temperature is 500 K, a modulation frequency of 900 Hz, and a reference bandwidth of 1 Hz. If the illuminating signal has intensity $I$ (W m$^{-2}$) and falls on a detector of area $A$ then we can write

$$\mathrm{NEP} = \frac{IA}{\sqrt{\Delta f}} \frac{V_\mathrm{N}}{V_\mathrm{S}}, \tag{21.37}$$

where $V_\mathrm{S}$ and $V_\mathrm{N}$ are the signal and noise voltages, respectively, measured with a bandwidth $\Delta f$.

We can rearrange Eq. (21.37) to give the equivalent intensity needed to generate an $S/N$ ratio of 1 in a bandwidth of 1 Hz as

$$I(S/N = 1, \Delta f = 1\ \mathrm{Hz}) = \frac{\mathrm{NEP}}{A}. \tag{21.38}$$

For example a detector with an NEP of $10^{-12}$ W Hz$^{-1/2}$ needs a total power of $10^{-12}$ W of black-body power to fall onto its sensitive surface to generate a signal equal to the *detector noise*. The photon-noise limit for such a detector would correspond to a received intensity of

$$I(\text{photon-noise-limited}, S/N = 1, \Delta f = 1\ \mathrm{Hz}) = 2h\nu.$$

So the relative magnitudes of the detector-noise-limited minimum detectable light signal and that set by photon noise is

$$\frac{I(\text{detector-noise-limited})}{I(\text{photon-noise-limited})} \simeq \frac{\mathrm{NEP}d}{2Ah\overline{\nu}}. \tag{21.39}$$

Strict equality does not hold in Eq. (21.39), since the NEP is generally defined for black-body radiation and the quantum $h\overline{\nu}$ implies an appropriate average frequency for the incoming radiation.

### 21.3.2 Detectivity

Many detectors exhibit an NEP that is proportional to the square root of the detector area; so a detector-area-independent parameter, namely the *detectivity* $D^*$, is frequently used. It is given by

$$D^* = \frac{\sqrt{A}}{\text{NEP}}, \tag{21.40}$$

where $A$ is the area of the detector, and $D^*$ is specified in the same way as the NEP; for example, $D^*(500\text{ K}, 900, 1)$. To specify the variation in response of a detector with wavelength, the *spectral detectivity* is often used. Thus, $D^*(\lambda, 900, 1)$ would specify the response of the detector to radiation of wavelength $\lambda$, modulated at 900 Hz, and detected with a bandwidth of 1 Hz; $D^*$ is measured in units of cm $\text{Hz}^{1/2}\ \text{W}^{-1}$.

The performance of photodiodes used in optical communication systems is frequently specified in terms of their responsivity $\mathcal{R}$, usually specified in units of A $\text{W}^{-1}$, which characterizes the response of the detector to unit irradiance. At high data rates the actual $S/N$ performance of these detectors will depend on the wide-band amplification electronics with which they are used. Specification of their performance in terms of the NEP or $D^*$ becomes less relevant.

### 21.3.3 The frequency response and the time constant

The frequency response of a detector is the variation of the responsivity or radiant sensitivity as a function of the modulation frequency of the incident radiation. The frequency variation of $\mathcal{R}$ and the time constant of the detector are generally related according to

$$\mathcal{R}(f) = \frac{\mathcal{R}(0)}{(1 + 4\pi^2 f^2 \tau^2)^{1/2}}. \tag{21.41}$$

For frequencies above $1/(2\pi\tau)$ the response is falling off significantly and at high enough modulation frequencies the detector will provide a DC output proportional to the average intensity.

## 21.4 Practical characteristics of optical detectors

The development of optical detectors has occurred, in common with various branches of electronics, by means of a series of advances from the use of gas-filled tubes and vacuum tubes to various semiconductor devices. However, whereas in general electronics the vacuum tube is now reserved for specialized applications, vacuum-tube optical detectors such as the photomultiplier are still in widespread use.

### 21.4.1 Photoemissive detectors

Photoemissive detectors include gas-filled and vacuum photodiodes, photomultiplier tubes, and photochanneltrons. These are all photon detectors that utilize the photoelectric effect. When radiation of frequency $\nu$ falls upon a metal surface, electrons are emitted, provided that the photon energy $h\nu$ is greater than a minimum critical value $\phi$, called the *work*

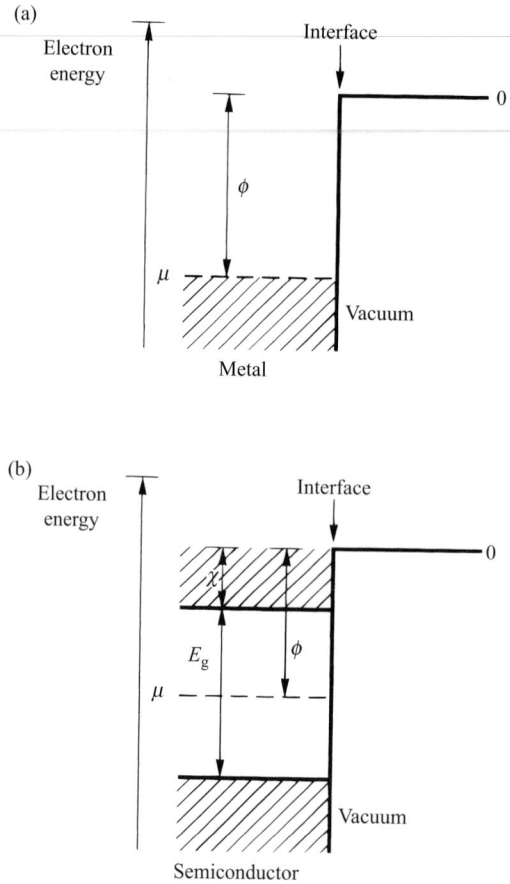


**Fig. 21.5** The band structure of (a) a metal–vacuum interface and (b) a pure-semiconductor–vacuum interface ($\phi$ is the work function, $\chi$ is the electron affinity, $E_g$ is the band-gap energy, and $\mu$ is the Fermi level).

*function*, which is characteristic of the material being irradiated. A simplified energy-level diagram illustrating this effect for a metal–vacuum interface is shown in Fig. 21.5(a). For most metals, $\phi$ is in the range 4–5 eV(1 eV $\equiv \lambda = 1.24$ μm), although for the alkali metals it is lower, for example 2.4 eV for sodium and 1.8 eV for cesium. Pure metals or alloys, particularly beryllium–copper, are used as photoemissive surfaces in ultraviolet and vacuum-ultraviolet detectors.

Lower work functions, and consequently sensitivities that can be extended into the infrared, can be obtained with special semiconductor materials. Figure 21.5(b) shows a schematic energy-level diagram of a semiconductor–vacuum interface. In this case the work function is defined as $\phi = E_{vac} - E_F$, where $E_F$ is the energy of the Fermi level. In a pure semiconductor, the Fermi level is in the middle of the band gap, as illustrated in Fig. 21.5(b). In a p-type doped semiconductor, $E_F$ moves down towards the top of the valence band, whereas in n-type material it moves up towards the bottom of the conduction band. Consequently, $\phi$ is not as useful a measure of the minimum photon energy for photoemission as it is for a metal. The electron affinity $\chi$ is a more useful measure of this minimum energy for a semiconductor. Except at absolute zero, photons with energy $h\nu > \chi$ cause photoemission. Photons with energy $h\nu > E_g$ lead to the production of carriers in the conduction band; this leads to intrinsic photoconductivity, which is the operative detection mechanism in various infrared detectors, such as InSb.

*(a) Vacuum photodiodes.* Once electrons have been liberated from a photoemissive surface (a photocathode), they can be accelerated to an electrode that is positively charged with respect to the cathode – the anode – and generate a signal current. If the acceleration of photoelectrons is directly from cathode to anode through a vacuum, the device is a vacuum photodiode. Because the electrons in such a device take a very direct path from anode to cathode and can be accelerated by high voltages – up to several kilovolts in a small device – vacuum photodiodes have the fastest responses of all photoemissive detectors. Risetimes of 100 ps or less can be achieved. External connections and electronics are generally the limiting factors in obtaining short risetimes from such devices. However, vacuum photodiodes are not very sensitive, since at most one electron can be obtained for each photon absorbed at the photocathode. In principle, of course, the limiting sensitivity of the device is set by its quantum efficiency. Practical quantum efficiencies for photoemissive materials range up to about 0.4.

If the space between the photocathode and the anode is filled with a noble gas, photoelectrons will collide with gas atoms and ionize them, yielding secondary electrons. Thus, an electron multiplication effect occurs. However, because the motion of the electrons from cathode to anode through the gas is slow, these devices have a long response time, typically about 1 ms. Gas-filled photocells are no longer competitive with solid-state detectors in practical applications.

*(b) Photomultipliers.* If photoelectrons are accelerated *in vacuo* from the photocathode and allowed to strike a series of secondary-electron emitting surfaces, called *dynodes*, held at progressively more positive voltages, a considerable electron multiplication can be achieved and a substantial current can be collected at the anode. Such devices are called *photomultipliers* (PMTs). Practical gains of $10^9$ (anode electrons per photoelectron) can be achieved from these devices for short light pulses. Because of their very high gain, photomultipliers can generate substantial signals when only a single photon is detected. For example, an anode pulse of duration 2 ns containing $10^9$ secondary photoelectrons produced from a single photoelectron will generate a voltage pulse of 4 V across 50 Ω. This, coupled with their low noise, makes photomultipliers effective single-photon detectors. Photomultiplier $D^*$ values can range up to $10^{16}$ cm $\mathrm{Hz}^{1/2}$ $\mathrm{W}^{-1}$. The dark-adapted human eye, which can detect bursts of about ten photons in the blue, comes close to this sensitivity.

The time-response characteristics of photomultiplier tubes depend to a considerable degree on their internal dynode arrangement. The response of a given device can be specified in terms of the output signal at the anode which results from emission of a single photoelectron at the photocathode. This is illustrated in Fig. 21.6. Because electrons passing through the dynode structure can generally take slightly different paths, secondary electrons arrive at the anode at different times. The anode pulse has a characteristic width called the *transit-time spread*, which usually ranges from 0.1 to 20 ns. The time interval between photoemission at the cathode and the appearance of an anode pulse is called

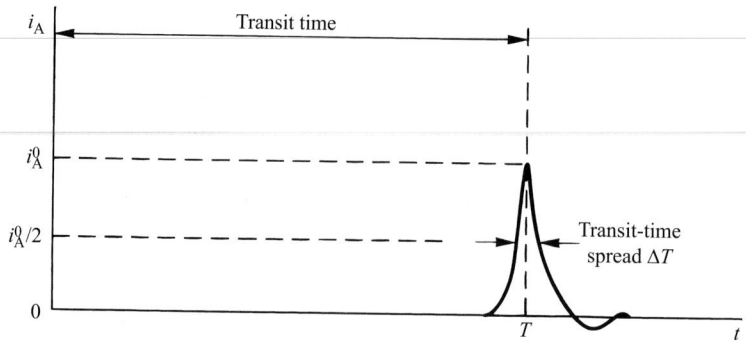


**Fig. 21.6** A typical anode pulse produced by a single photoelectron emission at the cathode of a photomultiplier tube ($t$ is the time following photoemission). The transit time $T$, the transit-time spread $\Delta T$, and the peak anode current $i_A^0$ all fluctuate from pulse to pulse.

the *transit time*, and is usually a few tens of nanoseconds. The transit time and transit-time spread also fluctuate slightly from one single-photoelectron-produced anode pulse to the next.

There are four main types of dynode structure in common usage in photomultiplier tubes; these are illustrated in Fig. 21.7. The circular cage structure is compact and can be designed for good electron-collection efficiency and small transit-time spread. This dynode structure works well with opaque photocathodes, but is not very suitable for high-amplification requirements, for which a larger number of dynodes is required. The box-and-grid and Venetian-blind structures offer very good electron-collection efficiency. Because they collect multiplied electrons independently of their path through the dynode structure, a wide range of secondary-electron trajectories is possible, leading to a large transit-time spread and slow response. Typical response times of these types of tube are 10–20 ns.

The focused dynode structure is designed so that electrons follow paths of similar length through the dynode structure. To accomplish this, electrons that deviate too much from a specified range of trajectories are not collected at the next dynode. These types of tube offer short response times, typically 1–2 ns.

Venetian-blind tubes can easily be extended to many dynode stages and have a very stable gain in the presence of small power-supply fluctuations. These tubes also have an optically opaque dynode structure, which contributes to their exhibiting very low dark-current noise when operated under appropriate conditions.

Closely related to these fundamental photomultiplier dynode structures is the *microchannel plate* (MCP). Instead of traditional dynodes, photoelectrons enter a bundle of capillary tubes fused together in an array. Each tube has a resistive coating on its inner surface that acts as a continuous electron multiplier [6]. Each capillary individually can be referred to as a channel electron multiplier (CEM) (or *channeltron* (a trademark of Burle Industries)). In these capillaries electrons undergo multiple reflections from the tube walls, generating secondary electrons. The MCPs have fast time responses, good immunity from external magnetic fields, and position-sensitive capability when provided with multiple anodes.

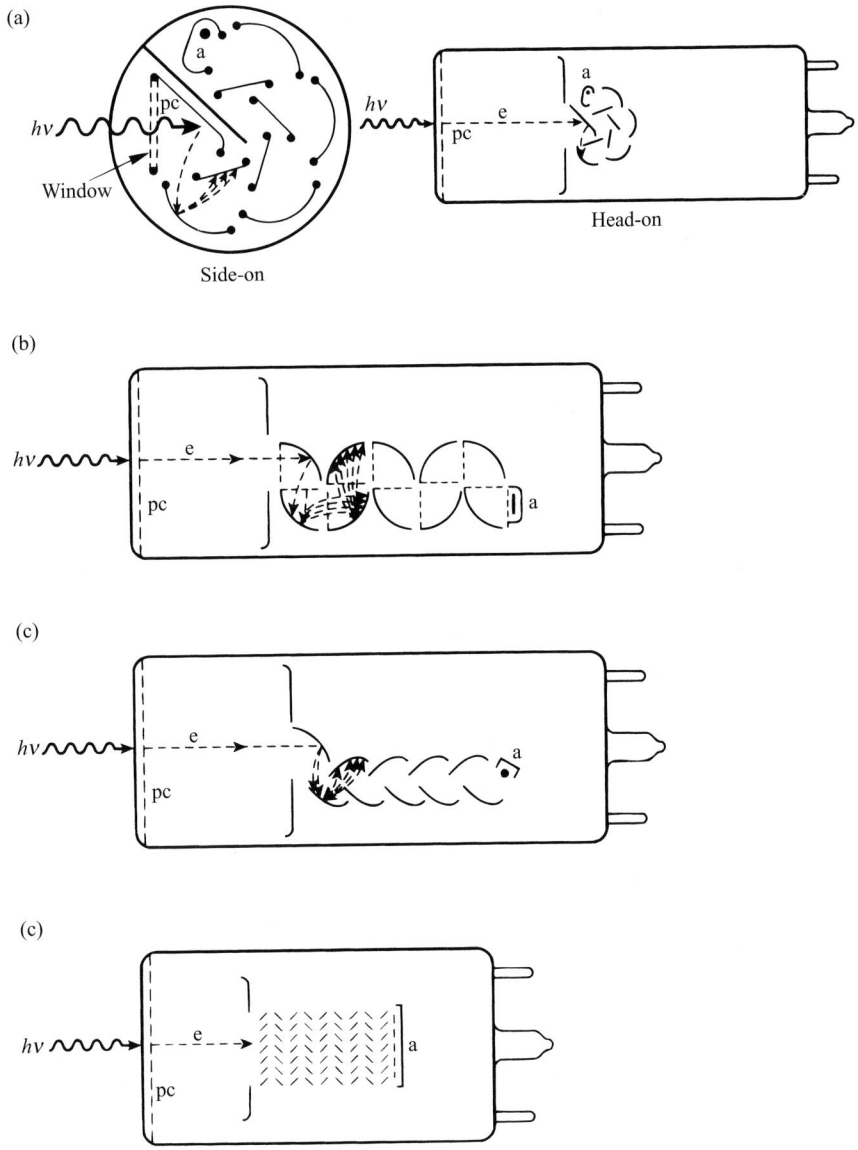


**Fig. 21.7** A schematic diagram of the internal structure of various types of photomultiplier tube: (a) a circular cage; (b) a box and a grid; (c) a focused dynode; and (d) a Venetian blind (a, anode; pc, photocathode).

*(c) Photocathode and dynode materials.* The performance of a photomultiplier depends not only on its internal structure, but also on the photoemissive material of its photocathode and on the secondary-electron-emitting material of its dynodes. The wavelength dependences of various commercially available photocathode materials are shown in Fig. 21.8; some radiant sensitivities and quantum efficiencies are given in Table 21.1. The short-wavelength cutoff of a given material depends on its work function. This cutoff is not sharp because, except at absolute zero, there are always a few electrons high up in the

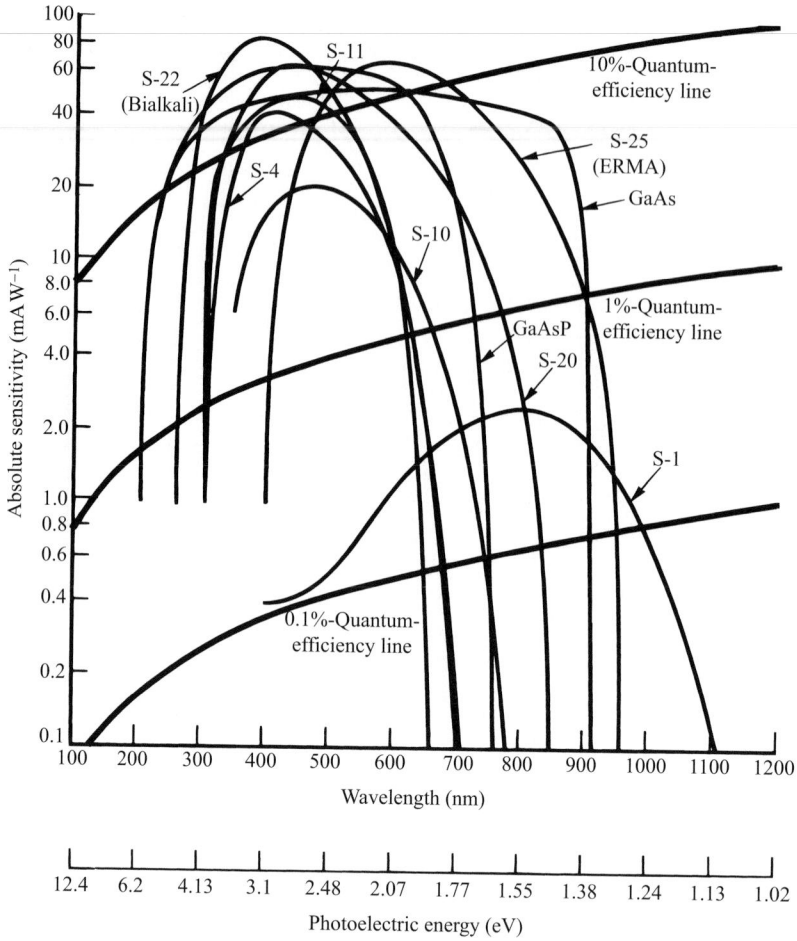


**Fig. 21.8** The wavelength dependence of the radiant sensitivity of several commercially available photocathode materials.

conduction band available for photoexcitation by low-energy photons. A few of these electrons, because of their thermal excitation, will be emitted without any photostimulation. This contributes the major part of the dark current observed from the photocathode. Materials that have low work functions, and are consequently more red- and infrared-sensitive, have higher – often much higher – dark currents than do materials that are optimized for visible and ultraviolet sensitivity.

Photocathodes are available both in opaque and in semitransparent forms, depending on the model of phototube. In the semitransparent form the photoelectrons are ejected from the thin photoemissive layer on the opposite side from the incident light. In both types the photocathode has to perform two important functions: it must absorb incident photons and allow the emitted photoelectron to escape. The latter event is inhibited if photons are absorbed too deep in a thick photoemissive layer, or if the photoelectron suffers energy loss from scattering in this layer. If the emitted photoelectron has too much energy, it can excite a further electron across the band gap. This pair-production phenomenon inhibits

**Table 21.1** Characteristics of photoemissive surfaces

| Cathode | Radiant sensitivity ($mA\ W^{-1}$) 515 nm | 694 nm | 1.06 μm | Peak quantum efficiency (%) | $\lambda_{peak}$ (μm) | Dark current[a] ($A\ cm^{-2}$) |
|---|---|---|---|---|---|---|
| S-1 Cs–O–Ag | 0.6 | 2 | 0.4 | 0.08 | 800 | $9 \times 10^{-13}$ |
| S-10 Cs–O–Ag–Bi[b] | 20 | 2.7 | 0 | 5 | 470 | $9 \times 10^{-16}$ |
| S-11 $Cs_3Sb$ on MnO[c] | 39 | 0.2 | 0 | 13 | 440 | $10^{-16}$ |
| S-20 (Cs)$Na_2KSb$ (trialkali) | 53 | 20 | 0 | 18 | 470 | $3 \times 10^{-16}$ |
| S-22 (bialkali) | 42 | 0 | 0 | 26 | 390 | $1.6 \times 10^{-18}$ |
| GaAs[b] | 48 | 28 | 0 | 14 | 560[e] | $10^{-16}$ |
| GaAsP[d] | 60 | 30 | 0 | 19 | 400[e] | $3 \times 10^{-15}$ |
| InGaAs[d] | – | – | 4.3 | – | –[e] | $3 \times 10^{-14}$ |
| InGaAsP[d] | – | – | – | 47 | 300[e] | $2.5 \times 10^{-13}$ |
| S-25 (ERMA)[f] | 53 | 26 | 0 | 25 | 430 | $10^{-15}$ |
| Cs–Te (solar blind) | – | – | – | 15 | 254 | $2.5 \times 10^{-17}$ |

Note that the table shows typical values, but these can vary greatly from one manufacturer to another.

[a] At room temperature.

[b] The cathode designated S-3 is similar.

[c] Several types of $Cs_3Sb$ photocathodes exist. The $Cs_3Sb$ is deposited onto different opaque and semitransparent substrates, and various window materials are used. These photocathodes have the designations S-4, S-5, S-13, S-17, and S-19, as well as S-11.

[d] Negative-electron-affinity (NEA) photoemitters.

[e] May show no wavelength of maximum quantum efficiency; the quantum efficiency falls with increasing wavelength. However, the exact spectral characteristics will depend on the thickness of the photoemitter and on whether it is used in transmission or not.

[f] Extended-red S-20.

the release of photoelectrons from the photoemissive layer, and accounts for the ultraviolet cutoff characteristics of the different materials shown in Fig. 21.8. With reference to Fig. 21.5, it can be shown that for pair production to occur the incident photon energy must be greater than $2E_g$. Photoelectrons have the best chance of escaping, and the photocathode has its highest quantum efficiency, in materials for which $\chi < E_g$.

Practical photoemissive materials fall into two main categories: classical photoemitters and negative-electron-affinity (NEA) materials. Classical photoemitters generally involve an alkali metal or metals, a group-V element such as phosphorus, arsenic, antimony, or bismuth, and sometimes silver and/or oxygen. Examples are the Ag–O–Cs (S-1) photoemitter, which has the highest quantum efficiency beyond about 800 nm of any classical photoemitter, and $Na_2KSbCs$, the so-called trialkali (S-20) cathode.

The NEA photoemitters utilize a photoconductive single-crystal semiconductor substrate with a very thin surface coating of cesium and usually a small amount of oxygen.

The cesium (oxide) layer lowers the electron affinity below the value it would have in the pure semiconductor, achieving an effectively negative value. Examples of such NEA photoemitters are GaAs (CsO) and InP (CsO). The NEA emitters can offer very high quantum efficiency and extended infrared response. GaAs (CsO), for example, has a higher quantum efficiency in the near-infrared than that of an S-1 photocathode.

The performance of the dynode material in photomultiplier tubes is specified in terms of the secondary-emission ratio $\delta$ as a function of energy. For a phototube with $n$ dynodes the gain is $\delta^n$. In the past, the commonest dynode materials were CsSb, AgMgO, and BeCuO. The last is also used as the primary photoemitter in windowless photomultipliers that are operated *in vacuo* for the detection of vacuum-ultraviolet radiation. BeCuO has the desirable property that it can be reactivated after exposure to air. The above materials have $\delta$ values of 3–4. Newer NEA dynode materials have much higher $\delta$ values: in particular, that of GaP can range up to 40 for an incident-electron input energy of 800 eV. With such high $\delta$ values a photomultiplier tube needs fewer dynodes for a given gain, which means a more compact and faster-response tube can be built. In very many commercial photomultipliers, the first dynode at least is now frequently made of GaP. This offers improved characterization of the single-photoelectron response of the tube, which is important in designing a system for optimum signal-to-noise ratio.

The accelerating voltages are supplied to the dynodes of a photomultiplier by a resistive voltage divider called a *dynode chain*. The relative resistance values in the chain determine the distribution of voltages applied to the dynodes. The total chain resistance $R$ determines the chain current at a given total photocathode–anode applied voltage. Many photomultiplier tubes have one or more focusing electrodes between the photocathode and the first dynode. The voltage on these electrodes can be adjusted to optimize the collection of photoelectrons from the photocathode [6].

The actual response-time behavior of the photomultiplier can be determined by observing its single-photoelectron response. This is done by looking at the anode pulses with a fast oscilloscope. The photocathode does not need to be illuminated for this to be done; sufficient noise pulses will usually be observed. The pulses should appear as in Fig. 21.6. These anode pulses reflect the time distribution and number of secondary electrons reaching the anode following single (or multiple) photoelectron emissions from the photocathode. If the height distribution of anode pulses is measured, a distribution such as that shown in Fig. 21.9(a) will probably be seen. The idealized distribution shown in Fig. 21.9(b) may be seen from newer tubes with GaP dynodes, which have a very high $\delta$ value.

### 21.4.2 Photoconductive detectors

Photoconductive detectors can operate through either intrinsic or extrinsic photoconductivity. The physics of intrinsic photoconductivity is illustrated in Fig. 21.10(a). Photons with energy $h\nu > E_g$ excite electrons across the band gap. The electron–hole pair that is thereby created for each photon absorbed leads to an increase in conductivity – which comes mostly from the electrons. Semiconductors with small band gaps respond to long-wavelength radiation but must be cooled accordingly, otherwise thermally excited electrons swamp any small photoconductivity effects. Table 21.2 lists commonly available intrinsic

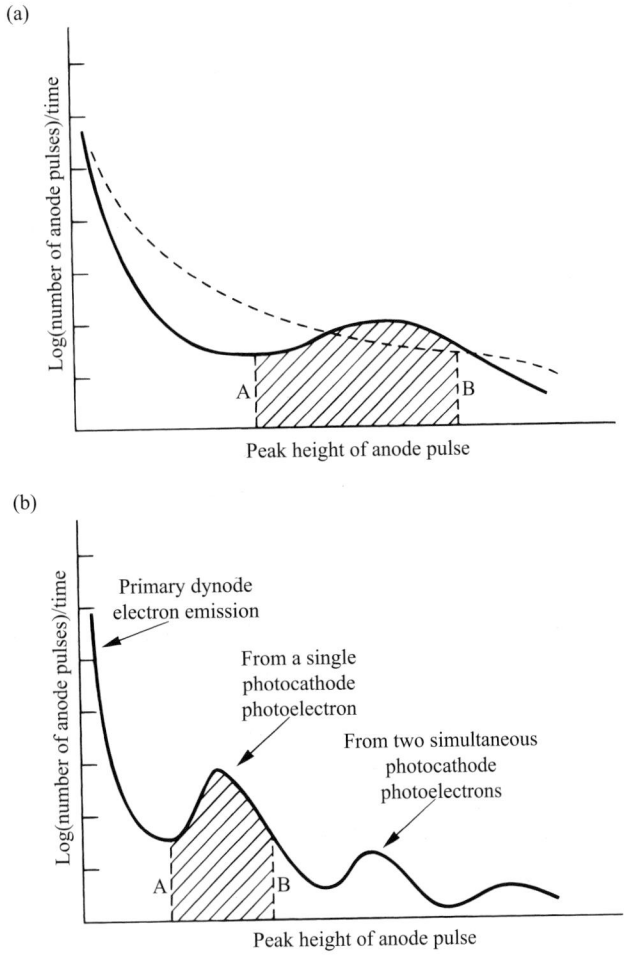


**Fig. 21.9** Photomultiplier anode pulse-height distributions: (a) two forms likely to be observed in practice; and (b) the idealized form from a tube with dynodes having a high and well-defined secondary emission coefficient. The best signal-to-noise ratio would be obtained in a photon-counting experiment by collecting only anode pulses in a height range roughly indicated by the shaded region AB.

photoconductive detectors together with their usual operating temperature and the limits of their long-wavelength response, $\lambda_0$, together with some representative figures for detectivities and time constants. Note that silicon and germanium are also operated both in photovoltaic and in avalanche modes.

If a semiconductor is doped with an appropriate material, impurity levels are produced between the valence and conduction bands as shown in Fig. 21.10(b). Impurity levels that are able to accept an electron excited from the conduction band are called *acceptor levels*, whereas impurity levels that can have an electron excited from them into the conduction band are called *donor levels*. Thus, in Fig. 21.10(b) photons with energy $h\nu > E_A$ excite an electron to the impurity level, leaving a hole in the valence band and thereby giving

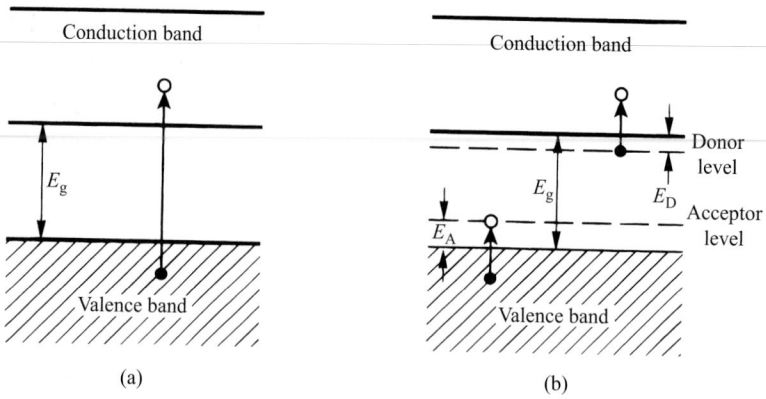


**Fig. 21.10** Mechanisms for (a) intrinsic and (b) extrinsic photoconductivity.

**Table 21.2** Intrinsic photoconductive detectors

| Semiconductor | $T$ (K) | $E_g$ (eV) | $\lambda_0$ (μm) | $D^*$(max) (cm Hz$^{1/2}$ W$^{-1}$) | $\tau$ |
|---|---|---|---|---|---|
| CdS | 295 | 2.4 | 0.52 | $3.5 \times 10^{14}$ | 50 ms |
| CdSe | 295 | 1.8 | 0.69 | $2.1 \times 10^{11}$ | ≃10 ms |
| Si | 295[a] | 1.12 | 1.1 | $\leq 2 \times 10^{12}$ | 50 ps |
| Ge | 295[a] | 0.67 | 1.8 | $\leq 10^{11}$ | 10 ns |
| PbS | 295 | 0.42 | 2.5 | $\leq 2 \times 10^{11}$ | 0.1–10 ms |
| | 195 | 0.35 | 3.0 | $\leq 4 \times 10^{11}$ | 0.1–10 ms |
| | 77 | 0.32 | 3.3 | $\leq 8 \times 10^{11}$ | 0.1–10 ms |
| PbSe | 295 | 0.25 | 4.2 | $(1–5) \times 10^{9}$ | 1 μs |
| | 195 | 0.22 | 5.4 | $(1.5–4) \times 10^{10}$ | 30–50 μs |
| | 77 | 0.21 | 5.8 | $\leq 3 \times 10^{10}$ | 50 μs |
| InSb[b] | 77 | ≃0.22 | 5.5–7.0 | $\leq 3 \times 10^{10}$ | 0.1–1 μs |
| $Hg_{0.8}Cd_{0.2}Te$ | 77 | ≃0.1 | 12–25 | $10^{9}–10^{11}$ | >1 ns |

[a] Increased sensitivity can be obtained by cooling.
[b] More commonly operated in a photovoltaic mode.

rise to p-type extrinsic photoconductivity. Photons with energy $h\nu > E_D$ will excite an electron into the conduction band, giving n-type extrinsic photoconductivity. For example, gold-doped germanium has an acceptor level 0.15 eV above the valence band and is an extrinsic p-type photoconductor, as is copper-doped germanium, which has an acceptor level 0.041 eV above the valence band. These are the two most commonly used extrinsic photoconductive detectors, responding out to about 9 μm and 30 μm, respectively. Curves showing the variation of their $D^*$ with wavelength are given in Fig. 21.11.

All photoconductive detectors, whether intrinsic or extrinsic, are operated in essentially the same way, although there are wide differences in packaging geometry. These differences arise from differing operating temperatures and speed-of-response considerations.

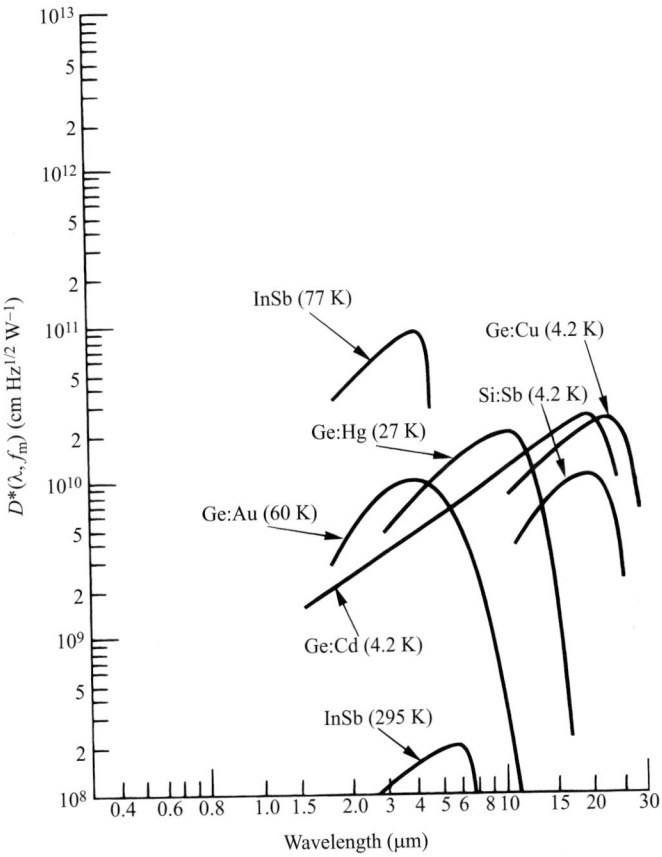


**Fig. 21.11** $D^*(\lambda)$ as a function of wavelength for various photoconductive detectors (courtesy of Hughes Aircraft Company).

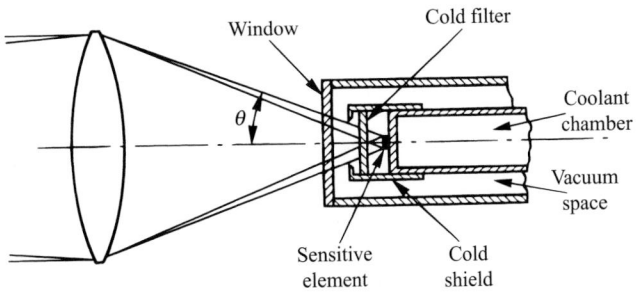


**Fig. 21.12** A radiation-shielded liquid-nitrogen-cooled photoconductive or photovoltaic infrared detector assembly.

A schematic diagram that shows the main construction features of a liquid-nitrogen-cooled photoconductive or photovoltaic detector is given in Fig. 21.12. Uncooled detectors can be of much simpler construction – for example, in a transistor or flat solar-cell package.

One feature of the cooled-detector design shown in Fig. 21.12 is worthy of note. The field of view of the detector is generally restricted by an aperture, which is kept at the

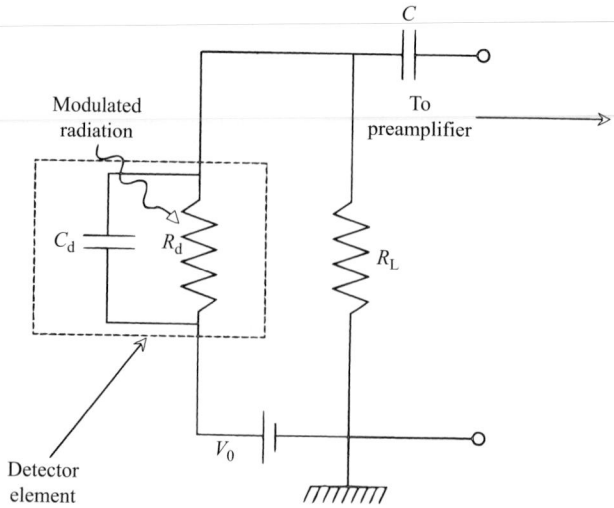


Fig. 21.13 A simple biasing circuit for operating a photoconductive detector with modulated radiation.

temperature of the detection element. This shields the detector from ambient infrared radiation, which peaks at 9.6 μm. For detection of low-level narrow-band infrared radiation the influence of background radiation can be further reduced by incorporating a cooled narrow-band filter in front of the detector element. The filter will radiate only beyond the cutoff wavelength of the detector, and it restricts transmitted ambient radiation to a narrow band.

Figure 21.13 shows a basic biasing circuit commonly used for operating photoconductive detectors. Here $R_d$ is the detector dark resistance. It is easy to see that the change in voltage, $\Delta V$, that appears across the load resistor, $R_L$, for a small change $\Delta R$ in the resistance of the detector is

$$\Delta V = -\frac{V_0 R_L \, \Delta R}{(R_d + R_L)^2}. \tag{21.42}$$

This is at a maximum when $R_L = R_d$. Thus it is common practice to bias the detector with a load resistance equal to the detector's dark resistance. The bias voltage is selected to give a bias current through the detector that gives optimum detectivity.

A few photoconductive detectors are worthy of brief extra comment. Silicon and germanium are much more commonly used for photodiodes, frequently in an avalanche mode. These devices are discussed in the next section. Lead sulfide detectors have high impedance, 0.5–100 MΩ, and slow response, but are sensitive detectors in the spectral region between 1.2 and 3 μm and can be used uncooled. The $D^*(\lambda)$ curves for these detectors and lead selenide are shown in Fig. 21.14.

Table 21.3 lists the characteristics of some extrinsic photoconductive detectors. Because of the small energy gaps involved in these detectors, they all operate at cryogenic temperatures. The use of extrinsic photoconductivity for the detection of far-infrared radiation requires the introduction of appropriate doping material into a semiconductor in order

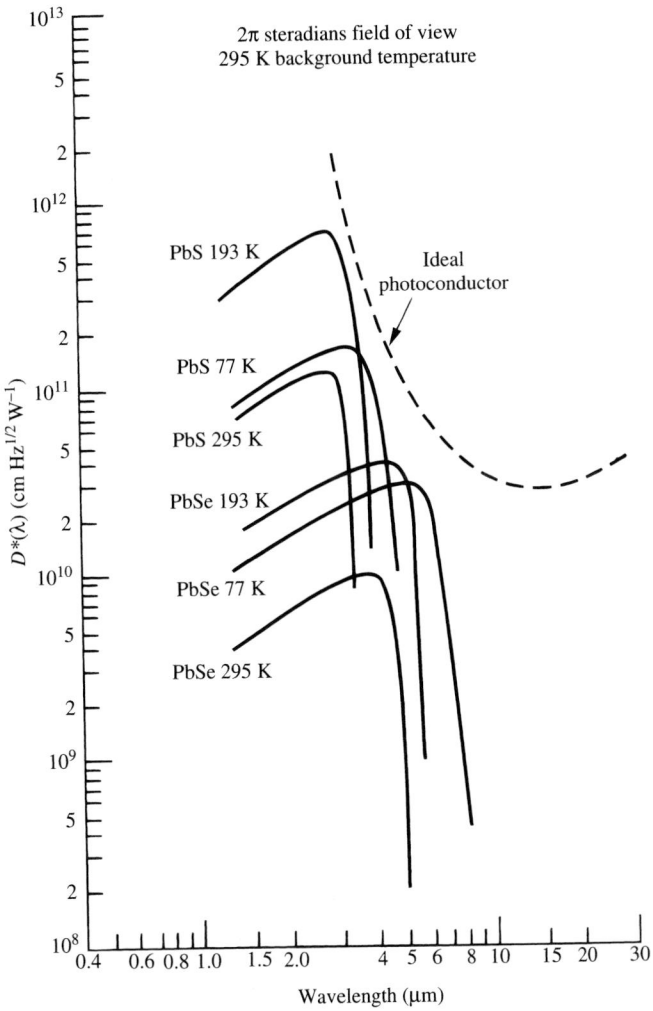


**Fig. 21.14** $D^*$ as a function of wavelength for various lead-salt photoconductive detectors (courtesy of Hughes Aircraft Company).

to generate an acceptor or donor impurity level that is extremely close to the valence or conduction band, respectively. Well-characterized impurity levels can be generated in germanium in this way using gallium, indium, boron, or beryllium doping, but the long-wavelength sensitivity limit is restricted to about 124 μm. Longer-wavelength-sensitive, extrinsic photoconductivity can be observed in appropriately doped InSb in a magnetic field. However, bulk InSb can be used more efficiently for infrared detection in a different photoconductive mode entirely [7–9]. Even at the low temperature at which far-infrared photoconductive detectors operate ($\leq$4 K), there are carriers in the conduction band. These free electrons can absorb far-infrared radiation efficiently and move into higher-energy states within the conduction band. This change in energy results in a change

**Table 21.3** Extrinsic photoconductive detectors

| Semiconductor | Impurity | $T$ (K) | $\lambda_0$ (μm) | $D^*$ | $\tau$ |
|---|---|---|---|---|---|
| Ge:Au[a] | p-type | 77 | 8.3 | $(3 \times 10^9)-10^{10}$ | 30 ns |
| Ge:Hg[b] | p-type | <28 | 14 | $(1-2) \times 10^{10}$ | >0.3 ns |
| Ge:Cd | p-type | <21 | 21 | $(2-3) \times 10^{10}$ | 10 ns |
| Ge:Cu[b] | p-type | <15 | 30 | $(1-3) \times 10^{10}$ | >0.4 ns |
| Ge:Zn[a,b] | p-type | <12 | 38 | $(1-2) \times 10^{10}$ | 10 ns |
| Ge:Ga | p-type | <3 | 115 | $2 \times 10^{10}$ | >1 μs |
| Ge:In | p-type | 4 | 111 | – | <1 μs |
| Si:Ga | p-type | 4 | 17 | $10^9-10^{10}$ | >1 μs |
| Si:As | n-type | <20 | 22 | $(1-3.5) \times 10^{10}$ | 0.1 μs |

[a] Sometimes also containing silicon.
[b] Sometimes also containing antimony.

of mobility of these free electrons, which can be detected as a change in conductivity. These *hot-carrier-effect* photoconductive detectors can be used successfully over a wavelength range extending from 50 to 10,000 μm; they have detectivities up to $2 \times 10^{12}$ cm $Hz^{1/2}$ $W^{-1}$ and response times down to 10 ns or less. They are frequently operated in a large magnetic field (several hundred kA $m^{-1}$ or more).

### 21.4.3 Photovoltaic detectors (photodiodes)

In a photovoltaic detector, photoexcitation of electron–hole pairs occurs near a junction when radiation of energy greater than the band gap is incident on the junction region. Extrinsic photoexcitation is rarely used in photovoltaic photodetectors. The internal energy barrier of the junction causes the electron and hole to separate, creating a potential difference across the junction. This effect is illustrated for a p–n junction in Fig. 21.1. Figure 21.15 shows how an actual p–n photodiode is constructed. Other types of structure are also used as illustrated in Fig. 21.16, such as p–i–n junctions, Schottky–barriers (a metal deposited onto a semiconductor surface), and heterojunctions. The p–n and p–i–n structures are the most commonly used. All these devices are commonly called photodiodes. The characteristics of some important photodiodes are listed in Table 21.4. Important photodiodes include silicon for detection of radiation between 0.1 and 1.1 μm and detectors based on the InGaAs(P) system for the region between 0.9 and 1.7 μm, which encompasses the important fiber-optical communication wavelengths of 1.3 μm and 1.55 μm (see Chapter 23). Typical spectral responsivities for some of the materials are shown in Fig. 21.17. Other photodiodes with more specialized applications include germanium between 0.4 and 1.8 μm, indium arsenide between 1 and 3.8 μm, indium antimonide between 1 and 7 μm, lead–tin telluride between 2 and 18 μm, and mercury–cadmium telluride between 1 and 12 μm. Some typical curves of $D^*(\lambda)$ for these photodiodes are shown in Fig. 21.18. These spectral-response regions are not all necessarily covered by a

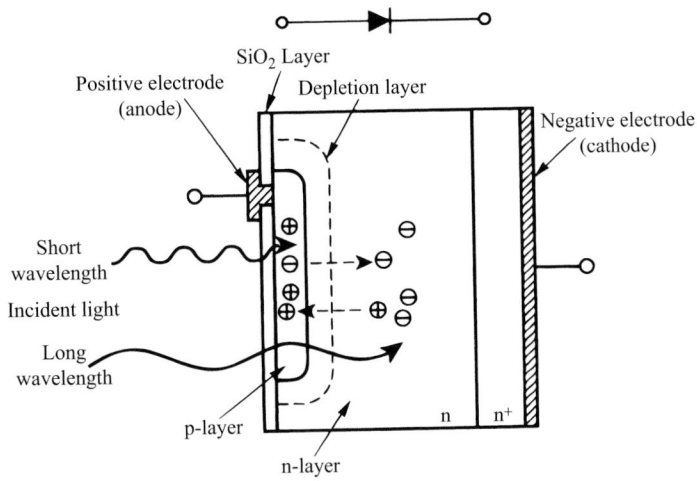


Fig. 21.15 Construction of a p–n junction photodiode (courtesy of Hamamatsu Corporation).

detector operating at the same temperature; for example, InSb responds to 7 μm at 300 K but to wavelengths no longer than 5.6 μm at 77 K. The wavelength response of PbSnTe and HgCdTe depends also on the stoichiometric composition of the crystal. All these photodiodes have very high quantum efficiency, defined in this case as the ratio of photons absorbed to mobile electron–hole pairs produced in the junction region. Values in excess of 90% have been observed in the case of silicon.

When a photodiode detector is illuminated with radiation of energy greater than the band gap, it will generate a voltage and can be operated in the very simple circuit illustrated in Fig. 21.19(a). However, it is much better to operate a photodiode detector in a reverse-biased mode, as shown in Fig. 21.19(b), where a positive voltage is applied to the n-type side of the junction and a negative voltage to the p-type side. In this case, the observed photosignal is seen as a change in current through the load resistor. The difference between the two modes of operation can easily be seen from Fig. 21.20(a), which shows the *I–V* characteristic of a photodiode in the dark and in the presence of illumination. In the forward and bias direction the light response of the photodiode is nonlinear. A photodiode responds much more linearly to changes in light intensity and has greater detectivity when operated in the reverse-biased mode. At increasing levels of reverse bias the photodiode first enters an *avalanche* region before it breaks down at sufficiently large reverse bias. Ideal operation is obtained when the diode is operated in the current mode with an operational amplifier that effectively holds the photodiode voltage at zero – its optimum bias point. A simple practical circuit that can be used to operate a photodiode in this way is shown in Fig. 21.20(b). This is called a *trans-impedance* amplifier circuit. In this circuit, the bias voltage $V_B$ is not necessary, but for many photodiodes it will improve the speed of response, albeit at the expense of an increase in noise. If the bias voltage is zero, then the photodiode is held at the short-circuit current point shown in Fig. 21.20(a). For DC operation, since no current can flow into an ideal op-amp, the output voltage will be $V_o = -I_{SC}R_F$, where $I_{SC}$ is the short-circuit photocurrent. The p–i–n structure is most

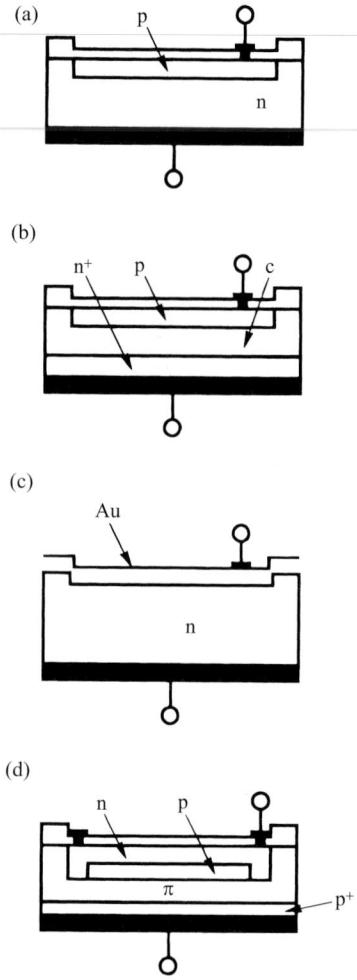


Fig. 21.16 Construction of various types of photodiode: (a) planar diffusion type; (b) p–i–n type; (c) Schottky type; and (d) avalanche type (courtesy of Hamamatsu Corporation).

commonly used in photodiodes because its performance, in terms of quantum efficiency (the number of useful carriers generated per photon absorbed) and frequency response, can be readily optimized. These devices have very low noise and fast response. In practice, the limiting sensitivity that can be obtained with them will be determined by the noise of the associated amplifier circuitry.

If the reverse-bias voltage on a photodiode is increased, photoinduced charge carriers can acquire sufficient energy traversing the junction region to produce additional electron–hole pairs. Such a photodiode exhibits current gain and is called an *avalanche photodiode* (APD). It is in some respects the solid-state analog of the photomultiplier. Avalanche photodiodes are noisier than p–i–n photodiodes, but because they have internal gain, the practical sensitivity that can be achieved with them is greater. Because of their importance p–i–n photodiodes and APDs are worthy of more detailed discussion.

**Table 21.4** Photovoltaic detectors (photodiodes)

| Semiconductor | $T$ (K) | Wavelength range (μm) | $D^*$(max) (or NEP)[b] | $\tau$ |
|---|---|---|---|---|
| Si | 300 | 0.2–1.1 | $\leq 2 \times 10^{13}$ | >6 ps |
| InGaAs[a] | 300 | 0.9–1.7 | $6 \times 10^{-14}$ (NEP, W Hz$^{-1/2}$) | 20 ps |
| GaAsP[a] | 300 | 0.3–0.76 | $2 \times 10^{-15}$ (NEP, W Hz$^{-1/2}$) | |
| Ge | 300 | 0.4–1.8 | $10^{11}$ | 0.3 ns |
| InAs | 300 | 1–3.8 | $\leq 4 \times 10^{9}$ | 5 ns–1 μs |
| InAs | 77 | 1–3.2 | $4 \times 10^{11}$ | 0.7 μs |
| InSb | 300 | 1–7 | $1.5 \times 10^{8}$ | 0.1 μs |
| InSb | 77 | 1–5.6 | $\leq 2 \times 10^{11}$ | >25 ns |
| PbSnTe | 77 | 2–18 | $\leq 10^{11}$ | 20 ns–1 μs |
| HgCdTe | 77 | 1–25 | $10^{9}$–$10^{11}$ | >1.6 ns |

[a] The precise performance depends on the stoichiometry. Quaternary versions of these detectors based on InGaAsP are also used.

[b] The $D^*$ or NEP of these detectors are typical values. The actual $S/N$ performance of these detectors in practical applications will depend on the inevitable additional noise added by amplification electronics [10].

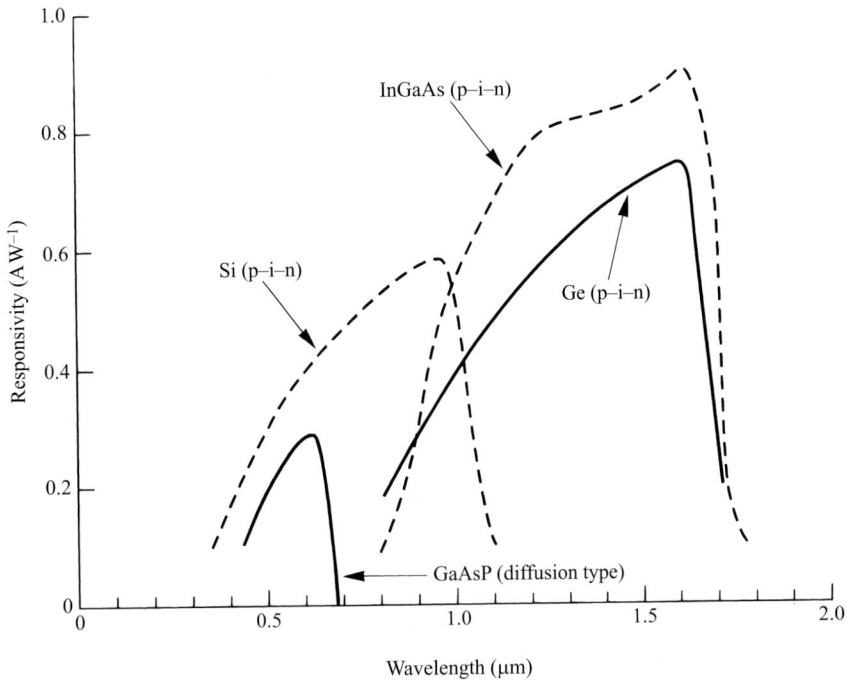


**Fig. 21.17** Spectral responsivities of some important near-infrared photodiodes.

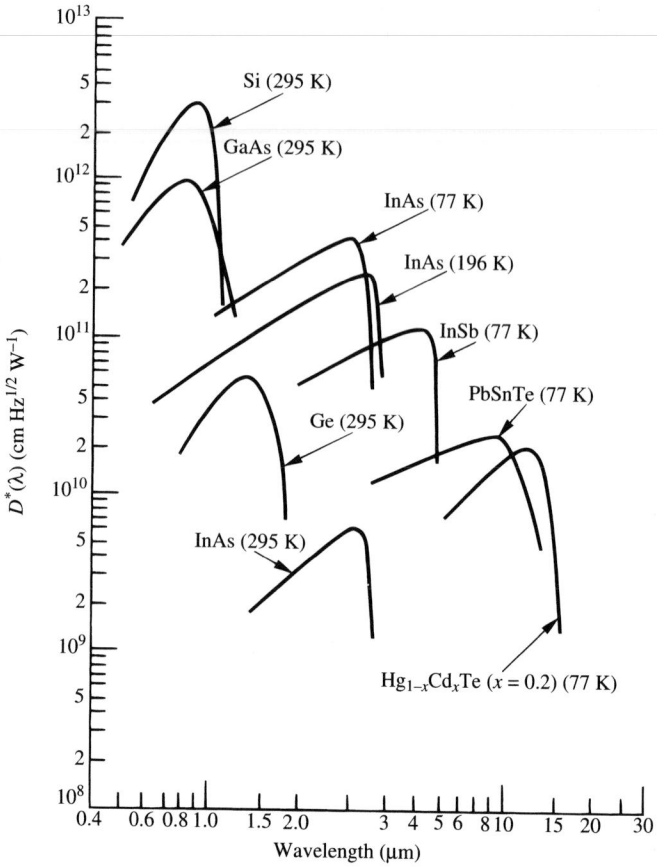


**Fig. 21.18** $D^*$ as a function of wavelength for various photovoltaic detectors (courtesy of Hughes Aircraft Company).

### 21.4.4 The p–i–n photodiodes

In a simple p–n-junction photodiode using the structure shown in Fig. 21.1 reverse bias leads to a current that increases linearly with incident optical power over up to nine orders of magnitude, say from 1 pW to 1 mW. In order for electron–hole pairs created by photon absorption to appear as useful current in the external circuit, they must be swept out of the junction region and collected at the electrodes before they have a chance to recombine. This is best accomplished if as much of the photon absorption as possible occurs in a thick depletion layer that is close to the electrodes, as shown in Fig. 21.21. Under reverse bias there are very few mobile charge carriers in the depleted i layer. There is a build-up of electrons on the heavily doped $p^+$ side of the device and of holes on the heavily doped $n^+$ side. Thus, the static electrical state is that of a parallel-plate capacitor of capacitance

$$C = \frac{\epsilon_0 \epsilon_r A}{d}. \tag{21.43}$$

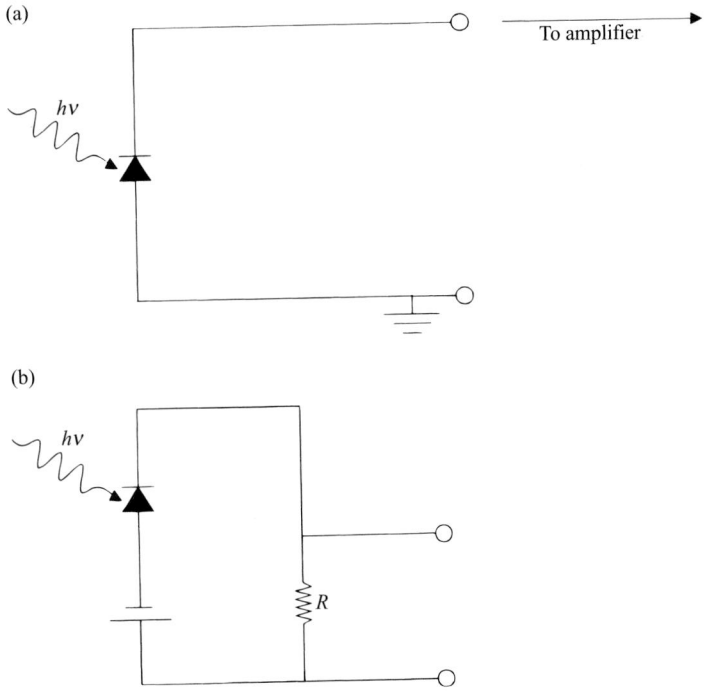


Fig. 21.19 A photovoltaic detector operated in (a) open-circuit and (b) reverse-biased mode.

For a typical device with $d = 20\ \mu\text{m}$, $A \simeq 10^{-8}\ \text{m}^2$, $\epsilon_r = 12$, and $C = 0.05$ pF. The intrinsic time constant of the diode driving a 50-Ω load is 2.7 ps. These are high-speed devices. There are very many variations in the detailed construction of p–i–n photodiodes [10–12]. It is possible to use heterostructures in these devices, for example, the $p^+$–i–$n^+$ layers can be GaAlAs/GaAs/GaAs or $(\text{InGaAsP})_1/(\text{InGaAsP})_2/\text{InP}$.[3] If the layer through which incident radiation enters has a larger band gap than that of the absorbing (intrinsic) layer, then long-wavelength photons will not be absorbed in the surface layer. As is common in these layered semiconductor structures (see also Chapter 12), additional highly doped layers may be included adjacent to contact (metal) electrodes. If a metal contact layer is placed on a layer that is not sufficiently heavily doped, a rectifying Schottky diode results. Such structures can actually be used as semiconductor detectors themselves, particularly for ultraviolet detection. A thin, transparent gold layer is placed on a substrate of GaAsP or GaP, as shown as the n-layer in Fig. 21.16(c).

### 21.4.5 Avalanche photodiodes

When the reverse bias of a photodiode is increased sufficiently, the internal electric field can accelerate photogenerated charge carriers to sufficient energy that they can excite additional electron–hole pairs. Both electrons and holes can contribute to this process, which is

[3] The subscripts indicate different stoichiometries. For example $(\text{InGaAsP})_1$ could be InP or InGaAs.

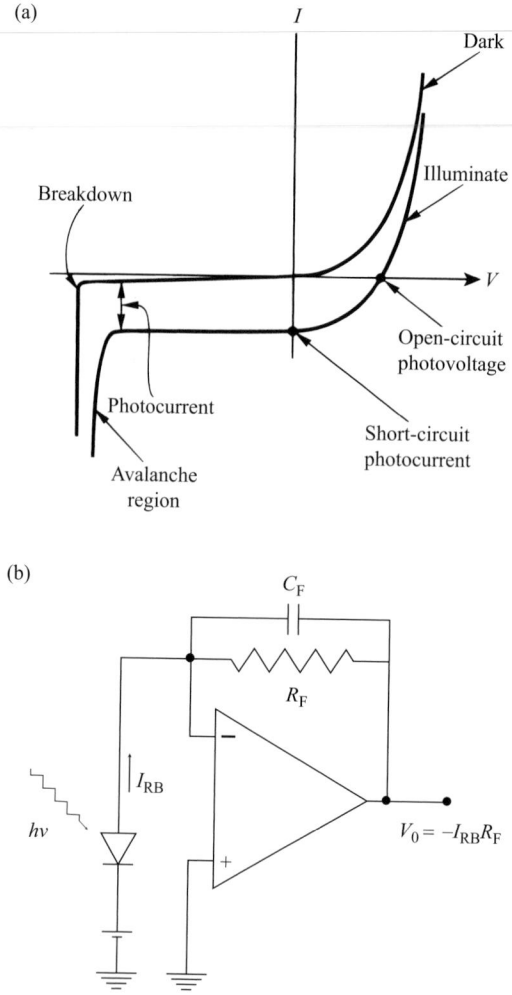


**Fig. 21.20** (a) Current–voltage characteristics of a photodiode; (b) current-mode-operated photodiode.

shown schematically for an electron in Fig. 21.22. The layout of a typical avalanche photodiode (APD) is shown schematically in Fig. 21.23. In this structure electron–hole pairs are created initially mostly in the lightly p-doped intrinsic layer (called a $\pi$ layer), because the $n^+$–p-junction region is very thin. There is sufficient voltage across the $\pi$ layer for photogenerated electrons and holes to drift rapidly across it. At the $n^+$–p junction near the positive electrode there is a large field gradient, and efficient avalanche multiplication occurs. The n-doped region at the edge of the p-doped region is called a *guard ring*. It prevents edge effects at the boundaries between the $n^+$, p, and $\pi$ layers. This keeps the electric field in the avalanche region uniform: there are no high fields at the edge where an avalanche could become destructive breakdown.

The performance of an APD is characterized by its multiplication factor $M$, which is the number of electron–hole pairs generated by the dominant carrier in the detector – electrons

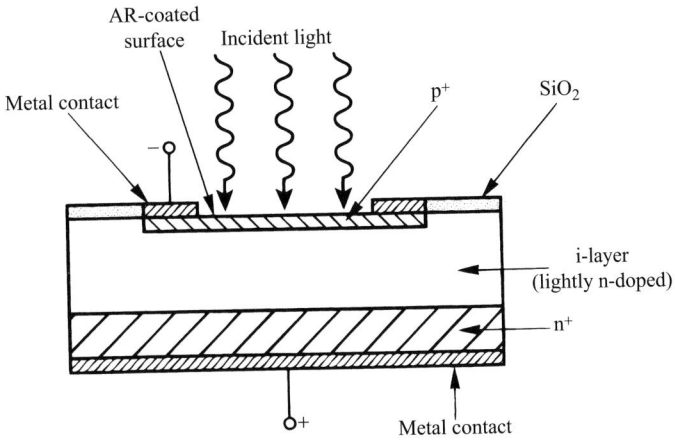


**Fig. 21.21** The construction of a p–i–n photodiode. The lightly n-doped intrinsic layer is frequently designated as a ν-layer.

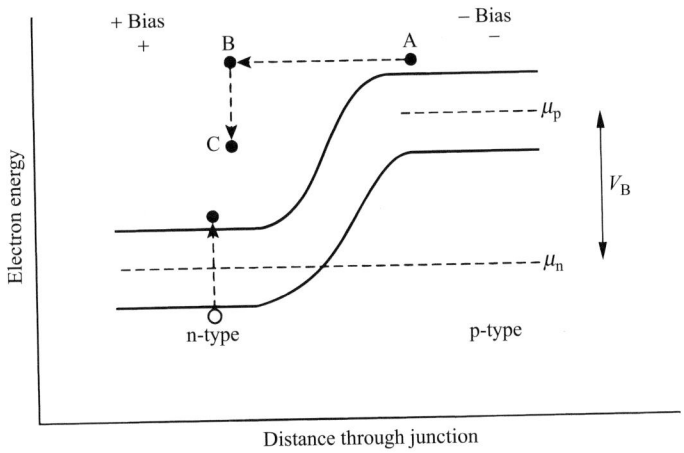


**Fig. 21.22** Illustration of avalanche electron–hole-pair production by an electron crossing a reverse-biased p–n junction. The Fermi energies are separated by the bias voltage $V_B$ across the junction. An electron crossing the junction from A to B has sufficient energy to excite an electron–hole pair and simultaneously fall to the lower energy at C.

in silicon, holes in InGaAs/InP devices. In the latter we see once again the use of a heterostructure, which allows control of where incident radiation is absorbed. Incident radiation can pass through the larger-band-gap InP and then be absorbed in the InGaAs. It is important both in APDs and in p–i–n photodiodes that, for the highest response speed, electron–hole pairs be produced in a fully depleted region of the device. If the charge carriers are generated in a region where electrons or holes are present in significant concentration, the motion of the photogenerated carriers is slowed by diffusion through these undepleted regions. The fastest response is obtained if the photogenerated carriers are pulled out of the depleted region by the bias field at the maximum possible speed. This occurs at the so-called *saturation velocity*, $v_s$, which is typically on the order of $10^5$ m s$^{-1}$.

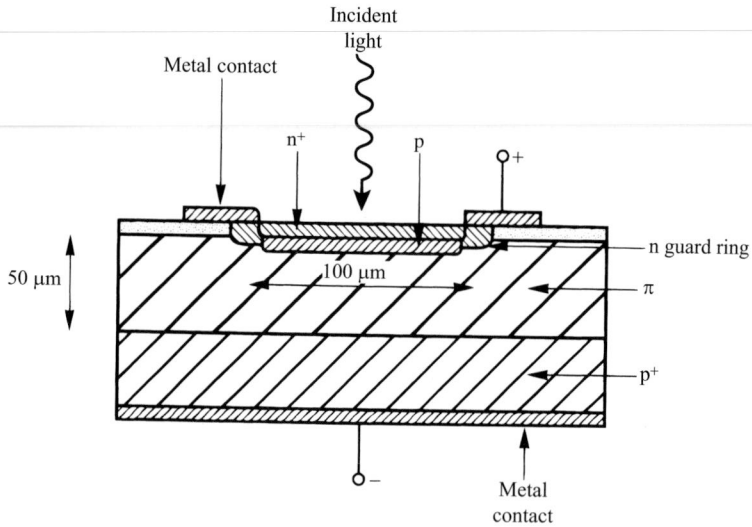


**Fig. 21.23** The construction of a planar $n^+-p-\pi-p^+$ APD with a guard ring, showing typical dimensions.

To achieve saturation velocity requires an applied field in the depleted region on the order of 1 MV $m^{-1}$. In a 50-μm-thick depleted π layer, as shown in Fig. 21.23, this requires a reverse bias of 50 V. A rough estimate for the speed of response will then be 0.5 ns. Of course, to determine the precise response of the detector, it would be necessary to include the actual spatial distribution of photogenerated carriers in the device [10]. The lateral dimensions of the device must also be kept small so as to minimize effects of the capacitance on the speed of response.

### 21.4.6 Geiger-mode avalanche photodetectors

If an APD with a very low dark current is operated in the *Geiger* mode [13–16], where the reverse bias voltage $V_R$ is set greater than the diode breakdown voltage $V_{BR}$, then a single photon can trigger a substantial circuit pulse. In other words, the current or voltage pulse produced from the detection of a single photon is above the noise level. The internal gain of the APD can reach $10^5-10^6$. An avalanche does not generally occur unless an actual photon is detected. The avalanche process generates a significant current pulse with peak current ~1 mA. When the gating voltage applied to the APD is removed, the avalanche is quenched because the APD is now below breakdown, and the current pulse terminates. To prevent the diode from suffering permanent damage it must be operated with a "quenching" circuit [17], which reduces the bias voltage to stop the avalanche process. Otherwise the diode will conduct a large current when it is biased above the breakdown voltage. It is desirable to reduce the dark current by cooling the APD. Avalanche photodetectors operated in this way are true "single-photon" detectors, and can replace PMTs in many photon-counting applications. Photodiodes using silcon, InGaAs/InP, InGaAs/Si, and HgCdTe have all been used successfully in this so-called Geiger mode. Dark counts below 1 count per second have been reported.

## 21.5 Thermal detectors

Thermal detectors, in principle, have a detectivity that is independent of wavelength from the vacuum-ultraviolet range upward as shown in Fig. 21.24, which also shows representative $D^*(\lambda)$ values for various types. In practice, the absorbing properties of the "black" surface of the detector will, in general, show some wavelength dependence, and the necessity for a protective window on some detector elements may limit the useful spectral bandwidth of the device. Most of the commonly available thermal detectors, although by no means as sensitive as various types of photodetectors, achieve spectral responses very far into the infrared – to the microwave region, in fact – while conveniently operating

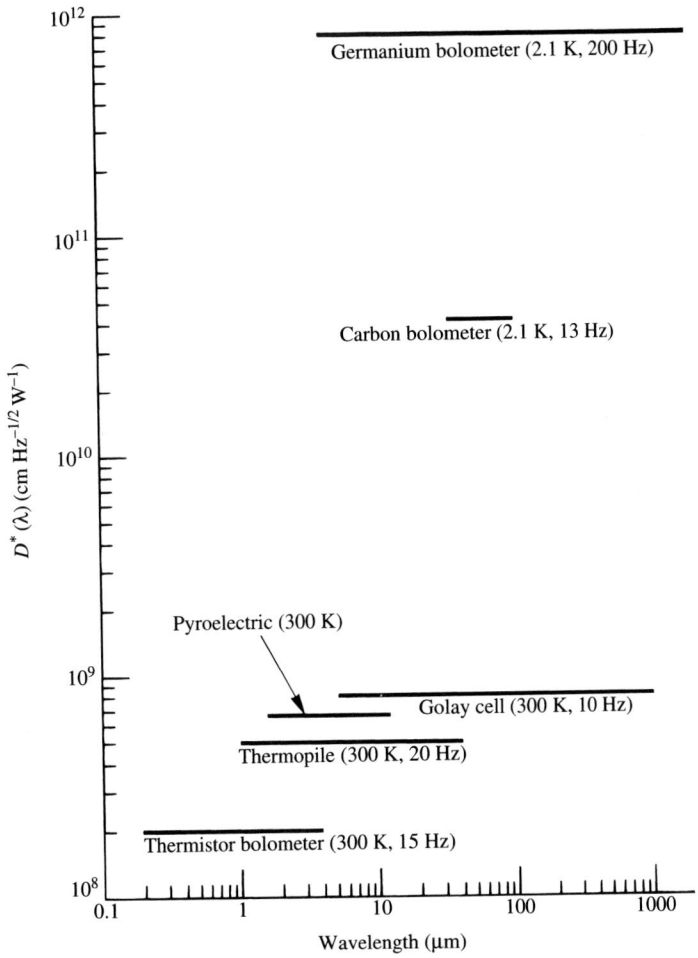


**Fig. 21.24** Representative $D^*(\lambda)$ response curves for various thermal detectors assuming total absorption of incident radiation. The operating temperature, modulation frequency, and typical useful wavelength range are shown for each detector. In each case the detector is assumed to view a hemispherical enclosure at a temperature of 300 K.

at room temperature. Each of these detectors is discussed briefly below. Putley [8] gives a more detailed discussion.

*(a) Thermopiles.* Thermopiles, although they are among the earliest forms of infrared detector, are still widely used. Their operation is based on the Seebeck effect, whereby heating the junction between two dissimilar conductors generates a potential difference across the junction. An ideal device should have a large Seebeck coefficient, low resistance (to minimize ohmic heating), and a low thermal conductivity (to minimize heat loss between the hot and cold junctions of the thermopile). These devices are usually operated with an equal number of hot (irradiated) and cold (dark) junctions, the latter serving as a reference to compensate for drifts in ambient temperature. Both metal (copper–constantan, bismuth–silver, antimony–bismuth) and semiconductor junctions are used as the active elements. The junctions can take the form of evaporated films, which improves the robustness of the devices and reduces their time constant, although this is still slow (0.1 ms at best). Because a thermopile has very low output impedance, it must be used with a specially designed low-noise amplifier, or with a step-up transformer [6].

*(b) Pyroelectric detectors.* These are detectors that utilize the change in surface charge that results when certain asymmetric crystals (ones that can possess an internal electric dipole moment) are heated. The crystalline material is fabricated as the dielectric in a small capacitor, and the change in charge is measured when the element is irradiated. Thus, these devices are inherently AC detectors. If the chopping frequency of the input radiation is slow compared with the thermal relaxation time of the crystal, the crystal remains close to thermal equilibrium and the current response is small. When the chopping period becomes shorter than the thermal relaxation time, much greater heating and current response result. The responsivity of the detector in this case can be written as

$$R = \frac{p(T)}{\rho C_p d} \quad (\mathrm{A\ W^{-1}}), \tag{21.44}$$

where $p(T)$ is the pyroelectric coefficient at temperature $T$, $d$ is the spacing of the capacitor electrodes, and $\rho$ and $C_p$ are the density and specific heat of the crystal, respectively.

The equivalent circuit of a pyroelectric detector is a current source in parallel with a capacitance, which can range from a few to several hundred picofarads. For optimum performance, the resultant high impedance must be matched to a high-input-impedance, low-output-impedance amplifier. These detectors can have response times as short as 2 ps. Their detectivities are comparable to those of thermopiles, and they also have a flat spectral response. Consequently, they can replace the thermopile in many applications as a convenient, room-temperature, wide-spectral-sensitivity detector of infrared and visible light.

*(c) Bolometers.* The resistance of a solid changes with temperature according to a relation of the form

$$R(T) = R_0[1 + \gamma(T - T_0)], \tag{21.45}$$

where $\gamma$ is the temperature coefficient of resistance, which is typically about 0.05 $\mathrm{K^{-1}}$ for a metal, and $R_0$ is the resistance at temperature $T_0$.

A *bolometer* is constructed from a material with a large temperature coefficient of resistance. Absorbed radiation heats the bolometer element and changes its resistance. Bolometers utilize metal, semiconductor, or almost-superconducting elements. Metal bolometers utilize fine wires (platinum or nickel) or metal films. The mass of the element must be kept small in order to maximize its temperature rise. Even so, the response time is fairly long ($\geq$1 ms). Semiconductor bolometer elements (thermistors) have larger absolute values of $\gamma$ and have largely replaced metals, except where very-long-term stability is required.

It is usual to operate bolometer elements in pairs in a bridge circuit, as shown in Fig. 21.25. One element is irradiated, while the second serves as a reference and compensates for changes in ambient temperature. Thermistors have a negative $I$–$V$ characteristic above a certain current and will exhibit destructive thermal runaway unless operated with a bias resistor. It is therefore usually best to operate the thermistor at currents below the negative-resistance part of its $I$–$V$ characteristic.

*(d) The Golay cell.* In a Golay cell (named for its inventor, M. J. E. Golay) [18], radiation is absorbed by a metal film that forms one side of a small sealed chamber containing xenon (which is used because of its low thermal conductivity). Another wall of the chamber is a flexible membrane, which moves as the xenon is heated. The motion of the membrane is used to change the amount of light reflected to a photodetector. The operating principle and essential design features of a modern Golay cell are shown in Fig. 21.26. Although these detectors are fragile, they are very sensitive and are still widely used for far-infrared spectroscopy.

## 21.6 Detection limits for optical detector systems

In a practical detection system the detector itself is coupled to various electronic devices such as amplifiers, filters, pulse counters, limiters, discriminators, and phase-locked loops. It is beyond our scope here to deal with all the consequent realities of how the optical detection limit is influenced by the additional noise contributions of these devices. Nor can we deal in detail with how various ways of encoding information onto a light signal can be used to enhance the signal-to-noise ratio of the overall detection system. For additional details the reader should consult the more specialized literature [10, 19–23]. However, in a well designed optical detection system the performance should be primarily limited only by the characteristics of the detector itself. Therefore, in our discussion of fundamental detector limits we will deal solely with the unavoidable noise sources of the detector and its associated resistors, as shown schematically in Fig. 21.3.

### 21.6.1 Noise in photomultipliers

Noise in photomultipliers comes from several sources.

(a) Thermionic emission from the photocathode.
(b) Thermionic emission from the dynodes.

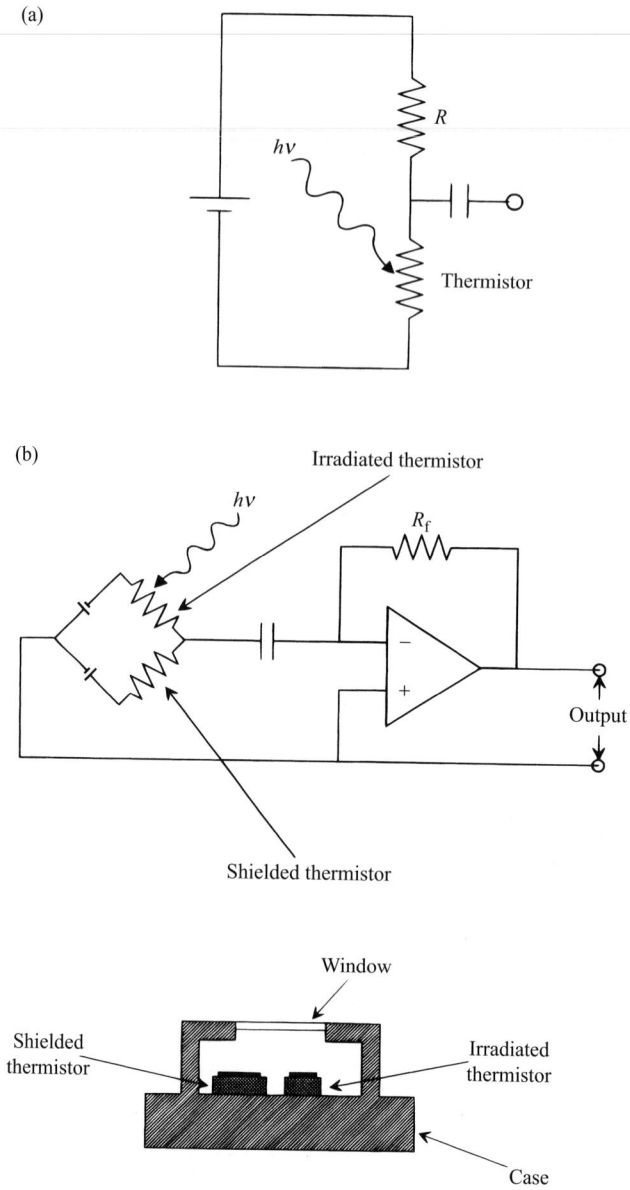


**Fig. 21.25** A bridge operating circuit for thermistor bolometers using a compensating shielded thermistor, with the device construction shown.

(c) Field emission from the dynodes (and from the photocathode) at high interdynode voltages. In this phenomenon the potential gradient near the emitting surface is sufficiently great to liberate electrons from the material.
(d) Radioactive materials in the tube envelope, for example $^{40}K$ in glass.
(e) Electrons striking the tube envelope and causing fluorescence.
(f) Electrons striking the dynodes and causing fluorescence.

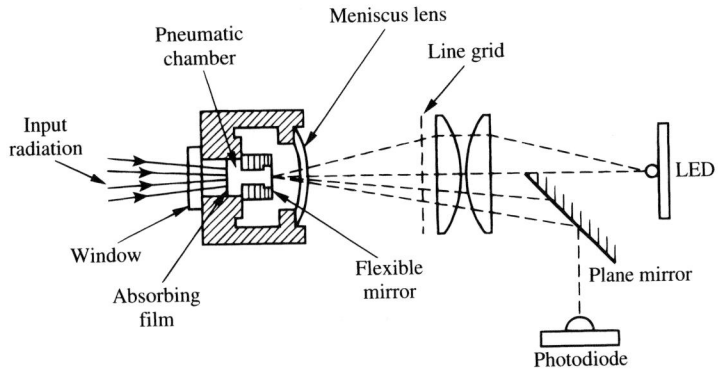


**Fig. 21.26** The design of the Golay detector. The top half of the line grid is illuminated by the LED and imaged back on the lower half of the grid by the flexible mirror and meniscus lens. Any radiation-induced deformation of the flexible mirror moves the image of the line grid and changes the illumination reaching the photodiode.

(g) Electrons colliding with residual atoms of vapor in the tube, cesium, for example, and causing fluorescence.
(h) Cosmic rays.

Of these sources, thermionic emission from the photocathode is generally the most important. It increases with the area of the photocathode. Thermionic emission from the dynodes leads to smaller anode pulses than for electrons originating at the photocathode. Therefore, when a photomultiplier is used in *photon counting*, where individual anode pulses are counted, discrimination against some noise can be effected by counting only within a specific range of anode pulse lengths, as shown in Fig. 21.9. Small anode pulses are likely to originate with noise source (b). Large anode pulses are likely to originate with noise sources (c)–(h).

## 21.6.2 Photon counting

Photomultipliers are *extremely* sensitive detectors of ultraviolet, visible, and near-infrared radiation. A photomultiplier has the ability to provide a macroscopic signal output from a single photoelectron liberated at its photocathode. In photon counting the anode pulses are counted individually, and the pulse-count rate in the case of illumination is compared with the dark count rate. The signal-to-noise ratio of this process can be further enhanced if the arrival time of signal photons can be isolated to a time window of width $\tau$, for example by synchronizing the pulse-counting electronics to the excitation of the phenomenon leading to the light.

Let us suppose that the (weak) light source to be detected emits $\overline{N}_s/\eta$ photons per second that are absorbed in the photocathode. Consequently, the average number of signal anode photoelectron counts per second is $\overline{N}_s$. These signal counts can be assumed to occur randomly within this time period.[4] The average number of signal counts in a sequence of time intervals each of duration $\tau$ is $\overline{N}_s\tau$, with a variance

[4] For a more detailed discussion of this point see Chapter 22.

$$\overline{(N_s - \overline{N_s})^2}\tau = \overline{N_s}\tau. \tag{21.46}$$

This can be interpreted as an associated shot noise $N_{SN} = \sqrt{\overline{N_s}}$.

If there are zero dark counts, the signal-to-noise ratio for determining $\overline{N_s}$ for observations during a time interval $\tau$ is

$$\frac{S}{N} = \frac{\overline{N_s}\tau}{\sqrt{\overline{N_s}\tau}}. \tag{21.47}$$

Therefore, by counting for a long time, a high accuracy can be obtained: for $\overline{N_s}\tau = 10^6$ counts the signal-to-noise ratio is 1,000.

If in addition there are dark counts that are detected randomly at a rate of $\overline{N_d}$ per second, then the variance in the number of dark counts in a time interval $\tau$ over a sequence of time intervals $\tau$ is

$$\overline{(N_d - \overline{N_d})^2}\tau = \overline{N_d}\tau. \tag{21.48}$$

The signal-to-noise ratio in this case for observations over a time interval $\tau$ is

$$\frac{S}{N} = \frac{\overline{N_s}\tau}{\sqrt{\overline{N_s} + \overline{N_d}}}. \tag{21.49}$$

An estimate of the limiting sensitivity of photon counting can be obtained by setting the signal-to-noise ratio to unity and choosing $\tau = 1$. The minimum sensitivity is then

$$(\overline{N_s})_{\min} = \frac{1}{2}\left[1 + \sqrt{1 + 4\overline{N_d}}\right], \tag{21.50}$$

and, since generally $\overline{N_s} \gg 1$,

$$(\overline{N_s})_{\min} = \sqrt{\overline{N_d}}. \tag{21.51}$$

So, for example, with a 500-nm source and a photomultiplier with a quantum efficiency of 20% and $10^2$ dark counts per second, the minimum detectable optical power for an observation lasting 1 s is

$$P_{\min} = 50h\nu = 2 \times 10^{-17}\ \text{W}. \tag{21.52}$$

Clearly the signal-to-noise ratio improves with integration of counts over long periods. It is entirely possible to detect signal counts that are much smaller than the noise by integrating for a sufficiently long time. For example, suppose that the dark count rate is 10 per second and the signal count rate is 1 per second. To obtain a signal-to-noise ratio of 10 requires $10 = \tau/\sqrt{\tau + 10\tau}$, which gives $\tau$ =1,100 s.

### 21.6.3 The signal-to-noise ratio in direct detection

In many applications of photomultiplier tubes the anode pulses are integrated to give a fluctuating analog current. The shot noise originating at the photocathode is, from Eq. (21.20),

$$\langle i_{\mathrm{N}}^2 \rangle_{\mathrm{c}} = 2e(\bar{i}_{\mathrm{c}} + \bar{i}_{\mathrm{d}})\Delta f, \tag{21.53}$$

where $\bar{i}_{\mathrm{c}}$ is the average photocathode current produced by a light source and $\bar{i}_{\mathrm{d}}$ is the average photocathode dark current. This noise is multiplied by the amplification of the electron number by interaction with the $N$ dynodes of the tube. If each dynode has a secondary emission multiplication efficiency of $\delta$ then the overall gain of the tube is[5]

$$G = \delta^N. \tag{21.54}$$

Because of statistical fluctuations in the secondary emission process and, in addition, because electrons can originate thermionically from the dynodes, the noise at the anode is further increased by a noise factor $F$. The overall noise appearing at the anode is

$$\langle i_{\mathrm{N}}^2 \rangle_{\mathrm{A}} = 2eG^2F(\bar{i}_{\mathrm{c}} + \bar{i}_{\mathrm{d}})\Delta f. \tag{21.55}$$

For $\delta = 4$ and a 14-stage tube, $G = 2.6 \times 10^8$, and typically $F \simeq \delta(\delta - 1) = 1.08$.

A photomultiplier is a current source: to convert this current into a detected voltage, the amplified photoelectron current is passed through an anode resistor of resistance $R$. This anode resistor may be part of the photomultiplier circuit, or may be provided in whole or in part by the input impedance of a following amplifier stage. The Johnson noise from the resistor is

$$\langle i_{\mathrm{N}}^2 \rangle_{\mathrm{R}} = \frac{4k_{\mathrm{B}}T\,\Delta f}{R}. \tag{21.56}$$

To optimize detection of a signal it is common practice to amplitude modulate the signal at some angular frequency $\omega_{\mathrm{m}}$ so as to permit synchronous detection at this frequency [6]. The optical power reaching the photocathode can be represented in this case as

$$P = P_0\,[1 + m\sin(\omega_{\mathrm{m}}t)]\,, \tag{21.57}$$

where $m$ is a modulation parameter.

The average photocathode current is

$$\bar{i}_{\mathrm{c}} = \frac{e\eta P_0}{h\nu}, \tag{21.58}$$

and

$$i_{\mathrm{c}}(t) = \bar{i}_{\mathrm{c}}\,[1 + m\sin(\omega_{\mathrm{m}}t)]\,. \tag{21.59}$$

At the anode the time-varying part of this current is

$$i_{\mathrm{s}}(t) = \bar{i}_{\mathrm{c}}Gm\sin(\omega_{\mathrm{m}}t). \tag{21.60}$$

[5] The first dynode often has a larger $\delta$ value than do the others, but we will assume that $\delta$ is an average value.

The signal-to-noise ratio ($S/N$) at the input to the electronics in Fig. 21.3 is therefore

$$\frac{\langle i_s^2 \rangle}{\langle i_N^2 \rangle_A + \langle i_N^2 \rangle_R} = \frac{\bar{i}_c^2 G^2 m^2/2}{2eG^2 F(\bar{i}_c + \bar{i}_d)\Delta f + 4k_B T \Delta f/R}. \tag{21.61}$$

The noise from the photomultiplier tube is usually sufficiently large that the Johnson noise can be neglected. If it is assumed that $\bar{i}_d \gg \bar{i}_c$, then for $m = 1$ and $S/N = 1$ we have, from Eqs. (21.58) and (21.61),

$$(P_0)_{\min} = \frac{2h\nu}{\eta}\sqrt{\frac{F\bar{i}_d\,\Delta f}{e}}. \tag{21.62}$$

**Example** Typical values for a good photomultiplier will be $\eta = 0.2, F \simeq 1$, and $\bar{i}_d \simeq 10^{-15}$ A. For a 530-nm source Eq. (21.62) gives $(P_0)_{\min} = 3 \times 10^{-16}$ W.

### 21.6.4 Direct detection with p–i–n photodiodes

The shot noise from a p–i–n photodiode is

$$\langle i_N^2 \rangle_1 = 2e(\bar{i}_s + i_d)\Delta f, \tag{21.63}$$

where $\bar{i}_s$ is the average signal current and $i_d$ is the dark current (which is usually specified for a low-noise photodiode in units of nA $\text{Hz}^{-1/2}$ or pA $\text{Hz}^{-1/2}$).

The average signal current is, in a similar way to Eq. (21.58),

$$\bar{i}_s = \frac{e\eta P_0}{h\nu}, \tag{21.64}$$

where $\eta$ is the quantum efficiency, which is much larger than for a photomultiplier: values of 0.7–0.8 are common. For the simple equivalent current discussed previously (Fig. 21.3) there is an additional Johnson-noise contribution of magnitude

$$(i_N^2)_2 = \frac{4k_B T \Delta f}{R}. \tag{21.65}$$

The overall signal-to-noise ratio for direct detection of an unmodulated signal is

$$\frac{\langle i_s^2 \rangle}{\langle i_N^2 \rangle_1 + \langle i_N^2 \rangle_2} = \frac{[e\eta P_0/(h\nu)]^2 R}{2eR[e\eta P_0/(h\nu) + i_d]\Delta f + 4k_B T \Delta f}. \tag{21.66}$$

This $S/N$ ratio would be reduced by a factor of $m^2/2$ for a modulated signal (cf. Eq. (21.61)).

If shot noise dominates over the dark current and the Johnson noise, then the shot-noise-limited signal-to-noise ratio is

$$\frac{S}{N} = \frac{\eta P_0}{2h\nu\,\Delta f}. \tag{21.67}$$

In a practical application using a photodiode there will be additional stages of electronic amplification that add noise. It is common to characterize the effect of an electronic circuit on the noise by its *noise figure*, $F_N$.

The noise figure can be defined conveniently as

$$F_N = \frac{\text{noise power at output of circuit}}{\text{amplified noise power at the input}}. \tag{21.68}$$

For input Johnson noise the mean-square, amplified, output noise current is

$$\langle i_N^2 \rangle_1 = \frac{4k_B TG\,\Delta f}{R}, \tag{21.69}$$

where $G$ is the power gain of the amplifier within the frequency band being considered. The actual output mean-square noise current is

$$\langle i_N^2 \rangle_2 = \frac{4k_B TGF_N\,\Delta f}{R}. \tag{21.70}$$

It is as if the amplifier were noiseless but the input Johnson noise is increased by the noise figure.

The noise figure is frequently quoted in decibels; a noise figure of 3 dB would represent a doubling of the output noise over the value expected from a noiseless amplifier. The term *noise temperature*, $T_i$, is also used. This is defined by

$$F_N = 1 + \frac{T_i}{T_{\text{ambient}}}. \tag{21.71}$$

**Example** We illuminate an InGaAs photodiode with a responsivity of 0.8 A $W^{-1}$ at 1.3 μm in an optical communication link in which there is 30 dB of loss between a 10-mW source and a receiver. The system bandwidth is 100 MHz, and the dark current is 5 nA (equivalent to 0.5 pA $Hz^{-1/2}$). We assume that the amplification electronics has a noise figure of 6 dB (relative to a 50-Ω input). We note the following:

received power $P_0 = 10^{-3} \times 10$ mW = 10 μW;
signal current $\bar{i}_s = 8$ μA;
dark current $\bar{i}_d = 5$ nA (dominated by $\bar{i}_s$);
signal power = $(\bar{i}_s)^2 R = 3.2$ nW;
shot-noise power = $2e(\bar{i}_s + \bar{i}_d)\Delta f R = 1.28 \times 10^{-14}$ W;
Johnson-noise power = $4k_B T\,\Delta f = 4(1.38 \times 10^{-23})(300)(10^8) = 1.66$ pW;
Effective Johnson-noise power including noise figure = $10^{0.6} \times 1.66$ pW = 6.6 pW.

In this case the Johnson noise is dominant, so the effective $S/N$ ratio is

$$\frac{S}{N} = \frac{3.2 \times 10^{-9}}{6.6 \times 10^{-12}} = 485.$$

### 21.6.5 Direct detection with APDs

In an APD there is a multiplication of the number of charge carriers by a factor of $M$. This multiplication can result from secondary ionizations produced by both electrons and holes. It is desirable that one or other of these charge carriers should have a significantly greater secondary ionization coefficient[6] than that of the other [24]. The current in an APD increases by the factor $M$, but the associated shot noise increases further because in a given avalanche process $M$ will fluctuate, taking values $M, M \pm 1, M \pm 2$, etc. The mean-square noise current thereby increases not by a factor of $M^2$, but by a factor of $FM^2$, where $F$ is called the *noise factor*. In silicon APDs $F$ typically lies in the range 2–20. Therefore, the shot noise becomes

$$\langle i_{\mathrm{N}}^2 \rangle_1 = 2eFM^2(\bar{i}_{\mathrm{s}} + i_{\mathrm{d}})\Delta f. \tag{21.72}$$

Both photogenerated and dark-generated carriers contribute to the shot noise. There are bulk charges that contribute an amount $i_{\mathrm{d1}}$ to the dark current and a surface leakage dark current $i_{\mathrm{d2}}$. Only the former is affected by multiplication and the noise factor. The overall signal-to-noise ratio is modified from Eq. (21.66) and becomes

$$\frac{\langle i_{\mathrm{s}}^2 \rangle}{\langle i_{\mathrm{N}}^2 \rangle_1 + \langle i_{\mathrm{N}}^2 \rangle_2} = \frac{M^2[e\eta P_0/(h\nu)]^2 R}{2eR\{FM^2[e\eta P_0/(h\nu) + i_{\mathrm{d1}}] + i_{\mathrm{d2}}\}\Delta f + 4k_{\mathrm{B}}T\Delta f}. \tag{21.73}$$

We can introduce a single effective dark current $i_{\mathrm{d}} = i_{\mathrm{d1}} + i_{\mathrm{d2}}/(M^2F)$ that simplifes the equations.

Equation (21.73) shows that the signal-to-noise ratio improves with increasing multiplication as the Johnson-noise contribution becomes less important until the avalanche shot noise becomes dominant. The avalanche-limited signal-to-noise ratio is

$$\frac{S}{N} = \frac{\eta P_0}{2Fh\nu\,\Delta f}, \tag{21.74}$$

a reduction of the signal-to-noise ratio from the quantum-limited value by the noise factor.

The NEP can be computed both for p–i–n photodiodes and for APDs by setting $S/N = 1$ in equations like (21.66) and (21.73), and thereby the minimum input power for $S/N = 1$ is determined. For $P_0 = P_{\mathrm{min}}$, we expect the effective dark current $i_{\mathrm{d}}$ to be dominant, so Eq. (21.73) becomes

$$\frac{S}{N} = \frac{[Me\eta P_0/(h\nu)]^2}{(2eFM^2 i_{\mathrm{d}} + 4k_{\mathrm{B}}T/R)\Delta f}. \tag{21.75}$$

For $S/N = 1$ and a bandwidth of 1 Hz, $P_0 = P_{\mathrm{min}} = \mathrm{NEP}$, which gives for an APD

$$\mathrm{NEP}(\mathrm{W\,Hz^{-1/2}}) = \frac{h\nu}{Me\eta}\sqrt{2eFM^2 i_{\mathrm{d}} + 4k_{\mathrm{B}}T/R}. \tag{21.76}$$

The equivalent result for a p–i–n diode can be obtained by setting $M = F = 1$.

[6] The secondary ionization coefficient is the number of secondary electron–hole pairs produced per unit length by a specific charge carrier (electron or hole) traveling through the material.

**Example** For an InGaAs APD with $F = 10$, $M = 100$, $R = 50\,\Omega$, $i_d = 2$ nA, and responsivity 0.8 A $W^{-1}$, the quantum efficiency is

$$\eta = \frac{0.8(6.626 \times 10^{-34})(3 \times 10^{8})}{(1.6 \times 10^{-19})(1.55 \times 10^{-6})} = 0.64. \tag{21.77}$$

From Eq. (21.73), the NEP is

$$\begin{aligned} &\text{NEP (W Hz}^{-1/2}) \\ &= \frac{(2 \times 1.6 \times 10^{-19} \times 10^{5} \times 2 \times 10^{-9} + 4 \times 1.38 \times 10^{-22} \times 300/50)^{1/2}}{0.8 \times 100} \\ &= 4.7 \times 10^{-13}\ \text{W Hz}^{-1/2}. \end{aligned}$$

Note that in this case the thermal noise is dominant.

## 21.7 Coherent detection

We have seen several examples in this chapter of how the signal-to-noise ratio of an optical detection system is limited by noise, in particular by Johnson noise and by the electronic noise of the amplification stages that follow an optical detector. Even if these sources of noise were not present, the signal-to-noise ratio would be limited by shot noise. In the direct detection schemes discussed so far the quantum noise limit set by shot noise is rarely attainable. However, it is possible to achieve the quantum limit for detection by the use of *coherent detection*. In this scheme the detector is illuminated simultaneously by the *signal* light and by a second source of light called the *local oscillator*, which must be phase-coherent with the signal. The degree of phase coherency that is required in order to make this scheme work well has been discussed in detail by Salz [25].

A useful rule of thumb is that the local oscillator must be phase-coherent over a time long enough to receive the information being transmitted. In an optical communication system in which binary information is being detected, the required phase coherence time is on the order of the pulse duration being detected.

The way in which coherent detection is carried out is shown schematically in Fig. 21.27. For optimum performance not only must the signal and the local oscillator be phase-coherent, but also their phase fronts and polarization states must be matched at the detector surface. This corresponds, for example, to a situation in which two linearly polarized $TEM_{00}$ Gaussian laser beams, which are coaxial and of equal spot size, coincide in a beam waist at the detector. Deviations from this ideal geometry, either through spot-size mismatch or arising from angular or lateral misalignment, decrease the efficiency of the detection process.

We represent the electric field of the signal beam at the detector surface as

$$E_s(t) = E_1 \cos(\omega_1 t + \phi_1) \tag{21.78}$$

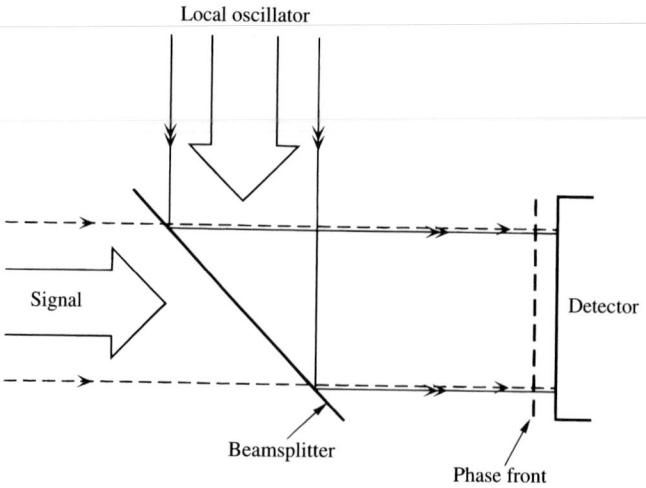


**Fig. 21.27** The arrangement for optical mixing.

and that of the local oscillator as

$$E_{\mathrm{lo}}(t) = E_2 \cos(\omega_2 t + \phi_2). \tag{21.79}$$

The detector responds to the intensity of the light falling on it. Therefore, the detector current is

$$i(t) = \frac{\mathcal{R}A}{Z}[E_{\mathrm{s}}(t) + E_{\mathrm{lo}}(t)]^2, \tag{21.80}$$

where $\mathcal{R}$ is the detector responsivity, $A$ is the effective detector area, and $Z$ is the characteristic impedance of the medium in front of the detector. Substitution from Eqs. (21.78) and (21.79) into (21.80) gives

$$i(t) = \frac{\mathcal{R}A}{Z}\Big[E_1^2 \cos^2(\omega_1 + \phi_1) + E_2^2 \cos^2(\omega_2 t + \phi_2) + 2E_1E_2 \cos(\omega_1 t + \phi_1)\cos(\omega_2 t + \phi_2)\Big]. \tag{21.81}$$

By the use of well-known trigonometrical identities,[7] the detector current can be written as

$$i(t) = \frac{\mathcal{R}A}{Z}\left\{\frac{E_1^2}{2} + \frac{E_1^2}{2}\cos[2(\omega_1 t + \phi_1)] + \frac{E_2^2}{2} + \frac{E_2^2}{2}\cos[2(\omega_2 t + \phi_2)] + E_1E_2\cos[(\omega_1 + \omega_2)t + \phi_1 + \phi_2] + E_1E_2\cos[(\omega_1 - \omega_2)t + \phi_1 - \phi_2]\right\}. \tag{21.82}$$

[7] $\cos^2 x = \frac{1}{2}[1 + \cos(2x)]$ and $\cos X + \cos Y = 2\cos[(X+Y)/2]\cos[(X-Y)/2]$.

The detector does not respond to the rapidly oscillating terms at frequencies $\omega_1, \omega_2$, and $\omega_1 + \omega_2$ in Eq. (21.82),[8] so the detector current is

$$i(t) = \frac{\mathcal{R}A}{Z}\left\{\frac{E_1^2}{2} + \frac{E_2^2}{2} + E_1E_2\cos[(\omega_1 - \omega_2)t + \phi_1 - \phi_2]\right\}. \tag{21.83}$$

We are assuming that the difference frequency $(\omega_1 - \omega_2)$ is within the response range of the detector. The generation of this difference frequency is called *optical mixing*. If $\omega_1 = \omega_2$, it is a *homodyne* mixing process; otherwise it is a *heterodyne* process. Similar processes occur in the operation of FM radio receivers. Note that the first two terms in Eq. (21.83) correspond to the detector current resulting from the signal and the local oscillator independently.

The detector shot noise is

$$\langle i_N^2\rangle_1 = \left[\frac{2e\mathcal{R}A}{Z}\left(\frac{E_1^2}{2} + \frac{E_2^2}{2}\right) + 2ei_d\right]\Delta f, \tag{21.84}$$

where $i_d$ is the detector dark current. In addition, we expect a Johnson-noise contribution of

$$\langle i_N^2\rangle_2 = \frac{4k_BT\,\Delta f}{R}. \tag{21.85}$$

It is usual to operate a coherent detection scheme in which $E_2 \gg E_1$: a weak signal is mixed with a strong local oscillator. Therefore, the effective signal current is

$$i_s(t) = \frac{\mathcal{R}AE_1E_2}{Z}\cos[(\omega_1 - \omega_2)t + \phi_1 - \phi_2]. \tag{21.86}$$

Equation (21.86) provides essential insight into the desirability of coherent detection for detection of a *weak* coherent optical field $E_1$. The signal current generated when this weak optical field is mixed with a local oscillator field $E_2$ is multiplied by the magnitude of the local oscillator field.

If $\omega_1 \neq \omega_2$, Eqs. (21.84)–(21.86) give a signal-to-noise ratio

$$\frac{\langle i_s^2(t)\rangle}{\langle i_N^2\rangle_1 + \langle i_N^2\rangle_2} = \frac{\mathcal{R}^2A^2E_1^2E_2^2}{2Z^2}\bigg/\left[\frac{e\mathcal{R}A}{Z}(E_1^2 + E_2^2) + 2ei_d + \frac{4k_BT}{R}\right]\Delta f. \tag{21.87}$$

With a sufficiently powerful local oscillator its shot noise dominates and the dark-current contribution and Johnson noise can be neglected. In this case, for $E_2 \gg E_1$, Eq. (21.87) reduces to

$$\frac{S}{N} = \frac{\mathcal{R}AE_1^2}{2eZ\,\Delta f}. \tag{21.88}$$

Noting that the average signal power is

$$P_s = \frac{AE_1^2}{2Z} \tag{21.89}$$

[8] In a more advanced analysis that reflects the fact that optical detectors respond to a light signal by absorbing photons, these high-frequency terms do not even appear. See Chapter 22 for further discussion.

and that $\mathcal{R} = \eta e/(h\nu)$, Eq. (21.88) becomes

$$\frac{S}{N} = \frac{\eta P_s}{h\nu\,\Delta f}. \tag{21.90}$$

It might appear from a comparison of Eqs. (21.67) and (21.90) that ideal heterodyne detection has a quantum-limited performance twice that of direct detection. However, to make such a comparison one must assume that it is possible to use the same signal-detection bandwidth $\Delta f$ in both schemes. If an unmodulated CW laser beam of power $P_s$ is to be detected then Eq. (21.67) would predict that the direct detection photon-noise-limited signal-to-noise ratio would be

$$\left(\frac{S}{N}\right)_{\mathrm{dd}} = \frac{\eta P_s}{2h\nu\,\Delta f}. \tag{21.91}$$

However, in practice one could not hope to achieve this signal-to-noise ratio because the resultant DC detector signal would be overwhelmed by $1/f$ noise.

It is always necessary to modulate the signal in some way. For example, if the signal power is

$$P = P_0[1 + m\sin(\omega_m t)], \tag{21.92}$$

the detected signal at frequency $\omega_m$ could be bandpass filtered. In this case the *effective* signal-to-noise ratio is

$$\left(\frac{S}{N}\right)_{\mathrm{dd}} = \frac{m^2\eta P_s}{4h\nu\,\Delta f}. \tag{21.93}$$

Particular care must be taken in comparing a homodyne receiver with a heterodyne receiver. If both the signal and the local oscillator are CW signals at the same frequency, it is meaningless to talk of detection of the signal, since the signal and the local oscillator are essentially indistinguishable. To discuss the signal-to-noise ratio, it is essential to consider how the signal beam is being modulated, since it is through modulation that information is transferred. To illustrate this we consider a situation in which the signal beam is phase modulated so that

$$E_s(t) = E_1\cos[\omega_1 t + m\sin(\omega_m t)]. \tag{21.94}$$

In this case the information being transferred is the sinusoidal phase modulation at frequency $\omega_m$. For small values of $m$, which is called the *modulation depth*, Eq. (21.94) can be rewritten as[9]

$$E_s(t) = E_1[J_0(m)\cos(\omega_1 t) + J_1(m)\cos(\omega_1 + \omega_m)t - J_1(m)\cos(\omega_1 - \omega_m)t]. \tag{21.95}$$

The signal beam has acquired *side-bands* at frequencies separated by $\pm\omega_m$ from the carrier frequency $\omega_1$.

The mixing process, in this case, can be written as

$$i(t) = \frac{\mathcal{R}A}{Z}\{E_1\cos[\omega_1 + m\sin(\omega_m t)] + E_2\cos(\omega_2 + \phi_2)\}^2. \tag{21.96}$$

[9] See Section 18.7.

By the use of Eq. (21.95), and with the assumption that $m \gg 1$, so that $[J_1(m)]^2 \gg J_1(m)$, Eq. (21.95) reduces to

$$i(t) = \frac{\mathcal{R}A}{Z}\left\{\frac{E_2^2}{2} + \frac{E_1^2 J_0^2(m)}{2} - 2J_1(m)E_1E_2\sin[(\omega_1 - \omega_2 - \phi_2)t]\sin(\omega_m t)\right\} \tag{21.97}$$

after the (nonexistent) high-frequency terms have been eliminated. The useful signal current contained in Eq. (21.97) is

$$i_s(t) = -\frac{2\mathcal{R}AJ_1(m)}{Z}E_1E_2\sin[(\omega_1 - \omega_2 - \phi_2)t]\sin(\omega_m t). \tag{21.98}$$

The information term $\sin(\omega_m t)$ is present as a low-frequency modulation of the intermediate frequency $(\omega_1 - \omega_2)$. In a similar way to Eq. (21.87), the signal-to-noise ratio is

$$\frac{\langle i_s^2(t)\rangle}{\langle i_N^2\rangle_1 + \langle i_N^2\rangle_2} = \frac{\mathcal{R}^2A^2J_1^2(m)E_1^2E_2^2}{Z^2\left(e\mathcal{R}AE_2^2/Z + 2ei_d + 4k_BT/R\right)\Delta f}, \tag{21.99}$$

where we have made the assumption that $E_1 \ll E_2$, as is usually the case in a heterodyne receiver.

For a sufficiently powerful local oscillator, Eq. (21.99) gives

$$\frac{S}{N} = \frac{\mathcal{R}AJ_1^2(m)E_1^2}{eZ\,\Delta f}. \tag{21.100}$$

For homodyne detection, we would optimize performance by setting $\phi_2 = \pi/2$ in Eq. (21.98). In this case the signal-to-noise ratio is

$$\left(\frac{S}{N}\right)_{\text{homodyne}} = \frac{2\mathcal{R}AJ_1^2(m)E_1^2}{eZ\,\Delta f}. \tag{21.101}$$

In this situation the homodyne $S/N$ is twice the value achieved by heterodyne detection. It should be stressed, however, that the $S/N$ in coherent detection schemes depends on the precise modulation scheme, and also on the contribution to the signal-to-noise ratio of the demodulation scheme used to extract the information term ($\sin(\omega_m t)$ in Eq. (21.98)), from the intermediate-frequency signal at $\omega_1 - \omega_2$ [22, 26]. A discussion of the merits of various demodulators, involving bandpass filters, discriminations, limiters, and phase-locked loops, is beyond our scope here.

We can note that the signal power in the two side-bands in Eq. (21.95) can be written as

$$P_s = \frac{E_1^2J_1^2(m)}{Z}, \tag{21.102}$$

so that, in terms of the side-band power, the heterodyne $S/N$ is, as before,

$$\frac{S}{N} = \frac{\eta P_s}{h\nu\,\Delta f}. \tag{21.103}$$

**Example** An interesting example of the coherent detection process is provided under conditions of *suppressed carrier* operation. If the modulation depth in Eq. (21.94) is set to a value $m = 2.405$, then $J_0(m) = 0$ and the carrier is suppressed. For a detector with

$\eta = 0.6$, and bandwidth 100 MHz operating at 1.55 μm, the minimum detectable signal power – defined to correspond to $S/N = 1$ – is

$$P_s = \frac{h\nu\,\Delta f}{1}, \tag{21.104}$$

which in this case gives

$$P_s = \frac{(6.626 \times 10^{-34})(3 \times 10^8)(10^8)}{0.6(1.55 \times 10^{-6})} = 21 \text{ pW}.$$

## 21.8 The bit-error rate

We conclude this chapter with a discussion of a final performance parameter that is widely used to characterize the performance of optical communication systems. This is the *bit-error rate* (BER) used to describe the performance of an optical detection system receiving binary-encoded signals. In its simplest form, such a scheme has a binary "one" represented by the detection of an optical pulse by the detector. A binary "zero" is represented by the failure to detect such a pulse. The "ones" and "zeros" constitute a bit stream at a prescribed clock rate, as shown schematically in Figs. 21.28(a) and (b). The detection of a "one" is characterized by the detector signal rising above a threshold level $\ell_1$. The detection of a "zero" is characterized by the detector signal failing to rise above a lower threshold level $\ell_2$. In the presence of additive noise, represented schematically by Fig. 21.28(c), the overall detector signal will appear as shown schematically in Fig. 21.28(d). The questions that then arise are the following: when will the noise reduce the detector signal below $\ell_1$ when a "one" is present, and when will the noise be above level $\ell_2$ when a "zero" is being detected? Both of these possibilities lead to a bit-detection error, as shown schematically in Fig. 21.28(e). One simple way to model this process is to assume that the detector gives rise to a noise current that fluctuates in a Gaussian fashion – more sophisticated analyses would examine more closely the statistical nature both of the detection and of the noise processes and the different ways in which the binary signal is encoded [22, 26].

For a detector noise current with variance $\langle i_N^2 \rangle$ that is Gaussian distributed about zero, the probability that the noise is *below* a set level is

$$P(i_N \le \ell) = \frac{1}{\sigma\sqrt{2\pi}} \int_{-\infty}^{\ell} e^{-x^2/(2\sigma^2)}\, dx, \tag{21.105}$$

where $\sigma^2 = \langle i_N^2 \rangle$.

On average a simple binary stream of "ones" and "zeros" will have equal numbers of ones and zeros. Therefore, for a large number $N$ of transmitted bits there will be $N/2$ "ones" and $N/2$ "zeros." There will be a bit error whenever a "one" is present but the noise current is below $-\ell_1$; there will also be a bit error whenever a "zero" is present but the noise is *above* level $\ell_2$. Therefore, we can write the probability of error as

$$p_e = \frac{1}{2}[P(i_N < -\ell_1) + P(i_N > \ell_2)]. \tag{21.106}$$

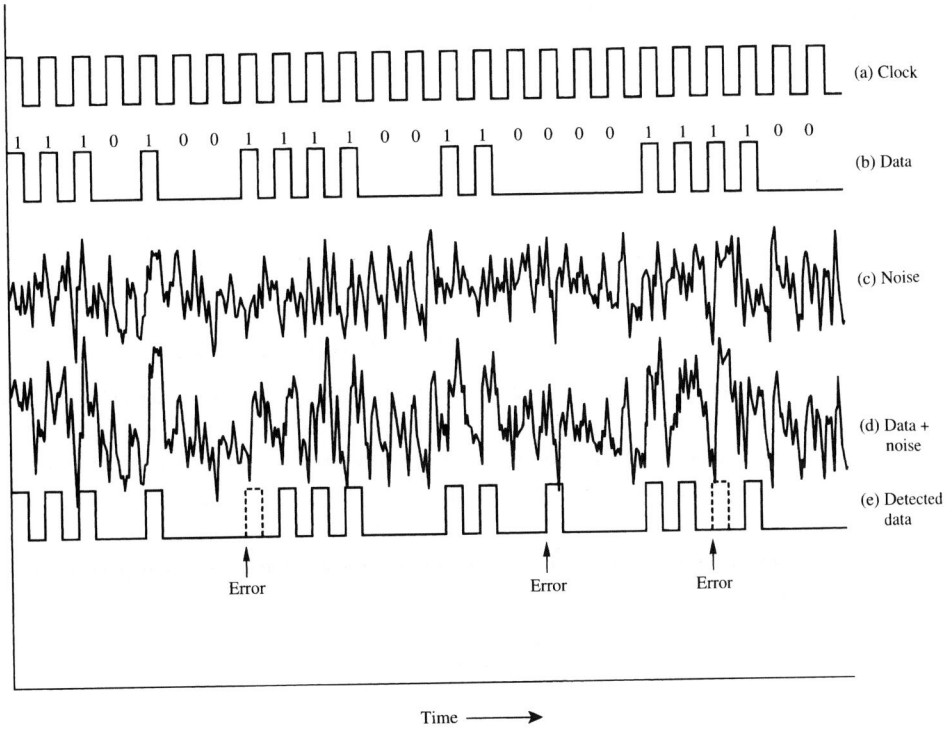


**Fig. 21.28** Binary signal transmission, showing the clock, data, the effect of added noise, and the resultant detection of "ones" and "zeros."

A simple result can be obtained if $\ell_1 = \ell_2 = i_s/2$, where $i_s$ is the detector current corresponding to a "one." In this case

$$p_e = \frac{1}{2\sigma\sqrt{2\pi}}\left(\int_{-\infty}^{-i_s/2} e^{-x^2/(2\sigma^2)}\,dx + \int_{i_s/2}^{\infty} e^{-x^2/(2\sigma^2)}\,dx\right), \tag{21.107}$$

which becomes

$$p_e = \frac{1}{\sigma\sqrt{2\pi}}\int_{i_s/2}^{\infty} e^{-x^2/(2\sigma^2)}dx = \frac{1}{\sqrt{\pi}}\int_{i_s/(2\sqrt{2}\sigma)}^{\infty} e^{-t^2}\,dt. \tag{21.108}$$

Equation (21.108) can be written in terms of the error function, erf($z$), which is defined by the relation [27]

$$\mathrm{erf}(z) = \frac{2}{\sqrt{\pi}}\int_0^z e^{-t^2}\,dt \tag{21.109}$$

where erf($z$)$\rightarrow$ 1 as $z \rightarrow \infty$. Therefore,

$$p_e = \frac{1}{2}\left[1 - \mathrm{erf}\left(\frac{i_s}{2\sqrt{2}\sigma}\right)\right] = \frac{1}{2}\mathrm{erfc}\left(\frac{i_s}{2\sqrt{2}\sigma}\right) \tag{21.110}$$

and, finally,

$$p_e = \frac{1}{2}\,\mathrm{erfc}\left(\frac{i_s}{2\sqrt{2}\langle i_N^2\rangle^{1/2}}\right). \tag{21.111}$$

The signal-to-noise ratio can be characterized by

$$\frac{S}{N} = \left(\frac{i_s}{\text{RMS noise current}}\right)^2, \tag{21.112}$$

so

$$p_e = \frac{1}{2}\,\mathrm{erfc}\left(\frac{1}{2\sqrt{2}}\sqrt{\frac{S}{N}}\right). \tag{21.113}$$

Figure 21.29 shows a plot of this probability of error, or bit-error rate, as a function of $S/N$. A frequently used performance measure is $P_e < 10^{-9}$, which implies $S/N = 141$, or 21.5 dB. This corresponds to $i_s$/RMS noise current = 11.89, so an optical signal only 11.89 times greater than the dark-current level is needed in order to achieve a bit-error rate of $10^{-9}$.

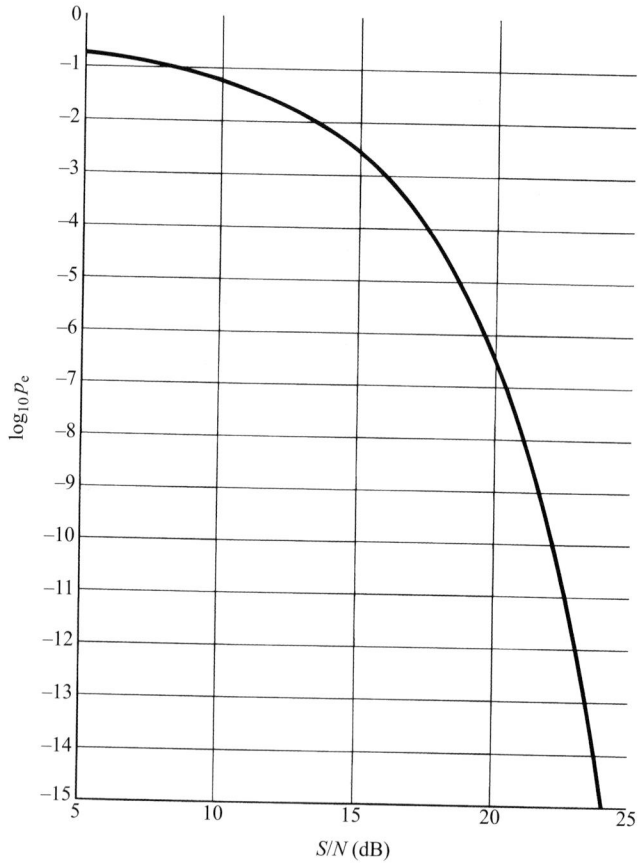


**Fig. 21.29** The bit-error rate as a function of the signal-to-noise ratio in a simple binary detection system.

In a practical test of bit-error rate a random, long sequence of binary pulses ($10^{15}$–$10^{21}$) would be transmitted from source to receiver and the received bit sequence compared with a record of the transmitted sequence.

## 21.9 Problems

21.1 A photomultiplier with a responsivity of 0.1 A $W^{-1}$ at 600 nm is being used to measure a very weak star spectral feature at 600 nm, and the average rate of signal counts is only 0.1 per second. The rate of dark counts is 10 per second. What integration time would be required to achieve a signal-to-noise ratio of 10? What minimum optical power is being detected in this case?

21.2 A photomultiplier (PMT) has a dark noise level of 100 pulses per second. Calculate the minimum optical power ($S/N = 1$) at 500 nm that can be detected with a 10-s integration time. Assume that the quantum efficiency of the PMT is 0.2 at 500 nm.

21.3 A p–i–n photodiode has a responsivity of 0.8 A $W^{-1}$ at 1.55 μm. The effective resistance of the detector circuit is 50 Ω. The photodiode dark current is 1 pA $Hz^{-1/2}$. Calculate the NEP at 300 K of the detector. Include both shot noise and Johnson noise in your calculation. If the area of the detector is 1 $mm^2$, what is its detectivity?

21.4 How would the answer to Problem (21.3) be modified if the detector were feeding an amplifier with a noise temperature of 600 K?

21.5 A p–i–n photodetector has a responsivity of 0.5 A $W^{-1}$ and 2 nA of dark current. The load resistance is 2,000 Ω and the system bandwidth is 50 MHz. The temperature is 40°C. (a) At what value of received optical power is the thermal noise equal to the shot noise? (b) What is the signal-to-noise ratio at this power level? (c) What is the value of the shot-noise power at this value of the received optical power?

21.6 Repeat Problem (21.5) for the case of an APD detector with a gain of 100 and a noise temperature of 1,000 K.

21.7 The optical power reaching a photodetector is 1 μW. The detector's responsivity is 0.5 A W, and its dark current is 4 nA, the temperature is 27 °C, the bandwidth is 500 MHz, and the load resistance is 50 Ω.

(a) Compute the signal-to-noise ratio.
(b) Compute the thermal-noise-limited signal-to-noise ratio.
(c) Compute the shot-noise-limited signal-to-noise ratio.
(d) What value of detector gain is required in order to make the actual signal-to-noise ratio just 5 dB less than the quantum limit? Assume that the excess noise factor of the system is 1.

21.8 An amplifier has a power gain of 8 and a noise figure of 3 dB. The amplifier follows a photodetector with a responsivity of 0.5 A $W^{-1}$. The load resistor has a resistance of 100 Ω, the temperature is 300K, the received optical power is 0.5μW, and the bandwidth is 1 MHz.

(a) Compute the signal power flowing through the load resistor.
(b) Compute the signal power exiting the amplifier.
(c) Compute the thermal noise power generated by the load resistor.
(d) Compute the thermal noise power exiting the amplifier.
(e) Compute the amplifier's noise temperature.
(f) Compute the signal-to-noise ratio at the input to the amplifier.
(g) Compute the signal-to-noise ratio at the output of the amplifier.

## References

[1] A. van der Ziel, *Noise*, New York: Prentice-Hall, 1954.

[2] H. Nyquist, "Thermal agitation of electric charge in conductors," *Phys. Rev.* **32**, 110–113, 1928.

[3] S. Ramo, J. R. Whinney, and T. Van Duzer, *Fields and Waves in Communication Electronics,* 3rd edn., New York: Wiley, 1994.

[4] B. I. Bleaney and B. Bleaney, *Electricity and Magnetism*, 3rd edn., Oxford: Oxford University Press, 1976.

[5] P. W. Kruse, L. D. McGlanchlin, and R. B. McQuistan, *Elements of Infrared Technology: Generation, Transmission, and Detection*, New York: Wiley, 1962.

[6] J. H. Moore, C. C. Davis, and M. A. Coplan, *Building Scientific Apparatus*, 4th edn., Cambridge: Cambridge University Press, 2009.

[7] E. H. Putley, "Indium antimonide submillimeter photoconductive detectors," *Appl. Opt.* **4**, 649–656, 1969.

[8] E. H. Putley, "Thermal detectors," in *Optical and Infrared Detectors*, 2nd edn., ed. R. J. Keyes, Berlin: Springer-Verlag, 1980, pp. 71–100.

[9] E. H. Putley and D. H. Martin, "Detectors," in *Spectroscopic Techniques for Infrared, Submillimeter and Millimeter Waves*, ed. D. H. Martin, Amsterdam: North-Holland, 1967, pp. 111–151.

[10] J. Gowar, *Optical Communication Systems*, Englewood Cliffs, NJ: Prentice-Hall, 1984.

[11] H. Kressel, ed., *Semiconductor Devices for Optical Communication*, Berlin: Springer-Verlag, 1982.

[12] H. Melchior, M. B. Fisher, and F. R. Arams, "Photodetectors for optical communication systems," *Proc. IEEE* **58**, 1466–1486, 1970.

[13] J. Gowar, *Optical Communication Systems*, 2nd edn., Englewood Cliffs, NJ: Prentice-Hall, 1993.

[14] H. Dautet, P. Deschamps, B. Dion, A. D. MacGregor, D. MacSween, R. J. McIntyre, C. Trottier, and P. P. Webb, "Photon-counting techniques with silicon avalanche photodiode," *Appl. Opt.* **32**, 3894–3900, 1993.

[15] S. Cova, A. Lacaita, F. Zappa, and P. Lovati, "Avalanche photodiodes for near-infrared photon counting," in *Advances in Fluorescence Sensing Technology II*, ed. J. Lakowicz, San Jose, CA: SPIE 1995, pp. 56–66.

[16] A. Lacaita, F. Zappa, S. Cova, and P. Lovati, "Single-photon detection beyond 1 μm: performance of commercially available InGaAs/InP detectors," *Appl. Opt.* **35**, 2986–2996, 1996.

[17] S. Cova, M. Ghioni, A. Lacaita, C. Samori, and F. Zappa, "Avalanche photodiodes and quenching circuits for single-photon detection," *Appl. Opt.* **35**, 1956–1976, 1996.

[18] M. J. E. Golay, "A pneumatic infra-red detector," *Rev. Sci. Instrum.* **18**, 357–362, 1947.

[19] H. Taub and D. L. Schilling, *Principles of Communication Systems*, 2nd edn., New York: McGraw-Hill, 1986.

[20] H. Kressel, ed., *Semiconductor Devices for Optical Communication*, Berlin: Springer-Verlag, 1982.

[21] P. K. Cheo, *Fiber Optics and Optoelectronics*, 2nd edn., Englewood Cliffs, NJ: Prentice-Hall, 1990.

[22] R. M. Gagliardi and S. Karp, *Optical Communications*, New York: Wiley, 1976.

[23] H. B. Killen, *Fiber Optic Communications*, Englewood Cliffs, NJ: Prentice-Hall, 1991.

[24] R. J. McIntyre, "Multiplication noise in uniform avalanche diodes," *IEEE Trans. Electron. Devices* **13**, 164–168, 1966.

[25] J. Salz, "Coherent lightwave communications," *AT&T Technol. J.* **64**, 2153–2209, 1985.

[26] P. E. Green, Jr., *Fiber Optic Networks*, Englewood Cliffs, NJ: Prentice-Hall, 1993.

[27] M. Abramowitz and I. A. Stegun, *Handbook of Mathematical Functions*, New York: Dover, 1965.

# 22 Coherence theory

## 22.1 Introduction

In this chapter we will put the concept of *coherence* on a more mathematical basis.[1] This will involve the formal definition of a number of functions that are used to describe the coherence properties of optical fields. These include the analytic signal, various correlation functions, and the degree of coherence for describing both temporal and spatial coherence. We shall see that the degree of temporal coherence is quantitatively related to the lineshape function and that the degree of spatial coherence between two points is determined by the size, intensity distribution, and location of an illuminating source. We will use the wave equation, and special solutions to the wave equation called Green's functions, to show how spatial coherence varies from point to point.

The chapter will conclude with a discussion of how the intensity fluctuations of a source depend on its coherence properties, and we will examine a specific scheme, namely the Hanbury Brown–Twiss experiment, by means of which this relationship is studied. This discussion will involve a discussion of "photon statistics," namely the time variation of the "detection" of *photons* from a source. In a quantum-mechanical context, square-law detectors respond to these quantized excitations of the optical field, which we call photons.

In classical coherence theory it is advantageous to represent the real electromagnetic field by a complex quantity, both for its mathematical simplicity and because it serves to emphasize that coherence theory deals with phenomena that are sensitive to the "envelope" or average intensity of the field. In spite of the fact that such a complex representation is rather artificial in the classical theory, it provides the best correspondence to the approach of quantum theory, wherein an equivalent representation is intimately connected with the process by which an electromagnetic field is detected.

## 22.2 Square-law detectors

Suppose that we illuminate a photodetector with a plane, monochromatic light wave containing two closely spaced frequencies, for example one for which the electric field can be written as

$$\mathbf{E} = \mathbf{A}\cos(\omega t) + \mathbf{B}\cos[(\omega + \Delta\omega)t], \tag{22.1}$$

[1] For a more detailed, extended, and definitive discussion of this material the reader should consult reference [1].

where **A** and **B** are parallel.

The detector responds to the intensity of this field and therefore to the square of the electric field; its output is

$$i \propto A^2 \cos^2(\omega t) + B^2 \cos^2[(\omega + \Delta\omega)t] + 2AB \cos(\omega t)\cos[(\omega + \Delta\omega)T], \tag{22.2}$$

which can be rewritten as

$$i \propto \frac{A^2}{2}[1 + \cos(2\omega t)] + \frac{B^2}{2}\{1 + \cos[2(\omega + \Delta\omega)t]\} + AB\cos[(2\omega + \Delta\omega)t] + AB\cos(\Delta\omega t). \tag{22.3}$$

If we make the practical assumption that typical optical detectors cannot follow the high-frequency terms at $2\omega$, $2(\omega + \Delta\omega)$, and $(2\omega + \Delta\omega)$ in the above expression, then the actual observed response can be calculated, as we have seen previously, to be

$$i \propto \frac{A^2}{2} + \frac{B^2}{2} + AB\cos(\Delta\omega t), \tag{22.4}$$

where we have assumed that the detector can respond to the beat signal at frequency $\Delta\omega$. If we use complex notation and write Eq. (22.1) as

$$|E| = Ae^{-j\omega t} + Be^{j(\omega+\Delta\omega)t} \tag{22.5}$$

then the detector response is $i \propto EE^* = |E|^2$, which gives

$$\begin{aligned} i &\propto [Ae^{j\omega t} + Be^{j(\omega+\Delta\omega)t}][Ae^{-j\omega t} + Be^{-j(\omega+\Delta\omega)t}] \\ &\propto A^2 + B^2 + ABe^{j\Delta\omega t} + ABe^{-j\Delta\omega t} \\ &\propto A^2 + B^2 + 2AB\cos(\Delta\omega t), \end{aligned} \tag{22.6}$$

and the use of the complex notation gives the same frequency dependence as in Eq. (22.4). In other words, the apparent averaging effect of the detector is automatically included by writing the fields in complex form.[2] This emphasizes the critical physical reality of the detection process. The detector does *not* average the high-frequency terms which appear in Eq. (22.3). These terms do not, in fact, exist for a photon detector. The detector responds to quanta of the electromagnetic field – photons. The conversion of these photons into charge carriers in the photodetector is well represented by the use of the complex notation in Eq. (22.6).

## 22.3 The analytic signal

We can generalize the use of complex fields to describe real fields by introducing the *complex analytic signal*.

[2] In general, if $a(t)$ and $b(t)$ are two real fields with *complex* amplitudes $A$ and $B$ then

$$\overline{a(t)b(t)} = \frac{1}{2}\mathcal{R}(AB^*).$$

If we have a real scalar quantity $V^{(\mathrm{r})}(t)$, where the superscript (r) indicates the reality of the quantity, which may, for example, represent a Cartesian component of the electric field **E**, then $V^{(\mathrm{r})}(t)$ possesses a Fourier transform,

$$V^{(\mathrm{r})}(t) = \int_{-\infty}^{\infty} V^{(\mathrm{r})}(\nu) e^{-j2\pi\nu t}\, d\nu, \tag{22.7}$$

where

$$V^{(\mathrm{r})}(\nu) = \int_{-\infty}^{\infty} e^{j2\pi\nu t} V^{(\mathrm{r})}(t) dt. \tag{22.8}$$

For positive $t$ this integral contains terms

$$[\cos(j2\pi\nu t) + j\sin(2\pi\nu t)]V^{(\mathrm{r})}(t).$$

For negative $t$ the integral contains terms

$$[\cos(j2\pi\nu t) - j\sin(2\pi\nu t)]V^{(\mathrm{r})}(-t).$$

So we can write

$$V^{(\mathrm{r})}(-\nu) = \int_{-\infty}^{\infty} e^{-j2\pi\nu t} V^{(\mathrm{r})}(t) dt \tag{22.9}$$

and

$$[V^{(\mathrm{r})}(-\nu)]^* = \int_{-\infty}^{\infty} e^{j2\pi\nu t} V^{(\mathrm{r})}(t) dt. \tag{22.10}$$

Therefore, $[V^{(\mathrm{r})}(-\nu)]^* = V^{(\mathrm{r})}(\nu)$ and $[V^{(\mathrm{r})}(\nu)]^* = V^{(\mathrm{r})}(-\nu)$.

Thus all the information in $V^{(\mathrm{r})}(\nu)$ is contained in the positive frequencies.

The complex analytic signal corresponding to $V^{(\mathrm{r})}(t)$ is defined by

$$V(t) = \int_{0}^{\infty} e^{-j2\pi\nu t} V^{(r)}(\nu) d\nu \tag{22.11}$$

and includes only contributions from positive frequencies. The positive-frequency Fourier amplitudes can be written as

$$V^{(\mathrm{r})}(\nu) = \begin{cases} \int_{-\infty}^{\infty} V(t) e^{j2\pi\nu t}\, dt \text{ for } \nu \geq 0, \\ 0 \text{ for } \nu < 0. \end{cases} \tag{22.12}$$

A real field, $V^{(\mathrm{r})}(t)$ for example, could be

$$A\cos(2\pi\nu t) = \frac{A}{2}(e^{j2\pi\nu t} + e^{-j2\pi\nu t}). \tag{22.13}$$

For such a field $V(t)$ can be written in terms of its real and imaginary (denoted by superscript (i)) parts as

$$V(t) = \frac{1}{2}[V^{(\mathrm{r})}(t) + jV^{(\mathrm{i})}(t)]. \tag{22.14}$$

A simple example of such an analytic signal could be $Ae^{-j2\pi\nu t}$.

If $V^{(\mathrm{r})}(t)$ satisfies the wave equation

$$\nabla^2 V^{(\mathrm{r})}(t) - \frac{1}{c^2}\frac{\partial^2 V^{(\mathrm{r})}(t)}{\partial t^2} = 0,$$

then both $V^{(\mathrm{i})}(t)$ and the analytic signal $V(t)$ do so also.

From Eq. (22.11)

$$\begin{aligned} V(t+j\tau) &= \int_0^\infty e^{-2\pi\nu(it-\tau)} V^{(\mathrm{r})}(\nu)d\nu \\ &= \int_0^\infty e^{2\pi\nu\tau} e^{-j2\pi\nu t} V^{(\mathrm{r})}(\nu)d\nu. \end{aligned} \tag{22.15}$$

The integrand $\to \infty$ for $\tau > 0$, so the derivative of $V(t+j\tau) = V(z)$ is not single-valued. Thus $V(t+j\tau)$ is analytic[3] only in the lower half plane, and $V(t)$ is the boundary value as $\tau \to 0^-$.

### 22.3.1 Hilbert transforms

A fundamental theorem of complex variables states that a function $f(z)$ that has no poles in the lower (or upper) $z$ plane has real and imaginary parts that are related by Hilbert transforms [2, 3].

When

$$V(t) = \frac{1}{2}[V^{(\mathrm{r})}(t) + jV^{(\mathrm{i})}(t)], \tag{22.16}$$

we have

$$V^{(\mathrm{i})}(t) = \frac{1}{\pi}\mathrm{PV}\int_{-\infty}^{\infty} \frac{V^{(\mathrm{r})}(t')}{t'-t}\,dt' \tag{22.17}$$

and

$$V^{(\mathrm{r})}(t) = -\frac{1}{\pi}\mathrm{PV}\int_{-\infty}^{\infty} \frac{V^{(\mathrm{i})}(t')}{t'-t}\,dt', \tag{22.18}$$

where PV indicates the Cauchy principal value[4] of the integral at the singular point $t' = t$.

[3] Possesses a finite derivative.

[4] Let $q(x) \to \infty$ as $x \to a$, then the Cauchy principal value of

$$\int_b^c q(x)dx$$

written

$$\mathrm{PV}\int_b^c q(x)dx \quad (b < a < c)$$

is

$$\lim_{\delta\to 0}\left\{\int_b^{a-\delta} q(x)dx + \int_{a+\delta}^c q(x)dx\right\}.$$

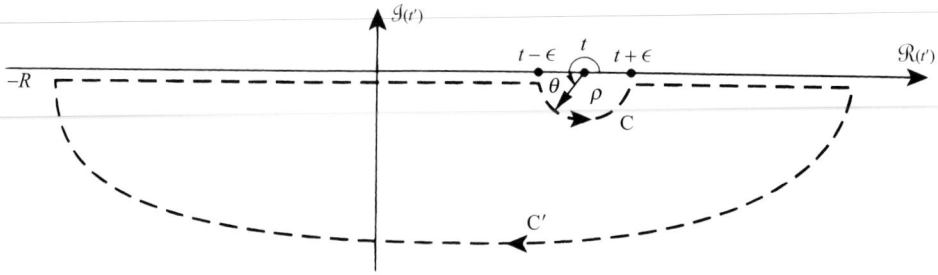


**Fig. 22.1** The complex plane, showing the contour over which the integral in Eq. (22.19) is evaluated.

Let us evaluate the integral

$$\int_{\text{contour}} \frac{V(t')}{t'-t}\,dt' = \int_{C'} \frac{V(t')}{t'-t}\,dt' + \int_{-R}^{t-\epsilon} \frac{V(t')}{t'-t}\,dt' + \int_{t+\epsilon}^{R} \frac{V(t')}{t'-t}\,dt' + \int_{C} \frac{V(t')}{t'-t}\,dt' = 0 \tag{22.19}$$

over the closed contour shown in Fig. 22.1. Now, the integral

$$\int_{C'} \frac{V(t')}{t'-t}\,dt'$$

vanishes as $|t'| \to \infty$. We integrate $\int_C \left[V(t')/(t'-t)\right] dt'$ over the infinitesimally small half circle of radius $\rho$ surrounding the point $t$:

$$\int_C \frac{V(t')}{t'-t}\,dt' = V(t)\int_C \frac{dt'}{t'-t} = V(t)\int_C \frac{d(\rho e^{j\theta})}{\rho e^{j\theta}}. \tag{22.20}$$

Over this half circle $\rho$ is constant, $\theta$ goes from $\pi$ to $2\pi$, and the integral becomes

$$V(t)j\int_{\pi}^{2\pi} d\theta = j\pi V(t). \tag{22.21}$$

Hence,

$$\int_C \frac{V(t')}{t'-t}\,dt' = \pi j V(t). \tag{22.22}$$

So

$$V(t) = \frac{1}{\pi}\left[\int_{-R}^{t-\epsilon} \frac{V(t')}{t'-t}\,dt' + \int_{t+\epsilon}^{R} \frac{V(t')}{t'-t}\,dt'\right]. \tag{22.23}$$

As $R \to \infty$ and $\epsilon \to 0$ the quantity in the square brackets becomes the Cauchy principal value, i.e.,

$$V(t) = \frac{j}{\pi}\text{PV}\int_{-\infty}^{\infty} \frac{V(t')}{t'-t}\,dt'. \tag{22.24}$$

Thus,

$$V^{(\mathrm{r})}(t) + jV^{(\mathrm{i})}(t) = \frac{j}{\pi}\mathrm{PV}\int_{-\infty}^{\infty}\left\{\frac{V^{(\mathrm{r})}(t)}{t'-t} + \frac{jV^{(\mathrm{i})}(t)}{t'-t}\right\}dt'. \tag{22.25}$$

On comparing the real and imaginary parts of Eq. (22.25), we have

$$V^{(\mathrm{r})}(t) = -\frac{1}{\pi}\mathrm{PV}\int_{-\infty}^{\infty}\frac{V^{(i)}(t')}{t'-t}\,dt', \tag{22.26}$$

$$V^{(\mathrm{i})}(t) = \frac{1}{\pi}\mathrm{PV}\int_{-\infty}^{\infty}\frac{V^{(r)}(t')}{t'-t}\,dt'. \tag{22.27}$$

## 22.4 Correlation functions

Let the (complex) electromagnetic field with polarization $\mu$ at a point $\mathbf{r}$ at time $t$ be $V_\mu(\mathbf{r}, t)$. The $(m+n)$th-order correlation function is an ensemble average of the products of these field functions considered at different space-time points and polarizations, defined as follows:

$$\Gamma^{(m,n)}_{\mu_1\ldots\mu_{m+n}}(\mathbf{r}_1\ldots\mathbf{r}_{m+n}, t_1\ldots t_{m+n}) = \left\langle\prod_{j=1}^{m}V^*_{\mu_j}(\mathbf{r}_j, t_j)\prod_{k=m+1}^{m+n}V_{\mu_k}(\mathbf{r}_k, t_k)\right\rangle, \tag{22.28}$$

where the $\langle\rangle$ brackets indicate an ensemble average.[5]

An important class of optical fields consists of those that are *stationary* and *ergodic*. For a *stationary* field the correlation functions are independent of the choice of time origin. For an *ergodic* field the ensemble average can be replaced by the time average, i.e.,

$$\langle\ldots\rangle = \lim_{T\to\infty}\frac{1}{2T}\int_{-T}^{T}\ldots dt. \tag{22.29}$$

So, leaving the polarization indices to be understood,

$$\begin{aligned}\Gamma^{(m,n)}(\mathbf{r}_1\ldots\mathbf{r}_{m+n}, t_1\ldots t_{m+n}) &\equiv \Gamma^{m,n}(\mathbf{r}_1\ldots\mathbf{r}_{m+n}, \tau_2\ldots\tau_{m+n})\\ &= \lim_{T\to\infty}\frac{1}{2T}\int_{-T}^{T}\prod_{j=1}^{m}V^*(\mathbf{r}_j, t+\tau_j)\prod_{k=m+1}^{m+n}V(\mathbf{r}_k, t+\tau_k)dt,\end{aligned} \tag{22.30}$$

where $\tau_1$ has been set to zero, and $\tau_j = t_j - t_1$ (the stationarity condition).

It is to be understood that $\mathbf{r}_j$ implies polarization $\mu_j$, although we are primarily concerned with linearly polarized fields where all the $\mu_j$ are the same. In all that follows we will be

[5] An ensemble average would be obtained by measuring the fields $V_{\mu_i}(\mathbf{r}_i, t_i)$ in a large number of systems that, as closely as possible, are identical. Such systems would contain identical sources, boundaries, temperatures, and detector configurations, but statistical fluctuations from one system to the other would cause individual fields to be different.

assuming linearly polarized fields all polarized in the same direction. If this were not the case, we could resolve the field into components along orthogonal axes.

The correlation function of second order $m = n = 1$ for stationary and ergodic fields is called the *mutual coherence function*, and is written

$$\Gamma(\mathbf{r}_1, \mathbf{r}_2, \tau) = \lim_{T\to\infty} \frac{1}{2T} \int_{-T}^{T} V^*(\mathbf{r}_1, t) V(\mathbf{r}_2, t+\tau) dt. \tag{22.31}$$

$\Gamma(\mathbf{r}_1, \mathbf{r}_2, \tau)$ is frequently written $\Gamma_{12}(\tau)$. Clearly

$$\Gamma_{12}^*(\tau) = \Gamma_{21}(-\tau). \tag{22.32}$$

We can also write

$$\Gamma(\mathbf{r}_1, \mathbf{r}_2, \tau) = \Gamma(\mathbf{r}_1, t_1, \mathbf{r}_2, t_2) = \overline{V^*(\mathbf{r}_1, t) V(\mathbf{r}_2, t+\tau)}. \tag{22.33}$$

This function is important in the mathematical description of effects such as interference and diffraction. When $\mathbf{r}_1 = \mathbf{r}_2 = \mathbf{r}$

$$\Gamma(\mathbf{r}, \mathbf{r}, \tau) = \overline{V^*(\mathbf{r}, t) V(\mathbf{r}, t+\tau)} = \Gamma_{11}(\tau). \tag{22.34}$$

This is the *autocorrelation function*. We define the *intensity* at point $\mathbf{r}$ at time $t$ as

$$I(\mathbf{r}, t) = V^*(\mathbf{r}, t) V(\mathbf{r}, t). \tag{22.35}$$

We have already illustrated this in a simple case for two different frequencies illuminating a photodetector.

It can be shown that [4, 5]

$$\overline{V_1^{(\mathrm{r})}(t) V_2^{(\mathrm{i})}(t+\tau)} = \overline{-V_1^{(\mathrm{i})}(t) V_2^{(\mathrm{r})}(t+\tau)}, \tag{22.36}$$

$$\overline{V_1^{(\mathrm{r})}(t) V_2^{(\mathrm{r})}(t+\tau)} = \overline{V_1^{(\mathrm{i})}(t) V_2^{(\mathrm{i})}(t+\tau)}, \tag{22.37}$$

and of course

$$\begin{aligned}\Gamma_{12}(\tau) &= \overline{V_1^*(t) V_2(t+\tau)} \\ &= \frac{1}{4}\left[\overline{V_1^{(\mathrm{r})}(t) V_2^{(\mathrm{r})}(t+\tau) + V_1^{(\mathrm{i})}(t) V_2^{(\mathrm{i})}(t+\tau)}\right] \\ &\quad - \frac{1}{4}\left[\overline{V_1^{(\mathrm{i})}(t) V_2^{(\mathrm{r})}(t+\tau) - V_1^{(\mathrm{r})}(t) V_2^{(\mathrm{i})}(t+\tau)}\right],\end{aligned} \tag{22.38}$$

where for convenience we write $V(\mathbf{r}_1, t_1) = V_1(t_1)$.

Since

$$V(t) = \int_0^\infty e^{-j2\pi\nu t} V^{(\mathrm{r})}(\nu) d\nu, \tag{22.39}$$

we can write

$$V(t) = \int_{-\infty}^\infty e^{-j2\pi\nu t} V(\nu) d\nu, \tag{22.40}$$

since, for $\nu \geq 0$, $V(\nu) = V^{(\mathrm{r})}(\nu)$, and, for $\nu < 0$, $V(\nu) = 0$.

Note that by definition $V(t)$ contains only positive frequencies. Similarly, for the correlation function

$$\Gamma(\mathbf{r}_1, \mathbf{r}_2, \tau) = \int_0^\infty \Gamma(\mathbf{r}_1, \mathbf{r}_2, \nu) e^{-j2\pi\nu t}\, d\nu. \tag{22.41}$$

This is equivalent to the Wiener–Khintchine theorem [6, 7] in the theory of stationary stochastic processes.

The *mutual spectral density is*

$$\Gamma(\mathbf{r}_1, \mathbf{r}_2, \nu) = \begin{cases} \int_{-\infty}^{\infty} \Gamma(\mathbf{r}_1, \mathbf{r}_2, \tau) e^{j2\pi\nu\tau}\, d\tau & \text{for } \nu \geq 0, \\ 0 & \text{for } \nu < 0. \end{cases} \tag{22.42}$$

$V(\mathbf{r}, t)$ obeys the wave equation

$$\left(\nabla^2 - \frac{1}{c^2}\frac{\partial^2}{\partial t^2}\right) V(\mathbf{r}, t) = 0, \tag{22.43}$$

where

$$\nabla^2 = \frac{\partial^2}{\partial x^2} + \frac{\partial^2}{\partial y^2} + \frac{\partial^2}{\partial z^2} \tag{22.44}$$

is an operator that acts in the vector space of the vector $\mathbf{r} = x\hat{\mathbf{i}} + y\hat{\mathbf{j}} + z\hat{\mathbf{x}}$. Multiplying Eq. (22.43) by the complex conjugate of the analytic field at the space-time point $(\mathbf{r}', t')$ gives

$$\left(\nabla^2 - \frac{1}{c^2}\frac{\partial^2}{\partial t^2}\right) V(\mathbf{r}, t) V^*(\mathbf{r}', t') = 0. \tag{22.45}$$

This equation holds for all the waves $V(\mathbf{r}, t)$ in the ensemble, so it holds for the ensemble average:

$$\left(\nabla^2 - \frac{1}{c^2}\frac{\partial^2}{\partial t^2}\right) \Gamma(\mathbf{r}, t, \mathbf{r}', t') = 0. \tag{22.46}$$

In a similar way, since $V(\mathbf{r}, t) V^*(\mathbf{r}', t') = [V(\mathbf{r}, t)^* V(\mathbf{r}', t')]^*$,

$$\left(\nabla'^2 - \frac{1}{c^2}\frac{\partial^2}{\partial t'^2}\right) \Gamma(\mathbf{r}, t, \mathbf{r}', t') = 0, \tag{22.47}$$

where

$$\nabla'^2 = \frac{\partial^2}{\partial x'^2} + \frac{\partial^2}{\partial y'^2} + \frac{\partial^2}{\partial z'^2} \tag{22.48}$$

is an operator that acts in the vector space of the vector $\mathbf{r}' = x'\hat{\mathbf{i}} + y'\hat{\mathbf{j}} + z'\hat{\mathbf{k}}$.

Now, since

$$\Gamma(\mathbf{r}, t, \mathbf{r}', t') = \int_0^\infty \Gamma(\mathbf{r}, \mathbf{r}', \nu) e^{-j2\pi\nu\tau}\, d\nu, \tag{22.49}$$

where $\tau = t' - t$,

$$\frac{\partial^2}{\partial t^2}(\Gamma(\mathbf{r}, t, \mathbf{r}', t')) = \frac{\partial^2}{\partial t^2}\int_0^\infty \Gamma(\mathbf{r}, \mathbf{r}', \nu)e^{-j2\pi\nu(t'-t)}\,d\nu$$
$$= -4\pi\nu^2\int_0^\infty \Gamma(\mathbf{r}, \mathbf{r}', \nu)e^{-2\pi j\nu(t'-t)}\,d\nu. \tag{22.50}$$

Hence,

$$\nabla^2\int_0^\infty \Gamma(\mathbf{r}, \mathbf{r}', \nu)e^{-j2\pi\nu\tau}\,d\nu + \frac{4\pi}{c^2}\int_0^\infty \nu^2\Gamma(\mathbf{r}, \mathbf{r}', \nu)e^{-j2\pi\nu(t'-t)}\,d\nu = 0, \tag{22.51}$$

so

$$\nabla^2\Gamma(\mathbf{r}, \mathbf{r}', \nu) + k^2\Gamma(\mathbf{r}, \mathbf{r}', \nu) = 0, \tag{22.52}$$

where

$$k = \frac{2\pi\nu}{c} = \frac{\omega}{c}. \tag{22.53}$$

Similarly,

$$\nabla'^2\Gamma(\mathbf{r}, \mathbf{r}', \nu) + k^2\Gamma(\mathbf{r}, \mathbf{r}', \nu) = 0, \tag{22.54}$$

and $\Gamma(\mathbf{r}, \mathbf{r}', \nu)$ obeys the homogeneous Helmholtz equation.

## 22.5 Temporal coherence

*Temporal coherence* implies a fixed phase relationship between the field amplitudes at two points A and B spaced a distance apart in the direction of the wave vector as shown in Fig. 22.2. Temporal coherence can be studied with a Michelson interferometer as shown in Fig. 22.3.

The path difference between the two arms of the interferometer is $\Delta s$. Interference will be observed if $\Delta s < \ell_c$, the coherence length. The coherence length is $\ell_c = c\tau_c$, where $\tau_c$ is the coherence time. If the bandwidth of the source is $\Delta\nu$, then $\tau_c \sim 1/\Delta\nu$.

*Examples*. For a low-pressure mercury lamp $\tau_c$ is a few nanoseconds, therefore the coherence length is a few centimeters.

For a stable laser, $\Delta\nu \simeq 1$ kHz, $\tau_c \simeq 1$ ms, and the coherence length $\simeq 3 \times 10^5$ m.

If the analytic signals coming from the two arms of the interferometer in Fig. 22.3 are $V_1$ and $V_2$, then the total analytic signal at some observation point $\mathbf{x}$ is

$$V(\mathbf{x}, t) = V_1(\mathbf{x}, t) + V_2(\mathbf{x}, t + \tau) = V_1(t) + V_2(t + \tau), \tag{22.55}$$

where $\tau$ is the time difference caused by the path difference $\Delta s$: $\tau = \Delta s/c$.

Fig. 22.2 A diagram for explaining temporal coherence.

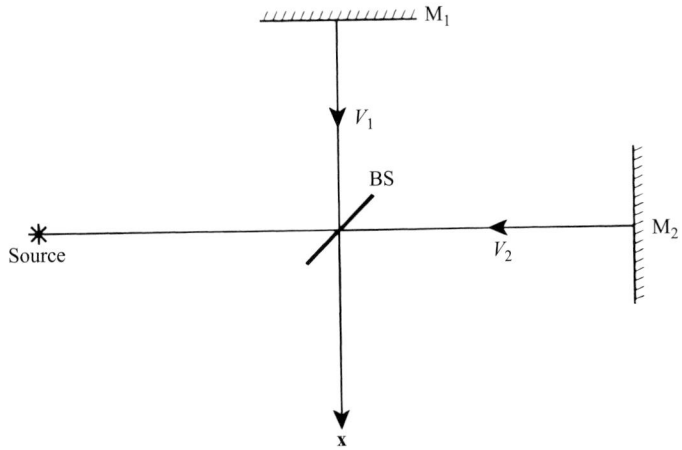


Fig. 22.3 The simple Michelson interferometer used in a discussion of the autocorrelation function of a light source. BS, beamsplitter.

The average intensity observed at $\mathbf{x}$ is

$$\begin{aligned} I(\mathbf{x}) &= \overline{V^*(\mathbf{x},t)V(\mathbf{x},t)} \\ &= \overline{V_1^*(t)V_1(t)} + \overline{V_2^*(t+\tau)V_2(t+\tau)} \\ &\quad + \overline{V_1^*(t)V_2(t+\tau)} + \overline{V_2^*(t+\tau)V_1(t)}. \end{aligned} \tag{22.56}$$

We can write $V_1 = a_1 V_0$ and $V_2 = a_2 V_0$, where $V_0(t)$ is related to the analytic field at the beamsplitter and the factors $a_1$ and $a_2$ take into account parameters such as different reflectivities at mirrors $M_1$ and $M_2$.

We can put

$$\overline{V_1^*(t)V_1(t)} = a_1^* a_1 |V_0(t)|^2 = |a_1|^2 |V_0(t)|^2 = |a_1|^2 I_0, \tag{22.57}$$

$$\overline{V_2^*(t+\tau)V_2(t+\tau)} = |a_2|^2 |V_0(t)|^2 = |a_2|^2 I_0, \tag{22.58}$$

and

$$\overline{V_1^*(t)V_2(t+\tau)} = a_1^* a_2 \Gamma_{11}(\tau), \tag{22.59}$$

where

$$\Gamma_{11}(\tau) = V_0(\mathbf{x},t)V_0(\mathbf{x},t+\tau) \tag{22.60}$$

and

$$\overline{V_2^*(t+\tau)V_1(t)} = a_2^* a_1 \Gamma_{11}(-\tau) = a_2^* a_1 \Gamma_{11}^*(\tau). \tag{22.61}$$

Thus,

$$\begin{aligned} I(\mathbf{x}) &= |a_1|^2 I_0 + |a_2|^2 I_0 + a_1^* a_2 \Gamma_{11}(\tau) + a_2^* a_1 \Gamma_{11}^*(\tau) \\ &= |a_1|^2 I_0 + |a_2|^2 I_0 + 2\,\mathcal{R}\,[a_1^* a_2 \Gamma_{11}(\tau)]. \end{aligned} \tag{22.62}$$

In the simplest case $|a_1|^2 = |a_2|^2 = K^2$ and

$$I(\mathbf{x}) = 2K^2[I_0 + \mathcal{R}\,\Gamma_{11}(\tau)], \tag{22.63}$$

and the variation in average intensity at $\mathbf{x}$ with time $\tau$ is contained in the autocorrelation function.

If $\tau > \tau_c$ then $\Gamma_{11}(\tau) = 0$: temporal coherence is measured by the autocorrelation function.

We write

$$|a_1|^2 I_0 = I_1, \tag{22.64}$$

$$|a_2|^2 I_0 = I_2. \tag{22.65}$$

The product $a_1^* a_2$ is generally real, so we can write

$$a_1^* a_2 = \sqrt{I_1 I_2}/I_0. \tag{22.66}$$

Introducing a normalized coherence function

$$\gamma(\mathbf{x}, \mathbf{x}, \tau) = \frac{\Gamma(\mathbf{x}, \mathbf{x}, \tau)}{\sqrt{I_1 I_2}} = \frac{\Gamma_{11}(\tau)}{\sqrt{I_1 I_2}}, \tag{22.67}$$

we can write

$$I(\mathbf{x}) = I_1 + I_2 + 2\sqrt{I_1 I_2}\mathcal{R}\,\gamma(\mathbf{x}, \mathbf{x}, \tau), \tag{22.68}$$

where $\gamma(\mathbf{x}, \mathbf{x}, \tau)$ is called the *degree of temporal coherence*. For the case in which $I_1 = I_2 = I$

$$\gamma(\mathbf{x}, \mathbf{x}, \tau) = \Gamma_{11}(\tau)/I. \tag{22.69}$$

If the source is quasi-monochromatic we use the slowly varying envelope (SVE) approximation and write

$$\gamma(\mathbf{x}, \mathbf{x}, \tau) = |\gamma(\mathbf{x}, \mathbf{x}, \tau)| e^{-j2\pi\nu_0\tau}, \tag{22.70}$$

where $|\gamma(\mathbf{x}, \mathbf{x}, \tau)|$ is a much more slowly varying function of time than $e^{-j2\pi\nu_0\tau}$. This separates a physically observable time variation $|\gamma(\mathbf{x}, \mathbf{x}, \tau)|$ from the rapid oscillation at the optical frequency $\nu$.

In this case

$$I(\mathbf{x}) = I_1 + I_2 + 2\sqrt{I_1 I_2}|\gamma(\mathbf{x}, \mathbf{x}, \tau)|\cos(2\pi\nu_0\tau). \tag{22.71}$$

For perfectly monochromatic light $|\gamma(\mathbf{x}, \mathbf{x}, \tau)| = 1$; $\gamma(\mathbf{x}, \mathbf{x}, \tau)$ is frequently written $\gamma(\tau)$.

We can define the coherence time exactly as

$$\tau_c^2 = \frac{1}{N}\int_{-\infty}^{\infty} \tau^2 |\gamma(\tau)|^2\, d\tau = \frac{2}{N}\int_0^{\infty} \tau^2 |\gamma(\tau)|^2\, d\tau, \tag{22.72}$$

where

$$N = \int_{-\infty}^{\infty} |\gamma(\tau)|^2 \, d\tau. \tag{22.73}$$

We can show that

$$\int_{-\infty}^{\infty} |\gamma(\tau)|^2 \, d\tau = \int_{0}^{\infty} g^2(\nu) d\nu, \tag{22.74}$$

where $g(\nu)$ is the normalized spectral density of the light or lineshape function. This is essentially Parseval's theorem [7].

Remember that

$$\Gamma(\mathbf{x}, \mathbf{x}, \nu) = \begin{cases} \int_{-\infty}^{\infty} \Gamma(\mathbf{x}, \mathbf{x}, \tau) e^{j2\pi\nu\tau} \, d\tau & \text{for } \nu \geq 0, \\ 0 \quad \text{for } \nu < 0, \end{cases} \tag{22.75}$$

$$\Gamma(\mathbf{x}, \mathbf{x}, \tau) = \int_{0}^{\infty} \Gamma(\mathbf{x}, \mathbf{x}, \nu) e^{-j2\pi\nu\tau} \, d\nu, \tag{22.76}$$

and

$$I(\mathbf{x}) = \Gamma(\mathbf{x}, \mathbf{x}, 0) = \int_{0}^{\infty} \Gamma(\mathbf{x}, \mathbf{x}, \nu) d\nu. \tag{22.77}$$

The lineshape function $g(\nu)$ is defined formally as

$$g(\nu) = \frac{\Gamma(\mathbf{x}, \mathbf{x}, \nu)}{\int_0^{\infty} \Gamma(\mathbf{x}, \mathbf{x}, \nu) d\nu} = \frac{\Gamma(\nu)}{\int_0^{\infty} \Gamma(\nu) d\nu}. \tag{22.78}$$

Clearly

$$\int_{0}^{\infty} g(\nu) d\nu = 1.$$

If we remember that $\Gamma(\mathbf{x}, \mathbf{x}, \nu)$ is 0 for $\nu < 0$, we can just as easily write

$$g(\nu) = \frac{\Gamma(\nu)}{\int_{-\infty}^{\infty} \Gamma(\nu) d\nu} = \frac{\Gamma(\nu)}{I(\mathbf{x})}, \tag{22.79}$$

in which case

$$\int_{-\infty}^{\infty} g(\nu) d\nu = 1, \tag{22.80}$$

where $g(\nu) = 0$ for $\nu < 0$.

Now

$$\begin{aligned} \gamma(\mathbf{x}, \mathbf{x}, \tau) = \frac{\Gamma(\mathbf{x}, \mathbf{x}, \tau)}{I(\mathbf{x})} &= \int_{0}^{\infty} \frac{\Gamma(\mathbf{x}, \mathbf{x}, \nu)}{I(\mathbf{x})} e^{-j2\pi\nu\tau} \, d\nu \\ &= \int_{0}^{\infty} g(\nu) e^{-j2\pi\nu\tau} \, d\nu. \end{aligned} \tag{22.81}$$

So, $\gamma(\tau)$ and $g(\nu)$ are Fourier transforms of each other. Since $g(\nu)$ is real,

$$\gamma^*(\mathbf{x}, \mathbf{x}, \tau) = \int_{0}^{\infty} g(\nu) e^{2\pi j\nu\tau} \, d\nu. \tag{22.82}$$

So

$$\int_{-\infty}^{\infty} |\gamma(\tau)|^2 \, d\tau = \int_{-\infty}^{\infty} \int_{0}^{\infty} \frac{\Gamma(\mathbf{x}, \mathbf{x}, \tau)}{I(\mathbf{x})} \frac{\Gamma^*(\mathbf{x}, \mathbf{x}, \nu)}{I(\mathbf{x})} e^{j2\pi\nu\tau} \, d\nu \, d\tau$$
$$= \int_{0}^{\infty} \frac{|\Gamma(\mathbf{x}, \mathbf{x}, \nu)|^2}{[I(\mathbf{x})]^2} \, d\nu = \int_{0}^{\infty} [g(\nu)]^2 \, d\nu. \tag{22.83}$$

## 22.6 Spatial coherence

*Spatial coherence* in the simplest case tells us about the phase relationship between the field amplitudes at two points in a plane normal to the wave vector, as shown in Fig. 22.4.

The existence, or not, of spatial coherence between the fields at two points can be demonstrated in an experiment employing Young's slits, as shown schematically in Fig. 22.5.

The intensity observed at point **r** is

$$I(\mathbf{r}) = \overline{I(\mathbf{r}, t)} = \overline{V^*(\mathbf{r}, t) V(\mathbf{r}, t)}, \tag{22.84}$$

where

$$V(\mathbf{r}, t) = a_1 V(\mathbf{r}_1, t - t_1) + a_2 V(\mathbf{r}_2, t - t_2). \tag{22.85}$$

$V(\mathbf{r}_1, t - t_1)$ is the analytic field at pinhole $S_1$, etc., and $a_1$ and $a_2$ are complex, time-independent transmission factors that depend on the mean frequency of the wave, the size

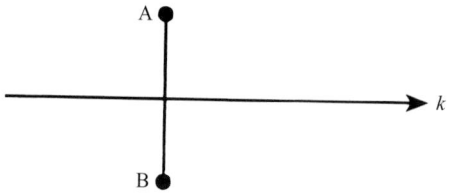


**Fig. 22.4** A diagram for explaining spatial coherence.

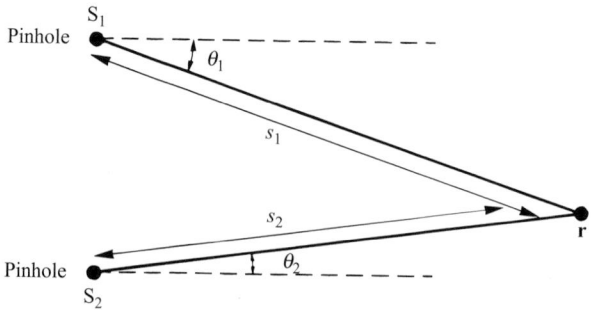


**Fig. 22.5** An interference experiment using Young's slits with two pinholes, $S_1$ and $S_2$.

of the pinholes, and the geometry of the experiment, for example the distances $s_1$ and $s_2$ from each of the pinholes to point $\mathbf{r}$.

In a similar way to our previous analysis, we get

$$I(\mathbf{r}) = \overline{I(\mathbf{r}, t)} = |a_1|^2 I(\mathbf{r}_1) + |a_2|^2 I(\mathbf{r}_2) + 2\mathcal{R}[a_1^* a_2 \Gamma(\mathbf{r}_1, \mathbf{r}_2, \tau)], \tag{22.86}$$

where

$$\Gamma_{12}(\tau) = \Gamma(\mathbf{r}_1, \mathbf{r}_2, \tau) = \overline{V^*(\mathbf{r}_1, t) V(\mathbf{r}_2, t + \tau)} \tag{22.87}$$

is the mutual correlation function, and $\tau = t_1 - t_2$.

We have used the stationarity condition

$$\overline{V^*(\mathbf{r}_1, t - t_1) V(\mathbf{r}_2, t - t_2)} = \overline{V^*(\mathbf{r}_1, t) V[\mathbf{r}_2, t + (t_1 - t_2)]}. \tag{22.88}$$

$|a_1|^2 I(\mathbf{r}_1)$ is the intensity arising at point $\mathbf{r}$ from pinhole $S_1$ alone, if pinhole $S_2$ is closed.

It can be shown that $a_1^* a_2$ is real. In fact

$$a_1 = \frac{jk \cos\theta_1}{2\pi s_1} = \frac{jks_1}{2\pi s_1^2} \cos\theta_1, \tag{22.89}$$

where $\theta_1$ is the angle shown in Fig. 22.5.

We can identify the $2\pi s_1^2$ factor in the denominator as the surface area of the half sphere illuminated through the pinhole, $jks_1$ is a complex amplitude, and $\cos\theta_1$ is an obliquity factor. The intensity at pinhole 1 is

$$I(\mathbf{r}_1) = \Gamma(\mathbf{r}_1, \mathbf{r}_1, 0) = I_1. \tag{22.90}$$

We can introduce the *degree of coherence*

$$\gamma(\mathbf{r}_1, \mathbf{r}_2, \tau) = \gamma_{12}(\tau) = \frac{\Gamma(\mathbf{r}_1, \mathbf{r}_2, \tau)}{\sqrt{\Gamma(\mathbf{r}_1, \mathbf{r}_1, 0)\Gamma(\mathbf{r}_2, \mathbf{r}_2, 0)}} = \frac{\Gamma_{12}(\tau)}{\sqrt{I_1 I_2}}. \tag{22.91}$$

If we write

$$|a_1|^2 I(\mathbf{r}_1) = I_1(\mathbf{r}), \tag{22.92}$$

$$|a_2|^2 I(\mathbf{r}_2) = I_2(\mathbf{r}), \tag{22.93}$$

then

$$I(\mathbf{r}) = I_1(\mathbf{r}) + I_2(\mathbf{r}) + 2\sqrt{I_1(\mathbf{r}) I_2(\mathbf{r})}\mathcal{R}\left[\gamma(\mathbf{r}_1, \mathbf{r}_2, \tau)\right]. \tag{22.94}$$

If the light illuminating both pinholes is monochromatic, then we can write

$$V(\mathbf{r}_1, t - t_1) = \sqrt{I_1} e^{-j2\pi\nu(t - t_1)}, \tag{22.95}$$

which gives

$$I(\mathbf{r}) = I_1(\mathbf{r}) + I_2(\mathbf{r}) + 2\sqrt{I_1(\mathbf{r}) I_2(\mathbf{r})} \cos[2\pi\nu(t_2 - t_1)]. \tag{22.96}$$

Equation (22.96) describes an ideal two-pinhole interference pattern.

From Schwarz's inequality[6] it follows that

$$\mathcal{R}\Gamma_{12}(\tau) \leq |\Gamma_{12}(\tau)| \leq \sqrt{\Gamma_{11}(\tau)\Gamma_{22}(\tau)} \leq \sqrt{I_1 I_2}, \tag{22.97}$$

and therefore,

$$\mathcal{R}\gamma_{12}(\tau) \leq |\gamma_{12}(\tau)| \leq 1. \tag{22.98}$$

If we are dealing with light that is not quite monochromatic, but is quasi-monochromatic, contained within a frequency band $\Delta\nu$ much less than its center frequency $\nu_0$, it is reasonable to write

$$\gamma_{12}(\tau) = |\gamma_{12}(\tau)| e^{-j2\pi\nu_0\tau}. \tag{22.99}$$

This is the SVE approximation again; $|\gamma_{12}(\tau)|$ is a slowly varying function that remains constant over many oscillations of the phase factor.

In this case

$$I(\mathbf{r}) = I_1(\mathbf{r}) + I_2(\mathbf{r}) + 2\sqrt{I_1(\mathbf{r})I_2(\mathbf{r})]}|\gamma_{12}(\tau)|\cos(2\pi\nu_0\tau). \tag{22.100}$$

Such an interference pattern would show the characteristic light and dark regions of a Young's-slits interference pattern in the center of the pattern, but the fringes would grow less distinct on moving away from the center. This is the effect of temporal coherence showing up. As we move away from the center of the interference pattern the time delay difference, $\tau$, for propagation from the pinholes grows larger. The time delay at the edges of the pattern approaches the value $\tau = 2d/c$, where $2d$ is the pinhole spacing.

Defining the visibility, $V$, of the fringes as

$$V = \frac{I_{\max} - I_{\min}}{I_{\max} + I_{\min}} \tag{22.101}$$

gives

$$V = \frac{2\sqrt{I_1 I_2}\,|\gamma_{12}(\tau)|}{I_1 + I_2}, \tag{22.102}$$

which, when $I_1(\mathbf{r}) = I_2(\mathbf{r})$, gives

$$V = |\gamma_{12}(t_1 - t_2)|. \tag{22.103}$$

For incoherent illumination of the two pinholes, $|\gamma_{12}(\tau)| = 0$ and no fringes are observed.

Perfect fringes are observed in the case of completely coherent illumination,

$$|\gamma_{12}(\tau)| = 1. \tag{22.104}$$

The intermediate case $0 < |\gamma_{12}(\tau)| < 1$ represents partial coherence.

[6] Schwarz's inequality states that for two functions, $f(x)$ and $g(x)$,

$$\int ff^*\,dx \int gg^*\,dx \geq \left|\int f^* g\,dx\right|^2.$$

To distinguish between the factors, $\gamma_{12}(\tau)$ is often called the *complex degree of coherence* and $|\gamma_{12}(\tau)|$ the *degree of coherence*.

## 22.7 Spatial coherence with an extended source

Consider the case of two pinholes illuminated by an extended source that takes the form of a square with sides of length $\Delta a$ as shown in Fig. 22.6.

If fringes are seen, then there exists at least some spatial coherence between light arriving at Q from $P_1$ and $P_2$. For this to happen, the overall field at $P_1$, which results from a superposition of the disturbances originating from all points on the source, must be phase-related to the overall field at $P_2$. If each point on the source is an independent emitter, the overall fields will have a stable phase relationship, provided that the extreme ends of the source, $A_1$ and $A_2$, are equidistant from $P_1$ (or $P_2$) within a half wavelength. It is easy to show that this leads to the condition $\theta\ \Delta a < \lambda_0$ (for quasi-monochromatic light of center wavelength $\lambda_0$), so the pinholes must be arranged within an area around O of size

$$\Delta A \approx (r\theta)^2 \simeq \frac{r^2\lambda_0^2}{\Delta a^2} = \frac{c^2 r^2}{\nu_0^2 S}, \tag{22.105}$$

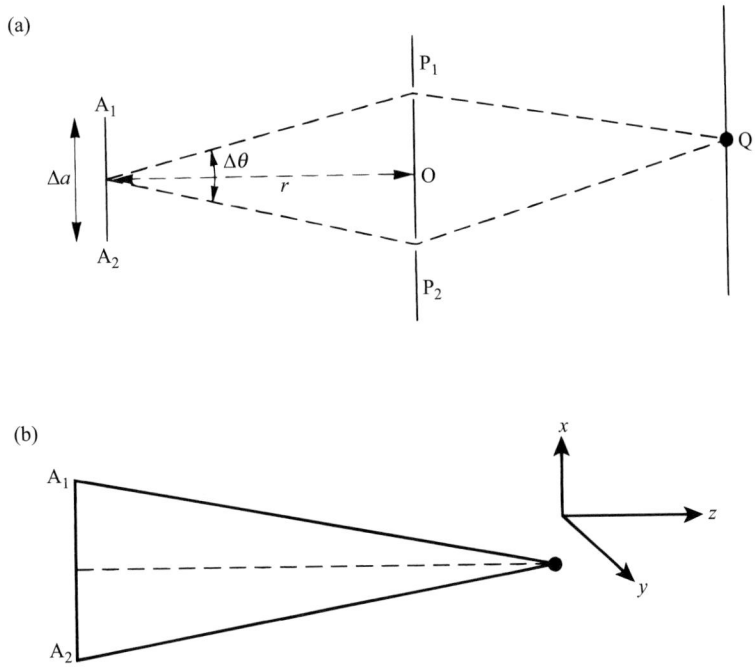


Fig. 22.6 (a) The geometry of a Young's-slits experiment where an extended source is used. (b) The coordinate system used in discussing uncertainties in photon momentum.

where $S$ is the area of the source. $\Delta A$ is called the coherence area in plane $A$. The volume of coherence is

$$\Delta V = \Delta A\, \ell_c = \frac{c}{\Delta\nu}\frac{c^2}{\nu_0^2}\frac{r^2}{S} = \frac{\lambda_0}{\Delta\lambda}\left(\frac{r}{\Delta a}\right)^2 \lambda_0^3. \tag{22.106}$$

The volume of coherence defined in this way has a length in the direction of propagation of the light that is equal to the coherence length, and a cross-section equal to the coherence area.

We can show that this volume is also the volume corresponding to one cell of phase space of the photons from the source.

The volume of an elementary cell of phase space is given by [8, 9]

$$\Delta p_x\, \Delta p_y\, \Delta p_z\, \Delta q_x\, \Delta q_y\, \Delta q_z = h^3, \tag{22.107}$$

where $\Delta p_x, \ldots, \Delta q_x, \ldots$ are the uncertainties in the components of the momentum and position, respectively, of the photon. We can estimate these uncertainties from the geometry of the double pinholes shown in Fig. 22.6(b).

The momentum of a photon is $h/\lambda_0$. The momentum in the $x$ direction for photons from $A_1$ is $-(h/\lambda_0)\sin(\theta/2) \simeq -h\theta/(2\lambda_0)$, and that for photons from $A_2$ is $h\theta/(2\lambda_0)$. Thus,

$$\Delta p_x = \Delta p_y = \frac{h\theta}{\lambda_0} = \frac{h}{\lambda_0}\frac{\Delta a}{r}. \tag{22.108}$$

$\Delta p_z$ comes mainly from uncertainties in the frequency of the source, i.e.,

$$\Delta p_z = \frac{h}{\lambda_0^2}\,\Delta\lambda, \tag{22.109}$$

so

$$\frac{h^2\,\Delta a^2}{\lambda_0^2 r^2}\cdot\frac{h\,\Delta\lambda}{\lambda_0^2}\,\Delta q_x\,\Delta q_y\,\Delta q_z = h^3, \tag{22.110}$$

which gives

$$\Delta q_x\,\Delta q_y\,\Delta q_z = \frac{\lambda_0}{\Delta\lambda}\left(\frac{r}{\Delta a}\right)^2 \lambda_0^3, \tag{22.111}$$

which is identical with the volume of coherence.

We can introduce the degeneracy parameter $\delta$ that represents the average number of photons in the same state of polarization in the volume of coherence. Thus, it represents the average number of photons in the same polarization state crossing the area of coherence per coherence time. If $N_\nu$ is the average number of photons emitted from unit area of the source, per unit frequency interval, per unit solid angle around the direction normal to the source, then it is easy to show that

$$\delta = \frac{1}{2}N_\nu S\,\Delta\nu\,\Delta\Omega\,\tau_c. \tag{22.112}$$

$S$ is again the area of the source, $\Delta\Omega$ is the solid angle subtended by the coherence area at the source, $\Delta\nu$ is the bandwidth of the source, and $\tau_c$ is its coherence time. The factor of

$\frac{1}{2}$ arises because the light coming from the source is assumed to be unpolarized. The solid angle is

$$\Delta\Omega = \frac{\Delta A}{r^2} = \frac{c^2}{\nu_0^2 S}, \tag{22.113}$$

and, because $\Delta\nu\ \tau_c \simeq 1$,

$$\delta \simeq \frac{c^2}{2\nu_0^2} N_\nu. \tag{22.114}$$

For black-body radiation the energy density $\rho(\nu)$ (J m$^{-3}$ Hz$^{-1}$) is

$$\rho(\nu) = \frac{8\pi h\nu^3}{c^3} \frac{1}{e^{h\nu/(k_B T)} - 1}, \tag{22.115}$$

so the photon flux $N_\nu$ is

$$\begin{aligned} N_\nu &= \frac{\rho(\nu)c}{4\pi h\nu} = \frac{c}{4\pi} \frac{8\pi h\nu^3}{c^3 h\nu} \frac{1}{e^{h\nu/(k_B T)} - 1} \\ &= \frac{2\nu^2}{c^2} \frac{1}{e^{h\nu/(k_B T)} - 1}. \end{aligned} \tag{22.116}$$

Therefore, from Eq. (22.114),

$$\delta \simeq 1/(e^{h\nu/(k_B T)} - 1). \tag{22.117}$$

This is identical with the mean number of photons in the mode – the occupation number $\overline{n}$, introduced in Chapter 1, which is generally very small, except at low frequencies and/or high temperatures.

## 22.8 Propagation laws of partial coherence

The classic interference experiment employing Young's slits is carried out with a point source illuminating the two pinholes. This ensures that the two pinholes will be within the area of coherence. One could imagine, in principle, this experiment being done using a star as the point source, with a suitable narrow-band filter placed over the pinholes to ensure quasi-monochromaticity. It is clear, however, that, if it were possible to place the two pinholes very close to the star, no fringes would be observed, because each pinhole would be illuminated by light from different points on the surface of the star – the pinholes would no longer be within the area of coherence. Therefore, the placement of the pinholes relative to an *extended* source is crucial in determining whether fringes will be seen. We can analyze this phenomenon mathematically by showing that the mutual coherence function varies as it propagates, so that the fields at laterally separated points go from no coherence very near the source to a high degree of coherence far away.

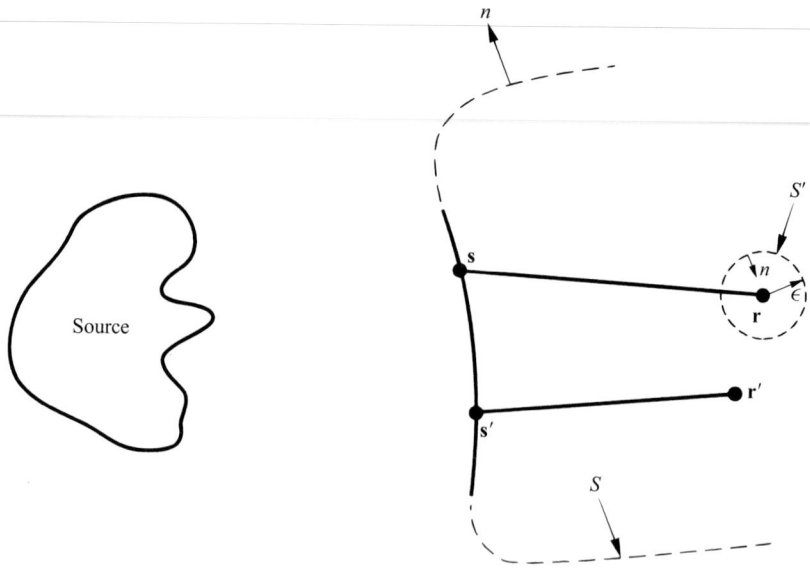


**Fig. 22.7** A diagram used in discussion of the propagation of the mutual coherence function from a surface $S$ to a surface $S'$.

We assume that we know the mutual coherence function $\Gamma_{12}(\tau) = \Gamma(\mathbf{r}, \mathbf{r}', \tau)$ over a certain surface $S$ of space. The problem is to determine $\Gamma_{12}(\tau) = \Gamma(\mathbf{r}, \mathbf{r}', \tau)$ in terms of its value $\Gamma(\mathbf{s}, \mathbf{s}', \tau)$ for points $\mathbf{s}$ and $\mathbf{s}'$ on the surface $S$, as shown in Fig. 22.7.

Now, we know that the mutual coherence function $\Gamma(\mathbf{r}, \mathbf{r}', \nu)$ satisfies the homogeneous Helmholtz equation

$$(\nabla^2 + k^2)\Gamma(\mathbf{r}, \mathbf{r}', \nu) = 0, \tag{22.118}$$

and

$$(\nabla'^2 + k^2)\Gamma(\mathbf{r}, \mathbf{r}', \nu) = 0. \tag{22.119}$$

We choose a Green's function $G(\mathbf{r}, \mathbf{s}, \nu)$ that satisfies

$$(\nabla^2 + k^2)G(\mathbf{r}, \mathbf{s}, \nu) = -\delta(\mathbf{r} - \mathbf{s}). \tag{22.120}$$

This Green's function is like a source of waves placed at $\mathbf{r}$. We have

$$G\nabla^2\Gamma + Gk^2\Gamma = 0,$$

where

$$\Gamma \equiv \Gamma(\mathbf{s}, \mathbf{r}', \nu) \text{ and } G = G(\mathbf{r}, \mathbf{s}, \nu), \tag{22.121}$$

and also

$$\Gamma\,\nabla^2 G + \Gamma k^2 G = -\delta(\mathbf{r} - \mathbf{s})\Gamma(\mathbf{s}, \mathbf{r}', \nu). \tag{22.122}$$

Subtracting Eq. (22.122) from Eq. (22.121) gives

$$G\,\nabla^2\Gamma - \Gamma\nabla^2 G = \delta(\mathbf{r} - \mathbf{s})\Gamma(\mathbf{s}, \mathbf{r}', \nu). \tag{22.123}$$

Using Green's theorem [8]

$$\int_V (G\,\nabla^2\Gamma - \Gamma\,\nabla^2 G)dV = \int_S (G\,\nabla\Gamma - \Gamma\,\nabla\mathbf{G})\cdot d\mathbf{S}, \tag{22.124}$$

where $V$ is the volume contained within the closed surface $S$, gives

$$\begin{aligned}\int_S (G\,\nabla\Gamma - \Gamma\,\nabla\mathbf{G})\cdot d\mathbf{S} &= \int_S \left(G\frac{\partial\Gamma}{\partial n} - \frac{\partial G}{\partial n}\Gamma\right) dS \\ &= \int_V \delta(\mathbf{r}-\mathbf{s})\Gamma(\mathbf{s},\mathbf{r}',\nu)dV \\ &= \Gamma(\mathbf{r},\mathbf{r}',\nu),\end{aligned} \tag{22.125}$$

where $\partial/\partial n$ is a derivative in the positive normal direction (pointing outward) on the surface $S$.

We must be careful about including the singular point $\mathbf{r}$ inside the integration volume in Eq. (22.125), but in this case it can be shown to be justified. If we use the expression $G = e^{jk|\mathbf{r}-\mathbf{s}|}/(4\pi|\mathbf{r}-\mathbf{s}|)$ and exclude the singular point at $\mathbf{s} = \mathbf{r}$ by taking a surface $S'$ of radius $\epsilon$ as shown in Fig. 22.7, then in this case the volume integral goes to zero and

$$\int_S \left(G\frac{\partial\Gamma}{\partial n} - \frac{\partial G}{\partial n}\Gamma\right) dS = -\int_{S'} \left(G\frac{\partial\Gamma}{\partial n} - \frac{\partial G}{\partial n}\Gamma\right) dS. \tag{22.126}$$

Writing $G = e^{jk\epsilon}/\epsilon$ gives

$$\int_S \left(G\frac{\partial\Gamma}{\partial n} - \frac{\partial G}{\partial n}\Gamma\right) dS = -\int_{S'} \frac{e^{jk\epsilon}}{4\pi\epsilon}\frac{\partial\Gamma}{\partial n}(\mathbf{r}) - \Gamma(\mathbf{r})\left(\frac{e^{jk\epsilon}}{4\pi\epsilon^2} - \frac{jke^{jk\epsilon}}{4\pi\epsilon}\right) dS. \tag{22.127}$$

It should be noted that we have used $\partial/\partial n = -\partial/\partial\epsilon$ because the normal $n$ points away from surface $S'$ into the small volume surrounding point $\mathbf{r}$. Thus

$$\begin{aligned}\int_S \left(G\frac{\partial\Gamma}{\partial n} - \frac{\partial G}{\partial n}\Gamma\right) dS &= -4\pi\epsilon^2\left[\frac{e^{jk\epsilon}}{4\pi\epsilon}\frac{\partial\Gamma}{\partial n}(\mathbf{r}) - \Gamma(\mathbf{r})\left(\frac{e^{jk\epsilon}}{4\pi\epsilon}\right)\left(\frac{1}{\epsilon} - jk\right)\right] \\ &= \Gamma(\mathbf{r}) \text{ in the limit as } \epsilon \to 0.\end{aligned} \tag{22.128}$$

This justifies Eq. (22.125). In summary,

$$\Gamma(\mathbf{r},\mathbf{r}',\nu) = \int_S \left\{G(\mathbf{r},\mathbf{s},\nu)\frac{\partial\Gamma}{\partial n}(\mathbf{s},\mathbf{r}',\nu) - \frac{\partial G}{\partial n}(\mathbf{r},\mathbf{s},\nu)\Gamma(\mathbf{s},\mathbf{r}',\nu)\right\} dS. \tag{22.129}$$

It is particularly convenient if we choose a Green's function such that $G(\mathbf{r},\mathbf{s},\nu) = 0$ on the surface $S$. Then

$$\Gamma(\mathbf{r},\mathbf{r}',\nu) = \int_S -\frac{\partial G}{\partial n}(\mathbf{r},\mathbf{s},\nu)\Gamma(\mathbf{s},\mathbf{r}',\nu)dS, \tag{22.130}$$

where $\partial/\partial n$ is a normal derivative at point $\mathbf{s}$. We can now use a second Green's function that satisfies

$$(\nabla'^2 + k^2)G'(\mathbf{r}',\mathbf{s}',\nu) = -\delta(\mathbf{r}'-\mathbf{s}'), \tag{22.131}$$

and which also vanishes on the surface $S$, to get in a similar way

$$\Gamma(\mathbf{s}, \mathbf{r}', \nu) = \int_S -\frac{\partial G'}{\partial n'}(\mathbf{r}', \mathbf{s}', \nu)\Gamma(\mathbf{s}, \mathbf{s}', \nu)(\mathbf{s}, \mathbf{s}', \nu)dS', \tag{22.132}$$

where $\partial/\partial n'$ is a normal derivative at point $\mathbf{s}'$. Combining Eqs. (22.130) and (22.132) gives

$$\Gamma(\mathbf{r}, \mathbf{r}', \nu) = \int_S \int_S \frac{\partial G}{\partial n}(\mathbf{r}, \mathbf{s}, \nu)\frac{\partial G'}{\partial n'}(\mathbf{r}', \mathbf{s}', \nu)\Gamma(\mathbf{s}, \mathbf{s}', \nu)dS\,dS', \tag{22.133}$$

and $\Gamma(\mathbf{r}, \mathbf{r}', \nu)$ can be determined from its value $\Gamma(\mathbf{s}, \mathbf{s}', \nu)$ on a surface.

Since we have already shown that $\Gamma(\mathbf{r}, \mathbf{r}, \nu) = \Gamma^*(\mathbf{r}, \mathbf{r}, \nu)$, it is straightforward to show that

$$\Gamma(\mathbf{r}, \mathbf{r}', \nu) = \Gamma^*(\mathbf{r}', \mathbf{r}, \nu). \tag{22.134}$$

Hence,

$$G(\mathbf{r}, \mathbf{s}, \nu) = G'^*(\mathbf{r}, \mathbf{s}, \nu). \tag{22.135}$$

If we define

$$K(\mathbf{r}', \mathbf{s}', \nu) = \frac{\partial G'}{\partial n'}(\mathbf{r}', \mathbf{s}', \nu) \tag{22.136}$$

then

$$\Gamma(\mathbf{r}, \mathbf{r}', \nu) = \int_S \int_S K^*(\mathbf{r}, \mathbf{s}, \nu)K(\mathbf{r}', \mathbf{s}', \nu)\Gamma(\mathbf{s}, \mathbf{s}', \nu)dS\,dS'. \tag{22.137}$$

$K(\mathbf{r}, \mathbf{s}, \nu)$ is called the diffraction function – this is the complex amplitude at $\mathbf{r}$ caused by a unit point source of frequency $\nu$ situated at $\mathbf{s}$. The mutual coherence function between the two points $\mathbf{r}$ and $\mathbf{r}'$ is

$$\Gamma(\mathbf{r}, \mathbf{r}', \tau) = \int_0^\infty \Gamma(\mathbf{r}, \mathbf{r}', \nu)e^{-j2\pi\nu\tau}\,d\nu. \tag{22.138}$$

## 22.9 Propagation from a finite plane surface

When we used Green's theorem in the preceding section we were converting a volume integral to an integral over a closed surface. However, if we are concerned with finite-area surfaces, which are not closed, we have to justify neglecting the integral over the rest of the closed surface that surrounds the finite region in which we are interested.

Now we have already shown that

$$\Gamma(\mathbf{r}, \mathbf{r}', \nu) = \int_S \left\{ G(\mathbf{r}, \mathbf{s}, \nu)\frac{\partial \Gamma}{\partial n}(\mathbf{s}, \mathbf{r}', \nu) - \frac{\partial G}{\partial n}(\mathbf{r}, \mathbf{s}, \nu)\Gamma(\mathbf{s}, \mathbf{r}', \nu) \right\} dS. \tag{22.139}$$

A typical basic Green's function that could be used in this equation is

$$G(\mathbf{r}, \mathbf{s}, \nu) = \frac{e^{jk|\mathbf{r}-\mathbf{s}|}}{4\pi|\mathbf{r}-\mathbf{s}|}. \tag{22.140}$$

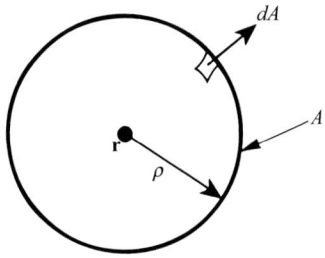


Fig. 22.8 The spherical volume surrounding a point $\mathbf{r}$ where a Green's function such as given in Eq. (22.140) becomes singular.

It is straightforward to show that this function satisfies the scalar Helmholtz equation

$$(\nabla_s^2 + k^2)G(\mathbf{r}, \mathbf{s}, \nu) = -\delta(\mathbf{r} - \mathbf{s}), \tag{22.141}$$

the $s$ subscript on $\nabla_s$ indicates differentiation with respect to the coordinates of point $\mathbf{s}$. Except at the point $\mathbf{r}$ the Green's function satisfies

$$(\nabla_s^2 + k^2)G(\mathbf{r}, \mathbf{s}, \nu) = 0. \tag{22.142}$$

It follows that

$$(\nabla_s^2 + k^2)\frac{e^{jk|\mathbf{r}-\mathbf{s}|}}{4\pi\,|\mathbf{r} - \mathbf{s}|} = \frac{e^{jk(|\mathbf{r}-\mathbf{s}|)}}{4\pi}\,\nabla_s^2\left(\frac{1}{|\mathbf{r} - \mathbf{s}|}\right). \tag{22.143}$$

We can show that the right-hand side of Eq. (22.143) is a $\delta$-function. $\nabla_s^2(1/|\mathbf{r} - \mathbf{s}|)$ is zero except when $|\mathbf{r} - \mathbf{s}| = 0$, and consequently

$$\int_{\text{all space}} \frac{e^{jk|\mathbf{r}-\mathbf{s}|}}{4\pi}\,\nabla_s^2\left(\frac{1}{|\mathbf{r} - \mathbf{s}|}\right) = \frac{1}{4\pi}\int_{\text{all space}} \nabla_s^2\left(\frac{1}{\mathbf{r} - \mathbf{s}}\right). \tag{22.144}$$

We can evaluate this integral by surrounding the point $\mathbf{r}$ by a small sphere $V$ of radius $\rho$ and surface $A$ as shown in Fig. 22.8.

Writing $|\mathbf{r} - \mathbf{s}| = R$,

$$\frac{1}{4\pi}\int_{\text{all space}} \nabla_s^2\left(\frac{1}{R}\right) = \frac{1}{4\pi}\int_V \nabla_s^2\left(\frac{1}{R}\right) = \frac{1}{4\pi}\int A\,\nabla_s\left(\frac{1}{R}\right)\cdot d\mathbf{A}, \tag{22.145}$$

where $d\mathbf{A}$ is outward-pointing. Finally,

$$\frac{1}{4\pi}\int_A \nabla_s\left(\frac{1}{R}\right)\cdot d\mathbf{A} = \frac{1}{4\pi}\int_A \frac{\rho\hat{\mathbf{r}}}{\rho^3}\cdot\rho^3\hat{\mathbf{r}}\,d\Omega = -1, \tag{22.146}$$

where $\hat{\mathbf{r}}$ is a unit vector in the direction of $\rho$ and $d\Omega$ is an elementary solid angle. This verifies Eq. (22.141).

Consider the finite aperture shown in Fig. 22.9, where the rest of the plane is, or can be imagined to be, blocked off with an opaque screen. The closed surface $S_0$ under

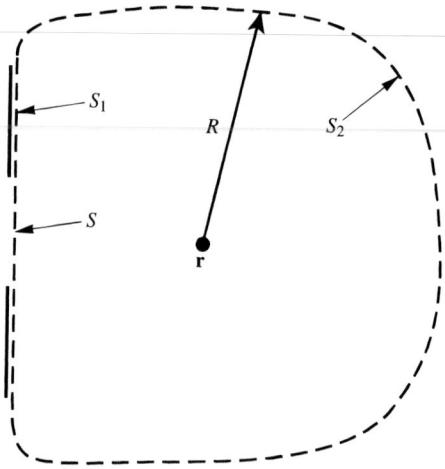


**Fig. 22.9** A closed surface $S_0$ consisting of three parts, namely an illuminated aperture $S$, an infinite opaque screen $S_1$ and an infinite half "sphere," $S_2$. In the limit the distance $R$ of a point $\mathbf{r}$ becomes infinite.

consideration consists of three parts, namely the aperture $S$, the infinite screen $S_1$, and the infinite half sphere $S_2$.

Now,

$$\Gamma(\mathbf{r}, \mathbf{r}', \nu) = \int_{S_0} \left[ G(\mathbf{r}, \mathbf{s}, \nu) \frac{\partial \Gamma(\mathbf{s}, \mathbf{r}', \nu)}{\partial n} - \frac{\partial G(\mathbf{r}, \mathbf{s}, \nu)}{\partial n} \Gamma(\mathbf{s}, \mathbf{r}', \nu) \right] dS$$
$$= \int_{S+S_1+S_2} \left[ G(\mathbf{r}, \mathbf{s}, \nu) \frac{\partial \Gamma(\mathbf{s}, \mathbf{r}', \nu)}{\partial n} - \frac{\partial G(\mathbf{r}, \mathbf{s}, \nu)}{\partial n} \Gamma(\mathbf{s}, \mathbf{r}', \nu) \right] dS. \quad (22.147)$$

Now, if $G = [1/(4\pi)]e^{-jkR}/R$ then, on $S_2$, $\mathbf{r}$ is essentially at the center of the sphere since on $S_2$, $R$ is large:

$$\frac{\partial G}{\partial n} = \frac{1}{4\pi} \left( jk - \frac{1}{R} \right) \frac{e^{jkR}}{R} \approx jkG \text{ as } R \to \infty. \quad (22.148)$$

Therefore, the surface integral over $S_2$ becomes

$$\int_{S_2} \left( G \frac{\partial \Gamma}{\partial n} - jkG\Gamma \right) dS = \int_{\Omega} G \left( \frac{\partial \Gamma}{\partial n} - jk\Gamma \right) R^2 \, d\omega, \quad (22.149)$$

where $\Omega$ is the solid angle subtended by $S_2$ at $\mathbf{r}$ and $\Gamma = \Gamma(\mathbf{s}, \mathbf{r}', \nu)$. Now $|RG|$ is finite on $S_2$; therefore the integral in Eq. (22.147) will vanish, provided that $\Gamma$ has the property

$$\lim R \to \infty, \quad R \left( \frac{\partial \Gamma}{\partial n} - jk\Gamma \right) = 0 \quad (22.150)$$

over the whole solid angle. This requirement is, in scalar diffraction theory, which involves similar considerations, known as the *Sommerfeld radiation condition* [10, 11], and is satisfied if $\Gamma$ vanishes at least as fast as a diverging spherical wave.

The rest of the surface integral can be reduced to an integral only over the illuminated aperture, provided that $\Gamma$ fulfills the *Kirchhoff boundary conditions*: (1) across the aperture

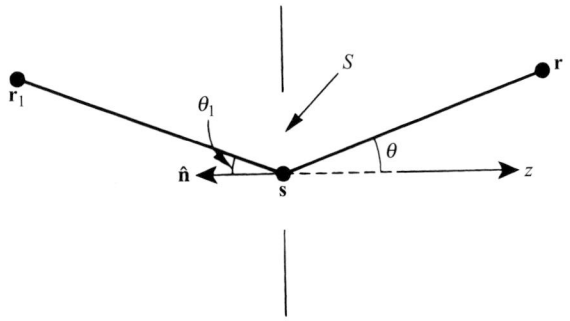


**Fig. 22.10** The geometry used in discussing a Green's function that vanishes over a plane. Point $\mathbf{r}_1$ is a mirror image of $\mathbf{r}$ in the plane $S$.

$S$ the field distribution $\Gamma$ and its derivative $\partial\Gamma/\partial n$ are exactly the same as they would be in the absence of the screen; and (2) over the portion of the surface obscured by the screen, $S_1$, the field distribution $\Gamma$ and $\partial\Gamma/\partial n$ are identically zero.

Thus, we have finally

$$\Gamma(\mathbf{r}, \mathbf{r}', \nu) = \int_{S(\text{aperture only})} \left( G\frac{\partial\Gamma}{\partial n} - \frac{\partial G}{\partial n}\Gamma \right) dS. \tag{22.151}$$

Unfortunately, the Kirchhoff boundary conditions are unnecessarily restrictive. The requirement that $\partial\Gamma/\partial n$ vanishes at the boundary between the aperture and screen is unreasonable. This restriction can be avoided by choosing a Green's function that vanishes over the plane surface comprising $S + S_1$. In studies of diffraction this is called the Rayleigh–Sommerfeld theory [3]. In this case

$$\Gamma(\mathbf{r}, \mathbf{r}', \nu) = \int_S -\frac{\partial G}{\partial n}\Gamma \, dS. \tag{22.152}$$

Green's functions that vanish over a plane do exist. For example, suppose that $G$ is generated by two identical point sources, one at $\mathbf{r}$ and one in the mirror image position $\mathbf{r}_1$ on the opposite side of the aperture, but, as shown schematically in Fig. 22.10, oscillating in antiphase. In this case

$$G(\mathbf{r}, \mathbf{s}, \nu) = \frac{1}{4\pi}\left( \frac{e^{jk|\mathbf{r}-\mathbf{s}|}}{|\mathbf{r}-\mathbf{s}|} - \frac{e^{jk|\mathbf{r}_1-\mathbf{s})}}{|\mathbf{r}_1-\mathbf{s}|} \right). \tag{22.153}$$

Clearly if $\mathbf{r}_1$ lies on $S$ this function vanishes, since then $\mathbf{r}$ lies on $S$ also. The Green's function $G^*(\mathbf{r}, \mathbf{s}, \nu)$ would be equally satisfactory in this regard.

If $\mathbf{s}$ is the point $(x_s, y_s, z_s)$, $\mathbf{r}$ is the point $(x, y, z)$, and $\mathbf{r}_1$ the point $(x_1, y_1, z_1)$, where

$$z_1 = z_s - (z - z_s) = 2z_s - z, \tag{22.154}$$

as shown in Fig. 22.10, then

$$|\mathbf{r}-\mathbf{s}| = \sqrt{(x-x_s)^2 + (y-y_s)^2 + (z-z_s)^2} \tag{22.155}$$

and

$$|\mathbf{r}_1 - \mathbf{s}| = \sqrt{(x - x_s)^2 + (y - y_s)^2 + (z + z_s)^2}. \tag{22.156}$$

$\hat{\mathbf{n}}$ is an outward pointing unit vector in Fig. 22.10.

The Green's function above can be written as

$$G(\mathbf{r}, \mathbf{s}, \nu) = \frac{1}{4\pi}\left(\frac{e^{jkR}}{R} - \frac{e^{jkR_1}}{R_1}\right) \tag{22.157}$$

and the normal derivative as

$$\frac{\partial}{\partial n} = +\frac{\partial}{\partial R}\cos\theta; \quad \frac{\partial}{\partial n} = -\frac{\partial}{\partial R_1}\cos\theta_1 = -\frac{\partial}{\partial R_1}\cos\theta. \tag{22.158}$$

Hence, for the above Green's function,

$$\frac{\partial G}{\partial n} = \frac{1}{4\pi}\left\{\cos\theta\left(jk - \frac{1}{R}\right)\frac{e^{jkR}}{R} + \cos\theta\left(jk - \frac{1}{R_1}\right)\frac{e^{jkR_1}}{R_1}\right\}. \tag{22.159}$$

Moreover, $R_1 = R$, and we have finally

$$\frac{\partial G}{\partial n} = 2\left(jk - \frac{1}{R}\right)\frac{e^{jkR}}{4\pi R}\cos\theta, \tag{22.160}$$

where $R = |\mathbf{r} - \mathbf{s}|$.

A corresponding Green's function for the variables $\mathbf{r}'$ and $\mathbf{s}'$ is

$$G(\mathbf{r}', \mathbf{s}', \nu) = \frac{1}{4\pi}\left(\frac{e^{jkR'}}{R'} - \frac{e^{jkR'_1}}{R'_1}\right), \tag{22.161}$$

with $R'_1 = R'$. So

$$\frac{\partial G'}{\partial n'} = 2\left(jk - \frac{1}{R'}\right)\frac{e^{jkR'}}{4\pi R'}\cos\theta'. \tag{22.162}$$

The angle between $\mathbf{r} - \mathbf{s}$ and the normal is $\theta$, and that between $\mathbf{r}' - \mathbf{s}'$ and the normal is $\theta'$. The primes on the variables in Eqs. (22.161) and (22.162) are not intended to indicate that the Green's functions $G'$ and $G$ are fundamentally different, but to distinguish the points over which the integration is being performed.

Restating Eq. (22.133), using $G = G'^*$,

$$\Gamma(\mathbf{r}, \mathbf{r}', \nu) = \int_S\int_S \frac{\partial G^*}{\partial n}(\mathbf{r}, \mathbf{s}, \nu)\frac{\partial G}{\partial n'}(\mathbf{r}', \mathbf{s}', \nu)\Gamma(\mathbf{s}, \mathbf{s}', \nu)dS\,dS', \tag{22.163}$$

which gives

$$\Gamma(\mathbf{r}, \mathbf{r}', \nu) = \frac{1}{4\pi^2}\int_S\int_S (1 + jkR)(1 - jkR') \times \frac{e^{-jk(R-R')}}{R^2R'^2}\cos\theta\cos\theta'\,\Gamma(\mathbf{s}, \mathbf{s}', \nu)dS\,dS'. \tag{22.164}$$

The mutual coherence function can be found by taking the Fourier transform of Eq. (22.164):

$$\Gamma(\mathbf{r},\mathbf{r}',\tau) = \frac{1}{4\pi^2}\int_0^\infty \int_S\int_S (1+jkR)(1-jkR') \times \frac{e^{-jk(R-R')-j2\pi\nu\tau}}{R^2R'^2}\cos\theta\cos\theta'\,\Gamma(\mathbf{s},\mathbf{s}',\nu)dS\,dS'\,d\nu. \tag{22.165}$$

If the observation points $\mathbf{r}$ and $\mathbf{r}'$ are not too close to the aperture, we can assume that $k = 2\pi/\lambda = 2\pi\nu/c \gg 1/R, 1/R'$. Consequently,

$$\Gamma(\mathbf{r},\mathbf{r}',\tau) = \frac{1}{4\pi^2}\int_0^\infty \int_S\int_S \frac{(-jkR)(jkR')\cos\theta\cos\theta'}{R^2R'^2}\Gamma(\mathbf{s},\mathbf{s}',\nu) \times e^{-j2\pi\nu[\tau+(R-R')/c]}\,dS\,dS'\,d\nu. \tag{22.166}$$

Writing $jk\cos\theta/(2\pi R) = K$ and $jk\cos\theta'/(2\pi R') = K'$ gives

$$\Gamma(\mathbf{r},\mathbf{r}',\tau) = \int_S\int_S \Gamma\left[\mathbf{s},\mathbf{s}',\left(\tau+\frac{R-R'}{c}\right)\right]K^*K'\,dS\,dS'. \tag{22.167}$$

The intensity at the point $\mathbf{r}$ is $\Gamma(\mathbf{r},\mathbf{r},0)$, which, from Eq. (22.167), gives

$$I(\mathbf{r}) = \int_S\int_S \sqrt{I(\mathbf{s})I(\mathbf{s}')}\gamma\left(\mathbf{s},\mathbf{s}',\frac{R-R'}{c}\right)K^*K'\,dS\,dS'. \tag{22.168}$$

Note that all the quantities on the right-hand side are measurable.

If the illuminated surface consists of two pinholes 1 and 2, then, from Eq. (22.168), performing the first integral over the surface $S$ gives

$$I(\mathbf{r}) = \int_S \sqrt{I(\mathbf{s})I_1}\gamma\left(\mathbf{s},1,\frac{R-R_1}{c}\right)K^*K_1 + \sqrt{I(\mathbf{s})I_2}\gamma\left(\mathbf{s},2,\frac{R-R_2}{c}\right)K^*K_2, \tag{22.169}$$

and integrating again gives

$$I(\mathbf{r}) = I_1K_1^*K_2\gamma(1,1,0) + \sqrt{I_2I_1}\gamma\left[2,1,\frac{R_2-R_1}{c}\right]K_2^*K_1 + I_2K_2^*K_2\gamma(2,2,0) + \sqrt{I_2I_2}\gamma\left[1,2,\frac{R_1-R_2}{c}\right]K_1^*K_2. \tag{22.170}$$

Now $\gamma(1,1,0) = \gamma(2,2,0) = 1$ and $\gamma(1,2,(R_1-R_2)/c) \equiv \gamma(1,2,\tau) = \gamma_{12}(\tau)$, where $\tau = (R_1-R_2)/c$ and $\gamma(2,1,(R_2-R_1)/c) \equiv \gamma(2,1,-\tau) = \gamma_{12}^*(\tau)$. Hence

$$I(\mathbf{r}) = |K_1|^2I_1 + |K_2|^2I_2 + 2\sqrt{I_1I_2}\mathcal{R}[K_1^*K_2\gamma_{12}(\tau)], \tag{22.171}$$

a similar result to Eq. (22.62). We can identify $K_1$ and $K_2$ with the complex propagation factors $a_1$ and $a_2$ introduced previously.

## 22.10 The van Cittert–Zernike theorem

An important feature of the mutual coherence function is the acquisition of partial coherence by the very process of propagation.

If the surface $S$ is that of the source itself, for example a star, then clearly $\Gamma(\mathbf{s}, \mathbf{s}', \nu)$ vanishes unless $\mathbf{s}$ and $\mathbf{s}'$ are sufficiently close together, say no more than a meter apart. Physically this simply means that the light sources are statistically independent over any macroscopic interval.

In the integral

$$\int_S \int_S \frac{\partial G}{\partial n}(\mathbf{r}, \mathbf{s}, \nu) \frac{\partial G^*}{\partial n'}(\mathbf{r}', \mathbf{s}', \nu) \Gamma(\mathbf{s}, \mathbf{s}', \nu) dS\, dS'$$

we can use

$$\Gamma(\mathbf{s}, \mathbf{s}', \nu) = \delta_s(\mathbf{s} - \mathbf{s}')[\Gamma(\mathbf{s}, \nu)], \tag{22.172}$$

where $\delta_s$ denotes a surface $\delta$-function. Thus the integral becomes

$$\Gamma(\mathbf{r}, \mathbf{r}', \nu) = \int_S \frac{\partial G}{\partial n}(\mathbf{r}, \mathbf{s}, \nu) \frac{\partial G^*}{\partial n}(\mathbf{r}', \mathbf{s}, \nu) \Gamma(\mathbf{s}, \nu) dS, \tag{22.173}$$

which need not vanish for $\mathbf{r} \neq \mathbf{r}'$, so the field is now partially coherent.

With our previous assumptions, i.e., that $k \gg 1/R$, and using the same "mirror-image" Green's function as before, Eq. (22.166) gives

$$\Gamma(\mathbf{r}, \mathbf{r}', \tau) = \int_0^\infty d\nu\, e^{-j2\pi\nu\tau} \int_S \int_S \Gamma(\mathbf{s}, \mathbf{s}', \nu) \times e^{-jk(R-R')} K^* K'\, dS\, dS'. \tag{22.174}$$

The intensity at the point $\mathbf{r}$ is $\Gamma(\mathbf{r}, \mathbf{r}, 0)$, which, from Eq. (22.170), gives

$$I(\mathbf{r}) = \int_0^\infty d\nu \int_S \int_S \Gamma(\mathbf{s}, \mathbf{s}', \nu) |K|^2\, dS\, dS', \tag{22.175}$$

and, if $\Gamma(\mathbf{s}, \mathbf{s}', \nu)$ is a surface $\delta$-function,

$$I(\mathbf{r}) = \int_0^\infty d\nu \int_S \Gamma(\mathbf{s}, \mathbf{s}, \nu) |K|^2\, dS = \int_S I(\mathbf{s}) |K|^2\, dS, \tag{22.176}$$

where we have used

$$\Gamma(\mathbf{s}, \mathbf{s}, \tau = 0) = I(\mathbf{s}) = \int_0^\infty \Gamma(\mathbf{s}, \mathbf{s}, \nu) d\nu. \tag{22.177}$$

Equation (22.176) is on examination an obvious result: the intensity observed from an extended, incoherent source is found by integrating the intensities produced by each infinitesimal surface element of the source over the area of the source.

If the extended, spatially incoherent source is quasi-monochromatic, then, provided that $[\tau + (R - R')/c] \gg 1/\Delta\nu$, Eq. (22.174) gives

$$\Gamma(\mathbf{r}, \mathbf{r}', \tau) = \int_0^\infty e^{-2j\pi\nu\tau}\, d\nu \int_S e^{-jk(R-R')} K^* K\, dS. \tag{22.178}$$

If the source is monochromatic $I(\mathbf{s}, \nu)$ is significant only for frequencies $\nu$ near $\nu_0$; therefore

$$\Gamma(\mathbf{r}, \mathbf{r}', \tau) = e^{-j2\pi\nu_0\tau} |\overline{K}|^2 \int_S I(\mathbf{s}) e^{-j\bar{k}(R-R')}\, dS, \tag{22.179}$$

where we have assumed small angles, such that $\cos\theta \approx 1$ and $\cos\theta' \approx 1$, and $\bar{k}$ is the value of $k$ appropriate to $\nu_0$, and similarly for $\overline{K}$.

The degree of coherence is

$$\gamma(\mathbf{r}, \mathbf{r}', \tau) = \gamma_{12}(\tau) = \frac{\Gamma(\mathbf{r}, \mathbf{r}', \tau)}{\sqrt{I(\mathbf{r})I(\mathbf{r}')}}. \tag{22.180}$$

On writing $\gamma_{12}(\tau) = \gamma_{12}(0)\exp(-j2\pi\nu_0\tau)$, where $\gamma_{12}(0)$ is the *degree of spatial coherence*,

$$\gamma_{12}(0) = \frac{|\overline{K}|^2}{\sqrt{I(\mathbf{r})I(\mathbf{r}')}} \int_S I(\mathbf{s}) e^{-j\bar{k}(R-R')}\, dS. \tag{22.181}$$

So, the degree of spatial coherence of an extended monochromatic source can be calculated from the intensity distribution over the source.

We can rewrite Eq. (22.179) for the mutual coherence function $\Gamma(\mathbf{r}, \mathbf{r}', \tau)$ as

$$\Gamma(\mathbf{r}, \mathbf{r}', \tau) = \Gamma(\mathbf{s}, \mathbf{s}, \tau) \int_S e^{-j\bar{k}(R-R')} K^* K'\, dS, \tag{22.182}$$

provided that $I(\mathbf{s}, \nu)$ does not depend on $\mathbf{s}$, where $\Gamma(\mathbf{s}, \mathbf{s}, \tau)$ is the Fourier transform of $I(\mathbf{s}, \nu)$.

For such a uniform source

$$\gamma(\mathbf{r}, \mathbf{r}', \tau) = \gamma(\tau)\gamma(\mathbf{r}, \mathbf{r}', 0), \tag{22.183}$$

$$\gamma(\tau) = \Gamma(\mathbf{s}, \mathbf{s}, \tau)/I(\mathbf{s}), \tag{22.184}$$

and

$$\gamma(\mathbf{r}, \mathbf{r}', 0) = \gamma_{12}(0), \tag{22.185}$$

where $\gamma(\tau)$ is the degree of temporal coherence of the source and $\gamma(\mathbf{r}, \mathbf{r}', 0)$ is the degree of spatial coherence. Here $\gamma(\tau)$ is a function of $\Delta\nu$, whilst $\gamma(\mathbf{r}, \mathbf{r}', 0)$ is a function of the geometrical properties of the source.

## 22.11 The spatial coherence of a quasi-monochromatic, uniform, spatially incoherent circular source

The geometry of the source and its location relative to the observation points $\mathbf{r}$ and $\mathbf{r}'$ are shown in Fig. 22.11.

The degree of spatial coherence between the points $\mathbf{r}$ and $\mathbf{r}'$ is

$$\gamma(\mathbf{r}, \mathbf{r}', 0) = \gamma_{12}(0) = \frac{1}{\sqrt{I(\mathbf{r})I(\mathbf{r}')}} \int_S I(\mathbf{s}) e^{-j\bar{k}(R-R')} K^* K'\, dS, \tag{22.186}$$

where

$$I(\mathbf{r}) = \int_S I(\mathbf{s})\, |K|^2\, dS. \tag{22.187}$$

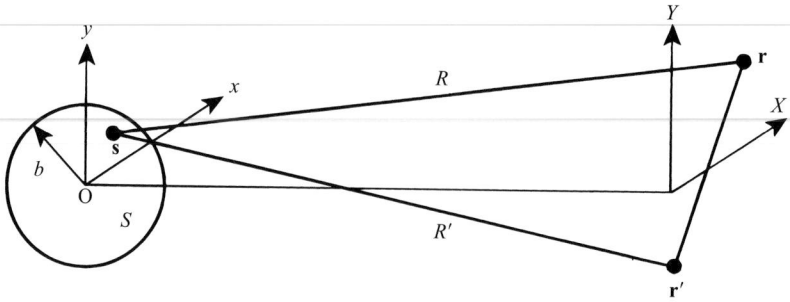


Fig. 22.11 The geometry of a circular source centered at O relative to two observation points $\mathbf{r}$ and $\mathbf{r}'$.

For a uniform source

$$I(\mathbf{r}) = I_S \int_S |K|^2 \, dS. \tag{22.188}$$

If we make the far-field approximation, namely $R^2 = R'^2 \gg S$ (where $S$ is the area of the source) and $R^2 \gg (R - R')^2$, then $I(\mathbf{r}) = I(\mathbf{r}') = I|K|^2$, where $I = \pi b^2 I_S$.

Further, in the far-field approximation, for a quasi-monochromatic source $K^*K'$ is constant, with $K^*K' = |\overline{K}|^2$, and

$$\gamma_{12}(0) = \frac{1}{\pi b^2} \int_S e^{-j\bar{k}(R-R')} \, dS. \tag{22.189}$$

It is left as an exercise for the reader to prove that

$$\bar{k}(R - R') = \bar{k}(|\mathbf{r} - \mathbf{s}| - |\mathbf{r}' - \mathbf{s}|) \simeq -\frac{\bar{k}}{R} \mathbf{s} \cdot (\mathbf{r} - \mathbf{r}'). \tag{22.190}$$

Integrating over the circular source gives

$$\gamma_{12}(0) = \frac{|K|^2 \int_0^b \int_0^{2\pi} I_S e^{(-j\bar{k}/R)sr\cos\phi} s \, ds \, d\phi}{I_S \pi b^2 |K|^2}. \tag{22.191}$$

The inner integral in Eq. (22.191) can be related to the Bessel function of zeroth order [12–15] giving

$$\gamma_{12}(0) = \frac{2}{I_S b^2} \int_0^b I_S J_0 \left( \frac{\bar{k}sr}{R} \right) s \, ds \tag{22.192}$$

and

$$\gamma_{12}(0) = \frac{2}{b^2} \left( \frac{Rb}{\bar{k}r} \right) J_1 \left( \frac{\bar{k}br}{R} \right), \tag{22.193}$$

which can be written as

$$\gamma_{12}(0) = \frac{2J_1(x)}{x}, \tag{22.194}$$

where $x = \bar{k}br/R$ and $r = |\mathbf{r} - \mathbf{r}'|$. This function is shown in Fig. 22.12.

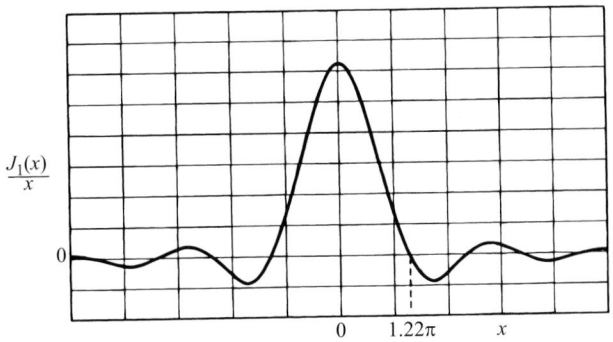


**Fig. 22.12** The spatial variation of the coherence of a circular source as predicted by Eq. (22.194) (the Airy function).

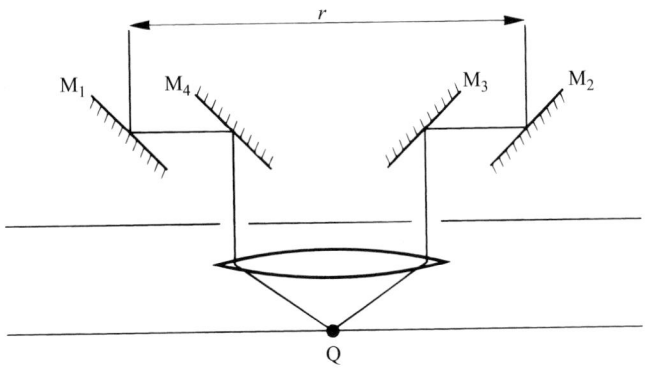


**Fig. 22.13** A schematic diagram of a Michelson stellar interferometer used to study the spatial coherence between two points separated by a distance *r*.

Equation (22.194) occurs in a similar form in the calculation of the far-field (Fraunhofer) diffraction pattern of a circular aperture – the Airy[7] diffraction pattern.

Equation (22.194) can be used to determine the angular diameters of distant objects, such as stars, in the Michelson stellar interferometer, which allows the visibility of interference fringes produced from laterally separated points in an incoming phase front from the star to be studied as a function of lateral separation, as shown in Fig. 22.13.

The visibility of the interference fringes first vanishes when $J_1(x) = 0$, i.e., when $x = 3.83 = 1.22\pi$. This occurs when

$$\frac{b}{R} = \alpha = \frac{0.61\bar{\lambda}}{r}, \tag{22.195}$$

where $\alpha$ is the angular diameter of the star. The drawback of this technique is that the visibility of the fringes is dependent also on the temporal coherence, which is affected by atmospheric turbulence, limiting the method to large stars.

[7] Named for the astronomer Sir George Airy.

For example, for a star the size of the Sun $b \approx 10^9$ m at a distance of 1 light year ($R \approx 10^{16}$ m), and, for a mean $\overline{k} \approx 10^7$ m$^{-1}$ ($\overline{\nu}_0 \approx 10^{15}$ Hz), the distance $r$ for the visibility first to vanish is somewhat less than 4 m.

## 22.12 Intensity-correlation interferometry

In the Michelson stellar interferometer the fundamental quantity that determines the visibility of the interference fringes is, just as it was in the case of Young's interference experiment considered previously, $\mathcal{R}\Gamma(\mathbf{r}, \mathbf{r}', \tau)$. If the optical paths are equal, then in theory the visibility is determined by spatial coherence alone, in other words by

$$\gamma_{12}(0) = \frac{\Gamma(\mathbf{r}, \mathbf{r}', 0)}{\sqrt{I(\mathbf{r})I(\mathbf{r}')}}. \tag{22.196}$$

However, in practice atmospheric turbulence degrades the degree of coherence between the two points $\mathbf{r}$ and $\mathbf{r}'$ below the value that would be expected if the waves propagated the entire distance from the source to $\mathbf{r}$ and $\mathbf{r}'$ through vacuum.

Now $\Gamma(\mathbf{r}, \mathbf{r}', \tau) = \overline{V^*(\mathbf{r}, t)V(\mathbf{r}', t')}$, where $t' - t = \tau$. The effect of atmospheric turbulence can be considered to modify the analytic signal at point $\mathbf{r}$ by a random, rapidly varying, phase factor, so $V(\mathbf{r}, t)$ changes according to

$$V(\mathbf{r}, t) \rightarrow V(\mathbf{r}, t)e^{j\phi(t)}, \tag{22.197}$$

where $\phi(t) = \phi(\mathbf{r}, t)$ is the phase factor at point $\mathbf{r}$. Consequently

$$\Gamma_{\text{obs}}(\mathbf{r}, \mathbf{r}', \tau) = e^{j[\phi(t) - \phi'(t')]}\Gamma_{\text{source}}(\mathbf{r}, \mathbf{r}', \tau). \tag{22.198}$$

Even when $t = t'$, the rapidly varying phase factor $e^{j[\phi(t) - \phi'(t')]}$ need not vanish, since the functions $\phi$ and $\phi'$ could be different and largely independent, as happens for a spatial interval greater than a few meters. Thus, amplitude-correlation interferometry breaks down because $\mathcal{R}\Gamma_{\text{obs}}$ is no longer equal to $\mathcal{R}\Gamma_{\text{source}}$.

The effect of random atmospherically induced fluctuations on the analytic signals can be largely eliminated by using *intensity-correlation interferometry*. In such experiments the quantity that is measured is $|\gamma(\mathbf{r}, \mathbf{r}', \tau)|^2$, generally for $\tau = 0$.

Since $\gamma(\mathbf{r}, \mathbf{r}', \tau) = \Gamma(\mathbf{r}, \mathbf{r}', \tau)/\sqrt{I(\mathbf{r})}$, when

$$\Gamma_{\text{obs}}(\mathbf{r}, \mathbf{r}', \tau) = e^{j[\phi(t) - \phi'(t')]}\Gamma_{\text{source}}(\mathbf{r}, \mathbf{r}', \tau) \tag{22.199}$$

it is clear that

$$|\Gamma_{\text{obs}}(\mathbf{r}, \mathbf{r}', \tau)| = |\Gamma_{\text{source}}(\mathbf{r}, \mathbf{r}', \tau)|, \tag{22.200}$$

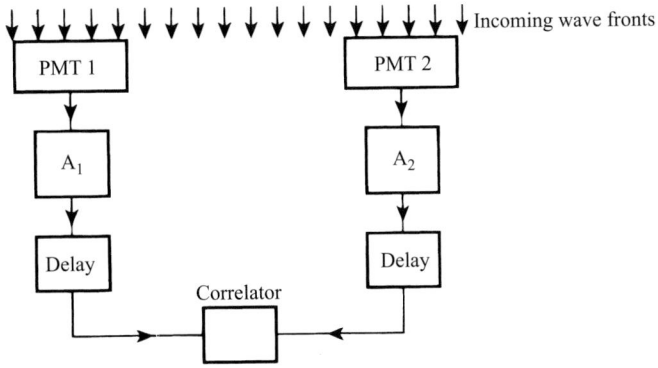


Fig. 22.14 An intensity-correlation (Hanbury Brown–Twiss) interferometer for use with visible light.

so that the phase distortions that plague amplitude-correlation interferometry are rendered harmless. The technique of intensity-correlation interferometry was developed by R. Hanbury Brown and R. Q. Twiss [12–15] and demonstrated in several interesting experiments beginning in the mid 1950s. Their first experiments were performed with radiowaves, and later ones used visible light. Using these techniques, they were able to extend the baseline of their stellar interferometer to 200 m, and thus were able to measure stars of much smaller angular diameter than could be done with amplitude-correlation interferometry.

The Hanbury Brown–Twiss (HBT) interferometer measures the angular diameter of stars by measuring $|\gamma_{12}(0)|^2$. As we have already shown $|\gamma_{12}|$ first vanishes for an $|\mathbf{r} - \mathbf{r}'|$ separation of $\bar{k}b\,|\mathbf{r} - \mathbf{r}'|/R = 3.83$, i.e., $\alpha = b/R = 0.61\bar{\lambda}/|\mathbf{r} - \mathbf{r}'|$, so, by determining the first vanishing point of $|\gamma_{12}(0)|^2$, the angular diameter can be determined.

The principle of the HBT interferometer is illustrated in Fig. 22.14. The two time delays allow compensation to be made for differences in the time of arrival of the wave trains at the two phototubes. In Hanbury Brown and Twiss's first optical experiments the correlations were determined from the analog signals from the photomultipliers, although in later experiments the ability of photomultipliers to detect individual photons was used in order to allow correlations to be measured from photoelectron-count fluctuations. We shall examine this situation in detail in Section 22.16.

## 22.13 Intensity fluctuations

In an experiment in which an optical signal is being detected by a photodetector, we expect the intensity falling on the photodetector to fluctuate. Fast photodetectors, such as photomultipliers, avalanche photodiodes, and p–i–n photodiodes, can follow such intensity fluctuations on quite a fast time scale, but not as fast as an optical period. If the source is quasi-monochromatic, it emits many wave trains of different frequencies lying within some frequency range $\Delta\nu$. Because there is normally no definite phase relationship between these different frequency components, the resultant real wave amplitude $V^{(\mathrm{r})}(t)$ must be regarded as a stochastic time function, which can be expanded as a Fourier series in a long interval $T$ in the form

$$V^{(\mathrm{r})}(t) = \sum_n a_n \cos\left(\frac{2\pi nt}{T}\right) + b_n \sin\left(\frac{2\pi nt}{T}\right). \tag{22.201}$$

The Fourier coefficients $a_n$ and $b_n$ can be regarded as statistically independent Gaussian variables with the same variance, provided that $V^{(\mathrm{r})}(t)$ is stationary and $T$ is sufficiently long. Such assumptions apply mainly to spontaneous emission from sources in thermal equilibrium rather than to lasers. We call the light from such sources *chaotic light*. If the field $V^{(\mathrm{r})}(t)$ is quasi-monochromatic, then in the slowly varying envelope approximation we can write

$$V(t) = U(t)e^{j2\pi\nu_0 t}, \tag{22.202}$$

where $\nu_0$ is the center frequency and $U(t)$ is the envelope function, which is nearly constant over any interval that is short compared with $1/\Delta\nu$ ($\approx\tau_c$); note that $|V(t)| = |U(t)|$. In other words, the intensity fluctuations of a quasi-monochromatic source occur on a time scale greater than the coherence time of the source. Such light can also be referred to as thermal or Gaussian light.

We had previously, from Eq. (22.41),

$$\begin{aligned}\Gamma(\mathbf{r},\mathbf{r}',\tau) &= \int_0^\infty \Gamma(\mathbf{r},\mathbf{r}',\nu)e^{-j2\pi\nu\tau}\,d\nu \\ &= e^{-j2\pi\nu_0\tau}\int_{-\nu_0}^\infty \Gamma(\mathbf{r},\mathbf{r}',\nu_0+\nu')e^{-j2\pi\nu'\tau}\,d\nu',\end{aligned} \tag{22.203}$$

where we have put $\nu = \nu_0 + \nu'$.

If the field is quasi-monochromatic, the integral in Eq. (22.203) vanishes for any $\nu'$ that is much outside the range $|\nu'| \lesssim \Delta\nu$. In this frequency range $2\pi\nu'\tau \ll 1$, provided that $\tau \ll \tau_c$.

Hence,

$$\begin{aligned}\Gamma(\mathbf{r},\mathbf{r}',\tau) &= e^{j2\pi\nu_0\tau}\int_{-\nu_0}^\infty \Gamma(\mathbf{r},\mathbf{r}',\nu_0+\nu')d\nu' \\ &= e^{j2\pi\nu_0\tau}\Gamma(\mathbf{r},\mathbf{r}',0) \text{ for } \tau \ll \tau_c.\end{aligned} \tag{22.204}$$

Now, the autocorrelation function of $U(t)$ is

$$\begin{aligned}\overline{U^*(t)U(t+\tau)} &= \overline{V^*(t)V(t+\tau)e^{j2\pi\nu_0\tau}} \\ &= e^{j2\pi\nu_0\tau}\Gamma(\mathbf{r},\mathbf{r},\tau) \\ &= \Gamma(\mathbf{r},\mathbf{r},0) = \Gamma_{11}(0) \text{ for } \tau \ll \frac{1}{\Delta\nu}.\end{aligned} \tag{22.205}$$

In other words, $U(t)$ is a slowly varying function over time intervals $\approx \tau \ll 1/\Delta\nu$. We can identify $|U(t)|^2$ as the instantaneous value of the intensity $I(t)$, a slowly varying function of $t$ for $t \ll \tau_c$.

Now, we previously quoted the result[8]

$$\overline{V_1^{(r)}(t)V_2^{(i)}(t+\tau)} = \overline{-V_1^{(i)}(t)V_2^{(r)}(t+\tau)}; \tag{22.206}$$

if we let $V_1 = V_2$ and $\tau = 0$ then

$$\overline{V_1^{(r)}(t)V_1^{(i)}(t)} = -\overline{V_1^{(i)}(t)V_1^{(r)}(t)} \tag{22.207}$$

and the values of $V_1^{(r)}(t)$ and $V_1^{(i)}(t)$ at the same time $t$ have no correlation.

If $V_1^{(r)}(t)$ is Gaussian then $V_1^{(i)}(t)$ is Gaussian also and both fluctuating quantities have the same variance $\sigma^2$. Thus, we can write the joint probability distribution of $V^{(r)}(t)$ and $V^{(i)}(t)$ as

$$p\left[V^{(r)}(t), V^{(i)}(t)\right] dV^{(r)}(t)dV^{(i)}(t) = \frac{1}{2\pi\sigma^2}\exp\left[\frac{-[V^{(r)2}(t) + V^{(i)2}(t)]}{2\sigma^2}\right] dV^{(r)}(t)dV^{(i)}(t). \tag{22.208}$$

Now

$$|U(t)|^2 = |V(t)|^2 = \frac{1}{4}\left[V^{(r)2}(t) + V^{(i)2}(t)\right]. \tag{22.209}$$

If we write $U(t)$ in the form

$$U(t) = |U(t)|e^{j\phi_u} \tag{22.210}$$

then

$$V(t) = U(t)e^{j2\pi\nu_0 t} = |U(t)|e^{j(\phi_u + 2\pi\nu_0 t)}. \tag{22.211}$$

Hence, $\phi_u = \phi_v - 2\pi\nu_0 t$, where $\phi_v$ is defined from

$$V(t) = |V(t)|e^{j\phi_v}. \tag{22.212}$$

We can transform the joint probability distribution Eq. (22.208) into polar coordinates as shown in Fig. 22.15 and get

$$p(|V(t)|, \phi_v)d|V(t)|d\phi_v = \frac{1}{2\pi\sigma^2}|V(t)|\exp\left(-\frac{|V(t)|^2}{2\sigma^2}\right) d|V(t)|d\phi_v, \tag{22.213}$$

which gives

$$p(|U(t)|, \phi_u + 2\pi\nu_0 t)d|U(t)|d\phi_u = \frac{1}{2\pi\sigma^2}|U(t)| \times \exp\left(-\frac{|U(t)|^2}{2\sigma^2}\right) d|U(t)|d\phi_u. \tag{22.214}$$

Note that the argument $\phi_u + 2\pi\nu_0 t$ does not appear as a variable in this probability distribution. Hence, all arguments are equally probable, and we can write

$$p(|U|, \phi_u)d|u|d\phi_u = \frac{1}{2\pi\sigma^2}|U|\exp\left(-\frac{|U|^2}{2\sigma^2}\right) d|U|d\phi_u, \tag{22.215}$$

[8] Results such as this are easily verified by, for example, setting $V^{(r)}(t) = \cos(\omega t)$ and $V^{(i)}(t) = \sin(\omega t)$.

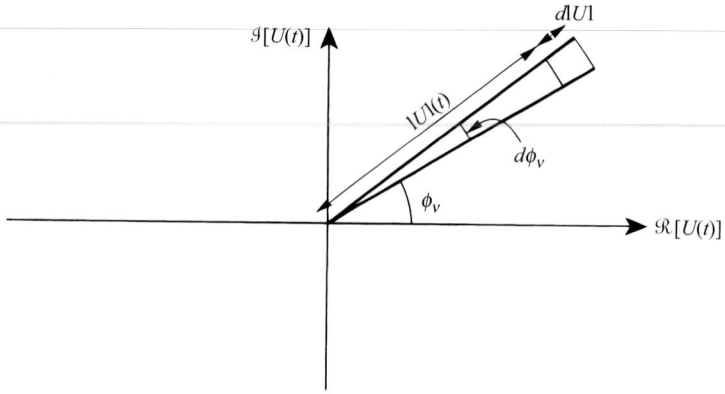


**Fig. 22.15** The coordinate system for transforming the probability density $V(t)$ into polar coordinates.

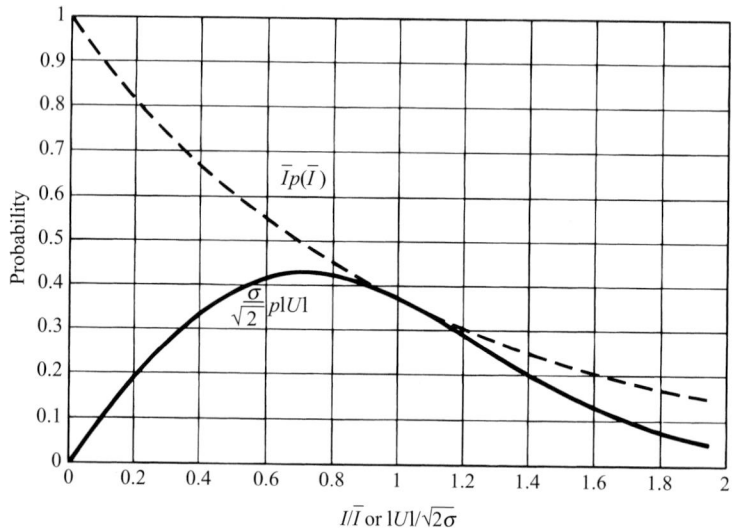


**Fig. 22.16** The probability distributions for the analytic signal amplitude $|u|$, called the Rayleigh distribution, and for the instantaneous intensity $I$, of chaotic light.

and, on integrating over all values of the argument ($0 \to 2\pi$),

$$p|U|d|U| = \frac{|U|}{\sigma^2} \exp\left(-\frac{|U|^2}{2\sigma^2}\right) d|U|, \tag{22.216}$$

which is called the Rayleigh distribution, and is shown in Fig. 22.16.

The distribution of $|U|^2$ is, from Eq. (22.216),

$$p|U|^2\, d|U|^2 = \frac{1}{2\sigma^2} \exp\left(-\frac{|U|^2}{2\sigma^2}\right) d|U|^2. \tag{22.217}$$

The average value of $|U|^2$ is

$$\overline{U^2} = \int_0^\infty \frac{|U|^2}{2\sigma^2} \exp\left(-\frac{|U|^2}{2\sigma^2}\right) d|U^2| = 2\sigma^2. \tag{22.218}$$

Consequently, since $|U|^2 = I$, on writing $\bar{I} = 2\sigma^2$, the distribution of the instantaneous intensity is

$$p(I)dI = \frac{1}{\bar{I}} \exp\left(-\frac{I}{\bar{I}}\right) dI, \tag{22.219}$$

which is also shown in Fig. 22.16.

## 22.14 Photon statistics

Since photomultiplier tubes have the ability to detect individual photons, albeit not with 100% quantum efficiency, it is important to determine how intensity fluctuations $I(t)$ influence the detection of photoelectron counts from an illuminated phototube.

Although the phenomenon of photoemission is fundamentally quantum mechanical in nature, we can still consider the statistics of the photoelectrons classically by making the plausible assumption that photoelectron counts are directly proportional to the incident intensity. Consequently, the statistical distribution of photoelectron counts reflects the statistics of the electromagnetic field illuminating the photocathode. Insofar as the full quantum-mechanical treatment of this problem is concerned, correlations between classical analytic fields become coincidences between photons detected at different points or delayed coincidences in time between photons detected at the same point.

The simplest type of photon statistics results when the source is of constant intensity over the observation period. For relatively long observation periods $T$, this implies temporal coherence of the source such that $\tau_c > T$. The photon statistics that result are those that would be observed for an ideal laser source.

### 22.14.1 Constant-intensity sources

If we make the plausible assumption that the rate of observed photoelectron counts is directly proportional to the instantaneous intensity, then the differential probability $dp$ that a count is observed in a time interval $dt$ when a phototube is illuminated with radiation intensity $I(t)$ is

$$dp(t) = \alpha I(t)dt = \alpha V^*(t)V(t)dt, \tag{22.220}$$

where the parameter $\alpha$ is a measure of the quantum efficiency of the detector and also depends on the area of the detector, the average spectral characteristics of the incident radiation, the operating conditions of the phototube and its associated pulse-counting electronics, etc.

If there are no random fluctuations in the intensity $I(t)$, then it is reasonable to assume that the probabilities of detecting photons in distinct time intervals are statistically independent. The probability of detecting a photon in a short time interval is independent of whether a photon was detected during the previous identical-length time interval. This is a characteristic of a *Markov* process [16]. The probability that no count occurs in the time interval $dt'$ at time $t'$ is $1 - dp(t')$. Thus the joint probability that no count occurs in an entire interval from $t$ to $t + T$ is

$$\prod_{k=0}^{n}\left[1 - dp\left(t + \frac{kT}{n}\right)\right] = P(0, T, t). \tag{22.221}$$

In the limit as $dp \to 0$ and $n \to \infty$, we can write

$$\ell\mathrm{n}\, P = \sum \ell\mathrm{n}[1 - dp(t')] = \sum_{\lim dp \to 0} -dp(t') = -\int_t^{t+T} dp(t'), \tag{22.222}$$

giving

$$P(0, T, t) = \exp\left[-\int_t^{t+T} dp(t')\right] = \exp\left[-\alpha \int_t^{t+T} I(t')dt'\right], \tag{22.223}$$

where it must be stressed that we have assumed that there are no intensity fluctuations.

In a similar way the probability that one count occurs between $t$ and $t + T$ is

$$\begin{aligned} P(1, T, t) &= \sum_{k''=0}^{\infty} dp\left(\frac{k''T}{n} + t\right) \prod_{k'=0}^{\infty} 1 - dp\left(t + \frac{k'T}{n}\right) \\ &= \int_t^{t+T} dp(t'')\exp\left[-\int_t^{t+T} dp(t')\right] \end{aligned} \tag{22.224}$$

in the limit as $k'', k' \to \infty, dp \to 0$, and $n \to \infty$. We have used the fact that the probability of obtaining a count in a time interval $dt''$ at time $t''$ and at no other times within the interval $t \to t + T$ is the same as the probability of a count at $t''$ in $dt''$ multiplied by the probability of there being no counts in the rest of the interval $t \to t + T$. Summing this probability over all infinitesimal time intervals $dt''$ within the range $t \to t + T$ gives the integral in Eq. (22.224).

Thus,

$$P(1, T, t) = \alpha \int_t^{t+T} I(t'')dt'' \exp\left[-\alpha \int_t^{t+T} I(t')dt'\right]. \tag{22.225}$$

In a similar way, we can show that the probability of obtaining $n$ counts in the interval $t \to t + T$ is

$$P(n, T, t) = \frac{1}{n!}\left(\alpha \int_t^{t+T} I(t'')dt''\right)^n \exp\left[-\alpha \int_t^{t+T} I(t')dt'\right], \tag{22.226}$$

where we have made use of the fact that $n$ counts observed at times $t_1, t_2, \ldots, t_n$ in intervals $dt_1, dt_2, \ldots, dt_n$ can be arranged among themselves in $n!$ different ways.

If we put $\mu = \alpha \int_t^{t+T} I(t')dt'$, where $\mu$ is the average number of photons detected in a time interval $T$, then

$$P(n, T, t) = \frac{\mu^n}{n!} e^{-\mu}. \tag{22.227}$$

This is a Poisson distribution, which we write as $P(n, \mu)$.

The mean number of counts observed in the interval $t \to t + T$ is

$$\overline{n} = \sum_{n=0}^{\infty} nP(n, \mu) = \sum_{n=0}^{\infty} \frac{\mu^n}{(n-1)!} e^{-\mu} = \mu. \tag{22.228}$$

The mean-square number is

$$\begin{aligned} \overline{n}^2 &= \sum_{n=0}^{\infty} n^2 P(n, \mu) \\ &= \sum_{n=0}^{\infty} \frac{n\mu^n}{(n-1)!} e^{-\mu} \\ &= \sum_{n=0}^{\infty} [(n-1)+1] \frac{\mu^n}{(n-1)!} e^{-\mu} = \mu^2 + \mu. \end{aligned} \tag{22.229}$$

The variance of the distribution of $n$ is

$$\sigma^2 = \overline{(n - \overline{n})^2} = \overline{n^2} - (\overline{n})^2. \tag{22.230}$$

This quantity is a measure of the fluctuations in $n$ about its mean value. For a Poisson distribution, from Eqs. (22.228) and (22.229),

$$\sigma^2 = \mu^2 + \mu - \mu^2 = \mu = \overline{n}. \tag{22.231}$$

### 22.14.2 Random intensities

We have shown that when the light intensity has no random fluctuations

$$P(n, T, t) = \frac{1}{n!} \left( \alpha \int_t^{t+T} I(t')dt' \right)^n \exp\left[ -\alpha \int_t^{t+T} I(t')dt' \right]. \tag{22.232}$$

If $I(t)$ is constant, $I(t) = I$, then

$$P(n, T, t) = \frac{1}{n!} (\alpha I T)^n e^{-\alpha I T}, \tag{22.233}$$

and it is clear that the choice of time $t$ is unimportant.

However, when $I(t)$ is fluctuating, it is apparent that $P(n, T, t)$ will depend on the choice of $t$. It can be shown, by dividing the time interval $T$ into small intervals, that even in this case $P(n, T, t)$ has the same form as Eq. (22.233). In the case of fluctuating intensity,

$$P(n, T, t) = \frac{1}{n!} (\alpha U_I)^n e^{-\alpha U_I}, \tag{22.234}$$

where

$$U_I = \int_t^{t+T} I(t')dt'. \tag{22.235}$$

In practice, the quantity $P(n, T, t)$ is not likely to be measured, since this would require many identical detectors being simultaneously irradiated coherently with the same radiation field at time $t$. What is usually measured is $P(n, T)$, which is an ensemble or time average of $P(n, T, t)$ that takes into account the distribution of $I(t)$, and consequently of $U_I$.

If $I(t)$ is a random variable, then $U_I$ will also be a random variable, with some distribution $p(U)$. Thus, if we average over the ensemble of $U_I$,

$$P(n, T) = \int_0^\infty \frac{(\alpha U_I)^n}{n!} e^{-\alpha U_I} p(U_I) du. \tag{22.236}$$

Even when $I(t)$ fluctuates, it remains essentially constant for time intervals $T \ll \tau_c$. In this case

$$U_I = \int_t^{t+T} I(t')dt' = I(t)T = IT \tag{22.237}$$

and $p(U_I) = p(I)$, which we previously derived. Therefore,

$$P(n, T) = \int_0^\infty \frac{(\alpha IT)^n}{n!} e^{-\alpha IP} p(I) dI, \tag{22.238}$$

where we stress that this applies to $T \ll \tau_c$.

If we substitute $p(I)$ from Eq. (22.219) into Eq. (22.238) and integrate, we get

$$P(n, T) = \frac{(\alpha \bar{I} T)^n}{(1 + \alpha \bar{I} T)^{n+1}}, \tag{22.239}$$

where

$$\bar{I} = \int_0^\infty I p(I) dI. \tag{22.240}$$

We can recognize $\alpha \bar{I} T$ as $\bar{n}$, the average number of counts during the period $T$, so

$$p(n, T) = \frac{\bar{n}^n}{(1 + \bar{n})^{n+1}}. \tag{22.241}$$

This is the Bose–Einstein distribution for $n$ identical particles in one quantum state [17]. This is the photon statistics that will be observed for an incoherent (thermal or Gaussian) source of light. The difference between the Poissonian statistics of an ideal laser (Eq. (22.227)) and those of a source of Gaussian light (Eq. (22.241)) is illustrated in Fig. 22.17.

Note that

$$\begin{aligned} \bar{n} = \sum_{n=0}^\infty n P(n, T) &= \int_0^\infty \sum_{n=0}^\infty \frac{n(\alpha IT)^n}{n!} e^{-\alpha IT} p(I) dI \\ &= \int_0^\infty \alpha IT p(I) dI = \alpha T \bar{I}. \end{aligned} \tag{22.242}$$

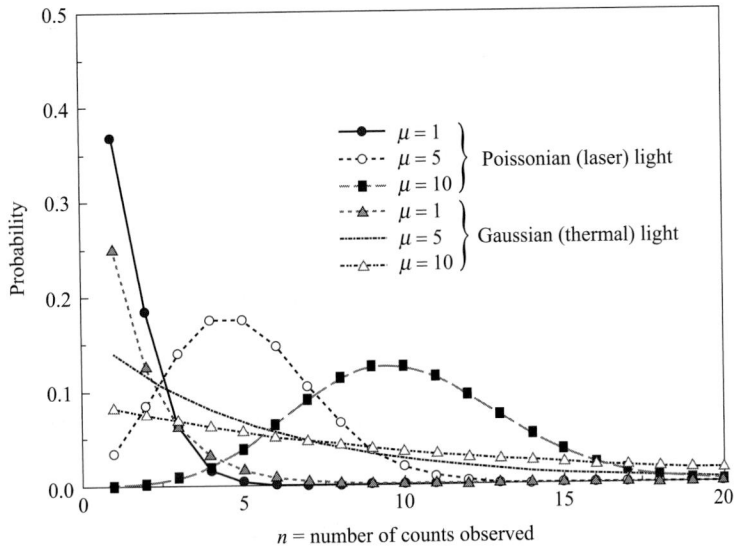


**Fig. 22.17** Photon-counting statistics for ideal coherent light and Gaussian light. The number of counts observed in a time interval is always an integer, so the lines shown in the graph are guides to the eye only.

The mean-square number of counts is

$$\overline{n^2} = \sum_{n=0}^{\infty} n^2 P(n,T) = \int_0^{\infty} \left(\alpha^2 T^2 I^2 + \alpha T I\right) p(I) dI$$
$$= \alpha^2 T^2 \overline{I^2} + \alpha T \bar{I}, \tag{22.243}$$

and the variance is

$$\sigma^2 = \overline{n^2} - (\bar{n})^2 = \alpha T \bar{I} + \alpha^2 T^2 [(\overline{I^2}) - (\bar{I})^2]. \tag{22.244}$$

This always exceeds $\bar{n}$ unless $p(I)$ is a $\delta$-function, $\delta(\bar{I} - I)$. For the thermal distribution $p(I) = e^{-I/\bar{I}}/\bar{I}$ and

$$\overline{I^n} = n!(\bar{I})^n. \tag{22.245}$$

Consequently,

$$\sigma^2 = \alpha T \bar{I} + \alpha^2 T^2 [2(\bar{I})^2] = \bar{n}(1 + \bar{n}), \tag{22.246}$$

which exceeds the value for a Poisson distribution. This implies that the fluctuations of the counts recorded by a phototube illuminated by Gaussian light depart from classical statistics in such a way that the variance always exceeds that given by a Poisson distribution. It follows that counts do not occur at random, but exhibit a bunching effect. Although this bunching effect is characteristic of Bose particles, it is not for this reason that it occurs. The bunching is correlated with intensity fluctuations of the source as illustrated in Fig. 22.18. Indeed, no bunching is observed in the output of an ideal laser, which is still a source of

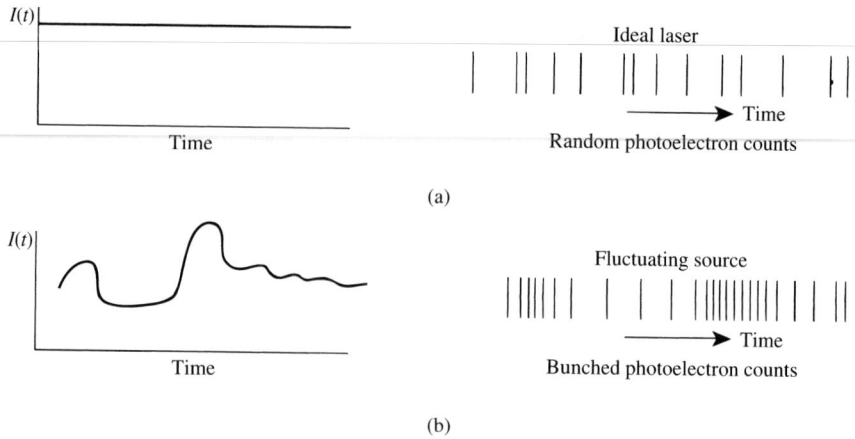


**Fig. 22.18** The variation of photoelectron counts from (a) a constant-intensity source, such as an ideal laser, where the distribution of counts is Poissonian, and (b) a source whose intensity is fluctuating, which gives rise to photoelectron "bunching."

Bose particles, as shown in Fig. 22.18. If two light beams are partially coherent, the bunching in the two beams will be correlated, and it is this correlation, that is responsible for the intensity correlations found by Hanbury Brown and Twiss.

When bunching occurs, photocounts are no longer independent; once a count is detected, there is an increased conditional probability of obtaining a second count within a short time $\tau \approx \tau_c$.

## 22.15 The Hanbury Brown–Twiss interferometer

This interferometer measures the intensity correlations between the waves illuminating two detectors. The quantity measured is

$$\overline{I_1(t)I_2(t+\tau)} = \frac{1}{16}\left\{\overline{V_1^{(r)2}(t)V_2^{(r)2}(t+\tau)} + \overline{V_1^{(i)2}(t)V_2^{(r)2}(t+\tau)} + \overline{V_1^{(r)2}(t)V_2^{(i)2}(t+\tau)} + \overline{V_1^{(i)2}(t)V_2^{(i)2}(t+\tau)}\right\}. \tag{22.247}$$

Now, it can be shown that, for two complex analytic signals $V_1(t)$ and $V_2(t)$,

$$\overline{V_1(t)V_2(t+\tau)} = 0, \tag{22.248}$$

$$\overline{V_1^{(r)}(t)V_2^{(r)}(t+\tau)} = \overline{V_1^{(i)}(t)V_2^{(i)}(t+\tau)}, \tag{22.249}$$

$$\overline{V_1^{(i)}(t)V_2^{(r)}(t+\tau)} = \overline{-V_1^{(r)}(t)V_2^{(i)}(t+\tau)}, \tag{22.250}$$

and, as we have mentioned previously,

$$\overline{V_1^{(i)}(t)V_2^{(r)}(t)} = 0. \tag{22.251}$$

From these relations, if $\tau = 0$ and $V_1(t) = V_2(t)$,

$$\overline{V_1^{(r)2}(t)} = \overline{V_1^{(i)2}(t)} = 2\overline{|V_1(t)|^2} = 2\overline{I_1}, \tag{22.252}$$

where $\overline{I_1}$ is the mean value of $I_1(t)$ (averaged over a time that is long compared with $\tau_c$).

Now,

$$\begin{aligned}\Gamma_{12}(\tau) &= \overline{V_1^*(t)V_2(t+\tau)} \\ &= \frac{1}{4}\left\{\overline{V_1^{(r)}(t)V_2^{(r)}(t+\tau)} + \overline{V_1^{(i)}(t)V_2^{(i)}(t+\tau)}\right. \\ &\qquad \left. + j\overline{V_1^{(r)}(t)V_2^{(i)}(t+\tau)} - j\overline{V_1^{(i)}(t)V_2^{(r)}(t+\tau)}\right\},\end{aligned} \tag{22.253}$$

which, using Eq. (22.250), gives

$$\Gamma_{12}(\tau) = \frac{1}{2}\overline{V_1^{(r)}(t)V_2^{(r)}(t+\tau)} - \frac{1}{2}j\overline{V_1^{(i)}(t)V_2^{(r)}(t+\tau)}. \tag{22.254}$$

Clearly,

$$\frac{1}{2}\overline{V_1^{(r)}(t)V_2^{(r)}(t+\tau)} = \mathcal{R}\Gamma_{12}(\tau) = \frac{1}{2}\overline{V_1^{(i)}(t)V_2^{(i)}(t+\tau)}, \tag{22.255}$$

$$-\frac{1}{2}\overline{V_1^{(i)}(t)V_2^{(r)}(t+\tau)} = \mathcal{I}\Gamma_{12}(\tau) = \frac{1}{2}\overline{V_1^{(r)}(t)V_2^{(i)}(t+\tau)}. \tag{22.256}$$

We now use the following relation that applies to Gaussian functions $X(t)$ and $Y(t)$, namely

$$\overline{X^2Y^2} = \overline{X^2}\,\overline{Y^2} + 2\overline{XY}^2. \tag{22.257}$$

If $X$ and $Y$ are independent, $\overline{XY} = 0$.

Applying this relation to the first of the four terms in Eq. (22.247) for $\overline{I_1(t)I_2(t+\tau)}$, gives

$$\begin{aligned}\overline{V_1^{(r)2}(t)V_2^{(r)2}(t+\tau)} &= \overline{V_1^{(r)2}(t)}\;\overline{V_2^{(r)2}(t+\tau)} + 2[\overline{V_1^{(r)}(t)V_2^{(r)}(t+\tau)}]^2 \\ &= 4\bar{I}_1\bar{I}_2 + 2[\overline{V_1^{(r)}(t)V_2^{(r)}(t+\tau)}]^2 \\ &= 4\bar{I}_1\bar{I}_2 + 8\{\mathcal{R}[\Gamma_{12}(\tau)]\}^2.\end{aligned} \tag{22.258}$$

Similarly, for the other three terms,

$$\overline{V_1^{(i)2}(t)V_2^{(r)2}(t+\tau)} = 4\bar{I}_1\bar{I}_2 + 8\{\mathcal{I}\Gamma_{12}(\tau)\}^2, \tag{22.259}$$

$$\overline{V_1^{(r)2}(t)V_2^{(i)2}(t+\tau)} = 4\bar{I}_1\bar{I}_2 + 8\{\mathcal{I}\Gamma_{12}(\tau)\}^2, \tag{22.260}$$

$$\overline{V_1^{(i)2}(t)V_2^{(i)2}(t+\tau)} = 4\bar{I}_1\bar{I}_2 + 8\{\mathcal{R}\Gamma_{12}(\tau)\}^2. \tag{22.261}$$

Therefore,

$$\overline{I_1(t)I_2(t+\tau)} = \bar{I}_1\bar{I}_2 + |\Gamma_{12}(\tau)|^2. \tag{22.262}$$

We can normalize Eq. (22.262) to give

$$\frac{\overline{I_1(t)I_2(t+\tau)}}{\bar{I}_1\bar{I}_2} = 1 + |\gamma_{12}(\tau)|^2. \tag{22.263}$$

The left-hand side of Eq. (22.263) defines the degree of second-order coherence $\gamma_{12}^{(2)}(\tau)$, so

$$\gamma_{12}^{(2)}(\tau) = 1 + |\gamma_{12}(\tau)|^2. \tag{22.264}$$

Clearly, for $\tau \gg \tau_c, \gamma_{12}^{(2)} \to 1$ and $\gamma_{12}^{(2)}(0) = 2$. Equation (22.264) holds for all kinds of chaotic light, even if the fields are not Gaussian variates. For an ideal laser $\tau_c \to \infty$ and the degree of second-order coherence $\gamma_{12}^{(2)}(\tau) = 2$. For special kinds of so-called *non-classical* light, Eq. (22.264) does not hold, and all that can be proven is that [18]

$$0 \leq \gamma_{12}^{(2)}(\tau) < \infty. \tag{22.265}$$

If we measure intensity fluctuations $\Delta I_1(t)$ and $\Delta I_2(t)$ from the mean values $\bar{I}_1$ and $\bar{I}_2$, then

$$\Delta I_1(t) = I_1(t) - \bar{I}_1, \qquad \Delta I_2(t) = I_2(t) - \bar{I}_2, \tag{22.266}$$

and

$$\overline{\Delta I_1(t)} = 0, \qquad \overline{\Delta I_2(t)} = 0. \tag{22.267}$$

Hence,

$$\begin{aligned}\overline{I_1(t)I_2(t+\tau)} &= \overline{[\bar{I}_1 + \Delta I_1(t)][\bar{I}_2(t+\tau) + \Delta I_2(t+\tau)]} \\ &= \bar{I}_1\bar{I}_2 + \overline{\Delta I_1(t)\Delta I_2(t+\tau)}.\end{aligned} \tag{22.268}$$

Consequently,

$$\overline{\Delta I_1(t)\Delta I_2(t+\tau)} = |\Gamma_{12}(\tau)|^2. \tag{22.269}$$

By normalizing,

$$\overline{\Delta I_1(t)\Delta I_2(t+\tau)} = \bar{I}_1\bar{I}_2|\gamma_{12}(\tau)|^2, \tag{22.270}$$

which therefore yields $|\gamma_{12}(\tau)|^2$, or $|\gamma_{12}(0)|^2$ if $\tau$ is set equal to zero. These fluctuations will be a function of the geometry and of the coherence time of the source.

The validity of the method of intensity-correlation interferometry was first demonstrated for radiowaves [12–15], and there was some doubt originally as to whether the method would still work in the optical region, where the best available detectors (photomultiplier tubes) were inherently quantum mechanical in nature and detected photons rather than classical waves.

## 22.16 The Hanbury Brown–Twiss experiment with photon-count correlations

When this experiment is performed by counting photons then the measured quantity can be written in terms of the mean number of electron counts as

$$\overline{I_1(t')I_2(t'+\tau)} \propto \overline{n_1 n_2}, \tag{22.271}$$

where $n_1$ and $n_2$ are the numbers of counts recorded when the two phototubes are illuminated for a time $T$ at the times $t'$ and $t' + \tau$, respectively. We have

$$\overline{n_1 n_2} = \sum_{n_1=0}^{\infty} \sum_{n_2=0}^{\infty} n_1 n_2 \overline{P_1(n_1, T, t') P_2(n_2, T, t'')}, \tag{22.272}$$

where $t'' = t' + \tau$ and $P(n, T, t)$ is given by Eq. (22.226).

Written out in full,

$$\begin{aligned}\overline{n_1 n_2} = \sum_{n_1=0}^{\infty} \sum_{n_2=0}^{\infty} \frac{n_1 n_2}{n_1! n_2!} \left[\int_0^T \alpha_1 I_1(t') dt'\right]^{n_1} \left[\exp\left(-\int_0^T \alpha_1 I_1(t') dt'\right)\right] \\ \times \left[\int_0^T \alpha_2 I_2(t'') dt''\right]^{n_2} \left[\exp\left(-\int_0^T \alpha_2 I_2(t'') dt''\right)\right],\end{aligned} \tag{22.273}$$

which, after some algebraic manipulation, gives

$$\begin{aligned}\overline{n_1 n_2} &= \alpha_1 \alpha_2 \int_0^T \int_0^T \overline{I_1(t') I_2(t'')} dt'\, dt'' \\ &= \alpha_1 \alpha_2 \int_0^T \int_0^T \left[\bar{I}_1 \bar{I}_2 + |\Gamma_{12}(t'' - t')|^2\right] dt'\, dt'' \\ &= \alpha_1 \alpha_2 \int_0^T \int_0^T \bar{I}_1 \bar{I}_2 \left[1 + |\gamma_{12}(\tau)|^2\right] dt'\, dt''.\end{aligned} \tag{22.274}$$

If we can write $\gamma_{12}(\tau) = \gamma_{12}(0)\gamma_{11}(\tau)$ – a situation of *cross-spectral purity*, then

$$\overline{n_1 n_2} = \alpha_1 \alpha_2 \bar{I}_1 \bar{I}_2 T^2 + \alpha_1 \alpha_2 \bar{I}_1 \bar{I}_2 |\gamma_{12}(0)|^2 \int_0^T \int_0^T |\gamma_{11}(t'' - t')|^2\, dt'\, dt'', \tag{22.275}$$

which can be rewritten as

$$\overline{n_1 n_2} = \bar{n}_1 \bar{n}_2 + \bar{n}_1 \bar{n}_2 [\xi(T)/T] |\gamma_{12}(0)|^2, \tag{22.276}$$

where we have written

$$[\xi(T)/T] = \frac{1}{T^2} \int_0^T \int_0^T |\gamma_{11}(t'' - t')|^2\, dt'\, dt''. \tag{22.277}$$

It follows that the fluctuations of $n_1$ and $n_2$ are correlated and that the correlation is given by

$$\overline{\Delta n_1\, \Delta n_2} = \overline{n_1 n_2} - \bar{n}_1 \bar{n}_2 = \bar{n}_1 \bar{n}_2 [\xi(T)/T] |\gamma_{12}(0)|^2. \tag{22.278}$$

This formula is analogous to Eq. (22.270) that we derived previously by considering classical waves.

It turns out that the correlations predicted by Eq. (22.278) are difficult to observe in practice unless $[\xi(T)/T]$, which is related to the coherence time of the source and the observation time $T$, is sufficiently large. This is equivalent also to the degeneracy parameter of the light being sufficiently large. If $\delta \ll 1$, bunching and correlation effects are not easily observed.

Note that

$$\xi(T) = \frac{1}{T}\int_0^T \int_0^T |\gamma_{11}(t' - t'')|^2 \, dt' \, dt''$$
$$= \frac{2}{T}\int_0^T (T - \tau)|\gamma_{11}(\tau)|^2 \, d\tau. \tag{22.279}$$

If $\gamma_{11}(\tau) = 1$ for the time interval $T$, then $\xi(T) = T$. The limiting value of $\xi(T)$ as $T \to \infty$ is

$$\xi(T) = 2\int_0^\infty |\gamma_{11}(\tau)|^2 \, d\tau, \tag{22.280}$$

which, from Eq. (22.67), is a finite quantity. Therefore, as $T \to \infty$, we have $\xi(T)/T \to 0$, and all correlations in the photoelectron counts disappear.

## 22.17 Squeezed light

Closely related to the correlations observed between photons detected in a Hanbury Brown–Twiss experiment are the correlations between photons generated in optical parametric oscillators (OPOs). Optical parametric oscillators operated at the degenerate operating point where $\omega_{\text{signal}} = \omega_{\text{idler}} = \omega_{\text{pump}}/2$ essentially convert each pump photon into two identical photons. These "two-photon" sources have very interesting correlation properties. In a quantum-mechanical description they can generate *squeezed* light [19] and *entangled* quantum states [20, 21]. Even operated away from the degeneracy point, OPOs still generate correlated photons at the signal and idler frequencies. The parametric process actually occurs at the single-photon level. One pump photon is "split" into two correlated photons at different frequencies. A comprehensive discussion has been provided by Mandel and Wolf [1].

A simple description of the distinction between light in an ideal *coherent state* [18, 22] and squeezed light can be given by reference to Fig. 22.19. If the amplitude and phase of light are viewed as two variables that characterize a state of light, then quantum mechanics

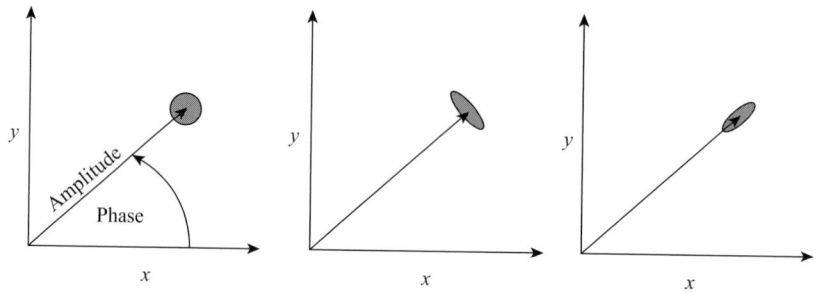


**Fig. 22.19** The behavior of the probability density for light in an ideal coherent state, amplitude-squeezed light, and phase-squeezed light.

tells us that there is an uncertainty involved in any measurement of two paired variables. This is the Heisenberg uncertainty principle, which in general says that an attempt to measure one of a pair of variables very precisely will render a simutaneous measurement of the paired variable less precise. The usual example of the Heisenberg uncertainty principle is that simultaneous measurements of the position $x$ and momentum of a particle $p_x$ have associated uncertainites $\Delta x$ and $\Delta P_x$ that obey the relation $\Delta x\,\Delta P_x \geq \hbar/2$. We can represent the state of light by a phasor vector, whose length describes the amplitude and whose polar angle describes its phase, as shown in Fig. 22.19. The uncertainties involved in measurements of amplitude and phase lie within a circle for an ideal coherent state and in ellipses for light that is squeezed in amplitude and phase. For example, amplitude-squeezed light has a better defined amplitude at the expense of a less well-defined phase.

## References


[1] L. Mandel and E. Wolf, *Optical Coherence and Quantum Optics*, Cambridge: Cambridge University Press, 1995.

[2] P. M. Morse and H. Feohbach, *Methods of Theoretical Physics*, New York: McGraw-Hill, 1953.

[3] W. R. LePage, *Complex Variables and the Laplace Transform for Engineers*, New York: Dover, 1961.

[4] J. Perina, *Coherence of Light*, London: Van Nostrand Reinhold, 1972.

[5] M. Françon, *Optical Interferometry*, New York: Academic Press, 1966.

[6] B. P. Lathi, *Linear Systems and Signals*, 2nd edn., Oxford: Oxford University Press, 2004.

[7] J. G. Proakis and M. Salehi, *Communication Systems Engineering*, Englewood Cliffs, NJ: Prentice-Hall, 1994.

[8] H. Margenau and G. M. Murphy, *The Mathematics of Physics and Chemistry*, 2nd edn., Princeton, MA: Van Nostrand, 1956.

[9] L. D. Landau and E. M. Lifshitz, *Statistical Physics*, 2nd revised edn., Oxford: Pergamon, 1969.

[10] M. Born and E. Wolf, *Principles of Optics*, 7th edn., Cambridge: Cambridge University Press, 1999.

[11] J. W. Goodman, *Introduction to Fourier Optics*, New York: McGraw-Hill, 1968.

[12] R. Hanbury Brown and R. Q. Twiss, "A new type of interferometer for use in radio astronomy," *Phil. Mag.* **45**, 663–682, 1954.

[13] R. Hanbury Brown and R. Q. Twiss, "A test of a new type of stellar interferometer on Sirius," *Nature* **178**, 1046–1048, 1956.

[14] R. Hanbury Brown and R. Q. Twiss, "Interferometry of the intensity fluctuations in light III. Applications to astronomy," *Proc. Roy. Soc.* **A248**, 199–221, 1958.

[15] R. Hanbury Brown and R. Q. Twiss, "Interferometry of the intensity fluctuations in light IV. A test of an intensity interferometer on Sirius A," *Proc. Roy. Soc.* **A248**, 222–237, 1958.

[16] D. T. Gillespie, *Markov Processes: An Introduction for Physical Scientists*, San Diego, CA: Academic Press, 1992.

[17] F. Reif, *Fundamentals of Statistical and Thermal Physics*, New York: McGraw-Hill, 1965.

[18] R. London, *The Quantum Theory of Light*, 3rd edn., Oxford: Oxford University Press, 2000.

[19] D. F. Walls, "Squeezed states of light," *Nature* **305**, 141–146, 1983.

[20] H. P. Yuen, "Two-photon coherent states of the radiation field," *Phys. Rev.* **13**, 2226–2243, 1976.

[21] H. P. Yuen and J. H. Shapiro, "Generation and detection of two-photon coherent states in degenerate four-wave mixing," *Opt. Lett.* **3**, 334–336, 1979.

[22] R. J. Glauber, "Coherent and incoherent states of radiation field," *Phys. Rev.* **131**, 2766–2767, 1963.

# 23 Laser applications

When the laser was first invented, it was described as "a solution looking for a problem." This comment did not long survive scrutiny, and nowadays lasers are ubiquitous in many aspects of daily life, with many technological, artistic, and educational applications. This chapter highlights some of the important application areas where lasers have become essential.

## 23.1 Optical communication systems

### 23.1.1 Introduction

Optical communications systems have a long history. Ancient man signaled with smoke and fire, often relaying messages from mountain top to mountain top. However, this optical communication scheme had limited transmission capacity. Such messages could serve as a warning, as Queen Elizabeth the First of England planned when she had a network of bonfires erected to be set in the event of a seaborne invasion from Spain. The smoke signals transmitted by native Americans had the capacity to transmit various messages. Since the end of the eighteenth century, messages have been passed by semaphore – the use of flags to indicate the transmission of one letter at a time. This form of communication could transmit information at a rate of about one letter per second over a direct line of sight, although messages could be relayed over long distances. Such means of communication were not very secure: anyone in the line of sight to the message sender could read the information (if he or she knew the code). The message could also be intercepted and altered during the relay process, as the Count of Monte Cristo did to his advantage in the eponymous novel by Alexandre Dumas [1].

Another historical use of optical communications involved the heliograph – a device to reflect the Sun's rays from a transmitting station to a receiving station using a code. This technique was widely used by the U.S. Cavalry in the desert south-west of the United States up until the early part of the twentieth century. For optical communication to progress past these early efforts, an information-carrying channel that was reliable, inexpensive, and could be used over long distances, preferably at high rates of data transmission, had to be developed. It was demonstrated as early as 1854 that light could be guided inside a transparent medium with a refractive-index discontinuity with its surroundings: John Tyndall [2] showed light being transmitted along a stream of water flowing from a container.

The development of optical fibers possessing extremely low attenuation has provided optical communication systems with the information-carrying channel that they need. But this did not come all at once. In the late 1950s and thereafter Charles Kao was an electrical engineer working for Standard Telephones and Cables (STC) (now Nortel Networks) at their research center in Harlow, England. While working at STC in 1966, Dr. Kao, together with his colleague George Hockham, made the seminal observation that a dielectric waveguide made of very pure glass could have a propagation loss below 20 dB $km^{-1}$ [3]. When he first proposed this, typical glasses had attenuations closer to 1000 dB $m^{-1}$,[1] so messages could realistically be transmitted over only a distance of a few meters. Kao recognized that this high attenuation was not a fundamental limitation of the glass itself but resulted from impurities such as copper, iron, cobalt, manganese, nickel, and chromium, which absorbed strongly in the 0.6–1.6-m spectral region. By purifying glasses so that these impurities were present only at concentrations below one part per million, Kao and Hockman demonstrated that a dielectric waveguide with attenuation below 20 dB $km^{-1}$ could be made. Such attenuations were achieved using fused silica by Maurer and his coworkers at Corning in 1970 [4]. By 1975 the attenuation was down to 4 dB $km^{-1}$. Attenuation in state-of-the-art fibers has declined steadily since then, reaching 0.5 dB $km^{-1}$ in 1976 and 0.2 dB $km^{-1}$ in 1979. This was achieved by further removal of impurities and reduction of the concentration of bound hydroxyl ions in the fiber to part-per-billion levels.

These reductions in attenuation have been accomplished by developing optical-fiber-fabrication techniques that eliminate impurities, particularly hydroxyl ions, when the lowest attenuations are required. It should be acknowledged that the lowest attentuation levels have not been achieved in the visible but at two near-infrared wavelengths near 1.3 μm and 1.55 μm. Figure 23.1 shows the variation of attentuation with wavelength of a typical modern fiber [5]. The attenuation in modern fibers can be as low as 0.17 dB $km^{-1}$ at wavelengths near 1,550 nm. The development of low-loss, low-dispersion, single-mode fibers, with the possibility of soliton transmission over long distances, may serve to satisfy the world's insatiable appetite for ever-increasing data rates. This is evidenced by Fig. 23.2, which shows how the data rates of communication channels have grown since the early days of the telegraph.[2]

The development of very-low-loss optical fibers has enabled a worldwide advanced telecommunications infrastructure. The Internet and a myriad of other digital data communication applications would not be possible without it. Technological developments in fiber design, optical sources (especially semiconductor lasers), modulation protocols, and high-speed detectors have moved data rates per wavelength to 40 Gb $s^{-1}$ and beyond. It is already two decades since Hitachi Ltd. announced that they had developed a 40-Gbit $s^{-1}$ link operating over a 40-km-long single-mode fiber.

[1] If the power emerging at the far end of a fiber of length $L$ km is $P_2$, for a power input $P_1$, the attenuation in dB $km^{-1}$ is $(10/L)\log(P_1/P_2)$.

[2] As of early 2012, the data rate along a single fiber had reached 109 Tb $s^{-1}$ [6].

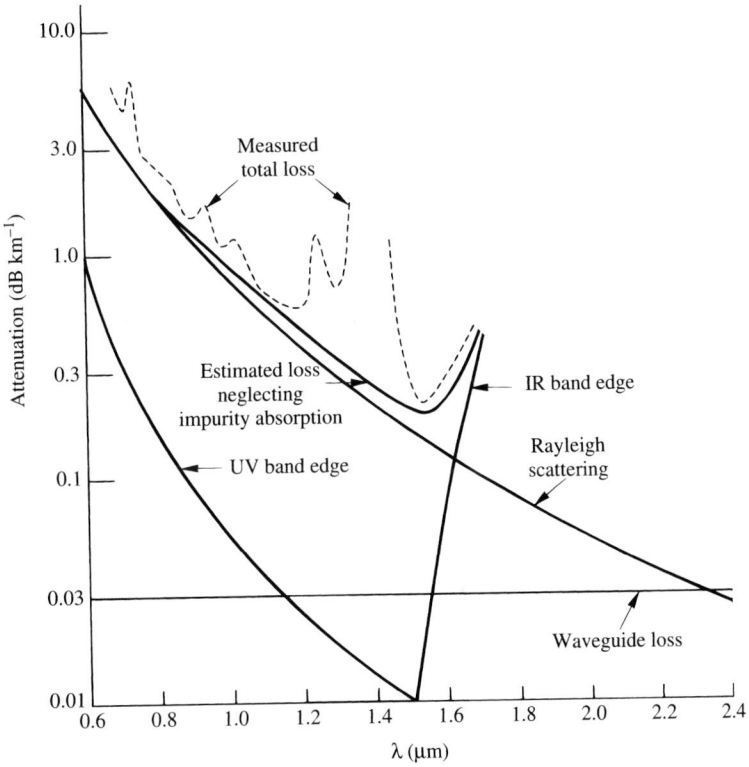


**Fig. 23.1** The variation of attenuation with wavelength of a modern, low-loss, germania-doped optical fiber.

## 23.1.2 Absorption in optical fibers

Residual absorption in low-loss modern optical fibers comes from several sources: residual ultraviolet absorption, residual infrared absorption, and Rayleigh scattering. Hydroxyl ions ($OH^-$) have a fundamental vibrational absorption at 2.73 μm, with overtone absorptions at 1.39, 0.95, and 0.72 μm. The hydroxyl content must be kept below 1 ppm in order that attenuation at 0.9 μm be less than 1 dB km$^{-1}$.

Even if the water content of the fiber is minimized, there remains residual absorption arising from the infrared absorption of the fundamental vibrations of the bonds that make up the glass, which are at 7.3 μm for B—O, 8.0 μm for P—O, 9.0 μm for Si—O, and 11.0 μm for Ge—O. Because these absorption features are very broad, their "wings" extend down to short wavelengths. Thus, the further the fundamental absorption is from the desired operating wavelength, the better. For this reason a germania-doped fiber achieves very low attenuation, as shown in Fig. 23.1 [5]. By the same token the "wings" of the ultraviolet absorption of the constituents of the fiber extend upwards in wavelength and

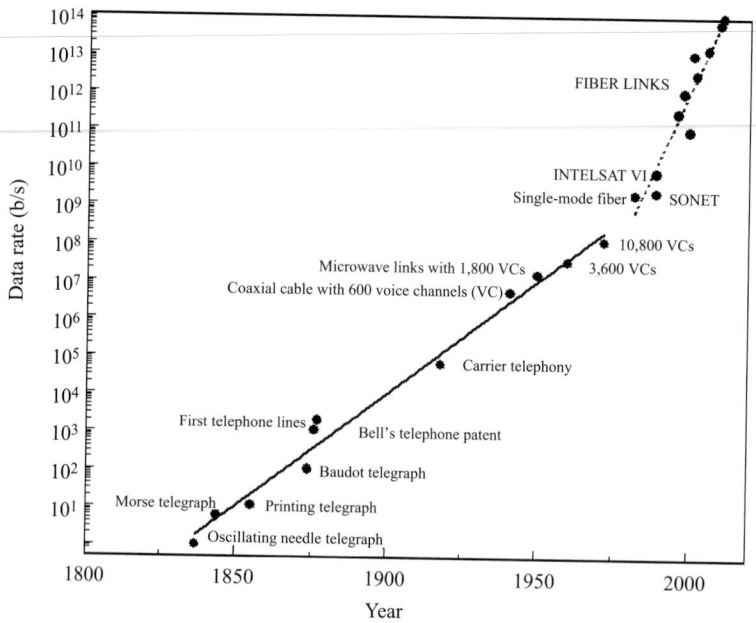


**Fig. 23.2** Historical variation of data transmission rates for various communication schemes since the early days of the telegraph.

contribute a small, but limiting, amount of absorption, as can also be seen in Fig. 23.1. The remaining significant, unavoidable attenuation mechanism in the fiber is Rayleigh scattering – the natural tendency for any atom or module to reradiate in all directions a part of the electromagnetic radiation incident on it. The magnitude of this scattering is proportional to $1/\lambda^4$, and is enhanced by the random structural character of glass. Microscopic variations in material density in a glass fiber lead to concomitant variations in refractive index that enhance Rayleigh scattering. This loss can be minimized by drawing the fiber during production so as to minimize compositional variations.

### 23.1.3 Optical data-transmission technology

The continuing improvement in the design and fabrication of optical fibers has made our modern communication infrastructure possible. Optical fibers themselves are cylindrical dielectric waveguides, which guide light by total internal reflection. In their simplest form they have a core doped with germania, phosphorus pentoxide, or alumina, which has a slightly higher refractive index than that of a surrounding cladding region. Some fiber designs have regions where the refractive index is locally reduced by doping with boron oxide. Single-mode fibers have a core diameter that is typically smaller than 10 μm, which ensures that only one mode (the $HE_{11}$ mode) can propagate. For long-distance communication, optical fibers must not only be low loss, but also have the modal and waveguide propagation characteristics required for long-distance, high-data-rate communication. A modulated optical signal will become impaired because of dispersion, when propagating

**Table 23.1** Principal telecommunication wavelength bands

| Band | Description | Wavelength range (μm) |
|---|---|---|
| O band | Original | 1.260–1.360 |
| E band | Extended | 1.360–1.460 |
| S band | Short wavelengths | 1.460–1.530 |
| C band | Conventional ("erbium window") | 1.530–1.565 |
| L band | Long wavelengths | 1.565–1.625 |
| U band | Ultralong wavelengths | 1.625–1.675 |

along an optical fiber. In its simplest sense, dispersion means that a short optical pulse will broaden as it travels. In a single-mode fiber several dispersion effects are still present. These include (1) chromatic dispersion (CD) with two contributors, namely (a) material dispersion ($dn/d\lambda$), the change in refractive index with wavelength, and (b) waveguide dispersion ($\beta(\omega)$), the change in waveguide propagation constant with frequency; and (2) polarization mode dispersion (PMD), which results from a lack of perfect cylindrical symmetry along the fiber. Owing to PMD the otherwise degenerate orthogonal polarization states of the $HE_{11}$ mode in the fiber have different propagation constants. The magnitude and direction of this anisotropy varies unpredictably along the length of the fiber, and causes energy mixing between the two polarization states. Advanced fiber designs, which involve engineering the radial refractive-index profile, and appropriate selection of dopants and operational wavelengths have reduced all these dispersion effects to manageable levels. There are three principal spectral regions where fiber-optic communication has been implemented extensively. Early on the 800–900-nm region was used, because GaAs/GaAlAs lasers and silicon detectors work well in this region, but loss and dispersion in fibers are much worse in this spectral region than is the case for the currently used regions near 1.3 m and 1.55 m. Near 1.3 m, fiber loss is low and chromatic dispersion is close to zero over a range of wavelengths. The region near 1.55 m dominates in long-haul applications, both because this region is where the attenuation is lowest, and also because of the availability of erbium-doped fiber amplifiers (EDFAs), which enable all-optical amplification without detection and retransmission. Table 23.1 summarizes the optical telecommunication bands currently in use for broadband optical communications. Chromatic dispersion effects can be minimized at 1.55 μm by moving the zero-dispersion point into this region with dispersion-shifted fiber, which has a core with a radial triangular, trapezoidal, or Gaussian refractive-index profile. Dispersion-flattened fibers have a relatively flat group velocity dispersion over some wavelength range, and can have near zero dispersion across the entire C band.

In these spectral regions, various III–V heterostructure lasers, such as InGaAs(N)(P)/InP, generally using single and multiple quantum wells, are used to transmit the data, and InGaAs photodetectors are used in the receivers. The first long-haul fiber-optic links used a single wavelength with regularly spaced repeaters, for example 65 km apart on the first transatlantic fiber cable (TAT-8), but this has been supplanted by the use of multiple wavelengths in a single fiber, with each wavelength carrying an independent stream of high-data-rate information. This is wavelength-division multiplexing (WDM). The C and L bands are in most widespread use because EDFAs operate in these spectral regions. Thulium-based fiber amplifiers have been developed for the S band and praseodymium-based amplifiers for the O band, but they are less well developed than the mature EDFA technology.

## 23.2 Optical amplification and wavelength-division multiplexing

Wavelength-division multiplexing (WDM) was made practical by the invention of the EDFA. The current most widely used International Telecommunications Union (ITU) standards for WDM separate the different wavelengths by 50 or 100 GHz. By using spectrally efficient modulation techniques, such as polarization-division-multiplexed (PDM) quaternary phase-shift keying (QPSK) paired with coherent detection, a spectral efficiency of 2 bits per Hz can be achieved, which allows 100 $\mathrm{Gb\,s^{-1}}$ channel data rates with 50 GHz spacing [7]. Wavelength-division multiplexing has allowed aggregate data rates to reach 69 $\mathrm{Tb\,s^{-1}}$ over 240 km using 432 wavelengths covering the C and extended L bands in a single fiber [8]. One challenge to using WDM in long-haul links is that dispersion effects are cumulative, which was not the case in single-wavelength, optical-to-electronic-optical, repeater-based systems, where pulse shapes could be corrected at each repeater before retransmission to the next repeater. Dispersion can be compensated for by inserting sections of special fiber with dispersion of the opposite sign. Polarization-division multiplexing is more difficult to manage since it is random along the length of the fiber, and also in time, but PMD compensation modules that can be placed at receivers in the network to compensate for its effects have been developed.

### 23.2.1 Optical communication networks

Fiber-optic networks have been installed extensively worldwide. They can be broadly divided into local-area networks, (LANs), metropolitan-area networks (MANs), and long-haul networks (LHNs). The LANs encompass short distances, and data rates that can cover a broad range – from $\mathrm{Mb\,s^{-1}}$ to $\mathrm{Gb\,s^{-1}}$. Because the data rates and distances involved are not very great, LANs use multimode fibers or even plastic optical fibers (POFs), since the latter offer simplicity of installation and interconnection. Such POFs are widely used in short-range (100-m) fiber-to-the-home applications, in home networks, and even as data buses in automobiles. The MANs provide the capacity to interconnect multiple LANs, and generally offer data capacities that range upwards from synchronous optical network (SONET) protocols, which operate at various data rates, for example SONET 2.5G (OC-48), to higher data capacities. The LHNs interconnect multiple MANs along trunk lines at ever increasing data capacities, and span the world's oceans for inter-continent connectivity. In high-data-rate networks many data streams must be multiplexed. It is

**Table 23.2** SONET/SDH data rates

| Designation | Payload Data Rate (Gb $s^{-1}$) | Overall data rate (Gb $s^{-1}$) |
|---|---|---|
| OC-1 | 0.05112 | 0.05184 |
| OC-3 | 0.150336 | 0.155520 |
| OC-12 | 0.601344 | 0.622080 |
| OC-24 | 1.202688 | 1.244160 |
| OC-48 | 2.4053766 | 2.488320 |
| OC-192 | 9.621504 | 9.953280 |
| OC-768 | 38.486016 | 39.813120 |
| OC-3072 | 153.944064 | 159.2524 |

The payload data rate corresponds to the useful data rate of transmitted information, which is smaller than the overall data rate, which includes additional data bits for "housekeeping" – the maintenance and measurement of the link.

beyond the scope of this chapter to describe in detail the various protocols that are used for the data transport. The SONET protocol (used in the United States and Canada) and the essentially identical Synchronous Digital Hierarchy (SDH) (used in the rest of the world) are standardized multiplexing protocols that are used to transfer multiple digital data streams over optical fibers using lasers or LEDs. Table 23.2 shows the various data rates used in SONET/SDH transmission. Especially in LANs, a parallel packet-switched data-transmission protocol named Ethernet is widespread, and standards exist for 10 Gig-E, 40 Gig-E, and 100 Gig-E. Gig-E data transport is not always compatible with SONET/SDH framing protocols. The 100 Gig-E connections can operate over short distances (at least 10 m) using copper cables, at least to 125 m using multimode fiber, and up to 40 km using single-mode fiber. At 100 Mbit $s^{-1}$ it is possible to use graded-index fibers, with typical modal dispersion of 0.2 ns $km^{-1}$ and either light-emitting diodes (LEDs) or laser sources.

Long-haul (trunk) lines use single-mode fiber and increasingly high data rates. A typical SONET optical fiber link will carry 32,000 simultaneous two-way voice or data channels on one pair of fibers. The SONET protocol uses single-mode fiber and different types of light source depending on the ranges involved. Short-reach links (up to 2 km) use LEDs or multimode lasers operating at 1.31 μm. Intermediate-reach links (up to 15 km) use 50-μW single-mode or multimode laser transmitters operating at 1.31 μm or 1.55 μm. Long-reach links (up to 40 km) use 500-μW laser transmitters operating at 1.31 μm or 1.55 μm.

In long-distance fiber links the signal decays with distance, so a repeater must be installed periodically in order to regenerate the signal. Early repeaters consisted essentially of an optical detector (usually an avalanche photodiode), pulse-reshaping electronics, and a laser transmitter. The ability to reshape and synchronize the bits in the data stream meant that the effects of dispersion are not cumulative from one repeater to the next. However, all-optical repeaters using EDFAs now dominate as a practical means for doing this without any optical → electronic → optical conversion. In the EDFA an optical pulse is amplified directly for transmission to the next repeater. In such a system the effects of dispersion must be compensated, since otherwise dispersion effects will be cumulative from one EDFA repeater to the next. In selecting repeater spacing the power transmitted into the fiber and system losses, including connectors and fiber splices, must be considered in order to calculate the acceptable distance.

*Example.* A fiber link has a laser transmitter that injects –7 dBm into the fiber,[3] where the fiber has an attenuation of 1 dB $km^{-1}$ and an additional loss of 6 dB from connectors and splices. The minimum received power for the required bit error rate (BER) is –30 dBm. The *power margin* for this system is defined as

$$P_m = -7\ \text{dBm} - (-30\ \text{dBm}) \equiv 23\ \text{dB}.$$

With 6 dB additional loss, the acceptable receiver spacing is 17 km.

The longest fiber links are those that span the world's oceans, including more than 20 across the Atlantic from North America to Europe and several linking the world's remaining continents. The first transatlantic cable, which was laid in 1988 (TAT-8), contained six fibers and could carry 40,000 telephone calls at one time. The TAT-15 cable laid in 2000 provides 64-10 Gb $s^{-1}$ channels. The next generation of cable will carry double this number of calls. The VSNL cable laid in 2001 has an overall capacity of 5.1 Tb $s^{-1}$. This cable system has two separate cables, which are laid along different routes over the ocean floor, but these two cables terminate near each other at the ends of the link. There are four fiber pairs per cable, each carrying 64-10 Gb $s^{-1}$ channels. The Fiber-Optic Link Around the Globe (FLAG) is a 28,000-km-long submarine fiber cable that connects the UK, Japan, and many places in between. There are currently about 300 submarine optical-fiber cables in service, under construction, or planned. They connect all the continents except Antarctica.

These fiber cables are heavily armored against shark bites, and in the shallow part of the ocean are buried under the ocean floor to prevent their being snagged by fishing boats. Optical–electrical–optical (OEO) repeaters were installed every 55–70 km. These repeaters have built-in redundancy, incorporating extra detectors and lasers so that if one device fails the channel remains functional. Because it is prohibitively expensive to dredge up cables to modify or repair repeaters, newer cables use highly reliable EDFAs. It should be pointed out, however, that even all-optical EDFA-based systems must still be provided with electric power to drive the pump lasers.

### 23.2.2 Advanced modulation formats

Because of this tremendous available bandwidth, there was very little incentive until recently for telecommunication providers to develop spectrally efficient modulation techniques for optical communication. By far the most prevalent modulation format used in optical communication is on–off keying (OOK): modulating the intensity of a laser between two states to represent a binary series of ones and zeros. Using fast modulators and detectors, such systems can easily reach data rates of 40 Gb $s^{-1}$. When more bandwidth was required, it was easy to simply add another, independently modulated, optical wavelength. Much of the development in the past 20 years was therefore focused on

[3] 0 dBm ≡ 1 mW, so –7 dBm is 7 dB below 1 mW, namely 0.2 mW.

the technologies needed to enable this type of WDM: optical bandpass filters, spectral multiplexers, and wide-band optical amplifiers. In contrast to the wireless domain, where spectral bandwidth is allocated and sold, often at great cost, fiber communications use an interference-free infrastructure. Wireless communication engineers never enjoyed the luxury of a bandwidth surplus, and they were forced to develop advanced modulation formats to more efficiently utilize the available bandwidth.

With the ever-growing demand for optical bandwidth, a similar trend is now emerging in optical communications. A key step in this process has been the development of optical receivers that simultaneously measure both the in-phase component and the quadrature component of the optical field in both polarization states. Although these coherent photoreceivers are larger and more costly than the earlier-generation intensity detectors, and they have to be widely deployed, they have paved the way for more sophisticated modulation formats including optical quadrature phase-shift keying (QPSK), which is equivalent to a four-quadrature amplitude-modulation (4-QAM) constellation modulation scheme, higher-order QAM modulation schemes, and polarization-multiplexed transmission. For more information on these modulation schemes the interested reader should consult the specialized literature [9, 10].

### 23.2.3 Optical-fiber-network architectures

Figure 23.3 shows three popular topologies of optical networks. In these arrangements of transmitters and receivers, each transmitter–receiver pair on the network, which is called a *node* or *subscriber*, can communicate with every other. The construction of such networks relies on the use of various passive optical components for their implementation. The most important of these is the fiber coupler, one form of which is shown schematically in Fig. 23.4, which either splits or combines optical signals. This is a four-port device. In the simplest sense such a device is fabricated by fusing together two single-mode fibers so that their cores come close together, allowing light to leak from one core to the other. Depending on the splitting ratio between the ports, the device can be described as either a *tap* or a *splitter*. For example, in Fig. 23.4(b), if intensity $I_1$ enters port 1 it is split on output between ports 3 and 4. Port 2 functions in the same way, although usually either port 1 or port 2 will be unused if it is desired that the device act as a Y (or T) coupler. If the intensities $I_3$ and $I_4$ leaving ports 3 and 4, respectively, on injection of $I_1$ are equal, the coupler is referred to as a 3-dB coupler or splitter, on the other hand if 99% of the input power passes from port 1 to port 3, with only 1% leaving at port 4, the device is described as a 20-dB coupler or tap, since $I_4 = I_1/100$, which is 20 dB below $I_1$.

Many of the current developments in fiber-optic communications are occurring in fiber-optic networks. Advanced fiber networks are moving towards all-optical routing, switching, and encryption. A key system development is the reconfigurable optical add–drop multiplexer (ROADM) [11]. An ROADM allows individual wavelengths to be added or dropped from a fiber without the need for optical to electronic to optical conversion using wavelength-selective filtering. This allows the routing of packets in the network in a wavelength-selected architecture. If necessary, data traveling on a channel at one wavelength can be transferred optically to a new wavelength. This provides great flexibility in

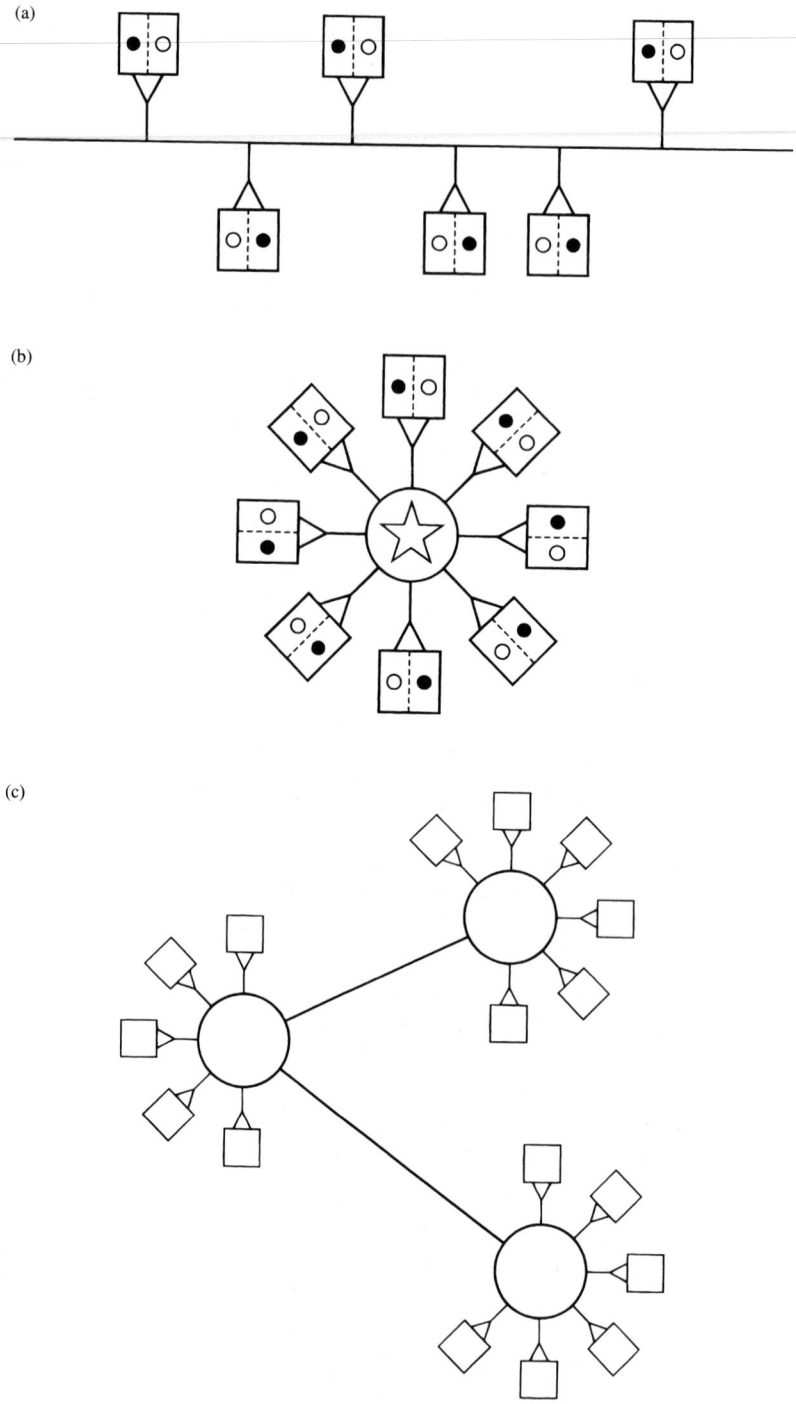


**Fig. 23.3** Optical network topologies (●, transmitter; ○, receiver) for (a) a bus, (b) a star, and (c) a multilevel star.

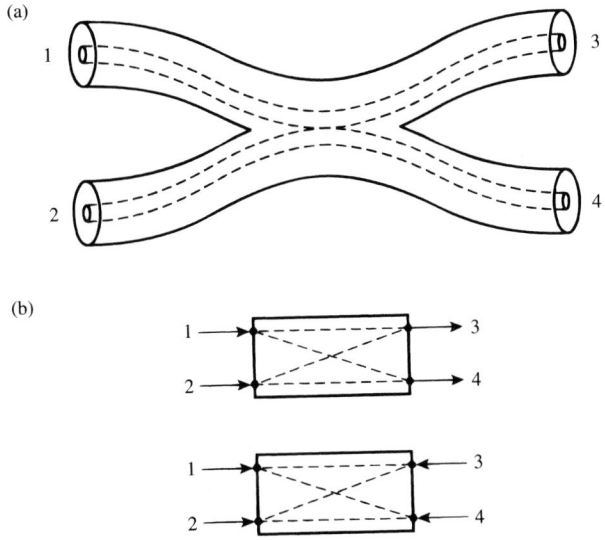


Fig. 23.4 Optical-fiber couplers: (a) a fused fiber-optic coupler and (b) bidirectional coupling pathways of a coupler.

the ability of the network to provide many different kinds of data services to subscribers with different bandwidth requirements [12].

### 23.2.4 Coding schemes in optical networks

As we have seen in previous chapters, a laser (or LED) source can be modulated in various ways, either by direct modulation of the source, for example by modulating the drive current of a semiconductor laser, or with an external amplitude or phase modulator. Almost all the practical implementations of optical communication networks utilize some form of binary modulation, which can amount simply to switching the laser on to transmit a "1" and off to transmit a "0." It is also possible to switch the laser between two discrete frequencies (frequency modulation), two distinct polarization states (polarization modulation), or two or more phases. It is beyond our scope here to go into detail regarding the precise implementation, advantages, and disadvantages of various schemes, for details the reader should consult the specialized literature [9, 10, 13–19]. However, a few details of how the binary information is most often encoded for transmission in the optical channel can be explained by reference to Fig. 23.5. These schemes are generally called pulse-code modulation (PCM).

The transmission of the digital data is synchronous with a clock rate. In the return-to-zero (RZ) format each "1" is emitted as a pulse during the first half of each clock cycle. In the non-return-to-zero (NRZ) format the pulse train stays "high" for an entire clock cycle when a "1" is to be transmitted, and stays "low" for the entire clock cycle when a "0" is to be transmitted. In the NRZI format the pulse train makes a transition from "high" or "low" only when the succeeding bit is a "0." The drawback of all these schemes is that the received data stream does not allow recovery of the clock rate, for example, in

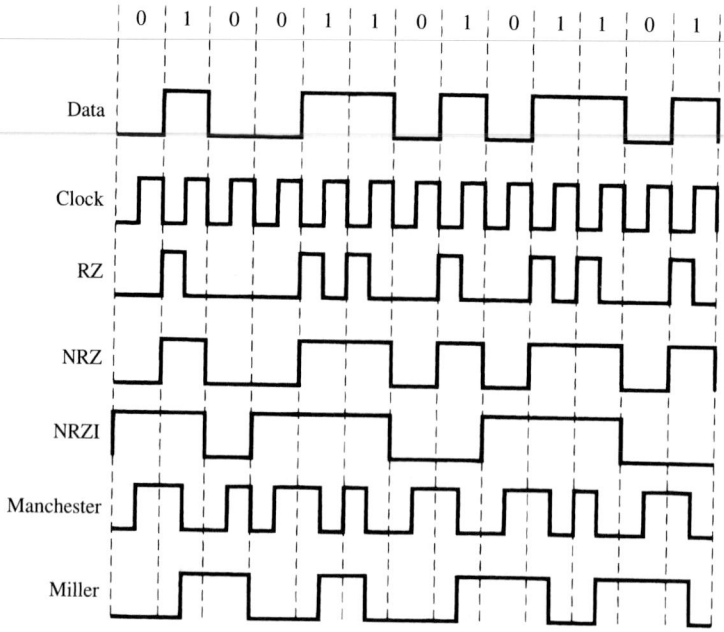


Fig. 23.5 Pulse-code-modulation formats.

the RZ scheme, if a long stream of zeros is transmitted. However, other schemes such as the Manchester and Miller encoding schemes allow recovery of the clock rate from the received data stream. The Manchester code is essentially a pulse-position modulation. For a "0" the signal makes a transition to the "high" state in the middle of each clock cycle, and each "1" is transmitted as a return from high to low in the middle of the clock cycle.[4] In the Miller scheme the transitions from high to low occur in the center of the clock cycle when a "1" is transmitted and at the beginning of the clock cycle when a "0" is transmitted after a "0."

### 23.2.5 Line-of-sight optical links

Line-of-sight optical links, which involve the free-space propagation of the encoded light signal from source to receiver, work best over short ranges in ground-based situations. This technology is often called *optical wireless* or *free-space optical* (FSO) communication and has a number of important emerging applications [20]. However, turbulence, scattering, and absorption by the atmosphere introduce amplitude and phase noise into the transmitted signal and cause significant signal-amplitude decay over moderate distances, typically over a few hundred meters near the ground [21]. These issues notwithstanding, there continues to be great interest in FSO communication networks, especially in military applications that require links with a low probability of interception (LPI) or detection (LPD)

[4] In logic terms the Manchester-encoded output is obtained as an exclusive OR operation between data and clock.

[22, 23]. However, ground-to-satellite, satellite-to-satellite, and deep-space communication links are important application areas for line-of-sight optical communications. When the communication channel must be guaranteed, as in a commercial satellite $\longleftrightarrow$ ground link, microwaves are generally preferred because their long wavelengths can penetrate clouds. An optical downlink, even to a generally cloudless location such as the mountains of Arizona or Hawaii, or the deserts of Australia, cannot provide 100%-reliable visibility conditions. Nonetheless, there are specialized applications of such links: a blue–green laser can be transmitted from an aircraft or satellite to provide a secure high-data-rate communication channel to a submerged submarine. Optical links between satellites in near-Earth orbit will be of increasing importance in the future for global communications. A series of satellites in synchronous Earth orbit (at a height of 38,600 km) can provide global communications coverage and can relay information from closer-in satellites used for Earth observation. The NASA Earth Observing System (EOS) program envisages the use of such satellites for continuous, long-term monitoring of Earth resources, climatic conditions, and atmospheric composition. The primary advantage of optical links for space communications is that a much more directional beam can be sent from source to receiver, so transmitter requirements for long-distance transmission are reduced. The most challenging application of such a link is in deep-space communications. In explorations of the outer planets in programs such as Pioneer and Voyager, data are relayed to and from the spacecraft with microwaves. Because the transmission distances are so long, high transmitter powers are required, which presents a problem for the spacecraft end of the system, which must rely on solar power. If a short-wavelength laser is used as the transmitter on the spacecraft, a substantial reduction in transmitted power is possible because a more directional beam can be employed.

The transmit/receive power analysis of such a link, and of any reasonably long-range line-of-sight link, can be carried out with reference to Fig. 23.6. The transmitter and receiver apertures are drawn schematically as lenses, although the transmitter and receiver will likely use reflective telescopes. We assume that the transmitter telescope sends a collimated Gaussian $\mathrm{TEM}_{00}$-mode beam of power $P_1$ towards the receiver situated at a distant range $R$. The diffraction angle of the transmitted beam is

$$\theta = \frac{\lambda}{\pi w_0}, \tag{23.1}$$

where we can safely assume that $\theta$ is a small angle. The transmitter aperture is chosen sufficiently large not to truncate this beam significantly: a commonly chosen condition for this is $r_1 = 1.4w_0$. At long range, $R$, the spot size is

$$w = R\theta, \tag{23.2}$$

and the power collected by the receiver is

$$P_2 = 2P_1\left(\frac{r_2}{w}\right)^2. \tag{23.3}$$

Combining Eqs. (23.1), (23.2), and (23.3) gives

$$P_2 = 2\left(\frac{\pi r_2 w_0}{R\lambda}\right)^2 P_1. \tag{23.4}$$

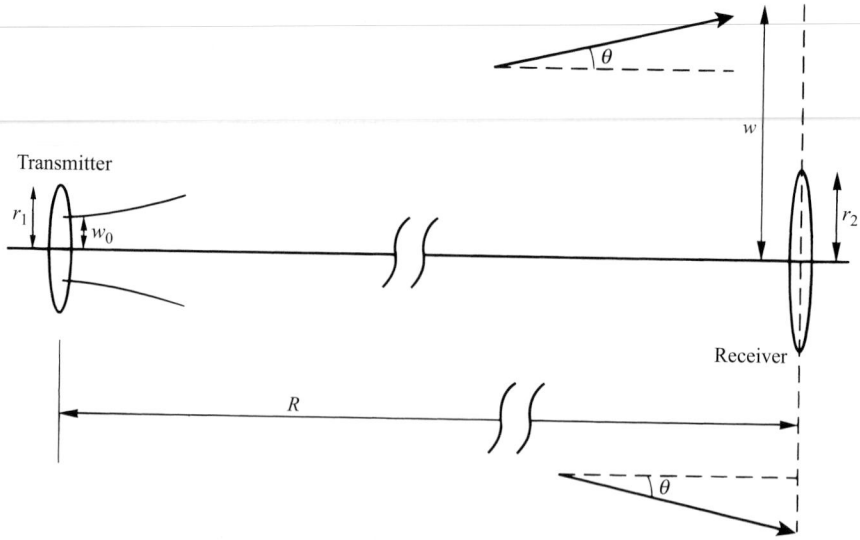


**Fig. 23.6** The geometry used in describing the performance of a line-of-sight optical link.

So, all other parameters being equal, the received power varies as $1/\lambda^2$. This highlights the diffraction advantage of an optical link. This advantage between 405 nm (a frequency-doubled GaAs laser) and 30-GHz microwaves is $6 \times 10^8$. Even allowing for the fact that larger microwave transmit/receive antennas than optical telescopes can be built, the advantage of the optical link is clear.

*Example.* We assume a spacecraft transmitter power of 0.1 W at 405 nm at a distance corresponding to the orbit of Jupiter, $780 \times 10^6$ km, a transmitted spot size $w_0 = 150$ mm, and a receiver aperture of 1.27 m (corresponding to a 100-inch telescope). The received power calculated from Eq. (23.4) is $1.9 \times 10^{-10}$ W. If we are using as our receiver an APD that has an excess noise factor, $F$, the minimum detectable received power in a bandwidth $\Delta f$ for a signal-to-noise ratio of unity is

$$P_{\min} = \frac{2eF\,\Delta f}{\mathcal{R}}, \tag{23.5}$$

where $\mathcal{R}$ is the responsivity.[5] With $F = 6$, and $\mathcal{R} = 0.5$ A W$^{-1}$,

$$P_{\min} = 3.84 \times 10^{-18}\,\Delta f \text{ W}.$$

For a BER of $10^{-6}$ (which would be acceptable in such a long-range link) the actual required signal-to-noise ratio is $S/N \simeq 19.6$ dB, or $S/N = 10$. Therefore, for a bandwidth of 200 kHz (100 kbit s$^{-1}$), the minimum required power is $P_{\min} = 3.84 \times 10^{-11}$ W: our proposed link would function at this data rate.

[5] See Chapter 21.

# 23.3 Holography

## 23.3.1 Wave-front reconstruction

Holography is lensless three-dimensional photography. A conventional photograph is only a flat record of a real image projected onto a photographic film. Information about the three-dimensional character of the object is almost entirely lost during the photographic recording process. Only the blurring of the images of objects that are not within the *depth of field* of the camera preserves any record of the location of the object relative to the camera. A *hologram*, on the other hand, is a special "photograph" of an object that retains information about the phase of waves coming from the object. Holography has a much longer history than that of the laser, although it was the invention of the laser that made the production of high-quality holograms a reality [24, 25].

In 1947, Denis Gabor was looking for a way to improve the quality of images that could be obtained with an electron microscope [26]. Although the electron microscopes of that time should have been capable of resolving down to the atomic scale, aberrations of the electron optics prevented them from doing so. Gabor conceived the idea of recording an electron hologram, which would contain amplitude and phase information, and then reconstructing a corrected image of the object by optical means. He did succeed in producing holograms and performing image reconstruction, but the optical sources available to him were not sufficiently temporally coherent to allow high-quality images to be achieved. For his pioneering work he was awarded the 1971 Nobel prize for physics [27–29].

The high degree of coherence obtainable with a laser has made the production of holograms and applications of holography into a large subject in its own right [30–37]. The principles underlying holography share much in common with the phenomena of interference and diffraction that we discussed in Chapter 5 and the concept of optical mixing discussed in Chapters 18 and 21. Taken together, these phenomena belong to the subject of *Fourier optics* [38–42].

In conventional photography the photographic emulsion becomes dark according to the local intensity of the real image falling on it. The emulsion is a mixture of silver halide crystals of different sizes, predominantly silver bromide, in a gelatin matrix. Incident light frees electrons from the halide, but the electrons become trapped. The addition of developer reduces the silver ions in an exposed crystal to silver atoms, leading to a strongly absorbing region where exposure levels were high. This is how a photographic negative is produced. If the incident field acting on the emulsion is

$$O(x,y) = |O(x,y)|e^{j\phi(x,y)}, \tag{23.6}$$

where $\phi(x,y)$ is the phase of the light at point $(x,y)$ in the image, then the incident intensity is

$$I(x,y) \propto |O(x,y)|^2, \tag{23.7}$$

and all phase information is lost. After exposure the field transmission coefficient of the film can be written as

$$t(x,y) = t_0 - \beta' I(x,y). \tag{23.8}$$

This linear relationship holds provided that the intensity at the location $(x,y)$ is not too high. Therefore, the field transmission coefficient of the developed film is

$$t(x,y) = t_0 - \beta |O(x,y)|^2. \tag{23.9}$$

If we illuminate the developed film uniformly, the transmitted intensity is

$$I_{\mathrm{T}}(x,y) \propto \left[t_0 - \beta |O(x,y)|^2\right]^2. \tag{23.10}$$

If the film is not overexposed, the dominant term of interest in Eq. (23.10) is

$$I_{\mathrm{T}}(x,y) \propto -2\beta t_0 |O(x,y)|^2, \tag{23.11}$$

which is a *negative* reconstruction of the original image intensity distribution but without any phase information.

We make a hologram, and preserve phase information in the photographic record, by allowing temporally and spatially coherent light to illuminate the object. Light reflected and scattered from the object falls on a photographic film together with a reference wave supplied directly to the film with the same laser as is used to illuminate the object. The way in which this is done is shown schematically in Fig. 23.7. No lenses are used to image the object onto the film. The relative location of all the components involved must be kept constant to within much less than the wavelength of the light being used during the exposure. Holograms of moving objects can be taken with pulsed lasers. The reference wave is incident on the film at angle $\theta_{\mathrm{R}}$.

The field from the object is

$$O(x,y) = |O(x,y)| e^{j\phi_{\mathrm{O}}(x,y)}, \tag{23.12}$$

where the amplitude $|O(x,y)|$ depends on the combined reflection plus scattering efficiency from each point on the object and $\phi_{\mathrm{O}}(x,y)$ is a composite phase factor that contains information about the location of each point on the object relative to the film. If an observer were to place his or her eye in the location of the film then he or she would *see* the object in full three-dimensionality. It does not matter precisely how the object is illuminated, whether this be directly by a collimated or expanding laser beam, or after sending a laser beam through a diffuser to illuminate the object uniformly as shown in Fig. 23.7. The illuminated object would be clearly visible to the observer in any case.

The field sent directly to the film from the laser which illuminates the object is called the reference field, which is written as

$$R(x,y) = |R(x,y)| e^{j\phi_{\mathrm{R}}(x,y)}. \tag{23.13}$$

In an actual exposure $|R(x,y)|$ might vary very little from point to point on the film. Furthermore, the phase factor $\phi_{\mathrm{R}}(x,y)$ will have only some smooth variation across the film depending on whether the reference wave is a plane wave or an expanding spherical wave.

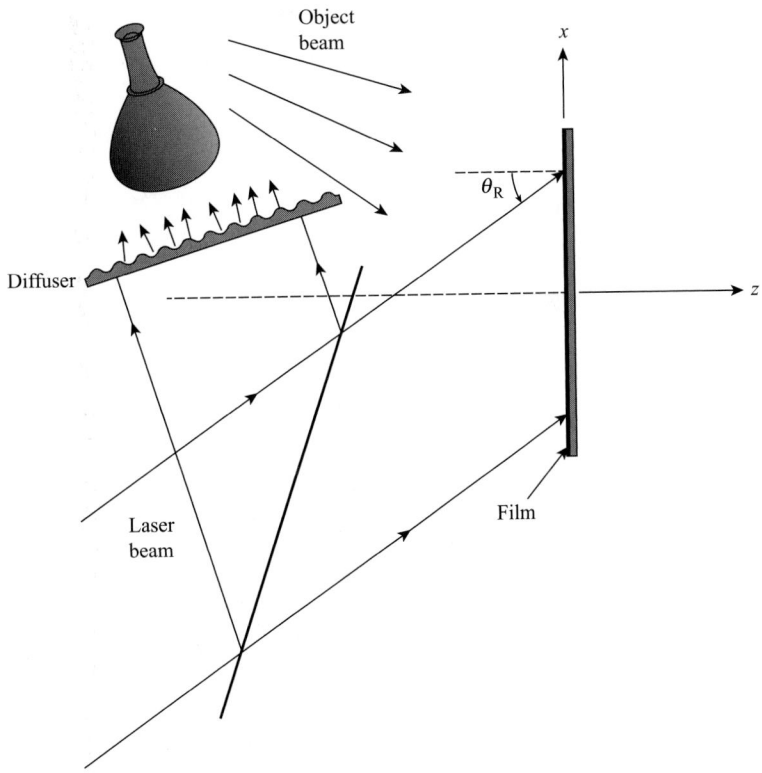


**Fig. 23.7** The schematic arrangement for making a simple hologram.

The transmission of the developed film in this case will depend on the total illuminating field $O(x,y) + R(x,y)$, and will be

$$t(x,y) = t_0 - \beta|O(x,y) + R(x,y)|^2, \tag{23.14}$$

which, in a simplified notation, can be written as

$$t(x,y) = t_0 - \beta OO^* - \beta(RR^*) - \beta(OR^*) - \beta(O^*R). \tag{23.15}$$

If the developed film is now illuminated by a laser that generates a reference wave just like the original $R$, then the transmitted field is

$$E_t = t_0R - \beta(OO^*)R - \beta(RR^*)R - \beta O(R^*R) - \beta O^*RR. \tag{23.16}$$

If the field from the object is much weaker than the reference field we can drop the second term in (23.16). The interesting remaining terms can be written out in full as

$$\begin{aligned} &-\beta|R(x,y)|^2R(x,y)e^{j\phi_R(x,y)} - \beta O(x,y)e^{j\phi_O(x,y)}|R(x,y)|^2 \\ &\quad - \beta O^*(x,y)e^{-j\phi_O(x,y)}|R(x,y)|^2e^{2j\phi_R(x,y)}. \end{aligned} \tag{23.17}$$

The first term in (23.17) is just the reference wave. The second term is the *primary reconstruction* of the original object. Apart from the (almost) spatially uniform illumination

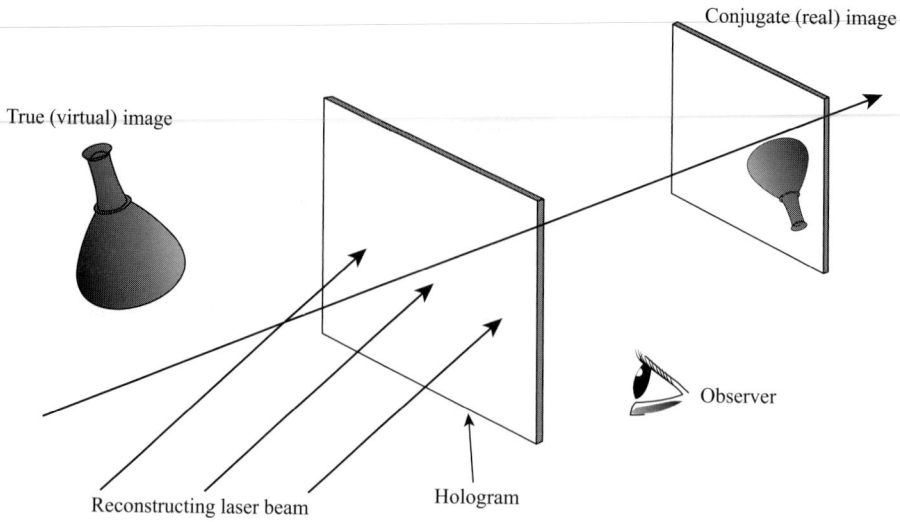


**Fig. 23.8** Reconstruction of the image stored in a hologram, showing the location of the real and virtual images.

factor $|R(x,y)|^2$ this field is proportional to $O(x,y)e^{j\phi_O(x,y)}$ – identical to the original field scattered by the illuminated object. This wave, if received in a geometry identical to the original exposure geometry, will appear exactly as it did when coming from the original object. Since the waves were diverging from the original object, the third term in (23.17) also corresponds to a diverging wave, and hence it will be viewed as a virtual image by an observer, as shown in Fig. 23.8. The final term in (23.17) is the complex-conjugate wave of the original object, except for the phase factor $e^{2j\phi_R(x,y)}$. Therefore, this corresponds to a set of waves that converge to form a real image of the original object, as shown in Fig. 23.8. Although this simple analysis does not show it, this image is inverted.

The developed hologram does not contain a recognizable image of the object as a conventional photograph does. Rather, the hologram generally appears as a collection of bright and dark lines with bright and dark patterns of concentric circles. The stored image can be regarded as a representation of the diffraction pattern of the object. We saw in Chapter 5 that the Fraunhofer diffraction pattern of an obstacle (or aperture) can be represented in terms of the Fourier transform of the amplitude transmission function of the obstacle. If the diffraction pattern is viewed close to the object so that the Fresnel diffraction pattern is being observed then the concept of the diffraction pattern as a simple Fourier transform must be modified. The object must be viewed as a series of point sources, each of which gives rise to a Huygenian secondary wavelet. The contributions of these secondary wavelets depend on a factor $e^{jkr}/r$, because each is a source of an expanding spherical wave. A detailed analysis has been given by Smith [33].

The hologram formed by exposure of a film and the production of a complex pattern of transmission is called an *amplitude hologram*. However, it is possible to process the film chemically by *bleaching* so that the dark silver atoms are converted into silver salts that are transparent but have a different refractive index from that of the surrounding film.

This converts the recorded hologram into a pattern of phase information and is called a *phase hologram*. As we shall see in the next section, the production of an image in holography can be viewed as a diffraction process. Diffraction will occur for structures that have variations either in amplitude or in phase transmission. However, the greater transmission through a bleached hologram provides much brighter reconstructions of the object [43, 44].

### 23.3.2 The hologram as a diffraction grating

To reinforce the interpretation of a hologram as a diffraction pattern, consider the simple situation shown in Fig. 23.9(a). The simplest kind of object is a distant point source, the waves from which can be treated as plane waves. The reference beam is also a plane wave. In the geometry of Fig. 23.9(a) the wave from the object is

$$O(x) = E_1 e^{j(\omega t + kx\sin\theta_1 - kz\cos\theta_1)}. \tag{23.18}$$

The reference wave is

$$R(x) = E_2 e^{j(\omega t - kx\sin\theta_2 - kz\cos\theta_2)}. \tag{23.19}$$

The transmission of the film placed in the plane $z = 0$ is

$$t(x) = t_0 - \beta|(E_1 e^{jkx\sin\theta_1} + E_2 e^{-jkx\sin\theta_2})|^2, \tag{23.20}$$

which gives

$$E(x) = t_0 - \beta E_1^2 - \beta E_2^2 - 2\beta E_1 E_2 \cos[k(\sin\theta_1 + \sin\theta_2)x]. \tag{23.21}$$

$E(x)$ clearly corresponds to a regular series of bright/dark bands on the film. The spacing $d$ of these bands is determined by the condition

$$k(\sin\theta_1 + \sin\theta_2)d = 2\pi, \tag{23.22}$$

which gives

$$d = \frac{\lambda}{\sin\theta_1 + \sin\theta_2}. \tag{23.23}$$

The interfering incident and reference waves have *written* a diffraction grating into the film.

In the language of Fourier theory $E(x)$ is a monochromatic function in $k$-space. As such it is a representation of a wave traveling at a well-defined angle – the incident wave from the object.

If we illuminate the film pattern corresponding to Eq. (23.21) with the reference wave, the transmitted field, in the plane $z = 0$, is

$$\begin{aligned} E_t(x) = {} & t_0 R(x) - \beta E_1^2 R(x) - \beta E_2^2 R(x) \\ & - 2\beta E_1 E_2^2 \cos[k(\sin\theta_1 + \sin\theta_2)x]\cos(\omega t - kx\sin\theta_2). \end{aligned} \tag{23.24}$$

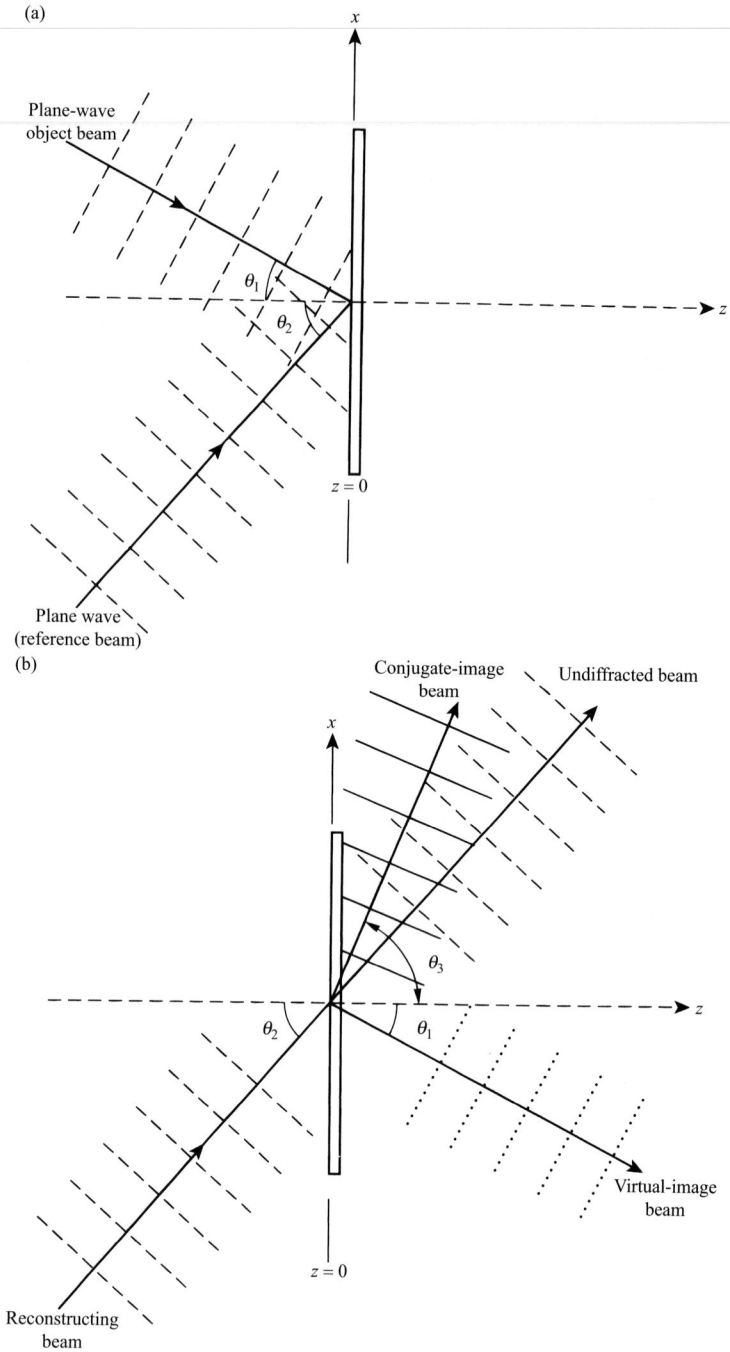


**Fig. 23.9** (a) The production of a hologram from a plane-wave object beam and a plane-wave reference beam. (b) The reconstruction of a simple plane-wave hologram showing diffracted waves corresponding to the production of real and virtual images.

The first three terms on the right-hand side of Eq. (23.24) correspond to the reference wave being transmitted through the film. However, the final term is the interesting one. It can be written as

$$E_t(x) = -\beta E_2^2[E_1 \cos(\omega t + kx \sin\theta_1) + E_1 \cos(\omega t - 2kx \sin\theta_2 - kx \sin\theta_1)]. \quad (23.25)$$

The first term in Eq. (23.25) is a reconstruction of the field of the incident wave *measured at the plane* $z = 0$. This field propagates away from the film as

$$E_1(x) = -\beta E_2^2 E_1 \cos(\omega t + kx \sin\theta_1 - kz \cos\theta_1), \quad (23.26)$$

which, apart from a proportionality constant, is equivalent to Eq. (23.18). The second term in Eq. (23.25) is a new plane-wave reconstruction of the original incident wave that propagates away from the film at angle $\theta_3$ as

$$E_3(x) = -\beta E_2^2 E_1 \cos(\omega t - kx \sin\theta_3 - kz \cos\theta_3), \quad (23.27)$$

where

$$\sin\theta_3 = 2\sin\theta_2 + \sin\theta_1. \quad (23.28)$$

Now the condition for diffraction at angle $\theta_3$ from a grating of spacing $d$ is

$$d(\sin\theta_3 - \sin\theta_2) = \lambda. \quad (23.29)$$

Substituting for $d$ from Eq. (23.23) gives $\sin\theta_3 = 2\sin\theta_2 + \sin\theta_1$, which is identical to Eq. (23.28).

We conclude, therefore, that the reference wave re-creates two versions of the original incident wave. One travels in the same direction as the original wave, for a more complex set of input waves this corresponds to the true or virtual image seen by the observer in Fig. 23.9(b). The second wave corresponds to the reconstruction of the conjugate or real image in Fig. 23.9(b).

### 23.3.3 Volume holograms

The holograms that we have been considering so far are called *plane holograms* because we have tacitly been assuming that interference of object and reference waves writes the hologram in a particular plane. We are neglecting the thickness of the photographic emulsion. However, since actual emulsions can be up to 20 μm thick, it is possible, and in many cases desirable, to write a *volume hologram* in which interference fringes are written throughout the emulsion. The hologram can be thought of as a continuous distribution of plane holograms at different values of $z$ through the emulsion thickness. Just as the simple plane hologram produced by the interference of two plane waves was a planar diffraction grating, so the volume hologram produced in this way will have a structure where the regions of bright and dark are distributed in a series of parallel planes in the recording medium. Figure 23.10 shows schematically how the transition from plane hologram to volume hologram changes the pattern of recorded information. Whether the volume hologram is a transmission or reflection hologram is determined by the way in which the hologram is recorded. A reflection hologram is recorded with the reference beam incident on the film

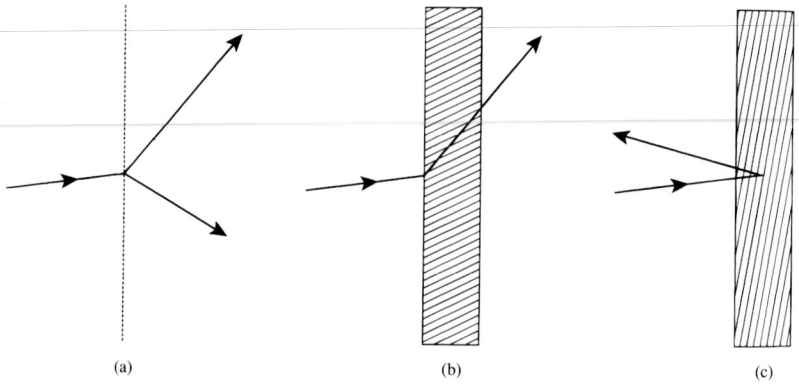


**Fig. 23.10** A schematic illustration of the transition from a plane hologram to a volume hologram. (a) A plane hologram in which the interference pattern written as the hologram is located in a plane. The real and virtual image waves are shown schematically as resulting from a reconstructing wave. (b) A volume hologram in which the interference of object and reference waves has written fringes almost perpendicular to the plane of the film. Only a single reconstructed wave results. (c) A volume hologram in which the interference fringes are almost parallel to the film. The direction of the reconstructed wave that results from illumination is shown.

from the opposite side from the object as shown in Fig. 23.11(a), so the resulting interference patterns are almost parallel to the film. A transmission hologram is recorded as shown in Fig. 23.12.

Because the interference patterns that constitute the volume hologram are families of planes, during reconstruction of the wave fronts not only must a diffraction condition like Eq. (23.29) be satisfied, but so also must the Bragg condition for reflection from the periodic planes in the hologram. Consequently, during wave-front reconstruction only one of the two images, either the true or the conjugate, will be reconstructed, depending on the direction of the reference beam used for reconstruction, as shown in Figs. 23.12(b) and (c). The diffraction and Bragg conditions cannot be satisfied for both at the same time. Reflection holograms are of particular interest, both aesthetic and practical, because they can be reconstructed with white light. These holograms are generally bleached because of the increased diffraction efficiency that this provides. The hologram picks out the appropriate narrow band of wavelengths that satisfy both the diffraction and the Bragg-reflection condition. The reconstructed image will be colored, and the color depends on the viewing angle. Full-color holograms can be produced by recording the hologram with the light from red, green, and blue lasers and then using these colors during reconstruction.

White-light holograms reconstruct images that lack the granularity, or *speckle*, that always appears when highly coherent light is used either to illuminate an object or to reconstruct an image. The primary cause of speckle is the microscopically rough surface of the illuminated object, or the diffuser if one is being used during hologram recording. In the case of a microscopically rough surface, each surface feature acts as a source of an expanding spherical wavelet. The many wavelets from the whole object add vectorially at each point in space and there will inevitably be a pattern of interference maxima and minima produced. This is the speckle pattern. It is, for example, the term $-\beta OO^*$ that

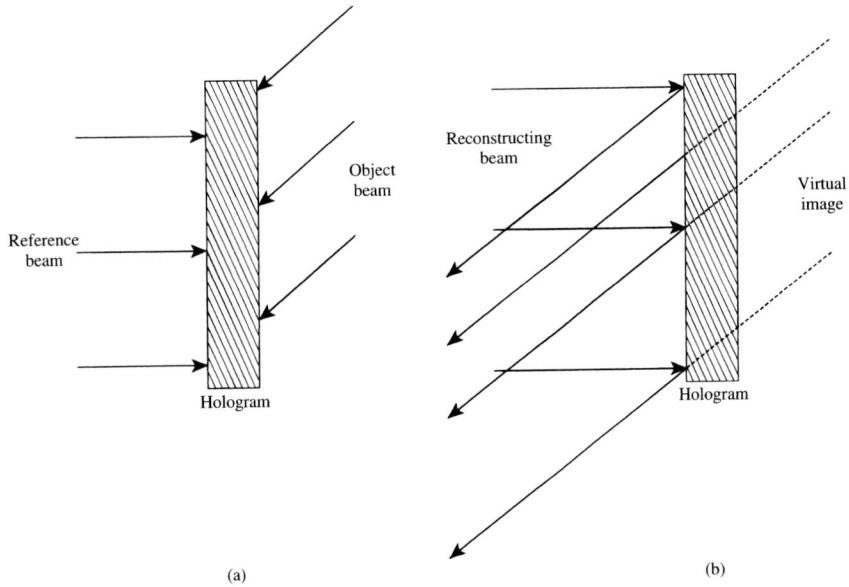


**Fig. 23.11** (a) The arrangement for writing a reflection volume hologram. (b) Reconstruction of the image.

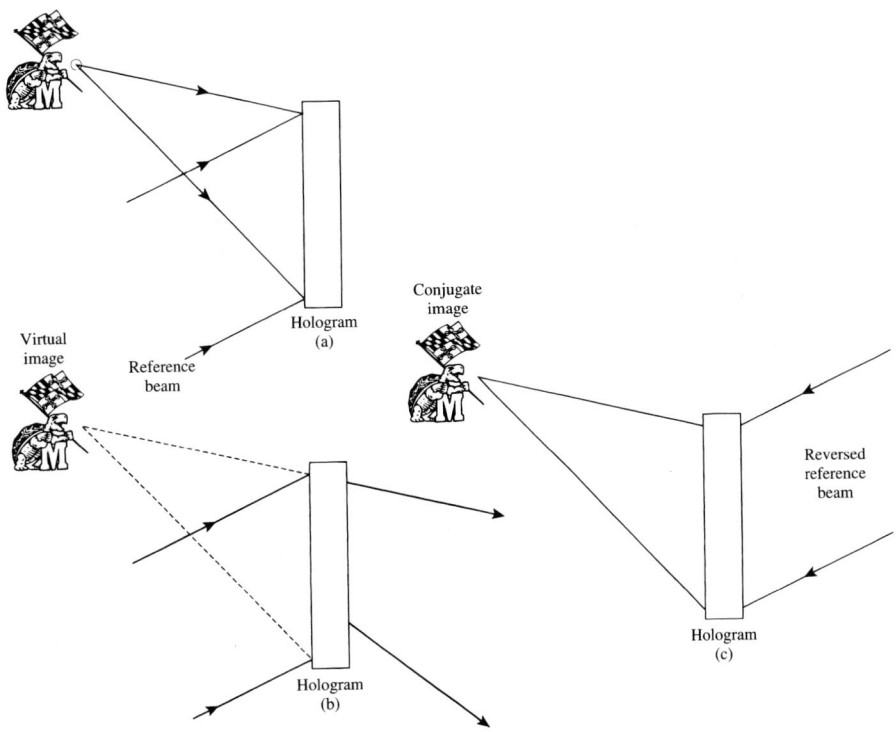


**Fig. 23.12** (a) The arrangement for writing a transmission volume hologram. (b) Reconstruction of the virtual image. (c) Reconstruction of the conjugate (real) image. The happy terrapin shown is Testudo, the mascot of the University of Maryland, College Park.

we neglected in Eq. (23.15). When a speckle pattern is viewed with the naked eye, the granularity appears to fluctuate. The speckle pattern itself is stationary in space, but the eye in its constant small motions moves through the various maxima and minima and leads to a subjective twinkling. The speckle pattern produced by an object contains useful information and can be used in various ways to study surface roughness and object motion [45–47].

Holography has found application in several areas. Of most practical importance is non-destructive testing. Extremely small deformations of an object can be detected by superimposing the reconstructed images of the object in its two different states: for example, before and after a stress is applied [48, 49]. Holography is also used in optical memories [38], and in this context is intimately related to the whole field of optical information processing [32, 34, 50–52], which is beyond our scope here. Practical details of recording holograms have been described by Iizuka [38] and Smith [33].

## 23.4 Medical applications of lasers

Almost from its inception the laser has found applications in medicine. Sometimes the laser has emerged as a means for carrying out a new surgical procedure, in other cases the laser presents an alternative means for carrying out an existing procedure; a comprehensive review has been given by Waynant [53]. Perhaps the best known of these surgical applications of the laser have been in the field of opthalmology, especially in vision correction using laser-assisted *in situ* keratomileusis (LASIK) and laser epithelial keratomileusis (LASEK), and in the reattachment of detached retinas.

### 23.4.1 Lasers in opthalmology

#### LASIK and LASEK

The correction of vision defects using LASIK and LASEK has become widespread [54]. Anesthetic drops are added to the patient's eye, which is immobilized with a corneal suction ring. In LASIK a short-pulsed laser then loosens a flap of the outer surface of the cornea, by creating a series of microbubbles, but leaves an unseparated region to act as a "hinge." The corneal flap is folded back and then an Ar–F (193-nm) exciplex laser ablates the exposed surface under computer control to produce the appropriate curvature modifications calculated to correct vision. The corneal flap is then returned to its place and healing begins. In a LASEK procedure an extremely thin epithelial layer is lifted from the eye's surface before laser energy is applied for ablation. After the LASEK procedure, the epithelium is replaced on the surface of the eye. These procedures have a very high success rate in producing significant vision improvement, although some people will experience dry eyes, "haloes" around lights at night, or double vision after the procedure.

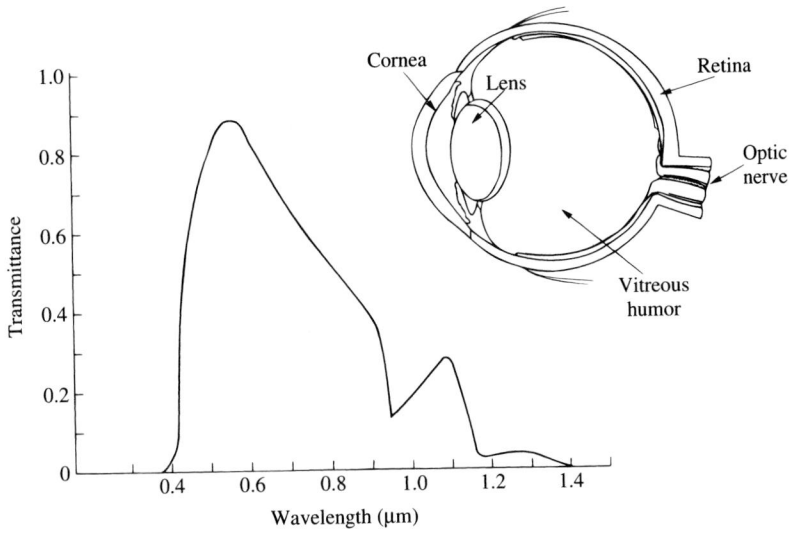


**Fig. 23.13** A simplified cross-section of the human eye, showing the spectral transmittance to the retina from the outside of the eye.

## Reattachment of detached retinas

If the detached retina will settle back into the correct position, it can literally be "welded" back into place by focusing a CW argon-ion laser, an 810-nm diode laser, or a pulsed or CW Nd:YAG laser through the lens of the eye onto the back of the eye. Bleeding inside the eye can also be stopped by laser heating the site of bleeding – a process called *photocoagulation*. The same lasers that present a serious eye-hazard, because they can penetrate to the back of the eye, can be used for these surgical procedures inside the eye. As can be seen from Fig. 23.13, wavelengths beyond about 1.5 μm are absorbed before reaching the retina. Lasers of these longer wavelengths can still be hazardous to the eye, but primarily as a means of causing burns to the outer surface of the cornea.

Ultraviolet, visible, and infrared lasers are readily absorbed by body tissue, so they have an ability to act as scalpels in specialized treatments. For treatment of certain skin conditions, including some cancers, the ability of the laser to burn, or photoablate,[6] and selectively remove thin layers near the surface without damaging underlying tissue is very valuable. Much of the ability of the laser to burn locally, and selectively, is based on the absorption of the laser wavelength by water. Consequently $CO_2$ lasers are more effective than argon-ion or Nd:YAG lasers in these scalpel-like applications. However, the absorption of water in the body is largest near 3 μm, so lasers operating near this wavelength, such as erbium:YAG lasers, are very useful. In dermatology lasers can be used to remove vascular malformations such as port-wine stains, unwanted body hair, and tattoos. Solid-state

[6] The rapid absorption by tissue of energy from a pulsed laser, leading to an ejection of material from the surface by shock without significant burning, is known as photoablation.

lasers based on Nd:YAG have largely supplanted argon-ion lasers in these applications. $CO_2$ and erbium lasers can be used to remove or reduce scars and re-surface skin in cosmetic procedures.

A clearly beneficial application of pulsed blue–green dye lasers is in the fragmentation of kidney stones that have left the kidney and become trapped in the urinary tract. Pulses of intense light from a holmium:YAG laser (operating at 2,150 nm) are directed along a quartz optical fiber at the kidney stone to successively break the stone into several pieces that are then able to leave the body [55].

For general application as scalpels lasers are attractive when used to cut very vascular tissue, which bleeds a great deal when cut. The laser cuts and cauterizes at the same time, the result being an incision with a layer of burnt tissue on both sides of the cut. Lasers are even finding application in "painless" dentistry. Lasers have many additional non-surgical uses in diagnostics and monitoring, for example in analytical procedures, and as optical sources in various types of sensors for temperature, pressure, and blood flow.

For performing surgery inside the body the laser beam must be directed inside an optical-fiber system that at the same time provides the surgeon with a TV image of the region at the end of the fiber. An exciting application of lasers in this way involves the burning away, or ablation, of the plaque inside blocked arteries, thereby opening them up once again to blood flow. This procedure is called laser angioplasty and is worthy of further discussion.

### 23.4.2 Laser angioplasty

Arteriosclerosis is the constriction, or blocking, of blood flow through arteries. The material that leads to this problem is called plaque, a composite material containing layers of fat, fibrous tissue, muscle tissue, and varying amounts of calcium (as $Ca_3(PO_4)_2$). There has been great interest, and some success, in using both CW and pulsed lasers to ablate plaque away. A laser beam is directed at the blockage in an artery along an optical fiber inserted through a catheter into the artery. The catheter is threaded over a previously placed guide wire. The use of lasers in this way is called *laser angioplasty*, in contrast with the earlier (and still more widely used and successful) technique of balloon angioplasty. In this technique a small balloon is threaded into the occluded region of the artery and then filled with fluid, thereby compressing the plaque against the wall of the artery.

The problems that have been faced in the use of lasers to remove plaque have involved the following issues.

(i) One must avoid causing damage to the wall of the artery, which is only about 1 mm thick. This problem has been alleviated to a large extent by placing a ball lens on the end of the fiber [56].
(ii) The opened region can reocclude within a few months.
(iii) Thermal damage to healthy tissue can occur. The use of pulsed lasers seems to avoid the thermal-damage problem, while short, intense pulses are effective in photoablating the plaque. Sub-microsecond ultraviolet pulses from XeCl lasers operating at 308 nm

are being investigated in this application [57], as is the use of holmium lasers operating at 2.1 μm. These lasers are used in a long-pulse mode delivering a few joules per pulse in pulses of duration 0.25 s at 1–10 pulses per second. Their wavelength of 2.1 μm is strongly absorbed by water. Both these ultraviolet and near-infrared wavelengths can be delivered to the occluded site along silica fibers. The treatment of the occluded artery is frequently followed up by an application of balloon angioplasty.

## 23.5 Laser weapons

Laser weapons are one example of what is called a *directed-energy* weapon. In principle, a laser weapon delivers its energy at the speed of light to exactly where it is aimed, although in terrestrial applications atmospheric effects must be taken into account. The path of the beam is affected by density (refractive-index) gradients in the atmosphere, and by heating of the path by the laser itself, which causes the beam to spread (thermal blooming). The use of lasers to cause permanent blindness is banned by a 1995 United Nations protocol on Blinding Laser Weapons. The use of visible laser dazzlers, which temporarily blind or disorient, is not forbidden by this protocol. Dazzlers can also be used to interfere with, or disable, electro-optic systems being used by an adversary.

The use of high-power lasers to disable or destroy vehicles such as missiles has met with some success in demonstrations, but has not led to deployment of these systems. The megawatt-class deuterium fluoride chemical laser (MIRACL) has successfully shot down Vandal missiles at the White Sands test range, but this required a large ground station, large amounts of the chemical fuels needed (see Section 10.7), and massive quantities of helium gas. The airborne laser (ABL) program involves a chemical oxygen–iodine laser (COIL) mounted on a modified Boeing 747 aircraft, which has successfully been used to destroy a liquid-fueled ballistic missile target. This success may be transient since the United States Air Force has not requested further funding for the program, and the Air Force Chief of Staff described this as "a magnificent technical achievement" but stated that the system "does not reflect something that is operationally viable." There is continuing work to make laser weapons more deployable by using very-high-power solid-state free-electron lasers [58] and high-power fiber lasers. A ship-borne system using six solid-state lasers has successfully shot down an unmanned aerial vehicle (UAV). Because a ship can provide large amounts of electrical power, it is a more viable platform than an aircraft for an electrically powered laser.

## 23.6 Laser isotope separation

The different isotopes of an element are chemically virtually indistinguishable and physically different to only a small extent. Consequently, it has always been extremely challenging to separate pure isotopes from the mixture of isotopes found for each element in nature. If very small quantities of the pure isotope are required, then it is possible to use a mass spectrometer to separate the ions of different isotopes. However, to produce larger quantities of commercially important isotopes such as heavy hydrogen (deuterium) or to enrich uranium from its naturally occurring 0.7% of uranium-235 to the 3.2% required for light-water nuclear reactors one must use different schemes. This problem is easiest to solve for the separation of hydrogen ($H_2$) and deuterium ($D_2$), since the deuterium is twice as heavy as the hydrogen. Electrolysis of water, which naturally contains about 0.015% of its hydrogen as heavy hydrogen, produces hydrogen and deuterium at different rates at the cathode, so sequential repeated electrolysis can enrich water in $D_2O$.[7]

The use of lasers for isotope separation is based on the spectroscopic difference between isotopes of an element [59–61]. The wavelengths of absorption differ for atoms or molecules that are chemically identical but contain different isotopes. Much of the stimulus for the development of laser isotope separation (LIS) arose from a need to improve the efficiency (and lower the cost) of increasing the concentration of $^{235}U$ relative to $^{238}U$. The traditional method for uranium enrichment is gaseous diffusion. The gas uranium hexafluoride ($UF_6$) is allowed to diffuse through a series of porous barriers. The $^{235}UF_6$ diffuses slightly faster, so a large number of successive diffusions, in which the slightly $^{235}UF_6$-enriched gas is passed to each diffusion stage, leads to the desired enrichment. Approximately 1,240 stages of diffusion are required in order to produce 3.2% $^{235}UF_6$. It is also possible to produce this enrichment in a gas centrifuge where the central forces acting on the different isotope species of $UF_6$ are used to produce enrichment. Both these techniques require gigantic amounts of electric power for their operation.

The way in which lasers can be used to produce isotopic enrichment is well illustrated by one of the schemes that has been used to enrich atomic uranium [62]. Figure 23.14 shows schematically the energy levels that are involved in the separation of $^{235}U$ and $^{238}U$. In one implementation of this scheme at the Lawrence Livermore National Laboratory (LLNL) in California a beam of uranium atoms emerges from an oven at 2,600 K. At this temperature 27% of the uranium atoms are in a metastable sub-level of the ground state. $^{235}U$ atoms are excited from this sub-level of energy $E_1$ (620 cm$^{-1}$) by a xenon-ion laser with $\lambda_1 = 378.1$ nm to an intermediate level of energy $E_2$. The wavelength of the transition from $E_1'$ to $E_2'$ in $^{238}U$ is at a slightly longer wavelength, so these atoms are not excited. Excited $^{235}U$ atoms that reach level $E_2$ are ionized with a krypton laser operating at 350.7 nm or 246.4 nm. By this two-step photodissociation process $^{235}U^+$ ions are selectively produced. The ions are collected by deflecting them to a collector with a strong electric field.[8] This isotope separation does not, in reality, proceed quite so straightforwardly as described. Excited uranium atoms can collide and exchange energy, thereby scrambling the process. No LIS uranium-enrichment plants are currently operating in the United States. However, this may change. The energy efficiency of LIS is better than is possible with gaseous diffusion or gas centrifuges. It remains to be seen whether there will be a resurgence of interest in

[7] Additional complexity arises from the presence of isotopically mixed water, HDO.

[8] The photoionizing processes in the currently used method of uranium isotope separation at the LLNL have not been revealed in detail, but the photons needed for the process are generated by copper-vapor-laser-pumped tunable dye lasers.

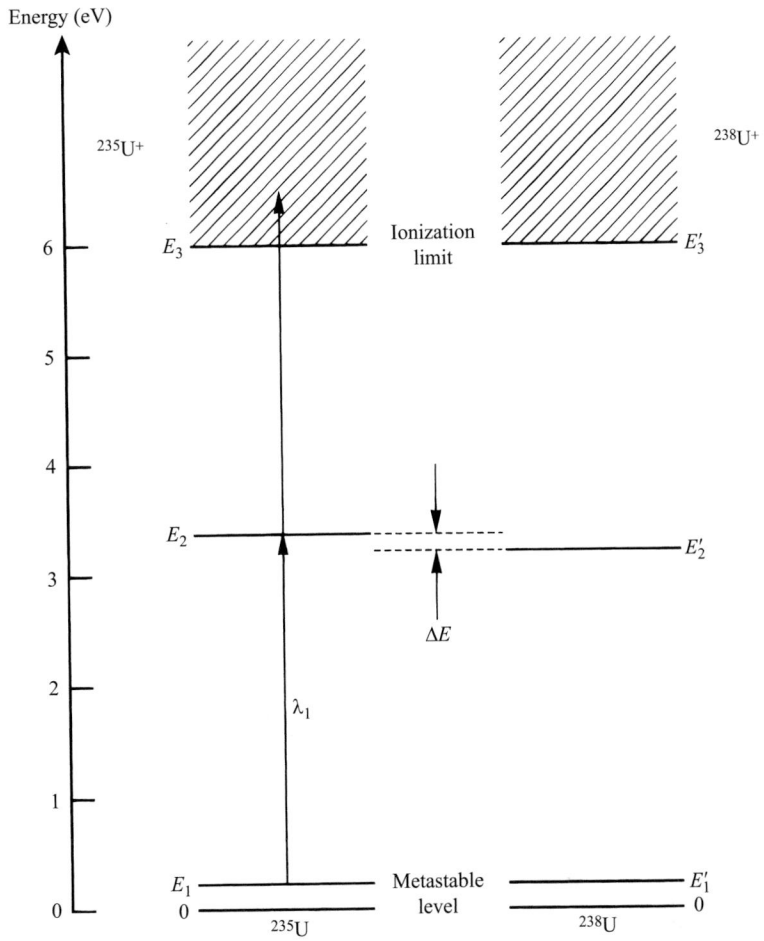


**Fig. 23.14** A schematic energy-level scheme of $^{235}U$ and $^{238}U$, showing how laser isotope separation can be performed.

nuclear fission for energy generation, which could rekindle interest in LIS. Also, LIS has the capability of producing many different isotopes for medical purposes.

There are also some interesting LIS processes that use molecular species. Two schemes that have been used are shown in Fig. 23.15. A tunable ultraviolet laser selectively excites one isotopic species of a molecule to a new electronic state, whereupon it dissociates. The fragments are thereby enriched in the isotopic species whose absorption line is coincident with the excitation laser. This scheme has been used to separate carbon isotopes through the predissociation of formaldehyde:

$$H_2\,{}^{13}CO \xrightarrow{h\nu} H_2 + {}^{13}CO_2. \tag{23.30}$$

Different isotopes of hydrogen have been separated from formaldehyde in a similar way:

$$HD^{12}CO \xrightarrow{h\nu} HD + {}^{12}CO. \tag{23.31}$$

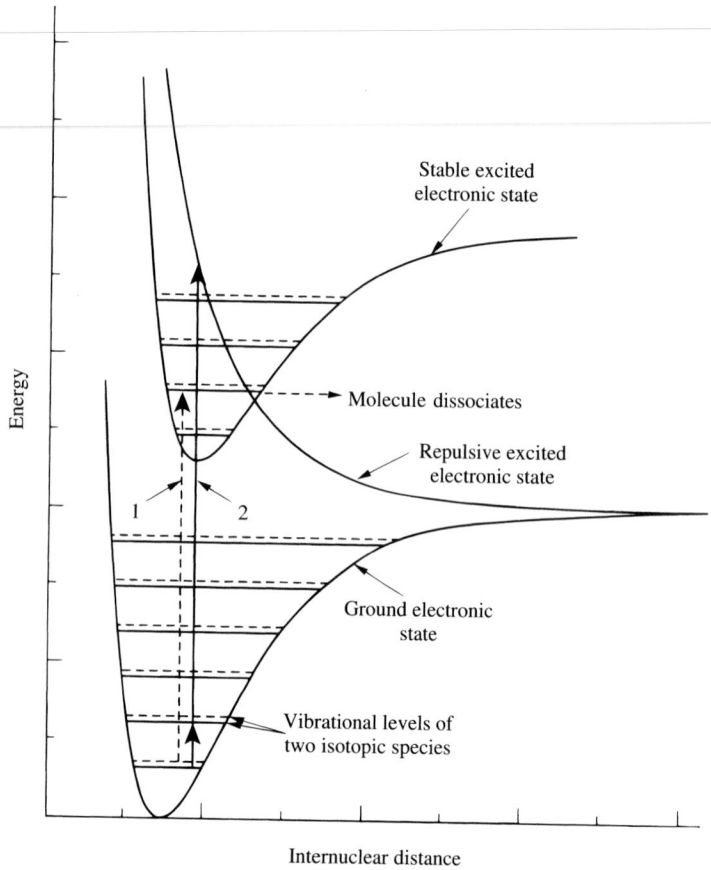


**Fig. 23.15** Laser isotope schemes for molecular species. In process 1 a single energetic photon excites a particular isotope to a level at which the molecule dissociates. In scheme 2 an infrared photon selectively excites a vibrational state of one isotope, which is then dissociated with a second, energetic photon.

The second scheme shown in Fig. 23.15 is one that has been actively studied as a means for separating $^{235}$U-rich fragments from $UF_6$. A laser that is resonant with the $\nu_3$ vibration–vibration absorption in the ground electronic state of $UF_6$ selectively excites $^{235}UF_6$. This vibrationally excited molecule is then dissociated with a second energetic photon to produce a $^{235}UF_5$ fragment and a fluorine atom. The first selective excitation in $UF_6$ is provided with a 16 μm photon, the second with a 200–300-nm photon from an excimer laser. The $^{235}UF_5$ fragments combine chemically and can be separated. In order that the 16-μm absorption occurs in $^{235}UF_6$ and not in $^{238}UF_6$, the $UF_6$ has to be cooled by expansion through a supersonic nozzle. This ensures that the absorption lines are sharp and higher vibrational levels are not populated thermally [63].

There has also been considerable success in separating isotopes through a process of multiphoton absorptions. In this process a molecule sequentially absorbs many identical photons and climbs up its ladder of vibrational energies to a sufficiently high energy that dissociation takes place. Although this process should not be efficient, because the vibrational energy levels of molecules are not equally spaced, it is in fact surprisingly so. For example, the various isotopes of sulfur in $SF_6$ have been separated by processes such as

$$^{32}SF_6 + h\nu \rightarrow {}^{32}SF_4 + 2F \tag{23.32}$$

followed by

$$SF_4 + H_2O \rightarrow SOF_2 + 2HF. \tag{23.33}$$

The $SOF_2$ is chemically relatively inert and can be separated from the other species [64].

A final isotope-separation scheme uses laser-enhanced chemical reactions. These schemes rely for their operation on the fact that excited states of atoms or molecules are frequently more reactive than those species are in the ground state. For example, the isotopes of chlorine can be separated by irradiating a mixture of iodine chloride and bromobenzene with red light from a dye laser. In order to obtain a sufficiently great photon flux, the reaction cell is placed inside the cavity of the dye laser:

$$I^{37}Cl + h\nu(\lambda = 605\ \text{nm}) \rightarrow (I^{37}Cl)^*, \tag{23.34}$$

followed by

$$(I^{37}Cl)^* + C_6H_5Br \rightarrow C_6H_5\,{}^{37}Cl + I + Br. \tag{23.35}$$

The resultant chlorobenzene ($C_6H_5{}^{37}Cl$) is enriched in $^{37}Cl$.

## 23.7 Laser plasma generation and fusion

Short-pulse, high-energy laser pulses when focused produce extremely high energy fluxes and electric and magnetic fields. For example, a 1-kJ, 1-ns rectangular laser pulse focused to a spot size of 1 μm produces an energy flux of $6.4 \times 10^{23}$ W m$^{-2}$, an electric field of $2.2 \times 10^{13}$ V m$^{-2}$, and a magnetic flux density in free space of 73 kT. Fields substantially smaller than this can strip outer- and inner-shell electrons from the atoms of a target material. Consequently, intensely focused lasers have found widespread use in the production of plasmas from the surface of target materials. These laser-produced plasmas can be copious producers of X-rays, which can be relatively coherent. Such X-rays have significant potential for high-resolution holography and lithography. Because diffraction limits the resolvable feature size in a microscope or lithography system, the shorter the wavelength being used the finer the resolution. Most laser-induced-plasma-generated X-radiation is coherent because of amplified spontaneous emission (ASE). The rapid ionization of a target material, frequently a thin metal foil, irradiated with an intense laser pulse [65–67], can lead to a short-lived population inversion on energy levels of the resultant multiply ionized ions. In many ways the process is an energetic analog of the excitation processes in self-terminating pulsed gas lasers such as copper-vapor and nitrogen lasers.

The irradiation of various materials at very high laser fluences (J $m^{-2}$) has, in and of itself, provided a fruitful means for studying matter at very high temperatures and pressures. However, much of the impetus for such studies has been provided by the desire to demonstrate laser-driven, inertial-confinement nuclear fusion. Nuclear fusion occurs in stars and hydrogen-bomb explosions, and there has been work for several decades on the controlled production of such thermonuclear reactions on a laboratory scale for nuclear power generation.

Most controlled thermonuclear reactions (CTRs) are initiated in magnetically contained plasmas heated to high temperatures in reactors such as Tokamaks and Stellerators [68]. In these reactors a plasma, an electrically neutral mixture of gaseous ions and free electrons, is confined by a complex arrangement of magnetic fields and heated. Mixtures of deuterium, D, and tritium, T, are used as fuels. At sufficiently high temperatures the following fusion reactions occur:

$$\mathrm{D}+\mathrm{D} \rightarrow {}^{3}\mathrm{He}(0.82\ \mathrm{MeV}) + \mathrm{n}(2.45\ \mathrm{MeV}), \tag{23.36}$$

$$\mathrm{D}+\mathrm{D} \rightarrow \mathrm{T}(1.01\ \mathrm{MeV}) + \mathrm{H}(3.02\ \mathrm{MeV}), \tag{23.37}$$

$$\mathrm{D}+\mathrm{T} \rightarrow {}^{4}\mathrm{He}(3.5\ \mathrm{MeV}) + \mathrm{n}(14.1\ \mathrm{MeV}), \tag{23.38}$$

$$\mathrm{D}+{}^{3}\mathrm{He} \rightarrow {}^{4}\mathrm{He}(3.6\ \mathrm{MeV}) + \mathrm{H}(14.7\ \mathrm{MeV}). \tag{23.39}$$

At temperatures of 100 million degrees the first two reactions occur at roughly equivalent rates. The sustainability of a nuclear fusion reactor is frequently characterized in terms of the *Lawson criterion*, which says that, at a temperature of 5 keV (1 eV = 11,600 K), the product of the ion number density, $n_i$, and the containment time, $\tau$, must satisfy

$$n_i \tau > 10^{20}\ \mathrm{m}^{-3}\ \mathrm{s}. \tag{23.40}$$

In a magnetically confined reactor containment times are typically in the millisecond-to-second range.

The approach taken in laser-fusion research is that of *inertial* confinement. The idea is to heat a small pellet of nuclear fuel (D–T mixtures are the most likely to be used) to high temperatures so fast that the mass of the fuel pellet prevents it from dispersing because of its own inertia. The Lawson criterion is thereby satisfied with a larger ion density and smaller containment time than for magnetically confined CTR. For a fuel pellet of radius $a$ the natural inertial containment time is

$$\tau_i \simeq a/v_s, \tag{23.41}$$

where $v_s$ is the velocity of sound in the fuel material at its elevated temperature. At a temperature of 10 keV the speed of sound is $\sim 10^6$ m $s^{-1}$, so, for a fuel pellet of radius 1 mm, the inertial-containment time is only 1 ns. To accomplish the heating of a fuel pellet in a short time, while simultaneously confining it and enhancing the achievement of the Lawson criterion, the fuel pellet is irradiated simultaneously by a large number of high-energy, short-pulse laser beams. The irradiation should be as spherically symmetric as possible. At the LLNL National Ignition Facility (NIF) 192 beams provide megajoule energies in 1-ns pulses at a wavelength of 351 nm [69, 70]. Each beam originates from an

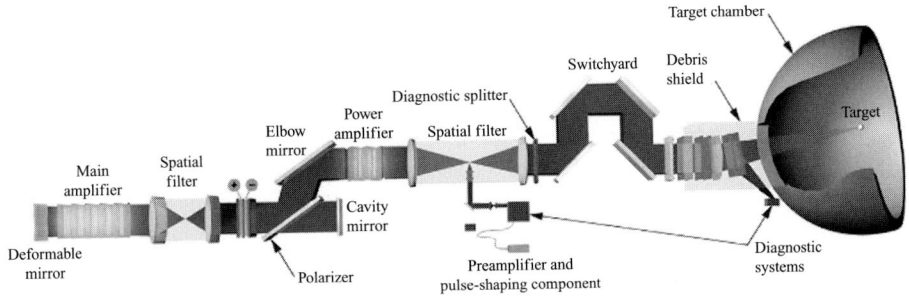


**Fig. 23.16** A schematic representation of the beam line of the National Ignition Facility (NIF) (from https://www.llnl.gov/str/JulAug 07/Remington.html).

ytterbium-doped optical-fiber laser, whose output is amplified by a series of Nd:phosphate-glass slab amplifiers operating at 1.053 μm. The final ultraviolet wavelength is obtained by third-harmonic generation of the beams from the oscillator–amplifier chains. Figure 23.16 shows how the NIF laser system is arranged.

The high-energy irradiation of the fuel pellet vaporizes its outer surface to form a plasma. The plasma allows the laser beam to penetrate a certain distance, to the *critical-density* surface, where substantial energy is absorbed. The hot plasma conducts thermal energy inwards to the solid boundary of the fuel pellet. The rapid heating of the solid boundary ablates material, which explodes away from the rest of the pellet like in a rocket engine. The equal and opposite reaction force of Newton's third law exerts an enormous pressure, perhaps $10^{12}$ atm, and compresses the remainder of the fuel pellet inwards. The core of the pellet is compressed to as much as 100 times solid density. The culmination of this process of compression and heating is that perhaps 10% of the original fuel pellet is heated and compressed sufficiently that it undergoes a fusion reaction. The aim is to generate 100 MJ of fusion energy (equivalent to the explosion energy of 48 lb of TNT)[9] in each explosion and heating process. Figure 23.17 shows schematically some of the processes that occur during the laser-fusion process. In early experiments the fuel pellets were cooled solid DT spheres [71], or small glass microspheres containing DT gas at high pressure. Current work is proceeding through a series of pellet designs beginning with pellets layered with beryllium doped with copper or hydrocarbon foam doped with germanium, and ultimately involving frozen DT pellets [70].

To heat the fuel pellet "smoothly," so that its core is not heated too rapidly before it is sufficiently compressed, a special shape of laser pulse appears to be necessary. Theoretically the ideal laser pulse is

$$I(t) = I_0(1 - t/t_c)^{-2}, \tag{23.42}$$

[9] The explosive force of the microfusion "bomb" is not as destructive as TNT since 80% of the energy is carried by neutrons. These neutrons can be absorbed in a surrounding blanket of $^6$Li to regenerate the tritium:

$$^6\mathrm{Li} + \mathrm{n} \rightarrow {}^4\mathrm{He} + \mathrm{T}(\simeq 4.6\,\mathrm{MeV}).$$

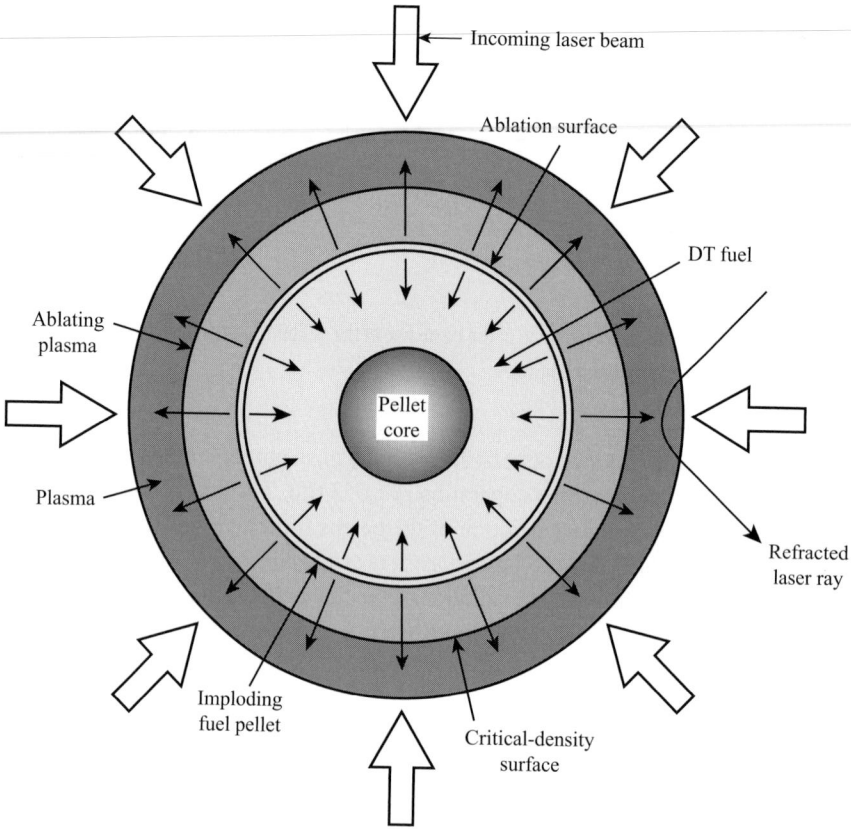


**Fig. 23.17** A schematic illustration of the laser fusion process whereby a spherical pellet of nuclear fuel is irradiated symmetrically with intense laser beams.

where $t_c$ is a characteristic time that depends on the precise characteristics of the fuel pellet. This pulse shape rises slowly at first and ultimately deposits a significant fraction of its total energy in a very short period at the end of the pulse.

Unfortunately, although the production of commercial quantities of electric power by laser fusion can be made to appear feasible [72–74] and complete reactor concepts have been developed, the "break-even" point, where the fusion reaction generates as much energy as is injected into the drive lasers, seems far away. However, this research has led to a substantial understanding of laser–matter interactions and significant developments in materials research.

## References

[1] A. Dumas, *The Count of Monte Cristo*, New York: Dodd Mead, 1979.

[2] J. Tyndall, "On some phenomena connected with the motion of liquids," *Royal Inst. Great Britain Proc.* **1**, 446–448, 1854.

[3] K. C. Kao and G. A. Hockham, "Dielectric-fiber surface waveguides for optical frequencies," *Proc. IEEE* **113**, 1151–1158, 1966.

[4] F. P. Kapron, D. B. Keck, and R. D. Maurer, "Radiation losses in glass optical waveguides," *Appl. Phys. Lett.* **17**, 423–425, 1970.

[5] T. Miya, Y. Terunuma, T. Hosaka, and T. Miyashita, "Ultimate low-loss single-mode fiber at 1.55 μm," *Electron. Lett.* **15**, 106–108, 1979.

[6] Y. Awaji, N. Wada, A. Kanno, T. Kawanishi, T. Hayashi, T. Taru, T. Kobayashi, and M. Watanabe, "Space division multiplexed transmission of 109-Tb/s data signals using homogeneous seven-core fiber," *J. Lightwave Technol.* **30**, 659–665, 2012.

[7] J. Renaudier, O. Bertran-Pardo, G. Charlet, M. Salsi, H. Mardoyan, P. Tran, and S. Bigo, "8 Tb/s long haul transmission over low dispersion fibers using 100 Gb/s PDM-QPSK channels paired with coherent detection," *Bell Labs Tech. J.* **14**, 27–45, 2010.

[8] A. Sano, H. Masuda, T. Kobayashi, M. Fujiwara, K. Horikoshi, E. Yoshida, Y. Miyamoto, M. Matsui, M. Mizoguchi, H. Yamazaki, Y. Sakamaki, and H. Ishii, "69.1 Tb/s (432 × 171 Gb/s) C and extended L band transmission over 240 km using PDM 16-QAM modulation and digital coherent detection," post-deadline paper presented at OFC 2010, San Diego, California, 2010.

[9] I. Djordjevic, W. Ryan, and B. Vasic, *Coding for Optical Channels*, Berlin: Springer-Verlag, 2010.

[10] Fuqin Xiong, *Digital Modulation Techniques*, 2nd edn., Boston, MA: Artech House, 2006.

[11] H. Zhu and B. Mukherjee, "Online connection provisioning in metro optical WDM networks using reconfigurable OADMs," *J. Lightwave Technol.* **23**, 2893–2901, 2005.

[12] H. Zang, J. P. Jue, and B. Mukherjee, "A review of routing and wavelength assignment approaches for wavelength-routed optical WDM networks," *Optical Networks Magazine* January, 47–60, 2000.

[13] G. P. Agrawal, *Fiber-Optic Communication Systems*, 4th edn., New York: Wiley, 2010.

[14] G. Keiser, *Optical Fiber Communications*, 4th edn., New York: McGraw-Hill, 2008.

[15] J. C. Palais, *Fiber Optic Communications*, 5th edn., Englewood Cliffs, NJ: Prentice-Hall, 2005.

[16] J. Gowar, *Optical Communication Systems*, 2nd edn., Englewood Cliffs, NJ: Prentice-Hall 1993.

[17] P. E. Green, Jr., *Fiber Optic Networks*, Englewood Cliffs, NJ: Prentice-Hall, 1993.

[18] H. B. Killen, *Fiber Optic Communications,* Englewood Cliffs, NJ: Prentice-Hall, 1991.

[19] D. J. Morris, *Pulse Code Formats for Fiber Optical Data Communication*, New York: Marcel Dekker, 1983.

[20] D. K. Borah, A. C. Boucouvalas, C. C. Davis, S. Hranilovic, and K. Yiannopoulos, "A review of communication-oriented optical wireless systems," *EURASIP J. Wireless Commun. Networking* **91**, April, 1–28, 2012.

[21] L. C. Andrews and R. L. Phillips, *Laser Beam Propagation through Random Media,* 2nd edn., Bellingham, WA: SPIE Press, 1998.

[22] C. C. Davis, I. I. Smolyaninov, and S. D. Milner, "Flexible optical high data rate wireless links and networks," *IEEE Commun. Mag.*, March, 51–57, 2003.

[23] J. C. Juarez, A. Dwivedi, A. R. Hammons, Jr., S. D. Jones, V. Weerackody, and R. A. Nichols, "Free-space optical communications for next-generation military networks," *IEEE Commun. Mag.*, November, 46–51, 2006.

[24] E. N. Leith and J. Upatnieks, "Wavefronts reconstruction with continuous-tone objects," *J. Opt. Soc. Am.* **53**, 1377–1381, 1963.

[25] E. N. Leith and J. Upatnieks, "Wavefront reconstruction with diffused illumination and three-dimensional objects," *J. Opt. Soc. Am.* **54,** 1295–1301, 1969.

[26] D. Gabor, "Holography, 1948–71," *Proc. IEEE* **60**, 655–668, 1972.

[27] D. Gabor, "A new microscopic principle," *Nature* **161**, 777–778, 1948.

[28] D. Gabor, "Microscopy by reconstructed wavefronts," *Proc. Roy. Soc.* **A197**, 454–487, 1949.

[29] D. Gabor, "Microscopy by reconstructed wavefronts: II," *Proc. Phys. Soc.* (*London*) **64**, 449–469,1951.

[30] V. Toal, *Introduction to Holography*, Boca Raton, FL: CRC Press, 2011.

[31] P. Hariharan, *Basics of Holography*, Cambridge: Cambridge University Press, 2002.

[32] P. Hariharan, *Optical Holography: Principles, Techniques and Applications*, Cambridge: Cambridge University Press, 1996.

[33] H. M. Smith, *Principles of Holography*, 2nd edn., New York: Wiley Interscience, 1975.

[34] G. W. Stroke, *An Introduction to Coherent Optics and Holography*, New York: Academic Press, 1969.

[35] W. T. Cathey, *Optical Information Processing and Holography*, New York: Wiley, 1974.

[36] H. J. Caulfield and S. Lu, *The Applications of Holography*, New York: Wiley-Interscience, 1970.

[37] B. J. Thompson, "Principles and applications of holography," in *Applied Optics and Optical Engineering*, Vol. 6, eds. R. Kingslake and B. J. Thompson, New York: Academic Press,1980.

[38] K. Iizuka, *Engineering Optics*, 3rd edn., Berlin: Springer-Verlag, 2010.

[39] J. W. Goodman, *Introduction to Fourier Optics,* 3rd edn., Greenwood Village, CO: Roberts and Co., 2005.

[40] A. Papoulis, *Systems and Transforms with Applications in Optics*, Malabar, FL: Krieger Publishing Co., 1981.

[41] P. M. Duffieux, *The Fourier Transform and Its Applications to Optics*, 2nd edn., New York: Wiley, 1983.

[42] J. D. Gaskill, *Linear Systems, Fourier Transforms, and Optics*, 2nd edn., New York: Wiley, 1978.

[43] J. Upatnieks and C. Leonard, "Diffraction efficiency of bleached photographically recorded interference patterns," *Appl. Opt.* **8**, 85–89, 1969.

[44] M. Young and F. H. Kittredge, "Amplitude and phase holograms exposed on Agfa-Gevaert 10E75 planes," *Appl. Opt.* **8**, 2453–2454, 1969.

[45] J. W. Goodman, *Speckle Phenomena in Optics: Theory and Applications* Greenwood Village, CO: Robert and Co., 2007.

[46] M. Francon, *Laser Speckle and Applications in Optics*, New York: Academic Press, 1979.

[47] J. C. Dainty, ed., *Laser Speckle and Related Phenomena*, 2nd edn., Berlin: Springer-Verlag, 1984.

[48] W. Schumann and M. Dubas, *Holographic Interferometry*, Berlin: Springer-Verlag, 1979.

[49] Y. I. Ostrovsky, M. M. Butusov, and G. V. Ostrovskaya, *Interferometry by Holography*, Berlin: Springer-Verlag, 1980.

[50] E. Marom, A. A. Friesem, and W. Wiener-Arenear, *Applications of Holography and Optical Data Processing*, New York: Pergamon, 1977.

[51] F. T. S. Yu, *Optical Information Processing*, New York: Wiley, 1983.

[52] S. H. Lee, ed., *Optical Information Processing*, Berlin: Springer-Verlag, 1981.

[53] R. W. Waynant, ed., *Lasers in Medicine*, Boca Raton, FL: CRC Press, 2001.

[54] E. Scerrati, "Laser *in situ* keratomileusis vs. laser epithelial keratomileusis (LASIK vs. LASEK)," *J. Refractive Surgery* **17**, S219–S221, 2001.

[55] D. H. Bagley, "Expanding role of ureteroscopy and laser lithotripsy for treatment of proximal ureteral and intrarenal calculi," *Current Opinion Urol.*, **12**, 277–280, 2002.

[56] G. S. Abela and G. R. Barbeau, "Laser angioplasty: potential effects and current limitations," in *Textbook of Interventional Cardiology*, ed. E. J. Topol, Philadelphia, PA: W. B. Saunders, 1990, pp. 724–737.

[57] D. L. Singleton, G. Paraskevopoulos, G. S. Jolly, R. S. Irwin, and D. J. McKenny, "Excimer lasers in cardiovascular surgery: ablation products and photoacoustic spectrum of arterial wall," *Appl. Phys. Lett.* **48**, 878–880, 1986.

[58] Committee on a Scientific Assessment of Free-Electron Laser Technology for Naval Applications, National Research Council, "Scientific Assessment of High-Power Free-Electron Laser Technology," 66 pages, 2008 (free download available at http://www.nap.edu/catalog/12484.html).

[59] V. S. Letokhov and C. B. Moore, "Laser isotope separation (review). I," *Sov. J. Quant. Electron.* **6**, 129–150, 1976.

[60] V. S. Letokhov and C. B. Moore, "Laser isotope separation (review). II," *Sov. J. Quant. Electron.* **6**, 249–276, 1976.

[61] C. D. Cantrell, S. M. Freund, and J. L. Lyman, "Laser-induced chemical reactions and isotope separation," in *Laser Handbook*, Vol. III, ed. M. L. Stitch, Amsterdam: North-Holland, 1979, pp. 485–576.

[62] P. T. Greenland, "Laser isotope separation," *Contemp. Phys.* **31**, 405–424, 1990.

[63] R. J. Jensen, O. P. Judd, and J. A. Sullivan, "Separating isotopes with lasers," *Los Alamos Sci.*, **3**, 2–33, 1982.

[64] R. V. Ambartsumyan, Yu. A. Gorokhov, V. S. Letokhov, G. N. Makarov, and A. A. Puretskii, "Explanation of the selective dissociation of the $SF_6$ molecule in a strong IR laser field," *JETP Lett.*, **23**, 22–25, 1976.

[65] D. L. Matthews, P. L. Hagelstein, M. D. Rosen, M. J. Eckart, N. M. Ceglio, A. U. Hazi, H. Medecki, B. J. MacGowan, J. E. Trebes, B. L. Whitten, E. M. Campell, C. W. Hatcher, A. M. Hawryluk, R. L. Kauffman, L. D. Pleasance, G. Rambach, J. H. Scofield, G. Stone, and T. A. Weaver, "Demonstration of a soft X-ray amplifier," *Phys. Rev. Lett.* **54**, 110–113, 1985.

[66] M. D. Rosen, P. L. Hagelstein, D. L. Matthews, E. M. Campbell, A. U. Hazi, B. l. Whitten, B. MacGowan, R. E. Turner, R. W. Lee, G. Charactis, G. E. Busch, C. L. Shepard, and P. D. Rockett, "Exploding-foil technique for achieving a soft X-ray laser," *Phys. Rev. Lett.* **54**, 106–109, 1985.

[67] S. Suckewer, C. H. Skinner, H. Milchberg, C. Keane, and D. Vorheas, "Amplification of stimulated soft X-ray emission in a confined plasma column," *Phys. Rev. Lett.* **55**, 1753–1756, 1985.

[68] R. W. Conn, "The engineering of magnetic fusion reactors," *Scient. Am.* **249**, October, 60–71, 1983.

[69] E. Moses, "The National Ignition Facility: an experimental platform for studying behavior of matter under extreme conditions," *Astrophys. Space Sci.* **336**, 37, 2011.

[70] E. I. Moses, "The National Ignition Facility and the National Ignition Campaign" *IEEE Trans. Plasma Sci.* **38**, 684–689, 2010.

[71] R. A. Sacks and D. H. Darling, "Direct drive cryogenic ICF capsules employing DT wetted foam," *Nucl. Fusion* **27**, 447–452, 1987.

[72] J. Nuckolls, L. Wood, A. Thiessen, and G. Zimmerman, "Laser compression of matter to super-high densities: thermonuclear (CTR) applications," *Nature* **249**, 139–142, 1972.

[73] L. A. Booth, D. A. Freiwald, T. G. Frank, and F. T. Finch, " Prospects of generating power with laser-driven fusion," *Proc. IEEE* **64**, 1460–1482, 1976.

[74] R. S. Craxton, R. C. McCrory, and J. M. Soures, "Progress in laser fusion," *Scient. Am.* **255**, August, 68–79, 1986.

# Appendix 1 Optical terminology

Light is electromagnetic radiation with a wavelength between 0.1 nm and 100 μm. Of course, this selection of wavelength range is somewhat arbitrary; short-wavelength vacuum-ultraviolet light between 0.1 nm and 10 nm (so called because these wavelengths – and a range above, up to about 200 nm – are absorbed by air and most gases) might just as easily be called soft X-radiation. At the other end of the wavelength scale, between 100 μm and 1000 μm, the sub-millimeter-wave region of the spectrum, lies a spectral region where conventional optical methods become difficult, as does the extension of microwave techniques that are easily applied to the centimeter- and millimeter-wave regions. The use of optical techniques in this region, and even in the millimeter region, is often called quasi-optics.

Table A1.1 provides a summary of the important parameters that are used to characterize light and the media through which it passes. A few comments on the table are appropriate. Although the velocity of light in a medium depends both on the relative magnetic permeability $\mu_r$ and on the dielectric constant $\epsilon_r$ of the medium, for all practical optical materials $\mu_r \simeq 1$, so the refractive index and dielectric constant are related by

$$n = \sqrt{\epsilon_r}. \tag{A1.1}$$

When light propagates in an anisotropic medium, such as a crystal of lower than cubic symmetry, $n$ and $\epsilon_r$ will, in general, depend on the direction of propagation of the wave and its polarization state. The velocity of light *in vacuo* is currently the most precisely known of all the physical constants – its value is known [1] to within 4 cm $s^{-1}$. A redefinition of the meter that is based on the cesium atomic-clock frequency standard [2] and a velocity of light of exactly $2.99792458 \times 10^8$ m $s^{-1}$ has been adopted [3]. The velocity, wavelength, and wavenumber of light traveling in the air have slightly different values from those of light traveling in *vacuo*. Tables of corresponding values of $\overline{\nu}$ *in vacuo* and in standard air are available [4].

A plane electromagnetic wave that is traveling in the $z$ direction can, in general, be decomposed into two independent, linearly polarized components.[1] The electric and magnetic field pairs associated with each of these components are themselves mutually orthogonal and transverse to the direction of propagation. They can be written as $(E_x, H_y)$ and $(E_y, H_x)$. The ratio of the mutually orthogonal **E** and **H** components is called the *impedance*, $Z$, of the medium:

$$\frac{E_x}{H_y} = -\frac{E_y}{H_x} = Z = \sqrt{\frac{\mu_r \mu_0}{\epsilon_r \epsilon_0}}. \tag{A1.2}$$

[1] The wave could also be decomposed into left- and right-hand circularly polarized components.

**Table A1.1** Fundamental parameters of electromagnetic radiation and optical media

| Parameter | Symbol | Value | Units |
|---|---|---|---|
| Velocity of light in free space *in vacuo* | $c_0 = \sqrt{\mu_0 \epsilon_0}$ | $2.99792458 \times 10^{8}$[a] | $\mathrm{m\,s^{-1}}$ |
| Permeability of free space | $\mu_0$ | $4\pi \times 10^{-7}$ | $\mathrm{H\,m^{-1}}$ |
| Permittivity of free space | $\epsilon_0$ | $8.85416 \times 10^{-12}$ | $\mathrm{F\,m^{-1}}$ |
| Velocity of light in a medium | $c = 1/\sqrt{\mu_r \mu_0 \epsilon_r \epsilon_0} = c_0/n$ | | $\mathrm{m\,s^{-1}}$ |
| Refractive index | $n = \sqrt{\mu_r \epsilon_r}$ | | Dimensionless |
| Relative permeability of a medium | $\mu_r$ | Usually 1 | Dimensionless |
| Dielectric constant of a medium | $\epsilon_r$ | | Dimensionless |
| Frequency | $\nu = c/\lambda$ | | Hz |
| Wavelength *in vacuo* | $\lambda_0 = c_0/\nu$ | | m |
| Wavelength in a medium | $\lambda = c/\nu = \lambda_0/n$ | | m |
| Wavenumber | $\overline{\nu} = 1/\lambda$ | | $\mathrm{cm^{-1}}$ (Kayser) |
| Wave vector | $\mathbf{k}, \lvert\mathbf{k}\rvert = 2\pi/\lambda$ | | $\mathrm{m^{-1}}$ |
| Photon energy | $E = h\nu$ | | J<br>1 eV (electron volt) $= 1.60202 \times 10^{-19}$ J |
| Electric field of wave | $\mathbf{E}$ | | $\mathrm{V\,m^{-1}}$ |
| Magnetic field of wave | $\mathbf{H}$ | | $\mathrm{A\,m^{-1}}$ |
| Poynting vector | $\mathbf{P} = \mathbf{E} \times \mathbf{H}$ | | $\mathrm{W\,m^{-2}}$ |
| Intensity | $I = \langle\lvert\mathbf{P}\rvert\rangle_{\mathrm{AV}} = \lvert\mathbf{E}\rvert^2/(2Z)$ | | $\mathrm{W\,m^{-2}}$ |
| Impedance of medium | $Z = E_x/H_y = -E_y/H_x$ | | $\Omega$ |
| Impedance of free space | $Z_0 = \sqrt{\mu_r \mu_0/\epsilon_r \epsilon_0} = \sqrt{\mu_0/\epsilon_0}$ | 376.7 | $\Omega$ |

[a]See the text.

The negative sign in Eq. (A1.2) arises because $(y, x, z)$ is not a right-handed coordinate system.

The Poynting vector, **S**, given by

$$\mathbf{S} = \mathbf{E} \times \mathbf{H}, \tag{A1.3}$$

is a vector that points in the direction of energy propagation of the wave. The average magnitude of the Poynting vector is called the intensity $I$,

$$I = \langle |\mathbf{S}| \rangle_{\mathrm{AV}} = \frac{|\mathbf{E}|^2}{2Z}. \tag{A1.4}$$

The factor of 2 comes from time-averaging the square of the sinusoidally varying electric field.

The wave vector **k** points in the direction perpendicular to the phase front of the wave (the surface of constant phase). In an isotropic medium **k** and **S** are always parallel.

The photon flux corresponding to an electromagnetic wave of average intensity, $I$, and frequency, $\nu$, is

$$N = I/(h\nu) = I\lambda/(hc), \tag{A1.5}$$

where $h$ is Planck's constant, $6.626 \times 10^{-34}$ J s.

For a wave of intensity 1 W m$^{-2}$ and wavelength 1 μm *in vacuo*, $N = 5.04 \times 10^{18}$ photons s$^{-1}$ m$^{-2}$. Photon energy is sometimes measured in electron volts (eV): 1 eV = $1.60202 \times 10^{-19}$ J. A photon of wavelength 1 μm has an energy of 1.24 eV. It is often important, particularly in the infrared, to know the correspondence between photon and thermal energies. The characteristic thermal energy at absolute temperature is $k_{\mathrm{B}}T$. At 300 K, $k_{\mathrm{B}}T = 4.14 \times 10^{-21}$ J = 208.6 cm$^{-1}$ = 0.026 eV.

## References

[1] K. M. Baird, D. S. Smith, and B. G. Whitford, "Conformation of the currently accepted value 299792458 meters per second for the speed of light," *Opt. Commun.* **31**, 367–368, 1979.

[2] G. W. C. Kaye and T. H. Laby, *Tables of Physical and Chemical Constants*, 14th edn., New York: Longman, 1972.

[3] E. R. Cohen and B. N. Taylor, Committee on Data for Science and Technology, CODATA Bulletin, No. 63, 1986.

[4] C. D. Coleman, W. R. Bozman, and W. F. Meggers, *Table of Wavenumbers*, Vols. 1 and 2, Washington, D.C.: National Bureau of Standards, 1960.

# Appendix 2 The $\delta$-function

The $\delta$-function[1] is a *pathological*[2] mathematical function with the following properties:

$$\delta(k-k_0)=0, \qquad k\neq k_0,$$
$$\delta(k-k_0)=\infty, \qquad k=k_0,$$
$$\int_{-\infty}^{\infty}\delta(k-k_0)dk=1. \tag{A2.1}$$

The $\delta$-function is the Fourier transform of a monochromatic complex exponential function, $f(x)$, where

$$f(x)=e^{-jk_0x}, \tag{A2.2}$$

and therefore,

$$\delta(k-k_0)=\frac{1}{2\pi}\int_{-\infty}^{\infty}e^{j(k-k_0)x}\,dx. \tag{A2.3}$$

We can simplify our notation by writing $\delta(k-k_0)$ as $\delta(k)$ when the function is centered at the origin, $k_0=0$. Additional properties of the $\delta$-function are

$$\delta(t)=\delta(-t), \tag{A2.4}$$
$$\int f(t)\delta(t-t_0)dt=f(t_0), \tag{A2.5}$$
$$\delta(at)=\frac{1}{a}\delta(t), \qquad a>0, \tag{A2.6}$$
$$\int\delta(t-t')\delta(t'-t'')dt'=\delta(t-t''), \tag{A2.7}$$
$$f(t)\delta(t-t_0)=f(t_0)\delta(t-t_0). \tag{A2.8}$$

The derivative of the $\delta$-function can also be defined:

$$\delta'(k)=\frac{1}{2\pi}\int_{-\infty}^{\infty}ike^{jkx}\,dx, \tag{A2.9}$$
$$\delta'(k)=-\delta'(-k), \tag{A2.10}$$
$$\int f(k)\delta'(k-k_0)dk=-f'(k_0). \tag{A2.11}$$

[1] Originally introduced in quantum theory, by Paul Dirac [1]. In engineering applications it is often called the *unit impulse function*.

[2] Strictly speaking, it is not a function at all.

Care must be taken in assigning values to the integrals above because of the singular behavior of $\delta(k)$. If, however, $\delta(k)$ is written as the limit of a rational function then such problems can be avoided.

For example, a normalized Gaussian function can be written as

$$g(x - x_0) = \frac{1}{\sigma\sqrt{2\pi}} e^{-(x-x_0)^2/(2\sigma^2)}, \tag{A2.12}$$

so that

$$\int_{-\infty}^{\infty} g(x - x_0)dx = \sigma \frac{1}{a\sqrt{2\pi}} \int_{\infty}^{\infty} e^{-(x-x_0)^2/(2\sigma^2)}\, dx = 1. \tag{A2.13}$$

This allows us to write the $\delta$-function as the limiting case of a normalized Gaussian function whose width shrinks to zero; that is,

$$\delta(x - x_0) = \lim_{\sigma \to 0} \frac{1}{a\sqrt{\sigma\pi}} e^{-(x-x_0)^2/(2\sigma^2)}. \tag{A2.14}$$

The $\delta$-function can be written in many different ways as the limit of an appropriately normalized function. For example, in terms of a normalized Lorentzian function

$$\delta(\nu - \nu_0) = \lim_{\Delta\nu \to 0} \frac{2/(\pi\ \Delta\nu)}{1 + [2(\nu - \nu_0)/\Delta\nu]^2}. \tag{A2.15}$$

We can extend the definition of the $\delta$-function to two or more dimensions. The three-dimensional $\delta$-function is

$$\begin{aligned} \delta(\mathbf{k}) &= \delta(k_x)\delta(k_y)\delta(k_z) \\ &= \frac{1}{(2\pi)^3} \int_{-\infty}^{\infty}\int_{-\infty}^{\infty}\int_{-\infty}^{\infty} e^{j(k_x x + k_y y + k_z z)}\, dx\, dy\, dz \\ &= \left(\frac{1}{2\pi}\right)^3 \int_{-\infty}^{\infty} e^{j\mathbf{k}\cdot\mathbf{r}}\, d\mathbf{r}. \end{aligned} \tag{A2.16}$$

The $\delta$-function is closely related to the unit-step or *Heaviside* function $u(t)$ defined by

$$u(t) = \begin{cases} 0, & t < 0, \\ 1, & t > 0, \end{cases} \tag{A2.17}$$

such that

$$\delta(t) = \frac{d}{dt} u(t). \tag{A2.18}$$

## Reference

[1] P. A. M. Dirac, *Principles of Quantum Mechanics*, 3rd edn., Oxford: Oxford University Press, 1947.

# Appendix 3 Black-body radiation formulas

The energy-density distribution with frequency of black-body radiation is

$$\rho(\nu) = \frac{8\pi h\nu^3}{c^3} \frac{1}{e^{h\nu/(k_B T)} - 1}, \tag{A3.1}$$

where $\rho(\nu)$ is measured in units of J m$^{-3}$ Hz$^{-1}$.

The corresponding energy distribution with wavelength is

$$\rho(\lambda) = \frac{8\pi hc}{\lambda^5} \frac{1}{e^{hc/(\lambda k_B T)} - 1}, \tag{A3.2}$$

where $\rho(\lambda)d\lambda$ is the energy stored per unit volume that lies within the region $\lambda \to \lambda + d\lambda$, measured in units of J m$^{-3}$ m$^{-1}$. From $\rho(\lambda)$ can be derived an expression for the *spectral emittance*, $M_{e\lambda}$, the total power emitted per unit wavelength interval into a solid angle $2\pi$ by unit area of the black body, where

$$M_{e\lambda} = \frac{c_1}{\lambda^5(e^{c_2/(\lambda T)} - 1)}, \tag{A3.3}$$

where $c_1 = 2\pi hc^2$ is called the *first radiation constant*, and has the value $3.7418 \times 10^{-16}$ W m$^2$; and $c_2 = ch/k_B$ is called the *second radiation constant*, and has the value $1.43877 \times 10^{-2}$ m K.

The *radiance*, $L_e$, of an extended source is the radiant flux that it emits per unit solid angle per unit area of the source:

$$L_e = \frac{\delta I_e}{\delta S_n}, \tag{A3.4}$$

where the area $\delta S_n$ is the projection of the surface element of the source in the direction being considered and $\delta I_e$ is the *radiant intensity*, the amount of power emitted by this surface element in a particular direction per solid angle, measured in watts per steradian. A black body has a radiance that is independent of the viewing angle – it is said to be a perfectly diffuse or *Lambertian* radiator. Clearly, for such a source the radiant intensity at an angle $\theta$ to the normal to its surface is

$$I_e(\theta) = I_e(0)\cos\theta. \tag{A3.5}$$

For a surface element $\delta S$, the projected area at angle $\theta$ to the normal is $\delta S \cos\theta$, so (A3.5) in conjunction with (A3.4) gives $L_e = I_e(0)$. The total flux emitted per unit area is its *radiant emittance*, $M_e$, where

$$M_e = \pi I_e(0). \tag{A3.6}$$

The relationship between the spectral emittance, $M_{e\lambda}$, and the spectral radiance, $L_{e\lambda}$ (W sr$^{-1}$ m$^{-2}$ m$^{-1}$), is

$$M_{e\lambda} = \pi L_{e\lambda}. \tag{A3.7}$$

For a black-body source the wavelength of maximum emittance, $\lambda_m$, obeys Wien's displacement law,

$$\lambda_m T = 2.898 \times 10^{-3} \text{ m K}. \tag{A3.8}$$

The total radiant emittance of a black body at temperature $T$ is

$$M_e = \int_0^\infty M_{e\lambda}\, d\lambda = \frac{2\pi^5 k_B^4}{15c^2h^3} T^4 = \sigma T^4. \tag{A3.9}$$

This is a statement of the Stefan–Boltzmann law. The coefficient $\sigma$, called the *Stefan–Boltzmann constant*, has a value of $5.6705 \times 10^{-8}$ W m$^3$ K$^{-4}$. The spectral emittance in terms of photons, $N_\lambda$, is

$$N_\lambda = \frac{M_{e\lambda}}{hc/\lambda}. \tag{A3.10}$$

# Appendix 4 *RLC* circuits

For an *RLC* circuit such as shown in Fig. 2.14 with a driving voltage $V(t)$, application of Kirchhoff's law to the circuit gives

$$L\frac{dI}{dt} + RI + \frac{Q}{C} = V(t), \tag{A4.1}$$

where $I$ is the current in the circuit and $Q$ is the charge on the capacitor. This can be more conveniently written as

$$L\frac{d^2I}{dt^2} + R\frac{dI}{dt} + \frac{I}{C} = \frac{dV}{dt}. \tag{A4.2}$$

Before considering the case of a sinusoidal driving voltage, we will consider the behavior of the circuit if the voltage source is shorted out. In this case

$$L\frac{d^2I}{dt^2} + R\frac{dI}{dt} + \frac{I}{C} = 0. \tag{A4.3}$$

For this linear, second-order differential equation we try the solution $I = I_0e^{-at}$, which gives

$$La^2 - Ra + \frac{1}{C} = 0, \tag{A4.4}$$

and, therefore,

$$a = \frac{R \pm \sqrt{R^2 - 4L/C}}{2L}. \tag{A4.5}$$

If $R > 4L/C$, $a$ is real and the current decays purely exponentially. In this case

$$a = \frac{R \pm R\sqrt{1 - 4L/(R^2C)}}{2L} \simeq \frac{R \pm R\sqrt{1 - 2L/(R^2C)}}{2L}, \tag{A4.6}$$

which can be written

$$a \simeq \frac{R \pm R}{2L} \mp \frac{1}{RC}. \tag{A4.7}$$

Clearly, we must choose the lower sign in order that the current decays as

$$I_0 = I_0e^{-t/(RC)}, \tag{A4.8}$$

which is the familiar result for the decay of a capacitor through a resistor.

If $R < 4L/C$ then

$$a = \frac{R - j\sqrt{4L/C - R^2}}{2L} = \alpha - j\omega', \tag{A4.9}$$

where $\alpha = R/(2L)$ and $\omega'$ is the natural oscillation frequency of the circuit,

$$\omega' = \sqrt{\frac{1}{LC} - \frac{R^2}{4L^2}}. \tag{A4.10}$$

The decay of the current is then given by the damped oscillation

$$I = I_0 e^{-R/(2L)} e^{j\omega' t}. \tag{A4.11}$$

If the circuit is driven with a sinusoidal voltage of the form $V_0 e^{j\omega t}$, then it is reasonable to assume that the current will be at the frequency $\omega$. Therefore, as a solution to Eq. (A4.2) we try $I = I_0 e^{j\omega t}$, where $I_0$ may be complex (it includes a phase factor). Substitution in Eq. (A4.2) gives

$$\left(-L\omega^2 + j\omega R + \frac{1}{C}\right) I_0 = j\omega V_0, \tag{A4.12}$$

which gives

$$I_0 = \frac{jV_0}{1/(\omega C) - L\omega + jR} = \frac{j[1/(\omega C) - L\omega - jR]}{[1/(\omega C) - L\omega]^2 + R^2} V_0. \tag{A4.13}$$

This can be written in the form

$$I_0 = \frac{V_0 e^{j\phi}}{\sqrt{[1/(\omega C) - L\omega]^2 + R^2}}, \tag{A4.14}$$

where $\tan\phi = [1/(\omega C) - L\omega]/R$.

The amplitude of the current varies with frequency according to

$$|I_0| = \frac{|V_0|}{\sqrt{[1/(\omega C) - L\omega]^2 + R^2}} = \frac{|V_0|/R}{\sqrt{1 + [1 - \omega^2 LC/(\omega RC)]^2}}. \tag{A4.15}$$

Maximum current results at the resonance frequency of the circuit, at which $1/(\omega_0 C) - L\omega_0 = 0$. Therefore

$$\omega_0 = \frac{1}{\sqrt{LC}}. \tag{A4.16}$$

Equation (A4.15) can be written in the form

$$|I_0| = \frac{|V_0|/R}{\sqrt{1 + \left(1 - \omega\sqrt{LC}\right)^2 \left(1 + \omega\sqrt{LC}\right)^2 / (\omega^2 R^2 C^2)}}, \tag{A4.17}$$

which near resonance, $\omega \simeq \omega_0$, becomes

$$|I_0| = \frac{|V_0|/R}{\sqrt{1 + [2(\omega - \omega_0)/(\omega_0^2 RC)]^2}}. \tag{A4.18}$$

The power $W(\omega)$ dissipated in the circuit is $R|I_0|^2$. Therefore

$$W(\omega) = \frac{|V_0|^2/R}{1 + [2(\omega - \omega_0)/(\omega_0^2 RC)]^2}, \tag{A4.19}$$

which is a Lorentzian-shaped resonance curve with FWHM

$$\Delta\omega = \omega_0^2 RC. \tag{A4.20}$$

The *quality factor*, $Q$, of the circuit is defined as

$$Q = \frac{\omega_0}{\Delta\omega} = \frac{1}{\omega_0 RC}. \tag{A4.21}$$

In terms of $Q$ the power spectrum of the circuit is

$$W(\omega) = \frac{|V_0|^2}{1 + [2(\omega - \omega_0)Q/\omega_0]^2}. \tag{A4.22}$$

# Appendix 5 Storage and transport of energy by electromagnetic fields

The total energy (per unit volume) stored by a system of electric and magnetic fields is [1], in an isotropic medium,

$$\rho = \frac{1}{2}(\mathbf{D}\cdot\mathbf{E} + \mathbf{B}\cdot\mathbf{H}) = \frac{1}{2}(\epsilon\epsilon_0 E^2 + \mu\mu_0 H^2). \tag{A5.1}$$

In an electromagnetic wave this energy is transported in the direction of energy flow, which in an isotropic medium is in the same direction as the wave vector (the normal to the wave front). The rate of energy flow across unit area per second is given by a vector **S** called the Poynting vector. If we consider a plane-polarized wave with fields $E_x$ and $H_y$, then **S** is in the direction of the $z$ axis and has magnitude

$$\begin{aligned} |\,\mathbf{S}\,| = c\rho &= \frac{1}{2\sqrt{\epsilon\epsilon_0\mu\mu_0}}(\epsilon\epsilon_0 E_x^2 + \mu\mu_0 H_y^2) \\ &= \frac{1}{2}\left(\sqrt{\frac{\epsilon\epsilon_0}{\mu\mu_0}}E_x^2 + \sqrt{\frac{\mu\mu_0}{\epsilon\epsilon_0}}H_y^2\right) = \frac{1}{2}\left(\frac{E_x^2}{Z_0} + Z_0 H_y^2\right) = E_x H_y \\ &= \frac{E_x^2}{Z_0} = Z_0 H_y^2, \end{aligned} \tag{A5.2}$$

so

$$S_z = E_x H_y. \tag{A5.3}$$

In general

$$\mathbf{S} = \mathbf{E}\times\mathbf{H}. \tag{A5.4}$$

The total energy per unit time that is carried by the wave into a volume $V$ bounded by a surface $A$, as shown in Fig. A5.1, is $\int_A -(\mathbf{S}\cdot\hat{\mathbf{A}})dA$; the minus sign is present because $\hat{\mathbf{A}}$ is a unit vector directed away from the surface. Now, from Gauss's theorem,

$$\int_A -(\mathbf{S}\cdot\hat{\mathbf{A}})dA = \int_V -\text{div}\,\mathbf{S}\,dV \tag{A5.5}$$

and

$$\int_V -\text{div}\,\mathbf{S}\,dV = \int_V -\text{div}(\mathbf{E}\times\mathbf{H})dV. \tag{A5.6}$$

Since $\text{div}(\mathbf{E}\times\mathbf{H}) = \mathbf{H}\cdot\text{curl}\,\mathbf{E} - \mathbf{E}\cdot\text{curl}\,\mathbf{H}$, using Maxwell's equations we get

$$\text{div}(\mathbf{E}\times\mathbf{H}) = \mathbf{H}\cdot\left(\frac{\partial\mathbf{B}}{\partial t}\right) - \mathbf{E}\cdot\left(\mathbf{j} + \frac{\partial\mathbf{D}}{\partial t}\right), \tag{A5.7}$$

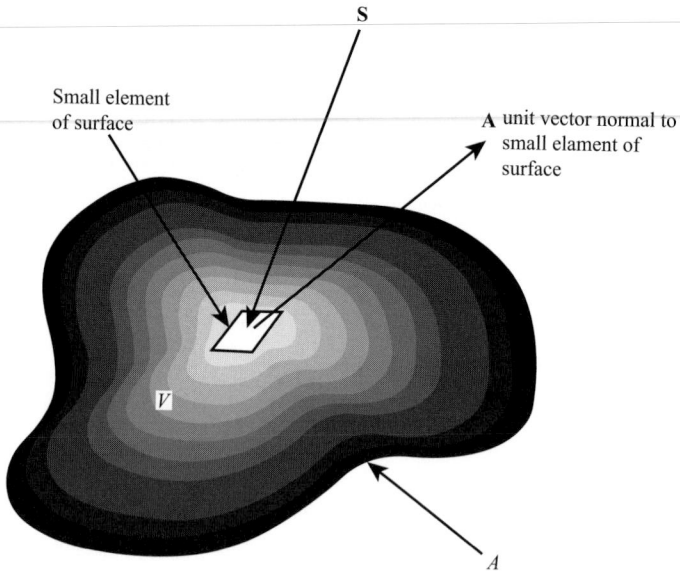


Fig. A5.1 Transport of electromagnetic energy into a region of space of volume $V$ and bounded by a closed surface $A$.

where $\mathbf{j}$ is the current density flowing in the medium. Now, remember that

$$\mathbf{D} = \epsilon_0 \mathbf{E} + \mathbf{P} = \epsilon\epsilon_0 \mathbf{E} \tag{A5.8}$$

and

$$\mathbf{B} = \mu_0(\mathbf{H} + \mathbf{M}) = \mu\mu_0 \mathbf{H}, \tag{A5.9}$$

where $\mathbf{M}$ is the magnetization (magnetic dipole moment per unit volume) in the medium. Therefore,

$$\int_V -\mathrm{div}(\mathbf{E} \times \mathbf{H}) dV = \int_V \left[ \mathbf{H} \cdot \left( \mu_0 \frac{\partial \mathbf{H}}{\partial t} \right) + \mu_0 \mathbf{H} \cdot \frac{\partial \mathbf{M}}{\partial t} + \mathbf{E} \cdot \mathbf{j} + \mathbf{E} \cdot \left( \epsilon_0 \frac{\partial \mathbf{E}}{\partial t} \right) + \mathbf{E} \cdot \frac{\partial \mathbf{P}}{\partial t} \right] dV. \tag{A5.10}$$

Using the equality

$$\frac{1}{2} \frac{\partial}{\partial t} (\mathbf{E} \cdot \mathbf{E}) = \mathbf{E} \cdot \frac{d\mathbf{E}}{dt} \tag{A5.11}$$

from Eq. (A5.10), the flow of energy into the volume $V$ becomes

$$\int_V -\mathrm{div}(\mathbf{E} \times \mathbf{H}) dV = \int_V \left[ \mathbf{E} \cdot \mathbf{j} + \frac{\partial}{\partial t} \left( \frac{\epsilon_0}{2\mathbf{E} \cdot \mathbf{E}} \right) + \frac{\partial}{\partial t} \left( \frac{\mu_0}{2\mathbf{H} \cdot \mathbf{H}} \right) + \mu_0 \mathbf{H} \cdot \frac{\partial \mathbf{M}}{\partial t} \right] + \mathbf{E} \cdot \frac{\partial \mathbf{P}}{\partial t} dV, \tag{A5.12}$$

where the terms on the right-hand side correspond, from left to right, to (i) the work done by the field in causing conventional current flow (ohmic heating), (ii) and (iii) the rate of increase of vacuum electromagnetic energy, (iv) the work done on magnetic dipoles in the volume, and (v) the final term, $\mathbf{E} \cdot \partial \mathbf{P}/\partial t$, represents work done by the electromagnetic field in polarizing the medium.

## Reference

[1] W. H. Hayt, Jr., *Engineering Electrodynamics*, 5th edn., New York: McGraw-Hill, 1989.

# Appendix 6 The reflection and refraction of a plane electromagnetic wave at a boundary between two isotropic media of different refractive indices

Figure A6.1 shows the directions of propagation and field directions for the incident, reflected, and transmitted electromagnetic waves at a planar boundary. The discussion is restricted to P waves, but a similar discussion is readily developed for S waves. For the incident wave the electric field varies with position as

$$\begin{aligned} E_1 &= E_{\mathrm{i}} e^{j(\omega t - k_1 s_1)} \\ &= E_{\mathrm{i}} e^{j(\omega t - k_1 x \sin\theta_1 - k_1 z \cos\theta_1)}, \end{aligned} \tag{A6.1}$$

where $s_1$ is the distance measured along the direction of propagation.

For the reflected wave

$$\begin{aligned} E_3 &= E_{\mathrm{r}} e^{j(\omega t - k_3 s_3)} \\ &= E_{\mathrm{r}} e^{j(\omega t - k_3 x \sin\theta_3 + k_3 z \cos\theta_3)}, \end{aligned} \tag{A6.2}$$

and for the transmitted wave

$$E_2 = E_{\mathrm{t}} e^{j(\omega t - k_2 x \sin\theta_2 - k_2 z \cos\theta_2)}. \tag{A6.3}$$

Similar expressions exist for the magnetic fields. At the boundary, $z = 0$, the tangential electric field must be continuous [1]. Therefore,

$$E_1 \cos\theta_1 + E_3 \cos\theta_3 = E_2 \cos\theta_2, \tag{A6.4}$$

which gives

$$E_{\mathrm{i}} \cos\theta_1 \, e^{-jk_1 x \sin\theta_1} + E_{\mathrm{r}} \cos\theta_3 \, e^{-jk_3 x \sin\theta_3} = E_{\mathrm{t}} \cos\theta_2 \, e^{-jk_2 x \sin\theta_2}. \tag{A6.5}$$

Equation (A6.5) must be true at all points along the boundary – that is, for all values of $x$. This can be true only if all the exponents are equal, which, since $k_1 = k_3 = 2\pi n_1/\lambda_0$ and $k_2 = 2\pi n_2/\lambda_0$, gives

$$\theta_1 = \theta_3. \tag{A6.6}$$

The angle of reflection equals the angle of incidence. Furthermore,

$$n_1 \sin\theta_1 = n_2 \sin\theta_2, \tag{A6.7}$$

which is Snell's law.

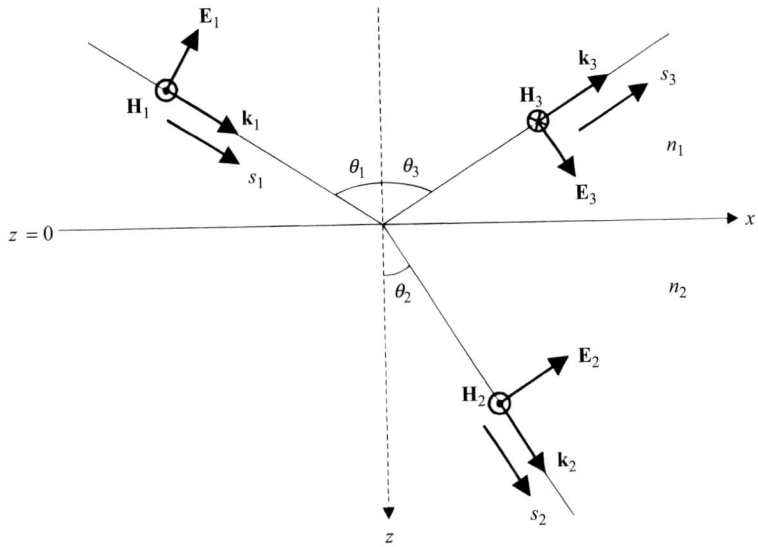


**Fig. A6.1** Field and propagation directions for incident, reflected, and transmitted waves at a planar boundary.

Tangential magnetic fields must also be continuous at the boundary, so

$$H_\mathrm{i} - H_\mathrm{r} = H_\mathrm{t}, \tag{A6.8}$$

which can be written in terms of impedances as

$$\frac{E_\mathrm{i}}{Z_1} - \frac{E_\mathrm{r}}{Z_1} = \frac{E_\mathrm{t}}{Z_2}. \tag{A6.9}$$

Equation (A6.5) can also be restated as

$$E_\mathrm{i} \cos\theta_1 + E_\mathrm{r} \cos\theta_1 = E_\mathrm{t} \cos\theta_2. \tag{A6.10}$$

Combination of Eqs. (A6.9) and (A6.10) gives

$$\frac{E_\mathrm{r}}{E_\mathrm{i}} = \frac{Z_2 \cos\theta_2 - Z_1 \cos\theta_1}{Z_2 \cos\theta_2 + Z_1 \cos\theta_1}, \tag{A6.11}$$

$$\frac{E_\mathrm{t}}{E_\mathrm{i}} = \frac{2Z_2 \cos\theta_2}{Z_2 \cos\theta_2 + Z_1 \cos\theta_1}. \tag{A6.12}$$

In terms of field components parallel to the boundary,

$$\frac{E_{\mathrm{r}x}}{E_{\mathrm{i}x}} = \rho = \frac{Z_2 \cos\theta_2 - Z_1 \cos\theta_1}{Z_2 \cos\theta_2 + Z_1 \cos\theta_1}, \tag{A6.13}$$

$$\frac{E_{\mathrm{t}x}}{E_{\mathrm{i}x}} = \frac{E_\mathrm{t} \cos\theta_2}{E_\mathrm{i} \cos\theta_1} = \tau = \frac{2Z_2 \cos\theta_2}{Z_2 \cos\theta_2 + Z_1 \cos\theta_1}. \tag{A6.14}$$

With the introduction of the effective impedance $Z' \equiv Z\cos\theta$ (for P waves),[1] we have the universal formulas

$$\rho = \frac{Z_2' - Z_1'}{Z_1' + Z_2'}, \tag{A6.15}$$

$$\tau = \frac{2Z_2'}{Z_1' + Z_2'}. \tag{A6.16}$$

Note that $\tau = 1 + \rho$.

For transmission in the opposite direction across the boundary we would have

$$\rho' = \frac{Z_1' - Z_2'}{Z_1' + Z_2'}, \tag{A6.17}$$

$$\tau' = \frac{2Z_1'}{Z_1' + Z_2'}. \tag{A6.18}$$

Note that $\rho' = -\rho$, so $|\rho'| = |\rho|$.

The fractional energy transmission across the boundary, the transmittance, is

$$T = 1 - |\rho|^2, \tag{A6.19}$$

which is independent of direction across the boundary. For lossless media, all the impedances are real and, from Eqs. (A6.15), (A6.16), and (A6.18),

$$T = 1 - |\rho|^2 = \frac{4Z_1'Z_2'}{(Z_1' + Z_2')^2}. \tag{A6.20}$$

Clearly then, from Eqs. (A6.16) and (A6.18),

$$T = \tau\tau'. \tag{A6.21}$$

## Reference

[1] S. Ramo, J. R. Whinnery, and T. Van Duzer, *Fields and Waves in Communication Electronics*, 2nd edn., New York: Wiley, 1984.

[1] For S waves the effective impendance is $Z/\cos\theta$.

# Appendix 7 The vector differential equation for light rays

The optical length $L$ in a uniform homogeneous medium is $n\ell$, where $n$ is the (constant) refractive index along a straight line of geometric path length $\ell$. In a situation where the refractive index varies from point to point, the optical path from point $P_1$ to point $P_2$ is

$$L = \int_{P_1}^{P_2} n(\mathbf{r})ds, \tag{A7.1}$$

where $s$ is the distance measured along the path of the ray. *Fermat's principle* states that the actual path taken by a ray will be the one for which $L$ is a minimum.

We can generalize the idea of the optical length by introducing a quantity called the *optical path* or *eikonal*, $\zeta$, defined as

$$\zeta = n\mathbf{r} \cdot \hat{\mathbf{s}}, \tag{A7.2}$$

where $\mathbf{r}$ is a point on the light ray measured with respect to an arbitrary origin, as shown in Fig. A7.1, and $\hat{\mathbf{s}}$ is a unit vector in the direction of the ray path at point $\mathbf{r}$. In an isotropic medium (one where the wave vector $\mathbf{k}$ and the Poynting vector are parallel) a typical monochromatic wave has field amplitudes that vary from point to point as $e^{j(\omega t - \mathbf{k} \cdot \mathbf{r})}$. The phase variation with distance depends on the factor $\mathbf{k} \cdot \mathbf{r}$, which can be written as

$$\mathbf{k} \cdot \mathbf{r} = \frac{2\pi}{\lambda_0} n\mathbf{r} \cdot \hat{\mathbf{s}} = \frac{2\pi}{\lambda_0}\zeta, \tag{A7.3}$$

so we can see that the eikonal is closely related to the spatial variation of the phase of the wave. Points where $\zeta$ = constant correspond to phase fronts.

From Eq. (A7.2)

$$\text{grad}\ \zeta = \text{grad}(n\mathbf{r} \cdot \hat{\mathbf{s}}). \tag{A7.4}$$

Provided that $n$ does not vary too rapidly with position, we can write Eq. (A7.4) as[1]

$$\text{grad}\ \zeta = n\ \text{grad}(\mathbf{r} \cdot \hat{\mathbf{s}}). \tag{A7.5}$$

By the use of a vector identity,

$$\text{grad}(\mathbf{u} \cdot \mathbf{v}) = (\mathbf{v} \cdot \nabla)\mathbf{u} + (\mathbf{u} \cdot \nabla)\mathbf{v} + \mathbf{v} \times \text{ curl } \mathbf{u} + \mathbf{u} \times \text{curl } \mathbf{v}, \tag{A7.6}$$

[1] For an alternative discussion of this point, see reference [1].

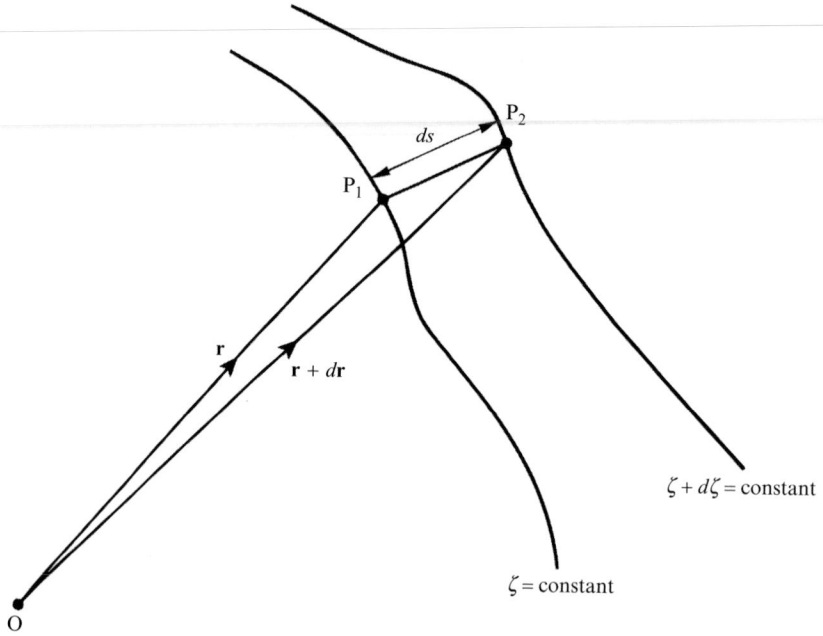


**Fig. A7.1** The geometry of the phase front, ray path, and position vectors used in discussing the ray path in a medium within which the refractive index varies with position.

we get

$$\text{grad}\,\zeta = n[(\hat{\mathbf{s}}\cdot\nabla)\mathbf{r} + (\mathbf{r}\cdot\nabla)\hat{\mathbf{s}}] + n[\mathbf{r}\times\text{curl}\,\hat{\mathbf{s}} + \hat{\mathbf{s}}\times\text{curl}\,\mathbf{r}]. \tag{A7.7}$$

Now, curl $\mathbf{r} = 0$ and curl $\hat{\mathbf{s}} = 0$, and, because $\hat{\mathbf{s}}$ is not explicitly a function of $x$, $y$, or $z$, $(\mathbf{r}\cdot\nabla)\,\hat{\mathbf{s}} = 0$. Furthermore,

$$(\hat{\mathbf{s}}\cdot\nabla)\mathbf{r} = \hat{\mathbf{s}}. \tag{A7.8}$$

Equation (A7.8) can be proved as follows. If $\hat{\mathbf{s}} \equiv a\hat{\mathbf{i}} + b\hat{\mathbf{j}} + c\hat{\mathbf{k}}$, and $\mathbf{r} = x\hat{\mathbf{i}} + y\hat{\mathbf{j}} + z\hat{\mathbf{k}}$, then

$$\begin{aligned}(\mathbf{s}\cdot\nabla)\mathbf{r} &= \left(a\frac{\partial}{\partial x} + b\frac{\partial}{\partial y} + c\frac{\partial}{\partial z}\right)(x\hat{\mathbf{i}} + y\hat{\mathbf{j}} + z\hat{\mathbf{k}})\\ &= a\hat{\mathbf{i}} + b\hat{\mathbf{j}} + c\hat{\mathbf{k}}\\ &= \hat{\mathbf{s}}.\end{aligned}$$

Consequently, from Eqs. (A7.6) and (A7.7),

$$\text{grad}\,\zeta = n\hat{\mathbf{s}}. \tag{A7.9}$$

In physical terms this is equivalent to saying that the ray path is perpendicular to the phase front.

In Fig. A7.1 an incremental ray path $ds$ between two closely spaced phase fronts is shown. Clearly,

$$\mathbf{r} + d\mathbf{r} = \mathbf{r} + ds\,\hat{\mathbf{s}}, \tag{A7.10}$$

so

$$\frac{d\mathbf{r}}{ds} = \hat{\mathbf{s}}. \tag{A7.11}$$

Therefore,

$$n\frac{d\mathbf{r}}{ds} = n\hat{\mathbf{s}} = \text{grad}\,\zeta \tag{A7.12}$$

and, consequently,

$$\frac{d}{ds}\left(n\frac{d\mathbf{r}}{ds}\right) = \frac{d}{ds}(\text{grad}\,\zeta). \tag{A7.13}$$

Now,

$$\frac{d}{ds}(\text{grad}\,\zeta) \equiv \left(\frac{dx}{ds}\frac{\partial}{\partial x} + \frac{dy}{ds}\frac{\partial}{\partial y} + \frac{dz}{ds}\frac{\partial}{\partial z}\right)\text{grad}\,\zeta. \tag{A7.14}$$

Therefore,

$$\frac{d}{ds}(\text{grad}\,\zeta) = \left(\frac{d\mathbf{r}}{ds}\cdot\nabla\right)\text{grad}\,\zeta, \tag{A7.15}$$

which, from Eq. (A7.12), can be written as

$$\frac{d}{ds}(\text{grad}\,\zeta) = \frac{1}{n}(\text{grad}\,\zeta\cdot\nabla)\,\text{grad}\,\zeta. \tag{A7.16}$$

Use of Eq. (A7.6) with $\mathbf{u} \equiv \mathbf{v} \equiv \text{grad}\,\zeta$ gives

$$\begin{aligned}\text{grad}(\text{grad}\,\zeta\cdot\text{grad}\,\zeta) &= (\text{grad}\,\zeta\cdot\nabla)\text{grad}\,\zeta + (\text{grad}\,\zeta\cdot\nabla)\text{grad}\,\zeta \\ &\quad + 2\,\text{grad}\,\zeta\times\text{curl}\,\text{grad}\,\zeta.\end{aligned} \tag{A7.17}$$

The last term on the right-hand side of Eq. (A7.17) is zero because curl grad $\zeta = 0$. Therefore,

$$(\text{grad}\,\zeta\cdot\nabla)\text{grad}\,\zeta = \frac{1}{2}\,\text{grad}(\text{grad}\,\zeta)^2. \tag{A7.18}$$

Combining Eqs. (A7.12), (A7.13), (A7.16), and (A7.18) gives

$$\frac{d}{ds}\left(n\frac{d\mathbf{r}}{ds}\right) = \frac{1}{2n}\,\text{grad}(n^2) \tag{A7.19}$$

and, since

$$\begin{aligned}\text{grad}(n^2) &= n\,\text{grad}\,n + (\text{grad}\,n)n \\ &= 2n\,\text{grad}\,n,\end{aligned} \tag{A7.20}$$

we have the final desired result,

$$\frac{d}{ds}\left(n\frac{d\mathbf{r}}{ds}\right) = \text{grad } n. \tag{A7.21}$$

## Reference

[1] M. Born and E. Wolf, *Principles of Optics*, 6th edn., Oxford: Pergamon Press, 1980.

# Appendix 8 Symmetry properties of crystals and the 32 crystal classes

The external form that crystals of a particular material adopt is a reflection of the symmetry of the internal arrangement of the atoms that make up the structure. The symmetry of the crystal is determined by observing what *symmetry operations* apply to it. These symmetry operations correspond to rotational symmetry, mirror symmetry, and inversion symmetry. The highest order of rotational symmetry possessed by a crystal determines to which of the seven *crystal systems* it belongs. Two-fold, three-fold, four-fold, and six-fold rotational symmetries are readily observed in nature.[1] If a crystal possesses a six-fold axis of symmetry, for example, then rotation of the crystal by $360°/6 = 60°$ brings the crystal into a configuration that is identical to the one in which it started. A mirror plane m is a self-explanatory symmetry element: if half of a crystal structure is a mirror image of the other half relative to a particular plane through the crystal then that plane is a mirror plane. The existence of a center of inversion symmetry implies that, if a face exists on the crystal perpendicular to a particular vector direction $\mathbf{r}$ from the origin, then a crystal face will also exist perpendicular to the vector $-\mathbf{r}$ from the origin. In terms of the internal structure an atom at coordinate $(x, y, z)$ has an identical counterpart at $(-x, -y, -z)$. Crystals can possess an axis of $n$-fold *inversion* symmetry: for example, in a crystal of six-fold inversion symmetry a rotation of 60° brings a crystal face on the bottom of the crystal into a position directly inverted from where an original face on the top of the crystal was in the 0° orientation. A crystal possessing a six-fold axis of inversion symmetry would be characterized as having a $\bar{6}$ axis.

The precise interplay between symmetry elements is best illustrated by giving specific examples of the 32 crystal classes that exist within the seven crystal systems. These seven systems are characterized by the highest order of rotational symmetry they possess, as shown in Table A8.1.

The 32 crystal classes are tabulated in Table A8.2, which gives their designation and lists briefly the symmetry elements that they possess. The highest-symmetry cubic class, designated m3m, provides a good example of how the various symmetry operations are related. This crystal class has three orthogonal four-fold axes, four three-fold axes (along the cube body diagonals), six two-fold axes, nine mirror planes, and a center of inversion symmetry, as shown schematically in Fig. A8.1. For details, and excellent drawings, of the crystal forms that occur in the 32 crystal classes the interested reader should consult reference [1]. However, a few examples might be helpful.

[1] Some examples of five-fold symmetry have been observed on a microscopic scale, but we will not deal with them further here.

**Table A8.1** The seven crystal systems

| System | Highest order of rotational symmetry |
|---|---|
| Triclinic | No axes of symmetry (one-fold axes) |
| Monoclinic | One two-fold axis |
| Orthorhombic | Three orthogonal two-fold axes |
| Trigonal | One three-fold axis |
| Hexagonal | One six-fold axis |
| Tetragonal | One four-fold axis |
| Cubic | Four three-fold axes |

**Table A8.2** The 32 crystal classes

| System | Designations of classes |
|---|---|
| Triclinic | 1, $\bar{1}$ |
| Monoclinic | 2, m, 2/m |
| Orthorhombic | mm, 222, mmm |
| Trigonal | 3, $\bar{3}$, 3m, $\bar{3}$ m, 32 |
| Hexagonal | 6, $\bar{6}$, 6/m, 6mm, $\bar{6}$m2, 62, 6/mmm |
| Tetragonal | 4, $\bar{4}$, 4/m, 4mm, $\bar{4}$2m, 42, 4/mmm |
| Cubic | 23, m3, $\bar{4}$3m, 43, m3m |

$\bar{2}$ indicates an axis of two-fold inversion symmetry.
m indicates a plane of mirror symmetry.
6/m indicates a six-fold axis orthogonal to a mirror plane.
4/mmm indicates a four-fold axis orthogonal to a mirror plane and additional mirror planes within which the four-fold axis lies.

## Class 6mm

This class has a six-fold symmetry axis at the intersection of two sets of three planes of mirror symmetry. Figure A8.2 shown as an example a crystal of zinc oxide that has this symmetry.

## Class $\bar{4}$2m

This class, represented by the important nonlinear crystals ADP and KDP, has an axis of four-fold inversion symmetry along the intersection of two orthogonal mirror planes, with two two-fold axes at 45° to these planes. Figure A8.3 shows these symmetry elements in a crystal of urea ($CO(NH_2)_2$).

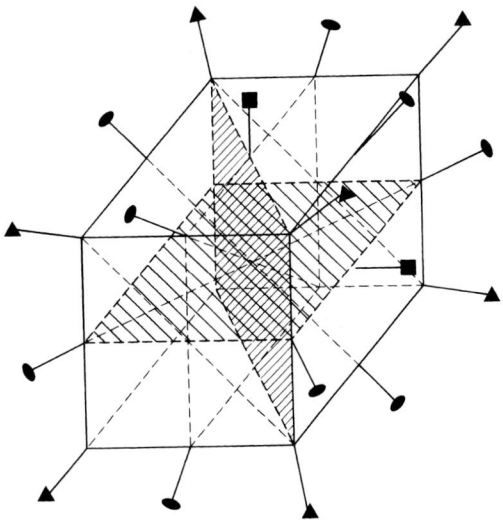

**Fig. A8.1** The symmetry elements of a cubic crystal of the highest symmetry class, m3m. The black squares indicate axes of four-fold symmetry, the triangles indicate axes of three-fold symmetry (triad axes), and the lenticular shapes indicate axes of two-fold symmetry (dyad axes). Various planes of mirror symmetry are also shown.

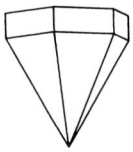

**Fig. A8.2** A zinc oxide crystal form with symmetry 6mm.

**Fig. A8.3** A urea crystal form with symmetry $\bar{4}2m$.

## Class 222

These crystals have three mutually orthogonal two-fold axes, but no mirror planes. The orthorhombic sphenoidal forms shown in Fig. A8.4 illustrate this symmetry. These crystals do not possess any planes of mirror symmetry.

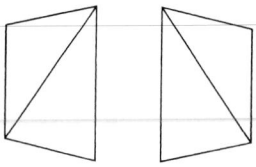

Fig. A8.4 Sphenoidal crystal forms with symmetry 222.

## Reference

[1] F. C. Phillps, *An Introduction to Crystallography*, 2nd edn., London: Longmans, Green and Co., 1956.

# Appendix 9 Tensors

One kind of tensor that is of particular interest to us here is a so-called tensor of the *second rank*,[1] denoted for example by $\bar{\bar{T}}$, which transforms one vector into another, for example,

$$\mathbf{B} = \bar{\bar{T}}\mathbf{A}. \tag{A9.1}$$

The tensor transforms the vector **A** into a new vector **B**, which is different in both magnitude *and direction* from **A**.

For example, the susceptibility tensor $\bar{\bar{\chi}}$ relates polarization and electric field by

$$\mathbf{P} = \bar{\bar{\chi}}(\epsilon_0\mathbf{E}). \tag{A9.2}$$

Clearly, the tensor can be written as an array of coefficients

$$\bar{\bar{\chi}} = \begin{pmatrix} \chi_{11} & \chi_{12} & \chi_{13} \\ \chi_{21} & \chi_{22} & \chi_{23} \\ \chi_{31} & \chi_{32} & \chi_{33} \end{pmatrix}. \tag{A9.3}$$

Physically, it does not matter what set of coordinate axes we choose; an electric field in a particular direction relative to specific crystal directions always gives the same polarization vector. However, the specific value of the coefficients $T_{ij}$ will depend on the coordinate system chosen. Thus, these coefficients must change, or *transform*, with a change in coordinate axes so as to preserve the specific relationship between **P** and **E**. Tensors are mathematically described as quantities that transform in a specific way under coordinate-axis rotations.

In a principal-axis system $(x, y, z)$ the tensor $\bar{\bar{\chi}}$ is diagonal,

$$\bar{\bar{\chi}} = \begin{pmatrix} \chi_{11} & 0 & 0 \\ 0 & \chi_{22} & 0 \\ 0 & 0 & \chi_{33} \end{pmatrix}. \tag{A9.4}$$

Let us choose a different set of axes $(x', y', z')$ in the following way.

(a) Rotate by angle $\phi$ about the $z$ axis: this produces axes $x''$, $y'$, and $z$, where

$$\begin{aligned} x'' &= x\cos\phi + y\sin\phi, \\ y' &= -x\sin\phi + y\cos\phi. \end{aligned} \tag{A9.5}$$

[1] A scalar is a tensor of rank zero; a vector is a tensor of rank 1.

(b) Rotate by angle $\theta$ about the new ($y'$) axis: this gives axes $x'$, $y'$, and $z'$, where

$$\begin{pmatrix} x' \\ y' \\ z' \end{pmatrix} = \begin{pmatrix} \cos\theta\cos\phi & \sin\phi\cos\theta & -\sin\theta \\ -\sin\phi & \cos\phi & 0 \\ \cos\phi\sin\theta & -\sin\phi\sin\theta & \cos\theta \end{pmatrix} \begin{pmatrix} x \\ y \\ z \end{pmatrix}. \tag{A9.6}$$

Equation (A9.6) can be used to demonstrate the transformation of a vector

$$\mathbf{r}' = \mathbf{R}\mathbf{r}, \tag{A9.7}$$

where $\mathbf{r}$ and $\mathbf{r}'$ are the representations of a vector in the two coordinate systems and $\mathbf{R}$ is the *rotation* matrix.

If we write Eq. (A9.4) in a non-principal-coordinate system that results from a rotation $\mathbf{R}$ of the principal coordinate system, then

$$\mathbf{P}' = \epsilon_0 \bar{\bar{\chi}}' \mathbf{E}', \tag{A9.8}$$

where $\bar{\bar{\chi}}'$ is the representation of the $\bar{\bar{\chi}}$ tensor in the primed coordinate system. Since $\mathbf{P}' = \mathbf{R}\mathbf{P}$ and $\mathbf{E}' = \mathbf{R}\mathbf{E}$, we have

$$\mathbf{R}\mathbf{P} = \epsilon_0 \bar{\bar{\chi}}' \mathbf{R}\mathbf{E} \tag{A9.9}$$

and

$$\mathbf{P} = \epsilon_0 \mathbf{R}^{-1} \bar{\bar{\chi}}' \mathbf{R}\mathbf{E} = \epsilon_0 \bar{\bar{\chi}} \mathbf{E}. \tag{A9.10}$$

Therefore, in terms of the tensor representation in the principal-axis system, $\bar{\bar{\chi}}$,

$$\mathbf{R}^{-1} \bar{\bar{\chi}}' \mathbf{R} = \bar{\bar{\chi}}' \tag{A9.11}$$

and

$$\bar{\bar{\chi}}' = \mathbf{R} \bar{\bar{\chi}}' \mathbf{R}^{-1}. \tag{A9.12}$$

Let us consider a simple case in which $\phi = 90°$ and $\theta = 90°$. This transforms $(x, y, z)$ to $(x', y', z')$, where, from the standpoint of directions, $x' \equiv -z$, $y' \equiv -x$, and $z' \equiv y$.

The transformation matrix is

$$\mathbf{R} = \begin{pmatrix} 0 & 0 & -1 \\ -1 & 0 & 0 \\ 0 & -1 & 0 \end{pmatrix}. \tag{A9.13}$$

It is easy to show that

$$\mathbf{R}^{-1} = \begin{pmatrix} 0 & -1 & 0 \\ 0 & 0 & -1 \\ -1 & 0 & 0 \end{pmatrix}, \tag{A9.14}$$

so a susceptibility tensor in a principal-axis system, such as Eq. (A9.4), would become

$$\bar{\bar{\chi}}' = \mathbf{R}\bar{\bar{\chi}}\mathbf{R}^{-1} = \begin{pmatrix} \chi_3 & 0 & 0 \\ 0 & \chi_1 & 0 \\ 0 & 0 & \chi_2 \end{pmatrix}, \tag{A9.15}$$

which correctly reflects the axis rearrangement that has occurred.

Note that $\mathbf{R}^{-1} = \tilde{\mathbf{R}}$, where $\tilde{\mathbf{R}}$ is the transpose of the matrix $\mathbf{R}$,[2] so we can write $(\mathbf{R}^{-1})_{ik} = \mathbf{R}_{ki}$. Therefore, Eq. (A9.12) can be written

$$\chi'_{ij} = \sum_k \sum_\ell R_{ik} R_{j\ell} \chi_{k\ell} \tag{A9.16}$$

or, in the repeated-index convention that is commonly used in tensor algebra,

$$\chi'_{ij} = R_{ik} R_{j\ell} \chi_{k\ell}. \tag{A9.17}$$

We can generalize this discussion to tensors of the *third rank*, such as the electro-optic tensor $\bar{\bar{d}}$, so that the coefficients of this tensor change under a coordinate transformation as

$$d'_{ijk} = R_{i\ell} R_{jm} R_{kn} d_{\ell mm}. \tag{A9.18}$$

The form of the $\bar{\bar{d}}$ tensor in Table 19.1 reflects the fact that $d'_{ijk}$ must be identical to $d_{\ell mn}$ if the coordinate transformation, or general symmetry transformation, takes the crystal to an identical form – such as a 60° rotation about the six-fold axis in a class-6 crystal.

[2] If all the elements of a matrix are real and $\mathbf{R}^{-1} = \tilde{\mathbf{R}}$, the matrix is said to be *orthogonal*.

# Appendix 10 Bessel-function relations

The Bessel-function relations used in deriving the unified boundary-condition equations, Eqs. (17.102)–(17.104), are

$$J_{-\nu}(u) = (-1)^{\nu} J_{\nu}(u), \tag{A10.1}$$

$$K_{-\nu}(w) = K_{\nu}(w), \tag{A10.2}$$

$$2J'_{\nu}(u) = J_{\nu-1}(u) - J_{\nu+1}(u), \tag{A10.3}$$

$$-2K'_{\nu}(w) = K_{\nu-1}(w) + K_{\nu+1}(w), \tag{A10.4}$$

$$J_{\nu+1}(u) + J_{\nu-1}(u) = 2\left(\frac{\nu}{u}\right) J_{\nu}(u), \tag{A10.5}$$

$$K_{\nu+1}(w) - K_{\nu-1}(w) = 2\left(\frac{\nu}{w}\right) K_{\nu}(w). \tag{A10.6}$$

The following equations result from (A10.1)–(A10.6):

$$\frac{J'_{\nu}(u)}{uJ_{\nu}(u)} = \frac{J_{\nu-1}(u)}{uJ_{\nu}(u)} - \frac{\nu}{u^2} = -\frac{J_{\nu+1}(u)}{uJ_{\nu}(u)} + \frac{\nu}{u^2}, \tag{A10.7}$$

$$\frac{K'_{\nu}(w)}{wK_{\nu}(w)} = -\frac{K_{\nu-1}(w)}{wK_{\nu}(w)} - \frac{\nu}{w^2} = -\frac{K_{\nu+1}(w)}{wK_{\nu}(w)} + \frac{\nu}{w^2}. \tag{A10.8}$$

# Appendix 11 Green's functions

In the most general sense, the use of Green's functions is a method for solving certain differential equations, *including* their boundary conditions [1]. For example, suppose it is desired to find the solution $g(x)$ for some inhomogeneous differential equation of the general form

$$L[g(x)] = f(x), \tag{A11.1}$$

where $L[g(x)]$ includes various differential operators acting on the function $g(x)$. If the solution to this equation, including its boundary conditions $G(x, x')$, can be determined for a function $f(x)$ that corresponds to a $\delta$-function at $x = x'$, then we could write

$$L[G(x, x')] = \delta(x - x'). \tag{A11.2}$$

A general function $f(x)$ can be imagined as being made up of a series of $\delta$-functions, since

$$f(x) = \int_{-\infty}^{\infty} f(x')\delta(x - x')dx'. \tag{A11.3}$$

Consequently, the solution to Eq. (A11.1) can be found in a similar way from the Green's function

$$g(x) = \int_{-\infty}^{\infty} G(x, x')g(x')dx'. \tag{A11.4}$$

To make this discussion more specific, we could choose examples from many different areas: the motion of damped vibrating systems subject to arbitrary forces, diffusion, the conduction of heat, or electromagnetic theory. Perhaps the simplest example comes from the theory of the electrostatic potential.

When a distribution of charge of charge density $\rho(\mathbf{r})$ is present in an unbounded region, the scalar potential $\phi$ obeys Poisson's equation

$$\nabla^2\phi = -\frac{\rho}{\epsilon}, \tag{A11.5}$$

where $\epsilon$ is the permittivity of the region being considered. The Green's function for this situation is the potential that results from a unit point charge situated at the origin. This is equivalent to choosing $x' = 0$ in our definition of the Green's function. For this function

$$\nabla^2 G(r, 0) = -\frac{\delta(r)}{\epsilon}. \tag{A11.6}$$

If we imagine replacing the point charge at the origin with a small spherical region of radius $a$, then the charge density within this region is

$$\rho_0 = \frac{3}{4\pi a^3}. \tag{A11.7}$$

Clearly, as $a \to 0$, $\rho_0 \to \infty$ and

$$\int \rho_0 \, dV = \int_0^a 4\pi r^2 \left( \frac{3}{4\pi a^3} \right) dv = 1. \tag{A11.8}$$

So, as $a \to 0$, $\rho_0$ is an appropriately normalized $\delta$-function.

Now, the potential produced by a unit point charge at the origin is well known. This is the Green's function for the problem,

$$G(r, 0) = \frac{1}{4\pi\epsilon r}. \tag{A11.9}$$

Writing Eq. (A11.6) in spherical coordinates,

$$\frac{1}{r^2}\frac{\partial}{\partial r}\left( r^2 \frac{\partial G}{\partial r} \right) = -\frac{\delta(r)}{\epsilon} \tag{A11.10}$$

and substituting from Eq. (A11.9) gives

$$\frac{1}{r^2}\frac{\partial}{\partial r}\left( r^2 \frac{\partial G}{\partial r} \right) = \frac{1}{r^2}\frac{\partial}{\partial r}\left( -\frac{1}{4\pi\epsilon} \right). \tag{A11.11}$$

Integrating the right-hand side of Eq. (A11.11) gives

$$\int \frac{1}{r^2}\frac{\partial}{\partial r}\left( -\frac{1}{4\pi\epsilon} \right) dV = 4\pi \int \frac{\partial}{\partial r}\left( -\frac{1}{4\pi\epsilon} \right) dv = -\frac{1}{\epsilon}, \tag{A11.12}$$

the same result as integrating the right-hand side of Eq. (A11.10). So the Green's function is indeed, a solution of Eq. (A11.6). Therefore, from Eq. (A11.4), we can see that the potential distribution of a distribution of charges is

$$\phi(\mathbf{r}) = \int \frac{\rho(\mathbf{r}')dV}{4\pi\epsilon|\mathbf{r} - \mathbf{r}'|}. \tag{A11.13}$$

In other words, the Green's function has allowed us to solve the inhomogeneous differential equation (Poisson's equation) by integrating over the solutions corresponding to individual point charges, namely $\delta$-functions, since their charge densities are infinite yet their total charge integrated over all space is finite.

If we solve the slightly more general problem of a point charge inside a region where a series of boundaries are at specified potentials we thereby determine the appropriate Green's function $G(\mathbf{r}, \mathbf{r}')$. The solution to this boundary-value problem for a distribution of charge is

$$\phi(\mathbf{r}) = \int G(\mathbf{r}, \mathbf{r}')\rho(\mathbf{r}')dV. \tag{A11.14}$$

Let us consider one more example. For monochromatic waves, the wave equation reduces to the Helmholtz equation

$$\nabla^2\psi + k^2\psi = 0. \tag{A11.15}$$

If we are considering spherical waves originating from a point source, then in spherical coordinates Eq. (A11.15) can be written as

$$\frac{1}{r^2}\frac{\partial}{\partial r}\left(r^2\frac{\partial\psi}{\partial r}\right)a + k^2\psi = 0, \tag{A11.16}$$

which clearly has the solution, except at $r = 0$,

$$\psi(\mathbf{r}) = G(\mathbf{r}, 0) = \frac{1}{4\pi r}e^{\pm jkr}. \tag{A11.17}$$

This is the appropriate Green's function for the problem, which satisfies the equation

$$\nabla^2 G(\mathbf{r}, 0) + k^2 G(\mathbf{r}, 0) = -\delta(\mathbf{r}). \tag{A11.18}$$

If we have a distribution of point sources of complex amplitude $g(\mathbf{r}')$ then the full solution to the Helmholtz equation can be obtained as an integral in a similar way to Eq. (A11.13):

$$\psi_{\text{Total}}(\mathbf{r}) = \frac{1}{4\pi}\int \frac{g(\mathbf{r}')}{|\mathbf{r} - \mathbf{r}'|}e^{\pm jk|\mathbf{r}-\mathbf{r}'|}\,dV. \tag{A11.19}$$

For further details of how this approach is extended to non-monochromatic sources the reader should consult reference [2].

## References

[1] H. Margenau and G. M. Murphy, *The Mathematics of Physics and Chemistry*, 2nd edn., Princeton, NJ: Van Nostrand, 1956.

[2] W. K. H. Panofsky and M. Philips, *Classical Electricity and Magnetism*, Reading, MA: Addison-Wesley, 1955.

# Appendix 12 Recommended values of some physical constants

| Quantity | Symbol | Value (with one-standard-deviation uncertainty) |
|---|---|---|
| Speed of light in vacuum | $c_0$ | $2.99792458 \times 10^8$ m s$^{-1}$ (exact) |
| Permeability of vacuum | $\mu_0$ | $4\pi \times 10^{-7}$ H m$^{-1}$ |
| Permittivity of vacuum, $1/\mu_0 c^2$ | $\epsilon_0$ | $8.854187817\ldots \times 10^{-12}$ F m$^{-1}$ |
| Elementary charge (of proton) | $e$ | $1.60217733(49) \times 10^{-19}$ C |
| Unified atomic mass constant | $m_u$ | $1.6605402(10) \times 10^{-27}$ kg |
| Rest mass | | |
| of electron | $m_e$ | $9.1093897(54) \times 10^{-31}$ kg |
| | $m_e/m_u$ | $5.48579903(13) \times 10^{-4}$ |
| of proton | $m_p$ | $1.6726231(10) \times 10^{-27}$ kg |
| | $m_p/m_u$ | $1.007276470(12)$ |
| | $m_p/m_e$ | $1836.152701(37)$ |
| of neutron | $m_n$ | $1.6749286(10) \times 10^{-27}$ kg |
| | $m_n/m_u$ | $1.008664904(14)$ |
| Planck's constant | $h$ | $6.6260755(40) \times 10^{-34}$ J s |
| $h/(2\pi)$ | $\hbar$ | $1.05457266(63) \times 10^{-34}$ J s |
| Avogadro's constant | $N_A$ | $6.0221367(36) \times 10^{23}$ mol$^{-1}$ |
| Boltzmann's constant | $k_B$ | $1.380658(12) \times 10^{-23}$ J K$^{-1}$ |
| Stefan–Boltzmann constant | $\sigma$ | $5.67051(19) \times 10^{-8}$ W m$^{-2}$ K$^{-4}$ |
| First radiation constant, $2\pi hc^2$ | $c_1$ | $3.7417749(22) \times 10^{-16}$ J m$^2$ s$^{-1}$ |
| Second radiation constant, $hc/k_B$ | $c_2$ | $1.438769(12) \times 10^{-2}$ m K |

# Index


1/*f* noise, 690

Abbé number, 390
aberration
  astigmatism, 393
  chromatic, 391
  coma, 393
  curvature of field, 393
  distortion, 393
  spherical, 393
aberrations, 389–398
  geometrical, 392–395
absorption
  stimulated, 18
absorption and fluorescence spectra of organic dye, 280
acceptor level, 701
acoustic Bragg diffraction, 609, 610
acousto-optic deflectors and scanners, 615, 633
acousto-optic interactions, 612
acousto-optic modulator, 608–614, 615, 633
  Debye–Sears, 611
acousto-optic modulators
  applications of, 614
  materials for fabrication, 617–620
adaptive optics, 671
additive pulse mode-locking, 146
ADP. ammonium dihydrogen phosphate, 588
airborne laser (ABL), 809
Airy, G.B., 115
ammonium dihydrogen phospate, ADP, 588
amplification at optical frequencies, 7, 11, 43–48
amplification at optical frequencies for homogeneously broadened transition, 43–46
amplification at optical frequencies for inhomogeneously broadened transition, 46–48
amplification in fiber communication system, 788
amplifier, 7
  electronic, 8
  transimpedance, 707
  voltage gain of, 9
amplitude condition for laser oscillation, 90
amplitude hologram, 800
amplitude modulation
  electro-optic, 595–597
analytic signal, 101, 737–739
anamorphic optics, 336
angioplasty by laser, 808
anharmonic potential, 622–626
anharmonic potential well, 237
anisotropic medium, 540
  field vector angles in, 547
  wave propagation in, 543
anisotropic meium
  stored electromagnetic energy in, 542
anode pulse height distribution in photomultiplier, 701
antisymmetric stretch mode of carbon dioxide, 238
APD
  direct detection with, 724
APD (avalanche photodiode), 708, 711–714
aperture stop, 386, 388
aplanatic imaging, 385
Arago, D., 111
arc, 223
ArF* laser, 267, 268
argon-ion laser, 211
asphere, 389
aspherical surface, 389
astigmatism, 393
astronomical telescope beam expander, 461
atmospheric turbulence, 15, 671
attosecond pulses, 149
autocorrelation function, 742
avalanche photodiode
  Geiger-mode, 714
avalanche photodiode (APD), 708, 711–714
axicon, 478

Babinet, J., 111
Babinet's principle, 111
band, 247, 301
  conduction, 11
  valence, 2, 11
band gap, 42, 302
band structures of insulator, metal, and semiconductor, 42, 302
band-gap
  indirect, 306
band-gap engineering, 322
band-stop filter, 412
band-tailing, 318
band-to-band transition, 313

Barker code, 605
Basov, N., 2, 122
beacons and guide stars, 671
beam deflector
electro-optic, 607
beam divergence angle, 119
beam divergence of Gaussian beam, 122
beam expander
astronomical telescope type, 461
Galilean, 461
beam parameter, 440
beam waist, 442
beam walk-off during nonlinear mixing, 653
beam waves
non-Gaussian, 477
beat frequency, 101
beats, 133
bending mode of carbon dioxide, 238
Bennett, W.R. Jr., 2
Bessel function relations, 848
Bessel functions, 505, 506
Bessel–Gauss beam, 478
biaxial crystals, 565
bisectrix
acute, 565
obtuse, 565
bistability, 674–678
bit-error rate, 730–733
black body, 20
black-body radiation, 20–25
spectral distribution of, 21
black-body radiation formulas, 826
Bloembergen, N., 2
bolometer, 716
Bose–Einstein distribution, 774
bouncing pulse interpretation of mode locking, 139
bound ray, 484, 488
bound state of molécule, 233
boundary conditions in cylindrical dielectric waveguide, 510
Bragg grating, 411
bremsstrahlung FEL, 293
Brewster angle window, 82
Brewster's angle, 402
brightness, 384
Brillouin scattering, 609
Brillouin zone, 422
broadening
inhomogeneous, 37–43
of energy level, 11
Burrus-type LED, 320

carbon dioxide
antisymmetric stretch vibration of, 238
bending mode vibration of, 238
normal mode of, 238
symmetric stretch vibration of, 238
carbon dioxide laser, 228, 244–247
carbon dioxide vibrational energy level diagram, 239
carbon monoxide laser, 249
carrier concentrations in semiconductor, 307
cataphoretic flow, 221
cavity round trip time, 142
centrosymmetric crystal, 585, 623, 634
cesium laser, 207
optical pumping of, 175, 207
chain reaction, 269
channeltron, 696
chaotic light, 768
charge transfer excitation, 220
chemical laser
CO, 271
chemical lasers, 269
chief ray, 388
chirped pulse, 148
chromatic aberration, 391
chromatic dispersion, 787
circular aperture
diffraction pattern of, 114
circularly polarized light, 563
left-handed, 563
right-handed, 563
Clausius-Mossotti equation, 631, 636
close-coupling in optical pumping, 174
CO chemical laser, 271
coding schemes in optical networks, 793
coherence
degree of, 749
partial, 750
spatial, 748–753
temporal, 125, 744–748
volume of, 752
coherence length, 125
coherence properties of radiation, 125
coherence time, 125
coherent detection, 725–730
colliding pulse mode-locking, 42, 145
collision
hard, 37
soft, 37
coma, 393
complex susceptibity, gain, and absorption, relation between, 63
Compton regime of FEL, 294
conduction band, 11
conductivity, 309
conjugate points in imaging system, 385
correlation functions, 741–744
Cotton-Mouton effect, 623, 634
Crick, F.H.C., 422
critical angle, 364

cross-section
stimulated emission, 159
crystal
biaxial, 565
centrosymmetric, 585, 623, 634
uniaxial, 550
crystal classes, 841
crystals
symmetry properties of, 841
crystals for nonlinear applications, 679
$CS_2$
phase conjugation in, 674
CTR (controlled thermonuclear reactions), 814
current–voltage characteristic of photodiode, 712
curvature of field, 393
cutoff frequencies in optical fiber, 518
cylindrical dielectric waveguide, 504
Czochralski method for ctrstal growth, 63, 170, 173

data transmission rate
historical variation over time, 786
DBR laser, 347, 348
de Broglie relation, 303
Debye–Sears effect, 609
deflector
acousto-optic, 615, 633
degenerate doping, 310
degree of coherence, 749
degree of spatial coherence, 763
degree of temporal coherence, 746
density of states, 308, 339, 340
detection limits for photodetectors, 717–725
detectivity, 692
detector
square-law, 736
detectors
photoconductive, 700–706
photoemissive, 693
DF laser, 269
DFB laser, 345
$\delta$-function, 824
diatomic molecule, 235
potential energu level diagram, 233
simple harmonic oscillator model of, 236
dielecric tensor, 539
dielectric constant, 6, 362
complex, 7
frequency-dependent, 409
dielectric slab waveguide
ray theory of, 490
wave theory of, 494–503
dielectric waveguide
boundary conditions in, 510
dielectric waveguides
cylindrical, 504
difference frequency generation, 660
diffraction, 109–111
diffraction efficiency of acousto-optic device, 614
diffraction grating
used for tuning, 282
diffraction losses in resonator, 474
diffraction pattern, 109
Fraunhofer, 112
Fresnel, 112
diffraction pattern of circular aperture, 114
diffraction pattern of single slit, 112
diffraction pattern of two dimensional aperture, 113
diffraction pattern of two narrow slits, 112
diffraction theory
Fresnel–Kirchoff, 474
diffusion length, 313
diode pumped solid state laser
frequency doubling, 40, 184
diode-pumped alkali vapor gas lasers, 208
diode-pumped solid state lasers, 183
direct band-gap transition, 305
direct detection, 687
signal-to-noise ratio, 721
direct detection with APD, 724
direct detection with p–i–n photodiode, 722
dispersion, 361, 787
chromatic, 787
material, 525, 787
polarization mode, 787
dispersion curves in slab waveguide, 498
dispersion diagram, 423
dispersion in optical fiber, 524–526
dispersion shifted optical fiber, 526
displacement vector, 2, 362
orientation in anisotropic medium, 545
distortion, 393
distributed feedback (DFB) laser, 345
donor level, 701
donut mode, 451
doping
degenerate, 310
Doppler broadening, 39
Dorsey, N.E., 4, 154
double heterostructure, 324–328
double-clad fiber design for fiber laser, 187
double-crucible method for optical fiber fabrication, 521, 523
doubly resonant parametric oscillator (DRO), 665
DRO (doubly resonant parametric oscillator), 665
duality of waves and particles, 2
dye cell
used for mode locking, 139
dye laser, 277–289
calculation of threshold pump power, 285–289
CW operation of, 284
end pumping, 281
excitation and relaxation pathways, 286

dye laser (cont.)
Hänsch-type, 283
Littman type, 283
side pumping, 281
dynode chain, 700
dynode materials, 697
dynodes, 693

E-wave, 551
EAM (electro-absorption modulator), 606
EDFA (erbium-doped fiber amplifier), 529, 787
effective lifetime, 153
effective mass, 303
eikonal (optical path), 837
Einstein A and B coefficients
relation between, 25
Einstein A coefficient for spontaneous emission, 12
Einstein B coefficient for stimulated absoprtion, 18
Einstein B coefficient for stimulated emission, 15
elasto-optic tensor, 585
electric circuit to drive a flashlamp, 77
electric field, 2
electric-optic effect
quadratic, 580, 587
electro-absorption modulator (EAM), 606
electro-optic amplitude modulation, 595–597
electro-optic beam deflector, 607
electro-optic effect
linear, 580, 581, 631, 637
electro-optic modulation
transverse, 590, 591, 594
electro-optic modulator
traveling wave, 600
waveguide, 599, 602
electro-optic modulators
high frequency, 599
electro-optic phase modulation, 597–598
electro-optic tensor, 581
electro-optic tensor values, 586
electro-optical modulation
linear, 588, 590
electromagetic radiation and optical media
fundamental parameters of, 822
electromagnetic field from accelerated electron, 291
electromagnetic fields
storage and transport of energy by, 831
electromagnetic theory, 3–4
electromagnetic waves, 6–7
electromagnetic waves and nonlinear polarization, 640–644
electron
electromagneic field resulting from acceleration, 291
electron energy distribution, 226
electron impact excitation, 210
electron oscillator model for radiative transition, 59–63
electron temperature, 226, 247
electron-beam preionization, 262
electron-beam pulser/sustainer laser, 264
electronic energy states, 207, 208, 232
ellipsoidal cavity for optical pumping, 159
elliptical reflector for optical pumping, 78, 174
enantiomorph, 406
energy band, 42
energy band diagram
parabolic, 304
energy barrier
rectification at, 324
energy density and intensity
relation between, 16
energy gap, 43
energy level diagram
rotational states, 241
ruby, 76
energy level diagram of organic dye molecule, 279
energy level diagram of three- and four-level lasers, 73
energy levels of molecules, 117, 232
energy levels, of atoms, molecules and condensed matter, 10–11
energy states
electronic, 232
rotational, 239
vibrational, 232–236
entrance pupil, 388
erbium-doped fiber amplifier (EDFA), 187, 529, 787
erector lens, 387
etalon
intracavity, 134
transfer matrix for, 415
evanescent wave, 365
excimer laser, 267
exciplex laser, 267
excitation transfer, 80
excited dimer (excimer), 265
excited state, 10
extended source
spatial coherence of, 751
external field in susceptibility analysis, 637
external quantum efficiency, 318, 319
extraordinary wave, 551
extrordinary wave, 365
eye, 807
structure of, 361
eyepiece lens, 387

f/number, 483
Fabrikant, V.A., 1
Fabry, M.P.A.C., 85
far-infrared lasers, 272–274

far-infrared waveguide laser, 273
Faraday dark space, 225
Faraday effect, 406, 634
feedback
  non-resonant, 85
feedback circuit, 8
feedback structure in laser, 85
FEL
  Compton regime, 294
  Raman regime, 294
  self-amplification spontaneous emission, 297
FEL bremsstrahlung, 293
FEL oscillator, 297
Fermat's principle, 837
Fermi–Dirac distribution, 307
Fermi–Dirac statistics, 307
fiber Bragg grating, 420
fiber lasers, 184–187
field stop, 387
field vector angular relationships in anisotropic medium, 547
filter
  band-stop, 412
filtering
  spatial, 462
finesse, 89
flashlamp
  coaxial, 281
  helical, 78
flashlamps
  output spectra, 158
flow-mixing laser, 253
fluorescence, 278
focal length, 374
  first focal length, 368
  second focal length, 368
focal points, 368
force constant, 235
four-wave mixing, 672
Fox–Smith interferometer, 135
Franklin, R., 422
Fraunhofer diffraction pattern, 112
free spectral range, 89
free-electron laser (FEL), 291–298
free-free transition, 293
free-space optical communications (FSO), 794
frequency doubling
  intracavity, 184
Fresnel condition, 117, 232
Fresnel diffraction condition, 228
Fresnel diffraction pattern, 112
Fresnel number, 474
Fresnel's equations, 545
Fresnel, A., 111
Fresnel–Kirchoff diffraction theory, 474
FSO (free-space optical communications), 794
full width at half maximum (FWHM), 36
fundamental Gaussian mode, 444
FWHM, 36

Gabor, D., 797
gain coefficient, 65
  threshold, 90, 207
gain in homogeneously broadened system, 51
gain saturation in homogeneouly broadened amplifer, 94
gain saturation in inhomogeneouly broadened amplifer, 96
gain-guided structure of semiconductor laser, 342, 343
Galilean beam expander, 461
gas discharge
  V-I characteristic of, 223
gas discharge lasers
  rate equations for, 225
gas discharges for exciting gas lasers, 222
gas dynamic lasers, 256–260
gas laser
  schematic of first one, 82
  schematic of modern version, 83
  TEA, 257, 260
gas laser action in a plasma, 206
gas laser excitation schemes, 206
gas lasers, 206
  high pressure pulsed, 260–264
  optical pumping of, 206
  pulsed, 214–216
  pumping saturation in, 213
  ultraviolet, 264
gas transport laser amplifier, 255
gas transport laser geometries, 254
gas transport lasers, 252–256
Gaussian beam
  focused spot size, 460
  fundamental solution, 441, 444
Gaussian beam angle, 442
Gaussian beam donut mode, 451
Gaussian beam in lens waveguide, 462
Gaussian beam in medium with parabolic gain profile, 465
Gaussian beam in medium with spatial gain or absorption, 464
Gaussian beam in quadratic index medium, 463
Gaussian beam in resonator, 467–474
Gaussian beam parameter, 440
Gaussian beam transfomation by optical system, 454
Gaussian beam waves, 118
Gaussian beams, 438
  optics of, 438–478
Gaussian beams with Cartesian symmetry, 445
Gaussian beams with cylindrical symmetry, 451
Gaussian bem transformation by lens, 459
Gaussian light, 768, 774

Gaussian line
narrowing of, 87
Gaussian modes
higher order, 445–454
Gaussian radial index variation
effect on ray propagation, 433
ray transfer matrix for, 434
Geiger-mode avalanche photodiode, 714
generation–recombination noise, 690
geometrical optics, 365
giant pulse lasers, 195–203
glass chart, 390
glow discharge, 224
Golay cell, 717
Goos-Hänchen shift, 491
graded index optical fiber, 486
Green's function in coherence theory, 754, 755, 759, 762
Green's functions, 849
Green's theorem, 755, 756
ground state, 10
group delay, 491
group velocity, 525
group velocity delay
negative, 148
group velocity dispersion, 146
growth of multilayer semiconductor structures, 322
guide stars, 671

Γ-valley, 306
half-wave plate, 563, 589
half-wave voltage, 593
Hanbury Brown, R., 767
Hanbury Brown–Twiss interferometer, 767, 776
Hanbury Brown–Twiss photon count correlatons, 778
hard collision, 37
helical flashlamp, 78
helium-cadmium laser, 221
helium-neon laser, 79–84
helium-selenium laser, 221
Helmholtz equation, 439
Hermite–Gaussian mode, 445
Herriott, D., 2
heterojunction, 321
heterostructure
double, 324–328
HF
vibrationally excited, 270
HF laser, 269
HgBr laser, 267
high pressure pulsed gas lasers, 260–264
higher order Gaussian beams, 445–454
Hilbert transforms, 739
Ho:YAG laser, 183
hole burning, 100
holes burnt in gain curve, 98
holey fiber, 527
hollow cathode structure, 221
hologram, 797
amplitude type, 800
phase type, 801
plane type, 803
volume type, 803
white light type, 804
hologram as a diffraction grating, 801
holography, 797–806
homogeneous line broadening, 6, 33–37
hot-carrier-effect detector, 706
human eye, 807
Humphreys C., 1
Huygenian constructions, 557
Huygens secondary wavelets, 116
Huygens, C., 116, 271
hydrogen chloride laser, 271

idler frequency in parametric process, 633
image hole, 98
imaging
aplanatic, 385
stigmatic, 385
imaging and magnification, 378
imaging and non-imaging optical systems, 383
imaging systems, 385
impedance
transformed, 403
use in multilayer systems, 403
impedance analysis of multilayer structure, 420
impedance of plane wave, 398
impedances
use in optics, 398–406
index ellipsoid, 548
index surfaces, 553
index-guided structure of semiconductor laser, 345
indicatrix, 548–550
change in electro-optic effect, 582
related surfaces, 555
uniaxial, 550
inertial confinement, 814
Inhomogeneous broadening, 37–43
injection efficiency, 317
injection luminescence, 313
inorganic liquid lasers, 289
input plane, 366
inside chemical vapor deposition (ICVD), 523
intensity, 14, 363
radiant, 13
intensity fluctuations of optical sources, 767
intensity-correlation interferometry, 766
interband quantum cascade laser, 354
interference, 127

interferometer
Fox-Smith, 135
Michelson, 125
internal quantum efficiency, 316, 321
intrinsic and extrinsic semiconductors, 309
intrinsic semiconductor, 307
inverse piezoelectric effect, 585
iodine chemical laser, 271, 272
iodine photodisocciation laser, 268
ion lasers, CW, 217–221
isolator
optical, 406
isotope separation by laser, 809–813
isotropic material
definition, 362
isotropic medium, 540

Javan, A., 2
Johnson noise, 688–690
Jones calculus, 570
Jones matrix, 572
Jones vector, 570
Joule-Thomson temperature, 256

k-space, 22
KDP
potassium dihydrogen phosphate, 588
Kerr effect, 580
Kerr lens mode-locking, 146
kidney stones
fragmentation of, 808
Kirchoff boundary conditions, 758
Kleinman's conjecture, 631, 636
Kohlrausch, R, 4

L-valley, 306
Laguerre–Gaussian mode, 451
Lamb dip, 99
Lamb, W.E. Jr., 100
Landenburg, R.W., 1
LASEK, 806
laser, 1,
carbon dioxide, 244–247
carbon monoxide, 249
chemical, 269
COIL, 271, 272
DF, 269
excimer, 267
exciplex, 267
far-infrared, 272–274
far-infrared waveguide, 273
flow-mixing, 253
free-electron, 291–298
gas dynamic, 256–260
gas transport, 252–256
helium-neon, 79–84
HF, 269
HgBr, 267
hydrogen chloride, 271
ideal photon statistics of, 774
inorganic liquid type, 289
iodine, 268
laser twisted mode operation of, 136
organic dye, 277–289
oscillation frequency of, 91–92
photodissociation, 268
power output of, 103–106
Q-switched, 116, 195–203
quantum well, 336–341
ruby, 74–79
semiconductor, 301–356
three- and four-level, 72–74
LASER -the acronym, 1
laser amplifier, 33
laser angioplasty, 808
laser cavity, 9, 154
laser cavity in semiconductor laser, 334
laser diode arrays, 350
laser isotope separation, 809–813
laser modes, 116–122
laser oscillation
amplitude condition for, 90
multimode, 93
phase condition for, 91
laser plasma generation , 813
laser pointer
construction of, 656
laser radiation
linewidth of, 122–125
laser resonator, 4, 9, 85
laser resonators
stable and unstable, 120
laser weapons, 809
lasers
medical applications of, 806–809
lasers as scalpels, 807
lasers in opthalmology, 806
LASIK, 806
lattic constants and band-gaps, 323
lattice
reciprocal, 422
lattice constant, 322
Lawson criterion, 814
LED
Burrus-type, 320
edge emitting, 327
lens
axicon, 478
effect on Gaussian beam, 459
erector, 387
objective, 387
lens sequence
periodic, 425–431

lens waveguide
Gaussian beam in, 462
identical thin lenses, 426
lens; eyepiece, 387
level degeneracy, 27–29
lifetime
effective, 296
natural, 13
light, 2–3
velocity of, 4
light absorption, 1
light emission, 1, 78
light ray, 4
light rays
differential equation of, 432, 837
light-emitting diode (LED), 314
line broadening, 33
homogeneous, 33–37
natural, 33–36
line-of-sight optical links, 794
lineshape function, 13, 747
Doppler broadened, 39
Lorentzian, 36
overall, 47
Voigt, 47
linewidth of laser radiation, 122–125
lithium niobate, 590
in transverse electro-optic modulator, 594
lithium niobate for parametric processes, 664
lithium niobate in parametric oscillator, 667
local field in susceptibility analysis, 631, 637
local oscillator, 725
longitudinal and transverse modes of semiconductor laser, 335
longitudinal electro-optic modulation, 588, 590
longitudinal mode, 93
Lorentz force, 293
Lorentzian and Gaussian lineshapes compared, 40
Lorentzian lineshape function, 36
loss line
in laser amplifier, 93

Mach number, 258, 262
Mach–Zehnder interferometer, 603
magnetic field, 2, 137–140
magnetic flux density, 2
magnetization, 3
magnification
angular, 379
linear, 379
Maiman, T.H., 2
Manchester code, 794
Manley-Rowe relations, 644, 658
marginal ray, 388
Markov process, 772
maser, 9
material dispersion, 525, 787
Maxwell, J.C., 4
Maxwell–Boltzmann distribution, 38
Maxwell–Boltzmann statistics, 23, 307
mean free path, 253
medical applications of lasers, 806–809
Meggers, W.F., 1
mercury laser
optically pumped, 208
meridional ray in step-index fiber, 481
metal film
transfer matrix of, 415
metal vapor lasers, 220
metastable state, 13, 119
methane cell
use in helium-neon laser, 84
Michelson interferometer, 125
Michelson stellar interferometer, 765
microchannel plate, 696
microlaser, 9
microring resonator, 605
Miller code, 794
minimum spot size, 442
MIRACL laser, 271, 809
mirror resonators
propagation of rays in, 428
mirrors
ray transfer matrix analysis of, 380–383
MISER laser, 185
mixing coefficients, 626–629
mobility, 309
mode
longitudinal, 93
mode beating, 100–102
mode patterns in optical fiber, 514–518
mode pulling, 92
mode-locked
mode of laser operation, 137
mode-locking, 132, 133
active, 141
additive pulse operation, 146
AM, 141
colliding pulse, 145
hard aperture, 146
Kerr lens operation, 146
methods for, 140–147
SESAM, 147
soft aperture, 146
with acousto-optic modulator, 141
modes
transverse, 118
modes and field patterns in optical fiber, 511
modulation formats for optical communication, 790
modulation transfer function (MTF), 395

modulator
acousto-optic, 608–615, 633
electro-absorption, 606
molecular beam epitaxy (MBE), 322
molecular energy state
bound, 233
unbound, 233
molecular gas lasers
optical pumping of, 209
molecule
diatomic, 235
monochromatic radiation, 17, 210
Mueller calculus, 576
Mueller matrix, 576
multilayer structure
analysis by impedance techniques, 420
multimode laser oscillation, 93, 132
multimode optical fiber, 520
multiphoton absorption for isotope separation, 812
multiple quantum well laser structure, 339, 340
mutual coherenece function, 742
mutual spectral density, 743

n-type semiconductor, 310
Nakamura, S., 352
National Ignition Facility, 170, 173, 814
natural line broadening, 33–36, 175
Nd:YAG laser, 160–169
CW operation, 168
threshold pump energy fo pulsed operation, 168
$Nd^{3+}$ glass laser, 169–173
absorption spectrum and energy levels, 172
$Nd^{3+}$ laser, 92, 160
oscillator-amplifier chain in, 173
$Nd^{3+}$:YAG
energy level diagram, 166
NEA photoemitter, 699
negative refractive index, 411
negative temperature, 46
NEP (noise equivalent power), 692
network architectures in optical fiber communication systems, 791
networks
optical communication, 788
noise, 684–691
noise equivalent power (NEP), 692
noise in photomultipliers, 717
non-imaging light collector, 384
nonlinear applications
crystals for, 679
nonlinear coefficient
effective, 646
nonlinear susceptibility, 626–629
normal glow, 222
normal modes of carbon dioxide, 238
nromal modes, 238
nuclear fusion using lasers, 813
numerical aperture, 385
numerical aperture of optical fiber, 483
Nyquist's theorem, 687

O-wave, 365, 551
objective lens, 387
occupation number, 25
opthalmology
lasers in, 806
optic angle, 565
optic axis
principal, 550
secondary, 567
optical bistability, 674–678
optical cavity
in laser, 85
optical communication
modulation formats for, 790
optical communication networks, 788
optical communication systems, 783–796
optical communications
history of, 783
optical data transmission technology, 786
optical fiber
coupling to, 533
cutoff frequency, 518
dispersion in, 524–526
dispersion shifted, 526
graded index, 486
meridional ray in, 481
multimode, 520
numerical aperture of, 483
ray theory of, 481–494
skew ray in, 481, 484
step index, 481, 484, 508
optical fiber connectors, 534
optical fiber fabrication, 521
optical fiber mode patterns, 514–518
optical fiber modes and field patterns, 511
optical fiber network architectures, 791
optical fibers
absorption in, 785
low loss, 784
optical fibers for communications, 784
optical isolator, 406
optical length, 132, 133
optical links
line-of-sight, 794
optical networks
coding schemes in, 793
optical path (eikonal), 837
optical pumping
axisymmetric, 174
close-coupling, 156
geometrical arrangements for, 174–181

optical pumping (cont.)
pulsed versus continuous operation, 157
threshold lamp power needed, 156
optical pumping in three- and four-level lasers, 153
optical pumping of molecular gas lasers, 209
optical pumping using elliptical reflector, 174
optical pumping with Rabl reflectors, 176
optical resonator containing amplifying medium, 87–91
optical sources
fluctuations of, 767
optimum coupling in laser, 106
optimum mirror reflectance, 247
ordinary wave, 365, 551
organic dye
absorption and fluorescence spectra, 280
organic dye laser, 277–289
organic dye molecule
energy level diagram of, 279
oscillation frequency of laser, 91–92
oscillation modes of laser, 116–122
oscillator
electronic, 8
parametric, 665–670
oscillator-amplifier chain in $Nd^{3+}$ laser, 173
output plane, 366
output spectra of semiconductor laser, 338
outside chemical vapor deposition (OCVD), 522

p–i–n photodiode, 710
direct detection with, 722
p–n junction, 311–316
photoexcitation at, 685
P-branch transition, 19, 243
P-polarization, 399
p-type semiconductor, 310
P-wave in slab waveguide, 495
parabolic gain profile medium
propagation of Gaussian beam in, 465
paramagnetic ion solid state lasers, 160
parametric amplifcation
optical, 661–665
parametric nonlinear processes, 615, 633
parametric oscillator, 665–670, 687
parametric oscillator (DRO), 665
parametric oscillator (SRO), 665
parametric oscillator tuning, 668
parametric process
idler, 633
pump, 617, 633
signal, 615, 633
paraxial ray analysis, 365–376
paraxial ray analysis sign convention, 370
paraxial ray tracing, 376
Parseval's theorem, 747
partial coherence, 750
partial coherence propagation laws, 753
partial population inversion, 242
passive mode-locking with nonlinear media, 145
passive mode-locking with saturable absorber, 19, 144
Patel, C.K.N, 245
Pauli exclusion principle, 42
Penning ionization, 220
periodic lens sequence, 425–431
periodic poling for phase matching, 656
permeability of free space, 3
permittivity of free space, 3
Perot, J.B.G.G.A., 66, 85
phase condition for laser oscillation, 91
phase conjugation, 670–674
phase conjugation in $CS_2$, 674
phase hologram, 801
phase matching
90 degree, 653
beam walk-off effects, 653
periodic poling in, 656
type-I, 652
type-II, 652
phase matching in second harmonic generation, 649
phase matching in sum frequency generation, 652
phase modulation
electro-optic, 597–598
phase space
elementary cell, 751
phase velocity
in anisotropic medium, 545
phase velocity of wave, 3
phase-matching, 646, 649
phonon, 37
phosphorescence, 278
photo-coagulation, 807
photocathode materials, 697
radiant sensivity of, 698
photoconductive detector
hot-carrier-effect, 706
photoconductive detectors, 700–706
extrinsic, 704, 706
intrinsic, 702
photodetector
pyroelectric, 716
photodetectors
thermal, 715
photodiode
p–i–n, 710
vacuum, 695
photodiodes, 706–708
spectral responsivities of, 709
photodissociation laser, 268
photoelastic effect, 585
photoelastic tensor, 585
photoemissive detectors, 693

photoemitter
NEA, 699
photomuliplier, 695–700
photomultiplier
anode pulse height distribution, 701
photon, 2
photon counting, 719
photon flux, 20
photon statistics, 771–776
photonic crystal, 411
photonic crystal lattice
two dimensional, 424
photonic crystals, 421–424
photonics, 8
photorefractive effect, 634
photovoltaic detectors, 706–708
physical constants
recommended values of, 852
piezo-optical coefficients, 587
piezoelecricity, 581
piezoelectric coefficients, 582
piezoelectric effect
inverse, 585
piezoelectric tensor, 581
Planck, M, 23
Planck's constant, 2
plane hologram, 803
plane surface
propagation from, 756
plane waves in complex media, 409
PMD (polarization mode dispersion), 787
Pockel's effect, 585
Poincaré, J.H., 2, 5
Poincaré sphere, 5
Poisson distribution, 773
Poisson's bright spot, 112
Poisson, S., 111, 148
polarization, 64, 362
circular, 2
elliptical, 2
left-hand circular, 5
linear, 2
nonlinear, 622–626, 640–644
power expended in inducing, 66
polarization mode dispersion (PMD), 787
polarization state
change on reflection, 404
polarization state of electromagnetic wave, 5
polarizer combinations
intensity trasmission through, 568
polyatomic molecule
vibrations of, 236
population inverison
partial, 242
population inversion, 46, 199
threshold, 91
positive column, 224
positive feedback, 9
potassium dihydrogen phosphate, KDP, 588
potential energy level diagram, 233
diatomic molecule, 233
potental well
harmonic, 237
potential
anharmonic, 622–626
potential well
anharmonic, 237
power output from laser amplifier, 58
power output of laser, 103–106
Poynting vector, 4, 363
preionization techniques, 258, 262
pressure broadening, 37
principal axes of material, 540
principal phase velocities, 545
principal planes, 368
principal points, 368
principal ray, 388
Prokhorov, A., 2
propagation from a finite plane surface, 756
propagation of spherical waves, 434
pulse compression, 147
pulse shortening in passive mode-locking, 144
pump frequency in parametric process, 617, 633
pumping power, 156
pumping saturation is gas lasers, 213
pupil
entrance, 388
pyroelectric detector, 716

Q (quality factor), 59, 830
$\theta_{\text{beam}}$, 442
Q-branch transition, 243
Q-switched lasers, 195–203
Q-switching, 195
theory of, 199–203
Q-switching with absorbing dye, 197
Q-switching with Kerr effect, 197
Q-switching with roof prism, 197
Q-switching with spinning wheel, 197
Q-switching with the electro-optic effect, 198
Q-switching with the Faraday effect, 197
quadratic index medium
Gaussian beam in, 463
quadratic index profile, 433
quality factor (Q), 59, 830
quantum, 8
quantum cascade laser, 353
quantum cascade laser energy level diagram, 355
quantum efficiency, 156
external, 318, 319
internal, 316, 321
quantum electronics, 8

quantum number
vibrational, 236
quantum well laser (MQW), 340
quantum well laser (SQW), 337
quantum well laser (SQW) energy band diagrem, 339
quantum well lasers, 336–341
quarter-wave plate, 563

R-branch transition, 243
Rabl reflectors, 176
radiant intensity, 13
radiation
monochromatic, 17
radiation resistance, 689
radiation traping, 83
radio-frequency spectrum analysis with acousto-optic device, 617, 633
radius of curvature of Gaussian beam, 442
Rama–Nath diffraction, 610
Raman–Nath regime in acousto-optic device, 609
Raman regime of FEL, 294
ray
bound, 484, 488
marginal, 388
principal or chief, 388
refracting, 484, 488
tunneling, 484, 488
ray and wave normal directions
in anisotropic medium, 552
ray direction, 363
ray of light, 4
ray propagation in media with spatially varying refractive index, 431
ray propagation in medium with radial Gaussian index variation, 433
ray propagation through isotropic media, 362
ray theory of optical fiber, 481–494
ray theory of slab waveguide, 490
ray tracing diagrams, 377
ray transfer matrix, 368
Gaussian medium, 434
mirrors, 380–383
planar interface, 370
slab, 371
thick lens, 372
thin lens, 374
two thin lenses, 375
uniform medium, 370
ray transfer matrix of unfolded system, 381
ray-velocity surface, 556
ray-velocity surface in biaxial crystal, 567
Rayleigh distribution, 770
rays
bound, refracting, and tunneling, 488
propagation in mirror resonators, 428
reciprocal lattice, 422
recombination and luminescence, 316
recombination radiation
spectrum of, 317
recombination velocity, 327
reconfigurable optical add-drop muliplexer (ROADM), 791
rectification at an energy barrier, 324
reduced mass, 48–58, 236
reflectance, 87, 400
reflectance at oblique incidence, 401
reflection
effect on polarization, 404
law of, 364
reflection and refraction at plane boundaries, 834
reflection coefficient, 87, 398
refractiive index
quadratic radial variation, 433
refracting ray, 484, 488
refractive index, 63, 362
negative, 411
relative permittivity, 64
relaxation oscillation
rate equations for, 190–194
relaxation oscillations, 188–190
undamped, 194
resonator
diffraction losses in, 474
Gaussian beam in, 467–474
microring, 603
passive, 89
symmetrical, 469
unstable, 475
resonator design, 471
resonator stability diagram, 430
resonators
stable and unstable configurations, 120
retardation, 560–565, 589
retardation in electro-optic modulation, 590
retina
reattachment of, 807
Rhodamine 6G, 287
RLC circuit, 60, 828
ROADM (reconfigurable optical add-drop multiplexer), 791
Rogowski profile, 258, 260
Rosa E.B., 4
rotational energy level, 10, 182
rotational energy level diagram, 241
rotational energy state, 239
rotational populations, 240
rotational temperature, 241
ruby
absorption spectrum, 74
energy level diagram, 76

ruby laser, 74–79, 188
threshold pumping for CW operation, 136, 158
threshold pumping for pulsed operation, 157, 188

S-polarization, 399
S-waves in slab waveguide, 502
sapphire
crystal structure, 75
saturation, 9
saturation in homogeneous system, 48
saturation in inhomogeneous system, 51–58
saturation intensity, 51
saturation velocity, 713
Schawlow, A., 2
Schottky diode, 324
second harmonic generation (SHG), 629, 645
second harmonic generation with Gaussian beams, 654
self-focusing, 146
self-mode-locking, 137
semconductor carrier concentrations, 307
semiconductor
intrinsic, 307
n-type, 310
p-type, 310
semiconductor laser, 301–356
blue–green, 352
distributed Bragg reflection, 347
distributed feedback, 345
edge-emitting, 342, 343
gain coefficient of, 331–334
gain-guided structure, 342, 343
index-guided structure, 345
interband quantum cascade, 354
longitudinal and transverse modes, 335
modulation of, 356
output spectra, 338
P–I characteristic, 335
quantum cascade, 353
surface-emitting, 348
threshold and power-voltage characteristics, 334
threshold current density, 334
VCSEL, 348
VECSEL, 349
semiconductor laser structures, 341
semiconductor lasers
broad area, 350, 352
DFB and DBR, 343–348
semiconductor materials, 302
ternary and quaternary, 323
semiconductor physics, 301
semiconductor structures
growth of mutlilayers, 322
semiconductors
intrinsic and extrinsic, 309
separate confinement heterostructure, 341
SESAM mode-locking, 147
SHG
effects of depletion during, 658–660
external, 655
intra-cavity, 655
SHG (second harmonic generation), 629, 645
shot noise, 684–687
signal frequency in parametric process, 615, 633
signal-to-noise ratio
direct detection, 721
silicon lasers, 355
simple harmonic oscillator model for diatomic molecule, 236
simple harmonic potential well, 237
sine condition for imaging systems, 385
single longitudinal mode operation, 133–137
single quantum well laser, 337
singlet state, 277
singly resonant parametric oscillator (SRO), 665
skew ray in optical fiber, 481, 484
slab
transfer matrix for, 412
slab waveguide
dispersion curves in, 498
fields in, 498
slab waveguide geometries, 503
Snell's law, 364
soft collision, 37, 63
solid state laser
face-pumped configuration, 182
solid state lasers
diode pumped, 183
Solid state lasers
high power pulsed, 182
solitons, 528
Sommerfeld radiation condition, 758
SONET, 789
SONET/SDH data rates, 789
spatial coherence, 127, 748–753
degree of, 763
extended source, 751
spatial coherence of a monochromatic circular source, 763
spatial cross-relaxation, 133
spatial filtering, 462
spatial hole-burning, 133
speckle, 804
spectrum of recombination radiation, 317, 319
spherical aberration, 393
spherical wave
focusing by lens, 435
spherical waves
propagation of, 434
spiking, 188–190
spontaneous and stimulated transitions, 7–8
spontaneous and stimulated transitions, ratio of, 29–30

Spontaneous emission, 12
spontaneous emission, 13
spot diagram, 395
spot size
  minimum, 442
spot size of focused Gaussian beam, 460
square law detector, 736
squeezed light, 780
SRO (singly resonant parametric oscillator), 665
stability diagram for resonators, 430
stagnation region in gas dynamic laser, 257, 260
Stark broadening, 37
Stark effect, 37
state
  metastable, 13
step indeix optical fiber, 481, 484, 508
steradian, 14
stigmatic imaging, 385
stimulated emission, 15
stimulated emission cross-section, 2, 159
stimulated emission rate in homogeneously broadened system, 46
stimulated emission, classical oscillator model, 66–69
stimulated transitions as photon-particle collision processes, 19
stochiometry, 302
Stokes parameters, 5
Stokes, G.G., 5
stop
  aperture, 386, 388
  field, 387
stop band, 419
stratified medium, 417
Strehl ratio, 396
stress, 581
striations, 225
structure with periodic dielecric properties, 411–421
superlattice, 323
surface plasmon polariton, 416
surface plasmon resonance, 416
surface-emitting laser
  grating coupled, 348
susceptibilities
  microscopic and macroscopic, 634–638
susceptibility
  complex, 6, 409
  electric, 4, 362
  electronic, 62
  magnetic, 4, 119
  nonlinear, 626–629
susceptibility analysis
  distinction between local and external fields, 631, 637
susceptibility tensor, 540
Sylvester's theorem, 426
symmetric stretch mode of carbon dioxide, 238
symmetrical resonator, 469
Symmetry and Kleinman's conjecture, 631, 636
symmetry properties of crystals, 841
synchronous pumping, 142

TEA laser, 257, 260
TEM modes of laser, 121
$TEM_{00}$ Gaussian mode, 444
temperature
  electron, 226, 247
  Joule-Thomson, 256
  rotational, 42, 241
  translational, 241, 258, 260
temporal and spatial coherence, 744
temporal coherence, 4, 125, 744–748
  degree of, 746
tensors, 845
ternary and quaternary matched materials, 323
thermal conduction time, 252
thermal detectors, 715
thermal lensing, 252
thermonuclear reactions
  controlled (CTR), 814
thermopile, 716
three- and four-level lasers, 72–74, 246
  energy-level diagram, 73
  threshold inversion in, 155
threshold current density in semiconductor laser, 334
threshold lamp power in optical pumping, 156
threshold pump power in dye laser, 285–289
TIR (total internal reflection), 365
titanium sapphire
  absorption and emission spectra, 166
titanium-sapphire laser, 160, 276
torr
  unit of pressure, 80
total internal reflection (TIR), 365
Townes, C.H., 2
Townsend dark discharge, 222
trans-impedance amplifer, 707
transfer matrix for metal film, 415
transfer matrix for single slab, 412
transformed impedance, 403
transition
  allowed, 13
  band-to-band, 313
  direct band-gap, 305
  fobidden, 13
  P-branch, 243
  Q-branch, 243
  R-branch, 243
translational temperature, 241
transmission coefficient, 87, 398
transmittance, 87, 400
transverse electro-optic modulation, 591, 594

transverse mode
  TEM, 121
transverse modes of laser, 118
triplet state, 277
tunneling ray, 484, 488
turbulence
  atmospheric, 15
Twiss R.Q., 767
twisted mode operation of laser, 136
type-I phase matching, 652
type-II phase matching, 652

ultraviolet catastrophe, 23
ultraviolet molecular gas lasers, 264
unbound state of molecule, 233
undulator, 293
unfolding
  in ray transfer matrix analysis, 381
uniaxial cryetal
  negative, 551
  positive, 551
uniaxial indicatrix, 550
unit cell, 422
unstable resonator, 475
up-conversion, 660
uranium isotope separation by laser, 810

V-number, 511
vacuum photodiode, 695
valence band, 11
van Cittert–Zernike theorem, 761
Van der Waals broadening, 37
vapor-phase axial-deposition (VAD), 523
VCSEL (vertical cavity surface emitting laser), 348
VECSEL (vertical-external-cavity surface-emitting laser), 349
velocity of light, 4
Verdet constant, 406
vibrational energy level diagram for carbon dioxide, 239
vibrational energy state, 10
vibrational energy states, 236
vibrational quantum number, 236
vibrations of polyatomic molecule, 236
vibronic state, 10
vibrot state, 10
vignetting, 389
Voigt profile, 47
volume hologram, 803
volume of an elementary cell of phase space, 752
volume of coherence, 752

Watson, J.D., 422
wave
  electromagnetic, 6–7
wave equation, 438
  beam solution of, 440
wave normal and ray directions, 552
wave propagation in anisotropic medium, 543
wave theory of slab waveguide, 494–503
wave vector, 3, 363
wave-particle duality, 2
wave-velocity surface, 555
waveguide dispersion, 526
waveguide structures
  integrated, 604
wavelength-division multiplexing (WDM), 531, 788
wavelengths important in optical design, 390
waveplate, 589
WDM (wavelength division multiplexing), 531
weakly guiding approximation, 512
Weber W.E., 4
Weber, J., 2
Weiner–Khintchine theorem, 743
whispering gallery modes, 605
wiggler, 293
  helical, 294
work function, 694

X-ray lasers, 206, 227
X-valley, 306
$Xe_2^+$ excimer laser, 265

Yablonovitch, E., 421
$Yb^{3+}$ in glass, energy levels, 186
Young's slits experiment, 127